Wastewater Treatment and Reuse

Theory and Design Examples

Volume 2: Post-Treatment, Reuse, and Disposal

Wastewater Treatment and Reuse

Theory and Design Examples

Volume 2: Post-Treatment, Reuse, and Disposal

Syed R. Qasim
The University of Texas at Arlington

Guang Zhu
CP&Y, Inc.

In Cooperation with

CP&Y, Inc.
Consulting Engineers · Planners · Project Managers
Dallas, Texas

CRC Press
Taylor & Francis Group
Boca Raton London New York

CRC Press is an imprint of the
Taylor & Francis Group, an **informa** business

First published 2018 by Taylor & Francis

Published 2023 by CRC Press
Taylor & Francis Group
6000 Broken Sound Parkway NW, Suite 300
Boca Raton, FL 33487-2742

ISBN-13: 978-1-138-30094-1 (hbk)

Visit the Taylor & Francis Web site at
http://www.taylorandfrancis.com

and the CRC Press Web site at
http://www.crcpress.com

For Product Safety Concerns and Information please contact our EU representative: GPSR@taylorandfrancis.com
Taylor & Francis Verlag GmbH, Kaufingerstraße 24, 80331 München, Germany

Contents

VOLUME 2 Post-Treatment, Reuse, and Disposal

Preface

Over the last decade there have been rapid developments and changes in the field of wastewater treatment. The emphasis has been on identification, detection, and removal of specific constituents; computer simulation and modeling; membrane processes; renovation and reuse of wastewater effluent; nutrients recovery, and reduction and utilization of biosolids; energy conservation; greater understanding of theory and principles of treatment processes; and application of these fundamentals into facility design. Environmental engineers have many responsibilities. One of the most demanding yet satisfying of these are the design of wastewater treatment and reuse facilities. There are several books that discuss the fundamentals, scientific principles, and concepts and methodologies of wastewater treatment. The actual design calculation steps in numerical examples with intense focus on practical application of theory and principles into process and facility design are not fully covered in these publications. The intent of the authors writing this book is threefold: *first*, to present briefly the theory involved in specific wastewater treatment processes; *second*, to define the important design parameters involved in the process, and provide typical design values of these parameters for ready reference; and *third*, to provide a design approach by providing numerical applications and step-by-step design calculation procedure in the solved examples. Over 700 illustrative example problems and solutions have been worked out to cover the complete spectrum of wastewater treatment and reuse from fundamentals through advanced technology applied to primary, secondary and advanced treatment, reuse of effluent, by-product recovery and reuse of biosolids. These examples and solutions enhance the readers' comprehension and deeper understanding of the basic concepts. They also serve as a good source of information for more experienced engineers, and also aid in the formal design training and instruction of engineering students. Equipment selection and design procedures are the key functions of engineers and should be emphasized in engineering curricula. Many practice problems with step-by-step solution provide skills to engineering students and professionals of diverse background for learning, and to master the problem-solving techniques needed for professional engineering (PE) exams. Also, these solved examples can be applied by the plant designers to design various components and select equipment for the treatment facilities. Thus, the book is a consolidated resource of valuable quick-and-easy access to a myriad of theory and practice information and solved examples on wastewater treatment processes and reuse.

This work is divided into two volumes. Principles and basic treatment processes are covered in Volume 1, which includes Chapters 1 through 10. Volume 2 contains Chapters 11 through 15 to cover post-treatment processes, reuse, and solids disposal.

Volume 1: Principles and Basic Treatment. Chapter 1 is an overview of wastewater treatment: past, present, and future directions. Chapters 2 and 3 cover the stoichiometry, reaction kinetics, mass balance, theory of reactors, and flow and mass equalization. Sources of municipal wastewater and flow rates and characteristics are provided in Chapters 4 and 5. Chapter 6 provides an in-depth coverage of wastewater treatment objectives, design considerations, and treatment processes and process diagrams. The preliminary treatment processes are covered in Chapters 7 and 8. These unit processes are screening and grit removal. Chapter 9 deals with primary treatment with plain and chemically enhanced sedimentation. Chapter 10 provides an in-depth coverage of biological waste treatment and nutrients removal processes.

Volume 2: Post-Treatment, Reuse, and Disposal. Chapter 11 covers major processes for effluent disinfection, while Chapter 12 deals with effluent disposal and reuse. Chapter 13 is devoted to residuals management, recovery of resources, and biosolids reuse. The plant layout, yard piping, plant hydraulics, and instrumentation and controls are covered in Chapter 14. Upgrading of secondary treatment facility, land application, wetlands, filtration, carbon adsorption, BNR and MBR; and advanced wastewater treatment processes such as ion exchange, membrane processes, and distillation for demineralization are covered in Chapter 15.

This book will serve the needs of students, teachers, consulting engineers, equipment manufacturers, and technical personnel in city, state, and federal organizations who are involved with the review of designs, permitting, and enforcement. To maximize the usefulness of the book, the technical information is summarized in many tables that have been developed from a variety of sources. To further increase the utility of this book six appendices have been included. These appendices contain (a) abbreviations and symbols, basic information about elements, useful constants, common chemicals used in water and wastewater treatment, and U.S. standard sieves and size of openings; (b) physical constants and properties of water, solubility of dissolved gases in water, and important constants for solubility and sodicity of water; (c) minor head loss coefficients for pressure conduits and open channels, normal commercial pipe sizes, and design information of Parshall fumes; (d) unit conversions; (e) design parameters for wastewater treatment processes; and (f) list of examples presented and solved in this book. These appendices are included in both volumes. The numerical examples are integrated with the key words in the subject index. This gives additional benefit to the users of this book to identify and locate the solved examples that deal with the step-by-step calculations on the specific subject matter.

Enough material is presented in this textbook that cover supplemental material for a water treatment course, and a variety of wastewater treatment courses that can be developed and taught from this title. The supplemental material for a water treatment course include components of municipal water demand (Section 4.3), rapid mix, coagulation, flocculation, and sedimentation (Sections 9.6, 9.7, and 10.9), filtration (Section 15.4.6), carbon adsorption (Section 15.4.8), chlorine and ozone disinfection (Sections 11.6 and 11.8), demineralization by ion exchange and membrane processes (Sections 15.4.9 and 15.4.10), and residuals management (Sections 13.4.1 through 13.4.3, 13.5 through 13.8, and 13.11.6). At least *three* one-semester, and *one* two-semester sequential wastewater treatment courses at undergraduate or graduate levels can be developed and taught from this book. The specific topics to be covered will depend on time available, depth of coverage, and the course objectives. The suggested wastewater treatment and reuse courses are:

Course A: A one-semester introductory course on wastewater treatment and reuse
Course B: A sequential two-semester advance course on wastewater treatment and reuse
Course C: A one-semester course on physical and chemical unit operations and processes
Course D: A one-semester course on biological wastewater treatment

The suggested course outlines of these courses are provided in the tables below. The information in these tables is organized under three columns: topic, chapter, and sections. The examples are not included in these tables. It is expected that the instructor of the course will select the examples to achieve the depth of coverage required.

Course A: Suggested course contents of a one-semester introductory course on wastewater treatment and reuse

Topic	Chapter	Section
Overview of wastewater treatment	1	All
Stoichiometry and reaction kinetics	2	2.1 and 2.2

Continued

Topic	Chapter	Section
Mass balance, reactors, and equalization	3	3.1 to 3.3, and 3.4.1 to 3.4.3
Sources and flow rates of wastewater	4	All
Characteristics of municipal wastewater	5	5.1 to 5.6, 5.7.1, 7.8, and 5.9
Treatment and design objectives, and processes	6	All
Screening	7	All
Grit removal	8	8.1 to 8.3, 8.4.1 to 8.4.5, 8.5, and 8.6
Conventional and chemically enhanced primary sedimentation	9	9.1 to 9.6, 9.7.1, and 9.7.2
Biological waste treatment: basics, oxygen transfer, fixed film attached growth processes, anaerobic treatment, biological nitrogen removal, and final clarifier	10	10.1, 10.2, 10.3.1, 3.3.2, 10.3.4 to 10.3.8, 10.3.10, 10.3.11, 10.4 to 10.6, 10.7.1 to 10.7.3, 10.8, and 10.9
Effluent disinfection	11	11.1 to 11.7
Effluent reuse and disposal	12	12.1, 12.2, 12.5, and 12.6
Residuals processing, reuse, and disposal	13	13.1 to 13.8, and 13.11
Plant layout, piping, hydraulics, and instrumentation and control	14	All
Advanced wastewater treatment and upgrading secondary treatment facility	15	15.1 to 15.3, 15.4.5, 15.4.6, and 15.4.8 to 15.4.10

Course B: Suggested course contents of a sequential two-semester advanced course on wastewater treatment and reuse

Topic	Chapter	Section
First Semester		
Overview of wastewater treatment	1	All
Stoichiometry and reaction kinetics	2	All
Mass balance, reactors, and equalization	3	All
Sources and flow rates of wastewater	4	All
Characteristics of municipal wastewater	5	All
Treatment objectives, design considerations, and treatment processes	6	All
Screening	7	All
Grit removal	8	All
Primary and enhanced sedimentation	9	All
Biological waste treatment: fundamentals and types	10	10.1 and 10.2
Second Semester		
Biological waste treatment (cont'd): suspended, attached, aerobic, anaerobic kinetics, oxygen transfer, biological nutrient removal (BNR), computer	10	10.3 to 10.10
application, and final clarifiers		
Disinfection and kinetics	11	All
Effluent reuse and disposal	12	All

Continued

Topic	Chapter	Section
Residuals processing, reuse, and disposal	13	All
Plant layout, piping, hydraulics, and instrumentation and control	14	All
Advanced wastewater treatment facilities	15	All

Course C: Suggested course contents of a one-semester course on physical and chemical unit operations and processes

Topic	Chapter	Sections
Overview of wastewater treatment	1	All
Stoichiometry and reaction kinetics	2	All
Mass balance, reactors, and equalization	3	All
Sources and flow rates of wastewater	4	4.4 and 4.5
Characteristics of municipal wastewater	5	5.1 to 5.4
Wastewater treatment processes	6	6.3.5
Screening: coarse and fine screens	7	7.1, and 7.2.1 to 7.2.4
Discrete settling and grit removal	8	8.1, 8.3, 8.4.2, and 8.4.4
Flocculant settling, rapid mixing, flocculation, and sedimentation	9	9.1, 9.2, 9.5.5, 9.6.5, 9.6.6, and 9.7.2
Zone or hindered settling	10	10.9.2
Disinfection kinetics, chlorination, dechlorination, ozonation, and UV radiation	11	11.4, 11.5, 11.6.1, 11.6.2, 11.7.1, 11.8.6, and 11.9.4 to 11.9.6
Compression settling, dissolved air flotation, anaerobic digestion, conditioning, and dewatering	13	13.4.1, 13.4.2, 13.5.1 to 13.5.3, 13.6.1, 13.6.2, 13.7.1, 13.8.1, and 13.8.2
Air stripping, filtration, carbon adsorption, ion exchange, and membrane processes	15	15.4.5, 15.4.6 , 15.4.8, 15.4.9, and 15.4.10

Course D: Suggested course contents of a one-semester course on biological wastewater treatment

Topic	Chapter	Section
Overview of wastewater treatment	1	All
Stoichiometry and reaction kinetics	2	All
Mass balance, reactors, and equalization	3	All
Sources and flow rates of wastewater	4	All
Characteristics of municipal wastewater	5	All
Wastewater treatment processes	6	6.3.5
Biological waste treatment, biological nutrient removal (BNR), and final clarifier	10	All
Pathogens reduction in treatment processes and natural die-off kinetics	11	11.2.1, and 11.5.1
Anaerobic and aerobic digestion of sludge, material mass balance, and composting	13	13.6.1, 13.6.2, 13.9, and 13.11.1
Aquatic treatment systems, and membrane processes	15	15.3.2, and 15.4.10

In the solutions of examples, full expressions are provided to demonstrate step-by-step calculations. Many process and hydraulic parameters are involved in these expressions. To be more efficient, these parameters are represented by symbols. Sometimes, in the same example, parameters are applied multiple times to different streams or reactors. Therefore, subscripted notations are also used to identify these parameters. Each symbol is fully defined when it appears for the first time in the solution of the example. After that this symbol is repeated in the entire solution. This approach is helpful in (1) saving space by replacing lengthy descriptions of a parameter, and (2) providing an identification of the numerical value used or obtained in the expression. Additionally, these symbols provide the designers a ready reference in their design calculations while using Mathcad or spreadsheet software.

The International System of Units (SI) is used in this book. This is consistent with the teaching practices in most universities in the United States and around the world. Most tables in the book have dual units and include conversion from SI to U.S. customary units in footnotes. Useful conversion data and major treatment process design parameters are provided in Appendices D and E.

Acknowledgment

A project of this magnitude requires the cooperation and collaboration of many people and organizations. We are indebted to many professionals, faculty members, students, and friends who have helped and provided constructive suggestions. We must acknowledge the support, encouragement, and stimulating discussion by Michael Morrison, W. Walter Chiang, and Pete K. Patel throughout this project. CP&Y, Inc., a multidisciplinary consulting engineering firm in Dallas, Texas provided the technical support. We gratefully appreciate the support and assistance provided by Michael F. Graves, Marisa T. Vergara, Gregory W. Johnson, Ellen C. Carpenter, Barbara E. Vincent, Megan E. Martin, Gil W. Barnett, and Dario B. Sanchez. Many students also assisted with typing, artwork, literature search, and proofreading. Among them are Bernard D'Souza, Rajeshwar Kamidi, Neelesh Sule, Richa Karanjekar, Gautam Eapi, and Olimatou Ceesay.

Kelcy Warren established Syed Qasim Endowed Professorship in Environmental Engineering in the Department of Civil Engineering at The University of Texas at Arlington. Funds from this endowment helped to support students. The support of the Department of Civil Engineering at The University of Texas at Arlington is greatly appreciated. In particular, we thank the support of Dr. Ali Abolmaali, and tireless support of Sara Ridenour in making departmental resources available.

Finally, we must acknowledge with deep appreciation the support, encouragement, and patience of our families.

Although the portions of this book have been reviewed by professionals and students, the real test will not come until this book is used in classes, and by professionals in design of wastewater treatment facilities. We shall appreciate it very much if all who use this book will let us know of any errors and changes they believe would improve its usefulness.

Syed R. Qasim and Guang Zhu
Arlington, Texas

Authors

Syed R. Qasim is a professor emeritus in the Department of Civil Engineering at the University of Texas at Arlington. Dr. Qasim earned PhD and MSCE from West Virginia University, and BSCE from India. He served on the faculty of Polytechnic University, Brooklyn, New York, and on the faculty of the University of Texas at Arlington, Texas, from 1973 till his retirement in 2007. Dr. Qasim has conducted full-time research with Battelle Memorial Institute, Columbus Laboratories, and has worked as a design engineer with a consulting engineering firm in Columbus, Ohio. He has over 47 years of experience as an educator, researcher, and practitioner in the related fields of environmental engineering. His principal research and teaching interests include water and wastewater treatment processes and plant design, industrial waste treatment, and solid and hazardous waste management. He served nationally and internationally as a consultant with governmental agencies and private concerns. Dr. Qasim has written 3 books, and he is the author or coauthor of over 150 technical papers, encyclopedia and book chapters, and research reports. His papers, seminars, and short courses have been presented nationally and internationally. He is a *life member* and *Fellow* of American Society of Civil Engineers, and a *life member* of Water Environment Federation; a member of Association of Environmental Engineering and Science Professors, American Water Works Association, and other professional and honor societies. He is a registered professional engineer in the state of Texas.

Guang Zhu is an associate and senior water and wastewater engineering director with CP&Y, Inc., a multidisciplinary consulting engineering firm headquartered in Dallas, Texas. Dr. Zhu has over 30 years of consulting experience in planning, process evaluation, pilot testing, design, and commissioning of numeral conventional and advanced water and wastewater treatment plants in the United States and China. He had 10 years of consulting experience with Beijing General Municipal Engineering Design and Research Institute, Beijing, China. Dr. Zhu has coauthored one textbook and many technical papers and has translated two water and wastewater books in Chinese. He has taught several design courses as an adjunct assistant professor at the University of Texas at Arlington. He is a registered professional engineer in the state of Texas.

11 Disinfection

11.1 Chapter Objectives

Wastewater contains many types of human enteric organisms that are associated with various waterborne diseases. Disinfection is the primary mechanism for inactivation and destruction of disease-causing organisms present in the wastewater, and it is one of the common steps required in wastewater treatment and reuse. Presently, chlorination of both water supply and wastewater effluent is a widespread practice for the control of waterborne diseases. Alternate methods of disinfection are receiving a great deal of attention because chlorination may result in the formation of disinfection by-products (DBPs); many of which are known carcinogens. The material presented in this chapter covers the following important issues related to the disinfection process and facility design:

- Objectives and requirements of disinfection
- Disinfection techniques to meet the disinfection goals
- Basic design considerations for chlorination and dechlorination
- Application of alternate methods of disinfection such as ozone and ultraviolet (UV) radiation

11.2 Objectives and Requirements of Disinfection

A disinfection process should be properly selected as the final defense after the routine nondisinfection treatment processes. The main objectives of wastewater disinfection are to protect public health from waterborne diseases, reduce adverse side effects to the aquatic life, eliminate threats to the environment, and comply with the regulatory requirements.

11.2.1 Pathogens and Indicator Organisms

The pathogens are normally divided into four major microbial groups: (1) bacteria, (2) virus, (3) protozoa, and (4) helminthes. Common pathogens that may be present in domestic wastewater are summarized in Table 11.1.[1–10] Because of the presence of a large number of organisms in wastewater samples, their number is usually expressed to the base 10 of the logarithmic scale (Equation 11.1).

$$\text{log concentration} = \log(N) \tag{11.1}$$

where N = number of organism present in a specified sample volume

It is important to specify the sample volume for the log scale. The volume of wastewater is typically 100 mL for bacteria and virus, and 100 L for protozoa and helminthes. A general discussion on the microbiological quality of wastewater, indicator organisms, and measurement techniques are presented in Chapter 5.

Traditional Indicator Organisms: The identification and measurement techniques for pathogens in wastewater are neither simple nor clear-cut. For this reason, a group of indicator organisms are used to

TABLE 11.1 Common Pathogens Carried in Raw Domestic Sewerage

Organism (Size Range)	Probable Concentration[a], log(N)	Probable Median Infectious Dose[b], log(N_{50})	Relative Infectivity	Associated Waterborne Disease
		Bacteria (0.2–10 μm)		
Escherichia coli (enteropathogenic)	5–8	6–10	Low	Gastroenteritis
Salmonella spp. (nontyphoid)	2–4	4–7	Low	Salmonellosis
Salmonella typhi	0–4	5–7	Low	Typhoid fever
Shigella spp.	0–3	1–2	Moderate	Shigellosis (dysentery)
Vibrio cholerae	1–4	3–8	Low	Cholera
		Viruses (0.02–0.3 μm)		
Astroviruses (7 types)	–	< 2		Gastroenteritis
Caliciviruses (Norwalk and Sapporo)	<6	1–2	High	Gastroenteritis
Coronavirus	–	–	–	Gastroenteritis
Enteric viruses (72 types)	6–7	0–3	High	Paralysis, gastroenteritis, fever, herpangina
Hepatitis A virus	–	0–2	High	Infectious hepatitis
Parvoviruses (3 types)	–	–	High	Gastroenteritis
Rotaviruses (4 types)	4–7	< 1	High	Gastroenteritis
		Protozoa (4–60 μm)		
Cryptosporidium parvum	2–6	0–2	High	Cryptosporidiosis
Entamoeba histolytica	2–8	0–2	High	Amebiasis
Giardia lamblia	2–7	0–2	High	Giardiasis
		Helminths (20–80 μm)		
Ascaris lumbricoides (roundworm)	2–6	0–1	High	Ascariasis
Schistosoma mansoni	1–2	0–1	High	Schistosomiasis
Trichuris trichiura (whipworm)	2–4	0–1	High	Trichuriasis

[a] The number of organisms (N) is typically expressed as colony forming unit (CFU) or most probable number (MPN) for bacteria, ova for helminthes, cysts or oocysts for protozoa, and plaque-forming unit (PFU) or MPN for viruses.
[b] The median infectious dose (N_{50}) indicates the number of organisms at which 50% of an exposed population exhibit a response.
Source: Adapted in part from References 1 through 10.

measure the potential for occurrence of pathogens in the water and wastewater. The traditional indicator organisms are the coliform bacteria of fecal origin. They have been used for a long time in establishing the bacterial surface water quality criteria. Federal water quality criteria for primary contact recreation waters were first established in 1968 based on fecal and total coliforms.[11] Under Section 304 (a) of the Clean Water Act of 1977, the criteria were revised by the U.S. Environmental Protection Agency in 1986 with the recommendations for establishing water quality standards based on *Escherichia coli* (*E. coli*) for fresh waters and *Enterococci* for fresh and marine waters. It was confirmed through time that *E. coli* and *Enterococci* were better indicator organisms than fecal and total coliforms for predicting the gastrointestinal illness-causing pathogens.[12] The characteristics of the traditional indicator organisms and the basic measurement techniques are covered in Sections 5.5.2 and 5.5.3.

Emerging Indicator Organisms: The recent practice of wastewater reclamation and reuse has indicated that use of only traditional indicator organisms may not be sufficient for predicting the safety of wastewater against most pathogens in the disinfected effluent.[13] Alternative emerging indicator organisms have been identified and recommended for microbial quality assessment.[2,5,14] The organisms most

TABLE 11.2 Indicator Organisms Used in Microbial Water Quality Criteria

Organism	Indication	Characteristics Shape and Size, μm	Probable Concentration in Raw Wastewater[a], $\log(N)/V$
		Traditional Bacterial Indicator Organisms	
Total coliforms	Bacteria	Rod 0.5–2	6–9
Fecal coliforms	Bacteria	Rod 0.5–2	5–8
Escherichia coli (*E. coli*)	Bacteria	Rod 1–2	5–8
Enterococci	Bacteria	Cocci 0.5–1	4–6
		Emerging Indicator Organisms	
Viruses			
Enteric viruses	Viral pathogen	Icosahedral capsid 0.025–0.03	6–7
Coliphages	Viral indicator	Icosahedral protein shell 0.025–0.2	5–7
Bacteria			
Clostridium perfringens	Protozoa surrogate	Rod spore 0.6–1.3 by 2.4–19	3–5
Protozoa			
Cryptosporidium parvum	Protozoan pathogen	Ovoid oocyst 4–6	2–6
Giardia lamblia	Protozoan pathogen	Ovoid cyst 8–12	2–7

[a] N = number of organisms, where CFU or MPN for bacteria and viruses, and cysts or oocysts for protozoan parasites. V indicates sample volume; 100 mL for bacteria and viruses, and 100 L for protozoa. For example, total coliform $\log N = 6$ means 10^6 organisms per 100 mL sample.

Source: Adapted in part from References 2, 5, 9, and 10.

commonly used in the recent microbial water quality criteria are summarized in Table 11.2. It is expected that the improved techniques for pathogens detection in reclaimed effluent for reuse may utilize multiple indicators rather than a single indicator.[5]

11.2.2 Microbial Reduction in Wastewater Treatment Processes

In wastewater treatment practice, the removal efficiency is usually expressed by percent removal of the most common contaminants. On account of high numbers, the efficiency as high as 99.99999% may be required for removal or inactivation of pathogens. It is, therefore, more convenient to express the log reduction or log inactivation for microbial removal or inactivation efficiency for wastewater treatment processes. The log reduction is expressed by Equation 11.2.

$$\text{log reduction} = \text{log inactivation} = \text{log removal} = -\text{log survival} \quad (11.2a)$$

$$\text{log reduction} = -\log(N/N_0) = -[\log(N) - \log(N_0)] \quad (11.2b)$$

$$\text{log reduction} = \log(N_0/N) = \log(N_0) - \log(N) \quad (11.2c)$$

where

N = number of organism present in the effluent from a treatment unit
N_0 = number of organism present initially in the influent

Wastewater treatment plants utilize many processes that can be selected for removal of different contaminants to meet the desired discharge permits or reuse criteria. Microbial reduction is achieved in most of these processes. The probable range of microbial log reductions through different wastewater treatment processes is developed and summarized in Table 11.3.[2,5,10–41] These ranges can be used only as

TABLE 11.3 Probable Microbial Log Reductions of Common and Advanced Wastewater Treatment Processes

Treatment Process	Viruses		Bacteria			Protozoa		Helminth
	Enteric Viruses	Coliphages	Total Coliform	*E. coli*, Fecal Coliform, *Enterococci*	*Clostridium perfringens*	*Cryptosporidium parvum*	*Giardia lamblia*	Ova, Cyst, Oocyst
Conventional Treatment Systems								
Primary treatment								
Conventional	0–0.5	0–1	0–0.5	0–0.5	0–0.5	0–0.5	0.5–1	0–2
CEPT	1–2	1–2	1–2.5	1–2.5	1–2.5	1–1.5	1–2	1–3
Secondary treatment								
Activated sludge	0.5–2	0–3	1–3	1–3	0.5–2	0.5–1.5	0.5–1.5	1–2
BNR facility	1–2	1.5–3	1–3	2–3	2–3	0.5–1.5	1.5–2.5	1–2
Extended aeration	1–2	1–3	1–3	1–3	0.5–2	0.5–1.5	0.5–1.5	1–2
Trickling filter	0–1.5	0–2	0–2	0–2	0–1	0–0.5	0–0.5	0.5–1
BAF	0.5–1.5	0.5–2	0.5–2	0.5–2	0–1	0–0.5	0–0.5	0.5–1
RBC	0–1	0–1.5	0.5–1.5	0.5–1.5	0–1	0–0.5	0–0.5	0.5–1
Tertiary treatment								
Coagulation/floc.	0.5–2	1–3	0–1	0–1	0–1	1–2	1–3	1–2
Slow sand filter	1–3	1–3	0–3	0–3	0–2	0–2.5	0–3	1–2
Monomedium filter	0.5–1	0–3	0.5–2	0–1	0–0.5	0–0.5	0–2	2–3
Dual media filter	0.5–2	0.5–1.5	0–3	0–2	0.5–1.5	1–1.5	1–2.5	2–3
Traveling bridge filter	0.5–1.5	0.5–1	2–2.5	1.5–2	1–1.5	0.5–1	0.5–1.5	2–3
Cloth media filter	0–0.5	0–0.5	2–3	1.5–2.5	0.5–1	0–0.5	0.5–1	2–3
MF	0.5–2	1–2	1–4	1–4	1–4	2–4	2–4	2–4
UF	2–4	3–4	3–4	3–4	3–4	3–> 4	3–> 4	> 4
Natural Treatment Systems								
Constructed wetlands	0.5–2	0.5–2	0.5–3	0.5–3	0.5–3	0.5–1	0.5–2	1–3
Stabilization pond	1–3	1–3	0.5–4	0.5–4	1–3	1–3	1–4	1–3
Storage pond/lagoon	1–4	1–4	1–5	1–5	1–5	1–3.5	3–4	1.5–3
Advanced Treatment Systems								
MBR	1.5–6	2.5–6	3.5–6	3.5–6	3.5–6	2.5–> 6	2.5–> 6	3–> 6
NF with MF or UF	4–7	4–7	4–7	4–7	4–7	> 6	> 6	> 6
RO with MF or UF	> 6	> 6	> 6	> 6	> 6	> 7	> 7	> 7

Note: BAF = biological aerated filter; BNR = biological nutrient removal; CEPT = chemically enhanced primary treatment; MBR = membrane bioreator; MF = microfiltration NF = nanofiltration; RBC = rotating biological contactor; UF = ultrafiltration.

Source: Adapted in part from References 2, 5, and 10 through 41.

a general guide. Actual reductions may be affected by the specific design and operating conditions at the plants. Normally, the primary treatment has limited microbial reductions. Secondary treatment achieves high reduction of pathogenic bacteria and enteric viruses. Removal increases with increasing mean cell residence time (MCRT) in biological treatment processes. Membrane processes are very efficient and reliable for control of pathogens. The performance of each and every predisinfection treatment process in a process train influences the overall removal of pathogens. As a result, the load on the disinfection process (which is the final step) can be greatly reduced by optimizing all components of the process train.

EXAMPLE 11.1: LOG CONCENTRATION OF ORGANISMS

A fecal coliforms count of 12,000 MPN/100 mL was found in the filtered effluent from a wastewater treatment plant. Calculate the log concentration of fecal coliforms in the filtered effluent.

Solution

Estimate from Equation 11.1, the log concentration of coliforms in filtered effluent

$$\text{log concentration} = \log(N) = \log(12{,}000) = 4.08 \text{ log}$$

EXAMPLE 11.2: LOG REMOVAL OF ORGANISMS

A tertiary filtration process was evaluated for removal of *Cryptosporidium parvum* oocysts from treated secondary effluent. The filter influent and effluent counts of oocysts were 100,000 and 300 per 100 L, respectively. Estimate the removal efficiency on (1) percentage basis and (2) log reduction basis.

Solution

1. Estimate the removal efficiency on percentage basis.

   The number or concentration of *Cryptosporidium parvum* oocysts in the filter influent

   $$N_0 = 100{,}000 \text{ oocysts/100 L}$$

   The number of concentration of *Cryptosporidium parvum* oocysts in the filtered effluent

   $$N = 300 \text{ oocysts/100 L}$$

   Calculate the removal efficiency on percentage basis.

   $$\text{Percent removal} = \frac{N_0 - N}{N_0} \times 100\% = \frac{100{,}000 \text{ oocysts/100 L} - 300 \text{ oocysts/100 L}}{100{,}000 \text{ oocysts/100 L}} \times 100\% = 99.7\%$$

   The removal efficiency of *Cryptosporidium parvum* oocysts by the filter is 99.7%.

2. Estimate the removal efficiency on log reduction basis.

   The log removal efficiency can be estimated from the following two methods.

   a. Direct method.

      The log removal (or reduction) is calculated directly from Equation 11.2b.

      $$\text{log removal} = -\log(N/N_0) = -\log\left(\frac{300}{100{,}000}\right) = 2.52 \text{ log}$$

b. Indirect method.

Calculate log concentrations in the filter influent and effluent from Equation 11.1.

$$\text{log concentration per 100 L in the filter influent} = \log(N_0) = \log(100{,}000) = 5 \log$$

$$\text{log concentration per 100 L in the filter effluent} = \log(N) = \log(300) = 2.48 \log$$

Calculate the log removal from Equation 11.2c.

$$\text{log removal} = \log(N_0) - \log(N) = (5 - 2.48) \log = 2.52 \log$$

The log removal efficiency of *Cryptosporidium parvum* oocysts by the tertiary filter is 2.52 log.

EXAMPLE 11.3: LOG CONCENTRATION AND MPN OF ORGANISMS AFTER TREATMENT PROCESS

A process assessment study was conducted to determine the microbial concentration remaining in the effluent from a tertiary plant. The process train includes primary and secondary treatment processes followed by dual media filtration. The monitoring records indicate that the total coliform count in the raw wastewater is 4×10^7 MPN/100 mL. The log reductions of total coliforms (MPN) by each treatment process are given below:

Primary treatment	0.25
Secondary treatment	2.00
Dual media filtration	2.50

Estimate the number of total coliforms (MPN) in the effluents from the primary, secondary, and tertiary treatment system.

Solution

1. Calculate the log concentration of total coliforms in the raw wastewater from Equation 11.1.

$$\log(N_0) = \log(4 \times 10^7) = 7.60 \log$$

2. Estimate the log concentration remaining after primary, secondary, and tertiary treatment processes.

a. Estimate the log concentration remaining in the primary effluent.

Rearrange Equation 11.2c and calculate the log concentration remaining after 0.25 log reduction during primary treatment.

$$\log(N_1) = \log(N_0) - \text{log reduction by primary treatment} = (7.60 - 0.25) \log = 7.35 \log$$

Rearrange Equation 11.1 and calculate the MPN (or concentration remaining) in the primary effluent.

$$N_1 = 10^{\log(N1)} = 10^{7.35} = 22{,}400{,}000 \text{ or } 2.24 \times 10^7 \text{ MPN/100 mL}$$

b. Estimate the log concentration remaining in the secondary and tertiary effluents.

Apply the procedure in Step 2.a to obtain the log concentration remaining and MPNs in the secondary and tertiary effluents, respectively.

$$\log(N_2) = \log(N_1) - \text{log reduction by secondary treatment} = (7.35 - 2.00)\ \log = 5.35\ \log$$

$$N_2 = 10^{\log(N2)} = 10^{5.35} = 224{,}000 \text{ or } 2.24 \times 10^5 \text{ MPN/100 mL}$$

$$\log(N_3) = \log(N_2) - \text{log reduction by tertiary treatment} = (5.35 - 2.50)\ \log = 2.85\ \log$$

$$N_3 = 10^{\log(N3)} = 10^{2.85} = 708 \text{ or } 7.08 \times 10^2 \text{ MPN/100 mL}$$

3. Summarize the estimation results for log concentration and MPN.

Location	log reduction	log concentration	Total coliforms concentration, MPN/100 mL
Raw wastewater influent	–	7.60	40,000,000 or 4×10^7
Primary effluent	0.25	7.35	22,400,000 or 2.24×10^7
Secondary effluent	2.00	5.35	224,000 or 2.24×10^5
Tertiary effluent from dual medial filtration	2.50	2.85	708 or 7.08×10^2

11.2.3 Regulatory Process and Requirements

Under the National Pollutant Discharge Elimination System (NPDES), a site-specific effluent discharge permit is required for all publically owned treatment works (POTWs). The discharge permit includes limitations on wastewater flow, and concentrations and mass loadings of specified pollutants. The U.S. Environmental Protection Agency has delegated the implementing authority of the NPDES to the individual states. The delegated states are responsible to establish the standards and compliance strategy for disinfection based on site-specific discharge limitations. Specific microbial criteria and monitoring requirements have also been established by the federal and many state governments for reuse of reclaimed effluent for many potential applications.

Discharge Limits: The individual states have variable and site-specific bacteria limitations based on coliforms for discharging effluent. The ranges can be from nondetectable to 5000 MPN/100 mL for fecal coliforms and from 2.2 to 10,000 MPN/100 mL for total coliforms. The most common standard is 200 and 1000 MPN/100 mL fecal coliforms for primary- and secondary-body contact recreation uses.

Criteria for Reuse of Reclaimed Effluent: The criteria for reuse of reclaimed effluent are much more stringent than those for effluent discharge. The guidance on desired microbial ranges as well as the minimum treatment requirements for reuse of reclaimed wastewater is covered in Chapter 12.

Design and Monitoring Requirements: The disinfection is the last barrier at a plant against microbial safety. For this reason, the design and monitoring requirements for effluent disinfection in the regulations are based on certain parameters that are properly specified and monitored. The parameters for chemical disinfection are based on disinfection properties of the chemical, and include the residual, contact time, initial mixing energy input, and the hydraulic characteristics of the reactor. The most important parameter for the UV disinfection process is the target dose. Detailed information about these parameters for major disinfection processes are covered in several examples later in this chapter.

11.3 Disinfection Techniques

The disinfection method broadly falls into three major categories: (1) chemical, (2) physical, and (3) radiation. Disinfection by radiation using gamma rays has been studied for a long time but is not practical for wastewater application. Chemical and physical processes have been successfully used for disinfection of wastewater effluent. Limited microbial removal is also achieved by natural die-off in a lagoon. These processes are described below.

11.3.1 Chemical Disinfection Processes

Chemical processes have been the traditional approach for disinfection of both drinking water and wastewater effluent. A chemical is fed into the effluent to maintain a desired residual for a desired contact time. Many chemicals are used for different disinfection applications. These chemicals include oxidants, alcohols, phenol and phenolic compounds, salts and heavy metals, quaternary ammonium compounds, soaps and detergents, and alkalis and acids.[42] The chemicals for disinfection of wastewater effluent are normally limited to strong oxidants that include halogen-based disinfectants, ozone (O_3), hydrogen peroxide (H_2O_2), potassium permanganate ($KMnO_4$), and peracetic acid (PAA). Hydrogen peroxide and $KMnO_4$ have been effectively used for odor control and color removal. They are rarely used for disinfection alone. Peracetic acid ($C_2H_4O_3$) is a mixture of acetic acid (CH_3COOH) and H_2O_2. It is a strong disinfectant with relatively high oxidation potential. One of the major advantages is that PAA does not react with TOC to form DBPs. The key concern of using PAA for wastewater disinfection is adding organics in the effluent.[43–45] The halogen-based disinfectants and ozone are most commonly used for chemical disinfection of wastewater. Recently, new technologies have been explored to use *mixed oxidants* for more effective disinfection than a single conventional disinfectant.[46,47]

Halogen-Based Disinfectants: The chlorine-based disinfectants include liquid/gas chlorine (Cl_2), sodium hypochlorite (NaOCl), calcium hypochlorite ($Ca(OCl)_2$), and chlorine dioxide (ClO_2). Chlorine dioxide is a stronger oxidant than chlorine and does not react with natural organic matters (NOMs) to produce trihalomethanes (THMs). However, application of chlorine dioxide is limited in wastewater disinfection because of its short residual life due to quick decomposition in sunlight and high costs associated with on-site generation. Other alternative halogen-based disinfectants include bromine, bromine chloride, and iodine. These chemicals are rarely used for general purpose of wastewater disinfection.[42,48,49] Liquid or gas chlorine and sodium hypochlorite are the most widely used halogen-based disinfectants.

Ozone: Ozone is a highly effective disinfectant that has been used for odor and color control as well as disinfection of water supply. Disinfection of wastewater effluent by ozone over chlorine is growing in recent years because of concerns over chlorinated organics, toxicity of chlorine residual to aquatic life in the receiving waters, and added cost of dechlorination.[42,48]

11.3.2 Physical Disinfection Processes

Common physical disinfection processes include heat (pasteurization), membrane, ultrasonic, and UV radiation. Heat or pasteurization is widely used in food processing, beverage and pharmaceutical industries, as well as medical practice. Heat is however cost prohibitive for wastewater disinfection. At a proper intensity, ultrasonic cavitation can effectively kill organisms by rupturing the cell wall. It is however limited to academic studies as well as small-scale applications. Membrane process for wastewater treatment is gaining popularity in recent years. It is effective for removal of larger organisms simply because of pore size. The application of membrane process is discussed in details in Chapter 15. UV irradiation is an excellent disinfectant and is now an effective and practical technology for disinfection of wastewater effluent. UV irradiation is discussed in Section 11.9.

11.3.3 Suitability of Disinfection Processes

The most commonly used technique for wastewater application is chlorination with or without dechlorination. Dechlorination is needed to reduce the toxicity of chlorine residual on aquatic life in the receiving water. Common chemicals used for dechlorination are sulfur dioxide and sodium bisulfite. Activated carbon adsorption may also be used as a physical dechlorination process. Ozone and UV irradiation are the effective emerging technologies that eliminate adverse effects of chlorine on the environment.[49] Suitability of chlorination with chlorine (Cl_2) or sodium hypochlorite (NaOCl), ozonation and UV irradiation are compared in Table 11.4.[2,42,48–57] The characteristics of an ideal disinfectant are availability, noncorrosive and nonscaling, nontoxic to higher forms of life, high solubility, homogeneity, stability, low side reactions, deep penetration, and safety.

Disinfection effectiveness is usually the most important consideration to assess the suitability of disinfectants. Chlorine, ozone, and UV irradiation are very effective for inactivation of bacterial pathogens and indicator organisms. Ozone is the best for inactivation of viruses, and UV irradiation for control of protozoa. A high UV irradiation target dose may be required for inactivation of some viruses and bacteria spores, including *Clostridium perfringens*. Chlorination is ineffective for control of viruses and protozoa. UV irradiation is ineffective for control of helminthes. Limited removal of helminth ova can be achieved by chlorination or ozonation processes. Tertiary treatment need to be considered seriously if removal of helminthes is a major concern. Probable log reduction of various organisms by chlorination, ozone, and UV irradiation are compared in Table 11.5.[5,10,13,44,58–63]

The predisinfection effluent quality is another major factor that may also affect the suitability of disinfection technologies, operation and maintenance (O&M), and costs. The predisinfection treatment and effluent quality criteria recommended for effective pathogens reduction by chlorination, ozone, and UV irradiation are summarized in Table 11.6.[9,55–57,64]

11.3.4 Chlorine and Ozone Doses for Required Disinfection

The disinfection efficiency for chlorination depends upon dosage, type and number of organisms, contact time, chlorine residual, and nature of liquid disinfected. Organic matter, ammonia, and reducing agents react with chlorine thus reducing its effectiveness. Suspended solids protect entrapped organisms. The typical chlorine and ozone dosages for reduction of total coliform in wastewater effluents from different processes are compared in Table 11.7.[2,42,48,52–54]

11.3.5 Disinfection By-products

The reaction of free chlorine with organic compounds in the effluent results in the production of many undesirable DBPs. These are halogenated organics or total organic halides (TOX). Among these are THMs, haloacetic acids (HAAs), haloaceticnitriles (HANs), chlorophenols, and chlorinated aldehydes, ketones, and others. Bromate concentration may also be elevated after ozonation if high bromide concentration is present in the predisinfection effluent. Many of these compounds are persistent in nature, and their reported toxicological effects in water supply are carcinogenic, mutagenic, genotoxic, heptotoxic, neurotoxic, and like.[63]

11.4 Inactivation Mechanisms

The inactivation mechanism for disinfection may be different depending upon the types of target organism as well as the chemical or physical agent selected. The impacts of these factors can be assessed by applicable microbial inactivation kinetics. For nondisinfection processes, microbial removal occurs simply through natural die-off. When disinfectants are used, different mechanisms may cause microbial inactivation. Some disinfectants have inhibition effect on special enzymes. Some may produce radicals that oxidize cellular material directly. Others may damage the cell wall, destroy proteins, or attack nucleic acid. The potential mechanisms that have been suggested for different disinfectants are summarized in Table 11.8.[2,42,49,55–57]

TABLE 11.4 Suitability of Common Disinfection Technologies for Wastewater Effluent

Application and Consideration	Chlorination		Ozone	UV Irradiation
	Cl_2	NaOCl		
Disinfection Effectiveness				
Oxidation potential at 25°C, volt	−1.36	1.49 (HOCl)	−2.07	–
Bacteria	Good	Good	Very good	Good
Viruses	Poor	Poor	Very good	Fair
Protozoa (cysts and oocysts)	Poor	Poor	Good	Very good
System Requirements				
Bulk supply	Common	Available	Not applicable	Not applicable
On-site generation	Not common	Common	Required	Required
Facility footprint	Large	Large	Medium to large	Small
Equipment reliability	Good	Good	Fair to good	Fair to good
Process Concerns				
Contact time	Long	Long	Moderate	Short
Stability of agent	Stable	Unstable	Unstable	Stable
Solubility	Moderate	High	High	Not applicable
Operational Concerns				
Process control	Well developed	Well developed	Developing	Developing
O&M intensity	Minimal	Minimal	High	Moderate
Energy consumption	Low	Low	High	Moderate
Capital	Low	Low to moderate	High	Moderate to high
O&M	Low	Low	High	Moderate
Impact of Predisinfection Effluent Quality				
BOD_5 and NOM	Increase demand	Increase demand	Increase demand	Increase UV absorption
TSS	Moderate	Moderate	Moderate	High
Hardness and TDS	Not significant	Not significant	Not significant	Fouling deposits
Alkalinity	Minor, consumed	Minor, produced	Minor	Affects UV absorbance
Fe, Mn, and H_2S	Increase demand	Increase demand	Increase demand	Decrease UV absorption
Ammonia	Reacts	Reacts	Reacts at high pH	Minor
Nitrite	Increases demand	Increases demand	Minor	Minor
Other Beneficial Applications				
Odor and color control	Common	Applicable	Good	No
Sludge bulking control	Common	Applicable	Not applicable	Not applicable
Removal of organics	Moderate	Moderate	Good	No
Potential Adverse Effects				
Corrosive	Yes	Yes	No	No
Explosive/flammable	Yes	No	No	No
Formation of DBPs	High	High	Medium	Negligible
U.S. EPA, RMP	Covered chemical	Not covered	Not covered	Not covered
Safety concern	High	Low	Moderate	Low
Overdosing concerns	Yes	Yes	Minor	No
Toxicity to aquatic life	High	High	None	None
Bacteria regrowth	Low	Low	Moderate to high	Moderate to high

(*Continued*)

TABLE 11.4 (*Continued*) Suitability of Common Disinfection Technologies for Wastewater Effluent

Application and Consideration	Chlorination		Ozone	UV Irradiation
	Cl_2	NaOCl		
Persistent by-products	Yes	Yes	None	None
Impact on TDS	Increase	Increase	No	No
Impact on pH	Reduce pH	Raise pH	None	None

Note: DBP = disinfection by-product; NOM = natural organic matter; O&M = operation and maintenance; RMP = Risk Management Program; TDS = total dissolved solids.

Source: Adapted from References 2, 42, and 48 through 57.

One of the most important concerns when using UV irradiation for wastewater disinfection is that certain indicator organisms and pathogens may have the capability of reversing the UV-induced DNA damages and regaining their infectivity during a critical period through the following two potential mechanisms:

- Photoreactivation using specific repairing enzymes after exposure to the visible light between 300 and 500 nm
- Photoreactivation or dark repair in the absence of light.

Due to lack of specific repairing enzymes in viral DNA, viruses may need to get the enzymes from the host cell to perform repair and therefore their repairing process is more complicated and slower than that of bacteria. Reactivation of pathogenic protozoa is unlikely after effective inactivation by UV irradiation. The amount of repairs required to regain their infectivity may not occur within the critical period.

11.5 Inactivation Kinetics

Microbial removal or inactivation kinetics is expressed by the rate of kill, that is, the change in the number of organisms with time. It is usually developed from a die-off coefficient or rate constant. For a nondisinfection treatment process, the microbial removal kinetics may be simply a function of time. When a disinfectant is involved, a residual or intensity is normally involved in the development of microbial inactivation kinetics. The residual-based kinetic is used to express microbial inactivation by chemical disinfection processes. UV irradiation is a physical disinfection process where the dosage is expressed by intensity-based kinetics.

11.5.1 Natural Die-Off Kinetics

Microbial inactivation can occur through natural die-off. The microbial die-off with time in natural water body can be expressed by Equation 11.3.[42,49]

$$N = \frac{1}{1 + k_d t_d} N_0 \qquad (11.3)$$

where

N = number of organisms present at time = t_d, number of organisms per unit sample volume
N_0 = number of organisms present initially at $t = 0$, number of organisms per unit sample volume
k_d = die-off coefficient or rate constant of a given type of organism, d^{-1}
t_d = average hydraulic retention time based on the flow through the water body, d

Equation 11.3 is also used to estimate the microbial reduction by enhanced die-off through a nondisinfection treatment processes such as natural treatment as well as primary or secondary treatment.[42] Many

TABLE 11.5 Probable Microbial Log Reductions in Effluent by Common Disinfection Technologies

Disinfection Technology	Log Reduction									
	Viruses		Bacteria					Protozoa		Helminth
	Enteric Viruses	Coliphages	Total Coliform	Fecal Coliform	*E. coli*	*Enterococci*	*Clostridium perfringens*	*Cryptosporidium parvum*	*Giardia lamblia*	Ova, Cyst, and Oocyst
Chlorination[a]	1–3	0–2.5	2–6	2–6	2–6	2–6	1–2	0–0.5	0.5–1.5	0–1
Ozone	3–6	2–6	2–6	2–6	2–6	2–6	0–0.5	1–2	2–4	0–2
UV irradiation	1– > 3	3–6	2–> 4	2–> 4	2–> 4	2–> 4	< 0.5	> 3	> 3	0

[a] Includes use of either Cl_2 or NaOCl.

Source: Adapted in part from References 5, 10, 13, 44, and 58 through 63.

TABLE 11.6 Predisinfection Treatment and Effluent Quality Criteria Recommended for Common Wastewater Disinfection Technologies

Disinfection Technology	Predisinfection Treatment	Predisinfection Effluent Quality			
		BOD_5, mg/L	TSS, mg/L	Turbidity, NTU	Other Parameter
Chlorination[a]	No requirement but prefer at least primary treatment	<15–20	<15–20	<5–10	Ammonia concentration must be assessed. High chlorine dose is required if free chlorine residual is desired.
Ozone	Usually requires secondary treatment with nitrification	<15–20	<10–15	<5	An ammonia concentration <1 mg/L as NH_3-N may be required to reduce initial ozone demand if effluent has high pH.
UV irradiation	Require tertiary treatment with filtration or membrane	<15–20	<5–10	<5	Water quality parameters required to ensure effective UV irradiation and to reduce the fouling potential due to scaling: • UV transmittance >50–70% or UV absorbance <0.15–0.3 cm^{-1} • Hardness <140 mg/L as $CaCO_3$, • Iron <0.1 mg/L as Fe, and • Hydrogen sulfide <0.2 mg/L as H_2S.

[a] Includes use of either Cl_2 or NaOCl.

Source: Adapted from References 9, 55 through 57, and 64.

environmental factors have impact on the enhanced pathogen die-off process. These factors may include temperature, retention time, algal toxins, predation, adsorption, sedimentation, filtration, photolysis, and solar UV irradiation.[17,18] The die-off rate constant may therefore vary significantly with actual conditions in a given treatment process. For instance, a rate constant of 0.5 and 0.03 d^{-1} has been reported for coliforms in a lagoon system under summer and winter conditions, respectively.[54]

TABLE 11.7 Typical Chlorine and Ozone Dosages Required to Achieve Total Coliform Reduction in Effluent from Various Treatment Processes[a,b]

Effluent from Treatment Process	Total Coliform, MPN/100 mL	Chlorine (Ozone) Dosage, mg/L			
		Effluent Standard, MPN/100 mL			
		1000	200	23	≤2.2
Septic tank	10^7–10^9	15–40 (15–40)	–	–	–
Raw wastewater	10^7–10^9	15–30 (15–40)	–	–	–
Primary	10^7–10^9	10–30 (10–40)	20–40	–	–
Trickling filter	10^5–10^6	6–10 (4–10)	12–20	20–30	–
Activated sludge	10^5–10^6	4–10 (4–10)	10–15 (4–8)	13–20 (16–30)	(30–40)
Filtered activated sludge	10^4–10^6	4–8 (6–10)	5–15 (4–8)	10–20 (16–25)	13–25 (30–40)
Filtered nitrified	10^4–10^6	4–10 (3–6)	6–12 (3–6)	8–14 (4–15)	8–16 (15–20)
Intermittent sand filter	10^2–10^4	1–5 (4–8)	2–8 (10–15)	5–10 (12–20)	8–15 (16–25)
Microfiltration	10–10^3	1–3 (2–6)	2–4 (2–6)	3–6 (3–8)	4–10 (4–8)

[a] The contact time for chlorination is 30 min and that for ozonation is 15 min.

[b] Chlorine dosages are based on the disinfection requirements of using chloramine residual.

Source: Adapted in part from References 2, 42, 48, and 52 through 54.

TABLE 11.8 Mechanisms of Common Wastewater Disinfection Technologies

Organism	Chlorination	Ozonation	UV Irradiation
Bacteria	Inhibits specific enzymes	Totally or partially destroys the cell membrane and causes lysis of the cell; and attacks enzymes to interrupt their activities	Causes photochemical damage to the cellular nucleic acids (deoxyribonucleic acid (DNA) or ribonucleic acid (RNA)) from the dimerization of pyrimidine molecules and destroys their replication abilities; and causes permanent damages to proteins, enzymes, coenzymes, and hormones through photochemical reactions
Viruses	Causes damages to either nucleic acids, the viral coat protein or both	Breaks viral capsid, particularly its proteins to liberate the nucleic acid and inactivate the DNA	Similar to that for bacteria
Protozoa	Not very effective	Totally or partially destroys the cyst wall and makes it more permeable; penetrates into the cyst and damages the plasma membrane; and attacks the nucleus ribosomes and other ultrastructural components	Similar to that for bacteria

Source: Adapted in part from References 2, 42, 49, and 55 through 57.

EXAMPLE 11.4: NATURAL DIE-OFF COLIFORMS AND REMOVAL EFFICIENCY

Wastewater collected from a residential park is treated in a stabilization pond system. Based on the results of a field study, a natural die-off rate constant of $0.4\ d^{-1}$ was obtained for coliforms in the lagoon system. The initial coliform count in the raw wastewater is 10^7 MPN/100 mL. Estimate the number of coliforms in the effluent from the lagoon system. Also calculate the coliform reduction efficiency of the pond system on both percentage and log bases. The basic information about the residential area and the stabilization pond system is summarized below:

Estimated population	600
Daily average wastewater generation rate	100 gallons per capita per day (gpcd)
Total surface area of the stabilization pond system	2 acres
Average water depth in the stabilization pond system	5 ft

Solution

1. Estimate daily average flow from the residential area.

$$\text{Daily average flow} = 100\ \text{gpcd} \times 600\ \text{persons} = 60{,}000\ \text{gpd or } 0.06\ \text{MGD}$$

2. Calculate total volume of the stabilization pond system.

$$\text{Total surface area of the pond system} = 43{,}560\ \text{ft}^2/\text{acre} \times 2\ \text{acres} = 87{,}120\ \text{ft}^2$$

$$\begin{aligned}\text{Total volume of the pond system} &= \text{total surface area} \times \text{average water depth} = 87{,}120\ \text{ft}^2 \times 5\ \text{ft}\\ &= 435{,}600\ \text{ft}^3 \text{ or } 435{,}600\ \text{ft}^3 \times 7.48\ \text{gal/ft}^3 = 3260{,}000\ \text{gal or } 3.26\ \text{Mgal}\end{aligned}$$

3. Calculate average hydraulic retention time in the pond system.

$$t_d = \frac{3.26 \text{ Mgal}}{0.06 \text{ MGD}} = 54 \text{ d}$$

4. Estimate the number of coliforms in the effluent.
   Apply Equation 11.3, the number of coliforms in the effluent,

$$N = \frac{1}{1 + kt_d} N_0 = \frac{1}{1 + 0.4 \text{ d}^{-1} \times 54 \text{ d}} \times 10^7 \text{ MPN/100 mL} = 4.4 \times 10^5 \text{ MPN/100 mL}$$

5. Calculate the coliform reduction efficiency of the pond system.
   a. Calculate the coliform percentage reduction efficiency.

$$\text{Percent reduction efficiency} = \frac{N_0 - N}{N_0} \times 100\%$$

$$= \frac{(10^7 - 4.4 \times 10^5) \text{ MPN/100 mL}}{10^7 \text{ MPN/100 mL}} \times 100\% = 96\%$$

   b. Calculate the coliform log reduction efficiency from Equation 11.2b.

$$\text{log reduction} = -\log(N/N_0) = -\log\left(\frac{4.4 \times 10^5 \text{ MPN/100 mL}}{10^7 \text{ MPN/100 mL}}\right) = 1.4 \log$$

Therefore, it is estimated that the coliform count in the effluent from the pond system is about 4.4×10^5 MPN/100 mL, and the coliform reduction efficiency is ~96% or 1.4 log.

11.5.2 Inactivation Kinetics for Chemical Disinfection Processes

Many important kinetic models have been developed to express the inactivation of organisms by chemical disinfectants. These kinetic models have been developed from chlorination studies, although models have been used for other oxidants, including ozone. These models are broadly divided into (1) first-order reaction-based models, (2) empirical models, (3) rationale models, and (4) mathematical models. These models are briefly presented below. Additional information for inactivation models is also presented for disinfection by ozonation in Section 11.8.6. The microbial inactivation equations for UV radiation are presented in Section 11.9.4.

First-Order Reaction-Based Models: During the early studies, the microbial inactivation by chlorination was assumed to follow the chemical reaction kinetics or the first-order reaction. This assumption was well applied in the Chick's law (Equation 11.4a). This earlier approach was then modified to include the effects of the chlorine residual on the microbial inactivation. On this premise, the Chick-Watson model was developed (Equation 11.4b). This equation is also recognized as the "mixed second-order" model that utilized the basic concept of "*Ct*" (residual concentration × contact time), also known as *Ct* or *CT* value. It was later confirmed that the actual microbial inactivation observed in disinfection practice deviated from the results obtained from the Chick-Watson model.[49,65] Many factors may be responsible for these deviations. Major factors are listed below:

- Disinfectant residual is not constant.
- Continuous and plug-flow conditions are more realistic than the batch and complete-mix conditions.
- Change in pH may affect reaction rate.

- Disinfectant may also be consumed by many competitive reactions.
- Resistance to disinfectant may develop.
- Initial microbial density may have impact on disinfection efficiency.

Two simplest deviations from Chick-Watson model were identified and known as *shoulder* (or *lag*) and *tailing off*. These phenomena are illustrated in Figure 11.1. The lag or shoulder effect is due to reaction and loss of disinfectant with many constituents in the liquid. The tailing effect is due to occulation, shielding, or shadowing of organisms by suspended solids. To express the nonheat-kill-related temperature dependence, the Arrhenius correction was later integrated into the Chick-Watson model (Equation 11.4c).

The Hom equation was developed to express these deviations and has been quite successful in describing the inactivation kinetics of a variety of pathogens, including viruses and protozoa.[49,55,65,66] Significant *tailing off* effect has been observed in disinfection by ozone in a batch mode because of rapid decay of ozone after application. The Hom model with $m < 1$ (Equation 10.4d) is considered better than other models to describe ozone inactivation of parasite cysts and oocysts, including *Giardia* and *Cryptosporidium*.[67,68] Haas equation (Equation 10.4e) further modified Hom model to include the effect of first-order decay of disinfectant with time on microbial inactivation.[69–72]

Chick's model $$N = N_0\, e^{-kt} \quad \text{or} \quad \frac{N}{N_0} = e^{-kt} \quad \text{or} \quad \ln\left(\frac{N}{N_0}\right) = -kt \tag{11.4a}$$

Chick-Watson $$N = N_0\, e^{-k'C^n t} \quad \text{or} \quad \frac{N}{N_0} = e^{-k'C^n t} \tag{11.4b}$$

Chick-Watson with temperature correction

$$N = N_0 \exp\left(-k'_{20}\beta^{(T-20)} C^n t\right) \tag{11.4c}$$

Hom $$N = N_0 \exp[-k'' C^n t^m] \quad \text{or} \quad \frac{N}{N_0} = \exp[-k'' C^n t^m] \tag{11.4d}$$

Haas $$N = N_0 \exp\left[-\left(\frac{m}{nk_c}\right)^m k'' C_0^n \times \left(1 - e^{\frac{-nk_c t}{m}}\right)^m\right] \tag{11.4e}$$

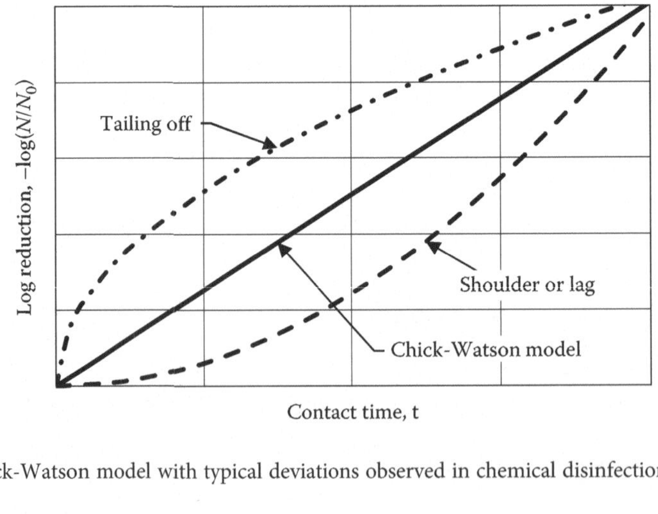

FIGURE 11.1 Chick-Watson model with typical deviations observed in chemical disinfection processes.

where

k = inactivation rate constant of a given type of organism due to a disinfectant, min^{-1}

k' = modified inactivation rate constant of a given type of organism due to a disinfectant, $(\text{L/mg})^n/\text{min}$

C = disinfectant residual, mg/L

n = coefficient obtained experimentally ($n > 0$, dimensionless. The significance of constant n in determining the activation efficiency is as follows. If $n = 1$, both C and t are equally important and the product Ct is actually more important than either C or t; and if $n > 1$, C is the dominant factor. If $n < 1$, t is the dominant factor.

t = contact time, min

k'_{20} = modified inactivation rate constant of a given type of organism due to a disinfectant at 20°C, $(\text{L/mg})^n/\text{min}$

β = empirical temperature-dependent constant found experimentally

T = temperature in the contact basin, °C

k'' = modified inactivation rate constant of a given type of organism and disinfectant, $(\text{L/mg})^n/\text{min}^m$

m = coefficient obtained experimentally, dimensionless. The significance of constant m on model behavior is as follows: (a) it is identical with the Chick-Watson model when $m = 1$, (b) it expresses the *shoulder* deviation when $m > 1$, and (c) it expresses the tailing off deviation when $m < 1$.

C_0 = initial disinfectant concentration, mg/L

k_c = first-order decay rate of disinfectant, min^{-1}. The disinfectant concentration with time is expressed by equation $C = C_0 e^{-k_c t}$ (Equation 2.15a).

Empirical Models: All first-order reaction-based models have been developed through academic development efforts. Experimental data for inactivation of different organisms by different disinfectants under predetermined experimental conditions have been used, including the "*Ct*" ranges. For a better expression of the actual microbial inactivation observed in wastewater disinfection practice, the original first-order model was heavily modified using actual experimental data. Developed on the basis of the "*Ct*" concept, three empirical models are given by Equation 11.5.[42,48,49,73–75] It should be noted that the similar empirical constants used in these equations may not be related to each other.[49] These models have been considered helpful design tools to estimate microbial inactivation by chlorine for disinfection of wastewater.[49,54,73] Among the empirical models, the Collins equation was developed from a study of disinfection of primary effluent. It can express a tailing phenomenon. The Selleck equation was developed for chlorine inactivation of coliform in wastewater.

Collins
$$N = N_0 \frac{1}{(1 + 0.23Ct)^3} \tag{11.5a}$$

Collins-Selleck
$$N = N_0 \quad \text{for } Ct \le b \tag{11.5b}$$

$$N = N_0 \left(\frac{Ct}{b}\right)^{-n} \quad \text{for } Ct > b \tag{11.5c}$$

Selleck
$$-\log(N/N_0) = a \log\left(1 + \frac{Ct}{b}\right) \tag{11.5d}$$

where

a = empirical constant determined experimentally, dimensionless

b = empirical constant determined experimentally, mg·min/L. The constant b is a threshold constant. Microbial inactivation can occur only until $Ct > b$. A b value of 4 and 3 is reported for total and fecal coliforms, respectively.
n = coefficient obtained experimentally, dimensionless. An n value of approximately 3 is typically used.
All other terms have been defined previously.

Rational Models: A rational approach has been used to develop specific kinetic models for disinfection by ozone. In these models, the effects of initial microbial density on disinfection are included.[76–79] The Power, and Hom-Power equations are expressed by Equation 11.6.

Power law $$\ln(N/N_0) = -\frac{1}{x-1}\ln\left[1 + N_0^{x-1}(x-1)k'C^n t\right] \tag{11.6a}$$

Hom-Power law $$\ln(N/N_0) = -\frac{1}{x-1}\ln\left[1 + N_0^{x-1}(x-1)k'C^n t^m\right] \tag{11.6b}$$

where
x = coefficient obtained experimentally ($x \neq 1$), dimensionless. The significance of constant x is as follows: it expresses the tailing off phenomenon when $x > 1$, and it expresses the shoulder phenomenon when $x < 1$.
n = coefficient obtained exponentially, dimensionless

Mathematical Models: Many nonideal factors may have either favorable or adverse impacts on the inactivation of organisms by UV irradiation. Mathematical models are suitable to deal with these nonideal factors since matching the data set is the primary goal of this method even though it may not require the model to be mechanistically approved.[80] Two mathematical expressions have been successfully established to account for the "shoulder" behavior in chemical disinfection processes based on available laboratory experimental or actual plant operating data.[81–83] The series-event model (Equation 11.7a) considers that a finite number of lethal events are required to occur for inactivation. The multiple-target model (Equation 11.7b) assumes that the microbial inactivation can only happen when multiple "target sites" are attacked by the disinfectant. A numerical method is usually required to solve these mathematical expressions as well as the rational models.

Series-event $$N = N_0 e^{-kCt}\left(\sum_{i=0}^{l-1}\frac{(kCt)^i}{i!}\right) \tag{11.7a}$$

Multiple target $$N = N_0\left(1 - (1 - e^{-kCt})^{n_c}\right) \tag{11.7b}$$

where
i = the event level
k = inactivation rate constant that is mathematically obtained for the model, L/min·mg
l = a threshold for effective inactivation for an organism
n_c = number of target sites that must be attacked by the disinfectant to inactivate the organisms

Inactivation Efficiency of Various Disinfectants: The inactivation efficiency of different disinfectants and target organisms can be compared on the basis of Ct value (product of disinfectant residual and contact time). Table 11.9 is developed from batch reactor data under controlled conditions to compare the effectiveness of various disinfectants on different organisms. For example, a 3-log inactivation of bacteria and viruses is achieved by free chlorine at Ct of 1.5–3 and 4–5 mg·min/L, respectively. This means that free chlorine is more effective for inactivation of bacteria than for viruses. Likewise, chlorine dioxide is a more effective disinfectant for protozoan cysts than chlorine.

TABLE 11.9 *Ct* Values Required for Inactivation of Organisms by Various Disinfectants

Organism	Disinfectant	*Ct* Value Required for Log Inactivation[a], mg·min/L			
		1-log	2-log	3-log	4-log
Bacteria (total coliform)	Chlorine (free)	0.4–0.6	0.8–1.2	1.2–1.8	1.6–2.4
	Chloramine	50–70	95–150	140–220	200–300
	Chlorine dioxide	0.4–0.6	0.8–1.2	1.2–1.8	1.6–2.4
	Ozone	0.005–0.01	0.01–0.02	0.015–0.02	0.02–0.04
	UV irradiation[b]	10–35	15–40	20–70	30–110
Viruses	Chlorine (free)	–	1.5–1.8	2.2–2.6	3–3.5
	Chloramine	–	370–400	550–600	750–800
	Chlorine dioxide	–	5–5.5	9–10	12.5–13.5
	Ozone	–	0.25–0.3	0.35–0.45	0.5–0.6
	UV irradiation[b]	20–60	25–70	30–120	50–190
Protozoan (*Giardia lamblia* or *Giardia lamblia* cysts)	Chlorine (free)	20–30	45–55	70–80	–
	Chloramine	400–450	800–900	1100–1300	–
	Chlorine dioxide	5–5.5	9–11	15–16	–
	Ozone	0.25–0.3	0.45–0.5	0.75–0.8	–
	UV irradiation[b]	5–10	10–15	10–20	15–30
Protozoan (*Cryptosporidium parvum*)	Chlorine (free)	2000–2600	4000–5000	–	–
	Chloramine	4000–5000	8000–10,000	–	–
	Chlorine dioxide	120–150	235–260	350–400	–
	Ozone	4–4.5	8–8.5	12–13	–
	UV irradiation[b]	10–15	10–15	15–25	20–40

[a] Applicable in filtered secondary effluent at pH 7 and temperature 20°C.
[b] UV dose (D_{UV}) = average UV irradiation intensity (I_{avg}) × exposure time (t) (Equation 11.39).
Source: Adapted in part from References 52 through 55, and 72.

EXAMPLE 11.5: CONTACT TIME FOR MICROBIAL INACTIVATION IN CHICK'S LAW

Batch tests were conducted to determine the contact time of an effluent disinfection facility at an activated sludge plant. At a desired chlorine residual, 99% kill of coliform organisms was achieved in 13 min. Calculate the contact time required to achieve 99.9%.

Solution

1. Determine the inactivation rate constant k.

   At 99% coliform kill, the organisms ratio N/N_0 remaining = 1 − 0.99 = 0.01 and t = 13 min. Use Chick's law (Equation 11.4a) to calculate k.

$$\frac{N}{N_0} = 0.01 = e^{-kt} \quad \text{or} \quad \ln\left(\frac{N}{N_0}\right) = \ln(0.01) = -kt$$

$$-4.6 = -kt \quad \text{or} \quad k = \frac{4.6}{t} = \frac{4.6}{13\text{ min}} = 0.35\text{ min}^{-1}$$

2. Determine the contact time for 99.9% kill.

   At 99.9% coliform kill, the organism ratio N/N_0 remaining $= 1 - 0.999 = 0.001$.

$$t = \frac{\ln(0.001)}{-k} = \frac{-6.9}{-0.35\ \text{min}^{-1}} = 19.7\ \text{min} \approx 20\ \text{min}$$

EXAMPLE 11.6: CHLORINE RESIDUAL AND CONTACT TIME FOR MICROBIAL INACTIVATION IN CHICK-WATSON MODEL

The Chick-Watson model applies to the product of chlorine residual and contact time Ct. Prove by using numerical data that log inactivation has a linear relationship with Ct. Also prove that log reduction decreases linearly as C increases. Assume $n = 1$ and $k' = 10$ L/mg·min.

Solution

1. Develop the linear relationship of log reduction $\log(N/N_0)$ with Ct using Equation 11.4b with $n = 1$.

$$\frac{N}{N_0} = e^{-k'Ct}$$

Take logarithm on both sides and rearrange the expression to have the following log reduction.

$$-\log\left(\frac{N}{N_0}\right) = k'Ct \times \log(e) = k'Ct \times 0.434 = (0.434\ k')Ct \quad \text{or} \quad -\log\left(\frac{N}{N_0}\right) = S_1 Ct$$

A linear relationship between $-\log(N/N_0)$ and Ct is developed with a slope $S_1 = 0.434\ k'$. At $k' =$ 10 L/mg·min the slope $S_1 = 0.434\ k' = 0.434 \times 10$ L/mg·min $= 4.34$ L/mg·min is obtained for the relationship.

Calculate $-\log(N/N_0)$ at $Ct = 0$, 0.125, 0.25, 0.5, and 1 mg·min/L.

Ct, mg·min/L	0	0.125	0.25	0.5	1
$-\log(N/N_0)$	0.00	0.54	1.09	2.17	4.34

Plot the liner relationship in Figure 11.2a.

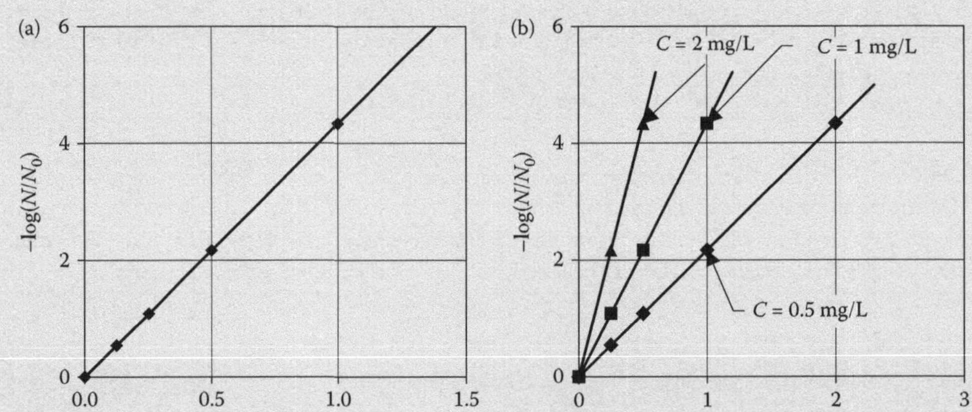

FIGURE 11.2 Log reduction expressions by Chick-Watson model using numerical data: (a) $-\log(N/N_0)$ versus Ct and (b) $-\log(N/N_0)$ versus t (Example 11.6).

2. Develop the linear expression of log reduction with t for disinfection concentration C of 0.5, 1.0, and 2.0.

Apply the Chick-Watson equation and rearrange the linear equation in Step 1.

$$-\log\left(\frac{N}{N_0}\right) = (0.434\,k'C)t \quad \text{or} \quad -\log\left(\frac{N}{N_0}\right) = S_2 t$$

A linear relationship is obtained at each given C. The slope of the line is $S_2 = 0.434\ k'C$ or $S_2 = 0.434 \times 10\ \text{L/mg·min} \times C = 4.34\ \text{L/mg·min} \times C$

Calculate S_2 for each assumed C and $-\log(N/N_0)$ values at different values of t.

Assume $C = 0.5$ mg/L at $t = 0$, 0.25, 0.5, 1, and 2 min.

$$S_2 = 4.34\ \text{L/mg·min} \times C = 4.34\ \text{L/mg·min} \times 0.5\ \text{mg/L} = 2.17\ \text{min}^{-1}$$

t, min	0	0.25	0.5	1	2
$-\log(N/N_0)$	0.00	0.54	1.09	2.17	4.34

Assume $C = 1$ mg/L at $t = 0$, 0.25, 0.5, and 1 min.

$$S_2 = 4.34\ \text{L/mg·min} \times C = 4.34\ \text{L/mg·min} \times 1\ \text{mg/L} = 4.34\ \text{min}^{-1}$$

t, min	0	0.25	0.5	1
$-\log(N/N_0)$	0.00	1.09	2.17	4.34

Assume $C = 2$ mg/L at $t = 0$, 0.25, and 0.5 min.

$$S_2 = 4.34\ \text{L/mg·min} \times C = 4.34\ \text{L/mg·min} \times 2\ \text{mg/L} = 8.68\ \text{min}^{-1}$$

t, min	0	0.25	0.5
$-\log(N/N_0)$	0.00	2.17	4.34

Plot the liner relationships in Figure 11.2b.

3. Discuss the results obtained from the linear plots using numerical data.
   a. The log reduction, expressed as $-\log(N/N_0)$ is a linear function of Ct with a slope $S_1 = 0.434\ k'$ (Figure 11.2a).
   b. At a given C, the log reduction increases with t at a constant rate of $S_2 = 0.434\ k'C$. As C increases the value of S_2 also increases. This means that log reduction at a given t increases rapidly as C is increased (Figure 11.2b).
   c. Both chlorine residual (C) and contact time (t) are equally important for an efficient disinfection. For a given log reduction, one-half of t is required when C is doubled.

EXAMPLE 11.7: CONTACT TIME AND CHLORINE RESIDUAL FOR A DESIRED INACTIVATION EFFICIENCY OBTAINED FROM THE CHICK-WATSON MODEL

The impact of contact time t and chlorine residual C on the inactivation efficiency is evaluated using experimental data from a bench-scale test. The experimental results indicate that the Chick-Watson model (Equation 11.4b) is good to express the *E. coli* inactivation kinetics. From the experimental results, the empirical constants were developed: $k' = 10.9\ (\text{L/mg})^{1.2}/\text{min}$, and $n = 1.2$. In order to achieve an inactivation efficiency of 99.9% or 3-log reduction, estimate (1) the contact time required at a constant chlorine residual of 0.2 mg/L and (2) the chlorine residual at a constant contact time of 10 min.

Solution

1. Determine the required contact time.

   Rearrange Equation 11.4b to obtain the following expression and calculate the contact time t required for a log inactivation efficiency $-\log(N/N_0) = 3$ at $C = 0.2$ mg/L.

$$t = \frac{1}{\log(e)k'C^n}[-\log(N/N_0)] = \frac{1}{0.434 \times 10.9\,(\text{L/mg})^{1.2}/\text{min} \times (0.2\text{ mg/L})^{1.2}} \times 3 = 4.4\text{ min}$$

   A contact time $t = 4.4$ min is required to achieve an inactivation efficiency of 99.9% or 3-log reduction at a chlorine residual $C = 0.2$ mg/L.

2. Determine the required chlorine residual.

   Rearrange Equation 11.4b to obtain the following expression and calculate the chlorine residual C required for a log inactivation efficiency $-\log(N/N_0) = 3$ at $t = 10$ min.

$$C = \left(\frac{1}{\log(e)k't}[-\log(N/N_0)]\right)^{\frac{1}{n}} = \left(\frac{1}{0.434 \times 10.9\,(\text{L/mg})^{1.2}/\text{min} \times 10\text{ min}} \times 3\right)^{\frac{1}{1.2}} = 0.1\text{ mg/L}$$

   A chlorine residual $C = 0.10$ mg/L is required to achieve an inactivation efficiency of 99.9% or 3-log reduction at a contact time $t = 10$ min.

EXAMPLE 11.8: TEMPERATURE-DEPENDENT CONSTANT FOR THE CHICK-WATSON MODEL

A virus disinfection study indicated that the contact time needs to be increased two fold when the temperature is lowered by 10°C to achieve the same inactivation level at a constant chlorine residual. Determine the applicable temperature-dependent constant for the Chick-Watson model based on the observations from the above study.

Solution

1. Develop the temperature-dependent β constant as an expression of contact time t and operating temperature T.

   Using Equation 11.4c to express the survival number of virus N_1 at a chlorine residual C_1, contact time t_1, and temperature T_1.

$$N_1 = N_0 \exp\left(-k'_{20}\beta^{(T_1-20)}C_1^n t_1\right)$$

   Similarly, express the survival number of virus N_2 at C_2, t_2, and T_2.

$$N_2 = N_0 \exp\left(-k'_{20}\beta^{(T_2-20)}C_2^n t_2\right)$$

   At the same inactivation level, $N_1 = N_2$, the following expression is obtained.

$$N_0 \exp\left(-k'_{20}\beta^{(T_1-20)}C_1^n t_1\right) = N_0 \exp\left(-k'_{20}\beta^{(T_2-20)}C_2^n t_2\right)$$

$$\exp\left(-k'_{20}\beta^{(T_1-20)}C_1^n t_1\right) = \exp\left(-k'_{20}\beta^{(T_2-20)}C_2^n t_2\right)$$

   Take natural logarithm on both sides of the equation.

$$-k'_{20}\beta^{(T_1-20)}C_1^n t_1 = -k'_{20}\beta^{(T_2-20)}C_2^n t_2 \quad \text{or} \quad \beta^{(T_1-20)}C_1^n t_1 = \beta^{(T_2-20)}C_2^n t_2$$

At the constant chlorine residual, $C_1 = C_2$, an expression of β as t and T is developed.

$$\beta^{(T_1-20)} t_1 = \beta^{(T_2-20)} t_2 \quad \text{or} \quad \beta^{(T_1-T_2)} = \frac{t_2}{t_1} \quad \text{or} \quad \beta = \left(\frac{t_2}{t_1}\right)^{\frac{1}{T_1-T_2}}$$

2. Estimate the temperature-dependent constant β.

Since the contact time is doubled when the temperature is lowered by 10°C, $\frac{t_2}{t_1} = 2$ and $T_1 - T_2 = 10$°C. Calculate the temperature-dependent constant β.

$$\beta = (2)^{\frac{1}{10}} = 1.072$$

Therefore, the applicable temperature-dependent constant β is 1.072. This will give twice the contact time to achieve the same level of virus deactivation if the reaction temperature is lowered by 10°C at a constant chlorine residual concentration.

EXAMPLE 11.9: DETERMINATION OF EMPIRICAL CONSTANTS IN HOM EQUATION FROM EXPERIMENTAL DATA

A chlorination study was conducted to determine the inactivation of total coliforms. The batch data at different chlorine residuals and contact times are given below:

Inactivation $(1 - N/N_0)$ at the residual concentration (%)						
Time (t), min	1	2	4	8	16	32
$C_1 = 0.05$ mg/L	57	65	75	85	91	96
$C_2 = 0.1$ mg/L	68	80	87	92.5	97.5	99
$C_3 = 0.2$ mg/L	81	88.5	94	97.5	99.3	–
$C_4 = 0.4$ mg/L	90	95	97.9	99.5	–	–

Determine the empirical constants of k'', n, and m in Hom equation (Equation 11.4a).

Solution

1. Develop the linear expression for determination of empirical constant m.

Take logarithm on both sides of Equation 11.4d.

$$-\log\left(\frac{N}{N_0}\right) = k''C^n t^m \times \log(e) = k''C^n t^m \times 0.434 = (0.434\, k''C^n) t^m$$

Assume $z = 0.434\, k''C^n$ at a given C to obtain the following expression.

$$-\log\left(\frac{N}{N_0}\right) = z\, t^m$$

Take logarithm on both sides to have the linear expression.

$$\log\left[-\log\left(\frac{N}{N_0}\right)\right] = m \log(t) + \log(z)$$

In this linear relationship, m is the slope and $\log(z)$ is the intercept on Y axis.

2. Prepare the values for linear plots from the experimental data.

At each chlorine residual, calculate $\log[-\log(N/N_0)]$ and $\log(t)$, and tabulate the calculation results below.

At $C_1 = 0.05$ mg/L	t, min	1	2	4	8	16	32
	$1 - N/N_0$	0.57	0.65	0.75	0.85	0.91	0.96
	N/N_0	0.43	0.35	0.25	0.15	0.09	0.04
	$\log(t)$	0	0.301	0.602	0.903	1.204	1.505
	$\log[-\log(N/N_0)]$	−0.436	−0.341	−0.220	−0.084	0.019	0.145
At $C_2 = 0.1$ mg/L	t, min	1	2	4	8	16	32
	$1 - N/N_0$	0.68	0.80	0.87	0.925	0.975	0.99
	N/N_0	0.32	0.20	0.13	0.075	0.025	0.01
	$\log(t)$	0	0.301	0.602	0.903	1.204	1.505
	$\log[-\log(N/N_0)]$	−0.306	−0.156	−0.053	0.051	0.205	0.301
At $C_3 = 0.2$ mg/L	t, min	1	2	4	8	16	
	$1 - N/N_0$	0.81	0.885	0.94	0.975	0.993	
	N/N_0	0.19	0.115	0.06	0.025	0.007	
	$\log(t)$	0	0.301	0.602	0.903	1.204	
	$\log[-\log(N/N_0)]$	−0.142	−0.027	0.087	0.205	0.333	
At $C_4 = 0.4$ mg/L	t, min	1	2	4	8		
	$1 - N/N_0$	0.9	0.95	0.98	0.995		
	N/N_0	0.1	0.05	0.02	0.005		
	$\log(t)$	0	0.301	0.602	0.903		
	$\log[-\log(N/N_0)]$	0	0.114	0.230	0.362		

3. Plot $\log[-\log(N/N_0)]$ versus $\log(t)$.

Plot $\log[-\log(N/N_0)]$ versus $\log(t)$ data for each chlorine residual in Figure 11.3.

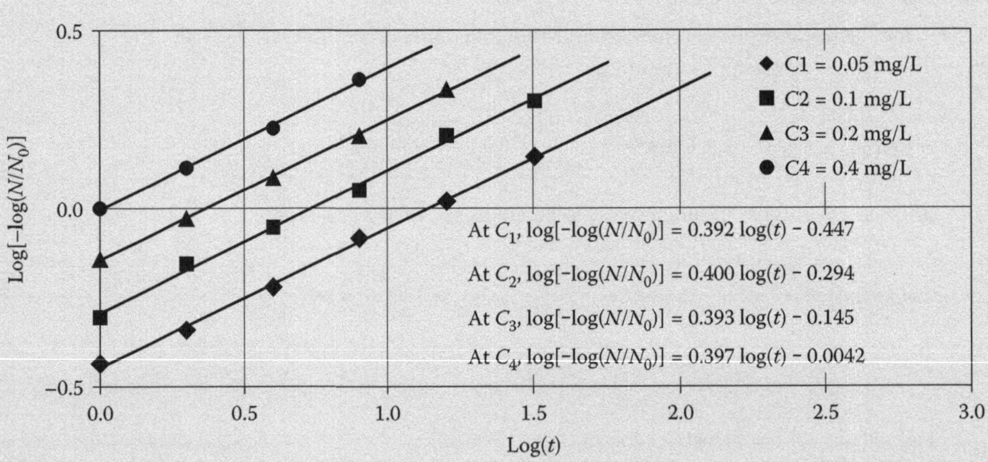

FIGURE 11.3 Plots of $\log[-\log(N/N_0)]$ versus $\log(t)$ at different chlorine residuals (Example 11.9).

From the plots, the following relationships are obtained.

C, mg/L	0.05	0.1	0.2	0.4
Slope, *m*	0.392	0.400	0.393	0.397
Intercept, log(*z*)	−0.447	−0.294	−0.145	−0.0042

Calculate the average slope or constant *m*.

$$m = 0.396 \approx 0.40$$

4. Develop the linear expression for determination of empirical constants k'' and n.
Take logarithm on both sides of the assumption $z = 0.434\, k''C^n$ to obtain the following expression.

$$\log(z) = n \log(C) + \log(0.434\, k'')$$

A plot of log(*z*) versus log(*C*) gives a linear relationship in which *n* is the slope and log(0.434 k'') is the intercept.
Tabulate log(*z*) and log(*C*) from the results obtained from Step 2.

C, mg/L	0.05	0.1	0.2	0.4
log(*C*)	−1.30	−1	−0.699	−0.398
log(*z*)	−0.447	−0.292	−0.145	−0.0042

5. Plot log(*z*) versus log(*C*).
The linear relationship between log(*z*) and log(*C*) is shown in Figure 11.4.

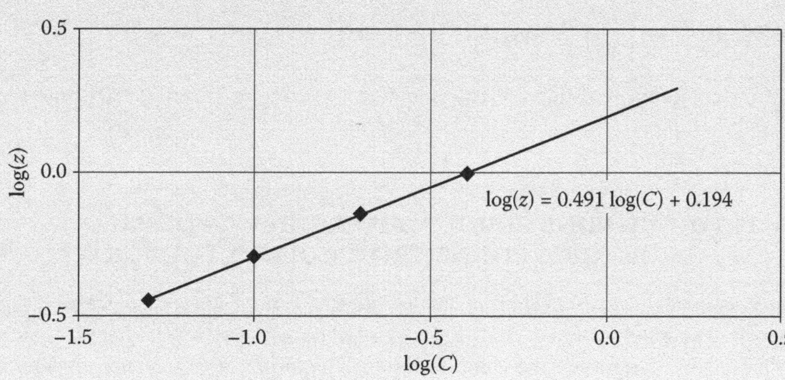

FIGURE 11.4 Plot of log(*z*) versus log(*C*) (Example 11.9).

From the plot, Slope, $n = 0.491 \approx 0.5$

Intercept, $\log(0.434\, k'') = 0.194$

Solve,

$$0.434\, k'' = 10^{0.194}$$

$$k'' = \frac{10^{0.194}}{0.434} = \frac{1.56}{0.434} = 3.59 \approx 3.6\ (\text{L/mg})^{0.5}/\text{min}^{0.4}$$

6. Summarize the developed kinetic coefficients.
The kinetic coefficients in Hom equation for inactivation of total coliforms with chlorine are: $m = 0.4$, $n = 0.5$, and $k'' = 3.6\ (\text{L/mg})^{0.5}/\text{min}^{0.4}$. The Hom equation is therefore expressed as $N = N_0 e^{-3.6C^{0.5}t^{0.4}}$.

EXAMPLE 11.10: MICROBIAL INACTIVATION FROM HAAS EQUATION

The Haas equation (Equation 11.4e) includes the effect of first-order decay of disinfectant. In a chlorination study of coliform kill, the following kinetic coefficients were obtained for a calibrated the Haas equation: $m = 0.4$, $n = 0.5$, $k'' = 3.0$, $k_c = 0.8\ \text{min}^{-1}$. Determine the log inactivation of coliform. The chlorine dose is 0.5 mg/L and contact time is 3 min.

Solution

Rearrange Equation 11.4e and substitute the kinetic coefficients in the Haas equation to determine the N/N_0 at chlorine dose of 0.5 mg/L and contact time of 3 min.

$$\frac{N}{N_0} = \exp\left[-\left(\frac{m}{nk_c}\right)^m k'' C_0^n \times \left(1 - e^{\frac{-nk_c t}{m}}\right)^m\right]$$

$$= \exp\left[-\left(\frac{0.4}{0.5 \times 0.8}\right)^{0.4} \times 3.0 \times 0.5^{0.5} \times \left(1 - e^{-\frac{0.5 \times 0.8 \times 3}{0.4}}\right)^{0.4}\right]$$

$$= \exp[-(1)^{0.4} \times 3.0 \times 0.707 \times (1 - e^{-3})^{0.4}] = \exp[-2.12 \times (0.950)^{0.4}] = \exp(-2.12 \times 0.980)$$

$$= e^{-2.08} = 0.125$$

Calculate the percent inactivation.

$$\text{Percent inactivation} = \left(1 - \frac{N}{N_0}\right) \times 100\% = (1 - 0.125) \times 100\% = 87.5\%$$

Calculate the log inactivation.

$$\text{log inactivation} = -\log\left(\frac{N}{N_0}\right) = -\log(0.125) = 0.9 \log$$

Therefore, the coliform inactivation at chlorine dose $C_0 = 0.5$ mg/L and contact time $t = 3$ min is 87.5% or 0.9 log.

EXAMPLE 11.11: CHLORINE RESIDUAL TO ACHIEVE A DESIRED COLIFORM COUNT FROM COLLINS EQUATION

Chlorination is evaluated for disinfection of treated effluent from a wastewater treatment plant. The filtered effluent had a coliform count of 10^4 MPN/100 mL. The discharge requirement for coliform count is ≤ 200 MPN/100 mL. Determine the chlorine residual required for disinfection of filtered effluent to meet the coliforms requirement at an average contact time of 20 min. Assume that the Collins equation applies.

Solution

Rearrange Equation 11.5a and estimate the chlorine residual (C) required to meet the coliforms requirement in the disinfected filtered effluent.

$$C = \frac{1}{0.23t}\left[\left(\frac{N_0}{N}\right)^{\frac{1}{3}} - 1\right] = \frac{1}{0.23\ \text{L/mg·min} \times 20\ \text{min}}\left[\left(\frac{10^4\ \text{MPN/100 mL}}{200\ \text{MPN/100 mL}}\right)^{\frac{1}{3}} - 1\right]$$

$$= \frac{1}{4.6\ \text{L/mg}} \times (50^{\frac{1}{3}} - 1) = \frac{3.68 - 1}{4.6\ \text{L/mg}}$$

$$= 0.58\ \text{mg/L as}\ Cl_2$$

EXAMPLE 11.12: DETERMINATION OF EMPIRICAL CONSTANTS IN COLLINS–SELLECK EQUATION FROM EXPERIMENTAL DATA

In an environmental laboratory class, a disinfection kinetic experiment was performed at a constant chlorine residual of 1 mg/L. The geometric mean density of fecal coliforms is 2×10^4 MPN/100 mL in the treated effluent sample. The experimental results are summarized below.

Time (t), min	5	10	20	30	40
Coliform count (N), MPN/100 mL	3200	390	50	14	5

Determine the empirical constants n and b, and validate the Collins–Selleck equation for disinfection of treated effluent using chlorine.

Solution

1. Develop the linear expression.

Assume that the Collins–Selleck equation (Equation 11.5c) is applicable ($Ct > b$). Take the logarithm on both sides of the equation and rearrange the expression.

$$-\log\left(\frac{N}{N_0}\right) = n\log(Ct) - n\log(b)$$

A plot of $-\log(N/N_0)$ versus $\log(Ct)$ gives a linear relationship in which n is the slope and $-n\log(b)$ is the intercept.

Tabulate calculation results at $C = 1$ mg/L and $N_0 = 2\times10^4$ MPN/100 mL.

t, min	5	10	20	30	40
Ct, mg·min/L	5	10	20	30	40
$\log(Ct)$	0.699	1.00	1.30	1.48	1.60
N/100 mL	3200	390	50	14	5
N/N_0	0.16	0.0195	0.0025	0.00070	0.00025
$-\log(N/N_0)$	0.796	1.71	2.60	3.15	3.60

2. Plot $-\log(N/N_0)$ versus $\log(Ct)$.

The linear relationship between $-\log(N/N_0)$ and $\log(Ct)$ is shown in Figure 11.5.

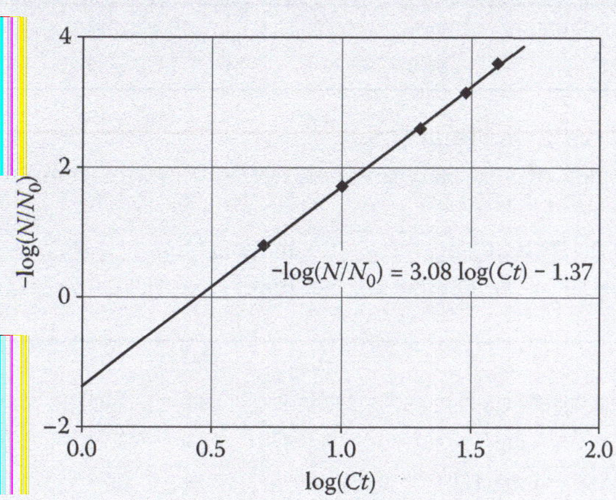

FIGURE 11.5 Plot of $-\log(N/N_0)$ versus $\log(Ct)$ using Collins-Selleck equation (Example 11.12).

3. Determine n and b from the plot.

From the plot, Slope, $n = 3.08$

Intercept, $-n\log(b) = -1.37$ or $n\log(b) = 1.37$

Solve, $b = 10^{\frac{1.37}{n}} = 10^{\frac{1.37}{3.08}} = 10^{0.445} = 2.79$ mg·min/L

4. Validate the equation.

Calculate the minimum value of Ct_{min} at $C = 1$ mg/L and $t = 5$ min (the shortest contact time used during the experiment).

$Ct_{min} = 1\text{ mg/L} \times 5\text{ min} = 5\text{ mg·min/L} > b = 2.79\text{ mg·min/L}$

Therefore, the experimental results indicate that the Collins-Selleck equation is applicable to express the kinetics of disinfection. The empirical form of Collins-Selleck equation for $Ct > 2.79$ mg·min/L is given below:

$$N = N_0\left(\frac{Ct}{2.79}\right)^{-3.08} \quad \text{or} \quad N = 23.6N_0(Ct)^{-3.08}$$

EXAMPLE 11.13: CONTACT TIME AND CHLORINE RESIDUAL ON MICROBIAL SURVIVAL FROM SELLECK EQUATION

The experimental results indicate that the Selleck equation applies to express the total coliform inactivation kinetics. The empirical constants are: $a = 2.1$ and $b = 0.1$ mg·min/L. Describe the effect of chlorine residual (C) and contact time (t) on the log reduction of total coliforms ($-\log(N/N_0)$).

Solution

1. Develop the log reduction of total coliforms ($-\log(N/N_0)$) as a function of Ct.

Apply Equation 11.5d $-\log(N/N_0) = a\log\left(1 + \frac{Ct}{b}\right)$ at $a = 2.1$ and $b = 0.1$ mg·min/L to calculate $-\log(N/N_0)$ at several assumed Ct values.

Ct, mg·min/L	0	0.125	0.25	0.5	1	2
$-\log(N/N_0)$	0.00	0.74	1.14	1.63	2.19	2.78

Plot $-\log(N/N_0)$ versus Ct in Figure 11.6.

2. Develop the log reduction of total coliforms ($-\log(N/N_0)$) as a function of t for a given chlorine concentration C.

At $C = 0.5$ mg/L, 1 mg/L, and 2 mg/L, apply Equation 11.5d at $a = 2.1$ and $b = 0.1$ mg·min/L to prepare a table of $-\log(N/N_0)$ at different contact times $t = 0$, 0.25, 0.5, 1, 2, and 4 min.

Contact time t, min	0	0.25	0.5	1.0	2.0	4
$-\log(N/N_0)$ at $C = 0.5$ mg/L	0	0.74	1.14	1.63	2.19	2.78
$-\log(N/N_0)$ at $C = 1$ mg/L	0	1.14	1.63	2.19	2.78	–
$-\log(N/N_0)$ at $C = 2$ mg/L	0	1.63	2.19	2.78	–	–

Plot $-\log(N/N_0)$ versus t at different C in Figure 11.7.

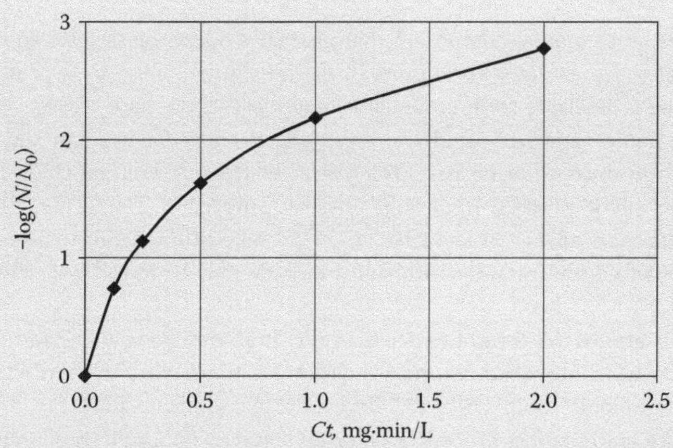

FIGURE 11.6 Plot of $-\log(N/N_0)$ versus Ct using Selleck equation (Example 11.13).

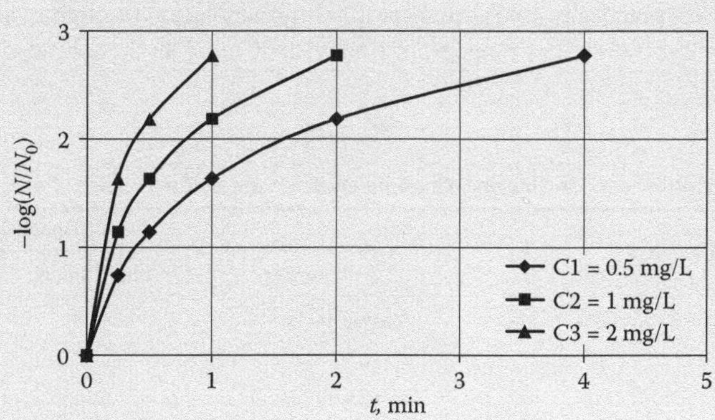

FIGURE 11.7 Plot of $-\log(N/N_0)$ versus t at different C values using Selleck equation (Example 11.13).

3. Describe the effect of Ct, C, and t on log reduction of organisms.
   a. The log reduction, that is, $-\log(N/N_0)$ is a nonlinear function of Ct in a *tailing off* trend. The increasing rate of log reduction is reduced with an increase in Ct.
   b. C and t are equally important for an efficient disinfection. For a given log reduction, C should be doubled if the contact time t is halved, and vice versa.
   c. When C is a constant, the log reduction shows also a *tailing off* trend with t.
   d. In general, the log reduction efficiency is improved when C is elevated at a given contact time t.

11.6 Chlorination

Chlorination is achieved by using liquid or gaseous chlorine (Cl_2), sodium hypochlorite (NaOCl), calcium hypochlorite ($Ca(OCl)_2$), or chlorine dioxide (ClO_2). *Dechlorination* is used to trim chlorine residual to meet the residual limit in the effluent for final discharge. The most common chemicals for dechlorination are sulfur dioxide (SO_2) and sodium bisulfite ($NaHSO_3$). The physical, chemical, and disinfection properties of Cl_2 and NaOCl are presented below.

11.6.1 Physical Properties of Chlorine and Hypochlorite

Chlorine: Chlorine is an elemental compound that normally occurs as Cl_2. It can be present in either gaseous or liquid states. At standard conditions (0°C and 1 atm), chlorine is present as an elemental gas. Liquefied chlorine is available from commercial suppliers in pressure vessels. When its pressure is reduced enough, the liquid vaporizes rapidly to a gas that is ready for use. On-site generation systems are also available to produce chlorine gas (99%) directly from a sodium chloride solution using a membrane-based electrolytic process. Important physical properties of chlorine, chlorine dioxide, and sulfur dioxide are summarized in Table 11.10.[42,48–55,84,85] Additional information about disinfection with chlorine dioxide and dechlorination with sulfur dioxide may be found in Sections 11.6.5 and 11.7, respectively.

Gaseous chlorine is almost 2.5 times heavier than air. It accumulates at the lowest point after being released to the atmosphere. Liquefied chlorine is about 1.5 times heavier than water. One liter (0.26 gal) of liquid chlorine weights about 1.5 kg (3.2 lbs) and can produce ~460 L (16 ft^3) of chlorine gas at 0°C and 1 atm. In the chlorination process, a chlorine solution is usually prepared from the chlorine gas and then fed at the application point. Chlorine is soluble in water with a maximum solubility of ~10,000 mg/L or 1% near 10°C (50°F) and 1 atm. Its solubility decreases rapidly with increase in temperature. To avoid excessive emission of gaseous chlorine from the solution, a maximum concentration limit of 3500 mg/L or 0.35% is practically used to prepare the chlorine solution through a vacuum injector. Normally, the chlorine concentration may range from 1000 to 2500 mg/L in the solution. Chlorine is a highly

TABLE 11.10 Physical Properties of Chlorine, Chlorine Dioxide, and Sulfur Dioxide

Property	Value for Chemical		
	Chlorine	Chlorine Dioxide	Sulfur Dioxide
General			
Molecular weight, g	70.9	67.5	64.1
Chemical form in use	Gas, liquid, or solution	Gas or solution	Gas, liquid, or solution
Odor	Distinct and irritating	Acrid	Pungent
Color of gas	Greenish-yellow	Yellow to reddish	Colorless
Vapor			
Specific gravity at 0°C (32°F), 1 atm and dry air = 1, dimensionless	2.49	1.86	2.26
Density at 0°C (32°F) and 1 atm, kg/m^3 (lb/ft^3)	3.21 (0.201)	2.40 (0.150)	2.93 (0.183)
Specific volume at 0°C (32°F) and 1 atm, m^3/kg (ft^3/lb)	0.311 (4.99)	0.417 (6.67)	0.342 (5.47)
Solubility at 20°C (68°F) and 1 atm, kg/m^3 (lb/gal)	7.3 (0.061)	70 (0.58)	100 (0.83)
Liquid			
Specific gravity at 0°C (32°F) and water = 1, dimensionless	1.47	-	1.44
Density at 20°C (68°F), kg/m^3 (lb/gal)	1410 (11.8)	-	1380 (11.5)
Specific volume at 20°C (68°F), L/kg (gal/lb)	0.71 (0.085)	-	0.72 (0.087)
Gas yield at 0°C (32°F) and 1 atm			
Yield per unit liquid volume, m^3/L (ft^3/gal)	0.46 (61)	-	0.49 (65)
Yield per unit liquid weight, m^3/kg (ft^3/lb)	0.31 (5.0)	-	0.34 (5.4)

Note: 1 atmosphere (atm) = 101.325 kPa = 760 mm Hg = 14.696 psi 1 kg/m^3 = 0.06243 lb/ft^3 = 0.008346 lb/gal.
Source: Adapted in part from References 42, 48 through 55, 84, and 85.

toxic substance that poses potential health risks to the plant personnel and off-site general publics. Therefore, a special risk management program (RMP) must be implemented under the federal regulations to ensure safe design and operation of chlorination process when the usage of liquid chlorine is exceeding a threshold quantity at the plant.

Sodium Hypochlorites: Sodium and calcium hypochlorites are hypochlorite salts. Their chemical formulas are NaOCl and $Ca(OCl)_2$. Calcium hypochlorite is available as powder, granules, and pellets. Calcium is the hardness causing component and tends to clog pipes and filters. It reacts with carbon dioxide and produces calcium carbonate precipitate. Also, it is not highly soluble in water. For all these reasons, its use is limited to on-site disinfection and disposal of effluent, in swimming pools, and as bleaching powder. On account of safety reasons, there has been a trend of replacing chlorine gas with sodium hypochlorite solution for disinfection at many larger plants in the populated urban areas.

Chlorine Content of Sodium Hypochlorite: Sodium hypochlorite is a strong oxidant and is obtained from commercial suppliers in solutions of many strengths. The chlorine content of the chemical is expressed as either actual or available chlorine.

The *weight percent actual chlorine* in a chlorine-containing disinfectant is obtained from Equation 11.8a. It is a simply measurement of chlorine content of the compound by weight.[85]

$$(\text{Weight \% Cl})_{\text{chemical}} = \frac{w_{\text{Cl}}}{mw_{\text{chemical}}} \times 100\% \tag{11.8a}$$

where

$(\text{Weight \% Cl})_{\text{chemical}}$ = weight percent actual chlorine (Cl) of the pure chemical, % by wt
w_{Cl} = weight of chlorine per mole chemical, g Cl/mole
mw_{chemical} = molecular weight of the pure chemical, g/mole

Since the measurement of actual chlorine does not provide direct information about the effectiveness in the disinfection process, the strength of the compound is typically measured by the *available chlorine.* It is a measure of the equivalent *oxidizing* or *disinfection* power of the chemical in comparing with that of Cl_2. The half oxidation reaction for Cl_2 to chloride ion (Cl^-) is shown in Equation 11.8b. In this reaction, the chlorine valance changes from 0 to −1 by accepting *one* electron (e^-). Therefore, the each chlorine element (Cl) in Cl_2 has an oxidizing power of accepting *one* (1) electron.

$$Cl_2 + 2\,e^- \rightarrow 2\,Cl^- \tag{11.8b}$$

$$NaOCl + 2\,e^- + H^+ \rightarrow Na^+ + Cl^- + HO^- \tag{11.8c}$$

The half oxidation reaction for NaOCl to chloride ion (Cl^-) is shown in Equation 11.8c. In this reaction, the chlorine valance changes from +1 to −1. As a result, each chlorine element in NaOCl molecule has an oxidizing power of accepting *two* (2) electrons. Therefore, the oxidizing power of chlorine in NaOCl is double that of chlorine. The chlorine equivalent of NaOCl, $eq_{\text{Cl}} = 2$ g Cl_2/g NaOCl. The *weight percent available chlorine* of the chemical is expressed by Equation 11.8d or 11.8e.[85]

$$(\text{Weight \% Cl}_2)_{\text{chemical}} = eq_{\text{Cl}} \times (\text{Weight \% Cl})_{\text{chemical}} \tag{11.8d}$$

$$(\text{Weight \% Cl}_2)_{\text{chemical}} = \frac{eq_{\text{Cl}}w_{\text{Cl}}}{mw_{\text{chemical}}} \times 100\% \tag{11.8e}$$

where

$(\text{Weight \% Cl}_2)_{\text{chemical}}$ = weight percent available chlorine (Cl_2) of the pure NaOCl, % by wt
eq_{Cl} = chlorine equivalent of the pure chemical, g Cl_2/g NaOCl

Therefore, the weight percent actual and available chlorine of pure NaOCl are:

$$(\text{Weight \% Cl})_{NaOCl} \text{ (from Equation 11.8a)} = \frac{35.5 \text{ g Cl/mole NaOCl}}{74.4 \text{ g/mole NaOCl}} \times 100\%$$
$$= 47.7\% \text{ or } 0.477 \text{ g Cl/g NaOCl}$$

$$(\text{Weight \% Cl}_2)_{NaOCl} = 2 \times 47.7\% \text{ (from Equation 11.8d)} = 95.4\% \text{ or } 0.954 \text{ g Cl}_2\text{/g NaOCl}$$

$$(\text{Weight \% Cl}_2)_{NaOCl} \text{ (from Equation 11.8e)} = \frac{2 \text{ g Cl}_2\text{/g Cl} \times 35.5 \text{ g Cl/mole NaOCl}}{74.4 \text{ g/mole NaOCl}} \times 100\%$$
$$= 95.4\% \text{ or } 0.954 \text{ g Cl}_2\text{/g NaOCl}$$

The molecular weight, chlorine equivalent, and weight percent actual and available chlorine in different chlorine-containing chemicals are given in Table 11.11.[48,49,51–54,84,85]

Concentration of Sodium Hypochlorite: The concentration of sodium hypochlorite solution may be expressed as either available chlorine or sodium hypochlorite.

As a common practice, the available chlorine concentration in a chemical solution is typically expressed by (a) *weight percent*, (b) *trade percent*, or (c) *grams per liter (gpl) available chlorine.* The weight percent available chlorine is the grams of available chlorine per gram of the chemical solution. The trade percent available chlorine is the weight of available chlorine (g/L as available Cl_2) over the weight of water per same unit volume (1000 g/L as H_2O). The gpl available chlorine is expressed as the grams of available chlorine per liter of the chemical solution. The relationships between these available chlorine concentrations are given by Equation 11.9a when the specific gravity of a chemical solution is known.[86]

$$(\text{Weight \% Cl}_2)_{\text{solution}} = \frac{(\text{Trade \% Cl}_2)_{\text{solution}}}{\text{sp. gr.}_{\text{solution}}} = \frac{C_{\text{Cl2,solution}}}{\rho_{\text{water}} \times \text{sp. gr.}_{\text{solution}}} \times 100\% \qquad (11.9a)$$

where

$(\text{Weight \% Cl}_2)_{\text{solution}}$ = weight percent available chlorine of the chemical solution, % by wt
$(\text{Trade \% Cl}_2)_{\text{solution}}$ = trade percent available chlorine of the chemical solution, % by wt
$\text{sp. gr.}_{\text{solution}}$ = specific gravity of the chemical solution at water = 1, unitless
$C_{\text{Cl2,solution}}$ = gpl available chlorine of the chemical solution, g/L (kg/m^3)
ρ_{water} = density of water, g/L (kg/m^3)

Similar to the available chlorine, the strength of sodium hypochlorite solution may also be directly expressed by the concentration of NaOCl in the solution. The relationships between sodium hypochlorite

TABLE 11.11 Actual and Available Chlorine Contents of Chlorine-Containing Chemicals

Chlorine-Containing Chemical	Molecular Weight (*mw*), g/mole	Chlorine Equivalent (eq_{Cl}), g Cl_2/g Cl	Weight Percent Actual Chlorine ($(\text{Weight \% Cl})_{\text{chemical}}$), %	Weight Percent Available Chlorine ($(\text{Weight \% Cl}_2)_{\text{chemical}}$), %
Chlorine, Cl_2	70.9	1	100	100
Sodium hypochlorite, NaOCl	74.4	2	47.7	95.3
Calcium hypochlorite, $Ca(OCl)_2$	143	2	49.6	99.2
Hypochlorous acid, HOCl	52.5	2	67.5	135
Chlorine dioxide, ClO_2	67.45	5	52.6	263
Monochloramine, NH_2Cl	51.5	2	68.9	138
Dichloramine, $NHCl_2$	85.9	2	82.5	165

Source: Adapted in part from References 48, 49, 51–54, 84, and 85.

and available chlorine concentrations are expressed by Equations 11.9b and 11.9c.[86]

$$(\text{Weight \% NaOCl})_{\text{solution}} = \frac{(\text{Weight \% Cl}_2)_{\text{solution}}}{(\text{Weight \% Cl}_2)_{\text{NaOCl}}} \times 100\% \tag{11.9b}$$

$$C_{\text{NaOCl,solution}} = \frac{C_{\text{Cl2,solution}}}{(\text{Weight \% Cl}_2)_{\text{NaOCl}}} \times 100\% \tag{11.9c}$$

$$C_{\text{NaOCl,solution}} = \rho_{\text{water}} \times \text{sp. gr.}_{\text{solution}} \times (\text{Weight \% NaOCl})_{\text{solution}} \tag{11.9d}$$

where

$(\text{Weight \% NaOCl})_{\text{solution}}$ = weight percent NaOCl of the solution, % by wt
$(\text{Weight \% Cl}_2)_{\text{NaOCl}}$ = weight percent available chlorine (Cl_2) of the pure NaOCl, 95.3 % by wt
$C_{\text{NaOCl,solution}}$ = gpl NaOCl concentration of the solution, g/L (kg/m^3)

Sodium hypochlorite solution is heavier than water and can be obtained from commercial supplies at a *trade percent* or *trade content* between 4% and 20% for available chlorine. At a trade percent of 12%, the specific gravity of the solution is about 1.17. The density of the solution is therefore 1170 kg/m^3 (9.8 lbs/gal) and the weight percent available chlorine is 10.3%. The trade percent is a very convenient practice to use in wastewater disinfection practice since 1 gal of 12% trade solution provides ~1 lb of available chlorine (9.8 lbs/gal × 0.103 lb Cl_2/lb solution = 1.01 lb Cl_2/gal ≈ 1 lb Cl_2/gal solution). See Example 11.15 for conversion calculations between different concentrations of a NaOCl solution.

Sodium hypochlorite is a strong oxidant, and is very reactive with many compounds and reducing substances. As an unstable weak base, an increase in acidity can cause a rapid breakdown of the hypochlorite in the presence of metallic ions. To improve its stability, excess caustic soda (NaOH) up to 10 g/L is normally added into a commercial sodium hypochlorite solution to raise the pH in the range of 11–13.5. Decomposition of sodium hypochlorite solution is accelerated significantly upon exposure to heat and UV from sunlight. These characteristics need to be considered thoroughly and properly during transportation, storage, and use of sodium hypochlorite solution. Important physical properties of NaOCl at typical trade percent available chlorine are summarized in Table 11.12.[48,49,51–54,84,86]

TABLE 11.12 Physical Properties of Sodium Hypochlorite at Typical Trade Percent Available Chlorine

Parameter	Typical Value at Different Trade Percent Available Chlorine ($(\text{Trade \% Cl}_2)_{\text{solution}}$)			
	4%	8%	12%	16%
$\text{sp. gr.}_{\text{solution}}$ at water = 1 at 20°C (68°F), unitless	1.06	1.12	1.17	1.21
Available chlorine concentration				
$(\text{Weight \% Cl}_2)_{\text{solution}}$, %	3.8	7.1	10.3	13.2
$C_{\text{Cl2,solution}}$, g/L or kg/m^3 (lb/gal) as available Cl_2	40 (0.33)	80 (0.67)	120 (1.00)	160 (1.33)
NaOCl concentration				
$(\text{Weight \% NaOCl})_{\text{solution}}$, %	4.0	7.5	10.8	13.9
$C_{\text{NaOCl,solution}}$, g/L or kg/$m^3$ (lb/gal) as NaOCl	42 (0.35)	84 (0.70)	126 (1.05)	168 (1.40)
Density, kg/m^3 (lb/gal) as solution	1058 (8.8)	1118 (9.3)	1168 (9.7)	1208 (10.1)
Specific volume				
L/kg (gal/lb) as available Cl_2	25 (3.0)	12.5 (1.5)	8.3 (1.0)	6.3 (0.75)
L/kg (gal/lb) as solution	0.95 (0.113)	0.89 (0.107)	0.86 (0.103)	0.83 (0.099)

Note: A water density of 998.2 g/L at 20°C is used (Table B.2 in Appendix B).
Source: Adapted in part from References 48, 49, 51–54, 84, and 86.

EXAMPLE 11.14: VOLUME OF LIQUID CHLORINE STORAGE FACILITY

A chlorine storage facility is designed for storage of 100,000 lbs (45,400 kg) of liquid chlorine. Estimate the bulk storage tank volume required at the facility.

Solution

1. Determine the volume requirement for the liquid chlorine.

   The Chlorine Institute guidelines suggest that the chlorine tank volume shall be at least 192.2 gal for each ton of liquid chlorine stored.[87] Calculate the Cl_2 storage volume requirement (req_{Cl2}) using a conservative value of 200 gal/ton.

$$req_{Cl2} = \frac{1 \text{ ton} \times 2000 \text{ lb/ton}}{200 \text{ gal/ton}} = 10 \text{ lb/gal } (1200 \text{ kg/m}^3)$$

   Note: The guidelines of Chlorine Institute suggest that bulk tank should be able to hold the design quantity of liquid chlorine with only ~95% full at a temperature of 122°F (50°C). There should be no release of chlorine from a pressure relief device set at 255 psig.[87]

2. Calculate the total liquid chlorine volume and select the tank volume.

   Calculate the total volume required to store the design quantity $w_{Cl2} = 100{,}000$ lbs.

$$V_{Cl2} = \frac{w_{Cl2}}{req_{Cl2}} = \frac{100{,}000 \text{ lbs}}{10 \text{ lb/gal}} = 10{,}000 \text{ gal } (38 \text{ m}^3)$$

   Provide two bulk storage tanks, each tank of capacity $V_{tank} = 5000$ gal (approx. 19 m^3).

3. Determine the limit for bulk shipment.

   The maximum storage capacity of each tank (w_{tank}).

$$w_{tank} = req_{Cl2} V_{tank} = 10 \text{ lb/gal} \times 5000 \text{ gal} = 50{,}000 \text{ lbs or } 25 \text{ tons } (22{,}700 \text{ kg})$$

   Calculate the maximum shipment limit at a safety factor of 1.25.

$$w_{shipment} = \frac{w_{tank}}{\text{S.F.}} = \frac{25 \text{ tons}}{1.25} = 20 \text{ tons } (18{,}100 \text{ kg})$$

   Note: A safety factor of 1.2 is provided in the guidelines of the Chlorine Institute, Inc.[87] For safe operation the tank truck should be less than 20 tons to receive and hold the entire shipment. This limit is within the typical range of 15–22 tons for the chlorine tank truck capacities in North America.[87] A larger storage tank volume may be required if the shipment size used by the local chemical suppliers exceeds this limit.

EXAMPLE 11.15: AVAILABLE CHLORINE IN SODIUM HYPOCHLORITE SOLUTION

Commercial sodium hypochlorite solution is used for effluent disinfection at a wastewater treatment facility. The solution has a trade percent available chlorine of 12%. At a solution temperature of 20°C, estimate (a) the weight percent available Cl_2, (b) gpl available chlorine concentration, (c) weight percent NaOCl, (d) gpl NaOCl concentration, (e) density of the solution, and (f) specific volume of the solution.

Solution

1. Estimate the weight percent available chlorine, $(\text{Weight \% } Cl_2)_{solution}$ in the solution.

From Table 11.12, a typical specific gravity sp. $gr._{solution} = 1.17$ is obtained for the sodium hypochlorite solution at $(\text{Trade } \% \ Cl_2)_{solution} = 12\%$. Apply Equation 11.9a to calculate $(\text{Weight } \% \ Cl_2)_{solution}$.

$$(\text{Weight } \% \ Cl_2)_{solution} = \frac{(\text{Trade } \% \ Cl_2)_{solution}}{sp.\ gr._{solution}} = \frac{12\%}{1.17} = 10.3\%$$

2. Estimate the gpl available chlorine concentration ($C_{Cl_2,solution}$) in the solution.

   The water density ρ_{water} at $20°C = 998.2$ g/L (Table B.2 in Appendix B). Rearrange Equation 11.9a and calculate the gpl available chlorine concentration $C_{Cl2,solution}$.

$$C_{Cl2,solution} = \frac{(\text{Trade } \% \ Cl_2)_{solution}}{100\%} \times \rho_{water} = \frac{12\%}{100\%} \times 998.2\,\text{g/L} = 120\,\text{g/L} \quad \text{or} \quad 120\,\text{kg/m}^3$$

3. Estimate the weight percent NaOCl $(\text{Weight } \% \ NaOCl)_{solution}$ in the solution from Equation 11.9b.

$$(\text{Weight } \% \ NaOCl)_{solution} = \frac{(\text{Weight } \% \ Cl_2)_{solution}}{(\text{Weight } \% \ Cl_2)_{NaOCl}} \times 100\% = \frac{10.3\%}{95.3\%} \times 100\% = 10.8\%$$

4. Estimate the gpl NaOCl concentration ($C_{NaOCl,solution}$) in the solution from Equation 11.9c.

$$C_{NaOCl,solution} = \frac{C_{Cl2,solution}}{(\text{Weight } \% \ Cl_2)_{NaOCl}} \times 100\% = \frac{120\,\text{g/L}}{95.3\%} \times 100\% = 126\ \text{g/L}$$

5. Estimate the density of the solution ($\rho_{solution}$).

   The density of the solution can be calculated from sp. $gr._{solution} = 1.17$ and $\rho_{water} = 998.2$ g/L.

$$\rho_{solution} = sp.\ gr._{solution} \times \rho_{water} = 1.17 \times 998.2\ \text{g/L} = 1168\ \text{g/L} \quad \text{or} \quad 1168\ \text{kg/m}^3$$

6. Estimate the specific volume of the solution (sp. $vol._{solution}$).

   The specific volume is the invert of the density of the solution.

$$sp.\ vol._{solution} = \frac{1}{\rho_{solution}} = \frac{1}{1168\ \text{g/L}} \times 10^3\text{g/kg} = 0.86\ \text{L/kg}$$

EXAMPLE 11.16: PERCENT LOSS OF AVAILABLE CHLORINE DUE TO DEGRADATION OF SODIUM HYPOCHLORITE SOLUTION

The available chlorine content of sodium hypochlorite solution is continuously monitored in a bulk storage tank. The results indicate that the available chlorine concentration is 107 g/L as Cl_2 in the solution after a storage period of 2 weeks. What is the percent loss of available chlorine due to decomposition if the initial trade percent available chlorine was 12%? Ignore the change in density with temperature.

Solution

1. Determine the initial weight percent available chlorine concentration of the NaOCl solution as g/L.

   The initial gpl available chlorine concentration $C_{Cl2,solution,initial} = 120$ g/L is obtained from Example 11.15, Step 2.

2. Estimate the percent loss of available chlorine at the end of storage period.

Calculate the percent loss of Cl_2 at the end gpl available chlorine concentration $C_{Cl2,solution,end} =$ 107 g/L after 2-week storage.

$$\text{Percent loss } Cl_2 = \frac{C_{Cl2,solution,initial} - C_{Cl2,solution,end}}{C_{Cl2,solution,initial}} \times 100\% = \frac{(120-107)\,\text{g/L}}{120\,\text{g/L}} \times 100\% = 11\%$$

Comments: Considerable decomposition of high strength NaOCl solution may occur during storage. For this reason, the storage period is typically <15 days for NaOCl solution with a strength >10%. For high strength NaOH solution, residual chlorine analyzer or oxidation-reduction potential (ORP) monitor can be provided in the tank. An automatic feed control system can also be used to compensate for the loss of strength due to decomposition.

EXAMPLE 11.17: SODIUM HYPOCHLORITE SOLUTION FEED RATE REQUIRED AT A DESIRED CHLORINE CONSUMPTION RATE

A NaOCl solution with 12% trade percent available chlorine concentration is used to disinfect a secondary effluent. Estimate (a) the daily quantity of NaOCl solution required to meet a chlorine consumption rate of 454 kg/d (1000 lbs/d) and (b) the set point for each pump if four identical chemical feed pumps are provided to deliver the desired chemical feed rate.

Solution

1. Determine the initial weight percent available chlorine concentration from feed of NaOCl solution as g/L.

   The gpl available chlorine concentration $C_{Cl2,solution} = 120$ g/L or kg/m^3 (1.0 lb/gal) is obtained from Step 2 of Example 11.15 for a trade percent Cl_2 of 12%.
2. Determine the daily total quantity of solution required.

$$Q_{total} = \frac{454\,\text{kg/d}}{120\,\text{kg/m}^3} = 3.78\,\text{m}^3/\text{d}\,(1000\,\text{gpd}) \text{ or } Q_{total} = 3.78\,\text{m}^3/\text{d} \times 10^3\,\text{L/m}^3 \times \frac{\text{d}}{24\,\text{h}} = 158\,\text{L/h}\,(42\,\text{gph})$$

   To meet a chlorine consumption rate of 454 kg/d (1000 lbs/d), a total flow of 158 L/h (42 gph) NaOCl solution is required at the plant.
3. Determine the set point for each pump.

$$Q_{pump} = \frac{Q_{total}}{N_{pump}} = \frac{158\text{ L/h}}{4\text{ pumps}} \approx 40\text{ L/h (10.5 gph) per pump}$$

   To meet the desired chlorine consumption rate of 454 kg/d (1000 lbs/d), the operators need to set each chemical feed pump at a flow of ~40 L/h (10.5 gph).

11.6.2 Basic Chlorine Chemistry

Many important reactions occur during chlorination or hypochlorination. The elemental chlorine (Cl_2) dissolves in water to make a chlorine solution. Hydrolysis reaction will then occur to form hypochlorous acid (HOCl) that is the desired disinfectant. HOCl will partially dissociate to hypochlorite ion (OCl^-). After the chlorine solution is fed into wastewater, the following reactions may occur: (1) HOCl reacts with inorganic and organic compounds causing an initial chlorine demand; (2) in the presence of ammonia or organic nitrogen, further reactions of HOCl occur to form chloramines or other nitrogen-containing compounds; and (3) HOCl may also react with other reducing contaminants in the wastewater or process chemicals added during treatment processes.

Dissolution of Gaseous Chlorine: When gaseous chlorine is mixed in water, the dissolution of chlorine occurs first. The dissolution of molecular chlorine in water is driven by the equilibrium between gaseous phase and solution phase, as given by Equation 11.10a.

$$Cl_{2(g)} \leftrightarrow Cl_{2(aq)} \quad (11.10a)$$

The dissolution phenomenon of gaseous chlorine can be described by the Henry's law. The empirical expression for mole fraction based Henry's law constant (H) is presented by Equation 10.84b.[53] Two practical expressions are also given derived by Equations 11.10b and 11.10c.

$$H_m = \frac{[Cl_2]}{P_{Cl2}} \quad \text{or} \quad [Cl_2] = H_m P_{Cl2} \quad (11.10b)$$

$$H_c = \frac{C_{Cl2}}{P_{Cl2}} \quad \text{or} \quad C_{Cl2} = H_c P_{Cl2} \quad (11.10c)$$

where

H_m = mole-concentration-based Henry's law constant, mole/L·atm

$[Cl_2]$ = dissolved gaseous chlorine concentration, mole/L

P_{Cl2} = partial pressure of chlorine or total pressure in gas phase, atm. When a nearly pure chlorine gas is applied, the mole fraction of chlorine is close to 1 so the partial pressure of chlorine is the same as the total pressure in the gaseous phase.

H_c = mass-concentration-based Henry's law constant, mg/L·atm

C_{Cl2} = dissolved aqueous chloride concentration, mg/L

Useful empirical equations are further derived for H_m and H_c by Equations 11.10d and 11.10e.

$$H_m = 10^{\left(\log(n_{H2O}) - B + \frac{A}{T}\right)} \quad \text{or} \quad H_m = 10^{\left(-B' + \frac{A}{T}\right)} \quad (11.10d)$$

$$H_c = 1000\, mw_{Cl2} H_m \quad \text{or} \quad H_c = 1000\, mw_{Cl2}\, 10^{\left(-B' + \frac{A}{T}\right)} \quad (11.10e)$$

where

n_{H2O} = mole concentration of water, 55.6 mole H_2O/L

B' = constant, $B' = B - \log(n_{H2O})$

mw_{Cl2} = molecular weight of chlorine, 70.9 g/mole

T = temperature, °K (T (°K) = 273.15 + T (°C))

A and B are defined in Equation 10.84b. See Example 11.18 for derivation of expressions of H_m and H_c. Equations 11.10b through 11.10e are applicable for chlorine as well as other gaseous chemicals used for disinfection of wastewater. Estimated constants for these chemicals are summarized in Table 11.13. See Example 11.19 for calculating H, H_m, and H_c for chlorine at 20°C. Also, see Section 15.4.5 and Examples 15.10 through 15.14 for applications of Henry's law in air-stripping process design.

Hydrolysis Reactions: The generalized hydrolysis reactions of gaseous chlorine and sodium hypochlorite are given by Equations 11.11a and 11.11b.

$$Cl_2 + H_2O \rightarrow HOCl + H^+ + Cl^- \quad \text{or} \quad Cl_2 + OH^- \rightarrow HOCl + Cl^- \quad (11.11a)$$

$$NaOCl + H_2O \rightarrow HOCl + Na^+ + OH^- \quad (11.11b)$$

TABLE 11.13 Henry's Law Constants for Gaseous Chemicals Used for Disinfection

Chemical	Empirical Constant[a]			Henry's Law Constant[b]		
	A	B	B′	H, atm	H_m, mole/L·atm	H_c, mg/L·atm
Ammonia, NH_3	1887.12	6.315	4.570	0.754	73.7	1,300,000
Chlorine, Cl_2	875.69	5.75	4.005	579	0.0960	6800
Chlorine dioxide, ClO_2	1041.77	6.73	4.985	1500	0.0370	2500
Ozone, O_3	1268.24	8.05	6.305	5290	0.0105	504
Sulfur dioxide, SO_2	1207.85	5.68	3.935	36.3	1.53	98,000

[a] See Reference 53 for constants A and B for other dissolved gases.
[b] The constants are estimated at a gas pressure of 1 atm and solution temperature of 20°C (or 293.15°K).
Source: Adapted in part from Reference 53.

The most important product of these reactions is hypochlorous acid, HOCl. The hydrolysis reaction of gaseous chlorine is described by the hydrolysis constant K_h expressed by Equation 11.11c.

$$K_h = \frac{[HOCl][H^+][Cl^-]}{[Cl_2]} \tag{11.11c}$$

where
K_h = hydrolysis constant, $mole^2/L^2$
[HOCl] = hypochlorous acid concentration (mole/L) (Equation 11.12a) for dissociation reaction of HOCl
$[H^+]$ = proton concentration, mole/L
$[Cl^-]$ = chloride ion concentration, mole/L

An experimental expression of hydrolysis constant has been developed for chlorine and is expressed by Equations 11.11d and 11.11e where T is in °K.[53] At a water temperature of 20°C, K_h is about 3.8×10^{-4} $mole^2/L$ and $pK_h = 3.4$.[2]

$$K_h = 10^{\left(0.4118 - \frac{1121.32}{T}\right)} \tag{11.11d}$$

$$pK_h = \frac{1121.32}{T} - 0.4118 \tag{11.11e}$$

where $pK_h = -\log(K_h)$

The hydrolysis reactions convert the dissolved gaseous chlorine into hypochlorous acid and chloride ion. The dissolved gaseous chlorine concentration is reduced which favors dissolution of additional gaseous chlorine. As a result, the overall chlorine solubility may be increased. See Example 11.21 for estimating the effect of hydrolysis on chlorine solubility.

In an aqueous solution, the pH is depressed by adding chlorine gas (Equation 11.11a) while the pH is raised by feeding sodium hypochlorite (Equation 11.11b). To form 1 g of hypochlorous acid as available chlorine, the expected changes in alkalinity are given below:

1. An alkalinity consumption of 0.70 g as $CaCO_3$ occurs when 1 g of gaseous chlorine as available chlorine is hydrolyzed (Equation 11.11a).
2. An alkalinity addition of 0.70 g as $CaCO_3$ occurs when 1 g of sodium hypochlorite as available chlorine is hydrolyzed (Equation 11.11b).

The ratio of 0.70 g $CaCO_3$/g Cl_2 is calculated below:

$$\frac{\text{Eq. wt. of } CaCO_3}{\text{Eq. wt. of } Cl_2} = \frac{50 \text{ g/eq. as } CaCO_3}{71 \text{ g/eq. as } Cl_2} = 0.70 \text{ g } CaCO_3/\text{g } Cl_2$$

Dissociation Reaction of Hypochlorous Acid: As a weak acid, hypochlorous acid has a tendency to form hypochlorite ion and hydrogen ion due to partial dissociation (or ionization). The dissociation of HOCl and the dissociation constant are expressed by Equations 11.12a and 11.12b.

$$HOCl \leftrightarrow H^+ + OCl^- \quad (11.12a)$$

$$K_a = \frac{[H^+][OCl^-]}{[HOCl]} \quad (11.12b)$$

where

K_a = dissociation constant for weak acid (proton reaction), mole/L

$[OCl^-]$ = hypochlorite ion concentration, mole/L

The hydrolysis constant is primarily a function of water temperature. An empirical expression of this constant for hypochlorous acid is expressed by Equations 11.12c or 11.12d.[48]

$$K_a = 10^{10.0686 - \frac{3000}{T}} - 0.0253\,T \quad (11.12c)$$

$$pK_a = \frac{3000}{T} + 0.0253\,T - 10.0686 \quad (11.12d)$$

where $pK_a = -\log(K_a)$

In dissociation reaction of HOCl, K_a is about 2.6×10^{-8} mole/L and a pK_a of 7.6 at a water temperature of 20°C. The chlorine residual created by hypochlorous acid is defined as the *free chlorine residual.* The free chlorine residual may be in the form of HOCl, OCl^-, as well as a small amount of Cl_2. HOCl is a stronger oxidant and more desirable disinfectant than OCl^-. At a given free chlorine residual, the effectiveness of disinfection is therefore affected by the percentage of HOCl at the equilibrium of the dissociation reaction. The percentage of HOCl can be predicted from the degree of dissociation reaction as a function of pH and K_a. An expression of percent HOCl in equilibrium is given by Equations 11.12e and 11.12f.

$$\%\,HOCl = \frac{[HOCl]}{[HOCl] + [OCl^-]} \times 100\% \quad \text{or} \quad \%\,HOCl = \frac{1}{1 + K_a \times 10^{pH}} \times 100\% \quad (11.12e)$$

$$\%\,HOCl = \frac{1}{1 + 10^{(pH - pK_a)}} \times 100\% \quad (11.12f)$$

where

% HOCl = molar percentage of HOCl in water, %

pH = pH of water

The percent HOCl in solution varies with pH and temperature. The percent HOCl concentration with respect to pH at three temperatures is shown in Figure 11.8. HOCl is normally the predominant species after hydrolysis reaction when pH is below 7.5. A noticeable decrease of HOCl percent may be experienced when the total dissolved solid (TDS) concentration in the wastewater is higher than 300–500 mg/L. Since protons are produced during dissociation reaction (Equation 11.12a), 0.70 g of alkalinity as $CaCO_3$ is consumed when 1 g of HOCl as available chlorine is dissociated into OCl^-.

Reactions with Ammonia Nitrogen: Ammonia nitrogen concentration in raw domestic wastewater is typically in a range of 20–40 mg/L as NH_3-N. Total ammonia nitrogen consists of ammonium ion (NH_4^+) and unionized ammonia (NH_3) in an equilibrium reaction. Partial dissociation of ammonium ion is expressed by Equation 11.13a.

$$NH_4^+ \leftrightarrow NH_3 + H^+ \quad (11.13a)$$

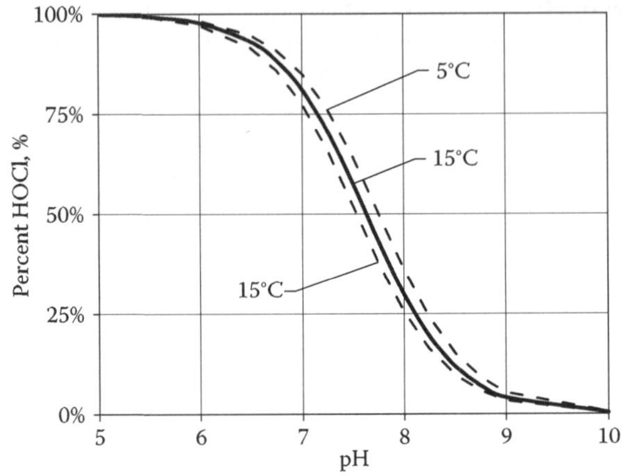

FIGURE 11.8 Percent HOCl versus pH.

The dissociation constant for the equilibrium reaction is expressed by Equations 11.13b through 11.13d.[48]

$$K_a = \frac{[H^+][NH_3]}{[NH_4^+]} \tag{11.13b}$$

$$K_a = 10^{-0.09018-\frac{2730}{T}} \tag{11.13c}$$

$$pK_a = 0.09018 + \frac{2730}{T} \tag{11.13d}$$

where

$[NH_4^+]$ = ammonium ion concentration, mole/L

$[NH_3]$ = undissociated ammonia concentration, mole/L

The percent NH_4^+ in equilibrium is determined from Equation 11.13e.

$$\%\,NH_4^+ = \frac{[NH_4^+]}{[NH_4^+]+[NH_3]} \times 100\% \quad \text{or} \quad \%\,NH_4^+ = \frac{1}{1+K_a \times 10^{pH}} \times 100\%$$
$$\text{or} \quad \%\,NH_4^+ = \frac{1}{1+10^{(pH-pK_a)}} \times 100\% \tag{11.13e}$$

where $\%\,NH_4^+$ = molar percentage of NH_4^+ in water, %

For dissociation reaction of NH_4^+ at a water temperature of 20°C, K_a is about 4.0×10^{-10} mole/L and $pK_a = 9.4$. At a pH <8, NH_4^+ is normally the predominant species (>90%) in water. At pH below 7, <1% of ammonia nitrogen is present in the form of NH_3.

Hypochlorous acid reacts with ammonia nitrogen to form various chloramines: *monochloramine* (NH_2Cl), *dichloramine* ($NHCl_2$), and *trichloramine* (NCl_3). Generalized chloramines formation reactions are expressed by Equations 11.14a through 11.14d.

Monochloramine:

$$HOCl + NH_4^+ \rightarrow NH_2Cl + H_2O + H^+ \tag{11.14a}$$

$$HOCl + NH_3 \rightarrow NH_2Cl + H_2O \tag{11.14b}$$

Dichloramine:

$$HOCl + NH_2Cl \rightarrow NHCl_2 + H_2O \tag{11.14c}$$

Trichloramine:

$$HOCl + NHCl_2 \rightarrow NCl_3 + H_2O \tag{11.14d}$$

The chloramines are usually formed by step reactions. The formation of different chloramine species is affected by factors such as the chlorine-to-ammonia nitrogen (Cl_2:NH_3-N) weight ratio, pH, contact time, and temperature. Under the normal wastewater disinfection conditions, the Cl_2:NH_3-N weight ratio is the most significant factor. At a low to medium ratio, monochloramine formation is predominant from initial reaction of HOCl with ammonia nitrogen (Equation 11.14a or 11.14b). The stoichiometric consumption rate of HOCl and ammonia nitrogen is 5.07 g (or approx. 5) as available CL_2 per g of ammonia nitrogen as NH_3-N (see Equation 11.26).

As the Cl_2:NH_3-N weight ratio increases, the extra HOCl provides favorable conditions for other side reactions. These reactions are affected by many factors such as the initial ammonia nitrogen concentration, initial Cl_2:NH_3-N weight ratio, pH, alkalinity, temperature, as well as other wastewater quality parameters. The nitrogenous end products from these side reactions may include dichloramine (Equation 11.14c), trichloramine (Equation 11.14d), and nitrate (NO_3^-). Generalized nitrate formation reactions from ammonia is given by Equation 11.15a or 11.15b. Nitrate may also be formed from oxidation of nitrite by HOCl.[48]

$$4\ HOCl + NH_4^+ \rightarrow NO_3^- + H_2O + 6\ H^+ + 4\ Cl^- \tag{11.15a}$$

$$4\ HOCl + NH_3 \rightarrow NO_3^- + H_2O + 5\ H^+ + 4\ Cl^- \tag{11.15b}$$

At a high weight ratio, the formation of significant amount of dichloramine leads to a *breakpoint phenomenon*; that is a breakdown or destruction of chloramines. The mechanism for this phenomenon is not clearly understood and defined. A combination of main *breakpoint reaction* and other *side reactions* is considered the major phenomenon. The main reaction is considered a direct breakdown of ammonia nitrogen to nitrogen gas (N_2). Generalized overall breakpoint reaction is given by Equations 11.15c or 11.15d.

$$3\ HOCl + 2\ NH_4^+ \rightarrow N_{2(g)} \uparrow + 3\ H_2O + 5\ H^+ + 3\ Cl^- \tag{11.15c}$$

$$3\ HOCl + 2\ NH_3 \rightarrow N_{2(g)} \uparrow + 3\ H_2O + 3\ H^+ + 3\ Cl^- \tag{11.15d}$$

The chlorine residual curve obtained from the reactions of hypochlorous acid with ammonia nitrogen is illustrated in Figure 11.9. Three typical zones are distinguished in the curve.

Zone I: Combined Chlorine Residual: NH_2Cl and $NHCl_2$ are the major species, and NH_2Cl is typically the predominant species in the pH range of 6.5–7.5 at Cl_2:NH_3-N weight ratio below 3:1. The chlorine residual due to chloramines is called the *combined chlorine residual*. Being a weaker oxidant than free chlorine, chloramines have lower disinfection power than HOCl. This property allows combined chlorine residual to last much longer than HOCl residual. Typically, the combined chlorine residual increases linearly with increase of the applied Cl_2:NH_3-N weight ratio in Zone I operation. Stoichiometrically, the peak combined chlorine residual is reached at a ratio of 5:1. At peak combined chlorine residual, usually NH_2Cl is about 60–90%.

During initial formation of NH_2Cl, proton reduction occurs (Equation 11.14a), and 0.70 g of alkalinity as $CaCO_3$ is consumed when HOCl reacts with ammonia nitrogen to form 1 g of NH_2Cl as residual Cl_2. No additional change in alkalinity is expected for further formation of either $NHCl_2$ or NCl_3.

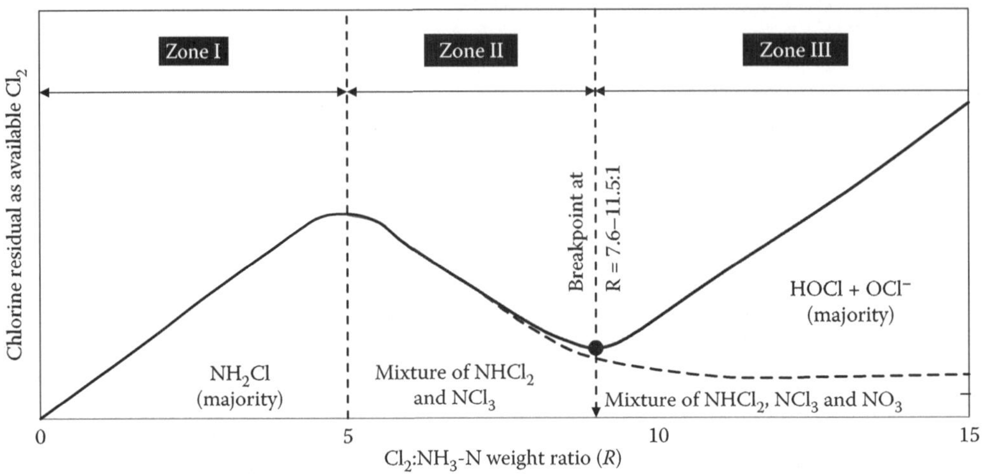

FIGURE 11.9 Example chlorine residual curves due to reactions of hypochlorous acid with ammonia nitrogen.

Zone II: Destruction of Combined Chlorine Residual: The combined chlorine residual is destroyed when the applied Cl_2:NH_3-N weight ratio exceeds 5:1. Usually, a linear decrease in the combined chlorine residual is observed with increase in the weight ratio until the *breakpoint* is reached at the dip. Zone II operation should be avoid in wastewater disinfection practice. Stoichiometrically, the breakpoint is reached at a Cl_2:NH_3-N weight ratio of 7.6:1 if nitrogen formation is the only reaction and other side reactions are negligible (Example 11.28 for determination of this ratio). In the presence of side reactions, the breakpoint may shift to a weight ratio in a range of 7.6:1–10:1. Trichloroamine (NCl_3) is the predominant species of the combined chlorine residual at the breakpoint.

Zone III: Free and Combined Chlorine Residual: Once the breakpoint is reached, the free chlorine residual begins to appear. Beyond this point, the free chlorine residual increases as the applied chlorine dose increases. The sum of free and combined chlorine residuals is the *total chlorine*. In wastewater disinfection practice, Zone III operation is used when free chlorine residual is desired.

Generalized Oxidation Reactions and Initial Chlorine Demand: HOCl is a very reactive oxidant. It quickly reacts with many inorganic and organic compounds in the wastewater. These compounds may include H_2S, organic contaminants, Fe^{2+}, Mn^{2+}, and other easily oxidizable organic compounds. The chlorine consumption in these reactions is called the *initial chlorine demand*. A generalized reaction due to initial chlorine demand is expressed by Equation 11.16. In this reaction, HOCl is reduced to Cl^-. Many other beneficial applications of chlorination process are covered in Section 11.6.4.

$$\text{HOCl} + \text{Chlorine demand} \rightarrow \text{Oxidized compound} + \text{H}^+ + \text{Cl}^- \qquad (11.16)$$

Typical chlorine residual curves in wastewater treatment are illustrated in Figure 11.10. The combined chlorine in the presence of ammonia nitrogen appears after the initial demand is met. The combined chlorine residual is about 0.7–0.8 mg/L as available Cl_2 per mg/L of excess chlorine dosage fed. Free chlorine residual is obtained after the breakpoint is reached at Cl_2:NH_3-N weight ratio of 8:1–12:1.

An alkalinity consumption of 0.70 g as $CaCO_3$ occurs when 1 g of HOCl as available Cl_2 is reduced to Cl^-. An overall alkalinity usage of 1.41 g as $CaCO_3$ is required when 1 g of gaseous chlorine (Cl_2) is hydrolyzed and then reduced to Cl^-. There is no change in alkalinity when NaOCl is hydrolyzed and used completely. For the disinfection of the effluent from a secondary or tertiary treatment, the initial chlorine demand is relatively low. The initial alkalinity of effluent is 150–250 mg/L as $CaCO_3$. The change in alkalinity and pH after chlorination or hypochlorination is small and is normally ignored.

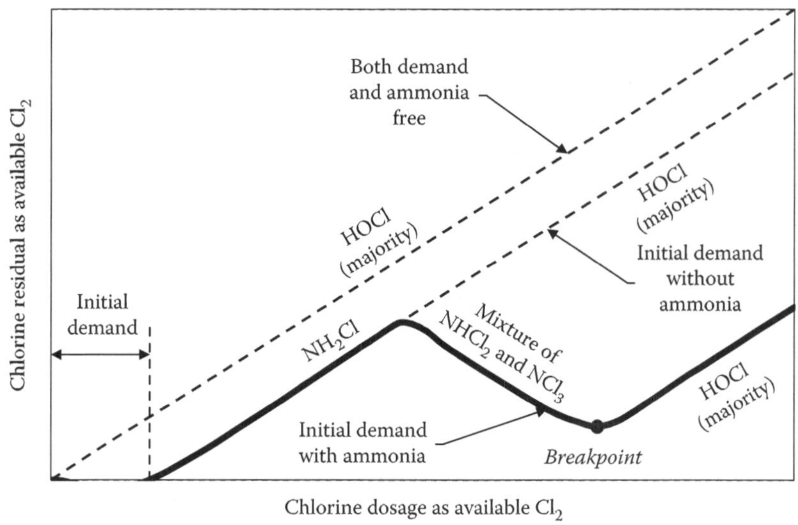

FIGURE 11.10 Typical curves of chlorine residual versus dosage.

TABLE 11.14 Increase in Total Dissolved Solids during Breakpoint Chlorination

Chemical	Ratio of TDS Increased to NH_4-N Consumed, g TDS/g NH_4-N
Chlorine gas	6.2
Sodium hypochlorite	7.1
Chlorine gas with lime (CaO) for neutralization	12.2
Chlorine gas with caustic (NaOH) for neutralization	14.8

Source: Adapted in part from Reference 53.

In wastewater treatment practice, alkaline chemicals may also be used for pH adjustment or other treatment purposes. A generalized reaction of HOCl with alkaline chemical in the form of HCO_3^- is expressed by Equation 11.17. These reactions do not change the total amount of available Cl_2 but may cause dissociation of HOCl to OCl^- when pH is raised. This may influence the effectiveness of the disinfection process.

$$HOCl + HCO_3^- \rightarrow OCl^- + CO_2 + H_2O \tag{11.17}$$

Dissolved Solids Buildup at Breakpoint Chlorination: Destruction of ammonia and chlorine at breakpoint chlorination adds significant amounts of dissolved solids. Depending upon the chemicals used, the major anions and cations are Cl^-, NO_3^-, SO_4^{2-}, Ca^{2+}, Na^+, and K^+. The ratios ofTDS per unit weight of NH_4-N destructed by different chemicals as breakpoint chlorination are provided in Table 11.14.

EXAMPLE 11.18: DERIVATION OF HENRY'S LAW CONSTANTS FOR DISSOLUTION OF GASEOUS CHLORINE

Derive the mole- and mass-concentration-based Henry's law constants H_m and H_c expressed by Equations 11.10d and 11.10e. Apply the empirical expression of mole-fraction-based constant (H) given by Equation 10.84b.

Solution

1. Develop the basic relationship of the mole-fraction-based Henry's law constant (H).

   Based on the generalized relationship given by Equation 10.84a, a simplified equation is obtained for mole fraction of gaseous chlorine in solution. The mole fraction of gas is close to 1 since nearly pure chlorine gas is applied in the gas phase (Equation 11.18a).

$$x_{Cl2} = \frac{P_{Cl2}}{H} \quad \text{or} \quad H = \frac{P_{Cl2}}{x_{Cl2}} \tag{11.18a}$$

   where

   x_{Cl2} = mole fraction of gaseous chlorine in the solution, mole chlorine/mole solution

   P_{Cl2} = total pressure above the liquid surface, atm

   The mole-fraction-based Henry's law constant (H) in atm has been defined by Equation 10.84a. The expression $H = 10^{\left(-\frac{A}{T}+B\right)}$ obtained from Equation 10.84b is also applicable to the dissolution of pure gaseous chlorine.

2. Derive the mole concentration chlorine gas based on Henry's law constants (H_m).

   The generalized relationship given by Equation 10.84c also applies to the mole fraction of gaseous chlorine in the solution. This equation can be simplified to Equation 11.18b since $\sum n_g + n_{H2O} \approx n_{H2O}$ as explained in Example 10.119.

$$x_{Cl2} = \frac{[Cl_2]}{\sum n_g + n_{H2O}} \approx \frac{[Cl_2]}{n_{H2O}} \tag{11.18b}$$

   Substitute x_{Cl2} from Equation 11.18b in Equation 11.18a to obtain Equation 11.18c.

$$\frac{n_{H2O}}{H} = \frac{[Cl_2]}{P_{Cl2}} \tag{11.18c}$$

   Assume $H_m = \frac{n_{H2O}}{H}$ and obtain the relationship $H_m = \frac{[Cl_2]}{P_{Cl2}}$ (Equation 11.10b) from Equation 11.18c.

   Apply $H = 10^{\left(-\frac{A}{T}+B\right)}$ (Equation 10.84b) to the assumption of $H_m = \frac{n_{H2O}}{H}$ and develop below expression that is also the first relationship Equation 11.10d.

$$H_m = \frac{n_{H2O}}{10^{\left(-\frac{A}{T}+B\right)}} = 10^{\log(n_{H2O})} \times 10^{\left(\frac{A}{T}-B\right)} = 10^{\left(\log(n_{H2O})-B+\frac{A}{T}\right)}$$

   Assume $B' = B - \log(n_{H2O})$ and substitute B′ into the above expression to obtain below expression that is also the second relationship in Equation 11.10d.

$$H_m = 10^{\left(-(B-\log(n_{H2O}))+\frac{A}{T}\right)} = 10^{\left(-B'+\frac{A}{T}\right)}$$

3. Derive the mass concentration based on Henry's law constants (H_c).

   The mass concentration of gaseous chlorine in the solution is estimated from the mole concentration given by Equation 11.18d.

$$C_{Cl2} = 1000\, mw_{Cl2}[Cl_2] \quad \text{or} \quad [Cl_2] = \frac{C_{Cl2}}{1000 mw_{Cl2}} \tag{11.18d}$$

Substitute Equation 11.18d into Equation 11.10b to obtain a new relationship expressed by Equation 11.18e.

$$H_m = \frac{C_{Cl2}}{1000\ mw_{Cl2} P_{Cl2}} \quad \text{or} \quad \frac{C_{Cl2}}{P_{Cl2}} = 1000\ mw_{Cl2}\ H_m \tag{11.18e}$$

Assume $H_c = \frac{C_{Cl2}}{P_{Cl2}}$ to obtain below expression that is also the first relationship in Equation 11.10e.

$$H_c = 1000\ mw_{Cl2}\ H_m$$

Apply the second relationship in Equation 11.10d to obtain below expression that is also the second relationship in Equation 11.10e.

$$H_c = 1000\ mw_{Cl2}\ 10^{(B' + \frac{A}{T})}$$

EXAMPLE 11.19: APPLICATION OF HENRY'S LAW CONSTANTS FOR DISSOLUTION OF GASEOUS CHLORINE

Apply the Henry's law constants for dissolution of gaseous chlorine in water at a gas pressure of 1 atm and solution temperature of 20°C. Determine (a) the mole-fraction-based constant (H), (b) mole-concentration-based constant (H_m), and (c) mass-concentration-based constant (H_c).

Solution

1. Estimate the value of H.

   Obtain from Table 10.13 the constants A and B for Cl_2: A = 875.69 and B = 5.75.

   Estimate the value of H for Cl_2 from Equation 10.84b at $T = (273.15 + 20)°K = 293.15°K$.

$$H = 10^{\left(-\frac{A}{T}+B\right)} = 10^{\left(-\frac{875.69}{293.15}+5.75\right)} = 579 \text{ atm}$$

2. Estimate the value of H_m.

   Calculate B′ from $n_{H2O} = 55.6$ g/mole for water.

$$B' = B - \log(n_{H2O}) = 5.75 - \log(55.6) = 4.005$$

   Calculate H_m from Equation 11.10d.

$$H_m = 10^{\left(-B' + \frac{A}{T}\right)} = 10^{\left(-4.005 + \frac{875.69}{293.15}\right)} = 0.0960 \text{ mole/L·atm}$$

3. Estimate the value of H_c.

   H_c is calculated from Equation 11.10e. Calculate B′ from $mw_{Cl2} = 70.9$ g/mole for Cl_2.

$$H_c = 1000\ mw_{Cl2} 10^{\left(-B' + \frac{A}{T}\right)} = 1000 \text{ mg/g} \times 70.9 \text{ g/mole} \times 10^{\left(-4.005 + \frac{875.69}{293.15}\right)} \text{ mole/L·atm}$$
$$= 6800 \text{ mg/L·atm}$$

EXAMPLE 11.20: CONCENTRATION OF DISSOLVED GASEOUS CHLORINE

The gaseous chlorine is induced by an injector at an operating pressure of 0.5 atm. Estimate the dissolved gaseous chlorine concentration in the chlorine solution at a water temperature of 20°C. Ignore the initial chlorine demand and the effect of hydrolysis on chlorine concentration in the solution.

Solution

The chlorine concentration may be calculated from either Henry's law constant provided in Table 10.13. In this example, Equation 11.10c is used to estimate the chlorine concentration. $H_c = 6800$ mg/L·atm at $P_{Cl2} = 0.5$ atm and water temperature = 20°C.

$$C_{Cl2} = H_c\, P_{Cl2} = 6800 \text{ mg/L·atm} \times 0.5 \text{ atm} = 3400 \text{ mg/L as } Cl_2$$

EXAMPLE 11.21: EFFECT OF HYDROLYSIS ON SOLUBILITY OF CHLORINE IN WATER

A chlorine solution is prepared from gaseous chlorine. The gaseous chlorine is dispersed into deionized water with negligible initial chlorine demand, buffering capacity, and chloride ion. The solubility of chlorine with effect of hydrolysis can be estimated from an empirical relationship expressed by Equation 11.19.[53] Calculate the chlorine solubility at a pressure of 0.5 atm and water temperature of 20°C.

$$S_{Cl2,m} = P_{Cl2}H_m + (K_h P_{Cl2} H_m)^{\frac{1}{3}} \quad (11.19)$$

where $S_{Cl2,m}$ = mole-concentration-based chlorine solubility with effect of hydrolysis (mole/L) as available Cl_2

Other terms have been defined previously.

Solution

1. Determine the value of H_m.
   $H_m = 0.0960$ mole/L·atm at 1 atm and 20°C is obtained from Table 11.13. The calculation procedure of H_m is also presented in Example 11.19.
2. Determine the value of K_h from Equation 11.11d at $T = (273.15 + 20)$°K = 293.15°K.

$$K_h = 10^{\left(0.4118 - \frac{1121.32}{T}\right)} = 10^{\left(0.4118 - \frac{1121.32}{293.15}\right)} = 3.86 \times 10^{-4} \text{ mole}^2/\text{L}^2$$

3. Determine the mole-concentration-based solubility of chlorine ($S_{Cl2,m}$) from Equation 11.19 at a pressure $P_{Cl2} = 0.5$ atm.

$$\begin{aligned} S_{Cl2,m} &= P_{Cl2}H_m + (K_h P_{Cl2} H_m)^{\frac{1}{3}} \\ &= 0.5 \text{ atm} \times 0.0960 \text{ mole/L·atm} + (3.86 \times 10^{-4} \text{mole}^2/\text{L}^2 \times 0.5 \text{ atm} \times 0.0960 \text{ mole/L·atm})^{\frac{1}{3}} \\ &= 0.0745 \text{ mole/L} \end{aligned}$$

4. Determine the mass-concentration-based solubility of chlorine.
   Calculate the mass-based concentration S_{Cl2} from $mw_{Cl2} = 70.9$ g/mole for Cl_2.

$$\begin{aligned} S_{Cl2} &= 1000\, mw_{Cl2}\, S_{Cl2,m} = 1000 \text{ mg/g} \times 70.9 \text{ g/mole} \times 0.0745 \text{ mole/L} \\ &= 5280 \text{ mg/L as } Cl_2 \end{aligned}$$

Note: The chlorine concentration estimated in Example 11.20 is 3400 mg/L as Cl_2. The solubility of gaseous chlorine based on Henry's law constant is 5250 mg/L as Cl_2.

EXAMPLE 11.22: PERCENT OF HOCL IN FREE CHLORINE RESIDUAL

Free chlorine residual is applied for disinfection of wastewater effluent. Estimate the percent of HOCl at equilibrium, and at 15°C and pH 7.2.

Solution

1. Determine the value of pK_a for chlorine at 15°C.

   From Equation 11.12c, calculate the dissociation constant K_a at temperature $T = (273.15 + 15)°\text{K} = 288.15°\text{K}$.

$$K_a = 10^{10.0686 - \frac{3000}{T} - 0.0253\,T} = 10^{10.0686 - \frac{3000}{288.15} - 0.0253 \times 288.15} = 2.33 \times 10^{-8} \text{ mole/L}$$

   Calculate pK_a, $pK_a = -\log(K_a) = -\log(2.33 \times 10^{-8}) = 7.63$

   The value of pK_a can also be calculated directly from Equation 11.12d.

$$pK_a = \frac{3000}{T} + 0.0253\,T - 10.0686 = \frac{3000}{288.15} + 0.0253 \times 288.15 - 10.0686 = 7.63$$

2. Determine the percent of HOCl at pH 7.2.

   From Equation 11.12e, calculate the percent of HOCl.

$$\%\ \text{HOCl} = \frac{1}{1 + 10^{(pH - pK_a)}} \times 100\% = \frac{1}{1 + 10^{(7.2 - 7.63)}} \times 100\% = 73\%$$

   Therefore, at 15°C and pH 7.2, the equilibrium concentration of HOCl is ~73% of free chlorine residual (Cl_2). HOCl is the most effective species of chlorine for disinfection.

EXAMPLE 11.23: DETERMINATION OF UNIONIZED AMMONIA CONCENTRATION IN THE EFFLUENT

Unionized ammonia (NH_3) is much more toxic species to aquatic life than ionized ammonium ion (NH_4^+). The total ammonia nitrogen concentration in the effluent from an activated sludge system is 3 mg/L. Estimate the unionized ammonia concentration in the effluent at a pH of 8.2 and temperature of 15°C.

Solution

1. Determine the pK_a for ammonia nitrogen at 15°C.

   At the temperature $T = (273.15 + 15)°\text{K} = 288.15°\text{K}$, calculate the pK_a from Equation 11.13d.

$$pK_a = 0.09018 + \frac{2730}{T} = 0.09018 + \frac{2730}{288.15} = 9.57$$

2. Determine the NH_3-N concentration.

   The NH_4^+ concentration is calculated from Equation 11.13e at pH 8.2.

$$\%\ NH_4^+ = \frac{1}{1 + 10^{(pH - pK_a)}} \times 100\% = \frac{1}{1 + 10^{(8.2 - 9.57)}} \times 100\% = 95.9\%$$

$$\%\ NH_3 = 100\% - \%\ NH_4^+ = 100\% - 95.9\% = 4.1\%$$

   Calculate NH_3 concentration at the given total ammonia concentration $C_{\text{Total } NH_3\text{-N}} = 3$ mg/L.

$$C_{NH_3\text{-N}} = \%\ NH_3 \times C_{\text{Total } NH_3\text{-N}} = 4.1\ \% \times 3 \text{ mg/L } NH_3\text{-N} = 0.12 \text{ mg/Las } NH_3\text{-N}$$

   Therefore, the NH_3 concentration is 0.12 mg/L as NH_3-N under the field conditions.

EXAMPLE 11.24: EFFECT OF PH ON UNIONIZED AMMONIA CONCENTRATION

Determine the total ammonia concentration remaining in the effluent if the pH is lowered from 8.2 to 7.2. The total ammonia nitrogen concentration and temperature of the effluent are 3 mg/L and 15°C.

Solution

1. Determine the value of pK_a for total ammonia nitrogen.

   In Example 11.23, the $pK_a = 9.57$ was calculated for ammonia nitrogen at 15°C in Step 1.

2. Determine the unionized NH_3-N concentration.

   Calculate the NH_4^+ concentration from Equation 11.13e at pH 7.2.

$$\% \, NH_4^+ = \frac{1}{1 + 10^{(pH - pK_a)}} \times 100\% = \frac{1}{1 + 10^{(7.2 - 9.57)}} \times 100\% = 99.58\%$$

$$\% \, NH_3 = 100\% - \% \, NH_4^+ = 100\% - 99.58\% = 0.42\%$$

Calculate NH_3 concentration at $C_{Total,NH_3\text{-}N} = 3$ mg/L.

$$C_{NH_3\text{-}N} = \% \, NH_3 \times C_{Total,NH_3\text{-}N} = 0.42\% \times 3 \text{ mg/L } NH_3\text{-N} = 0.013 \text{ mg/L as } NH_3\text{-N}$$

Note: In comparison, the concentration of ammonia is 0.12 mg/L as NH_3-N at pH 8.2 (Example 11.23. The NH_3 concentration is reduced nearly 90% to 0.013 mg/L as NH_3-N when the pH is lowered to 7.2.

EXAMPLE 11.25: EFFECT OF TEMPERATURE ON UNIONIZED AMMONIA CONCENTRATION

The discharge limit of unionized ammonia concentration in the effluent is set at <0.025 mg/L as NH_3-N to avoid toxicity to the aquatic life. Determine the maximum allowable concentrations of total ammonia nitrogen in the effluent at pH of 7.6 under the critical temperatures of 25°C and 10°C in the summer and winter conditions, respectively. Comment on the calculated results.

Solution

1. Determine the unionized NH_3-N concentration under the summer conditions.

   Calculate the pK_a from Equation 11.13d at the temperature $T = (273.15 + 25)$°K = 298.15°K.

$$pK_a = 0.09018 + \frac{2730}{T} = 0.09018 + \frac{2730}{298.15} = 9.25$$

Calculate the NH_4^+ concentration from Equation 11.13e at pH 7.6.

$$\% \, NH_4^+ = \frac{1}{1 + 10^{(pH - pK_a)}} \times 100\% = \frac{1}{1 + 10^{(7.6 - 9.25)}} \times 100\% = 97.8\%$$

$$\% \, NH_3 = 100\% - \% \, NH_4^+ = 100\% - 97.8\% = 2.2\%$$

Calculate the allowable total ammonia concentration to meet the unionized NH_3-N limit $C_{NH_3\text{-}N} =$ 0.025 mg/L

$$C_{Total,NH_3\text{-}N} = \frac{C_{NH_3\text{-}N}}{\% \, NH_3} \times 100\% = \frac{0.025 \text{ mg/L as } NH_3\text{-N}}{2.2\%} \times 100\% = 1.14 \text{ mg/L as } NH_3\text{-N}$$

Under the summer conditions, the maximum discharge limit for total ammonia nitrogen should be below 1 mg/L as NH_3-N.

2. Determine the unionized NH_3-N concentration under the winter conditions.

   Following the similar calculation procedure, the allowable total ammonia concentration is calculated below.

$$T = (273.15 + 10)°K = 283.15°K$$

$$pK_a = 0.09018 + \frac{2730}{283.15} = 9.73$$

$$\%\ NH_4^+ = \frac{1}{1 + 10^{(7.6-9.73)}} \times 100\% = 99.26\%$$

$$\%\ NH_3 = 100\% - 99.26\% = 0.74\%$$

$$C_{Total\ NH_3\text{-}N} = \frac{0.025\ mg/L\ as\ NH_3\text{-}N}{0.74\%} \times 100\% = 3.38\ mg/L\ as\ NH_3\text{-}N$$

   The discharge limit should be set below 3 mg/L as NH_3-N. This includes proper safety factor for the winter conditions.

3. Summarize and comment on the results.

   Seasonal discharge limits are typically used for ammonia concentration in the final effluent. In this example, the maximum allowable total ammonia concentrations should be below 1 and 3 mg/L as NH_3-N under the summer and winter conditions, respectively. The dilution factor and background ammonia nitrogen concentration in the receiving stream may also be considered in evaluation of the allowable ammonia nitrogen concentration in the discharge limit.

EXAMPLE 11.26: MONOCHLORAMINE FORMATION STOICHIOMETRY

Monochloramine is the desired species for disinfection of wastewater effluent at a treatment facility. The total ammonia concentration in the treated effluent is 1 mg/L as NH_3-N. Determine the HOCl concentration required based on stoichiometric weight ratio of Cl_2:NH_3-N for monochloramine formation. Assume the pH is relatively stable in the effluent.

Solution

1. Determine the stoichiometric weight ratio of Cl_2:NH_3-N.

   Apply Equation 11.14a to calculate the stoichiometric weight ratio of Cl_2:NH_3-N for monochloramine formation using technical data provided in Table 11.11.

$$\underset{\substack{52.5\text{ as HOCl}\\70.9\text{ as }Cl_2}}{HOCl} + \underset{\substack{18\text{ as }NH_4^+\\14\text{ as }NH_3\text{-N}}}{NH_4^+} \rightarrow \underset{\substack{51.5\text{ as }NH_2Cl\\70.9\text{ as }Cl_2}}{NH_2Cl} + H_2O + H^+$$

$$R_{Cl2:NH_3\text{-}N} = \frac{1\text{ mole HOCl}}{1\text{ mole }NH_4^+} = \frac{70.9\text{ g HOCl as available }Cl_2}{14\text{ g }NH_4^+\text{ as }NH_3\text{-N}} \approx 5\text{ g }Cl_2/\text{g }NH_3\text{-N}$$

   The stoichiometric weight ratio of Cl_2:NH_3-N is approximately 5:1 for NH_2Cl formation. It means that a stoichiometric HOCl concentration of 5 mg/L as available Cl_2 is required per mg/L of total ammonia nitrogen as NH_3-N consumed.

2. Determine the HOCl concentration required based on the stoichiometric weight ratio.

   Partial dissociation of ammonium ion does not affect the ratio. In accordance with Equation 11.13a, the reversal reaction will shift toward left when NH_4^+ is consumed in reaction with HOCl. At a relatively stable pH, NH_3 will react continuously with H^+ to form NH_4^+ until all ammonia nitrogen is consumed in the reaction with HOCl. Calculate the HOCl concentration required at a total ammonia nitrogen concentration $C_{\text{Total NH}_3\text{-N}} = 1$ mg/L as NH_3-N.

$$C_{\text{HOCl}} = R_{\text{Cl2:NH}_3\text{-N}} \times C_{\text{Total NH}_3\text{-N}} = 5 \text{ mg Cl}_2/\text{mg NH}_3\text{-N} \times 1 \text{ mg/L as NH}_3\text{-N} = 5 \text{ mg/L as Cl}_2$$

   At a HOCL concentration of 5 mg/L as Cl_2, a 5 mg/L of NH_2Cl as available Cl_2 is also produced, $C_{\text{NH2Cl}} = C_{\text{HOCl}} = 5$ mg/L as Cl_2.

EXAMPLE 11.27: DEVELOP BREAKPOINT CHLORINATION CURVE

The residual chlorine and chlorination data from a batch test are given below. Plot the chlorination curve and obtain the breakpoint chlorination dose. Determine the following points from the bench test data: (a) initial demand, (b) maximum combined chlorine residual, (c) breakpoint, and (d) operating point to provide a free chlorine residual of 1.2 mg/L. Calculate also the daily chlorine usage and overall demand at a flow of 1800 L/s.

Chlorine dosage, mg/L as Cl_2	1	2	3	4	5	6	7	8
Chlorine residual, mg/L as Cl_2	0	1.05	1.95	0.75	1.5	2.3	3.2	4.0

Solution

1. Plot the chlorination curve in Figure 11.11.

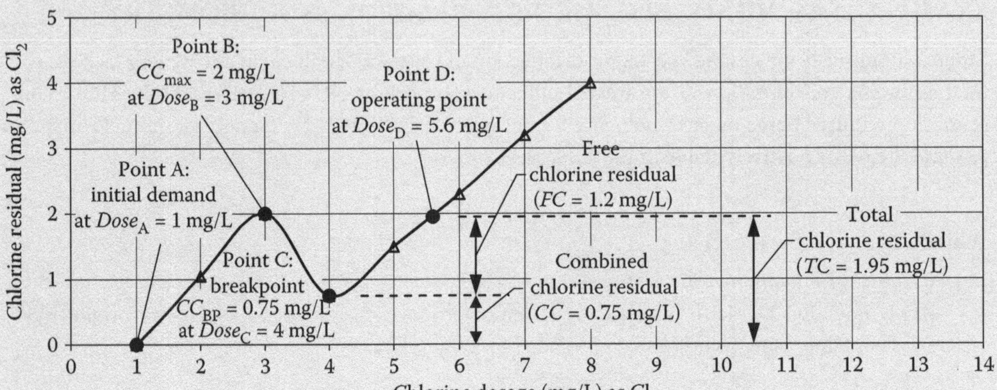

FIGURE 11.11 Plot of chlorine residual curve (Example 11.27).

2. Determine the initial chlorine demand.

   The initial chlorine demand is ~1 mg/L at $Dose_A = 1$ mg/L (Point A).

3. Determine the maximum combined chlorine residual.

   The maximum combined chlorine residual $CC_{max} = 2$ mg/L is achieved at $Dose_B = 3$ mg/L (Point B). From the difference in dosage between Points B and A, the ammonia concentration

($C_{NH_3\text{-}N}$) is estimated at the stoichiometric Cl_2:NH_3-N ratio of 5:1. (Other side reactions are ignored.)

$$C_{NH_3\text{-}N} = \frac{(Dose_B - Dose_A)_{max}}{R_{Cl2:NH_3\text{-}N}} = \frac{(3-1)\text{ mg/L as }Cl_2}{5\text{ mg }Cl_2/\text{mg }NH_3\text{-}N} = 0.4\text{ mg/L as }NH_3\text{-}N$$

Calculate the overall Cl_2:NH_3-N ratio at Point B.

$$R_B = \frac{Dose_B}{C_{NH_3\text{-}N}} = \frac{3\text{ mg/L as }Cl_2}{0.4\text{ mg/L as }NH_3\text{-}N} = 7.5\text{ mg }Cl_2/\text{mg }NH_3\text{-}N$$

4. Determine the breakpoint.
   The breakpoint is identified at Point C that has a combined chlorine residual $CC_{BP} = 0.75$ mg/L at $Dose_C = 4$ mg/L. Calculate the overall Cl_2:NH_3-N ratio at Point C.

$$R_C = \frac{Dose_C}{C_{NH_3\text{-}N}} = \frac{4\text{ mg/L as }Cl_2}{0.4\text{ mg/L as }NH_3\text{-}N} = 10\text{ mg }Cl_2/\text{mg }NH_3\text{-}N$$

   Note: The estimated ratio is within the typical range of 8–10:1.
5. Determine the operating point.
   Assume the combined chlorine residual $CC = CC_{BP} = 0.75$ mg/L when the breakpoint is exceeded. Calculate the total chlorine residual (TC) required to reach a desired free chlorine residual, $FC = 1.2$ mg/L.

$$TC = CC + FC = (0.75 + 1.2)\text{ mg/L} = 1.95\text{ mg/L as }Cl_2$$

   From the curve, the operating point is identified at Point D that gives a total chlorine residual $TC =$ 1.95 mg/L at $Dose_D = 5.6$ mg/L. Calculate the overall Cl_2:NH_3-N ratio at Point D.

$$R_D = \frac{Dose_D}{C_{NH_3\text{-}N}} = \frac{5.6\text{ mg/L as }Cl_2}{0.4\text{ mg/L as }NH_3\text{-}N} = 14\text{ mg }Cl_2/\text{mg }NH_3\text{-}N$$

6. Summarize the chlorination information.
   The chlorination information estimated from the bench test data is summarized below.

Parameter	Point A	Point B	Point C	Point D
	Initial Demand	Maximum Combined Chlorine Residual	Breakpoint	Operating Point
Dosage (mg/L) as Cl_2	1	3	4	5.6
Overall Cl_2:NH_3-N ratio, mg Cl_2/mg NH_3-N	N/A	7.5	10	14
Chlorine residual (mg/L) as Cl_2				
Combined (CC)	0	2	0.75	0.75
Free (FC)	0	0	0	1.2
Total (TC)	0	2	0.75	1.95

7. Determine the daily chlorine usage and demand.
   Calculate the total chlorine usage at the chlorine dosage $Dose_D = 5.6$ mg/L.

$$\text{Usage} = 5.6\text{ mg/L} \times 1800\text{ L/s} \times 86{,}400\text{ s/d} \times 10^{-6}\text{ kg/mg} = 871\text{ kg/d as } Cl_2$$

   Calculate the total chlorine residual TC at Point D by assuming $CC = CC_{BP} = 0.75$ mg/L.

$$TC = CC + FC = (0.75 + 1.2)\text{ mg/L} = 1.95\text{ mg/L as } Cl_2$$

   Calculate the overall chlorine demand concentration at Point D.

$$C_{\text{Demand}} = Dose_D - TC = (5.6 - 1.95)\text{ mg/L} = 3.65\text{ mg/L as } Cl_2$$

   Calculate the daily overall chlorine demand.

$$\text{Demand} = 3.65\text{ mg/L} \times 1800\text{ L/s} \times 86{,}400\text{ s/d} \times 10^{-6}\text{ kg/mg} = 568\text{ kg/d as } Cl_2$$

EXAMPLE 11.28: BREAKPOINT CHLORINATION STOICHIOMETRY

Ammonia nitrogen is destroyed at breakpoint chlorination. Determine the stoichiometric weight ratio of chlorine to ammonia nitrogen at the breakpoint chlorination and the HOCl concentration required if ammonia nitrogen concentration in the effluent is 1.4 mg/L. Also calculate the dosage required, and evaluate the alkalinity destruction if gaseous chlorine or NaOCl is used for ammonia nitrogen removal.

Solution

1. Determine the stoichiometric weight ratio of Cl_2:NH_3-N.
   Apply Equation 11.15c to calculate the stoichiometric weight ratio of Cl_2:NH_3-N for the breakpoint chlorination.

$$\underset{\substack{3\times52.5\text{ as HOCl}\\3\times70.9\text{ as }Cl_2}}{3\,HOCl} + \underset{\substack{2\times18\text{ as }NH_4^+\\2\times14\text{ as }NH_3\text{-N}}}{2\,NH_4^+} \rightarrow N_{2(g)}\uparrow + 3\,H_2O + 5\,H^+ + 3\,Cl^-$$

$$R_{Cl_2:NH_3\text{-N}} = \frac{3\text{ mole HOCl}}{2\text{ mole }NH_4^+} = \frac{3\times70.9\text{ g HOCl as available }Cl_2}{2\times14\text{ g }NH_4^+\text{ as }NH_3\text{-N}} \approx 7.6\text{ g }Cl_2/\text{g }NH_3\text{-N}$$

   The stoichiometric weight ratio of Cl_2:NH_3-N is ~7.6 for the breakpoint chlorination. In practice, the actual ratio is 8:1 to 10:1 due to side reactions.
2. Determine the HOCl concentration required for breakpoint chlorination.
   Calculate the HOCl concentration required at a total ammonia nitrogen concentration $C_{\text{Total }NH_3\text{-N}} =$ 1.4 mg/L as NH_3-N.

$$C_{HOCl} = R_{Cl_2:NH_3\text{-N}} \times C_{\text{Total }NH_3\text{-N}} = 7.6\text{ mg }Cl_2/\text{mg }NH_3\text{-N} \times 1.4\text{ mg/L as }NH_3\text{-N}$$
$$= 10.6\text{ mg/L as }Cl_2$$

3. Determine the chemical dosages.
   The required chemical dosages are calculated using the available chlorine information in Table 11.11. The HOCl concentration required for break point chlorination $C_{HOCl} = 10.6$ mg/L as Cl_2.

a. Chlorine dosage.

For chlorine (gas), the weight percent available chlorine $(\text{Weight \% } Cl_2)_{Cl2} = 100\%$ or 1 mg Cl_2/mg Cl_2.

$$Dose_{Cl2} = \frac{C_{HOCl}}{(\text{Weight \% } Cl_2)_{Cl2}} = \frac{10.6 \text{ mg/L as } Cl_2}{1 \text{ mg } Cl_2/\text{mg } Cl_2} = 10.6 \text{ mg/L as } Cl_2$$

b. NaOCl dosage.

For NaOCl $(\text{Weight \% } Cl_2)_{NaOCl} = 95.3\%$ or 0.953 mg Cl_2/mg NaOCl.

$$Dose_{NaOCl} = \frac{C_{HOCl}}{(\text{Weight \% } Cl_2)_{NaOCl}} = \frac{10.6 \text{ mg/L as } Cl_2}{0.953 \text{ mg } Cl_2/\text{mg } Cl_2} = 11.1 \text{ mg/L as NaOCl}$$

4. Evaluate the alkalinity destructions.

The overall reaction for alkalinity destruction is the hydrolysis and breakpoint reactions. The overall alkalinity destruction is calculated for each reaction.

a. Alkalinity destruction from gaseous chlorine.

A general expression for breakpoint reaction with gaseous chlorine is given by Equation 11.20a. This equation is developed by combining Equations 11.11a and 11.15c. Calculate the stoichiometric alkalinity destruction from Equation 11.20a.

$$\underset{3\times70.9 \text{ as } Cl_2}{3\,Cl_2} + \underset{\substack{2\times18 \text{ as } NH_4^+ \\ 2\times14 \text{ as } NH_3\text{-N}}}{2\,NH_4^+} \rightarrow N_{2(g)}\uparrow + \underset{8\times1 \text{ as } H^+}{8\,H^+} + 6\,Cl^- \qquad (11.20a)$$

8 moles of H^+ are produced when 2 moles of NH_4^+ are oxidized by 3 moles of available Cl_2. Calculate the weight ratio of H^+ to NH_3-N.

$$R_{H:NH_3\text{-N}} = \frac{8 \text{ mole } H^+}{2 \text{ mole } NH_3\text{-N}} = \frac{8 \times 1 \text{ g } H^+}{2 \times 14 \text{ g } NH_3\text{-N}} = 0.286 \text{ g } H^+/\text{g } NH_3\text{-N}$$

Calculate the weight ratio of alkalinity to NH_3-N from eq. wt. of $H^+ = 1$ g/eq. wt. and eq. wt. of $CaCO_3 = 50$ g/eq. wt.

$$R_{Alk:NH_3\text{-N}} = \frac{\text{eq. wt. of } CaCO_3}{\text{eq. wt. of } H^+} \times R_{H:NH_3\text{-N}} = \frac{50 \text{ g/eq. wt. as } CaCO_3}{1 \text{ g/eq. wt. as } H^+} \times 0.286 \text{ g } H^+/\text{g } NH_3\text{-N}$$
$$= 14.3 \text{ g alkalinity as } CaCO_3/\text{g } NH_3\text{-N}$$

Calculate the alkalinity destruction at $C_{Total\ NH_3\text{-N}} = 1.4$ mg/L as NH_3-N.

$$Alk_{NH_3\text{-N}} = R_{Alk:NH_3\text{-N}} \times C_{Total\ NH_3\text{-N}}$$
$$= 14.3 \text{ mg al kalinity as } CaCO_3/\text{mg } NH_3\text{-N} \times 1.4 \text{ mg/L as } NH_3\text{-N}$$
$$= 20 \text{ mg/L as } CaCO_3$$

b. Alkalinity destruction from NaOCl.

Equation 11.20b is developed for breakpoint reaction with NaOCl. This equation is developed by combining Equations 11.11b and 11.15c. Calculate the stoichiometric alkalinity destruction from Equation 11.20b.

$$\underset{\substack{3\times74.4 \text{ as NaOCl}\\ 3\times70.9 \text{ as } Cl_2}}{3\,NaOCl} + \underset{\substack{2\times18 \text{ as } NH_4^+\\ 2\times14 \text{ as } NH_3\text{-N}}}{2\,NH_4^+} \rightarrow N_{2(g)}\uparrow + \underset{\substack{2\times1 \text{ as } H^+\\ 2\times1 \text{ as } H^+}}{2\,H^+} + 3\,H_2O + 3\,Na^+ + 3\,Cl^- \qquad (11.20b)$$

2 moles of H^+ are produced when 2 moles of NH_4^+ are oxidized by 3 moles of NaOCl. Calculate the weight ratio of H^+ to NH_3-N.

$$R_{H:NH_3\text{-N}} = \frac{2 \text{ mole } H^+}{2 \text{ mole } NH_3\text{-N}} = \frac{2 \times 1 \text{ g } H^+}{2 \times 14 \text{ g } NH_3\text{-N}} = 0.0714 \text{ g}H^+/\text{g } NH_3\text{-N}$$

Calculate the weight ratio of alkalinity to NH_3-N.

$$R_{Alk:NH_3\text{-N}} = \frac{50 \text{ g/eq. wt. as } CaCO_3}{1 \text{ g/eq. wt. as } H^+} \times 0.0714 \text{ g}H^+/\text{g } NH_3\text{-N}$$
$$= 3.57 \text{ g alkalinity as } CaCO_3/\text{g } NH_3\text{-N}$$

Calculate the alkalinity destruction at $C_{\text{Total } NH_3\text{-N}} = 1.4$ mg/L as NH_3-N.

$$Alk_{NH_3\text{-N}} = R_{Alk:NH_3\text{-N}} \times C_{\text{Total } NH_3\text{-N}}$$
$$= 3.57 \text{ mg alkalinity as } CaCO_3/\text{mg } NH_3\text{-N} \times 1.4 \text{ mg/L as } NH_3\text{-N}$$
$$= 5.0 \text{ mg/L as } CaCO_3$$

c. Comment on alkalinity destruction by breakpoint chlorination.

The alkalinity destruction from gaseous chlorine is 20 mg/L as $CaCO_3$. The alkalinity destruction from NaOCl is 5 mg/L as $CaCO_3$. The alkalinity destruction with gaseous Cl_2 is significantly larger than that from NaOCl.

EXAMPLE 11.29: INCREASE IN TDS AND ALKALINITY FROM BREAKPOINT CHLORINATION

A small biological wastewater treatment facility is designed to provide well-nitrified effluent. The ammonia nitrogen limit is 1 mg/L. During peak sustained flow conditions, the effluent ammonia nitrogen concentration exceeds the limit. Breakpoint chlorination is proposed to control the nitrogen. Estimate the required chlorine dose and daily feed rate to reduce the NH_4-N concentration from a maximum of 6 mg/L to the required limit. Also calculate the alkalinity destruction as mg/L $CaCO_3$ and kg/d, and increase in TDS concentration if lime (CaO) is used to neutralize the acid formed. Assume the stoichiometric weight ratio of 7.6 mg Cl_2/mg NH_3-N applied. The peak sustained flow is three times the average design flow of 4000 m^3/d.

Solution

1. Determine the maximum Cl_2 requirements for ammonia destruction.

Calculate the maximum HOCl concentration required at $R_{Cl2:NH_3\text{-N}} = 7.6$ g Cl_2/g NH_3-N.

$$C_{HOCl} = R_{Cl2:NH_3\text{-N}} \times (6 - 1) \text{ mg/L as } NH_3\text{-N} = 7.6 \text{ mg } Cl_2/\text{mg } NH_3\text{-N} \times 5 \text{ mg/L as } NH_3\text{-N}$$
$$= 38 \text{ mg/L as } Cl_2$$

Since the weight percent available chlorine (Weight % $Cl_2)_{Cl2} = 100\%$, the gaseous chlorine dosage $Dose_{Cl2} = C_{HOCl} = 38$ mg/L or 38 g/m^3 as Cl_2. Calculate the maximum daily chlorine feed rate at the design flow $Q = 4000$ m^3/d and a peaking factor $PF = 3$.

$$w_{Cl2} = Dose_{Cl2}\, PF\, Q = 38 \text{ g/m}^3 \text{ as } Cl_2 \times 3 \times 4000 \text{ m}^3/\text{d} \times 10^{-3} \text{ kg/g} = 456 \text{ kg/d as } Cl_2$$

2. Determine the alkalinity destruction from ammonia oxidation.

   A weight ratio of alkalinity to NH_3-N, $R_{Alk:NH_3\text{-}N} = 14.3$ mg alkalinity as $CaCO_3$/mg NH_3-N for removal of ammonia nitrogen from gaseous chlorine (Example 11.28, Step 4.a). Calculate the alkalinity destruction from ammonia removal.

$$\begin{aligned} Alk_{NH_3\text{-}N} &= R_{Alk:NH_3\text{-}N} \times (6-1) \text{ mg/L as } NH_3\text{-N} \\ &= 14.3 \text{ mg alkalinity as } CaCO_3/\text{mg } NH_3\text{-N} \times 5 \text{ mg/L as } NH_3\text{-N} \\ &= 71.5 \text{ mg/L as } CaCO_3 \end{aligned}$$

   Calculate the maximum daily amount of alkalinity destruction from ammonia removal.

$$w_{Alk} = Alk_{Cl2} PF\, Q = 71.5 \text{ g/m}^3 \text{ as } CaCO_3 \times 3 \times 4000 \text{ m}^3\text{/d} \times 10^{-3} \text{ kg/g} = 858 \text{ kg/d as } CaCO_3$$

   Note: The destruction of alkalinity is significant, so effluent neutralization is required to produce stable effluent.

3. Determine the buildup of TDS concentration from ammonia removal followed by effluent neutralization.

   An TDS increase rate $R_{TDS:NH_3\text{-}N} = 12.2$ mg TDS/mg NH_3-N destroyed is obtained from Table 11.14, if CaO is used for neutralization of acid produced.

$$\begin{aligned} TDS_{NH_3\text{-}N} &= R_{TDS:NH_3\text{-}N} \times (6-1) \text{ mg/L as } NH_3\text{-N} = 12.2 \text{ mg TDS/mg } NH_3\text{-N} \times 5 \text{ mg/L as } NH_3\text{-N} \\ &= 61 \text{ mg/L as TDS} \end{aligned}$$

11.6.3 Components of Chlorination System

A chlorination system for efficient disinfection of wastewater effluent consists of four major components: (1) chlorine supply, storage and safety, (2) chlorine feed and application, (3) control systems, and (4) mixing and contact. These components are described below.

Chlorine Supply, Storage and Safety: For small- to medium-sized treatment facilities, the chlorine gas is typically supplied in pressure vessels that can be 45 and 70 kg (100 and 150 lb) cylinders, and 900 kg (1-ton) containers. The tank trucks with capacities between 13,600 and 20,000 kg (15 and 22 tons) are the most common method for bulk supplies to stationary storage tanks at medium to large plants. Rail cars with typical capacity in a range from 50,000 to 81,000 kg (55 to 90 tons) may also be transported via railroad to the site and parked near the application point at large facilities.

The chlorine supply and storage facilities should have scales, pipe headers, pressure gauges, pressure reducing valves, pressure relief valves, excess-flow valves, liquid chlorine expansion chambers, high- and low-pressure indicating switches with alarm, and shut-off valves.

Chlorine storage and chlorinator equipment must be housed in a separate building with adequate exhaust ventilation and leak detection device. Chlorine gas is heavier than the air. Therefore, the leak detection device should be at a maximum height of 0.3 m (1 ft) above the floor level. The facility should have temperature control and protections from fire hazards, containers from direct sunlight and equipped with appropriate emergency kit. Forced air mechanical ventilation fan should be 0.3 m (1 ft) below the ceiling (maximum), and the fan control should be located outside the room entrance. Personal protection equipment (PPE), including adequate type of self-contained breathing apparatus (SCBA) should be located outside chlorination buildings. The facilities should also meet the risk-based performance standards. Chlorine-handling equipment and safety procedures are covered in References 42, 48, 49, 51 through 54, and 85 through 91. All chlorination systems should conform to the standards of the Chlorine Institute (see References 85 through 90 for important guidance materials).

Chlorine Feed and Application: The chlorine feed and application system consists of three main process areas: (1) chlorine feed, (2) rate control, and (3) application. Brief descriptions about the major components in these three areas are provided in Table 11.15.[42,48,85,87,88,90]

Control Systems: Chlorine feed can be controlled either *manually* or *automatically*. Manual control may be used by operators during startup and shutdown, and in situations of almost constant flow and

TABLE 11.15 Major Components of Chlorine Feed and Application Systems

Component	Description
Chlorine feed	
Chlorine withdrawal	Cylinders are stored and used in upright position for gas withdrawal only. Containers are placed horizontally and two outlet valves must be aligned in vertical position for gas withdraw from top valve and liquid withdraw from the bottom valve. For gas withdraw, the ambient temperature should be maintained between 15°C and 20°C (60°F and 68°F). The maximum continuous gas withdraw rates are up to 19 kg/d (42 lb/d) from a 150-lb cylinder and 180 kg/d (400 lb/d) from a ton container. A manifold system with multiple ton containers can provide a chlorine gas withdrawal rate in a range of 180–700 kg/d (400–1500 lb/d). If the desired withdrawal rate is >700 kg/d (1500 lb/d), liquid withdraw with evaporator should be used. The maximum liquid withdraw rate is usually set at 4000 kg/d (9000 lb/d) for ton containers. Use of stationary storage tanks is more convenient for the facilities where the continuous withdrawal rates are >1200–1800 kg/d (2500–4000 lb/d).
Evaporator	The typical evaporator capacity ranges from 180 to 4500 kg/d (400 to 10,000 lb/d). A hot water bath, jacket, or kettle is typically used to provide heat of vaporization. The hot water can be supplied by (1) an external heat exchanger, (2) an electrical heater, or (3) direct steam injection. Proper pressure relief systems, including an expansion chamber on liquid side and rupture disc on gas side are essential for preventing piping from damages. An automatic shut-off valve on the evaporator outlet should close instantaneously when abnormal operating conditions occur in the evaporator. This is necessary to prevent the liquid chlorine from entering the vacuum regulator or chlorinator. Cathodic protection and effective insulation are also essential.
Vacuum regulator	For vacuum-operated feed rate controllers, a vacuum regulator is required. It reduces the chlorine gas from pressure to vacuum and allows the gas to flow through the device only if a vacuum condition is presented on the outlet side. It prevents pressured gas entering the controller. A drip leg with heater is usually required at the inlet of the device to prevent recondensed liquid chlorine from entering the vacuum regulator.
Automatic switchover	When multiple pressure vessels are manifolded in banks, an automatic switchover (or changeover) device is used to transfer the gas supply from the bank of empty vessels to the bank in full. The switchover system is of two types: vacuum and pressure. The automatic switchover is essential for unattended chlorination facilities.
Rate control	The feed rate can be controlled by (a) a simple manually adjusted rotameter or (b) a complex automatic chlorinator. The chlorinator receives chlorine gas from the storage container or evaporator and regulates the flow to the ejector. They work either under *pressure* or *vacuum*. The pressure-operated chlorinators provide chlorine gas to the application point directly. The vacuum-operated chlorinators control the chlorine feed flow rate by either (a) maintaining a constant vacuum differential or (b) achieving a sonic flow. They may also be installed in forms of wall panels or floor cabinets. The capacity categories of chlorinators cover a wide range from 45 to 4500 kg/d (100 to 10,000 lb/d). A typical remote vacuum chlorinator consists of an inlet pressure-reducing valve, a rotameter, a variable orifice, and a vacuum differential regulating valve.
Application	The application system consists of the following major components: (1) the operating water supply, (2) an injector (or ejector), (3) a pressure or vacuum chlorine gas line from the chlorinator, (4) a solution discharge line, and (5) a diffuser at the application point with a static, mechanical or hydraulic mixing device. The injectors are of two types: (a) pressure feed and (b) vacuum feed. The pressure injector poses the risk of gas escaping. In a vacuum feed injector, a vacuum is created by running pressurized water to draw chlorine gas from the supply source to the injector. The chlorine is mixed with water and is carried to the point of application as a solution of HOCl.

Source: Adapted in part from References 42, 48, 85, 87, 88, and 90.

demand. Automatic feed control is required under unattended situations with varying flows and demands. The automatic control system consists of (1) chlorine residual analyzer, (2) flow measurement device (meter or flume), (3) control system with programmable logic controller (PLC) to receive both flow and chlorine residual signals, (4) chlorinator with automatic chlorine feed function, and (5) compatible chlorine injection system. Two basic modes used for automatic chlorine feed control are (1) *residual control* and (2) *flow pacing*. The automatic residual control mode uses *closed-loop control* with a feedback element. It is more suitable for maintaining adequate residual under slow flow change conditions. On the other hand, the flow pacing control mode is an *open-* or *single-loop control* since no feedback is involved in the control system. It is capable of providing disinfection with rapid changes in flow rate. A *compound-loop-control*-based automatic chlorine feed control strategy is also available. This technique can be integrated with these two basic modes for maintaining a relatively constant chlorine residual near the preset value with both variable flow and demand. The details of control loops are presented in Section 14.5.[42,43] Alarms also serve as essential functions for automatic control system for timely response for safety and emergency situations. These situations may include high and low pressures in storage vessels, low quantity (or weight) in on-line vessels, chlorine leaks, high and low injector vacuum, high and low temperatures of evaporator water bath, low pressure of utility water supply, and high and low chlorine residuals.

Mixing and Contact: Adequate mixing of chlorine solution into effluent followed by a desired contact time in a basin is essential for effective disinfection. The chlorine solution may be mixed by (1) mechanical means, (2) baffle arrangement, or (3) hydraulic jump created by a downstream weir, venturi flume, or Parshall flume. To satisfy the mixing requirement, a velocity gradient G above 500 s^{-1} is essential. The velocity gradient may be calculated from Equation 9.30c.[42,53,84] Many examples for the determination of G from turbulence created by mixer, weir outfall, hydraulic jump, nozzle jet, baffled flocculators, and pneumatic mixer are discussed in Sections 9.5.5 and 9.5.6. Examples 9.32 through 9.35, 9.40 through 9.44, 11.32, and 11.33 cover the numerical aspects.

The purpose of the chlorine contact basin is to provide sufficient reaction time for chlorine to contact and reduce the number of living organisms to an acceptable level. The contact time may range from 15 to 30 min; and at least 15 min is typically required at peak design flow. The design objectives are to (1) minimize short circuiting and dead spaces, (2) maximize mixing for better contact, and (3) reduce solids settling in the basin. Chlorination improves the settling characteristics of the solids. These solids, if settled in the contact tank will exert chlorine demand. Anaerobic condition may also develop in the blanket and gasses may rise and carry the solids in the effluent. In a typical chlorine contact basin, there are normally two to four passes separated by longitudinal baffles connected by around-the-end turns. Such designs simulate a long narrow channel with a length-to-width ratio $\geq$ 20:1, and a depth-to-width ratio $\leq$ 1:1. A near- plug-flow regime is achieved in the channel.[42]

Chlorine Dosages: Proper chlorine dosage is essential for desired disinfection. The dosage is usually estimated from (a) initial chlorine demand, (b) chlorine decay during contact period, and (c) the required chlorine residual. Examples 11.35, 11.37, and 11.38 give the procedure for obtaining the desired chlorine dosage. As discussed previously in Section 11.6.2, an initial chlorine demand is typically required depending upon the organic and inorganic constituents in the effluent. A chlorine decay of 1–4 mg/L is normally expected when the contact time is in a range of 30–60 min. The chlorine residual required to achieve the desired effluent disinfection standards can be estimated from the disinfection inactivation kinetics (Section 11.5.2), and chlorine dosage required for desired disinfection is presented in Example 11.35. Typical dosages for disinfection of effluents from different processes are summarized in Table 11.7.[2,42,48,53,54]

Hypochlorination with Sodium Hypochlorite: Transportation, storage, and handling of liquid chlorine are associated with potential hazards. For this reason, hypochloration process using sodium hypochlorite (NaOCl) solution is used at many treatment plants in urban areas. Hypochlorination, however, is more expensive, high corrosive, and less stable in storage.

The most commonly used sodium hypochlorite solutions are in a range of 8–16 trade percent as available chlorine. It is commercially available in 200-L (55-gal) drums or 1100-L (300-gal) totes for industrial or small municipal applications. Large quantity is normally delivered by bulk tank trucks from 4 to 100 m^3 (1000 to 30,000 gal). The HOCl solution is stored in vertical or horizontal cylindrical tanks on the application sites. The storage tank is to have a volume at least 120% of the expected delivery truck capacity. The common construction materials for tank are rubber-line steel, fiberglass, or high density polyethylene (HDPE). An ORP probe is typically required to monitor the chemical decay in the tanks if high strength HOCl is stored. Commercial solution can be applied without further dilution. The HOCl solution may be withdrawn from the tank by (a) pumps, (b) gravity if the storage tanks are elevated, or (c) vacuum created by an educator system.

EXAMPLE 11.30: DESIGN OF A CHLORINATION FACILITY

Determine the capacities of chlorine feed and storage systems. The design data are given below. Peak 2-h and daily average design flows are 25,000 and 8000 m^3/d, respectively. The design maximum and average chlorine dosages are 12 and 5 mg/L, respectively. The chlorination feed equipment shall be capable of (a) delivering 1.25 times the maximum dosage under peak 2-h flow and (b) controlling separately the chlorine feed rate to maintain a chlorine residual >1 mg/L in each of two chlorine contact basins operated in parallel. The chlorine storage system shall provide a minimum of 15-day storage at 1.25 times the average dosage under average flow condition. The sustained low temperature in the chlorine feed room is 18°C. Verify the minimum turndown ratio required at the minimum flow of 5000 m^3/d. Also estimate the injector water flow rate.

Solution

1. Determine the capacity of chlorine feed system.

   The chlorine feed equipment is sized to meet the chlorine feed rate required at the maximum dosage at peak 2-hr flow.

   a. Determine the maximum chlorine feed rate.

   $$w_{max} = Dose_{max} SF\ Q_{2hpf} = 12\ g/m^3 \times 1.25 \times 25{,}000\ m^3/d \times 10^{-3}\ kg/g = 375\ kg/d \text{ as } Cl_2$$

   The maximum chlorine feed rate is much higher than the typical maximum continuous gas withdraw rate of 180 kg/d (400 lb/d) from a single ton container. A manifold system with multiple containers is therefore required.

   b. Determine the maximum chlorine withdrawal rate.

   The maximum continuous gas withdraw rate (WR_{max}) from a vessel (cylinder or container) can be estimated from Equation 11.21.[48]

   $$WR_{max} = (T - T_{th})WF_{max} \qquad (11.21)$$

   where

   WR_{max} = maximum continuous gas withdraw rate from a vessel, kg/d (lb/d)

   T = sustained low operating temperature, °C (°F)

   T_{th} = threshold temperature for gas withdraw, °C (°F). For chlorine gas withdraw, the typical values are −17.8°C (0°F) for vessel mounted vacuum regulator, and −12.2°C (10°F) for manifold system at 10–15 psig pressure, respectively.

   WF = maximum continuous gas withdraw factor, kg/d/°C (lb/d/°F). For chlorine gas withdraw, the typical ranges are 0.8–1.2 kg/d·°C (1–1.5 lb/d·°F) for a 150-lb cylinder, and 5–6.5 kg/d·°C (6–8 lb/d·°F) for a ton container.

Assume a conservative design value $WF = 5$ kg/d·°C for ton container. Calculate the WR_{max} from Equation 11.21 for a manifold system at $T_{th} = -12.2$°C and $T = 15$°C.

$$WR_{max} = (T - T_{th})WF_{max} = (18 - (-12.2))^\circ\text{C} \times 5\,\text{kg/d/}^\circ\text{C} = 151\,\text{kg/d} \approx 150\,\text{kg/d per container}$$

c. Determine the number of containers required in service.

$$N_{vessel} = \frac{w_{max}}{WR_{max}} = \frac{375\ \text{kg/d}}{150\ \text{kg/d/container}} = 2.5\ \text{containers} \approx 3\ \text{containers}$$

Provide two banks of containers for chlorine feed. Connect three containers to the manifold in each bank to prevent the maximum chlorine feed rate. The following components are needed in the manifold from each bank: (a) drip leg with heater, (b) a gas filter, (c) a pressure reducing valve (PRV), and (d) a vacuum regulator. A triple-container electronic scale with heavy-duty structure, trunnions for securing vessel, single load cell design, and digital LCD indicator are needed to monitor the total weigh of containers in each bank. An automatic vacuum switchover module is used to switch the operation from the empty to full bank.

d. Determine the capacity of chlorinators.

Provide three vacuum-operated chlorinators with identical capacity. Under the peak flow condition, two chlorinators will meet the peak demand, and the third unit will be a standby unit. This unit may also be used for miscellaneous applications. Each designated chlorinator will deliver the required amount of chlorine gas to the injector where the chlorine solution is prepared. The solution is transported to the point of application and then dispersed by a diffuser in the channel prior to a contact basin.

Assume that the peak flow is evenly split between two chlorine contact basins. Calculate the chlorine feed rate required for each chlorinator in service.

$$w_{chlorinator} = \frac{w_{max}}{N_{chlorinator}} = \frac{375\ \text{kg/d}}{2\ \text{chlorinators}} = 187.5\ \text{kg/d}$$

Provide two chlorinators each having a capacity of $w_{chlorinator}^{design} = 200$ kg/d. A conceptual process diagram of the chlorine feed system is shown in Figure 11.12.

e. Determine the minimum turndown ratio requirement.

Calculate the minimum chlorine feed rate at the minimum flow of 5000 m^3/d using the average dosage of 5 mg/L. To be conservative, assume both contact basins in service.

$$w_{chlorinator}^{min} = \frac{Dose_{avg}Q_{min}}{N_{chlorinator}} = \frac{5\ \text{g/m}^3 \times 5000\ \text{m}^3\text{/d} \times 10^{-3}\ \text{kg/g}}{2\ \text{chlorinators}} = 12.5\ \text{kg/d as Cl}_2$$

Calculate the minimum turndown ratio requirement when both chlorinators are in use.

$$R_{turndown} = \frac{w_{chlorinator}^{design}}{w_{chlorinator}^{min}} = \frac{200\ \text{kg/d}}{12.5\ \text{kg/d}} = 16$$

A maximum turndown ratio of 20:1 is typical for most chlorinators. Around this ratio, accurate control of chlorine feed rate under low flow condition is achieved at this facility. A maximum turndown ratio up to 100:1 is also commercially available in some newer models.

f. Estimate the injector utility water flow rate.

Provide three injectors, one for each chlorinator. Assume that the chlorine solution is prepared at a concentration $S_{Cl2} = 2000$ mg/L at the design feed rate capacity of $w_{chlorinator}^{design} = 200$ kg/d.

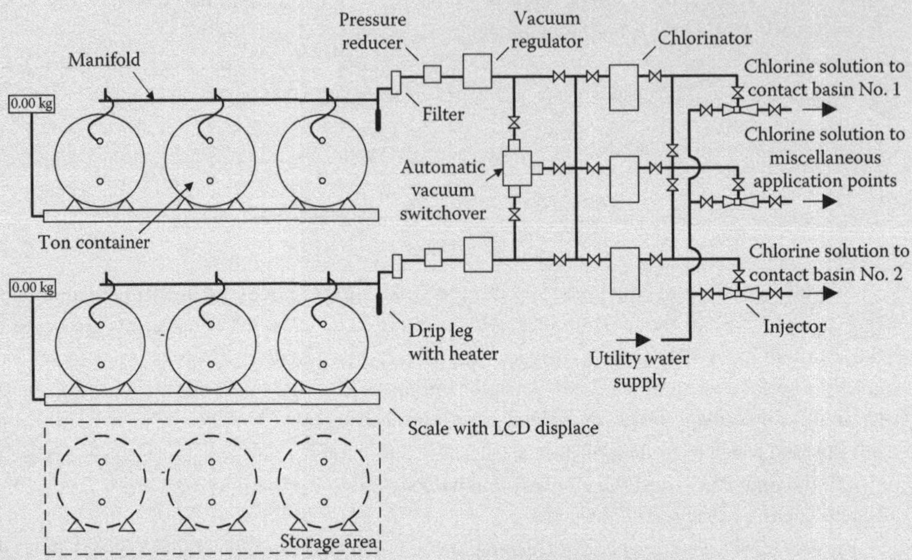

FIGURE 11.12 Conceptual process diagram of chlorine storage and feed systems (Example 11.30).

Estimate the water flow rate.

$$Q_{\text{injector}} = \frac{w^{\text{design}}_{\text{chlorinator}}}{S_{\text{Cl2}}} = \frac{200 \text{ kg/d}}{2 \text{ kg/m}^3} \times 10^3 \text{ L/m}^3 \times \frac{1 \text{ d}}{1440 \text{ min}} \approx 70 \text{ L/min}$$

A minimum utility water flow of 210 L/min may be required when all three injectors are used. Equipment manufacturers provide detailed information that can be used for injector sizing based on the chlorine feed rate, outlet back pressure, and injector nozzle/throat sizes.

2. Determine the storage requirement.

The chlorine storage system is sized to provide a minimum of 15-day storage. This storage does not include the containers connected to the bank in the standby position of the manifold system.

a. Determine the average chlorine feed rate.

$$w_{\text{avg}} = Dose_{\text{avg}}\ SF\ Q_{\text{avg}} = 5 \text{ g/m}^3 \times 1.25 \times 8000 \text{ m}^3\text{/d} \times 10^{-3} \text{ kg/g} = 50 \text{ kg/d as } Cl_2$$

b. Determine the chlorine quantity required for a 15-d storage.

$$w_{\text{storage}} = T_{\text{storage}}\ w_{\text{avg}} = 15 \text{ d} \times 50 \text{ kg/d} = 750 \text{ kg as } Cl_2$$

c. Determine the number of ton containers in storage.

One ton container contains ~900 kg liquid chlorine. Therefore, one container is theoretically sufficient to meet the storage requirement under average flow condition. Three containers are attached to each chlorinator with an automatic vacuum switchover system. All three containers from an empty bank should be replaced at the same time. Therefore, provide a storage space for three ton containers in the chlorine storage facility. A conceptual arrangement in storage area is also shown in Figure 11.12.

EXAMPLE 11.31: SIZING OF CHLORINE CONTACT BASIN

Two contact basins are provided for disinfection of secondary effluent at a wastewater treatment plant. The peak 2-h and average design flows are 0.66 and 0.22 m^3/s. The design contact time at peak design flow is 20 min. Determine the dimensions of a chlorine contact basin, and the influent and effluent structures.

Solution

1. Determine the contact basin configuration and dimensions.
   a. Describe the selected contact basin configuration and total volume requirement.
      Provide two identical and adjacent contact chambers. Each chamber shall have three passes around-the-end-baffled flow arrangement. A definition sketch of the chamber configuration is shown in Figure 11.13a.
      Calculate the total volume required in both contact basins.

$$V_{total} = Q_{2hpf}\,\theta_{contact} = 0.66\ m^3/s \times 20\ min \times 60\ s/min = 792\ m^3$$

$$\text{Volume required in each contact basin, } V = \frac{V_{total}}{N_{basin}} = \frac{792\ m^3}{2\ basins} = 396\ m^3$$

   b. Determine the contact basin dimensions.
      Provide a design water depth $H = 2$ m and calculate the surface area required in each contact basin.

$$A = \frac{V}{H} = \frac{396\ m^3}{2\,m} = 198\ m^2$$

      Provide a channel width $w = 2.5$ m, a total length $L_{total} = 84$ m in three passes, and a length of basin $L = 28$ m. The major dimensions of the contact basins are shown in Figure 11.13b.
   c. Verify the ratios and contact time requirements.

$$\text{Overall length-to-width ratio, } L/w \text{ Ratio} = \frac{L_{total}}{w} = \frac{84\ m}{2.5\,m} = 33.6 > 20$$

$$\text{Depth-to-width ratio, } H/w \text{ Ratio} = \frac{H}{w} = \frac{2\ m}{2.5\,m} = 0.8 < 1$$

      Actual volume of two contact basins, $V_{total} = N_{bsin}\,L_{total}\,Hw = 2 \times 84\,m \times 2\,m \times 2.5\ m = 840\ m^3$
      Compute the actual contact time when both basins are in operation at the peak design flow.

$$\theta_{contact} = \frac{V_{total}}{Q_{2hpf}} = \frac{840\ m^3}{0.66\ m^3/s \times 60\,s/min} = 21min > 20min$$

2. Determine the arrangement of the influent structure.
   a. Describe the selected influent structure and components.
      The secondary treated effluent is brought from the final clarifiers to the contact basins by a 1-m diameter pipe. The influent structure consists of (a) a 1.5 m × 1 m junction box, (b) a 1-m wide rectangular influent channel to carry the effluent to a contact basin after the flow split, and (c) a submerged 0.6 m × 0.6 m square orifice to feed the effluent into each basin. A sluice gate is provided over the opening to isolate one chamber from service for maintenance during the low flow period. Set the bottom of the junction box at 0.5 m below the basin bottom. The floor of the

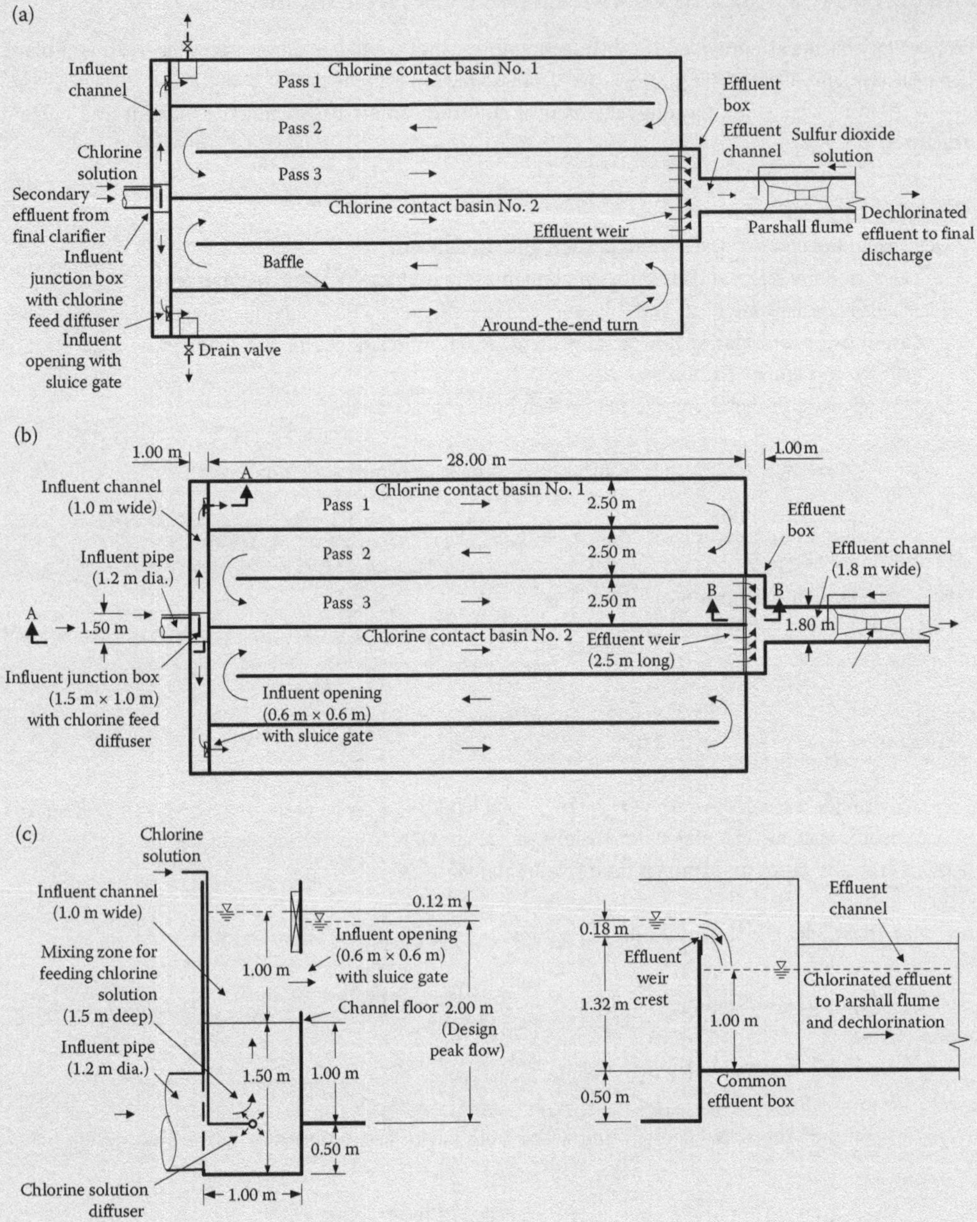

FIGURE 11.13 Design details of chlorine contact basins: (a) conceptual layout, (b) plan, (c) Section AA, and (d) Section BB (Example 11.31).

influent channel is 1 m above the bottom of the basin. A perforated diffuser is provided to apply the chlorine solution in the junction box before the flow splits into the influent channel. The design procedure of a perforated diffuser to provide energy for achieving a desired velocity gradient is given in Example 11.32. The design details of the influent structure are shown in Figure 11.13b and c.

b. Determine the head loss through the sluice gates.

The flow to each contact basin after an even flow split is Q_{basin}.

$$Q_{basin} = \frac{Q_{2hpf}}{N_{basin}} = \frac{0.66\ m^3/s}{2\ basin} = 0.33\ m^3/s\ \text{per basin}$$

The head loss across the gate at the peak design flow is calculated from orifice equation in Equation 7.4a. Assume $C_d = 0.6$.

$$h_{gate} = \frac{1}{2g}\left(\frac{Q_{basin}}{C_d A}\right)^2 = \frac{1}{2 \times 9.81\ m/s^2} \times \left(\frac{0.33\ m^3/s}{0.6 \times (0.6 \times 0.6)\ m^2}\right)^2 = 0.12\ m$$

Therefore, the water surface elevation in the influent channel is 0.12 m higher than that in the contact basins at the peak design flow.

3. Determine the arrangement of the effluent structure.

a. The effluent structure consists of (a) a rectangular weir for each basin, (b) a common effluent box, and (c) an exit channel with a Parshall flume. A standard Parshall flume with throat width of 1.5 ft (~0.45 m) is selected to measure the flow of chlorinated effluent and to provide a turbulent flow zone for mixing the chemicals. Sulfur dioxide is added in the mixing zone of the flume to dechlorinate the effluent. The dimensions of selected standard Parshall flume are given in Table C.3 in Appendix C. The design information about a Parshall flume and dechlorination facility are covered in Examples 9.38 and Section 11.7, respectively. The design details of the effluent structure are shown in Figure 11.13b and d.

b. Determine the head over the effluent weirs.

Calculated the head over the weir h_{weir} from Equation 8.10 at $Q_{basin} = 0.33\ m^3/s$, $C_d = 0.6$, $L_{weir} = w = 2.5$ m, and $n = 0$.

$$L' = L_{weir} - 0.1nH = L_{weir} = 2.5\ m$$

$$h_{weir} = \left(\frac{3}{2} \times \frac{Q_{basin}}{C_d L' \sqrt{2g}}\right)^{2/3} = \left(\frac{3}{2} \times \frac{0.33\ m^3/s}{0.6 \times 2.5\ m \times \sqrt{2 \times 9.81\ m/s^2}}\right)^{2/3} = 0.18\ m$$

Head over the effluent weir $h_{weir} = 0.18$ m at the peak design flow. The weir crest elevation will be 0.18 m below the design water surface in the basin. The weir crest height is 1.82 m above the bottom of the basin. The bottom of the common effluent box is set 0.5 m above the basin bottom, and the design water depth in the common effluent box is 1 m. Thus, a free fall of 0.32 m downstream of the weir is created at the design peak flow. The floor of effluent channel is same as that of the common effluent box (Figure 11.13c).

EXAMPLE 11.32: VELOCITY GRADIENT EXERTED BY PERFORATED DIFFUSER

In Example 11.31, a perforated diffuser is provided for feeding chlorine solution in the junction box before the flow is split into influent channels. Estimate the diameter and number of orifices required on the diffuser to achieve $G = 500\ s^{-1}$ at a sustained low temperature of 15°C and chlorine solution flow of 360 L/min.

Solution

1. Determine the mixing energy requirements.

The chlorine injection and mixing zone (D_{zone}) is near the floor of influent junction box (see Figure 11.13). The depth of $D_{zone} = 1.5$ m. The volume in the mixing zone (V_{zone})

$$V_{zone} = A_{box} D_{zone} = 1.5 \text{ m} \times 1 \text{ m} \times 1.5 \text{ m} = 2.25 \text{ m}^3$$

Calculate the mixing time at peak 2-hr flow of 0.66 m^3/s.

$$\theta_{zone} = \frac{V_{zone}}{Q_{2hpf}} = \frac{2.25 \text{ m}^3}{0.66 \text{ m}^3/\text{s}} = 3.4 \text{ s}$$

Typically adequate mixing of chlorine in the effluent is achieved within 2–15 s in the mixing zone. At 15°C, $\mu = 1.139 \times 10^{-3}$ N·s/m^2 is obtain from Table B.2 in Appendix B. The mixing water power imported in the junction box is calculate from Equation 9.27 at $G = 500 \text{ s}^{-1}$.

$$P_{water} = \mu\, VG^2 = 1.139 \times 10^{-3} \text{ N·s/m}^2 \times 2.25 \text{ m}^3 \times (500 \text{ s}^{-1})^2 = 640 \text{ N·m/s}$$

Assume the mixing efficiency of the diffuser is 50% of the hydraulic mixing power imported by the diffuser ($P_{diffuser}$).

$$P_{diffuser} = \frac{P_{water}}{E_{diffuser}} \times 100\% = \frac{640 \text{ N·m/s}}{50\%} \times 100\% = 1280 \text{ N·m/s} \quad \text{or} \quad 1280 \text{ kg·m}^2/\text{s}^3$$

2. Size the orifices.

Rearrange Equation 9.30a and calculate the water head loss needed through the orifices from the discharge of chlorine solution $q = 360$ L/min $= 6$ L/s $= 0.006$ m^3/s, and $\rho = 999.1$ kg/m^3 from Table B.2 in Appendix B. Ignore the mixing energy delivered by the influent.

$$h_{orifice} = \frac{P_{diffuser}}{\rho g q} = \frac{1280 \text{ kg·m}^2/\text{s}^3}{999.1 \text{ kg/m}^3 \times 9.81 \text{ m/s}^2 \times 0.006 \text{ m}^3/\text{s}} = 22 \text{ m}$$

Rearrange Equation 7.4a and calculate the total orifice area required at $C_d = 0.6$.

$$A = \frac{1}{C_d} \frac{q}{\sqrt{2gh_{orifice}}} = \frac{1}{0.6} \times \frac{0.006 \text{ m}^3/\text{s}}{\sqrt{2 \times 9.81 \text{ m/s}^2 \times 22 \text{ m}}} = 0.00048 \text{ m}^2$$

Velocity through the orifices, $v = \frac{q}{A} = \frac{0.006 \text{ m}^3/\text{s}}{0.00048 \text{ m}^2} = 12.5 \text{ m/s}$

High velocity through the orifices creates turbulence in the junction box for mixing the chlorine solution with the effluent. Select the orifice diameter of $d_{orifice} = 0.006$ m (~1/4 in) in the perforated diffuser. Calculate the area of each orifice and total number.

$$A_{orificee} = \frac{\pi}{4}(d_{orifice})^2 = \frac{\pi}{4}(0.006 \text{ m})^2 = 0.000028 \text{ m}^2 \text{ per orifice}$$

The number of orifices, $N_{orifice} = \frac{A}{N_{orifice}} = \frac{0.00048 \text{ m}^2}{0.000028 \text{ m}^2/\text{hole}} = 17 \text{ orifices}$

Provide a perforated diffuser with 16 orifices. The calculated results shall be validated with the diffuser performance based on the technical data obtained from the equipment manufacturer.

EXAMPLE 11.33: SELECT PUMP FOR MIXING CHLORINE SOLUTION BY JET NOZZLE

A jet mixing nozzle is use to feed chlorine solution into a 50-ft^3 mixing well prior to chlorine contact basins. The mixing intensity from the nozzle is such that a G of 600 s^{-1} is developed in the mixing well. Calculate (a) the mixing power output of the injector at a 60% efficiency and (b) the pump discharge pressure required at a solution pumping rate of 60 gpm. Assume the wastewater temperature is 60°F and the head loss in the pump discharge piping is 7 ft.

Solution

1. Calculate the mixing power output of the jet nozzle.
   At 60°F, $\mu = 2.36 \times 10^{-5}$ lb·s/ft^2 is obtain from Table B.3 in Appendix B.
   Calculate the power requirement in the mixing well from Equation 9.27 at $G = 600\ s^{-1}$.

$$P_{\text{water}} = \mu\ V\ G^2 = 2.36 \times 10^{-5}\ \text{lb·s/ft}^2 \times 50\ \text{ft}^3 \times (600\ \text{s}^{-1})^2 = 425\ \text{lb·ft/s}$$

Calculate the power required from a single nozzle at an efficiency of 60%.

$$P_{\text{nozzle}} = \frac{P_{\text{water}}}{E_{\text{nozzle}}} \times 100\% = \frac{425\ \text{lb·ft/s}}{60\%} \times 100\% = 708\ \text{lb·ft/s}$$

2. Calculate the required discharge pressure of the pump.

$$\text{Pumpflow}, Q = 60\ \text{gpm} \times \frac{1\ \text{ft}^3}{7.48\ \text{gal}} \times \frac{1\ \text{min}}{60\ \text{s}} = 0.134\ \text{ft}^3/\text{s (or cfs)}$$

At 60°F, the approximate water density $\gamma = 62.4$ lb/ft^3 is obtain from Table B.3 in Appendix B. Rearrange Equation 9.30a and calculate the head loss through the nozzle to provide the required energy for mixing.

$$h_{\text{nozzle}} = \frac{P_{\text{w}}}{\gamma Q} = \frac{708\ \text{lb·ft/s}}{62.4\ \text{lb/ft}^3 \times 0.134\ \text{ft}^3/\text{s}} = 84.7\ \text{ft}$$

Calculate the total head loss in the pump discharge pipings and nozzle.

$$h_{\text{discharge}} = h_{\text{piping}} + h_{\text{nozzle}} = (7 + 84.7)\ \text{ft} = 91.7\ \text{ft}$$

Convert the pump discharge head to pressure.

$$p_{\text{discharge}} = \gamma h_{\text{discharge}} = 62.4\ \text{lb/ft}^3 \times 91.7\ \text{ft} \times \frac{1\ \text{ft}^2}{144\ \text{in}^2} = 39.7\ \text{psi} \approx 40\ \text{psi}$$

Note: See also Examples 9.32 through 9.35 and 9.40 through 9.44 for additional information on calculating the velocity gradients from different mixing devices.

EXAMPLE 11.34: DISPERSION NUMBER OF A CHLORINE CONTACT BASIN

A chlorine contact basin is designed for a contact time of 30 min at a design peak flow of 6000 m^3/d. The dimensions of basin as initially proposed are: width $w = 1.5$ m and water depth $H = 3.0$ m. Validate if the required dispersion number is below the desired value of 0.015 at the operating temperature of 20°C.

Solution

1. Determine the horizontal velocity and length of the contact basin.

$$\text{Horizontal velocity, } v = \frac{Q}{Hw} = \frac{6000\ \text{m}^3/\text{d}}{3\ \text{m} \times 1.5\ \text{m}} \times \frac{\text{d}}{86{,}400\ \text{s}} = 0.0154\ \text{m/s}$$

$$\text{Length of contact basin, } L_{\text{total}} = \theta_{\text{contact}} v = 30\ \text{min} \times 0.0154\ \text{m/s} \times \frac{60\ \text{s}}{\text{min}} = 27.7\ \text{m}$$

Verify the ratios with the desired values.

$$\text{Overall length-to-width ratio, } L/w \text{ Ratio} = \frac{L_{\text{total}}}{w} = \frac{27.7\ \text{m}}{1.5\ \text{m}} = 18.5 < 20$$

$$\text{Depth-to-width ratio, } H/w \text{ Ratio} = \frac{H}{w} = \frac{3\ \text{m}}{1.5\ \text{m}} = 2 > 1$$

Both ratios are not meeting the typical guidance used in designing chlorine contact basins.

2. Select the equation for calculating the coefficient of dispersion.
The coefficient of dispersion is calculated from Equation 11.22.

$$D = 1.01 \nu N_{\text{R}}^{0.875} \tag{11.22a}$$

$$N_{\text{R}} = \frac{4\ vR}{\nu} \tag{11.22b}$$

where
D = coefficient of dispersion, m^2/s
ν = kinematic viscosity, m^2/s
N_{R} = Reynolds number
v = velocity in the channel, m/s
R = hydraulic mean radius (R = area/wetted perimeter), m

3. Determine the coefficient of dispersion.

$$R = \frac{Hw}{2H + w} = \frac{3\ \text{m} \times 1.5\ \text{m}}{2 \times 3\ \text{m} + 1.5\ \text{m}} = 0.6\ \text{m}$$

Obtain $\nu = 1.003 \times 10^{-6}\ \text{m}^2/\text{s}$ at 20°C from Table B.2 in Appendix B. Calculate N_{R} from Equation 11.22a.

$$N_{\text{R}} = \frac{4 \times 0.0154\ \text{m/s} \times 0.6\ \text{m}}{1.003 \times 10^{-6}\ \text{m}^2/\text{s}} = 36{,}800$$

Calculate D from Equation 11.22b.

$$D = 1.01 \times 1.003 \times 10^{-6}\ \text{m}^2/\text{s} \times (36800)^{0.875} = 0.0100\ \text{m}^2/\text{s}$$

4. Determine the dispersion number.
The dispersion number is calculated from Equation 3.41.

$$d = \frac{D}{vL_{\text{total}}} = \frac{0.0100\ \text{m}^2/\text{s}}{0.0154\ \text{m/s} \times 27.7\ \text{m}} = 0.023$$

The required dispersion number for chlorine contact basin at design peak flow is 0.015 or less. The calculated value of the dispersion number is 0.023, which exceeds the required value. The length-to-width and depth-to-width ratios are also out of the desired ranges. Therefore, an alternate design must be evaluated. Change the basin dimensions: $w = 1.75$ m, $H = 1.6$ m, and $L_{total} = 44.6$ m while maintaining the same design contact time of 30 min. Recalculate the dispersion number as well as the ratios. Follow the procedure presented in Steps 1, 3, and 4. The recalculated results are: $d = 0.0142$ (< 0.015), $L/w = 25.5$ (> 20), and $H/w = 0.91$ (< 1). The alternative design meets the typical design criteria as well as the required dispersion number.

EXAMPLE 11.35: CHLORINE DOSE FOR DISINFECTION OF FILTERED SECONDARY EFFLUENT

A filtered secondary effluent is disinfected by chlorine. The total coliform count before disinfection is 10^6 MPN/100 mL. The required summer and winter effluent total coliform counts are 23 and 240 MPN/100 mL, respectively. The initial chlorine demand of the effluent is 3.5 mg/L, and the demand due to decay in the contact basin is 2 mg/L after a required chlorine contact time of 40 min. Estimate the chlorine dose needed to disinfect the effluent. Assume that the shoulder effect exists.

Solution

1. State the applicable inactivation kinetic equation.

   Apply Collins–Selleck equation (Equation 11.5c). $N = N_0\left(\frac{Ct}{b}\right)^{-n}$ for $Ct > b$. Use the typical values of coefficients $n = 3$ and $b = 4$ for total coliforms (see Example 11.12 for determination of these coefficients).

2. Estimate the chlorine residual required during summer (C_{summer}).

   Rearrange Equation 11.5c and calculate C_{summer}.

$$C_{summer} = \frac{b}{t}\left(\frac{N_{summer}}{N_0}\right)^{-\frac{1}{n}} = \frac{4}{40} \times \left(\frac{23}{10^6}\right)^{-\frac{1}{3}} = 3.5 \text{ mg/L}$$

3. Estimate the chlorine residual required during winter (C_{winter}).

$$C_{winter} = \frac{b}{t}\left(\frac{N_{winter}}{N_0}\right)^{-\frac{1}{n}} = \frac{4}{40} \times \left(\frac{240}{10^6}\right)^{-\frac{1}{3}} = 1.6 \text{ mg/L}$$

4. Estimate the required chlorine dosages.

$$Dose_{summer} = Demand + Decay + C_{summer} = (3.5 + 2 + 3.5) \text{ mg/L} = 9 \text{ mg/L}$$

$$Dose_{winter} = Demand + Decay + C_{winter} = (3.5 + 2 + 1.6) \text{ mg/L} = 7.1 \text{ mg/L}$$

Note: Since the empirical constant b is a threshold constant, inactivation occurs during both summer and winter as long as it is significantly greater than b, and the Collins-Selleck equation is applicable. A contact time of 40 min applied in this example gives conservative chlorine dosages. A shorter contact time may be used if truly plug-flow condition can be achieved in the contact tank.

11.6.4 Other Beneficial Applications of Chlorine

Chlorine being a strong oxidant is used for many other beneficial applications. Among these are (a) control of sludge bulking, (b) odor control, (c) control of biological growth, (d) prevention of wastewater decomposition, (e) removal of organics, and (f) removal of inorganics. Each application is briefly described below.

Control of Sludge Bulking: Poor sludge settling or sludge bulking is a physical phenomenon caused by excessive growth of filamentous bacteria. These organisms are largely dispersed in the liquid, and therefore are more easily attacked by chlorine than are the healthy floc farming organisms that are protected within the floc structures. The disinfectant is typically applied into the return activated sludge (RAS). An average chlorine dose is 5–6 mg/L based on RAS flow, or daily feed rate of 0.4–0.7% by weight of VSS.[92] Also, biomass exposure to chlorine on the average should be no longer than 6–8 h during the bulking period.[5,92]

Odor Control: Offensive odors in wastewater systems are mainly caused by hydrogen sulfide (H_2S). Chlorine is used to destroy H_2S and inhibit growth of bacteria for effective odor control. Oxidation of H_2S by chlorine is a complex process. The factors affecting the process include pH, alkalinity, temperature, and the ratio of chlorine to sulfide.[92] The end products are a mixture of sulfate and sulfur. The generalized oxidation reactions of chlorine compounds with H_2S are given by Equations 11.23a through 11.23f.

$$4\ HOCl + H_2S + 6\ OH^- \rightarrow SO_4^{2-} + 4\ Cl^- + 6\ H_2O \tag{11.23a}$$

$$HOCl + H_2S + OH^- \rightarrow S\downarrow + Cl^- + 2\ H_2O \tag{11.23b}$$

$$4\ Cl_2 + H_2S + 10\ OH^- \rightarrow SO_4^{2-} + 8\ Cl^- + 6\ H_2O \tag{11.23c}$$

$$Cl_2 + H_2S + 2\ OH^- \rightarrow S\downarrow + 2\ Cl^- + 2\ H_2O \tag{11.23d}$$

$$4\ NaOCl + H_2S + 2\ OH^- \rightarrow SO_4^{2-} + 4\ Na^+ + 4\ Cl^- + 2\ H_2O \tag{11.23e}$$

$$NaOCl + H_2S \rightarrow S\downarrow + Na^+ + Cl^- + H_2O \tag{11.23f}$$

For odor control, chlorine can be fed directly into the wastewater, or used through a wet-air scrubber. Thorough assessment of chlorine dosage must be made to evaluate the buffering requirement, corrosion control, and explosion prevention issues that are potentially associated with odor problems. To determine the chlorine dosage requirement for oxidation of H_2S is the *chlorine demand test*.[92] The normal concentration of H_2S in municipal wastewater ranges 0–5 mg/L as S, and up to 30 ppm as S in the off-gases. Chlorine dosage may vary from 10 to 50 mg/L as Cl_2 for collection system and 10 to 20 mg/L for prechlorination at a treatment plant.[15] Theoretical chlorine dosage can be estimated using a ratio of 8.83 mg/L as Cl_2 per mg/L H_2S as S. Alkalinity consumption ratio is 15.6 mg/L alkalinity as $CaCO_3$ per mg/L of H_2S as S.

Control of Biological Growths: Excessive and desired attached biological growths over walls and other surfaces may affect the performance of many wastewater treatment units. Biofilm growth over effluent weirs and launders may cause uneven flow distribution, short circuiting through the final clarifiers, and may cause compliance problems with discharge permit. The undesirable growth can be burned off by periodical feeding of strong chlorine solution at proper locations. Examples are screens, effluent weirs and launders, division boxes, ahead of filters, and other locations. The chlorine concentration in washing solution may be in a range of 1–10 mg/L.[2]

Prevention of Wastewater Decomposition: High microbial population is associated with rapid decomposition rate and development of anaerobic activity and odors. Chlorine dose will reduce the microbial population and decomposition rate.

Removal of Organics: Chlorine oxidizes organic compounds and BOD_5 is redued. Example 11.39 gives a procedure to calculate BOD_5 reduction from oxidation of a specific organic compound. As a general rule for each mg/L of chlorine absorbed, there is 0.5 to 2 mg/L reduction of BOD_5. The generalized reactions of chlorine compounds with empirical expression $C_5H_7O_2N$ for organic matter (or VSS) are expressed by Equations 11.24a through 11.24c.

$$10\ HOCl + C_5H_7O_2N + 9\ OH^- \rightarrow 5\ CO_2 + NH_4^+ + 10\ Cl^- + 11\ H_2O \qquad (11.24a)$$

$$10\ Cl_2 + C_5H_7O_2N + 19\ OH^- \rightarrow 5\ CO_2 + NH_4^+ + 20\ Cl^- + 11\ H_2O \qquad (11.24b)$$

$$10\ NaOCl + C_5H_7O_2N \rightarrow 5\ CO_2 + NH_4^+ + 10\ Na^+ + 10\ Cl^- + OH^- + H_2O \qquad (11.24c)$$

Removal of Inorganic Compounds: Many inorganic compounds can be oxidized and removed from wastewater by chlorination or hypochlorination. These compounds are ammonia, hydrogen sulfide, cyanide, iron, manganese, and nitrite. The reactions with ammonia have been presented in Section 11.6.2. The reactions with H_2S are covered earlier under odor control (Equation 11.23). The reactions with other compounds are summarized below (Equations 11.25a through 11.28c).

Reactions with Cyanide (CN^-)

$$2\ HOCl + 2\ CN^- + 2\ OH^- \rightarrow 2\ CNO^- + 2\ Cl^- + 2\ H_2O \qquad (11.25a)$$

$$2\ Cl_2 + 2\ CN^- + 4\ OH^- \rightarrow 2\ CNO^- + 4\ C^- + 2\ H_2O \qquad (11.25b)$$

$$NaOCl + CN^- \rightarrow CNO^- + Na^+ + Cl^- \qquad (11.25c)$$

Note: The stoichiometric dosage is 2.7 mg/L as Cl_2 per mg/L of CN^- oxidized.

Reactions with Iron (Ferrous, Fe^{2+})

$$HOCl + 2\ Fe^{2+} + 5\ OH^- \rightarrow 2\ Fe(OH)_3 \downarrow + 5\ Cl^- \qquad (11.26a)$$

$$Cl_2 + 2\ Fe^{2+} + 6\ OH^- \rightarrow 2\ Fe(OH)_3 \downarrow + 2\ Cl^- \qquad (11.26b)$$

$$NaOCl + 2\ Fe^{2+} + 4\ OH^- + H_2O \rightarrow 2\ Fe(OH)_3 \downarrow + Na^+ + Cl^- \qquad (11.26c)$$

Note: The stoichiometric dosage is 0.63 mg/L as Cl_2 per mg/L of Fe^{2+} oxidized.

Reactions with Manganese (Mn^{2+})

$$HOCl + Mn^{2+} + 3\ OH^- \rightarrow MnO_2 \downarrow + Cl^- + 2\ H_2O \qquad (11.27a)$$

$$Cl_2 + Mn^{2+} + 4\ OH^- \rightarrow MnO_2 \downarrow + 2\ Cl^- + 2\ H_2O \qquad (11.27b)$$

$$NaOCl + Mn^{2+} + 2\ OH^- \rightarrow MnO_2 \downarrow + Na^+ + Cl^- + H_2O \qquad (11.27c)$$

Note: The stoichiometric dosage is 1.3 mg/L as Cl_2 per mg/L of Mn^{2+} oxidized.

Reactions with Nitrite (NO_2^-)

$$HOCl + NO_2^- + OH^- \rightarrow NO_3^- + Cl^- + H_2O \qquad (11.28a)$$

$$Cl_2 + NO_2^- + 2\ OH^- \rightarrow NO_3^- + 2\ Cl^- + H_2O \qquad (11.28b)$$

$$NaOCl + NO_2^- \rightarrow NO_3^- + Na^+ + Cl^- \qquad (11.28c)$$

Note: The NO_2^-–N is oxidized by Cl_2 at a stoichiometric dosage of 5.1 mg/L as Cl_2 per mg/L of NO_2^- as N.

EXAMPLE 11.36: CHLORINATION FOR CONTROL OF SLUDGE BULKING

An activated sludge plant experiences sludge bulking problems due to excessive growth of filamentous organisms. Chlorination of RAS is considered for control of sludge bulking. The operation conditions at the plant are summarized below:

Daily average influent flow to aeration basin $Q = 1.0$ MGD, ratio of RAS to influent flow $R_{ras} = 0.50$, total volume of aeration basin $V_{ab} = 0.25$ MG, total volume of final clarifier $V_{fc} = 0.125$ Mgal, MLSS concentration in aeration basins MLSS = 1500 mg/L, TSS concentration in the RAS flow $TSS_{ras} =$ 4500 mg/L, average TSS concentration in final clarifier $TSS_{fc} = 2000$ mg/L, and VSS/TSS ratio = 0.70. Chlorine dose in the RAS flow $Dose_{Cl2} = 5$ mg/L.

Determine the amount of chlorine supplied for bulking control based on the total VSS inventory in the biological system, and verify the exposure frequency of biomass to chlorine.

Solution

1. Evaluate the amount of chlorine applied into the RAS flow.

$$\text{Daily average RAS flow, } Q_{ras} = R_{ras}\,Q = 0.5 \times 1 \text{ MGD} = 0.5 \text{ MGD}$$

$$\text{Daily average amount of chlorine, } w_{Cl2} = Dose_{Cl2}\,Q_{ras} = 5 \text{ mg/L} \times 0.5 \text{ MGD} \times \frac{8.34 \text{ lb}}{\text{mg/L}\cdot\text{Mgal}}$$
$$= 21 \text{ lb/d as } Cl_2$$

2. Determine the total VSS inventory in the biological system.

$$\text{Total TSS inventory in the aeration basins, } \Delta TSS_{ab} = \text{MLSS}\,V_{ab} = 1500 \text{ mg/L} \times 0.25 \text{ Mgal} \times \frac{8.34 \text{ lb}}{\text{mg/L}\cdot\text{Mgal}}$$
$$= 3100 \text{ lbs as TSS}$$

$$\text{Total TSS inventory in the final clarifier, } \Delta TSS_{fc} = TSS_{fc}\,V_{fc} = 2000 \text{ mg/L} \times 0.125 \text{ Mgal} \times \frac{8.34 \text{ lb}}{\text{mg/L}\cdot\text{Mgal}}$$
$$= 2100 \text{ lbs as TSS}$$

$$\text{Total TSS inventory in the biological system, } \Delta TSS = \Delta TSS_{ab} + \Delta TSS_{fc} = (3100 + 2100) \text{ lbs}$$
$$= 5200 \text{ lbs as TSS}$$

$$\text{Total VSS inventory in the biological system, } \Delta VSS = \text{VSS/TSS ratio} \times \Delta TSS$$
$$= 0.7 \text{ lb VSS/lb TSS} \times 5200 \text{ lbs TSS}$$
$$= 3600 \text{ lbs as VSS}$$

3. Determine the amount of chlorine supplied based on the total VSS inventory.

$$\text{Daily amount of chlorine supplied, } f_{Cl2} = \frac{w_{Cl2}}{\Delta VSS} \times 100\% = \frac{21 \text{ lb/d as } Cl_2}{3600 \text{ lbs as VSS}} \times 100\% = 0.58\% \approx 0.6\%$$

Based on the total VSS inventory, the average daily chlorine dose for control of sludge bulking is about 0.7% VSS in the system. It is within the typical range of 0.4–0.7%.

4. Verify the average exposure frequency of biomass to chlorine.

Daily average mass of TSS in RAS flow, $m_{ras} = TSS_{ras}Q_{ras} = 4500\ \text{mg/L} \times 0.5\ \text{Mgal} \times \dfrac{8.34\ \text{lb}}{\text{mg/L·Mgal}}$

$$= 19{,}000\ \text{lbs as TSS}$$

$$\text{Average exposure frequency of biomass, } \theta_{ras/Cl2} = \frac{\Delta TSS}{m_{ras}} = \frac{5200\ \text{lbs as TSS}}{19{,}000\ \text{lb/d as TSS}} \times \frac{24\ \text{h}}{\text{d}} = 6.6\ \text{h}$$

(It is 6.6 h between exposures or 3.7 times per day.)

The average exposure frequency of biomass to chlorine is within the typical range of 6–8 h.

EXAMPLE 11.37: THEORETICAL CHLORINE REQUIREMENT AND ALKALINITY CONSUMPTION IN OXIDATION REACTIONS OF ORGANIC AND INORGANIC VOMPOUNDS

Chlorine oxidizes and removes numerous organic and inorganic compounds. Determine the theoretical chlorine requirements and associated alkalinity consumptions for oxidation of unit weight of organic and inorganic compounds. Use equations given for gaseous chlorine (Cl_2) in Section 11.6.4.

Solution

In Section 11.6.4, oxidation reactions of chlorine (in forms of HOCl, Cl_2, and NaOCl) with many inorganic and organic compounds are given. Determine the chlorine requirement as available Cl_2 and alkalinity consumption as $CaCO_3$ when unit mass of these compounds are oxidized. Tabulate the calculations and results in Table 11.16.

TABLE 11.16 Chlorine Requirements and Alkalinity Consumptions in Oxidation Reactions for Beneficial Applications (Example 11.37)

Oxidation Reaction	Applicable Eq.	Chlorine Required per Gram of Compound Oxidized	Alkalinity Consumption per Gram of Compound Oxidized
H_2S to SO_4^{2-}	11.23c	$(4 \times 70.9)/32.1 = 8.83$ g available Cl_2 required per g H_2S removed as S	$(10 \times 50.0)/32.1 = 15.6$ g alkalinity consumed as $CaCO_3$ per g H_2S removed as S
$C_5H_7O_2N$ to CO_2	11.24b	$(10 \times 70.9)/(5 \times 12{,}0) = 11.8$ g available Cl_2 required per g VSS oxidized as C or $(10 \times 70.9)/113 =$ 6.27 g available Cl_2 required per g VSS oxidized as $C_5H_7O_2N$	$(19 \times 50.0)/(5 \times 12.0) = 15.8$ g alkalinity consumed as $CaCO_3$ per g VSS oxidized as C or $(19 \times 50.0)/113 = 8.41$ g alkalinity consumed as $CaCO_3$ per g VSS oxidized as $C_5H_7O_2N$
CN^- to CNO^-	11.25b	$(2 \times 70.9)/(2 \times 26.0) = 2.73$ g available Cl_2 required per g CN^- oxidized	$(4 \times 50.0)/(2 \times 26.0) = 3.85$ g alkalinity consumed as $CaCO_3$ per g CN^- oxidized
Fe^{2+} to Fe^{3+}	11.26b	$70.9/(2 \times 55.8) = 0.63$ g available Cl_2 required per g Fe^{2+} oxidized as Fe	$(6 \times 50.0)/(2 \times 55.8) = 2.69$ g alkalinity consumed as $CaCO_3$ per g Fe^{2+} oxidized as Fe
Mn^{2+} to Mn^{3+}	11.27b	$70.9/54.9 = 1.29$ g available Cl_2 required per g Mn^{2+} oxidized as Mn	$(4 \times 50.0)/54.9 = 3.64$ g alkalinity consumed as $CaCO_3$ per g Mn^{2+} oxidized as Mn
NO_2^- to NO_3^-	11.28b	$70.9/14{,}0 = 5.06$ g available Cl_2 required per g NO_2-N oxidized as N	$(2 \times 50.0)/14.0 = 7.14$ g alkalinity consumed as $CaCO_3$ per g NO_2-N oxidized as N

EXAMPLE 11.38: SODIUM HYPOCHLORITE DOSAGE AND ALKALINITY CONSUMPTION FOR ODOR CONTROL

Feeding sodium hypochlorite (NaOCl) is recommended for odor control in the wet well of a lift station. The maximum concentration of H_2S in the wet well $C_{H2S} = 5$ mg/L as S. Determine the dosage of chlorine and consumption of alkalinity that may theoretically occur for odor control.

Solution

1. Identify the equation for odor control by NaOCl.
   The oxidation reaction of H_2S by NaOCl is expressed by Equation 11.23e.

$$\underset{4\times 74.4}{4\,\text{NaOCl}} + \underset{32.1}{\text{H}_2\text{S}} + \underset{\substack{2\times 17(\text{as OH}^-)\\ 2\times 50\ (\text{as CaCO}_3)}}{2\,\text{OH}^-} \rightarrow \text{SO}_4^{2-} + 4\,\text{Na}^+ + 4\,\text{Cl}^- + 2\,\text{H}_2\text{O}$$

2. Estimate the NaOCl dosage required for odor control.
   Calculate the NaOCl requirement ($R_{NaOCl:H2S}$) from the relationship identified in Step 1.

$$R_{NaOCl:H2S} = \frac{4 \times 74.4\ \text{g/mole of NaOCl}}{32.1\ \text{g/mole of H}_2\text{S as S}} = 9.27\ \text{g NaOCl/g H}_2\text{S as S or } 9.27\ \text{mg NaOCl/mg H}_2\text{S as S}$$

$$\text{NaOCl dosage,}\quad Dose_{NaOCl} = R_{NaOCl:H2S} \times C_{H2S} = 9.27\ \text{mg Cl}_2/\text{mg S} \times 5\ \text{mg/L as S}$$
$$= 46.4\ \text{mg/L as NaOCl}$$

Note: The available chlorine content of NaOCl is 95.3% (Table 11.11). The available Cl_2 requirement = 0.953 g available Cl2/g NaOCl × 9.27 g NaOCl/g H_2S as S = 8.83 g NaOCl/g H_2S as S. The available Cl_2 requirement is same as that calculated from Equation 11.23c in Example 11.37 (Table 11.16).

3. Estimate the alkalinity consumption during odor control.
   Calculate the alkalinity requirement ($R_{Alk:H2S}$) from the relationship identified in Step 1.

$$R_{Alk:H2S} = \frac{2 \times 17\ \text{g/mole of NaOH}}{32.1\ \text{g/mole of H}_2\text{S as S}} \times \frac{50\ \text{g/eq. of CaCO}_3}{17\ \text{g/eq. of NaOH}}$$
$$= 3.11\ \text{g Alk as CaCO}_3/\text{g H}_2\text{S as S or } 3.11\ \text{mg Alk as CaCO}_3/\text{mg H}_2\text{S as S}$$

$$\text{Alkalinity consumption, Comsumption}_{Alk} = R_{Alk:H2S} \times C_{H2S} = 3.11\ \text{mg CaCO}_3/\text{mg S} \times 5\ \text{mg/L as S}$$
$$= 15.6\ \text{mg/L as CaCO}_3$$

Note: When gaseous chlorine (Cl_2) is used, an alkalinity consumption of 15.6 g $CaCO_3$/g H_2S as S is calculated from Equation 11.23c in Example 11.37 (Table 11.16). Therefore, the alkalinity consumption from Cl_2 is five times that of NaOCl for odor control.

EXAMPLE 11.39: REMOVAL OF BOD_5 BY CHLORINE FOR STABILIZATION OF ORGANIC SUBSTANCES

Chlorine residual is consumed in reactions with a variety of organic substances. A combined chlorine consumption of 2.5 mg/L as Cl_2 is observed in a chlorine contact basin. It is estimated that 80% of chlorine consumed is utilized in the reaction with organic substances. Estimate the removal of BOD_5 by chlorine. The ratio $R_{BOD_5:BOD_L}$ of BOD_5 is typically 68% of BOD_L.

Solution

1. Determine the chlorine residual consumed in reaction with organic matter.
   Chlorine consumption from reaction with organic matter is 80% of total chlorine consumption.

$$C_{Cl2,VSS} = 80\% \times Dose_{Cl2} = 80\% \times 2.5 \text{ mg/L} = 2 \text{ mg/L as available } Cl_2$$

2. Estimate the organic substances (or VSS) oxidized by chlorine.
   The theoretical Cl_2 requirement, $R_{Cl2:VSS} = 6.27$ g Cl_2 per g VSS oxidized as $C_5H_7O_2N$. See Equation 11.24b in Example 11.37 and Table 11.16. The VSS destroyed by chlorine $= w_{VSS}$.

$$w_{VSS} = \frac{C_{Cl2,VSS}}{R_{Cl2:VSS}} = \frac{2 \text{ mg } Cl_2/L}{6.27 \text{ mg } Cl_2/\text{mg VSS as } C_5H_7O_2N} = 0.32 \text{ mg VSS/L}$$

3. Estimate the BOD_5 oxidized by chlorine.
   The ultimate oxygen demand (BOD_L) exerted by microbial cell (VSS), $R_{BODL:VSS} = 1.42$ g BOD_L/g VSS oxidized (Step 2 of Example 10.10). Calculate the BOD_5 oxidized by chlorine.

$$C_{BOD5} = R_{BOD5:BODL} R_{BODL:VSS} \Delta w_{VSS}$$
$$= 0.68 \text{ mg } BOD_5/\text{mg } BOD_L \times 1.42 \text{ mg } BOD_L/\text{mg VSS} \times 0.32 \text{ mg VSS/L} = 0.31 \text{ mg } BOD_5/L$$

The removal of BOD_5 by chlorine is about 0.31 mg/L in the chlorine contact basin.

Note: A significant portion of oxidizable organic compounds in the secondary effluent are rapidly oxidized by chlorine. A rule of thumb is for each mg/L chlorine consumed, BOD_5 reduction is 1–2 mg/L.

11.6.5 Disinfection with Chlorine Dioxide

Chlorine dioxide (ClO_2) has been used for years to bleach flour, paper, and textile. Its use in wastewater effluent disinfection is recent because of its powerful bactericide and viricide than chlorine. Its DBPs have much less adverse effects than those associated with chlorination. Also, ClO_2 does not react with ammonia. However, use of ClO_2 for wastewater disinfection may be very limited because it is more expensive than chlorine.

Properties of Chlorine Dioxide: ClO_2 is a greenish-yellow gas with an irritating odor stronger than chlorine, and is toxic to humans when inhaled. It makes an explosive mixture in air in concentrations above 10% by volume (corresponding to about 12 g/L concentration in a solution). Important physical, chemical, and biological properties are (a) it is a more powerful disinfectant than chlorine, although its oxidation potential is −0.95 volt (aq) lower than chlorine (see Table 11.4); (b) in the absence of free chlorine, it does not react with ammonia and does not produce THMs and other chlorinated by-products of concern; (c) in the presence of excess chlorine, many chlorinated and brominated organic compounds are produced; (d) the principal DBPs formed with ClO_2 are *chlorite* (ClO_2^-) and *chlorate* (ClO_3^-), both of which are toxic compounds, and are produced during generation of ClO_2; (e) free ClO_2 residuals have a short life and are less harmful to aquatic life than chlorine; and (f) the oxidizing power of ClO_2 is often referred to as *available chlorine*. The oxidizing power of ClO_2 has 263% of chlorine (Table 11.11).

Generation of Chlorine Dioxide: Chlorine dioxide is an unstable and explosive gas. It is typically generated on-site in a gaseous solution and must be used soon after preparation. Solutions up to 5000 mg/L may be stored for several days. There are three basic methods for ClO_2 generation:

(a) chlorite–chlorine, (b) chlorite–hypochlorite acid, and (c) chlorite acid. The overall reactions for these methods are given by Equations 11.29a through 11.29c.

$$2\ NaClO_2 + Cl_2(g) \rightarrow 2\ ClO_2(g) + 2\ Na^+ + 2\ Cl^- \quad (11.29a)$$

$$2\ NaClO_2 + NaOCl + 2\ HCl \rightarrow 2\ ClO_2(g) + 3\ Na^+ + 3\ Cl^- + H_2O \quad (11.29b)$$

$$5\ NaClO_2 + 4\ HCl \rightarrow 4\ ClO_2(g) + 5\ Na^+ + 5\ Cl^- + 2\ H_2O \quad (11.29c)$$

In the first method (Equation 11.29a), the stock solution contains sodium chlorite ($NaClO_2$) and chlorine gas Cl_2. Stoichiometrically, 1.34 g $NaClO_2$ reacts with 0.53 g Cl_2 to yield 1 g of ClO_2 (Example 11.40). The reaction is very efficient and has many operational flexibility, but there are serious safety concerns due to storage and application of Cl_2 gas at large plants. The second method (Equation 11.29b) is a viable alternative because the gaseous Cl_2 is not used. Sodium chlorite reacts with sodium hypochlorite (NaOCl) in presence of HCl. The stoichiometric ratios of these compounds are, respectively, 1.34, 0.55, and 0.54 g each to yield 1 g ClO_2. In the third method (Equation 11.29c), $NaClO_2$ and HCl are used directly in stoichiometric ratios of 1.68 and 0.54 g each to yield 1 g of Cl_2. This method is simple and easy to operate, and is suitable for small applications. The process has a maximum of only 80% efficiency of theoretical conversion from $NaClO_2$ to ClO_2.

Sodium chlorite ($NaClO_2$) is used in all three methods of chlorine dioxide generation. It is commercially available as an aqueous solution with 15–35% active $NaClO_2$, has specific gravity of 1.1–1.3 at 20–25°C, and a pH in the range of 12–13. The quantities of reactants in the stock solution are measured separately and then pumped, while vacuum is used for gaseous chlorine. As the combined solution flows upward through a reaction chamber, ClO_2 is generated when a desired reaction time is reached. The reaction time varies from 1 min for Cl_2 to 15 min for processes using HCl.

The source of undesired by-products chlorite ions (ClO_2^-) is the reduction of ClO_2 during generation process and escape of unreacted chlorite ion from the reactor. The undesired chlorate ions ClO_3^- are produced from the oxidation of chlorine dioxide from the impurities in sodium chlorite feedstock, and photolytic decomposition of ClO_2. The chlorite ion can be reduced to chloride ion by ferrous or sulfite ion (Fe^{2+} or SO_3^{2-}). The most effective way to control chlorite ion is to prevent its generation by careful management of feedstock and by using controlled excess chlorine feed beyond stoichiometric requirement.

Design Considerations: The basic design considerations for ClO_2 system are (a) ClO_2 dosage requirement, (b) ClO_2 generator, (c) supply, storage, and feed systems for reactants required for selected process, (d) temporary storage for generated ClO_2 solution, (e) transport of metered ClO_2 solution to the application point, and (f) dispersion of solution in wastewater. The *Ct* values of chlorine dioxide are compared with other disinfectants in Table 11.9. The ClO_2 generators are commercially available. The other components are similar to those of chlorination facility and may be partially adapted to chlorine dioxide system.

EXAMPLE 11.40: FEEDSTOCK SOLUTION FOR CHLORINE DIOXIDE GENERATION

The chlorine dioxide is generated from chlorine gas and sodium chlorite solution of 25% active content. The ClO_2 dosage is 5 mg/L as ClO_2 in an effluent flow of 8000 m^3/d. Determine (a) the quantity of Cl_2 gas and (b) the quantity and volume of sodium chlorite solution required per day. Assume the sodium chlorite solution has a specific gravity of ~1.2.

Solution

1. Determine the quantity of ClO_2 required for disinfection.

$$\Delta w_{ClO2} = C_{ClO2}Q = 5\ g/m^3 \times 8000\ m^3/d \times 10^{-3}\ kg/g = 40\ kg/d \text{ as } ClO_2$$

2. Determine the stoichiometric requirements of Cl_2 and $NaClO_2$ from Equation 1.29a.

$$\underset{2\times 90.5}{2\ NaClO_2} + \underset{70.9}{Cl_2(g)} \rightarrow \underset{2\times 67.5}{2\ ClO_2(g)} + 2\ Na^+ + 2\ Cl^-$$

The stoichiometric requirements of Cl_2,

$$R_{Cl2:ClO2} = \frac{70.9\ g/\text{mole of } Cl_2}{2 \times 67.5\ g/\text{mole of } ClO_2} = 0.53\ g\ Cl_2/g\ ClO_2$$

The stoichiometric requirements of $NaClO_2$, $R_{NaClO2:ClO2} = \dfrac{2 \times 90.5\ g/\text{mole of } NaClO_2}{2 \times 67.5\ g/\text{mole of } ClO_2}$

$$= 1.34\ g\ NaClO_2/g\ ClO_2$$

3. Determine the quantity of Cl_2 required.

Cl_2 required, $\Delta w_{Cl2} = R_{Cl2:ClO2}\Delta w_{ClO2} = 0.53\ kg\ Cl_2/kg\ ClO_2 \times 40\ kg\ ClO_2/d = 21\ kg/d \text{ as } Cl_2$

4. Determine the quantity and volume of $NaClO_2$ solution required.

$NaClO_2$ solution weight, $\Delta w_{NaClO2} = \dfrac{R_{NaClO2:ClO2}\Delta w_{Cl2}}{25\%} = \dfrac{1.34\ kg\ NaClO_2/kg\ ClO_2 \times 40\ kg\ ClO_2/d}{0.25\ kg\ NaClO_2/kg\ solution}$

$$= 214\ kg\ NaClO_2/d \text{ as solution}$$

$NaClO_2$ solution volume, $V_{NaClO2} = \dfrac{\Delta w_{NaClO2}}{1.2 \times 1000\ kg/m^3} = \dfrac{214\ kg\ NaClO_2/d \text{ as solution}}{1.2 \times 1000\ kg/m^3}$

$$= 0.18\ m^3/d \quad \text{or} \quad 7.5\ L/h$$

EXAMPLE 11.41: ACTUAL AND AVAILABLE CHLORINE IN CHLORINE DIOXIDE

Determine (a) the weight percent of actual chlorine and (b) weight percent of available chlorine in chlorine dioxide. Also determine g available chlorine per mole of ClO_2.

Solution

1. Determine the weight percent of actual chlorine in ClO_2.

Weight percent of actual chlorine, $(\text{Weight \% Cl})_{ClO2} = \dfrac{35.45\ g\ Cl/\text{mole of } Cl_2}{67.45\ g\ ClO_2/\text{mole of } ClO_2} \times 100\%$

$$= 52.6\%\ Cl \text{ in } ClO_2 \quad \text{or} \quad 0.526\ g\ Cl/g\ ClO_2$$

2. Determine the weight percent of available chlorine in ClO_2.
   Write the oxidation half reaction for ClO_2 in Equation 11.29d.

$$ClO_2 + 5e^- + 4\ H^+ \rightarrow Cl^- + 2\ H_2O \qquad (11.29d)$$

In the half oxidation reaction for ClO_2, the chlorine valance changes from +4 to −1; that is a net valence change of −5. It means that each chlorine element in ClO_2 molecule has an oxidizing power of accepting *five* (5) electrons. Therefore, the chlorine in ClO_2 has a chlorine equivalent $eq_{Cl} = 5$ g Cl_2/g Cl. The weight percent of available chlorine in ClO_2 is calculated from Equation 1.8d.

$$(\text{Weight \% } Cl_2)_{ClO2} = eq_{Cl} \times (\text{Weight \% actual})_{ClO2} = 5 \times 52.6\% = 263\% \text{ or } 2.63 \text{ g } Cl_2/\text{g } ClO_2$$

Therefore, 1 mg/L (or g/m^3) of ClO_2 is equivalent to 2.63 mg/L (or g/m^3) of Cl_2.

3. Determine g equivalent chlorine per mole of ClO_2.

   One mole of $ClO_2 = 67.45$ g.

$$\begin{aligned}\text{Equivalent chlorine per mole of } ClO_2, (\text{Weight \% } Cl_2)_{ClO2} &= (\text{Weight \% } Cl_2)_{ClO2} \times mw_{Cl2} \\ &= 2.63 \text{ g } Cl_2/\text{g } ClO_2 \times 67.45 \text{ g/mole of } ClO_2 \\ &= 177 \text{ g } Cl_2/\text{mole } ClO_2\end{aligned}$$

11.7 Dechlorination

Chlorine residual even at low level may have potential toxic effects on aquatic organisms. It may also interfere with the biomonitoring test required to determine the toxicity of an effluent. The purpose of dechlorination is to remove free and combined chlorine residuals from the effluent. It is achieved by reacting the residual chlorine with a reducing agent. The most commonly used reducing agents are sulfur dioxide (SO_2) and sodium bisulfite ($NaHSO_3$). Other compounds are sodium sulfite (Na_2SO_3), sodium metabisulfite ($Na_2S_2O_5$), and sodium thiosulfate ($Na_2S_2O_3$). As an alternative to a reducing agent, activated carbon will remove chlorine residual along with many other organic compounds.

11.7.1 Dechlorination with Sulfur Dioxide (SO_2)

Reactions: Sulfur dioxide is an effective reducing agent for removal of free and combined chlorine residuals. Useful physical properties of sulfur dioxide gas and liquid are provided in Table 11.10. When added to water, SO_2 hydrolyses to form sulfurous acid (H_2SO_3), a strong reducing agent. Sulfurous acid dissociates to produce HSO_3^- (bisulfite) ion which reacts with free and combined chlorine residuals producing chloride and sulfate ions. The reactions of SO_2 with free chlorine and chloramines are given by Equations 11.30a through 11.30d.

Reactions with Free Chlorine.

Hydrolysis of SO_2:

$$SO_2 + H_2O \rightarrow H^+ + HSO_3^- \quad (11.30a)$$

Reaction with HOCl:

$$HSO_3^- + HOCl \rightarrow 2\,H^+ + SO_4^{2-} + Cl^- \quad (11.30b)$$

Overall reaction:

$$SO_2 + HOCl + H_2O \rightarrow 3\,H^+ + SO_4^{2-} + Cl^- \quad (11.30c)$$

Reactions with Combined Chlorine Residual (Represented by Monochloramine NH_2Cl).

$$SO_2 + NH_2Cl + 2\,H_2O \rightarrow NH_4Cl + 2\,H^+ + SO_4^{2-} \quad (11.30d)$$

These stoichiometric reactions provide the following useful relationships:

1. The stoichiometric weight ratio of SO_2 to HOCl (expressed as available Cl_2) is 0.904 g SO_2/g available Cl_2 (64.1/70.9 in Equation 11.30c). In practice, a ratio of 1:1 to 1.2:1 may be required because of interference caused by the organic nitrogen.
2. The alkalinity consumption is 2.12 g as $CaCO_3$ when 1 g of free chlorine residual (HOCl) as available Cl_2 is removed in the reaction with SO_2 (3 × 50.0/70.9 in Equation 11.30c).
3. An alkalinity consumption of 1.41 g as $CaCO_3$ occurs when 1 g of combined chlorine residual (NH_2Cl) as available Cl_2 is removed in the reaction with SO_2 (2 × 50.0/70.9 in Equation 11.30d).

An excess dose of SO_2 may cause deoxygenation of effluent. The reaction is slow and expressed by Equation 11.30e.

$$2\,SO_2 + O_2 + 2\,H_2O \rightarrow 4\,H^+ + 2\,SO_4^{2-} \quad (11.30e)$$

In this reaction, an alkalinity consumption of 6.25 g as $CaCO_3$ occurs when 1 g of COD or BOD_L (expressed by O_2) is removed (4 × 50.0/(2 × 16.0) in Equation 11.30e).

Components of Sulfur Dioxide System: The sulfur dioxide supply, storage, and feed systems, and costs are very similar to those of chlorination facility as presented in details in Section 11.6.3. Contact tank is normally not required because the reaction of SO_2 is instantaneous. Rapid mixing at a velocity gradient G above 250 s^{-1} for 30–60 s at the application point is sufficient for dechlorination reaction.

11.7.2 Dechlorination with Sodium Bisulfite ($NaHSO_3$)

Sodium bisulfite is a stable reducing agent with 61.6% available SO_2 by weight. It is available from commercial suppliers at available $NaHSO_3$ content between 30% and 44%. The generalized dechlorination reactions are given by Equations 11.31a and 11.31b.

Reaction with Free Chlorine Residual.

$$NaHSO_3 + HOCl \rightarrow + NaHSO_4 + H^+ + Cl^- \quad (11.31a)$$

Reaction with Combined Chlorine Residual.

$$NaHSO_3 + NH_2Cl + H_2O \rightarrow NaHSO_4 + NH_4Cl \quad (11.31b)$$

The stoichiometric relationships of using sodium bisulfite for dechlorination are:

1. The stoichiometric requirement is 1.47 g as $NaHSO_3$ (104/70.9 in Equation 11.31a) or 0.904 g as SO_2 (64.1/70.9 in Equation 11.31a) for removal of 1 g of free chlorine residual (HOCl) as available Cl_2.
2. The alkalinity consumption is 0.705 g as $CaCO_3$ for removal of 1 g of free chlorine residual as available Cl_2 (50.0/70.9 in Equation 11.31a).
3. No change in alkalinity is expected when it is used for dechlorination of combined chlorine residual (Equation. 11.31b).

11.7.3 Dechlorination with Other Reducing Agents

Other less commonly used reducing agents for dechlorination are sodium sulfite (Na_2SO_3) and sodium metabisulfite ($Na_2S_2O_5$). The reactions with free chlorine residual and combined chlorine residual represented by monochloramine are expressed by Equations 11.32a through 11.33b.

Reactions with Sodium Sulfite.

$$Na_2SO_3 + HOCl \rightarrow Na_2SO_4 + H^+ + Cl^- \quad (11.32a)$$

$$Na_2SO_3 + NH_2Cl + 2\ H_2O \rightarrow Na_2SO_4 + NH_4Cl \quad (11.32b)$$

Reactions with Sodium Metabisulfite.

$$Na_2S_2O_5 + 2\ HOCl + H_2O \rightarrow 2\,NaHSO_4 + 2\ H^+ + 2\ Cl^- \quad (11.33a)$$

$$Na_2S_2O_5 + 2\ NH_2Cl + 3\ H_2O \rightarrow Na_2SO_4 + 2\ NH_4Cl + 2\ H^+ + SO_4^{2-} \quad (11.33b)$$

11.7.4 Dechlorination with Activated Carbon

Free and combined chlorine residuals are effectively removed by adsorption on the surface of activated carbon. The generalized reactions are expressed by Equations 11.34a and 11.34b.

Reaction with Free Chlorine Residual.

$$C + 2\ HOCl \rightarrow CO_2 + 2\ H^+ + 2\ Cl^- \quad (11.34a)$$

Reaction with Combined Chlorine Residuals.

$$C + 2\ NH_2Cl + 2\ H_2O \rightarrow CO_2 + 2\ NH_4Cl \quad (11.34b)$$

In these reactions, C represents the molecular carbon on the surface of activated carbon particles or granules to be actually involved in the reaction. In general, C first reacts with chlorine residual to form surface oxide that may then be partially or fully converted to CO_2. Adsorption and absorption are considered primary mechanisms for dechlorination by activated carbon. Gravity or pressure filter beds filled with granular activated carbon (GAC) are commonly used. Activated carbon bed removes chlorine residuals along with chlorinated and refractory organics. Therefore, activated carbon is cost-effective for dechlorination only if removal of other organic and inorganic contaminants is also required.

11.7.5 Dechlorination of Chlorine Dioxide

Dechlorination of chlorine dioxide is achieved by SO_2. The dechlorination reaction is given by Equation 11.35.

$$5\ SO_2 + 2\ ClO_2 + 6\ H_2O \rightarrow 12\ H^+ + 5\ SO_4^{2-} + 2\ Cl^- \quad (11.35)$$

The stoichiometric weight ratio of SO_2 and ClO_2 is 2.26 mg SO_2/mg ClO_2 as available Cl_2 ((5 × 64.1)/(2 × 70.9) in Equation 11.31a). Normally, ClO_2 residual is very unstable with a short life and dechlorination of ClO_2 is rarely needed. In case it is required, a practical ratio of 2.7 mg SO_2/mg ClO_2 may be used.

EXAMPLE 11.42: QUANTITY OF LIQUID SULFUR DIOXIDE REQUIRED TO DECHLORINATE EFFLUENT

A wastewater treatment plant is discharging 8000 m^3/d secondary treated effluent. The free chlorine residual is 1.8 mg/L. Estimate the quantity of SO_2 required to dechlorinate the effluent.

Solution

1. Determine the quantity of free chlorine residual in the effluent.

$$\text{The quantity of free chlorine residual in the effluent} = 1.8\,\text{g/m}^3 \times 8000\,\text{m}^3/\text{d} \times 10^{-3}\,\text{kg/g}$$
$$= 14.4\,\text{kg/d as available Cl}_2$$

2. Determine the stoichiometric requirement of sulfur dioxide to dechlorinate the effluent from Equation 1.30c.

$$\text{The SO}_2 \text{ consumption ratio} = \frac{mw_{\text{SO2}}}{mw_{\text{Cl2}}} = \frac{64.1\ \text{g/mole as SO}_2}{70.9\ \text{g/mole as Cl}_2} = 0.904\ \text{g SO}_2\text{/g as available Cl}_2$$

The stoichiometric consumption ratio is 0.904 kg of SO_2 per kg of available Cl_2.

3. Determine the actual quantity of liquid SO_2 required.
Assume that the actual SO_2 consumption ratio is 1.1 kg SO_2 per kg of Cl_2.

$$\text{The quantity of SO}_2 = \text{SO}_2 \text{ consumption ratio} \times \text{Quantity of free chlorine residual}$$
$$= 1.1\ \text{kg SO}_2\text{/kg of Cl}_2 \times 14.4\,\text{kg/d as available Cl}_2 = 15.8\,\text{kg/d as SO}_2$$

The quantity of liquid SO_2 required is 15.8 kg/d (or 34.8 lb/d) to meet the dechlorination requirement.

EXAMPLE 11.43: VOLUME OF LIQUID SULFUR DIOXIDE FOR DECHOLINATION NEED

Estimate the volume of liquid sulfur dioxide required to dechlorinate the effluent in Example 11.42.

Solution

1. Determine the specific volume of liquid SO_2.
From Table 11.10, the specific volume of liquid SO_2 is 0.72 L/kg (or 0.087 gal/lb) at 20°C.
2. Calculate the volume of liquid SO_2 required.

$$\text{The volume of SO}_2 \text{ required} = \text{Specific volume of liquid SO}_2 \times \text{Quantity of SO}_2$$
$$= 0.72\,\text{L/kg} \times 15.8\,\text{kg/d as SO}_2 = 11\,\text{L/d (or} \sim\!3\,\text{gpd)}$$

The volume of liquid SO_2 required for dechlorination is 11 L (or 3 gal) per day.

EXAMPLE 11.44: CONCENTRATION OF AVAILABLE SULFUR DIOXIDE IN SODIUM BISULFATE SOLUTION

Sodium bisulfate solution is used for dechlorination of effluent at a wastewater treatment plant. A commercial solution of $NaHSO_3$ is used at a $NaHSO_3$ content of 40% by weight. Estimate the weight of $NaHSO_3$ and the available sulfur dioxide per liter of the solution. Assume that the specific gravity of $NaHSO_3$ solution is 1.37.

Solution

1. Estimate the concentration of $NaHSO_3$ ($C_{\text{NaHSO3,solution}}$).

Equations 11.8 and 11.9 were presented for calculating the available Cl_2 in sodium hypochlorite (NaOCl) solution in Section 11.6.1. These equations can also be used for evaluating available SO_2 in $NaHSO_3$. Calculate the concentration of $NaHSO_3$ from Equation 11.9d. Assume that the density of water $\rho_{water} \approx 1$ kg/L or 1000 g/L.

$$C_{NaHSO3,solution} = \rho_{water} \times \text{sp. gr.}_{solution} \times \frac{(\text{Weight \% } NaHSO_3)_{solution}}{100\%}$$

$$= 1000 \text{ g/L} \times 1.37 \times \frac{40\%}{100\%} = 548 \text{ g/L or } 548 \text{ kg/m}^3$$

The concentration of $NaHSO_3$ is 548 g/L or 548 kg/m^3 (4.58 lb/gal) in the 40% $NaHSO_3$ solution.

2. Determine the available SO_2 in $NaHSO_3$ solution (Weight of SO_2).

   A comparison of Equations 11.30c and 11.31a indicates that 1 mole of either SO_2 or $NaHSO_3$ is required to react with 1 mole of HOCl. Calculate (Weight % of SO_2).

$$(\text{Weight \% } SO_2)_{NaHSO3} = \frac{mw_{SO2}}{mw_{NaHSO3}} \times 100\% = \frac{64.1 \text{ g/mole as } SO_2}{104 \text{ g/mole as } NaHSO_3} \times 100\% = 61.6\% \text{ as } NaHSO_3$$

   The available SO_2 in $NaHSO_3$ is 0.616 kg SO_2/kg $NaHSO_3$

3. Estimate the concentration of available SO_2 in the solution ($C_{SO2,solution}$).

   Rearrange Equation 1.9c to calculate $C_{SO2,solution}$.

$$C_{SO2,solution} = C_{NaHSO3,solution} \times \frac{(\text{Weight \% } SO_2)_{NsHSO3}}{100\%} = 548 \text{ g/L} \times \frac{61.6\%}{100\%}$$

$$= 338 \text{ g/L as } SO_2 \approx 340 \text{ g/L as available } SO_2$$

   The concentration of available SO_2 is ~340 g/L or 340 kg/m^3 (2.82 lb/gal) in the 40% $NaHSO_3$ solution.

EXAMPLE 11.45: VOLUME OF LIQUID SODIUM BISULFITE REQUIRED FOR DECHLORINATION

The chlorine consumption rate of 14.4 kg/d as available Cl_2 for effluent disinfection is calculated in Example 11.42. A 40% solution of $NaHSO_3$ by weight is used for dechlorination. Estimate the consumption rate of $NaHSO_3$ solution for dechlorination.

Solution

1. Determine the stoichiometric consumption of sodium bisulfite for dechlorination of liquid chlorine (Equation 11.31a).

$$\text{The } NaHSO_3 \text{ consumption ratio} = \frac{mw_{NaHSO3}}{mw_{Cl2}} = \frac{104 \text{ g/mole as } NaHSO_3}{70.9 \text{ g/mole as } Cl_2} = 1.47 \text{ g } NaHSO_3/\text{g as available } Cl_2$$

   The stoichiometric $NaHSO_3$ consumption ratio is 1.47 kg of $NaHSO_3$ per kg of available Cl_2.

2. Determine the quantity of $NaHSO_3$ required for dechlorination.

$$\text{The quantity of } NaHSO_3 = NaHSO_3 \text{ consumption ratio} \times Cl_2 \text{ consumption rate}$$

$$= 1.47 \text{ kg } NaHSO_3/\text{kg of } Cl_2 \times 14.4 \text{ kg/d as available } Cl_2 = 21.2 \text{ kg/d as } NaHSO_3$$

The quantity of $NaHSO_3$ required to meet the chlorine requirement for dechlorination is 21.2 kg/d (46.7 lb/d).

3. Estimate the consumption rate of $NaHSO_3$ solution.

The concentration of $NaHSO_3$ in a 40% by weight solution is 548 g/L or 548 kg/m^3 (Example 11.44, Step 1).

$$\text{The NaHSO3 solution consumption rate} = \frac{\text{Quantity of NaHSO}_3}{\text{Concentration of NaHSO}_3}$$

$$= \frac{21.2\ \text{kg/d as NaHSO}_3}{548\ \text{kg/m}^3\ \text{as 40\% NaHSO}_3\ \text{solution}}$$

$$= 0.039\ \text{m}^3/\text{d} \quad \text{or} \quad 39\ \text{L/d}$$

To meet the dechlorination requirement of 14.4 kg/d (31.8 lbs/d) as available Cl_2, a flow of 39 L/d (10 gpd) of 40% $NaHSO_3$ solution is required at the plant.

EXAMPLE 11.46: VOLUME OF LIQUID $NaHSO_3$ EQUIVALENT TO A REQUIRED SO_2 CONSUMPTION RATE

The actual amount of SO_2 is 15.8 kg/d for dechlorination which is calculated in Example 11.42. Estimate the volume of 40% $NaHSO_3$ by weight required for dechlorination.

Solution

1. Determine the available concentration of SO_2 in $NaHSO_3$.

The available SO_2 concentration in the 40% $NaHSO_3$ solution, $C_{SO2,solution}$ = 340 g/L or 340 kg/m^3 is calculated in Example 11.44, Step 3.

2. Determine the volume of 40% $NaHSO_3$ solution needed for dechlorination.

The volume of 40% $NaHSO_3$ solution required

$$= \frac{\text{Amount of SO}_2\ \text{required}}{\text{Available SO}_2\ \text{concentration in the 40\% NaHSO}_3\ \text{solution}}$$

$$= \frac{15.8\ \text{kg/d as SO}_2}{340\ \text{kg/m}^3\ \text{as available SO}_2} = 0.046\ \text{m}^3/\text{d} \quad \text{or} \quad 46\ \text{L/d}$$

The volume of 40% $NaHSO_3$ solution required for dechlorination is 46 L (or 12 gal) per day.

11.8 Disinfection with Ozone

In recent years, ozone (O_3) has received much interest for disinfection of municipal wastewater effluent. The reported benefits over chlorination are (1) no formation of persistent chlorinated organics, (2) less adverse DBPs, (3) no effluent toxicity, and (4) a powerful disinfectant. However, ozone being an unstable gas must be generated at site.

11.8.1 Ozone Chemistry

Ozone is a powerful oxidizing agent. The oxidation potential is −2.07 (Table 11.4). The decomposition reactions of ozone occur in a series of steps that are conceptually expressed by Equation 11.36.

$$O_3 + H_2O \rightarrow HO_3^+ + OH^- \qquad (11.36a)$$

$$HO_3^+ + OH^- \rightarrow 2\ HO_2^\bullet \quad (11.36b)$$

$$O_3 + HO_2^\bullet \rightarrow HO^\bullet + 2\ O_2 \quad (11.36c)$$

$$HO^\bullet + HO_2^\bullet \rightarrow H_2O + O_2 \quad (11.36d)$$

The free radicals $HO_2^\bullet$ and $HO^\bullet$ (free hydroxyl radical) or a mixture of O_3 and free hydroxyl radicals has great oxidizing power and carry out disinfection. The reaction is accelerated in neutral or alkaline environment (pH above 7), or exposure to UV light or hydrogen peroxide. The free hydroxyl radicals are more effective oxidizing agents than the molecular ozone, but they are extremely short-lived.

11.8.2 Properties of Ozone

Ozone is a clear to bluish in color, pungent smelling, and unstable gas. It has a density of 2.0 kg/m^3 (0.125 lb/ft^3) at 20°C and 1 atm, and is heavier than air. It is detected at low concentrations (0.01–0.05 ppm by volume). It may be toxic, but its odor is detected before health concerns develop. Gaseous ozone is explosive at concentrations above 240 g/m^3 (20% weight in air). Important properties of ozone are presented below.

1. Its solubility in water at 20°C is 12.07 mg/L. The solubility depends upon temperature and its concentration in feed gas as it enters the contactor. Its solubility in water increases with higher concentration of ozone in the feed gas. Increasing the pressure in the contactor and lowering the temperature also increases the solubility. A solubility of 570 mg/L at 20°C has been reported for pure O_3.[48] The solubility in water is governed by the Henry's law. The Henry's constants at 0°C, 10°C, and 20°C are 1940, 2480, and 3760 atm/mole fraction.[54]
2. The stability of ozone in air is greater than that in water, but in both cases, it is in the order of minutes. The half-life of ozone in water ranges from 8 min to 14 h depending on the level of ozone-demanding contaminants and temperature.
3. Ozone does not produce chlorinated DBPs such as THMs and HAAs. However, it will form other by-products in reactions with many organic and inorganic compounds. The inorganic compounds contain primarily ions in reduced state such as nitrite, ferrous, manganese, sulfide, and ammonium. Ozone also reacts with aliphatic and aromatic compounds, humic acids, and pesticides producing DBPs of lower molecular organic species. Among them are aldehydes, ketones, and acids that are biodegradable products. In the presence of bromide ions, hydrobromic acid is produced which may encourage formation of brominated DBPs. Some by-products of ozonation may have mutagenic or carcinogenic properties, but they may be short-lived. There is also a concern over the potential formation of N-nitrosodimethylamine (NDMA) and perfluoroalkyl acids (PFAAs) during ozonation of wastewater effluent and in reclaimed water.[93,94]
4. Ozone and free hydroxyl radicals being powerful oxidants are also very effective germicides. It is believed that bacteria and protozoa cysts are destroyed because of cell wall disintegration (cell lysis). It is also a very effective viricide.
5. The major ozone oxidation pathways include (a) direct oxidation by molecular O_3, which dominates primarily under acidic conditions and/or (b) hydroxyl oxidations under favoring conditions, such as high pH or in the advanced oxidation processes (AOPs) with O_3/UV or O_3/H_2O_2.[55]
6. Many other benefits of ozone include: (a) destruction of some harmful refractory organics and pesticides, (b) color removal, (c) increases biodegradability of organic compounds, (d) reduces initial chlorine demand, (e) does not add dissolved solids, (f) does not leave any residual, and (g) elevates dissolved oxygen level which may eliminate the need for postreaeration.

11.8.3 Ozone Dosage for Disinfection

The ozone dosage required for disinfection of wastewater effluent depends upon the quality of effluent, its constituents, and initial demand. Typical ozone dosage for effluent disinfection from various wastewater treatment processes after 15-min contact time is compared with those of chlorine in Table 11.7. The contact times for chlorine and ozone in this table are 30 and 15 min, respectively.

11.8.4 Ozone Generation

Ozone being a chemically unstable gas must be generated on-site and used quickly. It is generated by applying a high-energy electrical field to oxygen or dried air. The oxygen molecules are dissociated as expressed by Equation 11.37.

$$O_2 + 2\,e^- \leftrightarrow 2\,O^- \tag{11.37a}$$

$$2\,O^- + 2\,O_2 \leftrightarrow 2\,O_3 \tag{11.37b}$$

$$3\,O_2 \rightarrow 2\,O_3 \tag{11.37c}$$

The reaction is reversible and ozone decomposes to oxygen rapidly if temperature is above 35°C. Therefore, cooling is needed to dissipate heat produced during generation. The ozone generation system has three major components: (1) feed gas preparation; (2) power supply, and (3) ozone generation.

Feed Gas Preparation: The feed gas for ozone generation is (1) air, (2) oxygen generated on-site, or (3) liquid oxygen (LOX). The selection of feed gas is primarily determined by the availability and cost-effectiveness of local commercial oxygen supply. Air feed or oxygen from on-site generation is more suitable for small systems. Large applications typically use LOX.

Air Feed Systems: The oxygen content in air is ~21% by volume under ambient conditions. The feed air to ozone generator must be conditioned to remove moisture and particulate matter. The major steps involved are compression, cooling and drying, and filtration. Moisture in air reduces ozone production efficiency, causes fouling of dielectric tubes, and increases corrosion in the ozone generator and downstream equipment. The air must be dried to a maximum dew point of −65°C. Air filtration system should be capable of removing particles larger than 1 μm and oil droplets larger than 0.05 μm. Depending upon the operating pressure, the air feed systems are also classified as (a) ambient pressure, (b) low pressure lower than 2.1 kg/cm^2 (30 psig), (c) medium pressure in range of 2.1–4.2 kg/cm^2 (30–60 psig), or (d) high pressure >4.2 kg/cm^2 (60 psig).[55]

Oxygen Enriched Air from On-Site Generation Systems: High purity oxygen gas (>90% by volume) is generated on-site. These systems also require clean and dry air, and therefore utilize similar air preparatory equipment as those for air feed systems. Two common methods used for on-site oxygen generation are (1) pressure swing adsorption (PSA) and (2) cryogenic process with either vacuum pressure swing adsorption (VPSA) or vacuum swing adsorption (VSA).

LOX Feed Systems: LOX, 100% pure oxygen, is commercially available. Oxygen gas produced by evaporation of LOX is the most commonly used feed gas and a cost-effective method for ozone generation. These systems have many benefits over air feed system: (1) less complicated and without requirement of feed gas conditioning, (2) higher ozone production yield (more ozone per unit area of the dielectric), (3) high concentration of ozone in the feed gas, (4) less energy requirement, (5) smaller feed gas volume required for the same ozone output, (6) provides operational flexibility to meet peak ozone demands, and (7) less ancillary equipment.

One benefit of on-site oxygen generation is peak power load shedding. The oxygen-enriched feed gas may be generated during off-peak power demand. During the period of peak-power demand, the system may be switched to oxygen-enriched feed, thus reducing the power requirement for ozone generation during peak power demand.[42]

The typical components of a LOX feed system is illustrated in Figure 11.14. The pressure in the LOX bulk storage tank can be as high as 200 psi while the ozone generators are normally operated at a pressure of 15–30 psi. Therefore, a PRV/pressure regulating valve must be used. The production efficiency of ozone generator normally increases about 10–20% when the feed gas contains about 1–3% of nitrogen. An air

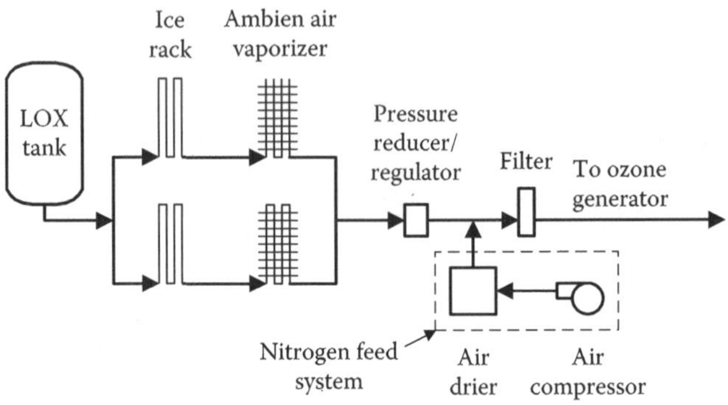

FIGURE 11.14 Conceptual process diagram of a liquid oxygen (LOX) feed system.

compressor with an air dryer is normally provided to blend air (containing 78% of nitrogen) with oxygen gas at a desired ratio to meet the nitrogen requirement.[52]

Power Supply: The voltage and frequency of the power must vary greatly to control the amount and rate of ozone generation. The normal power supply input to the ozone generator system is 480 V/3 phase/ 60 Hz. However, the power applied to the electrodes is typically medium to high frequency up to 6000 Hz and a high voltage over 20,000 V for ozone generation. For this reason, a specialized power supply unit (PSU), including transformer, power converter/inverter device, cooling system, and PLC are typically required.[55,95,96]

Ozone Generation: Ozone is generated by (1) UV radiation or (2) cold plasma or corona discharge. When UV light ($<$200 nm wavelength) passes through dry or oxygen-rich air, ozone is generated by photochemical reaction. The ozone concentration by this method is about 0.25%, and is suitable only for small systems. The most common and efficient method for ozone generation is corona discharge cell. A typical corona discharge cell is shown in Figure 11.15.[54,63] The feed gas is passed through the discharge gap between two electrodes while a desired voltage potential is maintained across a dielectric material.

Frequency and Voltage: Ideally, high ozone production yield by corona discharge would occur at a high voltage, a high frequency, a large dielectric constant, and a thin dielectric. Because of the practical limitations, a balance between these operating parameters is required to optimize the ozone production yield, operational reliability, and costs. Depending up on the applied frequency, the ozone generators are

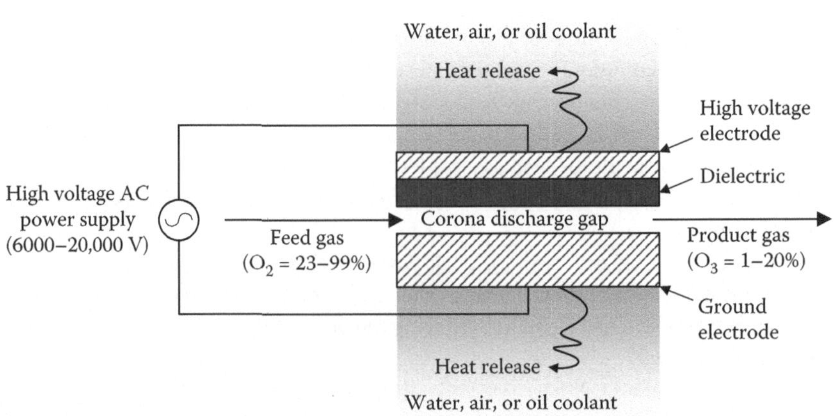

FIGURE 11.15 Corona discharge cell for ozone generation.

primarily classified as (a) low frequency (50 or 60 Hz at a high voltage >20,000 V), (b) medium frequency (up to 1000 Hz at a medium voltage 10,000–20,000 V), and (c) high frequency (>1000 Hz at low voltage <10,000 V).[52,95,96] Low-frequency, high-voltage generators are quite common, but recent improvements in electronic circuitry make high-frequency, low-voltage units more desirable.[30,54]

Ozone Concentration and Optimum Efficiency: The ozone concentration is typically 1–3% and 6–15% by weight in the product gas from air and high purity oxygen, respectively.[52] High ozone concentration up to 20% has also been reported using oxygen feed gas.[52,95–98]

Production Capacity and Optimum Efficiency: For low-frequency ozone generation, the ozone production capacity is normally below 10 kg O_3/h (20 lb O_3/h) per unit and the production at 60–70% of the maximum generation capacity is most cost-efficient. A high production capacity up to 250 kg O_3/h (115 lb O_3/h) can be provided by generators with a medium to high frequency. The optimum operating point is typically at 90–95% of the maximum capacity for these generators.[52,95–98]

Power Requirements: The theoretical specific energy requirement to break the covalent bonding (double bonds) in oxygen molecules and form ozone is only 0.82 kWh/kg O_3 produced (0.37 kWh/lb O_3 produced). However, the actual specific energy requirement is much higher. The overall electrical power requirement for feed gas preparation, ozone generation, and application may vary from 25–30 kWh/kg O_3 produced (10–13 kWh/lb O_3 produced) for air feed, and 15–20 kWh/kg O_3 produced (7–9 kWh/lb O_3 produced) for oxygen feed system.[52,95,96]

Cooling Requirements: Significant amount of input energy is converted into heat during ozone generation. In general, the higher the frequency is applied the higher amount of heat is produced. Therefore, an efficient cooling process is critically required to cool the production gas and prevent decomposition of ozone. The plates can be cooled by water, air, or oil. The cooling water requirement ranges from 4 to 12 L/kg O_3 produced (0.5 to 1.5 gal/lb O_3 produced) for an optimum cooling water temperature differential of 3–6°C (5–10°F).[55,95,96]

11.8.5 Ozone Application

The principal application of ozone is to disinfect the effluent prior to disposal or reuse. The major components for ozone application are (1) ozone contactor, (2) ozone destruction, and (3) other considerations for construction, safety, and O&M. The schematic of ozone disinfection system including generation and application is shown in Figure 11.16.

Ozone Contactor: The contact chambers are designed to maximize the transfer efficiency. The contactors may be tall vertical column, covered tank with multiple compartments, or inclined packed column. The ozone injection devices may be fine bubble diffusers, injector devices, or sidestream injection system with venture-type injector and degas vessel. The mixing energy can be provided by (1) mechanical mixing with static mixer or high-speed agitator at the base or (2) hydraulic mixing by baffles. A typical ozone contactor with the three covered chambers, fine bubble diffusers, and baffles is shown in Figure 11.17. The flow pattern can be counter current (Figure 11.17a), concurrent (Figure 11.17b), or combination of counter current and concurrent (Figure 11.17c). The production gas from an ozone generator is fed through the fine bubble

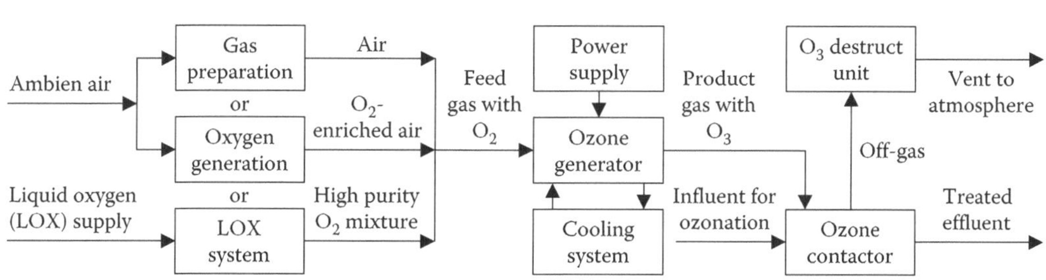

FIGURE 11.16 Schematic of ozone disinfection system.

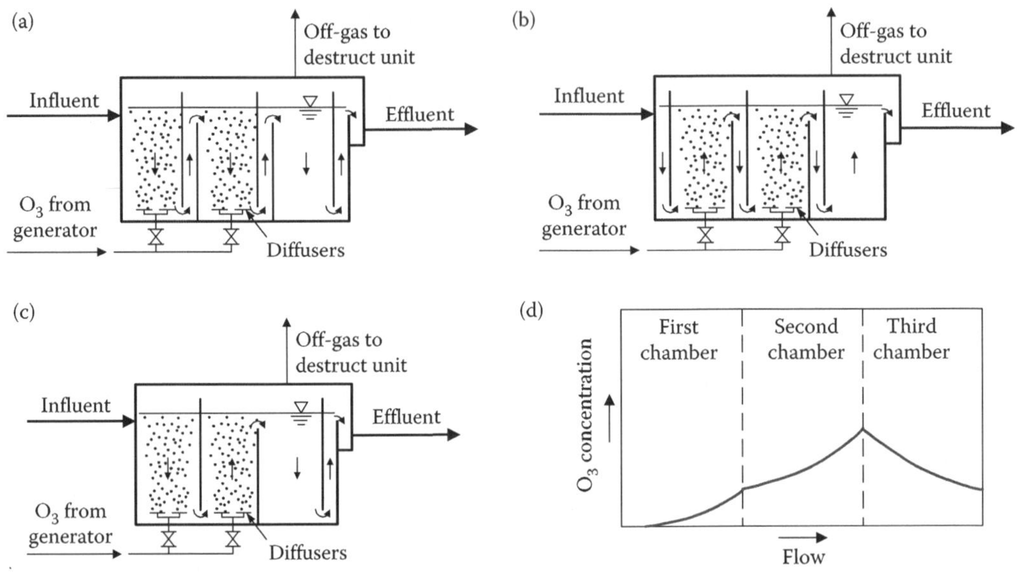

FIGURE 11.17 Ozone contactor flow patterns: (a) counter current, (b) cocurrent, (c) combination of counter current and concurrent, and (d) concentration profile.

diffusers installed in the first two chambers. Diffusers may also be installed in the third chamber for use only as necessary. The fine bubbles of ozone-containing gas cause mass transfer. The mass transfer coefficient (K_La) may lie in the range of 0.25–0.45 min^{-1} except for very low gas flow rate. The initial ozone demand is normally met in the first chamber and the residual concentration may reach in the range of 0.3–0.9 mg/L. The ozone residual in the combined wastewater–ozone mixture would reach the peak in the second chamber where microbial inactivation reaction starts. In the last chamber, the slow reaction is completed providing required disinfection while ozone is consumed. A typical ozone concentration profile is shown in Figure 11.17d. Typical ozone feed rate is 1–5 mg/L depending upon the purpose. Ozone transfer up to 90% may be achieved in a well-designed contactor. The supply pressure of production gas must be adequate to overcome the static pressure, diffuser exit pressure, and pipe losses. The desired ozone dose and contact time for effluent disinfection are best determined by pilot studies.

Ozone Destruction in Off-Gases: The ozone in the off-gases from contactor or degas vessel must be destroyed to a concentration < 0.1 ppm prior to release. Four methods may be used for ozone destruction: (1) thermal 300–380°C for 3 s, (2) heat/catalytic (~40°C), (3) catalytic (ambient temperature), and (4) moist GAC. Metal oxides such as granular manganese oxide are an effective catalyst for ozone destruction. Thermal destruction is more expensive but very reliable.

Other Considerations: Other important considerations for ozone generation and application include (1) material compatibility, (2) safety and O&M requirements, and (3) potential recycle of oxygen.

Material Compatibility: Materials of construction for ozone generation and application must be capable of resisting the strong oxidizing and corrosive effects of ozone. The construction materials recommended for ozone contactor, piping, and gaskets are reinforced concrete, Type 316L stainless steel, CPVC, HDPE, Teflon, and PTFE.[55]

Safety and O&M Requirements: Many safety and O&M requirements are needed for ozone generation and application. These are valves, gauges, metering, monitoring, and alarms. A list of important gauges, valves, and monitoring systems is given below.

1. *Feed Gas Preparation for Air Feed Systems:* Provide (a) monitors for power input, pressure, temperature, and dew point; (b) high dew point alarm; and (c) automatic generator shutdown.

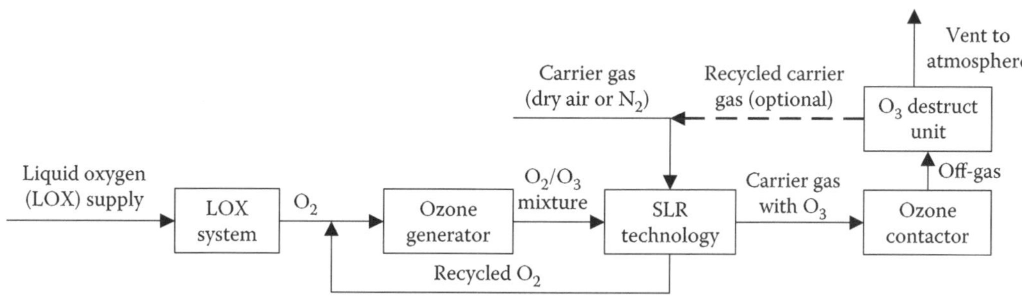

FIGURE 11.18 Oxygen recycling schematic.

2. *LOX Systems:* Provide (a) pressure sensors on the LOX storage tank and oxygen gas piping after vaporizer, (b) pressure reducing/regulating valve, and (c) pressure relief valve.
3. *Ozone Generator:* Provide (a) monitor for power input; (b) coolant flowmeter and temperature sensor; (c) feed gas flowmeter, (d) flowmeter, and sensors for ozone concentration, temperature, and pressure on discharge gas piping; and (e) automatic shutdown if coolant flow is interrupted, or the temperature or pressure exceeds a set value.
4. *Ozone Contactor:* Provide (a) sensors for monitoring ozone residuals, pressure and/or vacuum conditions; and (b) vacuum relief and pressure release valves.
5. *Ozone Destruct System:* Provide (a) sensors for inlet and outlet gas temperatures, (b) sensors to monitor the pressure drop across the catalyst chamber, (c) catalyst temperature, and (d) sensor with high-level alarm for ozone residual in the exhaust gas.
6. *Ambient Air Monitoring:* Provide (a) ozone leak detector with alarm to alert an ambient atmospheric ozone level above 0.1 ppm by volume and (b) automatic generator shutdown when the ambient ozone level exceeds 0.3 ppm by volume.

Oxygen Recycling: In oxygen feed ozone generation system, <20% ozone is actually used for disinfection in the ozone contactor. Over 80% of feed oxygen is wasted with the off-gas. Many processes have been investigated in recent years for recycling oxygen and ozone after disinfection. These concepts are generally categorized as either *direct* or *indirect* recycle of oxygen.[99]

One of the direct oxygen recycle processes is called "Short Loop Recycle (SLR)" technology. In this process (Figure 11.18), the O_2/O_3 mixture from the ozone generator is fed into an adsorption device where the O_3 is adsorbed while the O_2 passes through the device. The O_2 gas is directly recycled back to the ozone generator. The adsorbed O_3 is then stripped off by a carrier gas of the adsorption bed. The O_3/carrier gas mixture is sent into the ozone contactor where ozone is used and the remaining carrier gas is vented to the air. The carrier gas can be either nitrogen gas or air. It has been reported that the overall operating costs of ozone system was lowered by over 70% by using the SLR process.[99,100]

The most practical and cost-effective indirect recycle of spent oxygen is use of the off-gas (after ozone destruction) in the aeration system for an activated sludge process. The oxygen transfer efficiency (OTE) and aeration efficiency (AE) would be significantly improved since the oxygen content (75–85%) in the off-gas is much higher than that in the air.[99,101]

11.8.6 Kinetic Equations for Ozone Disinfection

Several kinetic models developed for disinfection by chlorine (Section 11.5.2) have also been applied to ozonation. The important ones are the Hom model (Equation 11.4d) and Hass–Karra model that is Chick-Watson equation in logarithm forms (Equations 11.38a and 11.38b). Also, empirical models (Equations 11.38c through 11.38f) have been developed for ozonation.[33,52–54,102]

Hass–Karra Model

$$\log\left(\frac{N}{N_0}\right) = -\Lambda_{10}C^n t \quad \text{or} \quad \ln\left(\frac{N}{N_0}\right) = -\Lambda_e C^n t \qquad (n \neq 1) \tag{11.38a}$$

$$\log\left(\frac{N}{N_0}\right) = -\Lambda_{10}(CT) \quad \text{or} \quad \ln\left(\frac{N}{N_0}\right) = -\Lambda_e(CT) \qquad (n = 1) \tag{11.38b}$$

Other empirical models

$$\frac{N}{N_0} = \left(\frac{U}{q}\right)^{-n} \quad \text{or} \quad \log\left(\frac{N}{N_0}\right) = -n\log\left(\frac{U}{q}\right) \qquad (U > q) \tag{11.38c}$$

$$\frac{N}{N_0} = 1 \quad \text{or} \quad \log\left(\frac{N}{N_0}\right) = 0 \qquad (U \leq q) \tag{11.38d}$$

$$D = \frac{100\%}{TE_{O_3}} U \tag{11.38e}$$

$$D = \frac{Q_{gas}}{Q}(C_{O_3,feed} - C_{O_3,off}) \tag{11.38f}$$

where

N = number of organisms remaining after disinfection, number of organisms per unit sample volume, organisms/100 mL

N_0 = number of organisms present initially at $t = 0$, number of organisms per unit sample volume, organisms/100 mL

C = disinfectant (O_3) residual, mg/L

n = coefficient obtained experimentally based on the slope of dose response curve, dimensionless

t = contact time, min

CT = product of ozone residual (C) and contact time (t), mg·min/L. It is an important design parameter used in the design on ozone contract. The typical ranges of CT required for inactivation of different organisms by ozone are provided in Table 11.9

Λ_{10} and Λ_e = coefficient of specific lethality (base 10 and e), L/mg·min. It is the slope of the linear portion of the dose (CT) response curve. The coefficients (base 10) for inactivation with ozone and other different disinfectants are summarized in Table 11.17.

U = utilized or transferred ozone dose, mg/L

q = estimated initial ozone demand, mg/L. It can be determined graphically from experimental data (Example 11.53)

D = total required ozone dosage, mg/L
D is obtained by applying the ozone transfer efficiency to U

TE_{O3} = ozone transfer efficiency, %
The ozone transfer efficiency may be in the range of 80–90%

Q_{gas} = flow rate of the feed gas from ozone generator, L/min

Q = flow rate of the disinfected effluent, L/min

$C_{O3,feed}$ = concentration of ozone in the feed gas, mg/L

$C_{O3,off}$ = concentration of ozone in the off-gas prior to destruction device, mg/L

TABLE 11.17 Typical Values of Coefficients of Specific Lethality for Inactivation of Different Organisms with Different Disinfectants[a]

Disinfectant	Unit	Coefficient of Specific Lethality (Λ_{10})[b]			
		Bacteria[c]	Virus	Protozoa *Cryptosporidium*	Protozoa *Giardia lamblia* Cysts
Chlorine, free	L/mg·min	2	1.2	0.00044	0.04
Chloramine	L/mg·min	0.016	0.0052	0.00022	0.0024
Chlorine dioxide	L/mg·min	2	0.38	0.008	0.2
Ozone	L/mg·min	44	7.27	0.24	4.21
UV radiation	cm^2/mJ	5.7	0.0215	0.31	0.33

[a] Values for inactivation in filtered secondary effluent at pH ≈ 7.5 and temperature ≈ 20°C.
[b] Coefficient base 10, $\Lambda_e = \ln(10)\Lambda_{10} = 2.303\Lambda_{10}$.
[c] Values are only applicable for total coliform. The values for fecal coliform and *E. coli* are quite different.
Source: Adapted in part from Reference 52.

EXAMPLE 11.47: SELECTION OF OZONE GENERATORS

Multiple ozone generators are selected to satisfy the average and peak demands as well as the necessary standby unit requirement. At a wastewater treatment plant, average and peak ozone requirements are 40 and 60 kg/d (88 and 132 lb/d). Determine the number of generators needed to satisfy the average and peak ozone demands, and standby unit requirement. The most cost-effective maximum ozone generation capacity is 67%.

Solution

Select three generators, each designed for 30 kg/d (66 lb/d) ozone generation capacity.

1. Determine the maximum capacity of two and three generators.

$$\text{Maximum most cost-effective capacity of two generators} = 2 \times 30\ \text{kg/d} \times 0.67 = 40\ \text{kg/d}$$
$$\text{Maximum most cost-effective capacity of three generators} = 3 \times 30\ \text{kg/d} \times 0.67 = 60\ \text{kg/d}$$

2. Select the generator arrangement.

   Two generators will provide the average ozone requirement of 40 kg/d. Three generators will provide the maximum ozone requirement of 60 kg/d. One standby unit will be available for maintenance during average demand. Two generators at 100% maximum generation capacity will provide maximum ozone requirement of 60 kg/d under emergency situation only in case one unit is out of service under peak condition.

EXAMPLE 11.48: COEFFICIENT OF SPECIFIC LETHALITY FROM EXPERIMENTAL DATA

At a wastewater treatment facility, inactivation of *Cryptosporidium parvum* in the filtered secondary effluent by ozonation was investigated in a bench-scale study. Determine the coefficient of specific lethality from Equation 11.38b. Apply the experimental data that are summarized below. Validate the applicability of the equation. The ratio of T_{10}/T is 0.4 for the experimental contactor.

Ozone Residual (C), mg/L	Retention Time (t), min	Number of Organisms (N), *parvums*/100 mL
0.0	0	21,000
3.1	2	5400
2.8	4	1900
3.0	6	410
2.9	8	160
3.1	10	38

Solution

1. Determine the coefficient of specific lethality (Λ_{10}) from tabulated data sets.
   a. Prepare the values of $\log(N/N_0)$ and CT from the experimental data and tabulate the calculation results below.

C, mg/L	t, min	T_{10}, min	CT, mg·min/L	N, *parvums*/100 mL	N/N_0	log, N/N_0	Λ_{10}, L/mg·min
0.0	0	0.0	0.00	21,000	1.000	0.000	–
3.1	2	0.8[a]	2.48[b]	5400	0.257[c]	−0.590[d]	0.24[e]
2.8	4	1.6	4.48	1900	0.0905	−1.04	0.23
3.0	6	2.4	7.20	410	0.0195	−1.71	0.24
2.9	8	3.2	9.28	160	0.00762	−2.12	0.23
3.1	10	4.0	12.4	38	0.00181	−2.74	0.22
						Average	0.23

[a] $T_{10} = 0.4 \times 2\text{ min} = 0.8\text{ min}$.

[b] $CT = 3.1\text{ mg/L} \times 0.8\text{ min} = 2.48\text{ mg·min/L}$.

[c] $N/N_0 = 5400/21{,}000 = 0.257$

[d] $\log(N/N_0) = \log(0.257) = -0.590$

[e] $\Lambda_{10} = -\dfrac{\log(N/N_0)}{CT} = -\dfrac{-0.590}{2.48\text{ mg·min/L}} = 0.24\text{ L/mg·min}$

   b. Rearrange Equation 11.38b and calculate the value of Λ_{10} from each set of $\log(N/N_0)$ and CT data. The calculated values are also summarized in the table above.
   c. An average value of coefficient of specific lethality $\Lambda_{10} = 0.23$ L/mg·min is obtained from the experimental data.
2. Determine the coefficient of specific lethality (Λ_{10}) from plot of $\log(N/N_0)$ versus CT.

The values of log (N/N_0) and CT are calculated from the experimental data in Step 1. Plot log (N/N_0) versus CT in Figure 11.19. A linear relationship between $\log(N/N_0)$ versus CT is obtained with $n = 1$. Equation 11.38b is valid for describing the inactivation of *Cryptosporidium parvums* by ozonation. The slope of the line is $\Lambda_{10} = -0.23$ L/mg·min is obtained from the slope of the line.

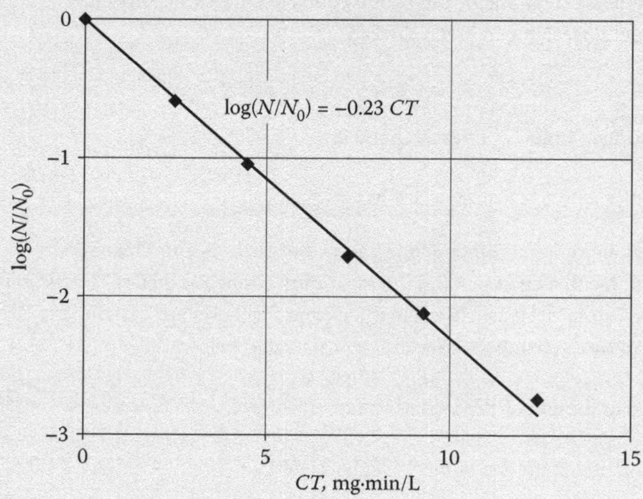

FIGURE 11.19 Plot of $\log(N/N_0)$ versus CT (Example 11.48).

EXAMPLE 11.49: CT REQUIREMENT TO ACHIEVE A DESIRED LOG REMOVAL

A 2-log inactivation of *Cryptosporidium parvums* is required in the filtered secondary effluent from a treatment facility. Ozonation process is one of the alternatives recommended for disinfection. Determine the CT requirement to meet the required inactivation. Assume that the coefficient of specific lethality determined in Example 11.48 is valid.

Solution

Rearrange Equation 11.38b and calculate the required CT_{req} for 2-log removal ($-\log(N/N_0) = 2$).

$$CT_{req} = -\frac{\log(N/N_0)}{\Lambda_{10}} = -\frac{-2}{0.23 \text{ L/mg·min}} = 8.7 \text{ mg·min/L}$$

Therefore, the ozone contactor must provide an overall $CT > 8.7$ mg·min/L.

EXAMPLE 11.50: INITIAL OZONE RESIDUAL REQUIRED TO MEET A DESIRED CT IN AN OZONE CONTRACTOR

A 4-chamber ozone contractor is designed to meet the disinfection requirement in Example 11.49. The hydraulic retention time (HRT) is 5 min in each chamber that has a baffling factor (BF) or T_{10}/T ratio of 0.6. The ozone is fed into the first chamber where the initial demand is met. A maximum ozone residual is reached at the end of first chamber. The ozone residual decreased gradually through the rest three chambers. Determine the (a) the maximum ozone residual required (C_1) at the end of first chamber, (b) the ozone dosage requirement at an initial ozone demand of 0.5 mg/L, (c) prepare an ozone residual profile through the ozone contactor, and (d) validate the performance of the ozonation process under design conditions. Assume that the decrease of ozone follows the following first-order decay curve (Equation 2.15a): $C = C_1 e^{-k_{O3}\, t}$, where $k_{O3} = 0.12\ \text{min}^{-1}$, and C_1 is the ozone residual at the end of first chamber, and t is the overall hydraulic retention time after the first chamber.

Note: The hydraulic calculations of the ozone contactor are similar to those used for the over-and-under baffled channel flocculator in Example 9.46.

Solution

1. Determine the contact time T_{10} in each chamber.

$$T_{10} = BF \times HRT = 0.6 \times 5 \text{ min} = 3 \text{ min}$$

2. Determine the ozone residual remaining at the end of each of four chambers.

Ozone is fed in the first chamber and the maximum ozone residual (C_1) is established at the end of the first chamber. Ozone decrease occurs in the second through fourth chambers. The ozone residual at the end of each of these chambers is separately calculated below:

Ozone residual at the end of the second chamber, $C_2 = C_1 e^{-k_{O_3} t_2} = C_1 e^{-0.12 \text{ min}^{-1} \times 5 \text{ min}} = 0.55\ C_1$

Overall reaction time at the end of the third chamber, $C_3 = C_1 e^{-k_{O_3} t_3} = C_1 e^{-0.12 \text{ min}^{-1} \times (2 \times 5 \text{ min})} = 0.30\ C_1$

Overall reaction time at the end of the fourth chamber,

$$C_4 = C_1 e^{-k_{O_3} t_4} = C_1 e^{-0.12 \text{ min}^{-1} \times (3 \times 5 \text{ min})} = 0.17\ C_1$$

3. Determine the overall CT value achieved in all four chambers.

Note: Ozone is fed into the first chamber. A maximum residual is only reached at the end of the chamber after the initial ozone demand is met. Therefore, it is common practice to exclude the CT value of the first chamber in the CT calculations ($CT_1 = 0$).

CT value in the second chamber, $CT_2 = C_2 \times T_{10} = 0.55 \times C_1 \times 3 \text{ min} = 1.65 \text{ min} \times C_1$

CT value in the third chamber, $CT_3 = C_3 \times T_{10} = 0.30 \times C_1 \times 3 \text{ min} = 0.90 \text{ min} \times C_1$

CT value in the fourth chamber, $CT_4 = C_4 \times T_{10} = 0.17 \times C_1 \times 3 \text{ min} = 0.51 \text{ min} \times C_1$

Calculate the overall CT value achieved in the entire ozone contactor.

$$\sum CT = CT_1 + CT_2 + CT_3 + CT_4 = (0 + 1.65 + 0.90 + 0.51) \text{ min} \times C_1 = 3.06 \text{ min} \times C_1$$

4. Determine the maximum ozone residual (C_1) required at the end of first chamber.

To meet the disinfection goal, the overall CT value achieved must be greater than the overall CT requirement of 8.7 mg·min/L (from Example 11.49).

$$\sum CT > CT_{\text{req}} = 8.7 \text{ mg·min/L} \quad \text{or} \quad 3.06 \text{ min} \times C_1 > 8.7 \text{ mg·min/L}$$

$$C_1 > \frac{8.7 \text{ mg·min/L}}{3.06 \text{ min}} = 2.84 \text{ mg/L}$$

Therefore, the ozone residual >2.84 mg/L is required at the end of the first chamber. Provide a design ozone residual $C_1 = 3$ mg/L.

5. Determine the ozone dosage required (D) in the first chamber.

At an initial ozone demand $q = 0.5$ mg/L, $D = C_1 + q = (3 + 0.5) \text{ mg/L} = 3.5 \text{ mg/L}$

6. Prepare the ozone residual profile through the ozone contactor.

Ozone residual at the end of the second chamber, $C_2 = 0.55\ C_1 = 0.55 \times 3$ mg/L = 1.65 mg/L

Overall reaction time at the end of the third chamber, $C_3 = 0.30\ C_1 = 0.30 \times 3$ mg/L = 0.90 mg/L

Overall reaction time at the end of the fourth chamber, $C_4 = 0.17\ C_1 = 0.17 \times 3$ mg/L = 0.51 mg/L

The design ozone residual profile is shown in Figure 11.20.

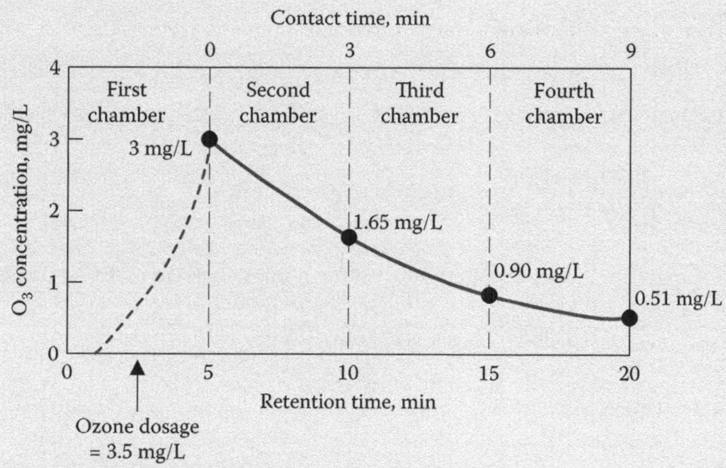

FIGURE 11.20 Ozone residual profile through the ozone contactor (Example 11.50).

7. Validate the overall CT value provided at the design ozone profile.

$$\sum CT = 3.06 \text{ min} \times 3 \text{ mg/L} = 9.2 \text{ mg·min/L} > 8.7 \text{ mg·min/L (overall CT value required)}$$

8. Validate the log inactivation provided at the design ozone profile at $\Lambda_{10} = 0.23$ L/mg·min (Example 11.49).

$$-\log(N/N_0) = \Lambda_{10} \sum CT = 0.23 \text{ L/mg·min} \times 9.2 \text{ mg·min}/L$$

$$= 2.1\text{–log} > 2\text{–log (overall inactivation required)}$$

EXAMPLE 11.51: OZONE GENERATOR PRODUCT GAS FEED RATE TO OZONE CONTACTOR

Product gas from a LOX ozone generation system is used for ozonation at a wastewater treatment plant. The ozone concentration is 14% by weight in the product gas from the generators. Determine the total product gas flow rate fed to the ozone contactors when the plant is operated at a flow of 10 MGD. Use the ozone dosage determined in Example 11.50. Assume that the ozone transfer efficiency is 90% in the ozone contactor. The feed gas contains 3% of nitrogen by weight.

Solution

1. Determine the mass of ozone required (m_{O_3}) at the ozone transfer efficiency of 90% and the ozone dosage $D = 3.5$ mg/L (from Example 11.50).

$$m_{O_3} = DQ \times \frac{100\%}{TE_{O_3}} = 3.5 \text{ mg/L} \times 10 \text{ MGD} \times \frac{8.34 \text{ lb}}{\text{mg/L·Mgal}} \times \frac{100\%}{90\%} = 325 \text{ lb/d as } O_3$$

2. Determine the ozone gas flow rate (q_{O3}) using the standard ozone density $\rho_{O3} = 0.125 \text{ lb/ft}^3$ (lb/scf) at 20°C and 1 atm.

$$q_{O3} = \frac{m_{O3}}{\rho_{O3}} = \frac{325 \text{ lb/d}}{0.125 \text{ lb/scf}} \times \frac{1 \text{ d}}{1440 \text{ min}} = 1.8 \text{ scfm as } O_3$$

3. Determine the total mass of feed gas required (m_{gas}) at the ozone concentration of 14% by weight.

$$m_{gas} = \frac{m_{O3}}{w_{O3}} = \frac{325 \text{ lb/d as } O_3}{0.14 \text{ lb } O_3\text{/lb gas}} = 2321 \text{ lb/d as gas mixture}$$

4. Determine the mass of nitrogen required (m_{N2}) at the nitrogen content of 3% by weight.

$$m_{N2} = w_{N2} \times m_{gas} = 0.03 \text{ lb } N_2\text{/lb gas} \times 2321 \text{ lb/d as gas} = 70 \text{ lb/d as } N_2$$

5. Determine the nitrogen gas flow rate (q_{N2}) using the standard nitrogen density $\rho_{N2} = 0.073 \text{ lb/ft}^3$ (lb/scf).

$$q_{N2} = \frac{m_{N2}}{\rho_{N2}} = \frac{70 \text{ lb/d}}{0.073 \text{ lb/scf}} \times \frac{1 \text{ d}}{1440 \text{ min}} = 0.7 \text{ scfm as } N_2$$

6. Determine the mass of oxygen remained (m_{O2}) in the product gas.
   a. Calculate the oxygen content in the product gas.

$$w_{O2} = 1 - (w_{O3} + w_{N2}) = 1 - (0.14 + 0.03) = 0.83 \text{ lb } O_2\text{/lb gas}$$

$$m_{O2} = w_{O2} \times m_{gas} = 0.83\text{lb } N_2\text{/lb gas} \times 2321 \text{ lb/d as gas} = 1926 \text{ lb/d as } O_2$$

7. Determine the oxygen gas flow rate (q_{O2}) using the standard oxygen density $\rho_{O2} = 0.083 \text{ lb/ft}^3$ (lb/scf).

$$q_{O2} = \frac{m_{O2}}{\rho_{O2}} = \frac{1926 \text{ lb/d}}{0.083 \text{ lb/scf}} \times \frac{1 \text{ d}}{1440 \text{ min}} = 16.1 \text{ scfm as } O_2$$

8. Determine the total product gas flow rate (q_{O2}) to the ozone contactors.

$$q_{gas} = q_{O3} + q_{N2} + q_{O2} = (1.8 + 0.7 + 16.1) \text{ scfm} = 18.6 \text{ scfm as gas mixture}$$

Calculate the standard product gas density at 20°C and 1 atm.

$$\rho_{gas} = \frac{m_{gas}}{q_{gas}} = \frac{2321 \text{ lb/d}}{18.6 \text{ scfm}} \times \frac{1 \text{ d}}{1440 \text{ min}} = 0.087 \text{ lb/ft}^3 \text{ as gas mixture}$$

EXAMPLE 11.52: GAS FEED RATE TO OZONE GENERATOR AND LOX USAGE

Determine (a) gas feed rate to ozone generators and (2) usage of LOX. Use the calculation results in Example 11.51.

Solution

1. Determine the mass of oxygen contained in the feed gas ($m_{O2,feed}$) to the ozone generators.

   During ozone generation, oxygen is converted to ozone without a change in mass. Therefore, the mass of oxygen in the feed gas to the ozone generators should equal the total mass of ozone and oxygen in the product gas $m_{O3} = 325$ lb/d as O_3 (Example 11.51, Step 1) and $m_{O2} = 1926$ lb/d as O_2 (Example 11.51, Step 6).

   $$m_{O2,feed} = m_{O3} + m_{O2} = (325 + 1926)\ \text{lb/d} = 2251\ \text{lb/d as } O_2$$

2. Determine the oxygen gas flow rate ($q_{O2,feed}$) using the standard oxygen density $\rho_{O2} = 0.083$ lb/ft^3 (lb/scf).

   $$q_{O2,feed} = \frac{m_{O2,feed}}{\rho_{O2}} = \frac{2251\ \text{lb/d}}{0.083\ \text{lb/scf}} \times \frac{1\ \text{d}}{1440\ \text{min}} = 18.8\ \text{scfm as } O_2$$

3. Determine the total feed gas mass ($m_{feed\ gas}$) and flow rate ($q_{O2,feed}$) to the ozone generators.

   There is no change in mass and flow rate for nitrogen during ozone generation.

   $q_{N2} = 0.7$ scfm (Example 11.51, Step 5) and $m_{N2} = 70$ lb/d as N_2 (Example 11.51, Step 4),

   $$q_{feed\ gas} = q_{O2,\ feed} + q_{N2} = (18.8 + 0.7)\ \text{scfm} = 19.5\ \text{scfm as gas mixture}$$

   $$m_{feed\ gas} = m_{O2,\ feed} + m_{N2} = (2251 + 70)\ \text{lb/d} = 2321\ \text{lb/d as gas mixture}$$

   or $m_{feed\ gas} = m_{gas} = 2321$ lb/d as gas mixture

   Calculate the standard feed gas density at 20°C and 1 atm.

   $$\rho_{feed\ gas} = \frac{m_{feed\ gas}}{q_{feed\ gas}} = \frac{2321\ \text{lb/d}}{19.5\ \text{scfm}} \times \frac{1\ \text{d}}{1440\ \text{min}} = 0.083\ \text{lb/ft}^3\ \text{as gas mixture}$$

   Note: The density of feed gas is close to that of oxygen.

4. Determine the usage of LOX (q_{LOX}) assuming the 100% LOX (cryogenic liquid) density $\rho_{LOX} =$ 9.5 lb/gal.

   $$m_{LOX} = m_{O2,feed} = 2251\ \text{lb/das LOX}$$

   $$q_{LOX} = \frac{m_{LOX}}{\rho_{LOX}} = \frac{2251\ \text{lb/d}}{9.5\ \text{lb/gal}} \times \frac{1\ \text{d}}{24\ \text{h}} = 10\ \text{gph as LOX}$$

EXAMPLE 11.53: OZONE DOSE FOR DESIRED COLIFORM COUNT IN THE EFFLUENT FROM EMPIRICAL EQUATION

A pilot plant was operated to determine the disinfection efficiency of ozonation for inactivation of coliforms on secondary treated effluent. The ozone dose was varied, and influent and effluent MPN number per 100 mL sample was measured after 15-min contact time. The ozone concentration utilized, MPN number per 100 mL sample before and after ozonation are tabulated below. The ozone transfer efficiency was 85%, and expected average MPN in the secondary treated plant effluent is $4 \times 10^5/100$ mL. Determine the coefficient (n) and initial demand (q) in Equation 11.38c from the experimental data. Also, estimate the ozone dose required to disinfect the secondary effluent to an MPN criteria of 240/100 mL.

U, mg/L	log(U)	N_0, MPN/100 mL	N, MPN/100 mL	N/N_0	log(N/N_0)
0.75	−0.12	2.0×10^4	2000	0.10	1.0
1.0	0.00	4.5×10^4	1800	0.040	1.4
2.0	0.30	2.1×10^5	1300	0.0062	2.2
3.0	0.48	3.2×10^5	800	0.0025	2.6
4.0	0.60	5.0×10^5	400	0.00080	3.1

Solution

1. Determine the log removal of coliforms based on MPN data.

   Rearrange Equation 11.38c to develop the following linear equation.

$$\log\left(\frac{N}{N_0}\right) = -n\log(U) + n\log(\text{q})$$

   The values of log(U) and log(N/N_0) are calculated for all data points and are summarized in the above table.

2. Plot the linear relationship and determine coefficients n and q.

   Plot log(N/N_0) versus log(U) in Figure 11.21. The plot gives a straight line. The slope of the line is $-n$ and the intercept on Y axis is n log(q).

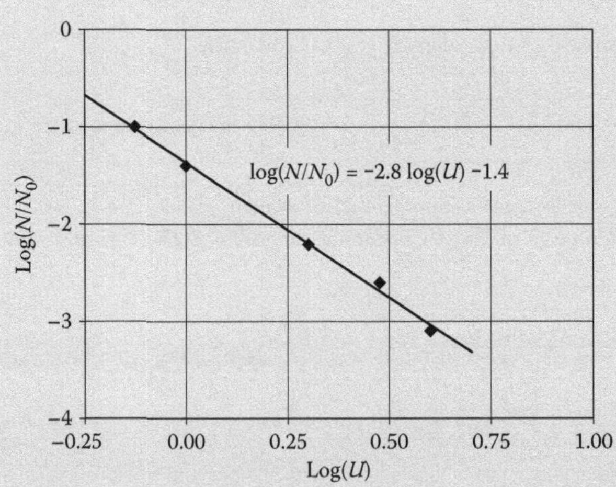

FIGURE 11.21 Plot of log(N/N_0) versus log(U) (Example 11.52).

The slope $-n = -2.8$, so $n = 2.8$.

n log(q) $= -1.4$ is determined from the intercept on Y axis.

$$q = 10^{\left(\frac{-1.4}{n}\right)} = 10^{\left(\frac{-1.4}{2.8}\right)} = 10^{-0.5} = 0.32 \text{ mg/L}$$

Note: The minimum U is 0.75 mg/L during the pilot test. Therefore, utilization of Equation 11.38c is valid since the condition $U > q$ is met.

3. Determine the ozone dose required to achieve a coliform count of 240 MPN/100 mL.
   Rearrange Equation 11.38c and substitute $N = 240$ MPN/100 mL and $N_0 = 4 \times 10^5$ MPN/100 mL. Solve for U.

$$U = q\left(\frac{N}{N_0}\right)^{-\frac{1}{n}} = 0.32\ \text{mg/L} \times \left(\frac{240\ \text{MPN/100 mL}}{4 \times 10^5\ \text{MPN/100 mL}}\right)^{-\frac{1}{2.8}} = 4.5\ \text{mg/L}$$

4. Determine the ozone dosage and loss.
   Apply Equation 11.38e to calculate ozone dosage (D) from $TE_{O_3} = 85\%$.

$$D = \frac{100\%}{TE_{O3}} U = \frac{100\%}{85\%} \times 4.5\ \text{mg/L} = 5.3\ \text{mg/L}$$

Ozone loss, $O_L = D - U = (5.3 - 4.5)\ \text{mg/L} = 0.8\ \text{mg/L}$

11.9 Disinfection with UV Radiation

UV radiation for disinfection of wastewater is an emerging technology. Being a physical process, it does not leave any chemical residuals. It is a safe and effective technology and is going through rapid development.

11.9.1 Mechanism of UV Disinfection

UV radiation is the transfer of electromagnetic energy from the source to the genetic material (DNA) of an organism. As the UV energy is absorbed by the DNA, the damage to the nucleic acid causes a change in cell structure that prevents the propagation of the organism. The amount of UV light that is absorbed by the organism is measured by *reflectance* or *transmittance*. Water and the constituents in water affect the transmission and absorption of UV rays. Suspended solids reduce the transmittance of UV radiation and also shield the organisms from exposure, particularly those that are encapsulated within the suspended particles. The absorbance of UV radiation by DNA molecule depends upon the wavelength of the radiation. The most effective spectral region for the germicidal effect is in the range of 250–265 nm with an optimum absorbance by DNA around 254 nm.[52,54]

11.9.2 Source of UV Radiation

UV radiation is produced from lamp that contains mercury vapors. The vapors are charged by striking an electric arc. Energy is generated by excitation of mercury vapors resulting in emission of UV light.

Types of Traditional UV Lamps: There are three types of traditional UV lamps based on their operating parameters. These are (1) *low-pressure low-intensity (LPLI)*, (2) *low-pressure high-intensity (LPHI)*, and (3) *medium-pressure high-intensity (MPHI)*. The typical operating parameters of these lamps are summarized in Table 11.18.[52,55] LPLI lamps are efficient and effective, and most widely used for UV radiation at small to medium facilities. Because of high output within small footprint, the MPHI lamps are more suitable for large applications for treatment of wastewater and combined sewer overflow (CSO). The typical lamp life ranges from 5000 to 15,000 h. However, the actual output intensity of lamps may decrease with age of the lamps, especially if notable scaling is developed on the surface due to high temperature.

TABLE 11.18 Typical Operating Parameters of Conventional UV Lamps

Parameter	Unit	Type of Lamp		
		LPLI	LPHI	MPHI
Pressure	mm Hg (in Hg)	0.007 (0.0003)	0.01–0.8 (0.004–0.03)	100–10,000 (4–400)
Total power draw	W (hp)	40–100 (0.05–0.13)	200–500 (0.27–0.67)	1000–13,000 (1.3–17)
Output at 254 nm	W (hp)	25–27 (0.03–0.04)	60–400 (0.08–0.5)	100–2000 (0.13–2.7)
Germicidal output/input	%	30–50	35–50	15–20
Operating temperature	°C (°F)	35–50 (95–120)	100–150 (200–300)	600–800 (1100–1500)
Dimension				
Length	m (ft)	0.75–1.5 (2.5–5)	1.8–2.5 (6–8)	0.3–1.2 (1–4)
Diameter	mm (in)	15–20 (0.6–0.8)	Variable	Variable
Estimated lamp life	h	8000–12,000	9000–15,000	3000–8000
Sleeve life	year	4–6	4–6	1–3
Ballast life	year	10–15	10–15	3–5
Decrease in output at estimated lamp life	% of output of new lamp	20–25	25–30	20–25

Note: LPLI: low-pressure low-intensity; LPHI: low-pressure high-intensity; MPHI: medium pressure high-intensity.
Source: Adapted in part from References 52 and 72.

New UV Lamp Technologies: UV lamp technologies have been improved continuously and many new developments are being seen in the research phase. New directions in future UV lamp market include: (1) the pulsed broadband energy xenon UV lamp, (2) the narrowband excimer UV lamp, (3) the microwave-powered high-intensity UV lamp, and (4) the UV light-emitting diodes (LED) lamp.

11.9.3 Types of UV Reactors and Lamp Arrangements

The UV reactors are primarily classified as either a *contact* or *noncontact* reactor. In a contract reactor, the UV lamps may be arranged in *parallel, perpendicular*, or *inclined* to the flow direction in an *open-channel* or *closed-vessel* application (Figure 11.22a).[103–106] The lamps are submerged in wastewater during operation. Each lamp is typically encased in a quartz sleeve that is slightly larger in diameter than the lamp. The lamps with sleeves are placed at a spacing to form a module or rack that may further be encased in a sealed shell. The lamp spacing varies from 75 mm (3 in) to 150 mm (6 in). The lamp module can be lifted up above the water surface or the lamps can be pulled out of a vessel for regular maintenance or replacement of lamps. Providing a proper mechanical or chemical–mechanical cleaning system is an essential requirement for preventing the quartz sleeves from fouling and reducing the performance of the UV disinfection system.[52] In a noncontact reactor, the lamps do not come in contact with the liquid. The lamps are suspended above the liquid or surround the conduits that carry the liquid (Figure 11.22b). These conduits are made of materials that are transparent to UV light, such as Teflon® or AFP^{TM}_{840}. Owing to the "dry" operating conditions, the encasement of lamps inside quartz sleeves is not required. The lamps are normally arranged in parallel to the flow direction in the noncontact systems. The turbulent hydraulic condition inside the conduits provides an automatic "self-cleaning" and "fouling-resistant" mechanism.[107]

11.9.4 Microbial Repair after UV Disinfection

Some organisms have the ability to repair UV-induced damage to DNA due to UV exposure. This happens when an injured organism is exposed to the visible light range (primarily blue spectrum). Two typical repair mechanisms are: (1) *photoreactivation* and (2) *dark repair*.

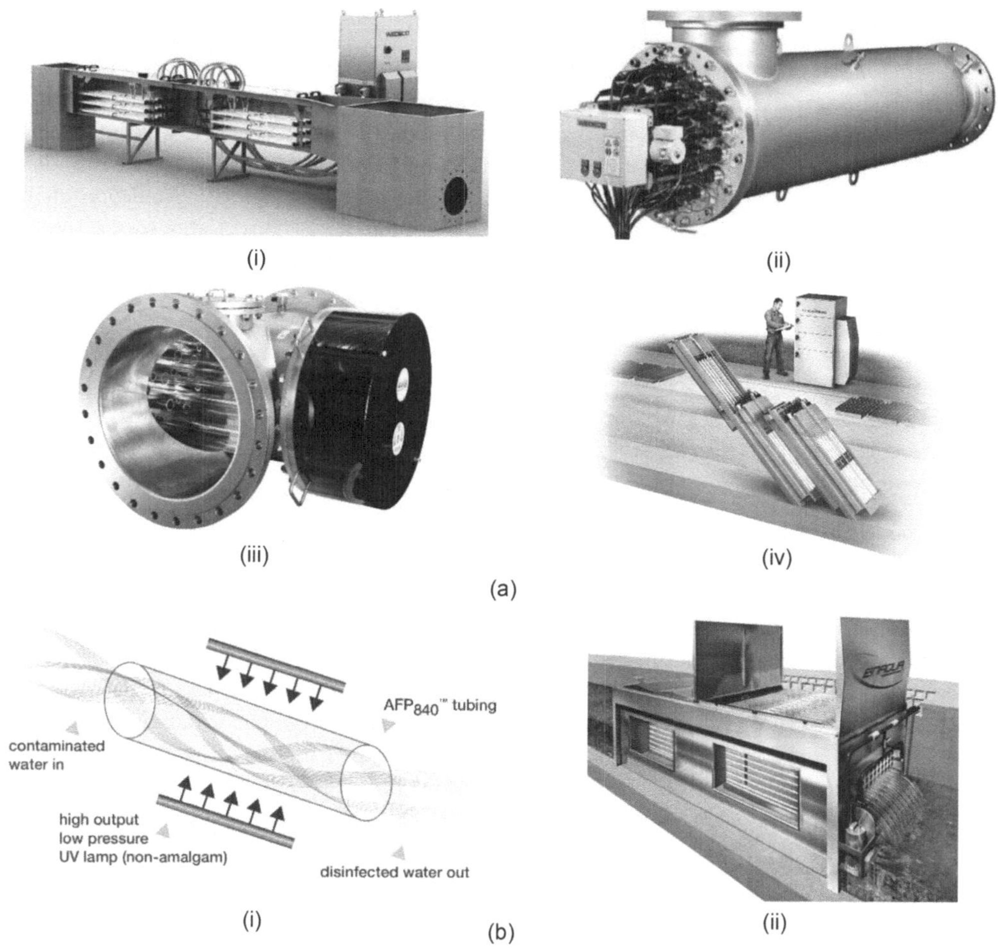

FIGURE 11.22 Types of UV Disinfection reactors and major system components: (a) contact UV systems: (i) horizontal lamps parallel to flow direction in channel (Courtesy Xylem, Inc.), (ii) horizontal lamps parallel to flow direction in closed vessel (Courtesy Xylem, Inc.), (iii) lamps perpendicular to flow direction in closed vessel (Courtesy Neptune Benson/Evoqua*), and (iv) inclined array of UV banks arranged in series (Courtesy TrojanUV); and (b) noncontact systems (Courtesy Enaqua/GRUNDFOS Pumps Manufacturing Corporation): (i) conceptual illustration and (ii) illustration of system operation.

Photoreactivation: The repair phenomenon of photoreactivation has been detected in many organisms.[108,109] There are two major steps involved in the photoreactivation: (1) a light-independent step to form photolyases, for example, enzyme–substrate complex and (2) light input-dependent process to repair the damage to the DNA by utilizing photo energy from visible light from the violet to blue end of the spectrum in wavelengths between 310 and 490 nm. The photoreactivation is limited to fix the damage to pyrimidine dimers only.[52]

Dark Repair: Two major steps in dark repair are: (1) excision repair to remove the damaged DNA section and (2) recombination repair to regenerate the damaged DNA section. In comparison with photoreactivation, it is a slower but more effective process for repairing a wider range of damage in the genome.

* Evoqua, ETS-UV, and Neptune-Benson are trademarks of Evoqua, its subsidiaries, and affiliates in some countries.

11.9.5 Inactivation Kinetics for UV Irradiation

Several kinetic models that have been discussed in Section 16.5.2 for chemical disinfection processes can also be used to express microbial inactivation by UV irradiation. Many important kinetic models for chemical disinfection are based on the concentration–time (*CT*) concept. For UV disinfection, this basic concept provides the basis of developing the *UV dose* concept. Known as the *Bunsen–Roscoe Law*, the UV dose (D_{UV}) or fluence is defined as the product of the average UV irradiation intensity (I_{avg}) and the exposure time t of the organism under UV irradiation.[2,62] This relationship is shown by Equation 11.39.

$$D_{UV} = I_{avg}t \tag{11.39}$$

where
- D_{UV} = UV dose to which organisms are exposed, mW·s/cm^2 (mJ/cm^2)
- I_{avg} = average UV irradiation intensity, mW/cm^2
- t = exposure time, s

Several models have been developed for disinfection by UV radiation. These models fall into three major groups: (1) first-order reaction-based models, (2) empirical models, and (3) mathematical models. These models are discussed below.

First-order Reaction-Based Model: The first-order Chick-Watson equation is a simple and common model. It assumes that the inactivation of bacteria, viruses, and protozoa by UV irradiation follows a first-order reaction.[2,42,54] This model has been used in comparative inactivation studies and validated by experimental and actual operating data. The results are reported in many recent publications.[51,58,110–112] The expression for the reduction of microorganisms by UV inactivation is given by Equation 11.40a. In this model, both UV intensity I_{avg} and time t are equally important, and the product of the parameters is actually a more meaningful factor than either parameter alone. The logarithmic expression of this equation yields a linear relationship (Equation 11.40b).

$$N = N_0 e^{-k_{UV} I_{avg} t} \quad \text{or} \quad N = N_0 e^{-k_{UV} D_{UV}} \tag{11.40a}$$

$$\ln\left(\frac{N}{N_0}\right) = -k_{UV} I_{avg} t \quad \text{or} \quad \ln\left(\frac{N}{N_0}\right) = -k_{UV} D_{UV} \tag{11.40b}$$

where
- k_{uv} = inactivation rate constant (base *e*) of a given type of organism receiving UV radiation, cm^2/mW·s (cm^2/mJ)
- All other terms have been defined previously

Equation 11.40b represents the ideal condition. In practice, many physical, chemical, and biological constituents in wastewater decrease the absorbance of UV light and may cause tailing and shouldering effects (Figure 11.1). The first-order Chick-Watson model is considered a reasonable inactivation kinetic when the UV dose and predisinfection water quality are both within the desirable ranges. However, significant deviations from the first-order behavior have been observed at either low or high UV doses as illustrated in Figure 11.23.[109,110,113–115] The typical *S-shape* curve shows a *shoulder*-type effect on microbial inactivation when the UV dose is low. Inadequate UV dose is the major attributor to this deviation. The *tailing off* response is a result of the interference of suspended particles on the effectiveness of UV irradiation by either shielding the organism from UV irradiation, scattering the UV light, or lowering the UV intensity due to decreased transmittance through the water.

Empirical Models: The first-order kinetic equations have been modified to include the tailing off and shoulder effects. Empirical models have been developed to incorporate the effects of suspended solids and photoreactivation. These procedures and models are presented below.

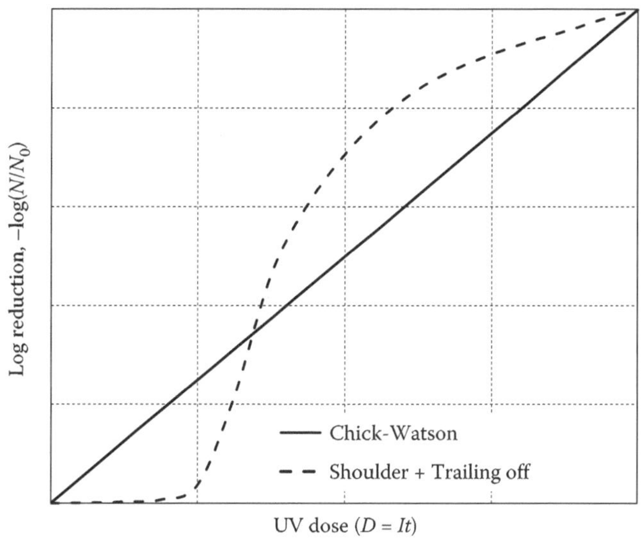

FIGURE 11.23 Typical deviations from Chick-Watson model observed in UV irradiation disinfection.

The suspended solids in the effluent reduce the disinfection efficiency of UV radiation. The Chick-Watson equation Equation 11.40a is modified to include this effect in Equation 11.41a.

$$N = N_0 e^{-k_{UV} I_{avg} t} + N_P \quad \text{or} \quad N = N_0 e^{-k_{UV} D_{UV}} + N_P \tag{11.41a}$$

where

N = total organism density remaining after UV exposure, organisms/100 mL

N_P = particulate organism density that is unaffected by UV light, organisms/100 mL. A relationship between N_P and total suspended solids (TSS) is given by Equation 11.41b. TSS is the most commonly regulated parameter in the effluent.

$$N_p = C_1 (\text{TSS})^{m_1} \tag{11.41b}$$

where

TSS = concentration of total suspended solids, mg/L

C_1 = proportionality constant

m_1 = constant

During UV disinfection process, various particles through the reactor have different exposure time around the ideal time. This is known as residence time distribution (RTD).[49,54,63,64] The RTD is a function of the dispersion characteristics of a reactor. A disinfection model that incorporates dispersive properties of a reactor was developed by Scheible.[116,117] The generalized expressions of this model for inactivation of coliform by UV are given by Equations 11.42a through 11.42f.

$$N = N' + N_P \quad \text{or} \quad N' = N - N_P \tag{11.42a}$$

$$N'/N_0 = \exp\left[\frac{ux}{2E}\left(1 - \left(1 + \frac{4kE}{u^2}\right)^{1/2}\right)\right] \tag{11.42b}$$

$$N' = N_0 \exp\left[\frac{ux}{2E}\left(1 - \left(1 + \frac{4kE}{u^2}\right)^{1/2}\right)\right] + N_P \tag{11.42c}$$

$$k = a(I_{avg})^b \quad \text{or} \quad \log\ k = \log\ a + b\ \log(I_{avg}) \tag{11.42d}$$

$$d = \frac{E}{ux} \quad \text{or} \quad E = dux \tag{11.42e}$$

$$N = N_0 \exp\left[\frac{ux}{2E}\left(1 - \left(1 + \frac{4Ea(I_{avg})^b}{u^2}\right)^{1/2}\right)\right] + C_1(\text{TSS})^{m_1} \tag{11.42f}$$

where

x = the length of the reactor or the average distance traveled by water under direct exposure to UV, cm

u = velocity of water, cm/s

E = dispersion coefficient, cm^2/s. It can be estimated from the RTD curve of a particular reactor.

d = dispersion number, dimensionless. The RTD curve and d is typically determined from the experimental data that are obtained through a tracer study using a conservative (nonreactive) dye. The practical range of d for UV reactor design is 0.03–0.05.

k = coliform inactivation rates, s^{-1}. The inactivation rate is expressed as a function of the average intensity of UV (I_{avg}) in Equation 11.42d. Term of I_{avg} has been defined in Equation 11.39. The unit of $I_{avg} = \mu W/cm^2$ in Equation 11.42d.

N' = nonparticulate effluent coliform density ($N' = N - N_P$), coliforms/100 mL

a and b = empirical constants. They are determined from the slope and intercept of the linear regression developed from log function of Equation 11.42d.

As a modification of first-order kinetics, the Collins-Selleck and Emerick equations (Equations 11.43a through 11.43c) were also developed. These equations describe the inactivation of organisms by either chemical disinfectants or UV irradiation with shoulder and moderate tailing effects.[70,91,118] Emerick model considers the effect of initial particles associated with microbial density on disinfection.[119] Specifically, it describes the inactivation kinetics of coliform bacteria by UV irradiation when both dispersed and particle-associated organisms are involved in the disinfection process.[2,119]

$$\text{Collins–Selleck} \quad N = N_0\left(\frac{I_{avg}t}{D_c}\right)^{-n_{uv}} \quad \text{or} \quad N = N_0\left(\frac{D_{UV}}{D_c}\right)^{-n_{uv}} \quad (I_{avg}t > D_c) \tag{11.43a}$$

$$N = N_0 \quad (I_{avg}t \leq D_c) \tag{11.43b}$$

$$\text{Emerick} \quad N = N_0 e^{-k_{UV} I_{avg} t} + \frac{N'_P}{k_{UV} I_{avg} t}\left(1 - e^{-k_{UV} I_{avg} t}\right) \tag{11.43c}$$

where

D_c = empirical threshold of UV dose determined experimentally, $mW{\cdot}s/cm^2$ (mJ/cm^2). Microbial inactivation can occur only when $I_{avg}\, t > D_c$.

n_{uv} = coefficient obtained experimentally, dimensionless

N'_P = number of particles per unit sample volume, particles/100 mL. Each particle contains at least one organism at $t = 0$.

Mathematical Models: Two mathematical expressions, that is, *series-event* and *multiple-target* models have been used to describe inactivation kinetics for chemical disinfection processes. After proper substitution of process variables, these two models have been applied for UV irradiation. Equations 11.44a and 11.44b express these models.[51,80,81,120]

$$\text{Series-event} \qquad N = N_0 e^{-k_{UV} I_{avg} t} \left(\sum_{i=0}^{l-1} \frac{(k_{UV} I_{avg} t)^i}{i!} \right) \tag{11.44a}$$

$$\text{Multiple target} \qquad N = N_0 \left[1 - \left(1 - e^{-k_{UV} I_{avg} t}\right)^{n_{cuv}} \right] \tag{11.44b}$$

where

i = the event level
l = threshold of effective inactivation for a organisms
n_{cuv} = coefficient obtained experimentally, dimensionless

In the multiple-target model, it is assumed that there are n_u target sites, and all of these sites must be attacked by the disinfectant to inactivate the organism.

EXAMPLE 11.54: INACTIVATION OF *E. COLI* FROM OLD AND NEW UV LAMPS

A UV lamp after 10,000-h operation is inactivating 99.9% *E. coli* at 72% intensity of a new lamp. The new lamp has a design intensity of 2 mW/cm^2 that is normally measured after 100 h of operation. The inactivation rate constant of *E. coli* is 0.13 cm^2/mW·s (cm^2/mJ). Determine the survival of *E. coli* when the lamps are new and the exposure time to organism is same as that of old lamps. Compare the results of new lamps with that of old lamps.

Solution

1. Determine the UV dose from a used lamp to which the organisms are exposed.
   Calculate the survival from the old lamps at an inactivation of 99.9% (3-log).

   $N_{used}/N_0 = (100 - 99.9)\% = 0.1\%$ or 0.001

   Rearrange Equation 11.40b and calculate $D_{UV,used}$, $k_{UV} = 0.13\ \text{cm}^2/\text{mW·s}$.

$$D_{UV,used} = -\frac{1}{k_{UV}} \times \ln\left(\frac{N_{used}}{N_0}\right) = -\frac{1}{0.13\ \text{cm}^2/\text{mW·s}} \times \ln(0.001) = -\frac{1}{0.13\ \text{cm}^2/\text{mW·s}} \times (-6.91)$$

$$= 53.2\ \text{mW·s/cm}^2 \quad \text{or} \quad 53.2\ \text{mJ/cm}^2$$

2. Determine the exposure time to achieve 99.9% inactivation.

   UV intensity from the used UV lamp, $I_{avg,used} = E_{used} \times I_{avg} = 0.72 \times 2\ \text{mW/cm}^2 = 1.44\ \text{mW/cm}^2$

   Rearrange Equation 1.39 and calculate the exposed time.

$$t = \frac{D_{UV,used}}{I_{avg.used}} = \frac{53.2\ \text{mW·s/cm}^2}{1.44\ \text{mW/cm}^2} = 37\ \text{s}$$

3. Determine the UV dose from Equation 1.39 for new lamp that has the same exposure time to organisms as that of old lamp.

$$D_{UV} = I_{avg} t = 2\ \text{mW/cm}^2 \times 37\ \text{s} = 74\ \text{mW·s/cm}^2 \text{or } 74\ \text{mJ/cm}^2$$

4. Determine the survival of *E. coli* after exposure from the new lamps.

$$\ln\left(\frac{N_{new}}{N_0}\right) = -k_{UV}\,D_{UV} = -0.13\ \text{cm}^2/\text{mW·s} \times 74\ \text{mW·s/cm}^2 = -9.62$$

Solve $N_{new}/N_0 = e^{-9.62} = 0.000066 = 6.6 \times 10^{-5}$ or 0.0066%

log inactivation, $-\log\left(\frac{N_{new}}{N_0}\right) = -\log(0.000066) = 4.2$

5. Compare the inactivation result from new and old lamps.
   log inactivation from new lamps, $-\log(N/N_0) = 4.2$ and log inactivation from old lamps, $-\log(N_{used}/N_0) = -\log\left(\frac{0.001}{1}\right) = 3.0$

   There is a decrease of 1.2-log in the inactivation of *E. coli* after 10,000-h operation of the UV lamps.

EXAMPLE 11.55: UV DOSE FOR A GIVEN REDUCTION TO OVERCOME THE EFFECTS BY PHOTOREACTIVATION

A UV inactivation study was conducted using a batch reactor. At a UV dose of 26 mW·s/cm^2, the survival of *E. coli* was 0.0001 (0.01%). If 1% of original organisms overcome the UV inactivation effect by photoreactivation, determine the total UV dose required to achieve total deactivation even after full recovery of 1% of original organisms from photoreactivation.

Solution

1. Determine the inactivation rate constant at the survival of = 0.0001.
   Rearrange Equation 11.40b and calculate k_{UV} using $N/N_0 = 0.0001$.

$$k_{UV} = -\frac{1}{D_{UV,initial}} \times \ln\left(\frac{N}{N_0}\right) = -\frac{1}{26\ \text{mW·s/cm}^2} \times \ln(0.0001) = -\frac{1}{26\ \text{mV·s/cm}^2} \times (-9.21)$$

$$= 0.35\ \text{cm}^2/\text{mW·s} \quad \text{or} \quad 0.35\ \text{cm}^2/\text{mJ}$$

2. Determine additional UV dose required to achieve complete deactivation even after full photoreactivation of 1% of original organisms.
   The survival after photoreactivation based on the original organisms ($N_{photoreactivation}/N_0$).

$$N_{photoreactivation}/N_0 = 0.01\ (1\%)$$

Calculate the survival after additional UV radiation based on the reactivated organisms ($N/N_{photoreactivation}$).

$$N/N_{photoreactivation} = \frac{N/N_0}{N_{photoreactivation}/N_0} = \frac{0.0001}{0.01} = 0.01$$

Rearrange Equation 11.40b and calculate the additional $D_{\text{UV,additional}}$ required after full recovery of 1% organisms.

$$D_{\text{UV,additional}} = -\frac{1}{k_{\text{UV}}} \times \ln\left(\frac{N}{N_{\text{photoreactivation}}}\right) = -\frac{1}{0.35\ \text{cm}^2/\text{mW}\cdot\text{s}} \times \ln(0.01)$$

$$= -\frac{1}{0.35\ \text{cm}^2/\text{mW}\cdot\text{s}} \times (-4.61) = 13.2\ \text{mW}\cdot\text{s/cm}^2 \quad \text{or} \quad 13.2\ \text{mJ/cm}^2$$

3. Determine the overall UV dose required to deactivate all organisms even 1% of original organisms that fully recover from photoreactivation.

The overall UV dose required, $D_{\text{UV,overall}} = D_{\text{UV,initial}} + D_{\text{UV,additional}} = (26 + 13.2)\ \text{mW}\cdot\text{s/cm}^2$
$= 39.2\ \text{mW}\cdot\text{s/cm}^2 \text{or } 39.2\ \text{mJ/cm}^2$

An overall UV dose of 39.2 $\text{mW}\cdot\text{s/cm}^2$ is needed to achieve deactivation of all organisms even after survival of 0.0001 and 1% of organisms that fully recover from photoreactivation.

EXAMPLE 11.56: DETERMINATION OF INACTIVATION RATE CONSTANT FOR CHICK-WATSON MODEL

A study was performed to assess the inactivation kinetics of *Cryptosporidium parvum* by UV irradiation. The laboratory data were analyzed and the following average log reduction results were obtained at different UV doses. Obtain the inactivation rate constant k_{uv} for *Cryptosporidium parvum.*

UV dose (D_{uv}), mJ/cm^2	0	1	2	3	4	5
log reduction ($-\log(N/N_0)$)	0	0.62	1.12	1.82	2.20	2.82

Solution

1. Develop the relationship of k_{uv} as a function of log reduction $-\log(N/N_0)$ and UV dose D_{uv}.
   Apply Equation 1.40a. $N/N_0 = e^{-k_{uv}D_{uv}}$
   Take logarithm on both sides and rearrange the expression to obtain the following expression.

$$-\log(N/N_0) = -\log(e^{-k_{uv}D_{uv}}) = -\log(e) \times (-k_{uv}D_{uv}) = \log(e) \times (k_{uv}D_{uv}) = 0.434\ k_{uv}D_{uv}$$

Rearrange the expression to obtain the desired relationship for k_{uv}.

$$k_{uv} = \frac{-\log(N/N_0)}{0.434\ D_{uv}}$$

2. Calculate the k_{uv} for inactivation of *Cryptosporidium parvam* at different UV doses.
   The calculated values of k_{uv} at different UV doses are tabulated below.

UV dose (D_{uv}), mJ/cm^2	0	1	2	3	4	5
log reduction ($-\log(N/N_0)$)	0	0.62	1.12	1.82	2.20	2.82
k_{UV}, cm^2/mJ	–	1.43	1.29	1.40	1.27	1.30

An average value $k_{uv} = 1.34\ \text{cm}^2/\text{mJ}$ for inactivation of *Cryptosporidium parvam*. The survival number of *Cryptosporidium parvum* by UV irradiation can be expresses as:

$$N = N_0 e^{-1.34\,D_{uv}} \quad \text{or} \quad N = N_0 e^{-1.34 I_{avg} t}$$

3. Plot the log reduction $-\log(N/N_0)$ versus UV dose D_{uv} in Figure 11.24.

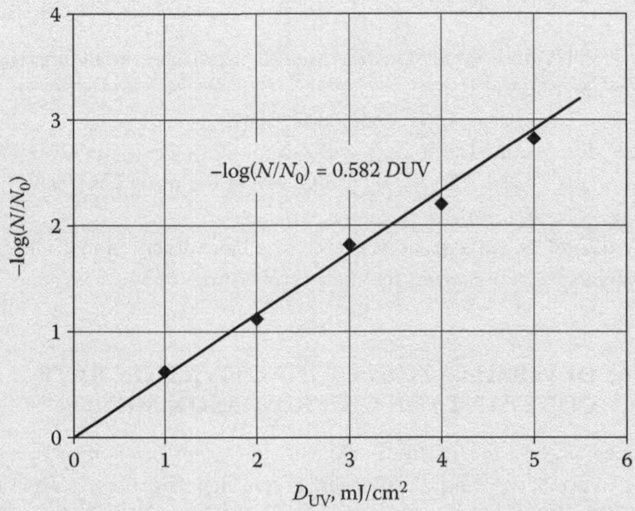

FIGURE 11.24 Plot of $-\log(N/N_0)$ versus D_{UV} (Example 11.56).

Note: A plot of log reduction of *Cryptosporidium parvam* ($-\log(N/N_0)$ vs. D_{uv}) is a linear relationship with a slope of $0.434\ k_{uv} = 0.434 \times 1.34\ \text{cm}^2/\text{mJ} = 0.582\ \text{cm}^2/\text{mJ}$. The expressions for log reduction are:

$$-\log(N/N_0) = 0.434\ k_{uv} D_{uv} \quad \text{or} \quad -\log(N/N_0) = 0.434\ k_{uv} I_{avg} t$$

$$-\log(N/N_0) = 0.582\ D_{uv} \quad \text{or} \quad -\log(N/N_0) = 0.582\ I_{avg} t$$

EXAMPLE 11.57: CHICK-WATSON MODEL USED TO DETERMINE THE UV DOSE TO ACHIEVE A DESIRED LOG REDUCTION

Estimate the UV dosage required to achieve a 6-log reduction of *Cryptosporidium parvum*. Apply Chick-Watson model and use $k_{uv} = 1.34\ \text{cm}^2/\text{mJ}$ that is obtained in Example 11.56.

Solution

1. Develop the relationship of UV dose D_{uv} as a function of log reduction $-\log(N/N_0)$.
   Apply the expression obtained in Example 11.56, Step 3. $-\log(N/N_0) = 0.434\ k_{uv} D_{uv}$
   Rearrange the above expression to obtain the following expression of D_{uv}.

$$D_{uv} = \frac{-\log(N/N_0)}{0.434\ k_{uv}}$$

Calculate the UV dose required for a 6-log reduction ($-\log(N/N_0) = 6$).

$$D_{uv} = \frac{-\log(N/N_0)}{0.434\ k_{uv}} = \frac{6}{0.434 \times 1.34\ \text{cm}^2/\text{mV·s}} = 10.3\ \text{mW·s/cm}^2 \quad \text{or} \quad 10.3\ \text{mJ/cm}^2$$

The UV dose required for 6-log inactivation of *Cryptosporidium* is 10.3 mJ/cm^2.

EXAMPLE 11.58: APPLICATION OF SCHEIBLE DISPERSION EQUATION WITH PARTICULATE BACTERIAL DENSITY

A UV inactivation study was conducted to determine the effectiveness of UV irradiation on the filtered secondary effluent. The initial coliform count $N_0 = 10^6$ MPN/100 mL and TSS = 10 mg/L. The UV reactor has uniform lamp array with the lamps parallel to flow direction. The reactor information is given below.

Effective lamp length $x = 147$ cm, velocity of water $u = 18.5$ cm/s, average UV intensity $I_{avg} = 8500$ $\mu\text{W/cm}^2$, and dispersion coefficient $E = 148\ \text{cm}^2/\text{s}$. Other constants are: $a = 1.5 \times 10^{-5}$, $b = 1.3$, $C_1 =$ 0.25, and $m_1 = 2.0$. Calculate (a) the nonparticulate effluent coliform density (N'), (b) the total coliform density remaining (N), and (c) the log inactivation after exposure to UV radiation.

Solution

1. Determine the coliform inactivation rate constant k from Equation 11.42d.

$$k = a(I_{avg})^b = 1.5 \times 10^{-5} \times (8500)^{1.3} = 1.92\ \text{s}^{-1}$$

2. Determine the nonparticulate effluent coliform density (N').

Apply Equation 1.42b, $N'/N_0 = \exp\left[\frac{ux}{2E}\left(1 - \left(1 + \frac{4kE}{u^2}\right)^{1/2}\right)\right]$

$$= \exp\left[\frac{18.5\ \text{cm/s} \times 147\ \text{cm}}{2 \times 148\ \text{cm}^2/\text{s}}\left(1 - \left(1 + \frac{4 \times 1.92\ \text{s}^{-1} \times 148\ \text{cm}^2/\text{s}}{(18.5\ \text{cm/s})^2}\right)^{1/2}\right)\right]$$

$$= \exp\left[9.19 \times \left(1 - (1 + 3.32)^{1/2}\right)\right]$$

$$= \exp\left[9.19 \times (1 - 2.08)\right]$$

$$= 4.9 \times 10^{-5}$$

$$N' = 4.9 \times 10^{-5}\ N_0 = 4.9 \times 10^{-5} \times 10^6\ \text{MPN/100 mL} = 49\ \text{MPN/100 mL}$$

3. Determine the coliform density remaining (N) after UV exposure.
   The particulate organism density is calculated from Equation 11.41b.

$$N_P = C_1(\text{TSS})^{m_1} = 0.25 \times (10\ \text{mg/L})^{2.0} = 25\ \text{MPN/100 mL}$$

   Calculate the total organism density remaining from Equation 11.42a.

$$N = N' + N_P = (49 + 25)\ \text{MPN/100 mL} = 74\ \text{MPN/100 mL}$$

   There are a total 74 MPN/100 mL remaining after UV exposure.

4. Determine the log inactivation after UV exposure.

$$-\log(N/N_0) = -\log\left(\frac{74 \text{ MPN/100 mL}}{10^6 \text{ MPN/100 mL}}\right) = 4.1\text{-log inactivation}$$

EXAMPLE 11.59: EMPIRICAL CONSTANTS IN COLLINS–SELLECK EQUATION FOR UV DISINFECTION UNDER THE INFLUENCE OF TURBIDITY

An experimental program was conducted to assess the impact of turbidity on inactivation of *E. coli* by UV radiation. The results are summarized below:

At turbidity of 2 NTU:

UV dose (D_{uv}), mJ/cm^2	0	2.5	5.0	7.5	10	15	20	30
log reduction ($-\log(N/N_0)$)	0	0.3	1.3	2.8	3.8	5.3	6.0	6.5

At turbidity of 20 NTU:

UV dose (D_{uv}), mJ/cm^2	0	2.5	5	7.5	10	15	20	30
log reduction ($-\log(N/N_0)$)	0	0.2	1.1	2.0	2.8	4.0	4.5	4.8

Discuss the impact of turbidity on the response of *E. coli* to UV irradiation. Also develop the empirical constants n_{uv} and D_c in Collins-Selleck model that can be used to express the general performance of UV inactivation at a constant turbidity level.

Solution

1. Plot the reduction of *E. coli* to UV irradiation for two turbidity levels.

Plot $-\log(N/N_0)$ versus D_{uv} for turbidity of 2 and 20 NTU, respectively, in Figure 11.25.

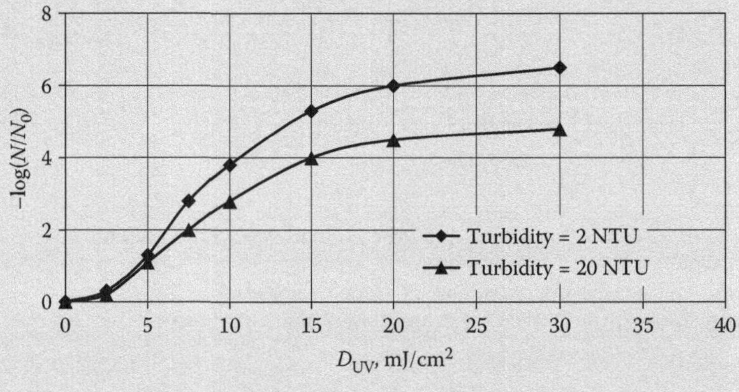

FIGURE 11.25 Plot of $-\log(N/N_0)$ versus D_{UV} (Example 11.59).

2. Discuss the response of *E. coli* to UV irradiation.

The following observations on the inactivation of *E. coli* by UV irradiation can be made from the plots:

a. There is a "shoulder" effect but almost no impact of turbidity at a low UV dose (below 5 mJ/cm^2).

b. There is a moderate "tailing off" effect with an increasing impact of turbidity at the UV doses in the range of 5–15 mJ/cm^2.

c. There is a significant "tailing off" effect at a UV dose higher than 15 mJ/cm^2.

3. Develop the inactivation kinetic model from the experimental data.

a. Plot $-\log(N/N_0)$ versus $\log(D_{uv})$.

Calculate the values of $\log(D_{uv})$ at different UV doses.

D_{uv}, mJ/cm^2	0	2.5	5.0	7.5	10	15	20	30
log (D_{uv})	–	0.40	0.70	0.88	1.0	1.2	1.3	1.5

Plot the values of $-\log(N/N_0)$ versus $\log(D_{uv})$ for turbidity of 2 and 20 NTU, respectively, in Figure 11.26. There is a linear portion on each curve in the UV dose (D_{uv}) range of 5–15 mJ/cm^2 or $\log(D_{uv})$ range of 0.7–1.2.

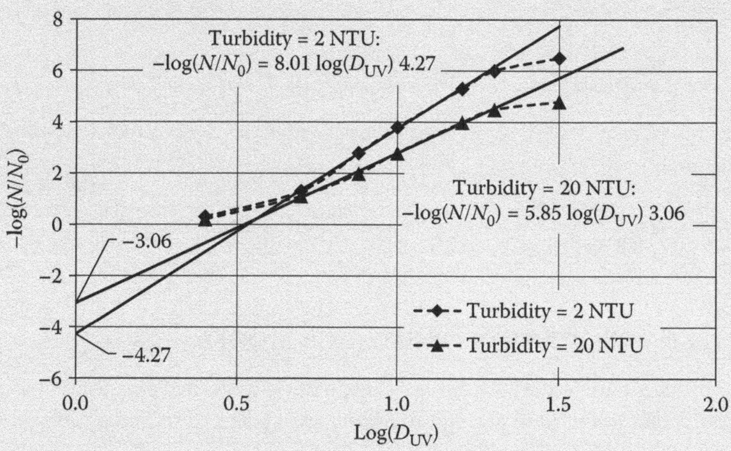

FIGURE 11.26 Plot of $-\log(N/N_0)$ versus D_{UV} (Example 11.59).

b. Develop the relationship between $-\log(N/N_0)$ and $\log(D_{uv})$.

Apply the Collins–Selleck model (Equation 11.43a). $N = N_0\left(\frac{D_{UV}}{D_c}\right)^{-n_{uv}}$

Take logarithm on both sides and rearrange the expression to obtain the following linear relationship that is the log function of Equation 11.43a.

$$-\log(N/N_0) = n_{UV}\log(D_{UV}) - n_{UV}\log(D_c)$$

The slope of the line is n_{uv} and the intercept on Y axis is $-n_{uv}\log(D_c)$.

The experimental data within the linear range are used to develop the linear relationship between $-\log(N/N_0)$ and $\log(D_{uv})$.

c. Develop the coefficient (n_{uv}) and empirical threshold (D_c) from the plot of 2-NTU turbidity.

The following linear relationship is obtained from experimental data:

$$n_{uv} = 8.01 \text{ and } -n_{uv}\log(D_c) = -4.27 \quad \text{or} \quad n_{uv}\log(D_c) = 4.27$$

Solve for D_c, $D_c = 10^{\frac{4.27}{n_{uv}}} = 10^{\frac{4.27}{8.01}} = 10^{0.533} = 3.41 \text{ mJ/cm}^2$

d. Develop the coefficient (n_{uv}) and empirical threshold (D_c) from the plot of 20-NTU turbidity.
   The following linear relationship is obtained from experimental data:

$$n_{uv} = 5.85 \text{ and } -n_{uv}\log(D_c) = -3.06 \quad \text{or} \quad n_{uv}\log(D_c) = 3.06$$

$$\text{Solve for } D_c,\ D_c = 10^{\frac{3.06}{n_{uv}}} = 10^{\frac{3.06}{5.85}} = 10^{0.523} = 3.33\ \text{mJ/cm}^2$$

e. Summarize the Collins–Selleck equation obtained from the experimental data for each turbidity level.
   The experimental results indicate that the Collins–Selleck equation is suitable to express the kinetics for inactivation of *E. coli* by UV irradiation under the influence of turbidity. The Collins–Selleck kinetic equations at a UV dose between 5 and 15 mJ/cm^2 are expressed below:
   a. At a turbidity of 2 NTU, 1- to 5-log reduction of *E. coli* can be predicted by the following equation:

$$N = N_0\left(\frac{D_{UV}}{3.41}\right)^{-8.01}$$

   b. At a turbidity of 20 NTU, 1- to 4-log reduction of *E. coli* can be predicted by the following equation:

$$N = N_0\left(\frac{D_{UV}}{3.33}\right)^{-5.85}$$

EXAMPLE 11.60: APPLICATION OF EMERICK MODEL

Emerick model includes the effect of both dispersed and particle-associated organisms of UV inactivation. A UV reactor is used for inactivation of coliform organisms in a filtered secondary effluent. Calculate log reduction. Use the following data.

$k_{uv} = 1.40$ cm^2/mJ, D_{uv} (or $I_{avg}\,t$) = 6 mJ/cm^2, $N_0 = 10^6$ MPN/100 mL, and $N'_P = 120$ particles containing at least one organism, particles/100 mL.

Solution

1. Determine the number of coliform organisms remaining after UV exposure.
   Apply Equation 1.43c.

$$\begin{aligned}
N &= N_0 e^{-k_{UV} I_{avg} t} + \frac{N'_P}{k_{UV} I_{avg} t}\left(1 - e^{-k_{UV} I_{avg} t}\right)\\
&= N_0 e^{-k_{UV} D_{UV}} + \frac{N'_P}{k_{UV} D_{UV}}\left(1 - e^{-k_{UV} D_{UV}}\right)\\
&= (10^6\ \text{MPN/100 mL}) \times e^{(-1.40\ \text{cm}^2/\text{mJ} \times 6\ \text{mJ/cm}^2)} + \frac{120\ \text{particles/100 mL}}{1.4\ \text{cm}^2/\text{mJ} \times 6\ \text{mJ/cm}^2} \times \left(1 - e^{(-1.40\ \text{cm}^2/\text{mJ} \times 6\ \text{mJ/cm}^2)}\right)\\
&= (10^6\ \text{MPN/100 mL}) \times e^{-8.4} + (14\ \text{particles/100 mL}) \times (1 - e^{-8.4})\\
&= (10^6\ \text{MPN/100 mL}) \times 2.25 \times 10^{-4} + (14\ \text{particles/100 mL}) \times (1 - 2.25 \times 10^{-4})\\
&= (225 + 14)\ \text{MPN/100 mL}\\
&= 239\ \text{MPN/100mL}
\end{aligned}$$

2. Determine the log reduction of coliform organisms.

$$\text{log reduction} = -\log(N/N_0) = -\log\left(\frac{239 \text{ MPN/100 mL}}{10^6 \text{ MPN/100 mL}}\right) = 3.6\text{-log}$$

11.9.6 UV Transmittance, Density, Intensity, and Dose

UV Transmittance: The initial *UV demand* in wastewater is analogous to the initial chlorine demand. It is expressed by the UV absorbance coefficient (A_{UV}) that is the absorption of energy or absorbance per unit depth (absorbance unit/cm or a.u./cm). The A_{UV} is related to UV *transmittance* (*UVT*) light as measured by a spectrophotometer and is expressed by Equation 11.45a. In most designs, a coefficient "α_{UV}" is used as the UV absorbance coefficient that is to the base *e* and is directly related to A_{UV}. This relationship is expressed by Equation 11.45.

$$UVT = 100\% \times 10^{-A_{UV}} \quad \text{or} \quad UVT = 100\% \times 10^{-(\alpha_{UV}/2.3)} \tag{11.45a}$$

$$\alpha_{UV} = 2.3\ A_{UV} \tag{11.45b}$$

$$A_{UV} = -\log\left(\frac{UVT}{100\%}\right) \quad \text{or} \quad \alpha_{UV} = -2.3\log\left(\frac{UVT}{100\%}\right) \tag{11.45c}$$

where

UVT = UV transmittance at wavelength of 254 nm and a path length of 1 cm, %
A_{UV} = UV absorbance coefficient to the base 10 at wavelength of 254 nm, a.u./cm
α_{UV} = UV absorbance coefficient to the base *e* at wavelength of 254 nm, a.u./cm

The UV absorbance and transmittance vary greatly in different wastewater effluents. At a wavelength of 254 nm, the range of UV absorbance and transmittance in different effluents is summarized in Table 11.19.

UV Density: The UV density (ρ_{UV}) is defined as the total nominal UV power at 253.7 nm available within a reactor divided by the liquid volume in the reactor (ρ_{UV} = total UV output/liquid volume). Determination of nominal UV density at any point in a reactor is not straight forward.

UV Intensity: Two UV intensities are used in UV disinfection practice: (1) nominal average UV intensity and (2) design UV intensity under the field operating conditions.

Nominal Average UV Intensity ($I_{avg,nom}$): The $I_{avg,nom}$ represents the lamps output at 100%. It is assumed that the quartz sleeves or Teflon tube transmits 100% of the energy emitted by the lamps. A mathematical model based on the *point source summation* (PSS) method is applied to determine the $I_{avg,nom}$ value from UV lamps. The model utilizes the basic characteristics of wastewater, spacing and physical properties of UV lamps, and configuration of multilamp chamber.[61,121] Typical lamp configurations

TABLE 11.19 Range of Typical UV Absorbance and Transmittance at a Wavelength of 254 nm in Various Treated Wastewater Effluents

Effluent from Treatment Process	Absorbance Coefficient (A_{UV}), a.u./cm	Transmittance (*UVT*), %
Primary	0.3–0.7	20–50
Secondary	0.15–0.35	45–70
Filtration secondary with or without nitrification	0.1–0.25	55–80
Microfiltration (MF)	0.05–0.1	80–90
Reverse osmosis (RO)	0.01–0.05	90–98

Source: Adapted in part from Reference 52.

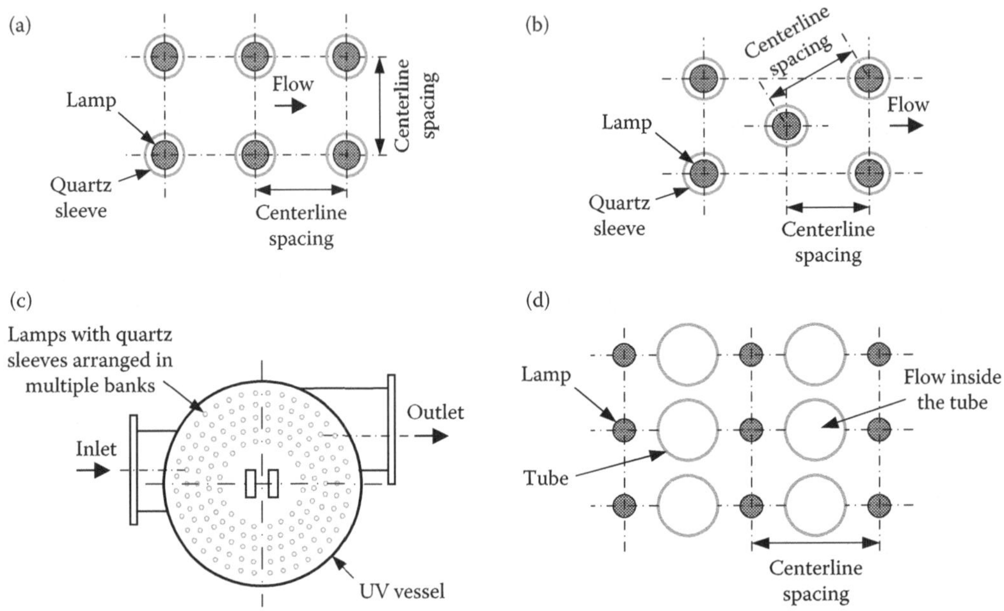

FIGURE 11.27 Schematic of different lamp arrangements: (a) uniform array, (b) staggered uniform array, (c) concentric array, and (d) tubular array.

are (1) uniform array, (2) staggered uniform array, (3) centric array, and (4) tubular array. These configurations are shown in Figure 11.27. The $I_{avg,nom}$ values for these four arrangements are presented graphically in Figure 11.28 with respect to UV absorbance coefficient (α_{UV}) and UV density (ρ_{UV}).

Average UV Intensity Under Field Conditions (I_{avg}): Under actual operation and for design purposes, the $I_{avg,nom}$ value is reduced to account for (a) aging of the lamps, (b) fouling of the inner side of the Teflon tubes or the outer surface of the quartz sleeves due to contact with wastewater, and (c) chemical constituents. The most influential chemical constituents to UV intensity are ferric and ferrous ions. Ozone may also be considered when an AOP with UV/ozone is used. For this reason, the nominal average UV intensity needs to be adjusted for the field conditions as given by Equation 11.46a.

$$I_{avg} = I_{avg,nom} \times F_a \times F_t \times F_c \tag{11.46a}$$

where

I_{avg} = design average UV intensity under field conditions, mW/cm^2 (μW/cm^2)

$I_{avg,nom}$ = nominal average UV intensity, mW/cm^2 (μW/cm^2). It may be obtained from UV lamp manufacturer or estimated for the lamp arrangement from Figure 11.28.

F_a = aging factor that is the ratio of the actual output of aged lamp to the nominal output of lamp, dimenssionless. $F_a = 0.8$ and 0.5, respectively, at the end of one-half and end of approximately the operating life (up to 8000–15,000 h).

F_t = material fouling factor that is the ratio of the actual transparency of the quartz sleeve or Teflon tubes to the nominal transparency (100%) of the enclosures, dimensionless. The minimum F_t for quartz and Teflon systems are 0.7 and 0.6, respectively.

F_c = water quality fouling factor that is the ratio of the actual output with chemical constituent to the nominal output of lamp, dimensionless. This factor may only be considered when the affecting chemical constituents are high. F_c can be expressed by Equation 11.46b.

$$F_c = 10^{-f_c C} \tag{11.46b}$$

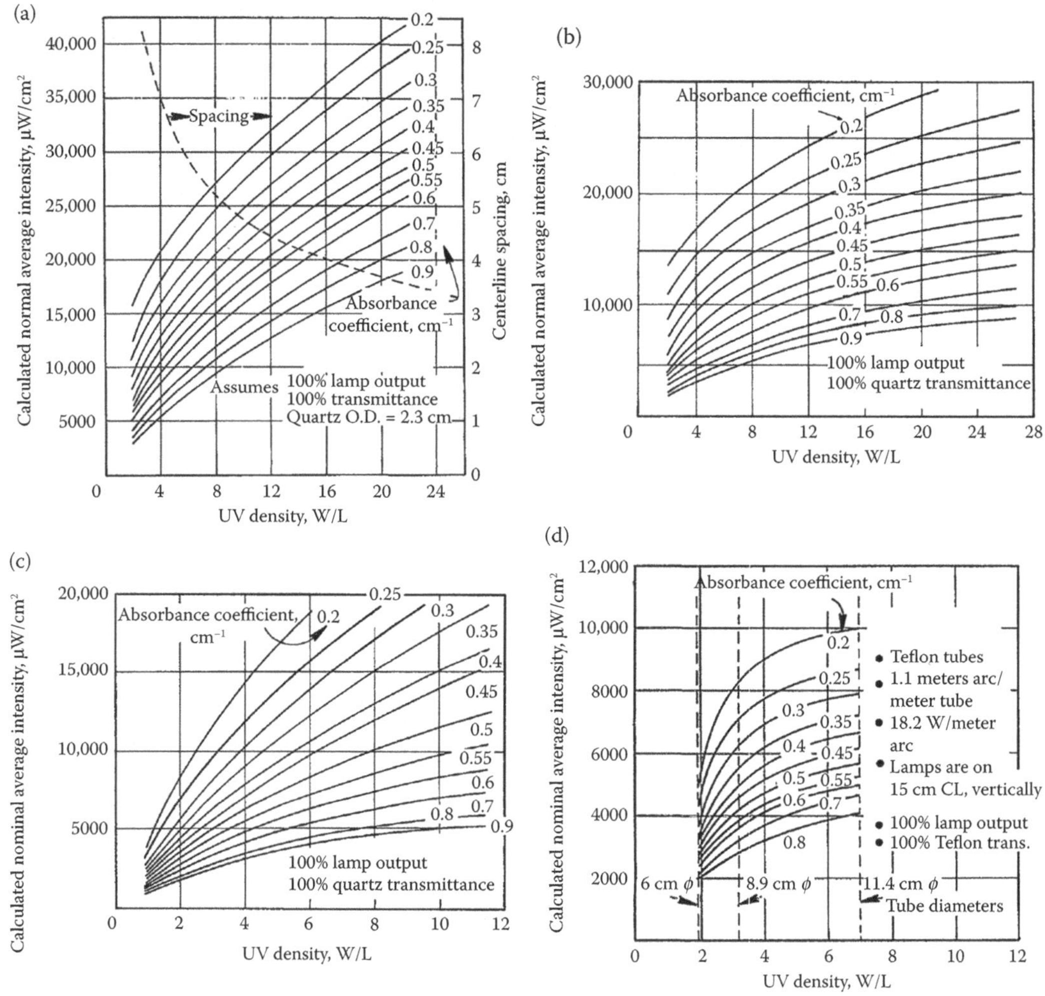

FIGURE 11.28 Nominal average intensity as a function of the reactor UV density and UV absorbance coefficient for different lamp arrangements: (a) uniform array, (b) staggered uniform array, (c) concentric array, and (d) tubular array. (Adapted in part from Reference 54.)

where

f_c = absorption coefficient for the chemical constituent in concern, L/mg. The value of f_c is 0.055 and 0.0083 L/mg for ferric and ferrous ion, respectively. For iron ion in general, the conservative value of 0.055 may be used. The value of f_c is 0.068 for ozone.

C = concentration of chemical constituent, mg/L. For iron ion, it is expressed in mg/L as Fe.

UV Dose: The UV dose is a product of UV intensity and exposure time (Equation 11.39). It can be further defined as (1) UV dose delivered (or available) by the UV lamps and (2) UV dose required (or demand) for meeting a desired microbial inactivation goal.

UV Dose Delivered: Three methods are considered for UV dose measurement. The first method involves determination of average UV intensity and exposure time; the second method uses computational fluid dynamics (CFD); and the third method uses bioassays.

The measurement of nominal UV intensity requires complex PSS method. The CFD methodology is available but has not been widely applied.[122,123] UV dose determination by collimated bean bioassay is commonly used. In this method, an empirical equation of UV dose delivered (mJ/cm^2) may be established as a function of the hydraulic loading applied to the lamp (L/min·lamp), and the UV transmittance (%) of the disinfected effluent. A generalized equation to deliver UV dose under field condition is expressed by Equation 11.47. Normally, the equation is only applicable to a specific UV bank arrangement within a UVT range. The equation needs to be specifically verified from pilot tests. The typical pilot test schedule and data analysis procedure are presented in detail with examples in References 52 and 124.

$$D_{\mathrm{UV,delivered}} = \frac{b}{10^a}\frac{(UVT)^m}{(q_{\mathrm{UV}})^n} \quad \text{or} \quad D_{\mathrm{UV,delivered}} = \frac{1}{B}\frac{(UVT)^m}{(q_{\mathrm{UV}})^n} \tag{11.47}$$

where

$D_{\mathrm{UV,delivered}}$ = UV dose delivered under the field conditions, mJ/cm^2
$D_{\mathrm{UV,delivered}}$ is normally a site-specific UV dose. It is developed from the experimental results for a given confidence level and adjusted to account for lamp aging and fouling effects.[52]

q_{UV} = average hydraulic loading applied to the UV lamp, L/min·lamp

a, b, m, n, and B = empirical constants ($B = \frac{10^a}{b}$). They are determined from pilot tests

UVT has been defined previously

UV Dose Required: UV dose requirement varies greatly to inactivate different organisms to various levels of inactivation. The typical UV doses for reduction of total coliform in wastewater effluents from different processes are summarized in Table 11.20. The estimated UV doses for inactivation of various organisms in the filtered secondary effluent are provided in Table 11.21.[49,52–54]

The upper and lower boundaries of inactivation of virus at a given UV dose have been established by the National Water Research Institute (NWRI) and the U.S. EPA based on the experimental data for the inactivation of Bacteriophage MS2.[52,124,125] These limits are expressed by Equation 11.48a through 11.48d and may be used as reference information for quality control purpose in design of UV disinfection systems.

Upper boundary:

$$-\log(N/N_0) = 0.040 \times D_{\mathrm{UV}} + 0.64 \quad \text{(NWRI)} \tag{11.48a}$$

$$-\log(N/N_0) = -9.6 \times 10^{-5} \times (D_{\mathrm{UV}})^2 + 4.5 \times 10^{-2} \times D_{\mathrm{UV}} \quad \text{(USEPA)} \tag{11.48b}$$

TABLE 11.20 Typical UV Dose Required to Achieve Total Coliform Reduction in Effluent from Various Treatment Processes

Effluent from Treatment Process	Total Coliform, MPN/100 mL	UV Dose, mJ/cm^2			
		Effluent Standard, MPN/100 mL			
		1000	200	23	≤2.2
Septic tank	10^7–10^9	20–40	25–50	–	–
Raw wastewater or primary	10^7–10^9	20–50	–	–	–
Trickling filter or activated sludge	10^5–10^6	20–35	25–40	40–60	90–110
Filtered activated sludge with or without nitrification	10^4–10^6	20–30	25–40	40–60	80–100
Intermittent sand filter	10^2–10^4	10–20	15–25	25–35	50–60
Microfiltration	10–10^3	5–10	10–15	15–30	40–50

Source: Adapted in part from Reference 52.

TABLE 11.21 Estimated Range of UV Dose for Various Levels of Inactivation of Bacteria, Viruses, and Protozoan Cysts in Filtered Secondary Effluent

Inactivation Level	UV Dose, mW·s/cm² (mJ/cm²)				
	Bacteria	Virus	*Cryptosporidium parvum*	*Cryptosporidium parvum* Oocyst	*Giardia lamblia* or *Giardia lamblia* Cysts
1-log	10–35	20–60	10–15	25–40	5–10
2-log	15–40	25–70	10–15	30–45	10–15
3-log	20–70	30–120	15–25	40–75	10–20
4-log	30–110	50–190	20–40	65–120	15–30

Source: Developed from the data in References 49, and 52 through 54.

Lower boundary:

$$-\log(N/N_0) = 0.033 \times D_{UV} + 0.20 \quad \text{(NWRI)} \quad (11.48c)$$

$$-\log(N/N_0) = -1.4 \times 10^{-4} \times (D_{UV})^2 + 7.6 \times 10^{-2} \times D_{UV} \quad \text{(USEPA)} \quad (11.48d)$$

All other terms have been defined previously.

11.9.7 Major Components of Open-Channel UV Disinfection Systems

The open-channel UV systems are commonly used for disinfection of filtered secondary effluent. The major components of an open-channel UV system include (1) UV channel, (2) UV banks, (3) power supply system, (4) monitoring and control system, and (5) level control system.

UV Channel: The UV channel is a long and narrow channel and houses the UV banks. The flow enters the influent structure ahead of UV channels. The influent structure has adjustable rectangular weir gates to regulate and distribute the flow into the UV channels. The channel simulates a plug-flow reactor with minimum dispersion and maximum contact time. Some turbulence axially in the direction of flow is desirable for uniform distribution of UV energy in the nonuniform intensity field of the reactor. A Reynolds number $N_R > 6000$ is normally required at one-half of the minimum flow. Excessive turbulence is associated with head loss, which is a function of velocity. Head loss is a controlling factor in the design of a reactor. The head loss calculations and hydraulic profile through the reactor are given in Example 11.62. The lamp modules should utilize maximum volume. Dead zones or short circuiting will cause ineffective use of lamp energy.

UV Banks: Depending upon the inactivation requirements, two or more UV banks are typically provided in series in each channel. Each UV bank contains many stainless steel UV modules, and each module holds the required number of UV lamps that may be installed horizontally or vertically. The UV modules are held in the support frame with the module handle above the water. The module is retrievable for easy access to the sleeves and lamps for maintenance. The lamps are enclosed individually in quartz sleeves and submerged in the channel. The number of lamps required should be calculated for different flow conditions and microbial quality of effluent to meet the permit requirements for discharge or reuse. These flow conditions may include (1) average daily, (2) maximum 7-day average, (3) maximum 30-day average, (4) peak dry weather, (5) peak wet weather, and (6) minimum dry weather. The procedure for determining the number of lamps required for peak wet weather flow is given in Example 11.61.

Power Supply System: Each UV bank is typically served by a power supply center that contains the ballasts to power UV lamps. The number and type of ballast, and the power requirement of the UV module depend on the number and type of lamps in the UV module. Each UV module is connected by power cable from a water tight connector to the power distribution center (PDC).

Monitoring and Control System: Each UV bank is normally equipped at least one submersible UV intensity sensor that measures the intensity of UV light at 254 nm in the UV bank. The UV intensity monitoring data are transmitted to the UV system control center (SCC) that will automatically turn on and off the UV banks as well as adjust the lamp power based on the flow and UV intensity. There are either microprocessor-based or PLC-based SCCs. The typical controls and displays include (1) electricity power supply data, (2) flow data, (3) UV intensity, transmittance, and dose, (4) control mode selection, (5) UV module run mode and fail status, and (6) alarms for low UV intensity, lamp failures, and breaker trip. These controls and displays may be provided by switches, buttons, message keypad, liquid crystal display (LCD), light-emitting diode (LED), and/or color/touchscreen human–machine interface (HMI).

Level Control System: The level control system is normally installed at the downstream end of the UV channel. It uses a special level control device to keep the effluent level at a proper depth through the UV channel even at variable flow. This keeps the UV lamps at a uniform submergence. These devices include (1) motorized adjustable rectangular weir gate, (2) automatic counter balanced gate, and (3) fixed finger weirs. The adjustable weir gate is continuously controlled by a controller based on the monitoring data from a level sensor installed in the UV channel. The weir plate moves vertically to maintain the water surface near the design set point. The counter balanced gate consists of a baffle plate balanced with weights. The baffle automatically swings partially open in proportion to the flow and releases the excess discharge to keep the liquid level near constant in the channel regardless of flow rate. Owing to a long weir length, the finger weirs may be used to maintain a relatively constant water level within a moderate flow range. This is a simple and cost-effective method for small to medium flow applications.

11.9.8 Hydraulics of UV Channel

A uniform velocity must be maintained in the channel and through the UV banks. Sufficient distance between the influent structure and lamp arrays is needed to attain uniform velocity in the channel. A minimum distance of 2 m (6 ft) is desired between the influent or effluent structure and the closest lamp array.[28]

The head loss through the UV banks may cause a drop in free water surface in the channel. This may cause serious operational problems in the disinfection process. If the required submergence is maintained at the downstream of the UV bank then the upper bank may have larger submergence and the liquid may pass through a region of lower intensity. Conversely, if the free water surface is set with respect to upstream of the lamps, then the portions of the lower end of the lamps may become exposed under diurnal fluctuations in the flow. Alternate immersion and dryness may cause fouling of the quartz sleeves, and irregular heat distribution may shorten the life of the lamps. For this reason, stepping down the lower banks may be needed.

The head loss through a UV bank depends upon the number and arrangement of UV lamps, system geometry, and velocity through the banks. Equation 11.49 gives a general expression of head loss as a function of velocity.

$$\frac{h}{L} = au + b\rho u^2 \quad \text{or} \quad h = (au + b\rho u^2)L \tag{11.49}$$

where

h = head loss, cm

L = length of chamber over which h is measured, cm

u = approach velocity, cm/s

ρ = liquid density, g/cm^3

a, b = equipment-specific empirical constants measured under the field operation

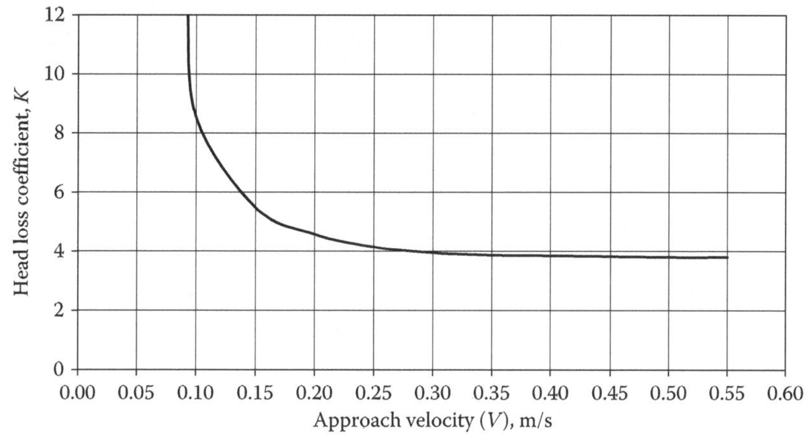

FIGURE 11.29 Head loss coefficient K through two banks arranged in series. (Developed from the information contained in Reference 42.)

The most commonly used equation for calculating the minor head loss through valves, fittings, elbows, contractions, and expansions is expressed by Equation 6.15b. The relationship between head loss coefficient (K) and approach velocity (V) for the head loss through two UV banks arranged in series is developed and is shown in Figure 11.29.[42]

EXAMPLE 11.61: DESIGN OF A UV REACTOR

Design a UV disinfection system to treat filtered effluent from the final clarifier of a BNR facility. The characteristics of filtered effluent, disinfection requirements, and UV reactor design considerations are given below:

Wastewater characteristics of the filtered final clarifier effluent

Peak wet weather flow = 57,000 m^3/d; peak wet weather fecal coliform density = 0.5×10^6 organisms/100 mL; TSS = 5 mg/L; iron = 0.2 mg/L as Fe; and UV transmittance at 253.7 nm wavelength of UV at daily average flow = 68%.

Disinfection requirements

Peak wet weather fecal coliform density required after UV disinfection $\leq$ 400 organisms/100 mL.

Basic requirements for the UV reactor design

Uniform lamp array arranged horizontally; lamps axially parallel to the flow direction; lamp length = 1.5 m; effective arc length = 1.47 m; the nominal unit arc length UV output at 254 nm = 18.2 W/m arc; the outer diameter of lamp's quartz sleeve = 2.3 cm; the aging factor $F_a = 0.8$; the fouling factor $F_t = 0.7$; and the site-specific coefficients that are developed experimentally from pilot tests for use of Equations 11.42d and 11.42f are: $a = 1.45 \times 10^{-5}$, $b = 1.3$, $C_1 = 0.25$ and $m_1 = 2.0$, and $E = 120–1300$ cm^3/s in proportion to the flow. Use the average nominal UV intensity data presented in Figure 11.29.

Provide two UV disinfection channels arranged in parallel. Each channel has two banks of UV lamps in series. The performance of UV disinfection system is high under plug-flow condition. In this case, low to moderation dispersion exists and dispersion number $d = 0.03$.

Solution

1. Determine the volume of liquid (V_{lamp}) exposed per lamp.

The lamp spacing $S_{lamp} = 6$ cm center to center, the effective arc length $L_{arc} = 147$ cm/lamp, diameter of quartz sleeve $d_{sleeve} = 2.3$ cm.

$$V_{lamp} = \left[(S_{lamp})^2 - \frac{\pi}{4}(d_{sleeve})^2\right] \times L_{arc} = \left[(6.0\ \text{cm})^2 - \frac{\pi}{4} \times (2.3\ \text{cm})^2\right] \times 147\ \text{cm/lamp}$$

$$= 4700\ \text{cm}^3/\text{lamp} \quad \text{or} \quad 4.7\ \text{L/lamp}$$

2. Determine the nominal UV density ($\rho_{UV,nom}$) of the lamp.

Calculate the total nominal UV output per lamp ($P_{lamp,nom}$) at the unit arc length UV output $P_{arc,nom} = 18.2$ W/m arc.

$$P_{lamp,nom} = P_{arc,nom} L_{arc} = 18.2\ \text{W/m} \times 1.47\ \text{m/amp} = 26.8\ \text{W/lamp}$$

$$\text{Nominal UV density,} \quad \rho_{UV,nom} = \frac{P_{lamp,nom}}{V_{lamp}} = \frac{26.8\ \text{W/lamp}}{4.7\ \text{L/lamp}} = 5.7\ \text{W/L}$$

3. Determine the average UV density (I_{avg}).

Calculate UV absorbance coefficient to the base e (α_{UV}) from Equation 11.45c at $UVT = 68\%$.

$$\alpha_{UV} = -2.3 \log\left(\frac{UVT}{100\%}\right) = -2.3 \log\left(\frac{68\%}{100\%}\right) = -2.3 \times 0.167 = 0.38\ \text{a.u./cm}$$

The nominal average UV intensity, $I_{avg,nom} = 16{,}000\ \mu\text{W/cm}^2$ is estimated from Figure 11.28a at $\rho_{UV} = 5.7$ W/L and $\alpha_{UV} = 0.38$ a.u./cm.

Note: A centerline spacing of ~6 cm is also estimated at $\rho_{UV,nom} = 5.7$ W/L from Figure 11.28a.

Calculate the water quality fouling factor F_c from Equation 11.46b at the iron concentration $C = 0.2$ mg/L; use a conservative average absorption coefficient $f_c = 0.055$ for ferric ion.

$$F_c = 10^{-f_c C} = 10^{-0.055 \times 0.2\ \text{mg/L}} = 0.975$$

Calculate I_{avg} from Equation 11.46a at $F_a = 0.8$, $F_t = 0.7$, and $F_c = 0.975$.

$$I_{avg} = I_{avg,nom} \times F_a \times F_t \times F_c = 16{,}000\ \mu\text{W/cm}^2 \times 0.8 \times 0.7 \times 0.975 = 8740\ \mu\text{W/cm}^2$$

4. Determine the coliform inactivation rate constant k from Equation 11.42d.

$$k = a(I_{avg})^b = 1.45 \times 10^{-5} \times (8740\ \mu\text{W/cm}^2)^{1.3} = 1.93\ \text{s}^{-1}$$

5. Develop the performance curve.

The performance curve is developed from UV volume loading rate ($UVL_{v,nom}$) and nominal exposure time.

a. Calculate the $UVL_{v,nom}$ that is the volume of water (V_{lamp}) exposed to the total nominal UV output per lamp ($P_{lamp,nom}$).

$$UVL_{v,nom} = \frac{V_{lamp}}{P_{lamp,nom}} = \frac{4.7\text{L/lamp}}{26.8\ \text{W/lamp}} = 0.175\ \text{L/W}$$

Note: UVL can also be obtained from reversing the nominal UV density (ρ_{UV}) of the lamp.

b. Calculate nominal exposure time (t_{nom}).

The value of t_{nom} is the nominal time the water is exposed to UV light. For each lamp, t_n is calculated from the volume of water exposed per lamp (V_{lamp}) and flow rate per min per lamp (q_{lamp}).

$$t_{nom} = \frac{V_{lamp}}{q_{lamp}}$$

Divide the numerator and denominator of the above expression by $P_{lamp,nom}$ to obtain Equation 11.50a.

$$t_{nom} = \frac{V_{lamp}/P_{lamp,nom}}{q_{lamp}/P_{lamp,nom}} = \frac{UVL_{v,nom}}{UVL_{f,nom}} \quad (11.50a)$$

where

t_{nom} = nominal time the water exposed to UV light, min

$UVL_{v,nom}$ = nominal UV loading in terms of the volume of water exposed per Watt, L/W

$UVL_{f,nom}$ = nominal UV loading in terms of the flow (or volume of water per unit time) exposed per Watt, L/min·W

q_{lamp} = flow (or volume of water per unit time) at the lamp, L/min

To develop the UV performance curve, assign different values to variable $UVL_{f,nom}$. For each assigned value of $UVL_{f,nom}$, calculate the corresponding value of t_{nom} from Equation 11.50a. As an example, calculate the t_{nom} at $UVL_{f,nom} = 0.5$ L/min·W below:

$$t_{nom} = \frac{UVL_{v,nom}}{UVL_{f,nom}} = \frac{0.175 \text{ L/W}}{0.5 \text{ L/min·W}} = 0.35 \text{ min} \quad \text{or} \quad 21 \text{ s}$$

Similarly, calculate t_{nom} for other assigned values of $UVL_{f,nom}$ ($UVL_{f,nom}$ between 1 and 5 L/min·W at an interval of 0.5 L/min·W). These values are summarized in Table 11.22.

TABLE 11.22 Calculated UV Performance Values $-\log(N'/N_0)$ for Assumed $UVL_{f,nom}$ at Peak Wet Weather Flow

$UVL_{f,nom}$, L/min·W	t_{nom}, s	u, cm/s	E, cm²/s	$-\log(N'/N_0)$
0.5	21.0	14.3	129	10.3
1.0	10.5	28.6	257	6.16
1.5	7.00	42.9	386	4.47
2.0	5.25	57.1	514	3.54
2.5	4.20	71.4	643	2.93
3.0	3.50	85.7	771	2.50
3.5	3.00	100	900	2.18
4.0	2.63	114	1026	1.94
4.5	2.33	129	1161	1.74
5.0	2.10	143	1287	1.58

c. Calculate the velocity through UV lamps in the channel.

There are two UV banks in the channel in series at the peak wet weather flow. The length of each lamp $L_{lamp} = 150$ cm. Total length of UV exposure in two banks $x = 2$ lamps × 150 cm/lamp = 300 cm. The nominal exposure time t_{nor} for $UVL_{f,nom} = 0.5$ L/min·W is 21 s. The corresponding velocity u for $UVL_{f,nom} = 0.5$ L/min·W is calculated below and those for other assumed values of $UVL_{f,nom}$ are also summarized in Table 11.22.

$$u = \frac{x}{t_{nom}} = \frac{300 \text{ cm}}{21 \text{ s}} = 14.3 \text{ cm/s}$$

d. Calculate the dispersion coefficient (E) from Equation 11.42e at $d = 0.03$ (given).

$$E = dux = 0.03 \times 14.3 \text{ cm/s} \times 300 \text{ cm} = 129 \text{ cm}^2\text{/s}$$

Similarly, the value of E for other assumed $UVL_{f,nom}$ values are calculated and summarized in Table 11.22.

e. Calculate the UV performance values of $-\log(N'/N_0)$ (log inactivation).

Take logarithm on both sides of Equation 11.42b and rearrange the equation to obtain the following expression.

$$-\log(N'/N_0) = -\log(e) \times \left[\frac{ux}{2E}\left(1 - \left(1 + \frac{4kE}{u^2}\right)^{1/2}\right)\right] = -0.434 \times \left[\frac{ux}{2E}\left(1 - \left(1 + \frac{4kE}{u^2}\right)^{1/2}\right)\right]$$

The UV performance values $-\log(N'/N_0)$ at the peak wet weather flow are calculated from the above expression at $UVL_{f,nom} = 0.5$ L/min·W.

$$-\log(N'/N_0) = -0.434 \times \left[\frac{14.3 \text{ cm/s} \times 300 \text{ cm}}{2 \times 129 \text{ cm}^2\text{/s}}\left(1 - \left(1 + \frac{4 \times 1.93 \text{ s}^{-1} \times 129 \text{ cm}^2\text{/s}}{(14.3 \text{ cm/s})^2}\right)^{1/2}\right)\right]$$

$$= -0.434 \times (-23.7) = 10.3$$

Similarly calculate the UV performance values $-\log(N'/N_0)$ for other assumed values of $UVL_{f,nom}$. These values are also summarized in Table 11.22.

6. Draw the UV inactivation performance curve.

The performance curve is shown in Figure 11.30. The following best fit equation (Equation 11.50b) is also obtained from the plot.

$$-\log(N'/N_0) = 6.08\ (UVL_{f,nom})^{-0.82} \qquad (11.50b)$$

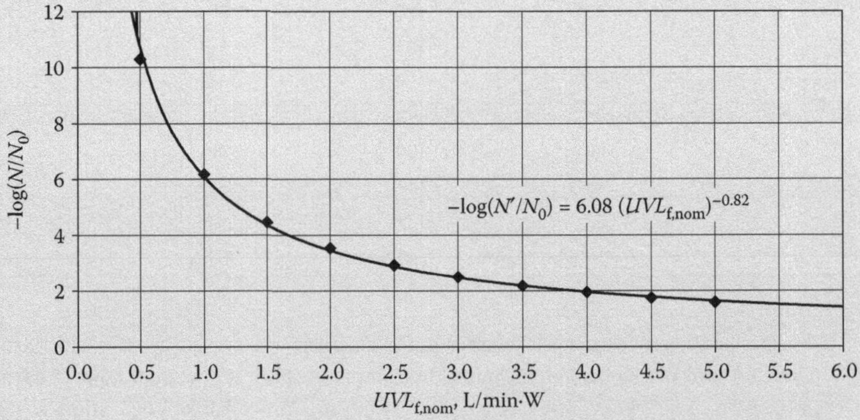

FIGURE 11.30 Performance curve for UV inactivation at peak wet weather flow (Example 11.61).

7. Establish the performance goal.

The performance goal of UV disinfection facility is developed from the permit requirements of fecal coliforms. The value of N_P is calculated from Equation 11.41b and subtracted from permit requirement.

a. Calculate N_P from Equation 11.41b at TSS = 5 mg/L, $C = 0.25$, and $m_1 = 2.0$.

$$N_P = C_1 \text{TSS}^{m_1} = 0.25 \times (5\,\text{mg/L})^2 = 6 \text{ orgnisms/100 mL}$$

b. Determine the performance goal N' at $N = 400$ organisms/100 mL from Equation 11.42a.

$$N' = N - N_P = (400 - 6) \text{ orgnisms/100 mL} = 394 \text{ orgnisms/100 mL}$$

c. Determine the performance value $-\log(N'/N_0)$.

$$-\log(N'/N_0) = -\log\left(\frac{394 \text{ organisms/100 mL}}{0.5 \times 10^6 \text{ organisms/100 mL}}\right) = 3.1 \log$$

8. Determine the design UV loading ($UVL_{\text{f,nom}}^{\text{design}}$) required to achieve the performance goal.

The design UV loading required to achieve the performance goal can be obtained from either of the two methods: (1) estimating $UVL_{\text{f,nom}}^{\text{design}}$ graphically from the curve in Figure 11.30 or (2) calculating $UVL_{\text{f,nom}}^{\text{design}}$ from Equation 11.50b. The second method is presented in this example. Rearrange Equation 11.50b and calculate $UVL_{\text{f,nom}}^{\text{design}}$ below.

$$UVL_{\text{f,nom}}^{\text{design}} = \left(\frac{-\log(N'/N_0)}{6.08}\right)^{-\frac{1}{0.82}} = \left(\frac{3.1}{6.08}\right)^{-\frac{1}{0.82}} = 2.27 \text{ L/min}\cdot\text{W}$$

The UV loading $UVL_{\text{f,nom}}^{\text{design}} = 2.27$ L/min·W is required to achieve the performance goal of 3.1-log inactivation.

9. Determine the required exposure time ($t_{\text{nom}}^{\text{design}}$).

The required exposure time $t_{\text{nom}}^{\text{design}}$ to achieve the performance goal is obtained from Equation 11.50a.

$$t_{\text{nom}}^{\text{design}} = \frac{UVL_{\text{v,nom}}}{UVL_{\text{f,nom}}^{\text{design}}} = \frac{0.175\text{L/W}}{2.27 \text{ L/min}\cdot\text{W}} = 0.077 \text{ min or } 4.6 \text{ s}$$

10. Determine the design UV doses to achieve the performance goal.

The UV dose (D_{UV}) is calculated from Equation 11.39.

$$D_{\text{UV}}^{\text{design}} = I_{\text{avg}} t_{\text{nom}}^{\text{design}} = 8740\,\mu\text{W/cm}^2 \times 4.6 \text{ s} \times \frac{\text{mW}}{10^3 \mu\text{W}} \times = 40\,\text{mW}\cdot\text{s/cm}^2 \quad \text{or} \quad 40\,\text{mJ/cm}^2$$

Note: The design UV dose $D_{\text{UV}}^{\text{design}}$ is within the typical range of 20–70 mJ/cm² for an approximately 3-log disinfection performance for reduction of bacteria (Table 11.21).

11. Calculate the number of lamps required.

The number of lamps required is calculated from Equation 11.50c.

$$N_{\text{lamp}} = \frac{Q}{P_{\text{lamp,nom}} UVL_{\text{f,nom}}} \tag{11.50c}$$

where

N_{lamp} = total number of UV lamps required at the flow Q, number of lamps

Q = total flow treated, L/min

Total flow at peak wet weather flow, $Q_{\text{pwwf,total}} = 57{,}000\,\text{m}^3/\text{d} \times \dfrac{\text{d}}{1440\,\text{min}} \times 10^3\,\text{L/m}^3 = 39{,}600\,\text{L/min}$

Calculate the number of total lamps ($N^{\text{design}}_{\text{lamp,total}}$) from Equation 1.50c.

$$N^{\text{design}}_{\text{lamp,total}} = \frac{Q_{\text{pwwf,total}}}{P_{\text{lamp,nor}}\,UVL^{\text{design}}_{\text{f,nom}}} = \frac{39{,}600\text{ L/min}}{26.8\text{ W/lamp} \times 2.27\text{ L/min}\cdot\text{W}} = 651\text{ lamps}$$

Provide total of 672 lamps for UV disinfection at the treatment facility. There are two UV disinfection channels and each channel has two UV banks.

Number of total lamps provided, $N^{\text{design}}_{\text{lamp,total}} = 672$ lamps

There are two banks in each channel, $N^{\text{design}}_{\text{channel}} = 2$ channels

Number of lamps per UV channel, $N^{\text{design}}_{\text{lamp,channel}} = \dfrac{N^{\text{design}}_{\text{lamp,total}}}{N^{\text{design}}_{\text{channel}}} = \dfrac{672\text{ lamps}}{2\text{ channels}} = 336$ lamps per channel

There are two banks in each channel, $N^{\text{design}}_{\text{bank,channel}} = 2$ banks/channel

Number of lamps per UV bank, $N^{\text{design}}_{\text{lamp,bank}} = \dfrac{N^{\text{design}}_{\text{lamp,channel}}}{N^{\text{design}}_{\text{bank,channel}}} = \dfrac{336\text{ lamps/channel}}{2\text{ banks/channel}} = 168$ lamps/bank

Select a UV module with 12 lamps per module, $N^{\text{design}}_{\text{lamp,module}} = 12$ lamps/module

Number of modules per bank, $N^{\text{design}}_{\text{module,bank}} = \dfrac{N^{\text{design}}_{\text{lamp,bank}}}{N^{\text{design}}_{\text{lamp,module}}} = \dfrac{168\text{ lamps/bank}}{12\text{ lamps/module}} = 14$ modules/bank

Total number of lamps provided.

$$\begin{aligned} N^{\text{design}}_{\text{lamp,total}} &= N^{\text{design}}_{\text{lamp,module}} \times N^{\text{design}}_{\text{module,bank}} \times N^{\text{design}}_{\text{bank,channel}} \times N^{\text{design}}_{\text{channel}} \\ &= 12\text{ lamps/module} \times 14\text{ modules/bank} \times 2\text{ banks/channel} \times 2\text{ channels} = 672\text{ lamps} \end{aligned}$$

12. Select the channel dimensions.

The lamps are 6 cm center to center in vertical and horizontal direction and 3 cm from center of lamps to channel bottom and sides, and 3 cm from the free water surface.

Channel width, $W = 13 \times 6\text{ cm} + 2 \times\ 3\text{ cm} = 84\text{ cm}$ or 0.84 m

Maximum water depth in the channel, $H_{\text{wd}} = 11 \times 6\text{ cm} + 2 \times 3\text{ cm} = 72\text{ cm}$ or 0.72 m

The layout and sectional views of UV channels and lamp arrangement are shown in Figure 11.31.

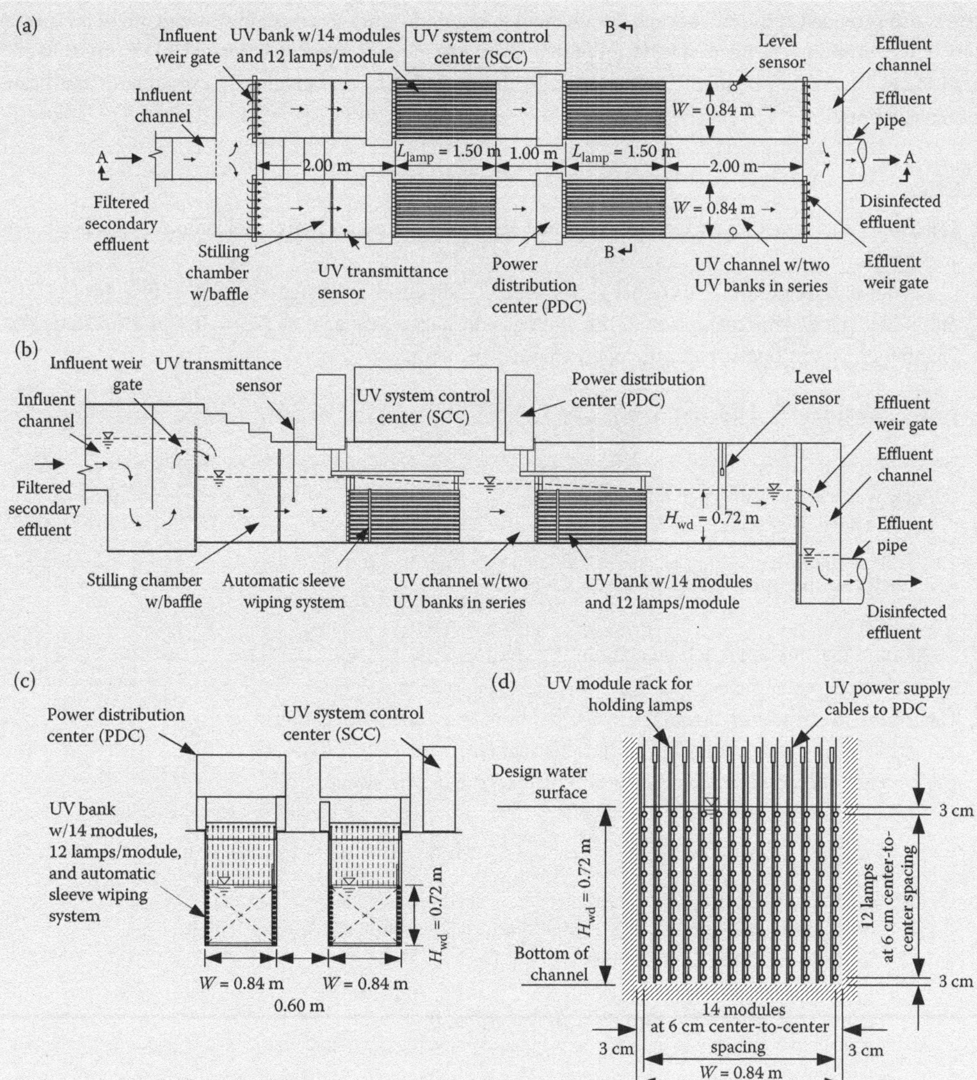

FIGURE 11.31 Design details of the UV disinfection system: (a) plan showing two UV channels in parallel operation, (b) Section AA, (c) Section BB, and (d) arrangement of modules and lamps (Example 11.61).

EXAMPLE 11.62: HYDRAULIC DESIGN OF UV CHANNEL

The hydraulic design of a UV channel is very important for proper operation of a facility. Develop the hydraulic design of the UV disinfection reactor provided in Example 11.61. There are two UV banks per channel, and each bank has 14 UV modules. Each module has 12 UV lamps. Diameter of quartz sleeve = 2.3 cm and the lamp spacing is 6 cm center to center. Peak wet weather flow per channel = 19,800 L/min, and average design flow is one-third of the peak wet weather flow. The details of UV channel and lamp arrangement are given in Figure 11.31. Calculate the head loss in the channel due to UV banks, and draw the hydraulic profile at average design and peak wet weather flows. An adjustable weir gate is provided at the beginning of each channel to isolate the channel for maintenance. An adjustable weir gate is also provided at the end of each UV channel to maintain a nearly constant water depth of 0.72 m at both flow conditions. See Figure 11.31b for both influent and effluent adjustable weir gates. Assume that the head losses in channel segments before and after the UV banks are small and ignorable in the head loss calculations.

Solution

1. Determine the velocities in the open-channel segments before and after UV banks at the peak wet weather flow.

   Layout and longitudinal sectional view of the UV disinfection system are shown in Figure 11.31a and b. Set the water depth prior to the effluent adjustable weir gate at $H_{wd} = 0.72$ m. Calculate the channel cross-sectional area in the open-channel segment.

   $$A_{open} = WH_{wd} = 0.84 \text{ m} \times 0.72 \text{ m} = 0.60 \text{ m}^2$$

   Peak wet weather flow per channel, $Q_{pwwf,channel} = 19{,}800 \text{ L/min} \times \dfrac{\text{min}}{60 \text{ s}} \times \dfrac{\text{m}^3}{1000 \text{ L}} = 0.33 \text{ m}^3/\text{s}$

   Velocity in the open channels, $v_{pwwf,open} = \dfrac{Q_{pwwf,channel}}{A_{open}} = \dfrac{0.33 \text{ m}^3/\text{s}}{0.60 \text{ m}^2} = 0.55 \text{ m/s}$

   Assume that the approach velocity in the channel segment with UV banks is equal to that in the open-channel segments, $v_{pwwf,approach} = v_{pwwf,open} = 0.55$ m/s.
2. Determine the head loss through the UV banks at the peak wet weather flow.
   a. Determine the velocities in the channel segment with the UV banks.

      The total cross-sectional area occupied by the quartz sleeve.

      $$A_{sleeve} = \frac{\pi}{4}(d_{sleeve})^2 \times N^{design}_{lamp,bank} = \frac{\pi}{4} \times (2.3 \text{ cm} \times \frac{\text{m}}{100 \text{ cm}})^2 \times 168 \text{ lamps/bank} = 0.070 \text{ m}^2$$

      The channel cross-sectional area upstream of this segment, $A_{open} = 0.60 \text{ m}^2$

      Net cross-sectional area through the UV banks,

      $$A_{bank} = A_{open} - A_{sleeve} = (0.60 - 0.070) \text{ m}^2 = 0.53 \text{ m}^2$$

      Velocity through the UV banks, $v_{pwwf,bank} = \dfrac{Q_{pwwf,channel}}{A_{bank}} = \dfrac{0.33 \text{ m}^3/\text{s}}{0.53 \text{ m}^2} = 0.62 \text{ m/s}$

      The head loss coefficient (K) in Equation 6.15b for the channel segment with the UV banks. $K_{pwwf} = 3.8$. This is obtained from Figure 11.29 for an approach velocity of 0.55 m/s.
   b. Calculate the head loss through the UV banks ($h_{m,pwwf,bank}$) from Equation 6.15b.

      Head loss per UV bank, $h_{m,pwwf,bank} = K_{pwwf} \dfrac{(v_{pwwf,bank})^2}{2g} = 3.8 \times \dfrac{(0.62 \text{ m/s})^2}{2 \times 9.81 \text{ m/s}^2}$

      $= 0.07$ m/bank or 7 cm/bank

      Calculate the head loss for two banks.

      $$h_{m,pwwf,channel} = N^{design}_{bank,channel} \times h_{m,pwwf,bank} = 2 \text{ banks/channel} \times 0.07 \text{ m/bank} = 0.14 \text{ m per channel}$$

c. Verify the actual total exposure time through two UV banks in series (t_{actual}).

$$t_{actual} = \frac{N^{design}_{bank,channel} L_{lamp}}{v_{pwwf,bank}} = \frac{2 \text{ banks/channel} \times 1.5 \text{ m/lamp (or bank)}}{0.62 \text{ m/s}} = 4.8 \text{ s}$$

The actual total exposure time is greater than $t^{design}_{nom} = 4.6$ s. Therefore, the design is valid.

3. Determine the head loss through the UV banks at the average design flow.

Average design flow per channel, $Q_{af,channel} = \frac{1}{3} \times Q_{pwwf,channel} = \frac{1}{3} \times 0.33 \text{ m}^3/\text{s} = 0.11 \text{ m}^3/\text{s}$

Velocity in the open channels, $v_{af,open} = \frac{Q_{af,channel}}{A_{open}} = \frac{0.11 \text{ m}^3/\text{s}}{0.60 \text{ m}^2} = 0.18 \text{ m/s}$

Assume $v_{approach,af} = v_{af,open} = 0.18$ m.

$K_{pwwf} = 4.7$ from Figure 11.29 for an approach velocity of 0.18 m/s.

Velocity through the UV banks, $v_{af,bank} = \frac{Q_{af,channel}}{A_{bank}} = \frac{0.11 \text{ m}^3/\text{s}}{0.53 \text{ m}^2} = 0.21 \text{ m/s}$

Head loss per UV bank, $h_{m,af,bank} = K_{af} \frac{(v_{af,bank})^2}{2g} = 4.7 \times \frac{(0.21 \text{ m/s})^2}{2 \times 9.81 \text{ m/s}^2}$

$= 0.01$ m/bank or 1 cm/bank

The head loss for two banks.

$$h_{m,af,channel} = N^{design}_{bank,channel} \times h_{m,af,bank} = 2 \text{ banks/channel} \times 0.01 \text{ m/bank} = 0.02 \text{ m per channel}$$

4. Determine the submergence of UV lamps and water depth in the channel at peak wet weather and average design flows.

The head loss in the channel segment between the 2nd UV bank and the effluent adjustable weir gate is ignored. The water depth at the lower end of the 2nd bank, $H_{wd} = 0.72$ m. the water surface elevation at the lower end of the 2nd bank is 3 cm above the center of the quartz sleeve. The submergence of lamps and water depths in the channel at the lower and upper ends of both banks are summarized in the following table.

Flow Condition	2nd (Downstream) Bank				1st (First) Bank			
	Lower End		Upper End		Lower End		Upper End	
	S_1, cm	D_1, m	S_2, cm	D_2, m	S_3, cm	D_3, m	S_4, cm	D_4, m
PWWF	1.85[a]	0.72[b]	8.85[c]	0.79[d]	8.85[e]	0.79[f]	15.85[g]	0.86[h]
AF	1.85	0.72	2.85	0.73	2.85	0.73	3.85	0.74

[a] $S_1 = (3-(2.3 \div 2))$ cm $= 1.85$ cm (Figure 11.31d).
[b] $D_1 = H_{wd} = 0.72$ m.
[c] $S_2 = S_1 + h_{m,wwf,bank} = (1.85 + 7)$ cm $= 8.85$ cm.
[d] $D_2 = D_1 + h_{m,pwwf,bank} = (0.72 + 0.07)$ m $= 0.79$ m.
[e] $S_3 = S_2 = 8.85$ cm.
[f] $D_3 = D_2 = 0.79$ m.
[g] $S_4 = S_3 + h_{m,pwwf,bank} = (8.85 + 7)$ cm $= 15.85$ cm.
[h] $D_4 = D_3 + h_{m,pwwf,bank} = (0.79 + 0.07)$ m $= 0.86$ m.
Note: AF = average design flow; D = depth; PWWF = peak wet weather flow; S = submergence

5. Determine the head over the influent and effluent adjustable weir gate at the peak wet weather and average design flows.

   Calculate the head over the rectangular weir from Equation 8.10 using $C_d = 0.6$. The length of the influent and effluent weirs are the same as the width of the UV channel, $L_{weir} = W = 0.84$ m and $n = 1$. Assume $L'_{weir,inf,pwwf} = 0.80$ m and $L'_{weir,inf,af} = 0.82$ m for the peak wet weather and average design conditions, respectively.

$$h_{weir,inf,pwwf} = \left(\frac{3}{2} \times \frac{Q_{pwwf,channel}}{C_d L'_{weir,inf,pwwf} \sqrt{2g}}\right)^{2/3} = \left(\frac{3}{2} \times \frac{0.33 \text{ m}^3/\text{s}}{0.6 \times 0.80 \text{ m} \times \sqrt{2 \times 9.81 \text{ m/s}^2}}\right)^{2/3} = 0.38 \text{ m}$$

$$h_{weir,inf,af} = \left(\frac{3}{2} \times \frac{Q_{af,channel}}{C_d L'_{weir,inf,af} \sqrt{2g}}\right)^{2/3} = \left(\frac{3}{2} \times \frac{0.11 \text{ m}^3/\text{s}}{0.6 \times 0.82 \text{ m} \times \sqrt{2 \times 9.81 \text{ m/s}^2}}\right)^{2/3} = 0.18 \text{ m}$$

$$\textit{Check: } L'_{weir,inf,pwwf} = L_{weir} - 0.1\, n\, h_{weir,inf,pwwf} = 0.84 \text{ m} - 0.1 \times 1 \times 0.38 \text{ m} = 0.80 \text{ m}$$

$$L'_{weir,inf,af} = L_{weir} - 0.1\, n\, h_{weir,inf,af} = 0.84 \text{ m} - 0.1 \times 1 \times 0.18 \text{ m} = 0.82 \text{ m}$$

   Both lengths are same as the initial assumptions. Since the influent and effluent weir gates are identical, the calculated heads over the weir for the influent weir are also applicable to the effluent weir gate: $h_{weir,inf,pwwf} = h_{weir,eff,pwwf} = 0.38$ m at PWWF, and $h_{weir,inf,af} = h_{weir,eff,af} = 0.18$ m at AF.

6. Determine the head loss at the effluent structure.

   An adjustable weir gate is provided at the end of each UV channel (Figure 11.31b. The gate is adjusted automatically by a controller to maintain a relatively constant water depth at 0.72 m in the channel upstream of the weir gate. Assume that the water surface in the effluent channel is 0.15 m below the floor of UV channel and a free fall is achieve after the weir at both flow conditions. Total head loss at the effluent structure = 0.72 m + 0.15 m = 0.87 m.

7. Determine the head loss at the influent structure.

   An adjustable weir gate is provided at the entrance of each UV channel (see Figure 11.31b). The top of weir is set at an elevation to provide a freefall of 0.15 m after the weir at the peak weather flow. The weir remains at this elevation under all flow conditions. The weir gate is raised to isolate the UV channel for maintenance. Total heal loss at the influent structure = 0.38 m + 0.15 m = 0.53 m under the peak wet weather flow.

8. Develop the hydraulic profile through the UV channel.

   a. Prepare major water surface elevations (WSELs) and elevations (ELs) of the major components.

      Assume the reference elevation is the bottom of the UV channel, $EL_{UV} = 0.00$ m. Prepare the elevations at both the peak wet weather and average design flows. The results are summarized in the table below:

Elevations and Reference Points	WSEL at Flow Condition, m		EL, m	Assumption or Remark
	PWWF	AF		
Floor EL of UV channel	–	–	0.00	Set as the datum.
WSEL in the effluent channel	–0.15	–0.15	–	Set WSEL in the effluent channel 0.15 m below the floor of UV channel.
Top EL of the effluent weir at PWWF	–	–	0.34	Top EL of weir at PWWF = (0.72 – 0.38) m = 0.34 m (Step 5). Freeboard = (0.34 + 0.15) m = 0.49 m
Top EL of the effluent weir at AF	–	–	0.54	Top EL of weir at AF = (0.72 – 0.18) m = 0.54 m (Step 5). Freeboard = (0.54 + 0.15) m = 0.69 m.

Continued

Elevations and Reference Points	WSEL at Flow Condition, m		EL, m	Assumption or Remark
	PWWF	AF		
WSEL between the adjustable weir gate and the lower end of the 2nd UV bank	0.72	0.72	–	Design requirement to keep submergence of the UV banks (Step 4). Total head loss at the weir = 0.87 m (Step 6).
WSEL between two UV banks	0.79	0.73	–	Head losses through the 2nd UV bank are 0.07 and 0.01 m at PWWF and AF, respectively (Steps 2 and 3).
WSEL between the upper end of the 1st UV bank and the influent weir gate	0.86	0.74	–	Head losses through a UV bank are 0.07 and 0.01 m at PWWF and AF, respectively (Steps 2 and 3).
Top EL of wall of UV channel	–	–	1.35	Select a design wall height above the channel floor = 1.35 m. Freeboard at PWWF = (1.35 − 0.86) m = 0.49 m. Freeboard at AF = (1.35 − 0.74) m = 0.61 m.
Top EL of the influent weir	–	–	1.01	Provide a free fall of 0.15 m below the weir gate at PWWF. Top EL of weir = (0.86 + 0.15) m = 1.01 m. Freeboard at AF = (1.01 − 0.74) m = 0.27 m.
WSEL in the UV influent channel upstream of the influent weir gate	1.39	1.19	–	WSEL at PWWF = (1.01 + 0.38) m = 1.39 m. WSEL at AF = (1.01 + 0.18) m = 1.19 m (Step 5).
Top wall EL of the UV influent channel	–	–	1.85	Select a design wall height of 1.85 m. Freeboard at PWWF = (1.85 − 1.39) m = 0.46 m. Freeboard at AF = (1.85 − 1.19) m = 0.66 m.

Note: AF = average design flow; EL = elevation; PWWF = peak wet weather flow; WSEL = water surface elevation.

b. Draw the hydraulic profile through the UV channel.

The hydraulic profile is prepared for the peak wet weather flow and is shown in Figure 11.32. The water surface profile at the average design flow will be similar but below that at the peak wet weather flow.

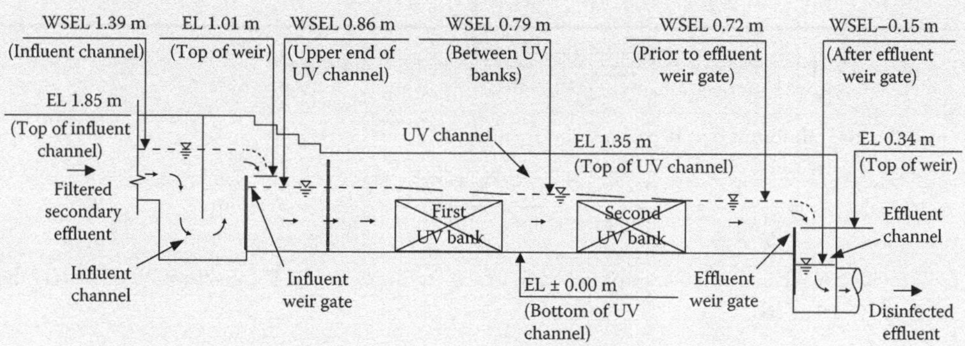

FIGURE 11.32 Hydraulic profile of the UV disinfection system at peak wet weather flow (Example 11.62).

c. Additional comments:

i. At the peak wet weather flow, the submergence at the upstream end of the upper UV bank is 15.85 cm. This is slightly deeper than the desired submergence. At average design flow, the submergence is only 3.85 cm. Stepping up of the upstream bank is not recommended as under low flows, some portions of the UV lamp may become exposed to air. At peak wet weather flows, some turbulence will occurs due to high velocity through the UV banks. This is desired for distribution of UV energy in regions of deeper submergences. Also, the peak wet weather flows occur infrequently and only for a shorter duration. Therefore, the head loss conditions, submergence, and water depth in the channel are acceptable.

ii. The number of UV lamps required under average design flow is less than one-third of peak wet weather flow. Therefore, one UV bank may be turned off during periods of low flows for energy savings and normal maintenance.

iii. An alternate method for head loss calculation is to apply energy equation (Equation 7.3a) at the channel sections upstream and downstream of the UV bank starting from upstream of the effluent structure. See Examples 7.3 and 7.4 for sample calculations.

EXAMPLE 11.63: APPLICATION OF MINIMUM UV DOSE

A pilot test was conducted for UV disinfection of filtered secondary effluent. A single bank unit was tested in a UVT range of 55–75%. The pilot test results were adjusted to account for an overall lamp aging and fouling. After calibration, the following site-specific coefficients were obtained for use in Equation 11.47: $m = 3.1$, $n = 0.65$, and $B = 420$. Estimate (a) the maximum allowable hydraulic loading applied to the UV lamp at a UVT = 70% and (b) the minimum number of UV lamps required for a single UV bank at a flow of 10,000 m^3/d. Use the NWRI equation for lower boundary (Equation 11.48c) to determine the minimum UV dose desired to achieve a 2-log reduction of virus.

Solution

1. Rearrange Equation 11.48c to estimate the minimum UV dose desired to achieve a 2-log reduction of virus.

$$D_{UV,min} = \frac{-\log(N/N_0) - 0.20}{0.033} = \frac{2 - 0.20}{0.033} = 55 \text{ mJ/cm}^2$$

2. Rearrange Equation 11.47 to estimate the maximum allowable hydraulic loading per lamp.

$$q_{UV,max} = \left(\frac{1}{B}\frac{(UVT)^m}{D_{UV,min}}\right)^{\frac{1}{n}} = \left(\frac{1}{420}\frac{(70)^{3.1}}{55 \text{ mJ/cm}^2}\right)^{\frac{1}{0.65}} = 122 \text{ L/min·lamp}$$

3. Estimate the minimum number of UV lamp required.

$$N_{UV,lamp,min} = \frac{Q}{q_{UV,max}} = \frac{10{,}000 \text{ m}^3/\text{d}}{122 \text{ L/min·lamp}} \times \frac{10^3 \text{ L}}{\text{m}^3} \times \frac{\text{d}}{1440 \text{ min}} = 57 \text{ lamps}$$

Provide a UV bank with 64 lamps per bank to achieve the desired 2-log inactivation of virus at the flow of 10,000 m^3/d.

EXAMPLE 11.64: CALIBRATION OF HEAD LOSS EQUATION FOR A UV BANK

A UV bank was tested for disinfection of filtered secondary effluent. The test unit has 147-cm effective lamps length horizontally arranged in a uniform array that is parallel to the flow direction. The head loss data are as follows:

The head loss through the lamps $h_L = 5$ cm at an approach velocity $u = 45$ cm/s, and $h_L = 1$ cm at approach velocity $u = 15$ cm/s, respectively. Determine the constants a and b in Equation 11.49.

Solution

1. Apply Equation 11.49 to obtain the following two expressions. The density of water is 1 g/cm^3 (Table B.1 in Appendix B).

$$\frac{h}{L} = au + b\,\rho u^2$$

$$\frac{5\text{ cm}}{147\text{ cm}} = (45\text{ cm/s}) \times a + (45\text{ cm/s})^2 \times 1.0\text{ g/cm}^3 \times b$$

$$\frac{1\text{ cm}}{147\text{ cm}} = (15\text{ cm/s}) \times a + (15\text{ cm/s})^2 \times 1.0\text{ g/cm}^3 \times b$$

2. Simplify the above expressions.

$$0.0756\text{ s/cm} = a + (0.45\text{ g/cm}^2\cdot\text{s}) \times b$$

$$0.0454\text{ s/cm} = a + (0.15\text{ g/cm}^2\cdot\text{s}) \times b$$

3. Solve for a and b from these expressions.

$$b = 1.01 \times 10^{-5}\text{ cm}\cdot\text{s}^2/\text{g}$$

$$a = 30.1 \times 10^{-5}\text{ s/cm}$$

4. Write the calibrated form of Equation 11.49.

$$\frac{h}{L} = (30.1\,u + 1.01\,\rho u^2) \times 10^{-5}$$

11.10 Recent Developments in Disinfection Reactor Design

The effectiveness of inactivation depends mainly on the disinfectant dose. The other factors of importance are (1) the influent and desired effluent quality; (2) disinfection process performance; and (3) the physical, chemical, and hydraulic characteristics of the reactor.

Many efforts to improve the design accuracy of the disinfection process have been made in recent years. One emerging design approach is to model the disinfection process using the concept of *integrated disinfection design framework* (IDDF) and utilize the *computational fluid dynamics* (CFD) analysis.

Integrated Disinfection Design Framework: The basic concept of IDDF approach is to incorporate the key elements of a disinfection process into a single framework. Based on this concept, an integrated model has been developed to predict disinfectant dose required for a disinfection process.[78,126] There are three basic components in this model: (a) a reactor hydraulics module, (b) a disinfectant demand and decay kinetics module, and (c) a pathogen inactivation module. A desired disinfectant dose can be predicted by solving the integrated model through complicated numerical analysis under the boundary conditions applied at the plant. This model can be used as a basis for developing the general regulatory requirements or site-specific disinfection design and operation criteria.[126] The advantage of applying IDDF concept is its enhanced accuracy for process prediction

TABLE 11.23 General Information about Commercially Available CFD Packages

Package	Developer	Information Source
ANSYS Fluent	ANSYS, Inc.	http://www.ansys.com
Autodesk® CFD	Autodesk, Inc.	http://www.autodesk.com
FLOW-3D	Flow Science, Inc.	http://www.flow3d.com
PHOENICS	CHAM	http://www.cham.co.uk
STAR-CD®	CD-adapco	http://www.cd-adapco.com

than that of the conventional approach. The IDDF approach was originally developed for evaluation of disinfection process in drinking water applications. It is also suitable for wastewater disinfection practice in protecting public health due to increased direct and indirect reuses of reclaimed effluent from wastewater treatment plants.

CFD Analysis: The CFD analysis is a computationally based modeling and simulation technique that can be used to study the dynamic behavior of a fluid. It has been used for many years in the modeling and simulation of major physicochemical treatment processes and hydraulic facilities. In recent years, it is used in modeling chlorine and ozone contactors as well as UV irradiation reactors.[123,127–136]

A special computer software package is usually required to perform CFD analysis. Commercially available CFD packages are summarized in Table 11.23. In the CFD analysis, a treatment process can be presented by a single or multiple mathematical models. These models contain information to describe the physical dimensions of the treatment units as well as the physical, chemical, and microbiological characteristics of the fluid and the objects that are studied. In modeling a disinfection process, it may provide information about chemical disinfectant concentration, UV dose (fluence) rate, UV intensity, residence or exposure time, and effects of particles or organisms. One simple example of using the CFD analysis is to identify the potential short circuiting or recirculation zones in a reactor. A sophisticated model is capable of simulating chemical reactions and decays, multiphase chemical mass transfer, and pathogen inactivation.

With further improvement, validation, and standardization, the CFD analysis will have the potential to provide engineers a unique and inexpensive tool to implement the IDDF concept for more adequate and complete design in the future. Utilities may also use a CFD-based site-specific model to optimize the disinfection process operation and meet the required pathogen inactivation goals without overdosing the disinfectant.

Discussion Topics and Review Problems

11.1 Municipal wastewater is treated in a primary and BNR treatment facility followed by a post polishing filter for reuse. The total coliform count in the effluent from the primary, BNR, and post polishing filters is 2.0×10^7, 2.5×10^5, and 8.0×10^2 per 100 mL, respectively. The total coliform count of raw wastewater is 5.8×10^7. Determine the log reduction in each treatment facility, and total log UV disinfection facility is designed to provide 1.9-log removal for water reuse. Determine the total coliform count in the reuse water.

11.2 Effluent from a secondary clarifier is stored in an earthen basin. The effluent from the basin is chlorinated then used for irrigation of golf course, highway medians, and landscaping areas. The retention time in the storage basin is 4 days, and natural die-off coefficient k for the coliform organisms is $2.5\ d^{-1}$. The total coliform count in the influent to the basin is 2.2×10^4 organism/100 mL. Determine the total coliform count in the effluent from the basin.

11.3 The reduction of organism in a chlorination process is expressed by Equation 1.4a. Using the midpoint chlorine of 3 mg/L and 130 coliform organisms per 100 mL remaining in the secondary effluent after 30 min contact time, calculate k. $N_0 = 10^6$ coliform per 100 mL. Also calculate the number of coliform organisms remaining after 20 min of contact time.

11.4 Chick-Watson equation is a widely used relationship for disinfection by chlorine. A bench-scale study was conducted to determine the constant k' when $n = 1$. The initial total coliform in the test sample was 10^6 organisms/100 mL. The chlorine residual was 1 mg/L. The experimental data is summarized below. Determine the inactivation rate constant k' for chlorine disinfection of total coliform organisms.

Time t, min	0	0.25	0.50	1.00
N, organism/100 mL	10^6	8.1×10^4	6.7×10^3	4.6×10^2
$-\log (N/N_0)$	0	1.09	2.17	4.34
Ct, mg·min/L	0	0.25	0.50	1.00

11.5 The Hom model expresses the tailing off deviation of chlorine disinfection of total coliform organism when $m < 1$. A bench-scale study was conducted to determine the exponents in Hom equation for deactivation of total coliform organisms by chlorine. The experimental values of exponents were: $n = 0.4$ and $k'' = 3.6\ (\text{L/mg})^{0.5}/\text{min}^{0.4}$. Determine the chlorine residual needed to achieve 99.99% inactivation of total coliform organisms. The contact time is 5 min.

11.6 The Collins–Selleck empirical model was developed for chlorine inactivation of coliform organisms. A bench-scale study was conducted on effluent from a polishing filter. The empirical constants n and b were 3.10 (dimensionless) and 2.1 mg·min/L, respectively. Determine the total coliform count in the filtered effluent after chlorination. A chlorine residual of 1.2 mg/L was maintained for 20 min in a contact basin. The initial coliform count of polishing filter effluent = 10^5 organism/100 mL.

11.7 Commercial NaOCl solution is used at a small wastewater treatment plant for disinfection of secondary treated effluent. The solution has a trade percent available chlorine of 8%. Estimate (a) weight percent available Cl_2, (b) weight percent NaOCl, (c) g/L available Cl_2 in solution, (d) g/L NaOCl concentration in solution, and (e) density and specific volume of solution. The temperature of the solution is 20°C.

11.8 Henry's law constants H, H_c, and H_m are, respectively, based on mole fraction, mass concentration, and mole concentration. Identify the numerical equations as they apply to chlorine chemistry and their units. List the value of each constant for dissolution of gaseous chlorine in water at 1 atm and at 20°C.

11.9 Determine the mass concentration based on solubility of chlorine in deionized water at 20°C and 1 atm.

11.10 Estimate the percent HOCl at equilibrium. The temperature and pH of the solution are 20°C and 7.5.

11.11 The secondary treated effluent has total ammonia nitrogen concentration of $C_{\text{total NH}_3\text{-N}}$ of 4 mg/L. The temperature and pH of the effluent are 20°C and 7.6. Estimate the unionized ammonia concentration.

11.12 Many factors and chemical reactions influence the disinfection efficiency of chlorine and many beneficial uses. Following statements and questions address too many properties of chlorine and other compounds. Give your answers to each statement.

a. What are the weight percent of chlorine and NaOCl in 16% trade chlorine solution?
b. Chlorine loss occurs during storage of NaOCl solution. How the loss is detected and compensated during the solution feed?
c. How much alkalinity consumption occurs when 1 g of gaseous chlorine as available chlorine is hydrolyzed?
d. What are different forms of chlorine residuals?
e. Chlorine reacts with ammonia to form chloramines. List the order in which various chloramines are formed. What is the significance of Cl_2 to ammonia ratio of 5:1?

f. At what stoichiometric Cl_2:NH_3-N ratio, the breakpoint chlorination is reached?
g. What is the effect of pH on unionized ammonia concentration?
h. What is the effect of temperature on unionized ammonia concentration?
i. Chlorine oxidizes many organic compounds in effluent. What is the typical BOD_5 reduction from each mg/L chlorine consumed?

11.13 A diffuser delivers chlorine solution into a mixing chamber. The flow rate of chlorine solution is 300 L/min. The mixing chamber is 1 m × 1 m × 2.5 m (deep). The diffuser has 22 orifices and each orifice has a diameter of 0.0051 m (1/5 in). Determine the velocity gradient *G* exerted by the diffuser. The coefficient of discharge $C_d = 0.6$ and water temperature is 15°C. Total flow to the contact basin is 0.48 m^2/s. The diffuser efficiency is 50%.

11.14 An ejector is used to dispense chlorine solution into a mixing well prior to chlorine contact basin. The operating water supply in the ejector creates required amount of vacuum into the line to the chlorinator and in all other components of the chlorinator system. The effluent pumping rate through the ejector is 300 L per min and the head loss in all connecting piping is 2.5 m. The mixing well is such that a *G* of 550 s^{-1} is developed. Calculate (a) the mixing power output of the ejector at 70% efficiency and (b) the pump discharge pressure. The field temperature is 15°C.

11.15 The residual chlorine and chlorination data are given below. Plot the chlorination curve. Obtain the break point chlorination dosage. What will be the initial chlorine demand, and breakpoint chlorination dosage, chlorine residual at break point, total chlorine residual, and total chlorine demand (kg/d) to give a free chlorine residual of 1.0 mg/L? The flow is 6000 m^3/d.

Chlorine dosage, mg/L	1	2	3	4	5	6	7
Chlorine residual, mg/L	0	0.8	1.4	1.0	1.1	2.0	3.0

11.16 Calculate the number of chlorinators, number of chlorine containers attached to the header, and the number of containers required for a 3-week chlorine supply. Use the following data: maximum and average flow = 2.00 and 0.67 m^3/s. Maximum chlorine feed rate = 9 mg/L. The chlorinators are 450 kg/d capacity, and the gaseous chlorine withdrawal rater per container = 180 kg/d.

11.17 Calculate the volume of a chlorine contact basin and the quantity of chlorine needed in kilograms per day. The average design flow is 0.2 m^3/s, contact time is 18 min, total chlorine demand is 17 mg/L, and chlorine residual maintained is 1.5 mg/L.

11.18 Calculate the dimensions of a chlorine contact basin that has a four-pass-around-the-end baffled arrangement. The contact time at 0.5 m^3/s flow is 20 min. The clear width of each pass and the openings at the baffles are 2.0 m and depth is 3 m. The total length of the basin is the centerline around the baffle.

11.19 A wastewater treatment plant is proposed to use chlorine dioxide ClO_2 to disinfect effluent. The expected average demand of ClO_2 is 65 kg/d. The chlorine dioxide is generated on-site using chlorine gas and sodium chlorite ($NaClO_2$) solution of 80% purity. Determine the weight of ClO_2 gas and sodium chlorite required per day in the feedstock. Also, calculate the equivalent chlorine supplied per day from chlorine dioxide.

11.20 A BNR facility is designed to treat 12,000 m^3/d average municipal wastewater. The free chlorine residual in the effluent is 1.5 mg/L. Liquid sulfur dioxide (SO_2) and sodium biosulfite solution ($NaHSO_3$) are evaluated for dechlorination. Estimate the following: (1) weight and volume of liquid SO_2 required and alkalinity destroyed, and (2) weight and volume of 40% $NaHSO_3$ solution required and alkalinity destroyed.

11.21 In recent years, ozone has received much interest for disinfection of municipal wastewater effluent. List the reported advantages and disadvantages of ozonation over disinfection. The disinfective power of ozone is attributed to what factors?

11.22 An experimental ozone contactor is tested to develop the coefficient of specific lethality and to obtain *CT* to achieve a desired log removal in the filtered effluent for reuse. The ratio of T_{10}/T

for the contactor is 0.45. The initial count of *Cryptosporidium parvum* in the filtered effluent is 20,000 organisms per 100 mL. After 8 min of residence time *t*, the ozone residual *C* reached 3.0 mg/L and the organisms count dropped to 120 per 100 mL. A 3.0-log inactivation of *Cryptosporidium parvum* is required in the reuse water. Determine the *CT* requirement to achieve the required inactivation and the number of organisms remaining in the effluent for reuse.

11.23 A 3-chamber counter current ozone contractor similar to that shown in Figure 11.17a is designed for ozonation of filtered effluent. The retention time in each chamber is 6 min. The baffling factor (*BF*) or T_{10}/T ratios is 0.65. In the first chamber, the initial ozone demand of 0.6 mg/L is met, and a maximum required ozone residual C_1 is reached at the end of the chamber before the ozone is diffused. The overall *CT* required to meet the disinfection goal is 9 mg·min/L. The ozone decay in chambers 2 and 3 follow a first-order reaction kinetics expressed by $C = C_0 e^{-k_{O3}\,t}$ (Equation 2.15a), where $k_{O3} = 0.15$/min. Determine (a) the ozone residual at the end of chamber and (b) log inactivation of *Cryptosporidium parvum*. The coefficient of specific lethality $\Lambda_{10} = 0.25$ L/mg·min.

11.24 An ozonation study was conducted in a bench-scale reactor. The filtered effluent sample was ozonated in the contact basin for 15 min. The MPN number per 100 mL sample was determined after two ozone doses were transferred. The MPN numbers after ozone doses *U* of 2.0 and 4.0 mg/L were 1500 and 450 per 100 mL. The initial MPN count before ozonation was 2.5×10^5 per 100 mL. Determine the coefficient *n* and initial demand *q* in Equation 1.38c. Also, estimate the ozone dose required to disinfect the secondary effluent to achieve a coliform count *N* of 240 MPN/100 mL.

11.25 A newly designed UV lamp is tested to achieve 99.9% inactivation of *E. coli* (3-log removal). The design intensity of the lamp is 2.5 mW/cm^2. The inactivation rate constant *k* of *E. coli* is 0.13 cm^2/mW·s (cm^2/mJ). Determine the survival of *E. coli* after the required UV exposure.

11.26 The Chick-Watson model utilizes the first-order reaction to assess the UV inactivation kinetics of bacteria, viruses, and protozoa. In a bench-scale reactor, *Cryptosporidium parvum* cysts were exposed to an average UV radiation intensity I_{avg} of 2 mW/cm^2 for 2 s. The organism count dropped to 1200 organism per 100 mL. Determine the inactivation rate constant k_{UV} (base *e*). Also calculate the UV dose needed to achieve 6-log reduction of *Cryptosporidium*.

11.27 Scheible dispersion equation is commonly used with particulate bacterial density. A UV disinfection facility is designed to reduce the fecal coliform number. The expected number of fecal coliform and TSS in the influence is 10^5 MPN/100 mL and 15 mg/L, respectively. The adjusted $I_{avg} = 9000$ μW/cm^2. The UV system has two banks; each bank uses 150 cm-long lamps. The flow velocity through the channel is 20 cm/s. The dispersion coefficient is 325 cm^2/s. The experimental constant $a = 1.45 \times 10^{-5}$, $b = 1.2$, $c_1 = 0.27$, and $m_1 = 2.1$. Calculate (a) the nonparticulate effluent coliform density N', (b) the total coliform density remaining, and (c) the log inactivation after exposure to UV radiation.

11.28 The effect of UV inactivation on dispersed and particle-associated organisms are assessed by the Emerick model. A UV reactor is designed for UV inactivation of coliform organisms in a filtered secondary effluent. The UV doses D_{UV} to which the organisms are exposed in the reactor and the inactivation rate constant k_{UV} are 5.5 mJ/cm^2 and 1.25 cm^2/mJ. The initial count of coliform organisms N_0 is 2.0×10^5 MPN/100 mL, and umber of particles N'_p is 150 particles/100 mL. Each particle contains at least one organism. Also, calculate the log reduction.

11.29 Some organisms have ability to repair UV-induced damage to DNA by photoreactivation. A UV batch reactor study was conducted to study the photoreactivation of *E. coli*. At a UV dose of 30 mW·s/cm^2, the survival of *E. coli* was 0.02%. It is estimated that an equivalent of total 0.5% original population before UV exposure fully recovers from photoreactivation. Calculate total UV exposure needed to inactivate all organisms even those that fully recover from photoreactivation.

11.30 A UV system is designed to disinfect the filtered secondary treated effluent. The design flow is peak wet weather flow of 25,000 m^3/d. The wet weather coliform density in filtered effluent $= 5 \times 10^5$ organisms/100 mL. TSS $= 8$ mg/L, and c_1 and m_1 are 0.25 and 2, respectively. The permit requirement of fecal coliform density N_T in the disinfected effluent $= 400$ organisms/100 mL. The total normal UV output per lamp ($P_{\text{lamp,nom}}$) $= 26.8$ W/lamp. There are two UV channels and one UV bank in each channel. Determine the following (a) total number of lamps and modules per bank and (b) width and depth of the UV channel.

References

1. Yates, M. V. and C. P. Gerba, Microbial considerations in wastewater reclamation and reuse, *Wastewater Reclamation and Reuse, Water Quality Management Library*, vol. 10, Asano, T. (Ed.), CRC Press, Boca Raton, FL, 1998.
2. Metcalf & Eddy|AECOM, Asano, T., F. L. Burton, H. L. Leverenz, R. Tsuchihashi, and G. Tchobanoglous, *Water Reuse: Issues, Technologies, and Applications*, McGraw-Hill Professional, New York, 2007.
3. Bitton, G. *Wastewater Microbiology*, 3rd ed., Wiley-Liss, New York, NY, 2005.
4. Mara, D. and N. J. Horan (Eds.), *Handbook of Water and Wastewater Microbiology*, Academic Press, London, 2003.
5. Rose, J. B., S. R. Farrah, V. J. Harwood, A. D. Levine, J. Lukasik, P. Menendez, and T. M. Scott, *Reduction of Pathogens, Indicator Bacteria, and Alternative Indicators by Wastewater Treatment and Reclamation Processes*, WREF Final Report, 00-PUM-2T, Water Environment Research Foundation, Alexandria, VA, 2004.
6. U.S. Environmental Protection Agency, *Guideline for Water Reuse*, EPA/625/R-04/108, U.S. Environmental Protection Agency, Office of Water, Washington, DC, September 2004.
7. World Health Organization (WHO), *Human Virus in Water, Wastewater and Soil*, WHO Technical Report Series 639, WHO, Geneva, Austria, 1979.
8. World Health Organization (WHO), *Guidelines for Drinking-Water Quality*, 4th ed., WHO, Geneva, Austria, 2011.
9. EPA Victoria, *Guidelines for Environmental Management: Disinfection of Treated Wastewater*, EPA Victoria, Victoria, Australia, 2002.
10. Environmental Protection Agency (EPA) of the State of Queensland, *Queensland Water Recycling Guidelines*, EPA of the State Queensland, Brisbane, Australia, 2005.
11. National Technical Advisory Committee, *Water Quality Criteria, Federal Water Pollution Control Administration*, Department of Interior, Washington, DC, 1968.
12. U.S. Environmental Protection Agency, *Ambient Water Quality Criteria for Bacteria—1986*, EPA440/5-84-002, U.S. Environmental Protection Agency, Office of Water, Washington, DC, January 1986.
13. Harwood, V. J., A. D. Levine, T. M. Scott, V. Chivukula, J. Lukasik, S. R. Farrah, and J. B. Rose, Validity of the indicator organism paradigm for pathogen reduction in reclaimed water and public health protection, *Applied and Environmental Microbiology*, 71(6), 2005, 3163–3170.
14. World Health Organization (WHO), *Assessing Microbial Safety of Drinking Water: Improving Approaches and Methods*, IWA Publishing, London, U.K., 2003.
15. Salvato, J. A., N. L. Nemerow, and F. Agardy, *Environmental Engineering*, 5th ed., John Wiley & Son, Inc., New York, NY, 2003.
16. Gray, N. F., *Water Technology: An Introduction for Scientists and Engineers*, 3rd ed., Taylor Francis Ltd, U.K., 2010.

17. Davies-Colley, R, Pond disinfection, *Pond Treatment Technology*, Shilton, A. (Ed.), IWA Publishing, London, U.K., 2006.
18. Gersberg, R. M., R. A. Gearheart, and M. Ives, Pathogen removal in constructed wetlands, *Constructed Wetlands for Wastewater Treatment: Municipal, Industrial, and Agricultural*, Hammer, D.A. (Ed.), CRC Press, Boca Raton, FL, 1989.
19. McCuin, R. M. and J. L. Clancy, Occurrence of *Cryptosporidium* oocysts in US wastewaters, *Journal of Water and Health*, 4(4), 2006, 437–452.
20. Rose, J. B., L. J. Dickson, S. R. Farrah, and R. P. Garnahan, Removal of pathogenic and indicator microorganisms by a full-scale water reclamation facility, *Water Research*, 30(11), November 1996, 2785–2797.
21. Rose, J. B., Water reclamation, reuse and public health, *Water Science & Technology*, 55(1–2), 2007, 275–282.
22. Zhang, K. and K. Farahbakhsh, Removal of native coliphages and coliform bacteria from municipal wastewater by various wastewater treatment processes: Implications to water reuse, *Water Research*, 41(12), June 2007, 2816–2824.
23. Garcia-Armisen, T., J. Prats, Y. Marrero, and P. Servais, Faecal bacterial indicators removal in various wastewater treatment plants located in Almendares River watershed (Cuba), *Water Science & Technology*, 58(4), 2008, 773–779.
24. Carducci, A., P. Morici, F. Pizzi, R. Battistini, E. Rovini, and M. Verani, Study of the viral removal efficiency in an urban wastewater treatment plant, *Water Science & Technology*, 58(4), 2008, 893–879.
25. U.S. Environmental Protection Agency, *Manual of Constructed Wetlands Treatment of Municipal Wastewaters*, EPA/625/R-99/010, U.S. Environmental Protection Agency, Office of Research and Development, Cincinnati, OH, September 2000.
26. Karim, M. R., F. D. Manshadi, M. M. Karpiscak, and C. P. Gerb, The persistence and removal of enteric pathogens in constructed wetlands, *Water Research*, 38(7), April 2004, 1831–1837.
27. Pundsack, J., R. Axler, R. Hicks, J. Henneck, D. Nordman, and B. McCarthy, Seasonal pathogen removal by alternative on-site wastewater treatment systems, *Water Environment Research*, 73 (2), March/April 2001, 204–212.
28. Campos, C., A. Guerrero, and M. Cárdenas, Removal of bacterial and viral faecal indicator organisms in a waste stabilization pond system in Choconta, Cundinamarca (Colombia), *Water Science & Technology*, 45(1), 2002, 61–66.
29. Bouhoum, K., O. Amahmid, and S. Asmama, Occurrence and removal of protozoan cysts and helminth eggs in waste stabilization ponds in Marrakech, *Water Science & Technology*, 42 (10–11), 2000, 159–164.
30. Jimenez, B., Helminth ova removal from wastewater for agriculture and aquaculture reuse, *Water Science & Technology*, 55(1–2), 2007, 485–493.
31. Payment, P., R. Plante, and P. Cejka, Removal of indicator bacteria, human enteric viruses, *Giardia* cysts, and *Cryptosporidium* oocysts at a large wastewater primary treatment facility, *Canadian Journal of Microbiology*, 47(3), March 2001, 188–193.
32. Sigmund, T., J. Siczka, T. Elliott, J. Awad, and R. Onderko, Operating chemically enhanced clarification for optimum disinfection performance, *Proceedings of the 79th Annual WEFTEC®*, Water Environment Federation, Dallas, TX, October 2006.
33. Comerton, A. M., R. C. Andrews, and D. M. Bagley, Evaluation of an MBR-RO system to produce high-quality reuse water: microbial control, DBP formation and nitrate, *Water Research*, 39(16), October 2005, 3982–3990.
34. Jimenez-Cisneros, B. E., C. Maya-Rendon, and G. Salgado-Velazquez, The elimination of helminth ova, faecal coliforms, *Salmonella* and protozoan cysts by various physicochemical processes in wastewater and sludge, *Water Science & Technology*, 43(12), 2001, 179–182.

35. Tawfik, A., H. Temmink, G. Zeeman, and B. Klapwijk, Sewage treatment in a rotating biological contactor (RBC) system, *Water, Air, & Soil Pollution*, 175(1–4), September 2006, 275–289.
36. Tawfik, A., B. Klapwijk, J. van Buuren, F. El-Gohary, and G. Lettinga, Physico-chemical factors affecting the *e.* coli removal in a rotating biological contactor (RBC) treating UASB effluent, *Water Research*, 38(5), 2004, 1081–1088.
37. Vaughn, S. D., W. Hurley, T. Madhanagopal, K. Kunihiro, and T. Slifko, Effect of hydraulic loading and chemical addition on filter performance and protozoa removal: A side-by-side comparison, *Proceedings of the 77th Annual WEFTEC®*, Water Environment Federation, New Orleans, LA, October 2004.
38. Knapp, T and D. Tucker, Side-by-side pilot testing of two disk filter manufacturers at the City of Merced wastewater treatment plant, *Proceedings of the 79th Annual WEFTEC®*, Water Environment Federation, Dallas, TX, October 2006.
39. Levine, A. D., V. J. Harwood, S. R. Farrah, T. M. Scott, and J. B. Rose, Pathogen and indicator organism reduction through secondary effluent filtration: Implications for reclaimed water production, *Water Environment Research*, 80(7), July 2008, 596–608.
40. Ottoson, J., A. Hansen, B. Bjorlenius, H. Norder, and T. A. Stenstrom, Removal of viruses, parasitic protozoa and microbial indicators in conventional and membrane processes in a wastewater pilot plant, *Water Research*, 40(7), April 2006, 1449–1457.
41. Lonigro, A., A. Pollice, R. Spinelli, F. Berrilli, D. Di Cave, C. D'Orazi, P. Cavallo, and O. Brandonisio, *Giardia* cysts and *Cryptosporidium* oocysts in membrane-filtered municipal wastewater used for irrigation, *Applied & Environmental Microbiology*, 72(12), 2006, 7916–1918.
42. Qasim, S. R., *Wastewater Treatment Plants: Planning, Design, and Operation*, 2nd ed., CRC Press, Boca Raton, FL, 1999.
43. Kitis, M., Disinfection of wastewater with peracetic acid: A review, *Environment International*, 30 (1), March 2004, 47–55.
44. Gehr, R., M. Wagner, P. Veerasubramanian, and P. Payment, Disinfection efficiency of peracetic acid, UV and ozone after enhanced primary treatment of municipal wastewater, *Water Research*, 37(19), November 2003, 4573–4586.
45. Wagner, M., D. Brumelis, and R. Gehr, Disinfection of wastewater by hydrogen peroxide or peracetic acid: Development of procedures for measurement of residual disinfectant and application to a physicochemically treated municipal effluent, *Water Environment Research*, 74(1), January–February 2002, 33–50.
46. Sobsey, M. D., M. J. Casteel, H. Chung, G. Lovelace, O. D., Simmons III, and J. S. Meschke, Innovative technologies for wastewater disinfection and pathogen detection, *Proceedings of Disinfection '98, the Latest Trends in Wastewater Disinfection: Chlorination versus UV Disinfection*, Water Environment Federation, Baltimore, MD, April 1998.
47. Son, H., M. Cho, J. Kim, B. Oh, H. Chung, and J. Yoon, Enhanced disinfection efficiency of mechanically mixed oxidants with free chlorine, *Water Research*, 39(4), February 2005, 721–727.
48. White, G. C., *Handbook of Chlorination and Alternative Disinfectants*, 4th ed., John Wiley & Sons, Inc., New York, NY, 1999.
49. Water Environment Federation, *Wastewater Disinfection*, Manual of Practice FD-10, WEF, Alexandria, VA, 1996.
50. U.S. Environmental Protection Agency, *Wastewater Technology Fact Sheet Ultraviolet Disinfection*, EPA 832-F-99-064, U.S. Environmental Protection Agency, Office of Water, Washington, DC, September 1999.
51. Task Force of the Water Environment Federation and the American Society of Civil Engineers/Environmental and Water Resources Institute, *Design of Municipal Wastewater Treatment Plant*,

WEF Manual of Practice No. 8 and ASCE Manuals and Reports on Engineering Practice No. 76, 5th ed., McGraw-Hill, New York, NY, 2009.

52. Metcalf & Eddy|AECOM, *Wastewater Engineering: Treatment and Resource Recovery*, 5th ed., McGraw-Hill, New York, NY, 2014.
53. Metcalf & Eddy, Inc., *Wastewater Engineering, Treatment and Reuse*, 4th ed., McGraw-Hill Companies, Inc., New York, NY, 2003.
54. U.S. Environmental Protection Agency, *Design Manual: Municipal Wastewater Disinfection*, EPA/625/1-86-021, U.S. Environmental Protection Agency, Office of Research and Development, Cincinnati, OH, October 1986.
55. U.S. Environmental Protection Agency, *Alternative Disinfectants and Oxidants Guidance Manual*, EPA 815-R-99-014, U.S. Environmental Protection Agency, Office of Water, Washington, DC, April 1999.
56. U.S. Environmental Protection Agency, *Ultraviolet Disinfection Guidance Manual for the Final Long Term 2 Enhanced Surface Water Treatment Rule*, EPA 815-R-06-007, U.S. Environmental Protection Agency, Office of Water, Washington, DC, November 2006.
57. Malley, J. P., N. Ballester, A. Margolin, K. Linden, and A. Mofidi, *Inactivation of Pathogens with Innovative UV Technologies*, American Water Works Association (AWWA) and the AWWA Research Foundation (AwwaRF), Denver, CO, 2004.
58. Shin, G., K. G. Linden, M. J. Arrowood, and M. D. Sobsey, Low-pressure UV inactivation and DNA repair potential of *Cryptosporidium parvum* oocysts, *Applied and Environmental Microbiology*, 67(7), July 2001, 3029–3032.
59. Jacangelo, J. G., P. Loughran, B. Petrik, D. Simpson, and C. McIlroy, Removal of enteric viruses and selected microbial indicators by UV irradiation of secondary effluent, *Water Science & Technology*, 47(9), 2003, 193–198.
60. Clancy, J. L., K. G. Linden, and R. M. McCuin, *Cryptosporidium* occurrence in wastewaters and control using UV disinfection, *International Ultraviolet Association (IUVA) News*, 6(3), September 2004, 10–14.
61. Keller, R., L. Passamani, L. Vaz, S. T. Cassini, and R. F. Goncalves, Inactivation of *Salmonella* spp. from secondary and tertiary effluents by UV irradiation, *Water Science & Technology*, 47(3), 2003, 147–150.
62. Finch, G. R. and M. Belosevic, Controlling *Giardia* spp. and *Cryptosporidium* spp. in drinking water by microbial reduction processes, *Journal of Environmental Engineering and Science*, 1, 2002, 17–31.
63. U.S. Environmental Protection Agency, *Technologies for Upgrading Existing or Designing New Drinking Water Treatment*, EPA 625/4-89/023, U.S. Environmental Protection Agency, Office of Drinking Water, Cincinnati, OH, March 1990.
64. U.S. Environmental Protection Agency, *Onsite Wastewater Treatment Systems Technology*, EPA/625/R-00/008, U.S. Environmental Protection Agency, Office of Water, Washington, DC, February 2002.
65. Haas, C.N., General kinetics of disinfection processes, *Control of Microorganisms in Drinking Water*, Lingireddy, S. (Ed.), ASCE, Reston, VA, 2002.
66. Hom, L. W., Kinetics of chlorine disinfection of an ecosystem, *Journal of the Sanitary Engineering Division, ASCE*, 98(1), January/February 1972, 183–194.
67. Finch, G. R., E. K. Black, L. L. Gyürék, and M. Belosevic, *Ozone Disinfection of Giardia and Cryptosporidium*, American Water Works Association (AWWA) and the AWWA Research Foundation (AwwaRF), Denver, CO., 1994.
68. Finch, G. R., C. N. Haas, J. A. Oppenheimer, G. Gordon, and R. R. Trussell, Design criteria for inactivation of *Cryptosporidium* by ozone in drinking water, *Ozone: Science & Engineering*, 23 (4), 2001, 259–284.
69. Hass, C. N. and J. Joffe, Disinfection under dynamic conditions: Modification of Hom model for decay, *Environmental Science & Technology*, 28(7), 1994, 1367–1369.

70. Haas, C. N., M. S. Heath, J. Jcangelo, J. Joffe, U. Anmangandla, J. C. Hornberger, and J. Glicker, *Development and Validation of Rational Design Methods of Disinfection*, American Water Works Association (AWWA) and the AWWA Research Foundation (AwwaRF), Denver, CO., 1995.
71. LeChevallier, M. W. and K. Au, *Water Treatment and Pathogen Control: Process Efficiency in Achieving Safe Drinking Water*, World Health Organization (WHO)/IWA Publishing, London, U.K., 2004.
72. Howe, K. J., D. W. Hand, J. C. Crittenden, R. R. Rrussell, and G. Tchobanoglous, *Montgomery Watson Harza, Principles of Water Treatment*, John Wiley & Sons, Inc., Hoboken, NJ, 2012.
73. Lin, S. D. and C. C. Lee, *Water and Wastewater Calculations Manual*, 2nd ed., McGraw-Hill Professional, New York, NY, 2007.
74. Collins, H. F., R. E. Selleck, and G. C. White, Problems in obtaining adequate sewage disinfection, *Journal of the Sanitary Engineering Division, ASCE*, 97(5), September/October 1971, 549–562.
75. Selleck, R. E., H. F. Collins, and B. M. Saunier, Kinetics of bacterial deactivation with chlorine, *Journal of the Environmental Engineering Division, ASCE*, 104(6), November/December 1978, 1197–1212.
76. Majumdar, S. B., W. H. Ceckler, and O. J. Sproul, Inactivation of poliovirus in water by ozonation, *Journal Water Pollution Control Federation*, 45(12), December 1973, 2433–2443.
77. Roy, D., R. S. Engelbrecht, and E. S. K. Chian, Kinetics of enteroviral inactivation by ozone, *Journal of the Environmental Engineering Division, ASCE*, 107(5), September/October 1981, 887–901.
78. Bellamy, W. D., G. R. Finch, and C. N. Haas, *Integrated Disinfection Design Framework*, American Water Works Association (AWWA) and the AWWA Research Foundation (AwwaRF), Denver, CO, 1998.
79. Haas, C. N. and B. Kaymak, *Effect of Initial Microbial Concentration on Disinfection Efficiency*, American Water Works Association (AWWA) and the AWWA Research Foundation (AwwaRF), Denver, CO, 2002.
80. Severin, B. F., M. T. Suidan, and R. S. Engelbrecht, Kinetic modeling of UV disinfection of water, *Water Research*, 17(11), 1983, 1669–1678.
81. Severin, B. F., M. T. Suidan, and R. S. Engelbrecht, Series-event kinetic model for chemical disinfection, *Journal of the Environmental Engineering, ASCE*, 110(2), March/April 1984, 430–439.
82. Gyürék, L. L. and G. R. Finch, Modeling water treatment chemical disinfection kinetics, *Journal of the Environmental Engineering, ASCE*, 124(9), September 1998, 783–793.
83. Gyürék, L. L., H. Li, M. Belosevic, and G. R. Finch, Ozone inactivation kinetics of *Cryptosporidium* in phosphate buffer, *Journal of Environmental Engineering*, 125(10), October 1999, 913–924.
84. American Water Works Association (AWWA), *Water Chlorination/Chloramination Practices and Principles, AWWA Manual M20*, 2nd ed., AWWA, Denver, CO, 2006.
85. The Chlorine Institute, Inc. (CI, Inc.), *Pamphlet 1: Chlorine Basics*, 7th ed., CI, Inc., Arlington, VA, 2008.
86. The Chlorine Institute, Inc. (CI, Inc.), *Pamphlet 96: Sodium Hypochlorite Manual*, 4th ed., CI, Inc., Arlington, VA, 2011.
87. The Chlorine Institute, Inc. (CI, Inc.), *Pamphlet 5: Bulk Storage of Liquid Chlorine*, 8th ed., CI, Inc., Arlington, VA, 2011.
88. The Chlorine Institute, Inc. (CI, Inc.), *Pamphlet 9: Chlorine Vaporizing Systems*, 7th ed., CI, Inc., Arlington, VA, 2011.
89. The Chlorine Institute, Inc. (CI, Inc.), *Pamphlet 85: Recommendations for Prevention of Personal Injuries for Chlorine Production and Use Facilities*, 5th ed., CI, Inc., Arlington, VA, 2010.
90. The Chlorine Institute, Inc. (CI, Inc.), *Pamphlet 155: Water Wastewater Operators Chlorine Handbook*, 2nd ed., CI, Inc., Arlington, VA, 2008.
91. Kawamura, S., *Integrated Design and Operation of Water Treatment Facilities*, 2nd ed., John Wiley and Sons, Hoboken, NJ, 2000.

92. Feachem, R. G., D. J. Bradley, H. Garelick, and D. D. Mara, *Sanitation and Disease: Health Aspects of Excreta and Wastewater Management*, John Wiley & Sons, Chichester, U.K., 1983.
93. Pisarenko, A. N., B. D. Stanford, D. Yan, D. Gerrity, and S. A. Snuder, Effects of ozone and ozone/peroxide on trace organic contaminants and NDMA in drinking water and water reuse applications, *Water Research, IWA*, 46(2), February 2012, 316–326.
94. WateReuse Research Foundation, Formation of nitrosamines and perfluoroalkyl acids during ozonation in water reuse applications, Webcast Presentation in *WateReuse-11-08*, Webcast, December 11, 2014.
95. Rakness, K. L., *Ozone in Drinking Water Treatment: Process Design, Operation, and Optimization*, AWWA, Denver, CO, 2005.
96. Shammas, N. K. and L. K. Wang, Ozonation, handbook of environmental engineering, Vol.3, *Physicochemical Treatment Processes*, Wang, L. K., Y. Hung, and N. K. Shammas (Eds.), Humana Press, Totowa, NJ, 2005.
97. Ozonia/Degremont Technologies Ltd., *Ozonia Disinfection Oxidation*, http://www.ozonia.com, 2011.
98. Primozone Production, *Primozone® GM-Series*, http://www.primozone.com, 2015.
99. Spartan Environmental Technologies, LLC, *Ozone Oxygen Recycling System*, http://www.spartanwatertreatment.com/ozone-oxygen-recovery.html
100. Air Product and Chemicals, Inc., *PRISM*™ *O_2/O_3 Short Loop Recycle (SLR) Technology for Ozonolysis Chemistry and Other Advanced Ozone Applications*, http://www.airproducts.com, 2015.
101. U.S. Environmental Protection Agency, *Wastewater Technology Fact Sheet Ozone Disinfection*, EPA 832-F-99-063, Office of Water, Washington, DC, September 1999.
102. Finch, G. R. and D. W. Smith, Ozone dose response of *Escherichia* coli in activated sludge effluent, *Water Research*, 23(8), 1989, 1017–1026.
103. Xylem, Inc., *TAK 55 Smart*™ *UV System, The Most Cost-Effective Solution for Wastewater Disinfection Using UV Light*, http://www.xylem.com, 2013.
104. Xylem, Inc., *WEDECO LBX Series, The Ideal UV Solution with Wastewater is under Pressure*, http://www.xylem.com, 2013.
105. TrojanUV, *TrojanUVSigna*™, *Wastewater Disinfection*, http://www.trojanuv.com, 2017.
106. Neptune Benson/Evoqua, *Multilamp, Medium Pressure UV Systems for Wastewater Applications*, http://www.neptunebenson.com, 2015.
107. Enaqua/GRUNDFOS Pumps Manufacturing Corporation, *Enagua Noncontact Ultraviolet Disinfection Systems, Municipal Wastewater*, http://www.enaqua.com, 2013.
108. Lindenauer, K. G. and J. L. Darby, Ultraviolet disinfection of wastewater: Effect of dose on subsequent photoreactivation, *Water Research*, 28(4), April 1994, 805–817.
109. Harris, G. D., V. D. Adams, D. L. Sorensen, and M. S. Curtis, Ultraviolet inactivation of selected bacteria and viruses with photoreactivation of the bacteria, *Water Research*, 21(6), June 1987, 687–692.
110. Thurston-Enriquez, J. A., C. N. Haas, J. Jacangelo, K. Riley, and C. P. Gerba, Inactivation of feline calicivirus and adenovirus Type 40 by UV radiation, *Applied and Environmental Microbiology*, 69 (1), January 2003, 577–582.
111. Nwachuku, N., C. P. Gerba, A. Oswald, and F. D. Mashadi, Comparative inactivation of adenovirus serotypes by UV light disinfection, *Applied and Environmental Microbiology*, 71(9), September 2005, 5633–5636.
112. Loge, F. J., K. Bourgeous, R. W. Emerick, and J. L. Darby, Variations in wastewater quality parameters influencing UV disinfection performance: Relative impact of filtration, *Journal of the Environmental Engineering Division, ASCE*, 127(9), September 2001, 823–837.
113. Wu, Y., T. Clevenger, and B. Deng, Impacts of goethite particles on UV disinfection of drinking water, *Applied and Environmental Microbiology*, 71(7), July 2005, 4140–4143.

114. Passantino, L., J. Malley Jr., M. Knudson, R. Ward, and J. Kim, Effect of low turbidity and algae on UV Disinfection performance, *Journal of American Water Works Association*, 96(6), June 2004, 128–137.
115. WEF Disinfection Committee, *Wastewater Disinfection: State-of-the-Art Report*, Water Pollution Control Federation, Alexandria, VA, 1984.
116. Scheible, O. K. and C. D. Bassell, *Ultraviolet Disinfection at a Secondary Wastewater Treatment Plant*, EPA-600/2-81-152, NTIS No. PB 81-242125, U.S. EPA, Cincinnati, OH, 1981.
117. Scheible, O. K., Development of a rationally based design protocol for the ultraviolet light disinfection process, *Journal Water Pollution Control Federation*, 59(1), January 1987, 25–31.
118. Collins, H. F. and R. E. Selleck, *Process Kinetics of Wastewater Chlorination*, Sanitary Engineering Research Laboratory (SERL) Report No. 72-5, University of California at Berkeley, Berkeley, CA, 1972.
119. Emerick, R. W., F. J. Loge, T. Ginn, and J. L. Darby, Modeling the inactivation of particle-associated coliform bacteria, *Water Environment Research*, 72(4), July–August 2000, 432–438.
120. Severin, B. F., M. T. Suidan, B. E. Rittmann, and R. S. Engelbrecht, Inactivation kinetics in a flow through UV reactor, *Journal Water Pollution Control Federation*, 56(2), February 1984, 164–169.
121. U.S. Environmental Protection Agency, *User's Manual for UV Dis, Version 3.1, UV Disinfection Process Design*, U.S. EPA G 0703, Risk Reduction Engineering Laboratory, Cincinnati, OH, 1992.
122. Blatchley, E. R., , III, W. L. Wood, and P. Pchuerch, UV pilot testing: Intensity distributions and hydrodynamics, *Journal of the Environmental Engineering, ASCE*, 121(3), March 1995, 258–262.
123. Do-Quang, Z., M. L. Janex, and D. Perrin, Predictive Tool for UV dose distribution assessment: Validation of CFD models by bioassays, *Proceedings of the American Water Works Association Annual Conference*, New Orleans, LA., 2002.
124. National Water Research Institute (NWRI), *Ultraviolet Disinfection Guidance for Drinking Water and Water Reuse*, 3rd ed., NWRI, Fountain Valley, CA, 2012.
125. U.S. Environmental Protection Agency, *Ultraviolet Disinfection Guidance Manual*, Draft, EPA 625/4-89/023, Office of Water, Washington, DC, 2003.
126. Carlson, K., D. Pier, W. Bellamy, M. Calson, J. Ducoste, G. Amy, and K. Rakness, *Implementation of the Integrated Disinfection Design Framework*, American Water Works Association (AWWA) and the AWWA Research Foundation (AwwaRF), Denver, CO, 2001.
127. Wang, H. and R. A. Falconer, Simulating disinfection processes in chlorine contact tanks under various turbulence models and high-order accurate difference schemes, *Water Research*, 32(5), March 1998, 1529–1543.
128. Greene, D. J., *Numerical Simulation of Chlorine Disinfection Processes in Nonideal Reactors*, PhD Dissertation, Drexel University, Philadelphia, PA., 2002.
129. Greene, D. J., B. Farouk, and C. N. Haas, CFD design approach for chlorine disinfection processes, *Journal of American Water Works Association*, 96(8), August 2004, 138–150.
130. Murrer, J., J. Gunstead, and S. Lo, The development of an ozone contact tank simulation model, *Ozone: Science and Engineering*, 17(6), 1995, 607–617.
131. Huang, T., C. J. Brouckaert, M. Docrat, and M. Pryor, A computational fluid dynamic and experimental study of an ozone contactor, *Water Science & Technology*, 46(9), 2002, 87–93.
132. Chiu, K., D. A. Lyn, P. Savoye, and E. R. Blatchley III, Integrated UV disinfection model based on particle tracking, *Journal of the Environmental Engineering Division, ASCE*, 125(1), January 1999, 7–16.
133. Lyn, D. A. and E. R. Blatchley III, Numerical computational fluid dynamics-based models of ultraviolet disinfection channels, *Journal of the Environmental Engineering Division, ASCE*, 131(6), June 2005, 838–849.

134. Munoz, A., S. Craik, and S. Kresta, Computational fluid dynamics for predicting performance of ultraviolet disinfection—Sensitivity to particle tracking input, *Journal of Environmental Engineering and Science*, 6(3), May 2007, 285–301.
135. Sozzi, D. A. and F. Taghipour, UV reactor performance modeling by Eulerian and Lagrangian Methods, *Environmental Science & Technology*, 40(5), March 2006, 1609–1615.
136. Greene, D. J., C. N. Haas, and B. Farouk, Computational fluid dynamics analysis of the effects of reactor configuration on disinfection efficiency, *Water Environment Research*, 78(9), September 2006, 909–919.

12

Effluent Reuse and Disposal

12.1 Chapter Objectives

High quality effluent from wastewater treatment plants is considered a reliable source of water supply. Increasing demand on fresh water for urban, industrial, irrigation, and many other needs has resulted in increased interest in effluent reuse. Even effluent discharged into natural waters is considered an indirect reuse for water supply. The objectives of this chapter are to present:

- Major issues related to effluent reuse
- Risk assessment
- Storage facility for reclaimed water
- Principal uses of wastewater effluent
- Effluent disposal considerations
- Design considerations for outfalls

12.2 Major Issues Related to Effluent Reuse

The public health and safety and environmental concerns are the major issues of wastewater reuse. Currently available wastewater treatment systems are well understood, effective, and reliable. These facilities are able to produce effluent of any desired quality. However, there are still concerns of long-term human health and safety, and environmental risks that are very difficult to establish. For these reasons, wastewater reuse in many cases has not received public acceptance.

12.2.1 Quality Parameters of Reclaimed Water

The major concerns of reclaimed water are the *constituents* remaining after treatment. These constituents are classified as conventional and nonconventional parameters and *emerging constituents*. The conventional parameters are pH, BOD, TSS, nitrogen, phosphorus, and pathogenic organisms. The nonconventional parameters are TDS, pesticides and refractory organics, surfactants, and metals.

The emerging constituents are an extraordinary group of chemicals extensively used by the general public. They include pharmaceutical and personnel care products (PPCPs). These products include: (1) prescription and nonprescription drugs, such as birth control hormones and other endocrine disruptors, antibiotics, steroid hormones, nonsteroid antiinflammatory drugs, and analgesics and (2) soaps, skin care products, insect repellants, sunscreens, and cosmetics.[1,2] Other emerging constituents include veterinary medicines, agricultural chemicals, and residential landscape-care products. These compounds are highly bioactive and occur at very low concentrations (μg/L or ng/L) in the environment.

As emerging constituents, PPCPs have raised serious concerns in recent years. They are carried into publicly owned treatment works (POTWs) via municipal wastewater and are not completely removed.[3,4] The health effects of PPCPs are poorly understood; the main reasons are: (1) They are used in large quantities in homes; (2) they have adverse biological effects, and certain PPCPs may have synergistic actions causing disproportionately large effects; and (3) they produce negative environmental effects on aquatic life even at very low concentrations. Some of these effects are feminized fish, that is, male fish have some female characteristics, and masculinization of fish.[1,2,5] The products of biggest concern are antibiotics and steroid hormones. Antibiotics lead to resistant pathogens, and steroid hormones may interfere with reproduction and growth and development of the living system.[1,5]

12.2.2 Treatment Technology for Reclaimed Water

The conventional parameters have traditionally been used to design and judge the performance of conventional wastewater treatment plants with or without nutrients removal functions. The ranges with typical values for the conventional parameters are summarized in Table 5.1. The performance of various unit operation and processes in terms of removal of the conventional parameters are listed in Table 6.8.

Many process trains and resulting effluent quality from combinations of these treatment processes are provided in Examples 6.13 through 6.18. Readers are referred to Section 6.3.5 to review and develop process trains for achieving the desired effluent quality. Additionally, References 6 through 9 provide background information on developing process train in the context of reclaimed water quality and reuse category.

EXAMPLE 12.1: PROCESS TRAIN FOR TREATMENT OF RECLAIMED WATER

A BNR facility is designed to produce reclaimed water for urban reuse. The process train of the facility includes primary sedimentation; biological reactor with combined anaerobic, anoxic, and aerobic zones; polishing filtration; and granular activated carbon (GAC) filter. The final effluent is disinfected with 1.0 mg/L free chlorine residual for a contact period of 30 min. Ammonia is then added to convert free chlorine residual into combined chlorine residual prior to clear well storage and distribution. Liquid ammonium sulfate (LAS) solution (40% $(NH_4)_2SO_4$ by weight and sp.gr. of 1.25) is used to provide the ammonia. The influent quality and percent removal by each unit process are summarized below.

Parameter	Raw Influent Concentration[a], mg/L	Typical Removal[b], %			
		Primary	BNR	Tertiary Filter	GAC Filter
BOD_5	210	30	90	50	70
TSS	250	60	90	70	60
TP	6	15	80	25	25
ON	15	25	85	60	40
AN	30	0	98.5	0	0

[a] See Table 5.1 for the characteristics of municipal wastewater.
[b] See Table 6.8 for the typical removal efficiencies of these parameters.

The influent has a total fecal coliform count of 10^6 organisms/100 mL. Draw the process diagram. Estimate (1) the effluent BOD_5, TSS, TP, ON, and AN; (2) total fecal coliform count in the unchlorinated effluent; (3) fecal coliform count after disinfection; (4) chlorine dosage required to establish a free chlorine residual of 1 mg/L; (5) ammonia dosage needed to convert 1.0 mg/L free chlorine residual into combined chlorine residual; and (6) LAS dosage and specific feed rate required. Fecal coliform removal in the treatment processes is 4.5-log prior to disinfection. Collins–Selleck equation is applicable to disinfection kinetics. Constants n and b are 3.0 and 2.8, respectively. Describe the quality of reclaimed water for intended reuse.

Solution

1. Draw the process diagram of proposed BNR facility in Figure 12.1.

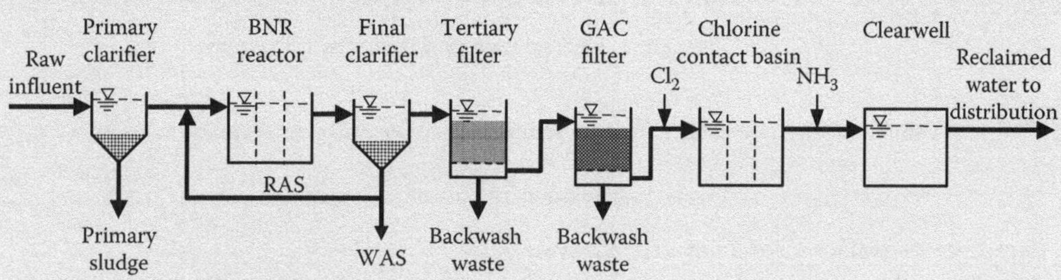

FIGURE 12.1 Process diagram of proposed BNR facility for reuse of reclaimed effluent (Example 12.1).

2. Determine the effluent quality.

$$\begin{aligned}
BOD_5 &= (1-0.30)\times(1-0.90)\times(1-0.50)\times(1-0.70)\times 210\,mg/L = 0.01\times 210\,mg/L = 2\,mg/L\\
TSS &= (1-0.60)\times(1-0.90)\times(1-0.70)\times(1-0.60)\times 250\,mg/L = 0.005\times 250\,mg/L = 1\,mg/L\\
TP &= (1-0.15)\times(1-0.80)\times(1-0.25)\times(1-0.25)\times 6\,mg/L = 0.1\times 6\,mg/L = 0.6\,mg/L\\
ON &= (1-0.25)\times(1-0.85)\times(1-0.60)\times(1-0.40)\times 15\,mg/L = 0.03\times 15\,mg/L = 0.5\,mg/L\\
AN &= (1-0)\times(1-0.985)\times(1-0)\times(1-0)\times 30\,mg/L = 0.015\times 30\,mg/L = 0.5\,mg/L
\end{aligned}$$

3. Determine the fecal coliform count in the reclaimed water prior to disinfection.
   The fecal coliform removal in the plant is 4.5-log.
   The influent fecal coliform count is 10^6 organisms/100 mL or 6.0-log.
   Fecal coliform count in the effluent = (6.0 − 4.5) log = 1.5-log or $10^{1.5}$ = 32 organisms/100 mL
4. Determine the fecal coliform count after disinfection.
   Apply Collins–Selleck equation (Equation 11.5c) for $Ct > b$.

$$N = N_0\left(\frac{Ct}{b}\right)^{-n} = 32\ \text{organisms/100 mL}\times\left(\frac{1.0\ \text{mg/L}\times 30\ \text{min}}{2.8\ \text{mg·min/L}}\right)^{-3.0} = 0.03\ \text{organisms/100 mL}$$

   The chlorinated reclaimed water has probably nondetectable fecal coliform count per 100 mL
5. Determine the chlorine dose required.
   Concentration of ammonia nitrogen in the effluent = 0.5 mg/L as N. The breakpoint chlorination is reached at stoichiometric Cl_2/NH_3-N ratio of 7.6:1 (see Example 11.28, Step 1).

   Stoichiometric dose of chlorine = 7.6 mg Cl_2/mg NH_3-N × 0.5 mg/L as NH_3-N = 3.8 mg/L as Cl_2

   The free chlorine residual is 1.0 mg/L.
   Stoichiometric chlorine dose required = (3.8 + 1.0) mg as Cl_2 = 4.8 mg/L as Cl_2
6. Determine the ammonia dosage required to convert free chlorine into combined chlorine residual.
   The stoichiometric Cl_2/NH_3-N ratio is 5:1 for monochlorine formation (see Example 11.26, Step 1).
   At the free chlorine residual of 1 mg/L,

$$\text{Stoichiometric } NH_3 \text{ dosage required} = \frac{1\,\text{mg/L as Free } Cl_2}{5\,\text{mg } Cl_2/\text{mg } NH_3\text{-N}}\times\frac{17\,\text{mg as } NH_3}{14\,\text{mg as } NH_3\text{-N}} = 0.24\,\text{mg/L as } NH_3$$

7. Determine the LAS dosage and specific feed rate (mL/m^3) required.

$$\text{LAS dosage required,} = 0.24\,\text{mg/L as } NH_3 \times \frac{132\,\text{mg as } (NH_4)_2SO_4}{2 \times 17\,\text{mg as } NH_3} \times \frac{100\,\text{mg as LAS solution}}{40\,\text{mg as } (NH_4)_2SO_4}$$

$$= 2.3\,\text{mg/L as LAS solution or } 2.3\,\text{g/m}^3\text{as LAS solution}$$

$$\text{LAS volume required,} = 2.3\,\text{g/m}^3\text{as LAS solution} \times \frac{1\,\text{mL}}{1.25\,\text{g}}.$$

$$= 1.8\,\text{mL LAS solution/m}^3\text{ reclaimed water}$$

8. Describe the quality of reclaimed water for urban reuse.

The quality of reclaimed water obtained from the above process train meets the quality guidelines for urban reuse. The U.S. EPA guidelines for reclaimed water reuse are summarized in Tables 12.1 and 12.2.

TABLE 12.1 Number of States with Regulations or Guidelines for Reuse of Reclaimed Water in Different Reuse Categories

Reuse Category	Description	No. of States with Regulations or Guidelines[a]
Urban reuse		
Unrestricted	In the areas where public access is not restricted, such as golf courses, parks, highway median, landscape, vehicle washing, toilet flush, and so on	32
Restricted	In the areas where public access is restricted by barriers, such as fencing, advisory signage, and so on	39
Agricultural irrigation		
Food crops	Irrigate crops intended for human consumption or eaten raw	27
Processed food and nonfood crops	Irrigate crops that are either processed before human consumption, including orchards and vineyards; or not consumed by humans, such as pasture land, fodder, fiber, and seed crops	42
Impoundment		
Unrestricted	Water recreational activities with no limits on body-contact, such as boating, fishing, and occasional full body contact	13
Restricted	No body-contact aesthetic landscaping	17
Environmental reuse	To create or enhance manmade and natural water bodies, including wetlands, marshes, aquatic habitats, or stream flow	17
Industrial reuse	Industrial applications and facilities, such as cooling tower makeup water, boiler feedwater, equipment washdown, oil and natural gas production, and so on	31
Groundwater recharge	Recharge to the aquifers that are not used as potable water sources	16
Potable reuse		
Indirect potable reuse (IPR)	Augment a surface or groundwater drinking water source with an environmental buffer prior to the treatment facility	9
Direct Potable Reuse (DPR)	Transfer directly into a water treatment plant with or without retention in an engineered storage buffer	0

[a] Applicable to 50 States and the District of Columbia by August 2012. The reuse categories may vary among state regulatory definitions.

Source: Adapted in part from Reference 11.

TABLE 12.2 U.S. EPA Guidelines for Reuse of Reclaimed Water in Different Reuse Categories

Reuse Category[a]	Treatment Requirement	Reclaimed Water Quality (and Monitoring) Requirement[a]						Setback Distance or Other Protection Requirement
		pH, s.u.	BOD_5, mg/L	TSS, mg/L	Turbidity[b], NTU	Fecal Coliform[c], organisms/100 mL	Min. Cl_2 Residual[d], mg/L	
Urban reuse								
Unrestricted	S, F, and D	6.0–9.0 (W)	≤10 (W)	–	≤2 (C)	Nondetectable (D)	≥1 (C)	15 m (50 ft) to PWSW
Restricted	S and D	6.0–9.0 (W)	≤30 (W)	≤30 (D)	–	≤200 (D)	≥1 (C)	90 m (300 ft) to PWSW, and 30 m (100 ft) to AATP for spray irrigation
Agricultural irrigation								
Food crops	S, F, and D	6.0–9.0 (W)	≤10 (W)	–	≤2 (C)	Nondetectable (D)	≥1 (C)	15 m (50 ft) to PWSW, and 30 m (100 ft) in area with porous media
Processed food and nonfood crops	S and D	6.0–9.0 (W)	≤30 (W)	≤30 (D)	–	≤200 (D)	≥1 (C)	90 m (300 ft) to PWSW, and 30 m (100 ft) to AATP for spray irrigation
Impoundment								
Unrestricted	S, F, and D	6.0–9.0 (W)	≤10 (W)	–	≤2 (C)	Nondetectable (D)	≥1 (C)	150 m (500 ft) to PWSW if bottom not sealed
Restricted	S and D	6.0–9.0 (W)	≤30 (W)	≤30 (D)	–	≤200 (D)	≥1 (C)	150 m (500 ft) to PWSW if bottom not sealed
Environmental reuse	Variable, or S and D	–	≤30 (W)	≤30 (D)	–	≤200 (D)	≥1 (C)	–
Industrial reuse								
Once-through cooling	S	6.0–9.0 (W)	≤30 (W)	≤30 (D)	–	≤200 (D)	≥1 (C)	90 m (300 ft) to public access area
Recirculating cooling tower	S and D, S, F, and D, or Site specific	6.0–9.0 (W)	≤30 (W)	≤30 (D)	–	≤200 (D)	≥1 (C)	90 m (300 ft) to AATP or reduced distance with high level disinfection
Others	Site specific and end use dependent			Site specific and end use dependent				Site specific and end use dependent
Groundwater recharge								
General	Site specific			Site specific and end use dependent				Site specific
Spreading	P, min			Site specific and end use dependent				Site specific
Injection	S, min			Site specific and end use dependent				Site specific

(Continued)

TABLE 12.2 (*Continued*) U.S. EPA Guidelines for Reuse of Reclaimed Water in Different Reuse Categories

Reuse Category[a]	Treatment Requirement	Reclaimed Water Quality (and Monitoring) Requirement[a]						Setback Distance or Other Protection Requirement
		pH, s.u.	BOD_5, mg/L	TSS, mg/L	Turbidity[b], NTU	Fecal Coliform[c], organisms/100 mL	Min. Cl_2 Residual[d], mg/L	
Indirect potable reuse (IPR)								
Spreading into potable aquifers[e]	S, F, D, and SAT	6.5–8.5 (D)	$\leq 2^f$ (W)	–	≤ 2 (C)	Nondetectable[g] (D)	≥ 1 (C)	Min. 2-month retention time in the ground to reach the nearest PWSW
Injection into potable aquifers[h]	S, F, D, and AWT	6.5–8.5 (D)	$\leq 2^f$ (W)	–	≤ 2 (C)	Nondetectable[g] (D)	≥ 1 (C)	Min. 2-month retention time in the ground to reach the nearest PWSW
Augment into surface water supply reservoirs[h]	S, F, D, and AWT	6.5–8.5 (D)	$\leq 2^f$ (W)	–	≤ 2 (C)	Nondetectable[g] (D)	≥ 1 (C)	Site specific

[a] Unless otherwise noted, the limits apply to the reclaimed water at the point of discharge from the treatment facility.

[b] The limit is the average turbidity based on a 24-h time period prior to disinfection. The turbidity should not exceed 5 NTU at any time.

[c] Unless otherwise noted, the limits are the median values determined from the bacteriological results of the last 7 days for which analyses have been completed.

[d] The limit is the total chlorine residual and applies only when chlorine is used as the primary disinfectant and the actual contact time is at least 30 min.

[e] Meet drinking water standards after percolation through vadose zone at least 2 m (6 ft) at maximum groundwater mounding.

[f] The limit is for TOC in mg/L.

[g] The limit is for total coliform organisms/100 mL.

[h] Meet drinking water standards at the point of injection.

Abbreviations:

Treatment Requirement

AWT = Advanced wastewater treatment processes that include chemical clarification, carbon adsorption, reverse osmosis and other membrane processes, advanced oxidation, air stripping, and ion exchange.

D = Disinfection that includes destruction, inactivation, or removal of pathogenic microorganisms by chemical, physical, or biological processes. These processes may be accomplished by chlorination, ozonation, oxidation by other chemical disinfectants, UV irradiation, or membrane processes.

F = Filtration processes that may include (a) natural undisturbed soils, (b) filter media such as sand and/or anthracite, and (c) microfiltration (MF) or other membrane processes.

P = Primary treatment process.

S = Secondary treatment process that may include activated sludge processes, trickling filters, rotating biological contractors, and stabilization pond. These processes should produce effluent concentrations

≤ 30 mg/L for both the BOD_5 and TSS.

SAT = Soil aquifer treatment.

Monitoring Requirement

C = Continuously, D = Daily, W = Weekly

Setback Distance Requirement

AATP = Area accessible to the public PWSW = Potable water supply well

Note: Meet drinking water standards at the point of discharge.

Source: Adapted in part from Reference 11.

12.2.3 Regulations or Guidelines Regarding Reuse of Reclaimed Water

There are no federal regulations governing water reclamation and reuse in the United States. Individual states have developed and implemented regulations at the state government level. U.S. EPA has developed guidelines and background information for water reuse. The first manual was published in 1980, and updated three times in 1992, 2004, and 2012.[10,11] For decades, many states have used these guidelines and other resources to develop regulations and guidelines for their states. U.S. EPA has been continuously monitoring and updating the status of the water reuse regulations and guidelines developed by these states in the updated guidelines. These publications also included the updated guidelines, technical advances, research data, and other information.

The reuse regulations developed by the individual states vary greatly. By August 2012, there were 43 states that have either regulations (29) or guidelines (14) for reuse of reclaimed water. The State of New Jersey has adopted reuse regulations as well as guidelines. Neither regulations nor guidelines have been developed in the remaining six States and the District of Columbia. The number of states that have developed regulations or guidelines for reclaimed water reuse are summarized in Table 12.1. The guidelines by the U.S. EPA include treatment levels, quality limits, monitoring frequencies, setback distances, and many other requirements for different reuse categories. Selected information from the latest guidelines is summarized in Table 12.2.[11] Readers are referred to Reference 11 to obtain the detail requirements in the guidelines.

12.3 Health Risk Analysis

Health risk analysis is often needed to determine the potential human health effects from exposure to a hazard. The exposure may occur from inhalation of air, ingestion of water or food, or absorption through the skin. Risk analysis involves (1) risk assessment, (2) risk management, and (3) risk communication.[12] The risk assessment is the qualitative or quantitative characterization of adverse effects. The risk management deals with controlling sources of exposure and risk. The risk communication involves exchange of information concerning risk assessment and management among risk assessors, risk managers, consumers, and regulators about nature, magnitude, significance, or control of a risk. In this section, the principles of toxicology and risk assessment in the context of water reuse are discussed.[13–15]

12.3.1 Quantitative Toxicology

Quantitative toxicology deals with quantifying the toxicity of hazardous chemicals in reclaimed water. The toxicity may be due to short-term exposure (*acute toxicity*) or due to long-term exposure (*chronic toxicity*).

Classification of Toxic Response: The acute toxicity occurs from a large dose that may cause an immediate reaction or sickness. The chronic toxicity has delayed reaction. A common chronic exposure to hazardous chemicals may occur over a lifetime from drinking or exposure from water contaminated with trace levels of hazardous chemicals. In quantifying toxic response, the terms *dosage* and *dose* are commonly used. The *dosage* is defined as the total mass of chemical which an organism is exposed. The *dose* is the chemical dosage normalized to body weight (mg/kg), skin surface (mg/cm^2), or air inhalation over a given period (mg/m^3 over 24 h). An example of conversion from dose to dosage is presented in Example 12.2.

Dose–Response Relationship: Toxicity is generally determined by administering the substance to a group of animals to study its effects. The objective is to select the organism that is (1) a good model of humans, (2) a response that is not subjective and can be consistently determined, and (3) a test period that is relatively short. A group of test animals is exposed to a measured substance under controlled condition, and the response (death, deformities, reproduction, etc.) is monitored. The end results are dose–response relationships that feature (1) an S-shape curve; (2) does not go through zero (this means effects of low dose may be unknown); (3) the effective dose (ED_{50}) that causes mild usually transitory response to

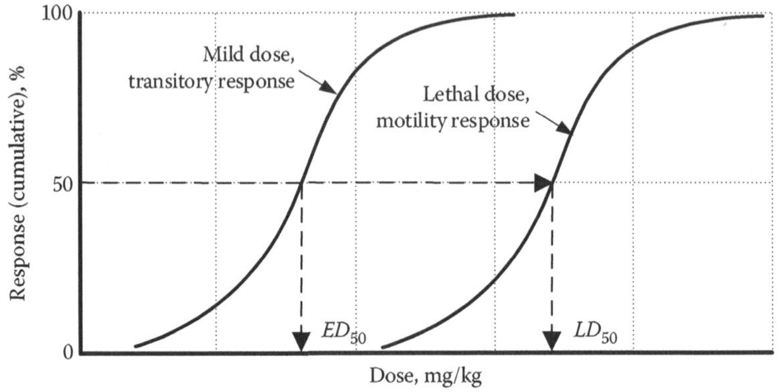

FIGURE 12.2 Dose–response curves showing ED_{50} and LD_{50}.

50% population; (4) the lethal dose (LD_{50}) that results in 50% mortality (death) of a population in a specified time* (it is a median lethal dose); and (5) the ratio of LD_{50}/ED_{50} is an important factor that gives an indication of the fractional lethal and nonharmful dose (see Example 12.4 for details). The typical dose–response curves are shown in Figure 12.2. The toxicity rating chart is given in Table 12.3.

Specific Contaminants and Mechanism of Toxic Action: The mechanism of toxic action for most hazardous chemicals occurs due to binding to receptor molecules on a cell, tissue, or organ. The most common receptors are enzymes, hemoglobin, membranes, deoxyribonucleic acid (DNA), and biological molecules of target organ or tissue. Some examples are *heptotoxicity* (liver), nephrotoxicity (kidney), genotoxicity (the chromosomes), and immunotoxicity (the immune system). The key mechanism for toxicity is where a toxic chemical locks into a specific receptor biomolecule.[11] Many specific hazardous chemicals of concern can be divided into the following three most important groups: *carcinogens*, *mutagens*, and *teratogens*.

Carcinogenicity: Cancer is a disease in which altered cells divide uncontrollably called *neoplastic growth*. Most chemically induced cancers have a *latency period* of up to 40 years. This may complicate the cause-and-effect relationships of carcinogenicity. It is suggested that mechanisms of carcinogenicity is a three-step process: (1) *Initiation*, which may be caused by genetic damage through a *mutation* to DNA; (2) *promotion*, in which the genetic damage that occurred previously goes through multiplication; and (3) *progression*, where the cancer spreads through uncontrolled growth. In 1986, the U.S. EPA published the first carcinogen risk assessment guidelines that categorize hazardous chemicals into several carcinogenic groups.[16] Examples of common chemicals in each of these groups are also listed in Table 12.4.

TABLE 12.3 Toxicity Rating Chart

Toxicity Rating or Class	Oral Acute LD_{50} for Rats, mg/kg	Examples of Toxic Chemicals
Extremely toxic	1 or less	Dioxin, botulinum toxin
Highly toxic	1–50	Strychnine
Moderately toxic	50–500	Dichlorodiphenyltrichloroethane (DDT)
Slightly toxic	500–5000	Morphine
Practically nontoxic	5000–15,000	Ethyl alcohol

Source: Adapted in part from References 13 and 14.

* *LD* is lethal dose by any route other than inhalation, and *LC* is lethal concentration of a substance in the air.

TABLE 12.4 U.S. EPA Categories for Carcinogenic Groups (1986)

Group	Description	Examples of Chemicals
A	Human carcinogen: sufficient evidence in humans	Arsenic, asbestos, benzene, chromium (VI)
B_1	Probable human carcinogen: limited human data available	Cadmium, formaldehyde
B_2	Probable human carcinogen: sufficient evidence in animals, and inadequate or no evidence in humans	Carbon tetrachloride, chlordane
C	Possible human carcinogen: limited evidence in animals in the absence of human data	Dibromochloromethane, 1, 1-Dichloroethane
D	Not classifiable as to human carcinogenicity: inadequate human and animal evidence or no data available	Acetone, n-Butanol, methyl ethyl ketone (MEK)
E	Evidence of noncarcinogenicity for humans	–

Source: Adapted in part from References 7, 13, and 16.

Since 2005, U.S. EPA has applied the new guidelines that characterize human carcinogenic potentials of hazardous chemicals more precisely based on more complete knowledge about the cancer assessment.[17] The new categories suggested by the 2005 guidelines are:

- Carcinogenic to humans
- Likely to be carcinogenic to humans
- Suggestive evidence of carcinogenic potential
- Inadequate information to assess carcinogenic potential
- Not likely to be carcinogenic to humans

The dose–response assessments of carcinogens under both guidelines are considered valid.

Mutagenicity: Mutation occurs within DNA, which causes a change in the genetic code. Mutations may occur naturally. Normally, organisms have special enzymes called repairases that can repair DNA. However, not all mutations are repairable, and some unrepaired mutations may cause the cell's metabolism to go out of control resulting in cancer. Carcinogenic chemicals are absorbed, bio activated, metabolized, and excreted like other noncarcinogenic chemicals. However, they are different because they cause cumulative and delayed effects.

Teratogenicity: Teratogenic chemicals cause toxicity to the fetus during pregnancy. This may cause developmental malformations from fertilization until birth resulting in deformities.

Discussions on priority pollutants, toxicity and biomonitoring, toxicity tests, and toxicity evaluation and reduction have been presented in Sections 5.6 and 5.7. Examples 5.51 through 5.56 cover a broad range of topics in inorganic and organic pollutants, lethal dose, effluent toxicity, and probit analysis. The following examples focus on the dose–response relationships.

EXAMPLE 12.2: DOSE AND DOSAGE RELATIONSHIP

A dose of 10 mg/kg was administered to a 3.5-g (7.7-lb) animal. Determine the dosage.

Solution

The dosage is the total mass of chemical that an organism is exposed. The dose is dosage per unit body measurement (weight, surface area, etc.). In this problem, dose is 10 mg chemical per unit weight of the animal. Therefore, it is weight specific.

$$\text{Dosage} = \text{Dose} \times \text{Body weight} = 10\,\text{mg/kg} \times 3.5\,\text{kg} = 35\,\text{kg}$$

EXAMPLE 12.3: LD_{50} OF A PESTICIDE ON MICE

A pesticide toxicity study was conducted on the mortality of mice. The average weight of mice is 25 g and 20 animals were tested in each group. The exposure period is 4 h. Determine LD_{50} for mice and that of humans. The safety factor is 10 and average weight of humans is 70 kg. The experimental data is tabulated below.

Dosage (amount injected), μg	No. of Mice That Died after 4-h Exposure	Mortality Fraction	Survival, %	log, Dosage
0	0	0	100	–
10	0	0	100	1.00
20	1	0.05	95	1.30
25	2	0.10	90	1.40
30	4	0.20	80	1.48
35	7	0.35	65	1.54
40	13	0.65	35	1.60
45	18	0.90	10	1.65
55	20	1.00	0	1.74

Solution

1. Determine $LD_{50,\ mice}$ for mice using arithmetic plot.

   The LD_{50} for mice can be assessed by two different graphical methods: arithmetic plot and semi-log plot methods. The procedures to determine the value of LD_{50} for mice from both methods are presented below.

   a. Arithmetic plot method.

      Prepare a plot between fractional mortality and dosage of pesticide in Figure 12.3a.

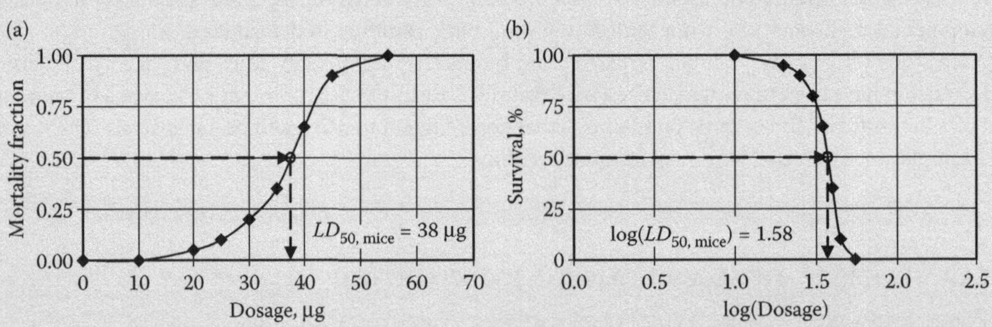

FIGURE 12.3 Dose–response curves for determining pesticide toxicity to mice: (a) arithmetic plot and (b) semi-log plot (Example 12.3).

      Draw a horizontal line from fractional mortality of 0.5, and read the corresponding value of LD_{50} on the curve. $LD_{50,\ mice} = 38$ μg pesticide for mice.

   b. Semi-log plot method.

      Prepare a semi-log plot between percent survival and log(Dosage) in Figure 12.2b.

Draw a horizontal line from 50% survival and read the corresponding $\log(LD_{50,\text{ mice}})$, and read $\log(LD_{50,\text{ mice}}) = 1.58$.

$LD_{50} = 10^{1.58} = 38$ μg pesticide for mice

c. Select the value of LD_{50} for mice.

An identical value of $LD_{50,\text{ mice}} = 38$ μg is obtained for pesticide exposure to mice from two different graphical methods.

2. Determine the lethal dose for mice ($Dose_{\text{mice}}$).

$$Dose_{\text{mice}} = \frac{LD_{50,\text{ mice}}}{Weight_{\text{mice}}} = \frac{38\ \mu\text{g}}{25\ \text{g}} \times \frac{10^3\ \text{g}}{\text{kg}} \times \frac{\text{mg}}{10^3\ \mu\text{g}} = 1.5\ \text{mg/kg}$$

3. Determine the equivalent lethal dosage for humans ($LD_{50,\text{ humans}}$) using a factor of safety, $SF = 10$.

$$LD_{50,\text{ humans}} = \frac{Dose_{\text{mice}} \times Weight_{\text{humans}}}{SF} = \frac{1.5\ \text{mg/kg} \times 70\ \text{kg}}{10} = 10.5\ \text{mg pesticide for humans}$$

EXAMPLE 12.4: ED_{50} AND LD_{50} OF A PESTICIDE ON MICE

A toxicity study was conducted on mice to obtain ED_{50} and LD_{50}. The exposure period is 48 h. The response for ED_{50} is sluggish movement or loss in agility in 48-h exposure, while the response for LD_{50} is death. The test data is summarized below. Determine the ED_{50} and LD_{50}, and LD_{50} ratio of the pesticide. Comment on the significance of the LD_{50}/ED_{50} ratio.

Determination of ED_{50}		Determination of LD_{50}	
Dosage (amount administered), μg	Fraction of mice affected	Dosage (amount injected), μg	Fraction of mice died
0	0	18	0
1	0	20	0
2	0.05	22	0.05
3	0.10	24	0.15
4	0.30	25	0.30
5	0.70	26	0.70
6	0.95	27	0.90
7	1.00	28	0.95

Solution

1. Plot the experimental data to obtain dose–response curves.

The dose–response curves for the effective dose and lethal dose are plotted in Figure 12.4.

2. Draw a horizontal line from 0.5 fraction of population affected by ingestion of pesticide.

The response for ED is sluggish movement or loss in agility. $ED_{50} = 4.5$ μg and $LD_{50} = 25.5$ μg.

3. Determine the LD_{50}/ED_{50} ratio.

$$LD_{50}/ED_{50}\ \text{ratio} = \frac{LD_{50}}{ED_{50}} = \frac{25.5\ \mu\text{g}}{4.5\ \mu\text{g}} = 5.7$$

4. Comment on the LD_{50}/ED_{50} ratio.

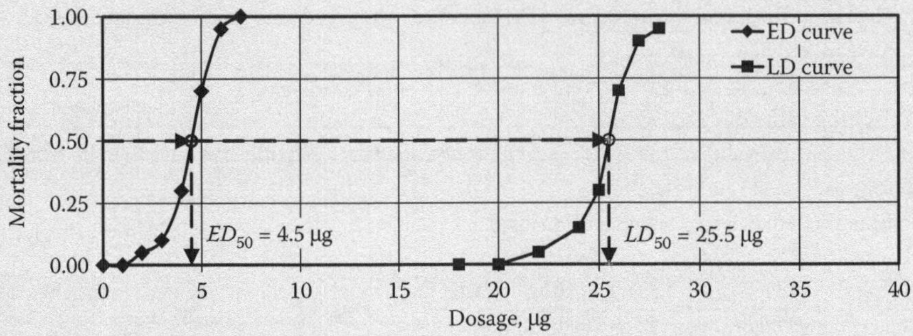

FIGURE 12.4 Dose-response curves for determining ED_{50} and LD_{50} (Example 12.4).

The LD_{50}/ED_{50} ratio represents the ratio of lethal dose and nonharmful dose. It can be used as an indicator for safety. A ratio of 5.7 provides sufficient safety warning before the lethal dose is administered. When the LD_{50}/ED_{50} ratio is small, the lethal dose may be administered before a sufficient distress warning is received.

12.3.2 Risk Assessment

Risk assessment is the quantitative or qualitative characterization and estimation of potential adverse health effects associated with exposure of individuals or population to hazardous materials and situations. The risk assessment is carried out in four discrete steps: (1) hazard assessment (or identification), (2) dose–response assessment, (3) exposure assessment, and (4) risk characterization. Each step is described below.[6–8,10,11]

Hazard Assessment: The purpose of hazard assessment is to review and evaluate whether an agent may pose carcinogenic hazards to humans and how it may be expressed. Hazard assessment involves source analysis, that is, chemicals present at the site or facility and their characteristics. Useful information needed is identity, nomenclature, concentration, and properties of all contaminants at the site or facility. Other specific tasks for hazard identification are sampling, monitoring wells, analytical work, and data analysis. Additionally, evidence is gathered on adverse health effects on humans and adverse environmental effects.

Dose–Response Assessment: The purpose of dose–response assessment is to estimate potential risks to humans at exposure levels of interest by establishing a relationship between the dose of a toxic agent and the adverse effect in humans. Such relationship requires extrapolation from animal test results to human population. To predict such dose–response relationships the data are fitted to a model that gives the probability of infection to the mean dose ingested. The typical dose–response models applied to human exposure are: (1) single-hit, (2) multistage, (3) linear multistage, (4) multihit, (5) beta-Poison, and (6) probit model. A discussion on these models and illustrative examples may be found in References 6–8 and 12–17.

Exposure Assessment: Exposure assessment is the determination of the magnitude, frequency, and duration of exposure. Exposure is contact of a subject with the hazard. Exposure may occur through different pathways such as inhalation, ingestion, or dermal. Steps involved with exposure assessment are identification of potential receptor population, evaluation of exposure pathways and routes, and quantification of exposures.

Risk Characterization: The risk characterization is to obtain quantitative estimates of both noncarcinogenic and carcinogenic risks to a *maximally exposed individual* (MEI) from all exposure routes. The quantitative risk assessments for noncarcinogenic and carcinogenic effects are different. The basis for the different approaches is that noncarcinogenic effects occur after exposure to the threshold dose but cancer initiation is not normally considered to have a threshold to cause a response. The typical dose–response relationships for noncarcinogenic and carcinogenic effects are shown in Figure 12.5.

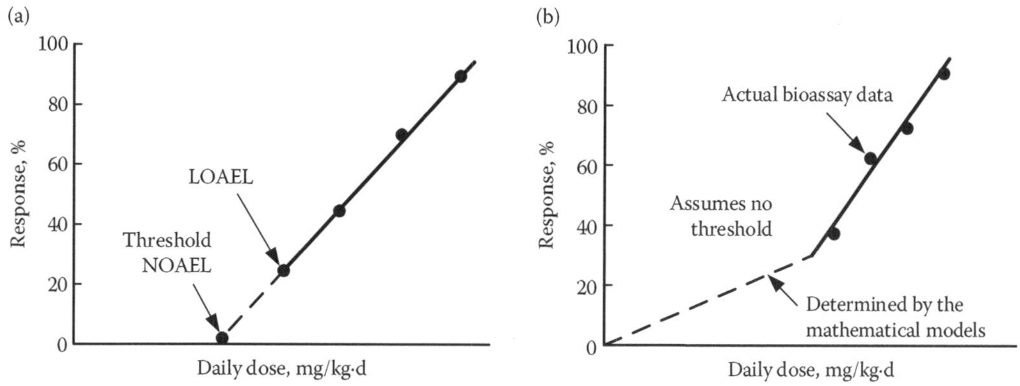

FIGURE 12.5 Dose–response relationships: (a) noncarcinogenic effects and (b) carcinogenic effects.

12.3.3 Assessment of Noncarcinogenic Effects

In a noncarcinogenic effect, the dose–response relationship is characterized by a *threshold value*, which is the concentration below which there is no measurable effect. The *no observed adverse effect level (NOAEL)* is the highest dose tested on animals that did not produce an effect. When a *NOAEL* is not available, a *lowest observed adverse effect level* (*LOAEL*) is obtained. It is the lowest dose tested that has an adverse effect.

Acceptable Daily Intake: It is the dose of toxic compound ingested daily that does not produce any adverse health effect upon test animals. *Acceptable daily intake* (*ADI*) is based on *NOAEL* and is lower than the threshold. It is expressed by Equation 12.1. A dose response for noncarcinogenic effects is illustrated in Figure 12.5a, which also identifies the threshold, NOAEL and *LOAEL*. Uncertainty factors of 10 are used to account for the confidence level of the data. If a well-designed and well-conducted study is the basis for the risk assessment, a lower safety factor may be needed. The safety factors for risk assessment of noncarcinogenic effects are given in Table 12.5.

$$ADI = \frac{NOAEL}{UF} \tag{12.1}$$

where

ADI = acceptable daily intake, mg/kg·d
$NOAEL$ = no observed adverse effect level, mg/kg·d
UF = uncertainty factor, dimensionless

TABLE 12.5 Values of Uncertainty and Modifying Factors

Uncertainty Factor (UF_1, UF_2, …)	Value
Extrapolation from animals to humans	10
Variation in the general population to protect most sensitive subpopulation (e.g., children and elderly)	10
Extrapolation from *LOAEL* to *NOAEL*	10
Extrapolation from subchronic to chronic	10
Reduction for short duration of study	10
Reduction for quality of epidemiological study	10
Modifying factor	0.1–10

Reference Dose: It is a contemporary surrogate used by the U.S. EPA instead of *ADI*. The *Reference Dose (RfD)* is developed using stricter procedures than that used for *ADI*, resulting in a lower value of acceptable intake. The standard procedure is to apply a series of uncertainty factors (UF_1, UF_2, ...) of 10 to account for variations in general population, animals to human, *LOAEL* to *NOAEL*, subchronic to chronic, uncertainties in data collection, and others. The U.S. EPA also applies a *modifying factor* (*MF*) which ranges from 1 to 10, to reflect a quantitative professional judgment of uncertainties. The *RfD* is derived from Equation 12.1 and expressed by Equation 12.2. The uncertainty and modifying factors are also given in Table 12.5.

$$RfD = \frac{NOAEL}{(UF_1 \times UF_2 \times \cdots) \times MF} \tag{12.2a}$$

$$RfD = \frac{LOAEL}{(UF_1 \times UF_2 \times \cdots) \times MF} \tag{12.2b}$$

where

RfD = reference dose, mg/kg·d
LOAEL = lowest observed adverse effect level, mg/kg·d
UF_1, UF_2, ... = uncertainty factors, dimensionless (mostly 10)
MF = modifying factor, dimensionless (1–10)

Exposure Assessment: The exposure assessment is the estimation of magnitude, frequency, duration, and route of exposure. The steps involved are: (1) identify potential population exposed, (2) identify potential exposure pathways, (3) estimate exposure concentration, and (4) estimate chemical intake. Sampling and analysis, and modeling effort may be needed to develop the data. The exposure data is normalized for contact rate and time, exposure concentration frequency, duration, and body weight. Four expressions for chemical intake in common use are Equations 12.3a through 12.3d.

$$I = \frac{C \times CR \times FD}{BW \times AT} \quad \text{or} \quad I = \frac{C \times CR \times EF \times ED}{BW \times AT} \tag{12.3a}$$

where

I = intake or amount of chemical at the exchange boundary, mg/kg·d
C = the average exposure concentration over the period, mg/L for water; mg/m^3 for air; and mg/kg for soil, food, or other ingestion sources
CR = contact rate, the amount of contaminated medium such as water, air, food, and soil contacted per unit time, L/d, m^3/d or kg/d
EFD = exposure frequency and duration (how often and how long). The *EFD* is normally divided into two terms, *EF* and *ED*
EF = exposure frequency, d/year. The maximum exposure is 350 d/year due to two week of travel and vacation schedule
ED = exposure duration, year
BW = the average body weight over the exposure period, kg
AT = average time period over which the exposure is averaged, d

For noncarcinogens, $AT = ED \times 365$ d/year, and intake is called *Chronic Daily Intake (CDI).*

For carcinogens, AT = lifetime exposure = 70 years × 365 d/year, and intake is called *Lifetime Average Daily Exposure (LADE).*

$$I_w = \frac{C_w \times IR_w \times EF \times ED}{BW \times AT} \tag{12.3b}$$

$$I_a = \frac{C_a \times IR_a \times EF \times ED}{BW \times AT} \tag{12.3c}$$

$$I_s = \frac{C_s \times IR_s \times CF \times FI \times EF \times ED}{BW \times AT} \tag{12.3d}$$

where

I_w, I_a, and I_s = intake by ingestion of water, inhalation of vapors or air, and ingestion of solids, mg/kg·d

C_w, C_a, and C_s = concentration of chemical(s) of concern, mg/L in water, mg/m^3 in vapor or air, and mg/kg in solids

IR_w = ingestion rate of water, L/d

IR_a = inhalation rate of air or vapor, m^3/d

IR_s = ingestion rate of solids, mg/d

CF = conversion factor = 10^{-6}, kg/mg

FI = fraction of ingested solids coming from the impacted source, dimensionless

All other terms have been defined previously.

The numerical values used in Equations 12.3a through 12.3d are provided in Table 12.6.

Hazard Index: The noncarcinogenic risk is characterized in terms of a *hazard index* (*HI*). An exposure typically involves multiple chemicals, and the index must be calculated for each chemical of concern

TABLE 12.6 U.S. EPA Maximum Exposure Assumptions

Parameter	Definition	Unit	Assumption	
			Adult	Child
BW	Body weight	kg	70	15
AT	Averaging time			
	For carcinogen	d	25,550[a]	–
	For non carcinogen	d	ED × 365/year	–
IR_w	Resident drinking water ingestion	L/d	2	1.4
IR_a	Air or vapor breathed	m^3/d	20	5
IR_s	Soil ingestion			
	Lifetime as resident	mg/d	100	200 during 1–6 years
	Occupational	mg/d	50	100 < 6 years
	Food consumption			
	Fish	g/d	54	–
	Beef	kg/meal	0.112	–
	Eggs	kg/meal	0.064	–
EF	Residential exposure frequency	d/year	350	350
ED	Maximum residential exposure duration	year	30	6
FI	Fraction of ingested soil coming from impacted source	dimensionless	1.0	1.0

[a] 70 years × 365 d/year.

Source: Adapted in part from References 16 and 18.

for all pathways and exposure routes. The index scores for all organ-specific noncarcinogens are added to obtain the total measure of risk to individual organs such as the liver, kidney, brain, and others. If the hazard index to each organ is <1.0, then the exposure level to the individual organ is acceptable even though the *HI* for all chemicals may exceed 1.0. These relationships are given by Equations 12.4a and 12.4b.

$$HQ = \frac{CDI}{RfD} \tag{12.4a}$$

$$HI = HQ_1 + HQ_2 + HQ_3 + \ldots \tag{12.4b}$$

where

HQ = hazard quotient, dimensionless. HQ_1, HQ_2, ... are for individual chemicals.
HI = hazard index, dimensionless. $HI = HQ$ if only one chemical is involved.
CDI = chronic daily intake, mg/kg·d. The notations I_w, I_a, and I_s in Equations 12.3b through 12.3d are changed to $I_{w,CDI}$, $I_{a,CDI}$, and $I_{s,CDI}$ for hazardous index.

12.3.4 Assessment of Carcinogenic Effects

Mathematical models are used to extrapolate from high dose to the lower dose for cancer assessment. The low-dose risk extrapolation assumes no threshold effect (see Figure 12.5b). Carcinogenic risk is defined as the lifetime average daily intake dose (developed in the exposure assessment) multiplied by the *carcinogenic slope factor* (*CSF*, determined by toxicity assessment). The product is the probability of lifetime cancer risk from exposure. Cancer risk is expressed by Equation 12.5.

$$\text{Cancer Risk} = LADE \times CSF \tag{12.5}$$

where

$LADE$ = lifetime average daily exposure, mg/kg·d. The notations I_w, I_a, and I_s in Equations 12.3b through 12.3d are changed to $I_{w,LADE}$, $I_{a,LADE}$, and $I_{s,LADE}$ for cancer risk.
CSF = carcinogen slope factor, kg·d/mg

CSF is also called *cancer potency factor* (*CPF*), and it is generally the 95% upper confidence limit of the slope of the dose–response curve at very low exposure. The slope factors and *RfD* for several carcinogens for oral route of exposure are summarized in Table 12.7.

TABLE 12.7 Carcinogenic Slope Factor (CSF) or Cancer Potency Factor (CPF) for Oral Route of Carcinogenic Chemicals Often Found in Drinking Water

Common Carcinogen	*CSF* or *CPF*, kg·d/mg	*RfD*, mg/kg·d
Arsenic (inorganic)	1.5	3.1×10^{-4}
Benzene	2.9×10^{-2}	4.1×10^{-3}
Bis(2-ethylhexyl) phthalate (BEHP)	1.4×10^{-2}	2.0×10^{-2}
Bromoform	7.9×10^{-3}	2.0×10^{-2}
Chloroform	6.1×10^{-3}	1.0×10^{-2}
Dieldrin	1.6×10	5.0×10^{-5}
Heptachlor	4.5	5.0×10^{-4}
N-nitrosodiethylamine (NDEA)	1.2×10^{2}	–
N-nitrosodimethylamine (NDMA)	5.1×10	–
Vinyl chloride	7.2×10^{-1}	–

Source: Adapted in part from References 9, 12, 13, and 19.

The U.S. EPA based on extensive modeling and data collection efforts has developed and maintains an extensive database on toxicological indices. It is known as the *Integrated Risk Information System* (IRIS). The IRIS database contains both qualitative and quantitative data regarding noncarcinogens as well as carcinogens. The database is updated continuously and is used extensively. It is available online at http://www.epa.gov/iris.19.

EXAMPLE 12.5: RECEPTOR, PATHWAY, AND LIFETIME AVERAGE DAILY EXPOSURE (LADE)

An adult drinks groundwater containing 60 μg/L trichloroethylene (TEC). Adult weighs 70 kg, and water consumption is 2 L/d over a lifetime of 70 years. Identify receptor, pathway, and exposure. Estimate the lifetime average daily exposure (LADE).

Solution

The receptor is an adult with $BW = 70$ kg; the pathway is drinking groundwater; the exposure is due to water containing TEC concentration $C_w = 60$ μg/L; the exposure quantification $IR_w = 2$ L/d; the exposure frequency $EF = 350$ d/year; exposure duration $ED =$ lifetime $= 70$ years; $AT =$ lifetime exposure $= 70$ years × 365 d/year = 25,550 d for a carcinogen. Estimate the *LADE* from Equation 12.3b.

$$I_{w,LADE} = \frac{C_w \times IR_w \times EF \times ED}{BW \times AT} = \frac{60\ \mu g/L \times 2\ L/d \times 350\ d/year \times 70\ years}{70\ kg \times 25{,}550\ d} = 1.64\ \mu g/kg{\cdot}d$$

EXAMPLE 12.6: DETERMINE RFD FROM HIGH QUALITY EPIDEMIOLOGIC STUDY

In a high quality epidemiologic study, $NOAEL = 50$ mg/kg·d. Calculate *RfD*.

Solution

Since the *NOAEL* is obtained from a high quality epidemiologic study, the result applies to humans. Only one uncertainty factor ($UF_1 = 10$) for variation within a human population will be applied while the modifying factor $MF = 1$. Determine the value of *RfD* from Equation 12.2a.

$$RfD = \frac{NOAEL}{UF_1 \times MF} = \frac{50\ mg/kg{\cdot}d}{10 \times 1} = 5\ mg/kg{\cdot}d$$

EXAMPLE 12.7: RFD FROM SHORT-TERM DATA BASED ON ANIMAL STUDY

A 90-day critical study with rats was conducted. The *LOAEL* was 100 mg/kg·d. Determine *RfD*.

Solution

This is a short-term animal study. The extrapolation is needed from using *LOAEL* instead of *NOAEL* data. Therefore, the following uncertainty factors will be used:

$UF_1 = 10$ as the uncertainty factor for animal data,
$UF_2 = 10$ as the uncertainty factor for the limited duration,
$UF_3 = 10$ as the uncertainty factor for human variability,
$UF_4 = 10$ as the uncertainty factor for *LOAEL* instead of *NOAEL*, and
$MF = 10$ as modifying factor for low quality data.

Determine the value of *RfD* from Equation 12.2b.

$$RfD = \frac{LOAEL}{UF_1 \times UF_2 \times UF_3 \times UF_4 \times MF} = \frac{100\ \text{mg/kg·d}}{10 \times 10 \times 10 \times 10 \times 10} = 0.001\ \text{mg/kg.d} = 1\ \mu\text{g/kg·d}$$

EXAMPLE 12.8: RFD FROM A SUBCHRONIC ORAL TOXICITY STUDY ON MICE

A subchronic oral toxicity study was conducted with mice. A *LOAEL* of 50 mg/kg·d was obtained for a pesticide. The quality of data was given a high rating by the experts. Determine *RfD*.

Solution

This is a highly rated study. The uncertainty factors are: (1) $UF_1 = 10$ for extrapolation from animals to humans, (2) $UF_2 = 10$ for extrapolation from subchronic to chronic, (3) $UF_3 = 10$ for extrapolation from *LOAEL* to *NOAEL*, and (4) $UF_4 = 10$ for variation within human population. The value of $MF = 1$ is used for high quality data.

Determine the value of *RfD* from Equation 12.2b.

$$RfD = \frac{LOAEL}{UF_1 \times UF_2 \times UF_3 \times UF_4 \times MF} = \frac{50\ \text{mg/kg·d}}{10 \times 10 \times 10 \times 10 \times 1} = 0.005\ \text{mg/kg·d} = 5\ \mu\text{g/kg·d}$$

EXAMPLE 12.9: ADULT EXPOSURE DOSE FROM DRINKING WATER

Groundwater is supplied as a residential water supply source to a community. Estimate the daily exposure dose to an adult exposed to two chemicals: (a) 0.05 mg/L benzene and (b) 0.5 mg/L toluene. Use the maximum exposure assumptions given in Table 12.6.

Solution

1. Identify the equation for estimating the daily exposure dose from drinking water source.

   Apply Equation 12.3b, $I_w = \dfrac{C_w \times IR_w \times EF \times ED}{BW \times AT}$

   The maximum exposure duration *ED* for adult is 30 years from Table 12.6.

2. Determine the intake of benzene as a carcinogen.

   For a carcinogen, *AT* for lifetime exposure = 70 years × 365 d/year = 25,550 d. Estimate the intake that is also called Lifetime Average Daily Exposure (*LADE*) for the exposure duration *ED* = 30 years.

$$I_{w,LADE} = \frac{0.05\ \text{mg/L} \times 2\ \text{L/d} \times 350\ \text{d/year} \times 30\ \text{years}}{70\ \text{kg} \times 25{,}550\ \text{d}} = 5.87 \times 10^{-4}\ \text{mg/kg·d}$$

3. Determine the intake of toluene as a noncarcinogen.

   For a noncarcinogen, *AT* = *ED* × 365 d/year = 30 years × 365 d/year = 10,950 d. Estimate the intake that is also called Chronic Daily Intake (CDI) for the exposure duration *ED* = 30 years.

$$I_{w,CDI} = \frac{0.5\ \text{mg/L} \times 2\ \text{L/d} \times 350\ \text{d/year} \times 30\ \text{years}}{70\ \text{kg} \times 10{,}950\ \text{d}} = 1.37 \times 10^{-2}\ \text{mg/kg·d}$$

EXAMPLE 12.10: CHILD DOSE FROM DRINKING WATER

Estimate the child's dose from drinking water for two cases: (a) 0.05 mg/L benzene (b) 0.10 mg/L toluene. Use the maximum exposure assumptions given in Table 12.6.

Solution

1. Determine the intake of benzene.

   The body weight $BW = 15$ kg and exposure duration $ED = 6$ years for a child are obtained from Table 12.6.

   For a carcinogen, $AT =$ lifetime exposure $= 70$ years $\times$ 365 d/year $=$ 25,550 d. Calculate the LADE from Equation 12.3b.

$$I_{\mathrm{w,LADE}} = \frac{0.05\ \mathrm{mg/L} \times 1.4\ \mathrm{L/d} \times 350\ \mathrm{d/year} \times 6\ \mathrm{years}}{15\ \mathrm{kg} \times 25{,}550\ \mathrm{d}} = 3.84 \times 10^{-4}\ \mathrm{mg/kg{\cdot}d}$$

2. Determine the intake of toluene.

   For a noncarcinogen, $AT = ED \times 365$ d/years $=$ 6 years $\times$ 365 d/year $=$ 2190 d. The *CDI* is calculated from Equation 12.3b.

$$I_{\mathrm{w,CDI}} = \frac{0.10\ \mathrm{mg/L} \times 1.4\ \mathrm{L/d} \times 350\ \mathrm{d/year} \times 6\ \mathrm{years}}{15\ \mathrm{kg} \times 2190\ \mathrm{d}} = 8.95 \times 10^{-3}\ \mathrm{mg/kg{\cdot}d}$$

EXAMPLE 12.11: ADULT EXPOSURE DOSE FROM INHALATION

An adult breathes outdoor air for 25 years. Calculate the daily exposure dose for two cases. The outdoor air contains (a) 0.01 mg/m^3 benzene and (b) 1.0 mg/m^3 toluene.

Solution

Calculate the average daily intake from Equation 12.3c,

$$I_{\mathrm{a}} = \frac{C_{\mathrm{a}} \times IR_{\mathrm{a}} \times EF \times ED}{BW \times AT}.$$

Obtain $IR_{\mathrm{a}} = 20\ \mathrm{m^3/d}$ for adult from Table 12.6. The estimated daily breathing doses for adult at $ED = 25$ years in two cases are:

$$I_{\mathrm{a,LADE}} = \frac{0.01\ \mathrm{mg/m^3} \times 20\ \mathrm{m^3} \times 350\ \mathrm{d/year} \times 25\ \mathrm{years}}{70\ \mathrm{kg} \times 25{,}550\ \mathrm{d}} = 9.78 \times 10^{-4}\ \mathrm{mg/kg{\cdot}d}\ \text{for benzene}$$

$$I_{\mathrm{a,CDI}} = \frac{1.0\ \mathrm{mg/m^3} \times 20\ \mathrm{m^3} \times 350\ \mathrm{d/year} \times 25\ \mathrm{years}}{70\ \mathrm{kg} \times 25\ \mathrm{years} \times 365\ \mathrm{d/year}} = 0.27\ \mathrm{mg/kg{\cdot}d}\ \text{for toluene}$$

EXAMPLE 12.12: CHILD EXPOSURE DOSE FROM SOIL INGESTION

A child ingests soil containing 1 mg/kg of benzene and 10 mg/kg of toluene in a residential situation. Estimate the dose of each chemical using appropriate intake equations and intake parameters.

Solution

1. Identify the equation for estimating the average daily intake from ingestion of soil.

Apply Equation 12.3d, $I_s = \dfrac{C_s \times IR_s \times CF \times FI \times EF \times ED}{BW \times AT}$

For a child, the exposure duration $ED = 6$ years, $IR_s = 200$ mg/d during 6 years, and $FI = 1.0$ from Table 12.6.

2. Determine the average daily intake of benzene as a carcinogen.

   For a carcinogen, $AT =$ lifetime exposure = 70 years × 365 d/year = 25,550 d. The intake as $I_{s,\text{LADE}}$ for exposure duration $ED = 6$ years at $C_s = 1$ mg/kg,

$$I_{s,\text{LADE}} = \frac{1\,\text{mg/kg} \times 200\,\text{mg/d} \times 10^{-6}\,\text{kg/mg} \times 1.0 \times 350\,\text{d/year} \times 6\,\text{years}}{15\,\text{kg} \times 25{,}550\,\text{d}} = 1.10 \times 10^{-6}\,\text{mg/kg·d}$$

3. Determine the average daily intake of toluene as a noncarcinogen.

   For a noncarcinogen, $AT = ED \times 365$ d/year = 6 years × 365 d/year = 2190 d. Estimate the intake as CDI for the exposure duration $ED = 6$ year at $C_s = 10$ mg/kg of toluene.

$$I_{s,\text{CDI}} = \frac{10\,\text{mg/kg} \times 200\,\text{mg/d} \times 10^{-6}\,\text{kg/mg} \times 1.0 \times 350\,\text{d/year} \times 6\,\text{years}}{15\,\text{kg} \times 2190\,\text{d}} = 1.28 \times 10^{-4}\,\text{mg/kg·d}$$

EXAMPLE 12.13: ADULT EXPOSURE DOSE FROM CONTAMINATED RECREATIONAL FISH

An adult consumes 120 g of recreational fish per meal for an average four meals per year over 30 years. Estimate the contaminant dose in the lifetime if the fish contains 2×10^{-6} mg dioxin/kg fish.

Solution

1. Determine the annual average daily toxin ingestion from consuming fish at $EF = 350$ d/year.

$$IR_s = \frac{120\,\text{g/meal} \times 4\,\text{meals/year}}{350\,\text{d/year}} = 1.37\,\text{g/d} \quad \text{or} \quad 1370\,\text{mg/d}$$

2. Determine the average daily intake of dioxin in the lifetime from Equation 12.3d for the exposure duration $ED = 30$ years at $C_s = 2 \times 10^{-6}$ mg dioxin/kg fish and $FI = 1.0$.

$$I_{s,\text{LADE}} = \frac{2 \times 10^{-6}\,\text{mg/kg} \times 1370\,\text{mg/d} \times 10^{-6}\,\text{kg/mg} \times 1.0 \times 350\,\text{d/year} \times 30\,\text{years}}{70\,\text{kg} \times 25{,}550\,\text{d}}$$
$$= 1.61 \times 10^{-11}\,\text{mg/kg·d}$$

EXAMPLE 12.14: RISK CHARACTERIZATION OF A NONCARCINOGENIC HAZARDOUS CHEMICAL

A drinking water supply is contaminated with 85 μg/L of *methyl ethyl ketone* (MEK). Determine the hazard index for exposure to an adult over 30 years. The *RfD* for oral exposure of $MEK =$ 0.6 mg/kg·d.

Solution

1. Determine from Equation 12.3b the oral *CDI* for MEK as a noncarcinogen for $ED = 30$ years.

$$I_{w,CDI} = \frac{85\ \mu g/L \times 10^{-3}\ mg/\mu g \times 2\ L/d \times 350\ d/year \times 30\ years}{70\ kg \times 30\ years \times 365\ d/years} = 2.33 \times 10^{-3}\ mg/k{\cdot}d$$

2. Determine the hazardous index (*HI*).

   Determine hazardous quotient (*HQ*) from Equation 12.4a.

$$HQ = \frac{CDI}{RfD} = \frac{2.33 \times 10^{-3}\ mg/kg \cdot d}{0.6\ mg/kg \cdot d} = 3.88 \times 10^{-3}$$

   Since there is only one chemical, $HI = HQ = 3.88 \times 10^{-3}$

EXAMPLE 12.15: RISK CHARACTERIZATION FROM EXPOSURE OF MULTIPLE NONCARCINOGENIC TOXIC CHEMICALS

An adult is exposed to five noncarcinogenic hazardous compounds by oral route. The names of the compounds, the chronic daily intake dose (*CDI*), and reference dose (*RfD*), are summarized below. Calculate the hazard index and check if the cumulative exposure is within acceptable level.

Noncarcinogenic Hazardous Compound	Chronic Daily Intake Dose (*CDI*), mg/kg·d	Reference Dose (*RfD*), mg/kg·d	Hazard Quotient (*HQ*)
n-Butanol	0.008	0.10	0.080
Chlorobenzene	0.0015	0.02	0.075
Ethylbenzene	0.010	0.10	0.10
Methyl ethyl ketone (MEK)	0.18	0.60	0.30
Phenol	0.12	0.60	0.20

Solution

1. Determine the hazard quotient (*HQ*) of each compound from Equation 12.4a.

   For n-Butanol, $HQ_1 = \frac{CDI_1}{RfD_1} = \frac{0.008\ mg/kg{\cdot}d}{0.1\ mg/kg{\cdot}d} = 0.08$

   Similarly calculate *HQ* for each compound. Summarize these values in the above summary table.

2. Determine the hazard index (*HI*) from Equation 12.4b.

$$HI = HQ_1 + HQ_2 + HQ_3 + HQ_4 + HQ_5 = 0.080 + 0.075 + 0.10 + 0.30 + 0.20 = 0.76$$

3. Comment on the hazardous risk of exposure of the chemicals.

   Since the value of *HI* is <1.0, the cumulative risk of exposure to an adult from these hazardous chemicals is acceptable.

EXAMPLE 12.16: HAZARDOUS RISK TO INDIVIDUAL ORGANS FROM NONCARCINOGENS

A receptor population is exposed to four chemicals by oral route. Two chemicals (A and B) are hepatotoxin and the other two (C and D) are nephrotoxins. The computed exposure and reference doses are tabulated below. Determine if the cumulative exposure to each target organ is within the acceptable level. Also, calculate total *HI* for all four chemicals.

Chemical	Target organ	Computed chronic daily intake (*CDI*), mg/kg·d	Reference dose (*RfD*), mg/kg·d	Hazard quotient (*HQ*)
A	Kidney	0.015	0.05	0.3
B	Liver	0.40	2.00	0.2
C	Kidney	0.16	0.40	0.4
D	Liver	0.12	0.20	0.6

Solution

1. Determine the hazard quotient (*HQ*) of each chemical.
   The *HQ* of each chemical is calculated from Equation 12.4a, $HQ = CDI/RfD$. The values of *HQ* for these four chemicals are provided in the above table.
2. Determine the *HI* of exposure to each target organ and overall *HI*.

   $HI = 0.3 + 0.4 = 0.7$ for kidney exposure to chemicals A and C

   $HI = 0.2 + 0.6 = 0.8$ for liver exposure to chemicals B and D

   $HI = 0.3 + 0.4 + 0.2 + 0.6 = 1.5$ for both organs exposure to all four chemicals

3. Describe the hazardous risk of chemicals exposure.
   Although the overall *HI* for all four chemicals in this example exceeds the acceptable level of 1.0, the hazard index score to each specific organ is within the acceptable level. This clearly means that the overall *HI* for all four chemicals may not be a viable measure of toxicity unless there is an evidence of synergistic effects of these toxic chemicals upon target organs.

EXAMPLE 12.17: CANCER RISK FROM DRINKING WATER CONTAINING CHLOROFORM

A water supply has 300 μg/L chloroform. An adult consumed water for 30 years. The cancer slope factor for oral chloroform intake is 6.1×10^{-3} kg·d/mg (see Table 12.7). Estimate the risk for cancer development in the lifetime.

Solution

1. Determine the *LADE* from Equation 12.3b for $ED = 30$ years.

$$I_{\mathrm{w,LADE}} = \frac{300\,\mu\mathrm{g/L} \times 10^{-3}\,\mathrm{mg}/\mu\mathrm{g} \times 2\,\mathrm{L/d} \times 350\,\mathrm{d/year} \times 30\,\mathrm{years}}{70\,\mathrm{kg} \times 25{,}550\,\mathrm{d}} = 3.52 \times 10^{-3}\,\mathrm{mg/kg{\cdot}d}$$

2. Determine the cancer risk from Equation 12.5 at $CSF = 6.1 \times 10^{-3}$ kg·d/mg.

$$\text{Cancer Risk} = I_{\mathrm{w,LADE}} \times CSF = 3.52 \times 10^{-3}\,\mathrm{mg/kg{\cdot}d} \times 6.1 \times 10^{-3}\,\mathrm{kg{\cdot}d/mg}$$
$$= 2.15 \times 10^{-5} \quad \text{or} \quad 2.15 \text{ per } 100{,}000$$

EXAMPLE 12.18: CANCER RISK FOR CHILD FROM BENZENE CONTAMINATED SOIL

A child ingests soil containing 2 mg/kg of benzene. The cancer slope factor (*CSF*) of 2.9×10^{-2} kg·d/mg for oral benzene intake is obtained from Table 12.7. Estimate the risk for cancer development in lifetime. The fraction of ingested soil coming from the impacted soil is 1.0.

Solution

1. Determine the *LADE* from Equation 12.3d.

   For a carcinogen, *AT* for lifetime exposure = 70 years × 365 d/year = 25,550 d. Calculate the *LADE* at $C_s = 2$ mg/kg, $IR_s = 200$ mg/d, $ED = 6$ years, and $FI = 1.0$.

$$I_{s,LADE} = \frac{C_s \times IR_s \times CF \times FI \times EF \times ED}{BW \times AT}$$

$$= \frac{2\,\text{mg/kg} \times 200\,\text{mg/d} \times 10^{-6}\,\text{kg/mg} \times 1.0 \times 350\,\text{d/year} \times 6\,\text{years}}{15\,\text{kg} \times 25{,}550\,\text{d}}$$

$$= 2.19 \times 10^{-6}\,\text{mg/kg·d}$$

2. Estimate the cancer risk from Equation 12.5 at $CSF = 2.9 \times 10^{-2}$ kg d/mg.

$$\text{Cancer Risk} = I_{s,LADE} \times CSF = 2.19 \times 10^{-6}\,\text{mg/kg·d} \times 2.9 \times 10^{-2}\,\text{kg·d/mg}$$

$$= 6.35 \times 10^{-8} \quad \text{or} \quad 6.35 \times 10^{-2} \text{ per million}$$

EXAMPLE 12.19: EXCESS LIFETIME CANCER CASES FOR A POPULATION

A population of 100,000 is drinking groundwater containing 0.05 μg/L of *N-Nitrosodimethylamine* (NDMA). What is the maximum number of excess lifetime cancer cases expected for lifetime exposure to NDMA?

Solution

1. Estimate the *LADE* from NDMA, $ED = 70$ years.

$$I_{w,LADE} \text{ from Equation 12.3b} = \frac{0.05\,\mu\text{g/L} \times 10^{-3}\,\text{mg/}\mu\text{g} \times 2\,\text{L/d} \times 350\,\text{d/year} \times 70\,\text{years}}{70\,\text{kg} \times 25{,}550\,\text{d}}$$

$$= 1.37 \times 10^{-6}\,\text{mg/kg·d}$$

2. Determine the individual lifetime cancer risk from drinking the groundwater. Apply Equation 12.5. The *CSF* of 5.1×10 kg d/mg is obtained for NDMA from Table 12.7.

$$\text{Cancer Risk} = I_{w,LADE} \times CSF = 1.37 \times 10^{-6}\,\text{mg/kg·d} \times 5.1 \times 10\,\text{kg·d/mg}$$

$$= 6.99 \times 10^{-5}$$

3. Estimate the maximum number of excess lifetime cases.

$$\text{Number of excess lifetime cancer cases} = \text{Cancer risk} \times \text{Population} = 6.99 \times 10^{-5} \times 100{,}000$$

$$= 7 \text{ persons}$$

   The estimated maximum number of excess lifetime cancer cases from exposure to drinking groundwater containing 0.05 μg/L of NDMA is 7 persons per 100,000 population.

EXAMPLE 12.20: MAXIMUM CONCENTRATION OF BENZENE IN DRINKING WATER FOR AN ACCEPTABLE EXCESS CANCER RISK

A population drinks groundwater containing carcinogenic benzene. Estimate the allowable concentration of benzene in drinking water from lifetime exposure. The acceptable excess cancer risk is one in a million. The cancer slope factor for oral exposure of benzene is 2.9×10^{-2} kg·d/mg (Table 12.7).

Solution

1. Compute the allowable *LADE* for a lifetime oral exposure to benzene from Equation 12.5.
   The acceptable excess cancer risk is one per million (1×10^{-6}), and $CSF = 2.9 \times 10^{-2}$ kg·d/mg.

$$\text{Cancer Risk} = I_{\text{w,LADE}} \times CSF$$

$$I_{\text{w,LADE}} \times 2.9 \times 10^{-2}\,\text{kg·d/mg} = 10^{-6}$$

$$\text{Solve for } LADE,\ I_{\text{w,LADE}} = \frac{10^{-6}}{2.9 \times 10^{-2}\,\text{kg·d/mg}} = 3.45 \times 10^{-5}\,\text{mg/kg·d}$$

2. Compute the allowable concentration of benzene in the drinking water from Equation 12.3b. The lifetime exposure ($ED = 70$ years) is at the acceptable *LADE* of 3.45×10^{-5} mg/kg·d.

$$I_{\text{w,LADE}} = \frac{C_{\text{w}} \times 10^{-3}\,\text{mg/μg} \times 2\,\text{L/d} \times 350\,\text{d/year} \times 70\,\text{years}}{70\,\text{kg} \times 25{,}550\,\text{d}} = C_{\text{w}} \times 2.74 \times 10^{-5}\,(\text{L/μg})(\text{mg/kg·d})$$

$$3.45 \times 10^{-5}\text{mg/kg·d} = C_{\text{w}} \times 2.74 \times 10^{-5}(\text{L/μg})(\text{mg/kg·d})$$

$$C_{\text{w}} = \frac{3.45 \times 10^{-5}\,\text{mg/kg·d}}{2.74 \times 10^{-5}\,(\text{L/μg})(\text{mg/kg·d})} = 1.26\,\text{μg/L}$$

The allowable benzene concentration of 1.26 μg/L in drinking water from lifetime exposure will result in one excess death due to cancer per million population.

12.4 Storage Facility for Reclaimed Water

Demand for reclaimed water varies according to the season of the year, weather conditions, day of the week, and time of the day. The demand is usually reaching the maximum value under dry weather conditions. Because of wide fluctuations in flows, a storage facility is typically required as an integral part of water reclamation and reuse system. The storage facilities are of two types: (1) enclose storage reservoir and (2) open earthen impoundment. Both types are described below.

12.4.1 Enclosed Storage Reservoirs

These are storage reservoirs made of concrete and are similar to a below ground Clearwater well used for finished water distribution at a drinking water treatment facility.

Design Considerations: The design of storage reservoir involves appropriate size, geometry, protection of the structure from excess hydraulic surge, and safety to maintenance workers. The hydraulic design is based on the difference between the reclaimed water production rate and the peak outflow into the distribution piping. Sufficient vents should be installed to expel and take in air as the water level rises and fall in the reservoir. To protect the structural integrity of the storage facility, overflow pipings and structures should be provided to release excess water beyond their full capacity. All openings to atmosphere should have screens or flap valves to prevent entry of insects and animals.

Water Quality Considerations: The quality of reclaimed water in a storage reservoir is of concern. Odors primarily due to H_2S are a common problem. H_2S also increases the chlorine demand. Color may be caused by humic materials. Color and turbidity may affect the aesthetic quality. Regrowth of bacteria may cause slime on the walls and may further affect the intended reuse. Provision should be made for diffused aeration and chemical treatment particularly with chlorine compounds.

High-Service Pumping: The high-service pumps are designed to supply peak demand of reclaimed water. In case distribution reservoirs are provided locally at the point of reuse, the high-service pumping rate can be reduced to average demand. The water distribution system is designed to transport reclaimed water from the production facility to the point of reuse. Basic design considerations and design examples may be found in References 20 and 21.

Storage Volume Calculation: The calculation procedure for storage volume is similar to that of equalization basin or clearwater well. The *mass diagram* or *cumulative change in volume deficiency* between the inflow and outflow may be used for determining the storage volume. These methods are presented in Section 3.6.3. Examples 3.53 through 3.56 show the calculation steps by both methods. Volume calculation procedure for clearwater well by both methods can be found in Reference 21.

12.4.2 Open Storage Reservoirs

These are large earthen basins. The storage volume calculations utilize precipitation, evaporation, and percolation data along with the inflow and outflow of reclaimed water. Such facilities are common for storage of reclaimed water for agricultural irrigation, and recreational and aesthetic reuse. Long-term storage may be needed for irrigation of crops or to supplement the natural rainfall and meet the needs during drought periods.

Water Quality: The most common concerns with open storage are significant changes in water quality that may occur seasonally. Common problems are: (1) excessive algae and plankton growth, (2) low dissolved oxygen and odors, (3) high turbidity, (4) rooted vegetation, (5) thermal stratification, (6) regrowth of microorganisms, and (7) bird and rodent infestation. Common operational strategies are: (a) aeration to increase dissolved oxygen (DO) and eliminate thermal stratification, (b) alum treatment to precipitate phosphorus and settle suspended solids, (c) copper sulfate treatment to control algae growth, (d) rock or slow sand filtration, (e) wetland or overland flow treatment, and (f) dredging to remove benthic deposits.

Storage Volume Calculation: The storage volume is calculated using mass diagram or change in volume deficiency methods. Water balance analysis is conducted over one year period using a drought-year data. The procedure is shown in Example 12.21. Additional examples on open storage reservoirs may be found in References 7 and 9.

EXAMPLE 12.21: STORAGE VOLUME OF OPEN STORAGE RESERVOIR BASED ON WATER BALANCE ANALYSIS

An open storage reservoir is designed to store reclaimed water. The surface area of the reservoir is 8 ha. The applicable percolation rate is 3.58 mm/d. The precipitation, evaporation, reclaimed water production, and reuse demand data are summarized below. Calculate the storage volume needed, average depth, and detention time in the reservoir.

Parameter	Month											
	January	February	March	April	May	June	July	August	September	October	November	December
Precipitation, mm	160	120	100	70	10	3	4	5	10	30	100	130
Evaporation, mm	30	40	60	80	140	180	210	180	140	80	50	30
Production of reclaimed water, m^3/d	5000	5000	4000	3500	3000	2500	2000	2500	3000	3500	4000	4500
Reclaimed water reuse demand, m^3/d	1463	1099	856	2584	4169	3880	4702	4892	3736	3729	4139	2306

Solution

1. Conduct the water balance analysis for the month of January.

The sample calculations for the month of January are provided below:

$$\text{Reclaimed water production (RWP)} = 5000\,m^3/d \times 31\,d = 155 \times 10^3\,m^3$$

$$\text{Precipitation (PREC)} = 160\,mm \times 10^{-3}\,m/mm \times 8\,ha \times 10{,}000\,m^2/ha = 12.8 \times 10^3\,m^3$$

$$\text{Total inflow} = \text{RWP} + \text{PREC} = (155 + 12.8) \times 10^3\,m^3/d \times 31\,d = 167.8 \times 10^3\,m^3$$

$$\text{Evaporation (EVAP)} = 30\,mm \times 10^{-3}\,m/mm \times 8\,ha \times 10{,}000\,m^2/ha = 2.4 \times 10^3\,m^3$$

$$\text{Percolation (PERC)} = 3.58\,mm/d \times 10^{-3}\,m/mm \times 31\,d \times 8\,ha \times 10{,}000\,m^2/ha = 8.88 \times 10^3\,m^3$$

$$\text{Reuse water demand (RWD)} = 1463\,m^3/d \times 31\,d = 45.35 \times 10^3\,m^3$$

$$\text{Total outflow} = \text{EVAP} + \text{PERC} + \text{RWD} = (2.4 + 8.88 + 45.35) \times 10^3\,m^3/d \times 31d = 56.6 \times 10^3\,m^3$$

$$\text{Change in volume} = \text{Total inflow} - \text{Total outflow} = (167.8 - 56.6) \times 10^3\,m^3 = 111 \times 10^3\,m^3$$

$$\text{Cumulative} = \text{Cumulative in December (last month)} + \text{Change in January (current month)} = (67 + 111) \times 10^3\,m^3 = 178 \times 10^3\,m^3$$

2. Conduct the water balance analysis for the remaining months.

Complete the inflow and outflow calculations to the reservoir for the remaining months, and summarize the 12-month data in the table below:

Month (d), days	Inflow to Reservoir, $10^3\,m^3$			Outflow from Reservoir, $10^3\,m^3$				Reservoir Volume, $10^3\,m^3$	
	RWP	PREC	Total Inflow	EVAP	PERC	RWD	Total Outflow	Change	Cumulative
Jan (31)	155	12.8	167.8	2.4	8.88	45.35	56.6	111	178
Feb (28)	140	9.6	149.6	3.2	8.02	30.77	42.0	108	286
Mar (31)	124	8.0	132.0	4.8	8.88	26.54	40.2	92	378
Apr (30)	105	5.6	110.6	6.4	8.59	77.52	92.5	18	396
May (31)	93	0.8	93.8	11.2	8.88	129.24	149.3	−56	340
Jun (30)	75	0.24	75.2	14.4	8.59	116.40	139.4	−64	276
July (31)	62	0.32	62.3	16.8	8.88	145.76	171.4	−109	167
Aug (31)	77.5	0.4	77.9	14.4	8.88	151.65	174.9	−97	70

Continued

Month (d), days	Inflow to Reservoir, 10^3 m^3			Outflow from Reservoir, 10^3 m^3				Reservoir Volume, 10^3 m^3	
	RWP	PREC	Total Inflow	EVAP	PERC	RWD	Total Outflow	Change	Cumulative
Sep (30)	90	0.8	90.8	11.2	8.59	112.08	131.9	−41	29
Oct (31)	108.5	2.4	110.9	6.4	8.88	115.60	130.9	−20	9
Nov (30)	120	8.0	128.0	4.0	8.59	124.17	136.8	−9	0
Dec (31)	139.5	10.4	149.9	2.4	8.88	71.49	82.8	67	67
Total (365)	1289.5	59.4	1349	97.6	104.5	1146.6	1349	0	-

RWP = reclaimed water production; PREC = precipitation; EVAP = evaporation; PERC = percolation; RWD = reuse water demand.

3. Determine the reservoir capacity.
   Identify the full and empty volumes from the calculated cumulative volume by month.
   The reservoir is full at the end of April, and the volume, $V_{full} = 396\ m^3$.
   The reservoir is empty at the end of November, and the maximum volume, $V_{empty} = 0\ m^3$.
   The reservoir capacity required is the difference between the cumulative volumes at full and empty.

$$V_{reservoir} = V_{full} - V_{empty} = (396 - 0) \times 10^3\ m^3 = 396{,}000\ m^3$$

   Provide reservoir volume of 400,000 m^3.
4. Determine the reservoir average depth.

$$D_{reservoir} = \frac{V_{reservoir}}{A_{reservoir}} = \frac{400{,}000\ m^3}{8\ ha \times 10{,}000\ m^2/ha} = 5\ m$$

5. Compute the average detention time in the reservoir.
   On annual basis, total inflow = total outflow = 1,349,000 m^3/year.

$$\text{Annual average daily flow, } Q_{aad} = \frac{1{,}349{,}000\ m^3/year}{365\ d/year} = 3696\ m^3/d$$

$$\text{Average detention time, } \theta_{reservoir} = \frac{V_{reservoir}}{Q_{aad}} = \frac{400{,}000\ m^3}{3696\ m^3/d} = 108\ d \quad \text{or} \quad 3.6\ \text{months} \quad \text{or} \quad 0.3\ \text{year}$$

12.5 Reclaimed Water Reuse

Reclaimed water reuse is on the rise. Even effluent discharged into natural water is being considered as an indirect reuse of effluent for water supply. The quality of reclaimed water is normally produced for intended reuse. The U.S. EPA has suggested guidelines for reclaimed water reuse (Table 12.2).[10,11] The common uses are: (1) urban reuse, (2) industrial reuse, (3) agricultural irrigation, (4) impoundments of recreational and aesthetic reuse, (5) environmental reuse for habitat restoration and enhancement, (6) groundwater recharge, and (7) augmentation of portable water supply. Each of these uses are briefly discussed below.[6–11,22]

12.5.1 Urban Water Reuse

Types of Reuse: The urban reuse of reclaimed water includes many nonpotable municipal uses within an urban area. These uses include: (1) irrigation of public parks and recreational centers, golf courses, athletic fields, school yards and playgrounds, highway medians and shoulders, and landscaped areas surrounding public buildings such as single-family and multifamily residences, and commercial and industrial developments[23]; (2) commercial uses such as vehicle washing facilities, window and street washing and general washdown; (3) ornamental landscapes and decorative water uses such as fountains, pools, and waterfalls; (4) dust control and concrete production on construction projects; (5) fire protection; and (6) toilet and urinal flushing in commercial and industrial buildings. The largest urban reuse however, constitutes irrigational uses within urban areas. Quality requirements of urban reuse water are summarized in Table 12.2.

Reliability, Public Health, and Safety: The reclaimed water is normally delivered to the consumers by a separate distribution network or dual system. The parallel network of distribution mains are separated from community potable water distribution system. The most important design and operation considerations for urban water reuse are reliability of service and protection of public health. The basic safeguards for urban water reuse systems are: (1) assurance that the reclaimed water delivered to the users meets the water quality requirements, (2) assurance that the treated effluent has no detectable fecal coliform count on daily monitoring, and desired level of chlorine residual is maintained at the point of use, (3) prevention of improper system operation, (4) prevention of cross-connection with potable water lines, and (5) prevention of improper use of reclaimed water. To avoid any cross-connection, all equipment associated with reclaimed water must be clearly marked. The national color standards have not been established, but the accepted practice by manufacturers and cities is to use purple color of all pipes carrying reclaimed water.[24,25]

Components of Reclamation System: The urban reclaimed water reuse systems have four major components: (1) production facility, (2) operational storage, (3) high service pumping, and (4) distribution system. The production facility is the municipal wastewater treatment plant. The process train should be designed to meet the desired effluent quality. In Example 12.1 a process train is developed to obtain reclaimed water for urban reuse. The operational storage is similar to that of the clearwell provided in a water production facility. It is an integral part of the production facility, and provides storage to balance the plant output with distribution system demand. Also, it provides the contact time for disinfection by free or combined chlorine residuals. The calculation procedure for sizing storage capacity is similar to that presented in Example 12.21.

The capacity of the high-service pumps is dependent upon the operations of the distribution, storage, and pumping systems. The high-service pumps are designed to supply peak demand. In case storage at the points of use is provided then the pumping capacity is significantly reduced because the peak demands can be supplemented by the local distribution reservoirs. The capacity of the high-service pump station should be equal to the sum of the combined capacities of all pumps with the largest unit out of service. This will assure that the pumping station will meet the emergency situation even in case of equipment failure. In case the distribution system serves an area that has high and low elevations, it is best to divide the areas into *pressure planes* or *pressure zones*. Multiple ground storage tanks with pumps are provided to reduce high energy cost. Numerous computer programs are commercially available to analyze the distribution system and simulate dynamic operation of pumps, pressure plans, pressure reducing valves (PRVs), check valves, reservoirs, and other appurtenances in the distribution system.[26–28] Excellent design examples of clearwell, high service pumps, and distribution system of a water production and distribution system are given in Reference 21. A design plan of clearwells with chlorine contact channels and high service pumps may also be found in this reference.

12.5.2 Industrial Water Reuse

U.S. industries use enormous amounts of water, and present very large market potential for reclaimed water. Major reclaimed water consuming industries are steel, chemicals, pulp and paper mills, petroleum refining, and power plants.[10,11] Most industries treat and recycle the wastewater from their own processes to conserve water and to avoid stringent regulatory standards for effluent discharge. Major industrial water reuses are: (1) cooling tower makeup water, (2) boiler feedwater, (3) process water, and (4) irrigation of landscaping around the plant.

The most common problems of using reclaimed water for cooling towers are scaling, corrosion, biological growth and fouling, and foaming. The boiler feedwater needs softening or demineralization. The industrial process water is used for manufacturing and is treated to meet the specific water quality requirements. The quality of irrigation water for landscaping is similar to that of urban water reuse (see Section 12.5.1 and Table 12.2).

Cooling Tower Makeup Water and Blowdowns: Recirculating cooling towers normally operate as a closed-loop system with blowdown and makeup water. Warm water from a heat exchanger is sprayed on the top of the internal packing of a tower. Cool air is pulled through the tower by large fans on the top. The water evaporates and takes the latent heat of vaporization from the warm water, and cool water is recirculated back to the heat exchanger from the bottom of the cooling tower. Water is lost by *evaporation* and also due to *drift* from wind. Salts in the cooling water concentrate. A portion of cooling water called *blowdown* is removed. Makeup water is added to maintain proper salt balance in the cooling water. A schematic process diagram of a cooling tower is shown in Figure 12.6.

From a recirculating cooling tower system water is lost by evaporation, drift, and blowdown. At a steady state, water flow and salt mass balances are expressed by Equations 12.6a and 12.6b.[7–9]

$$Q_m = Q_b + Q_d + Q_e \text{ (Water flow balance)} \tag{12.6a}$$

$$Q_m C_m = Q_b C_b + Q_d C_d + Q_e C_e \text{ (Salt mass balance)} \tag{12.6b}$$

where

Q_m = makeup water flow rate, L/min
Q_b = blowdown flow rate, L/min
Q_d = drift loss, L/min
Q_e = evaporation loss, L/min

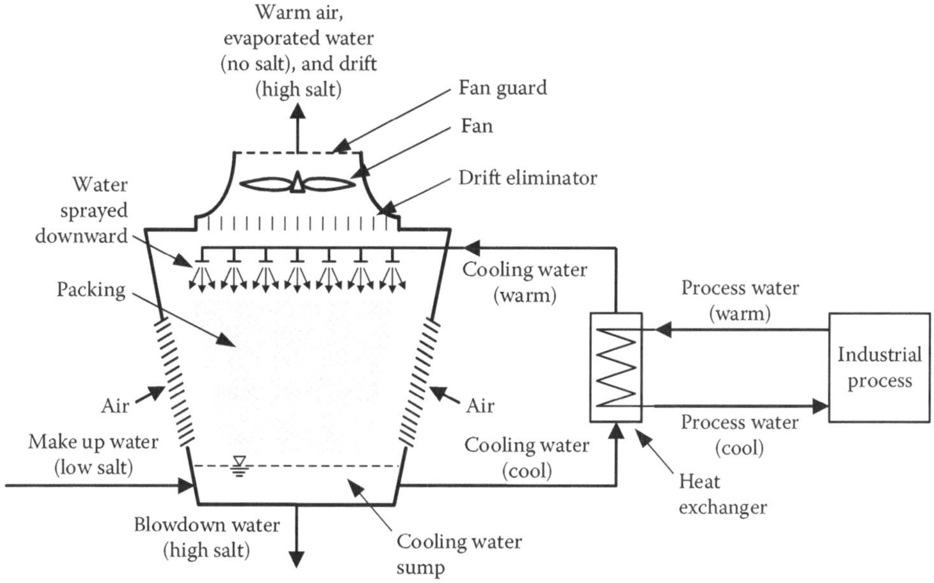

FIGURE 12.6 Schematic process diagram of a recirculating cooling tower.

C_m = salt concentration in makeup water, mg/L
C_b = salt concentration in blowdown, mg/L
C_d = salt concentration in drift loss, mg/L
C_e = salt concentration in evaporation loss, mg/L

The drift loss (Q_d), and salt concentration in the evaporation loss (C_e) are small. Therefore, the terms Q_dC_d and Q_eC_e are small and can be ignored. Equation 12.6b is simplified by Equations 12.6c and 12.6d.

$$Q_mC_m = Q_bC_b \tag{12.6c}$$

$$R_c = \frac{C_b}{C_m} = \frac{Q_m}{Q_b} \quad \text{or} \quad Q_m = R_cQ_b \quad \text{or} \quad C_b = R_cC_m \tag{12.6d}$$

where R_c = cycles of concentration, dimensionless

When the R_c value reaches 3–7, the solids in the recirculating liquid start to precipitate causing scale formation in the pipes and tower packing. Sulfuric acid is added to convert the calcium carbonate and magnesium hydroxide to more soluble sulfate compounds. As alkalinity is consumed, corrosion becomes a serious problem.

The makeup water and blowdown requirements may be calculated as a percent of the recirculation rate. Mass balance and heat balance relationships are used in these calculations. These relationships are developed in Reference 9, and are given by Equations 12.7a through 12.7d.

$$Q_m = MQ_r \tag{12.7a}$$

$$M = \frac{R_c}{R_c - 1}\frac{4.2\Delta T}{2300} = \frac{\Delta TR_c}{548(R_c - 1)} \tag{12.7b}$$

$$Q_b = NQ_r \tag{12.7c}$$

$$N = \frac{\Delta T}{548(R_c - 1)} - \frac{Q_d}{Q_r} \tag{12.7d}$$

where

Q_r = recirculation flow rate, L/min
ΔT = temperature change in cooling water through the cooling tower, °C
M = makeup water requirement ratio, dimensionless
N = blowdown requirement ratio, dimensionless
2300 = constant for heat loss to evaporate 1 kg (or 1 L) of water, kJ/kg
4.2 = constant for heat removal to cool one kg (or 1 L) of water by 1°C, kJ/kg°C
548* = constant, °C

Public Health Concerns of Cooling Water: Legionellosis is a severe and potentially fatal disease. It causes pneumonia, and has been linked to recirculating cooling water. The responsible organism is *Legionell spp.*; a heterotrophic bacterium that grows in the biofilms and within protozoan and algal cells that are found in the cooling tower.[9] The organism is sensitive to temperatures above 60°C. The human exposure is associated with aerosol from drifts. Control strategies include (1) thermal treatment (ability to intermittently increase the temperature over 60°C), (2) disinfection by chlorine or ozone, (3) use of high efficiency aerosol drift eliminator, and (4) control of solids and biofilm build up.

Boiler Feedwater and Blowdowns: Boilers produce hot water and/or steam for heating. Steam is also used in thermoelectric power generation plants. Boiler feedwater needs extensive treatment irrespective of

* $\frac{2300\ \text{kJ/kg}}{4.2\ \text{kJ/kg}^\circ\text{C}} = 548^\circ\text{C}$

its source. For this reason, reclaimed water is an attractive resource for boiler feedwater. It may be locally available and is a reliable water supply source.

The boilers are low, intermediate, and high pressure. The feedwater in these boilers must be of high quality because many impurities affect boiler performance. The constituents that may cause concerns in boiler feedwater are: (1) CO_2 causing corrosion; (2) O_2 causes corrosion and pitting of boiler tubes; (3) H_2S is odorous and corrosive; (4) Ca^{2+}, Mg^{2+}, Fe^{2+}, SiO_2, and SO_4^{2-} (in presence of Ca) cause scaling that inhibits heat transfer and may lead to boiler tube failure; (5) Cl^- causes uneven delivery of steam from boiler, lowers steam efficiency, and causes deposits on turbine blades; (6) Na^+ in the presence of alkalinity (OH^-, HCO_3^-, and CO_3^{2-}) causes foaming, corrosion, and embrittlement; (7) oil and grease and organic matter cause foaming, deposits (that can clog piping), and corrosion; and (8) sediments and turbidity cause sludge deposit and scale carryover.[9]

The reclaimed water for boiler feed is normally treated by ion exchange and membrane filtration at the industrial plant. The recommended quality of boiler feedwater may be found in References 9 and 10.

EXAMPLE 12.22: QUALITY OF COOLING TOWER RECIRCULATION WATER (OR BLOWDOWN)

Reclaimed water is used for cooling tower makeup water. The concentrations of major constituents in the reclaimed water are given below. Estimate the concentrations of major constituents in the recirculation water (or blowdown) after 6 cycles of concentration at 50°C. The solubility of $CaSO_4$ at 50°C is 2200 mg/L as $CaCO_3$.

Constituent	Concentration, mg/L
Ca^{2+}	35 mg/L
Mg^{2+}	8 mg/L
Total alkalinity	95 mg/L as $CaCO_3$
SO_4^{2-}	26 mg/L
Cl^-	20 mg/L
SiO_2	2 mg/L

Solution

1. Determine the breakdown of total hardness (TH) as $CaCO_3$ in the makeup (reclaimed) water.

$$\text{Calcium hardness (CaH)} = \frac{35\,\text{mg/L}\,Ca^{2+} \times 50\,\text{eq. wt. of}\,CaCO_3}{20\,\text{eq. wt. of}\,Ca^{2+}} = 87.5\,\text{mg/L as}\,CaCO_3$$

$$\text{Magnesium hardness (MgH)} = \frac{8\,\text{mg/L}\,Mg^{2+} \times 50\,\text{eq. wt. of}\,CaCO_3}{12.15\,\text{eq. wt. of}\,Mg^{2+}} = 33\,\text{mg/L as}\,CaCO_3$$

$$\text{Total hardness (TH)} = \text{CaH} + \text{MgH} = (87.54 + 33)\,\text{mg/L as}\,CaCO_3 = 120.5\,\text{mg/L as}\,CaCO_3$$

$$\text{Carbonate hardness (CH)} = \text{total alkalinity (TA)} = 95\,\text{mg/L as}\,CaCO_3$$

$$\text{Noncarbonate hardness (NCH)} = \text{TH} - \text{CH} = (120.5 - 95)\,\text{mg/L as}\,CaCO_3 = 25.5\,\text{mg/L as}\,CaCO_3$$

Note: NCH will combine with SO_4^{2-} first, and remaining with Cl^-. Since the CaH (87.5 mg/L as $CaCO_3$) is less than the TA (95 mg/L as $CaCO_3$), all CaH will be CaCH and there will be no CaNCH. The MgH (33 mg/L as $CaCO_3$) consists of MgCH ([95−87.5] mg/L as $CaCO_3$ = 7.5 mg/L as $CaCO_3$) and MgNCH ([33−7.5] mg/L as $CaCO_3$ = 25.5 mg/L as $CaCO_3$), respectively.

2. Determine the breakdown of TH as $CaCO_3$ in the recirculation water (or blowdown) after 6 cycles of concentration prior to adding H_2SO_4.

The hardness concentrations in the blowdown are calculated from Equation 12.6d at $R_c = 6$.

$$\begin{aligned}
&\text{CaH} &&= 6 \times 87.5\ \text{mg/L as CaCO}_3 = 525\ \text{mg/L as CaCO}_3\\
&\text{MgH} &&= 6 \times 33\ \text{mg/L as CaCO}_3 = 198\ \text{mg/L as CaCO}_3\\
&\text{TH} &&= (525 + 198)\ \text{mg/L as CaCO}_3 = 723\ \text{mg/L as CaCO}_3\\
&\text{CH} &&= 6 \times 95\ \text{mg/L as CaCO}_3 = 570\ \text{mg/L as CaCO}_3\\
&\text{NCH} &&= \text{TH} - \text{CH} = (723 - 570)\ \text{mg/L as CaCO}_3 = 153\ \text{mg/L as CaCO}_3
\end{aligned}$$

Note: TA = CH = 570 mg/L as $CaCO_3$. MgCH and MgNCH can also be calculated from Equation 12.6d at $R_c = 6$: MgCH = 6 × 7.5 mg/L as $CaCO_3$ = 45 mg/L as $CaCO_3$, and MgNCH = MgH − CH = (198 − 45) mg/L as $CaCO_3$ = 153 mg/L as $CaCO_3$.

3. Determine the concentration of H_2SO_4 added to the recirculation water in order to convert CH to NCH ($CaSO_4$ and $MgSO_4$) by destroying alkalinity to CO_2 and H_2O.

$$\underset{(100\ \text{as CaCO}_3)}{HCO_3^-\ (\text{alkalinity})} + \underset{\begin{pmatrix}98\ \text{as}\ H_2SO_4\\ 96\ \text{as}\ SO_4^{2-}\end{pmatrix}}{H_2SO_4} \rightarrow SO_4^{2-} + CO_2 + H_2O$$

Assume that 10% TA (or CH) is left to maintain pH and avoid corrosion.

TA (or CH) remaining = 0.1 × CH = 0.1 × 570 mg/L as $CaCO_3$ = 57 mg/L as $CaCO_3$

Note: All remaining CH (57 mg/L as $CaCO_3$) must be CaH since it is less than the CaH (525 mg/L as $CaCO_3$) in the recirculation water.

TA removed by H_2SO_4 = CH − TA remaining = (570 − 57) mg/L as $CaCO_3$ = 513 mg/L as $CaCO_3$

$$\begin{aligned}
\text{Concentration of } H_2SO_4 \text{ added} &= \text{TA removed} \times \frac{98\ \text{mol. wt. of}\ H_2SO_4}{100\ \text{mol. wt. of}\ CaCO_3}\\
&= 513\ \text{mg/L as}\ CaCO_3 \times \frac{98\ \text{mol. wt. of}\ H_2SO_4}{100\ \text{mol. wt. of}\ CaCO_3}\\
&= 503\ \text{mg/L as}\ H_2SO_4\ \text{or}\ 493\ \text{as}\ SO_4^{2-}
\end{aligned}$$

4. Determine the actual breakdown of TH in the recirculation water after adding H_2SO_4.

The concentrations of CaH, MgH, and TH will not change since precipitation of Ca^{2+} and Mg^{2+} is not expected after adding H_2SO_4. However, the split of TH between CH and NCH in the recirculation water will change.

From Step 2, CaH = 525 mg/L as $CaCO_3$, MgH = 198 mg/L as $CaCO_3$, and TH = 723 mg/L as $CaCO_3$.

CH remaining = TA remaining = 57 mg/L as $CaCO_3$

NCH = NCH in recirculation water after 6 cycles (Step 2) + TA removed by H_2SO_4 (Step 3) = (153 + 513) mg/L as $CaCO_3$ = 666 mg/L as $CaCO_3$

Note: Since there is no softening reaction after adding H_2SO_4 the TH will not change from 723 mg/L as $CaCO_3$. Due to effective conversion of hardness from CH to NCH, the risk of scaling in the recirculation water (or blowdown) will be reduced significantly.

5. Determine the $CaSO_4$ concentration in the recirculation water after adding H_2SO_4.

   The calculation results in Step 1 indicate that all CaH in the makeup water is as CaCH. The CaCH in the recirculation water after 6 cycles reaches 525 mg/L as $CaCO_3$. There is no CaNCH in the recirculation water before adding H_2SO_4 (see Step 2). Therefore, all CaNCH must be produced due to TA removal in the reaction of H_2SO_4 with alkalinity (see Step 3 for the reaction).

   CaCH converted to NCH after addition of H_2SO_4 = CaCH − TA remaining = (525 − 57) mg/L as $CaCO_3$ = 468 mg/L as $CaCO_3$

$$\begin{aligned}\text{CaSO}_4 \text{ produced} &= \text{CaCH converted} \times \frac{136 \text{ mol. wt. of CaSO}_4}{100 \text{ mol. wt. of CaCO}_3}\\ &= 468 \text{ mg/L as CaCO}_3 \times \frac{136 \text{ mol. wt. of CaSO}_4}{100 \text{ mol. wt. of CaCO}_3}\\ &= 636 \text{ mg/L as CaSO}_4\end{aligned}$$

   Note: The MgH in the recirculation water after 6 cycles is 198 mg/L as $CaCO_3$ (see Step 2). All MgH will be in the form of MgNCH after adding H_2SO_4. MgNCH after adding H_2SO_4 = MgH before adding H_2SO_4 = 198 mg/L as $CaCO_3$.

6. Evaluate if $CaSO_4$ remains in soluble state in the recirculation water.

$$\begin{aligned}\text{Solubility of CaSO}_4 \text{ at } 50^\circ\text{C} &= 2200 \text{ mg/L as CaCO}_3 \times \frac{136 \text{ mol. wt. of CaSO}_4}{100 \text{ mol. wt. of CaCO}_3}\\ &= 2992 \text{ mg/L as CaSO}_4\end{aligned}$$

   The $CaSO_4$ concentration of 636 mg/L in the recirculation water is significantly lower than the $CaSO_4$ solubility of 2992 mg/L. Therefore, no precipitation of $CaSO_4$ is expected after 6 cycles of concentration.

7. Determine the concentration of SO_4^{2-} in the recirculation water from Equation 12.6d at $R_c = 6$.

   The SO_4^{2-} concentration in the makeup water is 26 mg/L (given).

   Concentration of SO_4^{2-} in the recirculation water before adding H_2SO_4 = 6 × 26 mg/L = 156 mg/L

   Concentration of SO_4^{2-} concentration in the recirculation water after adding H_2SO_4

   = SO_4^{2-} concentration from makeup water + SO_4^{2-} concentration from H_2SO_4 addition (Step 3)
   = (156 + 493) mg/L = 649 mg/L

8. Determine the concentrations of Cl^- and SiO_2 in the recirculation water from Equation 12.6d at $R_c = 6$.

   Concentrations of Cl^- and SiO_2 in the recirculation water after 6 cycles are calculated from Equation 12.6d at $R_c = 6$.

   Concentration of Cl^- in the recirculation water = 6 × 20 mg/L (given) = 120 mg/L

   Concentration of SiO_2 in the recirculation water = 6 × 2 mg/L (given) = 12 mg/L

   Note: The concentrations of Cl^- and SiO_2 in the recirculation water will not be affected by adding H_2SO_4.

9. Summarize the concentrations of constituents in the makeup water as well as that in the recirculation water (or blowdown) before and after adding H_2SO_4.

Parameter	Concentration, mg/L		
	Makeup Water	Recirculation Water (or blowdown) after 6 Cycles	
		Before Adding H_2SO_4	After Adding H_2SO_4
Hardness (as $CaCO_3$)			
CaH	87.5	525 (see Step 2)	57 (see Step 3)
CaCH	87.5 (see Step 1)	525 (see Step 2)	57 (see Step 3)
CaNCH	0 (see Step 1)	0 (see Step 5)	468 (see Step 5)
MgH	33 (see Step 1)	198 (see Step 2)	198 (see Step 5)
MgCH	7.5 (see Step 1)	45 (see Step 2)	0 (see Step 5)
MgNCH	25.5 (see Step 1)	153 (see Step 2)	198 (see Step 5)
TH	120.5 (see Step 1)	723 (see Step 2)	723 (see Step 4)
CH	95 (see Step 1)	570 (see Step 2)	57 (see Step 4)
NCH	25.5 (see Step 1)	153 (see Step 2)	666 (see Step 4)
TA (as $CaCO_3$)	95	570 (see Step 2)	57 (see Step 3)
SO_4^{2-}	26	156 (see Step 7)	649 (see Step 7)
Cl^-	20	120 (see Step 8)	120 (see Step 8)
SiO_2	2	12 (see Step 8)	12 (see Step 8)

EXAMPLE 12.23: MAKEUP WATER AND BLOWDOWN REQUIREMENTS IN A COOLING TOWER

Makeup water and blowdown requirements as a ratio of the recirculation rate may be calculated from the energy balance. Calculate the makeup and blowdown flow rates if the recirculation rate in a cooling tower is 100,000 L/min. The cooling temperature drop is 10°C, and the recycle ratio is 6. The drift flow rate is 0.005% of recirculating water.

Solution

1. Calculate the makeup water flow rate (Q_m).

   Apply Equation 12.7b to obtain makeup water ratio M by substituting $R_c = 6$, and $\Delta T = 10°C$.

   $$M = \frac{\Delta T R_c}{548(R_c - 1)} = \frac{10°C \times 6}{548°C \times (6 - 1)} = 0.022$$

   Apply Equation 12.7a to calculate Q_m required at $Q_r = 100{,}000$ L/min.

   $$Q_m = MQ_r = 0.022 \times 100{,}000\ \text{L/min} = 2200\ \text{L/min}$$

2. Calculate the blowdown flow rate (Q_b).

   Apply Equation 12.7d to calculate N required at the drift rate $Q_d = 0.00005\ Q_r$.

   $$N = \frac{\Delta T}{548(R_c - 1)} - \frac{Q_d}{Q_r} = \frac{10°C}{548°C \times (6 - 1)} - 0.00005 = 0.0036$$

   Rearrange Equation 12.7c to calculate the Q_b required.

   $$Q_b = NQ_r = 0.0036 \times 100{,}000\ \text{L/min} = 360\ \text{L/min}$$

3. Check the cycles of concentration from Equation 12.6d.

   $$R_c = \frac{Q_m}{Q_b} = \frac{2200\ \text{L/min}}{360\ \text{L/min}} = 6.1$$ (The calculated R_c is close to the desired ratio of 6.0.)

12.5.3 Agricultural Irrigation Reuse

The agricultural irrigation offers a very large potential for reclaimed water reuse in the arid regions. Being a reliable water supply source, reclaimed water can be used to supplement the scarce natural water or sparse rainfall. The major considerations for agricultural irrigation reuse are: (1) demand, (2) quality of reclaimed water, (3) reliability, and (4) costs. The demand for agricultural irrigation may vary depending upon: (a) climatic conditions (temperature, precipitation, evapotranspiration, etc.), (b) crop type, (c) stage of plant growth, (d) storage requirements, and (e) method of irrigation. Proper crop selection is the most important part of the design process for reclaimed water irrigation system. The quality of reclaimed water is also important because some constituents have special significance in agricultural irrigation. Among these are salinity, ratio of cautions, trace elements, specific toxic ions, and nutrients. The impact of constituents are presented below. Quality requirements of reclaimed water for agricultural reuse are given in Table 12.2.

Salinity: The salinity is a measure of total dissolved solids (TDS). The electrical conductivity (EC_w) is a surrogate measure of TDS. At the standard temperature of 25°C, it may be expressed by Equations 12.8a and 12.8b.

$$\text{TDS} \approx 640\, EC_w \quad (\text{For} \quad EC_w < 5\,\text{ds/m}) \tag{12.8a}$$

$$\text{TDS} \approx 800\, EC_w \quad (\text{For} \quad EC_w > 5\,\text{ds/m}) \tag{12.8b}$$

where

TDS = total dissolved solids, mg/L

EC_w = electrical conductivity of water, ds/m (decisiems/m).* Equation 12.8a is normally applicable to potable water, reclaimed water, and sewage. This equation is not applicable to medium to high salinity waters such as brackish and saline waters and seawater.

High salinity affects the growth of a plant. The crop yield decreases with an increase in salinity above the threshold level. Good drainage is essential so that excess water can percolate down and salinity does not build up in the rooting zone to cause toxicity. Leaching fraction (*LF*) is an indication of the applied water that passes through and percolates below the entire root depth. It is expressed by Equation 12.9a. The salinity of the drainage water that percolates below the rooting zone is calculated from Equation 12.9b.

$$LF = \frac{D_r}{D_i} = \frac{D_i - ET_c}{D_i} \tag{12.9a}$$

$$EC_{dw} = \frac{EC_w}{LF} \tag{12.9b}$$

where

LF = leaching fraction, dimensionless. Higher *LF* value means less salt accumulation in the root zone.

D_r = depth of water below the root zone, mm (in)

D_i = depth of water applied at the surface, mm (in)

ET_c = crop evapotranspiration, mm (in)

EC_w = salinity of the reclaimed water used for irrigation reuse, mg/L

EC_{dw} = salinity of the drainage water percolating below the root zone, mg/L. It is equal to the salinity of the saturation extract of the soil sample.

Sodicity: Sodicity is a condition when the Na^+ is the dominant cation in the soil solution and in the irrigation water. Under sodic condition, the soil particles plug pore space in the soil matrix and reduce

* EC_w may also be expressed in units of μs/cm (microsiems/cm), ms/cm (millisiems/cm), ds/cm (decisiems/cm), mmho/cm (milliohms/cm), or μmho/cm (microohms/cm). The conversions among these units are: 1 ds/m = 10^3 μs/cm = 1 ms/cm = 1 mmho/cm = 10^3 μmho/cm.

the movement of water and air, causing water ponding at the surface (water logging). Sodicity is expressed by the SAR in Equation 12.10a.

$$SAR = \frac{[Na^+]}{\sqrt{([Ca^{2+}] + [Mg^{2+}])/2}} \tag{12.10a}$$

where

SAR = sodium adsorption ratio, dimensionless
$[Na^+]$ = concentration of sodium ion, meq/L
$[Ca^{2+}]$ = concentration of calcium ion, meq/L
$[Mg^{2+}]$ = concentration of magnesium ion, meq/L

In Equation 12.10a, it is assumed that the ratio of HCO_3^- to Ca^{2+} is at equilibrium for the given soil-water conditions. With reclaimed water the concentration of calcium ion needs to be adjusted to estimate the expected concentration that reaches equilibrium with the soil. The adjusted value of $[Ca_X^{2+}]$ is obtained from Equation 12.10b using HCO_3^-/Ca^{2+} ratio and salinity data given in Table B.10 of Appendix B.

$$SAR_{adj} = \frac{[Na^+]}{\sqrt{([Ca_X^{2+}] + [Mg^{2+}])/2}} \tag{12.10b}$$

where

SAR_{adj} = adjusted sodium adsorption ratio, dimensionless
$[Ca_X^{2+}]$ = adjusted concentration of calcium ion, meq/L. It is provided from Table B.10 in Appendix B.

Trace Elements: Many trace elements are essential for plant growth at low concentrations but may cause toxicity at high concentrations. Also, crops may accumulate trace elements in edible parts and present health risks to animals and humans. The recommended maximum concentrations of trace elements in the reclaimed water for irrigation reuse are given in Table 12.8. The potential toxicities of these elements at excessive concentrations on the plant growth are described in Reference 11.

Specific Ion Toxicity: Sodium, chloride, and boron ions in excessive concentrations in reclaimed water used for irrigation can cause toxicity to the plants. The common effects are leaf burn, chlorosis, and twig dieback of certain fruit trees. Minimal effects are expected on vegetable, grain, forage, or fiber crops at typical concentrations of these ions in the reclaimed water.

Nutrients: Important nutrients in agriculture are N, P, and K. Plants utilize nitrogen in the form of ammonium (NH_4^+) and nitrate (NO_3^-) ions. Organic nitrogen must be converted in these forms to be

TABLE 12.8 Recommended Maximum Concentrations of Trace Elements in Reclaimed Water for Irrigation Reuse

Element	Recommended Maximum Concentration, mg/L
Cd and Mo	0.01
Se	0.02
Co	0.05
As, Be, Cr, and V	0.1
Cu, Mn, and Ni	0.2
F	1.0
Zn	2.0
Li	2.5
Al, Fe, and Pb	5.0

Source: Adapted in part from Reference 11.

utilized by plants. Excessive nitrogen in the later part of the growing period may reduce the quality of some crops. Phosphorus in reclaimed water is primarily in inorganic form that may be taken up by the crop, accumulate in soil, or lost by leaching.

Guidelines for Interpretation of Water Quality for Irrigation: The reclaimed water quality required for irrigation may depend upon climate, soil type and characteristics, crops to be irrigated, and irrigation method. Additionally, the most important characteristics of irrigation water are TDS concentration or salinity, SAR, sodium, chloride, boron, nitrogen, bicarbonate, pH, and chlorine residual. A guideline for interpretation of water quality for irrigation is provided in Table 12.9. Readers are referred to Reference 11 for general guidelines for quality requirements of reclaimed water for irrigation use over processed food-crops and nonfood crops, and livestock consumption.

TABLE 12.9 Guidelines for Interpretation of Water Quality for Irrigation

Constituent (Problem)	Degree of Restriction on Irrigation Reuse		
	None	Slight to Moderate	Severe
Salinity (affects crop water availability and performance of irrigation system)			
EC_w, ds/m	<0.7	0.7–3.0	>3.0
TDS, mg/L	<450	450–2000	>2000
Iron (Fe) for drip irrigation, mg/L	<0.1	0.1–1.5	>1.5
Manganese (Mn) for drip irrigation, mg/L	<0.1	0.1–1.5	>1.5
Hydrogen sulfide (H_2S), mg/L	<0.5	0.5–2	>2
pH, standard unit	6.5–8.4		
TSS, mg/L	<50	50–100	>100
Permeability[a] *(affects infiltration rate of water into the soil)*			
EC_w, ds/m			
At $SAR_{adj} = 0–3$	>0.7	0.7–0.2	<0.2
At $SAR_{adj} = 3–6$	>1.2	1.2–0.3	<0.3
At $SAR_{adj} = 6–12$	>1.9	1.9–5.0	<0.5
At $SAR_{adj} = 12–20$	>2.9	2.9–1.3	<1.3
At $SAR_{adj} = 20–40$	>5.0	5.0–2.9	<2.9
Trace elements[b]			
Specific ion toxicity (affects sensitive crops)			
Sodium (Na^+)			
For surface irrigation, SAR	<3	3–9	>9
For sprinkler irrigation, mg/L	<70	>70	
Chloride (Cl^-), mg/L			
For surface irrigation	<140	140–350	>350
For sprinkler irrigation	<100	>100	
Residual chlorine for overhead sprinklers only, mg/L	<1	1–5	>5
Boron (B^{3+}), mg/L	<0.7	0.7–3	>3
Miscellaneous effects (affects susceptible crops)			
Nitrogen (total-N), mg/L	<5	5–30	>30
Bicarbonate HC,O_3^- for overhead sprinkling only, (mg/L)	<90	90–500	>500

[a] For wastewater irrigation, it is recommended that SAR be adjusted to include a more correct estimate of calcium in the soil water. A procedure is given in Example 12.25. The adjusted sodium adsorption ratio (SAR_{adj}) calculated by this procedure is to be substituted for the SAR value in this table for permissible criteria.

[b] See Table 12.8 for permeability criteria.

Source: Adapted in part from References 9, 11, 22, and 29 through 31.

EXAMPLE 12.24: SALINITY AND LEACHING FRACTION FROM CROP IRRIGATED BY RECLAIMED WATER

Reclaimed water is used to irrigate a crop. The electrical conductivity EC_w is 1.5 ds/m in the reclaimed water and the crop is irrigated to achieve a leaching fraction of 0.31. The allowable TDS for irrigation water is <1000 mg/L. Calculate (a) the salinity of reclaimed water and (b) the fraction of applied water used by the crop and lost by evapotranspiration.

Solution

1. Check the TDS of reclaimed water (TDS_w) from Equation 12.8a for $EC_w = 1.5$ ds/m <5 ds/m.

$$TDS_w \approx 640\,EC_w = 640 \times 1.5 = 960\text{ mg/L}$$

This is less than the allowable TDS limit of 1000 mg/L.

2. Determine the salinity of drainage water percolating below the root zone from Equation 12.9b.

$$EC_{dw} = \frac{EC_w}{LF} = \frac{1.5\text{ ds/m}}{0.31} = 4.8\text{ ds/m}$$

3. Determine the TDS of drainage water (TDS_{dw}) from Equation 12.8a for $EC_{dw} = 4.8$ ds/m <5 ds/m.

$$TDS_{dw} \approx 640\,EC_{dw} = 640 \times 4.8 = 3070\text{ mg/L}$$

4. Determine the fraction of applied water on the surface that is used by the crop and lost by evapotranspiration.

The leaching fraction is 0.31 and the fraction of applied water used by the crop and lost by evapotranspiration is $(1 - 0.31) = 0.69$.

EXAMPLE 12.25: ADJUSTED SODIUM ADSORPTION RATIO AND RESTRICTIONS FOR IRRIGATION REUSE

An open reservoir is used to store the reclaimed water. The water quality of stored water is given below. If this water is used for crop irrigation, determine (a) sodium adsorption ratio (SAR) and (b) adjusted sodium adsorption ratio (SAR_{adj}). Also, evaluate the potential restrictions for irrigation reuse.

Constituent	Value	Constituent	Value
pH (standard unit)	7.2	Chloride (Cl^-)	150
Calcium (Ca^{2+}), mg/L	40	Alkalinity (mg/L) as $CaCO_3$	260
Magnesium (Mg^{2+}), mg/L	18	TDS, mg/L	1450
Sodium (Na^+), mg/L	250		

Solution

1. Determine the concentrations of Ca^{2+}, Mg^{2+}, Na^+, and HCO_3^- in meq/L.

$$[Ca^{2+}] = \frac{40\text{ mg/L}}{20\text{ mg/meq}} = 2\text{ meq/L}$$

$$[Mg^{2+}] = \frac{18\text{ mg/L}}{12.15\text{ mg/meq}} = 1.5\text{ meq/L}$$

$$[Na^{+}] = \frac{250\text{ mg/L}}{23\text{ mg/meq}} = 10.9\text{ meq/L}$$

Since pH is below 8.3, the entire alkalinity is due to bicarbonate (HCO_3^-).

$$\text{Concentration of } HCO_3^- = 260\,\text{mg/L as } CaCO_3 \times \frac{61 \text{ eq. wt. of } HCO_3^-}{50 \text{ eq. wt. of } CaCO_3} = 317\,\text{mg/L as } HCO_3^-$$

$$[HCO_3^-] = \frac{317\,\text{mg/L}}{61\,\text{mg/meq}} = 5.2\,\text{meq/L}$$

2. Determine the sodium adsorption ratio (SAR) from Equation 12.10a.

$$\text{SAR} = \frac{[Na^+]}{\sqrt{([Ca^{2+}] + [Mg^{2+}])/2}} = \frac{10.9\,\text{meq/L}}{\sqrt{(2\,\text{meq/L} + 1.5\,\text{meq/L})/2}} = 8.2$$

3. Estimate the salinity of reclaimed water from Equation 12.8a.

$$EC_w \approx \frac{\text{TDS}}{640} = \frac{1450}{640} = 2.3\,\text{ds/m}$$

4. Determine the adjusted concentration of calcium ion ($[Ca_X^{2+}]$).

$$\text{Ratio of } HCO_3^-/Ca^{2+} = \frac{5.2\,\text{meq/L}}{2\,\text{meq/L}} = 2.6$$

$[Ca_X^{2+}] = 1.2\,\text{meq/L}$ is obtained from Table B.10 in Appendix B.

5. Determine the adjusted sodium adsorption ratio (SAR_{adj}) from Equation 12.10b.

$$\text{SAR}_{adj} = \frac{[Na^+]}{\sqrt{([Ca_X^{2+}] + [Mg^{2+}])/2}} \frac{10.9\,\text{meq/L}}{\sqrt{(1.2\,\text{meq/L} + 1.5\,\text{meq/L})/2}} = 9.4$$

6. Evaluate the potential restrictions for using the reclaimed water for crop irrigation.

The restrictions are obtained by comparing the available water quality data with the criteria provided in Table 12.9. Results are summarized below.

a. Check for the salinity criteria based on TDS and EC_w data.

The degree of restrictions is slight to moderate for irrigation reuse. The pH of the reclaimed water should not pose any restriction for crop irrigation.

b. Check for the permeability criteria based on SAR_{adj} and EC_w data.

The SAR_{adj} of 9.4 is within the range from 6 to 12 and EC_w of 2.3 ds/m is >1.9 ds/m. The degree of restriction is none to slight. Therefore, the permeability criteria are not restricted and the reclaimed water is suitable for surface irrigation. There will be no effect on the infiltration rate of the soil.

c. Check for the specific ion toxicity criteria based on SAR and chloride concentration data.

The SAR of 8.2 is within the range of 3–9. The degree of restriction is slight to moderate for surface irrigation. A chloride concentration of 150 mg/L suggests that the reclaimed water is more suitable for surface irrigation than sprinkler irrigation. The specific ion toxicity to sensitive crops and plants should be checked and identified.

d. Check for the miscellaneous effect.

At pH 7.2, the alkalinity consists of bicarbonate. The bicarbonate concentration of 260 mg/L is of slight to moderate concern for using the reclaimed water in an overhead sprinkler irrigation system.

e. Summarize the main findings.

In summary, the degrees of restrictions for irrigational reuse of reclaimed water in all categories are slight to moderate. Therefore, all sensitive plants and crops should be checked and evaluated for specific ion toxicity. There will be no effect on infiltration rate of soil, and salinity will not build up in the root zone to cause toxicity.

12.5.4 Impoundments for Recreational and Aesthetic Reuse

Reclaimed water is extensively used for recreational and aesthetic reuse, and beautification of urban landscape and environment. Common uses include: (1) maintenance of landscape and aesthetic impoundments; (2) ornamental fountains; (3) water-based recreational lakes for boating, fishing, and noncontact sports; (4) rearing of freshwater sport and commercial fisheries; (5) demonstration projects to emphasize the benefits of reusing high quality reclaimed water; and (6) snowmaking. The extent of reuse projects in this category will depend upon water demand and availability of reclaimed water of desired quality (see Table 12.2).

Impoundments of recreational and aesthetic reuse of reclaimed water help to beautify the environment in urban areas; serve the needs of noncontact uses and landscaping plan; and golf course and residential developments. These impoundments integrate water traps or ponds, which may also be used as storage facility for irrigation water. Artificial snow may be used to supplement natural snow for winter sports, mainly for skiing and snowboarding. Large-scale commercial fish farming in reclaimed water is increasing in the United States, and has potential for significant use of reclaimed water. However, the reclaimed water quality aspects should be thoroughly assessed for possible bioaccumulation of toxic contaminants through the food chain.

The required level of treatment varies with the intended use of reclaimed water. As the potential for human contact increases, the required treatment levels will also increase. For nonrestricted recreational uses, treatment of secondary effluent is to be followed by coagulation, filtration, and disinfection to reduce TSS, precipitate phosphorus, improve appearance, reduce total coliform count of <3 per 100 mL, and improve virus removal.[10,11,22]

12.5.5 Environmental Reuse for Habitat Restoration and Enhancement

Reclaimed water reuse may directly contribute to habitat restoration and enhancement of aquatic environment. Many other reuses include development and restoration of wetlands, habitat development, augmentation of existing stream flows, enhancement of water quality of ponds and lakes, and restoration of stream ecology. The water quality requirements for different water reuses are given in Table 12.2.

Over the past 200 years, ~50% of the wetlands in the United States have been destroyed for many diverse uses such as agriculture, mining, forestry, and urbanization. Wetlands serve many beneficial environmental functions. They may provide: (1) flood attenuation, (2) habitat restoration for wildlife and water fowl, (3) productivity to support the food chain, (4) aquifer recharge, (5) water quality enhancement, (6) nutrients assimilation from treated water, (7) aesthetic and environmental improvement, and (8) water quality enhancement of reclaimed water by cost-effective treatment through natural systems. Aquaculture has been practiced primarily for production of food, fiber, and fertilizer. In wetlands host of aquatic plants and animals such as water hyacinth, duckweed, seaweed, alligator weed, midge larvae, *Daphnia*, and others are cultured. These aquatic flora and fauna help to remove BOD, TSS, nutrients, and heavy metals. Water hyacinths and other aquatic plants are harvested for the production of compost, animal feed, fuel, and methane production. Many states have regulations that specifically address the use of reclaimed water in wetlands. *Natural wetlands* are protected under NPDES Permits and Water Quality Standards. Effluent quality equivalent of a secondary level or higher may be permitted. On the other hand, the *constructed wetlands* are built and operated for treatment only. The basic design considerations for constructed wetlands include types, propagation of appropriate plant and animal species, salinity reduction, providing a balance in hydraulic and constituents loadings, and ecosystem enhancement.[32–35] A discussion on natural and constructed wetlands may be found in Section 15.3.2.

Engineered augmentation of stream flow from wastewater effluent is done for environmental, aesthetic, ornamental, and recreational purposes, and propagation of aquatic life. In many arid regions the reclaimed water is the primary source of water to maintain the base flow for aquatic and riparian habitat enhancement. Riparian habitat consists of plants and animals occurring in a land area adjacent to a stream or river, between the wet and dry zones.

Reclaimed water may also be a source of water supply to ponds and lakes. These ponds and lakes may be ornamental and recreational impoundments. They may provide wildlife habitat, storage for irrigation water, and in some cases an indirect source of drinking water supply.

12.5.6 Groundwater Recharge

In many regions of the country excessive pumping of groundwater for municipal, industrial, and agricultural uses has resulted in lowering of groundwater table. As a result, subsidence, saltwater intrusion, and shortage of groundwater have occurred. There is a need for artificial recharge of these aquifers. Groundwater recharge by reclaimed water is a planned augmentation of groundwater aquifers that are not used as a potable water source (see Table 12.2). It has several advantages over surface storage and treatment. These are: (1) cost of artificial recharge of aquifer is less than the cost of surface storage; (2) no evaporation losses in subsurface storage; (3) no water quality deterioration due to associated algae, plankton, and aquatic growth; (4) no seasonal taste and odor problems; (5) no need for an environmentally acceptable reservoir site; (6) the aquifer may serve as a transport media eliminating the need for a conveyance system (pipelines, canals, aqueducts); and (7) transition between reclaimed water and groundwater may provide a more acceptable health, psychological, and aesthetic benefits than other direct pipe to pipe reuse.[7] There are two acceptable methods of groundwater recharge: (a) surface spreading and (b) direct injection.

Surface Spreading: Surface spreading is the simplest and the oldest method of groundwater recharge. It is similar to natural rainwater infiltration into the ground. The recharge may be by rapid infiltration, intermittent percolation, or slow rate infiltration. The surface spreading involves ridge and furrow, constructed wetlands, and infiltration basins. The depth of groundwater is in the range from 3 m to 15 m (10–50 ft.). Approximately 20–50% recharged water is recoverable on an annual basis.

Considerable improvement in the quality of reclaimed water may occur by percolation through the soil, unsaturated zone (vadose zone), and aquifer.[36] This is called soil-aquifer treatment (SAT) or geopurification. High reduction in BOD_5; nitrification–denitrification; and removal of synthetic organic compounds, microorganisms, and other chemical constituents may occur during their passage through suitable soils. The major removal processes are microbial decomposition, adsorption, filtration, ion exchange, vitalization, dilution, chemical oxidation and reduction, chemical precipitation, chelation, and complexation. The degree of contaminant removal is dependent upon the ability of the soil to remove the constituent. Heavy metal removal ranges from very small to >90%, depending upon soil properties and on speciation of the influent metals. There are indications that once heavy metals are adsorbed, they are not easily desorbed. Desorption may depend in part on the buffering capacity, salt concentration, and oxidation–reduction potential.[37,38] A discussion on land treatment of wastewater and quality of treated effluent may be found in Section 15.3.1.

The surface water infiltration is a function of vertical hydraulic conductivity of the soil and the depth of water in the recharge basin. It is given by Equation 12.11a.[9]

$$v_i = K\frac{H_w + L_f - H_c}{L_f} \tag{12.11a}$$

where
v_i = infiltration rate, m/d
K = hydraulic conductivity of the wetted zone, m/d
H_w = depth of water above the soil, m
L_f = depth of wetting front, m
H_c = critical pressure head (or water head) of soil for wetting, m. The range of H_c is from −0.1 m or greater for coarse soil to −1.0 m or less for fine soil.

The time for the wetting front to travel a given distance is calculated from Equation 12.11b.[9]

$$t = \frac{f}{K}\left(L_f - (H_w - H_c)\ln\left(\frac{H_w + L_f - H_c}{H_w - H_c}\right)\right) \tag{12.11b}$$

where

t = time since the start of filtration, d

f = fillable porosity, dimensionless. The value of f is normally 0.30–0.45.

Direct Injection: Direct injection involves pumping of high quality reclaimed water directly into the groundwater aquifer through injection wells. This method is practical in most cases where (1) groundwater is deep; (2) it is an unconfined aquifer; and (3) land use or topography makes surface spreading impractical or too expensive. Retention time in the ground is ~12 months, and ~20% of reclaimed water is recoverable on an annual basis. The direct injection is very effective in coastal regions to create a freshwater barrier against saltwater intrusion.

The injection wells vadose zone (or unsaturated zone) are normally used for direct recharge of the aquifer. The hydraulic conductivity is calculated from Equation 12.12.[9] The schematics of vadose zone injection well are shown in Figure 12.7.

For deep aquifer, when $L_d > 2L_w$

$$K = \frac{Q_r}{2\pi(L_w)^2}\left(\ln\left(\frac{L_w}{r_w} + \sqrt{\left(\frac{L_w}{r_w}\right)^2 - 1}\right) - 1\right) \tag{12.12a}$$

For shallow aquifers, when $L_d < 2L_w$

$$K = \frac{3Q_r}{\pi L_w(3L_w + 2L_d)}\ln\left(\frac{L_w}{r_w}\right) \tag{12.12b}$$

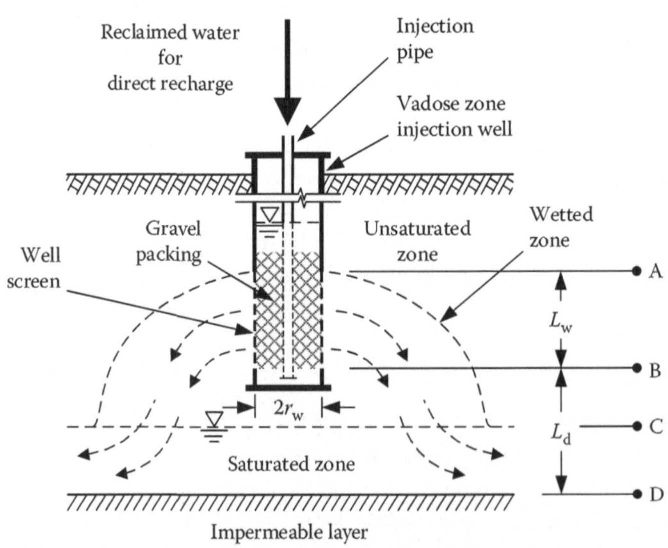

FIGURE 12.7 Schematic of vadose zone injection well and wetted zone.

where

Q_r = volumetric recharge rate, m^3/s

L_w = depth of water in the screened length of the injection well (AB), m

r_w = radius of the injection well, m

L_d = depth of permeable zone, m. It is the distance from the bottom of the well or below the screened section to the impervious layer ($L_d = BC + CD$).

Pretreatment Requirements for Groundwater Recharge: The common pretreatment prior to groundwater recharge includes primary and secondary treatment of wastewater followed by filtration and UV disinfection (see Table 12.2). The direct injection may require higher quality than that for surface spreading.

Problems Associated with Groundwater Recharge: While there are obvious advantages with groundwater recharge, there are numerous concerns too.[39–41] Some of these are: (1) extensive land areas may be needed for spreading basins; (2) the high energy cost of pumping required for direct injection; (3) not all injected water may be recoverable; (4) recharge may create the danger of aquifer contamination; and (5) aquifer remediation is difficult, expensive, and may take years to accomplish.

EXAMPLE 12.26: LAND AREA REQUIRED FOR RAPID INFILTRATION OF RECLAIMED WATER IN INFILTRATION BASIN

High-quality disinfected secondary effluent is stored in an open infiltration basin for recharging the groundwater aquifer. The recharge flow is 4000 m^3/d. The rapid infiltration rate is 0.38 $m^3/m^2{\cdot}d$. Determine the land area required for the infiltration basin.

Solution

Flow for aquifer recharge, $Q = 4000\ m^3/d$

Rapid infiltration rate, $r_f = 0.38\ m^3/m^2{\cdot}d$

$$\text{Area required, } A = \frac{Q}{r_f} = \frac{4000\ m^3/d}{0.38\ m^3/m^2{\cdot}d} = 10{,}500\ m^2 \quad \text{or} \quad 1.1\ \text{ha}$$

EXAMPLE 12.27: RECHARGE FROM SURFACE SPREADING IN INFILTRATION BASIN

A recharge basin receives reclaimed water. The clarified effluent infiltrates into an unconfined aquifer by direct recharge. The free water surface is 1-m deep. The hydraulic conductivity of wetted zone is 0.8 m/d and is assumed constant. The recharge area of the infiltration basin is 1350 m^2. The fillable porosity is 0.4 and the critical pressure head of soil for wetting is −0.4 m. Calculate (a) the time for wetting front to travel 10 m, (b) infiltration rate at a depth of 10 m, and daily percolation rate, and (c) volume of water infiltrated through 10 m.

Solution

1. Determine the time (t) required for wetting front to travel 10 m.

The travelling time is calculated from Equation 12.11b. $L_f = 10$ m, $H_w = 1$ m, $H_c = -0.4$ m, $f =$ 0.4 and $K = 0.8$ m/d.

$$t = \frac{f}{K}\left(L_f - (H_w - H_c)\ln\left(\frac{H_w + L_f - H_c}{H_w - H_c}\right)\right)$$

$$= \frac{0.4}{0.8\,\text{m/d}} \times \left(10\,\text{m} - (1\,\text{m} - (-0.4\,\text{m})) \times \ln\left(\frac{1\,\text{m} + 10\,\text{m} - (-0.4\,\text{m})}{1\,\text{m} - (-0.4\,\text{m})}\right)\right)$$

$$= 0.5\,\text{d/m} \times \left(10\,\text{m} - 1.4\text{m} \times \ln\left(\frac{11.4\,\text{m}}{1.4\,\text{m}}\right)\right)$$

$$= 0.5\,\text{d/m} \times (10\,\text{m} - 1.4\,\text{m} \times 2.1)$$

$$= 3.5\,\text{d}$$

2. Determine the infiltration rate (v_i) and the daily percolation rate (q_i) after the wetting front reaches the depth of 10 m.

   The infiltration rate (v_i) decreases as travelling through the soil. Calculate v_i from Equation 12.11a after reaching the depth $L_f = 10$ m.

$$v_i = K\frac{H_w + L_f - H_c}{L_f} = 0.8\,\text{m/d} \times \frac{1\,\text{m} + 10\,\text{m} - (-0.4\,\text{m})}{10\,\text{m}} = 0.8\,\text{m/d} \times \frac{11.4\,\text{m}}{10\,\text{m}} = 0.8\,\text{m/d} \times 1.14$$

$$= 0.91\,\text{m/d} \quad \text{or} \quad 0.91\,\text{m}^3/\text{m}^2{\cdot}\text{d}$$

$$\text{Dailypercolationrate, } q_i = Av_i = 1350\,\text{m}^2 \times 0.91\,\text{m}^3/\text{m}^2{\cdot}\text{d} = 1230\,\text{m}^3/\text{d}$$

3. Determine the volume of water (V_{inf}) that infiltrated through 10 m depth.

$$V_{inf} = fL_fA = 0.4 \times 10\,\text{m} \times 1350\,\text{m}^2 = 5400\,\text{m}^3$$

EXAMPLE 12.28: VOLUMETRIC RECHARGE RATE BY DIRECT INJECTION THROUGH VADOSE ZONE

Reclaimed water is recharged by direct injection into an unconfined aquifer that is 50-m deep. The hydraulic conductivity K of the wetted zone is 0.00068 m/s. The radius of the injection well is 1.2 m and the depth of water in the screened length is 18 m. The depth of vadose zone below the screened section is 20 m. Estimate the volumetric injection rate through vadose zone.

Solution

1. Determine the dimensions of L_w and L_d from Figure 12.7.

$L_w =$ depth of water in the screened length (between Points A and B in Figure 12.7) = 18 m

$L_d =$ depth of permeable or vadose zone below the screened depth (BC) + depth of unconfined aquifer (CD)

$= 20\,\text{m} + 50\,\text{m} = 70\,\text{m}$

2. Check the type of aquifer and identify the applicable equation.

$2L_w = 2 \times 18\,\text{m} = 36\,\text{m}$

Since L_d of 70 m $> 2L_w$, this is a deep aquifer, and Equation 12.12a is applicable.

3. Estimate the volumetric direct recharge (injection) rate (Q_r).
   Rearrange Equation 12.12a and calculate the flow rate Q_r.

$$Q_r = \frac{2\pi K(L_w)^2}{\ln\left(L_w/r_w + \sqrt{(L_w/r_w)^2 - 1}\right) - 1} = \frac{2\pi \times 0.00068\,\text{m/s} \times (18\,\text{m})^2}{\ln(18\text{m}/1.2\text{m} + \sqrt{(18\text{m}/1.2\text{m})^2 - 1}) - 1}$$

$$= \frac{1.38\,\text{m}^3/\text{s}}{\ln(15+15) - 1} = \frac{1.38\,\text{m}^3/\text{s}}{2.4} = 0.58\,\text{m}^3/\text{s} \quad \text{or} \quad 50{,}100\,\text{m}^3/\text{d}$$

EXAMPLE 12.29: PUMPING COST OF DIRECT INJECTION OF RECLAIMED WATER INTO AN AQUIFER

High quality reclaimed water is pumped directly into an aquifer. The average pumping rate is 4000 m^3/d. The total pressure needed for injunction is 2290 kN/m^2. The pump and motor efficiencies are 84% and 92%, respectively. Calculate the wire power needed, and annual pumping cost on the basis of 24 h/d and 365 d/year operation. The unit cost of electrical power is 25 cents per kWh.

Solution

1. Determine the water power required for direct injection.
   Pumping rate,

$$Q = 4000\,\text{m}^3/\text{d} \times \frac{\text{d}}{86{,}400\,\text{s}} = 0.0463\,\text{m}^3/\text{s}$$

   Apply Equation 6.17a and substitute $K' = 1$ kW/(kN·m/s), $Q = 0.0463$ m^3/s, and $(TDH)\ \gamma = p =$ 2290 kN/m^2.

$$P_w = K'Q(TDH)\gamma = 1\text{kW}/(\text{kN}\cdot\text{m/s}) \times 0.0463\,\text{m}^3/\text{s} \times 2290'\,\text{kN/m}^2 = 106\,\text{kW}$$

2. Calculate the wire power from combined Equations 6.17b and 6.17c.

$$P_m = \frac{P_w}{E_p E_m} = \frac{106\,\text{kW}}{0.84 \times 0.92} = 137\,\text{kW}$$

3. Determine the annual energy cost.

Daily energy consumption = 137 kW × 24 h/d = 3290 kWh/d
Annual energy consumption = 3290 kWh/d × 365 d/year
= 1,200,000 kWh/year or 1200 MWh/year (megawatt hour per year)
Annual energy cost = 1,200,000 kwh/year × \$0.25kwh = \$300,000/year or 0.3 million dollars per year

12.5.7 Augmentation of Potable Water Supply

Water is a renewable resource. Nature purifies it through an unending hydrological cycle. Currently, technology is available to treat wastewater to any extent the usage may require. The augmentation of potable water supply from reclaimed water is achieved by indirect or direct potable water reuse. Both methods are discussed below.

Indirect Potable Water Reuse: Indirect potable water reuse of reclaimed water is a common practice these days in the United States. Many public attitude surveys have indicated that the general public is willing to accept the indirect reuse of reclaimed water over direct reuse. Many surface water resources used for municipal water supply are also used for effluent disposal. Thus, the dilution effects, and natural assimilation processes provide further purification of wastewater effluent over the entire maturity period. The water treatment plants further treat the combined surface and reclaimed water or in rare cases only reclaimed water to meet the safe-drinking water quality standards. Similarly the groundwater recharge from reclaimed water provides natural purification and dilution over the maturation period prior to potable reuse.

The indirect potable water reuse is also achieved by percolating the polluted surface water or high quality reclaimed water into the groundwater zone from spreading basins, or from drain fields of porous pipes.[42,43] Partial treatment is achieved during passage through the soil until the groundwater is pumped out from an extraction well.

The infiltration gallery, riverbank, or dune filtration systems are a performed horizontal conduit or tunnel built along the bank of an impoundment or stream short distance away and typically in shallow depth. It intercepts and collects the subsurface groundwater. The water flows through perforated conduit into a tight chamber. From this chamber, the water is pumped into a treatment facility or into a distribution system. This method is distinctly different from groundwater recharge. The travel time is short, and the quality of water must be checked against the water quality requirements prior to indirect potable water reuse. A conceptual sketch of an infiltration gallery is shown in Figure 12.9. The unit flow rate collected from an infiltration gallery is expressed by Equation 12.13b that is derived in Example 12.30.

Direct Potable Water Reuse: Direct pipe-to-pipe reuse of reclaimed water involves blending of final product with other drinking water before distribution. The unit operations and processes used to produce potable water from wastewater are well understood, effective, and very reliable.[44] However, the practice of direct pipe-to-pipe reuse of reclaimed water is rare and has not gained public acceptance even though the quality may fully comply with the current safe drinking water standard.[44,45] The major concern however, is the chronic health effects that might result from ingesting the mixture of unregulated inorganic and organic constituents in trace amounts coming from both industrial chemicals and from home products such as pharmaceutical and personal care products (PPCPs). These constituents remain in minute concentration in reclaimed water, even after most advanced treatment methods such as microfiltration and reverse osmosis. The direct reuse of treated wastewater therefore is practicable only on a short-term emergency basis. The direct reuse of treated wastewater for human consumption over long periods will therefore depend upon economics of finding suitable natural water resource, government regulations, and public acceptance.[41,46]

EXAMPLE 12.30: DERIVE INFILTRATION GALLERY FLOW EQUATION FOR INDIRECT POTABLE REUSE OF RECLAIMED WATER

Infiltration galleries are constructed to treat in-situ the reclaimed water percolated from the bank of an impoundment. Develop the discharge equation for an infiltration gallery that is laid more or less at right angle to the direction of groundwater flow. The infiltration gallery will collect the treated reclaimed water for indirect potable reuse.

Solution

1. Draw a conceptual sketch of an infiltration gallery and flow lines.
   The conceptual sketch of infiltration gallery is shown in Figure 12.8.
2. Apply Darcy's law to express the discharge through the selected element.

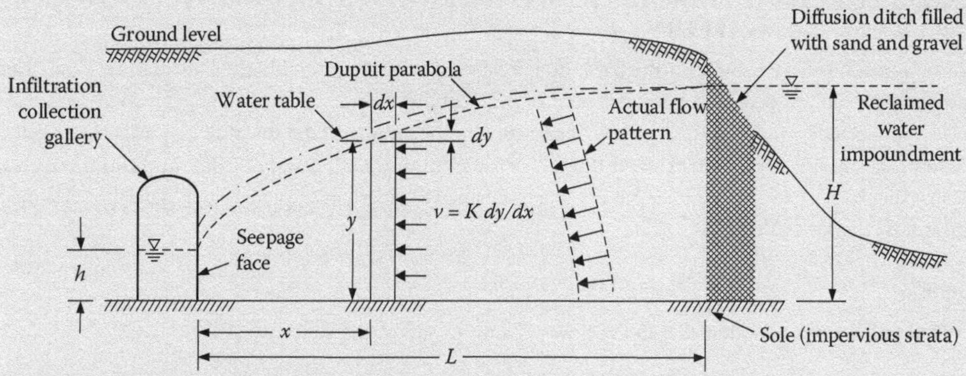

FIGURE 12.8 Conceptual sketch of an infiltration gallery (Example 12.30).

Flow per unit length of the gallery is expressed by Equation 12.13a.

$$q = K\frac{dy}{dx} \times (y \times 1) \tag{12.13a}$$

where

q = flow per unit length of the gallery, $m^3/d{\cdot}m$
K = coefficient of permeability (hydraulic conductivity), m/d
x = distance of the element from the seepage face of the gallery, m
$y \times 1$ = element area at the section, per unit length of the gallery, m^2/m
dy/dx = slope of the water surface, dimensionless
$K\,dy/dx$ = flow velocity through the section, m/d

3. Integrate Equation 12.13a over the limits to obtain Equation 12.13b that expresses the discharge equation per unit length of the gallery.

$$\int_0^L q dx = \int_h^H Ky dy$$

$$q = \frac{K}{2L}(H^2 - h^2) \tag{12.13b}$$

where

H = head of water in the diffusion ditch above the impervious strata, m
h = water depth in the infiltration collection gallery, m
L = total distance of the infiltration path through the porous stratum, m

The total discharge flow for an infiltration gallery is expressed by Equation 12.13c.

$$Q = Bq \quad \text{or} \quad Q = \frac{KB}{2L}(H^2 - h^2) \tag{12.13c}$$

where

Q = total discharge flow, m^3/d
B = length of infiltration gallery, m

EXAMPLE 12.31: WATER SUPPLY WITH DRAWN FROM AN INFILTRATION GALLERY

An infiltration gallery is built on the bank of a reclaimed water impoundment. The stratum is of clean sand and gravel 7-m deep and has coefficient of permeability $K = 10^{-1}$ cm/s. The percolated reclaimed water is collected in an infiltration gallery penetrating to the bottom of the stratum. The water level in the infiltration gallery is kept at 1.0 m above the sole of the stratum, and its distance from the diffusion ditch is 12 m. Calculate (a) flow per unit length of the gallery and (b) length of the gallery to obtain the reclaimed water flow rate of 4000 m^3/d.

Solution

1. Determine from Equation 12.13b the flow per unit length of the gallery.

$$K = 10^{-1}\,\text{cm/s} \times 10^{-2}\,\text{m/cm} \times \frac{86{,}400\,\text{s}}{\text{d}} = 86.4\,\text{m/d}$$

$$q = \frac{K}{2L}(H^2 - h^2) = \frac{86.4\,\text{m/d}}{2 \times 12\,\text{m}} \times ((7\,\text{m})^2 - (1\,\text{m})^2) = 173\,\text{m}^3/\text{d·m}$$

2. Determine the length of the gallery required to obtain the desired reuse water flow from rearranged Equation 12.13c.

$$B = \frac{Q}{q} = \frac{4000\,\text{m}^3/\text{d}}{173\,\text{m}^3/\text{d·m}} = 23\,\text{m}$$

Provide a 25-m long infiltration gallery.

12.6 Effluent Discharge into Natural Waters

The effluent discharge into natural waters is the most common disposal practice. The assimilation capacity of the natural waters is utilized to achieve further treatment and dilution. The 1972 Amendments to the Federal Water Pollutant Control Act (PL 92-500) and Clean Water Act of 1977 (PL 95-217) established the National Pollutant Discharge Elimination System (NPDES).[47,48] Under NPDES all municipalities and industries are required to obtain a permit to discharge a given quantity and quality of effluent into the natural waters. All states are required to adopt water quality standards. The interim goal is to restore and maintain the physical, chemical, and biological integrity of the nation's waters. Because of the importance of effluent disposal into natural waters, the remainder of this chapter is devoted to (1) requirements of outfall, (2) water quality models of surface water, and (3) design of outfall structures.

12.6.1 Requirements of Outfall

The outfall structures are designed to disperse the effluent into the receiving waters which may be a river, lake, estuary or marine ocean. In all cases, it is essential to provide (1) proper dispersion, (2) avoid localized pollution problem, (3) utilize the assimilative capacity of the natural water, and (4) protect the contamination of shores. In most cases the discharge is made below the water surface through one or multiple outlets to reduce the formation of foam and telltale signs, and achieve good dispersion. The requirements of river, lake, estuarine, and ocean outfalls are given below.

River Outfall: Shoreline releases of effluents into the rivers are poorly dispersed and hug the banks for long distances with little mixing. Installation of diffusers across the width of the river is preferred;

but any construction at the bottom of a navigable water is subject to approval by the U.S. Army Corps of Engineers. Sewage effluent is generally warmer than most receiving waters, and therefore it rises while being swept along by dominant currents. Formation of telltale signs such as sleek or detergent froth should be avoided.

Lake Outfall: In lake outfalls the best possible dispersion is needed to avoid shore contaminations. Shallow lakes and reservoirs are subject to significant mixing because of wind-induced currents, and effluents are dispersed well. In stratified lakes there is a possibility that the effluent may deplete the dissolved oxygen in the lower stratum. Warmer effluent may also destroy the thermocline, which is a zone of significant temperature change and is extremely resistant to mixing.

Estuarine Outfall: The effluents discharged into the tidal rivers and estuaries may oscillate due to freshwater flow, and tide- or wind-produced seiches. The mixing is caused by dispersion. If stratification occurs because of salinity or temperature, the effluent plume will rise to the surface.

Marine or Ocean Outfall: The marine outfall consists of long pipes to transport the effluent in long distance from the shore and then release it. The function of a diffuser is to mix the low-density effluent with high-density seawater at the bottom. The plume will rise under prevailing currents until the density of the mixture is equal to that of seawater below the surface. At this point the plume will spread horizontally and disperse with little or no visual effects. Some excellent references on marine outfalls are available in the literatures.[49–53]

12.6.2 Water Quality Models of Surface Water

Effluent discharged into rivers, lakes, estuaries, or ocean mixes with the surrounding water. The microbiological activity is enhanced to consume biodegradable organics under aerobic condition. The DO is lowered, which is replenished mainly by atmospheric reaeration, photosynthesis, and downstream dilution. A minimum DO of 4–5 mg/L is necessary for healthy aquatic life. The concentrations of conservative pollutants are reduced by dilution. The BOD discharged and DO response relationships in rivers, estuaries, and lakes are discussed below.[50,54–59]

Deoxygenation in Rivers: In rivers, the effluent is mixed as it flows downstream. The DO is supplied by reaeration and is proportional to the DO deficit. Several equations based on the Streeter and Phelps model have been developed to express DO with respect to time and many other relationships. These relationships are given by Equation 12.14.

$$C_0 = \frac{Q_r C_r + q_w C_w}{Q_r + q_w} \tag{12.14a}$$

$$D_t = \frac{K L_0}{K_2 - K}\left(e^{-Kt} - e^{-K_2 t}\right) + D_0 e^{-K_2 t} \tag{12.14b}$$

$$t_c = \frac{1}{K_2 - K}\ln\left(\frac{K_2}{K}\left(1 - \frac{(K_2 - K)D_0}{K L_0}\right)\right) \tag{12.14c}$$

$$D_c = \frac{K}{K_2} L_0 e^{-K t_c} \quad \text{or} \quad t_c = -\frac{1}{K}\ln\left(\frac{K_2 D_c}{K L_0}\right) \tag{12.14d}$$

$$x = tv \quad \text{or} \quad x_c = t_c v \tag{12.14e}$$

$$K = K_{20}\theta_T^{T-20} \quad \text{or} \quad K_2 = K_{2,20}\theta_T^{T-20} \tag{12.14f}$$

$$D = DO_{sat} - DO \quad \text{or} \quad DO = DO_{sat} - D \tag{12.14g}$$

where

C_0 = initial concentration of constituent in the mixture at the point of discharge (time $t = 0$), mg/L
C_r = concentration of constituent in the river before mixing, mg/L
C_w = concentration of constituent in the wastewater effluent, mg/L
D = DO deficit, mg/L
D_0 = initial DO deficit of the mixture at the point of discharge (time $t = 0$), mg/L
D_c = critical DO deficient at the critical distance x_c from the point of discharge or at the critical time t_c, mg/L
D_t = DO deficit at time t, mg/L
DO = DO concentration in the stream, mg/L
DO_{sat} = DO saturation concentration of the stream water, mg/L. See Table 10.16 for the DO saturation concentration at a given temperature, elevation (or pressure) and TDS concentration.
K_{20} = first-order reaction rate constant of constituent at 20°C, d^{-1} (base e). The value of K_{20} for domestic wastewater effluent may be in the range of 0.1–0.2 d^{-1}. More information on this topic may be found in Chapter 5.
$K_{2,20}$ = reaeration constant at 20°C, d^{-1} (base e). The value of $K_{2,20}$ depends upon the characteristics of stream for different types of stream conditions.
K = first-order reaction rate constant of constituent at temperature T (°C), d^{-1} (base e)
K_2 = reaeration constant at temperature T (°C), d^{-1} (base e). See additional information about K_2 below.
L_0 = initial ultimate BOD of the mixture at the point of discharge (time $t = 0$), mg/L. The value of L_0 can be estimated from Equation 5.7 based on K_{20}.
Q_r = river discharge flow rate, m^3/s
q_w = wastewater discharge flow rate, m^3/s
t = flow time in the river downstream from the point of discharge, d
t_c = critical time when the critical DO deficit D_c occurs, d. This is the time for flow to reach the critical distance x_c from the point of discharge, d
θ_T = temperature coefficient, dimensionless. For the first-order reaction rate constant K, the typical value of $\theta_T = 1.1$–1.2. For the reaeration constant K_2, the typical value of $\theta_T = 1.024$.
v = average velocity in the stream at the river discharge flow Q_r, m/d
x = distance downstream from the point of discharge to the point corresponding to flow time t or DO deficit D_t, m
x_c = critical distance downstream from the point of discharge to the point where the critical DO deficit D_c occurs, m

Note: Equation 12.14f may also be used for the following temperature corrections: (a) coefficient of oxygen dispersion (D_L) in Equation 12.15b, (b) reaction rate constant for nitrification oxygen demand (K_N) in Equation 12.17, and tidal dispersion coefficient (D_L) in Equations 12.18d and 12.18e.

Reaeration Constant: The reaeration constant (K_2) depends upon the characteristics of the stream. Several empirical equations have been proposed for estimating the constant. Three commonly used relationships are expressed by Equation 12.15.

Churchill $$K_2 = \frac{5.03v^{0.969}}{H^{1.637}} \quad (12.15a)$$

O'Connor and Dobbins $$K_2 = \frac{294(D_L v)^{0.5}}{H^{3/2}} \quad \text{or} \quad K_2 \approx \frac{3.9v^{0.5}}{H^{3/2}} \quad (12.15b)$$

Owens $$K_2 = \frac{5.34v^{0.67}}{H^{1.85}} \quad (12.15c)$$

where

D_L = coefficient of molecular diffusion of oxygen, m^2/d. The value of D_L is 1.76×10^{-4} m^2/d at 20°C, and the temperature coefficient $\theta_T = 1.037$ in Equation 12.14f.

H = average water depth in the stream, m

K_2 and v are defined previously. The value of v is in the unit of m/s in Equation 12.15.

The typical ranges of $K_{2,20}$ for different types of streams are: 0.1–0.23 d^{-1} in small ponds and backwaters; 0.23–0.35 d^{-1} in sluggish streams and large lakes; 0.35–0.45 d^{-1} in large streams at low velocity; 0.46–0.69 d^{-1} in large streams at normal velocity; 0.69–1.15 d^{-1} in swift streams; and >1.15 d^{-1} in rapid streams and waterfalls.[60]

Other Oxygen Sources and Sinks: There are other oxygen sources and sinks that may influence the oxygen sag equation (Equation 12.14b). These sources are: (1) oxygen supply from algal photosynthesis, (2) oxygen consumption due to respiration of photosynthetic organisms, and (3) benthic oxygen demand. The effects of these sources of oxygen supply and demand may be calculated from Equation 12.16a.

$$r_p = \frac{R_p}{H} \quad \text{or} \quad r_r = -\frac{R_r}{H} \quad \text{or} \quad r_b = -\frac{R_b}{H} \qquad (12.16a)$$

where

R_p = oxygen supply due to photosynthesis, g $O_2/m^2 \cdot d$. The value of R_p depends upon algal concentration and sunlight. It may range from 0.3 $g/m^2 \cdot d$ in deep slow-moving to 9 $g/m^2 \cdot d$ in shallow fast-moving streams.

R_r = respiration rate due to photosynthetic organisms, g $O_2/m^2 \cdot d$. The value may range from 0.5 $g/m^2 \cdot d$ in deep to 10 $g/m^2 \cdot d$ in shallow streams.

R_b = benthic oxygen demand, $g/m^2 \cdot d$. The value may range from 4 $g/m^2 \cdot d$ for fresh sediments near municipal wastewater effluent outfall to 1.5 $g/m^2 \cdot d$ for the aged sediments downstream of outfall.

r_p = oxygen supply rate due to algal photosynthesis ($r_p > 0$ or positive), $g/m^3 \cdot d$

r_r = oxygen consumption rate due to algal respiration ($r_r < 0$ or negative), $g/m^3 \cdot d$

r_b = oxygen consumption rate due to benthic oxygen demand ($r_b < 0$ or negative), $g/m^3 \cdot d$

The effects of photosynthesis, respiration, and benthic demand may be integrated in Equations 12.14b through 12.14d to get the combined relationships. These relationships are expressed by Equations 12.16b through 12.16d.

$$D_t = \frac{KL_0}{K_2 - K}(e^{-Kt} - e^{-K_2 t}) + D_0 e^{-K_2 t} - \frac{r_p + r_r + r_b}{K_2}(1 - e^{-K_2 t}) \qquad (12.16b)$$

$$t_c = \frac{1}{K_2 - K}\ln\left(\frac{K_2}{K} - \frac{K_2(K_2 - K)D_0 + (K_2 - K)(r_p + r_r + r_b)}{K^2 L_0}\right) \qquad (12.16c)$$

$$D_c = \frac{1}{K_2}(KL_0 e^{-Kt_c} - (r_p + r_r + r_b)) \quad \text{or} \quad t_c = -\frac{1}{K}\ln\left(\frac{K_2 D_c + (r_p + r_r + r_b)}{KL_0}\right) \qquad (12.16d)$$

All terms have been defined previously.

If the effects of photosynthesis, algal respiration and benthic oxygen demand are small then the term of $(r_p + r_r + r_b) \approx 0$ and the Equations 12.14b and 12.16b, Equations 12.14c and 12.16c, and Equations 12.14d and 12.16d become identical.

Effect of Nitrification on DO Sag Curve: Ammonia discharged in the effluent is oxidized to NO_2^- then to NO_3^- by the nitrifying bacteria. The oxygen is consumed due to both carbonaceous and nitrogenous oxygen demands. The DO deficit equation due to combined effect is given by

Equation 12.17.[61]

$$D_t = \frac{KL_0}{K_2 - K}\left(e^{-Kt} - e^{-K_2 t}\right) + D_0 e^{-K_2 t} + \frac{K_N L_N}{K_2 - K_N}\left(e^{-K_N(t-t_N)} - e^{-K_2(t-t_N)}\right) \quad (12.17)$$

where

K_N = reaction rate constant for nitrogenous oxygen demand, d^{-1}. The temperature coefficient for K_N is $\theta_T = 1.047$.

L_N = initial ultimate nitrogenous oxygen demand of the mixture at the point of discharge (time $t = 0$), mg/L. The value of L_N can be estimated from Equation 5.18c.

t_N = time when nitrification starts, d

All other terms have been defined previously.

Tidal Rivers and Estuaries: In tidal rivers and estuaries, the motion of water is caused not only by gravity but also by tidal action, density currents, and wind effects. The waste materials from the effluent discharged are mixed and diluted while carried back and forth over many cycles. Ultimately, the waste reaches the open sea. This is flushing time rather than travel time. The DO deficit and other relationships are expressed by Equation 12.18.

$$D_x = \frac{KL_0}{K_2 - K}(e^{J_1 x} - e^{J_2 x}) + D_0 e^{J_2 x} \quad (12.18a)$$

$$D_c = \frac{KL_0}{K_2 - K}(e^{J_1 x_c} - e^{J_2 x_c}) + D_0 e^{J_2 x_c} \quad (12.18b)$$

$$x_c = \frac{1}{J_1 - J_2}\ln\left(\frac{J_2}{J_1}\left(1 - \frac{(J_2 - J_1)D_0}{J_1 L_0}\right)\right) \quad (12.18c)$$

$$J_1 = \frac{U}{2D_L}\left(1 - \sqrt{1 + \frac{4KD_L}{U^2}}\right) \quad (12.18d)$$

$$J_2 = \frac{U}{2D_L}\left(1 - \sqrt{1 + \frac{4K_2 D_L}{U^2}}\right) \quad (12.18e)$$

where

D_x = DO deficit at distance x, mg/L

J_1 and J_2 = coefficients, m^{-1}

U = net downstream velocity calculated from freshwater flow, m/d

D_L = tidal dispersion coefficient (also called coefficient of dispersion, eddy coefficient, or molecular diffusion of oxygen), m^2/d. The temperature coefficient $\theta_T = 1.037$ for D_L.

x = longitudinal distance downstream from the point of discharge to the point corresponding to DO deficit D_x, m

x_c = longitudinal critical distance downstream from the point of discharge to the point where the critical DO deficit D_c occurs, m

All other terms have been defined previously.

The reported range of tidal dispersion coefficient D_L for different estuaries is 12–510 m^2/s with a typical value of 200 m^2/s. Several methods may be used to determine D_L based on (1) physical data, (2) salinity data, or (3) dye tracer data.

Physical Data. The D_L can be calculated from Equation 12.19a using the physical characteristics of the estuary.[50]

$$D_L = 63nU_T(R_H)^{5/6} \tag{12.19a}$$

where

n = Manning's roughness coefficient, dimensionless
U_T = maximum tidal velocity, m/d
R_H = hydraulic mean radius of the stream cross section, m

Salinity Data: Where salinity intrusion occurs, the salinity data is obtained with respect to distance upstream from the sea and used to calculate D_L from Equation 12.19b.

$$\ln\left(\frac{C_c}{C_0}\right) = -\frac{U}{D_L}x_s \quad \text{or} \quad D_L = \frac{U}{\ln(C_0/C_c)}x_s \quad \text{or} \quad C_c = C_0 e^{-U/D_L x_s} \tag{12.19b}$$

where

C_c = salinity concentration at a distance x upstream of the sea, mg/L
C_0 = salinity concentration at the sea, mg/L
x_s = distance upstream from the sea, m

U and D_L are defined previously.

Dye-Tracer Data: A slug of conservative dye tracer is released and the dye concentration is measured upstream along with the length of tidal river. The dye concentration distribution curve is prepared with respect to the distance upstream from the point the dye is released. Figure 12.9 shows a typical dye dispersion curve, and the relationships are given by Equations 12.19c and 12.19d.

$$\ln\left(\frac{C_x}{C_{max}}\right) = -\frac{(X - X_{max})^2}{4D_L t} \quad \text{or} \quad D_L = -\frac{(X - X_{max})^2}{4t\ln(C_x/C_{max})} \tag{12.19c}$$

$$X = tU \quad \text{or} \quad t = \frac{X}{U} \tag{12.19d}$$

where

C_x = the dye concentration monitored at point x, m
C_{max} = the maximum dye concentration monitored, m
X = the distance upstream from the point the dye is released, m

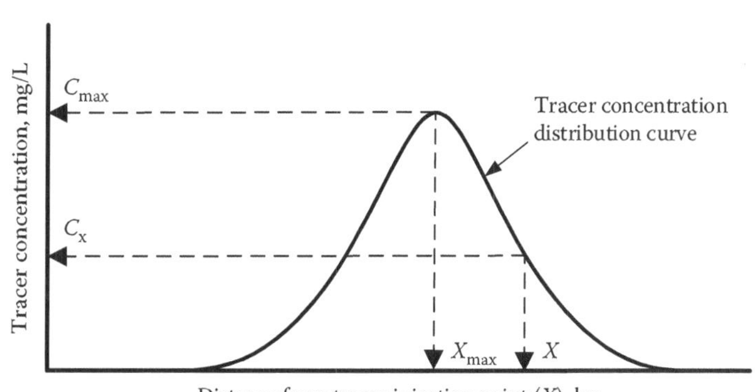

FIGURE 12.9 Definition sketch of a slug dye tracer profile for determination of tidal dispersion coefficient (D_L).

X_{max} = the distance between the location where the maximum dye concentration C_{max} is monitored and the point the dye is released, m
t = time for tide to reach point X, d

D_L and U are defined previously.

Lakes and Reservoirs: The pollutants discharged with the effluent into lakes and reservoirs also undergo dilution and decay with time. Most small natural bodies of water undergo significant mixing due to wind action, and are assumed completely mixed. Based on this assumption, a theoretical model has been developed to estimate the concentration of a contaminant that undergoes first-order decay. The mass balance relationships and equilibrium concentration of the contaminant are expressed by Equations 12.20a through 12.20d.[62]

$$W = Q_i C_i + Q_w C_w \tag{12.20a}$$

$$C_t = \frac{W}{\beta V}(1 - e^{-\beta t}) + C_0 e^{-\beta t} \tag{12.20b}$$

$$\theta = \frac{V}{Q_i + Q_w} \tag{12.20c}$$

$$\beta = \frac{1}{\theta} + K \tag{12.20d}$$

where

Q_i = normal flow to the reservoir, m^3/d
Q_w = wastewater effluent flow to the reservoir, m^3/d
C_i = concentration of pollutant in normal flow, mg/L (g/m^3)
C_w = concentration of pollutant in wastewater effluent, mg/L (g/m^3)
C_0 = concentration of pollutant initially present in the reservoir at $t = 0$, mg/L (g/m^3)
C_t = concentration of input pollutant at time t, mg/L (g/m^3)
W = total mass of pollutant reaching the reservoir, g/d
β = modified decay coefficient, d^{-1}
V = reservoir volume, m^3
θ = theoretical hydraulic detention time, d
K = first-order decay constant of the pollutant, d^{-1}
t = time after pollutant is discharged into the reservoir, d

The equilibrium concentration C_e is reached when $t = \infty$ and Equation 12.20b is reduced to Equation 12.20e.

$$C_e = \frac{W}{\beta V} \tag{12.20e}$$

where C_e = equilibrium concentration in the reservoir, mg/L (g/m^3)

All other terms have been defined previously.

Stratified lakes behave differently and complex models have been developed to model the physical phenomena that exist in different thermal zones.

Water Quality Modeling: Water quality modeling efforts have been steadily increasing over the years. Early models utilized only a few water quality parameters such as temperature, BOD and DO.[63–65] With time, models included additional parameters such as salinity; carbonaceous and nitrogenous BOD; benthic oxygen demand; total organic carbon; refractory organic carbon; ammonia, nitrite, nitrate and organic nitrogen; organic and inorganic phosphorus; algae as a chlorophyll; coliforms; phyto- and zooplanktons;

TABLE 12.10 General Information about Water Quality Modeling Packages

Package	Developer	Information Source
River and stream water quality models (QUAL2K and QUAL2E)	U.S. EPA	http://www.epa.gov
Water quality analysis simulation program (WASP)	U.S. EPA	http://www.epa.gov
Better assessment science integrating point and nonpoint sources (BASINS)	U.S. EPA	http://www.epa.gov
Watershed management optimization support tool (WMOST)	U.S. EPA	http://www.epa.gov
System water quality modeling (HEC-5Q)	U.S. Army Corps of Engineers	http://www.hec.usace.army.mil
Surface water quality modeling (SPARROW)	USGS	http://water.usgs.gov
2D and 3D water quality and ecological modelling (MIKE 21 or MIKE 21 in combination with ECO Lab)	DHI	http://www.mikepoweredbydhi.com
1D and 2D integrated catchment and river modeling (InfoWorks® ICM or InfoWorks® RS)	Innovyze	http://www.innovyze.com

and toxic compounds.[66,67] The commonly used water quality model packages are summarized in Table 12.10.[68–73] Readers may find many available models and tools for development and implementation of Total Maximum Daily Loads (TMDLs) from U.S. EPA on-line Center for Exposure Assessment Modeling (CEAM) at https://www.epa.gov/exposure-assessment-models/tmdl-models-and-tools.

EXAMPLE 12.32: DO PROFILE, CRITICAL DO AND DISTANCE FROM OUTFALL

An activated sludge plant is discharging effluent at a rate of 0.44 m^3/s. The effluent DO, temperature and total BOD_5 are 4.0 mg/L, 26°C and 25 mg/L, respectively. The reaction rate constant K_{20} for BOD at $20°C = 0.15\ d^{-1}$, and temperature correction factor $\theta_T = 1.12$. The 30-d mean low flow of the receiving stream and mean velocity are 3.4 m^3/s and 0.5 m/s (1.8 km/h). The average water temperature, DO and BOD_5 are 24°C, 8.22 mg/L and 1.2 mg/L, respectively. The reaeration rate constant $K_{2,20}$ at $20°C = 0.6\ d^{-1}$ and temperature correction factor $\theta_T = 1.024$. Calculate the critical DO, and develop a DO profile downstream from the outfall at the initial temperature of the mixture. Assume that TDS ≈ 0 in the freshwater stream and the elevation is near sea level.

Solution

1. Calculate the initial temperature (T), BOD_5 and DO (DO_0) of the mixture from Equation 12.14a.

$$\text{Initial temperature, } T = \frac{3.4\,\text{m}^3/\text{s} \times 24°\text{C} \times 0.44\,\text{m}^3/\text{s} \times 26°\text{C}}{(3.4 + 0.44)\,\text{m}^3/\text{s}} = 24.2°\text{C}$$

$$\text{Initial BOD}_5\text{, BOD}_5 = \frac{3.4\,\text{m}^3/\text{s} \times 1.2\,\text{mg/L} \times 0.44\,\text{m}^3/\text{s} \times 25\,\text{mg/L}}{(3.4 + 0.44)\,\text{m}^3/\text{s}} = 3.93\,\text{mg/L}$$

$$\text{Initial DO, } DO_0 = \frac{3.4\,\text{m}^3/\text{s} \times 8.22\,\text{mg/L} \times 0.44\,\text{m}^3/\text{s} \times 4.0\,\text{mg/L}}{(3.4 + 0.44)\,\text{m}^3/\text{s}} = 7.74\,\text{mg/L}$$

2. Calculate the initial DO deficit (D_0) of the mixture with initial DO concentration of 7.74 mg/L.

DO saturation $DO_{sat} = 8.39$ mg/L at the initial mixture temperature of 24.2°C is obtained from Table 10.16.

Calculate the D_0 from Equation 12.14g.

$$D_0 = DO_{sat} - DO_0 = (8.39 - 7.74)\,\text{mg/L} = 0.65\,\text{mg/L}$$

3. Calculate the ultimate BOD (L_0) of the mixture with initial BOD_5 concentration of 3.93 mg/L from Equation 5.7.

$$L_0 = \frac{BOD_5}{1 - e^{-K_{20}t}} = \frac{3.93\,\text{mg/L}}{1 - e^{-0.15\,d^{-1} \times 5\,d}} = 7.45\,\text{mg/L}$$

4. Correct the rate constants K and K_2 from Equation 12.14f at the initial mixture temperature of 24.2°C.

$$K = K_{20}\theta_T^{T-20} = 0.15\,d^{-1} \times (1.12)^{(24.2-20)^\circ C} = 0.24\,d^{-1}$$

$$K_2 = K_{2,20}\theta_T^{T-20} = 0.6\,d^{-1} \times (1.024)^{(24.2-20)^\circ C} = 0.66\,d^{-1}$$

5. Determine the critical time (t_c) from Equation 12.14c, critical distance (x_c) from Equation 12.14e, critical DO deficit (D_c) from Equation 12.14d, and critical DO (DO_c) from Equation 12.14g.

$$t_c = \frac{1}{K_2 - K}\ln\left(\left(\frac{K_2}{K}\left(1 - \frac{(K_2 - K)\,D_0}{KL_0}\right)\right)\right)$$

$$= \frac{1}{(0.66 - 0.24)\,d^{-1}} \times \ln\left(\frac{0.66\,d^{-1}}{0.24\,d^{-1}}\left(1 - \frac{(0.66 - 0.24)\,d^{-1} \times 0.65\,\text{mg/L}}{0.24\,d^{-1} \times 7.45\,\text{mg/L}}\right)\right)$$

$$= 2.01\,d$$

$$x_c = t_c v = 2.01\,d \times 1.8\,\text{km/h} \times 24\,\text{h/d} = 86.8\,\text{km}$$

$$D_c = \frac{K}{K_2}L_0 e^{-Kt_c} = \frac{0.24\,d^{-1}}{0.66\,d^{-1}} \times 7.45\,\text{mg/L} \times e^{-0.24\,d^{-1} \times 2.01\,d}$$

$$= 1.67\,\text{mg/L}$$

$$DO_c = DO_{sat} - D_c = (8.39 - 1.67)\,\text{mg/L} = 6.72\,\text{mg/L}$$

Note: The minimum DO concentration in the river under the low flow condition is high enough to support healthy aquatic life.

6. Develop the DO profile data for the river.

Based on $t_c = 2.01$ d, the DO profile up to 3 days is developed at a selected time increment of 0.3 d. Calculate x from Equation 12.14e, D_t from Equation 12.14b, and DO_t from Equation 12.14g at each time increment. The sample calculations for $t = 0.3$ d are given below.

$$x = tv = 0.3\,d \times 1.8\,\text{km/h} \times 24\,\text{h/d} = 13\,\text{km}$$

$$D_t = \frac{KL_0}{K_2 - K}\left(e^{-Kt} - e^{-K_2t}\right) + D_0 e^{-K_2t}$$

$$= \frac{0.24\,d^{-1} \times 7.45\,\text{mg/L}}{(0.66 - 0.24)\,d^{-1}}\left(e^{-0.24\,d^{-1} \times 0.3d} - e^{-0.66\,d^{-1} \times 0.3\,d}\right) + 0.65\,\text{mg/L} \times e^{-0.66d^{-1} \times 0.3\,d}$$

$$= 1.00\,\text{mg/L}$$

$$DO_t = DO_{sat} - D_t = (8.39 - 1.00)\,\text{mg/L} = 7.39\,\text{mg/L}$$

Similarly, calculate the values for other time increments and tabulate the results.

t, d	0.0	0.3	0.6	0.9	1.2	1.5	1.8	2.1	2.4	2.7	3.0
x, km	0	13	26	39	52	65	78	91	104	117	130
D_t, mg/L	0.65	1.00	1.26	1.44	1.56	1.63	1.66	1.67	1.65	1.62	1.57
DO_t, mg/L	7.74	7.39	7.13	6.95	6.83	6.76	6.73	6.72	6.74	6.77	6.82

7. Plot the DO profile (or sag curve) downstream from the effluent outfall in Figure 12.10.

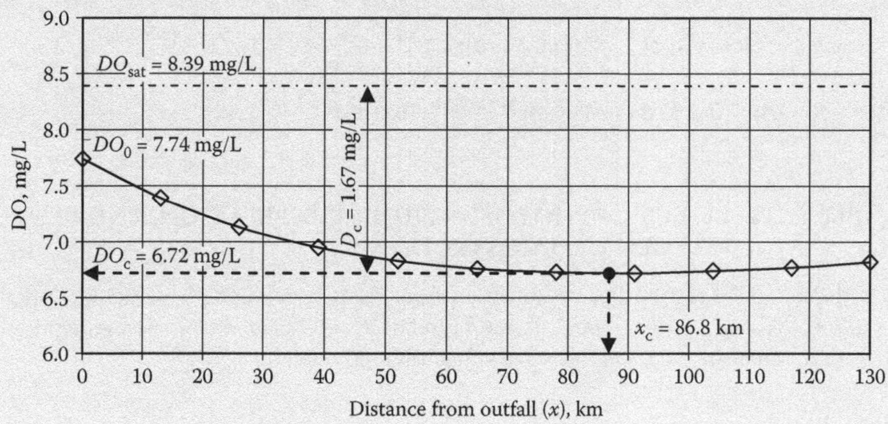

FIGURE 12.10 DO profile in the receiving stream from effluent bod discharged (Example 12.32).

Note: The DO profile is developed at a constant flow and velocity downstream of the outfall after initial mixing.

EXAMPLE 12.33: DETERMINATION OF REAERATION CONSTANT K_2

A stream has an average water depth of 2.1 m and velocity of 0.25 m/s. Determine the reaeration constants $K_{2,20}$ using three empirical relationships given by Equation 12.15. Also, calculate average K_2 at 25°C. The tidal dispersion coefficient of at 20°, $D_L = 1.76 \times 10^{-4}\,m^2/d$.

Solution

1. Calculate the $K_{2,20}$ using three empirical relationships.
   a. Churchill formula (Equation 12.15a).

$$K_{2,20} = \frac{5.03v^{0.969}}{H^{1.637}} = \frac{5.03 \times (0.25\,\text{m/s})^{0.969}}{(2.1\,\text{m})^{1.637}} = 0.39\,\text{d}^{-1}$$

   b. O'Conner and Dobbins formula (Equation 12.15b).

$$K_{2,20} = \frac{294(D_L v)^{0.5}}{H^{3/2}} = \frac{294 \times (1.76 \times 10^{-4}\,\text{m}^2/\text{d} \times 0.25\,\text{m/s})^{0.5}}{(2.1\,\text{m})^{3/2}} = \frac{3.9 \times (0.25\,\text{m/s})^{0.5}}{(2.1\,\text{m})^{3/2}} = 0.64\,\text{d}^{-1}$$

c. Owens formula (Equation 12.15c).

$$K_{2,20} = \frac{5.34\,v^{0.67}}{H^{1.85}} = \frac{5.34 \times (0.25\,\text{m/s})^{0.67}}{(2.1\,\text{m})^{1.85}} = 0.53\,\text{d}^{-1}$$

2. Determine the average value of three $K_{2,20}$.

$$\text{Average } K_{2,20} = \frac{1}{3} \times (0.39 + 0.64 + 0.53)\,\text{d}^{-1} = 0.52\,\text{d}^{-1}$$

3. Determine the value of K_2 at 25°C from Equation 12.14f using $\theta_T = 1.024$.

$$K_2 = K_{2,20}\theta_T^{T-20} = 0.52\,\text{d}^{-1} \times (1.024)^{(25-20)°\text{C}} = 0.59\,\text{d}^{-1}$$

EXAMPLE 12.34: EFFECT OF PHOTOSYNTHESIS, RESPIRATION, AND BENTHIC OXYGEN DEMAND ON DO SAG

Calculate the critical DO and the time it occurs from the outfall to critical DO. Also, check the critical DO in the stream. Assume $R_p = 0.6\,\text{g/m}^2\text{·d}$, $R_r = 0.4\,\text{g/m}^2\text{·d}$, $R_b = 0.7\,\text{g/m}^2\text{·d}$, and average depth of the stream is 2.5 m. Compare the results with those obtained in Example 12.32.

Solution

1. Calculate the rates of photosynthesis (r_p), respiration (r_r), and benthic oxygen demand (r_b) from Equation 12.16a.

$$r_p = \frac{R_p}{H} = \frac{0.6\,\text{g/m}^2\text{·d}}{2.5\,\text{m}} = 0.24\,\text{g/m}^3\text{·d}$$

$$r_r = -\frac{R_r}{H} = -\frac{0.4\,\text{g/m}^2\text{·d}}{2.5\,\text{m}} = -0.16\,\text{g/m}^3\text{·d}$$

$$r_b = -\frac{R_b}{H} = -\frac{0.7\,\text{g/m}^2\text{·d}}{2.5\,\text{m}} = -0.28\,\text{g/m}^3\text{·d}$$

$$r_p + r_r + r_b = (0.24 - 0.16 - 0.28)\,\text{g/m}^3\text{·d} = -0.20\,\text{g/m}^3\text{·d}$$

2. Determine the critical time (t_c) from Equation 12.16c.
$D_0 = 0.65$ mg/L (from Step 2 of Example 12.32), $L_0 = 7.45$ mg/L (from Step 3 of Example 12.32), $K = 0.24\,\text{d}^{-1}$ (from Step 4 of Example 12.32), and $K_2 = 0.66\,\text{d}^{-1}$ (from Step 4 of Example 12.32).

$$t_c = \frac{1}{K_2 - K}\ln\left(\frac{K_2}{K} - \frac{K_2(K_2 - K)D_0 + (K_2 - K)(r_p + r_r + r_b)}{K^2 L_0}\right)$$

$$= \frac{1}{(0.66 - 0.24)\,\text{d}^{-1}}\ln\left(\frac{0.66\,\text{d}^{-1}}{0.24\,\text{d}^{-1}} - \frac{\begin{array}{c}0.66\,\text{d}^{-1} \times (0.66 - 0.24)\,\text{d}^{-1} \times 0.65\,\text{g/m}^3 \\ +(0.66 - 0.24)\,\text{d}^{-1} \times (-0.20\,\text{g/m}^3\text{·d})\end{array}}{(0.24\,\text{d}^{-1})^2 \times 7.45\,\text{g/m}^3}\right)$$

$$= \frac{1}{0.42\,\text{d}^{-1}}\ln\left(2.75 - \frac{0.180\,\text{g/m}^3\text{·d} - 0.084\,\text{g/m}^3\text{·d}}{0.429\,\text{g/m}^3\text{·d}}\right) = \frac{1}{0.42\,\text{d}^{-1}}\ln(2.75 - 0.224) = \frac{0.927}{0.42\,\text{d}^{-1}}$$

$$= 2.21\,\text{d}$$

3. Determine the critical DO deficit (D_c) from Equation 12.16d and critical DO (DO_c) from Equation 12.14g.

$$D_c = \frac{1}{K_2}\left(KL_0 e^{-Kt_c} - (r_p + r_r + r_b)\right)$$
$$= \frac{1}{0.66\,\text{d}^{-1}}\left(0.24\,\text{d}^{-1} \times 7.45\,\text{g/m}^3 \times e^{-0.24\,\text{d}^{-1}\times 2.21\,\text{d}} - (-0.20\,\text{g/m}^3\cdot\text{d})\right)$$
$$= \frac{1}{0.66\,\text{d}^{-1}}\left(1.05\,\text{g/m}^3\cdot\text{d} + 0.20\,\text{g/m}^3\cdot\text{d}\right)$$
$$= 1.89\,\text{g/m}^3 \quad \text{or} \quad 1.89\,\text{mg/L}$$

$$DO_c = DO_{sat} - D_c = (8.39 - 1.89)\,\text{mg/L} = 6.50\,\text{mg/L}$$

4. Check the critical DO deficit (D_c) and DO (DO_c) calculated in Step 3.

D_c (from Equation 12.16b at $t = t_c = 2.21$ d)

$$= \frac{KL_0}{K_2 - K}\left(e^{-Kt_c} - e^{-K_2 t_c}\right) + D_0 e^{-K_2 t_c} - \frac{r_p + r_r + r_b}{K_2}\left(1 - e^{-K_2 t_c}\right)$$
$$= \frac{0.24\,\text{d}^{-1} \times 7.45\,\text{g/m}^3}{(0.66 - 0.24)\,\text{d}^{-1}}\left(e^{-0.24\,\text{d}^{-1}\times 2.21\,\text{d}} - e^{-0.66\,\text{d}^{-1}\times 2.21\,\text{d}}\right) + 0.65\,\text{g/m}^3 \times e^{-0.66\,\text{d}^{-1}\times 2.21\text{d}}$$
$$- \frac{-0.20\,\text{g/m}^3\cdot\text{d}}{0.66\,\text{d}^{-1}}\left(1 - e^{-0.66\,\text{d}^{-1}\times 2.21\,\text{d}}\right)$$
$$= 4.26\,\text{g/m}^3 \times (0.588 - 0.233) + 0.65\,\text{g/m}^3 \times 0.233 + 0.303\,\text{g/m}^3 \times (1 - 0.233)$$
$$= 1.51\,\text{g/m}^3 + 0.15\,\text{g/m}^3 + 0.23\,\text{g/m}^3$$
$$= 1.89\,\text{g/m}^3 \quad \text{or} \quad 1.89\,\text{mg/L}$$

DO_c (from Equation 12.14 g) $= DO_{sat} - D_c = (8.39 - 1.89)\,\text{mg/L} = 6.50\,\text{mg/L}$

5. Compare the results.
   a. The critical DO is calculated from Equation 12.16d in Step 3, and from Equation 12.16b in Step 4.
      Both equations give an identical result of $DO_c = 6.50$ mg/L.
   b. The critical t_c, D_c and DO_c are calculated in Examples 12.32 and 12.34. The results in Example 12.32 are $t_c = 2.01$ d, $D_c = 1.67$ mg/L, and $DO_c = 6.72$ mg/L. The results from Example 12.34 are $t_c = 2.21$ d, $D_c = 1.89$ mg/L, and $DO_c = 6.50$ mg/L. The respiration and benthic oxygen demands are sources of oxygen sinks and exceed the effect of photosynthesis. Larger critical DO deficit, lower critical DO, and longer time (or distances) to reach the critical DO are obtained in Example 12.34 than those in Example 12.32.

EXAMPLE 12.35: DO PROFILE IN A STREAM RECEIVING CARBONACEOUS AND NITROGENOUS WASTES

A stream receives effluent from a wastewater treatment plant. The water quality and physical data at the mixing zone of the stream are: 30-d mean flow plus effluent flow = 2.6 m^3/s, temperature 26°C, velocity = 0.2 m/s (0.72 km/h), $BOD_5 = 10$ mg/L, NH_3-N = 0.8 mg/L, $DO_0 = 6.1$ mg/L, K_{20} at 20° = 0.15 d^{-1} and $\theta_T = 1.12$; $K_{2,20}$ at 20°C = 0.54 d^{-1} and $\theta_T = 1.024$; and $K_N = 0.30$ d and $\theta_T = 1.047$. Nitrification in stream starts from day 2.8. Draw the DO profile in the stream with respect to distance from outfall. Comment on DO profile.

Solution

Since the nitrification starts from day 2.8, the DO sag curve has two parts. Part 1 is up to day 2.8 and is due to carbonaceous BOD demand only. Part 2 is from day 2.8 and onward; and the consumption of DO is due to combined effect of carbonaceous and nitrogenous oxygen demands.

1. Determine the rate constants K and K_2 at 26°C from Equation 12.14f.

$$K = K_{20}\theta_T^{T-20} = 0.15\,\text{d}^{-1} \times (1.12)^{(26-20)^\circ\text{C}} = 0.30\,\text{d}^{-1}$$

$$K_2 = K_{2,20}\theta_T^{T-20} = 0.54\,\text{d}^{-1} \times (1.024)^{(26-20)^\circ\text{C}} = 0.62\,\text{d}^{-1}$$

2. Determine the initial ultimate BOD (L_0) and initial DO deficit (D_0) in the mixing zone.

$$L_0 \text{ (from Equation 5.7)} = \frac{\text{BOD}_5}{1-\text{e}^{-K_{20}t}} = \frac{10\,\text{mg/L}}{1-\text{e}^{-0.15\,\text{d}^{-1}\times 5\,\text{d}}} = 19.0\,\text{mg/L}$$

$DO_{sat} = 8.11$ mg/L at 26°C is obtained from Table 10.16.

$$D_0 \text{ (from Equation 12.14g)} = DO_{sat} - DO_0 = (8.11 - 6.1)\,\text{mg/L} = 2.01\,\text{mg/L}$$

3. Check the critical time ($t_{c,BOD}$) for carbonaceous BOD.

$$t_{c,BOD}\,(\text{from Equation 12.14c}) = \frac{1}{K_2-K}\ln\left(\frac{K_2}{K}\left(1-\frac{(K_2-K)D_0}{KL_0}\right)\right)$$

$$= \frac{1}{(0.62-0.30)\text{d}^{-1}} \times \ln\left(\frac{0.62\text{d}^{-1}}{0.30\text{d}^{-1}}\left(1-\frac{(0.62-0.30)\text{d}^{-1}\times 2.01\,\text{mg/L}}{0.30\text{d}^{-1}\times 9.0\text{mg/L}}\right)\right)$$

$$= 1.89\text{d}$$

Since $t_{c,BOD}$ of 1.89 d is shorter than t_N of 2.8 d, the critical DO concentration for carbonaceous BOD ($DO_{c,BOD}$) will occur in Part 1 of the DO sag curve.

4. Determine the critical distance ($x_{c,BOD}$), critical DO deficit ($D_{c,BOD}$), and critical DO ($DO_{c,BOD}$) due to carbonaceous BOD demand.

$$x_{c,BOD}\,(\text{from Equation 12.14e}) = t_{c,BOD}v = 1.89\text{d} \times 0.72\text{km/h} \times 24\text{h/d} = 32.7\text{km}$$

$$D_{c,BOD}\,(\text{from Equation 12.14d}) = \frac{K}{K_2}L_0e^{-Kt_{c,BOD}} = \frac{0.30\text{d}^{-1}}{0.62\text{d}^{-1}} \times 19.0\text{mg/L} \times e^{-0.30\text{d}^{-1}\times 1.89\text{d}} = 5.21\,\text{mg/L}$$

$$\text{DO}_{c,BOD}\,(\text{from Equation 12.14g}) = DO_{sat} - D_{c,BOD} = (8.11-5.21)\text{mg/L} = 2.90\text{mg/L}$$

Note: The minimum DO concentration in the river may be a concern due to high BOD_5 load of 10 mg/L in the discharged effluent. Reduction in the effluent BOD_5 may be required to support healthy aquatic life.

5. Determine the D_t and DO_t for developing the Part 1 of DO sag curve at $t < 2.8$ d.

Select t values at an increment of 0.4 d and calculate x from Equation 12.14e, D_t from Equation 12.14b, and DO_t from Equation 12.14g at each time increment. $T = 26$°C. Calculation procedure similar to that presented in details in Example 12.32 is used.

6. Determine the rate constant K_N and initial ultimate nitrogenous oxygen demand (L_N).

$$K_N \text{ (from Equation 12.14f at 26°C)} = K_{N,20}\theta_T^{T-20} = 0.3\,\text{d}^{-1} \times (1.047)^{(26-20)^\circ\text{C}} = 0.40\,\text{d}^{-1}$$

$$L_N \text{ (from Equation 5.18c)} = 4.57(\text{NH}_3\text{-N}) = 4.57 \times 0.8\,\text{mg/L} = 3.66\,\text{mg/L}$$

Note: A discussion on ultimate nitrogenous oxygen demand L_N is given in Section 5.4.2 (see Equation 5.18 and Examples 5.26 and 5.27).

7. Calculate the distance where the nitrification starts at $t_N = 2.8$ d from Equation 12.14e.

   $x = tv = 2.8\text{ d } 0.72\text{ km/h } 24\text{ h/d} = 48.4\text{ km}$

8. Determine the D_t and DO_t for developing the Part 2 of DO sag curve at $t > 2.8$ d.

   When $t > 2.8$ days, the DO deficit is affected by both carbonaceous and nitrogenous oxygen demands. Select a time period up to 4.8 days. Use the same time increment of 0.4 d and calculate x from Equation 12.14e, D_t from Equation 12.17, and DO_t from Equation 12.14g at $t = 3.2$, 3.6, 4.0, 4.4, and 4.8 d. $T = 26°\text{C}$. The sample calculations for $t = 3.2$ d are given below.

   x (from Equation 12.14e) $= tv = 3.2\text{ d} \times 0.72\text{ km/h} \times 24\text{ h/d} = 55\text{ km}$

$$D_t \text{ (from Equation 12.17)} = \frac{KL_0}{K_2 - K}\left(e^{-Kt} - e^{-K_2t}\right) + D_0e^{-K_2t} + \frac{K_NL_N}{K_2 - K_N}\left(e^{-K_2(t-K_N)} - e^{-K_2(t-K_N)}\right)$$

$$= \frac{0.30\,\text{d}^{-1} \times 19.0\,\text{mg/L}}{(0.62 - 0.30)\,\text{d}^{-1}}\left(e^{-0.30\,\text{d}^{-1}\times 3.2\,\text{d}} - e^{-0.62\,\text{d}^{-1}\times 3.2\,\text{d}}\right) + 2.01\,\text{mg/L} \times e^{-0.62\,\text{d}^{-1}\times 3.2\,\text{d}}$$

$$+ \frac{0.40\,\text{d}^{-1} \times 3.66\,\text{mg/L}}{(0.62 - 0.40)\,\text{d}^{-1}}\left(e^{-0.40\,\text{d}^{-1}\times(3.2-2.8)\,\text{d}} - e^{-0.62\,\text{d}^{-1}\times(3.2-2.8)\,\text{d}}\right)$$

$$= 5.12\,\text{mg/L}$$

   DO_t (Equation 12.14g) $= DO_{sat} - D_t = (8.11\text{–}5.12)\text{ mg/L} = 2.99\text{ mg/L}$

   Similarly calculate D_t and DO_t for other given values of t.

9. Summarize the data developed for the DO sag curve.

   Tabulate the final results developed for Part 1 of DO sag curve in Step 5, and Part 2 of DO sag curve in Step 8.

t, d	0.0	0.4	0.8	1.2	1.6	2.0	2.4	2.8	3.2	3.6	4.0	4.4	4.8
x, km	0	7	14	21	28	35	41	48	55[a]	62	69	76	83
D_t, mg/L	2.01	3.47	4.39	4.92	5.16	5.20	5.10	4.91	5.12[b]	5.13	5.00	4.77	4.48
DO_t, mg/L	6.10	4.64	3.72	3.19	2.95	2.91	3.01	3.20	2.99[c]	2.98	3.11	3.34	3.63

10. Draw the DO profile (or sag curve).

    The DO sag curve in the stream with respect to distance from outfall is shown in Figure 12.11.

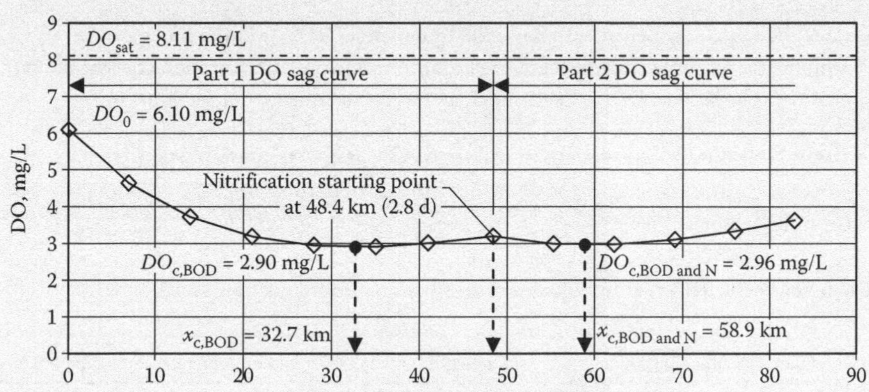

FIGURE 12.11 DO profile in the receiving stream due to carbonaceous and nitrogenous oxygen demands in the discharged effluent (Example 12.35).

11. Comment on the DO profile.
    a. The initial $DO_0 = 6.1$ mg/L after mixing at the outfall $x = 0$ km ($t = 0$ d). The saturation DO concentration $DO_{sat} = 8.11$ mg/L at 26°C and initial DO deficit $D_0 = 2.01$ mg/L.
    b. The oxygen consumption due to carbonaceous BOD occurs in Part 1 of DO sag curve. The first critical point with $DO_{c,BOD} = 2.90$ mg/L and $D_{c,BOD} = 5.21$ mg/L occur at a distance of $x_{c,BOD} = 32.7$ km from the outfall ($t_{c,BOD} = 1.89$ d).
    c. The DO starts to recover after the first critical point until the nitrification starts from a distance $x = 48.4$ km from the outfall ($t = 2.8$ d).
    d. The oxygen consumption due to combined carbonaceous and nitrogenous oxygen demands occurs in Part 2 of DO sag curve. The second critical point with $DO_{c,BOD\&N} = 2.96$ mg/L and $D_{c,BOD\&N} = 5.15$ mg/L occur at a distance $x_{c,BOD} = 58.9$ km from the outfall ($t_{c,BOD} = 3.41$ d). The second critical DO deficit is estimated from Equation 12.17 by a trial and error procedure.
    e. The DO recovery starts again after the second critical point.

EXAMPLE 12.36: CRITICAL DO IN A TIDAL RIVER OR ESTUARY

An estuary receives secondary treated effluent from a POTW. The downstream velocity in the estuary due to freshwater flow is 0.1 m/s (8640 m/d). The deoxygenation constant K_{20} at 20°C = 0.16 d^{-1} and $\theta_T =$ 1.048; and reaeration constant $K_{2,20}$ at 20°C = 0.53 d^{-1} and $\theta_T = 1.024$. The BOD_5 and DO at the outfall after mixing are 8.1 and 6.7 mg/L, respectively. The temperature in the estuary below outfall is 24°C, and tidal dispersion coefficient at 20°C, $D_{L,20} = 150\ m^2/s$ with $\theta_T = 1.037$. Calculate the critical DO and its distance from the outfall.

Solution

1. Determine the rate constants K, K_2, and D_L from Equation 12.14f under field conditions at 24°C.

$$K = K_{20}\theta_T^{T-20} = 0.16\,d^{-1} \times (1.048)^{(24-20)^\circ C} = 0.19\,d^{-1}$$

$$K_2 = K_{2,20}\theta_T^{T-20} = 0.53\,d^{-1} \times (1.024)^{(24-20)^\circ C} = 0.58\,d^{-1}$$

$$D_L = D_{L,20}\theta_T^{T-20} = 150\,m^2/s \times (1.037)^{(24-20)^\circ C}$$
$$= 173\,m^2/s \quad \text{or} \quad D_L = 173\,m^2/s \times 86400\,s/d = 1.49 \times 10^7\,m^2/d$$

2. Determine the initial DO deficit (D_0) and initial ultimate BOD (L_0) in the mixing zone.

   To simplify the example, the impact of TDS concentration on DO saturation concentration (DO_{sat}) is ignored the calculations. $DO_{sat} = 8.42$ mg/L at 24°C is obtained from Table 10.16.

$$D_0 \text{ (from Equation 12.14g)} = DO_{sat} - DO_0 = (8.42 - 6.7)\,mg/L = 1.72\,mg/L$$

$$L_0 \text{ (from Equation 5.7)} = \frac{BOD_5}{1 - e^{-K_{20}t}} = \frac{8.1\,mg/L}{1 - e^{-0.16\,d^{-1} \times 5\,d}} = 14.7\,mg/L$$

3. Calculate the coefficients J_1 from Equation 12.18d and J_2 from Equation 12.18e.

$$J_1 = \frac{U}{2D_L}\left(1 - \sqrt{1 + \frac{4KD_L}{U^2}}\right) = \frac{8640\,m/d}{2 \times 1.49 \times 10^7\,m^2/d}\left(1 - \sqrt{1 + \frac{4 \times 0.19\,d^{-1} \times 1.49 \times 10^7\,m^2/d}{(8640\,m/d)^2}}\right)$$
$$= 2.90 \times 10^{-4}\,m^{-1} \times (1 - \sqrt{1 + 0.152})$$
$$= -2.13 \times 10^{-5}\,m^{-1}$$

$$J_2 = \frac{U}{2D_L}\left(1-\sqrt{1+\frac{4K_2D_L}{U^2}}\right) = \frac{8640\,\text{m/d}}{2\times 1.49\times 10^7\,\text{m}^2/\text{d}}\left(1-\sqrt{1+\frac{4\times 0.58\,\text{d}^{-1}\times 1.49\times 10^7\,\text{m}^2/\text{d}}{(8640\,\text{m/d})^2}}\right)$$

$$= 2.90\times 10^{-4}\,\text{m}^{-1}\times(1-\sqrt{1+0.463})$$

$$= -6.08\times 10^{-5}\,\text{m}^{-1}$$

4. Determine the critical distance (x_c) from Equation 12.18c.

$$x_c = \frac{1}{J_1-J_2}\ln\left(\frac{J_2}{J_1}\left(1-\frac{(J_2-J_1)D_0}{J_1L_0}\right)\right)$$

$$= \frac{1}{(-2.13-(-6.08))\times 10^{-5}\,\text{m}^{-1}}$$

$$\times \ln\left(\frac{-6.08\times 10^{-5}\,\text{m}^{-1}}{-2.13\times 10^{-5}\,\text{m}^{-1}}\left(1-\frac{(-6.08-(-2.13))\times 10^{-5}\,\text{m}^{-1}\times 1.72\,\text{mg/L}}{-2.13\times 10^{-5}\,\text{m}^{-1}\times 14.7\,\text{mg/L}}\right)\right)$$

$$= 2.53\times 10^4\,\text{m}\times\ln(2.85\times(1-0.217))$$

$$= 2.03\times 10^4\,\text{m} \quad \text{or} \quad 20.3\,\text{km}$$

5. Determine the critical DO deficit (D_c) and critical DO (DO_c).
   Apply Equation 12.18b to obtain the D_c.

$$D_c = \frac{KL_0}{K_2-K}\left(e^{J_1x_c}-e^{J_2x_c}\right)+D_0e^{J_2x_c}$$

$$= \frac{0.19\text{d}^{-1}\times 14.7\,\text{mg/L}}{(0.58-0.19)\,\text{d}^{-1}}\times\left(e^{-2.13\times10^{-5}\,\text{m}^{-1}\times2.03\times10^4\,\text{m}}-e^{-6.08\times10^{-5}\,\text{m}^{-1}\times2.03\times10^4\,\text{m}}\right)$$

$$+1.72\,\text{mg/L}\times e^{-6.08\times10^{-5}\,\text{m}^{-1}\times2.03\times10^4\,\text{m}}$$

$$= 7.16\,\text{mg/L}\times\left(e^{-0.432}-e^{-1.23}\right)+1.72\,\text{mg/L}\times e^{-1.23}$$

$$= 7.16\,\text{mg/L}\times(0.649-0.292)+1.72\,\text{mg/L}\times 0.292$$

$$= 3.06\,\text{mg/L}$$

DO_c (from Equation 12.14g) $= DO_{sat} - D_c = (8.42-3.06)\,\text{mg/L} = 5.36\,\text{mg/L}$

EXAMPLE 12.37: TIDAL DISPERSION COEFFICIENT D_L OF AN ESTUARY

Determine the tidal dispersion coefficient D_L of an estuary from three methods: physical data, salinity data, and dye tracer data. The Manning coefficient $n = 0.0128$, minimum tidal velocity $= 6.5$ m/s (or 5.6×10^5 m/d), and the hydraulic mean radius at the cross section where maximum tidal velocity is measured $= 25$ m. At a distance of 1100 m upstream from the open sea the salinity is 2950 mg/L, and the salinity of sea water is 35,000 mg/L. The downstream average velocity is 0.15 m/s (or 1.3×10^4 m/d). The dye tracer study included release of a slug dye tracer slug at a certain time. The dye tracer concentration was measured at different distance intervals upstream from the released point. A dye tracer profile was similar to that in Figure 12.9 but skewed towards the left. $x_{max} = 800$ m and $C_{max} = 0.8$ mg/L. The dye concentration C_x at the distance $X = 2100$ m from the point of dye release was 0.5 mg/L.

Solution

1. Determine the D_L from Equation 12.19a based on physical data.

$$D_L = 63nU_T(R_H)^{5/6} = 63\times 0.0128\times 5.6\times 10^5\,\text{m/d}\times 25\,\text{m}^{5/6} = 6.6\times 10^6\,\text{m}^2/\text{d}$$

2. Determine the D_L from Equation 12.19b based on salinity data.

$$D_L = \frac{U}{\ln(C_0/C_c)}x = \frac{1.3 \times 10^4\,\text{m/d}}{\ln\left(\frac{35{,}000\,\text{mg/L}}{2950\,\text{mg/L}}\right)} \times 1100\,\text{m} = 5.8 \times 10^6\,\text{m}^2/\text{d}$$

3. Determine the D_L from Equation 12.19c based on dye tracer study data.
   The time t for tide to reach point x, is calculated from Equation 12.19d.

$$t = \frac{X}{U} = \frac{2100\,\text{m}}{0.15\,\text{m/s}} = 14{,}000\,\text{s} \quad \text{or} \quad 0.16\,\text{d}$$

$$D_L = -\frac{(X - X_{\max})^2}{4t\ln(C_x/C_{\max})} = -\frac{((2100 - 800)\,\text{m})^2}{4 \times 0.16\,\text{d} \times \ln\left(\frac{0.5\,\text{mg/L}}{0.8\,\text{mg/L}}\right)} = -\frac{1{,}690{,}000\,\text{m}^2}{0.64\,\text{d} \times (-0.470)} = 5.6 \times 10^6\,\text{m}^2/\text{d}$$

EXAMPLE 12.38: SALINITY INTRUSION IN A WATER INTAKE LOCATED ON AN ESTUARY

A city is located on an estuary. The water intake structure is in the estuary 32 km upstream from the ocean. The average cross section of the estuary is 460 m^2. The dispersion coefficient is 310 m^2/s (or 2.68×10^7 m/d), and the chloride concentration of the ocean is about 19,000 mg/L. The fresh water flow to the ocean is 28 m^3/s. To prevent the loss of fresh water a dam is proposed upstream of the intake structure. The fresh water flow from the dam will be reduced to 3 m^3/s. Determine the chloride concentration at the intake before and after dam construction. What course of action should the city take?

Solution

1. Determine the chloride concentration at the intake ($C_{c,1}$) before the dam construction.
   The average velocity of freshwater through the estuary,

$$U_1 = \frac{Q_1}{A} = \frac{28\,\text{m}^3/\text{s}}{460\,\text{m}^2} = 0.0609\,\text{m/s or } 5260\,\text{m/d}$$

   Apply Equation 12.19b to determine the concentration $C_{c,1}$.

$$C_{c,1} = C_0 e^{-\frac{U_1}{D_L}x_s} = 19{,}000\,\text{mg/L} \times e^{-\frac{5260\,\text{m/d}}{2.68\times10^7\,\text{m}^2/\text{d}}\times 32{,}000\,\text{m}} = 19{,}000\,\text{mg/L} \times 0.00187 = 36\,\text{mg/L}$$

2. Determine the chloride concentration at the intake ($C_{c,2}$) after dam construction

$$U_2 = \frac{Q_2}{A} = \frac{3\,\text{m}^3/\text{s}}{460\,\text{m}^2} = 0.00652\,\text{m/s or } 563\,\text{m/d}$$

$$C_{c,2} = C_0 e^{-\frac{U_2}{D_L}x_s} = 19{,}000\,\text{mg/L} \times e^{-\frac{563\,\text{m/d}}{2.68\times10^7\,\text{m}^2/\text{d}}\times 32{,}000\,\text{m}} = 19{,}000\,\text{mg/L} \times 0.511 = 9710\,\text{mg/L}$$

3. Describe the course of action the city must take.
   The maximum allowable chloride concentration in drinking water for aesthetic quality is 250 mg/L in the *Secondary Drinking Water Standards*. Current chloride concentration of 36 mg/L is well below the maximum allowable concentration. After the construction of the dam the chloride concentration of 9710 mg/L will exceed the limit for municipal water supply. It is recommended that the city build a dam and a pipeline to transport the raw water from the proposed upstream lake to the treatment plant.

EXAMPLE 12.39: DECAY OF CONTAMINANTS IN A RESERVOIR

The secondary treated effluent is discharged into a small lake for further treatment. The lake receives 10,000 m^3/d average drainage, while effluent discharge is 3800 m^3/d. The average lake volume is 26,500 m^3. The total BOD_5 in natural drainage is 3.2 mg/L and that in the effluent is 15 mg/L. The first-order decay coefficient for BOD_5 in the lake is 0.50 d^{-1}. Determine the BOD_5 in the reclaimed water withdrawn from the lake after discharge of reclaimed water.

Solution

1. Determine from Equation 12.20a the total BOD_5 load W reaching the lake.

$$W = Q_iC_i + Q_wC_w = 10{,}000\,\text{m}^3/\text{d} \times 3.2\,\text{g/m}^3 + 3800\,\text{m}^3/\text{d} \times 15\,\text{g/m}^3 = 89{,}000\,\text{g/d} \quad \text{or} \quad 89\,\text{kg/d}$$

2. Detention the detention time θ from Equation 12.20c.

$$\theta = \frac{V}{Q_i + Q_w} = \frac{26{,}500\,\text{m}^3}{(10{,}000 + 3800)\,\text{m}^3/\text{d}} = 1.92\,\text{d}$$

3. Determine the modified decay coefficient β from Equation 12.20d.

$$\beta = \frac{1}{\theta} + K = \frac{1}{1.92\,\text{d}} + 0.50\,\text{d}^{-1} = 1.02\,\text{d}^{-1}$$

4. Determine the equilibrium concentration of BOD_5 (C_e) from Equation 12.20e.

$$C_e = \frac{W}{\beta V} = \frac{89{,}000\,\text{g/d}}{1.02\,\text{d}^{-1} \times 26{,}500\,\text{m}^3} = 3.3\,\text{g/m}^3 \quad \text{or} \quad 3.3\,\text{mg/L}$$

The BOD_5 concentration in the lake with the reclaimed water = 3.3 g/m^3. This is close to the BOD_5 concentration of the receiving natural drainage water. The effect of reclaimed water discharge into the lake is a minimum.

12.6.3 Design of Outfall Structures

The outfall structures are designed to withstand the damaging forces such as erosion, uplift, and any natural forces due to velocity, turbulence, and impacts from debris in the receiving waters. The basic design considerations are: (1) flow velocity, (2) stratification due to salinity or temperature, (3) depth (shallow or deep), (4) shape (wide or narrow), (5) reversal of current (tidal or wind-produced seiches), (6) wind circulation, (7) temperature and salinity of effluent, and (8) proper mixing and dispersion of effluent.

Three principal design factors are: (1) construction material, (2) means of anchoring the structure in the body of water, and (3) the diffuser design. Each of these design factors are described below.

Construction Material: The typical construction material for outfall structure is concrete. The pipe may be of steel, concrete, plastic, or any similar material that is compatible with the waste stream, receiving water, site characteristics, and economics of fabrication and installation.

Installation and Anchoring: Installation and anchoring of outfall pipes and diffusers is especially important if strong currents prevail. Additional problems of concern are erosion and scour, and transport of bedload particularly objects such as sediments, tree trunks, and debris. Outfall pipes are secured by rock ballast, cement or sand bags, or chain and anchors. Often, pipes may be partly embedded in a concrete apron.[51,74]

Diffusers: Diffusers must disperse the effluent efficiently and evenly to achieve good mixing. Density and temperature gradients, as well as currents and winds influence the dispersion and mixing in the receiving waters. Knowledge of these conditions is necessary. Procedure for determining initial dilution and dispersion for ocean outfalls is given in Reference 51. Equations 12.21a through 12.21d are used for the design and analysis of the diffusers.[51] Design details of effluent outfall and diffuser system are shown in Figure 12.12.

$$E_1 = \frac{\Delta p}{\gamma} = HGL_1 - WSEL \tag{12.21a}$$

$$q_1 = C_{D,1} a_1 \sqrt{2gE_1} \quad \text{or} \quad q_1 = C_{D,1} \frac{\pi}{4}(d_1)^2 \sqrt{2gE_1} \tag{12.21b}$$

$$Q_i = q_1 + \cdots + q_{i-1} \quad (i = 2,3, \ldots, n) \tag{12.21c}$$

$$V_i = \frac{4Q_i}{\pi(D_i)^2} \quad \text{or} \quad V_i = \frac{Q_i}{A_i} \quad (i = 2, 3, \ldots, n) \tag{12.21d}$$

$$h_{f,i} = f \frac{L_i}{D_i} \frac{(V_i)^2}{2g} \quad (i = 2, 3, \ldots, n) \tag{12.21e}$$

$$E_i = E_{i-1} + h_{f,i} + \frac{\Delta S_g}{S_g} \Delta Z_i \quad (i = 2, 3, \ldots, n) \tag{12.21f}$$

$$q_i = C_{D,i} a_i \sqrt{2gE_i} \quad \text{or} \quad q_i = C_{D,i} \frac{\pi}{4}(d_i)^2 \sqrt{2gE_i} \quad (i = 2, 3, \ldots, n) \tag{12.21g}$$

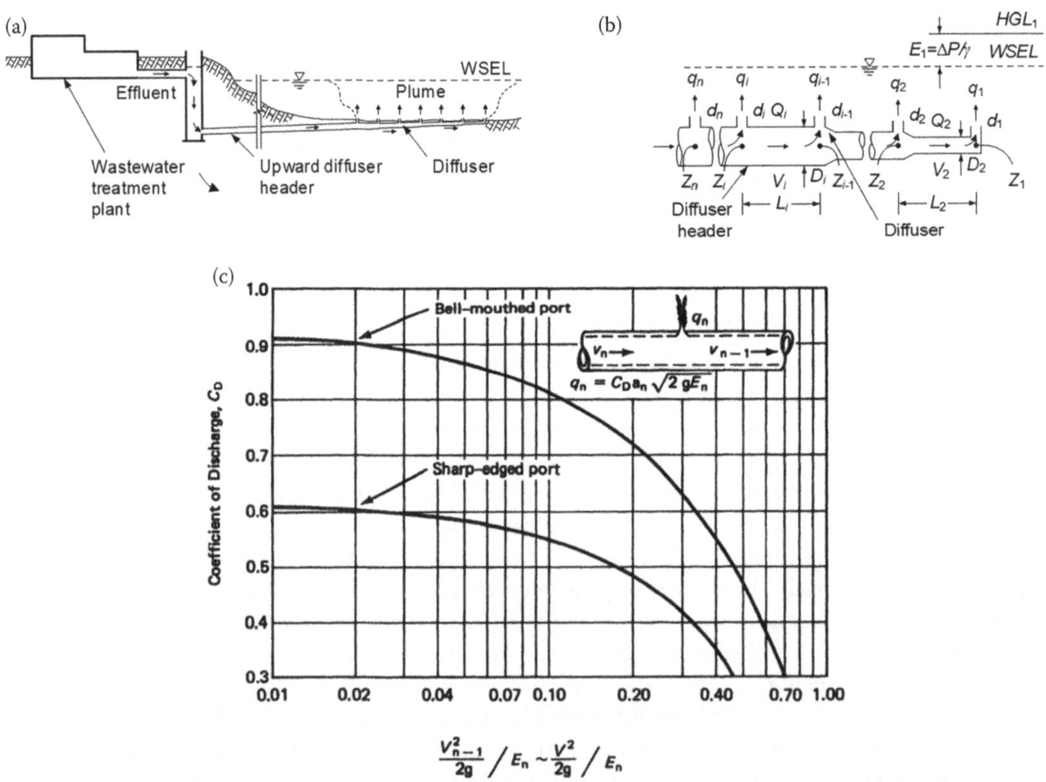

FIGURE 12.12 Effluent discharge outfall diffuser design details: (a) schematic of outfall pipe, (b) definition sketch of diffuser header, and (c) coefficient of discharge for small ports on the side of a pipe. (From Reference 51; used with permission of American Society of Civil Engineers.)

where

a_1 and d_1 = area and diameter of Port 1 which is the outermost port, m^2 and m

a_i and d_i = area and diameter of Port i, m^2 and m

A_i = area of diffuser pipe between Ports $(i-1)$ and i, m^2. Port i is upstream of Port $(i-1)$.

$C_{D,1}$ = coefficient of discharge of Port 1, dimensionless. The value of $C_{D,1}$ is 0.61 and 0.91 for sharp-edged and bell-mouthed ports, respectively.[51]

$C_{D,i}$ = coefficient of discharge of Port i, dimensionless. The value of $C_{D,i}$ is a function of the ratio of velocity to operating head $\left(h_{v,i}/E_i \text{ or } \frac{(V_i)^2}{2g}\frac{1}{E_i}\right)$. This relationship is graphically shown in Figure 12.12c.[51] The velocity head is calculated from Equation 6.15a.

D_i = diameter of the diffuser header between Ports $(i-1)$ and i, m

E_1 = assumed operating head for Port 1, m

E_i and E_{i-1} = total operating heads in the diffuser pipe at Ports i and $(i-1)$, m

f = Darcy friction factor, dimensionless

$h_{f,i}$ = head loss caused by friction in the diffuser header between Ports $(i-1)$ and i, m

HGL_1 = hydraulic grade line elevation in the diffuser pipe immediately upstream of Port 1, m

L_i = length of the diffuser header between Ports $(i-1)$ and i, m

$\Delta p/\gamma$ = difference in pressure head between inside and outside of diffuser Port 1, m

q_1 = discharge through Port 1, m^3/s

q_i = discharge through Port i, m^3/s

Q_i = total flow through the diffuser header between Ports $(i-1)$ and i, m^3/s

S_g = specific gravity of the receiving water, dimensionless. S_g for water = 1.

ΔS_g = difference in specific gravity between receiving water and effluent, dimensionless, dimensionless. ΔS_g for receiving water and effluent = 0 when effluent is discharged into freshwater.

V_i = velocity in the diffuser header between Ports $(i-1)$ and i, m/s

$WSEL$ = free water surface elevation in the receiving water, m

ΔZ_i = elevation differential of the centerline of diffuser header between Ports $(i-1)$ and i $(\Delta Z_i = Z_{i-1} - Z_i)$, m

Note: Ports 1 and n are the outermost and innermost ports, respectively.

The diffusers may have a total area approximately half the cross-sectional area of the diffuser pipe. The diameter of the port is governed by considerations of clogging, hydraulic performance, and spacing. Typical diffusers have port diameters of 10–30 cm (4–12 in), spaced over a desired length. As an alternative, two ports may be provided in the bulkhead at the end of the discharge pipe. One port is on the top and the other is at the bottom. These serve not only to discharge waste, but also to provide outlets for settleable and floatable materials. Other ports are normally alternated on the opposite sides of the pipe at the center line.[74] The minimum center-to-center spacing of the port is normally 10 times the port diameter.[51]

Design Procedure: The hydraulic analysis of multiport diffusers is basically a problem of manifolds and is somewhat complex. A discussion of this subject may be found in References 75 and 76. A simplified procedure is given in Reference 51. It is a process of hydraulic iteration using a trial-and-error solution. The procedure is illustrated in Example 12.41.

The diffusers are generally designed for maximum flow. The performance of the diffuser at lower flows must also be evaluated. In case low velocities are encountered under initial flow conditions, it is customary to plug the inboard ports during initial flow periods and open them later as the flow increases.

Another type of diffuser is a manifold and laterals normally used for flow distribution into a basin. These manifolds and laterals for influent distribution in a basin are presented in Section 9.4.5. Examples 9.13 has the derivation of various equations applied for flow distribution through manifold and laterals, and Example 9.14 covers the design calculations.[77–79]

EXAMPLE 12.40: OUTFALL PIPE DESIGN

A circular cast-iron outfall pipe discharges effluent into a creek. The outfall is totally submerged. The low and high water depths in the creek are 1.20 and 3.20 m, respectively. The pipe is anchored over a concrete pad that is flushed with the creek bed. The diameter of the outfall pipe is 92 cm (or 36 in) and the linear pipe length is 100 m (~330 ft). The upper end of the outfall pipe is connected to an effluent junction box in which the effluent is brought from a treatment facility via an open channel. A sluice gate is provided at the pipe entry. There are two 22.5° bends. The peak effluent discharge through the pipe is 1.321 m^3/s. Calculate the head loss in the pipe and determine the water surface elevation in the effluent junction box.

Solution

1. Calculate the velocity in the outfall pipe.

$$\text{Area of the pipe, } A = \frac{\pi}{4}D^2 = \frac{\pi}{4} \times (0.92\,\text{m})^2 = 0.665\,\text{m}^2$$

$$\text{Velocity at peak flow, } V = \frac{Q}{A} = \frac{1.321\,\text{m}^3/\text{s}}{0.665\,\text{m}^2} = 1.99\,\text{m/s}$$

2. Compute the head losses in the outfall pipe.

The friction and minor head losses in the outfall pipe are calculated separately. General discussions about hydraulic calculations may be found in Section 6.3.7.

a. Calculate the friction head loss from Darcy–Weisbach equation (Equation 6.13a) using coefficient of friction $f = 0.035$ and hydraulic mean diameter $D_h = D = 0.92$ m for circular pipe.

$$h_f = \frac{fLV^2}{2gD_h} = \frac{0.035 \times 100\,\text{m} \times (1.99\,\text{m/s})^2}{2 \times 9.81\,\text{m/s}^2 \times 0.92\,\text{m}} = 0.77\,\text{m}$$

b. Calculate the minor head loss from Equation 6.15b.

The minor head losses in the outfall pipe consist of those created by the open sluice gate (which is similar to an entrance loss, bends, and exit). The head loss coefficients are: $K_{exit} = 1.0$ for the exit loss, $K_{bend} = 0.15$ for each of two bend, and $K_{gate} = 0.5$ for the sluice gate (Appendix C).

$$\text{Head loss due to exit, } h_{m,exit} = K_{exit}\frac{V^2}{2g} = 1.0 \times \frac{(1.99\,\text{m/s})^2}{2 \times 9.81\,\text{m/s}^2} = 0.20\,\text{m}$$

$$\text{Head loss by two 22.5° bends, } h_{m,bend} = 2 \times K_{bend}\frac{V^2}{2g} = 2 \times 0.15 \times \frac{(1.99\,\text{m/s})^2}{2 \times 9.81\,\text{m/s}^2} = 0.06\,\text{m}$$

$$\text{Head loss by the sluice gate, } h_{m,gate} = K_{gate}\frac{V^2}{2g} = 0.5 \times \frac{(1.99\,\text{m/s})^2}{2 \times 9.81\,\text{m/s}^2} = 0.10\,\text{m}$$

$$\text{Total minor head loss, } h_m = h_{m,exit} + h_{m,bend} + h_{m,gate} = (0.20 + 0.06 + 0.10)\,\text{m} = 0.36\,\text{m}$$

c. Calculate the total head loss through the outfall pipe.

$$\Delta h = h_f + h_m = (0.77 + 0.36)\,\text{m} = 1.13\,\text{m}$$

3. Determine the water surface elevation in the effluent junction box.

   The maximum water surface elevation is reached when the peak flow is discharged in the creek at the high water depth of 3.20 m.

   $$\text{WSEL}_{max} = H_{high} + \Delta h = (3.20 + 1.13)\ \text{m} = 4.33\ \text{m}$$

4. Draw the sectional details of outfall structure and hydraulic profile in Figure 12.13.

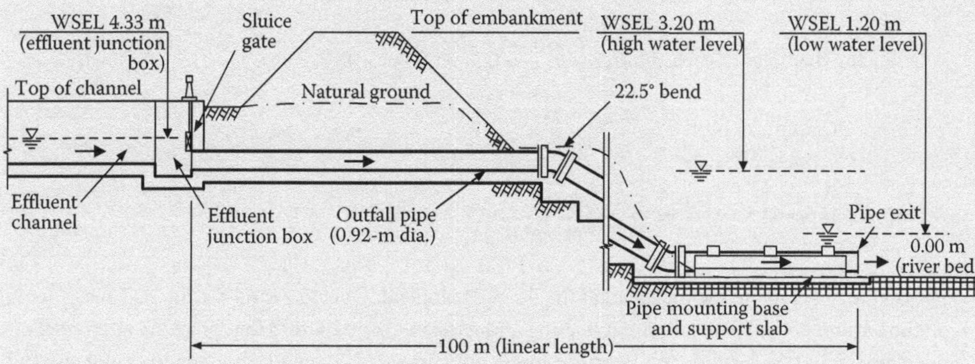

FIGURE 12.13 Conceptual side view of the outfall structure and hydraulic profile (Example 12.40).

EXAMPLE 12.41: DESIGN OF AN EFFLUENT DIFFUSER PIPE

The effluent should be dispersed in the receiving water to avoid localized pollution. Diffusers are most commonly designed for this purpose. Secondary effluent is discharged at a stream bed through a diffuser pipe. Design the diffuser pipe and connect it to the outfall pipe designed in Example 12.40. The diameter of the diffuser header is 92 cm, and it is connected with the 92 cm outfall pipe shown in Figure 12.13. The minimum center-to-center spacing of the ports is at least 10 times the port diameter. The peak design effluent flow is 1.321 m^3/s. Determine the head loss and water surface elevation in the effluent junction box provided at the entrance of the outfall pipe (see Example 12.40). Describe the design features and prepare the sketch of the diffuser pipe, and its layout.

Solution

1. *Select a procedure for diffuser pipe design*: The design of a diffuser header is basically a trial-and-error procedure.[51,74] The basic steps are: (a) select a tentative configuration, that is, number of ports, diameter, spacings, and layout; (b) select an operating head at the end of diffuser header; (c) calculate the flow through the outermost port; (d) calculate the head loss in the header pipe between the outermost port and next upstream port. These head losses are velocity head, the density differential head and the drop due to slope of the diffuser header; (e) calculate the operating head, velocity and flow at the next inboard port; and (f) continue the procedure over the next port, and then over the entire length of the diffuser header. The sum of discharges through all ports should equal the design effluent flow through the outfall pipe. If the total flow is not close, the operating head is adjusted and the calculations are repeated. The process can be tedious if done manually.

2. *Select the design equations*: Equations 12.21a through 12.21g are used for the design of the diffuser pipe. The design procedure and calculation steps are provided in Steps 4 and 5.
3. *Select a tentative diffuser configuration*: Provide a 92-cm diameter diffuser header pipe to connect it with the outfall pipe designed in Example 12.40, and shown in Figure 12.13. Provide total diffuser port area approximately half the area of the diffuser header pipe. The diffuser header area $A = 0.665\ \text{m}^2$ (Example 12.40, Step 1).

$$\text{Total diffuser area, } A_\text{d} = \frac{1}{2}A = \frac{1}{2} \times 0.665\ \text{m}^2 = 0.333\ \text{m}^2$$

Select the diameter of each diffuser port $d = 0.27$ m (10.5 in).

$$\text{Area of each diffuser port, } a = \frac{\pi}{4}D^2 = \frac{\pi}{4} \times (0.27\ \text{m})^2 = 0.0573\ \text{m}^2 \text{ per port}$$

$$\text{Number of ports, } N = \frac{A_\text{d}}{a} = \frac{0.333\ \text{m}^2}{0.0573\ \text{m}^2/\text{port}} = 5.8 \text{ ports} \quad \text{or} \quad 6 \text{ ports}$$

Provide two ports in the bulkhead at the downstream end. One port is on the top and the other is on the bottom of the bulkhead to exit floating and settleable solids. Other four ports are alternated on opposite sides of the header pipe along the center line. Provided an identical center-to-center spacing $L_\text{d} = 3$ m (which is greater than 10 times the diameter of the port). Therefore, total length of the diffuser header pipe is 13 m (~43 ft), including an additional segment of 1 m to connect the outfall pipe with the diffuser header. The design details, diffuser configuration, and diffuser header are shown in Figure 12.14. All diffuser ports have a divergent mouthpiece.[51]

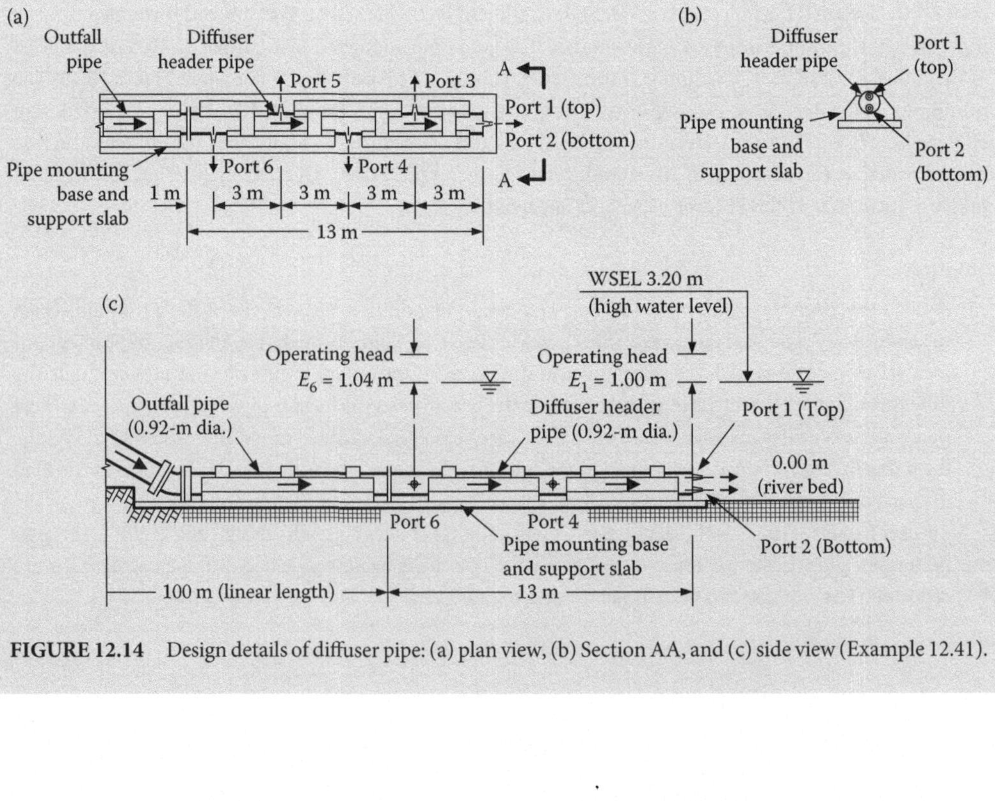

FIGURE 12.14 Design details of diffuser pipe: (a) plan view, (b) Section AA, and (c) side view (Example 12.41).

4. Estimate the approximate operating head (E_1) at Port 1.
   Assume that the total flow is evenly distributed among all six ports.
   Average flow through each port,

$$q_{avg} = \frac{Q}{6} = \frac{1.321\,\text{m}^3/\text{s}}{6\,\text{ports}} = 0.220\,\text{m}^3/\text{s per port}$$

For the outmost bell-mouthed port, $C_{D,1} = 0.91$. Rearrange Equation 12.21b and calculate the operating head required for Port 1 by assuming $q_1 = q_{avg} = 0.220\,\text{m}^3/\text{s}$.

$$E_1 = \frac{1}{2g}\left(\frac{q_1}{C_{D,1}a_1}\right)^2 = \frac{1}{2 \times 9.81\,\text{m/s}^2}\left(\frac{0.220\,\text{m}^3/\text{s}}{0.91 \times 0.0573\,\text{m}^2}\right)^2 = 0.91\,\text{m}$$

The initial operating head needs to be at least 0.91 m to start the iteration procedure. To accelerate iteration process, assume an operating head $E_1 = 1$ m.

5. *Calculate the discharges through Ports 1 and 2*: The head loss calculations and discharge through the diffuser ports is an iterative process. The diffuser Ports 1 and 2 are in the bulkhead. There is no connecting pipe between them. Ignore the vertical difference between the centerlines of two ports. $E_2 = E_1 = 1$ m, $C_{D,1} = 0.91$ for a bell-mouth port in the bulk head (Equation 12.21b and Figure 12.12). The average discharge through Ports 1 and 2 are equal, $q_2 = q_1$. Calculate the discharges q_1 and q_2 at Ports 1 and 2 from Equation 12.21b.

$$q_1 = C_{D,1}a\sqrt{2gE_1} = 0.91 \times 0.0573\,\text{m}^2 \times \sqrt{2 \times 9.81\,\text{m/s}^2 \times 1\,\text{m}} = 0.231\,\text{m}^3/\text{s} \quad \text{and}$$
$$q_2 = q_1 = 0.231\,\text{m}^3/\text{s}.$$

6. Calculate the discharge (Q_3), velocity (v_3), head loss ($h_{f,3}$), operating head (E_3), and velocity head ($h_{v,3}$) at diffuser header upstream of Ports 1 and 2.
   Discharge from Equation 12.21c, $Q_3 = q_1 + q_2 = (0.231 + 0.231)\,\text{m}^3/\text{s} = 0.462\,\text{m}^3/\text{s}$

   Velocity from Equation 12.21d, $V_3 = \dfrac{Q_3}{A} = \dfrac{0.462\,\text{m}^3/\text{s}}{0.665\,\text{m}^2} = 0.695\,\text{m/s}$

   Head loss from Equation 12.21e, $h_{f,3} = f\dfrac{L_d}{D}\dfrac{(V_3)^2}{2g} = 0.037 \times \dfrac{3\,\text{m}}{0.92\,\text{m}} \times \dfrac{(0.695\,\text{m/s})^2}{2 \times 9.81\,\text{m/s}^2} = 0.003\,\text{m}$

   For freshwater outfalls (rivers and lakes), the effect of density gradient expressed by the factor of $(\Delta S_g/S_g)\Delta Z_i$ in Equation 12.21f is near zero and is normally ignored.
   Operating head from Equation 12.21f, $E_3 = E_2 + h_{f,3} = (1 + 0.003)\,\text{m} = 1.003\,\text{m}$

   Velocity head from Equation 6.15a, $h_{v,3} = \dfrac{(V_3)^2}{2g} = \dfrac{(0.695\,\text{m/s})^2}{2 \times 9.81\,\text{m/s}^2} = 0.0246\,\text{m}$

7. Calculate the discharge through Port 3.

   Ratio of $V_3^2/2g$ (or) $h_{v,3}$ to E_3, $\dfrac{(V_3)^2}{2g}\dfrac{1}{E_3} = \dfrac{h_{v,3}}{E_3} = \dfrac{0.0246\,\text{m}}{1.003\,\text{m}} = 0.025$

$C_{D,3} = 0.90$ is obtained from Figure 12.12c at the ratio of $(V_3{}^2/2g)/E_3 = 0.025$.
Discharge through Port 3 from Equation 12.21g,
Calculate the discharges q_1 and q_2 at Ports 1 and 2 from Equation 12.21g.

$$q_3 = C_{D,3}a\sqrt{2gE_3} = 0.90 \times 0.0573\,\text{m}^2 \times \sqrt{2 \times 9.81\,\text{m/s}^2 \times 1.003\,\text{m}} = 0.229\,\text{m}^3/\text{s}$$

8. Calculate the discharge, velocity, head loss, and operating head in different diffuser header sections upstream of Port 3.

   Repeat Steps 6 and 7 to calculate Q, V, h_f, E, h_v (or $V^2/2g$), h_v/E (or $(V^2/2g)/E$), and C_D for diffuser header sections upstream of Port 3, and discharge (q) through Ports 4–6. The calculated values are summarized in Table 12.11.

TABLE 12.11 Design Calculations for Diffuser Pipe Used to Disperse Effluent into the Receiving Stream (Example 12.41)

Parameter	Port Number					
	1[a]	2[a]	3[b]	4	5	6
Distance from the end, m	0	0	3	6	9	12
Discharge through pipe (Q), m^3/s	–	–	0.462	0.691	0.910	1.118
Velocity in pipe (V), m/s	–	–	0.695	1.039	1.368	1.681
Head loss through pipe (h_f), m	–	–	0.003	0.007	0.012	0.017
Operating head (E), m	1	1	1.003	1.01	1.022	1.039
Velocity head (h_v or $V^2/2g$), m	–	–	0.0246	0.0550	0.0954	0.1440
Ratio of h_v to E (or $(V^2/2g)$ to E)	–	–	0.025	0.054	0.093	0.139
Coefficient of discharge (C_D)	0.91	0.91	0.90	0.86	0.81	0.78
Port discharge (q), m^3/s	0.231	0.231	0.229	0.219	0.208	0.202

[a] See Steps 4 and 5.
[b] See Steps 6 and 7.

9. Verify the total discharge through the diffuser pipe.

$$Q = \sum_{1}^{6} q_i = (0.231 + 0.231 + 0.229 + 0.219 + 0.208 + 0.202)\,\text{m}^3/\text{s} = 1.320\,\text{m}^3/\text{s}$$

   The calculated total flow through six ports is very close to the peak effluent discharge of 1.321 m^3/s. The results are acceptable. In the case the calculated total flow is significantly different from the peak design effluent flow, additional trials should be performed to achieve a closer agreement.

10. Determine the total head loss thought the diffuser pipe.

$$h_{\text{m, diffuser pipe}} = E_6 = 1.039\,\text{m} \approx 1.04\,\text{m}$$

11. Determine the total head loss in the outfall pipe, including the diffuser pipe.

   The head losses in the outfall pipe were calculated in Example 12.40. Since the outfall pipe is connected to the diffuser pipe, there is no exit loss. Calculate the total head loss in the outfall structure that include head losses in the outfall pipe (from Example 12.40) and in the diffuser pipe calculated above.

$$\Delta h = h_f + h_{\text{m,diffuser pipe}} + h_{\text{m,bend}} + h_{\text{m,gate}} = (0.77 + 1.04 + 0.06 + 0.10)\,\text{m} = 1.97\,\text{m}$$

12. Determine the water surface elevation in the effluent junction box.

The maximum water surface elevation in the junction box at the peak design effluent flow, and at high water depth of 3.20 m.

$$\mathrm{WSEL_{max}} = H_{\mathrm{high}} + \Delta h = (3.20 + 1.97)\,\mathrm{m} = 5.17\,\mathrm{m}$$

13. Describe the design features of the effluent structure.

The effluent structure consists of effluent junction box with sluice gate, cast iron outfall pipe, and steel diffuser pipe. All concrete, cast iron, and steel pipes and fittings shall conform to the American Society of Testing Materials (ASTM) and American Institute of Steel Construction (AISC) standards. The diffuser and outfall pipes shall be properly anchored deep in the channel bed and in the embankment to protect from erosion, scour, and uplift. The concrete apron shall be secured in place by poured concrete of sufficient depth below the channel bed. Construction of structures within navigable waters of the U.S. requires a permit from the U.S. Army Corps of Engineers. This permit, called Section 10 Permit, is applied for in conjunction with the Section 404 Permit. For information about these permits, contact the nearest U.S. Army Corps of Engineers district office. The design details of outfall and diffuser pipes are shown in Figures 12.13 and 12.14.

Note: The side section of outfall structure is also shown in Figure 12.13. The components shown are effluent junction box, sluice gate, outfall pipe with two 22.5° bends, and concrete-pipe-mounting base and support slab.

Discussion Topics and Review Problems

12.1 A municipal waste water treatment plant is producing effluent for industrial and urban reuse. The process train contains preliminary treatment, primary sedimentation, sequencing batch reactor (SBR), and polishing filter. The preliminary treatment facility removes screenings and grit; and no BOD_5, TSS and other constituents are removed. The influent quality and percent removal of traditional parameters are summarized below:

Parameter	Concentrations in Preliminary Treated Efficient, mg/L	Percent Removal, %		
		Primary	SBR	Polishing Filter
BOD_5	180	30	95	50
TSS	220	60	95	70
TP	6	15	85	25
ON	20	25	85	60
AN	25	0	97	0

Determine (a) Efficient quality from the polishing filter and (b) percent content of TP in the waste sludge. The biological kinetic coefficients are: $Y = 0.6$ g VSS/g BOD_5, $k_d = 0.5\ \mathrm{d}^{-1}$, and $\theta_c = 10$ d.

12.2 Municipal wastewater is treated to the highest quality for injection into a potable water aquifer. The reclaimed water meets the primary drinking water standards at the point of injection. Prior to injection, the reclaimed water is disinfected with a free chlorine residual of 1.1 mg/L at a contact time of 30 min. Ammonia is added to convert the free chlorine residual into combined chlorine residual. The total fecal coliform count in the influent to the treatment plant is 1.5×10^6 organisms/100 mL. A total of 4.8-log removal of total fecal coliform organisms is achieved in the water reclamation facility. Check if the reclaimed water meets the nondetectable microbiological quality requirement prior to injection. Chick-Watson equation is applicable to disinfection kinetics. The empirical constants k' and n are 0.31 $(L/mg)^n/min$ and 0.4, respectively.

12.3 Define LD_{50}. The toxicity of a pesticide was tested on the motility of mice. A group of 10 mice at each dose was used in the study. The average weight of mice was 24.5 g. Each animal in the group was administered the required pesticide dose. The motility was recorded after exposure period of 4 h. Determine the LD_{50} for mice and that for humans. The average weight of humans is 70 kg. The uncertainty factor for extrapolation from animal to humans is 10. The experimental data are summarized below:

Dosage, μg	0	10	20	25	30	35	40	45	50
No. of dead mice after 4 h of exposure	0	0	2	4	7	9	10	10	10

12.4 Define ED_{50} and the ratio LD_{50}/ED_{50}. A newly developed pesticide is tested to determine its toxicity and safety warning before the lethal dose is reached. A toxicity study was conducted with a group of 10 mice at each dose to obtain ED_{50} and LD_{50}. The response for ED_{50} was sluggish movement or loss in agility in 48 hours of exposure. The response for LD_{50} was death in 48 hours of exposure. Determine ED_{50}, LD_{50} and LD_{50}/ED_{50} ratio. Comment on safety warning. The experimental data are summarized below:

Low dose exposure to determine ED_{50}

Dosage, μg	0	1	2	3	4	5	5	5
No. of sluggish mice after 48 h of exposure	0	0	0	1	3	7	9	10

High dose exposure to determine LD_{50}

Dosage, μg	8	9	10	11	12	13	14	15
No. of dead mice after 48 h of exposure	0	1	4	6	8	9	10	10

12.5 Define NOAEL, LOAEL, ADI, *RfD*, UF, and MF. A 90-d high quality subchronic oral pesticide toxicity study with mice was conducted. A LOAEL of 100 mg/kg·d was obtained for the pesticide. List all uncertainty factors, and determine *RfD*.

12.6 Groundwater is the drinking water supply source for a small rural community. The water contains two chemicals of concern. Arsenic concentration is 0.1 mg/L, and toluene concentration is 0.5 mg/L. Estimate the daily exposure dose to an adult exposed to these chemicals. Use the U.S. EPA maximum exposure assumptions given in Table 12.6. Also, identify the receptor, pathway and exposure.

12.7 A child breaths outdoor air for a maximum exposure period of 6 years. Calculate the daily exposure dose for two cases: (a) 0.02 mg/m^3 benzene and (b) 1.0 mg/m^3 toluene.

12.8 A child consumes 40 g of recreational fish per meal. The average fish consumption is three meals per year over four years. The fish contains 2×10^{-6} mg dioxin/kg fish. Estimate the child's dose from eating contaminated fish. $FI = 1.0$.

12.9 A groundwater supply source is contaminated by seven noncarcinogenic hazardous organic compounds. The chronic daily intake (*CDI*) and reference dose (*RfD*) for oral intake of each compound are summarized below. Calculate the hazard index (*HI*) and check if the cumulative exposure is within the acceptable level to the consumers.

Noncarcinogenic Hazardous Compound	Oral Chronic Daily Intake (*CDI*), mg/kg·d	Reference Dose (*RfD*), mg/kg·d
Acetonitrile	4.2×10^{-4}	0.006
Benzoic acid	0.40	4.0
Bromomethane	1.02×10^{-4}	0.0014
Ethyl acetate	0.18	0.9
Ethyl ether	0.03	0.2
Formaldehyde	0.04	0.2
Hexabromobenzine	6.8×10^{-4}	0.002

12.10 Many noncarcinogenic hazardous compounds are organ specific. A receptor population is exposed to five chemicals by oral route. Three chemicals are hepatotoxins and two chemicals are nephrotoxins. The computed exposure and reference doses, and target organ of each chemical are tabulated below. Determine the cumulative exposure to each target organ, and total exposure from all five chemicals. Comment on the results.

Chemical	Target organ	Computed Chronic Daily Intake (*CDI*), mg/kg·d	Reference Dose (*RfD*), mg/kg·d
A	Kidney	0.016	0.08
B	Liver	0.210	0.70
C	Kidney	0.021	0.07
D	Kidney	0.281	0.70
E	Liver	0.128	0.21

12.11 The maximum contaminant level (*MCL*) for arsenic in primary drinking water standards is 10 μg/L. Determine the maximum number of excess lifetime cancer cases expected from consumption of water. The population exposed is 100,000 people. The carcinogenic slope factor (*CSF*) for oral arsenic intake is 1.5 kg·d/mg.

12.12 An open storage reservoir is built to store the reclaimed water. The average quarterly rate of reclaimed water production, precipitation, percolation, evaporation and reuse demand are tabulated below. The surface area of the storage reservoir is 8 ha. Determine the storage volume needed, depth and detention time in the reservoir.

Parameter	Monthly Data			
	January, February, and March	April, May, and June	July, August, and September	October, November, and December
Production of reclaimed water, m^3/d	4670	3000	2500	3974
Precipitation, mm	380	83	19	260
Reuse of reclaimed water, m^3/d	1140	3540	4443	3390
Evaporation, mm	130	400	530	160
Percolation, mm/d	3.6	3.6	3.6	3.6

12.13 Reclaimed water is used for makeup water in a cooling tower. The total hardness (TH) is due to Ca^{2+}, and total alkalinity (TA) is due to HCO_3^-. TH and TA concentrations in the reclaimed water are 135 and 95 mg/L as $CaCO_3$. Estimate the carbonate (CH) and noncarbonate hardness (NCH) of the blowdown water after six cycles of concentration at 50°C. Sulfuric acid is added in recirculation water to protect the cooling tower from $CaCO_3$ scaling. The solubility of $CaSO_4$ at 50°C is 2200 mg/L as $CaCO_3$. Also, estimate the concentration of H_2SO_4 to be added in the recirculation water, and total concentration of $CaSO_4$ as $CaCO_3$, and TA remaining in the blowdown. For pH control 10% CH is left in the recirculation water.

12.14 A cooling tower has a recirculation rate of 150,000 L/min. The recycle ratio is 6 and the drift flow rate is 0.007% of recirculating water. The temperature of cooling water is 12°C. Calculate the flow rates of makeup water and blowdown.

12.15 Reclaimed water is used for crop irrigation. The key water quality parameters of reclaimed water are: $Ca^{2+} = 50$ mg/L, $Mg^{2+} = 14$ mg/L, $Na^{+} = 200$ mg/L, $Cl^{-} = 180$ mg/L, TA = 200 mg/L as $CaCO_3$, TDS = 1380 mg/L, and pH = 7.5. Determine (a) Sodium adsorption ratio (*SAR*), (b) Adjusted sodium adsorption ratio (SAR_{ad}). Also, evaluate the potential restrictions of reclaimed water for crop irrigation reuse.

12.16 A 2-ha constructed wetland is designed as an infiltration basin, and to enhance the quality of reclaimed water. The average water depth is 1.2 m. The hydraulic conductivity of the wetted zone is 0.72 m/d and is assumed constant. The fillable porosity is 0.45, and the critical pressure head of soil for wetting is −0.5 m. Calculate (a) time for wetting front to travel 12 m, (b) infiltration rate at a depth of 12 m, and daily percolation rate, and (c) volume of water that infiltrated through 12 m depth.

12.17 An unconfined aquifer is 25-m deep. Reclaimed water is recharged by direct injection. The hydraulic conductivity K of the wetted zone is 0.008 m/s. The radius of injection well is 1.0 m and the depth of water in the screened length is 18 m. The depth of vadose zone below the screened section is 4 m. Estimate the volumetric injection rate through the vadose zone.

12.18 Reclaimed water is retained in a maturation basin. An infiltration gallery is constructed over the impervious strata 15 m from the bank of the impoundment, and 10-m deep. The diffusion ditch at the bank of impoundment is filled with sand and gravel. The stratum contains sand and silt, and has coefficient of permeability $K = 0.09$ cm/s. The water is pumped from the infiltration gallery to maintain a depth of 1.2 m. The length of the gallery is 30 m. Calculate the production rate of percolated reclaimed water.

12.19 Secondary effluent is discharged from a POTW into a stream. The wastewater flow is 10,000 m^3/d and has a BOD_5 of 20 mg/L. The summer temperature and DO in the effluent are 25°C and 6.5 mg/L. The water quality of stream above the effluent discharge point are $Q = 0.3$ m^3/s, DO = 7.9 mg/L, $BOD_5 = 3$ mg/L, and average water temperature is 21°C. The deoxygenation coefficient of effluent and reaeration coefficient of the stream after effluent mixing are 0.23 d^{-1} and 0.45 d^{-1} at 20°C. The average velocity of water below the outfall is 0.2 m/s. The temperature coefficients for deoxynation and reaeration coefficients are 1.12 and 1.024, respectively. Calculate critical DO and the distance from outfall where this value occurs. Draw the DO profile from the outfall to a distance where 90% DO saturation is resumed.

12.20 An outfall pipe 15 cm in diameter is discharging secondary treated effluent into a lake. The pipe is at the bottom of the lake at an elevation of 102.00 m. The water surface elevation is 105.50 m. The length of the pipe is 280 m, and it has four 45° elbows ($K = 0.2$). Calculate the water surface elevation in the upstream unit. $Q = 27$ L/s, $f = 0.024$, $K_{entrance} = 0.5$, and $K_{exit} = 1.0$.

12.21 A diffuser header is provided at an outfall structure to disperse the effluent. The diffuser header is designed in Example 12.41, and the design details and configuration are shown in Figure 12.14. During the initial years when the flow is small, some inboard ports are tightly plugged. Calculate the head loss through the diffuser header in Example 12.41 during the initial years. The peak design flow during the initial year is 966 L/s, and the inboard ports 5 and 6 in the diffuser header are tightly plugged.

References

1. Snyder, S. A., P. Westerhoff, Y. Yoon, and D. L. Sedlak, Pharmaceuticals, personal care products, and endocrine disruptors in water: Implications for the water industry, *Environmental Engineering Science*, 20 (5), 2003, 449–469.
2. Barceló, D. and M. Petrovic, Pharmaceutical and personal care products in the environment, *Analytical and Bioanalytical Chemistry*, 387, 2007, 1141–1142.
3. Yu, J. T., E. J. Bouwer, and M. Coelhan, Occurrence and biodegradability studies of selected pharmaceuticals and personal care products in sewage effluent, *Agricultural Water Management*, 86 (1–2), 2006, 72–80.
4. Hernando, M. D., M. Mezcua, A. R. Fernandez-Alba, and D. Barcelo, Environmental risk assessment of pharmaceutical residues in wastewater effluents, surface waters and sediments, *Talanta*, 69 (2), April 2006, 334–342.
5. Schwab, B. W. et al. Human pharmaceuticals in U.S. surface waters: A human health risk assessment, *Regulatory Toxicology and Pharmacology*, 42 (3), 2005, 296–312.
6. Qasim, S. R., *Wastewater Treatment Plants: Planning, Design, and Operation*, 2nd ed., CRC Press, Boca Raton, FL, 1999.
7. Metcalf & Eddy, Inc., *Wastewater Engineering, Treatment and Reuse*, 4th ed., McGraw-Hill Companies, Inc., New York, NY, 2003.
8. Metcalf & Eddy | AECOM, *Wastewater Engineering: Treatment and Resource Recovery*, 5th ed., McGraw-Hill, New York, NY, 2014.
9. Metcalf & Eddy AECOM, Asano, T., F. L. Burton, H. L. Leverenz, R. Tsuchihashi, and G. Tchobanoglous, *Water Reuse: Issues, Technologies, and Applications*, McGraw-Hill Professional, New York, NY, 2007.
10. U.S. Environmental Protection Agency and U.S. Agency for International Development, *Guidelines for Water Reuse*, EPA/625/R-04/108, U.S. Environmental Protection Agency and U.S. Agency for International Development, Washington DC, September 2004.
11. U.S. Environmental Protection Agency and U.S. Agency for International Development, *2012 Guidelines for Water Reuse*, EPA 600/R-12/618, U.S. Environmental Protection Agency and U.S. Agency for International Development, Washington, DC, September 2012.
12. Watts, R. J., *Hazardous Wastes: Sources, Pathways, Receptors*, John Wiley & Sons, Inc., New York, NY, 1998.
13. LaGrega, M. D., P. L. Buckingham, and J. C. Evans, *Hazardous Waste Management*, 2nd ed., Waveland Press, Inc., Long Grove, IL, 2010.
14. Cockerham, L. G., and B. S. Shane, *Basic Environmental Toxicology*, CRC Press, Boca Raton, FL, 1993.
15. Pepper, I. L., C. P. Gerba, and M. L. Brusseau (eds.), *Pollution Science*, Academic Press, San Diego, CA, 1996.
16. U.S. Environmental Protection Agency, *Guidelines for Carcinogen Risk Assessment*, EPA/630/R-00/004, U.S. Environmental Protection Agency, Risk Assessment Forum, Washington, DC, September 1986.
17. U.S. Environmental Protection Agency, *Guidelines for Carcinogen Risk Assessment*, EPA/630/P-03/001F, Risk Assessment Forum, Washington, DC, March 2005.
18. U.S. Environmental Protection Agency, *Risk Assessment and Management: Framework for Decision Making*, EPA 600/9-85-002, U.S. Environmental Protection Agency, Washington DC, December 1984.
19. U.S. Environmental Protection Agency, *Integrated Risk Information System (IRIS), Electronic On-Line Database of Summary Health Risk Assessment and Regulatory Information on Chemical Substances Assessment*, EPA 1630/R-92/001, U.S. Environmental Protection Agency, Washington, DC, 1996.

20. Walski, T. M., *Analysis of Water Distribution Systems*, Van Nostrand Reinhold, New York, NY, 1984.
21. Qasim, S. R., E. M. Motely, and G. Zhu, *Water Works Engineering: Planning, Design, and Operation*, Prentice Hall PTR, Upper Saddle River, NJ, 2000.
22. Miller, J. K., U.S. water reuse: Current status and future trends, *Water Environment and Technology*, 2 (12), 1990, 83–89.
23. Lazarova, V. and A. Bahri (eds.), *Water Reuse for Irrigation: Agriculture, Landscapes, and Turf Grass*, CRC Press, Boca Raton, FL, 2005.
24. AWWA, *Guidelines for Distribution of Nonpotable Water*, AWWA, California-Nevada Section, 1992.
25. AWWA, *Planning for the Distribution of Reclaimed Water, Manual of Water Supply Practices*, M24, 3rd ed., AWWA, Denver, CO, 2009.
26. Innovyze, *Software and Services Catalog 2016*, http://www.innovyze.com, 2016.
27. Bentley Systems, Inc., *Water Distribution Modeling and Management*, http://www.bently.com, 2014.
28. KYPipe, LLC, *KYPIPE, Pipe 2016*, http://kypipe.com, 2016.
29. Ayers, R. S. and D. W. Westcot, *Water Quality for Agriculture*, FAO Irrigation and Drainage Paper 29, Rev. 1, Food and Agricultural Organization of the United Nations, Rome, Italy, 1985.
30. Pettygrove, G. S. and T. Asano (eds.), *Irrigation with Reclaimed Municipal Wastewater—A Guidance Manual*, Lewis Publishers, Inc., Chelsea, MI, 1985.
31. Asano, T. (ed.), *Wastewater Reclamation and Reuse*, Water Quality Management Library, Vol. 10, CRC Press, Boca Raton, FL, 1998.
32. Hammer, D. A. (ed.), *Constructed Wetlands for Wastewater Treatment: Municipal, Industrial, and Agricultural*, Lewis Publishers, Inc., Chelsea, MI, 1989.
33. Gearheart, R. A., Use of constructed wetlands to treat domestic wastewater, city of Arcata, California, *Water Science and Technology*, 26 (7–8), 1992, 1625–1637.
34. U.S. Environmental Protection Agency, *Manual Constructed Wetlands Treatment of Municipal Wastewaters*, EPA/625/R-99/010, Office of Research and Development, Cincinnati, OH, September 2000.
35. U.S. Environmental Protection Agency, *Guiding Principles for Constructed Treatment Wetlands: Providing for Water Quality and Wildlife Habitat*, EPA 843-B-00-003, Office of Wetlands, Oceans and Watersheds, Washington, DC, October 2000.
36. Bouwer, H. and R. C. Rice, Effect of water depth in groundwater recharge basins on infiltration, *Journal of Irrigation and Drainage Engineering*, ASCE, 115 (4), 1989, 556–567.
37. Roberts, P. V., Water reuse for groundwater recharge: An overview, *Journal American Water Works Association*, 72 (7), 1980, 375–379.
38. Gerba, C. P. and S. M. Goyal, Pathogen removal from wastewater during groundwater recharge, *Artificial Recharge of Groundwater*, T. Asano (ed.), Butterworth Publishers, Stoneham, MA, 1985, pp. 283–318.
39. Oaksford, E. T., Artificial Recharge: Methods, Hydraulics, and Monitoring, *Artificial Recharge of Groundwater*, T. Asano (ed.), Butterworth Publishers, Stoneham, MA, 1985, pp. 69–128.
40. Bouwer, H., Role of groundwater recharge in treatment and storage of wastewater for reuse, *Water Science and Technology*, 24 (9), 1991, 295–302.
41. Bouwer, E. J., P. L. McCarty, H. Bouwer, and R. C. Rice, Organic contaminant behavior during rapid infiltration of secondary wastewater at the Phoenix 23rd avenue project, *Water Research*, 18 (4), 1984, 463–472.
42. Piet, G. J. and B. C. J. Zoetaman, Organic water quality changes during sand bank and dune filtration of surface waters in the Netherlands, *Journal American Water Works Association*, 72 (7), 1980, 400–404.

43. Sontheimer, H., Experience with riverbank filtration along the Rhine river, *Journal American Water Works Association*, 72 (7), 1980, 386–390.
44. Lauer, W. C., Denver's direct potable water reuse demonstration project, *Proceedings of the 21st Annual Conference on Water Policy and Management: Solving the Problems*, American Society of Civil Engineers, New York, NY , 1994, pp. 657–660.
45. Harhoff, J. and B. van der Merwe, Twenty-five years of wastewater reclamation in Windhoek, Namibia, *Water Science and Technology*, 33 (10–11), 1996, 25–35.
46. National Research Council, *Issues in Potable Reuse: The Viability of Augmenting Drinking Water Supply with Reclaimed Water*, National Research Council, National Academy Press, Washington, D C, 1998.
47. Federal Water Pollution Control Act Amendments of 1972 (P.L. 92-500), 92nd Congress, October 18th, 1972.
48. Clean Water Act (P.L. 95-217), 95th Congress, December 27, 1977.
49. Metcalf & Eddy, Inc., *Wastewater Engineering, Treatment, Disposal, and Reuse*, 3rd ed., McGraw-Hill, Inc., New York, NY, 1991.
50. Tchobanoglous, G. and E. D. Schroeder, *Water Quality: Characteristics, Modeling, Modification*, Addison-Wesley Publishing Co., Reading, MA, 1985.
51. Rawn, A. M., F. R. Bowerman, and N. H. Brooks, Diffusers for disposal of sewage in seawater, *Journal of the Sanitary Engineering Division, Proceedings of the ASCE*, ASCE, 86 (SA2), March 1960, 65–105.
52. Burchett, M. E., G. Tchobanoglous, and A. J. Burdoin, A practical approach to submarine outfall calculations, *Public Works*, 98 (5), 1967, 95–101.
53. Falk, L. L., Factors affecting outfall design, *Water and Sewage Works*, 113 (11), 1966, 233.
54. U.S. Environmental Protection Agency, *Characterization of Stream Reaeration Capacity, Ecological Research Series*, EPA-R3-72-012, Office of Reacher and Monitoring, Washington, DC, October 1972.
55. Streeter, H. W. and E. B. Phelps, *A study of the pollution and natural purification of the Ohio River, III, Factors Concerned in the Phenomena of Oxidation and Reaeration, Public Health Bulletin No. 146*, U.S. Public Health Services, Cincinnati, OH, 1925.
56. Thomann, R. V. and J. A. Mueller, *Principles of Surface Water Quality Modeling and Control*, Harper & Row, New York, NY, 1987.
57. Eckenfelder, W. W. and D. J. O'Connor, *Biological Waste Treatment*, Pergamon Press, New York, NY, 1961.
58. Wilcock, R. J., Study of river reaeration at different flow rates, *Journal of the Environmental Engineering Division*, ASCE, 114 (1), 1988, 91–105.
59. McCutcheon, S. C., Volume I transport and surface exchange in rivers, *Water Quality Modeling*, R. H. French (ed.), CRC Press, Boca Raton, FL, 1989.
60. Peavy, H. S., D. R. Rowe, and G. Tchobanoglous, *Environmental Engineering*, 3rd ed., McGraw-Hill, Inc., New York, NY, 1985.
61. Lin, S. D., *Water and Wastewater Calculations Manual*, 3rd ed., McGraw-Hill Professional, New York, NY, 2014.
62. Metcalf & Eddy, Inc., *Wastewater Engineering: Collection, Treatment, and Disposal*, McGraw-Hill Book Co., New York, NY, 1972.
63. Texas Water Development Board, *DOSAG-1 Simulation of Water Quality in Streams and Canals, Program Documentation and User's Manual*, Texas Water Development Board, Austin, TX, 1970.
64. Texas Water Development Board, *Simulation of Water Quality in Streams and Canals, Theory and Description of QUAL-I Mathematical Modeling System*, Report 128, Texas Water Development Board, Austin, TX, 1971.

65. Roesner, L. A., P. R. Giguere, and D. E. Evenson, *Computer Program Documentation for the Stream Quality Model QUAL-II*, EPA-600/9-81-014, Environmental Research Laboratory, Athens, GA, 1981.
66. U.S. Environmental Protection Agency, *Ecological Restoration: A Tool to Manage Stream Quality*, EPA 841-F-95-007, Office of Water, Washington, DC, November 1995.
67. Viessman, W., Jr., M. J. Hammer, E. M. Perez, and P. A. Chadik, *Water Supply and Pollution Control*, 8th ed., Prentice Hall PTR, Upper Saddle River, NJ, 2008.
68. IWA Task Group on River Water Quality Modeling, *River Water Quality Modeling No. 1*, IWA Publishing, London, UK, 2001.
69. U.S. Environmental Protection Agency, *TMDL Models and Tools*, website information at https://www.epa.gov/exposure-assessment-models/tmdl-models-and-tools, Accessed on March 16, 2016.
70. Hydrologic Engineering Center, *HEC-5Q: System Water Quality Modeling*, U.S. Army Corps of Engineers, Hydrologic Engineering Center, Davis, CA, 1986.
71. Schwarz, G. E., A. B. Hoos, R. B. Alexander, and R. A. Smith, *The SPARROW surface Water-Quality Model: Theory, Application and User Documentation*, U.S. Department of the Interior, USGS, Reston, VA, 2006.
72. DHI, *MIKE 2016 Embedding Knowledge in Technology*, Release Flyer, http://www.mikepowered-bydhi.com, 2016.
73. Innovyze, *InfoWorks® ICM Overview*, http://www.innovyze.com.
74. Rambow, C. A., Submarine disposal of industrial wastes, *Proceedings of the 24th Industrial Waste Conference*, Purdue University, Part 2, May 6–8, 1969, pp. 1486–1493.
75. McNown, J. S., Mechanics of Manifold Flow, *Transactions of American Society of Civil Engineers*, 119, 1954, 1103–1142.
76. Benefield, L. D., J. F. Judkins, and A. D. Parr, *Treatment Plant Hydraulics for Environmental Engineers*, Prentice-Hall, Inc., Englewood Cliffs, NJ, 1984.
77. Droste, R. D., *Theory and Practice of Water and Wastewater Treatment*, John Wiley & Sons, Inc., New York, NY, 1997.
78. Hudson, H. E., *Water Clarification Processes: Practical Design and Evaluation*, von Nostrand Reinhold Co., New York, NY, 1981.
79. Hudson, H. E., Jr., R. B. Uhler, and R. W. Bailey, Dividing-flow manifold with square-edged laterals, *Journal of the Environmental Engineering Division, ASCE*, 105 (EE4), 1979, 745–755.

13

Residuals Processing, Disposal, and Reuse

13.1 Chapter Objectives

Residuals are removed from wastewater, and they are processed through various treatment processes. Screenings and grit are separated at the preliminary treatment facilities. Settleable organic solids and scum are removed in the primary clarifier, and excess biomass is removed from the biological treatment in the secondary (or final) clarifier. The characteristics (quality and quantity) of screenings and grit, and disposal have been presented in Chapters 6 through 8. Discussion on scum, primary sludge, and waste activated sludge is presented in Chapters 6, 9, and 10. The objectives of this chapter are to present the following:

- Conventional and emerging technologies applied to residuals management
- Characteristics of municipal sludge
- Sludge processing systems, process diagrams, and regulatory requirements
- Methods of sludge-thickening and thickener design
- Methods of sludge stabilization, and theory and design of sludge digestion, and biogas utilization
- Methods of sludge conditioning and dewatering
- Material mass balance analysis and impact of sidestreams on influent characteristics of wastewater
- Emerging technologies for sludge processing
- Composting of sludge solids
- Heat drying, incineration, and other sludge destruction processes
- Land application of biosolids
- Disposal of sludge by landfilling

13.2 Conventional and Emerging Technologies for Sludge Processing

The residuals management facilities usually represent 40–60% of total construction costs of wastewater treatment plant, and consume a disproportionate share of operating costs. The conventional treatment facilities are thickening, digestion, conditioning, dewatering, and disposal. In recent years, many emerging technologies have been developed to recover resources such as nutrients and proteins, and to digest solids more completely and efficiently to reduce the quantity while recovering the energy. The conventional processes are presented in Sections 13.4 through 13.9. The emerging technologies are covered in Section 13.10. Finally, the biosolids reuse and disposal is presented in Section 13.11.

13.3 Characteristics of Municipal Sludge

13.3.1 Sources and Solids Content

The major sources of sludge production at a municipal wastewater treatment plant are the primary and final clarifiers. Additional sludge may result from chemical precipitation and tertiary filter backwash operation. The chemical precipitates are normally contained within either primary or secondary sludge. The filter backwash contains large volume of water. Often these sludge streams are recycled through the liquid treatment processes to concentrate the solids. Ultimately, a single concentrated sludge stream consisting of combined sludge is produced.

The sludge solids rapidly undergo putrification producing highly offensive odors. This creates serious and complex problems during processing and disposal. The physical properties of primary, secondary, and chemical sludges and combined digested sludge are summarized in Table 13.1.* The chemical characteristics of raw primary and waste activated sludge (WAS), and combined digested sludge are provided in Table 13.2.

13.3.2 Sludge Quantity Variations and Solids and Volume Relationship

Quantity Variation: There is a wide variation in solids generation rate. The peak day solids load over two to five consecutive days may be three to five times the average day load. The thickeners are normally sized for peak-day solid production rate. The digesters are designed for maximum day of solids production rate over the consecutive days of residence time. The mechanical dewatering facility is sized to handle solids production rate on the maximum day during the design operating schedule.

Relationships of Specific Gravity, Weight, and Volume: The sludge contains large volume of water. Also, the proportion of volatile and fixed solids varies depending upon the source of sludge and the degree of stabilization. The general equations used to calculate the specific gravity, weight, and volume of solids

TABLE 13.1 Physical Characteristics of Primary, Secondary, Combined Digested, and Chemical Sludge

Source of Sludge	Quantity, mg/L (g/m^3)	Typical Range			
		Solids Content (p_s), %	Ratio of VSS/TSS, kg-VSS/kg-TSS	Specific Gravity (water = 1)	
				Dry Solids (S_s)	Bulk Sludge (S_b)
PS	105–170	1–6	0.6–0.85	1.1–1.5	1.001–1.02
TFS	50–100	1–4	0.7–0.85	1.1–1.4	1.001–1.01
WAS	70–100	0.8–1.2 (from clarifier) 0.2–0.6 (from aeration basin)	0.7–0.85	1.1–1.4	1.000–1.003
PS and WAS	150–250	1–4	0.65–0.85	1.1–1.5	1.001–1.015
DGS	120–220	3–6	0.4–0.6	1.3–1.8	1.01–1.03
CHS					
Metal salts	200–300	0.5–4	0.5–0.6	1.3–1.6	1.001–1.015
Low lime	240–400	2–8	0.4–0.5	1.5–1.9	1.01–1.04
High lime	600–1300	4–10	0.3–0.45	1.6–2.1	1.02–1.06

Note: CHS = Chemical sludge from phosphorus removal in primary clarifier; DGS = Digested sludge; PS = Primary sludge; TFS = Trickling filter sludge; WAS = Waste activated sludge.

Source: Adapted in part from References 1 through 8.

* The distinction between total solids (TS) and total suspended solids (TSS) in municipal sludge is ignored since the total dissolved solids (TDS) is 250–700 mg/L, while TSS is 10,000–80,000 mg/L. In this book, the TSS-based parameters are used for liquid stream treatment processes. For sludge, it is also assumed that TS ≈ TSS.

TABLE 13.2 Typical Composition of Raw Primary and Secondary Sludge and Combined Digested Sludge

Parameter	Raw Primary Sludge	Raw Waste Activated Sludge	Combined Digested Sludge
pH, standard unit	5–8	6.5–8	6.5–7.5
Alkalinity, mg/L as $CaCO_3$	500–1500	200–1100	2500–3500
Organic acids, mg/L as HAc[a]	200–2000	1100–1700	100–600
Ratio of BOD_5/VSS	0.5–1.1	–	–
Ratio of COD/VSS	1.2–1.6	2–3	–
Ratio of Org-N/VSS	0.05–0.06	0.08–0.1	–
Cellulose, % of dry wt.	8–15	5–10	8–15
Hemicellulose, % of dry wt.	2–4	–	–
Lignin, % of dry wt	3–7	–	–
Grease and fats (ether soluble), % of dry wt.	5–8	5–12	5–20
Protein, % of dry wt.	20–30	30–40	15–20
Nitrogen as N, % of dry wt.	1.5–4	2.5–8.5	1.5–6
Phosphorus as P, % of dry wt.	0.3–1.2	1.2–9	1–5
Potassium as K, % of dry wt.	0–0.8	0.4–0.6	0.1–2
Heat content, kJ/kg VSS	20,000–29,000	16,000–23,000	6000–14,000
Lindane as $C_6H_6Cl_6$, mg/kg	0.6	1.0	–
Chlordane, mg/kg	2.6	4.4	–
Metal, mg/kg dry solids			
Cadmium	–	–	3–3400
Calcium	–	–	1000–250,000
Chromium	–	–	25–28,000
Copper	–	–	80–10,000
Iron	–	–	1000–150,000
Lead	–	–	15–20,000
Magnesium	–	–	1000–20,000
Manganese	–	–	20–7000
Mercury	–	–	0.5–10,000
Nickel	–	–	2–3500
Tin	–	–	2.5–300
Zinc	–	–	100–28,000

[a] HAc = acetic acid.
1 kJ/kg = 0.43 Btu/lb.
Source: Adapted in part from References 1 through 8.

and sludge are given by Equation 13.1.

$$\frac{w_s}{S_s\rho_w} = \frac{w_f}{S_f\rho_w} + \frac{w_v}{S_v\rho_w} \quad \text{or} \quad \frac{w_s}{S_s} = \frac{w_f}{S_f} + \frac{w_v}{S_v} \tag{13.1a}$$

$$\frac{1}{S_s} = \frac{1 - \text{VSS/TSS ratio}}{S_f} + \frac{\text{VSS/TSS ratio}}{S_v} \quad \text{or} \quad S_s = \frac{1}{\dfrac{1 - \text{VSS/TSS ratio}}{S_f} + \dfrac{\text{VSS/TSS ratio}}{S_v}} \tag{13.1b}$$

$$\frac{w_b}{S_b\rho_w} = \frac{w_s}{S_s\rho_w} + \frac{w_w}{S_w\rho_w} \quad \text{or} \quad \frac{w_b}{S_b} = \frac{w_s}{S_s} + w_w \tag{13.1c}$$

$$\frac{100\%}{S_b} = \frac{p_s}{S_s} + (100\% - p_s) \quad \text{or} \quad S_b = \frac{100\% \times S_s}{100\% \times S_s - p_s(S_s - 1)} \tag{13.1d}$$

$$V_b(\text{or } Q_b) = \frac{w_b}{S_b \rho_w} \quad \text{or} \quad V_b(\text{or } Q_b) = \frac{100\% \times w_s}{p_s S_b \rho_w} \tag{13.1e}$$

where

p_s = dry solids content in sludge, percent of wet sludge weight
ρ_w = density of water, kg/m^3
S_b = specific gravity of bulk wet sludge, dimensionless ($S_w = 1$)
S_f = specific gravity of fixed solids in sludge, dimensionless ($S_w = 1$). The typical ranges for S_f are 2.5–2.7 for primary sludge and WAS (with or without chemical precipitates from metal salts) and 2.8–3.1 for primary sludge with chemical precipitates from lime addition.
S_s = specific gravity of dry solids in sludge, dimensionless ($S_w = 1$)
S_v = specific gravity of volatile solids in sludge, dimensionless ($S_w = 1$). The typical ranges for S_v are 1.0–1.2 for VSS in the primary sludge and WAS.
S_w = specific gravity of water, dimensionless ($S_w = 1$)
V_b (or Q_b) = volume (or flow) of wet sludge, m^3 (m^3/d)
VSS/TSS ratio = ratio of volatile to total suspended solids, mg VSS/mg TSS
w_b = weight of wet bulk sludge, kg (kg/d)
w_f = weight of fixed solids, kg (kg/d)
w_s = weight of dry solids, kg (kg/d)
w_v = weight of volatile solids, kg (kg/d)
w_w = weight of water, kg (kg/d)

13.3.3 Sludge and Scum Pumping

Usually, sludge and scum are pumped from different liquid treatment process units to a designated area for on-site solids processing systems prior to final off-site disposal. Different types of pumps may be needed for different applications. A summary of pumps commonly used for different types of sludge applications is given in Table 13.3. Additional information about these pumps and their general applications is presented in Section 6.3.7 and Table 6.11.

TABLE 13.3 Sludge and Scum Pumps and Their Applications

Type of Pump[a]	Application in Sludge Transfer							
	Scum	Primary	Return	Waste Activated	Trickling Filter	Chemical	Thickened	Digested
Kinetic								
Centrifugal		x	x	x	x	x		x
Torque Flow		x	x	x	x	x		x
Positive displacement								
Diaphragm		x	x	x	x	x	x	x
High-pressure piston							x	x
Plunger	x	x				x	x	x
Progressive cavity		x	x	x	x	x	x	x
Rotary lobe		x				x	x	x

[a] Description of these pumps may be found in Section 6.3.7 and Table 6.11.

Source: Adapted in part from Reference 6.

TABLE 13.4 Hazen-Williams Roughness Coefficient for Sludge of Different Solids Contents

Sludge Solids Content, %	Roughness Coefficient *C*	
	Raw Sludge	Digested Sludge
0	100	100
1	83	100
2	71	91
3	60	83
4	53	78
5	47	73
6	42	69
7	37	65
8	33	60
9	29	55
10	25	48

Source: Adapted in part from References 3, 6, and 9.

Head Loss: The head loss in sludge piping depends upon pipe material, type of liner or coating, pumping velocity, and fluid properties of the sludge. Head loss normally increases with higher solid contents, higher volatile matter, and higher viscosity at lower temperature. For short-distance sludge pumping within the plant, Hazen–Williams equation Equation 6.13b is commonly used. The roughness coefficient *C* is modified for different sludge characteristics. Modified values of *C* for raw and digested sludges are given in Table 13.4. These values are developed for a laminar range of flow, which may extend up to a velocity of 1.1 m/s (3.5 ft/s).[3,6,9]

If the sludge is pumped over a longer distance, more accurate head loss calculations are needed. Friction losses are calculated using *thixotropic* behavior, *rheological concept*, and other flow properties of sludge. The procedure involves non-Newtonian fluid and turbulent condition. Both the Reynolds and Hedstom numbers are needed to calculate the Darcy-Weisbach friction factor.[3,6,7–12] An example to calculate the head loss in a pipe using sludge rheology is given in Reference 7.

Piping: The sludge has a tendency to settle and compact in pipelines' bends and connections. Also, grease has a tendency to coat the inside of the piping and pumping equipment. Special design considerations are given below when designing the sludge lines.

1. The pressure piping should not be <200 mm (8 in) in diameter unless they are glass lined.
2. Gravity piping should not be <150 mm (6 in) in diameter with a slope of 3% or larger.
3. Short radius elbows and sharp turns should be avoided. Sufficient clean outs in the forms of plugged tee, cross, or wye, and liberal number of hose connections should be provided so that the line could be cleaned and flushed.
4. Eccentric plug valves are typically used for isolation, and adjustment and control of flow in both pressure and gravity sludge piping.
5. The pump suction and discharge pipes should be at least 100 mm (4 in) in diameter. The velocity should be at least 0.75 m/s (2.5 ft/s) to prevent solids settling. If the sludge flow is small, large-capacity pump with a timer should be used to flush the line during pumping cycle. Also, the pump head should be sufficient to push the settled solids. A positive suction head of 24 in or higher is typically required for all sludge pumps.
6. Grease will coat the inside of piping. This will reduce the pipe diameter, and increase the velocity and head loss. Provision for melting the grease by hot water, steam, or digester supernatant should be made.
7. Provision for a high-pressure water jet, and pipe pigging and other cleaning devices should be made.

TABLE 13.5 Preliminary Sludge Preparation Operations

Preparation Operation	Description	Major Equipment and Design Consideration
Degritting	Degritting is need if separate grit removal facility before primary sedimentation is not provided.	Hydro-cyclone, centrifuge, and reciprocating rake (Sections 8.4.4 and 8.4.5, and Example 8.16).
Grinding	Grinding is needed to cut or shear stringy material, large particles, rags, and plastic. This is necessary to prevent clogging or wrapping around the rotating equipment, and rollers of sludge dewatering system and reduce wear on belts, and provide more uniform dewatering.	Slow-speed grinder, hardened steel cutters, overload sensor and reversible cutter rotation (Section 7.3.2).
Screening	Screening is also used an alternative to remove nuisance material. The screen opening is 3–6 mm (0.12–0.25 in). The captured screenings are removed by screw conveyer and dewatered. Recently, screening of return activated sludge has also gained interest for improving the biological activity of sludge.[13,14]	Step or rotary type fine screen with screw conveyor (Sections 7.2.3 and 7.2.4).
Blending	Blending is done to homogenize the sludge. Mostly primary, secondary, and advanced-process sludges are blended before or after thickenings. Blending may be done in primary clarifier, piping, sludge digester, or a designated blending tank.	Blending tank with slow mixing blending paddles (Examples 9.39 and 13.7).
Sludge storage	Sludge storage is needed to smooth out the fluctuations in flow and sustained solids loadings, and obtain a uniform feed rate to sludge processing processes. Short-term storage may be achieved in primary clarifier, blending tank, thickener and digester. For a storage period of 3–5 days or longer storage tanks with mixers are needed. Usually, the covered or open tanks must have provision for efficient ventilation, VOC emission, and odor controls.	Peaking factors need to be considered for storage (Example 13.7). Oxidizing chemicals such as Cl_2, Fe^{2+} salt, $KMnO_4$, $Mg(HO)_2$, and H_2O_2 are recommended for odor control. Biofiltration with or without activated carbon is also a valid option for odor control.[15,16]

Source: Adapted in part from References 6, 7, and 12 through 16.

8. Provision for air relief and blow off valves for the high and low points should be made.
9. Long sludge lines should have bypass lines around key sections for cleaning and maintenance. In special cases, two pipes in parallel should be provided so that one line can be shut down for cleaning periodically or repairing under emerging situations.

13.3.4 Preliminary Sludge Preparation Operations

Sludge contains stringy material and is nonhomogeneous. Preliminary sludge preparation operation may be needed for processing. Common preliminary operations are summarized in Table 13.5.

EXAMPLE 13.1: SPECIFIC GRAVITY OF DRY SOLIDS IN SLUDGE

A VSS/TSS ratio of 0.7 is obtained from the laboratory test results on a primary sludge sample. Estimate the specific gravity of dry solids of the sludge.

Solution

1. Select the specific gravity values for the fixed and volatile solids in the sludge.

   Use the $S_f = 2.5$ and $S_v = 1.1$. They are both in the typical ranges for primary sludge (see the definitions of these terms in Equation 13.1.

2. Calculate the specific gravity of the dry solids in the sludge (S_s) from Equation 13.1b using VSS/TSS ratio = 0.7.

$$S_s = \frac{1}{\frac{1 - \text{VSS/TSS ratio}}{S_f} + \frac{\text{VSS/TSS ratio}}{S_v}} = \frac{1}{\frac{1-0.7}{2.5} + \frac{0.7}{1.1}} = 1.32$$

Note: The calculated specific gravity of 1.32 is within the typical range for the dry solids in raw primary sludge given in Table 13.1.

EXAMPLE 13.2: SPECIFIC GRAVITY OF BULK WET SLUDGE

Estimate the specific gravity of bulk wet primary sludge with a solids content of 4%. Use the calculation results obtained in Example 13.1.

Solution

The specific gravity of dry solids obtained in Example 13.1 is 1.32. Calculate the bulk specific gravity of the wet primary sludge (S_b) from Equation 13.1d using $p_s = 4\%$.

$$S_b = \frac{100\% \times S_s}{100\% \times S_s - p_s(S_s - 1)} = \frac{100\% \times 1.32}{100\% \times 1.32 - 4\% \times (1.32 - 1)} = 1.010$$

Note: the calculated specific gravity of 1.010 is within the typical range for wet raw primary sludge given in Table 13.1.

EXAMPLE 13.3: MASS AND VOLUME OF PRIMARY, SECONDARY, AND COMBINED SLUDGE

A municipal wastewater treatment plant receives an average of 12,000 m^3/d flow. The average production rate of raw primary sludge and waste activated sludge (WAS) are 150 and 90 mg/L (or g/m^3), respectively. The solids content and bulk specific gravity values are: $p_{s,PS} = 4\%$ and $S_{b,PS} = 1.01$ for the primary sludge, and $p_{s,WAS} = 1.0\%$ and $S_{b,WAS} = 1.002$ for the WAS. Determine the daily total dry solids, weight, volume, specific gravity, solids content, and bulk density of combined raw sludge. Assume the water density $\rho_w = 1000\ kg/m^3$.

Solution

1. Determine the daily mass productions of dry solids in primary sludge (PS) and WAS.

$$\text{Mass of dry solids in primary sludge, } W_{s,PS} = \text{Sludge volume} \times \text{Sludge production rate}$$
$$= 12{,}000\ m^3/d \times 150\ g/m^3 \times 10^{-3}\ kg/g = 1800\ kg/d$$

$$\text{Mass of dry solids in WAS, } W_{s,WAS} = 12{,}000\ m^3/d \times 90\ g/m^3 \times 10^{-3}\ kg/g = 1080\ kg/d$$

2. Determine the daily volumes of primary sludge and WAS from Equation 13.1e.

$$\text{Volume of primary sludge, } Q_{b,PS} = \frac{100\% \times W_{s,PS}}{p_{s,PS} S_{b,PS} \rho_w} = \frac{100\% \times 1800\ kg/d}{4\% \times 1.01 \times 1000\ kg/m^3} = 44.6\ m^3/d$$

$$\text{Volume of WAS, } Q_{b,WAS} = \frac{100\% \times W_{s,WAS}}{p_{s,WAS} S_{b,WAS} \rho_w} = \frac{100\% \times 1080\ kg/d}{1.0\% \times 1.002 \times 1000\ kg/m^3} = 107.8\ m^3/d$$

3. Determine the daily total weight of primary sludge and WAS.
   Rearrange Equation 13.1e and calculate the total mass of the sludge.

$$\text{Total mass of wet primary sludge, } W_{b,PS} = S_{b,PS}\rho_w Q_{b,PS}$$
$$= 1.01 \times 1000\,\text{kg/m}^3 \times 44.6\,\text{m}^3/\text{d} = 45{,}000\,\text{kg/d}$$

$$\text{Total mass of wet WAS, } W_{b,WAS} = S_{b,WAS}\rho_w Q_{b,WAS}$$
$$= 1.002 \times 1000\,\text{kg/m}^3 \times 107.8\,\text{m}^3/\text{d} = 108{,}000\,\text{kg/d}$$

4. Determine the daily total weight of dry solids of combined sludge (CS).

$$\text{Total mass of dry solids in combined sludge, } W_{s,CS} = \text{Total mass of dry solids in primary sludge}$$
$$+ \text{Total mass of dry solids in WAS}$$
$$= W_{s,PS} + W_{s,WAS} = (1800 + 1080)\,\text{kg/d}$$
$$= 2880\,\text{kg/d}$$

5. Determine the daily total weight and volume of combined sludge.

$$\text{Total mass of wet combined sludge, } W_{b,CS} = \text{Total mass of wet primary sludge}$$
$$+ \text{Total mass of wet WAS}$$
$$= W_{b,PS} + W_{b,WAS} = (45{,}000 + 108{,}000)\,\text{kg/d}$$
$$= 153{,}000\,\text{kg/d}$$

$$\text{Total volume of combined sludge, } Q_{b,CS} = \text{Total volume of primary sludge}$$
$$+ \text{Total volume of WAS}$$
$$= Q_{b,CS} + Q_{b,CS} = (44.6 + 107.8)\,\text{m}^3/\text{d} = 152.4\,\text{m}^3/\text{d}$$

Note: The volume of combined sludge is 1.3% of the average wastewater flow.

6. Determine the overall specific gravity, solids content, and bulk density of combined wet sludge.
   Rearrange Equation 13.1e and calculate the overall specific gravity and solids content of combined wet sludge.

$$\text{Overall specific gravity of combined sludge, } S_{b,CS} = \frac{W_{b,CS}}{\rho_w Q_{b,CS}} = \frac{153{,}000\,\text{kg/d}}{1000\,\text{kg/m}^3 \times 152.4\,\text{m}^3/\text{d}} = 1.004$$

$$\text{Solids content of combined sludge, } p_{s,CS} = \frac{100\% \times W_{s,CS}}{S_{b,CS}\rho_w Q_{b,CS}} = \frac{100\% \times 2880\,\text{kg/d}}{1.004 \times 1000\,\text{kg/m}^3 \times 152.4\,\text{m}^3/\text{d}} = 1.9\%$$

$$\text{The density of combined sludge, } \rho_{CS} = S_{b,CS}\,\rho_w = 1.004 \times 1000\,\text{kg/m}^3 = 1004\,\text{kg/m}^3$$

7. Summarize the calculated results of the combined wet sludge.
   Daily total dry solids = 2880 kg/d, total wet sludge weight = 153,000 kg/d, total volume = 152.4 m^3/d, overall specific gravity = 1.004, solids content = 1.9%, and bulk density = 1004 kg/m^3.

EXAMPLE 13.4: BULK SPECIFIC GRAVITY OF DIGESTED SLUDGE

An anaerobically digested sludge contains 4% solids. The digested solids contain one-half mineral matter having specific gravity = 2.5. The remaining volatile solids has specific gravity = 1.0. Determine the specific gravity of total dry solids and the bulk specific gravity of digested sludge.

Solution

1. Determine the specific gravity of dry solids from Equation 13.1b.

$$S_s = \frac{1}{\dfrac{1 - \text{VSS/TSS ratio}}{S_f} + \dfrac{\text{VSS/TSS ratio}}{S_v}} = \frac{1}{\dfrac{1-0.5}{2.5} + \dfrac{0.5}{1.0}} = 1.43$$

2. Determine the bulk specific gravity of digested sludge from Equation 13.1d.

$$S_b = \frac{100\% \times S_s}{100\% \times S_s - p_s(S_s - 1)} = \frac{100\% \times 1.43}{100\% \times 1.43 - 4\% \times (1.43 - 1)} = 1.012$$

EXAMPLE 13.5: SPECIFIC GRAVITY AND VOLUME OF THICKENED AND ANAEROBICALLY DIGESTED SLUDGE

The thickened sludge has: total solids = 1000 kg/d at 6% solids, volatile solids = 70%, specific gravity of fixed solids = 2.5, and specific gravity of volatile solids = 1.0. The anaerobically digested sludge has: total solids = 10%, and 55% volatile solids are destroyed in the digester. The specific gravity of volatile and fixed solids remains the same as in the raw sludge. Determine the following: (1) specific gravity of dry solids and bulk thickened sludge, (2) volume of thickened sludge, (3) volatile solids in the digested sludge, (4) specific gravity of dry solids and bulk digested sludge, (5) volume of digested sludge, and (6) percent volume reduction in the digester. Assume the water density $\rho_w = 1000\,\text{kg/m}^3$.

Solution

1. Determine the specific gravity of dry solids and bulk thickened sludge (THS) from Equations 13.1b and 13.1d.

$$\text{Specific gravity of dry thickened solids, } S_{s,\text{THS}} = \frac{1}{\dfrac{1 - \text{VSS/TSS}_{\text{THS}}}{S_f} + \dfrac{\text{VSS/TSS}_{\text{THS}}}{S_v}} = \frac{1}{\dfrac{1-0.7}{2.5} + \dfrac{0.7}{1.0}} = 1.22$$

$$\text{Specific gravity of bulk thickened sludge, } S_{b,\text{THS}} = \frac{100\% \times S_{s,\text{THS}}}{100\% \times S_{s,\text{THS}} - p_{s,\text{THS}}(S_{s,\text{THS}} - 1)} = \frac{100\% \times 1.22}{100\% \times 1.22 - 6\% \times (1.22 - 1)} = 1.011$$

2. Determine the daily volume of thickened sludge from Equation 13.1e.

$$\text{Volume of thickened sludge, } Q_{b,\text{THS}} = \frac{100\% \times W_{s,\text{THS}}}{p_{s,\text{THS}} S_{b,\text{THS}} \rho_w} = \frac{100\% \times 1000\,\text{kg/d}}{6\% \times 1.011 \times 1000\,\text{kg/m}^3} = 16.5\,\text{m}^3/\text{d}$$

3. Determine the daily volatile solids and VSS/TSS ratio in the digested sludge (DGS).

$$\text{Volatile solids remaining after digestion, } W_{s,\text{VSS,DGS}} = W_{s,\text{THS}}\text{VSS/TSS}_{b,\text{THS}}(1 - \text{VSS}_{\text{destroyed}})$$
$$= 1000\,\text{kg TSS/d} \times 0.7\,\text{kg VSS/kg TSS} \times (1 - 0.55) = 315\,\text{kg VSS/d}$$

Calculate the fixed (or non-VSS) solids in the thickened sludge.

$$W_{s,\text{Non-VSS,DGS}} = \text{Weight of thickened sludge} \times (1 - \text{VSS/TSS ratio in the thickened sludge})$$
$$= W_{s,\text{THS}}(1 - \text{VSS/TSS}_{b,\text{THS}}) = 1000\,\text{kg TSS/d} \times (1 - 0.7\,\text{kg VSS/kg TSS})$$
$$= 1000\,\text{kg TSS/d} \times 0.3\,\text{kg Non-VSS/kg TSS} = 300\,\text{kg Non-VSS/d}$$

The fixed solids (or non-VSS) should remain the same in both the thickened and digested sludge.

$$W_{s,\text{Non-VSS,DGS}} = W_{s,\text{Non-VSS,THS}} = 300\,\text{kg Non-VSS/d}$$

Total solids after digestion, $W_{s,\text{DGS}} = \text{Weight of VSS} + \text{Weight fixed solids}$
$$= W_{s,\text{VSS,DGS}} + W_{s,\text{Non-VSS,DGS}} = (315 + 300)\,\text{kg/d}$$
$$= 615\,\text{kg TSS/d}$$

Volatile solids fraction remaining after digestion, $\text{VSS/TSS}_{\text{DGS}} = \dfrac{\text{Weight of VSS}}{\text{Weight of total solids}}$

$$= \frac{W_{s,\text{VSS,DGS}}}{W_{s,\text{DGS}}} = \frac{315\,\text{kg VSS/d}}{615\,\text{kg TSS/d}} = 0.512$$

4. Determine the specific gravity of dry solids and bulk digested sludge from Equations 13.1b and 13.1d.

Specific gravity of dry digested solids, $S_{s,\text{DGS}} = \dfrac{1}{\dfrac{1 - \text{VSS/TSS}_{\text{DGS}}}{S_f} + \dfrac{\text{VSS/TSS}_{\text{DGS}}}{S_v}}$

$$= \frac{1}{\dfrac{1 - 0.512}{2.5} + \dfrac{0.512}{1.0}} = 1.41$$

Specific gravity of bulk digested sludge, $S_{b,\text{DGS}} = \dfrac{100\% \times S_{s,\text{DGS}}}{100\% \times S_{s,\text{DGS}} - p_{s,\text{DGS}}(S_{s,\text{DGS}} - 1)}$

$$= \frac{100\% \times 1.41}{100\% \times 1.41 - 10\% \times (1.41 - 1)} = 1.030$$

5. Determine the daily volume of digested sludge from Equation 13.1e.

Volume of digested sludge, $Q_{b,\text{DGS}} = \dfrac{100\% \times W_{s,\text{DGS}}}{p_{s,\text{DGS}} S_{b,\text{DGS}} \rho_w} = \dfrac{100\% \times 615\,\text{kg/d}}{10\% \times 1.030 \times 1000\,\text{kg/m}^3} = 6.0\,\text{m}^3/\text{d}$

6. Determine the percent volume reduction in the digester.

Percent sludge volume reduction in digestion,

$$R_{b,\text{Reduction}} = \left(1 - \frac{\text{Volume of digested sludge}}{\text{Volume of thickened sludge}}\right) \times 100\%$$
$$= \left(1 - \frac{Q_{b,\text{DGS}}}{Q_{b,\text{THS}}}\right) \times 100\% = \left(1 - \frac{6.0\,\text{m}^3/\text{d}}{16.5\,\text{m}^3/\text{d}}\right) \times 100\% = 64\%$$

Note: Supernatant must be withdrawn from the digester to achieve high solids content of 10% at a 64% volume reduction.

EXAMPLE 13.6: HEAD LOSS IN A SLUDGE TRANSPORT PIPING

A sludge transport pipe is 50 m (164 ft) long and 150 mm (6 in) in diameter. Sludge is pumped from the sludge hopper of a primary clarifier to a sludge blending tank. The solids content in the sludge is 4% and pumping rate is 1000 L/min (264 gpm). Calculate the head loss and compare with pumping of water at the same flow rate through the same pipe.

Solution

1. Calculate the velocity in sludge pipe with a diameter $D = 0.15$ m.

$$\text{Pipe cross-sectional area,}\ A = \frac{\pi}{4}(\text{Diameter})^2 = \frac{\pi}{4}D^2 = \frac{\pi}{4} \times (0.15\ \text{m})^2 = 0.0177\ \text{m}^2$$

$$\text{Pumping rate,}\ Q = 1000\ \text{L/min} \times 10^{-3}\ \text{m}^3/\text{L} \times \frac{\text{min}}{60\text{s}} = 0.0167\ \text{m}^3/\text{s}$$

$$\text{Velocity,}\ v = \frac{\text{Flow}}{\text{Area}} = \frac{Q}{A} = \frac{0.0167\ \text{m}^3/\text{s}}{0.0177\ \text{m}^2} = 0.944\ \text{m/s}$$

2. Determine the head loss during sludge pumping ($h_{\text{f,sludge}}$) from Hazen-Williams equation Equation 6.13b.

The roughness coefficient $C = 53$ for raw sludge at 4% solids content is obtained from Table 13.4.

$$h_{\text{f,sludge}} = 6.82\left(\frac{v}{C}\right)^{1.85} \times \frac{L}{D^{1.167}} = 6.82\left(\frac{0.944\ \text{m/s}}{53}\right)^{1.85} \times \frac{50\ \text{m}}{(0.15\ \text{m})^{1.167}} = 1.81\ \text{m}$$

3. Determine the head loss if water is pumped ($h_{\text{f,water}}$).

Use $C = 100$ from Table 13.4 and calculate h_f from Equation 6.13b.

$$h_{\text{f,water}} = 6.82\left(\frac{0.944\ \text{m/s}}{100}\right)^{1.85} \times \frac{50\ \text{m}}{(0.15\ \text{m})^{1.167}} = 0.56\ \text{m}$$

4. Compare the head losses between pumping sludge and water.

The head loss from pumping sludge at 4% solids content is ($h_{\text{f,sludge}}/h_{\text{f,water}}$) = (1.81 m/0.56 m) = 3.2 times that from pumping water.

EXAMPLE 13.7: DESIGN OF A SLUDGE BLENDING TANK

Combined primary and biological sludge are blended in a tank prior to sludge thickening. The design blending and storage period is 2 h. The total peak flow of 1800 m^3/d to the blending tank includes combined sludge and dilution water. Describe different types of mixing arrangements and calculate (1) the dimensions of the blending tank and (2) the power requirements for mechanical mixing. The velocity gradient exerted by the mixer at the critical temperature of 5°C under the field condition is 60 s^{-1}, and the kinetic viscosity of sludge is 2 times that of water.

Solution

1. Describe the blending tank and mixing arrangement.

The blending tank is a cylindrical basin that receives the primary and biological sludge, and dilution water (normally effluent). Mixing may be achieved by (1) coarse air bubble diffusers, (2) liquid recirculation, or (3) mechanical mixers. The diffused air mixing has the advantage of freshening the sludge and may reduce odors in the thickener. Odor problem may be serious in the blending tank. The blending tank may be covered, and the captured air is scrubbed by chemical solutions of iron salts, chlorine, potassium permanganate, magnesium hydroxide, or hydrogen peroxide. Biofiltration is also be used effectively for odor control.

2. Calculate the dimensions of the blending tank.

$$\text{Volume of the blending tank,}\ V = \text{Flow} \times \text{Storage period} = Q\theta = 1800\ \text{m}^3/\text{d} \times 2\ \text{h} \times \frac{\text{d}}{24\ \text{h}} = 150\ \text{m}^3$$

Provide a design water depth, $H = 3$ m.

$$\text{Area of the tank, } A = \frac{\text{Tank volume}}{\text{Water depth}} = \frac{V}{H} = \frac{150\,\text{m}^3}{3\,\text{m}} = 50\,\text{m}^2$$

$$\text{Diameter of the tank, } D = \sqrt{\frac{4 \times \text{Area}}{\pi}} = \sqrt{\frac{4A}{\pi}} = \sqrt{\frac{4 \times 50\,\text{m}^2}{\pi}} = 8.0\,\text{m}^2$$

3. Determine the power requirements in blending tank.
   Kinetic viscosity $\mu = 1.519 \times 10^{-3}\,\text{N·s/m}^2$ for water at 5°C (Table B.2 in Appendix B).
   Kinetic viscosity for the blended sludge in the tank, $\mu = 2 \times 1.519 \times 10^{-3}\,\text{N·s/m}^2 = 3.038 \times 10^{-3}\,\text{N·s/m}^2$
   Calculate the power imparted to the sludge P from Equation 9.27 at $G = 60\,\text{s}^{-1}$ and $V = 150\,\text{m}^3$.

$$\text{Water power, } P_\text{w} = \mu VG^2 = 3.038 \times 10^{-3}\,\text{N·s/m}^2 \times 150\,\text{m}^3 \times (60\,\text{s}^{-1})^2$$
$$= 1640\,\text{N·m/s} = 1640\,\text{W or } 1.64\,\text{kW}$$

$$\text{Motor power at an overall efficiency of 75\%, } P_\text{m} = \frac{\text{Water power}}{\text{Efficiency}} = \frac{P_\text{w}}{E_\text{overall}} = \frac{1.64\,\text{kW}}{0.75} = 2.19\,\text{kW}$$

Note: Readers are referred to Section 9.5.6 for theory and design of flocculation process. The design procedure of blending paddles is similar to that of vertical flocculator in Example 9.39. The sludge blending tank and paddle details are shown in Figure 9.41.

EXAMPLE 13.8: VOLUME OF SLUDGE STORAGE TANK

Thickened sludge is stored for chemical conditioning and dewatering prior to trucking the sludge cake to a landfill for disposed. The dewatering facility runs five days a week while sludge production is seven days a week. A sludge storage facility, therefore, is needed for storage of the thickened sludge for a maximum of two consecutive days. The annual average daily solids production from the treatment facility is 10,000 kg/d. The solids concentration in the tank is 6% and the specific gravity of the bulk sludge is 1.01. Apply sustained peak solids mass loading data given in Example 5.63 to determine (a) the volume of sludge storage tank, and (b) the design capacity of dewatering facility. Apply a safety factor of 1.2 to maximum amount of solids stored in the tank, and for design of dewatering equipment.

Note: Dewatering of the thickened raw sludge is done for composting, incineration, and landfilling. Transportation of the dewatered sludge is also economical.

Solution

1. Develop the peaking factors from the sustained peak solids mass loading data given in Example 5.63.
   The dewatering facility, should meet the following two essential design requirements: (1) the sludge storage tank should hold the solids produced during two consecutive days and (2) the capacity of dewatering facility should be sufficient enough to process the total amount of solids produced during a week within a 5-day operating period. Therefore, the peaking factors for maximum sustained solids loading in two and seven consecutive days are needed.
   The following data are available from Table 5.16 (Example 5.63): (a) peaking factors of 2.13 and 2.31 for BOD_5 and TSS mass loadings in two consecutive days and (b) a peaking factor of 1.56 for both BOD_5 and TSS mass loadings in seven consecutive days.
   Assume that the primary sludge is ~ 60% of the combined sludge. Determine the peaking factors for the combined mass loading at the plant.

Peaking factor for two consecutive days, $PF_{M2D,CS} = (1-0.6) \times PF_{M2D,BOD5} + 0.6 \times PF_{M2D,TSS}$
$= (1-0.6) \times 2.13 + 0.6 \times 2.31 = 2.24$

Peaking factor for seven consecutive day factor, $PF_{M7D,CS} = PF_{M7D,BOD5} = PF_{M7D,TSS} = 1.56$

2. Determine the volume of sludge storage tank.

Calculate the maximum daily sludge production rate during two consecutive days from the annual average daily sludge production of combined sludge $W_{AD,CS} = 10{,}000$ kg/d.

$$W_{M2D,CS} = PF_{M2D,CS} \times \text{Annual average daily sludge production}$$
$$= PF_{M2D,CS} \times W_{AD,CS} = 2.24 \times 10{,}000\,\text{kg/d} = 22{,}400\,\text{kg/d}$$

Determine the maximum amount of solids production during two consecutive days.

$$\Delta W_{M2D,CS} = 2\,\text{d} \times \text{Maximum daily sludge production} = 2\,\text{d} \times W_{M2D,CS} = 2\,\text{d} \times 22{,}400\,\text{kg/d}$$
$$= 44{,}800\,\text{kg}$$

Apply the safety factor of 1.2 to calculate the maximum amount of solids stored in the tank.

$$\Delta W_{tank} = 1.2 \times \text{Maximum amount of sludge} = 1.2 \times \Delta W_{M2D,CS} = 1.2 \times 44{,}800\,\text{kg/d}$$
$$= 53{,}800\,\text{kg}$$

Total volume of storage tank Equation 13.1e, $V_{tank} = \dfrac{100\% \times \Delta W_{tank}}{p_s S_b \rho_w}$

$$= \frac{100\% \times 53{,}800\,\text{kg}}{6\% \times 1.01 \times 1000\,\text{kg/m}^3} = 888\,\text{m}^3$$

Provide two identical 450-m^3 storage tanks for operational flexibility. Also, provide the following accessories for the storage tank: (a) sludge recirculation for mixing, (b) sludge feed pumps and piping for dewatering facility, (c) a magnetic flow meter on the sludge force main to each dewatering unit, (d) chemical storage and feed systems for odor control, and (e) enclosure over the tank for odor control, if necessary.

Note: The volume determined above meets the worst case scenario because (a) it is assumed that the maximum mass loading for two consecutive days occurs exactly when the dewatering units are not in service, and (b) there are additional storage capacities in the primary and secondary clarifiers, and in gravity thickeners, are not utilized.

3. Determine the design capacity of dewatering facility.

Calculate the maximum daily sludge production rate during seven consecutive days.

$$W_{M7D,CS} = PF_{M7D,CS} \times W_{AD,CS} = 1.56 \times 10{,}000\,\text{kg/d} = 15{,}600\,\text{kg/d}$$

Determine the maximum amount of solids production during seven consecutive days.

$$\Delta W_{M7D,CS} = 7\,\text{d} \times W_{M7D,CS} = 7\,\text{d} \times 15{,}600\,\text{kg/d} = 109{,}200\,\text{kg}$$

Determine the daily solids reaching the dewatering units during five operating days.

$$W_{DS} = \frac{\Delta W_{M7D,CS}}{5\,\text{d}} = \frac{109{,}200\,\text{kg}}{5\,\text{d}} = 21{,}840\,\text{kg}$$

Determine the design daily solids processing capacity of the dewatering units with a safety factor of 1.2.

$$W_{DS}^{design} = 1.2 \times W_{DS} = 1.2 \times 21{,}840\,\text{kg/d} = 26{,}200\,\text{kg/d}$$

Note: The ratio of design capacity of dewatering facility to the annual average daily mass loading is 2.62. It is slightly higher than the maximum daily peaking factor of 2.58 for mass TSS loading in Table 5.16.

13.4 Conventional Sludge Processing Systems, Process Diagrams, and Regulatory Requirements

13.4.1 Sludge Processing Systems

The volume of combined sludge produced at a conventional secondary treatment plant is nearly 1% of the average wastewater flow. The sludge is odorous and many cause serious environmental problems. It requires complex processing equipment to concentrate, stabilize, condition and dewater. The cost is high and may exceed the cost of treating the entire wastewater stream. The major operations are: (1) thickening, (2) stabilization, (3) chemical conditioning, (4) dewatering, and (5) disposal and reuse. There are many variations with each operation. Table 6.10 has brief description of 21 unit operations and processes falling in five major sludge processing categories. Also, Figure 6.9 has generalized process diagrams for these major operations. Readers are referred to Section 6.3.5 for background information on sludge processing and disposal.

13.4.2 Sidestreams

The sludge-processing operations produce two streams (1) processed solids and (2) liquid stream. The liquid stream also called *sidestream* that contains high concentrations of BOD_5, suspended and dissolved solids, and nutrients. It is returned to the head of the plant for further treatment. Often an equalization basin is provided to collect the sidestreams and distribute the hydraulic and mass loading over 24-h operation. The quality of sidestreams from different sludge processing facilities is provided in Table 13.6. The returned sidestreams increase the mass loadings of BOD_5, TSS, and other constituents in the combined plant influent. To determine the incremental loadings from sidestreams a material mass balance analysis is usually performed at the average flow. Examples 6.22, 6.23, and 13.54 provide the generalized procedure for mass balance analysis to determine the impact of sidestreams upon the combined influent quality to be treated by the selected liquid process train.

13.4.3 Process Diagrams

Proper selection of sludge processing equipment is essential for plant operation and maintenance. Odor control factors are given serious consideration. The plant operation and environmental factors considered for process selection are: (1) public health and safety; (2) land requirement; (3) adverse effects of climatic conditions such as freezing and flooding; (4) ability to handle flow and loading variations; (5) ability to handle industrial wastes; (6) process reliability; (7) ease of operation and maintenance; (8) occupational hazardous; (9) quantity and quality of sidestreams; (10) environmental impacts, such as odors, VOCs, greenhouse gases, ground water contamination, etc.; (11) energy recovery such as methane, heat, and electricity; (12) resource recovery such as nutrients and lime; and (13) beneficial reuses such as biosolids for land application. Several process diagrams for producing Class A and B biosolids are given and discussed in Examples 13.9 and 13.10.

13.4.4 Regulatory Requirements for Biosolids Reuse and Disposal

To encourage the beneficial uses of sewage sludge for land application, the U.S. Environmental Protection Agency established the "Standards for the Use and Disposal of Sewage Sludge." These standards are under the 40 CFR Part 503 regulations and are commonly referred as the Part 503 Rule.[17,18] To protect the public health from reuse and disposal of biosolids over land, the numerical limits for pollutants are established and regulations impose significant monitoring, reporting, and record keeping requirements on POTWs.[17] The biosolids are considered clean and safe for land application if the regulatory requirements are met. In general, these requirements fall into three major groups: (a) pollutant limitations, (b) pathogen reduction, and (c) vector attraction reduction.[17–20] Brief discussions about these requirements are presented below.

TABLE 13.6 Typical Performance and Sidestream Quality Parameters from Various Sludge Processing Unit Operations and Processes

Sludge Processing System	Typical Feed Sludge	Feed Sludge Solids Content, %	Solids Capture Rate, %	Solids Content in Sludge, %	Concentration in Sidestream, mg/L	
					BOD_5	TSS
Thickener						
Gravity[a]	PS	1–6	85–92	4–10	500–3000	1000–5000
	WAS	0.2–1.2	60–85	2–4	200–1500	500–2500
	PS + WAS	1–4	80–90	3–6	300–2000	800–3500
Dissolved air flotation (DAF)	WAS	0.3–1.2	80–98	3–6	200–1000	300–2000
Centrifuge	WAS	0.5–1.2	80–98	3–8	300–500	500–1000
Gravity belt	WAS	0.5–1.2	90–98	4–8	150–300	300–600
Rotary drum	WAS	0.5–1.2	93–99	4–9	100–250	200–500
Membrane	WAS[b]	0.2–1.2	>99	3–4[c]	<1	<1
Stabilization						
Aerobic	WAS or THS	1–4	–	1–3	200–1500	1000–10,000
Anaerobic	THS	2–6	–	3–5	500–10,000	2000–15,000
Dewatering[d]						
Sludge drying beds[e]	WAS or THS	1–4	>90	15–40	100–500	200–1000
Solar drying bed[f,g]	THS or PDS	2–30	>95	50–90	100–500	150–1000
Belt-filter press	Various	1–8	85–98	15–30	300–1500	500–4000
Centrifuge	Various	1–8	85–98	10–35	500–2000	1000–10,000
Plate and frame filter press	Various	1–8	90–98	20–55	200–1000	300–3000
Vacuum filter	Various	3–8	90–98	15–25	300–1500	500–10,000
Rotary press	Various	2–6	90–98	20–30	–	–
Screw press	Various	2–6	90–98	20–30	–	–
Electro-dewatering (EDW)[f,g]	PDS	10–20	90–98	30–50	–	–
Drying						
Heat or thermal	DS	>15	–	>90	–	–

[a] The BOD_5 and TSS concentrations may be reduced when dilution water is used to meet the hydraulic loading requirement.
[b] Biological sludge from activated sludge processes with modifications, including MBR.
[c] No requirement for chemical conditioning by polymer or coagulant.
[d] Mechanical sludge dewatering processes are normally associated with chemical conditioning.
[e] Including conventional, paved, artificial-media, and vacuum assisted drying beds.
[f] Predewatering may be required for these processes (see Sections 13.8.1 and 13.8.2 for descriptions of these processes).
[g] These dewatering processes have partial drying function (see Section 13.11.2 for descriptions of solids drying processes).

Note: DGS = Digested sludge with chemical conditioning; DS = dewatered sludge; PDS = predewatered sludge; PS = Primary sludge; THS = thickened sludge; WAS = Waste activated sludge.

Source: Adapted in part from References 6 through 8.

Readers are referred to Section 13.11.5 for additional information on these requirements, and land application of biosolids in four designated groups.

Pollutant Limitations: In Part 503 Rule, the limits for land application of the biosolids are established in concentrations (mg/kg) and loading rates (kg/ha or kg/ha·year) for 10 heavy metals. These pollutants limitations are covered in Section 13.11.5.[17–20]

TABLE 13.7 General Information on Biosolids Classifications for Land Application

Parameter	Information by Class	
	Class A	Class B
Designation	The dewatered sewage sludge that is suitable for land application with no restrictions.	The treated sewage sludge that is only suitable for specific use as fertilizer with site restrictions.
Application	It is intended to have no detectable pathogens after treatment so that the biosolids can be applied to farm lands, nurseries, parks, golf courses, gardens, and home lawns, and sold or given away in bags or other containers for application over land without restrictions.	It is intended to reduce the pathogens to a level that it is not a threat to public health and environment so that the biosolids can be applied to agricultural lands, forests, or reclamation sites.
Qualification alternative	• Alt. 1: thermally treated • Alt. 2: treated in a high-pH/high-temperature process • Alt. 3: treated in an know process • Alt. 4: treated in other unknown processes • Alt. 5: treated in one of seven PFRPs • Alt. 6: treated in a process equivalent to a PFRP	• Alt. 1: monitoring of indicator organisms • Alt. 2: treated in one of five PSRPs • Alt. 3: treated in a process equivalent to a PSRP
Pathogen reduction criteria		
Bacterial[a]	• Fecal coliform < 1000 MPN/g total dry solids, or • *Salmonella* sp. < 3 MPN/4 g total dry solids	Fecal coliform < 2.0×10^6 MPN or CFU/g total solids
Viral and Parasitic[b,c]	• Enteric virus < 1 PFU/4 g total solids • Viable Helminth ova < 1 ova/4 g total solids	–
pH and temperature criteria[d]	pH > 12 for at least 72 h; temperature >50°C for at least 12 h; and then air dry sludge to ≥50% total solids	–

[a] For Class A classification, the criteria are applicable to all six alternatives (Alt. 1–6). For Class B classification, the criteria are applicable to Alt. 1.

[b] For Class A classification, the criteria are applicable to Alt. 3 when (a) testing prior to pathogen treatment, or (b) testing after the treatment if the criteria are not met in testing prior to the pathogen treatment.

[c] For Class A classification, the criteria are applicable to Alt. 4 after treatment and ready for distribution, etc.

[d] For Class A classification, the criteria are applicable to Alt. 2.

Note: PFRP = Processes to further reduce pathogens; PSRP = Processes to significantly reduce pathogens.

Source: Summarized from information contained in References 17 through 20.

Pathogen Reduction Requirements: The requirements for pathogen reduction vary depending upon the classification of the biosolids. Under the Part 503 Rule, the biosolids are divided into two classes, *Class A* and *Class B*. These classifications are based on the bacterial, viral, and parasitic pathogen reduction levels reached through a specific monitoring program, the degree of treatment received, and the specific operating parameters (temperature, pH, period of time, or solids content) maintained in the sludge processing processes. The basic information about the classifications of biosolids for land application is summarized in Table 13.7.[18]

Processes to Further Reduce Pathogens: The following processes are considered processes to further reduce pathogens (PFRPs) for treating the sewage sludge to qualify for Class A biosolids:[18]

1. *Composting* of the biosolids at 55°C or higher for (a) 3 days in a within-vessel or static aerated pile and (b) 15 days using windrow composting method with turning at least 5 times.
2. *Heat drying* of the biosolids to reduce the moisture content to 10% or lower at or over 80°C in a direct or indirect contact dryer.
3. *Heat treatment* of the liquid biosolids at a temperature of 180°C or higher for 30 min.

4. *Thermophilic aerobic digestion* (TAD) of the biosolids at temperatures 55–60°C for a mean cell residence time (MCRT) of 10 days.
5. *Beta ray irradiation* of the biosolids at dosages of at least 1.0 megarad (Mrad) at the room temperature (20°C).
6. *Gamma ray irradiation* of the biosolids at dosages of at least 1.0 Mrad at 20°C.
7. *Pasteurization* of the biosolids at a temperature of 70°C or higher for at least 30 min.

Processes to Significantly Reduce Pathogens: The biosolids treated by processes to significantly reduce pathogens (PSRP) contain reduced levels of pathogens that may still have a potential to transmit disease. Therefore, these solids are only qualified for Class B biosolid. The acceptable treatment processes for PSRP are given below:[18]

1. *Aerobic digestion* of the biosolids with MCRT and temperature between 40 days at 20°C and 60 days at 15°C.
2. *Air drying* of the biosolids on sand beds or in storage basins for a minimum of 3 months.
3. *Anaerobic digestion* of the biosolids with MCRT and temperature between 15 days at 35–55°C and 60 days at 20°C.
4. *Composting* of the biosolids for 5 days at 40°C or higher with the temperature in the compost pile exceeding 55°C for 4 h.
5. *Lime stabilization* to raise the pH of the biosolids up to 12 after 2 h of contact.

Vector Attraction Reduction Requirements: Twelve options are established in the Part 503 Rule for effectively reducing vector attraction. These options include reducing the volatile solids of biosolids by aerobic or anaerobic digestion (>38%); meeting a specific oxygen uptake rate (SOUR) for aerobic digestion (<1.5 mg O_2/g dry total solids·h at 20°C); adding sufficient alkaline agent to raise and maintain the pH above 12 for 2 h; reducing the moisture content (<10% with or <25% without unstabilized solids); and mixing the biosolids with other materials, injecting/incorporating the biosolids into soil within specified periods, and covering the biosolids on surface disposal sites. The biosolids product that meets or exceeds all requirements for Class A pathogen reduction, metals and VAR is also termed as Class A Exceptional Quality (EQ) biosolids.[17–20]

EXAMPLE 13.9: DEVELOP CONCEPTUAL PROCESS TRAINS FOR PROCESSING COMBINED SLUDGE

Draw conceptual process trains for processing the combined sludge. The following four commonly used processes are applied: (a) gravity belt thickener (GBT) and anaerobic digester, (b) aerobic digester and drying beds, (c) gravity thickener and dissolved air flotation (DAF) followed by dual digestion process, and (d) chemical conditioning of two-stage lime flocculation and incineration with lime recalcination. Give a brief description of each train.

Solution

1. Draw four conceptual process trains in Figure 13.1.
2. Describe each process train.
   a. Process Train 1—GBT and anaerobic digester:

      The process train is shown in Figure 13.1a. The combined sludge is thickened by GBT to solids content up to 6%. The high-rate mesophilic anaerobic digester has a solids retention time (SRT) of 15–20 d, and a VSS reduction of 40–60% is achieved. The digested sludge is stored in a sludge storage tank for dewatering. The belt-filter press (BFP) provides 90–95% solid capture, and solids content of sludge cake will be 20–30%. Class B biosolids criteria will be met if the sludge is maintained in the digester at a temperature of 35–55°C for at

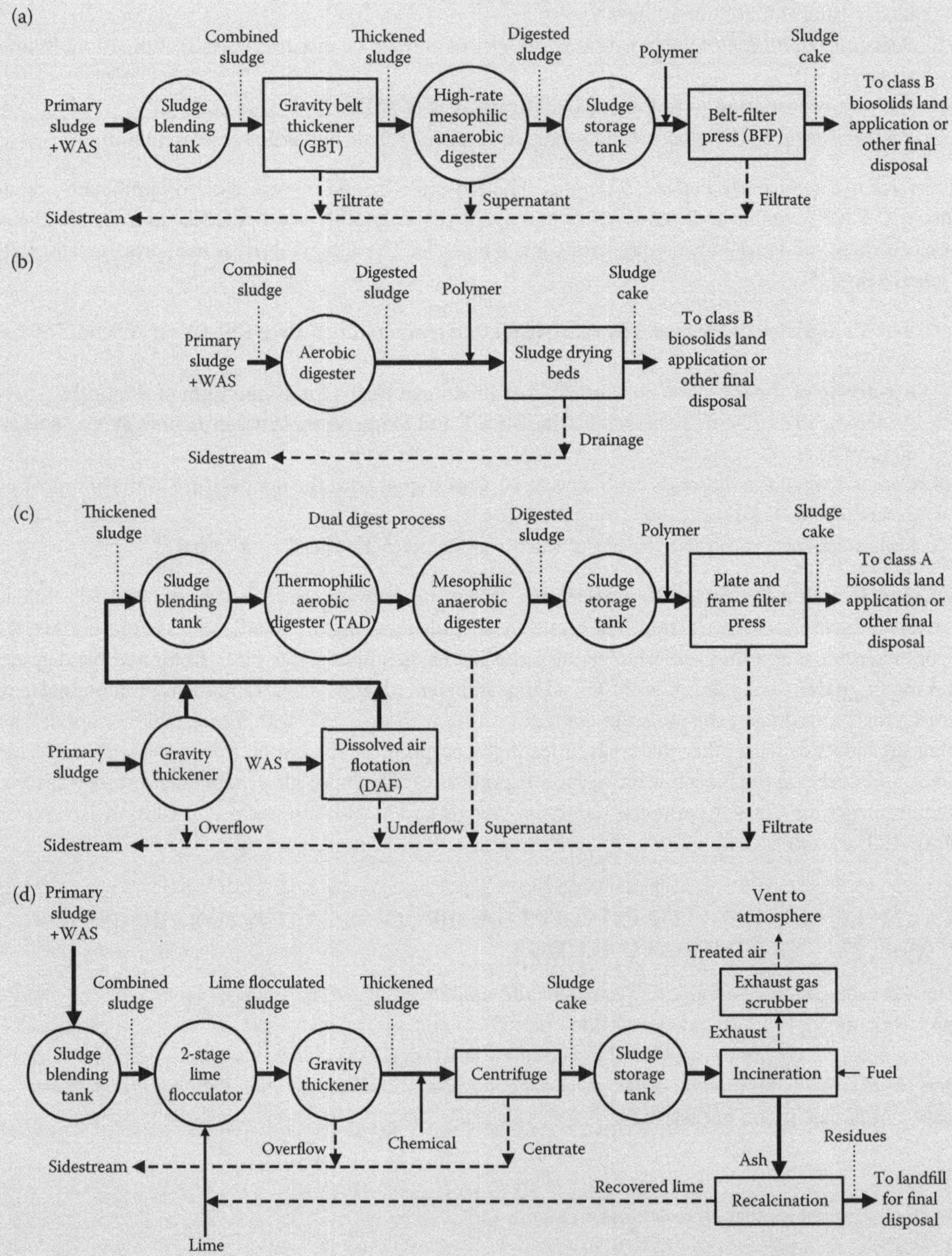

FIGURE 13.1 Conceptual process diagrams for sludge treatment: (a) Train 1: gravity belt thickener and anaerobic digester, (b) Train 2: aerobic digester and drying beds, (c) Train 3: gravity thickener and DAF followed by dual digestion process, and (d) Train 4: chemical conditioning of two-stage lime sludge and incineration with lime recalcination (Example 13.9).

least 15 d (Class B, Alt. 2, PSRP No. 3). See Section 13.6.1, Table 13.7, and Figure 13.8 for additional information.

b. Process Train 2—Anaerobic digester and drying beds:

The process train in Figure 13.1b consists of an aerobic digester and sludge drying beds. These processes are normally applied at small facilities with or without primary treatment. The aerobic digester provides a detention time of 40 d at a temperature over 20°C. The volatile solids reduction may reach over 60% resulting in significant odor reduction in the drying beds. The dewatered sludge will meet Class B biosolids criteria (Class B, Alt. 2, PSRP No. 1). See Section 13.6.2, and Table 13.7 for additional information.

c. Process Train 3—Gravity thickener and DAF followed by dual digestion process:

The process train in Figure 13.1c uses a gravity thickener for primary sludge and a DAF unit for WAS. These thickened sludges are combined. The combined sludge at 5–6% solids is digested in a dual digestion process. The thermophilic aerobic digester (TAD) operates at 55–60°C for a hydraulic retention time (HRT) of 24 h and a SRT over 10 d. The mesophilic anaerobic digester achieves enhanced VSS reduction. The plate and frame filter presses (PFFP) produces sludge cake at 30–40% solids. The solids meet Class A biosolids criteria (Class A, Alt 5, PFRP No. 4). See Section 13.6.2, and Table 13.7 for additional information.

d. Process Train 4—Chemical conditioning of two-stage lime sludge and incineration with lime recalcination:

In this process train (Figure 13.1d), the combined sludge is flocculated in a two-stage lime flocculator. The lime-flocculated sludge is thickened to about 6–7% solids in a gravity thickener. The thickened sludge is chemically conditioned and dewatered by centrifuge process. The dewatered sludge at 20–35% solids is stored in a sludge storage tank then incinerated. The ash is reclaimed for recovery of lime. The residues from recalcination are landfilled. The exhaust gas is scrubbed and released. See Sections 13.11.3 and 13.11.4 and Table 13.7 for additional information about the lime recalcination process.

EXAMPLE 13.10: MODIFICATIONS OF EXISTING PROCESS TRAIN TO PRODUCE CLASS A BIOSOLIDS

At a POTW the thickened primary sludge and WAS are combined and digested in a mesophilic anaerobic digester (35–55°C and SRT $>$ 15 days) to meet Class B biosolids criteria. The digested sludge is then dewatered by belt-filter press for land application. Develop five options to modify the existing process train to produce Class A biosolids.

Solution

1. Draw the existing process train.

The existing process train is shown in Figure 13.2a. The thickened primary sludge and WAS are blended in a sludge blending tank. The combined sludge is digested in a mesophilic anaerobic digester. The digested sludge is dewatered by a belt-filter press (BFP) system. The sludge cake is used as class B biosolids.

2. Modify the existing process train to produce class A biosolids.

The following five process trains are developed to modify the existing process train to produce Class A biosolids.

a. Modification 1—Alkaline stabilization process:

This modification is a continuation of existing process train (Figure 13.2a). Post treatment processes are added after the existing process train to achieve lime stabilization (Figure 13.2b). The sludge cake from BFP is mixed with lime slurry in a pug mill. The lime dosage ranges from

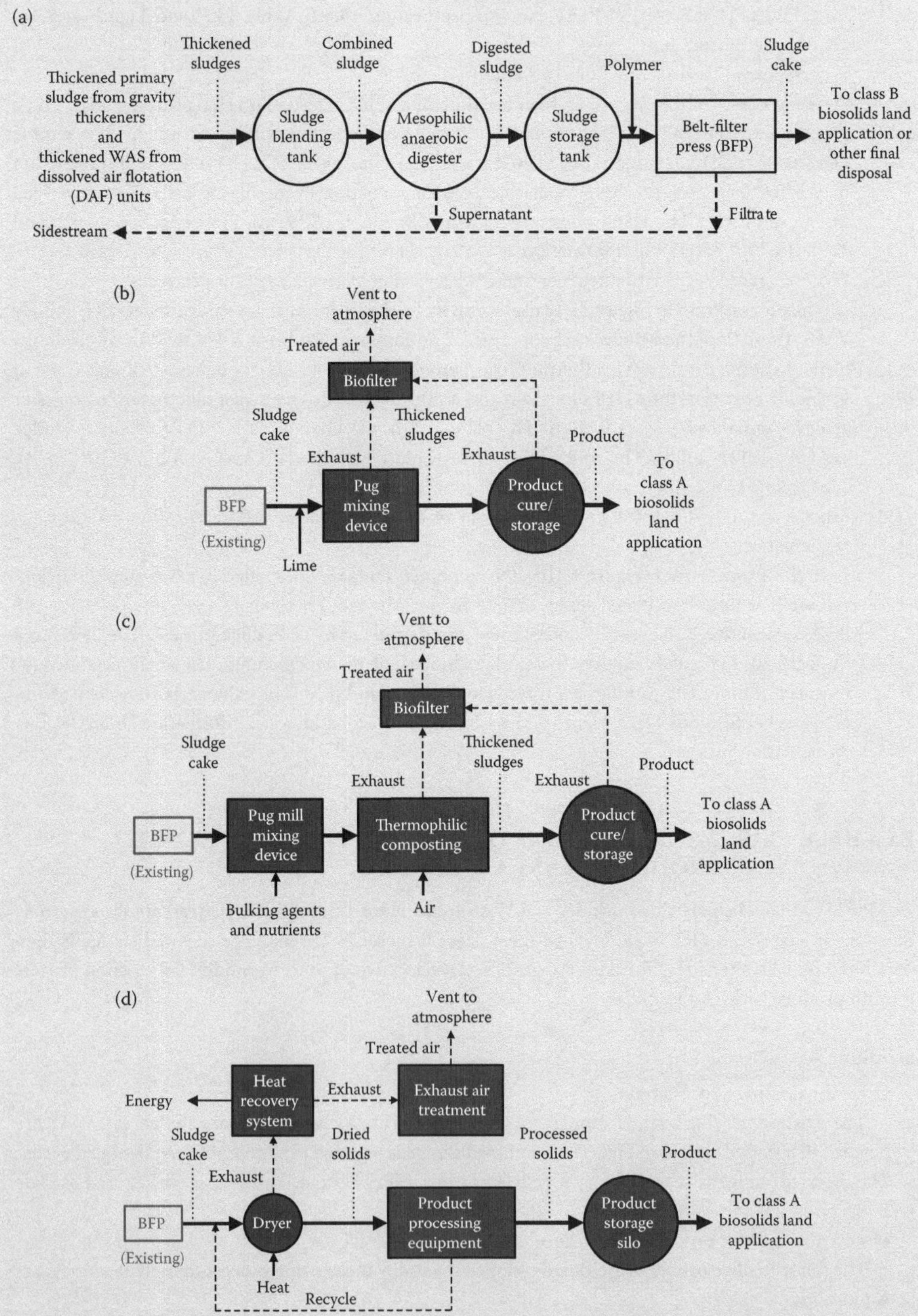

FIGURE 13.2 Conceptual process diagrams for existing sludge treatment process and modification options for producing Class A Biosolids: (a) existing process train, (b) alkaline stabilization, (c) thermophilic composting, (d) heat drying (Example 13.10). *(Continued)*

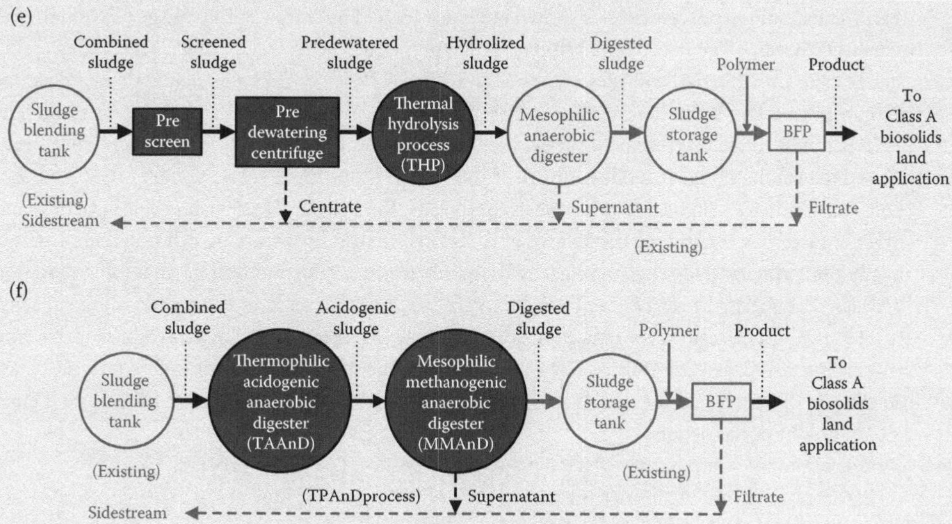

FIGURE 13.2 (Continued) Conceptual process diagrams for existing sludge treatment process and modification options for producing Class A Biosolids: (e) thermal hydrolysis process (THP), and (f) temperature-phased anaerobic digestion (TPAnD) (Example 13.10).

0.15 to 0.3 kg lime ($Ca(OH)_2$) per kg of dry solids. The treated biosolids is qualified for Class A biosolids (Class A, Alt. 5, PFRP No. 7) when (a) the pH of mixture is maintained at or above 12 for at least 72 h and at a temperature of 52°C for at least 12 h; or (b) the process is kept at a temperature above 70°C and pH of 12 for at least 30 min.[21,22] The exhaust gas is treated in a biofilter and released. The biosolids after lime addition is stored in a product storage silo. See Section 13.6.3, and Table 13.7 for additional information.

b. Modification 2—Thermophilic composting process:

This modification is also a continuation of existing process train (Figure 13.2a). Additional processes are provided after BFP to carry out thermophilic composting of combined sludge to achieve Class A biosolids (Figure 13.2c). The sludge cake is blended with bulking agent in a plug mill. The moisture is adjusted and nutrients and finished compost for seed are added. The thermophilic composting is then carried out in windows or in vessel. The temperature is sustained over 55°C for more than 3 days. The exhaust gasses from in vessel composting may be treated in a biofilter. The finished compost is stored in product cure/storage facility for 21–28 days. The finished product meets Class A biosolids criteria (Class A, Alt. 5, PFRP No. 1).[21,23] See Section 13.11.1, and Table 13.7 for additional information.

c. Modification 3—Heat drying process:

This modification is also a continuation of existing process train in Figure 13.2a. Extra processes are added to the sludge cake from BFP to achieve thermal heat drying for production of Class A biosolids. The added processes are shown in Figure 13.2d. The sludge cake from BFP is fed into a heat dryer. The typical operating temperature is 90–500°C (195–930°F) for heat dryers. The solids content after drying is 90–95%. The dried solids go through the product processing equipment. The final product ranges from 1to 8 mm in size and is stable enough to be qualified as Class A biosolids (Class A, Alt. 5, PFRP No. 2).[21,24–27] The final product is stored in product storage silo before delivery for final applications. See Section 13.11.2, and Table 13.7 for additional information.

d. Modification 4—Thermal hydrolysis process (THP):

In this modification, three new processes are added between the sludge blending tank and mesophilic anaerobic digester in the existing process train (Figure 13.2a) to achieve hydrolysis in a

THP. The modified process train is shown in Figure 13.2e. The combined sludge goes through prescreens (<5 mm). The prescreened sludge is then predewatered by a centrifuge. The predewatered sludge cake with 15–18% solids content goes into the THP.[21,28,29] The thermal hydrolyzed sludge is then processed by the digester and other subsequent processes of existing process train. The THP is considered a pretreatment process for enhanced sludge stabilization. High-pressure steam is used to heat the sludge to a temperature of 140–180°C (284–356°F) at a pressure of 500–900 kPa (70–130 psi) for 20–30 min. Under these conditions, the heat breaks biomass cell structure and organic materials (mainly in the WAS), and dissolves naturally occurring cell polymers (mainly in the raw primary sludge) into an easily digestible feedstock for mesophilic anaerobic digestion process. The thermal pretreatment results in (i) 2–3 times increase in digester loading capacity, (ii) 10–15% increase in VSS reduction, (iii) 25–35% increase in in biogas production by, and (iv) an improved dewaterability. The pasteurized sludge meets Class A biosolids criteria (Class A, Alt. 5, PFRP No. 7). See Sections 13.10.1 and 13.10.2, Tables 13.7 and 13.25, and Figure 13.32 for additional details.

e. Modification 5—Temperature-phased anaerobic digestion (TPAnD) process:

   In this process modification, the existing mesophilic anaerobic digester in existing process train (Figure 13.2a) is retrofitted into TPAnD. The modified process train is shown in Figure 13.2f. The combined sludge is digested in a two-phase anaerobic digestion (2PAD) system. The 1st phase is thermophilic acidogenic anaerobic digester (TAAnD). In this phase, the thermophilic condition is maintained at 55°C for SRT greater than 2.1 d. The 2nd phase is a mesophilic methanogenic anaerobic digester (MMAnD). In this phase, the mesophilic condition is maintained at 37°C for SRT greater than 10.5 d. The 2PAD has been granted "PFRP Conditional National Equivalency" by the U.S. EPA for producing Class A biosolids (Class A, Alt. 1 or 6).[21,30,31] See Section 13.6.1, Tables 13.7 and 13.13, and Figure 13.12 for additional information.

13.5 Sludge Thickening

Raw sludge from the wastewater treatment processes contains large volume of water. Thickening is critically needed to concentrate the solids and reduce the volume. Sludge thickening reduces the digester capacity and chemical requirement for conditioning. Also, the piping and equipment for sludge transport are smaller. In small plants, thickening of secondary sludge may be achieved in "cosettling thickening" phenomenon by returning the dilute biological sludge (0.8–1.2% solids) into the primary clarifier. A sludge blanket is maintained in the sludge hopper to retain the solids for 10–12 h. The solids compact into the hopper, and the underflow concentration of combined sludge solids may exceed 5%. Common methods of sludge thickening used at medium to large plants are: (1) gravity, (2) dissolved air flotation, (3) centrifuge, (4) gravity belt, and (5) rotary drum. Applications of membrane technology for sludge thickening has also been developed and commercialized during recent years. These methods are presented below.

13.5.1 Gravity Thickening

Gravity thickening of combined sludge is the most commonly used method. It is accomplished in a circular sedimentation basin. The sludge is fed into a center feed well. Three distinct zones develop in the thickener: (1) clear supernatant zone on the top, (2) sedimentation zone in the middle, and (3) thickening zone near the bottom. The supernatant discharges over a weir into an effluent launder and returned to the plant headworks. In the sedimentation zone, a stream of denser sludge moves toward the thickening zone. In the thickening zone, the individual particles agglomerate and a sludge blanket is maintained. The sludge mass is compressed by materials continuously added to the top. Water is squeezed out of interstitial spaces and flows upward through channels. With effective removal of rags and fibrous materials deep trusses or vertical pickets may be provided to gently stir the sludge blanket and move the gases and liquid to the surface. The thickened sludge is withdrawn from the bottom. In this section, a discussion on compression or Type IV settling and basic design considerations of gravity thickeners are provided.*

Compression Settling (Type IV): The settling behavior of particles in the thickening zone is typical of compression or Type IV settling. This topic was briefly introduced in Sections 8.3.1 and 10.9.2. As shown in Figure 10.119, the thickening or compression settling of sludge starts when the critical height of interface H_c is reached at compression point (Point C) at the settling time t_c. In the thickening zone, the particles remain supported on top of each other and further settling occurs by compression from the weight of particles above. The rate of consolidation at any time is proportional to the difference in the depth of the blanket and the maximum consolidated depth. The consolidation rate in the compression zone is expressed by Equation 13.2a. The time required for sludge line to drop from H_c to height H is obtained by integrating Equation 13.2a. The resulting equations are expressed by Equations 13.2b and 13.2c.[7,32–34]

$$-\frac{dH}{dt} = i(H - H_\infty) \tag{13.2a}$$

$$\int_{H}^{H_c} \frac{dH}{(H - H_\infty)} = \int_{t}^{t_0} i\, dt \tag{13.2b}$$

$$i(t - t_c) = \ln\left(\frac{H_c - H_\infty}{H - H_\infty}\right) \quad \text{or} \quad \ln(H - H_\infty) = \ln(H_c - H_\infty) - i(t - t_c) \tag{13.2c}$$

where

H = height of interface at time t, m
H_c = height of interface at the compression point, m
H_∞ = final sludge height of compression solids, m
i = setting constant for a given suspension, min^{-1}
t = time at which the height of interface is at H, min
t_c = time at which the height of interface is at the compression point (H_c), min

A plot of $\ln(H - H_\infty)$ versus $(t - t_c)$ is a straight line having a negative slope of $-i$. The intercept on Y-axis is $\ln(H_c - H_\infty)$. The final sludge height H_∞ is primarily dependent upon the liquid surface film adhering to the particles. Gentle agitation facilitates compaction.

Basic Design Considerations of Gravity Thickener: The gravity thickeners serve two purposes: (a) sludge concentration and equalization and (b) sludge storage to enhance the operation of downstream processes. The solids content in thickened sludge may vary from 2 to 5 times of the incoming solids, but the maximum solids content remains less than 10%. The typical design loading rates and expected performances of gravity thickener are summarized in Table 13.8. The general design guidelines for gravity thickener are given below:[6,7,33,35–37]

1. The thickener hydraulic and solids loading rates are important parameters to minimize the carry-over of solids in the overflow, arching of solids, and septic condition.
2. The thickener surface area should be calculated from solids and hydraulic loading rates based on the incoming sludge mass and flow rate. The typical ranges of design hydraulic loading rates are summarized in Table 13.8. However, a minimum hydraulic loading of 9 $m^3/m^2{\cdot}d$ (220 gpd/ft^2) is also recommended for operation to prevent arching of sludge solids, septic conditions, and odors. To achieve acceptable hydraulic loading, dilution water which may be primary or secondary effluent may be blended with the incoming sludge (see Examples 13.7 and 13.12 for blending tank and gravity thickener design).

* Depending upon the concentration and the tendency of particles to interact, four types of settling can occur in aqueous solution (Table 8.1). These are (1) discrete or Type I settling (Section 8.3.2), (2) flocculent or Type II settling (Section 9.2), (3) zone, hindered, or Type III settling (Section 10.9.2), and (4) compression or Type IV settling.

TABLE 13.8 Typical Loading Rates and Performance Parameter Ranges of Gravity Thickener

Type of Sludge	Influent Solids Content, %	Solids Loading, $kg/m^2 \cdot d$ ($lb/ft^2 \cdot d$)	Hydraulic Loading, $m^3/m^2 \cdot d$ (gpd/ft^2)	Solids Capture, %	Thickened Sludge Solids Content, %
Primary sludge	1–6	100–150 (20–30)	15–30 (350–750)	85–98	4–10
Trickling filter sludge	1–4	40–50 (8–10)	4–8 (100–200)	80–92	3–6
Waste activated sludge (WAS)	0.2–1.2	20–40 (4–8)	4–8 (100–200)	60–85	2–4
Primary sludge and WAS	1–4	25–75 (5–15)	6–12 (150–300)	80–92	3–6

$1\ kg/m^2 \cdot d = 0.205\ lb/ft^2 \cdot d$; $1\ m^3/m^2 \cdot d = 24.5\ gpd/ft^2$.

Source: Adapted in part from References 6, 7, 32, 35, and 36.

3. The TSS concentration in the overflow may range from 1000 to 5000 mg/L, while the BOD_5 concentration may be 500–3000 mg/L (see Table 13.6).
4. To achieve high solids content (10% or more) a sludge blanket of sufficient depth needs to be maintained. The sludge volume ratio (SVR) is the volume of sludge blanket held in the thickener divided by the volume of thickened sludge removed per day. The accepted value of SVR is 1–2 d for primary sludge and 0.75–1.5 for combined sludge at a typical sludge blanket depth of 0.3–1.5 m (1–5 ft). Higher SVR provide thicker sludge, but excessive retention may lead to gasification, buoying of solids, and odors due to septic condition.
5. The thickeners are generally circular concrete tanks with bottom sloping toward a central sludge well. The equipment is similar to that of a circular clarifier. It includes: (a) bottom scraper on a rotating rake arm, (b) vertical pickets (optional), (c) an overflow weir and collection trough, and (d) constant speed pump with timer or variable speed sludge pump and other accessories. The peripheral speed of sludge scrapper mechanism is usually in a range from 4.5 to 6 m/min (15–20 ft/min).
6. The side-water depth (SWD) of a thickener is 3–4.5 m (10–15 ft) with a hydraulic retention time (HRT) of 6–24 h. The slope of bottom floor (V:H) is generally 2:12–3:12, which is steeper than the slope of standard clarifiers. Normally, the diameter of the thickener should not exceed 25 m (80 ft).
7. To enhance the performance and improve overflow quality, chemicals may be added prior to the gravity thickener. The typical dosages of different chemicals are: (a) 1–3 mg/L for $FeCl_3$, (b) 6–9 mg/L as CaO for lime, (c) 10–40 mg/L for $KMnO_4$, or (d) 2.5–6 g/kg (5–12 lb/ton) dry solids for polymer.[37]

EXAMPLE 13.11: HEIGHT OF COMPRESSION ZONE AT A GIVEN TIME

A settling test was conducted to develop compression or Type IV settling data. Well-blended sludge sample was settled in a column. The interface height of 1.5 m at the start of compression point was reached in 50 min. The final depth of the compressed solids after a long time of settling was 0.3 m from the bottom. Calculate the time when the interface after the start of compression reaches 0.9 m from the bottom. The settling constant for the given suspension is $0.02\ min^{-1}$.

Solution

1. Describe the compression settling.

The zone settling and start of compression zone are discussed in Section 10.9.2. A graphical procedure for determination of interface height at the start of compression point is shown in Figure 10.119. Point C in this figure is the start of compression settling at interface height $H_c = 1.5$ m above the bottom. This height is reached at a settling time $t_c = 50$ min, and the height of compressed solids $H_\infty = 0.3$ m is reached in time t_∞.

2. Calculate the required time from rearranged Equation 13.2c at the interface height $H = 0.9$ m.

$$t = \frac{1}{i}\ln\left(\frac{H_c - H_\infty}{H - H_\infty}\right) + t_c = \frac{1}{0.02\ \text{min}^{-1}} \times \ln\left(\frac{(1.5-0.3)\ \text{m}}{(0.9-0.3)\ \text{m}}\right) + 50\ \text{min}$$

$$= 50\ \text{min} \times \ln(2) + 50\ \text{min} = 50\ \text{min} \times 0.693 + 50\ \text{min} = 85\ \text{min or } 1.4\ \text{h}$$

EXAMPLE 13.12: DESIGN OF A GRAVITY THICKENER

A gravity thickener receives combined primary and waste activated sludge. Under the peak sustained loading conditions, the design total weight of dry solids reaching the thickener is 10,000 kg/d. The combined sludge contains 2% solids and specific gravity of bulk sludge is 1.003. The criteria for solids loading rate is not to exceed 50 kg/m^2·d (10 lb/ft^2·d), and minimum hydraulic loading rate is 9 m^3/m^2·d (220 gpd/ft^2). The thickened sludge is pumped into the anaerobic digester. Design the thickener and determine the characteristics of thickened sludge and thickener overflow. Secondary effluent is used for dilution. The TSS and BOD_5 in the secondary effluent are 10 and 5 mg/L, respectively. The solids concentration in thickened sludge is 5%; the bulk density of thickened sludge is 1010 kg/m^3; and the solids and BOD_5 capture efficiencies are both 85%. The accepted SVR < 2 d. Assume (a) the ratio of biodegradable VSS (VSS_{bd}) to TSS is 0.55 kg VSS_{bd}/kg TSS, (b) each g of VSS_{bd} exerts 1.42 g ultimate BOD (BOD_L), and (c) ratio of BOD_5 to BOD_L is 0.68. Use the water density $\rho_w = 1000$ kg/m^3.

Solution

1. Compute the surface area of the thickener based on solids loading rate (*SLR*).

$$\text{Total surface area, } A_{\text{total}} = \frac{\text{Total weight of dry solids}}{\text{Solids loading rate}} = \frac{W_{s,CS}}{SLR} = \frac{10{,}000\ \text{kg/d}}{50\ \text{kg/m}^2\cdot\text{d}} = 200\ \text{m}^2$$

2. Determine the diameter of gravity thickener.
   Provide two circular thickeners and the required area $A = 100$ m^2 for each thickener.

$$\text{Required diameter of each thickener, } D = \sqrt{\frac{4 \times \text{Area}}{\pi}} = \sqrt{\frac{4\,A}{\pi}} = \sqrt{\frac{4 \times 100\ \text{m}^2}{\pi}} = 11.3\ \text{m}$$

   Provide two thickeners each 12 m (39 ft) in diameter.

$$\text{Calculate the actual surface area of each thickener, } A = \frac{\pi}{4}(\text{Diameter})^2 = \frac{\pi}{4}D^2 = \frac{\pi}{4} \times (12\ \text{m})^2 = 113\ \text{m}^2$$

3. Verify the hydraulic loading rate or surface overflow rate (*SOR*).
   Total weight of dry solids in the combined sludge, $W_{s,CS}$ = 10,000 kg/d.

$$\text{Combined sludge flow from Equation 13.1e, } Q_{b,CS} = \frac{100\% \times W_{s,CS}}{p_{s,CS} S_{b,CS} \rho_w}$$

$$= \frac{100\% \times 10{,}000\ \text{kg/d}}{2\% \times 1.003 \times 1000\ \text{kg/m}^3} = 500\ \text{m}^3/\text{d}$$

$$\text{Hydraulic loading rate from combined sludge flow, } SOR_{CS} = \frac{\text{Flow}}{\text{Total area}} = \frac{Q_{b,CS}}{2A} = \frac{500\ \text{m}^3/\text{d}}{2 \times 113\ \text{m}^2}$$

$$= 2.2\ \text{m}^3/\text{m}^2\cdot\text{d}$$

   The hydraulic loading rate is significantly lower than the minimum recommended hydraulic loading rate of 9 m^3/m^2·d. Therefore dilution water is needed.

4. Compute the dilution water flow required at the minimum operating hydraulic loading rate $SOR_{min} = 9\ m^3/m^2{\cdot}d$.

$$\text{Daily total volume of blended sludge, } Q_{b,blend} = \text{Minimum surface loading rate} \times \text{Total thickener area}$$
$$= SOR_{min}(2A) = 9\ m^3/m^2{\cdot}d \times 2 \times 113\ m^2$$
$$= 2030\ m^3/d$$

$$\text{Dilution water needed, } Q_{b,dilution} = \text{Blended sludge volume} - \text{Combined sludge volume}$$
$$= Q_{b,blend} - Q_{b,CS} = (2030 - 500)\ m^3/d = 1530\ m^3/d$$

5. Calculate the total amount of TSS in the blended sludge to the thickeners.

$$\text{TSS in dilution water, } W_{s,dilution} = \text{Flow} \times \text{TSS concentration} = Q_{b,dilution} \times TSS_{eff}$$
$$= 1530\ m^3/d \times 10\ g/m^3 \times 10^{-3}\ kg/g = 15\ kg\ TSS/d$$

$$\text{TSS in blended sludge, } W_{s,blend} = \text{TSS in sludge} + \text{TSS in dilution water} = W_{s,CS} + W_{s,dilution}$$
$$= (10{,}000 + 15)\ kg/d = 10{,}015\ kg\ TSS/d$$

6. Check solids and hydraulic loading rates based on the blended sludge flow.

$$\text{Solids loading rate with both thickeners, } SLR_{2\ thickeners} = \frac{\text{Weight of TSS}}{\text{Total area}} = \frac{W_{s,blend}}{2A} = \frac{10{,}015\ kg/d}{2 \times 113\ m^2}$$
$$= 44\ kg/m^2{\cdot}d$$

$$\text{Solids loading rate with one thickener, } SLR_{thickener} = \frac{W_{s,blend}}{A} = \frac{10{,}015\ kg/d}{113\ m^2} = 89\ kg/m^2{\cdot}d$$

$$\text{Hydraulic loading rate with both thickeners, } SOR_{2\ thickeners} = \frac{\text{Sludge volume}}{\text{Total area}} = \frac{Q_{b,blend}}{2A} = \frac{2030\ m^3/d}{2 \times 113\ m^2}$$
$$= 9\ m^3/m^2{\cdot}d$$

$$\text{Hydraulic loading rate with one thickener, } SOR_{thickener} = \frac{Q_{b,blend}}{A} = \frac{2030\ m^3/d}{113\ m^2} = 18\ m^3/m^2{\cdot}d$$

7. Calculate the total amount of BOD_5 in the blended sludge to the thickeners.

$$\text{Biodegradable VSS } (VSS_{bd}) \text{ in combined sludge, } W_{VSSbd,CS} = 0.55\ kg\ VSS_{bd}/kg\ TSS \times W_{s,CS}$$
$$= 0.55\ kg\ VSS_{bd}/kg\ TSS \times 10{,}000\ kg\ TSS/d$$
$$= 5500\ kg\ VSS_{bd}/d$$

$$BOD_L \text{ exerted by } VSS_{bd} \text{ in combined sludge, } W_{BODL,CS} = 1.42\ kg\ BOD_L/kg\ VSS_{bd} \times W_{VSSbd,CS}$$
$$= 1.42\ kg\ BOD_L/kg\ VSS_{bd} \times 5500\ kg\ VSS_{bd}/d$$
$$= 7800\ kg\ BOD_L/d$$

$$BOD_L \text{ exerted by } VSS_{bd} \text{ in combined sludge, } W_{BOD_5,CS} = 0.68\ kg\ BOD_5/kg\ BOD_L \times W_{BOD_L,CS}$$
$$= 0.68\ kg\ BOD_L/kg\ BOD_L \times 7800\ kg\ BOD_L/d$$
$$= 5300\ kg\ BOD_5/d$$

$$BOD_5 \text{ in dilution water, } W_{BOD_5,dilution} = Q_{b,dilution}S = 1530\ m^3/d \times 5\ g/m^3 \times 10^{-3}\ kg/g$$
$$= 8\ kg\ BOD_5/d$$

$$BOD_5 \text{ in blended sludge, } W_{BOD_5,blend} = BOD_5 \text{ in sludge} + BOD_5 \text{ in dilution water}$$
$$= W_{BOD_5,CS} + W_{BOD_5,dilution} = (5300 + 8)\ kg/d$$
$$= 5308\ kg\ BOD_5/d$$

8. Compute the TSS and BOD_5 concentrations in the blended sludge.

$$\text{TSS concentration in blended sludge, } C_{TSS,blend} = \frac{W_{TSS,blend}}{Q_{b,blend}} = \frac{10{,}015\ \text{kg TSS/d}}{2030\ \text{m}^3/\text{d}} \times 10^3\ \text{g/kg}$$

$$= 4930\ \text{g TSS/m}^3 \quad \text{or} \quad 4930\ \text{mg TSS/L}$$

$$BOD_5 \text{ concentration in blended sludge, } C_{BOD_5,blend} = \frac{W_{BOD_5,blend}}{Q_{b,blend}} = \frac{5308\ \text{kg BOD}_5/\text{d}}{2030\ \text{m}^3/\text{d}} \times 10^3\ \text{g/kg}$$

$$= 2610\ \text{g BOD}_5/\text{m}^3 \quad \text{or} \quad 2610\ \text{mg BOD}_5/\text{L}$$

9. Determine the characteristics of thickened sludge (THS).
   Total TSS in thickened sludge at a capture rate of 85%.

$$W_{s,THS} = 0.85 \times W_{s,blend} = 0.85 \times 10{,}015\ \text{kg TSS/d} = 8510\ \text{kg TSS/d}$$

$$\text{Thickened sludge volume from Equation 13.1e, } Q_{b,THS} = \frac{100\% \times W_{s,THS}}{p_{s,THS} S_{b,THS} \rho_w} = \frac{100\% \times 8510\ \text{kg/d}}{5\% \times 1.01 \times 1000\ \text{kg/m}^3}$$

$$= 170\ \text{m}^3/\text{d}$$

The daily volume of thickened sludge withdrawn from each thickener = 85 m^2/d.
Total amount of BOD_5 in thickened sludge at a capture rate of 85%.

$$W_{BOD_5,THS} = 0.85 \times W_{BOD_5,blend} = 0.85 \times 5308\ \text{kg BOD}_5/\text{d} = 4510\ \text{kg BOD}_5/\text{d}$$

$$\text{TSS concentration in thickened sludge, } C_{TSS,THS} = \frac{\text{Weight of TSS}}{\text{Flow}} = \frac{W_{s,THS}}{Q_{b,THS}}$$

$$= \frac{8510\ \text{kg TSS/d}}{170\ \text{m}^3/\text{d}} \times 10^3\ \text{g/kg}$$

$$= 50{,}100\ \text{g TSS/m}^3 \quad \text{or} \quad 50{,}100\ \text{mg TSS/L}$$

$$BOD_5 \text{ concentration in thickened sludge, } C_{BOD_5,THS} = \frac{W_{BOD_5,THS}}{Q_{b,THS}} = \frac{4510\ \text{kg BOD}_5/\text{d}}{170\ \text{m}^3/\text{d}} \times 10^3\ \text{g/kg}$$

$$= 26{,}500\ \text{g BOD}_5/\text{m}^3 \quad \text{or} \quad 26{,}500\ \text{mg BOD}_5/\text{L}$$

10. Determine the characteristics of overflow from the thickeners.

$$\text{Overflow rate from the thickeners, } Q_{b,overflow} = \text{Total blended sludge flow} - \text{Thickened sludge flow}$$

$$= Q_{b,blend} - Q_{b,THS} = (2030 - 170)\ \text{m}^3/\text{d} = 1860\ \text{m}^3/\text{d}$$

Daily overflow rate from each thickener = 930 m^3/d.

$$\text{Total TSS in overflow, } W_{s,overflow} = (W_{s,blend} - W_{s,THS}) = (10{,}015 - 8510)\ \text{kg TSS/d} = 1505\ \text{kg TSS/d}$$

$$\text{Total BOD}_5 \text{ in overlfow, } W_{BOD_5,overflow} = (W_{BOD_5,blend} - W_{BOD_5,THS}) = (5308 - 4510)\ \text{kg BOD}_5/\text{d}$$

$$= 798\ \text{kg BOD}_5/\text{d}$$

$$\text{TSS concentration in overflow, } C_{TSS,overflow} = \frac{W_{TSS,overflow}}{Q_{b,overflow}} = \frac{1505\ \text{kg TSS/d}}{1860\ \text{m}^3/\text{d}} \times 10^3\ \text{g/kg}$$

$$= 810\ \text{g TSS/m}^3 \quad \text{or} \quad 810\ \text{mg TSS/L}$$

$$BOD_5 \text{ concentration in overflow, } C_{BOD_5,overflow} = \frac{W_{BOD_5,overflow}}{Q_{b,overflow}} = \frac{798\ \text{kg BOD}_5/\text{d}}{1860\ \text{m}^3/\text{d}} \times 10^3\ \text{g/kg}$$

$$= 430\ \text{g BOD}_5/\text{m}^3 \quad \text{or} \quad 430\ \text{mg BOD}_5/\text{L}$$

11. Determine the depth of thickener.

The thickener depth consists of four separate zones: (1) clear liquid zone, (2) settling zone, (3) thickening or compression zone, and (4) sludge storage zone. The procedure for obtaining the thickener depth is given below.

a. Select the overall depth of clear liquid and settling zones (H_{csz}).

The clear liquid and settling zones are respectively 0.5–1.5 and 1.5–2 m. Provide an overall depth of clear liquid and settling zones, $H_{csz} = 3$ m (9.8 ft).

Volume of clear liquid and settling zones, $V_{csz} = H_{csz}A = 3\text{ m} \times 113\text{ m}^2 = 339\text{ m}^3$ per thickener.

b. Determine the depth of thickening zone (H_{tz}).

i. Determine the average solids contents ($p_{s,tz}$) and bulk sp. gr. ($S_{b,tz}$).

The interface settles leaving clear liquid zone. The incoming solids continue to settle through settling zone below the interface until the thickening zone is reached. Assume that solids content of 2% of the raw combined sludge before dilution is resumed at the upper end, and the thickened sludge solids content of 5% is reached at the lower end of the thickening zone. The average solids content in the thickening ($p_{s,tz}$).

$$p_{s,tz} = \frac{(p_{s,CS} + p_{s,THS})}{2} = \frac{(2+5)\%}{2} = 3.5\%$$

Estimate also the average bulk sp. gr. of the sludge in the thickening zone ($S_{b,tz}$).

$$S_{b,tz} = \frac{(S_{b,CS} + S_{b,THS})}{2} = \frac{1.004 + 1.01}{2} = 1.007$$

ii. Determine the depth of thickening zone (H_{tz}).

The thickening zone is normally sized to allow a maximum detention time of 1 d for the thickener sludge. Provide a detention time $\theta_{tz} = 1$ d. Calculate the amount of solids in the thickening zone per thickener.

$$\Delta W_{s,tz} = \frac{W_{s,THS}\theta_{tz}}{2\text{ thickeners}} \times 1\text{ d} = \frac{8510\text{ kg TSS/d} \times 1\text{ d}}{2\text{ thickeners}} = 4255\text{ kg TSS per thickener}$$

$$\text{The thickening zone volume, } V_{tz} = \frac{100\% \times \Delta W_{s,tz}}{p_{s,tz}S_{b,tz}\rho_w} = \frac{100\% \times 4255\text{ kg}}{3.5\% \times 1.007 \times 1000\text{ kg/m}^3} = 120\text{ m}^3$$

$$\text{Depth required in thickening zone, } H_{tz} = \frac{V_{tz}}{A} = \frac{120\text{ m}^3}{113\text{ m}^2} = 1.06\text{ m}$$

Provide 1-m depth of thickening zone.

Actual volume of thickening zone, $V_{tz} = H_{tz}A = 1\text{ m} \times 113\text{ m}^2 = 113\text{ m}^3$ per thickener

iii. Determine the depth of sludge storage zone (H_{sz}).

The sludge storage is provided in the bottom central cone of the thickener.

Provide bottom slope of 2:12 (V:H).

$$\text{Total drop at central well, } h_{cone} = \frac{2\text{ m}}{12\text{ m}} \times \frac{D}{2} = \frac{2\text{ m}}{12\text{ m}} \times \frac{12\text{ m}}{2} = 1\text{ m}$$

Provide sludge storage depth H_{sz} = height of cone = 1 m.

$$\text{Sludge storage volume in sludge storage cone, } V_{sz} = \frac{1}{3}\frac{\pi D^2}{4}H_{sz} = \frac{1}{3} \times \frac{\pi \times (12\text{ m})^2}{4} \times 1\text{ m}$$

$$= 38\text{ m}^3 \text{ per thickener}$$

iv. Determine the total side water depth (SWD), side wall height (SWH), total water depth (TWD), and total basin height (TBH) of the thickener.

$$\text{SWD} = H_{csz} + H_{tz} = (3+1)\,\text{m} = 4.0\,\text{m}\ (13.1\,\text{ft})$$

Provide a freeboard $FB = 0.5$ m.

Total side wall height, SWH = SWD + FB = (4.0 + 0.5) m = 4.5 m (14.8 ft)

Total water depth (TWD) in thickener, TWD = SWD + H_{sz} = (4.0 + 1.0) m = 5.0 m (16.4 ft)

Total basin height in thickener, TBH = SWH + H_{sz} = (4.5 + 1.0) m = 5.5 m (18.0 ft)

12. Check the sludge volume ratio (SVR).
The total thickened sludge flow is 170 m^3/d from two thickeners (Step 9).
SVR based on sludge storage volume $V_{sz} = 38\ m^3$ per thickener (step 11.b.iii),

$$\text{SVR} = \frac{2V_{sz}}{Q_{b,THS}} = \frac{2\times 38\,\text{m}^3}{170\,\text{m}^3/\text{d}} \times \frac{24\,\text{h}}{\text{d}} = 11\,\text{h}$$

SVR when the thickening volume $V_{tz} = 113\ m^3$ per thickener (Step 11.b.ii) is also considered for storage of excess,

$$\text{SVR} = \frac{2\times(V_{sz}+V_{tz})}{Q_{b,THS}} = \frac{2\times(38+113)\,\text{m}^3}{170\,\text{m}^3/\text{d}} = 1.8\,\text{d}$$

Note: The available SVR is between 11 h and 1.8 d. It is below the acceptable SVR of 2 d. It is suggested that the excess sludge storage capacity in the settling zone may be used under emergency situations such as equipment downtime or abnormally high sustained solids loading. If septic condition develops in the thickening and storage zone, sludge withdrawal rate should be increased.

13. Check the hydraulic retention time (HRT) in the thickeners.

$$\begin{aligned}\text{Total basin volume, } V_{\text{thickener}} &= V_{csz}(\text{Step 11a}) + V_{tz}(\text{Step 11b.ii}) + V_{sz}(\text{Step 11b.iii})\\ &= (339+113+38)\,\text{m}^3 = 490\,\text{m}^3 \text{ per thickener}\end{aligned}$$

$$\text{Hydraulic retention time based on combined sludge, } \theta_{CS} = \frac{2V_{\text{basin}}}{Q_{b,CS}} = \frac{2\times 490\,\text{m}^3}{500\,\text{m}^3/\text{d}} = 2.0\,\text{d}$$

$$\text{Hydraulic retention time based on blended sludge, } \theta_{CS} = \frac{2V_{\text{basin}}}{Q_{b,blend}} = \frac{2\times 490\,\text{m}^3}{2030\,\text{m}^3/\text{d}} = 0.48\,\text{d} \approx 0.5\,\text{d}$$

Note: The design HRT of 12 h with blended sludge is effective to prevent septic conditions and odor issues. The HRT is 2 d with combined sludge without diluted water. It is nearly same as the acceptable SVR. This will provide high solids content in thickened sludge.

14. Describe the influent and effluent structures and sludge blending tank design.
The influent structure consists of a sludge feed pipe and a central well. The effluent structure has V-notch weirs around the periphery, effluent launder, and outlet pipe. The general arrangement, design procedure of major components, and hydraulic profile through the gravity thickener are similar to those of a final clarifier. The design procedure of a final clarifier is given in Example 10.166. The thickener overflow may be collected in a sump and pumped to the headworks of the plant. The design details and layout of the gravity thickener designed in this example are shown

in Figure 13.3. A sludge blending tank may be added to homogenize the combined sludge and dilution water prior to thickening. Design of a sludge blending tank is given in Example 13.7. A thickened sludge storage tank may also be included to accommodate the operating schedule of dewatering process (see Example 13.8). Reader may refer to Reference 6 for a layout of sludge blending tank and gravity thickener.

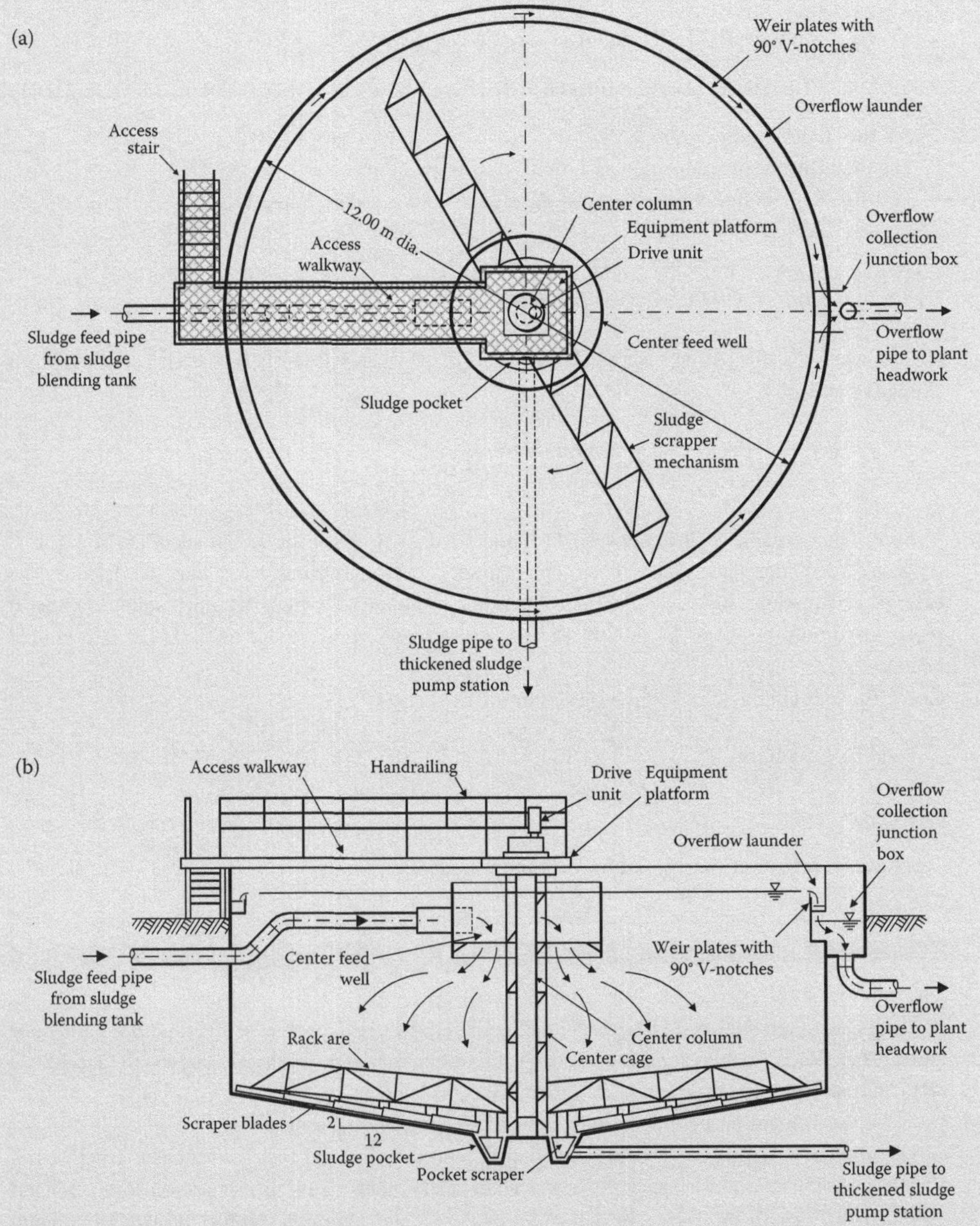

FIGURE 13.3 Design details of gravity thickener: (a) plan view (Example 13.12). (b) conceptual sectional view (Example 13.12).

15. Tabulate the characteristics of combined, blended and thickened sludge, and thickener overflow.

Parameter	Sludge Stream			Overflow (Step 10)
	Combined	Blended[a]	Thickened (Step 9)	
Daily volume, m^3/d	500	2030	170	1860
Solids[b]				
Solids content, %	2	≈ 0.5	5	≈ 0.08
Bulk specific gravity	1.004	1.001	1.01	1.000
TSS				
Mass, kg/d	10,000	10,015 (Step 5)	8510	1505
Concentration, mg/L	20,000[c]	4930 (Step 8)	50,100	810
BOD_5				
Mass, kg/d	5300 (Step 7)	5308 (Step 7)	4510	798
Concentration, mg/L	10,600[d]	2610 (Step 8)	26,500	430

[a] Dilution and blending is done with the secondary effluent. Volume of diluted water = 1530 m^3/d (Step 4), TSS = 15 kg (Step 5), and $BOD_5 = 8$ kg (Step 7).

[b] The calculation procedures for estimating solid contents and bulk specific gravity are given Examples 13.1 through 13.3.

[c] $C_{TSS,CS} = \frac{W_{s,CS}}{Q_{b,CS}} = \frac{10{,}000\,\text{kg TSS/d}}{500\,\text{m}^3/\text{d}} \times 10^3\,\text{g/kg} = 20{,}000\,\text{g TSS/m}^3 \quad \text{or} \quad 20{,}000\,\text{mg TSS/L}$

[d] $C_{BOD_5,CS} = \frac{W_{s,CS}}{Q_{b,CS}} = \frac{5300\,\text{kg BOD}_5/\text{d}}{500\,\text{m}^3/\text{d}} \times 10^3\,\text{g/kg} = 10,600\,\text{g BOD}_5/\text{m}^3 \quad \text{or} \quad 10,600\,\text{mg BOD}_5/\text{L}$

EXAMPLE 13.13: DESIGN SOLIDS LOADING (SOLIDS FLUX) AND AREA OF A GRAVITY THICKENER

Batch settling tests were conducted with combined sludge to determine the design solids loading. A series of solids concentrations were prepared and settled in batch columns. The interface subsidence rates at each concentration were obtained. The experimental data are provided below. Determine the design value of solids loading (solids flux) of a gravity thickener. Also, calculate the thickener area if total combined sludge solids input rate is 2500 kg/d and solids content in thickened sludge is 5%.

TSS concentration, kg/m^3	2	7	13	23	37	63	100
Subsidence rate, m/d	80	50	20	4	1	0.3	0.1

Note: The procedure to determine the design solids loading rate is discussed in Section 10.9.2, and Examples 10.163 through 10.165.

Solution

1. Plot the subsidence rate of interface versus TSS concentration on a log-log graph (Figure 13.4a).
2. Develop the computational data for plotting solid flux curve.

   Several solids loading concentrations are assumed between 2 and 100 kg/m^3, in Column (1) of the computational table below. Approximate solids content (%) is estimated for each TSS concentration using a suspension bulk density $\rho_b = 1000\,kg/m^3$ at $S_b = 1$. The estimated results are summarized in Column (2). The corresponding values of the subsidence rate of the interface are read from Figure 13.4a and recorded in Column (3). The solids flux ($kg/m^2{\cdot}d$) is calculated from the product of solids concentration (kg/m^3) and subsidence rate (m/d), and arranged in Column (4).

TSS Concentration[a], kg/m³	Solids Content (p)[b], %	Subsidence Rate[c], m/d	Solids Flux (G)[d], kg/m²·d
(1)	(2)	(3)	(4)
2	0.2	80	160
3	0.3	70	210
4	0.4	61	244
5	0.5	59	295
10	1	32	320
20	2	6	120
30	3	1.7	51
40	4	0.8	32
50	5	0.5	25
60	6	0.32	19
80	8	0.18	14
100	10	0.10	10

[a] Arbitrary data.
[b] Obtained by multiplying numbers in Column (1) by a factor of 0.1 (10 kg/m³ = 1%, and 1 kg/m³ = 1% × 1/10 = 0.1%).
[c] Obtained from Figure 13.4a.
[d] Product of numbers in Columns (1) and (3).

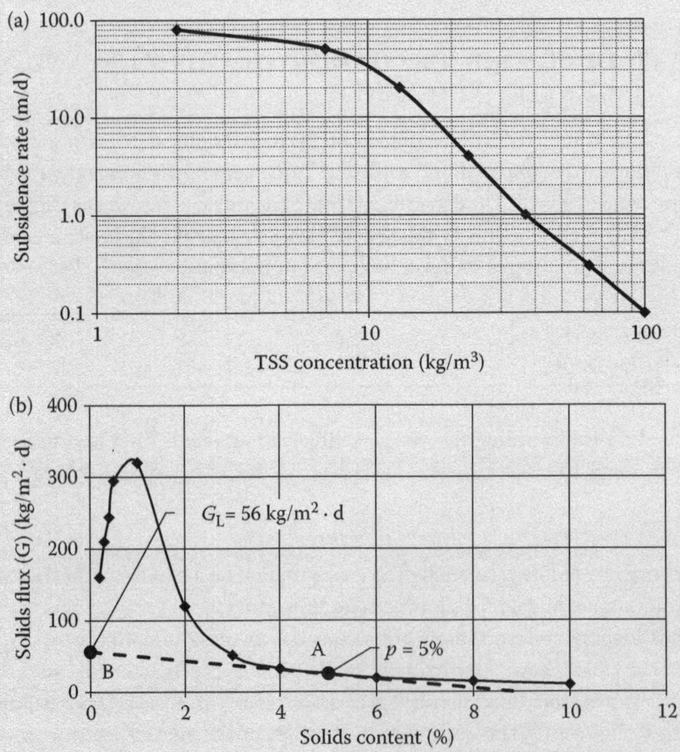

FIGURE 13.4 Determination of design solids loading rate: (a) plot of settling column data and (b) solids flux plot (Example 13.13).

3. Plot the solids flux data vs. solids content in Figure 13.4b.
4. Determine the design value of solids loading rate.
   Draw a tangent (dash line) at 5% solids content on the solids flux curve (Point A in Figure 13.4b). Read the design value of solids loading rate (*SLR*) from the intercept $G_L = 56\ \text{kg/m}^2\cdot\text{d}$ (Point B in Figure 13.4b).
5. Compute the surface area of gravity thickener.

$$\text{Thickener surface area, } A = \frac{\text{Combined sludge solids}}{\text{Solids loading rate}} = \frac{\Delta W_{TSS}}{SLR} = \frac{2500\ \text{kg/d}}{56\ \text{kg/m}^2\cdot\text{d}} = 45\ \text{m}^2$$

13.5.2 Dissolved Air Flotation Thickening

DAF process is effectively used to thicken WAS, coagulation-precipitation sludge, and suspend-growth nitrification sludge. The principal advantage of flotation over gravity thickening is the ability to thicken particles of small size and density close to that of fluid. These particles settle very slowly under gravity. The basic design features and general guidelines for the DAF process are given below:[7,32–36,38,39]

1. Flotation of solids is achieved by introducing fine bubbles into the liquid. The bubbles attach to the solid particles which then rise to the surface. In an DAF system, the air is dissolved into the liquid under a high pressure. The pressurized flow is released into a flotation tank at one atmosphere. Very fine bubbles rise that cause flotation.
2. Two variations of DAF process are commonly used: (1) pressurizing all or only small portion of incoming sludge (also called direct pressurizing) and (2) pressurizing the recycled flow from the flotation tank (also called indirect pressurizing). In the second method, high-pressure sludge pumps are not needed. The high-pressure sludge pumps may require high maintenance. The flow schematics of both systems are shown in Figure 13.5.
3. The solids content of thickened sludge from a DAF unit is ~3–8 times that of incoming sludge. The typical solids content in the float from thickening WAS ranges from 3% to 6% at a hydraulic loading rate of 1.25–6 $\text{m}^3/\text{m}^2\cdot\text{h}$ (0.5–2.5 gpm/ft^2). Polymer is usually needed to improve the performance of a DAF unit. The normal dosage is 1–5 g/kg dry solids (2–10 lb/ton) at SVI below 125 mL/g. A higher polymer dosage may be required at SVI > 200. The typical ranges of various design and performance parameters of a DAF thickener are summarized in Table 13.9.
4. The basic design and operation parameters of DAF systems are: (a) characteristics of incoming solids, (b) air/solids ratio, (c) solids loading rate, (d) hydraulic loading rate, (e) solids capture rate, (f) solids content in thickened sludge, (g) influent and recycled flows, (h) operating air pressure, and (i) chemical dosage and injection point. These parameters can be established from batch tests using a laboratory scale DAF apparatus.[35,38] The air/solids ratio for direct pressurization of influent and recycle flows are expressed by Equations 13.3a and 13.3b.[6,7,35]

$$A/S = \frac{1.3 s_a (fP - 1)}{S_a} \quad \text{(without recycle)} \tag{13.3a}$$

$$A/S = \frac{1.3 s_a (fP - 1) q}{S_a Q} \quad \text{or} \quad A/S = \frac{1.3 s_a (fP - 1) R_{rec}}{S_a} \quad \text{(with recycle)} \tag{13.3b}$$

where

A/S = air/solids ratio, kg air/kg solids (lb air/lb solids). The typical range of A/S ratio is 0.02–0.07.

s_a = solubility of air at the required temperature, mL/L. At the pressure of 1 atm, the values of s_a at temperatures 0°C, 10°C, 20°C, and 30°C are 29.2, 22.8, 18.7, and 15.7 mL/L, respectively.

S_a = solids concentration in incoming sludge, mg/L

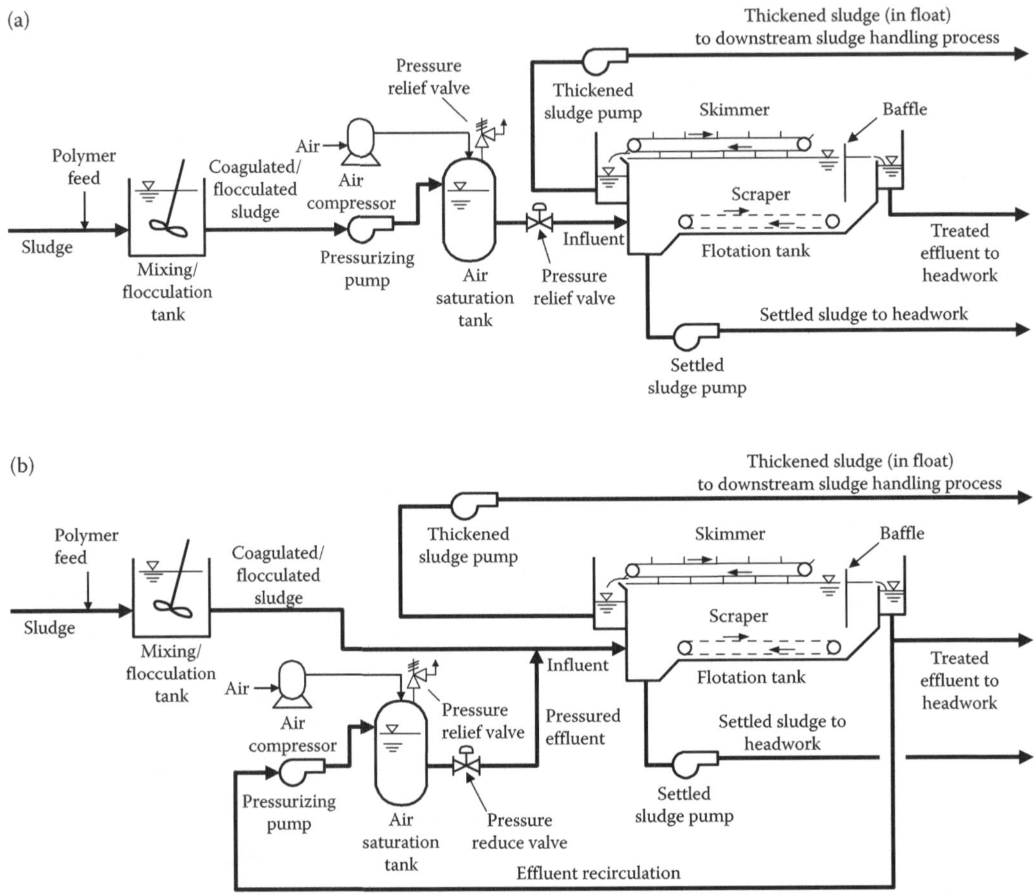

FIGURE 13.5 Schematics of DAF systems: (a) pressurized sludge without effluent recirculation and (b) pressurized effluent recirculation.

TABLE 13.9 Typical Design and Performance Parameter Ranges of Dissolved Air Flotation

Type of Sludge	Air/Solids Ratio	Solids Loading, kg/m^2·d (lb/ft^2·d)		Polymer Dose,[a] g/kg (lb/ton)	Solids Capture, %
		Without Polymer	With Polymer		
Primary sludge	0.04–0.07	100–150 (20–30)	Up to 300 (60)	2–5 (4–10)	80–95
Trickling filter sludge	0.02–0.06	70–100 (14–10)	240–270 (50–55)	1–3 (2–6)	90–98
Waste activated sludge (WAS)	0.02–0.05	30–70 (6–14)	220–240 (45–50)	1–3 (2–6)	80–95
Combined primary sludge and WAS	0.03–0.06	70–150 (14–30)	240–270 (50–55)	2–4 (4–8)	90–95

[a] Based on dry solids.
1 kg/m^2·d = 0.205 lb/ft^{2}d
Source: Adapted in part from References 1, 2, 6, 7, 32, 35, and 36.

f = fraction of air dissolved at pressure P, dimensionless. It is usually 0.5–0.95.
P = absolute pressure, atm (psia). It is calculated from Equation 13.3c in SI units, and Equation 13.3d in U.S. customary units.

$$P = \frac{p + 101.35\,\text{kPa}}{101.35\,\text{kPa}} \quad \text{(SI units)} \tag{13.3c}$$

$$P = \frac{p + 14.7\,\text{psi}}{14.7\,\text{psi}} \quad \text{(U.S. customary units)} \tag{13.3d}$$

p = gauge pressure, kPa (psig). The value of p is typically 2–5 atm (30–75 psig) in a DAF tank without recirculation.
q = recycled flow, or a portion of incoming flow pressurized, m^3/d (MGD)
Q = sludge flow to the DAF unit, m^3/d (MGD)
R_{rec} = sludge recirculation ratio ($R_{rec} = q/Q$), fraction of incoming sludge flow to the thickener. Typically it is in the range from 0 to 1 (0–100%).

In Equations 13.3a and 13.3b, the numerator is weight of air, and denominator is weight of solids. The factor 1.3 converts the mL of air to weight of air in mg.

5. The major equipment for a DAF thickener includes: (1) pressurizing pump, (2) air saturation tank with air compressor, (3) flotation tank (SWD = 2–4.5 m) with skimmer and scraper, (4) mixing/flocculation tank with chemical storage and feed system (mixing energy $G = 50\text{–}100\,\text{s}^{-1}$ and retention time $\theta = 1\text{–}2.5$ min), (5) polymer system with storage tanks and metering pumps, (6) thickened sludge pump, (7) settling sludge pump, (8) flow meters, and (9) pressure relief valves, pressure reducing valves, pressure switches, and pressure indicators.

EXAMPLE 13.14: DISSOLVED AIR FLOTATION FOR THICKENING OF WAS

A DAF thickener is designed to thicken WAS. The flow rate of WAS and TSS concentration are 500 m^3/d and 4000 g/m^3, respectively. The bench-scale data show that 4% solids content is reached in the float at A/S ratio of 0.02. The fraction of air saturation is 0.8. The operating temperature is 20°C and solubility of air at this temperature is 18.7 mL/L. The solids loading rate (SLR) in the floatation tank with polymer addition is 220 $kg/m^2{\cdot}d$, and maximum allowable surface overflow rate is 60 $L/m^2{\cdot}min$ (3.6 $m^3/m^2{\cdot}h$). The entire incoming WAS is pressurized and TSS capture efficiency of DAF unit is 85%. Determine (a) the applied pressure, (b) surface area of flotation tank, and (c) solids loading rate. Also, calculate the volume of thickened sludge at solids content of 4%. The bulk specific gravity of the thickened sludge is 1.006.

Solution

1. Compute the required operating gauge pressure.

Rearrange Equation 13.3a and calculated the absolute pressure (P).

$$P = \frac{1}{f}\left(\frac{S_a}{1.3s_a}(A/S) + 1\right) = \frac{1}{0.8} \times \left(\frac{4000\,\text{mg/L}}{1.3\,\text{mg/mL} \times 18.7\,\text{mL/L}} \times 0.02 + 1\right) = 5.4\,\text{atm}$$

Rearrange Equation 13.3c and calculated the gauge pressure (p).

$$p = 101.35\,\text{kPa/atm} \times (P - 1) = 101.35\,\text{kPa/atm} \times (5.4 - 1)\,\text{atm} = 450\,\text{kPa (65 psig)}$$

2. Compute the surface area of the flotation tank based on the solids loading rate.

Weight of TSS in incoming sludge, $W_{s,incoming} = \text{Flow} \times \text{TSS concentation} = QS_a$

$$= 500\,\text{m}^3/\text{d} \times 4000\,\text{g/m}^3 \times 10^{-3}\,\text{kg/g} = 2000\,\text{kg TSS/d}$$

$$\text{Total surface area of DAF, } A = \frac{\text{Weight of TSS}}{\text{Solids loading rate}} = \frac{W_{s,incoming}}{\text{SLR}} = \frac{2000\,\text{kg/d}}{220\,\text{kg/m}^2\cdot\text{d}} = 9.1\,\text{m}^2$$

3. Check the surface overflow rate.

$$\text{Surface overflow rate, } SOR = \frac{\text{Flow}}{\text{Area}} = \frac{Q}{A} = \frac{500\,\text{m}^3/\text{d}}{9.1\,\text{m}^2} = 55\,\text{m}^3/\text{m}^2\cdot\text{d} \text{ or } 2.3\,\text{m}^3/\text{m}^2\cdot\text{h} \text{ or } 38\,\text{L/m}^2\cdot\text{min}$$

Surface overflow rate is below the maximum allowable rate of 60 $\text{L/m}^2\cdot$min.

4. Determine the volume of thickened sludge.

Calculate the total amount of TSS in thickened sludge at a capture rate of 85%.

$$W_{s,THS} = 0.85 \times W_{s,incoming} = 0.85 \times 2000\text{ kg TSS/d} = 1700\text{ kg TSS/d}$$

Determine the daily volume of thickened sludge (THS) from Equation 13.1e.

$$\text{Flow of combined sludge, } Q_{b,THS} = \frac{100\% \times W_{s,THS}}{p_{s,THS} s_{b,THS} \rho_w} = \frac{100\% \times 1700\,\text{kg/d}}{4\% \times 1.006 \times 1000\,\text{kg/m}^3} = 42\,\text{m}^3/\text{d}$$

EXAMPLE 13.15: DAF THICKENING BY PRESSURIZING RECYCLE STREAM

Use the data given in Example 13.14 to determine the operating pressure, if the effluent is recycled and pressurized. The recycle ratio R_{rec} is 0.5. Also, determine the surface area of the flotation tank and solids loading rate.

Solution

1. Calculate the required operating gauge pressure.

Rearrange Equation 13.3b and calculated the absolute pressure (P).

$$P = \frac{1}{f}\left(\frac{S_a}{1.3s_a R_{rec}}(A/S) + 1\right) = \frac{1}{0.8} \times \left(\frac{4000\,\text{mg/L}}{1.3\,\text{mg/mL} \times 18.7\,\text{mL/L} \times 0.5} \times 0.02 + 1\right) = 9.5\,\text{atm}$$

Rearrange Equation 13.3c and calculated the gauge pressure (p).

$$p = 101.35\,\text{kPa/atm} \times (P - 1) = 101.35\,\text{kPa/atm} \times (9.5 - 1)\,\text{atm} = 860\,\text{kPa (125 psig)}$$

2. Determine the surface area of the flotation tank based on the solids loading rate.

Ignore the TSS in the recycled effluent. The TSS reaching the flotation tank is 2000 kg TSS/d and solids loading rate is 220 $\text{kg/m}^2\cdot$d. Therefore, the required surface area of floatation unit is 9.1 m^2 (see Example 13.14).

3. Check the hydraulic loading rate.

Surface overflow rate without recirculation = 38 $\text{L/m}^2\cdot$min (Example 13.14, Step 3).

$$\text{Surface overflow rate with recirculation, } SOR = \frac{(1 + R_{rec})Q}{A} = \frac{(1 + 0.5) \times 500\,\text{m}^3/\text{d}}{9.1\,\text{m}^2}$$

$$= 82\,\text{m}^3/\text{m}^2\cdot\text{d} \text{ or } 3.4\,\text{m}^3/\text{m}^2\cdot\text{h} \text{ or } 57\,\text{L/m}^2\cdot\text{min}$$

Surface overflow rate with recirculation is below the maximum allowable rate of 60 $\text{L/m}^2\cdot$min.

TABLE 13.10 Typical Design and Performance Parameter Range of Solid Bowl Centrifuge Thickener

Parameter	Range of Value
Bowl diameter, cm (in)	36–152 (14–60)
Capacity, L/min (gpm)	40–600 (10–150)
Rotational speed, rpm	1000–2500
Gravitational force, times of gravity force	1400–2300
Solids content in thickened sludge, %	5–8
Solids recovery rate, % by wt.	85–95
Polymer usage, g/kg dry solids (lb/ton dry solids)	0–6 (0–12)

1 L/min = 0.264 gpm.

Source: Adapted in part from References 1, 2, 6, 7, 32, 35, 39, and 40.

13.5.3 Centrifugal Thickening

Centrifugation creates a centrifugal field many times larger than the force of gravity. It can be used for thickening and dewatering of solids. Common types of centrifuges are: (1) imperforated basket, (2) disc nozzle, and (3) solid bowl (or scroll-type decanter). The solid-bowl centrifuges, in both concurrent and countercurrent configurations, have received widespread use in sludge thickening. Centrifuges require high power and high maintenance. For this reason, their use is limited to plants where space is limited, skilled operation is available, and sludge is difficult to thicken by other means.[6,7,32,40]

The principal design and operational variables in centrifugal thickening are: (1) characteristics of feed solids, (2) polymer use and injection point, (3) rotational speed and torque, (4) depth of liquid pool in the bowl, (5) solids and hydraulic loading rates (equipment specific), (6) solids capture rate, (7) solids content in thickened sludge, and (8) quantity of centrate. Major design and performance parameters of solid-bowl centrifuge are summarized in Table 13.10. The centrifuge thickeners may be available in unit capacities of 8–250 m^3/h (35–1100 gpm).[40–42]

The performance of a centrifuge is measured by solids capture rate, and is given by Equation 13.4.

$$R = \frac{C_t(C_f - C_c)}{C_f(C_t - C_c)} \times 100\% \quad \text{or} \quad C_t = \frac{RC_fC_c}{100\% \times C_c - (100\% - R)C_f} \tag{13.4}$$

where

R = solids capture efficiency rate, %
C_c = solids content in centrate, %
C_f = solids content in feed, %
C_t = solids content in thickened stream, %

EXAMPLE 13.16: SLUDGE THICKENING BY CENTRIFUGE

Centrifugal thickening is used for WAS. The manufacturer after bench-scale testing recommended that the solids capture efficiency of their unit is 90%. Determine the solids concentration in the thickened sludge if the solids concentration in the feed and centrate are 0.8% and 0.09%, respectively.

Solution

Determine the solids concentration in the thickened sludge from Equation 13.4.

$$C_t = \frac{RC_fC_c}{100\% \times C_c - (100\% - R)C_f} = \frac{90\% \times 0.8\% \times 0.09\%}{100\% \times 0.09\% - (100\% - 90\%) \times 0.8\%} = \frac{0.9 \times 0.8\% \times 0.09\%}{0.09\% - 0.1 \times 0.8\%} = 6.5\%$$

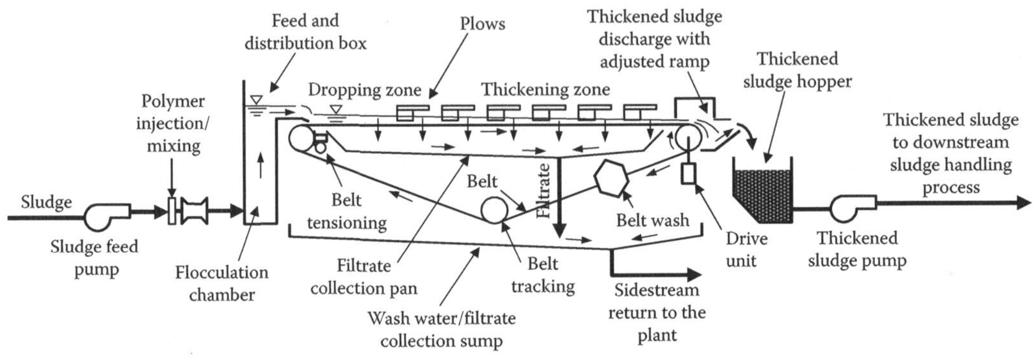

FIGURE 13.6 Schematic of gravity belt thickener.

TABLE 13.11 Design and Performance Parameters of Gravity Belt Thickener

Parameter	Range of Value
Solids content in thickened sludge, %	4–8
Polymer dosage, g/kg dry solids (lb/ton dry solids)	1–7 (2–14)
Solids capture rate, % by wt.	90–98
Hydraulic loading rate, L/min per m belt width (gpm per m belt width)	400–1000 (100–250)
Solids loading rate, kg/h per m belt width (lb/h per m belt width)	300–1350 (650–3000)
Washwater, L/min per m belt width (gpm per m belt width)	60–80 (15–20)

1 L/min·m = 0.264 gpm/m; 1 kg/h·m = 2.205 lb/h·m.
Source: Adapted in part from References 6, 7, 32, 35, and 44.

13.5.4 Gravity Belt Thickening

The gravity belt thickening (GBT) is accomplished over a porous horizontal belt similar to that of belt-filter press (BFP) used for dewatering sludge. The chemically conditioned sludge is applied and dispersed evenly over the moving belt. Water drains and a mat or solid layer builds over the belt. Rows of plows or vans expose the belt area for more drainage. Sludge pooling and backward rolling action may also be used to drain more free water from the sludge. The sludge travels over an adjustable ramp and into a hopper. The solids are scraped and fabric is washed by a water jet.[6,7,32,43–45] The schematic process diagram of GBT is shown in Figure 13.6.

The basic design and operation parameters of GBT are: (1) characteristics of feed solids, (2) polymer type and use, flocculation device and application point, (3) solids loading rate, (4) hydraulic loading rate, (5) solids capture rate, (6) solids content in thickened sludge, (7) belt width, (8) quantity of filtrate, and (9) wash water requirement. Some of these parameters are summarized in Table 13.11.

GBT units are commercially available in belt width between 0.4 and 4 m with an increment of 0.5–1 m. The belt size ranges from 2.5 to 20 m^2 (8–65 ft^2) per unit. Major components of a GBT system include polymer feed, polymer mixing device, belt thickener assembly with plows or vans, flocculation zone, thickening zone, thickened sludge hopper, belt scraper, and wash station.[43–45]

EXAMPLE 13.17: DESIGN OF A GRAVITY BELT THICKENER

The biological solids from an activated sludge plant are thickened over a gravity belt thickener (BGT). The maximum daily sludge production rate during seven consecutive days is 1500 kg/d with a solids content of 0.8% and a bulk specific gravity of 1.001. The thickener operates 5 days a week and 7 h per day. The hydraulic loading on the thickener belt is 600 L/min·m, and the optimum polymer dose is 4 g active content/kg dry solids. The active content of polymer emulsion is 29% by weight and density of solution is 1060 kg/m^3 (or 1.06 kg/L). Determine (a) the belt width requirement, (b) solids loading rate, (c) neat polymer feed rate (L/h), (d) the feed rate (L/h) of 1% diluted polymer solution, and (e) daily and weekly polymer usages.

Solution

1. Determine the daily total weight of sludge fed to the GBT on an operating day.
   Maximum daily sludge production rate during seven consecutive days, $W_{M7D} = 1500$ kg/d
   Total amount of sludge generated during a week,

   $$\Delta W_{M7D} = 7\,\text{d} \times W_{M7D} = 7\,\text{d/wk} \times 1500\,\text{kg/d} = 10{,}500\,\text{kg/wk}$$

   Total GBT operating hours during a week, $t_{GBT} = 5\,\text{d/w} \times 7\,\text{h/d} = 35\,\text{h/wk}$

   Calculate the average sludge loading rate over the GBT units during five days of operation in a week.

   $$W_{GBT} = \frac{\Delta W_{M7D}}{t_{GBT}} = \frac{10{,}500\,\text{kg/wk}}{35\,\text{h/wk}} = 300\,\text{kg/h}$$

   Apply a safety factor of 1.2 to obtain the design per h sludge processing capacity of the GBT.

   $$W_{GBT}^{design} = 1.2 \times W_{GBT} = 1.2 \times 300\,\text{kg/h} = 360\,\text{kg/h}\ (790\,\text{lb/h})$$

2. Determine the hourly total volume of sludge processed by the GBT.
   The design hourly volume of sludge from Equation 13.1e at the density of water $\rho_w = 1000$ kg/m^3,

   $$Q_{GBT}^{design} = \frac{100\% \times W_{GBT}^{design}}{p_s S_b \rho_w} = \frac{100\% \times 360\,\text{kg/h}}{0.8\% \times 1.001 \times 1000\,\text{kg/m}^3}$$
   $$= 45\,\text{m}^3/\text{h} \quad \text{or} \quad 750\,\text{L/min}\ (198\,\text{gpm})$$

3. Determine the belt width at the hydraulic loading rate, $HLR_{GBT} = 600$ L/min·m,

   $$B_{GBT} = \frac{Q_{GBT}^{design}}{HLR_{GBT}} = \frac{750\,\text{L/min}}{600\,\text{L/min·m}} = 1.25\,\text{m}$$

   Select the design belt width, $B_{GBT}^{design} = 1.5$ m (59 in).

4. Check the actual hydraulic and solids loading rates.

   Actual hydraulic loading rate, $HLR_{GBT}^{actual} = \dfrac{Q_{GBT}^{design}}{B_{GBT}^{design}} = \dfrac{750\,\text{L/min}}{1.5\,\text{m}} = 500\,\text{L/min·m}\ (132\,\text{gpm/m})$

   Note: The hydraulic loading rate is within the typical range of 400–1000 L/min·m.

   Actual solids loading rate, $SLR_{GBT}^{actual} = \dfrac{W_{GBT}^{design}}{B_{GBT}^{design}} = \dfrac{360\,\text{kg/h}}{1.5\,\text{m}} = 240\,\text{kg/h·m}\ (530\,\text{lb/h·m})$

   Note: The actual solids loading rate is below the typical range from 300 to 1350 kg/h·m (Table 13.11). It indicates that the design of GBT is controlled by the hydraulic loading rate in this application.

5. Determine the polymer feed rate.
   Calculate the daily polymer usage rate during five operating days at the polymer dosage of 4 g active content/kg dry solids.

   $W_{\text{polymer}} = Dose_{\text{polymer}} W_{\text{GBT}}^{\text{design}} = 4\,\text{g/kg} \times 360\,\text{kg/h} \times 10^{-3}\,\text{kg/g} = 1.44\,\text{kg/h}$

   $$\text{Polymer feed rate, } Q_{\text{polymer}} = \frac{100\% \times W_{\text{polymer}}}{p_{\text{s,polymer}}\rho_{\text{polymer}}} = \frac{100\% \times 1.44\,\text{kg/h}}{29\% \times 1.06\,\text{kg/L}} = 4.7\,\text{L/h (1.2 gph)}$$

6. Determine the diluted polymer solution feed rate at 1% concentration.
   The polymer solution feed rate is calculated from Equation 13.1e. Assume that the density of polymer solution $\rho_{\text{solution}} \approx \rho_{\text{water}} = 1000\,\text{kg/m}^3 = 1\,\text{kg/L}$.

   $$\text{Polymer solution feed rate, } Q_{\text{solution}} = \frac{100\% \times W_{\text{polymer}}}{p_{\text{s,solution}}\rho_{\text{solution}}} = \frac{100\% \times 1.44\,\text{kg/h}}{1\% \times 1\,\text{kg/L}} = 144\,\text{L/h (38 gph)}$$

   Dilution water flow, $Q_{\text{water}} = Q_{\text{solution}} - Q_{\text{polymer}} = (144 - 4.7)\,\text{L/h} = 139\,\text{L/h (36.8 gph)}$

7. Determine the polymer usages.

   Polymer daily usage, $\Delta Q_{\text{daily,polymer}} = 7\,\text{h/d} \times Q_{\text{polymer}} = 7\,\text{h/d} \times 4.7\,\text{L/h} = 33\,\text{L/d (8.7 gpd)}$

   $$\text{Polymer weekly usage, } \Delta Q_{\text{weekly,polymer}} = 35\,\text{h/wk} \times Q_{\text{polymer}} = 35\,\text{h/wk} \times 4.7\,\text{L/h}$$
   $$= 165\,\text{L/wk} \quad \text{or} \quad 0.165\,\text{m}^3/\text{wk (44 gal/wk)}$$

13.5.5 Rotary Drum Thickener

The rotary drum thickener (RDT) has a drum covered with stainless steel or polyester fabric, or a combination. The chemically conditioned sludge is conveyed inside along the drum length that rotates at a speed of 5–20 rpm. The free water drains through the screen while flocculated thickened solids roll along the length and exits through a discharge chute. Washwater periodically flushes the inside and outside of the drum to clean the screen openings.

The basic design and operation parameters for RDT are: (1) characteristics of feed solids, (2) polymer type and use, flocculation device, and application point, (3) solids and hydraulic loading rates per unit, (4) drum rotational speed, (5) solids capture rate, (6) solids content in thickened sludge, (7) drum dimensions, and (8) quantity of decant. Some of these parameters are summarized in Table 13.12. The rotary drum thickeners may be available in unit capacities up to 500 kg/h (1000 lb/h) and 1500 L/min (400 gpm). The typical range of polymer dosage for RDT is 2.5–5 g/kg dry solids (5–10 lb/ton dry solids).[46–48]

13.5.6 Membrane Thickener

Membrane bioreactor (MBR) is widely used for liquid treatment (see Section 10.3.7). Membrane process has also been investigated and applied for thickening sludge in recent years. The equipment used for membrane thickener (MBT) is similar to that for MBR. It consists of membrane cartridges in a tank. However,

TABLE 13.12 Design and Performance Parameters of Rotary Drum Thickener

Parameter	Range of Value		
	Primary Sludge	WAS	Combined Primary Sludge and WAS
Solids content in thickened sludge, %	7–9	4–9	5–9
Solids capture rate, % by wt.	93–98	93–99	93–98

1 L/min·m = 0.0805 gpm/ft 1 kg/h·m = 0.672 lb/h·ft.
Source: Adapted in part from References 6, 7, 32, and 35.

the operating flux is lower and transmembrane pressure (TMP) is higher for MBT than those for MBR process. The normal pore size of MBT membrane is 0.4 μ with an effective pore size of 0.1 μ. The membrane is scrubbed and cleaned continuously by air bubbles released from the diffuser incorporated in the lower part of the cartridge. The basic features of MBT are: (a) small footprint that is especially beneficial for use at both new and existing facility, (b) applicable to WAS from conventional activated sludge process and modifications, such as MBR, SBR, BNR, and enhanced biological phosphorus removal (EBPR), (c) the solids content in the thickened sludge ranges from 3% to 5% (typically 4%) at solids capture efficiency over 99%, (d) produces reusable quality permeate (BOD_5 and TSS < 1 mg/L) that can be combined with the secondary effluent, and (e) polymer or coagulant are not needed. In associated with aerobic digester process, MBT provides a cost-effective option to meet or exceed Class B biosolids in many cases.[49–51]

13.6 Sludge Stabilization

Sludge is stabilized to reduce offensive odors and pathogens, reduce biodegradable organic matter, and improve sludge dewatering property. Various methods of sludge stabilization fall in three categories: (1) biological: anaerobic and aerobic digestions, (2) chemical: chlorine and lime stabilizations, and (3) physical: thermal stabilization. Selection of any method depends largely on the ultimate sludge disposal method. Anaerobic digestion is normally a preferred method at medium and large sized plants because of energy recovery and little or no chemical requirement. Each of these stabilization methods is discussed below.

13.6.1 Anaerobic Digestion

Anaerobic digestion is achieved in air tight tanks in which anaerobic microorganisms breakdown organic matter and produce methane and carbon dioxide. The remaining sludge is stable, lack of offensive odors, low in pathogens count, and suitable for land application. The associated problems with anaerobic digestion are high capital cost, needs for skilled operation, and poor quality supernatant.

Anaerobic Process Description: Anaerobic process involves a series of complex biochemical reactions by several groups of facultative and anaerobic organisms. The overall process may be divided into three phases: (1) *hydrolysis phase*, (2) *acidogenic phase*, and (3) *methanogenic phase*. These phases have been discussed in Section 10.5.2. A simplified process diagram of anaerobic process is illustrated in Figure 10.82. Important aspects of each phase are briefly reviewed below.

1. *Hydrolysis phase*: The complex molecules of cellulose, carbohydrates, proteins, lipids and other complex organics contained in sludge are solubilized and broken down into simpler compounds such as monosaccharides, amino acids and long-chain fatty acids.
2. *Acidogenic phase, acidogenesis of fermentation*: Many acid-forming facultative organisms convert the solubilized organics into short-chain organic acids (acetic, propionic, butyric and others). In this phase some lowering of pH occurs, but there is little change in the total amount of organic matter in the system.
3. *Methanogenic phase or methanogenisis*: The volatile organic acids are converted into methane and carbon dioxide. Readers are referred to Section 10.5.3 for many important environmental factors for anaerobic decomposition of liquid wastes. Some basic features of methane phase are listed below.[3,6,7,32]
   a. The methane phase is essentially controlled by the methane forming bacteria.
   b. The methane formers are strict anaerobes and are very sensitive to pH, temperature, substrate composition, and toxic compounds.
   c. If pH drops below 6.0, methane formation essentially ceases, more acid accumulates and the digestion process reaches standstill. Thus, the pH and acid measurements constitute important operational parameters.
   d. The methane forming bacteria are highly active in the *mesophilic* 30–38°C (85–100°F), and *thermophilic* 50–57°C (122–135°F) temperature ranges. The anaerobic digesters are most

commonly operated in the mesophilic range. Thermophilic digestion enhances digestion, overall volatile solids reduction, and gas production; increases overall performance efficiency; and improves dewatering property of digested sludge. However, the energy requirement to maintain thermophilic range is high.

e. Alkalinity is important to provide buffering capacity. Bicarbonates of calcium, magnesium, and ammonium are main sources of alkalinity. Carbon dioxide produced during digestion consumes alkalinity. In a well-established digester, the total alkalinity is in the range from 2000 to 5000 mg/L, with a volatile acids to alkalinity (VA/Alk) ratio of 0.05–0.25. A VA/Alk ratio < 0.1 may be preferred.[7,32]

f. Sufficient HRT and SRT are needed in a digester to carry out hydrolysis, fermentation, and methanogenesis. There is a minimum SRT for each reaction. If SRT is less than these minimums, the bacteria will not fully establish and the overall digestion process will not reach equilibrium.

Types of Anaerobic Digesters: The anaerobic sludge digesters can be grouped into *single-* or *multiple-*stage systems. A single-stage digester may operate at a standard- or high-rate. The multiple-stage digesters are further categorized as *staged* or *phased* digestion. In the staged digestion, all reactors operate near neutral pH near same temperature (either mesophilic or thermophilic range). However, the SRT in the first digester is typically much longer than that in the second and third reactors. In the phased digestion, the pH and SRT in the first digester are favorable for the acid-forming facultative growth, while in the second digester these parameters are optimized for the growth of methane forming bacteria. Either mesophilic or thermophilic range may be separately selected for each reactor. Brief descriptions, important design and performance parameters, and simple definition sketches of different types of anaerobic digesters are provided in Table 13.13.

Process Design and Performance: The basic process design and performance factors are: (1) solids retention time (SRT), (2) volumetric solids loading (VSL), (3) per capita based capacity requirements (PCCRs), and (4) volatile solids reduction (VSR). Each of these factors are briefly discussed below. For different processes these factors are provided in Table 13.13.

Solids Retention Time: The value of solids retention time (SRT) represents the average time the solids are held in the digestion process. It may also be called the *digestion period, sludge age*, and *mean cell residence time (MCRT)*. The SRT and *hydraulic retention time (HRT)* are assumed same in a high-rate digester without recycle when the supernatant withdrawal volume is less than 5–10% of feed sludge volume. For mesophilic sludge digestion, the typical SRT in high-rate digester is 15–20 days (Table 13.13). The critical value of digestion period is 10 d. Below this value, the production of methane forming bacteria cannot keep up with the production of volatile acids and it begins to accumulate. Therefore, the digestion period must be checked against 7-d sustained peak flow and solids loading from the thickener, poor performance of thickening process, volume reduction due to grit accumulation and fluctuations in liquid depth due to difference in feed and withdrawal of sludge, and scum accumulation. Process analysis for anaerobic decomposition of liquid wastes has been presented in Section 10.5.4. The SRT (θ_c), observed biomass yield (Y_{obs}), COD stabilization rate (ΔS_M), and active biomass growth (P_x) for a continuous flow completely-mix, high-rate, anaerobic sludge digester without recycle can be estimated form Equations 10.16a, 10.16d, 10.78b, and 10.78d. The kinetic coefficients for anaerobic processes are provided in Table 10.44.

Volumetric Solids Loading: The volumetric solids loading (VSL) is normally based on the mass of volatile solids (VS) added per day per unit digester volume (kg VS/m^3·d (lb VSS/ft^3·d)). This parameter is also used in sizing an anaerobic suspended growth reactor for liquid waste treatment (see Section 10.5.5 and Table 10.46).

Per Capita Based Capacity Requirements (PCCRs): The capacity requirements for single-stage digesters are also established from TSS production rate (P_{TSS}) of 120 g per capita per day.* The per capita capacity requirements for stabilizing different types of sludge in various anaerobic digesters are given in Table 13.13.

* The concentration ranges and typical values for various constituents in municipal waste water are given in Table 5.1. The average loading for TSS is 113 g/capita·d. The TSS loading of 120 g/capita·d for digester capacity includes 6% sustained loading.

TABLE 13.13 Important Design and Performance Parameters of Various Anaerobic Digestion Processes

Process with Brief Description	Design and Performance Parameter		Definition Sketch
	Parameter	Value or Range	
1. Single-stage anaerobic digestion system			
a. Standard-rate digestion process These batch digesters are unheated, unmixed, fed intermittently, and supernatant and digested sludge are withdrawn periodically. It is normally stratified into four zones: (a) scum, (b) supernatant, (c) actively digesting solids, and (d) digested sludge layers (Figure 13.7). The biosolids may qualify for Class B biosolids under (a) Class B, Alt. 1 or (b) Class B, Alt. 2, PSRP No. 3 (see Table 13.7). These digesters are seldom used these days.[1,3,6,7,32,35]	Temperature, °C (°F)	20–30 (68–85)	
	pH.	6.8–7.2	
	SRT (θ_c), d	30–60	
	VSL[a], kg VSS/m^3·d (lb VSS/ft^3·d)	0.6–1.6 (0.04–0.1)	
	PCCR[b], m^3/capita (ft^3/capita)		
	Primary sludge	0.06–0.09 (2–3)	
	WAS	0.12–0.18 (4–6)	
	Primary sludge + WAS	0.12–0.15 (4–5)	
	Solids content, %		
	Feed sludge	2–4	
	Digested sludge	4–6	
	VSR[c], %	30–40	
	Biosolids classification potential	Class B	
b. High-rate digestion process The digester contents are heated and completely mixed in a CFSTR with optional decant (Figure 13.8). The solids loading is considerably higher than that of standard-rate digestion. The digester feed and sludge withdrawal are done on a regular schedule each day either feed before or after withdrawal of equal volumes. The biosolids may qualify for Class B biosolids under (a) Class B, Alt. 1 or (b) Class B, Alt. 2, PSRP No. 3 (see Table 13.7).[1,3,6,7,32,35]	Temperature, mesophilic range °C (°F)	30–38 (85–100)	
	pH	6.8–7.2	
	SRT (θ_c), d	15–20	
	VSL[a], kg VSS/m^3·d (lb VSS/ft^3·d)	1.6–4.8 (0.1–0.3)	
	PCCR[b], m^3/capita (ft^3/capita)		
	Primary sludge	0.03–0.06 (1–2)	
	Primary sludge + WAS	0.09–0.12 (3–4)	
	Solids content, %		
	Feed sludge	4–6	
	Digested sludge	3–5	
	VSR[c], %	40–60	
	Biosolids classification potential	Class B	

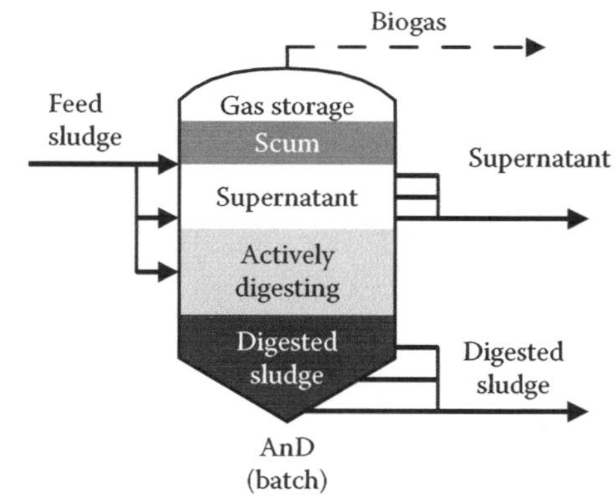

FIGURE 13.7 Standard-rate digestion process.

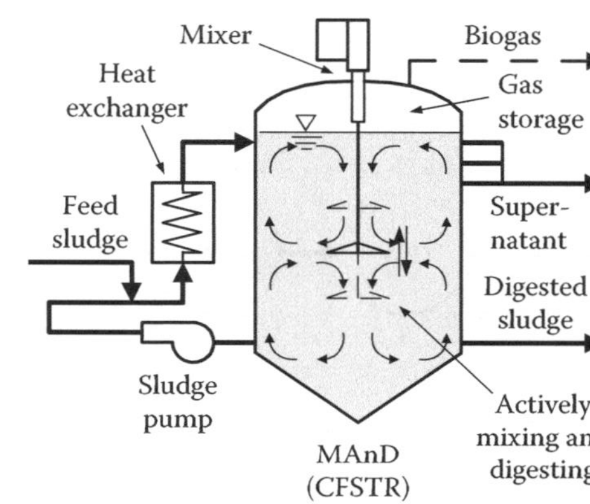

FIGURE 13.8 High-rate digestion process.

(Continued)

TABLE 13.13 (*Continued*) Important Design and Performance Parameters of Various Anaerobic Digestion Processes

Process with Brief Description	Design and Performance Parameter		Definition Sketch
	Parameter	Value or Range	
2. Staged anaerobic digestion system			
a. Staged mesophilic digestion process It is also commonly called two-stage digestion process. A high rate mesophilic anaerobic digester (MAnD) is arranged in series with a standard rate anaerobic digester (AnD) (Figure 13.9). The purpose of using the 2nd digester is to store digested sludge, clarify the supernatant, and provide additional digestion and gas recovery (~10% increase). The biosolids may qualify for Class B biosolids under (a) Class B, Alt. 1 or (b) Class B, Alt. 2, PSRP No. 3 (see Table 13.7). This process is not used in recent designs.[6,7,52]	Temperature, °C (°F)	30–38 (85–100)	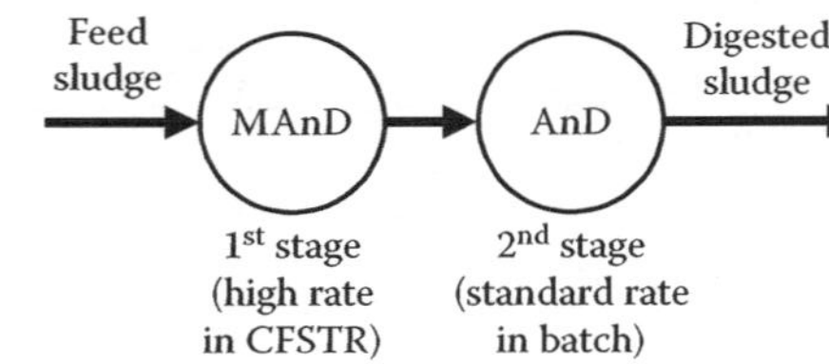 **FIGURE 13.9** Two-stage digestion process.
	pH	6.8–7.2	
	SRT (θ_c), d		
	1st reactor	7–10	
	2nd reactor	Variable	
	VSL[a], kg VSS/m^3·d (lb VSS/ft^3·d)	0.5–1.6 (0.03–0.1)	
	Solids content, %		
	Feed sludge	4–6	
	VSR[c], %	40–60	
	Biosolids classification potential	Class B	
b. Staged thermophilic anaerobic digestion process In this process, several thermophilic anaerobic digesters (TAnD) are arranged in series. A relatively large TAnD reactor is provided in the front to achieve efficient volatile solids reduction (VSR). It is followed by one or two smaller TAnD that operate in batch mode to prevent short circuiting of pathogens (Figure 13.10). The biosolids may qualify for Class A biosolids criteria under Class A, Alt. 1 (see Table 13.7).[7,32,52]	Temperature, °C (°F)	50–57 (122–135)	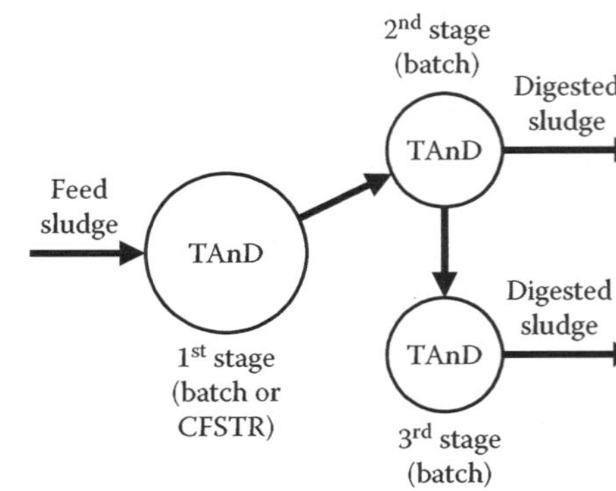 **FIGURE 13.10** Staged thermophilic digestion process.
	pH	6.8–7.2	
	SRT (θ_c), d		
	1st/2nd or 3rd reactor	17–22/1.5–2	
	VSL[a], kg VSS/m^3·d (lb VSS/ft^3·d)	4.8–6.4 (0.3–0.4)	
	Solids content, %		
	Feed sludge	5–6	
	VSR[c], %	50–65	
	Biosolids classification potential	Class A	

(*Continued*)

TABLE 13.13 (*Continued*) Important Design and Performance Parameters of Various Anaerobic Digestion Processes

Process with Brief Description	Design and Performance Parameter: Parameter	Value or Range	Definition Sketch
3. Phased anaerobic digestion system			
a. Acid/gas phased anaerobic digestion (A/GAnD) process In this process, the mesophilic acidogenic anaerobic digester (MAAnD) or the thermophilic acidogenic anaerobic digester (TAAnD) are arranged in series with mesophilic methanogenic anaerobic digester (MMAnD) or the thermophilic methanogenic anaerobic digester (TMAnD). The solids in the 1st reactor are hydrolyzed and acids are produced rapidly. The pH drops below 6 and short chain volatile fatty acids (SCVFAs) concentration may exceed 6000 mg/L. The 2nd digester utilizes methane producing bacteria to convert acid into CH_4 and CO_2. The SRT of methane phase is significantly larger than that for the acid phase. Two phases may be operated in either same or different temperature ranges (Figure 13.11). The VSS reduction and biogas production are notably improved. The digested sludge may meet the Class A biosolids criteria under Class A, Alt. 1 (see Table 13.7).[7,32,52]	Temperature, °C (°F)		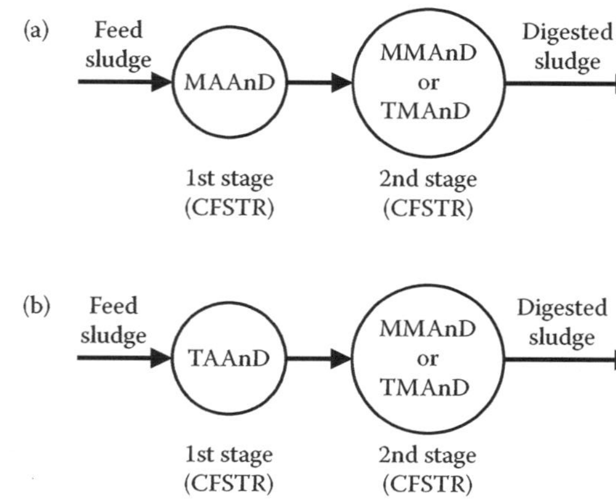 **FIGURE 13.11** Acid/gas phased (A/GAnD) digestion processes: (a) mesophilic acid phase and (b) thermophilic acid phase.
	Mesophilic	30–38 (85–100)	
	Thermophilic	50–57 (122–135)	
	pH		
	1st reactor	<6	
	2nd reactor	6.8–7.2	
	SRT (θ_c), d		
	1st/2nd reactor for mesophilic acid phase	1–3/> 10	
	1st/2nd reactor for thermophilic acid phase	1–2/> 10	
	VSL[a], kg VSS/m^3·d (lb VSS/ft^3·d)	4.8–6.4 (0.3–0.4)	
	Solids content, %		
	Feed sludge	5–6	
	VSR[c], %	50–60	
	Biosolids classification potential	Class A	

(*Continued*)

TABLE 13.13 (*Continued*) Important Design and Performance Parameters of Various Anaerobic Digestion Processes

Process with Brief Description	Design and Performance Parameter		Definition Sketch
	Parameter	Value or Range	
b. Temperate-phased anaerobic digestion (TPAnD) process Two digesters are operated in series at different temperatures. The first or second phases is either the thermophilic acidogenic anaerobic digester (TAAnD) or mesophilic acidogenic anaerobic digester (MAAnD) (Figure 13.12). At a high temperature the digestion rate is significantly higher, the SRT is much less, and the VSS destruction is 15–25% higher in comparison with that of a single-stage mesophilic high-rate digester. The process has capability of meeting Class A biosolids criteria under (a) Class A, Alt. 1 or (b) Class A, Alt. 6. The 2PAD[d] is an successful example for TPAnD process[e].[7,21,30–32,52]	Temperature, °C (°F)		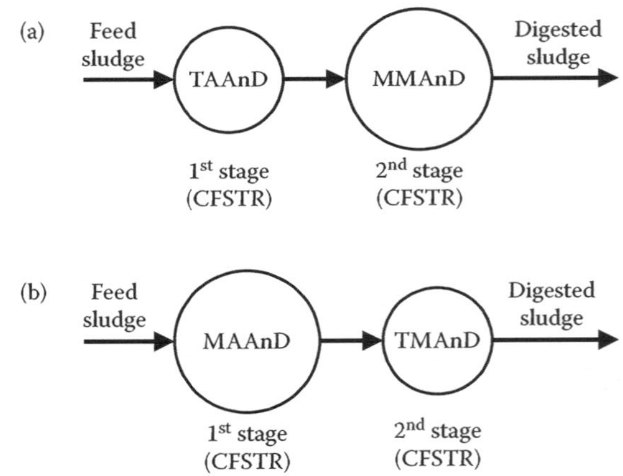 **FIGURE 13.12** Temperate-phased digestion (TPAnD) Process: (a) lead by a thermophilic acid phase and (b) lead by a mesophilic acid phase.
	Mesophilic	30–38 (85–100)	
	Thermophilic	50–57 (122–135)	
	pH		
	1st/2nd reactor	6.8–7.2	
	SRT (θ_c), d		
	1st/2nd reactor for thermophilic acid phase	3–5/7–15	
	1st/2nd reactor for mesophilic acid phase	7–10/> 5	
	VSL[a], kg VSS/m^3·d (lb VSS/ft^3·d)	4.8–6.4 (0.3–0.4)	
	Solids content, %		
	Feed sludge	4–6	
	VSR[c], %	50–65	
	Biosolids classification potential	Class A	

[a] VSL = volumetric solids loading rate
[b] PCCR = per capital based capacity requirement
[c] VSR = volatile solids reduction
[d] 2PAD = two-phased anaerobic digestion
[e] TPAnD = temperature-phased anaerobic digestion
1 kg/m^3·d = 0.0624 lb/ft^3·d.

A/GAnD = acid/gas phased digestion; AnD = anaerobic digester; CFSTR = continuous flow stirred tank reactor; MAAnD = mesophilic acidogenic anaerobic digester; MAnD = mesophilic anaerobic digester; MMAnD = mesophilic methanogenic anaerobic digester; SRT = solids retention time; TAAnD = thermophilic acidogenic anaerobic digester; TAnD = thermophilic anaerobic digester; and TMAnD = thermophilic methanogenic anaerobic digester.

Note: The 2PAD has been granted "PFRP Conditional National Equivalency" by the U.S. EPA for producing Class A biosolids (Class A, Alt. 1 or 6). In this system, the 1st stage is the thermophilic acidogenic anaerobic digestion with SRT ≥ 2.1 d at 55°C, and the 2nd stage is the mesophilic methanogenic anaerobic growth with SRT ≥ 10.5 d at 37°C.[21,31,32] PFRP is process to further reduce pathogens. See Sections 13.4.4 and 13.6.1, Tables 13.7, and Figure 13.2f, and Example 13.10 for additional information.

Source: Adapted in part from References 1, 3, 6, 7, 21, 30 through 32, 35, and 52.

Volatile Solids Reduction: The performance of anaerobic digester is also measured by the degree of destruction of volatile solids. Typically, the value of VSR depends on the SRT in the anaerobic digester. For a continuous flow completely-mixed, high-rate, anaerobic sludge digester, the VSR may be estimated from an empirical expression given by Equation 13.5a.[7]

$$VSR = 13.7\ln(\theta_c) + 18.9 \quad (13.5a)$$

where

VSR = volatile solids reduction, %

θ_c = solids retention time, d. The typical range is 15–20 d.

The VSR may also be calculated from Equations 13.5b through 13.5d. These equations are developed on mass balance basis using the VSS and TSS data on feed and digested sludges.[7] Calculation of VSR from different methods are shown in Example 13.22.

$$VSR = \frac{W_{VSS,FS} - (W_{VSS,DGS} + W_{VSS,SNT})}{W_{VSS,FS}} \times 100\% \quad (13.5b)$$

$$VSR = \frac{W_{VSS,FS} - W_{VSS,DGS}}{W_{VSS,FS}} \times 100\% \quad (W_{VSS,SPN} \approx 0) \quad (13.5c)$$

$$VSR = \frac{f_{VSS,FS} - f_{VSS,DGS}}{f_{VSS,FS}(1 - f_{VSS,DGS})} \times 100\% \quad (13.5d)$$

where

$W_{VSS,FS}$ = mass flow of VSS in the digester feed sludge, kg VSS/d

$W_{VSS,DGS}$ = mass flow of VSS in the digested sludge, kg VSS/d

$W_{VSS,SPN}$ = mass flow of VSS in the decanted supernatant from the digester, kg VSS/d

$f_{VSS,FS}$ = VSS/TSS ratio or weight fraction of volatile suspended solids in the digester feed on dry weight basis, g VSS/g TSS or dimensionless

$f_{VSS,DGS}$ = VSS/TSS ratio or weight fraction of volatile suspended solids in the digested sludge on dry weight basis, g VSS/g TSS or dimensionless

Other Design Considerations: Many other process and control factors need to be considered for design and operation of an anaerobic digester. These are: (1) tank geometry, (2) digester heating, (3) digester mixing, (4) digester covers, (5) biogas generation, collection, and utilization, (6) supernatant quality, and (7) codigestion. Each of these factors are discussed below.

Tank Geometry: The digesters are cylindrical or egg-shaped tanks. The cylindrical digesters are most common and range from 6 to 40 m (20–130 ft) in diameter. The floor is conical with slope of 1 vertical to 4–6 horizontal. The side water depth may range from 7.5 to 15 m (25–50 ft). The freeboard is at least 1.5 m (5 ft) to store scum buildup. Supernatant decanter pipes are provided for supernatant withdrawal if needed. Sufficient head space for gas collection and storage is also provided. The egg-shaped digesters when compared with cylindrical digesters provide the following advantages: (a) enhanced mixing efficiency and no dead zones, (b) little or no accumulation of grit, (c) reduced and better control of scum, (d) reduced cleaning frequency, (e) smaller land area, and (f) minimized foaming.

Digester Heating: Heat analysis is one of the most critical considerations in anaerobic sludge digester design. The energy required for heating the incoming sludge to reach the operating temperature is calculated from Equation 13.6a.[6,7]

$$H_{r,s} = C_{sh,s} W_s (T_d - T_s) \quad (13.6a)$$

where

$H_{r,s}$ = heat requirement for heating the sludge, J/d (Btu/d)

$C_{sh,s}$ = specific heat of the sludge, J/kg·°C (Btu/lb·°F). It is a common practice to assume that the specific heats of sludge and water are the same, $C_{sh,s} = C_{sh,w} = 4200$ J/kg·°C (1 Btu/lb·°F).

W_s = mass flow of wet feed sludge, kg/d (lb/d)
T_d = average digester operating temperature, (°F)
T_s = average temperature of the feed sludge, (°F)

The heat losses from digester sidewalls, bottom slab, and top cover are calculated from Equation 13.6b.[6,7] Example 13.23 shows these calculations.

$$H_{l,sf} = U_{sf}A(T_d - T_a) \tag{13.6b}$$

where

$H_{l,sf}$ = heat loss through the surface, J/s (Btu/h)
U_{sf} = overall heat transfer coefficient, J/s·m^2·°C (Btu/h·ft^2·°F).* The typical coefficients are: 0.68 J/s·m^2·°C (0.12 Btu/h·ft^2·°F) for floor, 1.02 J/s·m^2·°C (0.18 Btu/h·ft^2·°F) for concrete wall in dry earth, 1.42 J/s·m^2·°C (0.25 Btu/h·ft^2·°F) for concrete wall in wet earth, 1.99 J/s·m^2·°C (0.35 Btu/h·ft^2·°F) for concrete wall with air space insulation, 1.36 J/s·m^2·°C (0.24 Btu/h·ft^2·°F) for floating cover, and 2.84 J/s·m^2·°C (0.50 Btu/h·ft^2·°F) for fixed concrete roof.[32]
A = surface area through which the heat loss occurs, m^2 (ft^2)
T_a = average ambient temperature outside of the surface, (°F)

The mesophilic and thermophilic digesters are typically heated by recirculating part of the reactor sludge through external heat exchangers (see Figure 13.8). The external exchangers that are typically used in sludge digestion are: (a) tube-in-tube (or jacketed pipe) and (b) spiral-plate or water-bath. A heat exchanger design is given in Example 13.24.

Digester Mixing Devices: Sufficient mixing of digester contents is necessary to (a) maintain intimate contact between feed and active biomass, and mass and temperature uniformity throughout the digester, (b) disperse metabolic end products and toxic compounds entering the digester, and (c) prevent formation of surface scum. In anaerobic digesters, some natural mixing occurs due to rising of gas bubbles and thermal convection currents created by heated sludge entering the digester. The intensity of natural mixing is normally not sufficient. Therefore, additional mixing is needed for meeting the mixing requirements. The commonly used mixing methods are: (a) gas injection, (b) mechanical mixers, and (c) sludge circulations. A velocity gradient $G = 50–80\ s^{-1}$ and a digester content turnover time $T_{to} = 20–30$ min is typically required for all mixing systems. The types of mixing devices and typical design parameters for different mixing systems are summarized in Table 13.14. Schematics of different mixing devices are shown in Figure 13.13. Readers are referred to References, 1, 3, 6, 7, and 32 for detailed information on device schematics and design and advantages and disadvantages of these mixing systems.

Digester Covers: Properly designed digester covers are required to (a) prevent oxygen from entering, (b) contain odors, (c) collect and hold biogas, and (d) maintain desired operating temperature. The gas pressure under a digester cover is typically 0–3.7 kN/m^2 (0–15 in of water). The types of covers are: (a) *floating*, (b) *fixed*, and (c) *membrane*. These types of covers are shown in Figure 13.14. All three cover types are applicable to cylindrical tanks, while fixed cover is mostly used for egg-shaped digesters. Brief descriptions of each type of digester covers are presented below. Readers are referred to References 6 and 7 for design details, and information about typical appurtenances in digester covers.

1. The floating covers (Figure 13.14a) are more expensive but offer many advantages: (a) independent of sludge feed and withdraw, (b) reduce gas hazards and explosion, and (c) control the formation of scum mat. The covers float directly over liquid and have a maximum vertical travel of 2–3 m (6–10 ft). Floating cover designs are typically the Wiggins (Pontoon) type and the Downes type. A variation of the floating cover is the floating gas holder that has an extended skirt 3 m (10 ft) for additional gas storage when the biogas production is high.[6]
2. The fixed covers (Figure 13.14b) are less expensive but there are some operational problems. The raw sludge feed and digested sludge withdrawal must be precisely controlled. Rapid drop in liquid

* 1 J/s·m^2·°C = 0.176 Btu/h ft^2·°F.

TABLE 13.14 Types of Devices and Design Parameters for Digester Mixing Systems

Parameter	Design Information	
	Mixing Requirement	Types of Mixing Device
Gas injection		
Unconfined	0.0045–0.005 m^3/m^3·min (4.5–5 $ft^3/10^3$ ft^3·min)	• Cover-mounted lances (Figure 13.13a) • Bottom-mounted diffusers
Confined	0.005–0.007 m^3/m^3·min (5–7 $ft^3/10^3$ ft^3·min)	• Gas lifters (Figure 13.15b) • Gas bubble pistons (Figure 13.13b)
Mechanical mixing	0.005–0.008 kW/m^3 (0.2–0.3 hp/10^3 ft^3)	• Low-speed turbines (Figure 13.13c) • Low-speed mixers • Linear motion mixing devices (Figure 13.13d)
Hydraulic mixing[a]	Meeting the general requirements of $G = 50$–$80\ s^{-1}$ and $T_{to} = 20$–30 min	• Internal draft tubes (Figure 13.13e) • External draft tubes • Recirculation pumps with jet nozzles (Figure 13.13f)

1 m^3/m^3·min = 1000 $ft^3/10^3$ ft^3·min; 1 kW/m^3 = 38.0 hp/10^3 ft^3.
[a] Sludge recirculation may be used as an integrated part of both mixing and heating.
Source: Adapted in part from References 1, 3, 6, 7, and 32.

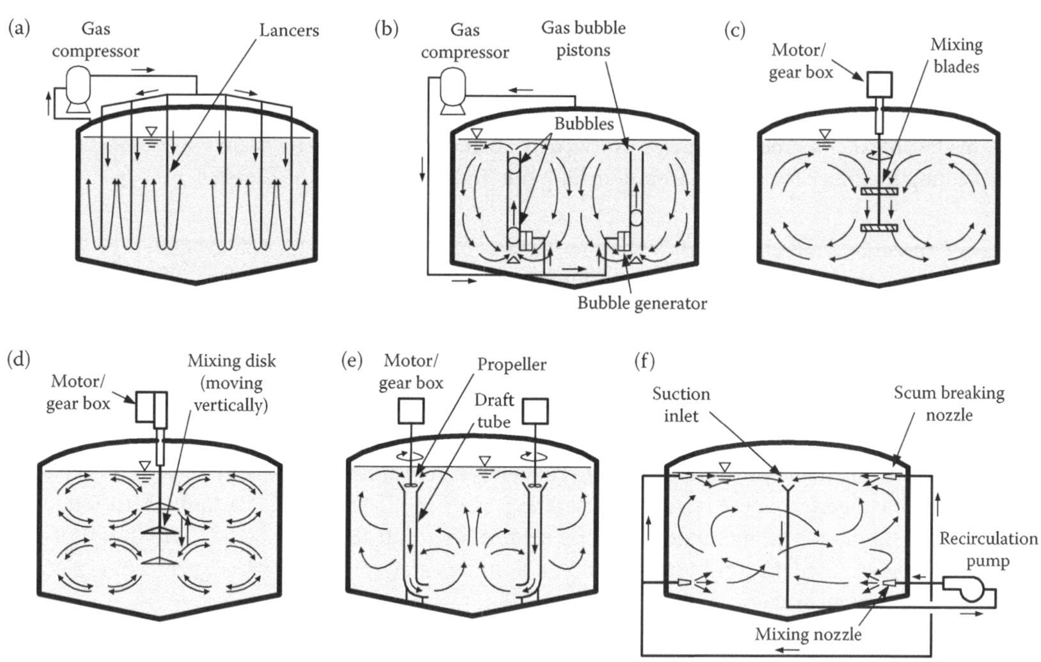

FIGURE 13.13 Schematics of mixing devices: (a) cover-mounted lances for unconfined gas injection, (b) gas pistons for confined gas injection, (c) low-speed turbine for mechanical mixing, (d) linear motion mixer for mechanical mixing, (e) hydraulic mixing with internal draft tubes for hydraulic mixing, and (f) recirculation pump with jet nozzles for hydraulic mixing.

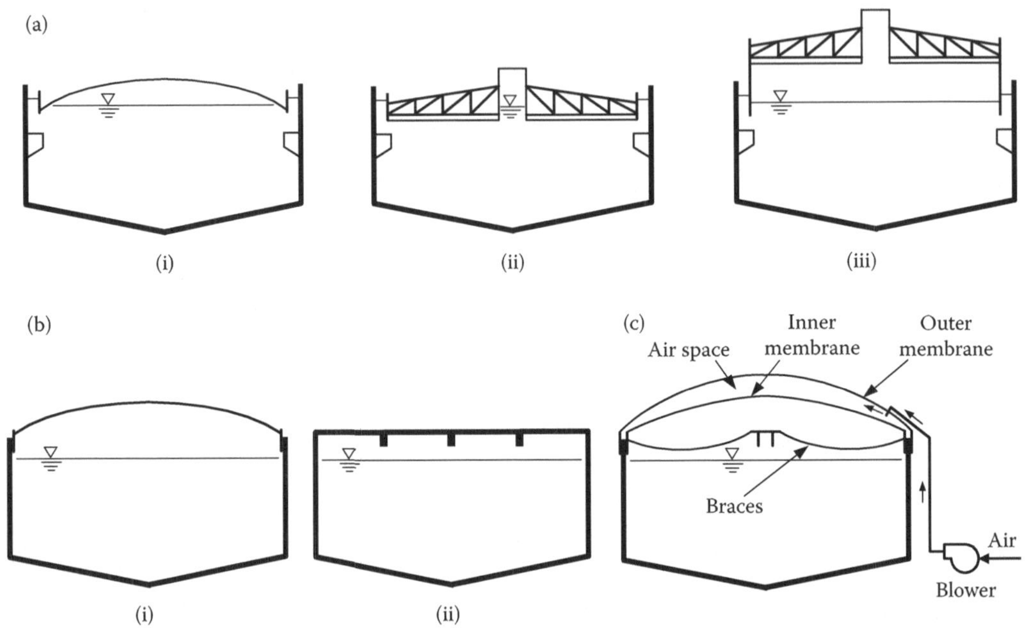

FIGURE 13.14 Sample digester covers: (a) floating types: (i) Wiggins, (ii) Downes, and (iii) gas holder; (b) fixed types: (i) domed, and (ii) flat slab; and (c) membrane type.

depth may draw air into the tank creating explosive mixture of biogas and oxygen since 5–15% CH_4 by volume in air is explosive. Also, rapid rise in liquid level may pressurize the tank, which may cause structural damage to the digester cover. Therefore, external storage tanks may be required for the digesters with fixed covers.[3,6,7]

3. The membrane cover or gas holder (Figure 13.14c) is a double-layer system with air space between the flexible membranes and are pressurized by an air blower. A brace system may be used for preventing the inner membrane from touching the sludge content in the digester.

Biogas Generation, Collection, and Utilization: The gas produced in an anaerobic digester is a valuable resource for energy recovery. Typically, the biogas contains ~65–70% of CH_4, 25–30% of CO_2, and trace amounts of H_2, N_2, H_2S, H_2O (vapor), and other gases. The density of digester gas is about 86% that of air (air density is ~1.162 kg/m^3 (0.07251 lb/ft^3)). The heating value of biogas is 21,000–24,000 kJ/m^3 (550–650 Btu/ft^3). For comparison, the heat contents of methane and natural gases are 35,800 and 37,300 kJ/m^3 (960 and 1000 Btu/ft^3), respectively.[6,7] The properties of methane gas are summarized in Table 10.45.

The volumetric methane generation rate (Q_M) may be estimated from the kinetic equations developed for the anaerobic digester. These relationships are expressed by Equations 10.78a through 10.78d.[3,6,7,32] Other rules of thumb for estimating digester gas volume are: (a) 0.75–1.12 m^3/kg (12–18 ft^3/lb) of VSS reduced, (b) 0.4–0.75 m^3/kg (8–12 ft^3/lb) of VSS loading, and (c) 0.03–0.04 m^3 per capita per day (1.1–1.4 ft^3 per capita per day).

The digester gas collection system includes digester covers, gas pipings and pressure relief valves, adequate flame traps, gas compressors, gas meters, and gas storage tank. The digester gas makes an explosive mixture with air, therefore, necessary safety precautions must be utilized to prevent explosion. Biogas flares are also provided to burn the excess biogas under normal, abnormal, and/or emergency conditions.

The biogas generated in anaerobic digester contains harmful impurities. Pretreatment is needed before utilization. Pretreatment mainly removes: (a) moisture, (b) H_2S, and (c) siloxanes (silicon containing

TABLE 13.15 Characteristics of Supernatant from Anaerobic Sludge Digestion

Parameter	Concentration, mg/L
TSS	2000–15,000
BOD_5	500–10,000
COD	1500–30,000
Ammonia nitrogen as N	400–1000
Total phosphorus as P	300–1000

Source: Adapted in part from References 1, 3, 6, and 32.

volatile organic compounds). The clean biogas is used for (a) cogeneration and (b) natural gas supply. At many POTWs, the biogas is used directly to heat the digesters, buildings/maintenance shops, and to fuel boilers and kilns to generate steam or heat. An increasing trend is to produce electricity using internal combustion engines, gas turbines, or micro-turbines for on-site use or off-site sale to the power grid. The overall energy efficiency can be further improved if a combined heat and power (CHP) system is also involved.* Electricity generated from biogas is considered renewable energy. It is also a cost-effective and sustainable approach to reduce greenhouse emissions.[7,53,54]

Digester Supernatant Quality: The supernatant from an anaerobic digester contains high levels of suspended and dissolved organic and inorganic solids. The concentration depends upon: (a) digestion period, (b) digestion temperature, (c) single- or two-stage digestion, (d) degree of mixing, and (e) method of solids separation from the liquor. These sidestreams are normally returned to the plant headworks for treatment. Return of these streams may significantly increase the loadings to the treatment facility. The characteristics of supernatant from an anaerobic digester treating thickened combined primary sludge and WAS are summarized in Table 13.15.

Codigestion: Traditionally, anaerobic digesters are designed to treat the thickened combined primary sludge and WAS produced on-site. However, there is an increasing trend of using the excess design capacity of an existing digester (typically, 15–30%) to codigest other organic wastes during the sustained low loading periods. The high strength organic substrates suitable for codigestion include: (a) sorted organic portion of municipal solid waste (MSW), (b) fats, oils and grease (FOG), and food waste collected from restaurants and meat processing industries, and (c) special industrial or agricultural organic wastes collected from beverage bottling facilities, cosmetic and beauty products, airport deicing operations, diaries, and others. It has been reported that codigestion benefits the digestion process. These benefits are: (a) more stable operation, (b) increased overall VSS reduction and gas production, (c) improved sludge dewatering, (d) reduced industrial discharges into collection systems that may upset the liquid treatment processes, (e) reduced carbon footprint of MSW management, and (f) extended operating life of MSW landfills.[7,55,56]

EXAMPLE 13.18: KINETIC COEFFICIENTS OF ANAEROBIC DIGESTION

A pilot plant study using a completely-mixing anaerobic batch reactor was conducted on sludge samples from a POTW. Determine the kinetic coefficients Y and k_d from the data given below:

Test Run	COD Uptake Rate of Sludge ($-\Delta S/\Delta t$), kg COD/d	Average Net Biomass Yield in the Digester ($\bar{X}$), kg VSS	Sludge Withdraw Rate from Digester ($\Delta X/\Delta t$), kg VSS/d
1	500	200	37
2	250	60	20

* The CHP system is an on-site electricity generation facility. It also converts the waste heat from power generation to other useful thermal energy, such as steam or hot water that can be used for other applications.

Solution

1. Identify the applicable equation.

   Microbial growth kinetics in a completely-mixed batch reactor is presented in Section 10.3.1. The relationship between $\frac{(-\Delta S/\Delta t)}{\bar{X}}$ and $\frac{(\Delta x/\Delta t)}{\bar{X}}$ is described by Equation 10.10a.

   $$\frac{1}{\bar{X}}\frac{\Delta X}{\Delta t} = Y\left(-\frac{1}{\bar{X}}\frac{\Delta S}{\Delta t}\right) - k_d$$

   Rearrange the equation to obtain the expression below.

   $$\frac{\Delta X}{\Delta t} = Y\left(-\frac{\Delta S}{\Delta t}\right) - k_d\bar{X}$$

2. Determine the kinetic coefficient Y and k_d from the given data.

   Substitute the experimental data into the above rearranged expression.

   $37\,\text{kg VSS/d} = Y \times 500\,\text{kg COD/d} - k_d \times 200\,\text{kg VSS}$ (use data from Test Run 1)

   $20\,\text{kg VSS/d} = Y \times 250\,\text{kg COD/d} - k_d \times 60\,\text{kg VSS}$ (use data from Test Run 2)

   Solve for Y and k_d.

   $Y = 0.09\,\text{kg VSS/kg COD}$ and $k_d = 0.04\,\text{d}^{-1}$

EXAMPLE 13.19: BIOMASS GROWTH AND BOD_L STABILIZATION KINETICS IN ANAEROBIC DIGESTION

Kinetic coefficients for the anaerobic digestion process are obtained from pilot test data in Example 13.18. Use these kinetic coefficients to determine: (a) the observed yield coefficient (Y_{obs}), (b) the active biomass growth (P_x), (c) COD stabilization rate (ΔS_M), and (d) observed COD stabilization efficiency (E_M) for an anaerobic sludge digester. The digester receives sludge solids that equivalent COD of 4500 kg/d. Assume that the SRT (θ_c) = 15 d and COD utilization efficiency $E = 0.75$.

Solution

1. Calculate the Y_{obs} from Equation 10.16d.

   $$Y_{obs} = \frac{Y}{1 + k_d\theta_c} = \frac{0.09\,\text{mg VSS/mg COD}}{1 + 0.04\,\text{d}^{-1} \times 15\,\text{d}} = 0.056\,\text{mg VSS/mg COD} = 0.056\,\text{kg VSS/kg COD}$$

2. Calculate the P_x from Equation 10.78d.

   $$P_x = Y_{obs}E\Delta S_0 = 0.056\,\text{kg VSS/kg COD} \times 0.75 \times 4500\,\text{kg COD/d} = 189\,\text{kg VSS/d}$$

3. Calculate the COD stabilization rate ΔS_M from Equation 10.78b.

   $$\Delta S_M = E\Delta S_0 - 1.42P_x = 0.75 \times 4500\,\text{kg COD/d} - 1.42\,\text{kg COD/kg VSS} \times 189\,\text{kg VSS/d} = 3110\,\text{kg COD/d}$$

4. Estimate the COD stabilization efficiency E_M from Equation 10.78e.

   $$E_M = \frac{\Delta S_M}{\Delta S_0} \times 100\% = \frac{3110\,\text{kg COD/d}}{4500\,\text{kg COD/d}} \times 100\% = 69\%$$

Note: The results indicate that about 69% of incoming COD loading is utilized for methane formation. The overall COD removal efficiency of the anaerobic sludge digester is 75%. The remaining incoming COD loading of 6% is used for synthesis of biomass.

EXAMPLE 13.20: DIGESTER CAPACITY FROM DIFFERENT METHODS

Calculate the digester capacity (V) based on (1) solids retention time (θ_c), (2) volumetric solids loading (VSL), (3) per capita based capacity requirement (PCCR), and (4) observed volume reduction. The thickened combined sludge flow $Q_{FS} = 150\ m^3/d$, total solids reaching the digester $W_{TSS} = 8200$ kg TSS/d, VSS/TSS ratio in the feed sludge $f_{VSS,FS} = 0.75$, $\theta_c = 15$ d, $VSL = 2.5$ kg VSS/m^3·d, $PCCR = 0.09\ m^3$ per capita, sludge solids (TSS) production $p_{TSS} = 120$ g TSS per capita per day, and average volume of digested sludge $Q_{DGS} = 132\ m^3/d$.

Solution

1. Compute the required digester volume without recycle V based on hydraulic retention time $(\theta) = \theta_c = 15$ d.

$$V = \text{Retention time} \times \text{Flow} = \theta Q = 15\,\text{d} \times 150\,\text{m}^3/\text{d} = 2250\,\text{m}^3$$

2. Compute the required V based on volumetric solids loading $VSL = 2.5$ kg VSS/m^3·d.
   The total volatile solids reaching the digester W_{VSS}.

$$W_{VSS} = f_{VSS,FS} \times \text{Weight of total solids} = f_{VSS,FS} \times W_{TSS}$$
$$= 0.75\,\text{kg VSS/kg TSS} \times 8200\,\text{kg TSS/d} = 6150\,\text{kg VSS/d}$$

$$V = \frac{\text{Weight of volatile solids}}{\text{Volumetric solids loading}} = \frac{W_{VSS}}{VSL} = \frac{6150\,\text{kg VSS/d}}{2.5\,\text{kg VSS/m}^3\cdot\text{d}} = 2460\,\text{m}^3$$

3. Compute the required V based on per capita capacity requirement.
   Estimate the population (P) served by the digester at $p_{TSS} = 120$ g TSS/capita·d = 0.12 kg TSS/capita·d.

$$P = \frac{W_{TSS}}{p_{TSS}} = \frac{8200\,\text{kg TSS/d}}{0.12\,\text{kg TSS/capita}\cdot\text{d}} = 68{,}300\ \text{persons}$$

$$V = \text{per capita capacity requirement (PCCR)} \times P = 0.09\,\text{m}^3/\text{capita} \times 68{,}300\ \text{persons} = 6150\,\text{m}^3$$

4. Compute the required V based on the observed volume reduction.
   During the digestion process, the volume of solids is slightly reduced and supernatant is withdrawn. Therefore, the volume remaining in the digester decreases exponentially. The digester volume may be approximated by Equation 13.7 based on the observed volume reduction.[1,3,32]

$$V = \theta_c\left[Q_{FS} - \frac{2}{3}(Q_{FS} - Q_{DGS})\right] \tag{13.7}$$

where
- V = required digester volume, m^3 (ft^3)
- Q_{FS} = sludge feed rate into the digester, m^3/d (ft^3/d)
- Q_{DGS} = sludge withdrawal rate from the digester, m^3/d (ft^3/d)
- θ_c = solids retention time, d. $\theta_c \approx \theta$ when the digester supernatant volume is relatively small.

Calculate required V from Equation 13.7.

$$V = 15\text{d} \times \left[150\,\text{m}^3/\text{d} - \frac{2}{3} \times ((150 - 132)\,\text{m}^3/\text{d})\right] = 2070\,\text{m}^3$$

5. Select the digester capacity
   The digester capacity calculated from four methods is in the range from 2070 m^3 to 6150 m^3. The digester capacity based on the observed volume reduction is the smallest among four values. This may be the minimum volume required for the digester. The volume based on per capita capacity requirement is the largest. Normally, this method is conservative and may be used when the solids loading data are not reliable. The volumes determined from SRT (θ_c) and VSL are close. Select an active digester capacity of 2500 m^3 which provides two digesters each with an active capacity of 1250 m^3.

EXAMPLE 13.21: DIMENSIONS OF AN ANAEROBIC DIGESTER

The design active capacity of 1250 m^3 is obtained for each anaerobic digester in Example 13.20. Develop the digester geometry, dimensions, and conceptual plan and sectional views. Provide excess volumes for grit accumulation at the bottom, scum collection near the liquid level, and clearance for floating cover. Also, check for the active volume ratio (AVR), digestion period (θ), and volumetric solids loading (VSL). The sludge flow and organic loading to each digester are $Q_{\text{FS,digester}} = 75\,\text{m}^3/\text{d}$ and $W_{\text{VSS,digester}} = 3075\,\text{kg VSS/d}$ (see Example 13.20).

Solution

1. Develop the digester geometry.
   a. Provide a cylindrical digester with a bottom cone. The floor of the digester is sloped at 1 vertical (V) to 3 horizontal (H).
   b. Provide a 1-m depth (H_{grit}) for the grit accumulation in the bottom of cone.
   c. Provide a 0.6-m (2 ft) water depth for the scum blanket (H_{scum}) below the maximum sludge surface level.
   d. Provide an active side water depth of $H_{\text{act,swd}} = 8.5$ m (~28 ft) below the scum layer.
   e. Provide a 0.6-m (2 ft) additional clearance space (H_{space}) between the floating cover and the maximum sludge surface level.
   f. Side water depth of the digester, $H_{\text{swd}} = H_{\text{act,swd}} + H_{\text{scum}} = (8.5 + 0.6)\text{ m} = 9.1\text{ m}$ (~30 ft) below the maximum sludge surface level.
   g. Total height of side wall of the cylindrical portion of the digester, $H_{\text{sw}} = H_{\text{swd}} + H_{\text{space}} = (9.1 + 0.6)\text{ m} = 9.7\text{ m}$ (~32 ft).
2. Determine the digester diameter.
   Estimate the surface area of the digester required at $H_{\text{act,swd}} = 8.5$ m.

$$A = \frac{\text{Volume of digester}}{\text{Active SWD}} = \frac{V}{H_{\text{act,swd}}} = \frac{1250\,\text{m}^3}{8.5\,\text{m}} = 147\,\text{m}^2$$

$$\text{Required digester diameter, } D = \sqrt{\frac{4}{\pi} \times (\text{Digester area})} = \sqrt{\frac{4}{\pi}A} = \sqrt{\frac{4}{\pi} \times 147\,\text{m}^2} = 13.7\,\text{m (45 ft)}$$

It is assumed that the digester of standard floating cover is commercially available at a 5-ft (~ 1.5 m) increment. Therefore, select a digester diameter $D = 45$ ft (13.7 m).

3. Determine the depth of bottom cone.

Depth of cone at a slope of 1 V to 3 H, $H_{cone} = \frac{V}{H} \times \frac{\text{Digester diameter}}{2} = \frac{V}{H} \times \frac{D}{2} = \frac{1}{3} \times \frac{13.7\,\text{m}}{2} = 2.3\,\text{m}$

Total height of the digester, H_{total} = Height of side wall + Depth of cone $= H_{sw} + H_{cone} = (9.7 + 2.3)\,\text{m}$
$= 12.0\,\text{m}$

4. Calculate the total active volume available in the digester.

Allowance for grit accumulation is given at the bottom of the cone. Most grit accumulation will occur at the sludge withdrawal well and away from the diffusers if they are used for mixing purpose.

Cone diameter for grit accumulation, $D_{grit} = 2 \times \frac{H}{V} \times (\text{Depth of grit}) = 2 \times \frac{H}{V} \times (H_{grit})$
$= 2 \times \frac{3}{1} \times 1\,\text{m} = 6\,\text{m}$

Volume provided for grit accumulation,

$$V_{grit} = \frac{\pi}{12}(\text{Diameter})^2 \times \text{Height} = \frac{\pi}{12}(D_{grit})^2 H_{grit} = \frac{\pi}{12} \times (6\,\text{m})^2 \times 1\,\text{m} = 9.4\,\text{m}^3$$

Total volume of the cone, $V_{cone} = \frac{\pi}{12}(D)^2 H_{cone} = \frac{\pi}{12} \times (13.7\,\text{m})^2 \times 2.3\,\text{m} = 113\,\text{m}^3$

Active volume available in the cone, $V_{act,cone}$ = Cone volume − Grit voume $= V_{cone} - V_{grit}$
$= (113 - 9.4)\,\text{m}^3 = 104\,\text{m}^3$

Active volume available in the cylindrical portion, $V_{act,cyl} = \frac{\pi}{4}(\text{Diameter})^2 \times \text{Height} = \frac{\pi}{4}(D)^2 H_{act,swd}$
$= \frac{\pi}{4} \times (13.7\,\text{m})^2 \times 8.5\,\text{m} = 1250\,\text{m}^3$

Total active volume available in the digester, V_{act} = Active cone volume + Active cylinder volume
$= V_{act,cyl} + V_{act,cone} = (1250 + 104)\,\text{m}^3 = 1354\,\text{m}^3$

Note: The available active volume of 1354 m^3 is about 8% larger than the required active capacity of 1250 m^3.

5. Check the active volume ratio, AVR.

Total volume provided in the cylindrical portion including scum layer,

$$V_{cyl} = \frac{\pi}{4}(\text{Diameter})^2 \times \text{Side water depth} = \frac{\pi}{4}(D)^2 H_{swd} = \frac{\pi}{4} \times (13.7\,\text{m})^2 \times 9.1\,\text{m} = 1340\,\text{m}^3$$

Total volume provided in the digester, $V_{digester}$ = Total cylinder volume + Total cone volume
$= V_{cyl} + V_{cone} = (1340 + 113)\,\text{m}^3 = 1453\,\text{m}^3$

$$\text{AVR provided in the digester, AVR} = \frac{\text{Active volume}}{\text{Total volume}} = \frac{V_{act}}{V_{digester}} = \frac{1354\,\text{m}^3}{1453\,\text{m}^3} = 0.93$$

Note: The AVR of 0.93 provided in the digester is above 0.85. It is considered efficient utilization of digester volume.

6. Check the SRT (θ_c) or the digestion period (θ).

$$\text{SRT achieved in the digester, } \theta_c \approx \theta = \frac{\text{Active volume}}{\text{Feed sludge flow}} = \frac{V_{act}}{Q_{FS,digester}} = \frac{1354\,\text{m}^3}{75\,\text{m}^3/\text{d}} = 18\text{ d}$$

Note: The actual digestion period is >15 d (given in Example 13.20) and within the typical range from 15to 20 d for the high-rate digestion process (see Table 13.13).

7. Check the volumetric solids loading, VSL.

$$\text{VSL maintained in the digester, VSL} = \frac{\text{Weight of feed volatile solids}}{\text{Active volume}} = \frac{W_{\text{VSS,digester}}}{V_{\text{act}}}$$

$$= \frac{3075\ \text{kg VSS/d}}{1354\ \text{m}^3} = 2.3\ \text{kg VSS/m}^3{\cdot}\text{d}$$

Note: The actual VSL is lower than the design VSL of 2.5 kg VSS/m^3·d (given in Example 13.20) and within the typical range from 1.6 to 4.8 kg VSS/m^3·d for the high-rate digestion process (see Table 13.13).

8. Draw the conceptual views and describe briefly the major components of digester.

The conceptual plan and sectional views of the digester are provided in Figure 13.15. Each anaerobic digester is mixed by a gas lift mixer that receives compressed biogas from a gas compressor. The digester content is recirculated through a heat exchanger to maintain the sludge temperature within the desired mesophilic range. The feed sludge is added into the sludge recirculation line prior to the heat exchanger. The digested sludge (biosolids) is withdrawn from the reactor for dewatering. Design details of sludge heating and exchanger, and gas mixing and storage, are given in several examples (see Examples 13.23, 13.24, 13.26, and 13.27). Readers are referred to References 6 and 7 for more design details and illustrations of anaerobic digester.

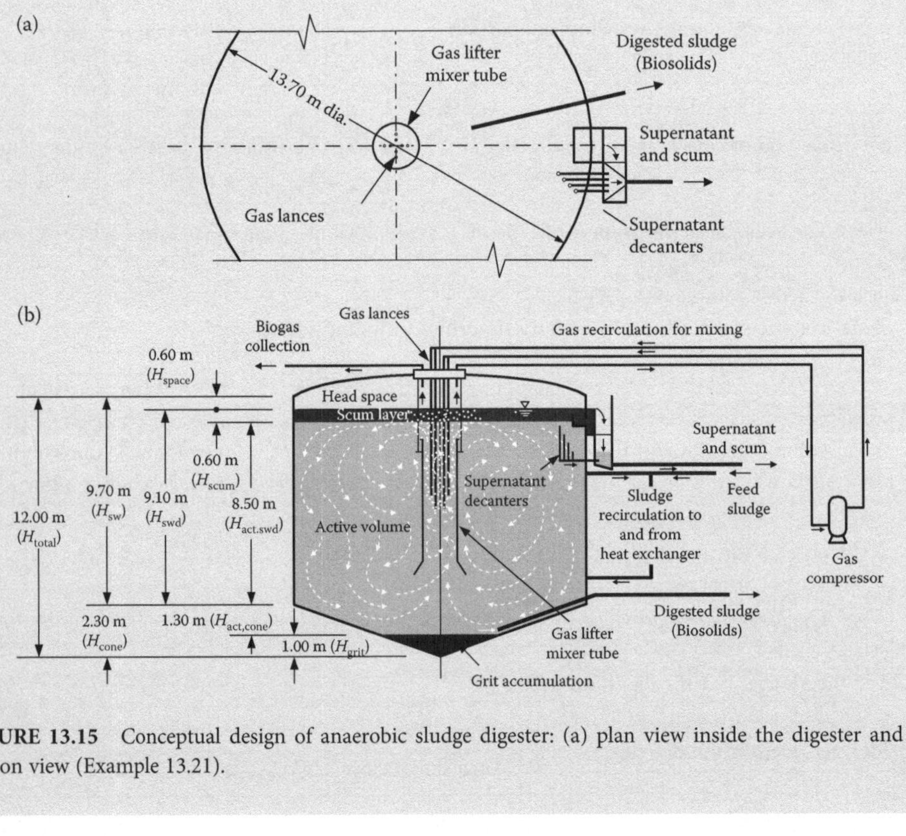

FIGURE 13.15 Conceptual design of anaerobic sludge digester: (a) plan view inside the digester and (b) section view (Example 13.21).

EXAMPLE 13.22: VOLATILE SOLIDS REDUCTION IN RAW AND DIGESTED SLUDGE

A high-rate digester operates at an average SRT (θ_c) of 15 d. The historical operating data were used to obtain the average VSS/TSS ratios of 0.75 and 0.57 for feed and digested sludges, respectively. Determine the percent volatile solids reduction (VSR) in the anaerobic digester based on (a) digester SRT, (b) mass balance analysis, and (c) simplified equations using the VSS/TSS ratios. Estimate also the total solids reduction (TSR). Assume that the weight of fixed solids in the feed and digested sludge remains unchanged. Only the volatile solids portion in the feed sludge undergoes reduction during the digestion process.

Solution

1. Determine the percent *VSR* based on the SRT in digester from Equation 13.5a.

$$VSR = 13.7\ln(\theta_c) + 18.9 = 13.7 \times \ln(15\,\text{d}) + 18.9 = 56\%$$

2. Determine the *VSR* from mass balance analysis.
   Assume that the mass rate of feed sludge $W_{\text{TSS.FS}} = 1$ kg/d.
   a. Determine the mass of volatile and fixed solids in the feed sludge.

$$\text{Mass of volatile solids, } W_{\text{VSS, FS}} = f_{\text{VSS, FS}} W_{\text{TSS, FS}} = 0.75\,\text{kg VSS/kg TSS} \times 1\,\text{kg TSS/d}$$
$$= 0.75\,\text{kg VSS/d}$$

$$\text{Mass of fixed solids, } W_{\text{FSS, FS}} = W_{\text{TSS, FS}} - W_{\text{TSS, FS}} = 1\,\text{kg TSS/d} - 0.75\,\text{kg VSS/d} = 0.25\,\text{kg FSS/d}$$

   b. Determine the mass of volatile solids in the digested sludge.
   It is given that the weight of fixed solids in the feed and digested sludge remains unchanged.
   Mass of fixed solids in the digested sludge, $W_{\text{FSS, DGS}} = W_{\text{FSS, FS}} = 0.25$ kg FSS/d
   Determine the FSS/TSS ratio or weight fraction of fixed solids in the digested sludge.

$$f_{\text{FSS, DGS}} = 1 - f_{\text{VSS, DGS}} = 1 - 0.57 = 0.43\,\text{kg FSS/kg TSS}$$

$$\text{Mass of TSS in the digested sludge, } W_{\text{TSS, DGS}} = \frac{W_{\text{FSS, DGS}}}{f_{\text{FSS, DGS}}} = \frac{0.25\,\text{kg FSS/d}}{0.43\,\text{kg FSS/kg TSS}} = 0.58\,\text{kg TSS/d}$$

   Mass of VSS in the digested sludge,

$$W_{\text{VSS, DGS}} = W_{\text{TSS, DGS}} - W_{\text{FSS, DGS}} = 0.58\,\text{kg TSS/d} - 0.25\,\text{kg/FSS/d} = 0.33\,\text{kg VSS/d}$$

   c. Determine the *VSR* from Equation 13.5b.

$$VSR = \frac{W_{\text{VSS, FS}} - W_{\text{VSS, DGS}}}{W_{\text{VSS, FS}}} \times 100\% = \frac{(0.75 - 0.33)\,\text{kg VSS/d}}{0.75\,\text{kg VSS/d}} \times 100\% = 56\%$$

3. Determine the *VSR* from Equation 13.5d.

$$VSR = \frac{f_{\text{VSS, FS}} - f_{\text{VSS, DGS}}}{f_{\text{VSS, FS}}(1 - f_{\text{VSS, DGS}})} \times 100\% = \frac{(0.75 - 0.57)}{0.75 \times (1 - 0.57)} \times 100\% = 56\%$$

Note: Same value of VSR should be obtained from three different methods when the reliable sludge data are available.

4. Determine the total solids reduction (*TSR*) from the calculated data.

$$TSR = \frac{W_{\text{TSS, FS}} - W_{\text{TSS, DGS}}}{W_{\text{TSS, FS}}} \times 100\% = \frac{(1 - 0.58)\,\text{kg TSS/d}}{1\,\text{kg TSS/d}} \times 100\% = 42\%$$

EXAMPLE 13.23: HEATING REQUIREMENTS FOR DIGESTER FEED AND RADIATION LOSSES

Two high-rate, anaerobic digesters are operating. The digester gas is used to heat the feed sludge and both digesters to maintain the mesophilic operating temperature of 35°C. The total thickened sludge pumped to the digesters under sustained loading condition is 150 m^3/d. The bulk density of sludge is 1010 kg/m^3, and the critical sludge temperature during winter is 12°C. Each digester is 13.7 m in diameter, and the height of side wall is 9.7 m. The height of side wall below ground is 5 m. The digester domed roof is elevated at a vertical rise of 0.5 m at the center. The digester floor is sloping with a total drop of 2.3 m at the center. The overall heat transfer coefficients U_{sf} as J/s·m^2·°C of various components of the digester are: fixed concrete roof $U_{sf} = 2.48$, exposed concrete walls with insulation $U_{sf} = 1.99$, concrete walls buried by dry earth $U_{sf} = 1.02$, and digester concrete bottom $U_{sf} = 0.68$. Compute total digester gas requirement to raise the temperature of incoming sludge and to maintain the desired digester operating temperature of 35°C. Assume that the critical sustained low temperatures surrounding the digesters are: −2°C for ambient air, 3°C for earth outside buried wall, and 6°C for earth below the floor. The groundwater table is below the slab of the digester.

Solution

1. Draw the schematic of digester heating system.

   The schematic process diagram of digester heating system is shown in Figure 13.16. Each digester has a heat exchanger which can also serve the other digester under emergency situation. The incoming feed sludge is metered and added into the recirculation line prior to the heat exchanger. A third heat exchanger may also be added as a redundant unit serving both digesters. The redundant unit is not shown in Figure 13.16. See Example 13.24 for design of the heat exchangers.

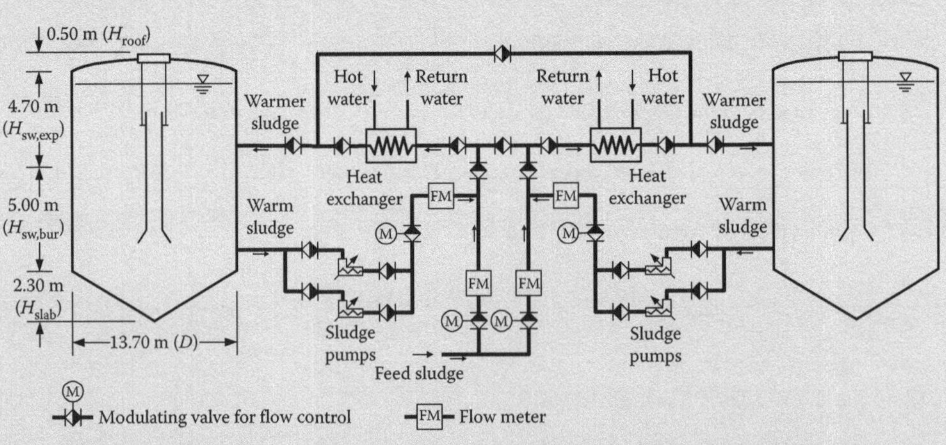

FIGURE 13.16 Schematic flow diagram of sludge heating and recirculation system (Example 13.23).

2. Calculate the heat requirement for incoming feed sludge to two digesters ($H_{r,s,adj}$).

   Mass of feed wet sludge, W_s = Density × Flow = $\rho_s Q_s = 1010\,\text{kg/m}^3 \times 150\,\text{m}^3/\text{d} = 151{,}500\,\text{kg/d}$

The total heat required to heat the feed sludge to the operating temperature $T_d = 35°C$ is calculated from Equation 13.6a at $T_s = 12°C$.

$$H_{r,s} = C_{sh,s} W_s (T_d - T_s) = 4200\,\text{J/kg·°C} \times 151{,}500\,\text{kg/d} \times (35 - 12)°\text{C}$$
$$= 1.46 \times 10^{10}\,\text{J/d} \quad \text{or} \quad 1.46 \times 10^4\,\text{MJ (megajoule)/d}$$

Provide additional 20% heat loss for uncertainties, and 25% heat loss for extreme situations. The adjusted total heat requirement for entire incoming feed sludge to two digesters is calculated below,

$$H_{r,s,adj} = (1 + 0.2 + 0.25) H_{r,s} = 1.45 \times 1.46 \times 10^{10}\,\text{J/d} = 2.12 \times 10^{10}\,\text{J/d} \quad \text{or} \quad 2.12 \times 10^4\,\text{MJ/d}$$

3. Compute the surface areas of each digester for heat loss.
   The heat loss occurs from the roof, side wall exposed to air, buried side wall, and bottom slab.
   a. Surface area of the doomed roof (A_{roof}).
      In each digester, the doomed roof surface area is calculated using the equation for spherical lateral surface area.

$$A_{roof} = \pi \times \left(\left(\frac{\text{Diameter of digester}}{2} \right)^2 + (\text{Height of roof})^2 \right)$$
$$= \pi \left(\left(\frac{D}{2} \right)^2 + (H_{roof})^2 \right) = \pi \left(\left(\frac{13.7\,\text{m}}{2} \right)^2 + (0.5)^2 \right) = 148\,\text{m}^2$$

   b. Surface area of the exposed side wall ($A_{sw,exp}$) at a height $H_{sw,epx} = 4.7$ m.

$$A_{sw,exp} = \pi D H_{sw,exp} = \pi \times 13.7\,\text{m} \times 4.7\,\text{m} = 202\,\text{m}^2$$

   c. Surface area of buried side wall ($A_{sw,bur}$) at a height $H_{sw,bur} = 5$ m.

$$A_{sw,bur} = \pi D H_{sw,bur} = \pi \times 13.7\,\text{m} \times 5.0\,\text{m} = 215\,\text{m}^2$$

   d. Surface area of bottom slab ($A_{sw,slab}$).
      The surface area is calculated using the equation for lateral surface area of an inverted cone.

$$\text{Lateral length of slab, } L_{slab} = \sqrt{\left(\frac{\text{Diameter of digester}}{2} \right)^2 + (\text{Height of cone})^2}$$
$$= \sqrt{\left(\frac{D}{2} \right)^2 + (H_{slab})^2} = \sqrt{\left(\frac{13.7\,\text{m}}{2} \right)^2 + (2.3\,\text{m})^2} = 7.23\,\text{m}$$

$$A_{slab} = \pi \frac{D}{2} L_{slab} = \pi \times \left(\frac{13.7\,\text{m}}{2} \right) \times 7.23 = 156\,\text{m}^2$$

4. Compute from Equation 13.6b the total heat loss from two digesters.
   a. Heat loss from the roof using $U_{sf,roof} = 2.84\,\text{J/s·m}^2\text{·°C}$ for fixed concrete roof at $T_{air} = -2°C$.

$$H_{l,sf,roof} = U_{sf,roof} A_{roof} (T_d - T_{air}) = 2.48\,\text{J/s·m}^2\text{·°C} \times 148\,\text{m}^2 \times (35 - (-2))°\text{C} \times 86{,}400\,\text{s/d}$$
$$= 1.17 \times 10^9\,\text{J/d}$$

b. Heat loss from the exposed wall using $U_{sf,sw,exp} = 1.99\ J/s{\cdot}m^2{\cdot}°C$ for concrete wall with air space insulation at $T_{air} = -2°C$.

$$H_{l,sf,sw,exp} = U_{sf,sw,exp}A_{sw,exp}(T_d - T_{air}) = 1.99\ J/s{\cdot}m^2{\cdot}°C \times 202\ m^2 \times (35 - (-2))°C \times 86{,}400\ s/d$$
$$= 1.29 \times 10^9\ J/d$$

c. Heat loss from the buried wall using $U_{sf,sw,bur} = 1.02\ J/s{\cdot}m^2{\cdot}°C$ for concrete wall in dry earth at $T_{earth,sw} = 3°C$.

$$H_{l,sf,sw,bur} = U_{sf,sw,bur}A_{sw,bur}(T_d - T_{earth,sw}) = 1.02\ J/s{\cdot}m^2{\cdot}°C \times 215\ m^2 \times (35 - 3)°C \times 86{,}400\ s/d$$
$$= 6.06 \times 10^8\ J/d$$

d. Heat loss from the bottom slab using $U_{sf,slab} = 0.68\ J/s{\cdot}m^2{\cdot}°C$ for concrete slab at $T_{earth,slab} = 6°C$.

$$H_{l,sf,slab} = U_{sf,slab}A_{sw,slab}(T_d - T_{earth,slab}) = 0.68\ J/s{\cdot}m^2{\cdot}°C \times 156\ m^2 \times (35 - 6)°C \times 86{,}400\ s/d$$
$$= 2.66 \times 10^8\ J/d$$

e. Total heat loss from each digester ($H_{l,sf,d}$).

$$H_{l,sf,d} = H_{l,sf,roof} + H_{l,sf,sw,exp} + H_{l,sf,sw,bur} + H_{l,sf,slab} = (11.7 + 12.9 + 6.06 + 2.66) \times 10^8\ J/d$$
$$= 3.33 \times 10^9\ J/d$$

Provide additional 20% heat losses for uncertainties and 25% heat loss for extreme situation. The adjusted total heat loss from both digesters.

$$H_{l,sf,adj} = 2 \times (1 + 0.2 + 0.25)H_{l,sf,d} = 2 \times 1.45 \times 3.33 \times 10^9\ J/d = 9.66 \times 10^9\ J/d$$

5. Compute the total heat requirement (H_r) to heat the incoming sludge and compensate the heat losses in two digesters.

$$H_r = \text{Adjusted total heat requirement for heating sludge} + \text{Adjusted total heat loss}$$
$$= H_{r,s,adj} + H_{l,sf,adj} = (21.2 + 9.66) \times 10^9\ J/d = 3.09 \times 10^{10}\ J/d \quad \text{or} \quad 3.09 \times 10^4\ MJ/d$$

$$\text{or} \quad H_r = 3.09 \times 10^{10}\ J/d \times 10^{-3}\ kJ/J \times \frac{d}{24\ h} = 1.29 \times 10^6\ kJ/h$$

$$\text{or} \quad H_r = 1.29 \times 10^6\ kJ/h \times \frac{kW}{3600\ kJ/h} = 358\ kW$$

Note: About 31% of the total heat requirement is caused by the heat losses at the sustained low temperature during winter.

6. Determine the digester gas required to meet the heating demand in two digesters ($Q_{r,biogas}$).
Assume that the heating value of biogas with about 65% CH_4 is 24,000 kJ/m^3 or 24 MJ/m^3 and calculate the digester gas requirement assuming the overall efficiency of heating units is 60%.

$$Q_{r,biogas} = \frac{\text{Total heat requirement}}{\text{Heat value of biogas} \times \text{Heating efficiency}} = \frac{H_r}{24\ MJ/m^3 \times 0.6} = \frac{3.09 \times 10^4\ MJ/d}{24\ MJ/m^3 \times 0.6}$$
$$= 2150\ m^3/d$$

EXAMPLE 13.24: DESIGN OF HEAT EXCHANGER FOR DIGESTER HEATING

Two external-jacket heat exchangers are provided for heating anaerobic digesters shown in Figure 13.16 and designed in Example 13.23. Each unit is sized for heating the recirculated sludge and feed sludge to one digester. Determine (a) hot water recirculation rate through the jacket, (b) sludge recirculation flow and mass rate, (c) temperature rise in the heat exchanger jacket, and (d) surface area and length of sludge pipe in the heat exchanger jacket. To avoid sludge caking inside the sludge pipe in the jacket, the temperatures of water entering and leaving the jacket are 65°C and 45°C, respectively. Assume that the heat transfer coefficient of external water jacketed heat exchanger $U_{hex} = 4000$ kJ/h·m²·°C (196 Btu/ft²·h·°F), specific heat of water $C_{sh,w} = 4200$ J/kg·°C, and heat transfer efficiency $E_{hex} = 0.8$. The sludge recirculating pump flow is such that a recirculation velocity of 1 m/s is maintained in the 20-cm (8-in) sludge recirculation lines. The density of recirculated sludge is 1010 kg/m³. The feed sludge flow and mass are 75 m³/d and 75,750 kg/d for each digester.

Solution

1. Calculate the hot water recirculation rate through the external heat exchanger.

   The hot water is pumped countercurrent to the sludge recirculation pipe (Figure 13.16). The total heat supplied by 1 kg recirculated hot water through the jacket.

$$h_{hw} = C_{sh,w}(T_{hw,in} - T_{hw,out}) = 4200\,\text{J/kg·°C} \times (65 - 45)\text{°C} = 8.40 \times 10^4\,\text{J/kg water}$$

   Total heat required to heat the incoming and recirculating sludge from both digesters, $H_r = 3.09 \times 10^{10}$ J/d (from Example 13.23, Step 5). One jacketed pipe heat exchanger is sized for each digester. Calculate the heat output requirement for each heat exchanger serving one digester.

$$H_{hex} = \frac{H_r}{2} = \frac{3.09 \times 10^{10}\,\text{J/d}}{2} = 1.55 \times 10^{10}\,\text{J/d}$$

   Calculate the hot water recirculation rate through each heat exchanger at $E_{hex} = 0.8$.

$$Q_{hw} = \frac{H_{hex}}{h_{hw}E_{hex}} = \frac{1.55 \times 10^{10}\,\text{J/d}}{8.40 \times 10^4\,\text{J/kg} \times 10^3\,\text{kg/m}^3 \times 0.8} = 230\,\text{m}^3\text{/d} \quad \text{or} \quad 160\,\text{L/min (42 gpm)}$$

   Note: This is the flow rate required for each heat exchanger at the sustained low temperature during winter. A lower flow rate will be required under warmer weather conditions. Therefore, hot water recirculation pump with variable frequency drive (VFD) is required. The flow rate from the pump will be adjusted by the digester automatic control system to provide the actual heating requirements.

2. Calculate the sludge recirculation flow and mass rate.

$$\text{Sludge recirculation pump flow rate, } Q_{rs} = \frac{\pi}{4}(d_{rs})^2 v_{rs} = \frac{\pi}{4} \times (0.2\,\text{m})^2 \times 1\,\text{m/s} \times 86{,}400\,\text{s/d} = 2710\,\text{m}^3\text{/d}$$

$$\text{Mass of recirculated sludge, } W_{rs} = \rho_{rs}Q_{rs} = 1010\,\text{kg/m}^3 \times 2710\,\text{m}^3\text{/d} = 2.74 \times 10^6\,\text{kg/d}$$

   Note: The feed sludge flow of 75 m³/d is much smaller than the recirculated sludge flow.

3. Calculate the required temperature rise in the heat exchanger.

   The recirculated sludge flow is designed to compensate for the heat losses in one digester. The total heat losses in the both digesters $H_{l,sf,adj} = 9.66 \times 10^9$ J/d (from Example 13.23, Step 4). Calculate the required heat supply in the recirculated sludge flow for each digester:

$$H_{rs} = \frac{H_{l,sf,adj}}{2} = \frac{9.66 \times 10^9\,\text{J/d}}{2} = 4.83 \times 10^9\,\text{J/d}$$

Rearrange Equation 13.6a to calculate the temperature rise required after the heat exchanger ($T_{s,out}$) to provide the required heat supply in the recirculated sludge.

$$T_{s,out} = T_d + \frac{H_{rs}}{C_{sh,s} W_{rs}} = 35°C + \frac{4.83 \times 10^9\ J/d}{4200\ J/kg{\cdot}°C \times 2.30 \times 10^6\ kg/d} = 35.4°C$$

The required temperature rise through the heat exchanger is 0.4°C. It is within the typical allowable temperature variation range of ($\pm$0.5°C) for optimum performance of methane forming bacteria. Select a design sludge temperature $T_{s,out} = 35.5°C$ at the outlet of the heat exchanger.

4. Calculate the required radiation surface area and length of sludge pipe in the heat exchanger jacket.

Average temperature of circulating water through the jacket,

$$T_{hw,avg} = \frac{T_{hw,in} + T_{hw,out}}{2} = \frac{(65+45)°C}{2} = 55°C$$

Calculate the average temperature of the combined sludge after mixing the feed and recirculated sludge flows prior to the heat exchanger.

$$T_{s,in} = \frac{W_{fs}T_s + W_{rs}T_d}{W_{fs} + W_{rs}} = \frac{75{,}750\ kg/d \times 12°C + 2{,}740{,}000\ kg/d \times 35°C}{(75{,}750 + 2{,}740{,}000)\ kg/d} = 34.4°C$$

Average temperature of sludge through the pipe in the heat exchanger jacket,

$$T_{s,avg} = \frac{T_{s,in} + T_{s,out}}{2} = \frac{(34.4 + 35.5)°C}{2} = 34.95°C \approx 35°C$$

Rearrange Equation 13.6b to calculate the radiation surface area required to meet the heat requirement $H_{hex} = 1.55 \times 10^{10}$ J/d at $U_{hex} = 4000$ kJ/h·m²·°C and $E_{hex} = 0.8$.

$$A_{s,pipe} = \frac{H_{hex}}{U_{hex}(T_{w,avg} - T_{fs,avg})E_{hex}}$$

$$= \frac{1.55 \times 10^{10}\ J/d}{4000\ kJ/h{\cdot}m^2{\cdot}°C \times 24\ h/d \times 10^3 \times J/kJ \times (55-35)°C \times 0.8}$$

$$= 10.1\ m^2 \quad \text{or} \quad 109\ ft^2$$

Calculate the required length of sludge pipe in the heat exchanger jacket.

$$L_{s,pipe} = \frac{A_{s,pipe}}{\pi d_{rs}} = \frac{10.1\ m^2}{\pi \times 0.2\ m} = 16.1\ m \quad \text{or} \quad 53\ ft$$

Note: This is the minimum area and length required for heat transfer from hot water to the sludge mixture at the sustained low temperature during winter. Additional safety factor may be considered while selecting the commercial heat exchanger.

EXAMPLE 13.25: DIGESTER BIOGAS GENERATION

A wastewater treatment plant has two identical high-rate anaerobic digesters. Each anaerobic digester receives 4100 kg/d TSS. Estimate the quantity of biogas produced from each digester and that produced at the plant. Apply the kinetic equations and rules of thumb to estimate gas quantity. The kinetic coefficients and digester operating parameters are: $Y = 0.08$ kg VSS/kg COD, $k_d = 0.03\ d^{-1}$, SRT (θ_c) = 15 d, COD utilization efficiency $E = 0.8$. Assume VSS/TSS ratio = 0.75, the ratio of biodegradable VSS (VSS_{bd})

to TSS is 0.56 kg VSS_{bd}/kg TSS, and each g of VSS_{bd} exerts 1.42 g COD. Other relationships of the biogas generation are: digester gas production rate = 0.55 m^3/kg VSS loading, 0.85 m^3/kg VSS reduced, and 0.035 m^3 per capita per day. The average sludge production rate is 0.12 kg TSS per capita per day.

Solution

1. Calculate the biogas production from each digester using the kinetic coefficients.
   a. Calculate COD in the feed sludge received by the digester.

$$\begin{aligned}\text{Biodegradable VSS }(\text{VSS}_{\text{bd}})\text{ in feed sludge}, W_{\text{VSSbd,FS}} &= \text{VSS}_{\text{bd}}/\text{TSS ratio} \times W_{\text{TSS, FS}}\\ &= 0.56\,\text{kg VSS}_{\text{bd}}/\text{kg TSS} \times 4100\,\text{kg TSS/d}\\ &= 2300\,\text{kg VSS/d}\end{aligned}$$

$$\begin{aligned}\text{COD exerted in feed sludge}, \Delta S_0 &= \text{COD/VSS}_{\text{bd}}\text{ ratio} \times W_{\text{VSSbd,FS}}\\ &= 1.42\,\text{kg COD/kg VSS}_{\text{bd}} \times 2300\,\text{kg VSS}_{\text{bd}}/\text{d}\\ &= 3270\,\text{kg COD/d}\end{aligned}$$

   b. Calculate the Y_{obs} from Equation 10.16d.

$$Y_{\text{obs}} = \frac{Y}{1+k_d\theta_c} = \frac{0.08\,\text{mg VSS/mg COD}}{1+0.03\,\text{d}^{-1}\times 15\,\text{d}} = 0.055\,\text{mg VSS/mg COD} = 0.055\,\text{kg VSS/kg COD}$$

   c. Calculate the active biomass production P_x from Equation 10.78d.

$$P_x = Y_{\text{obs}}E\Delta S_0 = 0.055\,\text{kg VSS/kg} \times 0.75 \times 3270\,\text{kg COD/d} = 135\,\text{kg VSS/d}$$

   d. Calculate the COD stabilization ΔS_M from Equation 10.78b.

$$\begin{aligned}\Delta S_M &= E\Delta S_0 - 1.42P_x = 0.75 \times 3270\,\text{kg COD/d} - 1.42\,\text{kg COD/kg VSS} \times 135\,\text{kg VSS/d}\\ &= 2260\,\text{kg COD/d}\end{aligned}$$

   e. Calculate the methane production (Q_M) from Equation 10.78c using $f_v = 0.35\,m^3\,CH_4$/kg COD (see Example 10.113).

$$Q_M = f_v\Delta S_M = 0.35\,\text{m}^3\,\text{CH}_3/\text{kg COD} \times 2260\,\text{kg COD/d} = 791\,\text{m}^3\,\text{CH}_3/\text{d}$$

   f. Calculate the biogas production (Q_{biogas}) by assuming that CH_4 content is 65% by volume in the biogas.

$$Q_{\text{biogas}} = \frac{Q_M}{f_{\text{CH4}}} = \frac{791\,\text{m}^3\,\text{CH}_3/\text{d}}{0.65} = 1220\,\text{m}^3\,\text{biogas/d}$$

2. Estimate the biogas production from each digester using the rules of thumb.
   a. Calculate the biogas production based on VSS loading.

$$\begin{aligned}\text{Mass of VSS in feed sludge}, W_{\text{VSS, FS}} &= \text{VSS/TSS ratio} \times W_{\text{TSS, FS}}\\ &= 0.75\,\text{kg VSS/kg TSS} \times 4100\,\text{kg TSS/d}\\ &= 3075\,\text{kg VSS/d}\end{aligned}$$

$$\begin{aligned}\text{Biogas production}, Q_{\text{biogas}} &= 0.55\,\text{m}^3\,\text{biogas/kg VSS} \times W_{\text{VSS, FS}}\\ &= 0.55\,\text{m}^3\,\text{biogas/kg VSS} \times 3075\,\text{kg VSS/d}\\ &= 1690\,\text{m}^3\,\text{biogas/d}\end{aligned}$$

b. Calculate the biogas production based on volatile solids reduction (VSR).

VSR from Equation 13.5a, $VSR = 13.7\ \ln(\theta_c) + 18.9 = 13.7 \times \ln(15\text{d}) + 18.9 = 56\%$

$$\text{Mass of VSS destroyed in digester, } \Delta W_{\text{VSS,reduced}} = \frac{\text{VSR}}{100\%} \times W_{\text{VSS, FS}}$$

$$= \frac{56\%}{100\%} \times 3075\,\text{kg VSS/d} = 1720\,\text{kg VSS/d}$$

$$\text{Biogas production, } Q_{\text{biogas}} = 0.85\,\text{m}^3\ \text{biogas/kg VSS reduced} \times \Delta W_{\text{VSS,reduced}}$$

$$= 0.85\,\text{m}^3\ \text{biogas/kg VSS reduced} \times 1720\,\text{kg VSS reduced/d}$$

$$= 1460\,\text{m}^3\ \text{biogas/d}$$

c. Calculate the biogas production based on population.
Estimate the population (P) served by the digester at $p_{\text{TSS}} = 0.12$ kg/capita·d.

$$P = \frac{W_{\text{TSS, FS}}}{p_{\text{TSS}}} = \frac{4100\,\text{kg TSS/d}}{0.12\,\text{kg TSS/capita·d}} = 34{,}200 \text{ persons}$$

$$\text{Biogas production, } Q_{\text{biogas}} = 0.035\,\text{m}^3\ \text{biogas/capita·d} \times P$$

$$= 0.035\,\text{m}^3\ \text{biogas/capita·d} \times 34{,}200 \text{ persons}$$

$$= 1200\,\text{m}^3\ \text{biogas/d}$$

3. Summarize the calculated gas production results, and total gas produced at the plant.

Kinetic equations = 1220 m^3/d
VSS loading = 1690 m^3/d
VSS reduction = 1460 m^3/d
Population basis = 1200 m^3/d

A conservative estimate for gas production = 1250 m^3/d per digester. The total biogas production is 2500 m^3/d from two digesters.

4. Comment on the gas utilization.
A biogas flow requirement of 2150 m^3/d is estimated for heating the sludge to two digesters (see Example 13.23. At the sustained low temperature during winter, the total biogas generated by the anaerobic digesters is sufficient to meet the heating requirements of the digesters. Excess biogas may be available for use by other on- or off-site users under warm weather conditions. To be energy efficient at the plant, uses of excess biogas should be evaluated and included in the comprehensive biosolids management plan.

EXAMPLE 13.26: DIGESTER GAS MIXING

Two high-rate anaerobic digesters are mixed with gas lifters (see Figure 13.15b). The volume of each digester is 1450 m^3. The acceptable velocity gradient G for mixing is 80 s^{-1}. Determine the power requirement and gas flow rate from the compressors. The operating temperature is 35°C. Assume that the dynamic viscosity μ of sludge is twice that of water.

Solution

1. Describe the gas lifter.

The gas lifter is a confined gas injection system for mixing the sludge content in anaerobic sludge digester. Effective gas mixing is achieved by introducing the recirculated biogas through lances inside a draft tube that is typically located in the center of the tank. The rising gas bubbles carry the sludge into an upward movement that is confined and enhanced by the draft tube. Effective circulation in the digester is created when the sludge flow is drawn from the bottom and released near the top of the tank. A conceptual arrangement of the gas lifter is shown in Figure 13.15.

2. Estimate the dynamic viscosity μ of digester sludge.

   Dynamic viscosity of water $\mu_w = 0.726 \times 10^{-3}\,\text{N·s/m}^2$ at 35°C (estimated from data in Table B.2 in Appendix B).

   Dynamic viscosity of sludge $\mu_s = 2\mu_s = 2 \times 0.726 \times 10^{-3}\,\text{N·s/m}^2 = 1.45\times\,10^{-3}\,\text{N·s/m}^2$

3. Compute the compressor power requirement.

   Calculate the required gas mixing power P_g from Equation 9.27 at $G = 80\,\text{s}^{-1}$ and $V = 1450\,\text{m}^3$.

$$P_g = \mu VG^2 = 1.45 \times 10^{-3}\,\text{N·s/m}^2 \times 1450\,\text{m}^3 \times (80\,\text{s}^{-1})^2 = 1.35 \times 10^4\,\text{N·m/s} = 1.35 \times 10^4\,\text{W or } 13.5\,\text{kW}$$

   Assume 65% overall efficiency of the gas mixing system to calculate the required compressor input or motor output power for each compressor.

$$P_m = \frac{P_g}{E_{\text{mixer}}} = \frac{13.5\,\text{kW}}{0.65} = 20.8\,\text{kW (28 hp)}$$

   Provide three compressors each with a 22.5-kW (30 hp) motor; one compressor for each digester. The third compressor will be a standby unit serving both digesters.

4. Compute the gas recirculation flow in each digester.

   In Section 10.3.8, the relationship between the air flow and shaft horsepower (SHP) of an air blower is given by Equation 10.45a. This equation is also applicable to a gas compressor.

   Assume that (a) the compressor inlet pressure $P_i = 1.03$ atm which is equal to the gas pressure inside the headspace, (b) the compressor outlet pressure $P_d = 2.4$ atm which is sufficient to overcome static head and recirculation pipe losses, (c) compressor inlet temperature $T_i = (273 + 35°\text{C}) = 308°\text{K}$, which is equal to the digester operating temperature, and (d) compressor efficiency $\eta_c = 75\%$. The density of digester gas $\rho_g = 1.03\,\text{kg/sm}^3$, which is typically 86% of air density $\rho_a = 1.20\,\text{kg/sm}^3$ $(0.075\,\text{lb/ft}^3)$. The compressor input power $P_{\text{compressor}} = 20.8$ kW, $C = 28.97$ kg/k mole, $(k-1)/k = 0.283$, and $R = 8.314$ kJ/k mole·°K.

   Rearrange Equation 10.45a and calculate the gas flow delivered by the compressor to a digester.

$$Q_g = \frac{C\left(\frac{k-1}{k}\right)\left(\frac{\eta_b}{100\%}\right)}{\left[\left(\frac{P_d}{P_i}\right)^{\frac{k-1}{k}} - 1\right]}\frac{60P_{\text{compressor}}}{\rho_g RT_i}$$

$$= \frac{28.97\,\text{kg/k mole} \times 0.283 \times \left(\frac{75\%}{100\%}\right)}{\left[\left(\frac{2.4\,\text{atm}}{1.03\,\text{atm}}\right)^{0.283} - 1\right]} \times \frac{60 \times 20.8\,\text{kW}}{1.03\,\text{kg/m}^3 \times 8.314\,\text{kJ/k mole·°K} \times 308°\text{K}}$$

$$= 10.8\,\text{sm}^3/\text{min}$$

   Check the volumetric gas recirculation (VGR) rate provided in the digester.

$VGR = \frac{Q_g}{V} = \frac{10.8\ sm^3/min}{1450\ m^3} = 0.007\ sm^3/m^3{\cdot}min$ (The VGR rate is at the upper end of the typical range from 0.005 to 0.007 $m^3/m^3{\cdot}min$ for the confined gas injection mixing in the sludge digester (see Table 13.14)).

EXAMPLE 13.27: GAS STORAGE SPHERE AND COMPRESSOR

Two high-rate anaerobic digesters are producing biogas at a daily volume of 2500 m^3/d at the standard condition (1 atm and 0°C). A separate spherical shaped gas holder is provided for 5-d gas storage to serve the needs of digester heating and other on-site uses. The biogas is compressed by the high-pressure gas compressors into the sphere. The storage pressure in the sphere is 5.5 atm at the critical temperature of 40°C during summer. The design total capacity of the compressors is 200% of the biogas production rate. The gas pressure inside the head space of the digester is 1.03 atm. Calculate the diameter of the sphere and power requirement of the compressor. Assume that the density of digester gas $\rho_g = 1.03\ kg/sm^3$ under standard conditions.

Solution

1. Compute the diameter of gas storage sphere.

   Apply the ideal gas law (Equation 10.82).

$$nR = \frac{P_1 V_1}{T_1} = \frac{P_2 V_2}{T_2}$$

Total volume of gas stored at the standard condition, $V_1 = t_{storage} Q_{biogas} = 5\ d \times 2500\ m^3/d = 12{,}500\ m^3$

Rearrange Equation 10.82 to calculate the total volume of gas stored in the sphere at 5.5 atm and 40°C.

$$V_2 = \frac{P_1 T_2}{P_2 T_1} V_1 = \frac{1\ atm \times (273 + 40)\ {}^\circ K}{5.5\ atm \times (273 + 0)\ {}^\circ K} \times 12{,}500\ m^3 = 2610\ m^3 \quad \text{or} \quad 92{,}200\ ft^3$$

Volume of a sphere, $V = \frac{\pi}{6} d^3$

Required diameter of the sphere, $d = \left(\frac{6}{\pi} V_2\right)^{1/3} = \left(\frac{6}{\pi} \times 2610\ m^3\right)^{1/3} = 17.1\ m$

Provide a standard 18.44 m (60.5 ft) diam sphere with capacity of 3280 m^3 (20,000 barrels or 116,000 ft^3).

2. Compute the total power requirement of the compressors.

   Total flow from the compressors at 200% of the biogas production rate,

$$Q_g = 2Q_{biogas} = 2 \times 2500\ m^3/d \times \frac{d}{1440\ min} = 3.5\ m^3/min$$

Calculate the total compressor power requirement from Equation 10.45a. Assume the machine efficiency $\eta_b = 75\%$. $\rho_g = 1.03\ kg/sm^3$, $R = 8.314\ kJ/k\ mole{\cdot}{}^\circ K$, $(k-1)/k = 0.283$, $C = 28.97\ kg/k\ mole$, $Q_{biogas} = 120\ sm^3/min$, $T_i = (35^\circ C + 273) = 308^\circ K$, and $P_i = 1.03$ atm, and $P_d = 5.5$ atm.

$$P_{\text{compressor}} = \frac{\left(\frac{Q_g}{60}\right)\rho_g R T_i}{C\left(\frac{k-1}{k}\right)\left(\frac{\eta_b}{100\%}\right)}\left[\left(\frac{P_d}{P_i}\right)^{\frac{k-1}{k}} - 1\right]$$

$$= \frac{\left(\frac{3.5\,\text{sm}^3/\text{min}}{60\,\text{s/min}}\right) \times 1.03\,\text{kg/m}^3 \times 8.314\,\text{kJ/k mole}\cdot{}^\circ\text{K} \times 308^\circ\text{K}}{28.97\,\text{kg/k mole} \times 0.283 \times \left(\frac{75\%}{100\%}\right)}\left[\left(\frac{5.5\,\text{atm}}{1.03\,\text{atm}}\right)^{0.283} - 1\right]$$

$$= 15.2\,\text{kW}\,(20.4\,\text{hp})$$

Provide two constant-speed compressors, each driven by a 10-kW (13.4-hp) electric motor.

EXAMPLE 13.28: DIGESTED SLUDGE AND SUPERNATANT PRODUCTIONS

An anaerobic digester receives 4100 kg/d total solids at a solids content of 5.4% and bulk specific gravity of 1.008. The digested sludge has 3.5% solids and bulk specific gravity of 1.009. The solid concentration in supernatant is 5000 mg/L. The VSS/TSS ratio is 0.75 in the feed sludge, and volatile solids reduction (VSR) in the digester is 56%. Determine the mass and volume of digested sludge and the supernatant. Assume that no liquid volume change occurs in the digester.

Solution

1. Compute from Equation 13.1e the daily volume of feed sludge reaching the digester.

$$\text{Volume of feed sludge, } Q_{FS} = \frac{100\% \times W_{TSS,FS}}{p_{s,FS} S_{b,FS} \rho_w} = \frac{100\% \times 4100\,\text{kg/d}}{5.4\% \times 1.008 \times 1000\,\text{kg/m}^3} = 75\,\text{m}^3/\text{d}$$

2. Compute the solids in digested sludge.

$$\text{Mass of volatile solids, } W_{VSS,FS} = f_{VSS,FS} W_{TSS,FS} = 0.75\,\text{kg VSS/kgTSS} \times 4100\,\text{kg TSS/d}$$
$$= 3075\,\text{kg VSS/d}$$

$$\text{Mass of volatile solids destroyed in the digester, } \Delta W_{VSS,reduced} = \frac{\text{VSR}}{100\%} \times W_{VSS,FS}$$
$$= \frac{56\%}{100\%} \times 3075\,\text{kg VSS/d} = 1720\,\text{kg VSS/d}$$

$$\text{Mass of total solids remaining after digestion, } W_{TSS,REM} = W_{TSS,FS} - \Delta W_{VSS,VSR}$$
$$= 4100\,\text{kg TSS/d} - 1720\,\text{kgVSS/d}$$
$$= 2380\,\text{kg FSS/d}$$

3. Compute the daily volume and solids in digested sludge and in the digester supernatant.

   Since there is no change in volume during digestion process, the following relationships are established from mass balance.

   a. Obtained a weight relationship from daily volume balance.

Volume of feed sludge (Q_{FS}) = Volume of digested sludge (Q_{DGS}) + Volume of digester supernatant (Q_{SPN})

$$Q_{FS} = Q_{DGS} + Q_{SPN} = \frac{100\% \times W_{TSS,DGS}}{p_{s,DGS} S_{b,DGS} \rho_w} + \frac{W_{TSS,SPN}}{TSS_{SPN}}$$

$$= \frac{100\% \times W_{TSS,DGS}}{3.5\% \times 1.009 \times 1000\,kg/m^3} + \frac{W_{TSS,SPN}}{5000\,g/m^3 \times 10^{-3}\,kg/g}$$

$$= \frac{W_{TSS,DGS}}{35.3\,kg/m^3} + \frac{W_{TSS,SPN}}{5\,kg/m^3}$$

$$\text{or} \quad \frac{W_{TSS,DGS}}{35.3\,kg/m^3} + \frac{W_{TSS,SPN}}{5\,kg/m^3} = 75\,m^3/d \quad \text{(from Step 1)} \qquad (13.8a)$$

The weight relationship from daily volume balance is given by Equation 13.8a.

b. Obtained a weight relationship from daily mass balance of total solids (TSS).

Weight of TSS remaining in the system ($W_{TSS,REM}$)

= Weight of TSS in digested sludge ($W_{TSS,DGS}$) + Weight of TSS in digester supernatant ($W_{TSS,SPN}$)

$$W_{TSS,REM} = W_{TSS,DGS} + W_{TSS,SPN}$$

$$2380\,kg\,TSS/d \text{ (from Step 2)} = W_{TSS,DGS} + W_{TSS,SPN} \quad \text{or} \quad W_{TSS,SPN} = 2380\,kg\,TSS/d - W_{TSS,DGS} \qquad (13.8b)$$

The weight relationship from total weight of TSS remaining in the system is given by Equation 13.8b.

Substitute $W_{TSS,SPN}$ from Equation 13.8b in Equation 13.8a and solve for $W_{TSS,DGS}$.

$$\frac{W_{TSS,DGS}}{35.3\,kg/m^3} + \frac{2380\,kg\,TSS/d - W_{TSS,DGS}}{5\,kg/m^3} = 75\,m^3/d$$

$$W_{TSS,DGS} = 2336\,kg\,TSS/d$$

$$Q_{DGS} = \frac{100\% \times 2336\,kg\,TSS/d}{3.5\% \times 1.009 \times 1000\,kg/m^3} = 66\,m^3/d \text{ (from Equation 13.1e)}$$

$$W_{TSS,SPN} = W_{TSS,REM} - W_{TSS,DGS} = (2380 - 2336)\,kg\,TSS/d = 44\,kg\,TSS/d \text{ (from Equation 13.8b)}$$

$$Q_{SPN} = Q_{FS} - Q_{DGS} = (75 - 66)\,m^3/d = 9\,m^3/d$$

Note: The volume of supernatant is relatively small. It is only about 12% of the feed sludge volume.

EXAMPLE 13.29: VSS AND COD COMPONENTS OF FEED AND DIGESTED SLUDGES

A high-rate anaerobic digester without recycle is treating thickened sludge. The waste characteristics of the feed sludge are given below.

Average daily sludge volume $Q_{FS} = 75\,m^3/d$	Solids concentration $TSS_{FS} = 54\,g/L$
$VSS/TSS = 0.75$	Biodegradable $(VSS_{bd})/VSS = 0.73\,g\,VSS_{bd}/g\,VSS$
Soluble COD in feed sludge (COD_{FS}) is negligible.	

The digester design and performance parameters are:

Kinetic coefficients: $Y = 0.08$ g VSS/g COD, $k_d = 0.03\ d^{-1}$, $k = 3.2\ d^{-1}$, and $K_s = 800$ mg/L.

Process parameters: SRT safety factor $SF = 1.5$, Operating temperature $T_d = 35°C$, and VSS reduction $(VSR) = 55\%$.

Quality of digested sludge quality: soluble COD in the digester feed sludge, $S_{DGS} = 1000$ mg/L.

Calculate (a) VSS and COD components of feed sludge, (b) design SRT, (c) VSS reduction and COD removal, and (d) VSS and COD components of the digested sludge. Assume that the ratio of COD and biodegradable VSS (VSS_{bd}), COD/VSS_{bd} ratio = 1.42 g COD/g VSS_{bd}.

Solution

1. Determine the VSS components of the feed sludge.

$$\text{VSS in the feed sludge, } VSS_{FS} = VSS/TSS \text{ ratio} \times TSS_{FS}$$
$$= 0.75\,\text{g VSS/gTSS} \times 54\,\text{g TSS/L} = 40.50\,\text{g VSS/L}$$

$$\text{Biodegradable } VSS_{bd} \text{ in the feed sludge, } VSS_{bd,FS} = VSS_{bd}/VSS \text{ ratio} \times VSS_{FS}$$
$$= 0.73\,\text{g } VSS_{bd}/\text{gVSS} \times 40.50\,\text{g VSS/L} = 29.57\,\text{g VSS/L}$$

$$\text{Nonbiodegradable VSS in the feed sludge, } VSS_{nd,FS} = VSS_{FS} - VSS_{bd,FS}$$
$$= (40.50 - 29.57)\,\text{g VSS/L} = 10.93\,\text{g VSS/L}$$

$$\text{Fixed suspended solids in the feed sludge, } FSS_{FS} = TSS_{FS} - VSS_{FS}$$
$$= 54\,\text{g TSS/L} - 40.50\,\text{g VSS/L} = 13.50\,\text{g FSS/L}$$

2. Determine the COD components of the feed sludge.
It is given that soluble COD in the feed sludge is ignorable, $S_{FS} = 0$.

$$\text{Particulate COD due to } VSS_{bd} \text{ in the feed slduge, } S_{p,FS} = COD/VSS_{bd} \text{ ratio} \times VSS_{bd,FS}$$
$$= 1.42\,\text{g COD/g } VSS_{bd} \times 29.57\,\text{g } VSS_{bd}/\text{L}$$
$$= 41.99\,\text{g COD/L}$$

$$\text{Total COD in the feed sludge, } S_{t,FS} = S_{FS} + S_{p,FS} = (0 + 41.99)\,\text{g COD/L} = 41.99\,\text{g COD/L}$$

3. Determined the VSS reduction and COD removal.

$$\text{Volatile solids destroyed in the digester, } \Delta VSS_{reduced} = \frac{VSR}{100\%} \times VSS_{FS} = \frac{56\%}{100\%} \times 40.50\,\text{g VSS/L}$$
$$= 22.68\,\text{g/L}$$

Assume that the COD removal in the digester $E_{COD} = 75\%$.

$$\text{COD removed in the digester, } \Delta S_{removed} = E_{COD} S_{t,FS} = 0.75 \times 41.99\,\text{g COD/L}$$
$$= 31.49\,\text{g COD/L}$$

4. Determined the total VSS in the digested sludge.

$$\text{The total volatile solids in the digested sludge, } VSS_{DGS} = VSS_{FS} - \Delta VSS_{reduced}$$
$$= (40.50 - 22.68)\,\text{g/L} = 17.82\,\text{g VSS/L}$$

5. Determine the Design SRT (or HRT).

$\theta = \theta_c$ for anaerobic digester without recycle. Apply Equation 10.16a to calculate the HRT (θ) required for anaerobic stabilization of soluble COD, $S_{DGS} = 1000$ mg/L in the feed sludge.

$$\frac{1}{\theta} = \frac{YkS_{DGS}}{K_s + S_{DGS}} - k_d = \frac{0.08 \times 3.2\,\text{d}^{-1} \times 1000\,\text{mg COD/L}}{800\,\text{mg COD/L} + 1000\,\text{mg/L COD/L}} - 0.03\,\text{d}^{-1} = (0.142 - 0.03)\,\text{d}^{-1}$$

$$= 0.112\,\text{d}^{-1}$$

$$\theta = \frac{1}{0.112\,\text{d}^{-1}} = 8.93\,\text{d}$$

Required design HRT, $\theta^{\text{design}} = SF\theta = 1.5 \times 8.93\,\text{d} = 13.4\,\text{d}$

Provide $\theta^{\text{design}} = \theta_c^{\text{design}} = 15$ d. The design SRT is within the typical range (see Table 13.13).

6. Determine the VSS components of the digested sludge.

The total volatile solids produced in anaerobic digestion (VSS_{DGS}) is estimated from the following relationship.

$$\text{Digester sludge } (VSS_{DGS}) = \text{Biomass } (VSS_{bm,DGS}) + \text{Cell debris } (VSS_{cd,DGS}) + \text{Nonbiodegradable } (VSS_{nb,DGS}) + \text{Biodegradable } (VSS_{bd,DGS})$$

This relationship is presented previously in Equation 10.85a.

$$p_{x,t} = p_{x,bm} + p_{x,cd} + p_{x,nb} + p_{x,bd}$$

The VSS components in the above expression are calculated below.

a. Calculate the biomass generated from COD removal ($VSS_{bm,DGS}$).

$$Y_{obs} \text{ from Equation 10.16d} = \frac{Y}{1 + k_d\theta^{\text{design}}} = \frac{0.08\,\text{mg VSS/mg COD}}{1 + 0.03\,\text{d}^{-1} \times 15\,\text{d}}$$

$$= 0.055\,\text{mg VSS/mg COD} = 0.055\,\text{g VSS/g COD}$$

Calculate $p_{x,bm}$ from Equation 10.16e.

$$VSS_{bm,DGS} = p_{x,bm} = Y_{obs}\Delta S_{removed} = 0.055\,\text{g VSS/g COD} \times 31.49\,\text{g COD/L} = 1.73\,\text{g VSS/L}$$

b. Calculate the cell debris produced ($VSS_{cd,DGS}$).

Use $f_{cd} = 0.15$ and apply Equation 10.85b to calculate $P_{x,cd}$.

$$VSS_{cd,DGS} = p_{x,cd} = f_d k_d \theta_c^{\text{design}} p_{x,bm} = 0.15 \times 0.03\,\text{d}^{-1} \times 15\,\text{d} \times 1.73\,\text{g VSS/L}$$
$$= 0.12\,\text{g VSS/L}$$

c. Calculate the nonbiodegradable VSS ($VSS_{nb,DGS}$).

$$VSS_{nb,DGS} = p_{x,nb} = VSS_{nd,FS} = 10.93\,\text{g VSS/L} \quad \text{(from Step 1)}$$

d. Calculate the biodegradable VSS remaining in the digested sludge ($VSS_{bd,DGS}$).

Rearrange Equation 10.85a to calculate the $VSS_{bd,DGS}$ using $VSS_{DGS} = 17.82$ g VSS/L (from Step 4)

$$VSS_{bd,DGS} = VSS_{DGS} - (VSS_{bm,DGS} + VSS_{cd,DGS} + VSS_{nb,DGS})$$
$$= (17.82 - (1.73 + 0.12 + 10.93))\,\text{g VSS/L}$$
$$= 5.04\,\text{g VSS/L}$$

Total VSS_{bd} in the digested sludge, $VSS_{t,bd,DGS} = VSS_{bm,DGS} + VSS_{bd,DGS}$

$$= (1.73 + 5.04)\ \text{g VSS/L} = 6.77\ \text{g VSS/L}$$

Total VSS_{bd} in the digested sludge, $VSS_{t,nd,DGS} = VSS_{cd,DGS} + VSS_{nd,DGS}$

$$= (0.12 + 10.93)\ \text{g VSS/L} = 11.05\ \text{g VSS/L}$$

FSS in the digested sludge, $FSS_{DGS} = FSS_{FS} = 13.50$ g FSS/L (from Step 1)

TSS in the digested sludge, $TSS_{DGS} = VSS_{DGS} + FSS_{DGS} = 17.82$ g VSS/L + 13.50 g FSS/L

$$= 31.32\ \text{g TSS/L}$$

VSS/TSS ratio in the digested sludge, $\text{VSS/TSS ratio} = \dfrac{VSS_{DGS}}{FSS_{DGS}} = \dfrac{17.82\ \text{g VSS/L}}{31.32\ \text{g TSS/L}} = 0.57$

7. Determine the COD components of the digested sludge.

   Soluble COD in the digested sludge is given in the problem statement, $S_{DGS} = 1000$ mg COD/L = 1 g COD/L. Particulate COD ($S_{P,DGS}$) due to volatile solids in the digested sludge,

   $S_{p,DGS} = COD/VSS_{bd}\ \text{ratio} \times VSS_{t,bd,DGS} = 1.42\ \text{g COD/g}\ VSS_{bd} \times 6.77\ \text{g}\ VSS_{bd}/\text{L} = 9.61\ \text{g COD/L}$

   Total COD in the digested sludge, $S_{t,DGS} = S_{DGS} + S_{p,DGS} = (1 + 9.61)$ g COD/L = 10.61 g COD/L

   Check the COD removal based on the total COD in the feed sludge, $S_{t,FS} = 41.99$ g COD/L (from Step 2).

   $$E_{COD} = \frac{S_{t,FS} - S_{t,DGS}}{S_{t,DGS}} \times 100\% = \frac{(41.99 - 10.61)\ \text{g COD/L}}{41.99\ \text{g COD/L}} \times 100\% = 75\%$$

   Note: The calculated COD removal efficiency is same as that assumed in Step 3.

8. Summarize the calculated results.

   The calculated results for VSS and COD components as well as other data in the feed and digested sludges are summarized below.

Parameter	Feed Sludge	Digested Sludge
Solids concentration, g/L (% of TSS)		
VSS		
Biodegradable	$VSS_{bd,\ FS}$, 29.57 (55) (Step 1)	$VSS_{bd,\ DGS}$, 5.04 (22) (Step 6.d)
Nonbiodegradable	$VSS_{nd,\ FS}$, 10.93 (20) (Step 1)	$VSS_{nd,\ DGS}$, 10.93 (35) (Step 6.c)
Total	VSS_{FS}, 40.50 (75) (Step 3)	VSS_{DGS}, 17.82 (57) (Step 4)[a]
FSS	FSS_{FS}, 13.50 (25) (Step 1)	FSS_{DGS}, 13.50 (43) (Step 6.d)
TSS	TSS_{FS}, 54.00 (100) (Problem statement)	FSS_{DGS}, 31.32 (100) (Step 6.d)
COD concentration, g/L (% of Total COD)		
Soluble	S_{FS}, 0.00 (0) (Problem statement)	S_{DGS}, 1.00 (9) (Problem statement)
Particulate	$S_{P,\ FS}$, 41.99 (100) (Step 2)	$S_{p,DGSS}$, 9.61 (91) (Step 7)
Total	$S_{t,FS}$, 41.99 (100) (Step 2)	$S_{p,DGS}$, 10.61 (100) (Step 7)

[a] $VSS_{DGS} = VSS_{bd,DGS} + VSS_{nd,DGS} + VSS_{bm,DGS} + VSS_{cd,DGS} = (5.04 + 10.93 + 1.73 + 0.12)$ g VSS/L = 17.82 g VSS/L.

13.6.2 Aerobic Digestion

Aerobic digestion is commonly used for small plants (20,000 m^3/d (5 MGD) or less). It is similar to the activated sludge process except that it operates at nearly zero food-to-microorganism (F/M) ratio. The oxidation of biodegradable organic matter occurs within the cell tissues resulting in endogenous respiration of biomass. The stabilization of organic matter continues over a long SRT (40–60 days) until the digested sludge is fully stabilized, and odorless, and humus like biologically stable solids remain. In aerobic digesters, about 75–80% of the cell tissues are oxidized. The remaining 20–25% biomass consists of fixed suspended solids (FSS), nonbiodegradable VSS, cell debris from endogenous respiration, and small amount of biodegradable VSS. The advantages of the aerobic digestion are: (1) simple to operate, (2) low capital cost, (3) the supernatant is low in BOD_5, (4) suitable for digesting nutrient-rich sludge, and (5) the digested sludge is dewatered readily on sand drying beds.[6,7] The reported disadvantages are: (1) poor dewaterability by mechanical equipment; (2) high operation cost due to high air supply rate; (3) performance is affected by temperature, solids concentration, and aeration and mixing equipment; and (4) does not produce methane for energy recovery.[6,7] Depending upon the process utilized, the aerobic digestion is capable of producing either Class A or B biosolids (see Section 13.4.4).

Stoichiometric Reactions: The simplified biochemical reactions of biomass ($C_5H_7O_2N$) under aerobic conditions are given by Equations 13.9a and 13.9b.[7]

Biomass stabilization with oxidation of organics when nitrification is inhibited:

$$C_5H_7O_2N + 5\ O_2 \rightarrow 4\ CO_2 + NH_4^+ + HCO_3^- + H_2O \tag{13.9a}$$

Overall biomass stabilization with complete nitrification at sufficient DO supply:

$$C_5H_7O_2N + 7\ O_2 + HCO_3^- \rightarrow NO_3^- + 6\ CO_2 + 4\ H_2O \tag{13.9b}$$

A variation of aerobic digestion is an alternation between aeration and mixing (typically, 75% aeration and 25% mixing). During mixing cycle, DO is absent and denitrification may occur using nitrate as an electron acceptor. The overall biomass stabilization with complete nitrification and denitrification is expressed by Equation 13.9c:

$$5\ C_5H_7O_2N + 28\ O_2 \rightarrow 2\ N_2 + 24\ CO_2 + NH_4^+ + HCO_3^- + 15\ H_2O \tag{13.9c}$$

Nitrification lowers pH and consumes alkalinity. Denitrification helps to recover part of alkalinity. The oxygen and alkalinity requirements for stabilization of biomass may be estimated from these stoichiometric equations (see Examples 13.30 and 13.36). When combined primary and biological sludge is applied to an aerobic digester, biomass stabilization occurs with simultaneous oxidation of biomass and raw organic substrates present in the primary sludge.

Aerobic Digestion Kinetics: The suspended growth biological treatment processes are presented in Section 10.3. Most aerobic microbial kinetics developed in this section is applicable to aerobic sludge digestion. Similar to Equation 10.10a, the biomass production and substrate utilization in a continuous flow completely-mixed aerobic digester without recycle can be expressed by Equation 13.10a.

$$\frac{dX}{dt} = Y\left(-\frac{dS}{dt}\right) - k_d X \tag{13.10a}$$

where

dX/dt = rate of change of volatile solids concentration per unit of time, mg VSS/L·d

dS/dt = rate of change of substrate (COD or BOD) utilized per unit of time (d), mg COD (or BOD)/L·d

Y = biomass yield coefficient, mg VSS/mg COD (or BOD)

k_d = specific endogenous decay coefficient, d^{-1}. This constant depends on the type of sludge, temperature, and solids concentration. The typical values of k_d range from 0.05 d^{-1} at 15°C to 0.14 d^{-1} at 25°C for conventional aerobic digestion of waste activated sludge (WAS). The average temperature correction coefficient θ_T for k_d in Equation 10.17b = 1.05. The value of k_d may also be seriously affected due to solubilization (hydrolysis) of biodegradable and nonbiodegradable VSS. These effects are shown in Example 13.32. A typical value $k_d = 0.12$ d^{-1} at 20°C with $\theta_T = 1.023$ may be used for thermophilic aerobic digestion (see Example 13.37).

X = VSS concentration remaining at time t in the aerobic digester, mg VSS/L

To meet the major objective of sludge reduction in aerobic digestion, the term $Y(-dS/dt)$ must be nearly zero, and the change in VSS concentration due to decay of biomass is determined from the first-order decay reaction expressed by the Equation 13.10b.

$$\frac{dX}{dt} = -k_d X \qquad (13.10b)$$

The decay of biomass is similar to that of BOD under substrate limited condition (see Equation 5.5). The time dependent reaction rate equations after integration are expressed by Equations 13.10c and 13.19d while the volatile solids reduction (VSR) can be calculated from Equation 13.10e.

$$X_t = X_0 e^{-k_d t} \qquad (13.10c)$$

$$X_r = X_0(1 - e^{-k_d t}) \qquad (13.10d)$$

$$\text{VSR} = \frac{X_r}{X_0} \times 100\% = (1 - e^{-k_d t}) \times 100\% \qquad (13.10e)$$

where

X_t = concentration of VSS remaining at time t, mg VSS/L

X_0 = concentration of VSS available in the feed sludge, mg VSS/L

X_r = concentration VSS reduced at time t, mg VSS/L

t = retention time in the digester, d

VSR = volatile solids reduction, %

The plots of X_t vs. t, and X_r vs. t are similar to those developed from Equations 5.6 and 5.7 (see Figure 5.3 in Example 5.15, and Figure 5.4 in Example 5.19). Determination of k_d from batch test results is covered in Example 13.31.

Two most commonly known aerobic digestion processes are: (1) conventional aerobic digestion and (2) thermophilic aerobic digestion (TAD). Brief descriptions of these processes are presented below.

Conventional Aerobic Digestion: The process involves aeration of waste activated sludge (WAS) or combined sludge in an aeration basin for an extended period. Air is the oxygen source for oxidation of organic matter in batch or continuous-flow reactors. Schematic flow diagrams of conventional aerobic digestion process are shown in Figure 13.17.[7] Batch-operated aerobic digesters are widely used as an integrated component of small package plants (less than 0.5 MGD). A batch-operated aerobic digester uses fill, settle and decant sequence (Figure 13.17a). The continuous-flow aerobic digestion process may have a gravity thickener for pre- or postthickening of sludge. The digestion of combined or thickened sludge is achieved in a conventional aerobic digester followed by storage in a sludge storage tank prior to dewatering. The conventional aerobic digestion may also be operated in series with a gravity thickener, and thickened biomass is recycled similar to that of an achieved sludge process. These process trains are shown in Figures 13.17b(i) and 13.17b(ii).

Process Parameters: The important process parameters for conventional aerobic digestion are: (a) sludge age (SRT) or aeration period, (b) volumetric solids loading (VSL), (c) air or oxygen requirements, (d) VSR, (e) sludge concentration and volatile solids fraction in the digester, (f) mixing energy

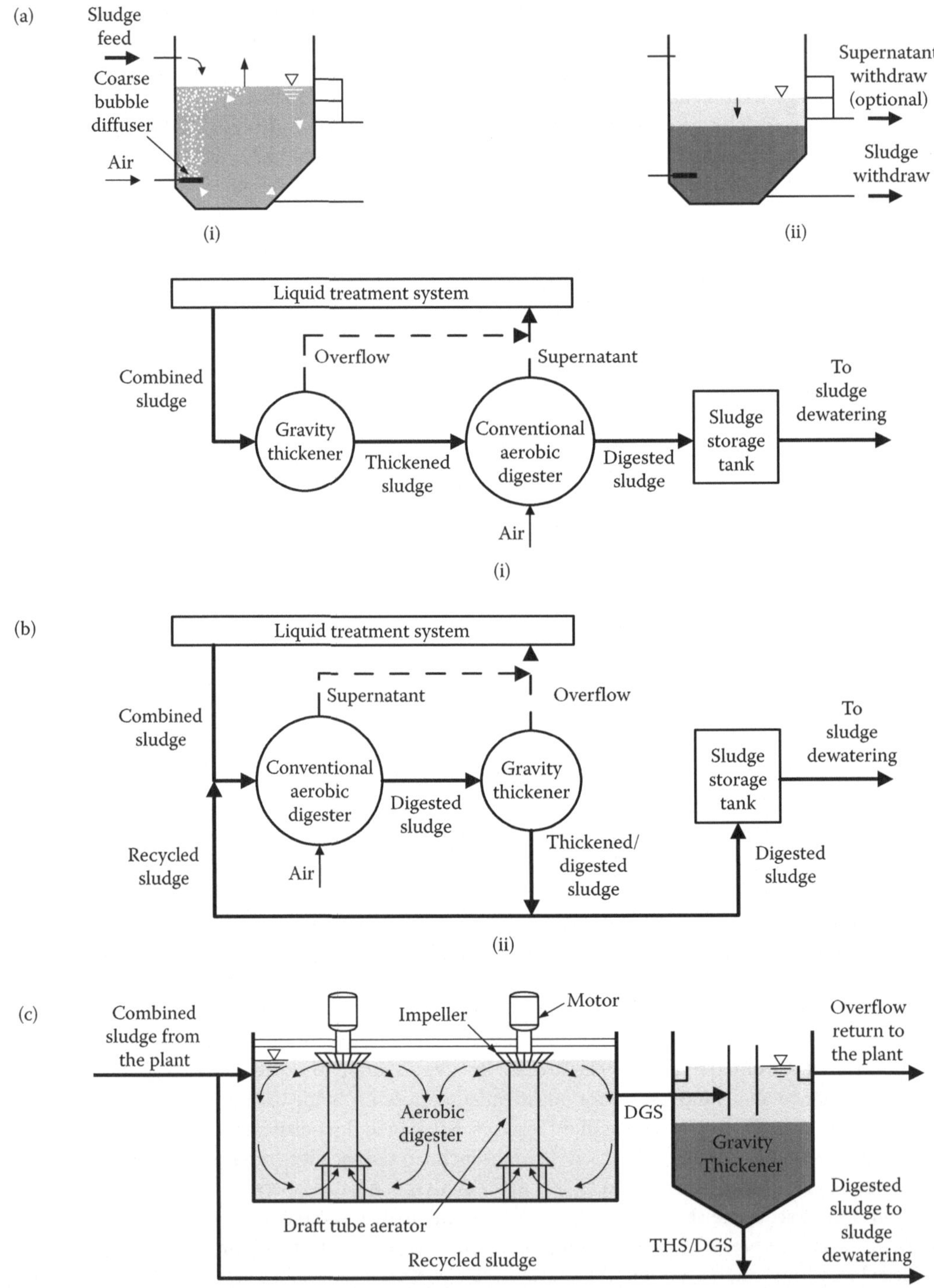

FIGURE 13.17 Conventional aerobic sludge digestion processes: (a) batch-operated aerobic digester: (i) aeration cycle with sludge feed and (ii) settling cycle with supernatant and digested sludge withdraw; (b) continuous-flow aerobic digester: (i) prethickening and (ii) postthickening; and (c) schematic of a continuous-flow rectangular aerobic digester with postthickening.

TABLE 13.16 Design and Performance Parameters of Conventional Aerobic Digestion

Parameter	Value or Range
SRT (θ_c), d	
Required for VSS reduction only	10–15
Required for meeting Class B biosolids requirement	40 at 20°C or 60 at 15°C
Volumetric solids loading (VSL), kg VSS/m^3·d (lb VSS/ft^3·d)	1.6–4.8 (0.1–0.3)
Per capita capacity requirement (PCCR), m^3/capita (ft^3/capita)	0.06–0.2 (2–7)
Oxygen requirement, kg O_2/kg VSS reduced (lb O_2/lb VSS reduced)	
Stabilization of cell tissues in waste activated sludge (WAS)	~ 2.3 (2.3)
Oxidation of BOD_5 in primary sludge	1.6–1.9 (1.6–1.9)
Volatile solids reduction (VSR), %	35–50
Energy requirement for mixing, minimum	
Diffused aeration, sm^3/m^3·min (ft^3/10^3 ft^3·min)	
WAS	0.02–0.04 (20–40)
Primary sludge and WAS	0.06 (60)
Mechanical aeration, kW/m^3 (hp/10^3 ft^3)	0.02–0.04 (0.75–1.5)
Power requirement per 10,000 population equivalent, kW (hp)	6–7.5 (8–10)
Solids content[a], %	
Feed sludge	1–4
Digested sludge	1–3
Minimum DO in the liquid[b], mg/L	1–2
Minimum sustained low temperature[c], °C (°F)	15 (60)
Minimum pH[d]	> 5.5

1 kg/m^3·d = 0.0624 lb/ft^3·d; 1 kW/m^3 = 38.0 hp/10^3 ft^3.

[a] Thickened solids content of 3–4% may require careful selection of mixing and aeration equipment.

[b] Alternation between aeration and mixing may be used to optimize nitrification and denitrification for alkalinity recovery and energy saving.

[c] High-purity oxygen may be used in the modified aerobic digestion process that is more suitable for use in cold climate.

[d] Supplemental alkalinity is considered if necessary to increase the sludge buffering capacity and maintain operating pH above 5.5 in the digester.

Source: Adapted in part from References 1 through 3, 6, 7, and 12.

requirement, (g) DO level, (h) temperature, (i) alternation of aeration and mixing periods, (j) biosolids requirement for land application, and (k) supernatant quality (if decanting). The basic design and performance parameters of conventional aerobic digestion are summarized in Table 13.16.

Digestion Tank and Equipment: The digestion tanks are open basins of rectangular, square, or circular shape. There is no heating or insulation requirement unless the minimum operating temperature of 15°C cannot be maintained in the tanks. The schematic of a rectangular aerobic digestion basin with draft-tube aerators and post thickening of digested biosolids is shown in Figure 13.17c.

Volume Requirements: The digester tank volume is determined from the design SRT or VSR. The per capita based volume requirement may be used when detailed design data is not available. The typical ranges of these design parameters are given in Table 13.16. The digester volume may also be calculated from Equation 13.11, which is a mass balance based equation.[7,32]

$$V = \frac{Q_{FS}(X_{FS} + f_{PS}S_{PI})}{X_{DG}(k_d f_{VSS/TSS,DGS} + (1/\theta_c))} \tag{13.11}$$

where

V = aerobic digester volume, m^3

Q_{FS} = daily volume of the feed sludge, m^3/d. It is equal to the daily volume of the digested sludge from a completely-mixed aerobic digester without recycle

X_{FS} = VSS in the feed sludge, mg/L (g/m^3)

X_{DG} = VSS concentration maintained in the digester, g/m^3. In a completely-mixed aerobic digester without recycle it is same as that in the digested sludge ($VSS_{DG} = VSS_{DGS}$), mg/L (g/m^3)

S_{PI} = total BOD_5 concentration in the primary influent, mg/L (g/m^3)

f_{PS} = volumetric fraction of raw primary sludge in the feed sludge, m^3 primary sludge/m^3 feed sludge

$f_{VSS/TSS,DGS}$ = VSS/TSS ratio of the digested sludge, mg VSS/mg TSS

θ_c = mean cell residence time or solids retention time (SRT), d. In a completely-mixed aerobic digester without recycle, SRT equals to the hydraulic retention time ($\theta_c = \theta$)

k_d = specific endogeneous decay coefficient, d^{-1}

Volatile Solids Reduction: The achievable VSR in a conventional aerobic digester is usually a function of digester liquid *temperature-time factor* (°C-d). This relationship may be found in References 7 and 57. Common values are summarized in Table 13.17. The aeration period is normally 15 days to achieve 40–50% VSR. Oxygen uptake rate exclusive of nitrification may vary from 3 to 30 mg/g VSS·h.

Digester Supernatant Quality: The characteristics of supernatant from the conventional aerobic sludge digesters are summarized in Table 13.18. In comparison, the organic and inorganic contaminants in the supernatant from anaerobic sludge digester (see Table 13.15) are much higher than those from aerobic digester. The contaminant levels in supernatant from aerobic digesters are closer to those in the raw wastewater influent, therefore, the impact of recycled sidestream on plant operation is significantly less.

Thermophilic Aerobic Digestion: The TAD process is the aerobic digestion operating in the thermophilic temperature range from 55°C to 70°C (131–158°F). At high pH > 8 and the operating temperature over 41°C (104°F), nitrification is inhibited and the ideal reaction in an TAD system may be expressed by Equation 13.9a. It is an exothermic microbial oxidation reaction that produces over 20,000 kJ/kg VSS reduced (or 9000 Btu/lb VSS reduced). The heat produced from the reaction is sufficient to maintain the process in the desired thermophilic temperature range. This is called autothermal (or auto-heating) thermophilic aerobic digestion (ATAD). Properly insulated reactors will maintain thermophilic temperature without external heating. Other main advantages of using ATAD process over the anaerobic and conventional aerobic digestion processes are: (a) capable of producing Class A biosolids, (b) reduced HRT or SRT to achieve the same VSR as that in a conventional aerobic digester, and (c) improved dryness of dewatered sludge cake after proper chemical conditioning. The schematic flow diagram of a two-stage ATAD system is shown in Figure 13.18a.[7,58–64]

TABLE 13.17 Typical Volatile Solids Reduction with Respect to Value of Temperature–Time Factor in Conventional Aerobic Digestion

Digester Liquid Temperature-Time Factor[a], °C-d	Volatile Solids Reduction (VSR), %
135	20
235	30
325	35
500	40
854	45
1400	50

[a] Digester temperature-time factor = Temperature (in °C) × SRT (in days).

Source: Developed based on information contained in References 7 and 57.

TABLE 13.18 Characteristics of Supernatant from Conventional Aerobic Digestion

Parameter	Range (Typical Value)
pH, standard unit	6–7.5 (7)
BOD_5, mg/L	200–1500 (500)
Soluble BOD_5, mg/L	5–200 (50)
COD, mg/L	200–8000 (2500)
TSS, mg/L	500–10,000 (3500)
Total Kjeldahl N (TKN), mg/L as N	10–400 (150)
Total P (TP), mg/L as P	20–250 (100)
Soluble P, mg/L as P	2–60 (25)

Source: Adapted in part from References 7 and 57.

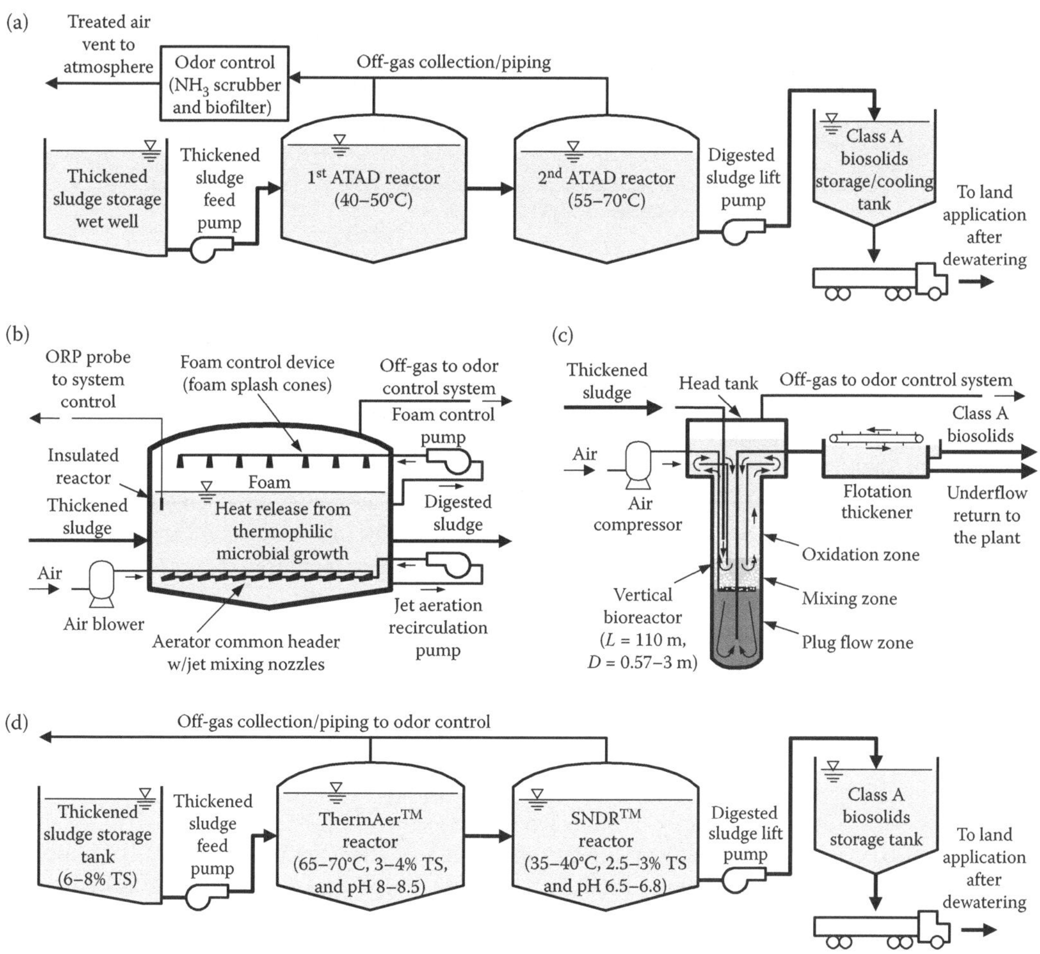

FIGURE 13.18 Autothermal thermophilic aerobic digestion (ATAD) processes: (a) schematic flow diagram of a two-stage ATAD system, (b) major components of the ATAD reactor, (c) VERTAD™ system, and (d) schematic flow diagram of ThermAer™/SNDR™ system.

TABLE 13.19 Design and Performance Parameters for ATAD Process

Parameter	Value or Range
SRT (θ_c)[a], d	
Required for VSS reduction only	4–10
Required for meeting Class A biosolids requirement	10–15
Volumetric solids loading (VSL), kg VSS/m^3·d (lb VSS/ft^3·d)	3–4 (0.2–0.25)
Operating temperature[b], °C (°F)	55–70 (131–158)
Volatile solids reduction (VSR), %	40–60
Aeration and mixing requirement[c]	
Oxygen requirement, kg O_2/kg VSS reduced (lb O_2/lb VSS reduced)	1.3–1.9
Air supply, m^3/m^3·min (ft^3/10^3 ft^3·min)	0.02–0.03 (20–30)
Power requirement, kW/m^3 (hp/10^3 ft^3)	0.1–0.12 (3.8–4.5)
Solids content in feed sludge, %	3–8
Control setting	
Target DO, mg/L	~0.5
Target ORP, mV	~0
Target pH, s. u.	8–8.5

1 kg/m^3·d = 0.0624 lb/ft^3·d; 1 kW/m^3 = 38.0 hp/10^3 ft^3.

[a] SRT = HRT in biological reactor without recycle.

[b] A temperature of 40–55°C (104–131°F) may be maintained in the 1st reactor of an ATAD system with multireactors in series.

[c] Jet aerators are operated with recirculation pump and blower (see Table 10.18 and Figure 10.52c).

Source: Adapted in part from References 7, and 58 through 64.

The basic design and performance parameters of ATAD process are summarized in Table 13.19. Other important design considerations include: (a) both single- and multi-stage ATAD systems are commercially available; (b) high solids content (6–8% TSS and > 4% VSS) in the feed sludge is required for generating sufficient heat for autothermal operation (see Example 13.34); (c) operating DO level must be low to maintain the efficient oxygen transfer since the DO saturation in sludge at operating temperatures is ~4 mg/L; (d) semi-batch operation with a holding time of at least 23 h at 55°C is preferred to prevent short circuiting and meet the Class A biosolids requirements; (e) continuous operating systems are under development (see Example 13.35); (f) ORP is used as a surrogate parameter for process control due to low DO level; (g) jet aerator operated with recirculation pump and blower is preferred aeration method over induced-air surface and aspirator aerator (see Table 10.18 and Figure 10.52); (h) VFD device is required on both pump and blower to match the oxygen uptake rate for operation near the target ORP; (i) air is typically supplied to the aerator but high-purity oxygen may also be considered; (j) proper odor control (typically a NH_3 scrubber followed by a biofilter) is required since high contents of ammonia (800–1500 ppm from protein degradation) and VFAs (from acid formation under low DO) in the off-gas are expected (see calculations of ammonia generation and dissociation in Example 13.36); and (k) a foam layer on top of the liquid must be properly controlled and maintained for effective insulation. Foam splash cones or foam cutters are typically used to keep foam layer in place. The major components of ATAD system are illustrated in Figure 13.18b.[7,58-64]

TAD has also been applied in many proprietary stabilization systems for producing Class A biosolids. The examples of these modified systems include: (a) dual digestion process, where the TAD is followed by a second stage of mesophilic anaerobic digestion (Figure 13.1c); (b) integration of ATAD into in-ground vertical reactor (or deep-shaft) technology with flotation thickener, similar to VERTAD™ system of NORAM Engineering and Constructors Ltd. (Figure 13.18c); and (c) integration of ATAD with simultaneous nitrification/denitrification process such as ThermAer™/SNDR™ system of Thermal Process Systems (Figure 13.18d).[65,66]

EXAMPLE 13.30: OXYGEN AND ALKALINITY REQUIREMENTS IN AEROBIC DIGESTION PROCESSES

Estimate the oxygen and alkalinity requirements based on stoichiometric oxidations in (a) conventional aerobic digestion with complete nitrification and (b) autothermal thermophilic aerobic digestion (ATAD) processes.

Solution

1. Estimate the oxygen and alkalinity requirements in a conventional aerobic digester with complete nitrification.
   a. Identify the stoichiometric reaction.
      The ideal biomass stabilization reaction with complete nitrification is given by Equation 13.9b.

$$C_5H_7O_2N + 7\ O_2 + HCO_3^- \rightarrow NO_3^- + 6\ CO_2 + 4\ H_2O$$

   b. Estimate the oxygen requirement.
      Seven moles of O_2 are theoretically needed per mole of $C_5H_7O_2N$ oxidized. The ratio of oxygen consumed to oxidize biomass (g O_2/g VSS reduced) with complete nitrification is determined from Equation 13.9b.

$$O_{2,VSS} = \frac{7\text{ moles } O_2 \times 32\text{ g } O_2/\text{mole } O_2}{1\text{ mole } C_5H_7O_2N \times 113\text{ g } C_5H_7O_2/\text{mole } C_5H_7O_2N\text{-oxidized}}$$

$$= 1.98\text{ g } O_2/\text{g}C_5H_7O_2N\text{-oxidized} \quad \text{or} \quad 1.98\text{ g } O_2/\text{g VSS-reduced}$$

   Note: The stoichiometric oxygen requirement of ~ 2 g O_2/g VSS reduced is obtained from the ideal oxidation reaction. It is close to the practical range from 1.6 to 2.3 g O_2/g VSS reduced (see Table 13.16). The actual oxygen requirement may be lower than that estimate, if nitrification is not complete.

   c. Estimate the alkalinity requirement.
      One mole of HCO_3^- is needed per mole of $C_5H_7O_2N$ oxidized (Equation 13.9b).

$$\text{Alkalinity consumed, } w_{CaCO3/HCO3} = \frac{1\text{ mole } HCO_3^- \times 61\text{ g } HCO_3^-/\text{mole } HCO_3^-}{61\text{ g } HCO_3^-/\text{eq. as } HCO_3^-} \times 50\text{ g/eq. as CaC}$$

$$= 1\text{ eq.} \times 50\text{ g/eq. as } CaCO_3 = 50\text{ g as } CaCO_3$$

$$\text{Alkalinity to VSS ratio, } Alk_{CaCO3/VSS} = \frac{50\text{ g as } CaCO_3}{1\text{ mole } C_5H_7O_2N \times 113\text{ g } C_5H_7O_2/\text{mole } C_5H_7O_2N\text{ oxidized}}$$

$$= 0.44\text{ g Alk as } CaCO_3/\text{g } C_5H_7O_2N\text{ oxidized}$$

$$\text{or} \quad 0.44\text{ g Alk as } CaCO_3/\text{g VSS reduced}$$

   Note: Alkalinity is consumed in biomass stabilization with complete nitrification. During aerobic digestion, partial nitrification may occur at low DO. This will consume alkalinity. A depression of pH will occur in most aerobic sludge digesters. The process should be monitored carefully to ensure that the operating pH does not drop below 5.5 (see Table 13.16).

2. Estimate the oxygen and alkalinity requirements in ATAD process.
   a. Identify the stoichiometric reaction.

Nitrification is inhibited under typical operating conditions in an ATAD system. Therefore, the digestion process maybe expressed by the stoichiometric reaction expressed by Equation 13.9a.

$$C_5H_7O_2N + 5\ O_2 \rightarrow 4\ CO_2 + NH_4^+ + HCO_3^- + H_2O$$

b. Estimate the oxygen requirement.

The ratio of oxygen requirement is calculated from Equation 13.9a.

$$O_{2,VSS} = \frac{5 \text{ moles } O_2 \times 32\,\text{g } O_2/\text{mole } O_2}{1 \text{ mole } C_5H_7O_2N \times 113\,\text{g } C_5H_7O_2/\text{mole } C_5H_7O_2N \text{ oxidized}}$$

$$= 1.42\,\text{g } O_2/\text{g}C_5H_7O_2N \text{ oxidized} \quad \text{or} \quad 1.42\,\text{g } O_2/\text{g VSS-reduced}$$

Note: The calculated oxygen requirement of 1.42 g O_2/g VSS reduced is within the typical range from 1.3 to 1.9 g O_2/g VSS reduced (see Table 13.19).

c. Estimate the increase in alkalinity.

One mole of HCO_3^- is created per mole of $C_5H_7O_2N$ oxidized. The alkalinity produced by the ATAD process is calculated from Equation 13.9a.

$$Alk_{CaCO3/VSS} = \frac{50\,\text{g as } CaCO_3}{1 \text{ mole } C_5H_7O_2N \times 113\text{ g } C_5H_7O_2/\text{mole } C_5H_7O_2N\text{-oxidized}}$$

$$= 0.44\,\text{g Alk as } CaCO_3/\text{g } C_5H_7O_2N\text{-oxidized}$$

$$\text{or} \quad 0.44\,\text{g Alk as } CaCO_3/\text{g VSS-reduced}$$

Note: Alkalinity produced by the aerobic digestion without nitrification will increase the pH. For this reason, the pH level in ATAD system is typically in the range from 8 to 8.5 (see Table 13.19).

3. Summarize the calculation results.

The calculation results of the oxygen and alkalinity requirements in the aerobic digestion processes are summarized below.

Parameter	Aerobic Sludge Digestion	
	Conventional	ATAD
Oxygen requirement, g O_2/g VSS reduced	1.98 (consumed)	1.42 (consumed)
Alkalinity requirement, g Alk as $CaCO_3$/g VSS reduced	0.44 (consumed)	0.44 (produced)

EXAMPLE 13.31: SPECIFIC ENDOGENOUS DECAY COEFFICIENT (K_d) IN AEROBIC DIGESTION

A batch aerobic digester was operated at 20°C. The VSS concentrations in two samples withdrawn after 10 and 20 days of digestion are 21 and 11 g VSS/L, respectively. Determine (a) the specific endogenous decay coefficient (k_d) and (b) the total biodegradable VSS (X_0) in the feed sludge.

Solution

1. Substitute the bench-scale data in Equation 13.10c to obtain the following simultaneous equations.

$$21\,\text{g VSS/L} = X_0 e^{-k_d \times 10\,\text{d}}$$

$$11\,\text{g VSS/L} = X_0 e^{-k_d \times 20\,\text{d}}$$

2. Solve for k_d.

   Divide two simultaneous equations.

$$\frac{21\text{ g VSS/L}}{11\text{ g VSS/L}} = \frac{X_0 e^{-k_d \times 10\text{ d}}}{X_0 e^{-k_d \times 20\text{ d}}}$$

$$1.91 = e^{-k_d \times (10-20)d} = e^{k_d \times 10\,d}$$

$$k_d = \frac{1}{10\text{ d}} \times \ln(1.91) = 0.065\text{ d}^{-1}$$

   The final value of $k_d = 0.065\text{ d}^{-1}$.

3. Determine the initial total biodegradable VSS (X_0).

   Rearrange Equation 13.10c to calculate X_0.

$$X_0 = X_t e^{k_d t} = 21\text{ g VSS/L} \times e^{0.065\text{ d}^{-1} \times 10\text{ d}} = 40\text{ g VSS/L}$$

EXAMPLE 13.32: OBSERVED ENDOGENOUS DECAY COEFFICIENT AND EFFECT OF SOLUBILIZATION ON VSS

An aerobic digester was operated to stabilize waste activated sludge from a POTW. The digestion period is 20 days, and VSS concentrations in the feed and digested sludge are 8000 and 4500 mg/L, respectively. The nonbiodegradable portion of VSS in the feed sludge (X_{nb}) is 25%. The COD of VSS is 1.42 g COD/g VSS_{bd} (biodegradable VSS). The soluble COD of the feed and digested sludge are 150 and 1000 mg/L, respectively. Determine the observed endogenous decay coefficient ($k_{d,obs}$) based on (a) VSS reduction only, (b) VSS reduction with solubilization of VSS, (c) biodegradable VSS reduction only, (d) biodegradable VSS reduction with solubilization of biodegradable portion of VSS, and (e) biodegradable VSS reduction with solubilization of nonbiodegradable portion of VSS. Ignore the cell debris from endogenous respiration.

Solution

1. Identify the expression used for estimating the observed endogenous decay coefficient.

   The observed k_d is estimated from Equation 13.12, which is approximately the same as Equation 13.10b.

$$k_{d,obs} = \frac{\Delta X}{\theta X_e} = \frac{X_0 - X_e}{\theta X_e} \tag{13.12}$$

   where,

   $k_{d,obs}$ = observed endogenous decay coefficient, d^{-1}

   ΔX = amount of VSS that undergoes endogenous decay, mg VSS/L

   X_0 = initial VSS concentration in the feed sludge, mg VSS/L

   X_e = VSS concentration in the digested sludge, mg VSS/L

   θ = solids retention time (digestion period) without recirculation, d

2. Compute the $k_{d,obs}$ based on VSS reduction only.

   VSS that undergoes endogenous decay, $\Delta X = X_0 - X_e = (8000 - 4500)$ mg VSS/L = 3500 mg VSS/L

$$k_{d,obs}\text{ from Equation 13.12, } k_{d,obs} = \frac{\Delta X}{\theta X_e} = \frac{3500\text{ mg VSS/L}}{20\text{ d} \times 4500\text{ mg VSS/L}} = 0.039\text{ d}^{-1}$$

3. Compute the $k_{d,obs}$ based on VSS reduction with solubilization of VSS.

$$\text{VSS solubilized, } X_s = \frac{S_e - S_0}{\text{COD/VSS}_{bd}\text{ ratio}} = \frac{(1000 - 150)\text{ mg/L}}{1.42\text{ mg COD/mg VSS}_{bd}} \approx 600\text{ mg VSS/L}$$

$$\text{Nonsolubilized VSS available for endogenous decay, } X_{ns} = X_0 - X_s = (8000 - 600)\text{ mg VSS/L} = 7400\text{ mg VSS/L}$$

$$\text{VSS that undergoes endogenous decay, } \Delta X = X_{ns} - X_e = (7400 - 4500)\text{ mg VSS/L} = 2900\text{ mg VSS/L}$$

$$k_{d,obs}\text{ from Equation 13.12, } k_{d,obs} = \frac{\Delta X}{\theta X_e} = \frac{2900\text{ mg VSS/L}}{20\text{ d} \times 4500\text{ mg VSS/L}} = 0.032\text{ d}^{-1}$$

4. Compute the $k_{d,obs}$ based on biodegradable VSS reduction only.

$$\text{Nonbiodegradable VSS in the feed sludge, } X_{nb} = f_{nb}X_0 = 0.25\text{ mg VSS/mg VSS} \times 8000\text{ mg VSS/L} = 2000\text{ mg VSS/L}$$

$$\text{Biodegradable VSS in the feed sludge, } X_{bd} = X_0 - X_{nb} = (8000 - 2000)\text{ mg/L} = 6000\text{ mg VSS/L}$$

Assume that the nonbiodegradable portion of VSS in the feed sludge remains unchanged in the aerobic digestion. Nonbiodegradable portion of VSS in the digested sludge, $X_{e,nb} = X_{nb}$.

$$\text{Biodegradable portion of VSS in the digested sludge, } X_{e,bd} = X_e - X_{e,nb} = (4500 - 2000)\text{ mg/L} = 2500\text{ mg VSS/L}$$

$$\text{VSS}_{bd}\text{ that undergoes endogenous decay, } \Delta X_{bd} = X_{bd} - X_{e,bd} = (6000 - 2500)\text{ mg VSS/L} = 3500\text{ mg VSS/L}$$

$$k_{d,obs}\text{ from Equation 13.12, } k_{d,obs} = \frac{\Delta X_{bd}}{\theta X_{e,bd}} = \frac{3500\text{ mg VSS/L}}{20\text{ d} \times 2500\text{ mg VSS/L}} = 0.070\text{ d}^{-1}$$

5. Compute the $k_{d,obs}$ based on biodegradable VSS reduction with solubilization of biodegradable portion of VSS.

$$\text{Nonsolubilized VSS}_{bd}\text{ available for endogenous decay, } X_{bd,ns} = X_{bd} - X_s = (6000 - 600)\text{ mg VSS/L} = 5400\text{ mg VSS/L}$$

$$\text{VSS}_{bd}\text{ that undergoes endogenous decay, } \Delta X_{bd} = X_{bd,ns} - X_{e,bd} = (5400 - 2500)\text{ mg VSS/L} = 2900\text{ mg VSS/L}$$

$$k_{d,obs}\text{ from Equation 13.12, } k_{d,obs} = \frac{\Delta X_{bd}}{\theta X_{e,bd}} = \frac{2900\text{ mg VSS/L}}{20\text{ d} \times 2500\text{ mg VSS/L}} = 0.058\text{ d}^{-1}$$

6. Compute the $k_{d,obs}$ based on biodegradable VSS reduction with solubilization of nonbiodegradable portion of VSS.

Nonbiodegradable portion of VSS in the digested sludge, $X_{e,nb} = X_{nb} - X_s = (2000 - 600)\text{ mg/L} = 1400\text{ mg VSS/L}$

Biodegradable portion of VSS in the digested sludge, $X_{e,bd} = X_e - X_{e,nb} = (4500 - 1400)\text{ mg/L} = 3100\text{ mg VSS/L}$

VSS_{bd} that undergoes endogenous decay, $\Delta X_{bd} = X_{bd} - X_{e,bd} = (6000 - 3100)\text{ mg VSS/L} = 2900\text{ mg VSS/L}$

$k_{d,obs}$ from Equation 13.12, $k_{d,obs} = \dfrac{\Delta X_{bd}}{\theta X_{e,bd}} = \dfrac{2900\text{ mg VSS/L}}{20\text{ d} \times 3100\text{ mg VSS/L}} = 0.047\text{ d}^{-1}$

7. Summarize the calculation results.

The range of observed endogenous decay coefficient $k_{d,obs}$ based on VSS reduction with and without solubilization is 0.039 and 0.032 d^{-1}. Based on the biodegradable VSS reductions with and without solubilization, the $k_{d,obs}$ values range from 0.047 to 0.070 d^{-1}. The effects of VSS solubilization on $k_{d,obs}$ is quite significant.

EXAMPLE 13.33: DESIGN OF AN AEROBIC DIGESTER

A conventional aerobic digester is designed to digest 1000 kg TSS/d combined sludge. The solids are concentrated in a gravity thickener to 3% solids content. The specific gravity is 1.005. To meet the Class B biosolids requirement the SRT (θ_c) is 60 days at the critical sludge temperature of 15°C in winter. The specific endogenous decay coefficient is 0.05 d^{-1} at 15°C. The sustained operating temperature in the digester is 25°C in summer. The VSS fraction in the feed sludge $f_{VSS/TSS,FS} = 0.75$, the fraction of primary sludge $f_{PS} = 0.3$, and total BOD_5 in plant influent $S_{PI} = 250$ mg/L. The operating VSS concentration in the digester $TSS_{DG} = 20{,}000$ mg/L. The water density $\rho_w = 1000\text{ kg/m}^3$.

Solution

1. Determine the daily average sludge flow to the digester from Equation 13.1e.

Feed sludge flow, $Q_{FS} = \dfrac{100\% \times W_{s,FS}}{p_{s,FS} S_{b,FS} \rho_w} = \dfrac{100\% \times 1000\text{ kg/d}}{3\% \times 1.005 \times 1000\text{ kg/m}^3} = 33.2\text{ m}^3\text{/d}$

2. Determine the volatile solids reductions (VSRs) in winter and summer.

In winter, the temperature-time factor = 15°C × 60 d = 900°C-d. The achievable VSR_{winter} from Table 13.17 = 45%.

In summer, the temperature-time factor = 25°C × 60 d = 1500°C-d. The achievable VSR_{summer} from Table 13.17 = 50%. Both VSRs are meeting the volatile solids reduction of 38% for vector attraction reduction requirements for Class B biosolids.

3. Determine the mass components of solids in the digested sludge in winter.

Mass of VSS reaching the digester, $W_{VSS,FS} = f_{VSS/TSS,FS} W_{s,FS} = 0.75\text{ kg VSS/kg TSS} \times 1000\text{ kg TSS/d} = 750\text{ kg VSS/d}$

Mass of fixed solids (FSS) reaching the digester, $W_{FSS,FS} = W_{s,FS} - W_{VSS,FS} = (1000 - 750)\,kg/d = 250\,kg\ FSS/d$

Mass of VSS reduced in winter, $\Delta W_{VSS,winter} = VSR_{winter} \times W_{VSS,FS} = 0.45 \times 750\,kg\ VSS/d = 338\,kg\ VSS/d$

Mass of VSS remained in the digested sludge, $W_{VSS,DGS} = W_{VSS,FS} - W_{VSS,winter} = (750 - 338)\,kg\ VSS/d = 412\,kg\ VSS/d$

Mass of TSS remained in the digested sludge, $W_{TSS,DGS} = W_{VSS,DGS} + W_{FSS,FS} = (412 + 250)\,kg/d = 662\,kg\ VSS/d$

$$\text{VSS fraction in the digester sludge, } f_{VSS/TSS,DGS} = \frac{W_{VSS,DGS}}{W_{TSS,DGS}} = \frac{412\,kg\ VSS/d}{662\,kg\ TSS/d} = 0.62\,kg\ VSS/kg\ TSS$$

4. Determine the required total digester volume.

$$\text{VSS concentration in the feed sludge, } X_{FS} = \frac{W_{s,FS}}{Q_{FS}} = \frac{750\,kg/d}{33.2\,m^3/d} \times 10^6\,mg/kg \times 10^{-3}\,m^3/L = 22{,}600\,mg/L$$

VSS concentration in the digester, $X_{DG} = f_{VSS/TSS,DGS}\,TSS_{DG} = 0.62\,mg\ VSS/mg\ TSS \times 20{,}000\,mg\ TSS/L = 12{,}400\,mg\ VSS/L$

The required total digester volume from Equation 13.11.

$$V_{DG} = \frac{Q_{FS}(X_{FS} + f_{PS}S_{PI})}{X_{DG}(k_d f_{VSS/TSS,DGS} + (1/\theta_c))}$$

$$= \frac{33.2\,m^3/d \times (22{,}600\,mg\ VSS/L + 0.3 \times 250\,mg\ BOD_5/L)}{12{,}400\,mg\ VSS/L \times (0.05\,d^{-1} \times 0.62 + (1/60\,d))} = 1270\,m^3$$

5. Select the dimensions of the aerobic digester.
Provide a total digester volume $V_{DG} = 1400\ m^3$. Provide two 700-m^3 aerobic digesters each having a post thickener and sludge recycle capability (see Figures 13.17b, ii and c). The dimensions of each digester are:

length $L = 16\,m$, width $W = 8\,m$, depth $H = 5.5\,m$, and actual volume $V = 700\,m^3$.

6. Verify the volumetric solids loading (*VSL*).

$$VSL = \frac{W_{VSS,FS}}{V_{DG}} = \frac{750\,kg\ VSS/d}{1400\,m^3} = 0.54\,kg\ VSS/m^3{\cdot}d$$

The *VSL* is below the typical range from 1.6 to 4.8 kg VSS/m^3·d (see Table 13.16). The design volume is considered conservative.

7. Determine the aeration requirement in summer.

Mass of VSS reduced in the digester, $\Delta W_{VSS,summer} = VSR_{summer} \times W_{VSS,FS} = 0.5 \times 750\,\text{kg TSS/d}$
$= 375\,\text{kg VSS/d}$

Provide the conservative oxygen requirement of 2.3 kg O_2/kg VSS reduced (Table 13.16).

Mass of oxygen required, $W_{O_2} = 2.3\,\text{kg } O_2/\text{kg VSS reduced} \times W_{VSS,summer}$
$= 2.3\,\text{kg } O_2/\text{kg VSS reduced} \times 375\,\text{kg VSS/d} = 863\,\text{kg } O_2/\text{d}$

At standard condition (20°C and 1 atm), the air contains 23.2% oxygen by weight and the density of air is 1.204 kg/m^3. Use a standard oxygen transfer efficiency $SOTE = 10\%$ for coarse bubble diffusers (see Table 10.18).

$$\text{Air requirement for aeration, } Q_{air} = \frac{W_{O_2} \times 100\%}{0.232\,\text{kg } O_2/\text{kg air} \times 1.204\,\text{kg/sm}^3\text{ air} \times \text{SOTE}}$$
$$= \frac{863\,\text{kg } O_2/\text{d} \times 100\%}{0.232\,\text{kg } O_2/\text{kg air} \times 1.204\,\text{kg/m}^3\text{ air} \times 10\%}$$
$$= 31{,}000\,\text{sm}^3/\text{d} \quad \text{or} \quad 21.5\,\text{sm}^3/\text{min}$$

8. Verify the air requirement for mixing.

$$\text{Air supply per unit digester volume, } Q_{air} = \frac{Q_{air}}{V_{DG}} = \frac{21.5\,\text{sm}^3/\text{min}}{1400\,\text{m}^3}$$
$$= 0.015\,\text{sm}^3/\text{m}^3\cdot\text{min}$$

The minimum air supply required for diffused aeration is 0.06 sm^3/m^3·min for mixing combined primary sludge and WAS (see Table 13.16). Therefore, provide the minimum air requirement for mixing.

$$Q_{air} = 0.06\,\text{sm}^3/\text{m}^3\cdot\text{min} \times 1400\,\text{m}^3 = 84\,\text{sm}^3/\text{min}$$

Provide a design air supply capacity of 125 sm^3/min. This gives a safety factor of 1.5.

EXAMPLE 13.34: HEAT GENERATION IN ATAD SYSTEM

Estimate the solids content required in the feed sludge to generate sufficient heat to maintain the operating temperature of 65°C in an ATAD system. The temperature and the VSS/TSS ratio of feed sludge are 20°C and 0.75. Assume that the volatile solids reduction $VSR = 50\%$. The heat production from the exothermic microbial oxidation reaction is 20,000 kJ/kg VSS reduced, sludge density $\rho_s \approx \rho_w = 1$ kg/L, and the specific heat of sludge $C_{sh,s} \approx C_{sh,w} = 4200$ J/kg·°C.

Solution

1. Calculate the overall heat requirement for heating the feed sludge to the operating temperature from Equation 13.6a.

$$H_{r,s} = C_{sh,s}\rho_s(T_d - T_s) = 4200\,\text{J/kg}\cdot°\text{C} \times 1\,\text{kg/L} \times (65 - 20)°\text{C}$$
$$= 1.89 \times 10^5\,\text{J/L} \quad \text{or} \quad 189\,\text{kJ/L}$$

Assume an overall efficiency of heat transfer and losses from the digester = 0.6. Calculate the overall heat requirement from the exothermic microbial oxidation reaction.

$$H_{r,reaction} = \frac{H_{r,s}}{E} = \frac{189\,\text{kJ/L}}{0.6} = 315\,\text{kJ/L}$$

2. Calculate the required VSS concentration in the feed sludge for providing the heat requirement.

$$\text{Required reduction of VSS concentration, } \Delta X = \frac{H_{r,\text{reaction}}}{C_{h,\text{VSS}}} = \frac{315\,\text{kJ/L}}{20{,}000\,\text{kJ/kg VSS reduced}} \times 10^3\,\text{g/kg}$$

$$= 15.75\,\text{g VSS/L}$$

$$\text{Required VSS concentration in the feed sludge, } X_{\text{FS}} = \frac{\Delta X}{VSR} = \frac{15.75\,\text{g VSS/L}}{0.5} = 31.5\,\text{g VSS/L}$$

$$\text{Required TSS concentration in the feed sludge, } \text{TSS}_{\text{FS}} = \frac{X_{\text{FS}}}{\text{VSS/TSS ratio}} = \frac{31.5\,\text{g VSS/L}}{0.75\,\text{g VSS/g TSS}}$$

$$= 42\,\text{g TSS/L}$$

Note: For meeting the autothermal operation of ATAD system, the TSS concentrations of 6–8, and VSS concentration greater than 4% in the feed are needed.

EXAMPLE 13.35: HOLDING TIME REQUIRED IN THERMOPHILIC AEROBIC DIGESTER (TAD)

Thermophilic aerobic digestion is evaluated for treating the combined sludge from a POTW to meet the Class A biosolids requirement. Assume that the solids content in the digester is 6%. Determine the minimum holding times required at operating temperatures of 55°C, 60°C, or 65°C prior to discharge from the digester.

Solution

1. Identify the applicable time–temperature regime.

   U.S. EPA has developed time–temperature regimes for achieving Class A pathogen reduction. For sludge solids less than 7% and temperature 50°C or greater, the relationship between time and temperature is expressed by Equation 13.13.[18]

$$t = \frac{50{,}070{,}000}{10^{0.14T}} \tag{13.13}$$

   where

   t = sludge holding time, d

   T = sludge temperature, °C

2. Determine the minimum time required from Equation 13.13.

$$\text{Holding time required at temperature of 55°C, } t_{55} = \frac{50{,}070{,}000}{10^{0.14\times 55°\text{C}}} = 1\,\text{d} \quad \text{or} \quad 24\,\text{h}$$

$$\text{Holding time required at temperature of 60°C, } t_{60} = \frac{50{,}070{,}000}{10^{0.14\times 60°\text{C}}} = 0.2\,\text{d} \quad \text{or} \quad 5\,\text{h}$$

$$\text{Holding time required at temperature of 65°C, } t_{65} = \frac{50{,}070{,}000}{10^{0.14\times 65°\text{C}}} = 0.04\,\text{d} \quad \text{or} \quad 1\,\text{h}$$

   The holding times are >30 min at all three temperatures, and Equation 13.13 condition is valid.

   Note: To ensure effective pathogen reduction, it is desired to operate the TAD in semi-batch mode on daily basis. The digested sludge is withdrawn once a day and the digester is fed within a short time (typically within 1 h). The digester is then held in batch operation at a temperature above 55°C for the rest of the day.

EXAMPLE 13.36: ALKALINITY AND AMMONIA GENERATIONS IN ATAD SYSTEM

Estimate (a) the increase in alkalinity and (b) ammonia generation due to endogenous decay of VSS in an ATAD system. Also estimate the initial equilibrium concentration of ammonia species at the typical operating temperature of 65°C and pH of 8.25 in the digester. The VSS concentration in the feed sludge $X_{FS} =$ 45 g/L and the volatile solids reduction VSR = 50%.

Solution

1. Identify the stoichiometric reaction.

The nitrification process is inhibited under the typical operating conditions in an ATAD system. Therefore, the digestion process is expressed by the stoichiometric reaction expressed by Equation 13.9a.

$$C_5H_7O_2N + 5\ O_2 \rightarrow 4\ CO_2 + NH_4^+ + HCO_3^- + H_2O$$

2. Estimate the alkalinity increase due to endogenous decay of VSS.

VSS concentration reduced in the digester, $\Delta X = VSR \times X_{FS} = 0.5 \times 45$ g VSS/L = 22.5 g VSS/L

An alkalinity production ratio $Alk_{CaCO_3/VSS} = 0.44$ g Alk as $CaCO_3$/g VSS is obtained in Step 2.c of Example 13.30. Calculate the alkalinity increase in the digester.

$$\begin{aligned} Alk_{generation} &= Alk_{CaCO_3/VSS} \times \Delta X = 0.44\,\text{g Alk as CaCO}_3/\text{g VSS-reduced} \times 22.5\,\text{g VSS/L} \\ &= 9.9\,\text{g/L as CaCO}_3 \quad \text{or} \quad 9900\,\text{mg/L as CaCO}_3 \end{aligned}$$

Note: A significant amount of alkalinity is produced from the endogenous decay of VSS. An elevated pH is expected in the ATAD digester.

3. Estimate the concentration of ammonia generated in the digester.

The ratio of ammonia generation and VSS reduced from Equation 13.9a,

$$\begin{aligned} N_{NH_4/VSS} &= \frac{1\text{ mole NH}_4^+ \times 14\,\text{g NH}_4^+\text{-N/mole NH}_4^+}{1\text{ mole C}_5\text{H}_7\text{O}_2\text{N} \times 113\,\text{g C}_5\text{H}_7\text{O}_2/\text{mole C}_5\text{H}_7\text{O}_2\text{ N-oxidized}} \\ &= 0.124\,\text{g NH}_4^+\text{-N/g C}_5\text{H}_7\text{O}_2\text{N-oxidized} \quad \text{or} \quad 0.124\,\text{g NH}_4^+\text{-N/g VSS-reduced} \end{aligned}$$

Increase in ammonia concentration from the endogenous decay in the digester,

$$\begin{aligned} N_{NH_4,generation} &= N_{NH_4/C_5H_7O_2N} \times \Delta X = 0.124\,\text{g NH}_4^+\text{-N/g VSS-reduced} \times 22.5\,\text{g VSS/L} \\ &= 2.79\,\text{g NH}_4^+\text{-N/L} \quad \text{or} \quad 2790\,\text{mg NH}_4^+\text{-N/L} \end{aligned}$$

Note: Ammonia nitrogen concentration in raw domestic wastewater is typically in a range from 20 to 40 mg/L as NH_4^+-N. A significant amount of ammonia nitrogen is generated in an ATAD system.

4. Estimate the initial equilibrium concentration of ammonia species in the ATAD system.

As discussed in Section 11.6.2, ammonia nitrogen generated from endogenous decay will remain in equilibrium as ammonium ion (NH_4^+) and unionized ammonia (NH_3) at equilibrium. The partial dissociation of NH_4^+ is expressed by Equation 11.13a.

$$NH_4^+ \leftrightarrow NH_3 + H^+$$

Determine the pK_a from the equilibrium reaction (Equation 11.13d) at the average operating temperature of $T = 273 + 65°C = 338°K$.

$$pK_a = 0.09018 + \frac{2730}{T} = 0.09018 + \frac{2730}{338} = 8.17$$

Note: The pK_a is within the typical operating pH range of 8–8.5 for ATAD systems. Ammonia nitrogen in the ATAD digester will reach an equilibrium between NH_4^+ and NH_3.

The percent NH_4^+ in equilibrium can be determined from Equation 11.13e at the average operating pH = 8.25.

$$\%NH_4^+ = \frac{1}{1+10^{(pH-pK_a)}} \times 100\% = \frac{1}{1+10^{(8.25-8.17)}} \times 100\% \approx 45\%$$

The equilibrium concentration of NH_4^+ initially reached in the digester content,

$$N_{NH4,equilibrium} = \%NH_4^+ \times N_{NH4,generation} = 0.45 \times 2.79\,g\ NH_4^+\text{-N/L}$$
$$= 1.26\,g\ NH_4^+\text{-N/L} \quad \text{or} \quad 1260\,mg\ NH_4^+\text{-N/L}$$

The equilibrium concentration of NH_3 initially formed in the digested content,

$$N_{NH3,equilibrium} = N_{NH4,generation} - N_{NH4,equilibrium} = (2.79 - 1.26)\,g\ NH_4^+\text{-N/L}$$
$$= 1.53\,g\ NH_3\text{-N/L} \quad \text{or} \quad 1530\,mg\ NH_3\text{-N/L}$$

Note: Unionized ammonia may be stripped out with other off-gases. Therefore, high ammonia concentration is expected in both liquid discharge and off-gases. A special ammonia scrubber is needed to treat the off-gases prior to the biofilter for odor control. Nitrification is also required if sidestream is returned to the plant. Integration of ATAD with simultaneous nitrification/denitrification process such as ThermAer™/SNDR™ system may also be considered to reduce the high ammonia concentration in the digested sludge and sidestream (see Figure 13.18d).

EXAMPLE 13.37: VOLATILE SOLIDS REDUCTION (VSR) IN MULTISTAGE THERMOPHILIC AEROBIC DIGESTION

A two-stage TAD system is operated at an average operating temperature of 60°C and a SRT of 6 d. Estimate the VSS reductions based on (a) biodegradable VSS and (b) overall VSS. Assume $k_{d,20} = 0.12\ d^{-1}$ and $\theta_T = 1.023$. The biodegradable portion of VSS in the feed sludge is 75%. Ignore the solubilization of biosolids.

Solution

1. Determine the specific endogenous decay coefficient (k_d) at 60°C from Equation 10.17b.

$$k_{d,60} = k_{d,20}\theta_T^{(T-20)} = 0.12\ d^{-1} \times 1.023^{(60-20)} = 0.298\ d^{-1}$$

2. Estimate the VSS reduction based on biodegradable (VSR_{bd}).

The biodegradable VSS concentration ($X_{bd,DGS}$) in the digested sludge and overall biodegradable VSS reduction (VSR_{bd}) in a multistage TAD may be estimated from Equation 13.14.[60]

$$X_{bd,DGS} = \left(\frac{1}{1+k_d\theta_c}\right)^n X_{bd,FS} \tag{13.14a}$$

Rearrange Equation 13.14a to obtain Equation 13.14b.

$$\frac{X_{bd,FS} - X_{bd,DGS}}{X_{bd,DGS}} = 1 - \left(\frac{1}{1+k_d\theta_c}\right)^n \quad \text{or} \quad VSR_{bd} = 1 - \left(\frac{1}{1+k_d\theta_c}\right)^n \tag{13.14b}$$

where

$X_{bd,DGS}$ = biodegradable VSS concentration in the nth digester, mg VSS/L

$X_{bd,FS}$ = biodegradable VSS concentration in the feed sludge, mg VSS/L

VSR_{bd} = overall biodegradable VSS based volatile solids reduction (VSR), dimensionless
n = number of stages in the TAD system, stage

Calculate the VSR_{bd} from Equation 13.14b.

$$VSR_{bd} = 1 - \left(\frac{1}{1 + k_d\theta_c}\right)^n = 1 - \left(\frac{1}{1 + 0.298\,d^{-1} \times 6\,d}\right)^2 = 0.87 \quad \text{or} \quad VSR_{bd} = 87\%$$

3. Estimate the VSS reduction (VSR).

$$VSR = VSS_{bd}/VSS \text{ ratio} \times VSR_{bd} = 0.75 \times 0.87 = 0.65 \quad \text{or} \quad VSR = 65\%$$

13.6.3 Chemical Stabilization

Chemical stabilization involves reducing odors and slowing down microbial activity. Most commonly used chemical is chlorine (either Cl_2 or NaOCl) for small operations. For larger operations, chlorine oxidation is not cost-effective because large amount of chlorine is required. Other chemicals that may also be considered are hydrogen peroxide (H_2O_2) and ozone (O_3).[1,3,32]

Lime stabilization of sludge with either quicklime (CaO) or hydrated lime ($Ca(OH)_2$) has been widely used at large plants. By adding lime, the pH of sludge is raised to 12 or higher. At such high pH and after sufficient contact time, a significant level of pathogens are killed (see Section 3.4.4). Lime-treated sludge does not petrify or create odors, or pose health hazards. Typically, lime stabilization does not destroy the organic matter. Instead, the organic matter is temporarily denaturized and petrification is halted. There is a 10–35% apparent reduction in VSS/TSS ratio due to increase in inorganic matters of lime sludge (mostly $CaCO_3$). Odor production is minimized and is reduced as long as pH remains high. At high pH, there is also loss of NH_4^+-N due to volatilization of NH_3. The sludge must be quickly disposed of before the sludge starts to petrify again. Lime may also react with many organic and inorganic constituents in sludge. Heavy metal will be immobilized from liquid phase into sludge. Lime reactions with alkalinity, hardness, and phosphorus have been presented in Sections 9.5.2 through 9.5.4. Many stoichiometric equations and examples for dose requirement and sludge production with lime reactions may be found in these sections. Some basic information is summarized below.[6,7,18,67–69]

1. Lime stabilization will require a designated lime facility for lime storage, dry-feeder, slaking, slurry preparation with mixing, and pug mill for blending lime slurry into the sludge. The process design and equipment selection are similar to that presented in Sections 9.5.2 through 9.5.6.
2. Lime can be applied either before or after sludge dewatering for pre- or postlime stabilization. The lime dosages for prelime stabilization are given in Table 13.20. Quicklime is typically used in postlime stabilization at a dosage of 0.15–0.3 kg $Ca(OH)_2$/kg dry solids in the dewatered sludge cake.

TABLE 13.20 Lime Dosage for Sludge Prelime Stabilization

Sludge Type	Solids Content, %		Lime Dosage, kg $Ca(OH)_2$/kg Dry Solids	
	Range	Typical	Range	Typical
Primary sludge	1–6	4	0.05–0.15	0.1
Waste activate sludge (WAS)	0.2–1.2	1	0.2–0.4	0.30
Combined primary sludge and WAS	1–4	2.5	0.15–0.35	0.25
Digested sludge	1–6	3.5	0.15–0.3	0.2

Source: Adapted in part from References 6, 7, 12, and 57.

3. If quicklime is used, considerable amount of heat is released and temperature rises. Pathogen destruction as high as 99% or more occurs. The process is capable of producing Class A biosolids by pasteurizing the biosolids at 70°C or higher for at least 30 min (see Section 13.4.4, Table 13.7, and Figure 13.2b for Class A, Alt. 5, PFRP No. 7).

13.6.4 Heat Treatment or Thermal Stabilization

Thickened sludge is heated to 140–250°C (285–480°F) for ~ 30 min at a pressure of 2760 kN/m^2 (400 psi). Heating coagulate solids, break the gel structure, and reduce the water affinity of solids. The heat-treated sludge is conditioned, stabilized, deodorized, and dewaters readily. This process is capable of producing Class A biosolids by heating the liquid biosolids at a temperature of 180°C or higher for 30 min (see Section 13.4.4 for Class A, Alt. 5, PFRP No. 3). The capital cost is high and sidestream contains high concentrations of dissolved organic solids, nitrogen, and phosphorus. The process is limited for large installations.[6,7,18] Thermal treatment of sludge may be an integral part of a comprehensive biosolids management system. See additional information in Sections 13.10 and 13.11.2 for enhanced sludge stabilization and heat drying.

13.7 Sludge Conditioning

Sludge conditioning is needed to improve the dewatering characteristics of the sludge solids. The solids in the sludge are extremely fine and carry electrostatic charges. This makes dewatering quite difficult. Sludge conditioning is used to stabilize the suspension and form large flocs for dewatering. The most common methods of sludge conditioning are chemical and physical. The chemical methods utilize coagulants, lime, and polymers. The physical methods include heat treatment, elutriation, freeze and thaw, irradiation, and ultrasonic vibration. However, chemical conditioning is the most commonly used method.

13.7.1 Chemical Conditioning

Most commonly used inorganic chemicals for sludge conditioning are ferric chloride, ferrous sulfate, alum, and lime. The important reactions involved in the process are: (1) the metal salts hydrolyze in water to form positively charged soluble metal complexes that neutralize the negatively charged sludge solids causing them to aggregate; (2) the metal salts also react with alkalinity forming hydroxides that cause flocculation; (3) increased aluminum, iron, and calcium contents in the sludge enhance dewaterability by counterbalancing the monovalent cations (Na^+, K^+, etc.) if they are high in sludge; (4) use of hydrated lime with iron salts controls pH to provide disinfection and reduces odors; (5) $CaCO_3$ produced in the reaction with lime produces granular structure, increases porosity of sludge, and decreases sludge compressibility, and (6) increased calcium content may help in optimizing the biosolids dewaterability by lowering the monovalent to divalent (M/D) ratio to <2.[6,7,32,70–72] In depth coverage of these reactions of iron salts, alum, and lime with municipal wastewater is provided in Section 9.5.2. In Examples 9.22 through 9.31, the dosages of these chemicals, alkalinity consumptions, phosphorus precipitations, and sludge production are extensively covered. Addition of these chemicals increases 20–30% sludge quantity on dry weight basis.

Power plant fly ash, cement kiln dust, and incinerator ash have also been used as sludge conditioning agents. They improve sludge dewaterability, but the quantity of sludge cake is increased considerably. Organic polymers (polyelectrolytes) are also used for sludge conditioning. Since sludge solids mostly carry a negative charge, the cationic polymers are most commonly used. The basic advantages of these polymers in comparison with metal coagulants are: (1) do not appreciably increase sludge quantity, (2) do not lower the fuel valve of the cake, and (3) safer and easier to handle than the inorganic chemicals. The dosage of polymer may vary depending upon the type of sludge, solids concentration, pH, and alkalinity. The dosages of all chemicals are normally determined by laboratory studies. The common laboratory tests include:

(1) *Büchner funnel*, (2) *leaf filter*, (3) *capillary suction time test (CST)*, and (4) *standard jar test*. These tests are used to determine the chemical dosages, filter yield, and suitability of various filtering media. The chemical dosages based on dry sludge weight for different chemicals are: (1) iron salts: 2–6%, (2) lime: 7–15%, (3) polymers: up to 1%, and (4) fly ash: 25–50%.[7,32,73–78]

Intimate mixing of these chemicals with sludge solids is necessary for proper conditioning. The topics of chemical mixing and flocculation are covered in Sections 9.5.5 and 9.5.6. Examples 9.33 through 9.47 contain design information on rapid mixing and flocculation. The Büchner funnel test procedure is described in Example 13.39. The leaf test was developed closely to simulate the vacuum filter dewatering. There is a decline in use of vacuum filtration for dewatering of municipal wastewater sludge. See References 77 and 78 for detailed testing procedures for CST and standard jar tests. Readers may also find discussions and many examples in References 79 and 80.

EXAMPLE 13.38: FERRIC CHLORIDE AND LIME DOSAGE AND EXCESS SLUDGE PRODUCTION

Ferric chloride ($FeCl_3$) and hydrated lime ($Ca(OH)_2$) are used to condition the anaerobically digested sludge. The digested sludge contains 2500 kg/d dry solids at 3.5%. The ferric chloride and hydrated lime dosages are 4% and 8% of dry solids, respectively. At an average 80% chemicals are incorporated into the sludge solids. The remaining 20% chemicals react with the alkalinity. The pH and alkalinity are 7.2 and 3500 mg/L as $CaCO_3$, respectively. Calculate the percent increase in the solids mass in the conditioned sludge. The specific gravity of digested sludge is 1.010.

Solution

1. Calculate the mass of $FeCl_3$ and $Ca(OH)_2$ added for conditioning.

$$FeCl_3 \text{ added} = 0.04\,\text{kg } FeCl_3/\text{kg dry solids} \times 2500\,\text{kg/d as dry solids} = 100\,\text{kg/d}$$

$$Ca(OH)_2\text{added} = 0.08\,\text{kg } Ca(OH)_2/\text{kg dry solids} \times 2500\,\text{kg/d as dry solids} = 200\,\text{kg/d}$$

2. Calculate the mass of $FeCl_3$ and $Ca(OH)_2$ incorporated in the conditioned sludge solids.

$$FeCl_3 \text{ in sludge solids} = 0.8 \times 100\,\text{kg/d} = 80\,\text{kg/d}$$

$$Ca(OH)_2 \text{ in sludge solids} = 0.8 \times 200\,\text{kg/d} = 160\,\text{kg/d}$$

3. Calculate the mass of $FeCl_3$ and $Ca(OH)_2$ available to react with the liquid phase.

$$FeCl_3 \text{ in liquid phase} = (100 - 80)\,\text{kg/d} = 20\,\text{kg/d}$$

$$Ca(OH)_2 \text{ in liquid phase} = (200 - 160)\,\text{kg/d} = 40\,\text{kg/d}$$

4. Calculate the amount of total alkalinity in the digested sludge.

$$\text{Sludge volume from Equation 13.1e, } V_s = \frac{100\% \times W_s}{p_s S_{b,s} \rho_w} = \frac{100\% \times 2500\,\text{kg/d}}{3.5\% \times 1.010 \times 1000\,\text{kg/m}^3} = 71\,\text{m}^3/\text{d}$$

Amount of total alkalinity in the digested sludge,

$$W_{\text{Alk}} = Alk \times V_s = 3500\,\text{g/m}^3 \text{ as } CaCO_3 \times 71\,\text{m}^3/\text{d} \times 10^{-3}\,\text{kg/g} = 249\,\text{kg/d as } CaCO_3$$

At pH < 8.3, the entire alkalinity is due to bicarbonate (HCO_3^-).

5. Calculate the mass of $Fe(OH)_3$ precipitated.

The $Fe(OH)_3$ precipitation reaction is given by Equation 9.18a.

$$\underset{2\times162}{2\ Fe(Cl)_3} + \underset{\substack{3\times100\\(\text{as } CaCO_3)}}{3\ Ca(HCO_3)_2} \rightarrow \underset{2\times107}{2\ Fe(OH)_3(s)} + 3\ CaCl_2 + 6\ CO_2$$

$Fe(OH)_3$ precipitated from Equation 9.18a at 20 kg/d of $FeCl_3$ in the liquid phase (from Step 3)

$$= \frac{2\times107}{2\times162} \times 20\,\text{kg/d} = 13\,\text{kg/d}$$

6. Calculate the mass of $CaCO_3$ precipitated.
   The reaction between $Ca(OH)_2$ and bicarbonate alkalinity (HCO_3^-) is given by Equation 9.16b.

$$\underset{74}{Ca(OH)_2} + \underset{100}{Ca(HCO_3)_2} \rightarrow \underset{2\times100}{2\ CaCO_3(s)} + 2\ H_2O$$

$CaCO_3$ precipitated from Equation 9.16b at 40 kg/d of $Ca(OH)_2$ in the liquid phase (from Step 3)

$$= \frac{2\times100}{74} \times 40\,\text{kg/d} = 108\,\text{kg/d}$$

7. Verify the availability of alkalinity for the precipitation reactions.

$$\text{Alkalinity consumption from Equation 9.18a} = \frac{3\times100}{2\times162} \times 20\,\text{kg/d} = 19\,\text{kg/d as } CaCO_3$$

$$\text{Alkalinity consumption from Equation 9.16b} = \frac{100}{74} \times 40\,\text{kg/d} = 54\,\text{kg/d as } CaCO_3$$

$$\text{Total alkalinity consumption} = (19+54)\,\text{kg/d} = 73\,\text{kg/d as } CaCO_3$$

The alkalinity consumed in precipitation reactions (73 kg/d as $CaCO_3$) is much lower than the alkalinity available in the digested sludge (249 kg/d as $CaCO_3$). Therefore, reactions of $FeCl_3$ and $Ca(OH)_2$ with alkalinity are valid.

8. Determine the total mass of solids generated due to chemical conditioning.

$$\text{Total sludge from conditioning} = (80+160+13+108)\,\text{kg/d} = 361\,\text{kg/d}$$

9. Determine the percent increase in sludge mass due to chemical conditioning.

$$\text{Percent increase in sludge mass due to chemical conditioning} = \frac{361\,\text{kg/d}}{2500\,\text{kg/d}} \times 100\% = 14.4\%$$

EXAMPLE 13.39: BÜCHNER FUNNEL TEST FOR OPTIMUM CHEMICAL DOSE

Büchner funnel test is conducted to obtain optimum chemical dose for dewatering. Describe the Büchner funnel test and develop the equations that characterize the sludge filterability.

Solution

1. Describe Büchner funnel apparatus and test procedure.
   The Büchner funnel test measures the filtration rate of conditioned sludge through a porous media with respect to time under a vacuum. The procedure utilizes vacuum filtration apparatus (Figure 13.19). A sludge sample of known moisture content is filtered through a predried and weighted filter paper.

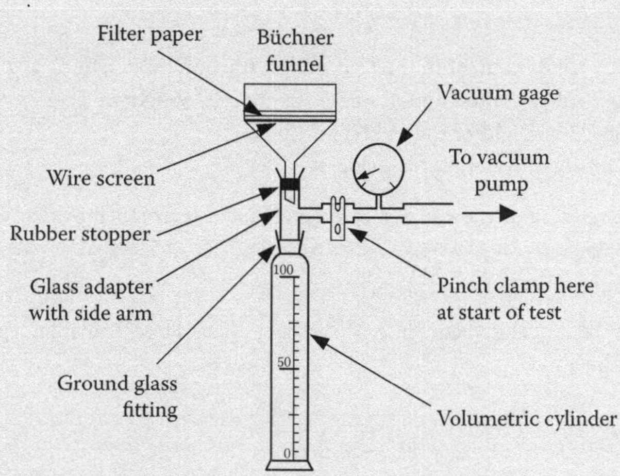

FIGURE 13.19 Büchner funnel test apparatus for determination of specific resistance (r) of sludge (Example 13.39).

Vacuum is applied, and a known volume of conditioned sludge is poured into the Büchner funnel. Volume of filtrate is measured with respect to time until the vacuum breaks. This is determined by an abrupt change in vacuum gauge reading, and may be difficult to ascertain at low vacuum setting. The moisture content of the sludge cake deposited at the filter paper is determined. The filtration rate obtained by this test is used to calculate the sludge filterability.

2. Develop the specific resistance equations.

The rate of filtration is expressed by Equation 13.15a.

$$\frac{dV}{dt} = \frac{PA^2}{\mu(rwV + R_m A)} \qquad (13.15a)$$

where

V = volume of filtrate collected at time t, m^3 (ft^3)
t = filtration time, s
P = applied pressure (or vacuum), N/m^2 or $kg/s^2{\cdot}m$ (lb_f/ft^2 or $lb/s^2{\cdot}ft$)
A = filter area, m^2 (ft^2)
μ = dynamic viscosity of filtrate, $N{\cdot}s/m^2$ or $kg/s{\cdot}m$ ($lb_f{\cdot}s/ft^2$ or $lb/s{\cdot}ft$). It is assumed equal to that of water.
r = specific resistance of the sludge cake, m/kg (ft/lb)
R_m = initial resistance of the filter medium, m^{-1} (ft^{-1})
w = mass of dry solids per unit of volume of filtrate, kg/m^3 (lb/ft^3). The value of w is calculated from Equation 13.15b.

In SI Units

$$w = \frac{\gamma}{g} \frac{10^3}{\left(\frac{1-p_f}{p_f}\right) - \left(\frac{1-p_c}{p_c}\right)} \quad \text{or} \quad w = \frac{\rho_w}{\left(\frac{1-p_f}{p_f}\right) - \left(\frac{1-p_c}{p_c}\right)} \qquad (13.15b)$$

In U.S. customary units

$$w = \frac{\gamma}{\left(\frac{1-p_f}{p_f}\right) - \left(\frac{1-p_c}{p_c}\right)} \quad \text{or} \quad w = \frac{\rho_w\, g}{\left(\frac{1-p_f}{p_f}\right) - \left(\frac{1-p_c}{p_c}\right)} \qquad (13.15c)$$

where

p_f = dry solids content in the feed sludge, fraction by weight or g/g
p_c = dry solids content in the sludge, fraction by weight or g/g
γ = specific weight of water, kN/m^3 (lb/ft^3)
ρ_w = density of water, kg/m^3 (slug/ft^3)

At a constant pressure, Equation 13.15a is integrated and rearranged to yield Equation 13.15d.

$$\frac{t}{V} = \frac{\mu r w}{2PA^2} V + \frac{\mu R_m}{PA} \quad \text{or} \quad \frac{t}{V} = SV + I \qquad (13.15d)$$

This is a linear relationship between t/V versus V. The slope and intercept are expressed by Equations 13.15e and 13.15f.

$$\text{Slope: } S = \frac{\mu r w}{2PA^2} \qquad (13.15e)$$

$$\text{Intercept: } I = \frac{\mu R_m}{PA} \qquad (13.15f)$$

where

S = slope of the line, s/m^6 (s/ft^6)
I = intercept of the line, s/m^3 (s/ft^3)

The specific resistance of a given sludge r and the initial resistance R_m of the filtration medium are determined from Equations 13.15g and 13.15h.

$$r = \frac{2PA^2 S}{\mu w} \qquad (13.15g)$$

$$R_m = \frac{IPA}{\mu} \qquad (13.15h)$$

The Büchner funnel test data are used to determine these design parameters.

Note: The unit of specific resistance r is often expressed as s^2/g (s^2/lb). In SI units, the value is divided by (9.81 m/s^2 10^3 g/kg) to convert the unit from m/kg to s^2/g. In U.S. customary units, the value is divided by 32.2 ft/s^2 to convert the unit from ft/lb to s^2/lb.

EXAMPLE 13.40: SPECIFIC RESISTANCE AND INITIAL RESISTANCE OBTAINED FROM BÜCHNER FUNNEL TEST

The specific resistance of an anaerobically digested chemically conditioned sludge sample was measured by Büchner funnel test. The filtration data are summarized below. The solids contents in the feed sludge and cake are 0.044 and 0.20 g/mL (or g/g) sample, respectively. The filter diameter used in Büchner funnel = 11 cm. The applied vacuum = 381 mm Hg. The operating temperature is 25°C.

Time (t), s	15	30	45	60	75	90	105	120
Filtrate volume (V), mL (or cm^3)	64	91	112	131	148	163	174	186
t/V, s/mL (or s/cm^3)	0.23	0.33	0.40	0.46	0.51	0.55	0.60	0.65

Develop the linear relationship and calculate (a) the specific resistance of the conditioned sludge (r) and (b) the initial resistance of the filter medium (R_m). Also, compare the specific resistance of this sludge with the typical values of the different sludges.

Solution

1. Plot the calculated data of t/V versus filtrate volume (V) to determine slop (S) and intercept (I) in Figure 13.20.

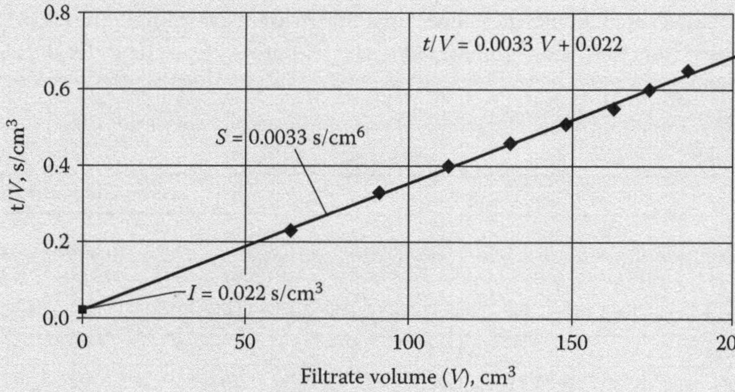

FIGURE 13.20 Plot of Büchner funnel test data t/V versus V (Example 13.40).

2. Determine the slope (S) and intercept (I) of the linear relationship in Figure 13.20.
   Slope $S = 0.0033\ \text{s/cm}^6$ or $3.310^9\ \text{s/m}^6$, and intercept $I = 0.022\ \text{s/cm}^3$ or $2.210^4\ \text{s/m}^3$.
3. Prepare the input parameters for calculating r and R_m.
   a. Determine the applied vacuum (P).

$$760\ \text{mm Hg} = 101.325\ \text{kN/m}^2\ \text{at 1 atm}$$

$$P = \frac{101.325\ \text{kN/m}^2}{760\ \text{mm Hg}} \times 381\ \text{mm Hg} = 50.8\ \text{kN/m}^2 \quad \text{or} \quad 5.08 \times 10^4\ \text{N/m}^2$$

   b. Calculate the filter area (A).

$$A = \frac{\pi}{4} d^2 = \frac{\pi}{4} \times (11\ \text{cm})^2 = 95\ \text{cm}^2 \quad \text{or} \quad 9.5 \times 10^{-3} \text{m}^2$$

   c. Calculate the mass of dry solids per unit of volume of filtrate (w).
   Density of water $\rho_w = 997\ \text{kg/m}^3$ at 25°C is obtained from Table B-2 in Appendix B. Calculate the w from Equation 13.15b using $p_f = 0.044$ g/g, and $p_c = 0.2$ g/g.

$$w = \frac{\rho_w}{\left(\frac{1-p_f}{p_f}\right) - \left(\frac{1-p_c}{p_c}\right)} = \frac{0.997\ \text{kg/m}^3}{\left(\frac{1-0.044}{0.044}\right) - \left(\frac{1-0.2}{0.2}\right)} = 56\ \text{kg/m}^3$$

4. Calculate the specific resistance of the chemically conditioned sludge (r) from Equation 13.15g.
   Dynamic viscosity of water $\mu = 0.89 \times 10^{-3}\ \text{N·s/m}^2$ (or kg/s·m) at 25°C is obtained from Table B-2 in Appendix B.

$$r = \frac{2PA^2S}{\mu w} = \frac{2 \times 5.08 \times 10^4\ \text{N/m}^2 \times (9.5 \times 10^{-3}\ \text{m}^2)^2 \times 3.3 \times 10^{-9}\ \text{s/m}^6}{0.89 \times 10^{-3}\ \text{N·s/m}^2 \times 56\ \text{kg/m}^3} = 6.1 \times 10^{11}\ \text{m/kg}$$

$$\text{or} \quad r = \frac{6.1 \times 10^{11}\ \text{m/kg}}{9.81\ \text{m/s}^2 \times 10^3\ \text{g/kg}} = 6.2 \times 10^7\ \text{s}^2/\text{g}$$

5. Calculate the initial resistance of the filter medium (R_m) from Equation 13.15h.

$$R_m = \frac{IPA}{\mu} = \frac{2.2 \times 10^4\ \text{s/m}^3 \times 5.08 \times 10^4\ \text{N/m}^2 \times 9.5 \times 10^{-3}\ \text{m}^2}{0.89 \times 10^{-3}\ \text{N·s/m}^2} = 1.2 \times 10^{10}\ \text{m}^{-1}$$

6. Compare the calculation result with the typical values of specific resistance of different sludges.
   The typical values of r for different sludges are summarized in Table 13.21. The $r = 6.1 \times 10^{11}$ m/kg that is calculated from the Büchner funnel test data is within the typical range for the digested sludge after conditioning.

TABLE 13.21 Typical Values of Specific Resistance of Municipal Sludge from Different Sources (Example 13.40)

Sludge Source	Specific Resistance (r)	
	m/kg	s^2/g
Primary sludge	$1.5–5 \times 10^{14}$	$1.5–5 \times 10^{10}$
Waste activated sludge (WAS)	$1–10 \times 10^{13}$	$1–10 \times 10^9$
Digested sludge	$1–6 \times 10^{14}$	$1–6 \times 10^{10}$
Digested sludge after conditioning by coagulation	$3–40 \times 10^{11}$	$3–40 \times 10^7$

Note: The unit of r in s^2/g is obtained by dividing the r value in m/kg by ($9.81\ \text{m/s}^2 \times 10^3$ g/kg).

EXAMPLE 13.41: DETERMINE OPTIMUM DOSAGE FROM BÜCHNER FUNNEL TEST

Combined digested sludge samples were conditioned by using different doses of alum. Each sample was filtered through the Büchner funnel apparatus to determine the specific resistance r. The alum dosages were 0%, 3%, 6%, 9%, 12% and 15% dry weight of sludge solids. The linear t/V versus V plots of each conditioned sludge sample were developed. A similar plot is shown in Figure 13.20. The slopes and intercepts of lines at different alum dosages are determined using the procedures presented in Steps 1 and 2 of Example 13.40. The results are tabulated below. Calculate the specific resistance of each conditioned sludge sample, and determine optimum alum dose.

Alum Dose, % Dry Solids	0	3	6	9	12	15
Slope (S), s/cm^6	0.0040	0.0028	0.0023	0.0017	0.0014	0.0021
Intercept (I), s/cm^3	0.10	0.10	0.10	0.10	0.11	0.11

Additional experimental information is provided below:

The filter area = 38.5 cm^2, applied pressure $P = 8.78 \times 10^4\ \text{N/m}^2$, and dynamic viscosity of water $\mu = 0.89 \times 10^{-3}\ \text{N·s/m}^2$. Assume that the mass of dry solids per unit of volume of filtrate $w = 0.003\ \text{g/cm}^3$.

Solution

1. Calculate from Equation 13.15g the specific resistance (r) for each of the sludge samples at different alum dosage and tabulate the calculated values in the table below. See sample calculation of r in Example 13.40, Step 4.

Alum Dose, % Dry Solids	0	3	6	9	12	15
Specific resistance (r)						
m/kg	3.9×10^{12}	2.7×10^{12}	2.2×10^{12}	1.7×10^{12}	1.4×10^{12}	2.0×10^{12}
s^3/g	4.0×10^8	2.8×10^8	2.2×10^8	1.7×10^8	1.4×10^8	2.0×10^8

2. Plot the specific resistance (r) versus alum dose in Figure 13.21.

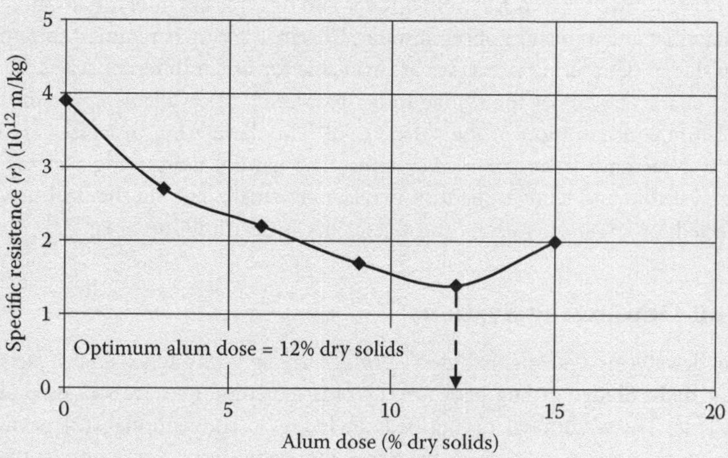

FIGURE 13.21 Optimum dose of alum for sludge conditioning (Example 13.41).

3. Determine the optimum dose of alum for conditioning of sludge.
   Read the optimum dose from the plot in Figure 13.21. The optimum dose of alum for sludge conditioning is 12% of dry weight of solids.

13.7.2 Physical Conditioning

The most common method of physical sludge conditioning are heat treatment, freeze and thaw, and elutriation. These methods are briefly presented below.

Heat Conditioning: Heat conditioning is used for stabilization of sludge. It breaks the gel structure, coagulates solids, and improves dewaterability. Heat treatment is covered in Section 13.6.4.

Freeze and Thaw: Freeze and thaw breaks the gel structure and solids are dewatered readily. This process is also used for demineralization and is covered in Section 15.4.11.

Elutriation: Large quantities of inorganic coagulants and lime are needed for sludge conditioning if alkalinity is high. Elutriation is washing of sludge with effluent to remove soluble organic and inorganic compounds particularly bicarbonate alkalinity. The elutriation can be achieved in (a) single-tank, multiple elutriation stages or (b) multiple-tank countercurrent elutriation. The elutriation tanks are typically designed as gravity thickeners with solids loading of 40–50 $kg/m^2{\cdot}d$ (8–10 $lb/ft^2{\cdot}d$). The washwater to sludge ratio is about 2–5 for multiple-tank countercurrent elutriation process. The cost of additional tanks and equipment for washing and solid separation may not justify the savings in the cost of chemicals. Also, elutriation generates washwater waste two to six times the volume of sludge. This liquid contains high concentrations of solids. It is returned to the plant that increases organic and solids loadings. Additionally, polymer dosage for sludge conditioning is increased substantially. For these reasons, use of elutriation is rapidly decreasing. Additional information on sludge elutriation may be found in References 3, 32, 73, and 74.

13.8 Sludge Dewatering

Sludge dewatering is needed to remove free water so that the reduced volume of sludge cake can be transported by truck, applied as biosolids over farms or nonagricultural lands, composted, incinerated or disposed of by landfilling. The water contained in the sludge are normally in three forms: *free*, *interstitial*, and *bond* water. The bond water consists of surface and chemically bonding water to the sludge particles. Sludge dewatering is effective for removal of free water and only partial interstitial water. Newer technologies remove interstitial and a portion of bond water. Thermal drying is required to remove bond water effectively. Many sludge-dewatering processes are available for use at different scales. The selection may depend upon: (1) characteristics of the sludge to be dewatered, (2) available space for dewatering, and (3) required moisture concentration of the sludge cake. The dewatering processes are broadly divided into (a) *natural* systems and (b) *mechanical* systems. The natural dewatering systems are used where land is remotely available and sludge quantity is relatively small. The mechanical dewatering systems are generally selected for larger operations and the facility is near built up areas.[6,7,21,74]

13.8.1 Natural Dewatering Systems

Common natural dewatering systems include: (1) *sludge drying beds* and (2) *sludge lagoons*.

Sludge Drying Beds: Sludge drying beds are the oldest method and are still used at many small-to medium-sized plants. The traditional drying beds include: (1) conventional, (2) paving, (3) artificial-media, and (4) vacuum. In recent years, there have been innovative improvements over the traditional drying beds. These drying beds are discussed below.

Conventional Sand Beds: The conventional sand beds consist of perforated-pipe-underdrain system underneath the gravel and sand layers. The free water in the sludge filters through the bed to leave the sludge cake over sand surface. The typical plan and section of conventional drying beds are shown in Figure 13.22a. Basic design features are given below.[6,7,32,74]

1. The drying bed is divided into sections. Each section is normally 6–7.5 m (20–25 ft) wide and 30–60 m (100–200 ft) long.
2. The sand layer is typically 20–45 cm (8–18 in) in depth and the uniformity coefficient and effective size are ≤ 4 and 0.3–0.75 mm, respectively.
3. The graded gravel layer is 20–30 cm (8–12 in) deep and supports the sand.
4. The perforated plastic or vitrified clay drainage pipes are typically 150–200 mm (6–8 in) in diameter and spaced 2.5–6 m (8–20 ft) center-to-center.
5. The bottom subsoil liner needs to have a thickness at least 0.3 m (12 in), permeability $< 1 \times 10^7$ cm/s, and minimum liner slope of 1% toward the drain. An impermeable concrete liner is required in the drying beds where the groundwater table is within 1.2 m (4 ft) of the bottom of the liner.
6. A watertight exterior wall is normally extended from the gravel layer up and has a freeboard of 0.3–0.9 m (12–36 in) above the sand layer or ground level, whichever is higher.
7. Each drying bed section is flooded by the sludge to a depth of 20–30 cm (8–12 in) and allowed to dry for a period of 10–20 days. The liquid percolates to underdrain system and is returned to the plant. Normally, the dewatering occurs by drainage for about 2 days and the drying is done by evaporation for 2–3 weeks.
8. The sludge cake builds up over the sand. It is then removed (a) manually by shoveling into the wheelbarrow or truck or (b) mechanically by a scraper or front-end loader. The sludge cake contains 20–30% solids, and solids capture rate is over 90%.
9. Poorly digested sludge may cause odor problems. The drying bed may be open or covered depending upon the climatic condition and odor control requirements.
10. Design criteria: The dry solids based sludge loading rates are 50–125 $kg/m^2 \cdot year$ (10–25 $lb/ft^2 \cdot year$) for open beds and 60–200 $kg/m^2 \cdot year$ (12–40 $lb/ft^2 \cdot year$) for covered beds. The per capita required bed areas are: (a) 0.14–0.23 m^2 per capita (1.5–2.5 ft^2/capita) for open beds and (b) 0.09–0.2 m^2 per capita (1–2 ft^2/capita) for covered beds.
11. For anaerobically digested sludge, the BOD_5 and COD of filtrate from the drying beds may be 40 mg/L and 350 mg/L, respectively.

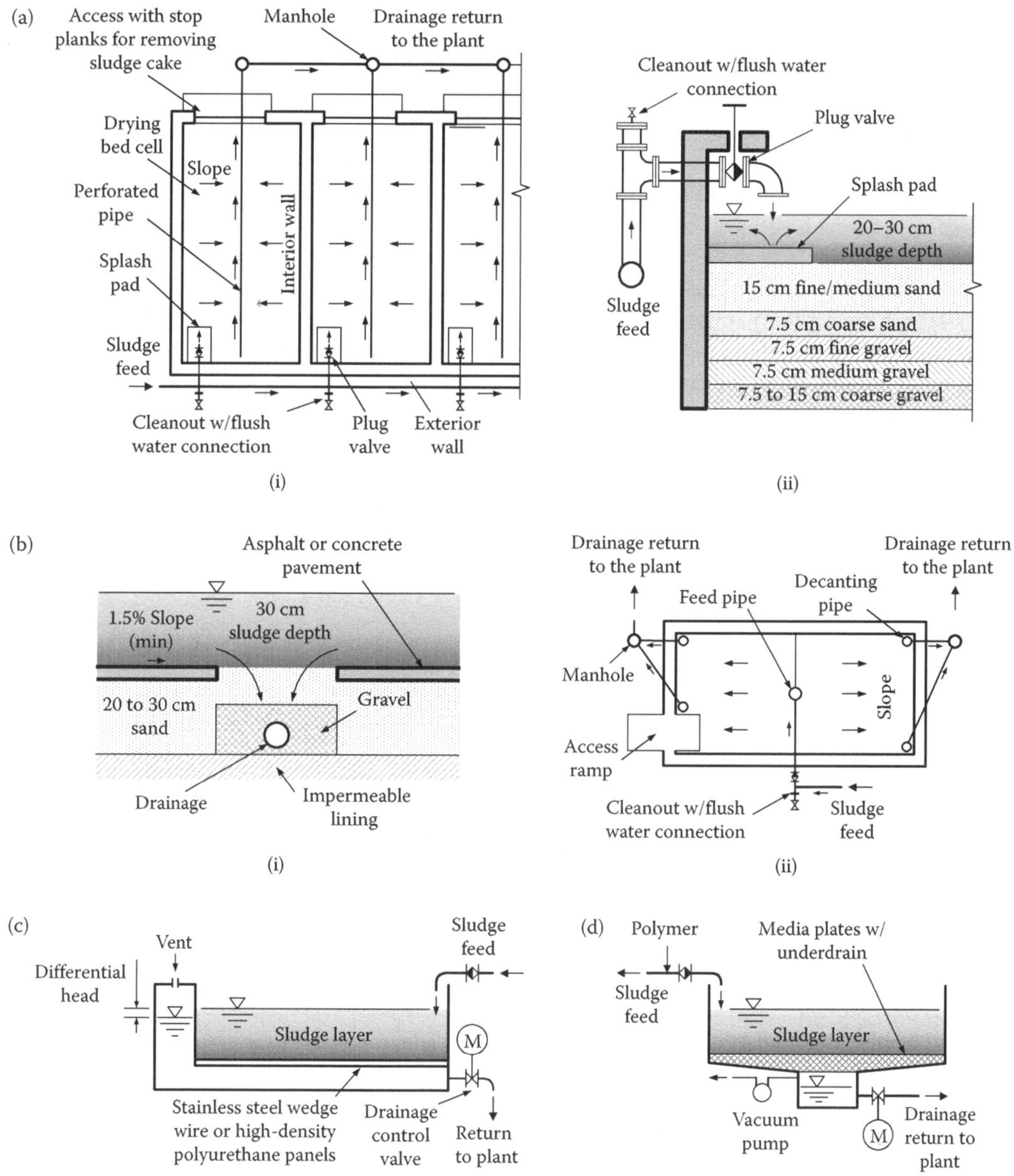

FIGURE 13.22 Sludge drying bed dewatering facilities: (a) conventional drying bed: (i) conceptual layout plan and (ii) typical section at sludge feed; (b) paved drying bed: (i) construction of drainage type bed and (ii) conceptual layout of decanting type bed; (c) the components of artificial-media drying bed; and (d) the components of vacuum assisted drying bed.

The drying bed area required per capita for combined digested sludge may be estimated from an empirical equation given by Equation 13.16.[81]

$$A = K(C_1 R + C_2) \tag{13.16}$$

where

A = area per capita, m^2/capita (ft^2/capita)

K = factor depending on type of digestion, dimensionless. $K = 1.0$ for anaerobic digestion, and 1.6 for aerobic digestion.
$C_1 = 0.0366$ (m^2/capita)/(m/year) for SI units and 0.01 (ft^2/capita)/(in/year) for U.S. customary units
$C_2 = 0.0929$ m^2/capita for SI units and 1 ft^2/capita for U.S. customary units
R = annual rainfall, m/year (in/year)

Paved Drying Beds: The paved drying beds are: (a) *drainage* type and (b) *decanting* type.[6,7,32,75] The drainage-type drying beds have concrete or bituminous concrete paved tracks for a front-ended loader. The paved tracks are laid over 20–30 cm (8–12 in) sand or gravel base and sloped 1.5% towards the unpaved drainage area in the middle. The unpaved beds are 0.6 –1 m wide, and their operation and design are similar to that of conventional drying beds. The percolation area may be agitated by mobile equipment for improved drainages. The rectangular beds are 6–15 m (20–50 ft) wide and 20–45 m (65–150 ft) long. The solids concentration in the sludge cake may range from 20% to 40%. The sectional view of drainage type drying bed is shown in Figure 13.22b(i).

The decanting type drying beds have draw-off pipes for decanting the supernatant. These beds have paved impermeable base of soil-cement mixture. The influent pipe is a vertical pipe located at the center. The decanting pipes are also vertical pipes and placed at the four corners (Figure 13.22b(ii)). The applied liquid depth is typically 30 cm (12 in). The decanting achieves about 20–30% of water after efficient sludge setting. The remaining moisture is evaporated in 30–40 days to achieve solids concentration of 40–50% in arid climate. Mixing of drying sludge is also practiced for enhanced evaporation of moisture. This type of drying bed is more efficient for use under warm and arid climates. Equations 13.17a through 13.17e are used for designing the decanting type paved drying beds.[6,74,82,83]

$$W_0 = W_s\left(\frac{1-p_0}{p_0}\right) \quad \text{or} \quad W_d = W_s\left(\frac{1-p_d}{p_d}\right) \quad \text{or} \quad W_e = W_s\left(\frac{1-p_e}{p_e}\right) \tag{13.17a}$$

$$R_e = \rho_w k_e E_p \tag{13.17b}$$

$$R_p = \rho_w P \tag{13.17c}$$

$$\Delta W_e = W_d - W_e + R_p A \quad \text{or} \quad \Delta W_e = \left(\frac{1-p_d}{p_d} - \frac{1-p_e}{p_e}\right)W_s + \rho_w PA \tag{13.17d}$$

$$A = \frac{W_d - W_e}{R_e - R_p} \quad \text{or} \quad A = \frac{W_s}{\rho_w(k_e E_p - R_p)}\left(\frac{1-p_d}{p_d} - \frac{1-p_e}{p_e}\right) \tag{13.17e}$$

where
W_0 = mass of water content in the applied sludge, kg/year (lb/year)
W_s = annual sludge application over the drying beds, kg/year (lb/year) dry solids
W_d = mass of water content in the sludge after decanting, kg/year (lb/year). The solids lost in the decant flow are ignored.
W_e = mass of water content in the sludge after evaporation, kg/year (lb/year)
p_0 = solids content in applied sludge, decimal fraction of dry solids
p_d = solids content in the sludge after decanting, decimal fraction of dry solids
p_e = solids content in the sludge after evaporation, decimal fraction of dry solids
R_e = annual evaporation rate, kg/m^2·year (lb/ft^2·year)
k_e = efficiency of water evaporation rate from sludge when compared with that from free water surface, decimal fraction. Typical value of $k_e = 0.6$.
ρ_w = density of water, kg/m^3 (lb/ft^3). It is typically using $\rho_w = 1000$ kg/m^3 in IS units, or 62.4 lb/ft^3 in U.S. customary units.
E_p = evaporation rate from free water surface in the pan, m/year (ft/year)
R_p = annual precipitation rate, kg/m^2·year (lb/ft^2·year)

P = annual precipitation, m/year (ft/year)
ΔW_e = annual total amount of water to be evaporated after decanting, kg/year (lb/year)
A = bottom area of paved bed, m^2 (ft^2)

Artificial-Media Drying Beds: Artificial-media used for drying beds include: (a) *stainless steel wedge wire* panels and (b) *high-density polyurethane* panels (see Figure 13.22c). The wedge wire drying bed consists of a shallow rectangular basin fitted with a false floor of panels made of stainless-steel wedge-shaped bars with slotted openings of 0.25 mm (0.01 in). The high-density polyurethane media system has 30 cm (12 in) square interlocking panels with slotted area. Both systems are operated similarly. An outlet valve controls the rate of the drainage. Initially, the outlet valve is closed, and the space below and ~2–3 cm above the wire-wedge is filled with wastewater. This water serves as a cushion when sludge is applied on the top of the false floor. After the bed is filled to the desired depth by the sludge, the outlet valve is opened carefully to drain the initially applied wastewater. The water from the sludge is drained and the sludge cake concentrates by drainage and evaporation. When the desired solids concentration is reached, the sludge cake is removed by a front-end loader and the panels are washed with high water jet. The artificial-media drying beds have the following advantages in comparison with the sand drying beds: (1) low media clogging, (2) higher through put rate, (3) constant rapid drainage, (4) easier sludge cake removal, (5) easier to maintain, and (6) the ability to handle dilute and difficult-to-dewater sludge, such as aerobically digested sludge. The typical solids loading rates are 2–5 kg/m^2 (0.4–1 lb/ft^2) per cycle or 750–1800 kg/m^2·year (150–360 lb/ft^2·year). A cycle of 24 h typically achieves dry sludge solids of 8–12%.[6,7,81]

Vacuum Assisted Drying Beds: In this type of drying beds, vacuum is applied to accelerate dewatering process. The major components of vacuum assisted drying beds are shown in Figure 13.22d. The construction and operation features are listed below.[6,7,32,79,84]

1. The principal components are: (a) concrete housing with porous media plates over a support underdrain layer, (b) polymer addition and mixing equipment, (c) a vacuum pump, (d) sludge cake removal equipment, and (e) high-pressure surface washing device.
2. The solids loading rate range from 5 to 20 kg/m^2 (1–4 lb/ft^2) per application. Typical application cycle time is 1–2 days.
3. The polymer dose may range from 1 to 20 g/kg (2–40 lb/ton).
4. The conditioned sludge is applied to a depth of 30–75 cm (12–30 in). The liquid drains for 1 h, then vacuum is applied and maintained at 30–48 kPa (10–25 in of Hg).
5. The suction is stopped when the cake is cracked and vacuum is lost. At this time, a solids level of 15–25% is reached.
6. The cake is removed by a front-end loader and the plates are washed by a high-pressure hose to remove the residues.

Innovative Drying Beds: Innovative sludge drying beds have been developed in recent years. They include: (a) solar drying beds and (b) reed beds. The solar drying beds are essentially large greenhouse with sensors, air louvers, ventilation fans, and a mobile sludge agitator (Figure 13.23a). The system is capable of producing dry pelletized sludge with solids content of 50–90%, meeting Class A biosolids requirements. The design solids loading rates without drainage floor are typically in the range from 500 to 900 kg/m^2·year (100–180 lb/ft^2·year) for predewatered sludge feed at 15–30% solids. The design solids loading rate for thickened sludge feed at 2–6% solids with drainage floor is 50–400 kg/m^2·year (10–80 lb/ft^2·year) for small applications.[6,7,21,85–88] See Section 15.4.11 for additional information about solar evaporation for disposal of waste brine.

The reed drying beds are also called planted drying beds and are similar to constructed wetlands. From the bottom up, a typical bed consists of three layers: (1) 250 mm (10 in) deep coarse/medium gravel, (2) 250 mm (10 in) deep pea gravel, and (3) 100–150 mm (4–6 in) layer of coarse sand. Above these layers, a depth of 1 m (3 ft) is provided for storing accumulated sludge. Phragmites and other reeds are planted. The plants provide continuous drainage of water from sludge to underdrains. Plants also absorb and

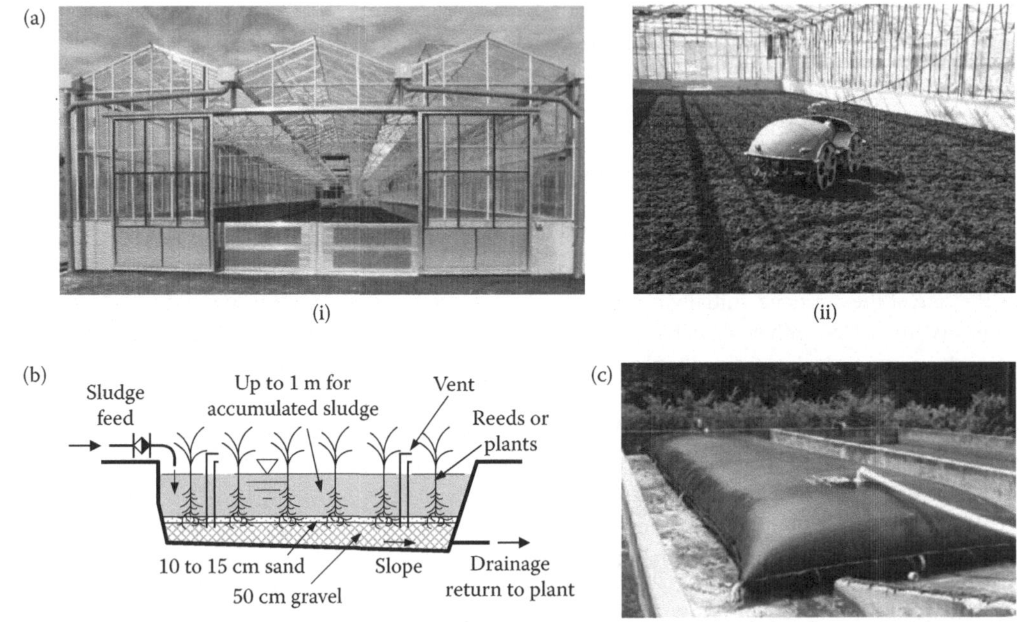

FIGURE 13.23 Innovative alternatives of sludge drying beds: (a) solar drying bed (Courtesy Thermo-System): (i) greenhouse and (ii) sludge agitator (or called "electric mole"); (b) reed drying bed; and (c) geotextile tube container in a sludge drying bed cell (Courtesy TenCate Geosynthetics Americas).

evaporate water, and roots assist in aerobic stabilization and mineralization of sludge. The design loading rate is 30–100 kg/m^2·year (6–20 lb/ft^2·year). The applied depth of sludge is 75–100 mm (3–4 in) once every 1–2 weeks (see Figure 13.23b).[6,7,21]

Another innovative technology is the geotextile tube containers (see Figure 13.23c). This technology may be cost-effective to retrofit or replace existing drying beds. This process has been used for years to contain and dewater dredging and mining materials. At little to no operation and maintenance requirements, it achieves high solids capture rate, reduces odor from the dewatering operation, and produces filtrate with low solids, BOD, and nutrients.[21,89]

Sludge Drying Lagoons: Drying lagoons are relatively low-cost substitute for drying beds. These are shallow earthen basins. The sludge is applied 0.75–1.25 m (2.5–4 ft) in depth and the supernatant is decanted from the surface. The loss of water is mainly due to evaporation and percolation. Normally, 3–6 months are required to reach 20–40% solids in the sludge cake. The sludge cake is removed by mechanical equipment. The typical solids-loading rate is 32–40 kg/m^3·year (2–2.5 lb/ft^3·year). The lagoon capacity is typically 0.3–0.4 m^2/capita (3.2–4.3 ft^2/capita). The design of lagoon is similar to that of storage reservoirs and drying beds (see Examples 12.21 and 13.42).

The major concerns associated with sludge drying lagoons are serious odors resulting from poorly digested sludge, insect breeding, and potential groundwater contamination. The major design considerations include climate, subsoil permeability, sludge characteristics, lagoon depth, groundwater table, and area management practices.[90,91]

EXAMPLE 13.42: CONVENTIONAL SLUDGE DRYING BED AREA BASED ON EVAPORATION DATA

Conventional sludge drying beds are designed to dewater digested combined sludge at the maximum monthly sludge production of 250 kg/d dry solids, solids content of 3.5%, and specific gravity of 1.01.

The sludge application is 25 cm/cycle. Laboratory data indicates that solids content of 18 % is reached after drainage in 24 h. Sludge cake is removed from drying beds at 25% solids. The experimental data also show that 43% rainfall drains through the cake and 57% is evaporated. The average evaporation from wet sludge is 75% of that of free water surface. The monthly total rainfall and evaporation data of the area are tabulated below. To be conservative, assume that monthly total rainfall occurs during the drying period. Calculate the required total area of drying beds. Ignore loss of solids with the initial drainage. Water density $\rho_w = 1000\ kg/m^3$.

Parameter	Monthly Total Value in Each Month, cm											
	January	February	March	April	May	June	July	August	September	October	November	December
Precipitation	10	8	8	10	10	9	5	8	8	8	7	7
Evaporation	15	17	19	22	29	28	34	29	23	17	12	8

Solution

1. Calculate the weight of sludge solids applied per m^2 of the conventional drying beds per cycle (or the sludge application rate, SAR_s).

   Apply Equation 13.1e, the weight of sludge solids applied to the drying beds are developed in Equation 13.18a through 13.18f.

$$w_s = \frac{Q_s p_s S_b \rho_w}{100\%} \quad \text{(It is derived from Equation13.1e.)} \tag{13.18a}$$

$$Q_s = D_s A \tag{13.18b}$$

$$SAR_s = \frac{w_s}{A} \quad \text{or} \quad SAR_s = \frac{Q_s p_s S_b \rho_w}{100\% \times A} \quad \text{or} \quad SAR_s = \frac{D_s p_s S_b \rho_w}{100\%} \quad \text{or} \quad SAR_s = \frac{W_s \theta_{cycle}}{A} \tag{13.18c}$$

$$A = \frac{w_s}{SAR_s} \quad \text{or} \quad A = \frac{W_s \theta_{cycle}}{SAR_s} \tag{13.18d}$$

$$w_s = \frac{SAR_s A}{\theta_{cycle}} \quad \text{or} \quad w_s = \frac{W_s}{\theta_{cycle}} \tag{13.18e}$$

$$w_{s,stage} = \frac{100\%}{p_{s,stage}} SAR_s \tag{13.18f}$$

where

w_s = weight of sludge solids applied to the drying beds per cycle (kg/cycle) on basis of dry solids

Q_s = volumetric application rate of wet sludge, m^3/cycle

D_s = application depth of wet sludge, m/cycle

A = sludge drying bed area, m^2

SAR_s = solids application rate, kg/m^2·cycle on basis of dry solids

W_s = weight of sludge solids applied to the drying beds per day, kg/d on basis of dry solids

$W_{s,stage}$ = weight of wet sludge remaining in unit bed area after a solids content of ps, operation is reached in a stage during a drying cycle, kg/m^2·cycle

$p_{s,stage}$ = solids content of wet sludge in the bed after a stage during a drying cycle, percent of wet sludge weight

θ_{cycle} = drying period, d/cycle

Other variables are defined previously in Equation 13.1.

SAR_s from Equation 13.18c, $SAR_s = \dfrac{D_s p_s S_b \rho_w}{100\%} = \dfrac{25\,\text{cm/cycle} \times 10^{-2}\,\text{m/cm} \times 3.5\% \times 1.01 \times 1000\,\text{kg/m}^3}{100\%}$

$= 8.8\,\text{kg/m}^2\text{·cycle}$

2. Calculate the water evaporation from the sludge drying bed in the month of January.
   a. Depth of water evaporated from sludge ($D_{s,evap}$) from mass balance of the sludge solids in cm/cycle.

      Weight of wet solids per m^2 bed area at 18% solids content after drainage from Equation 13.18f,

$$W_{s,drainage} = \frac{100\%}{p_{s,drainage}} \times SAR_s = \frac{100\%}{18\%} \times 8.8\,\text{kg/m}^2\text{·cycle} = 48.9\,\text{kg/m}^2\text{·cycle}$$

      Weight of sludge cake per m^2 bed area at 25% solids content after drying from Equation 13.18f,

$$W_{s,drying} = \frac{100\%}{p_{s,drainage}} \times SAR_s = \frac{100\%}{25\%} \times 8.8\,\text{kg/m}^2\text{·cycle} = 35.2\,\text{kg/m}^2\text{·cycle}$$

      Water content evaporated from the sludge,

$$W_{s,evap} = \text{Wet Sludge weight after drainage} - \text{Wet sludge weight after drying}$$
$$= W_{s,drainage} - W_{s,drying} = (48.9 - 35.2)\,\text{kg/kg/m}^2\text{·cycle} = 13.7\,\text{kg/m}^2\text{·cycle}$$

      Water depth evaporated from the sludge per m^2 of the drying bed,

$$D_{s,evap} = \frac{\text{Weight of water evaporated}}{\text{Water density}} = \frac{W_{s,evap}}{\rho_w} = \frac{13.7\,\text{kg/m}^2\text{·cycle}}{1000\,\text{kg/m}^3} \times 10^2\,\text{cm/m} = 1.4\,\text{cm/cycle}$$

   b. Depth of water evaporated from precipitation ($D_{p,evap}$).

      It is given that monthly total precipitation depth is 10 cm in January, and 57% of it is evaporated. Assume that the evaporation is completed during the drying cycle as a worst case scenario.

      Average precipitation depth evaporated,

$$D_{p,evap} = 0.57 \times 10\,\text{cm/mo} = 5.7\,\text{cm/mo} \approx 5.7\,\text{cm/cycle}$$

   c. Total depth of water evaporated from the sludge drying beds ($D_{t,evap}$).

$$D_{t,evap} = D_{s,evap} + D_{p,evap} = (1.4 + 5.7)\,\text{cm/cycle} = 7.1\,\text{cm/cycle}$$

   d. Average depth of evaporation from the sludge drying bed (R_{evap}).

      In January, the monthly total evaporation rate is 15 cm from a free water surface, and the evaporation from sludge is 75% of that from the free water surface.

      Average evaporation depth evaporated,

$$R_{epav} = 0.75 \times 15\,\text{cm/mo} = 11.3\,\text{cm/mo}$$

   e. The number of days required to evaporate water from the sludge drying bed (θ_{cycle}).

$$\theta_{cycle} = \frac{D_{t,evap}}{R_{evap}} = \frac{7.1\,\text{cm/cycle}}{11.3\,\text{cm/mo}} \times 31\,\text{d/mo} = 19.5\,\text{d/cycle}$$

      Note: It is less than the total days in January.

TABLE 13.22 Required Drying Periods to Evaporate Water from Sludge and Precipitation (Example 13.42)

Parameter	Calculation Result											
	January	February	March	April	May	June	July	August	September	October	November	December
Depth of water to evaporate, cm/cycle												
From sludge ($D_{s,evap}$)	1.4	1.4	1.4	1.4	1.4	1.4	1.4	1.4	1.4	1.4	1.4	1.4
From precipitation ($D_{p,evap}$)	5.7	4.6	4.6	5.7	5.7	5.1	2.9	4.6	4.6	4.6	4.0	4.0
Total ($D_{t,evap}$)	7.1	6.0	6.0	7.1	7.1	6.5	4.3	6.0	6.0	6.0	5.4	5.4
Average evaporation rate (R_{evap}), cm/mo	11.3	12.8	14.3	16.5	21.8	21.0	25.5	21.8	17.3	12.8	9.0	6.0
Required drying period (θ_{cycle}), d/cycle	19.5	13.1	13.0	12.9	10.1	9.3	5.2	8.5	10.4	14.5	18.0	27.9

3. Summarize the water evaporation data for the month of January.
   The water evaporation data and drying period for the month of January are calculated in Step 2. These values are summarized in Table 13.22.
4. Calculate the water evaporation data for the rest of the year.
   Repeat the calculation procedure in Step 2 for the rest of the year (February to December), and summarize the results in Table 13.22.
5. Determine the design drying period (θ_{cycle}).
   The design is determined from the maximum monthly drying period required. It is 27.9 d/cycle from Table 13.22. Provide a design drying period of 30 d.
6. Calculate the area of drying beds (A).
   The required area of drying bed at the m, is expressed by Equation 13.18d at the maximum monthly sludge production $W_s = 250$ kg/d.
   Required total drying bed area from Equation 13.18d.

$$A = \frac{W_s\theta_{cycle}}{SAR_s} = \frac{250\,\text{kg/d} \times 30\,\text{d/cycle}}{8.8\,\text{kg/m}^2\cdot\text{cycle}} = 850\,\text{m}^2$$

   Provide six (6) drying bed cells each having the dimensions of 6 m width by 30 m long. The total area of each bed cell is 180 m^2. The total area of all drying bed cells is 1080 m^2. It will take about 6 days to fill one cell. Different cells will be at different operational stages, and one to two cells will be ready to receive the sludge at all times. This will give the desired operational flexibility.

EXAMPLE 13.43: DESIGN OF PAVED DRYING BEDS

Decanting type paved drying beds are used for dewatering aerobically digested sludge from a small community. The sludge production is 91,000 kg/year dry solids. The sludge after decanting has solids concentrations of 0.09%. The percent dry solids required for final disposal is 25%. The annual precipitation and free surface evaporation rate from pan are 100 and 250 cm, respectively. Assume $k_e = 0.6$, and water density $\rho_w = 1000$ kg/m^3. Calculate the total area of paved drying beds. Ignore any loss of solids during the process (See Example 13.42 for additional data).

Solution

1. Calculate from Equation 13.17a the water content in the sludge after decanting and drying by evaporation.
   Water content of sludge after decanting,

$$W_d = W_s\left(\frac{1-p_d}{p_d}\right) = 91{,}000\,\text{kg/year} \times \left(\frac{1-0.09}{0.09}\right) = 920{,}000\,\text{kg/year}$$

   Water content of sludge after drying,

$$W_e = W_s\left(\frac{1-p_e}{p_e}\right) = 91{,}000\,\text{kg/year} \times \left(\frac{1-0.25}{0.25}\right) = 273{,}000\,\text{kg/year}$$

2. Calculate from Equation 13.17b the annual evaporation rate from the sludge drying beds at $E_p = 250$ cm/year (2.5 m/year) and $k_e = 0.6$.

$$R_e = \rho_w k_e E_p = 1000\,\text{kg/m}^3 \times 0.6 \times 2.5\,\text{m/year} = 1500\,\text{kg/m}^2\cdot\text{year}$$

3. Calculate from Equation 13.17c the annual precipitation rate at $P = 100$ cm/year $= 1$ m/year.

$$R_p = \rho_w P = 1000\,\text{kg/m}^3 \times 1\,\text{m/year} = 1000\,\text{kg/m}^2\cdot\text{year}$$

4. Calculate the total area of the paved drying beds from Equation 13.17e.

$$A = \frac{W_d - W_e}{R_e - R_p} = \frac{(920{,}000 - 273{,}000)\,\text{kg/year}}{(1500 - 1000)\,\text{kg/m}^2\cdot\text{year}} = 1290\,\text{m}^2$$

Provide six (6) paved drying beds for operational flexibility. Each bed has dimensions of 8 m × 30 m. The total drying bed area is 1440 m^2.

EXAMPLE 13.44: ARTIFICIAL-MEDIA AND VACUUM ASSISTED DRYING BEDS

A POTW produces aerobically digested sludge at an average rate of 250 kg dry solids per day. The chemically conditioned sludge is dewatered by an artificial-media and vacuum assisted drying beds. The average solids loading rate and cycle time for artificial-media drying beds are 4 kg/m^2·cycle, and 2 days. The solids loading rate and cycle time for vacuum assisted drying beds are 10 kg/m^2·cycle and 1.5 days. Calculate and compare the drying bed areas for both types of sludge drying beds tested.

Solution

1. Estimate from Equation 13.18d the required total bed area of artificial-media beds (A_{amdb}).

$$A_{amdb} = \frac{W_s\theta_{cycle,amdb}}{SAR_{s,amdb}} = \frac{250\,\text{kg/d} \times 2\,\text{d/cycle}}{4\,\text{kg/m}^2\cdot\text{cycle}} = 125\,\text{m}^2$$

Provide three (3) beds each bed 5 m × 10 m.

2. Estimate the required total bed area of vacuum assisted beds (A_{vadb}) from Equation 13.18d.

$$A_{vadb} = \frac{W_s\theta_{cycle,vadb}}{SAR_{s,vadb}} = \frac{250\,\text{kg/d} \times 1.5\,\text{d/cycle}}{10\,\text{kg/m}^2\cdot\text{cycle}} = 37.5\,\text{m}^2$$

Provide two (2) beds each bed 5 m × 5 m.

3. Compare the required total bed areas of artificial-media and vacuum assisted sludge drying bed.

The required vacuum assisted bed area is only 30% of that of artificial-media beds. It should be noted that the energy requirements for vacuum assisted drying beds are significantly higher than that for artificial media drying beds.

EXAMPLE 13.45: AREAS NEEDED FOR SOLAR DRYING BEDS

Solar drying beds may be considered an improvement alternative to replace the conventional sludge drying beds. Estimate the following for the solar drying beds: (a) area required and (b) percent reduction in biosolids volume in comparison with that from conventional sludge drying beds. The solar drying beds concentrate the sludge cake to 85% dry solids. The specific gravity of concentrated sludge cake is 1.5. The solids loading rate for design of solar drying beds is 200 kg/m^2·year. Total solids production at the plant is 250 kg/d dry solids. The solids content in the sludge cake from a conventional drying bed is 25% and specific gravity of 1.1. Ignore any loss of solids during the process.

Solution

1. Estimate the annual total sludge production at the plant (W_s).

$$W_s = 250\,\text{kg/d} \times 365\,\text{d/year} = 91{,}000\,\text{kg/year}$$

2. Estimate the required total bed area of solar drying beds (A).

$$A = \frac{W_s}{SAR_s} = \frac{91{,}000\,\text{kg/year}}{200\,\text{kg/m}^2\cdot\text{year}} = 455\,\text{m}^2$$

Provide two solar drying beds, each bed of area 230 m^2.

3. Estimate from Equation 13.1e the volume of biosolids produced from conventional and solar drying beds.

a. Volume of biosolids from conventional drying beds ($V_{s,conv}$).

$$V_{s,convl} = \frac{100\% \times W_s}{p_{s,conv} S_{b,conv} \rho_w} = \frac{100\% \times 91{,}0000\,\text{kg/year}}{25\% \times 1.1 \times 1000\,\text{kg/m}^3} = 331\,\text{m}^3/\text{year}$$

b. Volume of biosolids from solar drying beds ($V_{s,solar}$).

$$V_{s,solar} = \frac{100\% \times W_s}{p_{s,solar} S_{b,s,solar} \rho_w} = \frac{100\% \times 91{,}0000\,\text{kg/year}}{85\% \times 1.5 \times 1000\,\text{kg/m}^3} = 71\,\text{m}^3/\text{year}$$

4. Estimate the percent reduction of volume of biosolids before and after concentration in drying beds, and volume reduction (V_{red}).

$$V_{red} = \left(\frac{V_{s,conv} - V_{s,solar}}{V_{s,conv}}\right) \times 100\% = \left(\frac{331 - 71\,\text{m}^3/\text{yr}}{331\,\text{m}^3/\text{yr}}\right) \times 100\% = 79\% \approx 80\%$$

EXAMPLE 13.46: COMPARISON OF AREA AND SOLIDS CONTENT OBTAINED FROM DIFFERENT DRYING BEDS

The dewatering areas of drying solids, and solids contents in the biosolids are calculated for five different types of drying beds in Examples 13.42 through 13.45. Compare the total area required and expected solids contents in biosolids from these drying beds. The combined digested sludge is dewatered by all drying beds. The maximum monthly sludge production rate is 250 kg/d dry solids.

Solution

Five different types of drying beds are evaluated in Examples 13.42 through 13.45. These are conventional, paved, artificial-media, vacuum assisted, and solar drying beds. The average dry solids production rate is 250 kg/d or 91,000 kg/year at the plant. The total bed areas estimated for different drying beds are summarized below with the expected solids contents of the biosolids.

Parameter	Drying Bed Types					Comparison Comment
	C	P	AM	VA	S	
Total area required, m^2	850	1290	125	37.5	455	Large areas will be required for conventional or pave drying beds. Significantly smaller areas are expected with the artificial-media or vacuum assisted drying beds.[a] The

Continued

Parameter	Drying Bed Types					Comparison Comment
	C	P	AM	VA	S	
						required area for solar drying beds is in the middle of conventional and vacuum assisted drying beds.[b]
Expected solids content, % dry solids	25	25	25	25	85	The solids content around 25% is expected in the first four types of drying beds. In comparison with these, a much higher dryness at nearly 80% volume reduction is achievable by the solar drying beds.[c]

[a] Polymer addition may be required in vacuum assisted drying beds.

[b] The required area of solar drying beds can be reduced by 50–90% if the feed sludge is predewatered to a solids content of 25%.

[c] Solar drying beds are used to concentrate dewatered sludge cake to produce Class A biosolids. The biosolids will also be available for beneficial reuse.

Note: AM = artificial-media; C = conventional; P = paved; S = solar; VA = vacuum assisted.

13.8.2 Mechanical Dewatering Systems

The mechanical dewatering systems are used for large operations and are installed inside a building along with chemical feed system. The most commonly used systems are: (a) belt-filter press (BFP), (b) centrifuge, (c) plate and frame filter press, and (d) vacuum filter. Several emerging technologies have also gained interest in recent years. The mechanical dewatering systems and emerging technologies are discussed below.[6,7,21]

Belt-Filter Press: The belt-filter presses are currently very popular in the United States. The main advantages are: (a) simple operation and maintenance; (b) relatively fast start-up and shutdown; (c) continuous operation, low power requirement, and high solids content in the cake; and (d) low noise level during operation. The major disadvantages are due to requirements for (a) large volume of wash water, (b) screening or grinding of sludge to protect the belt from damages by sharp-edged objects, and (c) chemical addition and an enclosure for odor control. It is also not suitable for oily solids. The typical solids contents in the sludge cake are between 15% and 35% at a solids capture rate of 95–98%. A continuous wash water supply of 150–300 L/m·min (40–80 gpm/m) at 12–17 kPa (80–120 psig) is typically required. The typical design and performance parameters of BFP systems are summarized in Table 13.23.[6,7,32,74,92]

The major process components of a typical BFP system are shown in Figure 13.24a. Since the BFP process is sensitive to characteristics of incoming sludge, adequate polymer dosing followed by efficient polymer/sludge mixing for 15–45 s is required. Other important auxiliary systems are: (a) belt wash water supply, (b) compressed air for pneumatic or hydraulic belt tensioning systems, and (c) belt or screw conveyor for loading the trucks with biosolids.

The process details of a typical BFP are shown in Figure 13.24b. The mechanical components are: (1) dewatering belts, (2) rollers, (3) drive unit with control, (4) belt tracking (or steering) system, (5) belt tensioning system, (6) belt washing system, and (7) automatic shutdown system that will be activated under abnormal conditions such as (a) belt misalignment, insufficient belt tension, or belt drive failure; (b) sludge conditioning tank failure; (c) loss of pneumatic or hydraulic system pressure; (d) low belt wash water pressure; and (e) emergency stop.[6,7,92–96]

The available belt widths are between 0.5 and 3.4 m (20–134 in); most common width is 2-m. Currently, 2-belt and 3-belt BFP units are available to dewater thickened and unthickened sludges with solids contents of 2–6% and 1.5% or less, respectively. Pressure rollers are provides for pressuring (or sqeezing), driving, tracking (or steering), and tensioning the belt and sludge cake. The number of rollers is 6–14 that are arranged in series. These pressure rollers and belt arrangements also categorize BFP as a horizontal or vertical unit. The details of belt and roller arrangements are shown in Figures 13.24b through d. The BFP operation involves (a) gravity drainage zone, (b) low-pressure zone, also called wedge zone, and

TABLE 13.23 Typical Design and Performance Parameters of Belt-Filter Presses for Sludge Dewatering[a]

Type of Sludge	Feed Sludge Solids Content, %	Solids Loading[b], kg/m·h	Hydraulic Loading[b], L/m·min	Polymer Dose[c], g/kg (lb/ton)	Cake Solids Content, %
Raw or thickened sludge					
PS	4–8	1000–1500	250–600	1.5–2.5 (3–5)	25–35
WAS	1–4	200–350	200–350	3–7.5 (6–15)	15–20
PS and WAS	3–6	350–800	350–800	5–10 (10–20)	15–25
SBR/MBR sludge[c]	1–2	250–400	250–350	5–10 (10–20)	10–20
Anaerobically digested sludge					
PS	2–5	700–900	250–600	2–5 (4–10)	25–35
WAS	2–3	250–400	100–350	4–10 (8–20)	15–20
PS and WAS	2–4	300–550	150–450	4–8 (8–16)	15–25
Aerobically digested					
WAS	1–3	250–400	150–350	6–10 (12–20)	10–20
TAD[c] biosolids	2–5	350–600	100–500	2–12.5 (4–25)	10–20

[a] Based on dry solids.
[b] Based on belt width in meter.
[c] Based on weight of dry solids and polymer dose.
Note: TAD = thermophilic aerobic digestion; MBR = membrane bioreactor; PS = primary sludge; SBR = sequencing batch reactor; WAS = waste activated sludge.
Source: Adapted in part from References 6, 7, 32, 74, 75, and 92 through 96.

(c) high pressure zone. The important process control parameters are sludge feed rate, polymer dosage, and belt speed and belt tension. These parameters are monitored for optimum performance. Readers are referred to References 93 through 96 for in depth equipment details of BFP systems.

Centrifugal Dewatering: Besides sludge thickening (see Section 13.5.3 and Example 13.16), the centrifuges are also widely used for sludge dewatering. The main advantages are: (a) continuous operation, (b) relatively high cake solids, (c) totally enclosed, clean appearance and minimal odors, (d) no or low wash water requirement, (e) suitable for direct dewatering of thin sludge like WAS (<0.5% solids), and (f) easy installation, small footprint, and low initial capital costs. Major concerns of centrifuges are: (a) high power consumption, (b) relatively high TSS in the centrate, (c) high operation and maintenance (O&M) costs, and (d) potential noise and vibration.[6,7,40,74,97]

The polymer-conditioned sludge is fed into the horizontal or cylindrical bowl. At a high rotational speed of 1600–2000 rpm, the cake is formed on the outer wall and scraped. The liquid is collected as centrate and returned to the plant. The basic design of centrifuges is either *countercurrent* or *cocurrent.* The major components and general arrangement of centrifuges are shown in Figure 13.25. The centrifuges are classified into: (a) standard solid-bowl and (b) high-solids centrifuges. These centrifuges produce sludge cake of ~20–30% solids at a solids capture rate 85–95%. The available unit capacity is 5–280 m^3/h (20–1250 gpm). The polymer dosage is in the range from 1 to 25 g/kg (2–50 lb/ton) of dry solids. The important operating parameters are sludge feed rate, rotational speed, differential speed of the scroll, depth of settling zone, polymer used, and physiochemical properties of suspended solids (size, shape, density, temperature, and liquid viscosity).[6,7,98–101]

Plate and Frame Filter Press: They are also called *filter presses* or *recessed plate filter presses.* It consists of round or rectangular recessed plates covered with filter cloth. When pressed together and bolted, hallow chambers are formed and sealed to withstand the pressure applied during filtration. The typical filtration area ranges from 0.3 to 6 m^2 (from 3 to 65 ft^2) per chamber with a plate size of 0.5–2 m (20–80 in). A filter unit may consists of 20–200 chambers. The two types of filter presses are: (a) fixed volume and (b) variable volume. The sequence of operation includes: (1) fill the chamber with conditional sludge, (2) apply desired pressure and maintain for required period, (3) the filtrate passes through the filter cloth and plate to the

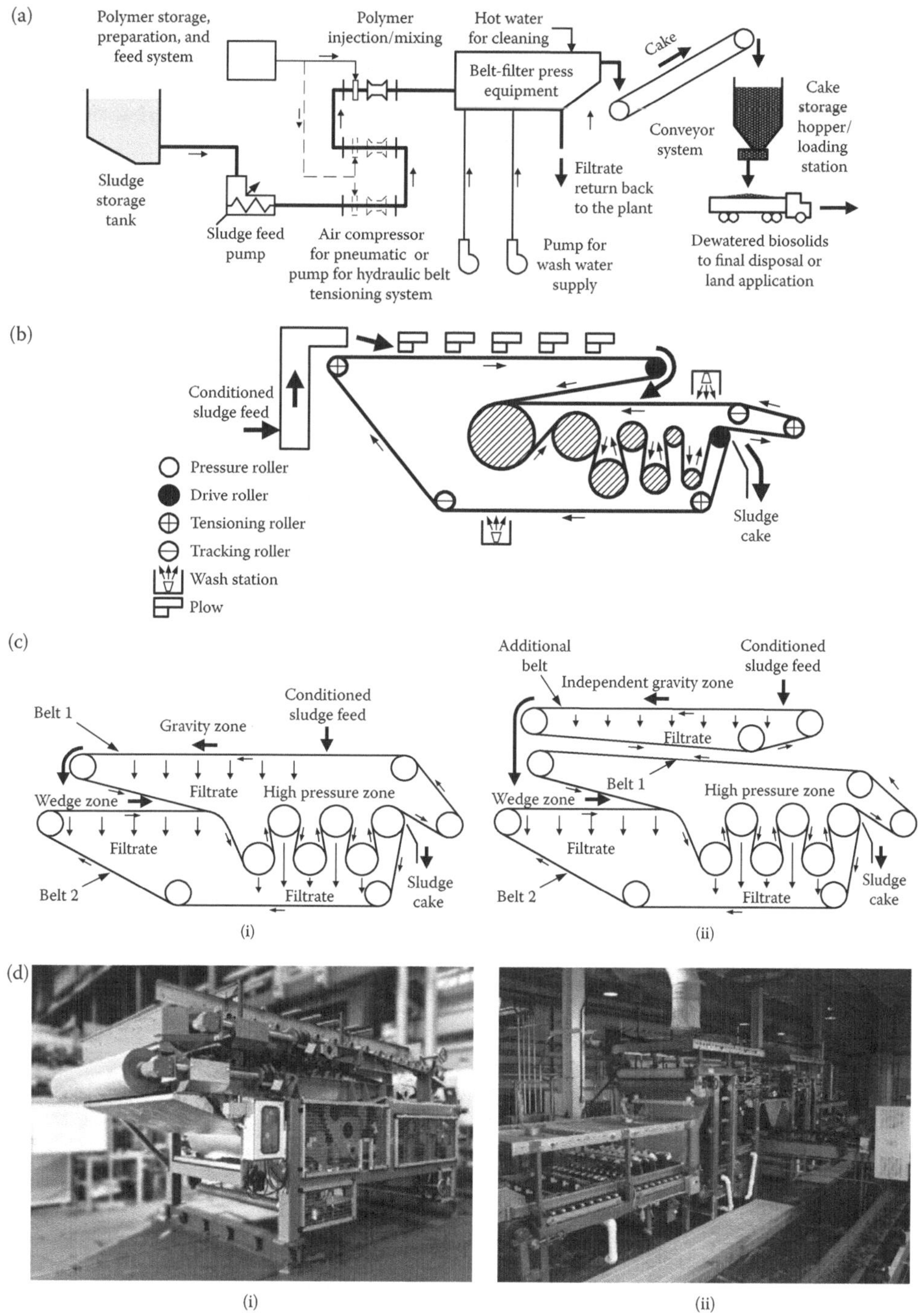

FIGURE 13.24 Belt-filter press for sludge dewatering: (a) schematic process diagram; (b) major mechanical components; (c) belt alternatives with operational stages: (i) 2-belt and (ii) 3-belt; and (d) belt arrangements: (i) horizontal* (Courtesy Andritz (USA) Inc.) and (ii) vertical (Courtesy BDP Industries).

* ANDRITZ Heavy-Duty Belt Press CPF 2200 S8 for tailings dewatering.

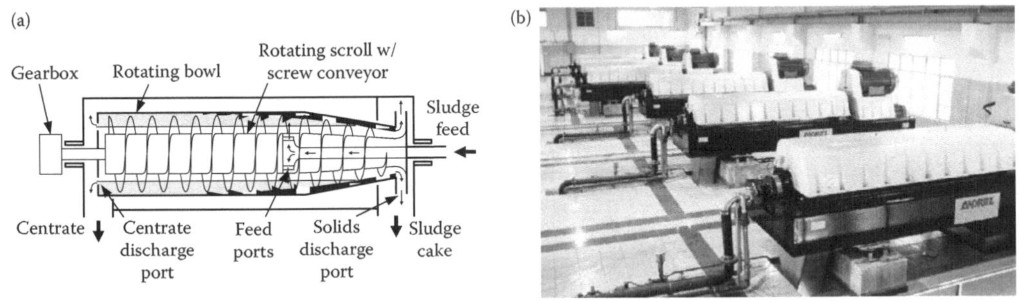

FIGURE 13.25 Decanter solid-bowl centrifugal dewatering: (a) components of a typical centrifuge and (b) general arrangement of centrifuges for dewatering (Courtesy Andritz (USA) Inc.).

outlet ports, (4) separate the plates and remove the cake, and (5) wash the cloth and close the press for the next operating cycle. The solids content of sludge cake is 25–55%. and solids loading rate is 5 $kg/m^2 \cdot h$ (1 $lb/ft^2 \cdot h$). The typical chemical dosages are 50 g/kg (100 lb/ton) for ferric chloride and 100 g/kg (200 lb/ton) for CaO on dry solids basis. In a fixed-volume recessed-plate filter press, a pressure of 700–1550 kPa (100–225 psig) is applied by the feed pumps and maintained for 1–3 h. The cycle time is 2–5 h. In a variable-volume recessed-plate filter press, a membrane diaphragm is placed behind the filter cloth. When expanded under air or hydraulic pressure, the membrane squeezes the sludge in the chamber and presses more filtrate through the filter cloth (Figure 13.26). The total cycle time is about 60–180 min, including

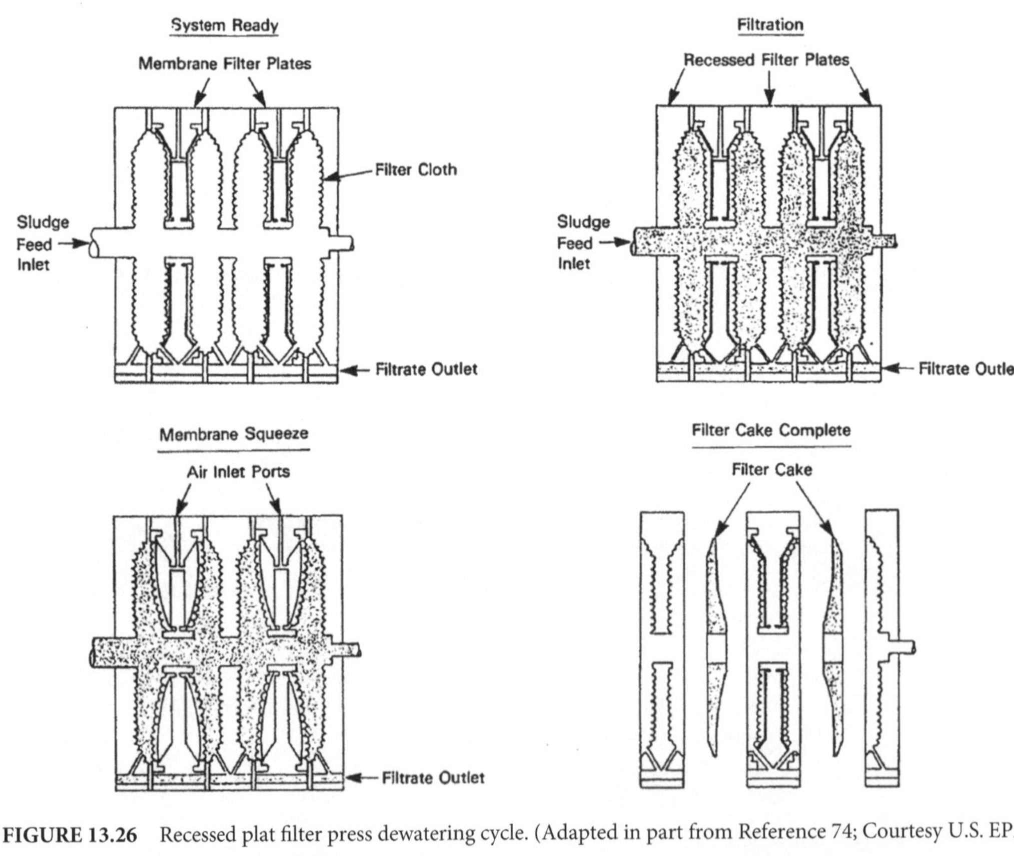

FIGURE 13.26 Recessed plat filter press dewatering cycle. (Adapted in part from Reference 74; Courtesy U.S. EPA.)

15–30 min additional time for cleaning and reset. The typical pressure is 700–900 kPa (100–130 psig) for pre- and intermediate-compressions and 1400–1550 kPa (200–225 psig) for final compression.[6,8,74,102–106]

The advantages of plate and frame filter press include: (a) high cake dryness, (b) high solids capture rate, (c) low TSS concentration in filtrate, and (d) simple operation. The major disadvantages are: (a) batch operation only, (b) high in equipment and O&M costs, (c) labor intensive, (d) large floor area requirement, and (e) high chemical solids associated with high inorganic chemical dosage. The filter presses may be more suitable for dewatering water treatment sludge where metal salt and/or lime are already added as coagulants in the treatment processes. This process is also a valid method for many industrial applications where a high cake dryness is desired.[6–8,74,102]

Vacuum Filter: In recent years, the use of vacuum filters in wastewater application has declined. Vacuum filters consist of a cylindrical drum covered with cloth of natural or synthetic fabric. The drum remains partly submerged in a vat of sludge and rotates slowly. The belt submergence time in the vat is *form time* and that in the air is the *dewatering time*. The form time is normally 40–60% of the cycle time. A vacuum of 70 kPa (10 psi) is applied inside the drum, while a cake layer with 20–35% solids is formed on the outside of the drum. The filter solids loading rate is 10–25 $kg/m^2{\cdot}h$ (2–5 $lb/ft^2{\cdot}h$) at a chemical dose of 20–40 g/kg (40–80 lb/ton) for ferric chloride and 100–150 g/kg (200–300 lb/ton) for CaO on dry solids basis.[6,8]

Other Emerging Mechanical Dewatering Systems: Three emerging mechanical dewatering systems have been tested in full-scale applications in recent years. These are: (a) rotary press, (b) screw press, and (c) electro-dewatering (EDW) systems. New developments that have been reported in bench- or pilot-scale testing stages may include electroosmotic, electroacoustic, membrane filter press (MFP), and tubular filter press (TFP).[7,21,107–114] A brief description of each of these systems is given below.

Rotary Press: The rotary press has a rectangular channel formed by two parallel slow-rotating stainless steel screens in a containment vessel. The polymer-conditioned sludge is fed at a low pressure into the channel and dewatered while travelling with the slow-rotating screens. Filtrate flows through the screens and collected by the vessel. High solids sludge cake is formed because (a) solids accumulate after free water is drained through the screens, (b) a pushing force is created by the friction of slow-moving screens at 1–3 rpm, and (c) a backpressure is created at the restricted outlet. The cross section of typical module is shown in Figure 13.27a(i). The typical design parameters are: (a) feed sludge 2–6% solids, (b) design hydraulic loading 2.5–3.5 $m^3/m^2{\cdot}h$ (1–1.5 gpm/ft^2), (c) solids loading rate ~250 $kg/m^2{\cdot}h$ (50 $lb/ft^2{\cdot}h$), and (d) polymer dosage 5–15 g/kg (10–30 lb/ton). At a typical solids capture rate of 90–98%, the rotary press is capable of achieving a solids contents of 25–30% for dewatering raw and anaerobically digested combined sludge, and 20–25% for dewatering anaerobically digested combined sludge. The wash water requirement is about 200 L/min (50 gpm) for 5 min per module per day. The rotary press is supplied by the manufacturer in standard modules of 1-, 2-, 4- and 6-unit configurations. The total effective dewatering area ranges from 0.2 to 7 m^2 per unit based on these modular configurations. A complete packaged system include rotary press module, pumps, polymer feed equipment, mixing device, power supply, process monitoring, and operational control. A 4-module rotary press unit is illustrated in Figure 13.27a(ii).[7,107–109]

Screw Press: The screw press is a long cylindrical enclosed vessel that contains a stationary stainless steel wedge-wire-screen basket with a slow-rotating screw inside. The polymer-flocculated sludge is fed at low pressure into the basket from one end of the vessel and the filtrate is leached from the sludge through the screen under the pressure applied by the screw at low speed of 0.5 rpm. The dewatering backpressure may be elevated by an adjustable restriction outlet that is located on the opposite end from the sludge feeding. The screw press may be available at a throughput of 7–45 L/min (2–12 gpm) at an operating pressure of 280–560 kPa (40–80 psig). The polymer dosage is 4–20 g/kg (8–40 lb/ton). At a typical solids capture rate of over 90%, the screw press is capable of achieving a solids contents of 25–35% for dewatering raw and anaerobically digested combined sludge or 20–25% for dewatering anaerobically digested combined sludge. The screen is washed for ~15 s every 10 min operation. The screw press is available in either horizontal or inclined design (see Figure 13.27b). The *volute dewatering press* is a modification of screw press that uses moving and fixed rings to improve dewatering performance while preventing the clogging problem.[7,110–113]

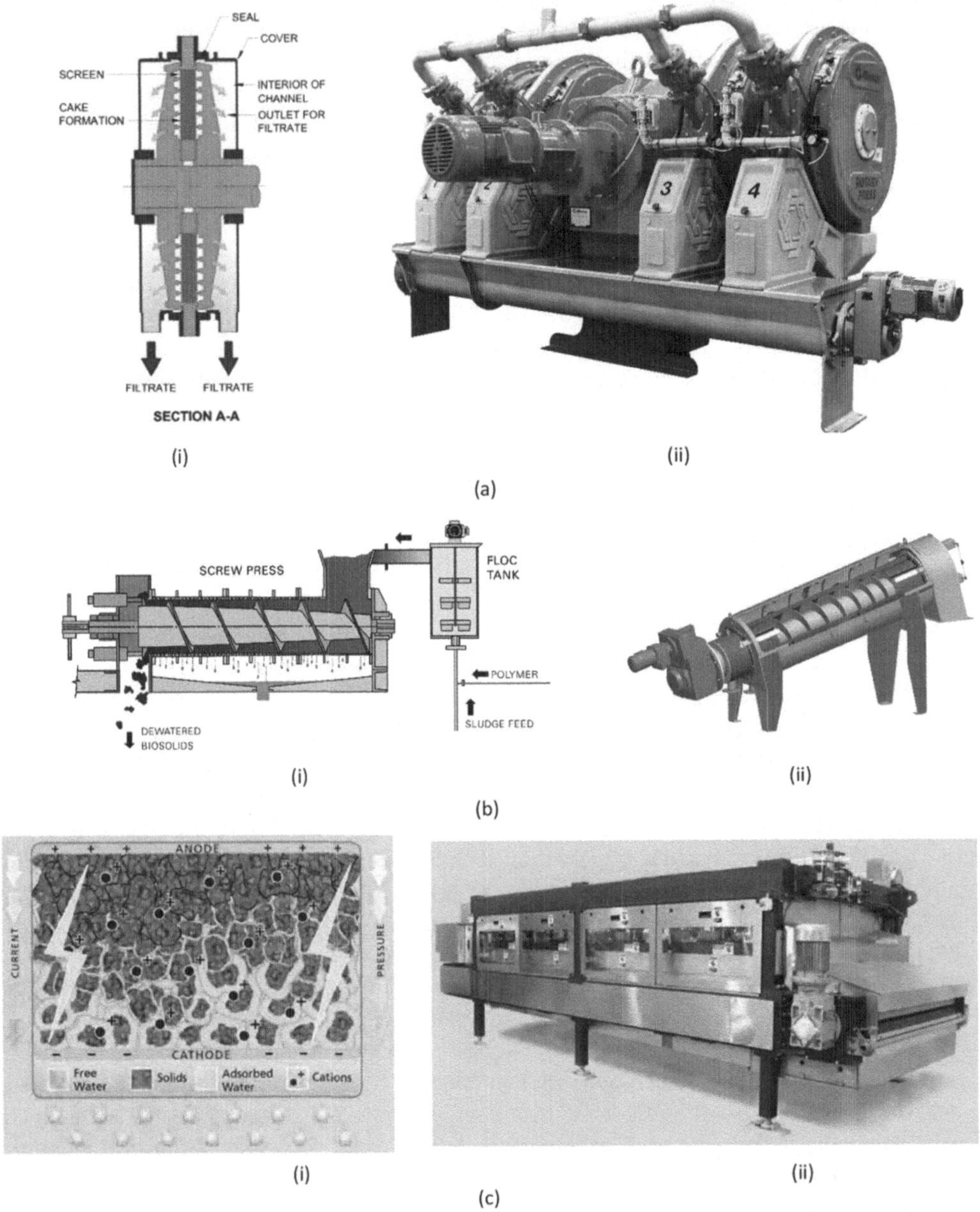

FIGURE 13.27 Selected Emerging Mechanical Dewatering Systems: (a) rotary press (Courtesy Fournier Industries Inc.): (i) cross sectional view and (ii) 4-module configuration; (b) screw press: (i) horizontal design (Courtesy FKC Co., Ltd.) and (ii) inclined design (Courtesy Huber Technology, Inc.); and (c) electro-dewatering (EDW) system (Courtesy Ovivo): (i) definition sketch of process and (ii) equipment.

Since both the rotary and screw process are low-pressure and low-speed devices, they require low power, low maintenance, and low O&M costs. Other advantages of both devices include: enclosed vessel for containing odor, relatively low vibration and noise levels, compact and expandable configuration, and low washwater and operator attention requirements. The major disadvantages of both devices are relatively sensitive to sludge quality, high polymer requirements, limited throughput, and high capital costs.[7,21,107–113] It is recommended that pilot test be conducted to establish the design parameter.

Electro-dewatering system: They are also called *electrical field-assisted dewatering* systems. The electrical field is created by the electrodes at a direct voltage of about 50–70 V. The predewatered sludge with 10–20% solids is fed between the electrodes. The sludge particles (negatively charged) and water molecules (positively charged) move in opposite directions (see Figure 13.27c(i)). Filter cloth or membrane is provided over the cathode to allow the filtrate passing through. Extra mechanical pressure is also applied to accelerate the filtration rate. The capacity of available EDW system may be in a range from 300 to 2400 kg/h per unit (600–5000 lb/h) at a power consumption of 0.2–0.3 kWh/kg (180–270 kWh/ton). The process is capable of achieving solids contents 30–50%. Wash water requirement is about 15–20 L/min (4–5 gpm) per unit since the negatively charged particles are rejected by the cathode to prevent clogging of filter cloth. The EDW process removes free, interstitial and portion of bond water. Its advantages are less sensitive to sludge characteristics, simple in operation, low wash water requirement, pathogen reduction capability, and capable of automatic operation. The disadvantages include predewatering requirement, low throughput, and high capital costs. With limited experience of full-scale installation, unpredictable performance is one of the main concern. A commercial EDW unit is shown in Figure 13.27c(ii).[7,114]

EXAMPLE 13.47: DESIGN OF A BELT-FILTER PRESS (BFP) SLUDGE DEWATERING SYSTEM

A wastewater treatment facility produces digested combined sludge. The sustained weekly peak sludge solids production = 5000 kg/d, solids content = 4%, and specific gravity of liquid sludge = 1.01. The design data for the BFP system are: total solids in the sludge cake = 20%; specific gravity = 1.05; solids concentration in the waste stream = 850 mg/L; the maximum throughput loading (or hydraulic loading) less than 200 L/m·min, solids loading = 300 kg/m·h; the BFP operation = 8 h/d and 5 d/week; and the washwater is plant effluent wtih 10 mg/L TSS at a flow rate = 180 L/m·min. The active polymer dose is 5 g/kg dry solids and 100% active polymer is retained in the sludge. Calculate (a) the number and size of BFPs required; (b) daily volumes of sludge cake, total waste from the BFPs, and filtrate; (c) weights of solids in sludge cake, total waste from the BFPs, and filtrate; and (d) solids capture rate. Also, describe the major components and prepare a conceptual layout of the BFP system.

Solution

1. Calculate the total solids processed by the BFPs per hour of filter operation.

$$\text{Weight of digested sludge during operating day, } W_s = \frac{5000\,\text{kg/d} \times 7\,\text{d/week}}{5\,\text{d/week} \times 8\,\text{h/d operation}} = 875\,\text{kg/h}$$

Weight of polymer in processed solids, $W_{\text{polymer}} = 5\,\text{g/kg} \times 875\,\text{kg/h} \times 10^{-3}\,\text{kg/g} = 4.4\,\text{kg/h}$

Weight of total solids processed, $W_{\text{solids}} = \text{W}_s + \text{W}_{\text{polymer}} = (875 + 4.4)\,\text{kg/h} = 879\,\text{kg/h}$

2. Calculate the BFP size and number based on the solids loading rate.

$$\text{BFP effective belt width required, } L_{\text{BFP}} = \frac{W_{\text{solids}}}{\text{SLR}_{\text{BFP}}} = \frac{879\,\text{kg/h}}{300\,\text{kg/m·h}} = 2.93\,\text{m}$$

Provide three BFPs, each having a belt width of 1.5 m.

Under design operating condition, two units will be in service while one unit will be a standby unit.

$$\text{Actual design solids loading rate, } SLR_{\text{BFP}}^{\text{Design}} = \frac{W_{\text{solids}}}{2 \times L_{\text{BFP}}^{\text{Design}}} = \frac{879\,\text{kg/h}}{2 \times 1.5\,\text{m}} = 293\,\text{kg/m·h}$$

3. Check the hydraulic loading rate on the belt.

Digested sludge volume from Equation 13.1e, $V_s = \dfrac{100\% \times W_s}{p_s S_{b,s} \rho_w} = \dfrac{100\% \times 875\,\text{kg/h}}{4\% \times 1.01 \times 1000\,\text{kg/m}^3} = 21.7\,\text{m}^3/\text{h}$

Assume the polymer solution strength is 0.5%, and specific gravity is ~1.

Polymer solution volume from Equation 13.1e, $V_{\text{polymer}} = \dfrac{100\% \times 4.4\,\text{kg/h}}{0.5\% \times 1 \times 1000\,\text{kg/m}^3} = 0.9\,\text{m}^3/\text{h}$

Total sludge volume applied, $V_{\text{solids}} = V_s + V_{\text{polymer}} = (21.7 + 0.9)\,\text{m}^3/\text{h} = 22.6\,\text{m}^3/\text{h}$

Actual design hydraulic loading rate, $HLR_{\text{BFP}}^{\text{Design}} = \dfrac{V_{\text{solids}}}{2 \times L_{\text{BFP}}^{\text{Design}}} = \dfrac{22.6\,\text{m}^3/\text{h} \times 10^3\,\text{L/m}^3}{2 \times 1.5\,\text{m} \times 60\,\text{min/h}} = 126\,\text{L/m·min}$

This is below the maximum allowed hydraulic loading of 200 L/m·min.

4. Calculate the volume and solids in wash water.

Volume of wash water, $V_{\text{wash water}} = 180\,\text{L/m·min} \times 2 \times 1.5\,\text{m} \times 60\,\text{min/h} \times 10^{-3}\,\text{m}^3/\text{L} = 32.4\,\text{m}^3/\text{h}$

Weight of solids contained in wash water, $W_{\text{wash water}} = 10\,\text{g/m}^3 \times 32.4\,\text{m}^3/\text{h} \times 10^{-3}\,\text{kg/g} = 0.3\,\text{kg/h}$

5. Calculate the volumes of sludge cake, total waste from the BFPs, and filtrate.

a. Prepare flow balance around the BFP units (Equation 13.19a).

$$V_{\text{solids}} + V_{\text{wash water}} = V_{\text{cake}} + V_{\text{waste}}$$

$$(22.6 + 32.4)\,\text{m}^3/\text{h} = V_{\text{cake}} + V_{\text{waste}} \quad \text{or} \quad V_{\text{waste}} = 55\,\text{m}^3/\text{h} - V_{\text{cake}} \qquad (13.19\text{a})$$

b. Prepare mass balance around the BFP units (Equation 13.19b).

$$W_{\text{solids}} + W_{\text{wash water}} = W_{\text{cake}} + W_{\text{waste}}$$

$$(879 + 0.3)\,\text{kg/h} = W_{\text{cake}} + W_{\text{waste}} \quad \text{or} \quad 879.3\,\text{kg/h} = W_{\text{cake}} + W_{\text{waste}} \qquad (13.19\text{b})$$

c. Determine the volume of sludge cake.

Rearranged Equation 13.1e to calculate the weight of solids in sludge cake (Equation 13.19c).

$$W_{\text{cake}} = \frac{p_{\text{cake}} S_{\text{cake,s}} \rho_w V_{\text{cake}}}{100\%} = \frac{20\% \times 1.05 \times 1000\,\text{kg/m}^3}{100\%} \times V_{\text{cake}} \qquad (13.19\text{c})$$
$$= 210\,\text{kg/m}^3 \times V_{\text{cake}}$$

It is assumed that the solids concentration in the waste stream from BFPs, $C_{\text{waste}} = 1500$ mg/L. Weight of solids in total waste from BFPs (Equation 13.19d),

$$W_{\text{waste}} = C_{\text{waste}} V_{\text{waste}} = 850\,\text{g/m}^3 \times 10^{-3}\,\text{kg/g} \times V_{\text{waste}} = 0.85\,\text{kg/m}^3 \times V_{\text{waste}} \qquad (13.19\text{d})$$

Substitute the V_{waste} in Equation 13.19d from Equation 13.19a to obtain (Equation 13.19e).

$$W_{\text{waste}} = 0.85\,\text{kg/m}^3 \times (55\,\text{m}^3/\text{h} - V_{\text{cake}}) = 46.8\,\text{kg/h} - 0.85\,\text{kg/m}^3 \times V_{\text{cake}} \qquad (13.19\text{e})$$

Substitute the W_{cake} from Equation 13.19c and W_{waste} from Equation 13.19e to obtain Equation 13.19f.

$$879.3\,\text{kg/h} = 210\,\text{kg/m}^3 \times V_{\text{cake}} + 46.8\,\text{kg/h} - 0.85\,\text{kg/m}^3 \times V_{\text{cake}} \qquad (13.19\text{f})$$

Solve V_{cake} from Equation 13.19f.

$$V_{cake} = \frac{(879.3 - 46.8)\ \text{kg/h}}{(210 - 0.85)\ \text{kg/m}^3} = 3.98\ \text{m}^3/\text{h}$$

d. Determine the volume of total waste from BFPs from Equation 13.19a.

$$V_{waste} = 55\ \text{m}^3/\text{h} - V_{cake} = (55 - 3.98)\ \text{m}^3/\text{h} = 51\ \text{m}^3/\text{h}$$

e. Determine the volume of filtrate.

$$V_{filtrate} = V_{waste} - V_{wash\ water} = (51 - 32.4)\ \text{m}^3/\text{h} = 18.6\ \text{m}^3/\text{h}$$

6. Calculate the weights of solids in sludge cake, total waste from the BFPs, and filtrate.

Weight of solids contained in sludge cake from Equation 13.19c, $W_{cake} = 210\ \text{kg/m}^3 \times 3.98\ \text{m}^3/\text{h} = 836\ \text{kg/h}$

Weight of solids contained in filtrate, $W_{filtrate} = W_{solids} - W_{cake} = (879 - 836)\ \text{kg/h} = 43\ \text{kg/h}$

Weight of solids contained in total waste from BFPs,

$$W_{waste} = W_{filtrate} + W_{wash\ water} = (43 + 0.3)\ \text{kg/h} = 43.3\ \text{kg/h}$$

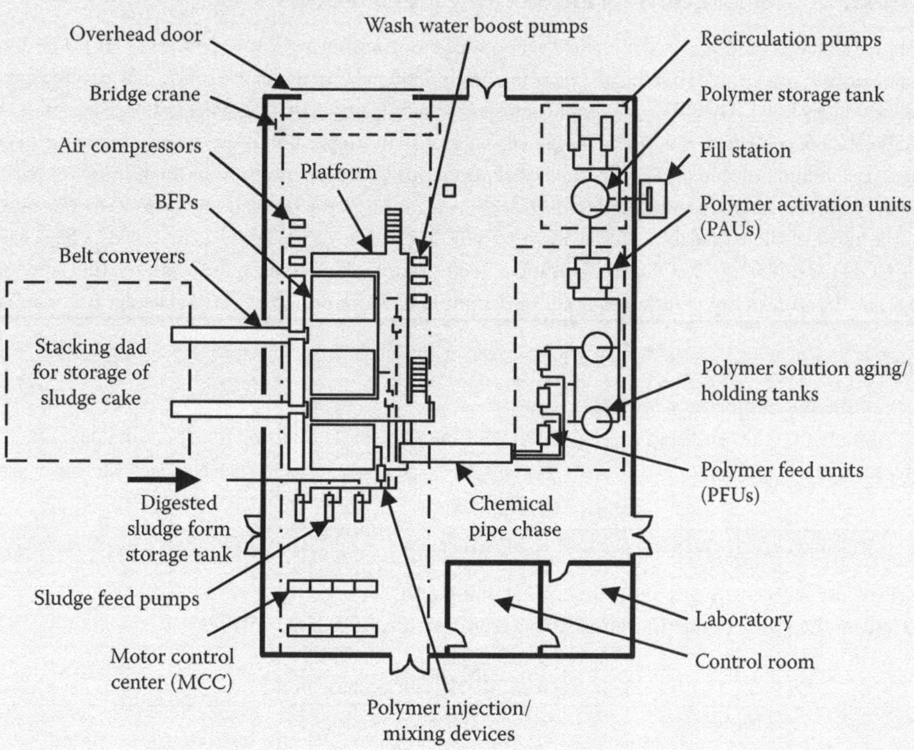

FIGURE 13.28 Conceptual layout of belt-filter press sludge dewatering system and biosolids dewatering building (Example 13.47).

7. Verify the solids concentration in the waste stream from BFPs.

$$C_{\text{waste}} = \frac{W_{\text{waste}}}{V_{\text{waste}}} = \frac{43.1\,\text{kg/h}}{29.4\,\text{m}^3/\text{h}} = 1.46\,\text{kg/m}^3 \text{ or } 1460\,\text{mg/L}$$

The calculated solids concentration is close to the assumed value of 1500 mg/L in Step 5.c.

8. Calculate the solids capture rate of BFPs.

Solids capture rate,

$$E_{\text{solids}} = \frac{W_{\text{cake}}}{W_{\text{solids}}} \times 100\% = \frac{836\,\text{kg/h}}{879\,\text{m}^3/\text{h}} \times 100\% = 95\%$$

9. Describe the BFP facility components and prepare a conceptual layout.

The BFP system is located inside a Biosolids Dewatering Building. It consists of the following equipment: (a) three BFP units, (b) sludge feed pumps, (c) polymer injection/mixing devices, (d) polymer storage, preparation, and feed system, (e) air compressors, (f) wash water booster pumps, (g) BFP waste collection and drainage system, (h) belt conveyer system, (i) bridge crane, (j) power supply and motor control center (MCC), and (k) process control room with operation laboratory. A schematic process diagram is shown in Figure 13.24a. A conceptual layout of the Biosolids Dewatering Building is provided in Figure 13.28.

EXAMPLE 13.48: PREDEWATERING BY CENTRIFUGES

Combined sludge is predewatered by centrifuges prior to the thermal hydrolysis process (THP). The design flow rate of combined sludge during an operating day is 3300 m^3/d at about 2% solids. The specific gravity of bulk sludge is 1.004. The solids content of sludge cake is 16% and specific gravity is 1.04. The solids capture efficiency of centrifuge is 90% at a design capacity of 50 m^3/h per unit. Determine (a) number of centrifuges, (b) volume of sludge cake, (c) flow of centrate, (d) TSS concentration in the filtrate, (e) mass of COD in the predewatered sludge, and (f) BOD_5 concentration in the centrate. Assume VSS/TSS ratio = 0.75, the ratio of biodegradable VSS (VSS_{bd}) to VSS is 0.73 kg VSS_{bd}/kg VSS, each g of VSS_{bd} exerts 1.42 g COD (or BOD_L), and BOD_5/BOD_L ratio = 0.68. Soluble COD is negligible. Assume that a polymer dosage less than 1g/kg dry solids is required. See Example 13.55 for design of THP for sludge pretreatment.

Solution

1. Determine the number of centrifuges.

Assume that the centrifuges are operated continuously for 24 h during the operating day. Calculate the number of centrifuges.

$$N_{\text{req}} = \frac{V_{\text{WAS}}}{q_{\text{centrifuge}}} = \frac{3300\,\text{m}^3/\text{d}}{50\,\text{m}^3/\text{h per unit} \times 24\,\text{h/d}} = 2.75 \text{ units}$$

Provide four centrifuges, including one standby unit.

2. Calculate the solids in the thickened sludge from rearranged Equation 13.1e.

$$W_{\text{sludge}} = \frac{p_s S_{b,s} \rho_w V_{\text{WAS}}}{100\%} = \frac{2\% \times 1.004 \times 1000\,\text{kg/m}^3 \times 3300\,\text{m}^3/\text{d}}{100\%} = 66{,}300\,\text{kg/d}$$

Note: The polymer dose is small (< 0.1% of the dry solids weight), so its weight retained in the sludge is ignored.

3. Calculate the volume of predewatered sludge at a solids capture efficiency of 90%.

Weight of predewatered sludge, $W_{\text{pds}} = 0.9 \times W_{\text{sludge}} = 0.9 \times 66{,}300\,\text{kg/d} = 59{,}700\,\text{kg/d}$

Predewatered sludge volume from Equation 13.1e.

$$V_{pds} = \frac{100\% \times W_{pds}}{p_{pds} S_{b,pds} \rho_w} = \frac{100\% \times 59{,}700\,\text{kg/d}}{16\% \times 1.04 \times 1000\,\text{kg/m}^3} = 360\,\text{m}^3/\text{d}$$

4. Determine the flow of centrate.
   Volume of centrate, $V_{centrate} = V_{sludge} - V_{pds} = (3300 - 360)\,\text{m}^3/\text{d} = 2940\,\text{m}^3/\text{d}$
5. Estimate the solids concentration in the centrate.
   Weight of solids in centrate, $W_{centrate} = W_{sludge} - W_{pds} = (66{,}300 - 59{,}700)\,\text{kg/d} = 6600\,\text{kg/d}$

$$\text{Solids concentration in centrate, } C_{centrate} = \frac{W_{centrate}}{V_{centrate}} = \frac{6600\,\text{kg/d}}{2940\,\text{m}^3/\text{d}} \times 10^3\,\text{g/kg} = 2240\,\text{g/m}^3 \quad \text{or} \quad 2240\,\text{mg/L}$$

6. Estimate the mass of COD in the dewatered sludge.
   Biodegradable VSS (VSS_{bd}) in the predewatered sludge.

$$\begin{aligned} W_{VSSbd,pds} &= VSS_{bd}/VSS \text{ ratio} \times VSS/TSS \text{ ratio} \times W_{pds} \\ &= 0.73\ VSS_{bd}/\text{kg VSS} \times 0.75\,\text{kg VSS/kg TSS} \times 59{,}700\,\text{kg TSS/d} \\ &= 32{,}700\,\text{kg VSS/d} \end{aligned}$$

COD exerted in the predewatered sludge.

$$\begin{aligned} \Delta S_0 &= COD/VSS_{bd} \text{ ratio} \times W_{VSSbd,pds} = 1.42\,\text{kg COD/kg } VSS_{bd} \times 32{,}700\,\text{kg } VSS_{bd}/\text{d} \\ &= 46{,}400\,\text{kg COD/d} \end{aligned}$$

7. Estimate the BOD_5 concentration in the centrate.

$$\begin{aligned} C_{BOD_5,centrate} &= BOD_5/BOD_L \text{ ratio} \times BOD_L/VSS_{bd} \text{ ratio} \times VSS_{bd}/VSS \text{ ratio} \times VSS/TSS \text{ ratio} \\ &\quad \times C_{centrate} = 0.68\,\text{mg } BOD_5/\text{mg } BOD_L \times 1.42\,\text{mg } BOD_L/\text{mg } VSS_{bd} \\ &\quad \times 0.73\,\text{mg } VSS_{bd}/\text{mg VSS} \times 0.75\,\text{mg VSS/mg TSS} \times 2240\,\text{mg TSS/L} \\ &= 1180\,\text{mg } BOD_5/\text{L} \end{aligned}$$

EXAMPLE 13.49: DESIGN OF A PLATE AND FRAME FILTER PRESS

Design a variable-volume recessed plate filter press facility to dewater 5000 kg/d anaerobically digested combined sludge. The sludge characteristics are: volatile solids = 55%, solids concentration = 3.5%, and specific gravity of bulk sludge and sludge cake are 1.01 and 1.07. The following design data apply: maximum solids loading rate = 5 kg/m^2·h (1 lb/ft^2·h); minimum solids content of sludge cake = 25%; optimum ferric chloride ($FeCl_3$) and lime (CaO) dosages = 50 and 100 g/kg of dry solids; and operating time = 8 h/d, and 5 d/week. Assume that 95% of sludge solids are captured and 80% flocculants are retained in the sludge cake. Determine (a) number of variable-volume recessed plate filter press assemblies required, (b) the volumes of dewatered cake and filtrate, and (c) solids concentration in the filtrate. Also develop a simple process diagram and describe briefly the four functional stages in the operating cycle of a filter press.

Solution

1. Calculate the total solids processed by dewatering operation.

$$\text{The daily weight of digested sludge processed, } W_s = \frac{5000\,\text{kg/d} \times 7\,\text{d/week}}{5\,\text{d/week}} = 7000\,\text{kg/d}$$

Weight of $FeCl_3$ added in sludge solids, $W_{FeCL3} = 50$ g/kg × 7000 kg/d × 10^{-3} kg/g = 350 kg/d
Weight of CAO added in sludge solids, $W_{CaO} = 100$ g/kg × 7000 kg/d × 10^{-3} kg/g = 700 kg/d
Weight of total sludge solids, $W_{solids} = W_s + W_{FeCl3} + W_{CaO} = (7000 + 350 + 700)$kg/d = 8050 kg/d

2. Obtain the manufacturer's technical data on the selected filter press.

   The plate and frame filter press units are available in many sizes. Select a filter press that has the following specifications: (a) plate size = 1200 mm × 1200 mm (47¼ in × 47¼ in); (b) cake thickness $B_c = 33$ mm (1¼ in); (c) number of chambers per unit $N_c = 80$ chambers/unit·cycle; (d) volume per chamber $V_c = 0.035\ m^3$/chamber (1.24 ft^3/chamber); and (e) filter area per chamber $A_c = 2.5\ m^2$/chamber (26.9 ft^2/chamber).

3. Determine the number of filter presses based on total solids volume applied.

   The solids loss through the filtrate is normally ignored in selecting the number of filter presses. Volume of dewatered sludge from Equation 13.1e,

$$V_{dewatered} = \frac{100\% \times W_{solids}}{p_s S_{b,s} \rho_w} = \frac{100\% \times 8050\,\text{kg/d}}{25\% \times 1.07 \times 1000\,\text{kg/m}^3} = 30.1\,\text{m}^3/\text{d}$$

   Total cake volume per filter press per cycle per unit,

$$V_u = N_c V_c = 80\ \text{chambers/unit·cycle} \times 0.035\,\text{m}^3/\text{chamber} = 2.8\,\text{m}^3/\text{unit·cycle}$$

   Assume that a total operating cycle of 2 h is required to achieve the desired solids content in the sludge cake. There are a total of four cycles per unit during 8 h operation time.

$$\text{Number of units required, } N_u = \frac{V_{dewatered}}{N_{cycle} V_u} = \frac{30.1\,\text{m}^3/\text{d}}{4\ \text{cycles/d} \times 2.8\,\text{m}^3/\text{unit·cycle}} = 2.7\ \text{units}$$

   Provide four (4) identical units, including one standby unit.

4. Check the dewatering capacity based on total volume of solids applied.

   The available dewatering capacity is validated for: (a) firm capacity, when the largest unit is out of service and (b) maximum capacity, when all units are in service.

   a. Check the firm capacity available.

      Firm volumetric capacity with 3 units.

$$V_{firm} = (N_u^{design} - 1) N_{cycle} V_u = (4 - 1)\ \text{units} \times 4\ \text{cycles/d} \times 2.8\,\text{m}^3/\text{unit·cycle} = 33.6\,\text{m}^3/\text{d}$$

$$\text{Firm capacity availability, } Av_{firm} = \frac{V_{firm}}{V_{dewatered}} \times 100\% = \frac{33.6\,\text{m}^3/\text{d}}{30.1\,\text{m}^3/\text{d}} \times 100\% \approx 110\% > 100\%$$

      Note: It is required that the firm capacity when the largest unit is out of service is larger than the design total solids volume. The current design meets this requirement.

   b. Check the maximum capacity availability.

      Maximum volumetric capacity with 4 units,

$$V_{max} = N_u^{design} N_{cycle} V_u = 4\ \text{units} \times 4\ \text{cycles/d} \times 2.8\,\text{m}^3/\text{unit·cycle} = 44.8\,\text{m}^3/\text{d}$$

$$\text{Maximum capacity availability, } Av_{max} = \frac{V_{max}}{V_{dewatered}} \times 100\% = \frac{44.8\,\text{m}^3/\text{cycle}}{30.1\,\text{m}^3/\text{cycle}} \times 100\% \approx 150\% > 125\%$$

      Note: It is required that the maximum capacity, when all units are in service is at least 125% of the design total solids volume dewatered. The current design meets this requirement.

5. Check the solids loading rate based on total weight of solids applied.

   Filter area per cycle per filter press unit,

   $A_u = N_c A_c = 80$ chambers/unit·cycle × 2.5 m^2/chamber = 200 m^2/unit·cycle

   Firm filter area with 3 units,

   $A_{firm} = (N_u^{design} - 1)N_{cycle}A_u = (4-1)$ units × 4 cycles/d × 200 m^2/unit·cycle = 2400 m^2/d

   Solids loading rate per cycle, $SLR_{cycle} = \dfrac{W_{solids}}{A_{firm}} = \dfrac{7000\text{ kg/d}}{2400\text{ m}^2\text{/d}} = 2.9\text{ kg/m}^2\cdot\text{cycle}$

   Average solids loading rate during 2-h cycle time, $SLR = \dfrac{SLR_{cycle}}{t_{cycle}} = \dfrac{2.9\text{ kg/m}^2\cdot\text{cycle}}{2\text{ h/cycle}} = 1.5\text{ kg/m}^2\cdot\text{h}$

   Note: It is below the maximum solids loading rate of 5 kg/m^2·h.

6. Calculate the total volume of solids processed during the operating day.

   Digested sludge volume from Equation 13.1e,

   $$V_s = \frac{100\% \times W_s}{p_s S_{b,s} \rho_w} = \frac{100\% \times 7000\text{ kg/d}}{3.5\% \times 1.01 \times 1000\text{ kg/m}^3} = 198\text{ m}^3\text{/d}$$

   The ferric chloride solution is prepared at 10% strength with a specific gravity of 1.04.

   Volume of $FeCl_3$ solution from Equation 13.1e, $V_{FeCl3} = \dfrac{100\% \times 350\text{ kg/d}}{10\% \times 1.04 \times 1000\text{ kg/m}^3} = 3.4\text{ m}^3\text{/d}$

   Lime slurry is fed at 10% content with the specific gravity of 1.08.

   Volume of lime slurry from Equation 13.1e, $V_{CaO} = \dfrac{100\% \times 700\text{ kg/d}}{10\% \times 1.08 \times 1000\text{ kg/m}^3} = 6.5\text{ m}^3\text{/d}$

   Total volume of solids processed, $V_{solids} = V_s + V_{FeCl3} + V_{CaO} = (198 + 3.4 + 6.5)\text{m}^3\text{/d} = 208\text{ m}^3\text{/d}$

7. Calculate the volume of sludge cake during the operating day.

   Weight of sludge solids captured in the cake, $W_{s,cake} = 0.95 \times W_s = 0.95 \times 7000$ kg/d = 6650 kg/d

   Weight of solids added from $FeCl_3$ conditioning and retained in the cake,

   $W_{FeCl3,cake} = 0.8 \times W_{FeCl3} = 0.8 \times 350$ kg/d = 280 kg/d

   Weight of solids added from lime conditioning and retained in the cake,

   $W_{CaO,cake} = 0.8 \times W_{CaO} = 0.8 \times 700$ kg/d = 560 kg/d

   Total weight of solids in the sludge cake,

   $W_{cake} = W_{s,cake} + W_{FeCl3,cake} + V_{CaO,cake} = (6650 + 280 + 560)$ kg/d = 7490 kg/d

   Sludge cake volume from Equation 13.1e,

   $$V_{cake} = \frac{100\% \times W_{cake}}{p_{cake} S_{b,cake} \rho_w} = \frac{100\% \times 7490\text{ kg/d}}{25\% \times 1.07 \times 1000\text{ kg/m}^3} = 28\text{ m}^3\text{/d}$$

   Firm capacity availability based on sludge cake volume,

   $$Av_{firm} = \frac{V_{firm}}{V_{cake}} \times 100\% = \frac{33.6\text{ m}^3\text{/cycle}}{28\text{ m}^3\text{/cycle}} \times 100\% = 120\%$$

   Note: The designed filter press units provide 20% additional capacity. This gives desired operational flexibility.

8. Calculate the volume of waste generated during the operating day.

   a. Calculate the filtrate from the difference between the volume of solids processed (Step 6) and the volume of sludge cake (Step 7).

      Volume of filtrate, $V_{filtrate} = V_{solids} - V_{cake} = (208 - 28)\text{ m}^3\text{/d} = 180\text{ m}^3\text{/d}$

b. Determine the volume of wash water.

The unit wash water requirements per cycle shall be determined based on manufacturer's recommendations. The typical wash water requirements for a filter press unit per operating cycle are: (a) wash water = 2000 L/uint·cycle (~500 gal/uint·cycle), (b) cake extraction = 200 L/uint·cycle (~50 gal/uint·cycle), and (c) diaphragm losses = 100 L/uint·cycle (~25 gal/uint·cycle). Ignore the solids contained in the wash water ($W_{\text{wash water}} \approx 0$).

Total wash water cycle per filter press unit,

$$V_{\text{u,washwater}} = (2000 + 200 + 100)\ \text{L/unit·cycle} \times 10^{-3}\ \text{m}^3/\text{L} = 2.3\ \text{m}^3/\text{unit·cycle}$$

Total wash water with 3 units,

$$V_{\text{washwater}} = \left(N_{\text{u}}^{\text{design}} - 1\right) N_{\text{cycle}} V_{\text{u,washwater}} = (4-1)\ \text{units} \times 4\ \text{cycles/d} \times 2.3\ \text{m}^3/\text{unit·cycle} = 28\ \text{m}^3/\text{d}$$

c. Calculate the volume of total waste from dewatering operation.

Volume of total waste, $V_{\text{waste}} = V_{\text{filtrate}} + V_{\text{wash water}} = (180 + 28)\ \text{m}^3/\text{d} = 208\ \text{m}^3/\text{d}$

Percent wash water in total waste volume,

$$R_{\text{washwater}} = \frac{V_{\text{washwater}}}{V_{\text{waste}}} \times 100\% = \frac{28\ \text{m}^3/\text{d}}{208\ \text{m}^3/\text{d}} \times 100\% = 13.5\%$$

Note: Percent wash water is 13.5% of total waste flow from dewatering operation. It is within the typical range from 10% to 20%.

9. Estimate the solids concentration in the waste flow from dewatering operation.

Weight of solids in the waste, $W_{\text{waste}} = W_{\text{solids}} - W_{\text{cake}} = (8050 - 7490)\ \text{kg/d} = 560\ \text{kg/d}$

$$\text{Solids concentration in the waste flow, } C_{\text{waste}} = \frac{W_{\text{waste}}}{V_{\text{waste}}} = \frac{560\ \text{kg/d}}{208\ \text{m}^3/\text{d}} \times 10^3\ \text{g/kg} = 2690\ \text{g/m}^3 \text{ or } 2690\ \text{mg/L}$$

10. Develop the process diagram of plate and frame filter press facility for dewatering anaerobically digested combined sludge.

A simplified process diagram is shown in Figure 13.29a.

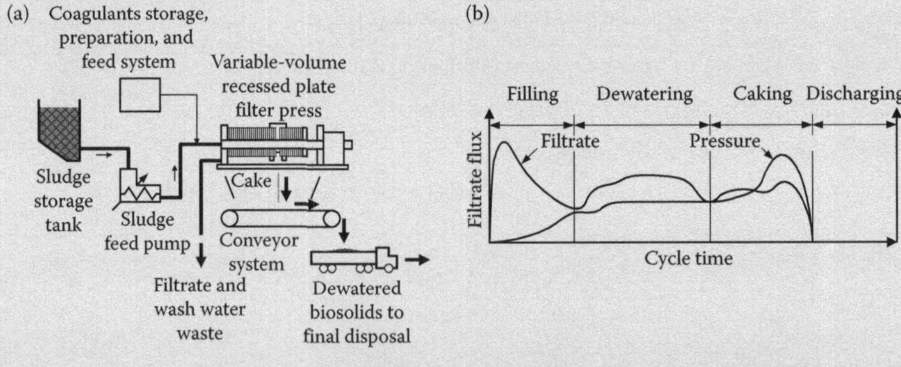

FIGURE 13.29 Design of variable-volume recessed plate filter press: (a) process flow diagram and (b) definition sketch of functional stages in operating cycle (Example 13.49).

11. Describe the functional stages of variable-volume recessed plate filter press in the operating cycle.

There are four basic functional stages in an operating cycle: (a) filling, (b) dewatering, (c) caking, and (d) discharging. A brief description of each stage is given below.

a. Filling Stage (10–30 min). Upon closing the clean press completely, the feed pump is turned on to fill the chambers and form an initial solids layer over the filter clothes while filtrate is released. The pumping is continued until the required initial pressure is reached.

b. Dewatering Stage (20–75 min): The feed pump is turned off and the diaphragm membranes are inflated by compressed air for pre- and intermediate-compressions of the solids to optimize solids dewatering and cake-formation at low pressure.
c. Caking Stage (15–45 min). The final compression is performed at high pressure to achieve the desired solids content of the cake. Air-drying may also be used to remove additional moisture.
d. Discharging Stage (15–30 min). The press is opened to discharge the sludge cake. The filter cloth is cleaned by the automatic wash water system, and the unit is ready for the next cycle of filter operation. A definition sketch showing the filtration rate and applied pressure during these functional stages is shown in Figure 13.29b.

EXAMPLE 13.50: VACUUM FILTER YIELD

A wastewater treatment plant is using vacuum filter for dewatering of digested-combined chemically-conditioned sludge. The specific resistance r of sludge $= 2.4 \times 10^7\ s^2/g$. The applied vacuum $= 650$ mm of Hg (26 in). The solids in conditioned feed and sludge cake are 3.5% and 20% dry weight. The temperature of sludge is 20°C. The cycle time is 6 min and the cake forming time is 40% of the cycle time. Calculate the filter yield in kg/m^2·h.

Solution

1. Identify the filter yield equation.
   The filter yield equation is expressed by Equation 13.20.

$$L = \left(\frac{2\ Pwx}{\mu r \theta g}\right)^{1/2} \tag{13.20}$$

where
L = filter yield as dry solids, kg /m^2·h (lb/ft^2·h)
P = applied vacuum, N/m^2 (lb/ft^2)
w = mass of dry solids per unit of volume of filtrate, kg/m^3 (lb/ft^3). w is calculated from Equation 13.15b.
r = specific resistance, s^2/kg (ft^2/lb)
μ = viscosity of filtrate, N·s/m^2 (lb$_f$·s/ft^2)
θ = cycle time (time of one revolution of the drum), s
x = ratio of cake forming time to cycle time, dimensionless
g = acceleration due to gravity, 9.81 m/s^2 (32.2 ft/s^2)

2. Calculate the dry solids per unit volume of filtrate from Equation 13.15b.
   Density of water $\rho_w = 998.2$ kg/m^3 at 20°C is obtained from Table B-2 in Appendix B. Calculate the w from Equation 13.15b using $p_f = 0.035$ g/g, and $p_c = 0.2$ g/g.

$$w = \frac{\rho_w}{\left(\frac{1-p_f}{p_f}\right) - \left(\frac{1-p_c}{p_c}\right)} = \frac{998.2\,\text{kg/m}^3}{\left(\frac{1-0.035}{0.035}\right) - \left(\frac{1-0.2}{0.2}\right)} = 0.0423\,\text{g/cm}^3$$

3. Determine the applied vacuum (P).
   760 mm Hg = 101.325 kN/m^2 at 1 atm.

$$P = \frac{650\,\text{mm Hg}}{760\,\text{mm Hg}} \times 101.325\,\text{kN/m}^2 = 86.7\,\text{kN/m}^2 \quad \text{or} \quad 8.67 \times 10^4\,\text{N/m}^2$$

4. Calculate the filter yield.

Dynamic viscosity of water $\mu = 1.002\ 10^{-3}$ N·s/m^2 (or kg/s·m) at 20°C is obtained from Table B-2 in Appendix B.

The value of $x = 0.4$, specific resistance $r = 2.4 \times 10^7$ s^2/g $\times 10^3$ g/kg $= 2.4 \times 10^{10}$ s^2/kg, and the cycle time $t = 6$ min $\times$ 60 s/min = 360 s. Calculate the filter yield (L) from Equation 13.20.

$$L = \left(\frac{2\,Pwx}{\mu r \theta g}\right)^{1/2} = \left(\frac{2 \times 8.67 \times 10^4\ \text{N/m}^2 \times 42.3\ \text{kg/m}^3}{1.002 \times 10^{-3}\ \text{N·s/m}^2 \times 2.4 \times 10^{10}\ \text{s}^2\text{/kg} \times 360\ \text{s} \times 9.81\ \text{m/s}^2}\right)^{1/2}$$

$$= 9.29 \times 10^{-3}\ \text{kg/m}^2\text{·s} \times 3600\ \text{s/h} = 33.4\ \text{kg/m}^2\text{·h}$$

EXAMPLE 13.51: VACUUM FILTER AREA

A wastewater treatment plant produces 175,000 L/d digested combined sludge at 4% solids, and specific gravity of 1.01. The conditioning chemicals dosage is 40 g/kg of dry solids. The vacuum filter loading is 20 kg/m^2·h. Calculate the drum area required. Assume that the conditioning chemicals are retained at 100% in the sludge. The dewatering facility is operated for 8 h/d.

Solution

1. Calculate the total solids in the conditioned sludge from rearranged Equation 13.1e.

$$W_s = \frac{p_s S_{b,s} \rho_w V_s}{100\%} = \frac{4\% \times 1.01 \times 1000\ \text{kg/m}^3 \times 175{,}000\ \text{L/d} \times 10^{-3}\ \text{m}^3\text{/L}}{100\%} = 7070\ \text{kg/d}$$

Solids from conditioning chemicals, $W_{chemicals} = 40$ g/kg $\times$ 7070 kg/d $\times 10^{-3}$ kg/g = 283 kg/d

Weight of total solids, $W_{solids} = W_s + W_{chemical} = (7070 + 283)$ kg/d = 7353 kg/d

2. Calculate the vacuum filter drum area for an operating period of 8 h/d.

$$\text{Filter drum area required, } A_{drum} = \frac{W_{solids}}{\text{SLR}t_{operating}} = \frac{7353\ \text{kg/d}}{20\ \text{kg/m}^2\text{·h} \times 8\ \text{h/d}} = 46\ \text{m}^2$$

EXAMPLE 13.52: ROTARY PRESS

Rotary press is piloted at the treatment facility described in Example 13.51. To achieve the desired solids in the sludge cake, the following performance criteria were developed during the pilot test: (a) solids loading rate = 200 kg/m^2·h (40 lb/ft^2·h), (b) hydraulic loading rate = 2.4 m^3/m^2·h (1.25 gpm/ft^2), and (c) active polymer dose = 10 g/kg dry solids. Select the number of filter presses. Assume that the polymer solution is prepared at 0.5%, specific gravity of ~1, and 100% active polymer is retained in the sludge.

Solution

1. Calculate the total solids in the conditioned sludge per operating hour.

$W_s = 7070$ kg/d is calculated in Step 1 of Example 13.51.

$$\text{Hourly weight of sludge, } W_s = \frac{7070\ \text{kg/d}}{8\ \text{h/d}} = 884\ \text{kg/h}$$

Solids from conditioning chemicals, $W_{polymer} = 10$ g/kg $\times$ 884 kg/h $\times 10^{-3}$ kg/g = 8.8 kg/h

Weight of total solids, $W_{solids} = W_s + W_{polymer} = (884 + 8.8)$ kg/h = 893 kg/h

2. Calculate the total volume of the conditioned sludge per operating hour.

$V_s = 175{,}000$ L/d or 175 m^3/d is given in problem statement of Example 13.51.

Hourly volume of sludge, $V_s = \dfrac{175\,m^3/d}{8\,h/d} = 21.9\,m^3/h$

Polymer solution volume from Equation 13.1e, $V_{polymer} = \dfrac{100\% \times 8.8\,kg/h}{0.5\% \times 1 \times 1000\,kg/m^3} = 1.8\,m^3/h$

Volume of total solids, $V_{solids} = V_s + V_{polymer} = (21.9 + 1.8)\,m^3/h = 23.7\,m^3/h$

3. Determine the total area of the rotary presses required.
   a. Determine the area based on solids loading rate.

$$A_{SLR} = \frac{W_{solids}}{SLR} = \frac{893\,kg/h}{200\,kg/m^2{\cdot}h} = 4.5\,m^2$$

   b. Determine the area based on hydraulic loading rate.

$$A_{HLR} = \frac{W_{solids}}{HLR} = \frac{23.7\,m^3/h}{2.4\,m^3/m^2{\cdot}h} = 9.9\,m^2$$

   c. Determine the area required.
   Provide a total area of $A = 10\,m^2$ for equipment selection.
4. Determine the technical data of the selected rotary press.
   The technical data are available from equipment manufacturers. The typical values of different parameters are obtained for the selected rotary press unit: (a) rotary diameter $D = 915$ mm (36 in); (b) number of modules $N_m = 4$ modules/unit; and (c) effective dewatering area $A_u = 3.9\,m^2$/unit (42 ft^2/unit).
5. Select the number of rotary presses based on total area applied.

Number of units required, $N_u = \dfrac{A}{A_u} = \dfrac{10\,m^3}{3.9\,m^2/unit} = 2.6$ units

Provide four (4) identical units, including one standby unit.

Note: See the procedure for checking the firm and maximum capacity availabilities of the selected dewatering equipment in Example 13.49, Step 4.

13.9 Material Mass Balance Analysis

The sludge processing facilities produce liquids that are returned to the plant. The returned flows (also called sidestreams) are produced from sludge thickener overflows, digester supernatants, and waste flows from dewatering facilities. These sidestreams contain significant amounts of solids and contaminants that contribute extra BOD_5, TSS, nitrogen, and phosphorus loadings to the wastewater treatment units after blending with the process streams. Equalization basins are often provided to dampen the surge from the sidestreams. Many designers return these sidestreams back to the headwork and blend them with the incoming raw wastewater. In many cases, however, these flows may also be returned to the primary sedimentation or secondary treatment units. Each location of returned stream must be carefully evaluated. Considerations should be given to (1) the effect of incremental flow and mass loadings, (2) geometry and location of treatment units, (3) treatment processes and available treatment capacity, and (4) odor problems that may result from the returned flows.

A material mass balance analysis is used to estimate the concentrations of various contaminants in the resultant flows. If the sidestreams are produced intermittently, an equalization basin may be necessary. The general concepts and procedures of mass balance analysis have been introduced in Chapters 3 and 6 (see Examples 3.1 through 3.13, and 6.21 through 6.23). Complete mass balance analysis of a secondary and BNR facilities are also available in References 6 and 7, respectively.

EXAMPLE 13.53: CHARACTERISTICS OF SIDESTREAM FROM A GRAVITY THICKENER

A gravity thickener receives blended combined sludge. The characteristics of blended sludge are: volume = 500 m^3/d, TSS = 10,000 kg/d, total BOD_5 = 4960 kg/d, soluble BOD_5 = 35 kg/d, Org-N = 720 kg/d, NH_4-N = 8 kg/d, TP = 73 kg/d, Ortho-P (soluble) = 7% of TP, and VSS/TSS ratio = 0.7. Determine the characteristics of thickened sludge and sidestream. It is assumed that a capture rate of 85% is applicable for solids, particulate BOD_5, Org-N, and Org-P in the gravity thickener. The solids content of the thickened sludge = 5% and the specific gravity is 1.009.

Solution

1. Calculate the solids and volume of thickened sludge.

TSS in thickened sludge, $W_{TSS,THS} = 0.85 \times W_{TSS,CS} = 0.85 \times 10,000\ \text{kg/d} = 8500\ \text{kg/d}$

VSS in thickened sludge, $W_{VSS,THS} = \text{VSS/TSS ratio} \times W_{TSS,THS} = 0.7 \times 8500\ \text{kg/d} = 5950\ \text{kg/d}$

$$\text{Volume of thickened sludge, } V_{THS} = \frac{100\% \times W_{THS}}{p_{THS} S_{b,THS} \rho_w} = \frac{100\% \times 8500\ \text{kg/d}}{5\% \times 1.009 \times 1000\ \text{kg/m}^3} = 168\ \text{m}^3/\text{d}$$

2. Determine the BOD_5, TN, and TP in the thickened sludge.

a. BOD_5.

Particulate BOD_5 in combined sludge,

$$W_{PBOD5,CS} = W_{BOD_5,CS} - W_{SBOD5,CS} = (4960 - 35)\ \text{kg/d} = 4925\ \text{kg/d}$$

Particulate BOD_5 in thickened sludge,

$$W_{PBOD5,THS} = 0.85 \times W_{PBOD5,CS} = 0.85 \times 4925\ \text{kg/d} = 4186\ \text{kg/d}$$

$$\text{Soluble BOD}_5 \text{ in thickened sludge, } W_{SBOD5,THS} = W_{SBOD5,CS} \times \frac{V_{THS}}{V_{CS}} = 35\ \text{kg/d} \times \frac{168\ \text{m}^3/\text{d}}{500\ \text{m}^3/\text{d}} = 12\ \text{kg/d}$$

Total BOD_5 in thickened sludge,

$$W_{BOD_5,THS} = W_{PBOD5,THS} + W_{SBOD5,THS} = (4186 + 12)\ \text{kg/d} = 4198\ \text{kg/d}$$

b. TKN.

Org-N in thickened sludge, $W_{Org\text{-}N,THS} = 0.85 \times W_{Org\text{-}N,CS} = 0.85 \times 720\ \text{kg/d} = 612\ \text{kg/d}$

$$\text{NH}_4\text{-N in thickened sludge, } W_{NH4\text{-}N,THS} = W_{NH4\text{-}N,CS} \times \frac{V_{THS}}{V_{CS}} = 8\ \text{kg/d} \times \frac{168\ \text{m}^3/\text{d}}{500\ \text{m}^3/\text{d}} = 3\ \text{kg/d}$$

TKN in thickened sludge, $W_{TKN,THS} = W_{Org\text{-}N,THS} + W_{NH4\text{-}N,THS} = (612 + 3)\ \text{kg/d} = 615\ \text{kg/d}$

c. TP.

Ortho-P in combined sludge, $W_{Ortho\text{-}P,CS} = 0.07 \times W_{TP,CS} = 0.07 \times 73\ \text{kg/d} = 5\ \text{kg/d}$

Org-P in combined sludge, $W_{Org\text{-}P,CS} = W_{TP,CS} - W_{Ortho\text{-}P,CS} = (73 - 5)\ \text{kg/d} = 68\ \text{kg/d}$

Org-P in thickened sludge, $W_{Org\text{-}P,THS} = 0.85 \times W_{Org\text{-}P,CS} = 0.85 \times 68\ \text{kg/d} = 58\ \text{kg/d}$

$$\text{Ortho-P in thickened sludge, } W_{Ortho\text{-}P,THS} = W_{Ortho\text{-}P,CS} \times \frac{V_{THS}}{V_{CS}} = 5\ \text{kg/d} \times \frac{168\ \text{m}^3/\text{d}}{500\ \text{m}^3/\text{d}} = 2\ \text{kg/d}$$

TP in thickened sludge, $W_{TP,THS} = W_{Org\text{-}P,THS} + W_{Ortho\text{-}P,THS} = (58 + 2)\ \text{kg/d} = 60\ \text{kg/d}$

3. Determine the characteristics of overflow from thickener (or sidestream).

a. Volume.

Volume of overflow, $V_{OF} = V_{CS} - V_{THS} = (500 - 168)\ \text{m}^3/\text{d} = 332\ \text{m}^3/\text{d}$

b. Solids.

TSS in overflow, $W_{TSS,OF} = W_{TSS,CS} - W_{TSS,THS} = (10,000 - 8500)\ \text{kg/d} = 1500\ \text{kg/d}\ (4518\ \text{mg/L})$

VSS in overflow, $W_{VSS,OF} = \text{VSS/TSS ratio} \times W_{TSS,THS} = 0.7 \times 1500\ \text{kg/d} = 1050\ \text{kg/d}\ (3163\ \text{mg/L})$

c. BOD_5.

Particulate BOD_5 in overflow, $W_{PBOD5,OF} = W_{PBOD5,CS} - W_{PBOD5,THS} = (4925 - 4186)$ kg/d
$= 739$ kg/d (2223 mg/L)

Soluble BOD_5 in overflow, $W_{SBOD5,THS} = W_{SBOD5,CS} - W_{SBOD5,THSS} = (35 - 12)$kg/d
$= 23$ kg/d (69 mg/L)

Total BOD_5 in overflow, $W_{BOD5,OF} = W_{PBOD5,OF} + W_{SBOD5,OF} = (739 + 23)$ kg/d
$= 762$ kg/d (2295 mg/L)

d. TKN.

Org-N in overflow, $W_{Org\text{-}N,OF} = W_{Org\text{-}N,CS} - W_{Org\text{-}N,THS} = (720 - 612)$ kg/d
$= 108$ kg/d (325 mg/L)

NH_4-N in overflow, $W_{NH4\text{-}N,OF} = W_{NH4\text{-}N,CS} - W_{NH4\text{-}N,THS} = (8 - 3)$ kg/d $= 5$ kg/d (15 mg/L)

TKN in overflow, $W_{TKN,OF} = W_{Org\text{-}N,OF} + W_{NH4\text{-}N,OF} = (108 + 5)$ kg/d $= 113$ kg/d (340 mg/L)

e. TP.

Org-P in overflow, $W_{Org\text{-}P,OF} = W_{Org\text{-}P,CS} - W_{Org\text{-}P,THS} = (68 - 58)$ kg/d $= 10$ kg/d (30 mg/L)

Ortho-P in overflow, $W_{Ortho\text{-}P,OF} = W_{Ortho\text{-}P,CS} - W_{Ortho\text{-}P,THS} = (5 - 2)$ kg/d $= 3$ kg/d (9 mg/L)

TP in overflow, $W_{TP,OF} = W_{Org\text{-}P,OF} + W_{Ortho\text{-}P,OF} = (10 + 3)$ kg/d $= 13$ kg/d (39 mg/L)

EXAMPLE 13.54: SIDESTREAMS FROM A CONVENTIONAL SECONDARY TREATMENT PLANT

A wastewater treatment plant receives an average flow of 38,000 m^3/d. The average BOD_5, TSS, and VSS concentrations in the raw influent are: $S_0 = 250$ mg/L, $TSS_0 = 210$ mg/L, and $VSS_0 = 160$ mg/L. The treatment processes include headworks (bar screens, influent pump station, and degritting units), primary clarifiers, aeration basins, final clarifiers, and chlorine contact basins. Both BOD_5 and TSS concentrations in the effluent are 10 mg/L. The sludge process train includes combined sludge blending tank, gravity thickeners, anaerobic digesters, and belt-filter press units. The sidestreams from sludge handling facilities are collected in a sump and then pumped into a junction box prior to the aeration basins. A conceptual process diagram is shown in Figure 13.30. Conduct the mass balance analysis under the average operating conditions to establish the flow and mass loadings of BOD_5 and TSS of the combined flow before mixing with return activated sludge (RAS) at the entrance of the aeration basins.

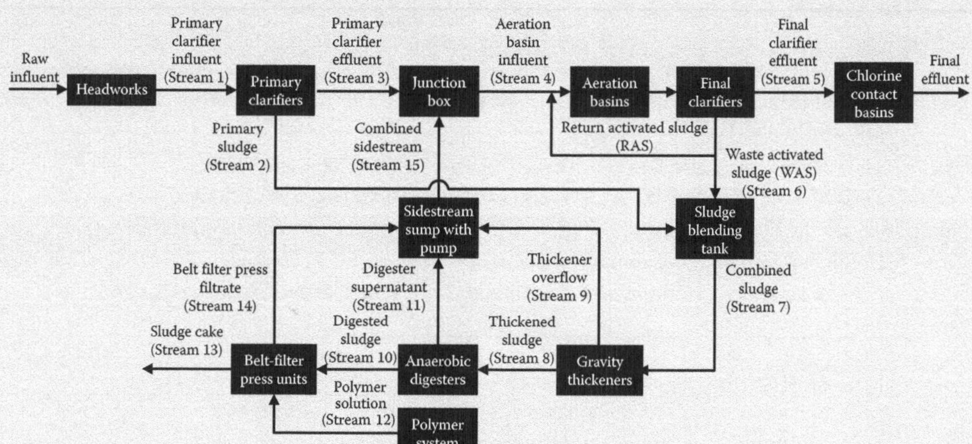

FIGURE 13.30 Process diagram of a conventional secondary municipal wastewater treatment plant (Example 13.54).

Solution

1. Describe the material mass balance strategy.

   The sidestreams from sludge handling areas are normally returned to the head of the plant. In most cases, the sidestreams contain mainly colloidal particles that are normally not removed in the primary clarifiers. For this reason, the combined sidestream is returned to the aeration basins for bioflocculation and removal of colloidal particles. The mass balance process involves iterative process. In the first iteration, only the influent data is applied to generate sidestreams. From the second iteration, the combined sidestream is blended with the primary effluent, the flow and loadings that are applied to the aeration basins are increased. As a result, the sidestream flows and loadings are also increased accordingly. The mass balance analysis is repeated using the updated sidestream data until the change from the previous iteration is insignificant. The iterative process is given below.

2. First iteration.

   In mass balance analysis, a simple numbering system is used to identify the stream instead of stream descriptions as shown in Figure 13.30. The stream numbers are sequential in order by which the stream is calculated in the first iteration. A total of fifteen (15) streams are involved in the first iteration (see Figure 13.30).

   a. Calculate the mass loadings in the primary clarifier influent (Stream 1).

      The changes in volume and solids through screening and degritting units are relatively small. Both processes may be ignored in mass balance analysis.

      Mass of BOD_5, $W_{BOD_5,1} = S_0Q = 250\ \text{g/m}^3 \times 38{,}000\ \text{m}^3/\text{d} \times 10^{-3}\ \text{kg/g} = 9500\ \text{kg/d}$

      Mass of TSS, $W_{TSS,1} = TSS_0Q = 210\ \text{g/m}^3 \times 38{,}000\ \text{m}^3/\text{d} \times 10^{-3}\ \text{kg/g} = 7980\ \text{kg/d}$

      Mass of VSS, $W_{VSS,1} = VSS_0Q = 160\ \text{g/m}^3 \times 38{,}000\ \text{m}^3/\text{d} \times 10^{-3}\ \text{kg/g} = 6080\ \text{kg/d}$

   b. Calculate the primary sludge characteristics (Stream 2).

      Assume that the removals of BOD_5 and TSS are 30% and 60%, respectively in the primary clarifiers. Also the solids content in the primary sludge $p_{s,2} = 4\%$ and specific gravity $S_{s,2} = 1.01$. The VSS/TSS ratio is 0.7.

      Mass of BOD_5, $W_{BOD5,2} = 0.3 \times W_{BOD5,1} = 0.3 \times 9500\ \text{kg/d} = 2850\ \text{kg/d}$

      Mass of TSS, $W_{TSS,2} = 0.6 \times W_{TSS,1} = 0.6 \times 7980\ \text{kg/d} = 4788\ \text{kg/d}$

      Mass of VSS, $W_{VSS,2} = 0.7 \times W_{TSS,2} = 0.7 \times 4788\ \text{kg/d} = 3352\ \text{kg/d}$

      $$\text{Volume of sludge from Equation 13.1e, } V_2 = \frac{100\% \times W_{TSS,2}}{p_{s,2}S_{b,2}\rho_w} = \frac{100\% \times 4788\ \text{kg/d}}{4\% \times 1.01 \times 1000\ \text{kg/m}^3} = 119\ \text{m}^3/\text{d}$$

   c. Calculate the flow and mass of BOD_5 and TSS in the primary settled effluent (Stream 3).

      Flow of primary effluent, $V_3 = V_1 - V_2 = (38{,}000 - 119)\ \text{m}^3/\text{d} = 37{,}881\ \text{m}^3/\text{d}$

      Mass of BOD_5, $W_{BOD5,3} = W_{BOD5,1} - W_{BOD5,2} = (9500 - 2850)\ \text{kg/d} = 6650\ \text{kg/d}$

      Mass of TSS, $W_{TSS,3} = W_{TSS,1} - W_{TSS,2} = (7980 - 4788)\ \text{kg/d} = 3192\ \text{kg/d}$

      Mass of VSS, $W_{VSS,3} = W_{VSS,1} - W_{VSS,2} = (6080 - 3352)\ \text{kg/d} = 2728\ \text{kg/d}$

   d. Calculate the flow and mass of BOD_5 and TSS applied to the aeration basin (Stream 4).

      In the first iteration, the combined sidestream (Stream 14) is ignored.

      Flow of aeration basin influent before mixing with return activated sludge (RAS),

      $$V_4 = V_3 + V_{14} = (37{,}881 + 0)\ \text{m}^3/\text{d} = 37{,}881\ \text{m}^3/\text{d}$$

      Mass of BOD_5, $W_{BOD5,4} = W_{BOD5,3} + W_{BOD5,14} = (6650 + 0)\ \text{kg/d} = 6650\ \text{kg/d}$

      Mass of TSS, $W_{TSS,4} = W_{TSS,3} + W_{TSS,14} = (3192 + 0)\ \text{kg/d} = 3192\ \text{kg/d}$

      Mass of VSS, $W_{VSS,4} = W_{VSS,3} + W_{VSS,14} = (2728 + 0)\ \text{kg/d} = 2728\ \text{kg/d}$

   e. Estimate the mass of total solids generated by the secondary treatment process.

      i. Estimate the soluble BOD_5 in the effluent from final clarifiers.

         Assume VSS/TSS ratio = 0.75, the ratio of biodegradable VSS (VSS_{bd}) to TSS is 0.55 kg

VSS_{bd}/kg TSS, each g of VSS_{bd} exerts 1.42 g BOD_L, and BOD_5/BOD_L ratio = 0.68. Particulate BOD_5 concentration in the final clarifier effluent,

$$\begin{aligned} S_{PBOD5,e} &= BOD_5/BOD_L \text{ ratio} \times BOD_L/VSS_{bd} \text{ ratio} \times VSS_{bd}/TSS \text{ ratio} \times TSS_e \\ &= 0.68 \text{ mg } BOD_5/\text{mg } BOD_L \times 1.42 \text{ mg } BOD_L/\text{mg } VSS_{bd} \\ &\quad \times 0.55 \text{ mg } VSS_{bd}/\text{mg TSS} \times 10 \text{ mg TSS/L} \\ &= 5.3 \text{ mg } BOD_5/L \end{aligned}$$

Soluble BOD_5 concentration in the final clarifier effluent,

$$S = S_{BOD_5,e} - S_{PBOD5,e} = (10 - 5.3) \text{ mg/L} = 4.7 \text{ mg/L} \quad \text{or} \quad 4.7 \text{ g/m}^3$$

ii. Estimate the mass of BOD_5 removed by the secondary treatment process.

$$\begin{aligned} \Delta W_{BOD_5,AB} &= W_{BOD_5,4} - V_4 = 6650 \text{ kg/d} - 4.7 \text{ g/m}^3 \times 37{,}881 \text{ m}^3/\text{d} \times 10^{-3} \text{ kg/g} \\ &= (6650 - 178) \text{ kg/d} = 6472 \text{ kg/d} \end{aligned}$$

iii. Estimate the mass of total solids after the secondary treatment process.

Assume that biomass yield coefficient $Y = 0.5$ mg VSS/mg BOD_5, specific endogenous decay coefficient $k_d = 0.06\ d^{-1}$, and sludge retention time $\theta_c = 10$ d. Calculate observed yield coefficient (Y_{obs}) from Equation 10.15l.

$$Y_{obs} = \frac{Y}{1 + k_d\theta_c} = \frac{0.50 \text{ mg VSS/mg } BOD_5}{1 + 0.06 \text{ d}^{-1} \times 10 \text{ d}} = 0.31 \text{ mg VSS/mg } BOD_5 = 0.31 \text{ g VSS/g } BOD_5$$

Calculate biomass growth or VSS generation rate (P_x) in the aeration basin from Equation 10.15m

$$P_x = Y_{obs}\Delta W_{BOD_5,AB} = 0.31 \text{ g VSS/g } BOD_5 \times 6472 \text{ kg } BOD_5/\text{d} = 2006 \text{ kg VSS/d}$$

Increase in mass of total solids after the secondary treatment process,

$$\Delta W_{MLSS} = \frac{P_x}{\text{VSS/TSS ratio}} = \frac{2006 \text{ kg VSS/d}}{0.75 \text{ kg VSS/kg TSS}} = 2675 \text{ kg/d}$$

f. Calculate the flow and mass of BOD_5 and TSS in the final clarifier effluent (Stream 5).

To maintain a balance of biomass in the secondary treatment, the solids lost in the final clarifier effluent and waste activated sludge (WAS) must be equal to the increase in mass of total solids. Assume that the solids concentration in the WAS $TSS_{WAS} = 10{,}000$ mg/L, the following expression is obtained.

$$\Delta W_{MLSS} = TSS_{WAS}\,(V_4 - V_5) + TSS_e V_5$$

Rearrange the expression to solve the total flow from the final clarifiers.

$$V_5 = \frac{TSS_{WAS} V_4 - \Delta W_{MLSS}}{TSS_{WAS} - TSS_e} = \frac{10{,}000 \text{ g/m}^3 \times 37{,}881 \text{ m}^3/\text{d} - 2675 \text{ kg/d} \times 10^3 \text{ g/kg}}{(10{,}000 - 10) \text{ g/m}^3} = 37{,}651 \text{ m}^3/\text{d}$$

Mass of BOD_5, $W_{BOD_5,5} = S_e V_5 = 10 \text{ g/m}^3 \times 37{,}651 \text{ m}^3/\text{d} \times 10^{-3} \text{ kg/g} = 377 \text{ kg/d}$

Mass of TSS, $W_{TSS,5} = TSS_e V_5 = 10 \text{ g/m}^3 \times 37{,}651 \text{ m}^3/\text{d} \times 10^{-3} \text{ kg/g} = 377 \text{ kg/d}$

g. Calculate the characteristics of WAS (Stream 6).

Volume of WAS, $V_6 = V_4 - V_5 = (37{,}881 - 37{,}651) \text{ m}^3/\text{d} = 230 \text{ m}^3/\text{d}$

Mass of TSS, $W_{TSS,6} = \Delta W_{MLSS} - W_{TSS,5} = (2675 - 377) \text{ kg/d} = 2298 \text{ kg/d}$

Mass of VSS, $W_{VSS,6} = \text{VSS/TSS ratio} \times W_{TSS,6} = 0.75 \times 2298 \text{ kg/d} = 1724 \text{ kg/d}$

The total BOD_5 in the WAS consists of particulate and soluble BOD_5. The soluble portion of BOD_5 is too small to be considered in mass balance analysis. Therefore, the mass of BOD_5 in

WAS is estimated from the particulate BOD_5 only.

$$W_{BOD_5,6} = BOD_5/BOD_L \text{ ratio} \times BOD_L/VSS_{bd} \text{ ratio} \times VSS_{bd}/TSS \text{ ratio} \times W_{TSS,6}$$
$$= 0.68\,kg\,BOD_5/kg\,BOD_L \times 1.42\,kg\,BOD_L/kg\,VSS_{bd} \times 0.55\,kg\,VSS_{bd}/kg\,TSS \times 2298\,kg\,TSS/d$$
$$= 1220\,kg/d$$

h. Calculate the volume and mass of BOD_5 and TSS in the combined sludge (Stream 7).
Volume of sludge, $V_7 = V_2 + V_6 = (119 + 230)\,m^3/d = 349\,m^3/d$
Mass of BOD_5, $W_{BOD5,7} = W_{BOD5,2} + W_{BOD5,6} = (2850 + 1220)\,kg/d = 4070\,kg/d$
Mass of TSS, $W_{TSS,7} = W_{TSS,2} + W_{TSS,6} = (4788 + 2298)\,kg/d = 7086\,kg/d$
Mass of VSS, $W_{VSS,7} = W_{VSS,2} + W_{VSS,6} = (3352 + 1724)\,kg/d = 5076\,kg/d$

i. Calculate the thickened sludge characteristics (Stream 8).
Assume that a capture rate of 90% is achieved for TSS, VSS, or BOD_5 in the gravity thickeners. The thickened sludge solids content $p_{s,8} = 5\%$ and specific gravity $S_{s,8} = 1.01$.
Mass of BOD_5, $W_{BOD5,8} = 0.9 \times W_{BOD5,7} = 0.9 \times 4070\,kg/d = 3663\,kg/d$
Mass of TSS, $W_{TSS,8} = 0.9 \times W_{TSS,7} = 0.9 \times 7086\,kg/d = 6377\,kg/d$
Mass of VSS, $W_{VSS,8} = 0.9 \times W_{VSS,7} = 0.9 \times 5076\,kg/d = 4568\,kg/d$

Volume of sludge from Equation 13.1e, $$V_8 = \frac{100\% \times W_{TSS,8}}{p_{s,8} S_{b,8} \rho_w} = \frac{100\% \times 6377\,kg/d}{5\% \times 1.01 \times 1000\,kg/m^3} = 126\,m^3/d$$

j. Calculate the flow and mass of BOD_5 and TSS in the thickener overflow (Stream 9).
Flow of thickener overflow, $V_9 = V_7 - V_8 = (349 - 126)\,m^3/d = 223\,m^3/d$
Mass of BOD_5, $W_{BOD5,9} = W_{BOD5,7} - W_{BOD5,8} = (4070 - 3663)\,kg/d = 407\,kg/d$
Mass of TSS, $W_{TSS,9} = W_{TSS,7} - W_{TSS,8} = (7086 - 6377)\,kg/d = 709\,kg/d$
Mass of VSS, $W_{VSS,9} = W_{VSS,7} - W_{VSS,8} = (5076 - 4568)\,kg/d = 508\,kg/d$
Note: Treated effluent may be used for dilution water to meet the hydraulic loading requirement of the thickener. The thickener overflow will be increased significantly. However, the solids in the dilution water are not high enough to cause any effect on the mass loading to the thickener.

k. Calculate the solids remaining after digestion.
Assume that volatile solids reduction is 50% in the anaerobic digesters.

Mass of VSS destroyed, $$\Delta W_{VSS,DG} = \frac{VSR}{100\%} \times W_{VSS,8} = \frac{50\%}{100\%} \times 4568\,kg/d = 2284\,kg/d$$

Mass of TSS remaining, $W_{TSS,REM} = W_{TSS,8} - \Delta W_{VSS,DG} = (6377 - 2512)\,kg/d = 4093\,kg/d$
Mass of VSS remaining, $W_{VSS,REM} = W_{VSS,8} - \Delta W_{VSS,DG} = (4568 - 2284)\,kg/d = 2284\,kg/d$

Ratio of VSS to TSS, $$VSS/TSS\ ratio_{REM} = \frac{W_{VSS,REM}}{W_{TSS,REM}} = \frac{2284\,kg/d}{4093\,kg/d} = 0.558$$

l. Calculate the digested sludge characteristics (Stream 10).
Assume that the solid concentration in supernatant is 5000 mg/L. The digested sludge solids content $p_{s,10} = 4\%$ and specific gravity $S_{s,8} = 1.01$. A general procedure of solving the digested sludge and supernatant production is presented in Example 13.28. Using the same procedure, a simplified approach is given below. Readers are referred to Example 13.28 to find step-by-step calculation procedure.

$$V_8 = \frac{100\% \times W_{TSS,10}}{p_{s,10}S_{b,10}\rho_w} + \frac{W_{TSS,REM} - W_{TSS,10}}{TSS_{SPN}}$$

$$= \frac{100\% \times W_{TSS,10}}{4\% \times 1.01 \times 1000\,kg/m^3} + \frac{4093\,kg\ TSS/d - W_{TSS,10}}{5000\,g/m^3 \times 10^{-3}\,kg/g}$$

$$= \frac{W_{TSS,10}}{40.4\,kg/m^3} + \frac{4093\,kg\ TSS/d - W_{TSS,10}}{5\,kg/m^3}$$

or $$\frac{W_{TSS,10}}{40.4\,kg/m^3} + \frac{4093\,kg\ TSS/d - W_{TSS,10}}{5\,kg/m^3} = 126\,m^3/d$$

Solve for $W_{TSS,10}$.

Mass of TSS, $$W_{TSS,10} = \frac{4093\,kg\ TSS/d - 5\,kg/m^3 \times 126\,m^3/d}{40.4\,kg/m^3 - 5\,kg/m^3} \times 40.4\,kg/m^3 = 3952\,kg/d$$

Mass of VSS, $W_{VSS,10} = VSS/TSS\ ratio_{REM} \times W_{TSS,10} = 0.558 \times 3952\,kg/d = 2205\,kg/d$

Assume that the VSS_{bd}/TSS ratio is reduced to 0.3 kg VSS_{bd}/kg TSS after anaerobic decomposition of biodegradable biosolids and calculate the mass of BOD_5,

$$W_{BOD_5,10} = BOD_5/BOD_L\ ratio \times BOD_L/VSS_{bd}\ ratio \times VSS_{bd}/TSS\ ratio \times W_{TSS,10}$$

$$= 0.68\,kg\ BOD_5/kg\ BOD_L \times 1.42\,kg\ BOD_L/kg\ VSS_{bd} \times 0.3\,kg\ VSS_{bd}/kg\ TSS \times 3952\,kg\ TSS/d$$

$$= 0.29\,kg\ BOD_5/kg\ TSS \times 3952\,kg\ TSS/d = 1146\,kg/d$$

Volume of sludge from Equation 13.1e, $$V_{10} = \frac{100\% \times W_{TSS,10}}{p_{s,10}S_{b,10}\rho_w} = \frac{100\% \times 3952\,kg/d}{4\% \times 1.01 \times 1000\,kg/m^3} = 98\,m^3/d$$

m. Calculate the flow and mass of BOD_5 and TSS in the digester supernatant (Stream 11).

Flow of digester supernatant, $V_{11} = V_8 - V_{10} = (126 - 98)\,m^3/d = 28\,m^3/d$

Mass of TSS, $W_{TSS,11} = W_{TSS,REM} - W_{TSS,10} = (4093 - 3952)\,kg/d = 141\,kg/d$

Mass of VSS, $W_{VSS,11} = W_{VSS,REM} - W_{VSS,10} = (2284 - 2205)\,kg/d = 79\,kg/d$

Mass of BOD_5, $$W_{BOD_5,11} = 0.29\,kg\ BOD_5/kg\ TSS \times W_{TSS,11} = 0.29\,kg\ BOD_5/kg\ TSS \times 141\,kg/d = 41\,kg/d$$

n. Calculate the polymer solution characteristics (Stream 12).

Assume that the active polymer dose = 5 g/kg dry solids, polymer concentration $p_{polymer}$ = 0.5%, specific gravity of solution $S_{polymer} \approx 1$, and 100% polymer is no-biodegradable volatile solids.

Mass of TSS, $W_{TSS,12} = 5\,g/kg \times W_{TSS,10} = 5\,g/kg \times 3952\,kg/d \times 10^{-3}\,kg/g = 20\,kg/d$

Mass of VSS, $W_{VSS,12} = 1 \times W_{TSS,12} = 1 \times 20\,kg/d = 20\,kg/d$

Mass of BOD_5, $W_{BOD5,12} = 0\,kg/d$

Calculate volume of polymer solution from Equation 13.1e.

$$V_{12} = \frac{100\% \times W_{TSS,12}}{p_{12}S_{12}\rho_w} = \frac{100\% \times 20\,kg/d}{0.5\% \times 1 \times 1000\,kg/m^3} = 4\,m^3/d$$

o. Calculate the sludge cake characteristics (Stream 13).

Make the following assumptions for the sludge dewatering by belt-filter press: solids capture efficiency of belt-filter press = 95%; 75% polymer added is retained in the sludge cake; solids content of sludge cake $p_{s,13}$ = 20%; and specific gravity of sludge cake $S_{s,13}$ = 1.06.

Mass of BOD_5, $W_{BOD5,13} = 0.95 \times W_{BOD5,10} = 0.95 \times 1146\,kg/d = 1089\,kg/d$

Mass of TSS, $W_{TSS,13} = 0.95 \times W_{TSS,10} + 0.75 \times W_{TSS,12} = (0.95 \times 3952 + 0.75 \times 20)\,kg/d = 3769\,kg/d$

Mass of VSS, $W_{VSS,13} = 0.95 \times W_{VSS,10} + 0.75 \times W_{VSS,12} = (0.95 \times 2205 + 0.75 \times 20)\,kg/d = 2110\,kg/d$

Volume of sludge from Equation 13.1e, $V_{13} = \dfrac{100\% \times W_{TSS,13}}{p_{s,13} S_{b,13} \rho_w} = \dfrac{100\% \times 3769\,kg/d}{20\% \times 1.06 \times 1000\,kg/m^3} = 18\,m^3/d$

p. Calculate the flow and mass of BOD_5 and TSS in the filtrate from BFP units (Stream 14).
   Flow of waste flow, $V_{14} = V_{10} + V_{12} - V_{13} = (98 + 4 - 18)\,m^3/d = 84\,m^3/d$
   Mass of BOD_5, $W_{BOD5,14} = W_{BOD5,10} - W_{BOD5,13} = (1146 - 1089)\,kg/d = 57\,kg/d$
   Mass of TSS, $W_{TSS,14} = W_{TSS,10} + W_{TSS,12} - W_{TSS,13} = (3952 + 20 - 3769)\,kg/d = 203\,kg/d$
   Mass of VSS, $W_{VSS,14} = W_{VSS,10} + W_{VSS,12} - W_{VSS,13} = (2205 + 20 - 2110)\,kg/d = 115\,kg/d$
   Note: Treated effluent is used for belt wash. Ignore the mass and solids in the wash water.

q. Calculate the combined sidestream characteristics (Stream 15).
   Flow of combined sidestream, $V_{15} = V_9 + V_{11} + V_{14} = (223 + 28 + 84)\,m^3/d = 335\,m^3/d$
   Mass of BOD_5, $W_{BOD5,15} = W_{BOD5,9} + W_{BOD5,11} + W_{BOD5,14} = (407 + 41 + 57)\,kg/d = 505\,kg/d$
   Mass of TSS, $W_{TSS,15} = W_{TSS,9} + W_{TSS,11} + W_{TSS,14} = (709 + 141 + 203)\,kg/d = 1053\,kg/d$
   Mass of VSS, $W_{VSS,15} = W_{VSS,9} + W_{VSS,11} + W_{VSS,14} = (508 + 79 + 115)\,kg/d = 702\,kg/d$

TABLE 13.24 Summary of Final Results of Material Mass Balance Analysis after the Fourth Iteration (Example 13.54)

Stream		Volume[a], m^3/d	Mass Loading, kg/d			VSS/TSS Ratio[b]
No	Description		BOD_5	TSS	VSS	
1	Primary clarifier influent	38,000	9500	7980	6080	0.76
2	Primary sludge	119	2850	4788	3352	0.70
3	Primary clarifier effluent	37,881	6650	3192	2728	0.85
4	Aeration basin influent[c]	38,237	7170	4277	3453	0.81
5	Final clarifier effluent	37,986	380	380	285	0.75
6	Waste activated sludge (WAS)	251	1332	2509	1882	0.75
7	Combined sludge	370	4182	7297	5234	0.72
8	Thickened sludge	130	3764	6567	4711	0.72
9	Thickener overflow	240	418	730	523	0.72
10	Digested sludge	101	1179	4064	2272	0.56
11	Digester supernatant	29	43	147	83	0.56
12	Polymer solution	4	0	20	20	1.00
13	Sludge cake	18	1120	3876	2173	0.56
14	Belt-filter press filtrate	87	59	208	119	0.57
15	Combined sidestream	356	520	1085	725	0.67

[a] Total system inflow = primary clarifier influent (V_3) + polymer solution (V_{12}) = (38,000 + 4) m^3/d = 38,004 m^3/d (see Step 2. n for the polymer solution volume). Total system outflow = final clarifier effluent (V_5) + sludge cake (V_{13}) = (37,986 + 18) m^3/d = 38,004 m^3/d. The loss of water due to evaporation and gas production is ignored.

[b] The ratio is calculated from VSS and TSS data for each stream.

Note: The results of mass balance analysis indicate that flow and mass loadings in combined stream (Stream 4) going to aeration basin have Increased as follows: flow 1%, BOD_5 loading 8%, TSS loading 34%, and VSS loading 27%. The flow of Stream 4 may be increased by up to 4–5% if the dilution water (Step 2.j) and belt wash water flows (Step 2.p) are considered in the mass balance calculations. However, the mass loadings in Stream 4 will not be affected. Readers are referred to Examples 13.12 and 13.47 for calculations of these additional flows.

3. Second iteration.

   Since the combined sidestream is returned to the aeration basins the calculations in the first iteration will remain the same for Streams 1 through 3. Calculations need to be revised for Streams 4 through 15 (see Figure 13.30).

   a. Calculate the flow and mass of BOD_5 and TSS in Stream 4.

      The combined sidestream characteristics obtained for Stream 15 in the first iteration is used in the calculations below.

$$V_4 = V_3 + V_{15} = (37{,}881 + 335)\ \text{m}^3/\text{d} = 38216\ \text{m}^3/\text{d}$$

      Mass of BOD_5, $W_{BOD5,4} = W_{BOD5,3} + W_{BOD5,15} = (6650 + 505)\ \text{kg/d} = 7155\ \text{kg/d}$
      Mass of TSS, $W_{TSS,4} = W_{TSS,3} + W_{TSS,15} = (3192 + 1053)\ \text{kg/d} = 4245\ \text{kg/d}$
      Mass of VSS, $W_{VSS,4} = W_{VSS,3} + W_{VSS,15} = (2728 + 702)\ \text{kg/d} = 3430\ \text{kg/d}$

   b. Calculate the characteristics of Streams 5 through 15.

      Repeat the calculation steps presented in Step 2 to obtain the following characteristics of Stream 15: $V_{15} = 355\ \text{m}^3/\text{d}$, $W_{BOD5,15} = 520\ \text{kg/d}$, $W_{TSS,15} = 1084\ \text{kg/d}$, and $W_{VSS,15} = 724\ \text{kg/d}$.

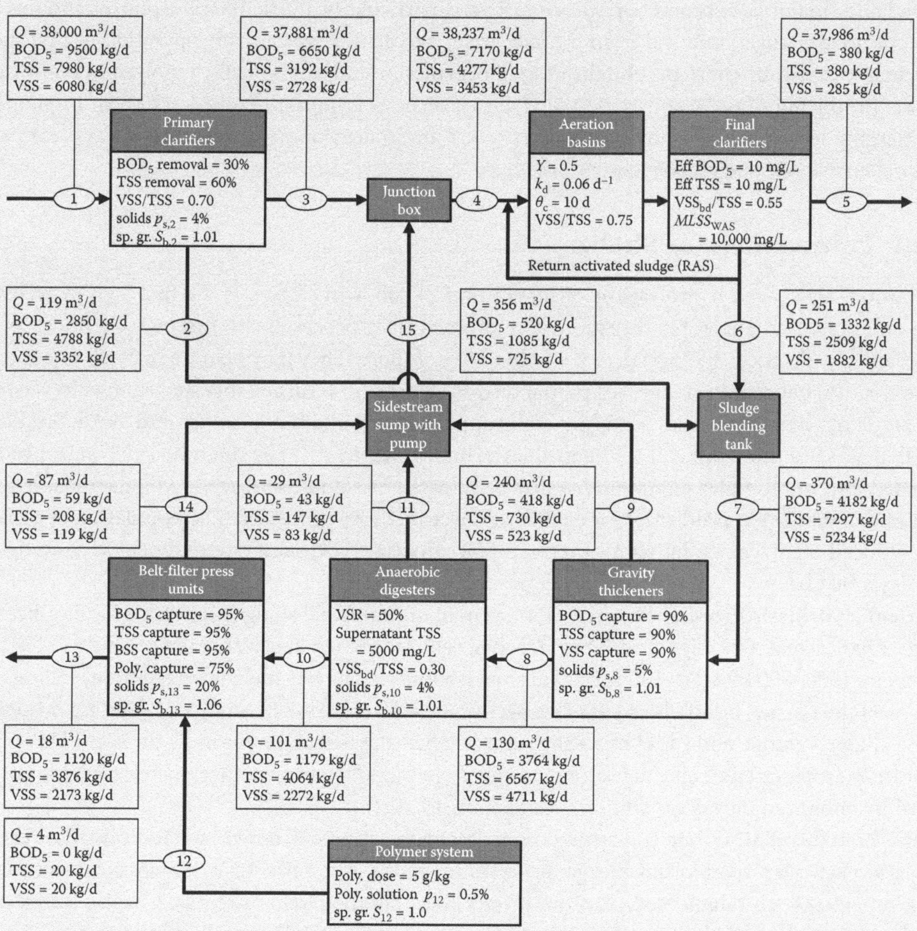

FIGURE 13.31 Final results of material mass balance analysis after the fourth iteration (Example 13.54).

4. Third and fourth iterations.

The above computational procedure was repeated two more iterations. After the fourth iteration stable flow and mass loadings to the aeration basin were obtained in Stream 4. The final results of the fourth iteration are summarized in Table 13.24 and shown in Figure 13.31.

It is important to note that the general procedures presented in this example can also include the material mass balance analysis for nutrients (nitrogen and phosphorus). Complete mass balance analysis for conventional parameters and nutrients may be found in Reference 6.

13.10 Emerging Technologies for Enhanced Sludge Stabilization and Resource Recovery

The sludge handling and disposal processes are expensive and may often be the overriding factor in process selection during design of wastewater treatment facilities. Recent technical developments in this area have focused on the following: (a) enhanced anaerobic digestion to increase biogas generation and reduce biodegradable organics; (b) promote resource recovery, particularly, nutrients (phosphorus and nitrogen); and (c) optimize nitrogen removal from the sidestreams to achieve sustainable operation of BNR facility. As a necessary pretreatment step, solubilization of organics in sludge is typically involved in many of these processes. Producing Class A and/or EQ biosolids is also an important consideration of implementing these emerging technologies. Many of these processes are in development stages but offer great promise in comprehensive biosolids management practices.

13.10.1 Pretreatment of Sludge

Solubilization, also called hydrolysis or disintegration of solids in sludge, is the first step in recovery of protein and enhanced anaerobic digestion. The pretreatment processes are thermal, chemical, physical (or mechanical), electrical, biological, or a combination of these. They may operate in batch or continuous mode to treat (a) entire feed sludge, (b) partial feed sludge, or (c) returned digested sludge. In wastewater treatment plants, anaerobic digestion is generally applied to a mixture of primary and WAS. Studies have shown that WAS is more difficult to digest than primary sludge.[7,115] The microbial cell walls need to be fragmented and intracellular compounds must be released as soluble and readily biodegradable organic substrates to enhance gas production or enable resource recovery processes. The population of pathogens is also reduced significantly by using pretreatment processes. Selected pretreatment technologies are briefly described below.

Thermal Hydrolysis Process: Thermal pretreatment of municipal sludge is effective for liquification of solids to enhance anaerobic digestion. The microbial cells are broken at elevated temperature (140–180°C) and pressure (500–800 kPa (70–120 psi)). Thermal treatment also has the advantage of (a) killing pathogens to meet the Class A biosolids criteria (see Section 13.4.4), (b) improve sludge dewatering property, (c) reducing sludge volume, and (d) eliminating odors. When the thermal hydrolysis process (THP) is integrated with mesophilic anaerobic digestion, the energy required to perform thermal treatment is positively balanced by enhanced biogas generation (see Section 13.10.2).[7,21,28,29,116–118]

Ozone Pretreatment: Ozone is a strong oxidizing agent. It breaks down into radicals that react with soluble and particular organic and inorganic solids in sludge. The refractory organic structures and cell walls of organisms are fragmented, and the intracellular compounds are released in the forms of low-molecular and biodegradable organic compounds with a sBOD/sCOD ratio higher than 60%. The optimum ozone consumption may be in a range from 0.05 to 0.5 g O_3 per g total solids. Besides enhanced anaerobic digestion, ozone pretreatment also provides the benefits of effective degradation of micro-pollutants and significant reduction of foaming.[116,119–123]

Peroxidation Pretreatment: Use of peroxidants for disintegration of biosolids has also been reported in many recent studies for enhancing biogas production. The mechanisms are similar to that for ozone pretreatment. The preoxidants produce hydroxyl radicals that attack the cell walls and improve the sBOD/sCOD ratio in the biosolids. The commonly used peroxidants include Fenton's reagent (a mixture of hydrogen peroxide (H_2O_2) and ferrous iron), peracetic acid (PAA), peroxymonosulfate (POMS), and dimethyldioxirane (DMDO).[124–126]

Ultrasonic Destruction: Ultrasonic pretreatment (or called sonication) of WAS is very effective in rupturing the cellular wall. At the optimum frequency range from 20 to 40 kHz, it involves cavitation and formation of microbubbles in the liquid phase. These bubbles grow and then violently collapse when they reach a critical size. Cavitation collapse produces (a) intense local heating and high pressure at the liquid-gas interface, (b) turbulence and high shearing phenomena in the liquid phase, and (c) formation of radicals. The cell wall is ruptured and intracellular contents are released into the liquid. The released substrates are readily biodegradable that improve the performance of anaerobic digesters. Hydrolysis of WAS by alkaline treatment followed by ultrasonic treatment improves the subsequent biotransformation of organics into fatty acids. The applicable ultrasonic dose may be in a range from 500 to 5000 kJ/kg total solids at a wide irradiation time of 1–60 min.[116,118,127–136]

Enzyme Hydrolysis Process: The enzyme hydrolysis of solids is also an effective pretreatment step for liquidification of sludge. The enzymic reactions break down the lipoprotein structure of cell wall and release the cell contents that enhances protein recovery and anaerobic digestion processes.[52,128] The extracellular polymeric substances (EPSs) extracted from activated sludge have the ability to hydrolyze organic solids in sludge and wastewater.[52] The endogenous amylase and protease extracted from bacterial fermentation of mixed culture increase the solubility of the solids in sludge and increase gas production in anaerobic digestion.[137–139]

High Pressure Homogenization: The high pressure homogenization (HPH) process may also be efficient for solubilization of sludge. Typically, the process involves the following functional steps: (a) chemical conditioning with alkaline (caustic or lime) to soften the cell walls and lower the viscosity of the biosolids, (b) grinding (if desired) and fine-screening the chemically conditioned solids to remove large particles, (c) pressurization under a high pressure (1.2–80 MPa (175–12,000 psi)) in a homogenizer, and (d) destruction (or lysis) of cell membranes to release the intracellular contents. The effective lysis is created by the sudden pressure drop, violent turbulent eddies, and strong shear forces when the pressurized solids pass through an outlet valve after the homogenizer. The HPH device may be applied for (a) *pretreatment* of a portion of feed WAS to the anaerobic digester or (b) *posttreatment* of digested or post-thickened sludge that is recycled back to the anaerobic digesters. The HPH process may also lower the overall sludge quantity and polymer requirement for dewatering.[7,140–142]

Electrical Pretreatment: In this process, biosolids pass through an electrical field that causes (a) separation of hydrogen bonds that creates pores on the cell membranes to release the cell contents, and (b) breakup of ionic bonds in the WAS flocs to reduce the particles size. These reactions increase soluble volatile acids and colloidal particles that accelerate the anaerobic digestion. The pretreatment may be achieved by (a) electrokinetic (EK) disintegration, and (b) focused pulses (FPs) technology. In the EK process, a high-voltage typically in a range from 10 to 100 kV is applied to the microbial cells under the electrical field created by electrodes. The energy input is in a range from 200 to 300 kJ/kg total solids and the residence time is about 3 s per electrode. The FPs process is derived from electroporation that is typically used to kill pathogens in food processing for over 40 years.[7,118,143–145]

Microwave Pretreatment: The microwave (MW) technologies for sludge pretreatment have been recently demonstrated by pilot studies. At a frequency of 2.45 GHz, both thermal and athermal effects may be involved for sludge disintegration by the MW irradiation. The operating temperature range is 70–175°C, and irradiation time is 1–30 min. Similar to the ultrasonic or electrokinetic pretreatment, it has the capability to (a) break the cell walls and release the soluble cell matter, and (b) disturb the ionic bonds to disperse the sludge floc. It is also highly effective for pathogens destruction and improving the sludge dewaterability. The soluble COD concentration may increase by 10–25% after MW pretreatment.

TABLE 13.25 Sludge Pretreatment Processes for Enhanced Anaerobic Digestion Performance

Type of Drying	Potentially Enhanced Performance[a], % Increase		Sample Commercial System
	Biogas Production	Volatile Solids Reduction (VSR)	
Thermal hydrolysis process (THP)	10–50	10–50	*Cambi*®, *Bio Thelys*™, and *Exelys*™
Ozone pretreatment	30–40	30–40	*Lyso*™
Peroxidation pretreatment[b]	20–75	–	N/A
Ultrasonic destruction	15–50	10–55	*Sonolizer*™, *Sonix*™, and *VTA GSD*
Enzyme hydrolysis process (EHP)[c]	10–25	5–10	*Enhanced Enzymic Hydrolysis (EEH)*
High pressure homogenization (HPH)	10–45	5–50	*MicroSludge* and *Crown*
Electrical pretreatment	20–60	20–60	Bio*Crack* and *OpenCel*
Microwave (MW) irradiation[d]	10–30	5–25	N/A

[a] The enhanced performances of anaerobic digestion are estimated from published data. Readers may consult the equipment manufacturers for the expected performances of their commercial systems.

[b] System integration with advanced oxidation equipment and anaerobic digestion is required.

[c] The enzymatic solutions may be obtained from different suppliers for use in enhanced anaerobic digestion process.

[d] There has been widespread application of MW technologies in many industries. Currently, there is no full-scale application of MW irradiation for pretreatment of municipal sludge.

Source: Summarized from information contained in References 7, 21, 28, 29, and 115 through 161.

MW technologies may be integrated into a hybrid process with either ultrasonic, H_2O_2, or acid/alkali. It may also be a valid method for enhanced recovery of nutrients (nitrogen and phosphorus) and heavy metals, and improve sludge drying.[21,146–151]

13.10.2 Enhanced Anaerobic Digestion Performance

The sludge pretreatment technologies enhance the anaerobic digestion to improve biogas generation, VSR, and quality of biosolids for land application and for final disposal. The increased biogas productions and VSRs, and commercially available systems for sludge pretreatment are summarized in Table 13.25.[7,21,28,29,115–161]

Currently, the THP process is one of the most successful systems demonstrated at full-scale application. Both batch and continuous systems are commercially available. CAMBI® process is a promising example of batch THP process. A simplified process diagram of batch THP is shown in Figure 13.32. In this process,

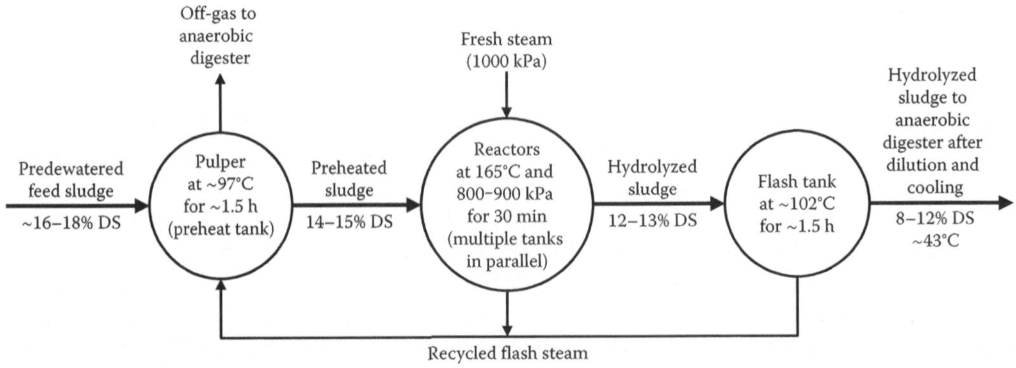

FIGURE 13.32 A simplified process diagram of batch thermal hydrolysis process (THP).

the feed sludge is prescreened (5 mm) and predewatered to a solids content of 15–18% dry solids (DS). The sludge is first preheated by recycled stream and diluted to about 15% DS in the pulper. The THP typically goes through the following five cycles (or steps): (1) *Filling*: The preheated sludge is added into the reactor; (2) *Steam Injection*: Fresh steam is injected to reach a target temperature of 165°C (329°F) and a pressure of 800–900 kPa (120—130 psi); (3) *Reaction*: The thermal disintegration is then carried out for 30 min and the solids content is lowered to about 12% DS; (4) *Steam Out*: Steam is released and recycled to the pulper; and (5) *Empty*: The hydrolyzed sludge is depressurized and transferred to the flash tank. The total operating period is at least 90 min to complete all five cycles. The hydrolyzed sludge is finally cooled to about 43°C (110°F) and diluted to about 8–12% DS prior to being pumped into the anaerobic digesters. The digestion period is reduced to less than half of that of the conventional digester. Significant increases in both volatile solids destruction and biogas production are achieved. The biosolids has little odor and meets Class A criteria.[7,28]

13.10.3 Resource Recovery from Municipal Sludge

The resource recovery from municipal biosolids includes nutrients (nitrogen and phosphorus) and protein. Three steps that are involved in complete nutrients recovery are: (a) accumulation of high concentrations of phosphorus (>100 mg/L) and ammonia (>1000 mg/L), (b) release of large amount of nutrients in a small volume, and (c) extraction of a product high in nutrient contents. The EBPR process (see Section 10.7), membrane filtration (Section 15.4.10), and algal farming are good examples of accumulation step. Many sludge pretreatment technologies (see Section 13.10.2), anaerobic digestion process (see Section 13.6.1), and P-release from WAS are effective methods for nutrients release. The extraction technologies include chemical precipitation, electrodialysis, gas permeable membrane adsorption, gas stripping, ion exchange, and solvent extraction.[162] Brief descriptions of nutrients recovery processes are presented below.

Phosphorus Recovery: Phosphorus is a nonrenewable resource, and there is a danger of running out of it. Nitrogen, potassium, and phosphorus are key components of commercial fertilizers. It is unlikely that we could sustain the food supply required under the current global population growth without readily available inorganic fertilizers. Nitrogen can be produced from atmospheric nitrogen resource which is plentiful. Potassium is mined and potassium reserves are also plentiful. Phosphorus is also mined from phosphate rock deposits, however reserves are relatively limited. It is estimated that the expected peak of global phosphorus production will reach in around 2030, and the current global reserves may be depleted within 50–100 years after the expected peak production. Phosphorus removal from wastewater is aggressively enforced by the regulatory agencies. This has resulted in high phosphorus content in biosolids from EBPR processes, causing an imbalance between phosphorus and nitrogen ratio for agricultural use. Mandatory removal and recovery of excess phosphorus from residuals will be required for sustainable land application of biosolids.[163,164]

Precipitation of struvite crystals (also known as magnesium ammonium phosphate (MAP), or $MgNH_4PO_4{\cdot}6H_2O$) on equipment, pipe surfaces, pump impellers and flow meters is a nuisance, and is a significant maintenance concern for the treatment plants with anaerobic digestion. Treatment plants that employ EBPR face severe struvite scaling. The phosphorus accumulating organisms (PAOs) concentrate phosphate in their cells in aerobic environment resulting in phosphorus removal from the liquid stream. Under anaerobic condition, PAOs release stored phosphate back into solution (see Section 10.7 and Figure 10.105a). Thus, a buildup of high concentration of soluble phosphorus occurs when the WAS undergoes anaerobic condition in digesters or WAS holding tanks at the EBPR facilities. The sidestreams are rich in phosphorus that is trapped in a vicious cycle of uptake, release and return back to the plant. In addition to achieving resource recovery, there are many other benefits of recovering phosphorus from biosolids: (1) break the vicious cycle of uptake, release and return, (2) reduce the load of phosphorus and ammonia in the sidestreams returned to the plant, (3) reduce or eliminate the struvite scaling on solids handling equipment, (4) reduce quantity of biosolids, and (5) reduce phosphorus content to optimize the N:P ratio in the biosolids.[6,7,165]

Phosphorus is effectively recovered by precipitation of struvite from stream with high phosphorus content. The generalized reaction is expressed by Equation 13.21.[7]

$$Mg^{2+} + NH_4^+ + PO_4^{3-} + 6\ H_2O \rightarrow MgNH_4PO_4{\cdot}6\ H_2O \quad (13.21)$$

Basic steps for struvite recovery include:

1. *Addition of magnesium*: Magnesium chloride ($MgCl_2$), magnesium hydroxide ($Mg(OH)_2$) or magnesium oxide (MgO) may be added in the stream containing high concentrations of phosphate and ammonia.
2. *Elevated pH*: The optimum pH range is 8.5–9.5 for precipitation of struvite. Sodium hydroxide (NaOH) is added to adjust the pH.
3. *Excess ammonia*: The N:P ratio is 0.45 g N/g P in the struvite. An excess N:P ratio of 2.4–2.6 is required in the feed stream. When the ratio is low, aqueous ammonia (NH_4OH) solution may be added to raise the N:P ratio.

Another critical requirement for effective recovery of phosphorus is proper P-release in the WAS holding tank. Readily biodegradable organics (or SCVFAs) are required for P-release. A fermenter or an acidogenic anaerobic phase reactor may be used to produce SCVFAS (see Figure 10.107). An external carbon source, such as acetic acid, or methanol may be applied if SCVFAS cannot be generated onsite. A propriety system, the WASSTRIP® reactor may also be used for releasing Mg, K, and P from WAS prior to thickening. Conceptual process diagram of phosphorus recovery by struvite precipitation is illustrated in Figure 13.33. Performance data of several propriety reactors for phosphorus recovery systems are summarized in Table 13.26.[7,162,165–171] Example process diagrams for phosphorus recovery from a BNR facility are shown in Example 13.57.

During recent years, the electrochemical processes (ECPs) or bioelectrochemical systems (BESs) have also been investigated for nutrient recovery from sludge and wastewater. The ECPs may include (a) electrolysis (EL), (b) conventional electrodialysis (CED), and (c) electrodialysis with bipolar membranes (EDBM). The BESs may include (a) microbial fuel cells (MFCs), (b) microbial electrolysis cells (MECs), and (c) microbial desalination cells (MDCs). Currently, most of these processes are limited to lab-scale studies. One small-scale electrolysis reactor (ePhos®) by Fraunhofer IGB is ready for market. In this system, an electrolyte cell consisting of an inert cathode and a sacrificial magnesium anode is the catalyst for operation. The water molecule is split at the cathode into hydrogen and hydroxide ions. The pH is raised to around nine which is the desired level for struvite recovery. Oxidation takes place

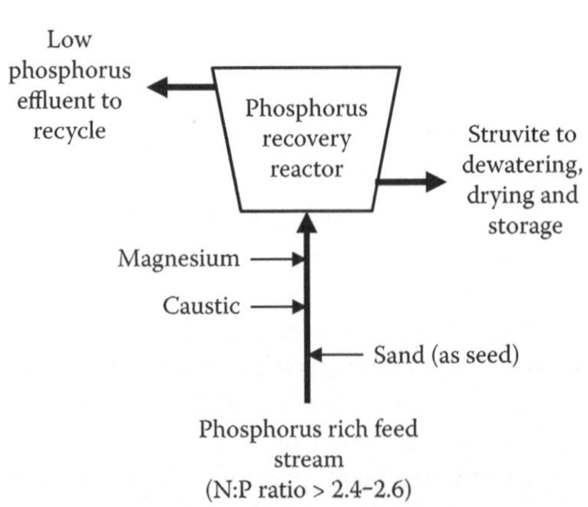

FIGURE 13.33 Conceptual process flow diagram of a phosphorus recovery system.

TABLE 13.26 Performance Data of Several Propriety Reactors for Phosphorus Recovery Systems

Process	Type of Reactor	Chemical Addition	Feed Stream	Removal Efficiency, %		Product
				P	NH_4-N	
Airpex™	CSTR with CO_2 stripping	$MgCl_2$	Digested sludge	80–90[a]	10–40[a]	Struvite
Crystalactor™	UFBR	$Mg(OH)_2$	Sidestream	85–95	10–40	Struvite, $Ca_3(PO4)_2$, and $Mg_3(PO4)_2$
NuReSys™	CSTR after air stripping	$MgCl_2$ NaOH	Sidestream	>85	5–20	Struvite
Multiform™	UFBR	$MgCl_2$ NaOH	Sidestream	80–90	10–40	Struvite
Pearl®[b]	UFBR	$MgCl_2$ NaOH	Sidestream	80–90	10–50	Struvite
Phospaq™	CSTR with aeration	MgO	Sidestream	80	10–40	Struvite
Phosnix®	UFBR with aeration	$Mg(OH)_2$ NaOH	Sidestream	>90	-	Struvite

[a] Removals are based on soluble P and NH_4-N. Removal is about 7% of total P in the feed sludge.
[b] Pearl® reactor may also be complemented by the WASSTRIP® reactor for phosphorus removal system.
Note: CSTR = continuous stirred-tank reactor; UFBR = upflow fluidized bed reactor.
Source: Adapted in part from References 7, 162, and 165 through 171.

at the anode to release magnesium ions into the solution, which reacts with phosphorus and nitrogen to form struvite. The energy consumption is relatively small at around 70 Wh/m^3 of wastewater. The phosphorus recovery is about 85%.[172–176]

Nitrogen Recovery: Nitrogen is recovered by three different processes: (a) combined recovery of phosphorus and nitrogen, (b) recovery of protein, and (c) recovery of ammonia only. The combined recovery of phosphorus and nitrogen is achieved by struvite precipitation. This process has been covered under phosphorus recovery, presented above. Ammonia is recovered as bound nitrogen with protein recovery. A brief discussion of protein recovery is presented below. Ammonia is recovered from sidestream by *gas stripping* or *ion exchange* processes (see Sections 15.4.5 and 15.4.9).

Protein Recovery: The WAS contains polysaccharides, nucleic acids, enzymes and proteins. The protein content of primary sludge and WAS are 20–30% and 32–41%, respectively.[6] The protein in sludge can be solubilized by sodium hydroxide, sodium chloride, or combination of these. The solubilization may also be achieved by sonication after alkaline treatment. The solubilized protein is precipitated by hydrochloric acid, sodium lingosulfoxide, sulfuric acid, or acetic acid and ammonium sulfate. The recovery of protein is up to 91%.[177,178] The protein contains all essential amino acids and has been used on trial basis as an alternate animal feed. The rat toxicity tests showed no effects on mortality. Another use of recovered protein is biobased wood adhesive. The sludge quantity is reduced significantly while useful byproducts are produced.[179]

13.10.4 Nitrogen Removal by Partial Nitrification/Anammox (PN/A) Process

The traditional method for nitrogen removal from wastewater is achieved by nitrification of ammonia by ammonia-oxidizing bacteria (AOB) to nitrite then by nitrite-oxidizing bacteria (NOB) to nitrate. In presence of a suitable carbon source, heterotrophic organisms carry out denitrification of nitrate into nitrogen (see Section 10.6 and Figure 10.87). In a breakthrough process known as shortcut nitrogen removal the nitrification and denitrification processes are replaced by a single process called *deammonification*. In this process, nitrogen gas is formed in a two-step biological pathway (a) partial nitritation that converts 56% ammonia nitrogen (NH_4-N) into nitrite (NO_2-N) by AOB, and (b) anaerobic ammonia oxidation that converts most NH_4-N into N_2 using NO_2-N as an electron acceptor by a different class of bacteria called *anammox*. This process may also be called partial nitritation/anammox (PN/A) process. An overall stoichiometric reaction, with cell synthesis is given by Equation 13.22.[7]

$$NH_4^+ + 0.799\ O_2 + 1.109\ HCO_3^- \rightarrow 0.436\ N_2 + 0.111\ NO_3^- + 1.034\ CO_2 + 2.496\ H_2O + \text{new cell mass} \quad (13.22)$$

From the stoichiometric reactions Equation 13.22, the following useful relationships about the PN/A process are observed and summarized below. Readers may refer to Reference 7 for detailed stoichiometric reactions for each step of the deammonification process.

1. The potential nitrogen removal efficiency is 87% stoichiometrically after deammonification. Thirteen percent of NH_4-N remains in the sidestream, including 11% as NO_3-N and nearly 2% of NH_4-N assimilated into the new cell mass. A simplified nitrogen balance for PN/A process is illustrated in Figure 13.34.
2. The overall oxygen requirement for conversion of ammonia to nitrogen gas and nitrate is 1.83 g O_2/g NH_4-N converted (see Example 13.60). It is about 60% lower than that of 4.57 g O_2/g NH_4-N oxidized in nitrification (see Example 10.124).
3. A stoichiometric alkalinity consumption of 3.97 g as $CaCO_3$ is required per gram of NH_4-N converted (see Example 13.60). This is significantly lower than that of 7.14 g alkalinity as $CaCO_3$/g NH_4-N oxidized by nitrification (see Example 10.124).
4. The overall biomass synthesis is 0.12 g VSS/g NH_4-N converted by the PN/A process (0.07 and 0.05 g VSS/g NH_4-N converted for synthesis of AOB and anammox bacteria, respectively). It is about 30% lower than the overall biomass yield of 0.17 g VSS/g NH_4-N utilized in nitrification (see Examples 13.60 and 10.125).

For sidestream treatment, the use of PN/A process has many advantages over the conventional nitrification process. Some of these are: (1) 50–60% energy saving due to reducing aeration requirement, (2) 90% net CO_2 consumption within the process rather than emission from carbon oxidation into atmosphere, (3) about 45% reduction in alkalinity requirement, (4) reduced sludge production, and (5) 50% reduction in land area requirement. The process is typically designed and operated at a low DO concentration (~0.3 mg/L or less) for a SRT > 20 d. The optimal temperature range is 30–37°C for anammox bacteria, while the optimal pH range is between 7 and 8.5. The main challenges in operation of a PN/A process include: (a) slow startup due to slow growth of anammox bacteria and (b) potentially irreversible loss of activity of anammox bacteria, since they are sensitive to nitrite.[7]

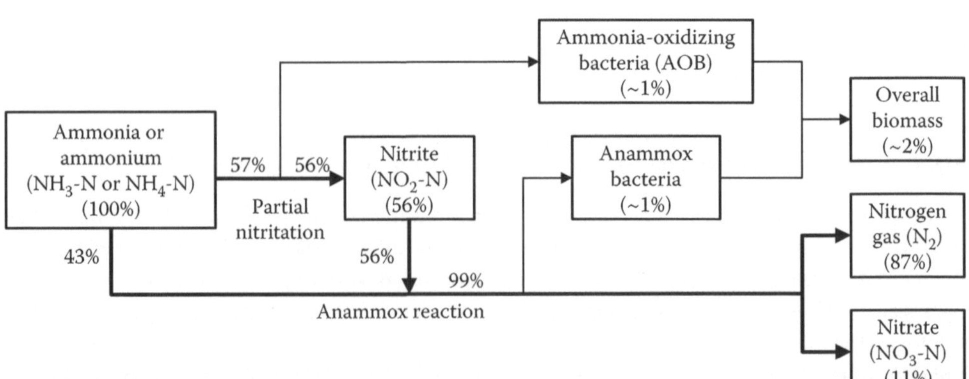

FIGURE 13.34 Simplified nitrogen balance for deammonification process.

Since the initial discovery of anammox bacteria in 1995, several propriety processes have been developed and applied to treatment of high nitrogen containing sidestreams from BNR and other processes. The PN/A process has been applied to suspended, attached, or hybrid growth reactors in batch and continuous operations. Examples of these commercial configurations using the PN/A process are: ANITA™ in single-stage moving-bed biofilm reactor (MBBR), DeAmmon® in multistage MBBRs, and DEMON® using sequencing batch reactor (SBR). Another hybrid system is Phospaq®-Anammox® process that combines nitrogen removal with phosphorus recovery. A recent survey indicated that by 2014, over 100 full-scale PN/A plants have been installed worldwide. Among these installations 75% are for sidestream treatment, and over 50% are with SBRs.[7,171,180–185]

EXAMPLE 13.55: DESIGN OF THP FOR SLUDGE PRETREATMENT

A wastewater treatment plant is planning to improve anaerobic digestion performance by implementing batch thermal hydrolysis (THP) reactors for sludge pretreatment. The facility produces 50,000 kg/d dry solids in predewatered sludge. Determine (a) the unit capacity of each reactor, (b) number of reactors, (c) pumping rate for reactor feed/pulper recirculation, (d) pulper and flash tank volumes, and (e) steam requirement. Also, prepare a conceptual chart for operating the THP reactors.

Solution

1. Determine the design parameters of batch THP system.

   The following design information is provided by a THP manufacturer: (a) total reactor volume = 12 m^3, (b) filling volume = 7.5 m^3 for each batch, (c) cycle time = 90 min, including a filing time = 15 min, steam in time = 15 min, reaction time = 30 min, steam out time = 15 min, and empty time = 15 min, (d) desired feed sludge solids content = 14.5%, (e) pulper and flash tank volume = twice of total reactor volume, and (f) steam requirement = 1 kg steam/kg dry solids.

2. Calculate the unit capacity of THP reactor.

   Total number of cycles per reactor per day, $N_{\text{cycle}} = \dfrac{24\,\text{h/d} \times 60\,\text{min/h}}{90\,\text{min/cycle}} = 16\,\text{cycles/d}$

   Note: To be energy efficient, frequent shutdowns shall be avoided between batch operating cycles. Total volume of sludge handled by a reactor per day,

   $V_{\text{treated}} = N_{\text{cycle}} V_{\text{fill}} = 16\,\text{cycles/d} \times 7.5\,\text{m}^3/\text{cycle} = 120\,\text{m}^3/\text{d}$

   Assume specific gravity of feed sludge 1.03, calculate the total solids handled by a reactor per day from rearranged Equation 13.1e,

   $$W_{\text{treated}} = \frac{p_s S_{b,s} \rho_w V_{\text{WAS}}}{100\%} = \frac{14.5\% \times 1.03 \times 1000\,\text{kg/m}^3 \times 120\,\text{m}^3/\text{d}}{100\%} = 17{,}900\,\text{kg/d per reactor}$$

3. Determine the number of THP reactors.

   $$\text{Number of reactors required, } N_{\text{reactor}} = \frac{W_s}{W_{\text{treated}}} = \frac{50{,}000\,\text{kg/d}}{17{,}900\,\text{kg/d per reactor}} = 2.8\,\text{reactors} \approx 3\,\text{reactors}$$

   Provide four reactors, including one redundant unit.

4. Determine the pumping rate for reactor feed/pulper recirculation.

   $$\text{The pumping rate during 15 min filling time, } q_{\text{feed}} = \frac{V_{\text{fill}}}{t_{\text{fill}}} = \frac{7.5\,\text{m}^3/\text{cycle} \times 10^3\,\text{L/m}^3}{15\,\text{min/cycle}} = 500\,\text{L/min}$$

   Provide three 250 L/min progressive cavity pumps, including one stand-by unit.

5. Determine the volumes of pulper and flash tank.

$$V_{\text{pulper}} = V_{\text{flash}} = 2V_{\text{reactor}} = 2 \times 12\,\text{m}^3 = 24\,\text{m}^3$$

Provide two 25-m^3 pulper tanks with one stand-by unit, and two 25-m^3 flash tanks with one stand-by unit.

Note: See Figure 13.32 for arrangement of pulper, reactor and flash tanks.

6. Determine the steam requirement.

$$W_{\text{steam}} = 1\,\text{kg steam/kg solids} \times W_{\text{treated}} = 1\,\text{kg steam/kg solids} \times 17{,}900\,\text{kg/d}$$
$$= 17{,}900\,\text{kg steam/d per reactor}$$

$$\text{Steam requirement per cycle, } W_{\text{steam,cycle}} = \frac{W_{\text{steam,cycle}}}{N_{\text{cycle}}} = \frac{17{,}900\,\text{kg steam/d}}{16\,\text{cycles/d}} = 1120\,\text{kg steam/cycle}$$

Provide a steam supply capacity of 1600 kg/cycle to encounter the unaccounted for heat losses during the process.

7. Prepare the reactor sequence operation schedule.

The sequence operation schedule for three reactors during the 90-min cycle time is shown in Figure 13.35.

<table>
<tr><th>Reactor No.</th><th>15 min</th><th>15 min</th><th>15 min</th><th>15 min</th><th>15 min</th><th>15 min</th></tr>
<tr><td>No. 1</td><td>Fill</td><td>Steam in</td><td colspan="2">Reaction</td><td>Steam out</td><td>Empty</td></tr>
<tr><td>No. 2</td><td>Steam out</td><td>Empty</td><td>Fill</td><td>Steam in</td><td colspan="2">Reaction</td></tr>
<tr><td>No. 3</td><td colspan="2">Reaction</td><td>Steam out</td><td>Empty</td><td>Fill</td><td>Steam in</td></tr>
</table>

FIGURE 13.35 Operation sequence schedule of batch thermal hydrolysis process (THP) reactors (Example 13.55).

EXAMPLE 13.56: ENHANCED VOLATILE SOLIDS REDUCTION BY THP

A conventional biological treatment plant produces 50,000 kg/d dry solids of combined sludge with a VSS/TSS = 0.7. The sludge is anaerobically digested after the sludge is disintegrated by a THP system. The overall volatile solids reduction (VSR) is 65%, including THP and anaerobic digestion. Compare the percent reduction in quantity of digested sludge with conventional anaerobic digestion which achieves a 50% VSR.

Solution

1. Determine the quantities of VSS and fixed suspended solids (FSS) in the combined sludge.

Mass of VSS, $W_{\text{VSS,CS}} = \text{VSS/TSS ratio} \times W_{\text{TSS,CS}} = 0.7 \times 50{,}000\,\text{kg/d} = 35{,}000\,\text{kg/d}$

Mass of FSS, $W_{\text{FSS,CS}} = W_{\text{TSS,CS}} - W_{\text{VSS,CS}} = (50{,}000 - 35{,}000)\,\text{kg/d} = 15{,}000\,\text{kg/d}$

2. Determine the quantities of VSS and TSS remaining after the anaerobic digestion with disintegration by THP system.

$$\text{Mass of VSS destroyed, } \Delta W_{\text{VSS,DG,THP}} = \frac{VSR_{\text{THP}}}{100\%} \times W_{\text{VSS,CS}} = \frac{65\%}{100\%} \times 35{,}000\,\text{kg/d}$$
$$= 22{,}750\,\text{kg/d}$$

$$\text{Mass of VSS remaining, } W_{\text{VSS,REM,THP}} = W_{\text{VSS,CS}} - \Delta W_{\text{VSS,DG,THP}} = (35{,}000 - 22{,}750)\,\text{kg/d}$$
$$= 12{,}250\,\text{kg/d}$$

Mass of TSS remaining, $W_{TSS,REM,THP} = W_{TSS,CS} - \Delta W_{VSS,DG,THP} = (50{,}000 - 22{,}7500)$ kg/d
$= 27{,}250$ kg/d

Ratio of VSS to TSS, $\text{VSS/TSS ratio}_{REM,THP} = \dfrac{W_{VSS,REM,THP}}{W_{TSS,REM,THP}} = \dfrac{12{,}250\ \text{kg/d}}{27{,}250\ \text{kg/d}} = 0.45$

3. Determine the quantities of VSS and TSS remaining after the conventional anaerobic digestion.

Mass of VSS destroyed, $\Delta W_{VSS,DG} = \dfrac{VSR}{100\%} \times W_{VSS,CS} = \dfrac{50\%}{100\%} \times 35{,}000\ \text{kg/d} = 17{,}500\ \text{kg/d}$

Mass of VSS remaining, $W_{VSS,REM} = W_{VSS,CS} - \Delta W_{VSS,DG} = (35{,}000 - 17{,}500)\ \text{kg/d} = 17{,}500\ \text{kg/d}$
Mass of TSS remaining, $W_{TSS,REM} = W_{TSS,CS} - \Delta W_{VSS,DG} = (50{,}000 - 17{,}500)\ \text{kg/d} = 32{,}500\ \text{kg/d}$

Ratio of VSS to TSS, $\text{VSS/TSS ratio}_{REM} = \dfrac{W_{VSS,REM}}{W_{TSS,REM}} = \dfrac{17{,}500\ \text{kg/d}}{32{,}500\ \text{kg/d}} = 0.54$

4. Compare the percent reduction in digested sludge after sludge pretreatment by THP.
Reduction in sludge remaining after enhanced anaerobic digestion.

$$R_{REM} = \frac{W_{TSS,REM} - W_{TSS,REM,THP}}{W_{TSS,REM}} \times 100\% = \frac{(32{,}500 - 27{,}250)\ \text{kg/d}}{32{,}500\ \text{kg/d}} = 16\%$$

EXAMPLE 13.57: DEVELOP CONCEPTUAL PROCESS TRAINS FOR PHOSPHOROUS RECOVERY FROM AN EBPR FACILITY

Draw conceptual process diagrams with brief descriptions for phosphorus recovery from a EBPR facility. Assume that the following three alternatives are used to provide carbon source for WAS phosphorus release prior to the struvite precipitation: (a) addition of VFA-containing chemical, (b) overflow from thickener after fermenter, and (c) supernatant from the first reactor of acid/gas phased anaerobic digestion (A/GAnD) process.

Solution

1. Draw the conceptual process diagrams in Figure 13.36.
2. Describe each process alternative.
   a. Process Train 1—Phosphorus recovery using external carbon source:
      This is a basic process train in which SCVFAs from external carbon source are mixed with P-rich WAS wasted from EBPR process (Figure 13.36a). The mixture is stirred in a phosphorus release tank. The WAS is thickened in a gravity belt thickener. The filtrate has high nutrient concentration. The high nitrogen and phosphorus recycled streams from mesophilic anaerobic digester and belt-filter press are mixed with high phosphorus containing filtrate. The combined stream goes into phosphorus recovery system. Magnesium and NaOH are added to precipitate struvite (magnesium ammonium phosphate (MAP)). The clarified liquid from phosphorus recovery system is low in phosphorus and high in nitrogen. It is recycled to gravity thickener. In this unit the primary sludge is mixed and overflow is returned to the EBFR system. The thickened sludge is mixed with phosphorus stripped WAS into a sludge blending tank. The thickened combined primary and WAS solids are digested in a mesophilic anaerobic digester. The digested sludge is dewatered in a belt-filter press and sludge cake is used as Class B biosolids.
   b. Process Train 2—Phosphorus recovery using overflow from fermenter/ thickener:
      In this process train, the SCVFA is generated by fermenting primary sludge in a fermenter (Figure 13.36b). The fermented primary sludge is thickened in a gravity thickener. The overflow

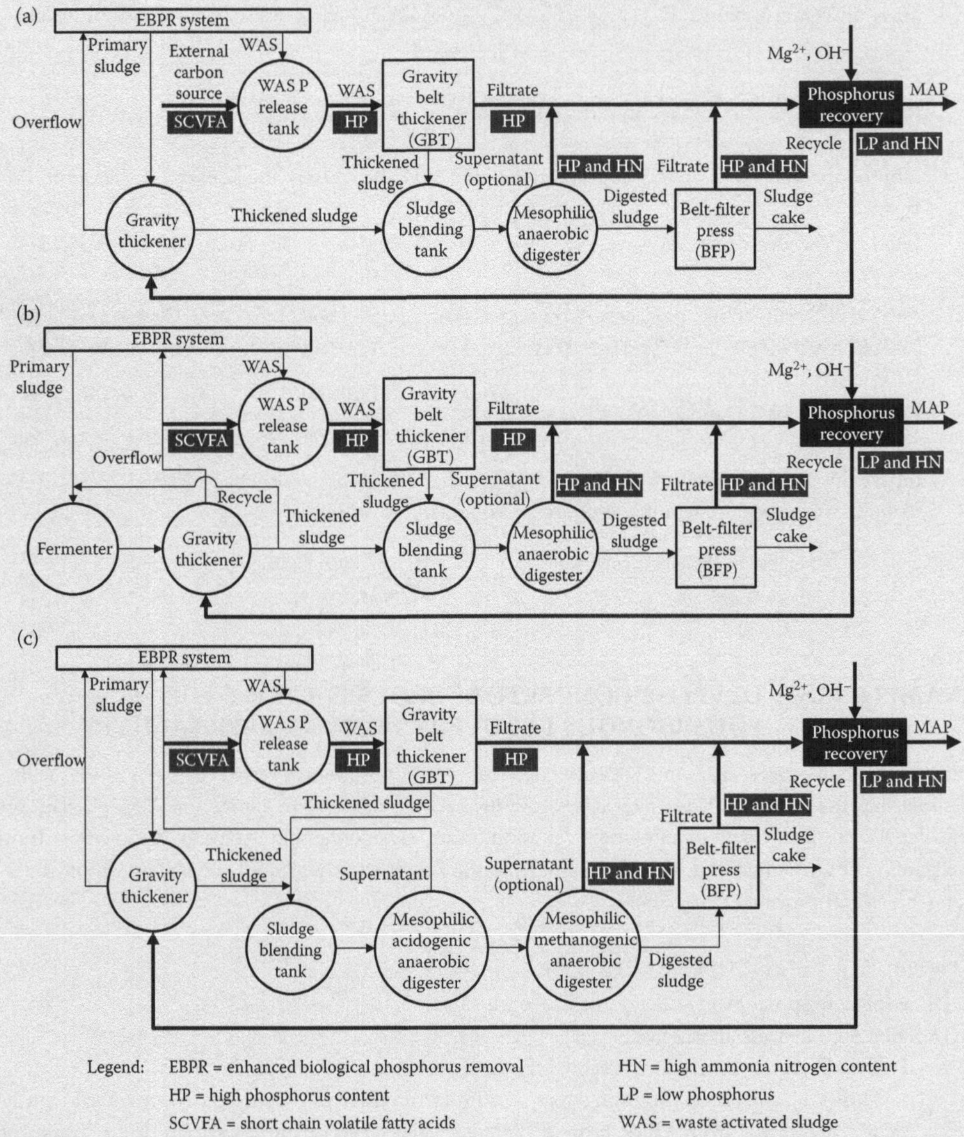

FIGURE 13.36 Conceptual process diagrams for phosphorus recovery using different carbon source alternatives: (a) Train 1: external carbon source, (b) Train 2: overflow from fermenter/thickener, and (c) Train 3: supernatant from acid phase reactor of A/GAnD process (Example 13.57).

from the thickener provides the needed SCVFA, which is mixed with P-rich WAS from EBPR system in a phosphorus release tank. The remainder of the Process Train 2 is similar to that in Process Train 1 for phosphorus recovery as struvite. In this process train, Class B biosolids production is identical to the Process Train 1.

c. Process Train 3—Phosphorus recovery using supernatant from acid phase reactor of A/GAnD process:

This process train is very similar to Process Train 2. The major difference is application of acid/gas phased anaerobic digestion (A/GAnD) process to improve the anaerobic digestion

(Figure 13.36c). The SCVFA produced in mesophilic acidogenic anaerobic digester is partly diverted to phosphorus release tank to supply needed SCVFA for phosphorus release from P-rich WAS. The remainder of Process Train 3 from mesophilic methanogenic anaerobic digester to phosphorus recovery as struvite is similar to that in Process Train 1 or 2. Class B biosolids production is identical to Process Train 1 or 2.

Note: In all three process trains, ammonia concentration in the recycled streams from the phosphorus recovery systems may remain high. Effective removal of ammonia by a PN/A process may be necessary for reducing the ammonia loading from the recycled sidestream to the EBPR system (see the conceptual process diagram in Example 13.61).

EXAMPLE 13.58: PHOSPHOROUS RECOVERY FROM AN EBPR FACILITY

An EBPR facility produces 4000 kg/d WAS (as dry solids) with VSS/TSS ratio = 0.8. The overall P content in the WAS is 0.08 g P/g VSS. Cothickening of WAS with primary sludge in a gravity thickener under anaerobic condition produces the volatile fatty acids (VFAs) needed for P-release. The efficiency of P-release is 40% of total P in the WAS. After phosphorus release, the combined sludge is separated in a thickener and the overflow rich in phosphorus is processed for the phosphorus recovery. $Mg(OH)_2$ and NaOH are added, and needed ammonia for struvite formation is recycled through digester overflow and waste flows from BFPs. Estimate the quantity of struvite precipitated if the recovery is 85% in the reactor. Ignore the P contented in the filtrate from dewatering operation.

Solution

1. Calculate the quantity of P released from WAS.

   Total quantity of P in the WAS,

$$W_{\text{P,WAS}} = f_{\text{P,WAS}} \times \text{VSS/TSS Ratio} \times W_{\text{WAS}} = 0.08\,\text{kg P/kg VSS} \times 0.8\,\text{kg VSS/kg TSS} \times 4000\,\text{kg TSS/d}$$
$$= 256\,\text{kg P/d}$$

   Quantity of Org-P released from WAS at a release rate of 40% ($f_{\text{P,release}} = 0.4$),

$$W_{\text{P,release}} = f_{\text{P,release}} \times W_{\text{P,WAS}} = 0.4 \times 256\,\text{kg P/d} = 102\,\text{kg P/d}$$

2. Calculate the quantity of struvite precipitated at phosphorous recovery, $f_{\text{P,recovery}} = 85\%$.

$$W_{\text{P,recovery}} = f_{\text{P,recovery}} \times W_{\text{P,release}} = 0.85 \times 102\,\text{kg P/d} = 87\,\text{kg P/d}$$

   Apply Equation 13.21 and calculate the quantity of struvite (magnesium ammonium phosphate (MAP)) precipitated.

$$W_{\text{MAP}} = W_{\text{P,recovery}} \times \frac{\text{molar wt. of MAP}}{\text{molar wt. of P}} = 87\,\text{kg P/d} \times \frac{245\,\text{g/mole MAP}}{31\,\text{g/mole P}} = 688\,\text{kg MAP/d}$$

   Note: The molecular weight of MAP ($MgNH_4PO_4{\cdot}6H_2O$) = 245 g/mole.

EXAMPLE 13.59: PROTEIN RECOVERY FROM COMBINED SLUDGE

An activated sludge plant produces 2500 kg/d combined sludge. The average protein content of the sludge is 30% of VSS. The sludge solids are solubilized by alkaline treatment followed by ultra-sonication. The protein precipitation is achieved by acetic acid and ammonium sulfate treatment. The recovery of protein is 90%. Calculate the quantity of protein precipitated. Assume the VSS/TSS ratio = 0.7.

Solution

1. Calculate the quantity of soluble protein in the combined sludge.

$$\text{Quantity of protein} = 0.3\,\text{kg protein/kg VSS} \times 0.7\,\text{kg VSS/kg TSS} \times W_{cs}$$
$$= 0.3\,\text{kg protein/kg VSS} \times 0.7\,\text{kg VSS/kg TSS} \times 2500\,\text{kg TSS/d}$$
$$= 525\,\text{kg protein/d}$$

2. Calculate the quantity of protein precipitated.

$$\text{Protein precipitated} = 0.9 \times 525\,\text{kg protein/d} = 473\,\text{kg protein/d}$$

EXAMPLE 13.60: OXYGEN REQUIREMENT, ALKALINITY CONSUMPTION, AND NEW CELL MASS GENERATION FACTORS FOR PARTIAL NITRITATION/ANAMMOX (PN/A) PROCESS

The factors for oxygen requirement, alkalinity consumption, and new cell mass generation in the PN/A process are 1.83 g O_2, 3.97 g alkalinity as $CaCO_3$, and 0.12 g VSS per gram of NH_4-N converted, respectively. Derive these factors from Equation 13.22.

Solution

1. Identify the overall reaction for PN/A process.

   The overall stoichiometric reaction for PN/A process is given by Equation 13.22.

$$NH_4^+ + 0.799\,O_2 + 1.109\,HCO_3^- \rightarrow 0.436\,N_2 + 0.111\,NO_3^- + 1.034\,CO_2 + 2.496\,H_2O + \text{New cell mass}$$

2. Determine the oxygen requirement for ammonia oxidation in the PN/A process.

   The ratio of g O_2/g NH_4-N for converting ammonia to nitrite is 0.799 mole of O_2 theoretically needed per mole of NH_4-N converted.

$$O_2 = \frac{0.799\,\text{moles}\,O_2 \times 32\,\text{g}\,O_2/\text{mole}\,O_2}{1\,\text{mole}\,NH_4^+\text{-N} \times 14\,\text{g}\,NH_4^+\text{-N/mole}\,NH_4^+\text{-N}} = 1.83\,\text{g}\,O_2/\text{g}\,NH_4^+\text{-N converted}$$

3. Determine the alkalinity consumed by the PN/A process.

   1.109 mole of HCO_3^- is theoretically consumed per mole of NH_4-N converted.

$$\text{Bicarbonate} = \frac{1.109\,\text{mole}\,HCO_3^- \times 61\,\text{g}\,HCO_3^-/\text{mole}\,HCO_3^-}{1\,\text{mole}\,NH_4^+\text{-N} \times 14\,\text{g}\,NH_4^+\text{-N/mole}\,NH_4^+\text{-N}} = 4.84\,\text{g}\,HCO_3^-/\text{g}\,NH_4^+\text{-N converted}$$

   Alkalinity recovery as $CaCO_3$ is calculated from the bicarbonate generation.

$$\text{Alkalinity} = 4.84\,\text{g}\,HCO_3^-/\text{g}\,NH_4^+\text{-N converted} \times \frac{50\,\text{g/eq. as}\,CaCO_3}{61\,\text{g/eq. as}\,HCO_3^-}$$
$$= 3.96\,\text{g}\,CaCO_3/\text{g}\,NH_4^+\text{-N converted}$$

4. Estimate the new cell mass generated in the PN/A process.

   The molecular weight of new cell generation (mw_{VSS}) as g VSS/mole new cell mass is estimated from Equation 13.22.

$$mw_{VSS} = (NH_4^+ + 0.799\,O_2 + 1.109\,HCO_3^-) - (0.436\,N_2 + 0.111\,NO_3^- + 1.034\,CO_2 + 2.496\,H_2O)$$
$$= (18 + 0.799 \times 32 + 1.109 \times 61) - (0.436 \times 28 + 0.111 \times 62 + 1.034 \times 44 + 2.496 \times 18)$$
$$= 111.217 - 109.514 = 1.703\,\text{g VSS/mole new cell mass}$$

The amount of new cell mass produced (VSS) as g VSS/g NH_4-N converted,

$$VSS = \frac{1 \text{ mole new cell mass} \times 1.703\,\text{g VSS/mole new cell mass}}{1 \text{ mole } NH_4^+\text{-N} \times 14\,\text{g } NH_4^+\text{-N/mole } NH_4^+\text{-N}}$$
$$= 0.12\,\text{g VSS/g } NH_4^+\text{-N converted}$$

Note: Cell mass generation factors for AOB and anammox bacteria can also be separately estimated from stoichiometric reactions for partial nitritation and anammox reactions. These reactions may be found in Reference 7.

EXAMPLE 13.61: CONCEPTUAL PN/A PROCESS ADDED FOR REMOVAL OF HIGH AMMONIA IN SIDESTREAM FROM A PHOSPHORUS RECOVERY PROCESS TRAIN

High ammonia concentration in the sidestream from a phosphorus recovery system is indicated in all three alternative process trains developed in Example 13.57. Add PN/A process in Process Train 3 (see Figure 13.36c) for ammonia removal from the high ammonia containing sidestream.

Solution

As described in Section 13.10.4, the PN/A process is a viable process for removal of ammonia from sidestream after phosphorus recovery system. A conceptual process diagram for ammonia removal from sidestream of Process Train 3 in Example 13.57 is shown in Figure 13.37.

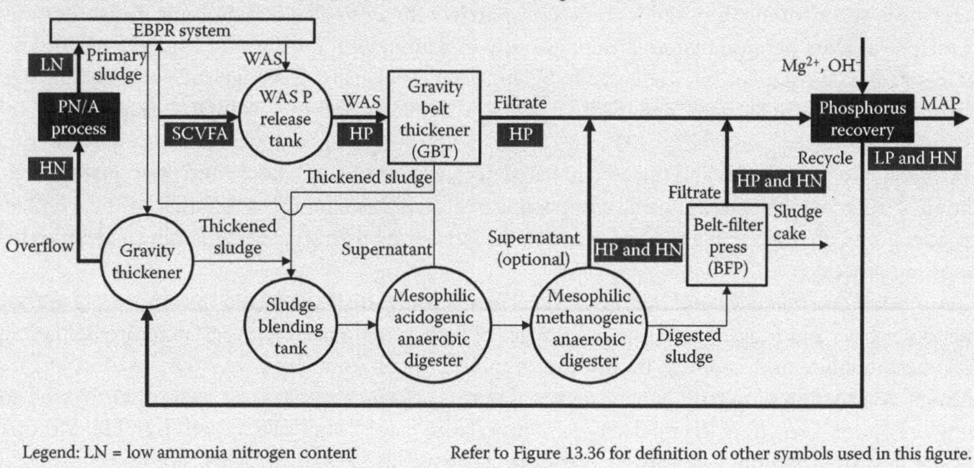

FIGURE 13.37 Conceptual PN/A process diagram for removal of ammonia from sidestream of a phosphorus recovery process (Example 13.61).

Note: The Phospaq®-Anammox® process, a hybrid propriety system is also a good example of combined phosphorus recovery and ammonia removal.[185]

13.11 Sludge Disposal and Biosolids Reuse

Safe disposal and reuse of residuals generated at a wastewater treatment plant are an integral part of planning, design and management of a treatment facility. Residuals such as screenings, grit and skimmings are mostly landfilled. The sludge disposal processes and biosolids reuse options may include (1) composting,

(2) heating drying, (3) thermal oxidations, (4) recalcination, (5) land application of biosolids, (6) dedicated land disposal, and (7) residuals disposal by landfilling. These common practices are briefly discussed below.

13.11.1 Composting

In this process the organic matter undergoes microbial decomposition to produce a stable product that is usually acceptable as a soil conditioner. Normally, the dewatered sludge cake is blended with a bulking agent such as wood chips, sawdust, chipped brush, leaves and yard wastes. Moisture is controlled and aerobic decomposition is carried out. Approximately 20–30% of volatile solids are converted to carbon dioxide and water. The heat generated by composting organic materials is typically 21 MJ/kg of organics converted. The temperature rises to the thermophilic range from 50°C to 70°C (120−160°F) and is maintained for several days. The temperature rise is an indication of active microbial decomposition of organic matter. The enteric pathogenic organisms are destroyed, and the compost may be used as soil conditioned in agriculture and horticulture. Cocomposting with MSW may also be a valid approach in an integrated waste-disposal management program.[7,186–188]

Design and Operational Considerations: Composting should be carried out in a specially designed facility. Many design and operational parameters must be monitored and controlled to achieve optimum results. These parameters are discussed below.[7,20,186–188]

Moisture Content and Proportion of Bulking Agent: The optimum moisture content of the composting mixture should be in the range from 40% to 60%. The optimum moisture content is obtained by blending the sludge and bulking agent in the right proportion. The moisture content of sludge cake is 75–80%, and that of bulking agent is 30–40%. The blending proportion of bulking agent and sludge cake by volume is typically 3:1 to 4:1 to achieve the optimum moisture content. The bulking agent serves as a carbon source, reduces the moisture of mixture, and increases the porosity for air to circulate. Volume measurements are used instead of mass because materials are generally measured with a front-end loader, truck of known volume, or piles. Weight conversion is made by applying the density of sludge cake and bulking agents which may be 1050–1070 kg/m^3 (66–68 lb/ft^3) and 400–600 kg/m^3 (25–37 lb/ft^3), respectively. If the moisture content of the blend falls <40%, the rate of compositing will be slower. Also, if moisture content is >60%, water logging occurs and the air circulated decreases resulting in odors and poor quality product. Addition of water may be needed during composting operation to maintain optimum moisture content. In many operations, after screening the end product, the large-size bulking agent remaining is recycled in the composting process.

Temperature, Microbiology and Pathogen Kill: The microorganisms involved in composting are bacteria, actinomycetes and fungi. There are three stages of activity associated with temperature: initial mesophilic, thermophilic, and cooling. In the thermophilic range from 50°C to 70°C (120−160°F), the maximum degradation of organic matter occurs. Decrease in temperature is an indication that the composting operation is reaching the final stages. Temperature plays a vital role in pathogen kill. Most pathogens are destroyed rapidly when the temperature is above 55°C (130°F). For example, *Salmonella* sp. bacteria are destroyed at 60°C (140°F) in 15–20 min and in 1 h at 55°C. These conditions are easily met in well operated composting system producing Class A biosolids.

pH Control: pH changes during the composting operation are good indication of the process performance. The initial pH of sludge/bulking agent is in the range from 6 to 8.5. Like temperature, pH also varies with time. In the first few days of composting operation, the pH drops to 5 or less. At this stage, active decomposition of organic matter occurs and temperature starts to rise rapidly. Organic acids are broken down and pH begins to rise to ~ 8 for the remainder of aerobic process. The pH starts to fall slightly during the cooling phase and reaches to a value in the range from 7 to 7.5 in the mature compost. If aeration is not adequate, pH will not rise. This is an indication of anaerobic condition and organic acids production. A drop in pH is an indication that the composting process is occurring very slowly.

Seeding Requirements: Seeding involves addition of a small portion of finished compost in the mixture of sludge and bulking agent to increase the initial microbial population. Seeding accelerates the composting operation. The amount of seeding is typically 20% of dry solids in the sludge. A 55% or higher organic content in the composting mixture is usually desired. Carbon should also be biodegradable, and is lost as CO_2 during composting.

Nutrient Ratios: Proper nutrient ratios for composting operation are also quite important. The initial carbon to nitrogen (C/N) ratio should be 20:1 to 35:1 by weight. The C/N ratio of finished product should be less than 10:1. The desired nitrogen to phosphorus (N/P) ratio may range from 75:1 to 100:1 by weight.

Organic and Oxygen Requirements: The source of oxygen is from the diffusion of air into porous structure of the compost material. Turning of pile stimulates the oxygen demand due to exposure of anaerobic or low oxygen containing material, and increases the flow of oxygen into the pile. A 5–15% oxygen is required throughout the pile to ensure aerobic condition, while the CO_2 content should not be >8%.

Odor Control: Majority of odor problems are associated with development of anaerobic condition. Because sufficient oxygen is often not available in the center of the pile, anaerobic condition develops. Many organic acids are produced and many of them are extremely odorous. Turning of pile is often accompanied by the release of offensive odors.

Process Description: The composting process takes place in many interrelated steps that are shown in Figure 13.38. These steps are: (1) *mixing and preparation*: The dewatered sludge is mechanically mixed with the bulking agent to create a feedstock that optimizes the porosity, carbon content, nutrients, and moisture content; and the mixture is well seeded; (2) *high-rate decomposition*: The temperature rises under aerobic condition by aeration or mechanical turning in this step that may take 3–4 weeks to complete; (3) *curing*: It takes about 30 days to further stabilize and reduce pathogens, cool and degas the compost, and make the finished compost more marketable; (4) *drying*: This step may vary from several days to a couple of weeks and is especially necessary if the *bulking agent recovery* from the finished compost is desired for recycling; (5) *postprocessing*: It may include screening for removal of undesired material or grinding for size reduction; and (6) *product storage*: It is normal to provide a 30- to 60-d storage capacity.[7,187]

Types of Composting Operations: Composting systems are generally divided into three categories: *windrow, aerated static pile* and *in-vessel*. These methods are described below.[7,20,23,187–191]

Windrow: In the windrow composting the sludge, bulking agent, and small amount of finished compost in proper proportion are mixed by mechanical equipment and windrows are constructed. Initial mixing helps to distribute moisture, nutrients and seeding microorganisms. The windrows are from 1 to 2 m (3–6 ft) high and 2–4.5 m (6–14 ft) wide at the base. The composting period is about 3–4 weeks. Convective air movement and diffusion moves oxygen into and releases CO_2 from the windrows. The windrows are turned and mixed periodically. Turning increases porosity for air movement, distributes anaerobic areas, and helps to maintain aerobic condition. The frequency of turning depends upon the moisture content, sludge characteristics and air requirement. It is more during initial stage and tapers off towards the completion of composting (starting daily then 3 times per week). Turning of the windrows may be accompanied by the release of offensive odors.

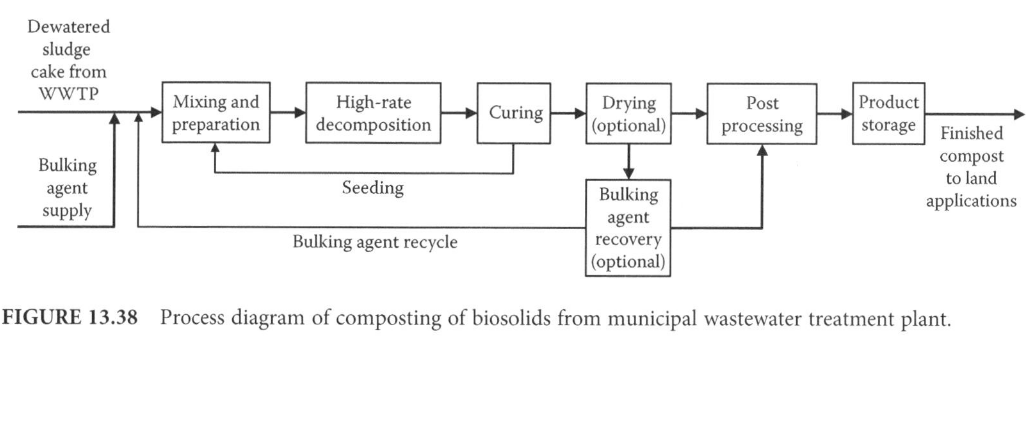

FIGURE 13.38 Process diagram of composting of biosolids from municipal wastewater treatment plant.

TABLE 13.27 Suggested Parameters with Monitoring and Testing Frequencies of Composting Operation

Activity and Parameter	Monitoring and Testing Frequency
Sludge	
Moisture content	Initial and periodic (monthly), depending upon process changes at the facility
pH	Initially and as processes change
Heavy metals and toxic organics	Initially to establish range and periodic sampling thereafter
Pathogens	As required by local and state regulations
Bulking agent	
Moisture content	Seasonally and change in climatic condition
Composting feedstock	
Moisture content	Initially to adjust bulking material ratio, and periodic as related to sludge consistency and moisture content of bulking agent
Composting process	
Temperature	Initially and once a day; daily for 3–4 consecutive days above 55°C; and weekly thereafter
Oxygen	Initially and after every turn in windrow composting, and periodic depending on the temperature change in the pile
Odors	Daily to identify and correct and sources of odors such as improper cover, anaerobic condition, high moisture content, or stagnant pool of leachate
Heavy metals	Initially, and as required by local and state regulations
Site	
Meteorological	Daily records of precipitation, sunshine, overcast, humidity, prevailing wind direction
Leachate and runoff	Daily records of quality and quantity and action taken
Applications of finished product	
Nutrients	Initially to establish the range
Pathogens	As required by health authorities

Source: Adapted in part from References 23, and 187 through 190.

Aerated Static Pile: In the static pile composting, the aeration system consists of a series of perforated pipes running underneath each pile and connected to a pump or blower that draws or blows air through the pile. The pile is 2–2.5 m (6–8 ft) high. The composting period is 3–4 weeks.

In-vessel System: In-vessel composting (also known as mechanical or enclosed reactor system) is accomplished inside an enclosed container or vessel. These systems may be either mixed or unmixed (or plug-flow) process in a horizontal or vertical reactor. The main advantage of in-vessel composting is better process and odor control, higher throughput, reduced composting time, lower labor cost, and smaller area requirement.

Monitoring and Testing Requirements: Monitoring and testing of operational parameters for sludge, bulking agents, composting process, product are essential for operational, and marketing and public relations. Key operational parameters and their sequence of sampling for composting operation are summarized in Table 13.27.

EXAMPLE 13.62: VOLUME AND WEIGHT RATIOS OF BULKING AGENT AND SLUDGE FOR COMPOSTING

A POTW uses belt filter press for sludge dewatering. The sludge cake contains 20% solids and density of sludge cake is 1.05. Total quantity of dry dewatered solids is 2800 kg/d. The bulking agent contains 65% solids by weight and bulk density is 500 kg/m^3. Calculate the volume and weight ratios of bulking agent to sludge in the composting feedstock so that the final moisture content is 50%.

Solution

1. Calculate the volume of sludge cake from Equation 13.1e.

$$\text{Volume of sludge, } V_s = \frac{100\% \times W_s}{p_s S_b \rho_w} = \frac{100\% \times 2800\,\text{kg/d}}{20\% \times 1.05 \times 1000\,\text{kg/m}^3} = 13.3\,\text{m}^3/\text{d}$$

2. Calculate the wet weight of bulking agent mixed with the sludge to give an overall moisture content of 50%.

$$\text{Total wet weight of dewatered sludge, } W_{s,wet} = \frac{W_s}{p_s} = \frac{100\% \times 2800\,\text{kg/d}}{20\%} = 14{,}000\,\text{kg/d}$$

Assume wet weight of bulking agent is $W_{ba,wet}$ in kg/d.
Total wet weight of the mixture, $W_{mix,wet} = W_{s,wet} + W_{ba,wet} = 14{,}000\,\text{kg/d} + W_{ba,wet}$

$$\text{Total weight of moisture in the mixture, } W_{mix,water} = \left(1 - \frac{p_s}{100\%}\right) W_{s,wet} + \left(1 - \frac{p_{ba}}{100\%}\right) W_{ba,wet}$$

$$= \left(1 - \frac{20\%}{100\%}\right) \times 14{,}000\,\text{kg/d} + \left(1 - \frac{65\%}{100\%}\right) W_{ba,wet}$$

$$= 11{,}200\,\text{kg/d} + 0.35\,W_{ba,wet}$$

$$\text{Ratio of moisture content in the mixture, } \frac{W_{mix,water}}{W_{mix,wet}} = \frac{11{,}200\,\text{kg/d} + 0.35 W_{ba,wet}}{14{,}000\,\text{kg/d} + W_{ba,wet}} = \frac{p_{mix}}{100\%} = \frac{50\%}{100\%} = 0.5$$

Solve for $W_{ba,wet}$.

$$W_{ba,wet} = \frac{11{,}200\,\text{kg/d} - 0.5 \times 14{,}000\,\text{kg/d}}{0.5 - 0.35} = \frac{4200\,\text{kg/d}}{0.15} = 28{,}000\,\text{kg/d}$$

3. Calculate the volume of bulking agent.

$$\text{Volume of bulking agent, } V_{ba} = \frac{W_{ba,wet}}{\rho_{ba}} = \frac{28{,}000\,\text{kg/d}}{500\,\text{kg/m}^3} = 56\,\text{m}^3/\text{d}$$

4. Calculate the ratios of bulking agent to sludge cake.

$$\text{Volume ratio, } R_v = \frac{V_{ba}}{V_s} = \frac{56\,\text{m}^3/\text{d}}{13.3\,\text{m}^3/\text{d}} = 4.2 \quad \text{or} \quad 4.2{:}1 \text{ by volume}$$

$$\text{Weight ratio, } R_w = \frac{W_{ba,wet}}{W_{s,wet}} = \frac{28{,}000\,\text{kg/d}}{14{,}000\,\text{kg/d}} = 2 \quad \text{or} \quad 2{:}1 \text{ by weight}$$

EXAMPLE 13.63: COCOMPOSTING OF DIGESTED SLUDGE WITH MUNICIPAL SOLID WASTE

Feedstock is prepared by mixing digested liquid sludge into mulch. The mulch separated from municipal solids waste at a rate of 8000 lb/d. The moisture content of mulch is 20%. Digested liquid sludge at 5% solids is added in the mulch to bring the moisture content of the feedstock at the optimum value of 55%. Calculate the quantity of liquid sludge needed per day.

Solution

1. Determine the dry solids in mulch at a moisture $p_{w,m} = 20\%$.

$$\text{Weight of dry solids in mulch, } W_{m,s} = \left(1 - \frac{p_{w,m}}{100\%}\right) W_{m,wet} = \left(1 - \frac{20\%}{100\%}\right) \times 8000\,\text{lb/d} = 6400\,\text{lb/d}$$

2. Develop the relationship between the moisture content of composting feedstock ($p_{w,fs}$) and the weight of wet sludge ($W_{s,wet}$).

$$\text{Moisture in feedstock} = \frac{\text{Water weight in feedstock}}{\text{Weight of feedstock}} = \frac{\text{Weight of feedstock} - \text{Dry solids in feedstock}}{\text{Weight of feedstock}}$$

$$\frac{p_{w,fs}}{100\%} = \frac{\text{Weight of wet feedstock } (W_{fs,wet}) - \text{weight of solids in feedstock } (W_{fs,s})}{W_{fs,wet}}$$

$$= \frac{(W_{m,wet} + W_{s,wet}) - (W_{m,s} + W_{s,s})}{(W_{m,wet} + W_{s,wet})}$$

where

$W_{m,wet}$ = weight of wet mulch, kg/d (lb/d)
$W_{s,wet}$ = weight of wet sludge, kg/d (lb/d)
$W_{s,s}$ = weight of dry solids in sludge, kg/d (lb/d)

3. Rearrange the above expression and solve for $W_{s,wet}$ required to achieve optimum moisture in feedstock $p_{w,fs} = 55\%$.

$W_{m,wet} = 8000$ kg/d, $W_{m,s} = 6400$ kg/d, $W_{s,s} = \frac{5\%}{100\%} \times W_{s,wet} = 0.05\ W_{s,wet}$, and $p_{w.fs} = 55\%$.

$$\frac{55\%}{100\%} = \frac{(8000\,\text{lb/d} + W_{s,wet}) - (6400\,\text{lb/d} + 0.05\ W_{s,wet})}{8000\,\text{lb/d} + W_{s,wet}}$$

$$\text{Solve for } W_{s,wet} = \frac{6400\,\text{lb/d} - (1 - 0.05\ W_{s,wet}) \times 8000\,\text{lb/d}}{1 - 0.05 - 0.55} = 7000\,\text{lb/d}$$

4. Check the moisture in the feedstock.

$$p_{w,fs} = \frac{(8000\,\text{lb/d} + 7000\,\text{lb/d}) - (6400\,\text{lb/d} + 0.05 \times 7000\,\text{lb/d})}{8000\,\text{lb/d} + 7000\,\text{lb/d}} \times 100\%$$

$$= \frac{15{,}000\,\text{lb/d} - (6400\,\text{lb/d} + 350\,\text{lb/d})}{15{,}000\,\text{lb/d}} \times 100\% = \frac{8250\,\text{lb/d}}{15{,}000\,\text{lb/d}} \times 100\% = 55\%$$

Therefore, 7000 kg/d liquid sludge is added in 8000 kg/d mulch to obtain the moisture content of 55% in 15,000 kg/d feedstock.

EXAMPLE 13.64: LENGTH OF WINDROW FOR COMPOSTING

A conventional secondary wastewater treatment plant produces 2000 kg/d dry dewatered sludge at 20% solids. The specific weight is 1050 kg/m^3. Garden mulch is added at a ratio of 4:1 by volume. The windrow dimensions are 3 m at the base and 1 m on the top. The height of the windrow is 1.5 m. Calculate the length of windrow needed each operating day.

Solution

1. Determine from Equation 13.1e the volume of sludge cake per operating day.

$$\text{Volume of sludge, } V_s = \frac{100\% \times W_s}{p_s S_b \rho_w} = \frac{100\% \times 2000\,\text{kg/d}}{20\% \times 1.05 \times 1000\,\text{kg/m}^3} = 9.5\,\text{m}^3/\text{d}$$

2. Determine the volume of mulch needed each operating day.
   Volume of mulch, $V_m = 4V_s = 4 \times 9.5\,\text{m}^3/\text{d} = 38\,\text{m}^3/\text{d}$
3. Determine the total volume of composting feedstock each operating day.
   Volume of feedstock, $V_{fs} = V_s + V_m = (9.5 + 38)\,\text{m}^3/\text{d} = 47.5\,\text{m}^3/\text{d}$

4. Determine the cross-sectional area of the windrow.

$$\text{Cross-sectional area of windrow, } A = \frac{1}{2} \times (\text{Width of top} + \text{Width of base}) \times \text{Hieght}$$
$$= \frac{1}{2} \times (1+3)\ \text{m} \times 1.5\,\text{m} = 3\,\text{m}^2$$

5. Calculate the length of windrow per operating day.

$$\text{Length of windrow, } L = \frac{\text{Volume of feedstock}}{\text{Cross-section area}} = \frac{V_{fs}}{A} = \frac{47.5\,\text{m}^3/\text{d}}{3\,\text{m}^2} \approx 16\,\text{m/d}$$

EXAMPLE 13.65: BLENDING WASTES FOR OPTIMUM C/N RATIO

Mulch from yard waste is blended with sludge cake from belt-filter press. The C/N ratios, moisture contents and nitrogen concentrations of the mulch and sludge cake are summarized below. Determine (a) the weight ratio between two components to achieve a C/N blend ratio of 25 in the feedstock, and (b) the moisture content in the feedstock. If loss of carbon as CO_2 is 60% during composting process, determine also the C/N ratio in the finished compost.

	Mulch	Sludge Cake
Moisture Content	45%	80%
Nitrogen Content	0.7% of dry solids	5.8% of dry solids
C/N Ratio	50:1	6.1:1

Solution

1. Summarize the water content, dry matter, nitrogen and carbon in 1 kg of mulch and sludge.

Component	Moisture, kg	Dry Solids, kg	Nitrogen (N), kg	Carbon (C), kg
Mulch	0.45	0.55	0.00385[a]	0.1925[b]
Sludge cake	0.80	0.20	0.01160[c]	0.07076[d]

[a] 0.55 × 0.7% ÷ 100% = 0.00385.
[b] 0.00385 × 50 =0.1925.
[c] 0.2 × 5.8% ÷ 100% = 0.0116.
[d] 0.0116 × 6.1 = 0.07076.

2. Determine the ratio of sludge cake to mulch to achieve a C/N ratio of 25.
 Assume W_s is the amount of sludge cake added to 1 kg of mulch. Determine the W_s required to achieve a C/N ratio of 25.

$$C/N_{\text{mix}} = \frac{\text{C in mixture}}{\text{N in mixture}} = \frac{\text{C in 1 kg of mulch} + (\text{C in 1 kg of cake}) \times W_s}{\text{N in 1 kg of mulch} + (\text{N in 1 kg of cake}) \times W_s}$$
$$= \frac{0.1925\,\text{kgC/kg mulck} + 0.07076\,\text{kg C/kg cake} \times W_s}{0.00385\,\text{kgN/kg mulch} + 0.0116\,\text{kgN/kg cake} \times W_s} = 25\,\text{kg C/kgN}$$

Solve for W_s.

$$W_s = \frac{0.1925\,\text{kgC/kg mulck} - 25\,\text{kg C/kgN} \times 0.00385\,\text{kgN/kg mulch}}{25\,\text{kg C/kgN} \times 0.0116\,\text{kgN/kg cake} - 0.07076\,\text{kg C/kg cake}} = \frac{0.09625}{0.21924} = 0.44\,\text{kg cake/kg mulch}$$

The required weight ratio of sludge cake to mulch is 0.44.

3. Check the C/N ratio (C/N_{fs}) in the feedstock of 1 kg mulch and 0.44 kg sludge cake.

$$C/N_{fs} = \frac{0.1925\ \text{kg C/kg mulck} + 0.07076\ \text{kg C/kg cake} \times 0.44\ \text{kg cake/kg mulch}}{0.00385\ \text{kg N/kg mulch} + 0.0116\ \text{kg N/kg cake} \times 0.44\ \text{kg cake/kg mulch}}$$
$$= \frac{0.224\ \text{kg C}}{0.00895\ \text{kg N}} = 25\ \text{kg C/kg N in the mixture}$$

This is the desired ratio for the feedstock. Therefore, a C/N ratio of 25 is achieved by adding 0.44 kg sludge cake into 1 kg mulch.

4. Determine the moisture content ($p_{w,fs}$) in the feedstock of 1 kg mulch and 0.44 kg sludge cake.

$$p_{w,fs} = \frac{0.45\ \text{kg moisture/kg mulch} + 0.8\ \text{kg moisture/kg cake} \times 0.44\ \text{kg cake/kg mulch}}{1\ \text{kg mulch/kg mulch} + 0.44\ \text{kg cake/kg mulch}} \times 100\%$$
$$= \frac{0.802\ \text{kg moisture}}{1.44\ \text{kg mixture}} \times 100\% = 56\%$$

Therefore, 0.44 kg sludge cake is added to 1 kg mulch to obtain a moisture content of 56% in the feedstock to composting process.

5. Determine C/N ratio in the finished compost ($C/N_{product}$) after a 60% loss of carbon.

$$C/N_{product} = (1 - \text{Loss of C}) \times C/N_{fs} = (1 - 0.6) \times C/N_{fs} = 0.4 \times 25\ \text{kg C/kg N} = 10\ \text{kg C/kg N}$$

Therefore, the C/N ratio is 10 in the finished compost.

EXAMPLE 13.66: OXYGEN REQUIREMENT AND HEAT-GENERATION IN COMPOSTING

A composting plant feedstock is prepared at 1000 kg/d organics as dry weight from the wastewater sludge cake and garden mulch. The feedstock has optimum moisture content, seed and nutrients. Upon completion the finished compost is 200 kg/d organics as dry weight. Assume that the chemical formulas of organics in the initial feedstock and finished product are $C_{31}H_{50}O_{26}N$ and $C_{11}H_{14}O_4N$, respectively. Determine (a) oxygen requirement in kg/d, and (b) heat energy released in kJ/d.

Solution

1. State the generalized stoichiometric equation, and define the terms in Equation 13.23a.

$$C_aH_bO_cN_d + q\,O_2 \rightarrow n\,C_wH_xO_yN_z + s\,CO_2 + r\,H_2O + m\,NH_3 \qquad (13.23a)$$

where

$C_aH_bO_cN_d$ = composition of organics in the feedstock entering the process
$C_wH_xO_yN_z$ = composition of organics in the finished compost leaving the process
q = moles of oxygen consumed, $q = 0.5\,(ny + 2s + r - c)$
n = molar ratio of organics in the compost leaving the process to that in the feedstock entering the process,
s = moles of CO_2 produced, $s = (a - nw)$
r = moles of water produced, $r = 0.5[b - nx - 3\,(d - nz)]$
m = moles of NH_3 produced, $m = d - nz$

2. Calculate the molecular weights of feedstock and compost.

Element	Atomic Weight, g/atom	Molecular Weight, g/mole	
		Feedstock ($C_{31}H_{50}O_{26}N$)	Compost ($C_{11}H_{14}O_4N$)
C	$aw_C = 12$	a $aw_C = 31 \times 12 = 372$	w $aw_C = 11 \times 12 = 132$
H	$aw_H = 1$	b $aw_H = 50 \times 1 = 50$	x $aw_H = 14 \times 1 = 14$
O	$aw_O = 16$	c $aw_O = 26 \times 16 = 416$	y $aw_O = 4 \times 16 = 64$
N	$aw_N = 14$	d $aw_N = 1 \times 14 = 14$	z $aw_N = 1 \times 14 = 14$
Total	-	$mw_{feedstock} = 372 + 50 + 416 + 14 = 852$	$mw_{compost} = 132 + 14 + 64 + 14 = 224$

3. Calculate the moles of feedstock and compost, and molar ratio n.

Moles of feedstock, $m_{fs} = \dfrac{W_{fs}}{mw_{fs}} = \dfrac{1000\,\text{kg/d} \times 10^3\,\text{g/kg}}{852\,\text{g/mole}} = 1174\ \text{moles/d}$

Moles of finished compost, $m_{compost} = \dfrac{W_{compost}}{mw_{compost}} = \dfrac{200\,\text{kg/d} \times 10^3\,\text{g/kg}}{224\,\text{g/mole}} = 893\ \text{moles/d}$

Molar ratio of compost to feedstock,

$$n = \frac{m_{compost}}{m_{fs}} = \frac{893\ \text{moles/d}}{1174\ \text{moles/d}} = 0.761\ \text{mole of compost/mole of feedstock}$$

4. Calculate the moles of O_2 required (q), and CO_2, H_2O, and NH_3 produced per mole of feedstock.

Moles of CO_2 produced, $s = \text{a} - n\text{w} = 31 - 0.761 \times 11 = 22.6$ mole CO_2/mole of feedstock

Moles of H_2O produced, $r = 0.5[\text{b} - n\text{x} - 3(\text{d} - n\text{z})]$

$= 0.5 \times [50 - 0.761 \times 14 - 3 \times (1 - 0.761 \times 1)]$

$= 19.3$ mole H_2O/mole of feedstock.

Moles of O_2 required, $q = 0.5\,(n\,\text{y} + 2\,s + r - \text{c}) = 0.5 \times (0.761 \times 4 + 2 \times 22.6 + 19.3 - 26)$

$= 20.8$ mole O_2/mole of feedstock

Moles of NH_3 produced, $m = \text{d} - n\,\text{z} = 1 - 0.761 \times 1 = 0.239$ mole NH_3/mole of feedstock

The composting process is expressed by Equation 13.23b.

$$C_{31}H_{50}O_{26}N + 20.8\ O_2 \rightarrow 0.761\ C_{11}H_{14}O_4N + 22.6\ CO_2 + 19.3\ H_2O + 0.239\ NH_3 \qquad (13.23b)$$

5. Calculate the amount of O_2 required (W_{O2}) from Equation 13.23b.

$W_{O2} = q\,m_{fs}\,mw_{O2} = 20.8$ mole O_2/mole feedstock $\times$ 1174 moles/d $\times$ 32 g/mole $\times 10^{-3}$ kg/g

$= 781$ kg O_2/d

6. Identify the general equation for calculating the amount of heat energy released during composting.

Based on the chemical composition the total chemical energy content in an organic compound is expressed by Equations 13.23c through 13.23f.

$$\Delta h = h_{feedstock} - h_{compost} \qquad (13.23c)$$

$$h = E\,W \qquad (13.23d)$$

$$E = 127\,R + 400 \qquad (13.23e)$$

$$R = \frac{100 \times [(2.66 \times \%C) + (7.94 \times \%H) - \%O]}{398.9} \qquad (13.23f)$$

where

Δh = heat released during composting, calorie

$h_{feedstock}$ = total energy in feedstock, calorie

$h_{compost}$ = total energy in compost, calorie
h = total energy in organic matter, calorie
E = specific chemical energy content in organic matter, calories/g dry weight ash free
W = total weight of organic matter, g dry weight ash free
R = degree of reduction expressed by Equation 13.23f, dimensionless
%C = carbon content, % based on dry weight ash free
%H = hydrogen content, % based on dry weight ash free
%O = oxygen content, % based on dry weight ash free

Note: The conversion factor from calorie to kJ is 4.18×10^{-3} kJ/calorie.

7. Calculate the element contents in the feedstock and compost.

Element	Atomic Weight, g/atom	Ratio of Element to Molecular Weight, %	
		Feedstock ($C_{31}H_{50}O_{26}N$)	Compost ($C_{11}H_{14}O_4N$)
C	$aw_C = 12$	$\%C = \frac{a\ am_C \times 100}{mw_{feedstock}} = \frac{31 \times 12 \times 100}{852} = 43.66$	$\%C = \frac{w\ am_C \times 100}{mw_{compost}} = \frac{11 \times 12 \times 100}{224} = 58.93$
H	$aw_H = 1$	$\%H = \frac{b\ am_H \times 100}{mw_{feedstock}} = \frac{50 \times 1 \times 100}{852} = 5.87$	$\%H = \frac{x\ am_H \times 100}{mw_{feedstock}} = \frac{14 \times 1 \times 100}{224} = 6.25$
O	$aw_O = 16$	$\%O = \frac{c\ am_O \times 100}{mw_{feedstock}} = \frac{26 \times 16 \times 100}{852} = 48.83$	$\%O = \frac{y\ am_O \times 100}{mw_{feedstock}} = \frac{4 \times 16 \times 100}{224} = 28.57$
Overall	-	$43.66 + 5.87 + 48.83 = 98.36$	$58.93 + 6.25 + 28.57 = 93.75$

8. Calculate the total energy in feedstock on dry weight ash free basis.
   Degree of reduction from Equation 13.23f,

$$R_{feedstock} = \frac{100 \times [(2.66 \times 43.66) + (7.94 \times 5.87) - 48.83]}{398.9} = 28.6$$

Specific chemical energy content from Equation 13.23e, $E_{feedstock} = 127 \times 28.6 + 400$
$= 4032$ calorie/g feedstock

Total energy from Equation 13.23d, $h_{feedstock} = 4032 \text{ calorie/g} \times 1000 \text{ kg/d} \times 10^3 \text{ g/kg}$
$= 4.032 \times 10^9$ calorie/d

or $h_{feedstock} = 4.032 \times 10^9 \text{ calorie/d} \times 4.18 \times 10^{-3} \text{ kJ/calorie} = 1.69 \times 10^7$ kJ/d

9. Calculate the total energy in compost on dry weight ash free basis.

Degree of reduction from Equation 13.23f, $R_{compost} = \frac{100 \times [(2.66 \times 58.93) + (7.94 \times 6.25) - 28.57]}{398.9} = 44.6$

Specific chemical energy content from Equation 13.23e, $E_{compost} = 127 \times 44.6 + 400$
$= 6064$ calorie/g feedstock

Total energy from Equation 13.23d, $h_{compost} = 6064\ \text{calorie/g} \times 200\ \text{kg/d} \times 10^3\ \text{g/kg}$
$= 1.213 \times 10^9\ \text{calorie/d}$

or $h_{compost} = 1.213 \times 10^9\ \text{calorie/d} \times 4.18 \times 10^{-3}\ \text{kJ/calorie} = 5.07 \times 10^6\ \text{kJ/d}$

10. Calculate the total heat energy released on dry weight ash free basis from Equation 13.23c.

$$\Delta h = h_{feedstock} - h_{compost} = (1.69 \times 10^7 - 5.07 \times 10^6)\ \text{kJ/d} = 1.18 \times 10^7\ \text{kJ} \quad \text{or} \quad 1.18 \times 10^4\ \text{MJ/d}$$

Comments:

a. Released heat energy heats up the compost, and may also result in evaporating a large part of the moisture in the composting process.
b. Released heat causes a temperature rise of composting material, and the temperature is sustained for several days.

11. Calculate the specific heat released.
The heat released by organics converted during composting is calculated from Equation 13.23g.

$$h' = \frac{h_{feedstock} - h_{compost}}{W_{feedstock} - W_{compost}} \quad \text{or} \quad h' = \frac{\Delta h}{\Delta W} \qquad (13.23g)$$

where
h' = heat released per unit weight of organics destroyed, MJ/kg organics destroyed
$W_{feedstock}$ = total weight of organics in feedstock, kg dry weight ash free
$W_{compost}$ = total weight of organics in compost, kg dry weight ash free
ΔW = total organics converted during composting, kg dry weight ash free

$$h' = \frac{\Delta h}{\Delta W} = \frac{h_{feedstock} - h_{compost}}{W_{feedstock} - W_{compost}} = \frac{1.18 \times 10^4\ \text{kJ/d}}{(1000 - 200)\ \text{kg/d}} \approx 15\ \text{MJ/kg organics converted}$$

Note: The heat released from composting may vary with the characteristics of organic materials.

EXAMPLE 13.67: AIR REQUIREMENT FOR IN-VESSEL COMPOSTING

A continuous plug flow in-vessel composting system is designed to compost 2500 kg/d feedstock as dry solids. The in-vessel composting equipment receives forced aeration at variable rates along the length; and moisture and nutrients are injected at several locations. The total in-vessel composting period is 6 days for each daily load. The biodegradable volatile solids (VS_{bd}) portion of the feedstock is represented by a chemical equation $C_{60}H_{95}O_{38}N$. The initial moisture content is 40%, the organic fraction (VS) is 90% of total feedstock, and VS_{bd} is 60% of VS. The VS_{bd} conversion efficiency is 95%. Determine the air requirement to compost the daily feed over the composting period. The oxygen demand for each day of composting period over 6 days are 18%, 33%, 21%, 15%, 10%, and 3%, respectively. Determine the air supply over the 6-d composting period. The safety factor for aeration equipment is 1.5.

Solution

1. Estimate the dry mass of VS_{bd} and expected VS_{bd} conversion.
Daily amount of VS_{bd}, $W_{VSbd} = (1 - \text{Initial moisture content}) \times \text{Dry weight of feedstock}$
$\times\ \text{VS fraction} \times VS_{bd}\ \text{fraction}$
$= (1 - 0.4) \times 2500\ \text{kg/d} \times 0.9 \times 0.6 = 810\ \text{kg/d}$
Daily amount of VS_{bd} conversion, $W_{VSbd,converted} = VS_{bd}\ \text{fraction} \times \text{Daily amount}\ VS_{bd}$
$= 0.95 \times 810\ \text{kg/d} = 770\ \text{kg/d}$

The amount of VS_{bd} fed into the plug-flow in-vessel reactor (IVR) for the 6-d composting period is $W_{VSbd,IVR} = 770$ kg.

2. Identify the stoichiometric equation for aerobic decomposition of VS_{bd}.
   The aerobic decomposition of VS_{bd} is given by Equation 13.23h.

$$C_aH_bO_cN_d + \left(\frac{4a + b - 2c - 3d}{4}\right)O_2 \rightarrow a\ CO_2 + \left(\frac{b - 3d}{2}\right)H_2O + d\ NH_3 \quad (13.23h)$$

3. Calculate the molecular weight of VS_{bd} and moles of oxygen consumed for biochemical oxidation of VS_{bd}.
   It is given that a = 60, b = 95, c = 38, and d = 1.
   Molecular weight of $C_{60}H_{95}O_{38}N$, $mw_{org} = 60 \times 12 + 95 \times 1 + 38 \times 16 + 1 \times 14 = 1437$ g/mole
   Moles of O_2 consumed per mole of VS_{bd} oxidized,

$$\text{Moles of } O_2 \text{ consumed per mole of } VS_{bd} \text{ oxidized, } \frac{4 \times 60 + 95 - 2 \times 38 - 3 \times 1}{4} = 64 \text{ mole } O_2/\text{mole } VS_{bd}$$

4. Calculate the total amount of O_2 requirement during 6-d composting period.

$$O_2 \text{ required per kg of } VS_{bd} \text{ oxidized, } W_{O2/VSbd} = \frac{64 \text{ mole } O_2/\text{mole } VS_{bd} \times 32 \text{ g/mole of } O_2}{1437 \text{ g/mole of } VS_{bd}}$$

$$= 1.43 \text{ g } O_2/\text{g } VS_{bd} \text{ oxidized or } 1.43 \text{ kg } O_2/\text{kg } VS_{bd} \text{ oxidized}$$

$$\text{Total } O_2 \text{ required in the IVR over the 6-d period, } W_{O2,IVR} = W_{O2} W_{VSbd,IVR}$$
$$= 1.43 \text{ kg } O_2/\text{kg } VS_{bd} \times 770 \text{ kg } VS_{bd}$$
$$= 1100 \text{ kg } O_2$$

5. Calculate the air supply requirement.
   Assume that air contains 23.2% oxygen by weight and the density of air is 1.23 kg/m^3 at 15°C.

$$\text{Theoretical air requirement over the 6-d period, } V_{air,IVR} = \frac{1100 \text{ kg } O_2}{0.232 \text{ kg } O^2/\text{kg air} \times 1.23 \text{ kg/m}^3} = 3855 \text{ m}^3$$

   Total air supply with a safety factor of 1.5 over the 6-d period, $V_{air,supply} = 1.5 \times 3855 \text{ m}^3 = 5800 \text{ m}^3$
   Provide a total air volume of 6000 m^3 to the in-vessel composting equipment.

6. Describe the air supply system.
   The in-vessel composting equipment provides a plug-flow regime by moving each daily load through 6 equal volume segments along its length. The average retention time in each segment is a day for each daily load during the 6-d composting period. The air is supplied at a variable rate as follows.

Parameter	Segment No.						Total
	1	2	3	4	5	6	
Retention time, d	1	1	1	1	1	1	6
Oxygen demand distribution, %	18	33	21	15	10	3	100
Daily air supply, m^3/d	1080[a]	1980	1260	900	600	180	6000
Air flow, L/min	750[b]	1375	875	625	417	125	4167

[a] 0.18×6000 m^3/d = 1080 m^3/d.
[b] 1080 m^3/d × (1d/1440 min) × 10^3 L/m^3 = 750 L/min.

13.11.2 Heat Drying

Heat drying is a physical process where the sludge is heated by a drying equipment (or *dryer*) to evaporate moisture from the biosolids. It is also a proven technology for significant pathogen reduction to meet requirements for Class A EQ biosolids that is readily marketable for beneficial reuses. Additionally, the biosolids after drying has high heating value, and reduced storage and hauling costs. The major disadvantages include high capital and operating costs, high energy consumption, increased fire and explosion hazards, and odor and dust related issues.

Heat drying may be categorized as: (a) *direct*, (b) *indirect*, (c) *combined*, (d) *infrared*, and (e) *MW*. The direct and indirect dryers are the most commonly used devices for biosolids drying. In direct (convective) dryers the biosolids are directly heated by hot air in contact with the biosolids. Common direct dryers are: (a) *rotary drum*, (b) *fluidized-bed*, and (c) *belt*. In indirect (conductive) drying, the biosolids are heated by contact with hot surfaces that separate the heat transfer medium (steam, thermal oil, or other hot fluid) from the biosolids. The indirect dryers are either horizontal or vertical. Screw, paddles, flights, trays, or disks are used to move the biosolids through the dryers.

The typical operating temperature ranges are: (a) 90–500°C (195–930°F) for direct air dryers, (b) 150–175°C (300–350°F) for indirect steam dryers, and (c) ~400°C (750°F) for indirect oil dryers. The pellet-type products (<8 mm in dia.) from heat drying processes have solids contents of 90–98%. A simplified process diagram for heat drying system is shown in Figure 13.2d. Commercial available sludge drying systems may include: (a) belt-, drum-, or fluidized-bed drying systems by Andritz Separation, Inc., (b) K-S paddle dryer by Komline-sanderson, (c) Bio-SCRU® screw dryer system by Therma-Flite, Inc., and (d) BioCon® rotary drum dryer by Veolia Water Technologies.[7,24–27,191]

13.11.3 Thermal Oxidations

Thermal oxidation processes used for energy recovery from municipal sludge may include (a) incineration, (b) gasification, (c) wet air oxidation, and (d) pyrolysis. Brief description and application of each process are presented below.

Incineration: Incineration (or advanced thermal oxidation, ATO) of sludge cake involves drying followed by complete oxidation of organic matter into carbon dioxide, water, and ash. The minimum temperature is 700°C (~1300°F). The excess air is 20–100% over the stoichiometric air requirement. The heating value of digested and raw sludge is in the range from 12,000 to 25,000 kJ/kg (5000–11,000 Btu/lb). Normally, the feed sludge must contain at least 15–25% solids for self-sustaining incineration. Natural gas or fuel oil as auxiliary fuel is needed for ignition and to maintain proper burning temperature. The by-products of combustion are CO_2, H_2O, and ash. Air pollution control and carbon footprint reduction are important considerations in design and operation of sludge incineration. Advanced heat recovery system (HRS) for incineration process may be required for improving energy efficiency.

Two types of incinerators are in common use: (a) multiple-hearth furnace and (b) fluidized-bed. The multiple-hearth incinerator consists of several hearths arranged in a vertical stack. The cake enters the top hearths where water is vaporized and solids are dried at about 450–650°C (850–1200°F). The solids are ignited at 650–820°C (1200–1500°F). In the lower hearths, the slow-burning material is burned, and ash is cooled at a temperature of about 315°C (600°F). The excess air required is 50–100% of the theoretical amount. The efficiency of multiple-hearth incinerators is about 55%. The fluidized-bed incinerator utilizes a hot sand reservoir in which hot air is blown from below to expand and fluidize the bed. A temperature of 760–820°C (1400–1500°F) is maintained. The excess air is 20–25%. Air pollution control equipment is needed to remove the ash.[7,33,192]

Gasification: Gasification process produces a fuel gas (or called syngas) from organic materials. It operates in a temperature range from 590°C to 980°C (1100–1800°F) with restricted air supply that is below the stoichiometric requirement. The syngas consists of N_2, CO_2, CO, H_2, and CH_4. The heating value of syngas is about 4500–5500 kJ/m^3 (120–150 Btu/ft^3) which is about 10–15% of that of the natural gas. Gasifiers are commercially available in either fixed-bed (upflow or downflow), fluidized-bed, or entrained-flow reactor. The feedstock may be prepared from dried biosolids (45–95% dry solids) with or without blending with municipal solids waste. Commercial biomass gasification systems include *Kopf* by Sulzle Kopf Anlagenbau GmbH and *Waste-To-Energy* by PHG Energy.[7,194,195]

Wet Air Oxidation: Wet air oxidation (WAO) also called Zimmerman process is used to oxidize organic matter in liquid phase.[2,196] Dewatering of sludge is not required and the organic matter is actually broken down to improve the dewaterability. The sludge is pumped into a reactor where high temperature of 150–320°C (300–610°F) and high pressure of 5–225 bar (70–3300 psig) are maintained. The product residues may be further dewatered and dried to over 90% dry solids. This process may be considered as an alternative to incineration. Zimpro® system is a commercially available from Siemens.[197,198]

Pyrolysis: Pyrolysis or destructive distillation is heating of organic matter in oxygen-free or oxygen-starved atmosphere. At a temperatures of 200–590°C (390–1100°F) the organic matter undergoes cracking to produce combustible gases, oil and tar, and charcoal. These by-products are used as an energy source. Currently, pyrolysis of MSW is considered an effective approach for waste-to-energy (WTE) but the environmental impacts must be thoroughly evaluated. Copyrolysis of biosolids with MSW is a viable alternative. Either gasification or pyrolysis can be an integral component of a comprehensive combined heat and power (CHP) system.[7,199,200]

EXAMPLE 13.68: DEWATERING REQUIREMENT FOR INCINERATION

Raw combined sludge is thickened and dewatered prior to incineration. Calculate the moisture content of the dewatered sludge cake to achieve self-sustaining burning (no auxiliary fuel) at the winter temperature of 5°C. Assume that the fuel value of the sludge $H_{f,s} = 20{,}000$ kJ/kg dry solids, and the furnace efficiency $E_{furnace} = 60\%$ (fuel value available to evaporate water).

Solution

1. Compute the heat required to evaporate 1 kg of water.

   Rearrange Equation 13.6a and calculate the heat required to raise water temperature from 5°C to 100°C.

   $$H_{r,w} = C_{sh,w}\,(T_d - T_w) = 4200\,\text{J/kg·°C} \times (100 - 5)\text{°C} = 399{,}000\,\text{J/kg} = 399\,\text{kJ/kg} \approx 400\,\text{kJ/kg}$$

   Heat of vaporization of water (to convert 1 kg water into steam at 100°C), $H_{v,w} = 2260$ kJ/kg.
   Total heat required to evaporate 1 kg of water.

   $$H_{t,w} = H_{r,w} + H_{v,w} = (400 + 2260)\,\text{kJ/kg} = 2660\,\text{kJ/kg}$$

2. Compute the heat required to evaporate water from 1 kg of sludge cake.

   Assume that the solids content of sludge cake is p_s (fraction).
   Heat required to evaporate water from 1 kg of sludge cake

   $$H_{r,s} = (1 - p_s)H_{t,w} = (1 - p_s) \times 2660\,\text{kJ/kg}$$

3. Compute the heat available to evaporate water from 1 kg of sludge cake ($H_{a,s}$) at the furnace efficiency of 60%.

   $$H_{a,s} = H_{f,s}p_sE_{furnace} = 20{,}000\,\text{kJ/kg} \times p_s \times 0.6 = 12{,}000\,\text{kJ/kg} \times p_s$$

4. Compute the moisture content required in the sludge cake.

   Based on heat balance, $H_{r,s} = H_{a,s}$.

   $$(1 - p_s) \times 2660\,\text{kJ/kg} = 12{,}000\,\text{kJ/kg} \times p_s$$

Solve for p_s.

$$p_s = \frac{2660\ \text{kJ/kg}}{(12{,}000 + 2660)\ \text{kJ/kg}} = 0.18$$

Therefore, the sludge cake must be dewatered to at least 18% solids, to achieve incineration without auxiliary fuel.

EXAMPLE 13.69: HEATING VALUE OF ACTIVATED SLUDGE

The commonly used empirical formula to express the organic fraction of cells is $C_5H_7O_2N$. Estimate the heating value of the activated sludge. Assume VSS/TSS ratio = 0.8.

Solution

1. Identify the empirical equation for estimating heating value of sludge.

   As a general practice, the heating value of organic matter may be estimated from empirical expressions based on the fractions of elements. One expression that has been developed for biosolids is given by Equation 13.24.[7,32]

$$\text{HHV}_s = 33.83\, f_C + 144.70\, f_H - 18.09\, f_O + 9.42\, f_S \tag{13.24}$$

where

HHV_s = heating value of sludge, MJ/kg dry solids

f_C, f_H, f_O, and f_S = fractions of carbon, hydrogen, oxygen, and sulfur of the sludge, g/g based on dry weight

2. Calculate the fractions of elements in the biomass.

   Molecular weight of organic portion of sludge.

$$mw_{VSS} = 5 \times aw_C + 7 \times aw_H + 2 \times aw_O + 1 \times aw_N = 5 \times 12 + 7 \times 1 + 2 \times 16 + 1 \times 14 = 113\ \text{g/mole}$$

   Molecular weight of dry solids in the activated sludge based on VSS/TSS ratio of 0.8.

$$mw_s = \frac{mw_{VSS}}{\text{VSS/TSS ratio}} = \frac{113\ \text{g/mole}}{0.8} = 141\ \text{g/mole dry solids}$$

$$\text{Fraction of carbon, } f_C = \frac{5 \times am_C}{mw_s} = \frac{5 \times 12}{141} = 0.43$$

$$\text{Fraction of hydrogen, } f_H = \frac{7 \times am_H}{mw_s} = \frac{7 \times 1}{141} = 0.050$$

$$\text{Fraction of oxygen, } f_O = \frac{2 \times am_O}{mw_s} = \frac{2 \times 16}{141} = 0.23$$

   Ignore sulfur content, $f_S \approx 0$.

3. Estimate the heating value of the activated sludge from Equation 13.24.

$$\text{HV}_{VSS} = 33.83 \times 0.43 + 144.70 \times 0.050 - 18.09 \times 0.23 + 9.42 \times 0 \approx 18\ \text{MJ/kg dry solids}$$

Note: The typical value of activated sludge ranges 20–23 MJ/kg. The heating value of 18 MJ/kg estimated from the general expression of $C_5H_7O_2N$ is slightly below the typical range.

EXAMPLE 13.70: THEORETICAL AND EXCESS AIR

Determine the theoretical air requirement of 50 kg of organic waste (VSS) expressed by the chemical formula $C_5H_7O_2N$. Also, calculate 100% excess air.

Solution

1. Write the balanced oxidation equation and calculate theoretical amount of oxygen required.
   The stoichiometric oxidation equation is given by Equation 10.4.

$$\underset{113}{C_5H_7O_2N} + \underset{5\times 32}{5\,O_2} \rightarrow 5\,CO_2 + NH_3 + 2\,H_2O$$

Theoretical amount of oxygen required to combust 50 kg of organic waste.

$$W_{O2} = \frac{5 \times mw_{O2}}{mw_{VSS}} \times W_{VSS} = \frac{5 \times 32\,\text{g/mole}}{113\,\text{g/mole}} \times 50\,\text{kg} = 71\,\text{kg}\ O_2$$

2. Compute the theoretical amount of air required for burning 50 kg of organic waste.
   At 15°C air contains 23.2% oxygen by weight and the density of air is 1.23 kg/m^3.

$$\text{Theoretical air requirement, } V_{air} = \frac{71\,\text{kg}\ O_2}{0.232\,\text{kg}\ O^2/\text{kg air} \times 1.23\,\text{kg/m}^3} = 249\,\text{m}^3 \approx 250\,\text{m}^3$$

3. Calculate the air requirement with 100% excess amount.

$$V_{air,100\%;\,ex} = \left(1 + \frac{Ex}{100\%}\right) V_{air} = \left(1 + \frac{100\%}{100\%}\right) \times 250\,\text{m}^3 = 500\,\text{m}^3$$

EXAMPLE 13.71: BURNING TEMPERATURES AT DIFFERENT EXCESS AIR

An incinerator is burning undigested sludge mixed with municipal solid waste. The moisture content of the feedstock is 30%. Ambient air temperature is 85°F. The incineration rate is 2400 lb feedstock per day. Calculate the burning temperatures at theoretical air, and at 100%, 200%, and 300% excess air. The product of combustion (including gaseous and ash) at these air supplies are respectively 6.2, 11.3, 16.5, and 21.7 lbs per lb of dry feedstock burned. Specific heat of the combustion product is 0.26 Btu/lb·°F. Net heat available during combustion is 400,000 Btu/h. Assume that the initial moisture content in the feedstock is excluded in the combustion product.

Solution

1. Determine the moisture content and dry solids in feedstock burning at hourly rate.

$$\text{Hourly burning rate of feedstock, } W_{s,wet} = 2400\ \text{lb/d} \times \frac{1\text{d}}{24\text{h}} = 100\ \text{lb/h}$$

   Hourly moisture content in feedstock, $W_w = p_w W_{s,wet} = 0.3 \times 100\,\text{lb/h} = 30\,\text{lb/h}$
   Hourly dry solids in feedstock, $W_s = (1 - p_w) W_{s,wet} = (1 - 0.3) \times 100\,\text{lb/h} = 70\,\text{lb/h}$

2. Determine the total product of combustion at different excess air.
   Product of combustion at theoretical air, $W_p = 6.2\,\text{lb/lb} \times W_s = 6.2\,\text{lb/lb} \times 70\,\text{lb/h} = 434\,\text{lb/h}$
   Total product of combustion to include the initial moisture, $W_t = W_w + W_p = (30 + 434)\,\text{lb/h} = 464\,\text{lb/h}$ Similarly, calculate the total products of combustion at different amounts of excess air, and summarize the results below.

Excess Air	Moisture (W_w), lb/h	Product of Combustion (W_p), lb/h	Total Product of Combustion (W_t), lb/h
Theoretical	30	434	464
100%	30	791	821
200%	30	1155	1185
300%	30	1519	1549

3. Determine the burning temperatures at different excess air.

The temperature rise for the net heat available during combustion is calculated from Equation 13.25.

$$\Delta T = \frac{H_{ava}}{C_{sh,t} \times W_t} \quad (13.25)$$

where

ΔT = temperature rise during combustion, °C (°F)

H_{avi} = net heat available during combustion, kJ/h (Btu/h)

$C_{sh,t}$ = specific heat of combustion product, kJ/kg·°C (Btu/lb·°F)

W_t = total combustion product, kg/h

Calculate the temperature rise at theoretical air from Equation 13.25.

$$\Delta T = \frac{H_{ava}}{C_{sh,t} \times W_t} = \frac{400{,}000\ \text{Btu/h}}{0.26\ \text{Btu/lb·°F} \times 464\ \text{lb/h}} = 3316\text{°F}$$

Burning temperature at theoretical air, $T_b = T_{amb} + \Delta T = (85 + 3316)\text{°F} = 3401\text{°F}$ Similarly, calculate the burning temperatures at different amounts of excess air and summarize the results below.

Excess Air	Ambient Temperature (T_{amb}), °F	Temperature Rise (ΔT), °F	Burning Temperature (T_b), °F
Theoretical	85	3316	3401
100%	85	1874	1959
200%	85	1298	1383
300%	85	993	1078

EXAMPLE 13.72: PRODUCTS OF PYROLYSIS: COMBUSTIBLE GASES, LIQUIDS AND CHAR

500 kg of organic material of chemical formula $C_6H_{10}O_5$ is pyrolized in a reactor. Calculate the quantity of combustible gases, liquid and char produced.

Solution

1. Write the balanced equation of pyrolysis process in Equation 13.26.

$$\underset{3\times162}{3\,C_6H_{10}O_5} \rightarrow \underset{96}{C_6H_8O} + \underset{16}{CH_4} + \underset{2}{H_2} + \underset{2\times28}{2CO} + \underset{2\times44}{2\,CO_2} + \underset{8\times18}{8\,H_2O} + \underset{7\times12}{7\,C} \quad (13.26)$$

2. Determine the weight of combustible liquid produced as C_6H_8O from Equation 13.26.

$$W_{liquid} = \frac{mw_{C6H8O}}{3mw_s} \times W_s = \frac{96\ \text{g/mole}}{3 \times 162\ \text{g/mole}} \times 500\ \text{kg} = 99\ \text{kg as } C_6H_8O\ \text{(liquid)}$$

3. Determine the weight of combustible gases (CH_4, H_2 and CO) produced Equation 13.26.

$$W_{gas} = \frac{(mw_{CH4} + mw_{H2} + 2\,mw_{CO})}{3mw_s} \times W_s = \frac{(16 + 2 + 2 \times 28)\,\text{g/mole}}{3 \times 162\,\text{g/mole}} \times 500\,\text{kg}$$

$$= 76\,\text{kg of combustable gases } (CH_4, H_2, \text{ and CO})$$

4. Determine the weight of char produced as C from Equation 13.26.

$$W_{char} = \frac{7\,aw_C}{3\,mw_s} \times W_s = \frac{7 \times 12\,\text{g/mole}}{3 \times 162\,\text{g/mole}} \times 500\,\text{kg} = 86\,\text{kg as C (char)}$$

5. Determine the total energy recovered as combustible products.

$$W_{energy} = W_{liquid} + W_{gas} + W_{char} = (99 + 76 + 86)\,\text{kg} = 261\,\text{kg}$$

$$\text{Recovery efficiency, } E_{recovery} = \frac{W_{energy}}{W_s} \times 100\% = \frac{261\,\text{d}}{500\,\text{kg}} \times 100\% = 52\%$$

13.11.4 Recalcination

Coagulation and precipitation with lime requires a dosage of 100–500 mg/L as CaO in one or two stages. The amount of sludge produced is also high. Recalcining may be used to recover and reuse the lime, and reduce the quantity of sludge for disposal. The process train includes thickening, dewatering and recalcining (see Figure 13.1d). The recalcining involves heating the sludge in a multiple-hearth furnace to evaporate water and burn the organic matter. Recalcining occur at around 1000°C. Large pieces of recalcined lime (CaO) are collected from the bottom outlet of the furnace. These CaO granules are ground and reused. The impurities are nitrogen and phosphorus. A portion of lime sludge is wasted to reduce buildup of impurities. The cost of recalcining must be weighed against the cost of new lime, savings in cost from reclaimed lime, and reduction in sludge quantities for disposal.[7,38,71]

13.11.5 Land Application of Biosolids

Land application of biosolids involves spreading of solids over or just below the soil surface. It is estimated that in the U.S., the quantity of biosolids marketed or distributed free for land application may reach 40% of total sludge production.[4]

Factors Affecting Land Application of Biosolids: There are many factors that must be considered for land application of biosolids. Some of these factors are: (1) regulatory requirements for sludge quality, (2) site evaluation and selection, and (3) monitoring requirements. These factors are briefly discussed below.

Regulatory Requirements for Sludge Quality: The U.S. Environmental Protection Agency under 40 CFR Part 503 Sludge Regulations imposes significant monitoring, reporting, and recordkeeping requirements on land application of biosolids. These requirements have been discussed in Section 13.4.4. The biosolids are considered clean, if the concentration limits are not exceeded for many pollutants. These limits are summarized in Table 13.28.[7,17,18]

Site Evaluation and Selection: Proper site selection is essential for different land application options such as agriculture, forest, land reclamation, and dedicated disposal use. Proper site selection will reduce future environmental problems, monitoring requirements, overall program costs, and public reaction. A two-phase approach for site evaluation and selection is needed. Phase I is the preliminary screening, in which the physical characteristics, land availability, and rough estimate of land area required for each option are determined. In Phase II, the final site selection is made using careful evaluation of topography, soil type

TABLE 13.28 Concentration Limits of Pollutants for Land Application[a]

Pollutant	Ceiling Concentration[b], mg/kg (lb/ton)	Monthly Average Concentration[c], mg/kg (lb/ton)	Annual Pollutant Loading Rate (APLR)[d], kg/ha·year (lb/ac·year)	Cumulative Pollutant Loading Rate (CPLR)[e], kg/ha (lb/ac)
Arsenic (As)	75 (0.15)	41 (0.082)	2.0 (1.8)	41 (37)
Cadmium (Cd)	85 (0.17)	39 (0.078)	1.9 (1.7)	39 (35)
Chromium (Cr)	–[f]	–[f]	–[f]	–[f]
Copper (Cu)	4300 (8.60)	1500 (3.0)	75 (67)	1500 (1338)
Lead (Pb)	840 (1.68)	300 (0.60)	15 (13)	300 (268)
Mercury (Hg)	57 (0.11)	17 (0.034)	0.85 (0.76)	17 (15)
Molybdenum (Mo)	75 (0.15)	–[f]	–[f]	–[f]
Nickel (Ni)	420 (0.84)	420 (0.84)	21 (18)	420 (375)
Selenium (Se)	100 (0.20)	100 (0.20)[f]	5.0 (4.5)	100 (89.2)
Zinc (Zn)	7500 (15.0)	2800 (5.60)	140 (125)	2800 (2498)

[a] All concentrations are on dry weight bases.

[b] All biosolids sold or given away shall not be applied to the land if the concentration exceeds ceiling concentration for any pollutant.

[c] If bulk (Class A or B) or bagged (Class A only) biosolids are applied to a lawn or a home garden, the concentration of each pollutant shall not exceed the given monthly average concentration.

[d] If bagged (Class A only) biosolids are sold or given away for land application, either the concentration of pollutant should not exceed the monthly average concentration, or the annual pollutant loading rate (APLR) shall not exceed the given value.

[e] If bulk (Class A or B) biosolids are applied on agricultural land, forest, or a public contact site, either the monthly average concentration of each pollutant shall not exceed, or the cumulative pollutant loading rate (CPLR) for each pollutant shall not exceed the given value.

[f] Deleted or revised in the later Amendment to the 40 CFR Part 503 Sludge Regulations.

1 mg/kg = 0.002 lb/ton; 1 kg/ha = 0.892 lb/ac.

Source: Adapted in part from References 7, 17, and 18.

and permeability, site drainage, depth of ground water, subsurface geology, proximity to critical areas, and accessibility.[4]

Monitoring Requirements: Specific regulatory requirements for every biosolids application projects are imposed by the regulatory agencies. These requirements include monitoring of soils, crops, and surface water and groundwater. The monitoring requirements are less extensive if biosolids are applied at agronomic rates as a substitute for commercial fertilizer. If the pH of the soil is maintained above 6.5, groundwater monitoring may not be required. The following parameters may require monitoring. The frequency may depend upon the amounts applied, depth of groundwater, and climatological conditions.

a. Soils: pH, K, P, and some heavy metals
b. Groundwater: nitrate and fecal coliforms
c. Surface water: BOD, nitrogen, phosphorous, and fecal coliforms
d. Crops: cadmium

Potential Options for Land Application of Biosolids: The direct disposal of liquid sludge over cropland may eliminate the cost of dewatering, but the transportation cost may be high. Small cities utilize truck hauling, while the large cities pump liquid sludge. The liquid sludge is incorporated into the soil directly by the injection. The land application of biosolids in large scales include four potential options: (1) agricultural utilization to enhance production of food-chain and nonfood chain crops, (2) forest land utilization to enhance forest productivity, (3) land reclamation utilization to revegetate for the primary purpose of sludge disposal while crop production (if any) is of secondary importance and landfilling is excluded, and (4) dedicated land disposal (DLD) where disposal of biosolids is the primary purpose of land application. A combination of two or more options may also be used.

Agricultural Land Utilization: The biosolids application over agricultural land offers many benefits: (1) reduces or eliminates the need of commercial fertilizer; (2) may improve crop production; and

TABLE 13.29 Soil Limitations for Agricultural Land Application of Biosolids

Soil Feature	Degree of Soil Limitation[a]		
	Slight	Moderate	Severe
Slope[b], %	<6	6–12	>12
Depth to seasonal water table, m (ft)	>1.2 (4)	0.6–1.2 (2–4)	<0.6 (< 2)
Flooding and ponding	None	None	Occasional to frequent
Depth to bedrock, m (ft)	>1.2 (4)	0.6–1.2 (2–4)	<0.6 (<2)
Permeability of the most restricting layer above a 1-m (3-ft) depth[c], cm/h (in/h)	0.24–0.8 (0.1–0.3)	0.08–0.24 (0.03–0.1) or 0.8–2.4 (0.3–1.0)	<0.08 (<0.03) or >2.4 (>1.0)
Available water capacity, cm/h (in/h)	>2.4 (>1.0)	1.2–2.4 (0.5–1.0)	<1.2 (<0.5)

[a] The degree of soil limitation is determined based on the nitrogen fertilizer utilization rate.

[b] Slope is an important factor in determining the runoff that is likely to occur: 0–3% is ideal with no concern for runoff; 3–6% is acceptable with slight risk for erosion; 6–12% medium runoff expected, surface application of dewatered sludge is usually acceptable; and >12% may have rapid to very rapid runoff and runoff control measures may be necessary.

[c] Permeability rating (cm/h) is as follows: <0.15 very slow, 0.15–0.5 slow, 0.5–1.5 moderately slow, 1.5–5 moderate, 5–15.2 moderately rapid, 15.2–51 rapid, and >51 very rapid.

1 ft = 0.3048 m; 1 in = 2.54 cm.

Source: Adapted in part from References 4 and 73.

(3) with proper site selection and application rates, the potential harmful effects on phytotoxicity, ground and surface water contamination, and public health or nuisance many be avoided. The crop utilization option is designed for nitrogen and/or phosphorus need for the crop. The application is at agronomic rates on an annual basis to ensure acceptance by the farmers and to minimize environmental monitoring. The average biosolids application rate is the amount of biosolids applied annually between harvesting and planting. This rate may be 2–70 mt/ha·year (1–30 ton/ac·year)* on dry solids basis. Soil limitations for agricultural land application of bisolids with respect to nitrogen fertilizer rate are given in Table 13.29. The agricultural land utilization is generally controlled by regulatory guidelines or nutrients loading rate to meet the regulation requirements. Nitrogen is mostly the nutrient of concern. The pollutant loading, nitrogen and phosphorus limitations, and land requirements are expressed by several equations. These equations and design considerations are presented below.

1. Biosolids Application Rate Based on Metal Pollutant Loading: The amount of biosolids on dry solids basis that can be applied annually or over the useful life of the site is given by Equation 13.27. The limiting values of various contaminants are summarized in Table 13.28.

$$S_m = \frac{L_m}{C_m C_f} \tag{13.27}$$

where

S_m = maximum amount of biosolids that can be applied annually or over the useful life of the site, mt/ha·year [ton/ac·year] or mt/ha [ton/ac] on dry solids basis

L_m = maximum rate of pollutant that can be applied annually or over the useful life of the site as established by regulations, kg/ha·year [lb/ac·year] (kg/ha [lb/ac]). It should not exceed either the annual pollutant loading rate (APLR) for the annual amount applied, or the cumulative pollutant loading rate (CPLR) for the cumulative amount applied. Both limits are given in Table 13.28.

C_m = content of concerned pollutant in biosolids applied (kg/kg; lb/lb) of dry solids

C_f = conversion factor, 1000 kg/mt for SI units and 2000 lb/ton for U.S. customary units

2. Plant-available Nitrogen: Organic nitrogen in biosolids is slowly mineralized for several years, while nitrate and ammonia nitrogen becomes available within a year. When biosolids are applied at a

* Unit conversions: mt = metric ton, tonne or 1000 kg; 1 mt/ha·year = 0.446 ton/ha·year.

constant annual application rate, the overall plant-available nitrogen in the biosolids can be estimated from Equations 13.28a through 13.28d.

$$N_{a,n} = N_{a,NO3\text{-}N} + N_{a,NH4\text{-}N} + N_{a,Org\text{-}N,n} \quad (13.28a)$$

$$N_{a,NO3\text{-}N} = C_f(NO_3\text{-}N) \quad (13.28b)$$

$$N_{a,NH4\text{-}N} = C_f K_v(NH_4\text{-}N) \quad (13.28c)$$

$$N_{a,Org\text{-}N,n} = N_{a,Org\text{-}N,n-1} + C_f f_n(Org\text{-}N)_{n-1} \quad (13.28d)$$

where

$N_{a,n}$ = overall plant-available nitrogen in the *n*th year, (kg N/mt; lb N/ton) on dry solids basis. The value of *n* represents the year under consideration. It is the year since the biosolids application began at the site.

$N_{a,NO3\text{-}N}$ = plant-available nitrogen from nitrate nitrogen (NO_3-N) in biosolids that are placed during the *n*th year, (kg N/mt; lb N/ton) on dry solids basis

$N_{a,NH4\text{-}N}$ = plant-available nitrogen from ammonia nitrogen (NH_4-N) in biosolids that are placed during the *n*th year (kg N/mt; lb N/ton) on dry solids basis

$N_{a,Org\text{-}N,n}$ = overall organic nitrogen (Org-N) mineralization in the *n*th year (kg N/mt; lb N/ton) on dry solids basis

NO_3-N = NO_3-N in biosolids applied (kg N/kg; lb N/lb) of dry solids

NH_4-N = NH_4-N in biosolids applied (kg N/kg; lb N/lb) of dry solids

Org-N = Org-N in biosolids applied (kg N/kg; lb N/lb) of dry solids

K_v = volatilization factor for NH_4-N, dimensionless. It is 0.5 for surface or sprinkler application of liquid sludge, and 1.0 for dewatered sludge or incorporated liquid sludge.

$N_{a,Org\text{-}N,n-1}$ = overall organic nitrogen (Org-N) mineralization in the $(n-1)$th year (kg N/mt; lb N/ton) on dry solids basis

f_n = fraction factor for Org-N mineralization of biosolids that have been stayed on the site for *n* years, dimensionless.

$(Org\text{-}N)_{n-1}$ = Org-N content remaining in the biosolids after Org-N mineralization on the site for $(n-1)$ years, kg N/kg (lb N/lb) of dry solids. It is calculated from Equation 13.28e.

$$(Org\text{-}N)_{n-1} = (1-f_1)(1-f_2)\ldots(1-f_{n-1})(Org\text{-}N) \quad (13.28e)$$

The values of $N_{a,NO3\text{-}N}$ and $N_{a,NH4\text{-}N}$ calculated from Equations 13.28b and 13.28c are the nitrogen available from the biosolids that are applied in the concerned year. Additional nitrogen will also be available from mineralization of Org-N in all biosolids applied during the years since the biosolids application began. The overall Org-N mineralization of biosolids that is available in the *n*th year may be calculated by either one of the following two methods: (1) *stepwise method* from Equation 13.28d or (2) *alternate method* from Equation 13.28f. Both methods are illustrated in Example 13.73.

$$N_{a,Org\text{-}N,n} = N_{a,Org\text{-}N,n-1} + (\%Org\text{-}N)_0 K_n \quad (13.28f)$$

where

K_n = mineralization factor for the biosolids that have been stayed on the site for *n* years, (kg N/mt·%; lb N/ton·%) on dry solids basis. The values of K_n are given in Table 13.30.

$(\%Org\text{-}N)_0$ = initial Org-N content in biosolids, percent of dry solids

3. Biosolids Application Rate Based on Nutrient Limitations: The annual biosolids application rate may also be evaluated based on the nutrient (N, P, and K) requirements of the crop grown. Soil

TABLE 13.30 Estimated Mineralization of Org-N from Sludge Applied to Soils

Year Since Biosolids Application Began (n), Year	Type of Sludge							
	Unstabilized Combined Primary and Waste Activated Sludges		Aerobically Digested		Anaerobically Digested		Composted	
	f_n	K_n, kg/mt·% (lb/ton·%)	f_n	K_n, kg/mt·% (lb/ton·%)	f_n	K_n, kg/mt·% (lb/ton·%)	f_n	K_n, kg/mt·% (lb/ton·%)
1	0.4	4 (8)	0.3	3 (6)	0.2	2 (4)	0.10	1 (2)
2	0.2	1.2 (2.4)	0.15	1.05 (2.1)	0.1	0.8 (1.6)	0.05	0.45 (0.9)
3	0.1	0.48 (0.96)	0.08	0.45 (0.9)	0.05	0.36 (0.72)	0.03	0.25 (0.5)
4	0.05	0.22 (0.44)	0.04	0.21 (0.42)	0.03	0.21 (0.42)	0.03	0.25 (0.5)
5	0.03	0.12 (0.24)	0.03	0.16 (0.32)	0.03	0.2 (0.4)	0.03	0.24 (0.48)
6	0.03	0.12 (0.24)	0.03	0.15 (0.3)	0.03	0.19 (0.38)	0.03	0.23 (0.46)
7	0.03	0.12 (0.24)	0.03	0.15 (0.3)	0.03	0.19 (0.38)	0.03	0.23 (0.46)
8	0.03	0.11 (0.22)	0.03	0.15 (0.3)	0.03	0.18 (0.36)	0.03	0.22 (0.44)
9	0.03	0.11 (0.22)	0.03	0.15 (0.3)	0.03	0.18 (0.36)	0.03	0.21 (0.42)
10	0.03	0.11 (0.22)	0.03	0.15 (0.3)	0.03	0.17 (0.34)	0.03	0.21 (0.42)

1 kg/mt = 2 lb/ton.

Source: Adapted from References 4 and 73.

tests are conducted to determine the available nutrients and the requirements of the crops grown and their yield. The biosolids application rate to meet the nutrient requirements is calculated from Equation 13.29.

$$S_{\mathrm{Nu}} = \frac{R_{\mathrm{Nu}}}{C_{\mathrm{Nu}} C_{\mathrm{f}}} \quad \text{or} \quad S_{\mathrm{Nu}} = \frac{R_{\mathrm{Nu}}}{N_{\mathrm{Nu,a}}} \tag{13.29}$$

where

S_{Nu} = biosolids application rate to satisfy the specific nutrient requirement of plants, mt/ha·year (ton/ac·year) on dry solids basis

R_{Nu} = nutrient requirement of plants, kg/ha·year (lb/ac·year)

C_{Nu} = concentration of available nutrient in biosolids applied (kg/kg; lb/lb) of dry solids

$N_{\mathrm{Nu,a}}$ = overall plant-available nutrient in biosolids, kg/mt (lb/ton)

4. Land Requirement Based on the Biosolids Application Rate: The land requirement is calculated from the design biosolids application rate. The design loading rate is obtained from the critical concentration of nitrogen, phosphorus, cadmium, or any other constituent of concern. The land area is calculated from Equation 13.30.

$$A = \frac{Q_{\mathrm{s}}}{S_{\mathrm{d}}} \tag{13.30}$$

where

A = land area required at the design biosolids application rate, ha (ac)

Q_{s} = total sludge production (mt/year; ton/year) on dry solids basis

S_{d} = design biosolids application rate (mt/ha·year; ton/ac·year) on dry solids basis

5. Design Considerations. The basic design considerations for agricultural land application are given below.
   a. Determine the amounts of nutrients (N, P and K) present in the biosolids.

b. Determine the nutrients needs of the crops to achieve the desired yields. Common crop grown are wheat, oats, barley, soybean, cotton, legume, forage (alfalfa, clover, trefoil, etc.), and grass (orchard, timothy, brome, reed, canary, etc.) See Example 13.75 for sample calculations for two crops having similar nutrients requirements and cropping patterns.
c. Compare the nutrient availability with the needs. If comparison shows that one or more nutrients are suboptimal, the appropriate amount of commercial fertilizer is added.
d. Determine the long term accumulation of Cd in the soil to protect the productivity of crops. The biosolids application is terminated when the cumulative amount of Cd or sometimes also Pb, Zn, Cu, and Ni exceed the specific limit based on cation exchange capacity (CEC) of the soil. The recommended limits are given in Table 13.28.
e. Conduct the monitoring requirements as required. These may include soil pH, concentrations of P and K in soil, nitrate in groundwater, and Cd in crops.
f. Follow other operational requirements such as biosolids application methods, storage and scheduling based on climatic conditions and crop needs.

Forest Land Utilization: Land application over forest land constitutes a major utilization/disposal option of biosolids. Major benefits are: (1) availability of land throughout the country and some around the city and metropolitan areas, (2) recycling nutrients over nonfood crop, and (3) no cost to acquiring land. Three categories of forested lands may be available for biosolids application: (1) recently cleared land prior to planting, (2) newly established plantation (about 3 to 10 years old), and (3) established forests. The basic features of forest land utilization are given below.[4,6,73]

1. The normal biosolids application range is 10–225 mt/ha·year (4–100 ton/ac·year) depending upon soil, tree species and sludge quality.
2. The minimum distance to drinking water sources and excluded aquifers are 2 m and 1 m, respectively.
3. The sludge application may be one time over an interval of several years. Dramatic improvement in vegetative growth has been reported.
4. Potential harmful impacts on vegetation, surface and groundwater, and human health may be avoided by proper selection of biosolids, application rate and management practices.
5. Proper site selection, calculations for heavy metals accumulation, and nitrogen and phosphorus loading rates may be performed using procedures discussed under the previous section.

Land Reclamation Utilization: The strip mining of coal and other minerals has drastically disturbed the lands. These lands are unable to support vegetation because of nutrient deficiencies, lack of organic matter, physical and chemical properties, and topography. The biosolids application range from 7 to 450 mt/ha·year (3–200 ton/ac·year) depending upon the soil and site conditions. Application is normally one time. Additional applications are at 5- to 10-year intervals. The minimum distances to drinking water sources and excluded aquifers are 1 and 0.5 m, respectively. The benefits of using biosolids are: (1) the organic material and nutrients are beneficially recycled; (2) use of commercial fertilizer is reduced or not needed; and (3) it allows soil to support vegetation, and retards erosion. Detailed site investigations for topography, soil chemistry, climate, sludge nutrients, heavy metals, and biosolids loading rates are normally conducted on case-by-case basis.

Dedicated Land Disposal: The primary purpose of dedicated land disposal site is to provide long-term disposal of biosolids by spreading. The production of crops, improvement of soil characteristics, or use of nutrients are secondary to sludge disposal. The application rate ranges widely from 12 to 2250 mt/ha·year (from 5 to 1000 ton/ac·year) depending upon the soil characteristics and climate. The application rate may be 5–20 days on a routine basis. The danger for surface water and groundwater contamination exists. The minimum distance to drinking water sources and excluded aquifers are 1 and 0.5 m, respectively. Evaluation of environmental constraints, nutrients, heavy metals, climate, vegetation, and biosolids loading are all considered in the selection process and site development.[4,6,7]

EXAMPLE 13.73: MINERALIZATION RATE OF ORG-N IN BIOSOLIDS

Organic nitrogen is mineralized slowly. Anaerobically digested biosolids has Org-N = 3%. Calculate the overall Org-N mineralization in the second year after the biosolids application began on the site. Use both the stepwise and alternate methods.

Solution

1. Determine the Org-N mineralization by the stepwise method.
   From Table 13.30, $f_1 = 0.2$ for the first year, and $f_2 = 0.1$ for the second year.
   Org-N mineralization of biosolids in the first year from Equation 13.28d using $f_1 = 0.2$.

$$N_{a,\text{Org-N},1} = N_{a,\text{Org-N},0} + C_f f_1 (\text{Org-N})_0 = 0 + 1000\,\text{kg/mt} \times 0.2 \times 0.03\,\text{kg N/kg}$$
$$= 6\,\text{kg N/mt of dry solids}$$

   Org-N remaining after the mineralization during the first year from Equation 13.28e.

$$(\text{Org-N})_1 = (1 - f_1)(\text{Org-N})_0 = (1 - 0.2) \times 0.03\,\text{kg N/kg} = 0.024\,\text{kg N/kg}$$

   Org-N mineralization of biosolids in the second year from Equation 13.28d using $f_2 = 0.1$.

$$N_{a,\text{Org-N},2} = N_{a,\text{Org-N},1} + C_f f_2 (\text{Org-N})_1 = 6\,\text{kg N/mt} + 1000\,\text{kg/mt} \times 0.1 \times 0.024\,\text{kg N/kg}$$
$$= 8.4\,\text{kg N/mt of dry solids}$$

2. Determine the Org-N mineralization by the alternate method.
   From Table 13.30, $K_1 = 2$ kg/mt·% for the first year, and $K_2 = 0.8$ kg/mt·% for the second year.
   Calculate from Equation 13.28f the mineralization of Org-N in biosolids in the first year.

$$N_{a,\text{Org-N},1} = N_{a,\text{Org-N},0} + (\%\text{Org-N})_0 K_1 = 0 + 3\% \times 2\,\text{kg/mt}\cdot\% = 6\,\text{kg N/mt of dry solids}$$

   Calculate from Equation 13.28f the mineralization of Org-N in biosolids in the second year.

$$N_{a,\text{Org-N},2} = N_{a,\text{Org-N},1} + (\%\text{Org-N})_0 K_2 = 6\,\text{kg N/mt} + 3\% \times 0.8\,\text{kg/mt}\cdot\%$$
$$= (6 + 2.4)\,\text{kg N/mt} = 8.4\,\text{kg N/mt of dry solids}$$

   Note: Same Org-N mineraliztion rate of 8.4 kg N/mt is obtained using two different methods.

EXAMPLE 13.74: LIMITING METAL LOADING RATES FOR LAND APPLICATION

Biosolids are used over a farmland for crops production. The concentrations of metals as mg/kg in biosolids are given below:

$C_{As} = 45$ $C_{Cd} = 60$ $C_{Cu} = 2500$ $C_{Pb} = 650$ $C_{Hg} = 20$ $C_{Ni} = 400$ $C_{Se} = 25$ $C_{Zn} = 3800$

Calculate the allowable annual biosolids loading rate for metals. If nitrogen loading rate is 20 mt/ha·year, determine the limiting biosolids loading rate.

Solution

1. Compare the metal concentrations against ceiling concentrations.
   The ceiling concentrations of metals are given in Table 13.28. A comparison shows that all metal concentrations are below the ceiling concentration. Therefore, the biosolids are suitable for application over agricultural land.
2. Calculate the maximum amount of biosolids that can be applied annually for each metal.

The maximum annual amount of biosolids applied is calculated for each metal from Equation 13.27 based on the annual pollutant loading rate (APLR) provided in Table 13.28 ($L_m = APLR_m$).

$$C_{As} = 45\,\text{mg/kg} = 0.000045\,\text{kg/kg},\ S_{As} = \frac{APLR_{As}}{C_{As}C_f} = \frac{2.0\,\text{kg/ha·year}}{0.000045\,\text{kg/kg} \times 1000\,\text{kg/mt}} = 44\,\text{mt/ha·year}$$

$$C_{Cd} = 60\,\text{mg/kg} = 0.00006\,\text{kg/kg},\ S_{Cd} = \frac{APLR_{Cd}}{C_{Cd}C_f} = \frac{1.9\,\text{kg/ha·year}}{0.00006\,\text{kg/kg} \times 1000\,\text{kg/mt}} = 32\,\text{mt/ha·year}$$

$$C_{Cu} = 2500\,\text{mg/kg} = 0.0025\,\text{kg/kg},\ S_{Cu} = \frac{APLR_{Cu}}{C_{Cu}C_f} = \frac{75\,\text{kg/ha·year}}{0.0025\,\text{kg/kg} \times 1000\,\text{kg/mt}} = 30\,\text{mt/ha·year}$$

$$C_{Pb} = 650\,\text{mg/kg} = 0.00065\,\text{kg/kg},\ S_{Pb} = \frac{APLR_{Pb}}{C_{Pb}C_f} = \frac{15\,\text{kg/ha·year}}{0.00065\,\text{kg/kg} \times 1000\,\text{kg/mt}} = 23\,\text{mt/ha·year}$$

$$C_{Hg} = 20\,\text{mg/kg} = 0.00002\,\text{kg/kg},\ S_{Hg} = \frac{APLR_{Hg}}{C_{Hg}C_f} = \frac{0.85\,\text{kg/ha·year}}{0.00002\,\text{kg/kg} \times 1000\,\text{kg/mt}} = 43\,\text{mt/ha·year}$$

$$C_{Ni} = 400\,\text{mg/kg} = 0.0004\,\text{kg/kg},\ S_{Ni} = \frac{APLR_{Ni}}{C_{Ni}C_f} = \frac{21\,\text{kg/ha·year}}{0.0004\,\text{kg/kg} \times 1000\,\text{kg/mt}} = 53\,\text{mt/ha·year}$$

$$C_{Se} = 25\,\text{mg/kg} = 0.000025\,\text{kg/kg},\ S_{Se} = \frac{APLR_{Se}}{C_{Se}C_f} = \frac{5.0\,\text{kg/ha·year}}{0.000025\,\text{kg/kg} \times 1000\,\text{kg/mt}} = 200\,\text{mt/ha·year}$$

$$C_{Zn} = 3800\,\text{mg/kg} = 0.0038\,\text{kg/kg},\ S_{Zn} = \frac{APLR_{Zn}}{C_{Zn}C_f} = \frac{140\,\text{kg/ha·year}}{0.0038\,\text{kg/kg} \times 1000\,\text{kg/mt}} = 37\,\text{mt/ha·year}$$

The limiting metal loading of lead is 23 mt/ha·year. This loading rate controls the annual biosolids application over farm land for metals.

3. Compare with the nitrogen loading.

Nitrogen loading at agronomical loading rate is 20 mt/ha·year. This loading is less than that of metals. Therefore, the limiting value of nitrogen will control the annual biosolids application. Excess nitrogen may cause serious nitrate pollution of groundwater.

EXAMPLE 13.75: DESIGN OF BIOSOLIDS APPLICATION SYSTEM OVER FARMLAND

A municipal district is applying at an average 1500 kg/d (548 mt/year) dry solids produced from anaerobically digested sludge after dewatering. The biosolids, soil, crops and climate data are provided below. Design a land application system for using biosolids over the farmland.

A. Biosolid Data

Solids Content, $p_s = 20\%$	sp. gr., $S_b = 1.06$	VSS/TSS ratio = 55%		
NO_3-N = 0	NH_4-N = 0.2%	Org-N = 6.4%	P = 2%	K = 0.5%
Arsenic = 41 mg/kg	Cadmium = 42 mg/kg	Copper = 1480 mg/kg	Lead = 218 mg/kg	
Mercury = 14 mg/kg	Nickel = 380 mg/kg	Selenium = 35 mg/kg	Zinc = 1800 mg/kg	

B. Soil Data

Farmland soil is sandy loam. pH (water) = 6.0 CEC = 18 meq./100 g

Lime needed to raise the pH to 6.5, $S_{lime} = 4.6$ mt/ha (2.1 ton/ac) as $Ca(OH)_2$

C. Selected Crops Data

The crops grown in the area are corn, oat, soybeans, grain sorghum, wheat, barley, and forages for hay and pasture. The crop yield (mt/ha) and fertilizer requirements (N, P, and K) for each crop are different. This makes calculations repetitive and complex. Select two crops : corn and grain sorghum. Both crops have same nutrients requirements and cropping patterns. Therefore, the calculations can be combined for one crop and the cropping period is extended. The combined cropping requirements are: Crop yield = 8.4–10.1 mt/ha, N requirement = 190 kg/ha·year, P requirement = 60 kg/ha·year, K requirement = 125 kg/ha·year. Ignore the nutrients available in the soil.

Solution

1. Calculate the biosolids application rate based on heavy metal concentrations.
   a. Compare the metal concentrations against ceiling concentrations.
      In comparison with the ceiling concentrations of metals in Table 13.28, the biosolids are suitable for land application over farmland.
   b. Compute the maximum amount of biosolids applied based on the heavy metal loading limits.
      The maximum amount of biosolids applied is separately calculated from Equation 13.27 for (a) the annual amount applied based on the APLR ($L_m = APLR_m$), and (b) the cumulative amount over the lifetime based on the CPLR ($L_m = CPLR_m$). Values of APLR and CPLR for each metal are given in Table 13.28. The calculation results are tabulated below. See Example 13.74 for detailed procedure of calculating these values.
   c. Determine the biosolids application rates.
      The calculation results indicate that the maximum amount of biosolids that can be applied is governed by cadmium. The annual limiting application rate is $S_m = 45.2$ mt/ha·year, and the maximum amount of biosolids applied over the lifetime is limited to $S_{m,life} = 929$ mt/ha.

Metal	Concentration (C_m)		Maximum Annual Amount		Maximum Cumulative Amount	
	mg/kg	kg/kg	$L_{m,APLR}$ (from Table 13.28), kg/ha·year	$S_{m,APLR}$, mt/ha·year	$L_{m,CPLR}$ (from Table 13.28), kg/ha	$S_{m,CPLR}$, mt/ha
As	41	0.000041	2.0	48.8	41	1000
Cd	42	0.000042	1.9	**45.2**	39	**929**
Cu	1480	0.00148	75	50.7	1500	1014
Pb	218	0.000218	15	68.8	300	1376
Hg	14	0.000014	0.85	60.7	17	1214
Ni	380	0.00038	21	55.3	420	1105
Se	35	0.000035	5.0	142.9	100	2857
Zn	1800	0.0018	140	77.8	2800	1556

2. Calculate the biosolids application rate to satisfy the nutrient requirements.
   a. Calculate the biosolids application rate to meet the N requirement.
      i. Determine the overall plant-available N in biosolids in the first year.
         Available N from NO_3-N. From Equation 13.28b, $N_{a,NO3\text{-}N} = 0$, since $NO_3\text{-}N = 0$.
         Available N from NH_4-N. From Equation 13.28c at $NH_4\text{-}N = 0.2\% = 0.002$ kg N/kg, and $K_v = 1.0$.

$$N_{a,NH4\text{-}N} = C_f K_v(NH_4\text{-}N) = 1000\,\text{kg/mt} \times 1.0 \times 0.002\,\text{kg N/kg} = 2\,\text{kg N/mt of dry solids}$$

From Table 13.30, $K_1 = 2$ kg/mt·% for the first year. Calculate from Equation 13.28f the mineralization of Org-N in biosolids in the first year, using $(\%\text{Org-N})_0 = 6.4\%$.

$$N_{\text{a,Org-N,1}} = N_{\text{a,Org-N,0}} + (\%\text{Org-N})_0 K_1 = 0 + 6.4\% \times 2\,\text{kg/mt·\%} = 12.8\,\text{kgN/mt of dry solids}$$

Calculate from Equation 13.28a the overall N available in the first year.

$$N_{\text{a,1}} = N_{\text{a,NO3-N}} + N_{\text{a,NH4-N}} + N_{\text{a,Org-N,1}} = (0 + 2 + 12.8)\,\text{kgN/mt} = 14.8\,\text{kgN/mt of dry solids}$$

ii. Determine the plant-available N due to mineralization of Org-N in subsequent years.
The overall Org-N mineralization in each of the subsequent years is calculated from Equation 13.28f. Calculated results for a 10-year period are summarized below. See Example 13.73 for additional information about the calculation procedure for Org-N mineralization from biosolids.

Year Since Biosolids Application Began, (n), year	$N_{\text{a,NO3-N}}$ (from Equation 13.28b), kg/mt	$N_{\text{NH4-N}}$ (from Equation 13.28c), kg/mt	Org-N Mineralization		Plant-Available N (N_a) (from Equation 13.28a), kg/mt
			K_n (from Table 13.30), kg/mt·%	$N_{\text{a,Org-N},n}$ (from Equation 13.28f), g/mt	
1	0	2	2	12.8	14.8
2	0	2	0.8	17.9[a]	19.9[b]
3	0	2	0.36	20.2	22.2
4	0	2	0.21	21.5	23.5
5	0	2	0.2	22.8	24.8
6	0	2	0.19	24.0	26.0
7	0	2	0.19	25.2	27.2
8	0	2	0.18	26.4	28.4
9	0	2	0.18	27.6	29.6
10	0	2	0.17	28.7	30.7

[a] $N_{\text{a,Org-N,2}} = 12.8\,\text{kg/mt} + 6.4\% \times 0.8\,\text{kg/mt·\%} = 17.9\,\text{kgN/mt}$.
[b] $N_{\text{a,2}} = (0 + 2 + 17.9)\,\text{kg/mt} = 19.9\,\text{kgN/mt}$.

iii. Determine the biosolids application rate.
Assume that the plant-available N from biosolids reaches the steady-state in the 10th year and the Org-N mineralization after that is ignorable. Therefore, the peak plant-available N is 30.7 kg N/mt as dry solids, or a concentration $C_N = 0.0307$ kg N/kg of dry solids.
The N requirement for crops $R_N = 190$ kg/ha·year.

$$\text{Biosolids application rate from Equation 13.29, } S_N = \frac{R_N}{N_a} = \frac{190\,\text{kg/ha·year}}{30.7\,\text{kg/mt}}$$
$$= 6.2\,\text{mt/ha·year as dry solids}$$

b. Calculate the biosolids application rate to meet P requirement.
Assume that 50% P in biosolids is available for crop production.

Plant-available P concentration in biosolids, $C_P = 0.5 \times 2\% = 1\% = 0.01$ kg P/kg

The P requirement for crops is $R_P = 60$ kg/ha·year.

$$\text{Biosolids application rate from Equation 13.29, } S_P = \frac{R_P}{C_P C_f} = \frac{60\,\text{kg/ha·year}}{0.01\,\text{kg P/kg} \times 1000\,\text{kg/mt}}$$
$$= 6\,\text{mt/ha·year as dry solids}$$

c. Calculate the biosolids application rate to meet K requirement.

Assume that 100% K in biosolids is available for crop production.

Plant-available K concentration in biosolids, $C_K = 0.5\% = 0.005$ kg K/kg The K requirement for crops is $R_K = 125$ kg/ha·year.

$$\text{Biosolids application rate from Equation 13.29, } S_K = \frac{R_K}{C_K C_f} = \frac{125\ \text{kg/ha·year}}{0.005\ \text{kg K/kg} \times 1000\ \text{kg/mt}}$$

$$= 25\ \text{mt/ha·year as dry solids}$$

3. Select the design biosolids application rate (S_d).

The biosolids application rates calculated for limitations by metals, and nutrients (N, P, and K) are summarized below.

Concern	Biosolids Application Rate, mt/ha·year as Dry Solids	Comment
Metals		
Cd (limiting factor)	$S_m = 45.2$ (Step 1c)	This is the limiting loading rate ($S_d \leq S_m$).
Nutrients		
N (governing factor)	$S_N = 6.2$ (Step 2a,iii)	Select a design biosolids application rate $S_d = S_N = 6.2$ mt/ha·year to satisfy the agronomic rate of N.
P	$S_P = 6$ (Step 2b)	P loading will exceed the crops requirements.
K	$S_K = 25$ (Step 2c)	K needs will not be met by applying biosolids.

Note: In most designs, the biosolids application rate is normally selected based on agronomic loading rate of nitrogen, because migration of excess nitrogen may cause serious nitrate contamination of groundwater. At biosolids application rate of 6.2 mt/ha·year the Cd loading will be significantly lower than the limiting loading rate of 45.2 mt/ha·year for metals. The N requirement will be met but P loading will slightly exceed the crops requirement causing slow accumulation in soils, and subsequent minor release in the surface runoff. The K needs for crops will not be met, and supplemental chemical fertilizer will be needed.

4. Compute the farmland needed for cultivation

$$\text{Apply Equation 13.30 to calculate the area requirement, } A = \frac{Q_s}{S_d} = \frac{548\ \text{mt/year}}{6.2\ \text{mt/ha·year}} = 88\ \text{ha}$$

Add 25% land for buffer. Total land area 110 ha.

5. Estimate the potential life of farmland for biosolids application.

Maximum amount of biosolids applied over the lifetime, $S_{m,life} = 929$ mt/ha (Step 1c).

$$\text{Life of farmland for biosolids application, } \theta = \frac{S_{m,life}}{S_d} = \frac{929\ \text{mt/ha}}{6.2\ \text{mt/ha·year}} = 150\ \text{years}$$

Consider a safety factor, the life of land application is at least 120 years.

6. Check the nutrient balance and determine the chemical fertilizer need.

a. Check the N balance.

The nitrogen release from biosolids gradually increases over time. Therefore, chemical fertilizer needs decreases with time. The chemical fertilizer need to meet the N requirement is summarized in the table below.

Year Since Biosolids Application Began, Year	N Requirement by Crops (R_N), kg N/ha·year	N Available from Biosolids Application			N Supplement from Chemical Fertilizer	
		Design Application Rate (S_d), mt/ha·year	Plant-Available N (N_a), kg N/mt	Annual Application Rate, kg N/ha·year	Requirement, kg N/ha·year	Quantity, mt N/year
1	190	6.2[a]	14.8[b]	91.8[c]	98.2[d]	8.6[e]
2	190	6.2	19.9	123.4	66.6	5.9
3	190	6.2	22.2	137.6	52.4	4.6
4	190	6.2	23.5	145.7	44.3	3.9
5	190	6.2	24.8	153.8	36.2	3.2
6	190	6.2	26.0	161.2	28.8	2.5
7	190	6.2	27.2	168.6	21.4	1.9
8	190	6.2	28.4	176.1	13.9	1.2
9	190	6.2	29.6	183.5	6.5	0.6
10	190	6.2	30.7	190.3	−0.3	0.0

[a] Selected in Step 3 based on steady-state plant-available N from biosolids applied.
[b] Calculated in Step 2.a.ii based on overall plant-available N from biosolids applied.
[c] 6.2 mt/ha·year × 14.8 kg/mt = 91.8 kg/ha·year.
[d] (190 − 91.8) kg/ha·year = 98.2 kg/ha·year.
[e] 98.2 kg/ha·year × 88 ha × 10^{-3} mt/kg = 8.6 mt/year as N.

b. Check the P balance.

The amount of P available in biosolids $C_P = 0.01$ kg P/kg dry solids (Step 2.b) and the P need for crops, $R_P = 60$ kg/ha·year.

P provided by biosolids, $W_P = C_P C_f S_d = 0.01\,\text{kg P/kg} \times 1000\,\text{kg/mt} \times 6.2\,\text{mt/ha·year} = 62\,\text{kg/ha·year}$

Calculate the annual excess P remained in soil and released in surface runoff.

$$R_{\text{release,P}} = W_P - R_P = (62 - 60)\,\text{kg/ha·yr} = 2\,\text{kg/ha·year}$$

c. Check the K balance.

The amount of K available in biosolids $C_K = 0.005$ kg K/kg dry solids (Step 2.c) and the K need for crops, $R_K = 125$ kg/ha·year.

K provided by biosolids, $W_K = C_K C_f S_d = 0.005\,\text{kg P/kg} \times 1000\,\text{kg/mt} \times 6.2\,\text{mt/ha·year} = 31\,\text{kg/ha·year}$

$$\text{K supplement required from chemical fertilizer, } R_{\text{fertilizer,K}} = R_K - P_K = (125 - 31)\,\text{kg/ha·year} = 94\,\text{kg/ha·year}$$

Quantity of K in fertilizer, $W_{\text{fertilizer,K}} = R_K A = 94\,\text{kg/ha·year} \times 88\,\text{ha} \times 10^{-3}\,\text{mt/kg} = 8.3\,\text{mt/year as K}$

7. Calculate the quantity of lime required.

Lime is needed to raise the pH of the soil. The required lime application rate is $S_{\text{lime}} = 4.6$ mt/ha as $Ca(OH)_2$ to maintain a pH above 6.5. At this pH, the uptake of heavy metals particularly Cd in corn crop is reduced.

$$\text{Quantity of lime needed, } W_{\text{lime}} = S_{\text{lime}}\, A = 4.6\,\text{mt/ha} \times 88\,\text{ha} = 405\,\text{mt as Ca(OH)}_2$$

8. Describe other design considerations.

a. Biosolids transport equipment.

The options for transporting biosolids include truck, pipeline, railroad and barge. Select 11-m^3 dump or hopper truck.

b. Biosolids storage and application schedule.

Biosolids storage is needed because of fluctuations in production rate, equipment breakdown, cropping patterns, adverse weather conditions, and seasonal variation in demands. Provide 5-d storage at the plant and 30-d storage at the distribution centers. Biosolids application corresponds to farming operation is influenced by climate and soil properties. Application is not recommended (a) during and after rains, and frozen ground, and (b) crop tillage, planting and harvesting periods.

c. Biosolids application equipment.

The biosolids application is similar to spreading semisolid fertilizer, lime or animal manure. The truck-mounted or tractor-powered box spreaders are most commonly used.

d. Monitoring requirements.

The environmental monitoring requirements are briefly covered in Section 13.11.5.

13.11.6 Residuals Disposal by Landfilling

Many types of residuals are produced at a wastewater treatment facility that are normally landfilled. These are screenings, grits, skimmings, ashes, and leftover or unused dewatered biosolids. These residuals may be *disposed off directly* in a monofill or *codisposed* with MSW. Planning, design and operation of all landfills require detailed investigation. Common problems are soil erosion, dust, vectors, noise, odors and continued regeneration of leachate and gases for many years. Important factors considered in landfill planning are: (1) regulations and permits, (2) site selection, (3) residuals characteristics, (4) methods of landfilling, (5) design considerations, (6) operation and maintenance, and (7) monitoring and post closure care. Each of these factors is briefly discussed below.

Regulations and Permits: Direct disposal of sludge in a monofill is regulated under 40 CFR Part 503. However, codisposal of municipal biosolids with MSW in a municipal landfill will not have to meet the numerical pollutants limits for land application of biosolids in 40 CFR Part 503 regulations.[17,18] Codisposal practice is regulated by the U.S. Environmental Protection Agency under 40 CFR Part 258 rule.[201] This rule established minimum federal criteria for MSW Landfills to address location restrictions, facility design criteria, operation criteria, groundwater monitoring, and requirements for corrective actions, financial assurance, and closure and post closure care. Many regulatory agencies require permit for construction and operation of sludge landfills. The design must follow the permit requirements which are generally available from the federal, state and local agencies. The permit requirements address to environmentally sensitive areas, surface water and groundwater, odors, safety, and public participation program.[201,202] Recently, mandatary monitoring of landfill gas (LFG) is also required under 40 CFR Part 98 (also referred as the Greenhouse Gas Reporting Program, GHGRP).[203]

Site Selection: Site selection of a sanitary landfill is perhaps the most difficult and controversial issue of the planning process. The major factors are listed below:

1. *Useful Life*: The selected site should last several decades to justify the cost of permanent structures.
2. *Topography*: Flat sites (slopes 1%) cause ponding, and steeper site (slope >20%) may cause erosion.
3. *Surface Water*: Runoffs from landfill areas must comply with NPDES permit.
4. *Groundwater*: Depth of aquifer, hydraulic gradient, quality of groundwater and projected water use, and location of primary recharge zone should be considered.
5. *Soil and Geology*: Fine-textured soils such as clays are desirable because of low permeability and high attenuation capacity against many pollutants.
6. *Site Access*: Existing and future all weather access roads and traffic impacts on neighborhoods should be evaluated.

7. *Land Use*: Current and future land use patterns around the site should be coordinated with local planning agencies.
8. *Social and Political Issues*: The extent of public resistance and social and political implications should be evaluated.
9. *Ecology*: Environmentally sensitive areas such as wetlands, flood plains, permafrost, critical habitats of endangered species, and recharge zones of sole source aquifers should be evaluated. Vegetation should be utilized as a buffer.
10. *Economics*: Early in the sites evaluation the relative costs of different sites should be developed and compared.

Residuals Characteristics: In determining the suitability of sludge for landfilling the following factors should be evaluated: (1) raw wastewater characteristics, (2) wastewater treatment processes, (3) sludge handling and disposal methods, (4) organic and nutrients content, (5) heavy metals contents, and (6) characteristics of chemical precipitates.

The preferable conditions of biosolids for landfilling include (1) dewatered sludge with high solids content ($\geq$20%, for monofill), (2) high pH ($\geq$10, optimum) to prevent heavy metals from leaching out, (3) low metal contents, (4) low organic content for less gas production and low risk of explosion, and (5) low nutrients to reduced risk of groundwater pollution.

Methods of Sludge Landfilling: Common methods of landfilling are: (1) *trenching* and (2) *area fill*. In both methods the base is designed to collect leachate for treatment. Also, gas collection system is installed to stop migration of gases. Typically, the cover material is excavated soil. The daily cover is up to 15 cm and top cover is 0.6–1 m (2–3 ft). Sludge with a solids content of 15–20% cannot support cover. Bulking agent (usually dry soil or MSW) is mixed with sludge to increase workability and soil bearing capacity. Both landfilling methods are briefly described below.

Trenching: There is a *narrow* or *wide* trench. The width of narrow trench is 0.6–1 m (2–3 ft) for low solids sludge (15–20%), and 1–3 m (3–10 ft) for sludge of 20–30% solids. Instead of sinking into the sludge, the cover soil bridges over each side of the trench walls. Backhoe, excavator, trenching machine, truck loader, and doser dragline and scraper are common equipment. The width of trench is 3–15 m (10–50 ft). If sludge is incapable of supporting equipment, cover is applied from solid ground. A front-end loader is suitable for trench width up to 3 m, and a dragline up to a width of 15 m. If sludge can support equipment, a truck dozer should apply the cover from within the trench. The haul vehicles should directly dump the sludge from the top, or by entering the trench. Dikes are used to confine sludge to a specific area in a continuous trench.

Area Fill: This is a sludge disposal operation in which sludge is placed over a constructed impervious base above the ground surface and subsequently covered with soil. To achieve stability and soil-bearing capacity, sludge is mixed with a bulking agent (dry soil). The soil absorbs moisture and increases workability. The imperious base normally has clay or flexible membrane liner (FML) with leachate collection system, surface drainage control, and gas collection.

The area-filled landfilling has three variations: *mound*, *layer*, and *diked containment*. In the area filled mound method the sludge with bulking agent is stacked in mounds ~2 m (6 ft) high. The cover is about 1 m (3 ft) thick. Cover thickness is increased if additional mounds are applied over the top of earlier lift. Earthen containments increase the stability. The area fill layer method utilizes even layers of sludge and bulking agent mixture 0.6–1 m (2–3 ft) thick covered by intermediate layer of soils 15–30 cm (6–12 in) thick. Lightweight equipment with swamp pad tracks is normally used. The layering is continued until the final cover is applied. The diked containment method utilizes containment walls. The haul vehicles dump the sludge from the top of the dike. The intermediate cover is 0.3–1 m (1–3 ft) and final cover is 1–2 m (3–6 ft). Typical dimensions are 15–30 m (50–100 ft) wide, 30–60 m (100–200 ft) long, and 3–10 m (10–30 ft) deep. The details of these landfilling methods may be found in References 2, 3, 6 and 191.

Design Considerations: Basic design considerations for sludge landfills include cost-effective utilization of site, storage volume, soil, and equipment. The important factors are: (1) leachate control,

(2) gas control, (3) stormwater management, and (4) environmental factors and safety. These factors are briefly presented below.

Leachate Control: Leachate containment, removal and disposal is essential for groundwater protection. Leachate control involves (1) reduction of rainwater entering the fill and (2) containment and removal of leachate from the fill. Diversion of rainwater around the fill by sloping the landfill surface, or providing an impervious liner below the top cover with a sand drainage layer are the common methods to reduce the rainwater from entering the fill. The leachate collection and removal system utilize impervious base, drainage blanket and leachate collection pipe. The impervious liner may be single or composite, natural soil or clay, admixed or spray-on-lining, synthetic flexible membrane liner (FML) or geomembranes. Typical liners used for municipal landfill are summarized in Table 13.31. The details of classification and construction of these liners may be found in References 6, 202, and 204–207.

Landfill Gas Control: The landfill gas contains 50–65% methane and 30–45% carbon dioxide. The remaining portion contains nitrogen, oxygen, water vapors, ammonia, and hydrogen sulfide. Gas production from undigested sludge is 0.5–10 m^3/kg (8–16 ft^3/lb) of dry solids. Digested sludge produces much less gas, and it is generated over a long period. Methane is an important greenhouse gas and landfills are ranked the third largest source of methane emissions in the United States.[208] Therefore, effective collection of landfill gas is essential in design, and operation of a municipal landfill. The gas collection system is either *passive* or *active*. In a passive system, the gas movement is driven by internal pressure, while in an active system, the gas movement is created by induced negative pressure or vacuum. See design details and operating criteria of landfill gas collection in References 6, 201, 209, and 210.

Stormwater Management: The surface drainage from upland areas should be collected and directed around the landfill. The channels are earthen, concrete, stone-lined and corrugated metal pipes. The surface of the landfill should be well-graded outward. The desirable slope is 2–5%.

Environmental Factors and Safety: The environmental factors include odors, dust, vector control, and aesthetics. No sludge should be left uncovered for extended period. Coverage with well compacted soil will reduce odors and control flies. Sludge spillage, erosion, and dust and mud formation should be controlled. The entire area should be well lighted. Provide facilities for employees and needed utilities such as water, sewer, electrical, and telephone. The typical landfill equipment needed for safe operation may be found in References 2 and 6.

TABLE 13.31 Typical Liners Used for Municipal Landfill

Type	Description
Natural soil and clay	Native clays, engineered soils formed on site
Admixed liners	Formed on site
Asphalt concrete	Hot mixture of asphalt cement and mineral aggregate
Soil cement	Mixture of Portland cement, water, and selected soil
Soil asphalt	Mixture of soil and liquid asphalt
Sprayed-on linings	Liners are formed in the field by spraying the material on a prepared surface. These are air blown asphalt, emulsified asphalt, urethane-modified asphalt, and rubber and plastic latexes.
Soil sealants	Permeability of soils is reduced by application of sealants by spraying or pressure injection.
Flexible membrane liners (FMLs) or geomembrane liners[a]	These are prefabricated liners made by different polymeric material used to serve special functions.
Geosynthetic clay liners	These liners are fabricated from sodium bentonite in between a woven and a nonwoven geotextile. They may include (1) geotextile, (2) geogrids, (3) geonets, and (4) geocomposites.

[a] See Example 10.91 for a list of commercial FMLs that is used for lining stabilization ponds. Similar materials may also be used for municipal landfills.

Source: Adapted in part from References 6 and 202.

Operation and Maintenance: Colandfilling sludge operation is an ongoing process. Special considerations for efficient operation and maintenance may include compatibility of equipment with sludge characteristics, site conditions, landfilling method, and operating schedule.

Monitoring and Post Closure Care: All completed landfills are monitored for many years for any adverse environmental effects. The monitoring includes (1) groundwater, (2) surface water, and (3) gas movement. The water quality parameters are pH, electrical conductivity, nitrate, total organic carbon (TOC) or COD, and heavy metals. Gas monitoring is done around the property line. The gas monitoring is done by methane probes that are installed in a hole at different depths.

The landfill will continue to settle. The area should be regraded for maintaining proper drainage. Also, on a timetable check the settlement, grading and vegetation cover, sediment and erosion control, leachate and gas control, and groundwater and surface runoff quality.

EXAMPLE 13.76: DESIGN OF LANDFILL SLUDGE IN A MONOFILL

A POTW is designed to treat a flow of 30,000 m^3/d. The residuals include screenings, grit, scum, and anaerobically digested and dewatered sludge cake that meets Class B biosolids criteria. The residuals quality, production rates and quantity produced are given below:

Residual	Moisture Content, %	Density, kg/m^3	Production Rate, m^3/m^3 Raw Influent	Quantity Produced, m^3/d
Screenings	80	960	20×10^{-6}	0.60
Grit	25	1600	30×10^{-6}	0.90
Scum (or skimming)	60	950	16×10^{-6}	0.48
Dewatered sludge cake	75	1060	4.7×10^{-4}	14.1

At an average 65% of the biosolids is used over highway embankments and medians, and land reclamation. The selected sludge-only landfill site is a 12 ha wooded land area located on the north of the treatment plant. The landfilling area is flat and nearly rectangular. Four test borings were performed to determine the soil condition and hydrology. The soil is silt loam, medium texture, permeability is 2×10^{-4} cm/s (moderately slow), pH = 6.5, and CEC = 18 meq./100 g. The depth of groundwater table is 6 m. The narrow-trench landfilling method is selected. Determine the landfill area required per year and life of landfill. Show the layout and details of the trenches.

Solution

1. Select the depth of excavated trench and base construction.

   The depth of excavated trench or cell is established initially to meet the minimum distance requirement for groundwater table. If 4-m deep cells are excavated, a separation depth of the soil from the water table will be 2 m. At CEC of 18 meq./100 g and permeability classified as "moderately slow," the containment and attenuation of contaminants will be acceptable. Provide 0.5 m compacted clay on the bottom and sides to meet the regulatory requirement. Provide a minimum of 0.6-m drainage layer at the base, and a central pipe for leachate collection. These provisions will give a minimum distance of 3.1 m from the bottom of the filled material to groundwater table. The cross section of the trench is shown in Figure 13.39a.
2. Select the dimensions of the trench.

   The selected narrow trench dimensions are: bottom dimensions = 2 m × 40 m, side slope = 1:1, and top dimensions = 10 m × 48 m.
3. Calculate the volume available for residuals, bulking agents and intermediate cover.

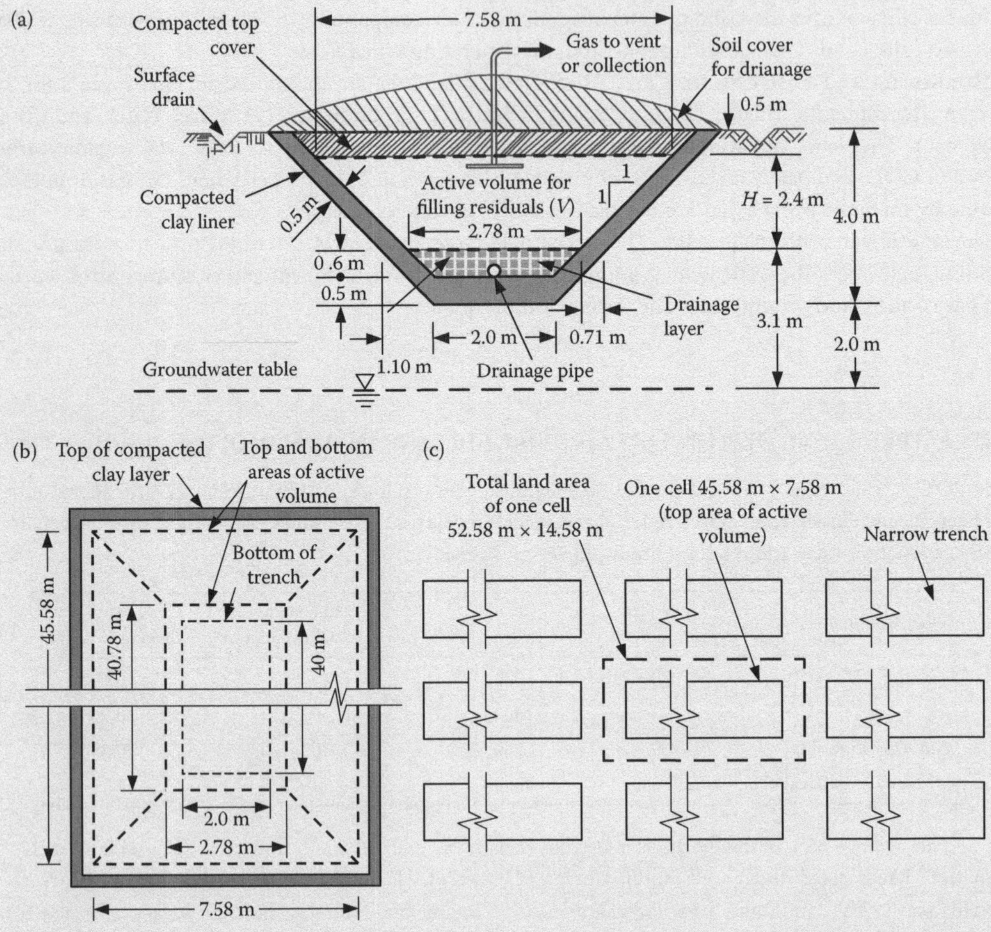

FIGURE 13.39 Design details and layout of narrow trench sludge landfill: (a) cross section, (b) plan of a single cell, and (c) layout plan and outer boundary of a cell in a cluster of trenches (Example 13.76).

The plan view of a cell is shown in Figure 13.39b.

The horizontal projection of 0.5 m compacted clay layer on the slope = 0.71 m.

The dimensions of the area available for residuals disposal are calculated below.

Bottom width $= 2\,\text{m} + 2 \times 1.1\,\text{m} - 2 \times 0.71\,\text{m} = 2.78\,\text{m}$

Bottom length $= 40\,\text{m} + 2 \times 1.1\,\text{m} - 2 \times 0.71\,\text{m} = 40.78\,\text{m}$

Top width $= 2.78\,\text{m} + 2 \times 2.4\,\text{m} = 7.58\,\text{m}$

Top length $= 40.78\,\text{m} + 2 \times 2.4\,\text{m} = 45.58\,\text{m}$

Total active volume in each cell is given by Equation 13.31.

$$V = \frac{1}{3} \times H \times \left(A_1 + A_2 + \sqrt{A_1 A_1}\right) \tag{13.31}$$

where

V = total active volume, m^3

A_1 and A_2 = bottom and top areas of active volume available for residuals disposal, m^2

H = depth of the volume, m
Bottom area $A_1 = 2.78\text{ m} \times 40.78\text{ m} = 113\text{ m}^2$
Top area $A_2 = 7.58\text{ m} \times 45.58\text{ m} = 345\text{ m}^2$
$H = 2.4\text{ m}$

$$\text{Active volume from Equation 13.31, } V = \frac{1}{3} \times 2.4\,\text{m} \times \left(113\,\text{m}^2 + 345\,\text{m}^2 + \sqrt{113\,\text{m}^2 \times 345\,\text{m}^2}\right) = 524\,\text{m}^3$$

4. Calculate the disposal capacity of each cell.
   If 20% volume is occupied by the bulking agent and intermediate cover, the total volume available for residuals disposal is about 420 m^3.
5. Calculate the operating period to fill one cell.
   Volume of residuals $= (0.60 + 0.90 + 0.48 + (1 - 0.65) \times 14.1)\ \text{m}^3/\text{d} \approx 7\ \text{m}^3/\text{d}$

$$\text{Time to fill one cell } \theta = \frac{420\,\text{m}^3}{7\,\text{m}^3/\text{d}} = 60\,\text{d}$$

6. Calculate the total land area of one cell.
   The conceptual layout of a cluster of trenches is shown in Figure 13.39c.
   Provide 7 m clear spacing between adjacent cells on all sides.
   The centerline boundary dimensions of each cell $= (7.58 + 7)\ \text{m} \times (45.58 + 7)\ \text{m} = 14.58\ \text{m} \times 52.58\ \text{m}$

$$A = 14.58\,\text{m} \times 52.58\,\text{m} \times \frac{1\,\text{ha}}{10{,}000\,\text{m}^2} = 0.077\,\text{ha/cell}$$

7. Calculate the total number of cells over the life of the landfill.

$$N = \frac{12\,\text{ha}}{0.077\,\text{ha/cell}} = 156\ \text{cells}$$

   Use a conservative number of 150 cells.
8. Calculate the total life of landfill site.

$$\text{Life of the landfill site, } A = 150\ \text{cells} \times \frac{60\,\text{d}}{\text{cell}} \times \frac{1\ \text{year}}{365\,\text{d}} \approx 25\ \text{years}$$

9. Describe the operational details.
   The narrow trench is excavated and soil is stockpiled on the side. After base construction the landfilling operation will begin. In each cell, the screenings, grit and skimmings shall be placed on one end while the biosolids are placed on the opposite end. This way the biosolids will remain segregated from other residues, and can be excavated at a later date and used as topsoil, soil conditioner or biosolids. After completion of each cell the sodded surface drains are constructed on both sides of the trench. The surface drains shall carry the surface runoff to a detention pond located at the lowest elevation on the residuals disposal area. The landfilling operation shall be performed by equipment operating outside the cell. Commonly used equipment is backhoe with loader and trenching machine. An asphalt-paved road shall be constructed all around the trench filled area.

Discussion Topics and Review Problems

13.1 The primary sludge and WAS are combined then blended. The total solids content of primary sludge = 1850 kg/d at 4% solids and specific gravity = 1.02. The total solids content, solids concentration and specific gravity of WAS are respectively 1200 kg/d, 1% and 1.01. Sludge after blending is pumped into a gravity thickener. The solids capture efficiency of thickener is 85%. Solids concentration and specific gravity of thickened sludge are 6% and 1.03. Determine (a) the volume and specific gravity of blended sludge, and (b) volume of thickened sludge.

13.2 An anaerobic sludge digester produces 4000 kg/d dry solids. The conditioning chemicals are added at a rate of 6 g/kg dry solids prior to dewatering by belt filter press. Determine the volume of sludge cake. The solids capture efficiency of belt filter press is 90%, and 80% conditioning chemicals are incorporated into the sludge cake. The specific gravity of sludge cake is 1.05.

13.3 What is the purpose of sludge blending and dilution water? A sludge blending tank is designed to blend primary sludge, WAS and dilution water. The volume of combined sludge and dilution water are 800 m^3/d and 1080 m^3/d. The blending time is 1.5 h. The critical temperature is 50°C and design velocity gradient under field condition is 58 s^{-1}. The kinetic viscosity of blended sludge is twice that of water. Calculate the water power and wire power. The blender and motor efficiencies are 80% and 90%, respectively.

13.4 Calculate the dimensions of a gravity thickener to thicken combined sludge. The sludge volume is 500 m^3/d and has a solids concentration of 1.0% and sp. gr. of 1.008. The solids loading must not exceed 50 kg/m^2·d, and hydraulic loading must not be less than 6 m^3/m^2·d. The underflow concentration of thickened sludge is 6% (sp. gr. = 1.03). Calculate the volume of thickened sludge, volume of dilution water, volume of supernatant, and TSS concentration in the supernatant. Assume the solids capture efficiency of 85%, one-day solids retention time, clear liquid zone of 1.0 m, settling zone of 2.0 m, and slope of thickener slab is at 17% towards the hopper. Provide a freeboard of 0.5 m.

13.5 Determine the power requirement, and the paddle area required to achieve $G = 60\ s^{-1}$ in a tank that has a volume = 400 m^3, water temperature = 20°C, and $C_D = 1.8$. The paddle tip velocity is 1.6 m/s. The kinematic viscosity of paddle relative to the fluid is 0.75 times the paddle tip speed.

13.6 Determine the motor power required for a sludge-blending tank using vertical redwood paddles. There are12 paddles, six on each side of the central shaft. Each paddle is 0.10 m wide × 2.75 m high. The distances to the middle of each paddle from the center of the shaft are 0.5, 1.0, 2.0, 2.5, and 3.0 m. Assume rotational speed of the shaft = 0.06 rps, $C_D = 1.8$, motor efficiency = 75%, and specific gravity of blended sludge = 1.01.

13.7 Calculate air to solids ratio and dimensions of the floatation tank. Use the following data:

Combined sludge flow	= 760 m^3/d
Solids concentration	= 1.2%, sp. gr. = 1.01
Operating temperature	= 30°C
Solids loading rate not to exceed	= 75 kg/m^2·d
Hydraulic loading rate not to exceed	= 8 m^3/m^2·d
Length to width ratio	= 5:1
Depth	= 2.5
Operating pressure	= 19 atm
Fraction of air dissolved at P	= 0.65

The entire influent sludge stream is pressurized.

13.8 A bench-scale anaerobic digester study was conducted to develop the biological kinetic coefficients. The average biomass yield coefficient Y and endogenous decay coefficient k_d at 35°C were 0.08 kg

VSS/kg COD and 0.05 d^{-1} respectively. The average COD utilization efficiency E was 75%. A high-rate anaerobic digester is designed to stabilize combined sludge containing solids that has equivalent COD of 5000 kg/d and the average digestion period of 15 days. Estimate the following (a) the active biomass growth P_x, (b) COD utilization rate ΔS_m, and COD stabilization efficiency E_m.

13.9 An anaerobic digester receives combined thickened sludge at a rate of 100 m^3/d, solids content of 5% and specific gravity of 1.08. The VSS/TSS ratio ($f_{VSS.FS}$) = 0.75, and $\theta_c = 15$ d. Determine the digester capacity based on (a) volatile solids loading VSL, (b) per capita capacity requirement (PCCR), and (c) average volume of digested sludge. Use the following data: VSL = 2.4 kg VSS/m^3.d, PCCR = 0.04 m^3 per capita TSS, solids production = 120 g TSS per capita per day, and the observed sludge volume reduction = 0.9. Summarize the results and select the design capacity.

13.10 An anaerobic sludge digester receives 2000 kg/d TSS. Estimate the quantity produced by kinetic equation and rule of thumb. The kinetic coefficients are $Y = 0.08$ kg VSS/kg COD, $k_d = 0.03\ d^{-1}$, $SRT = 15$ d, COD utilization efficiency $E = 0.8$, VSS/TSS ratio = 0.75, the ratio of biodegradable VSS (VSS_{bd}) to TSS = 0.56 kg VSS_{bd}/kgTSS, and each g of VSS_{bd} exerts 1.42 g COD. Other relationships for biogas generation are 0.52 m^3/kg VSS loading, 0.82 m^3/kg VSS reduced, and 0.043 m^3 per capita per day. The average TSS production is 0.12 kg per capita per day. Compute the gas generation from different methods and give your recommendation.

13.11 A high-rate anaerobic digester has 12 m diameter and vertical side depth of 8.5 m. The floor of the digester is sloped 1 vertical to 3 horizontal. Grit accumulation allowance is 1-m depth at the bottom of the cone. Scum accumulation allowance is 0.6 m above sludge level, and clearance between the floating cover above the scum layer is 0.5 m. Determine the active volume ratio AVR and digestion period. The average flow blended and thickened sludge to the digester is 50 m^3/d.

13.12 An anaerobic sludge digester has an average liquid depth of 7 m and a diameter of 10 m. It receives 1600 kg/d thickened sludge at 6% solids. The specific gravity of the thickened sludge is 1.03. Calculate (a) volatile solids loading factor, (b) digestion period, and (c) capacity per capita if the population served is 19,000. Assume VS in sludge solids is 79%.

13.13 An anaerobic digester receives 2000 kg total dry solids per day. The sludge contains 5.8% solids at a specific gravity of 1.03. The digested sludge contains 4.2% solids and has a specific gravity of 1.02. Calculate the BOD_5 and TS of the digester supernatant. Use the following data:

TVS in raw sludge	= 75%
TVS stabilized	= 56%
Digested sludge withdrawal rate	= 24 m^3/d
BOD_L of digested sludge	= 0.65 × TVS in digested sludge
BOD_5/BOD_L	= 0.68

13.14 Compute the power requirements for gas mixing in a digester. The volume of the digester is 2000 m^3, velocity gradient for proper mixing is 88/s, and viscosity of sludge at operating temperature is 1.53×10^{-3} N·s/m^2.

13.15 Thickened sludge is heated from 15 to 36°C before reaching an anaerobic digester. The average quantity of sludge is 100 m^3/d, the specific heat of sludge is 4200 J/kg·°C (same as for water), and the heat exchange capacity of jacketed heat exchanger is 3060 kJ/h·m^2·°C. Calculate the following:

a. Length of sludge pipe in the jacket.
b. Volume of hot water recirculated through the heat exchanger if inlet and outlet recirculating water temperatures are 85°C and 40°C. Assume heat transfer efficiency of 80%.
c. Digester gas needed in the boiler (m^3/d). Assume 75% boiler efficiency, and heating value of the digester gas is 24,500 kJ/m^3.

13.16 Discuss different factors that affect the performance of an anaerobic digester. List various tests that must be performed daily for control of the digestion process. What is the significance of each test in relation to the normal digester operation?

13.17 An anaerobic digester is designed to receive 4000 kg solids at 7% by weight. The digester roof, bottom, and side wall areas are 450, 470, and 550 m^2. The overall coefficients of heat transfer are 0.9, 0.62, and 0.68 $J/s{\cdot}m^2{\cdot}°C$, respectively. Assume that the temperatures of incoming sludge, digester operating, and average outside are 25°C, 38°C, and 21°C, respectively. Minor heat losses are 25% of total heat loss. Calculate the total heat requirement of the digester. The specific gravity of the incoming sludge is 1.03.

13.18 Describe and compare the operation of staged mesophilic anaerobic digester (MAnD), and staged thermophilic anaerobic digester (TAnD). List key operational parameters, typical values, and advantages and disadvantages of both digestion processes.

13.19 Describe and compare the operation of acid/gas phased anaerobic digester (A/GAnD) and temperature-phased anaerobic digester (TPAnD). List key operational parameters, typical values, and advantages and disadvantages of both digestion processes.

13.20 An aerobic digester is designed to digest 30 m^3/d combined blended and thickened sludge. The solids content in the feed is 3% and specific gravity is 1.01. The sludge is digested to meet the Class B biosolids criteria under the critical temperature of 15°C and SRT of 60 days. The VSS/TSS ratio is 0.8, the fraction of primary sludge $f_{ps} = 0.25$, and total BOD_5 in plant influent $S_{PI} = 210$ mg/L. The operating TSS concentration in the digester = 20,000 mg/L. The endogenous decay coefficient k_d is 0.05 d^{-1} at 15°C. Determine the volume of anaerobic digester and the volumetric solids loading.

13.21 An aerobic digester is designed to digest combine sludge. The digester volume is 400 m^3. The volatile solids reduction (VSR) during the critical summer condition is 58% and VSS reaching the digester is 180 kg VSS/d. Determine the air required for stabilization of organic matter during summer, and that for mixing the digester contents. The oxygen requirement for oxidation and mixing are obtained from Table 13.16. The oxygen transfer efficiency of coarse diffusers (SOTE) is 6%. Oxygen content of air and density under standard conditions are 23.2% and 1.204 kg/m^3.

13.22 Describe autothermal aerobic digestion (ATAD) process. An ATAD is operating at a temperature of 60°C. The ratio of VSS/TSS in the feed sludge is 0.8, and VSS reduction in the digester is 0.52%. The heat generation from microbial oxidation reaction is 20,000 kJ/kg VSS reduced. The density ρ_s and specific heat $C_{sh,s}$ of sludge are same as those for water, (1 kg/L and 4200 $J/kg{\cdot}°C$). The overall efficiency of heat transfer is 0.55. Estimate the solids content in the feed sludge to generate sufficient heat to maintain the operating temperature of the digester. The temperature of feed sludge is 20°C.

13.23 An ATAD is digesting combined sludge that has feed volatile suspended solids concentration of 50 g/L. The volatile solids reduced in the digester VSR = 55%. The alkalinity production ratio is 0.44 g Alk as $CaCO_3$ per g VSS reduced. Calculate the alkalinity increase and ammonia generated due to endogenous decay of VSS in the aerobic digester.

13.24 Hydrated lime ($Ca(OH)_2$) is used to condition the anaerobically digested sludge. The digested feed sludge contains 3000 kg/d dry solids at 4% solids. The specific gravity is 1.010. The lime dosage is 8.0% of dry solids. At an average, 80% lime is incorporated in the sludge solids. The remaining 20% lime reacts with the alkalinity in the digested sludge to produce excess sludge. The pH and alkalinity of digested sludge are 7.2 and 2500 mg/L as $CaCO_3$. Calculate the percent increase of solids in the conditioned sludge.

13.25 A BNR facility is designed to achieve nutrients removal and phosphorus recovery. The WAS stream contains 2500 kg/d solids at VSS/TSS ratio of 0.8. The phosphorus content of WAS is 0.085 g P/g VSS. The WAS is mixed with gravity thickener overflow in a basin to provide volatile fatty acids (VAF) for P-release. Anaerobic condition is maintained in the basin in which 65% total P released occurs. The solids are settled in another basin and phosphorus rich supernatant is processed for P recovery. Needed amounts of $Mg(OH)_2$, NaOH and ammonia are mixed for formation of magnesium ammonium sulfate phosphate (MAP) or struvite formation ($MgNH_4PO_4.6H_2O$). Estimate the quantity of struvate precipitated. The recovery of phosphorus in the struvite is 80%. Molecular weight of MAP = 245 g/mole.

13.26 Combined sludge from a conventional activated sludge plant is processed for recovery of protein. The quantity of sludge produced is 3400 kg/d. The VSS/TSS ratio = 0.8. The sludge solids are solubilized by alkaline treatment followed by ultra-sonication. The protein from the processed sludge is precipitated by acetic acid and ammonium sulfate treatment. The protein precipitation efficiency is 45%. Calculate the protein recovery.

13.27 An autothernal thermophilic aerobic digestion (ATAD) process is used for sludge digestion. Ammonia nitrogen concentration in the sidestream from ATAD process is 1500 mg/L and flow is 20 m^3/d. The sidestream is treated in a partial nitrification/Anammox (PN/A) process. The factors for oxygen requirement, alkalinity consumption, and new cell mass generation in the PN/A process are respectively 1.83 g O_2, 3.97 g alkalinity as $CaCO_3$, and 0.12 g VSS per gram of NH_4-N converted. The ammonia removal efficiency of the process is 65%. The alkalinity of combined sludge is 300 mg/L as $CaCO_3$. Calculate the oxygen and alkalinity requirements of the process, and total biomass yield.

13.28 The combined blended sludge is thickened in a gravity thickener. The characteristics of blended sludge are: flow = 400 m^3/d, TSS = 8000 kg/d, total BOD_5 = 4000 kg/d, soluble BOD_5 = 30 kg/d, Org-N = 550 kg/d, NH_4-N = 6 kg/d, TP = 56 kg/d, Orth-P (soluble) = 4 kg/d. VSS/TSS = 0.75 and biodegradable biosolids/TSS = 0.55. The solids capture efficiency of gravity thickener = 85% and TS in the thickened sludge = 5% and sp.gr = 1.01. Determine the characteristics of the sidestream and thickened sludge.

13.29 Conventional sludge drying beds are designed at a municipal wastewater treatment plant. The digested sludge production rate is 350 kg/d at a solids constant of 4.0%, and specific gravity of 1.015. The application rate of sludge is 30 cm/cycle. The cycle time is 25 d/cycle. Calculate the total area of sludge drying beds and total number of cells. The dimensions of each cell is 6 m × 15 m.

13.30 At a conventional drying bed 40% rainfall drains through the cake and 60% is evaporated. The average evaporation from the wet sludge is 70% of that of free water surface. The rainfall and evaporation in the month of May are 12 and 25 cm. The average solids loading L_s = 10.5 kg/m^2·cycle. After drainage of the beds, the solids content of 20% is reached in 24 h. The sludge cake is removed from the beds at 28% solids. Calculate the number of days required per cycle to evaporate water from the sludge drying beds.

13.31 A small wastewater treatment plant is using paved drying beds to dewater aerobically digested sludge. The sludge contains 220 kg/d dry solids. After decanting the sludge contain 0.08% solids, and the percent dry solids required for final disposal is 25%. The average annual precipitation and evaporation rates are 120 and 280 cm, respectively. The efficiency factor for water evaporation k_c = 0.6. Calculate the total area of paved drying beds.

13.32 A small municipal wastewater treatment plant produces 300 kg/d Class A biosolids at 25% dry solids content, and specific gravity of 1.08. The biosolids are further dehydrated in solar drying beds to produce pelles at 90% solids with specific gravity of 1.5. The design solids loading rate to the solar drying beds is 180 kg/m^2.year. Estimate (a) the area of solar drying beds, (b) volumes of biosolids before and after concentration in solar drying beds, and percent volume reduction, and (c) volume of water evaporated in solar drying beds.

13.33 Calculate the number of variables volume recessed plate filter units to dewater 5000 kg/d digested sludge under average flow conditions. The filter operation is 8 h/d and 5 d/week. The lime, ferric chloride, and polymer dosages are 6%, 1.3%, and 0.5% of dry solids, respectively. The filter cycle time is 80 min, and the filtration rate is 15 kg/m^2.h. The filter area of each chamber is 0.5 m^2. Also calculate how many chambers in each filter unit shall be provided so that 100% filtration can be achieved with one unit out of service, and 124% filtration capability is available when all units are in operation.

13.34 A belt filter is designed for dewatering sludge from a municipal wastewater treatment plant. The conditioned digested sludge that reaches the belt filter is 3360 kg/d at 6% solids, and has a specific gravity of 1.03. The design solids loading rate is 200 kg per meter of belt width per hour. The sludge cake is at 22% solids and has a specific gravity of 1.06. Assume that filter operation is 5 d/week. Calculate (a) the effective belt width and number of belt filters if available belt is 1.0 m wide; (b) the volume of filter cake if solids capture efficiency is 95%; (c) the volume of filtrate, assuming belt wash water is 30 L/kg TSS in sludge and TSS concentration in washwater is 15 mg/L; and (d) the concentration of TSS in filtrate.

13.35 The cumulative Cd loading rate for agricultural land is 39 kg/ha. The average concentration of Cd in biodolids is 80 mg/kg. Calculate the maximum amount of dry biosolids that can be applied over the useful life of the site (mt/ha). If the biosolids application rate is 10 mt dry solids per ha per year, and there is no loss of Cd in the crop, how many years will it take to reach the limiting value?

13.36 Anaerobically digested and dewatered biosolids containing NH_4-N = 0.5%, NO_3-N = 0.2%, and Org-N = 4.8% is applied over a farm land. The solids application rate is 9 mt/ha·year. Calculate the plant-available nitrogen in the fifth and sixth years. How much increase in plant-available N will occur between the fifth and sixth years?

13.37 A biosolids disposal project is using digested dewatered solids with a nutrient supplement. The nutrient contents in biosolids are NH_4-N = 0.8%, NO_3-N = 0.1%, Org-N = 3.8%, P = 4.5%, and K = 0.4%. The nutrient requirements for the crop are N = 30 kg/ha·year, P = 40 kg/ha·year, and K = 180 kg/ha·year. Estimate the quantities of nutrient supplements if the biosolids application rate is 10 kg/ha·year.

13.38 Calculate the useful life of a site in which sludge will be landfilled by an area fill mound operation. The site is ~3 ha, and after removing the cover and bedding material, the average depth of the disposal site is 3.5 m. The finished grade must be kept in line with the ground surface above the depression. Approximately 27 mt of residues will be landfilled each day. The bedding material and intermediate covers constitute 18% volume. The top cover is 1.3. Assume that the specific gravity of the mixed residue is 1.2. Also determine the equipment needed.

References

1. U.S. Environmental Protection Agency, *Process Design Manual for Sludge Treatment and Disposal*, EPA 625/1-74-006, Technology Transfer, Washington, DC, October 1974.
2. U.S. Environmental Protection Agency, *Process Design Manual: Municipal Sludge Landfills*, EPA-625/1-78-010, SW-705, Environmental Research Information Center, Technology Transfer, Office of Solid Waste, Cincinnati, OH, October 1978.
3. U.S. Environmental Protection Agency, *Process Design Manual for Sludge Treatment and Disposal*, EPA 625/1-79-011, Center for Environmental Research Information, Cincinnati, OH, September 1979.
4. U.S. Environmental Protection Agency, *Process Design Manual for Land Application of Municipal Sludge*, EPA-625/1-83-016, Center for Environmental Research Information, Cincinnati, OH, October 1983.
5. U.S. Environmental Protection Agency, *Environmental Regulations and Technology, Use and Disposal of Municipal Wastewater Sludge*, EPA 625/10-84-003, Intra-Agency Sludge Task Force, September 1984.
6. Qasim, S. R., *Wastewater Treatment Plants: Planning, Design, and Operation*, 2nd ed., CRC Press, Boca Raton, FL, 1999.
7. Metcalf & Eddy | AECOM, *Wastewater Engineering: Treatment and Resource Recovery*, 5th ed., McGraw-Hill, New York, NY, 2014.

8. Metcalf & Eddy, Inc., *Wastewater Engineering, Treatment and Reuse*, 4th ed., McGraw-Hill Companies, Inc., New York, NY, 2003.
9. Metcalf & Eddy, Inc., *Wastewater Engineering: Collection and Pumping of Wastewater*, McGraw-Hill Book Co., New York, NY, 1981.
10. Carthew, G. A., C. A. Goehring, and J. E. Van Teylingen, Development of dynamic head loss criteria of raw sludge pumping, *Journal of Water Pollution Control Federation*, 55(5), 1983, 472–483.
11. Mulbarger, M. C., S. R. Cropas, J. R. Kordic, and F. M. Cash, Pipeline friction losses for wastewater sludge, *Journal of Water Pollution Control Federation*, 53(8), 1981, 1303–1313.
12. Sanks, R. L., G. Tchobanoglous, D. Newton, B. E. Bosserman, and G. M. Jones (eds.), *Pumping Station Design*, Butterworths, Stoneham, MA, 1989.
13. Novak, J. T., D. H. Chon, B. Curtis, and M. Doyle, Biological solids reduction using the cannibal process, *Water Environment Research*, 79(12), 2007, 2380–2386.
14. Siemens Water Technologies Corp., *Cannibal® Solids Reduction Process*, http://www.siemens.com, 2007.
15. U.S. Environmental Protection Agency, *Biosolids and Residuals Management Fact Sheet: Odor Control in Biosolids Management*, EPA 832-F-00-067, Office of Water, Washington, DC, September 2000.
16. Evoqua Water Technologies, *Zabocs® Biological Odor Control Systems*, http://www.evoqua.com, 2014.
17. U.S. Environmental Protection Agency, 40 CFR Part 503 – Standards for the Use or Disposal of Sewage Sludge, 58 *FR* 9247, Feb. 19, 1993, and as amended later.
18. U.S. Environmental Protection Agency, *A Plain English Guide to the EPA Part 503 Biosolids Rule*, EPA/832/R-93/003, Office of Wastewater Management, Washington, DC, September 1994.
19. U.S. Environmental Protection Agency, *Process Design Manual: Land Application of Sewage Sludge and Domestic Septage*, EPA/625/R-95/001, Office of Research and Development, Washington, DC, September 1995.
20. U.S. Environmental Protection Agency, *Environmental Regulations and Technology: Control of Pathogens and Vector Attraction in Sewage Sludge*, EPA/625/R-92/013, Office of Research and Development, Washington, DC, Revised July 2003.
21. U.S. Environmental Protection Agency, *Emerging Technologies for Biosolids Management*, EPA 832-R-06-005, Office of Wastewater Management, Washington, DC, September 2006.
22. U.S. Environmental Protection Agency, *Biosolids Technology Fact Sheet: Alkaline Stabilization of Biosolids*, EPA 832-F-00-052, Office of Water, Washington, DC, September 2000.
23. U.S. Environmental Protection Agency, *Biosolids Technology Fact Sheet: Use of Composting for Biosolids Management*, EPA 832-F-02-024, Office of Water, Washington, DC, September 2002.
24. U.S. Environmental Protection Agency, *Biosolids Technology Fact Sheet: Heat Drying*, EPA 832-F-06-029, Office of Water, Washington, DC, September 2006.
25. Andritz Separation, Inc., *Drying Technologies for Sewage Sludge*, http://www.andritz.com, 2012.
26. Komline-Sanderson, *Komline-Sanderson Paddle Dryer, Drying Technology for Biosolids, Sludges and By-Products*, http://www.komline.com, 2008.
27. Therma-Flite, Inc., *Therma-Flite IC Series BIO-SCRU® Dryer System*, http://www.Therma-Flite.com, 2010.
28. Cambi Group AS, *CambiTHP™*, website information at http://www.cambi.com/Products/CambiTHP
29. Veolia Water Technologies, *Thermal Hydrolysis Processes*, http://www.veoliawatertechnologies.com, 2016.
30. U.S. Environmental Protection Agency, *Examples of Equivalent Processes: PFRP and PSRP*, website information at https://www.epa.gov/biosolids/examples-equivalent-processes-pfrp-and-psrp
31. SUEZ Treatment Solutions, Inc., *Suez 2PAD, Two-Phase Anaerobic Digestion System*, http://www.suez-na.com, 2012.

32. Task Force of the Water Environment Federation and the American Society of Civil Engineers/Environmental and Water Resources Institute, *Design of Municipal Wastewater Treatment Plant*, WEF Manual of Practice No. 8 and ASCE Manuals and Reports on Engineering Practice No. 76, 5th ed., McGraw-Hill, New York, NY, 2009.
33. Reynolds, T. D. and P. A. Richards, *Unit Operations and Processes in Environmental Engineering*, 2nd ed., PWS Publishing Co., Boston, MA, 1996.
34. Schroeder, E. D., *Water and Wastewater Treatment*, McGraw-Hill Book Co., New York, 1977.
35. Water Environment Federation, *Sludge Thickening*, Manual of Practice No. FD-1, Water Environment Federation, Alexandria, VA, 1980.
36. Andreoli, C. V., M. von Sperling, and F. Fernandes (eds.), *Biological Wastewater Treatment Series*, Vol. 6, Sludge Treatment and Disposal, IWA Publishing, London, UK, 2007.
37. U.S. Environmental Protection Agency, *Biosolids Technology Fact Sheet: Gravity Thickening*, EPA 832-F-03-022, Office of Water, Washington, DC, September 2003.
38. Suidan, M. T. (ed.), *Environmental Engineering Unit Operations and Unit Processes Laboratory Manual*, 4th ed., Association of Environmental Engineering Professors, University of Texas at Austin, Austin TX, 1988.
39. WesTech Engineering, Inc., *Dissolved Air Flotation for Municipal Clarification and Thickening*, http://www.westech-inc.com, 2012.
40. U.S. Environmental Protection Agency, *Biosolids Technology Fact Sheet: Centrifuge Thickening and Dewatering*, EPA 832-F-00-053, Office of Water, Washington, DC, September 2000.
41. Centrisys, Inc., *THK Series – Thickening Centrifuge: The Most Efficient Solution to Thicken Waste Activated Sludge (WAS)*, http://www.centrisys.us
42. Flottweg Separation Technology, *Flottweg OSE Thickening Centrifuges, Optimizing the Digestion Process in Wastewater Treatment Plants*, http://www.flottweg.com, 2013.
43. Andritz Separation, Inc., *PowerDrain, Gravity Belt Thickener for Efficient Sludge Treatment*, http://www.andritz.com, 2013.
44. Alfa Laval Inc., *Alfa Laval AS-H Belt Thickener*, http://www.alfalaval.com, 2014.
45. Komline-Sanderson, *The Komline-Sanderson Gravity Belt Thickener for Sludge/Slurry Volume Reduction/Gravabelt®*, http://www.komline.com, 1996.
46. Andritz Separation, Inc., *Drum Thickeners – PDR Series for Sludge Thickening Large Flow*, http://www.andritz.com, 2012.
47. Parkson Corporation, *ThickTech™ Rotary Drum Thickener*, http://www.parkson.com, 2010.
48. Kruger/Veolia Water Technologies, *Rotary Drum Thickener Treats All Types of Sludge*, http://www.kruger.dk
49. Woo, B., Membrane thickening aerobic digestion processes, *Water Science & Technology*, 69(7), 2014, 1509–1517.
50. Ovivo, *Ovivo ECO-MAT™ Technology*, http://www.ovivowater.com
51. Ovivo, *Ovivo Mem-TAD™ Process, Membrane Thickened Aerobic Digestion*, http://www.ovivowater.com, 2014.
52. U.S. Environmental Protection Agency, *Biosolids Technology Fact Sheet: Multi-Stage Anaerobic Digestion*, EPA 832-F-06-031, Office of Water, Washington, DC, September 2006.
53. U.S. Department of Agriculture, U.S. Environmental Protection Agency, and U.S. Department of Energy, *Biogas Opportunities Roadmap, Voluntary Actions to Reduce Methane Emissions and Increase Energy Independence*, U.S. Department of Agriculture, U.S. Environmental Protection Agency, and U.S. Department of Energy, Washington, DC, August 2014.
54. Energy and Environmental Analysis, Inc., and Eastern Research Group, Inc., *Biomass Combined Heat and Power Catalog of Technologies*, Vol. 1.1, U.S. Environmental Protection Agency, Combined Heat and Power Partnership, September 2007.
55. U.S. Environmental Protection Agency, *Food Waste to Energy: How Six Water Resource Recovery Facilities are Boosting Biogas Production and the Bottom Line*, EPA 600/R-14/240, Office of Research and Development, Cincinnati, OH, September 2014.

56. Crawford, G. and J. Sandino, *Energy Efficiency in Wastewater Treatment in North America: A Compendium of Best Practices and Case Studies of Novel Approaches*, WERF Report OWSO4R07e, Water Environmental Research Foundation (WERF) and IWA Publishing, 2010.
57. Water Environment Federation, *Wastewater Residuals Stabilization*, Manual of Practice No. FD-9, Water Environment Federation, Alexandria, VA, 1995.
58. Scisson, Jr. J. P., ATAD, the next generation: Design, construction, start-up and operation of the first municipal 2nd generation ATAD, *Proceedings of the WEF/AWWA/CWEA Joint Residuals and Biosolids Management Conference and Exhibition*, Baltimore, MD, February 2003.
59. Staton, K. L., J. E. Alleman, R. L. Pressley, and J. Eloff, 2nd generation autothermal thermophilic aerobic digestion: Conceptual issues and process advancements, *Proceedings of the WEF/AWWA/CWEA Joint Residuals and Biosolids Management Conference*, San Diego, CA, February 2001.
60. Gyger, R. F., J. R. McWhirter, and P. Balan, Continuous multi-stage autothermal aerobic digestion process, *Proceedings of the WEF/AWWA/CWEA Joint Residuals and Biosolids Management Conference and Exhibition*, Denver, CO, April 2007.
61. Layden, N. M., An evaluation of Autothermal Thermophilic Aerobic Digestion (ATAD) of municipal sludge in Ireland, *Journal Of Environmental Engineering Science*, 6(1), 2007, 19–29.
62. Layden, N. M., D. S. Mavinic, H. G. Kelly, R. Moles, and J. Bartlett, Autothermal Thermophilic Aerobic Digestion (ATAD) – Part I: Review of origins, design, and process operation, *Journal of Environmental Engineering and Science*, 6, 2007, 665–678.
63. Layden, N. M., H. G. Kelly, D. S. Mavinic, R. Moles, and J. Bartlett, Autothermal Thermophilic Aerobic Digestion (ATAD) – Part II: Review of research and full-Scale operating experiences, *Journal of Environmental Engineering and Science*, 6, 2007, 679–690.
64. Fuchs Gas- und Wassertechnik GmbH, *Fuchs ATAD Process, Advanced Integrated Concept from Sewage Sludge to Class A Biosolids*, www.fuchs-germany.com, 2011.
65. Thermal Process Systems, *TPS ThermAer™ Process, Your Class A Solution for Biosolids Management*, www.thermalprocess.com
66. NORAM Engineering and Constructors Ltd., *VERTAD™ Class A Biosolids Production*, http://www.noram-eng.com/
67. National Lime Association, *Lime: Handling, Application, and Storage in Treatment Processes*, Bulletin 213, 5th ed., National Lime Association, Arlington, VA, 1988.
68. U.S. Environmental Protection Agency, *Full Scale Demonstration of Lime Stabilization*, EPA 600/2-78-171, Municipal Environmental Research Laboratory, Cincinnati, OH, September 1978.
69. U.S. Environmental Protection Agency, *Lime Stabilized Sludge: Its Stability and Effect on Agricultural Land*, EPA 670/2-75-012, National Environmental Research Center, Cincinnati, OH, April 1975.
70. Park, C., C. D. Muller, M. M. Abu-Orf, and J. T. Novak, The effect of wastewater cations on activated sludge characteristics: Effects of aluminum and iron in floc, *Water Environment Research*, 78(1), 2006, 31–40.
71. Abu-Orf, M. M., C. D. Muller, C. Park, and J. T. Novak, Innovative technologies to reduce water content of dewatered municipal residuals, *Journal of Residuals Science & Technology*, 1(2), 2004, 83–91.
72. Higgins, M. J. and J. T. Novak, The effect of cations on the setting and dewatering of activated sludges: Laboratory results, *Water Environment Research*, 69, 1997, 215.
73. Metcalf and Eddy, Inc., *Wastewater Engineering: Treatment, Disposal and Reuse*, 3rd ed., McGraw-Hill Inc., New York, NY, 1991.
74. U.S. Environmental Protection Agency, *Design Manual, Dewatering Municipal Wastewater Sludges*, EPA/625/1-87/014, Center for Environmental Research Information, Cincinnati, OH, September 1987.
75. Water Pollution Control Federation, *Sludge Conditioning*, Manual of Practice No. FD-14, Water Pollution Control Federation, Alexandria, VA, 1991.

76. U.S. Environmental Protection Agency, *Innovations in Sludge Drying Beds: A Practical Technology*, EPA-832R87101, Center for Environmental Research Information, Cincinnati, OH, October 1987.
77. American Public Health Association (APHA), American Water Works Association (AWWA) and Water Environmental Federation (WEF), *Standard Methods for the Examination of Water and Wastewater*, 22nd ed., Rice, E. W., R. B. Baird, A. D. Eaton, and L. S. Clesceri (eds.), APHA, Washington, DC, 2012.
78. ASTM, *Standard Practice for Coagulation-Flocculation Jar Test of Water*, ASTM D2035-08, ASTM International, West Conshohocken, PA, 2008, http://www.astm.org
79. Eckenfelder, W. W., Jr. and D. L. Ford, *Water Pollution Control: Experimental Procedures for Process Design*, Pemberton Press, Jenkins Publishing Co., Austin and New York, 1970.
80. Ramalho, R. S., *Introduction to Wastewater Treatment Processes*, Academic Press, New York, NY, 1977.
81. Texas Department of Health, *Design Criteria for Sewage System*, Austin, TX, 1976.
82. Water Pollution Control Federation, *Sludge Dewatering*, Manual of Practice No. 20, Alexandria, VA, 1983.
83. U.S. Environmental Protection Agency, Design information report: Sidestreams in wastewater treatment plants, *Journal Water Pollution Control Federation*, 59(1), 1987, 54–59.
84. U.S. Environmental Protection Agency, Design information report: Design, operational, and cost considerations for vacuum associated sludge dewatering bed systems, *Journal Water Pollution Control Federation*, 59(4), 1987, 228–234.
85. Thermo-System, *Solar Drying of Sludge*, http://www.thermo-system.com, 2015.
86. Thermo-System, *Solar Storage Dryer*, http://www.thermo-system.com, 2015.
87. Veolia Water Technologies, *Solia™ The New Generation of Solar Sludge Drying*, http://www.krugerusa.com, 2016.
88. SUEZ Treatment Solutions, Inc., *Heliantis™ Solar Sludge Drying*, http://www.suez-na.com.
89. TenCate Geosynthetics Americas, *Geotube® Industrial Fabrics Dewatering Systems*, www.geotube.com, 2011.
90. Crites, R. W. and G. Tchobanoglous, *Small and Decentralized Wastewater Management Systems*, McGraw-Hill, New York, NY, 1998.
91. Jeffrey, E. A., Dewatering rates of digested sludge in lagoons, *Journal Water Pollution Control Federation*, 32(11), 1960, 1153–1160.
92. U.S. Environmental Protection Agency, *Biosolids Technology Fact Sheet: Belt Filter Press*, EPA 832-F-00-057, Office of Water, Washington, DC, September 2000.
93. Andritz Separation, Inc., *Heavy-Duty Belt Press CPF SMX-Q, A New Standard in Flexible, Effluent, and High-Throughput Dewatering*, http://www.andritz.com, 2016.
94. Alfa Laval Inc., *Alfa Laval AS-H Belt Press G2, Sludge Dewatering Machine*, http://www.alfalaval.com, 2014.
95. BDP Industries, *Model 3DP Belt Filter Press*, http://www.bdpindustries.com
96. Charter Machine Company, *Sentry Press, The Next Generation of 3-Belt Press from Charter Machine Company*, http://www.chartermachine.com
97. U.S. Environmental Protection Agency, *Design Information Report: Centrifuges*, EPA/600/M-86/023, Water Engineering Research Laboratory, Cincinnati, OH, September 1986.
98. Alfa Laval Inc., *ALDEC High-Performance Decanter Centrifuge*, http://www.alfalaval.com, 2013.
99. Andritz Separation, Inc., *Decanter Centrifuges, D-Series: High-Performance*, http://www.andritz.com, 2012.
100. Flottweg Separation Technology, *Flottweg Decanters for Water Treatment and Wastewater Facilities, Optimizing the Digestion Process in Wastewater Treatment Plants*, http://www.flottweg.com, 2016.

101. Centrisys, Inc., *Hydraulic Scroll Drive based on Rotodiff® Technology vs. Gearbox Drive: A Technical Clarification*, http://www.centrisys.us
102. U.S. Environmental Protection Agency, *Biosolids Technology Fact Sheet: Recessed-Plate Filter Press*, EPA 832-F-00-058, Office of Water, Washington, DC, September 2000.
103. Andritz Separation, Inc., *Overhead Filter Presses for Demanding Applications*, http://www.andritz.com, 2009.
104. Alfa Laval Inc., *Alfa Laval Plate Press Sludge Dewatering Machine*, http://www.alfalaval.com, 2014.
105. Putsch® FST, *Fully Automated Filter Presses, Series PKF*, http://www.putschusa.com, 2010.
106. FLSmidth Salt Lake City, Inc., *FLSmidth Shriver® Filter Press Products*, http://www.flsmidth.com, 2012.
107. Fournier Industries Inc., *Optimum-CV Rotary Press Dewatering…, Municipal and Industrial Waste Water Applications*, http://www.rotary-press.com, 2013.
108. Prime Solution, Inc., *The Prime Rotary Fan Press Family*, http://www.psirotary.com
109. Scullion, M., K. R. Tsang, L. Tunnell, W. Hurley, and M. Burgess, Biosolids dewatering alternatives for orange county, *Florida Water Resources Journal*, http://fwrj.com/?page_id=163, 2012, 18–26.
110. FKC Co., Ltd., *Biosolids Dewatering*, http://www.fkcscrewpress.com
111. Huber Technology, Inc., *Screw Press ROTAMAT® RoS 3Q*, http://www.huber-technology.com, 2013.
112. BDP Industries, *Model DSP Screw Press*, http://www.bdpindustries.com, 2016.
113. Process Wastewater Technologies LLC, *Volute Dewatering Press*, http://www.PWTech.us, 2016.
114. Ovivo, *CINETIK® Linear Electro-Dewatering, Sustainable Biosolids Solutions for the Wastewater Industry*, http://www.PWTech.us, 2010.
115. Lafitte-Trouque, S. and C. F. Forster, The use of ultrasound and γ-irradiation as pretreatments for the anaerobic digestion of waste activated sludge at mesophilic and thermophilic temperatures, *Bioresource Technology*, 84(2), 2002, 113–118.
116. Bougrier, C., C. Albasi, J. P. Delgenes, and H. Carrere, Effect of ultrasonic, thermal and ozone pretreatments on waste activated sludge solubilisation and anaerobic biodegradability, *Chemical Engineering and Processing*, 45(8), 2006, 711–718.
117. Haug, R. T., D. C. Stuckey, J. M. Gossett, P. L. McCarty, Effect of thermal pretreatment on digestibility and dewaterability of organic sludge, *Journal Water Pollution Control Federation*, 50(1), 1978, 73–85.
118. Long, J. H. and C. M. Bullard, Waste activated sludge pretreatment to boost volatile solids reduction and digester gas production: Market and technology assessment, *Florida Water Resources Journal*, http://fwrj.com/?page_id=163, 2014, 44–50.
119. Yeom, I. T., K. R. Lee, K. H. Ahn, and S. H. Lee, Effects of ozone treatment on the biodegradability of sludge from municipal wastewater treatment plants, *Water Science and Technology*, 46(4–5), 2002, 421–425.
120. Weemaes, M., H. Grootaerd, F. Simoens, and W. Verstraete, Anaerobic digestion of ozonated biosolids, *Water Research*, 34(8), 2000, 2330–2336.
121. Battimelli, A., C. Millet, J. P. Delgenes, and R. Moletta, Anaerobic digestion of waste activated sludge combined with ozone post-treatment and recycling, *Water Science and Technology*, 48(4), 2003, 61–68.
122. Vranitzky, R. and J. Lahnsteiner, *Sewage Sludge Disintegration Using Ozone – A Method of Enhancing the Anaerobic Stabilization of Sewage Sludge*, VA TECH WABAG, R&D Process Engineering, Vienna, Austria, http://www.wabag.com/wp-content/uploads/2012/04/Sewage-sludge.pdf, 2012.
123. Vranitzky, R. and J. Lahnsteiner, Ozone treatment of organic micro-pollutants in sewage sludge, *Proceedings of the 12th IWA Sludge Conference – Sustainable Management of Water and Wastewater Sludges*, Harbin, China, August 2009.

124. Dewil, R., L. Appels, J. Baeyens, and J. Degreve, Peroxidation enchances the biogas production in the anaerobic digestion of biosolids, *Journal of Hazardous Materials*, 146(3), 2007, 557–581.
125. Shang, M. and H. Hou, Studies on effects of peracetic acid pretreatment on anaerobic fermentation biogas production from sludge, *Proceedings of Asia-Pacific Power and Energy Engineering Conference (APPEEC 2009)*, Wuhan, China, March 2009.
126. Appels, L., A. Van Assche, K. Willems, J. Degreve, J. Van Impe, and R. Dewil, Peracetic acid oxidation as an alternative pre-treatment for the anaerobic digestion of waste activated sludge, *Bioresource Technology*, 102(5), 2011, 4124–4130.
127. Atchley, A. A. and L. A. Crum, Acoustics cavitation and bubble dynamics, *Ultrasound: Its Chemical, Physical and Biological Effects*, K. S. Suslick (ed.), VCH Publishers, Weinheim, 1988.
128. Kelly, H. G., Emerging processes in biosolids treatment 2003, Paper Presented at the 2nd Canadian Organic Residuals Conference, Penticton, British Columbia, April 2003.
129. Chiu, Y., C. Chang, J. Lin, and S. Huang, Alkaline and ultrasonic pretreatment of sludge before anaerobic digestion, *Water Science and Technology*, 36(11), 1997, 155–162.
130. Tiehm, A., K. Nickel, M. Zellhorn, and U. Neis, Ultrasonic waste activated sludge disintegration for improving anaerobic stabilization, *Water Research*, 35(8), 2001, 2003–2009.
131. Bougrier, C., H. Carrere, and J. P. Delgenes, Solubilisation of waste-activated sludge by ultrasonic treatment, *Chemical Engineering Journal*, 106(2), 2005, 163–169.
132. Show, K. Y., T. Mao, and D. J. Lee, Optimisation of sludge disruption by sonication, *Water Research*, 41(20), 2007, 4741–4747.
133. Braguglia, C. M., G. Mininni, and A. Gianico, Is sonication effective to imporve biogas production and solids reduction in excess sludge digestion? *Water Science and Technology*, 57(4), 2008, 479–483.
134. Kidak, R., A. Wilhelm, and H. Delmas, Effect of process parameters on energy requirement in ultrasonical treatment of waste sludge, *Chem. Eng. and Processing*, 48(8), 2009, 1346–1352.
135. Perez-Elvira, S., M. Fdz-Polanco, F. I. Plaza, G. Garralon, and F. Fdz-Polanco, Ultrasound pretreatment for anaerobic digestion improvement, *Water Science and Technology*, 60(6), 2009, 1525–1532.
136. Carrere, H., C. Dumas, A. Battimelli, D. J. Batstone, J. P. Delgenes, J. P. Steyer, and I. Ferrer, Pretreatment methods to improve sludge anaerobic degradability: A review, *Journal of Hazardous Materials*, 183(1–3), 2010, 1–15.
137. Guellil, A., M. Boualam, H. Quiquampoix, P. Ginestet, J. M. Audic, and J. C. Block, Hydrolysis of wastewater colloidal organic matter by extracellular enzymes extracted from activated sludge flocs, *Water Science and Technology*, 43(6), 2001, 33–40.
138. Werker, A. G., M. Carlsson, F. Morgan-Sagastume, M. S. Le, and D. Harrison, Full scale demonstration and assessment of enzymic hydrolysis pre-treatment for mesophilic anaerobic digestion of municipal wastewater treatment sludge, *Proceedings of the 80th Annual WEFTEC®*, Water Environment Federation, San Diego, CA, October 2007.
139. Yu, S., G. Zhang, J. Li, Z. Zhao, and X. Kang, Effect of endogenous hydrolytic enzymes pretreatment on the anaerobic digestion of sludge, *Bioresource Technology*, 146, 2013, 758–761.
140. Onyeche, T., O. Schlafer, and M. Sievers, Advanced anaerobic digestion of sludge through high pressure homogenization, *Journal of Solid Waste Technology and Management*, 29(1), 2003, 56–61.
141. Zhang, Y., P. Zhang, J. Guo, W. Ma, W. Fang, B. Ma, and X. Xu, Sewage sludge solubilization by high-pressure homogenization, *Water Science and Technology*, 67(11), 2013, 2399–2405.
142. Fang, W., P. Zhang, J. Ye, Y. Wu, H. Zhang, J. Liu, Y. Zhu, and G. Zeng, Physicochemical properties of sewage sludge disintegrated with high pressure homogenization, *International Biodeterioration & Biodegradation*, 102(5), 2015, 126–130.
143. Rittmann, B., H. S. Lee, H. Zhang, J. Alder, J. E. Banaszak, and R. Lopez, Full-scale application of focused-pulse pre-treatment for improving biosolids digestion and conversion to methane, *Water Science and Technology*, 58(10), 2008, 1895–1901.

144. Simmon, H., Sludge pretreatment system holds promise for increased biogas production, *WaterWorld*, 28(2), http://www.waterworld.com, 2012.
145. Montgomery, L. F. R. and G. Bochmann, *Pretreatment of Feedstock for Enhanced Biogas Production*, IEA Bioenergy, http://www.iea-biogas.net, 2014.
146. Park, B., J. H. Ahn, J. Kim, and S. Hwang, Use of microwave pretreatment for enhanced anaerobiosis of secondary sludge, *Water Science and Technology*, 50(9), 2004, 17–23.
147. Eskicioglu, C., N. Terzian, K. J. Kennedy, R. L. Droste, and M. Hamoda, Athermal microwave effects for enhancing digestibility of waste activated sludge, *Water Research*, 41(11), 2007, 2457–2466.
148. Yu, Q., H. Lei, Z. Li, H. Li, K. Chen, X. Zhang, and R. Liang, Physical and chemical properties of waste-activated sludge after microwave treatment, *Water Research*, 44(9), 2010, 2841–2849.
149. Elagroudy, S. and F. El-Gohary, Microwave pretreatment of mixed sludge for anaerobic digestion enhancement, *International Journal of Thermal & Environmental Engineering*, 5(2), 2013, 105–111.
150. Tyagi, V. K. and S. L. Lo, Microwave irradiation: A sustainable way for sludge treatment and resource recovery, *Renewable and Sustainable Energy Reviews*, 18, 2013, 288–305.
151. Wang, Y., Q. Xiao, J. Liu, H. Yan, and Y. Wei, Pilot-scale study of sludge pretreatment by microwave and sludge reduction based on lysis-cryptic growth, *Bioresource Technology*, 190, 2015, 140–147.
152. Veolia Water Technologies, *Bio Thelys™ Batch Thermal Hydrolysis*, http://www.veoliawater technologies.com, 2016.
153. Veolia Water Technologies, *Exelys™ Continuous Thermal Hydrolysis*, http://www.veoliawater technologies.com, 2016.
154. Praxair Technoloby, Inc., *Lyso™ Ozone Sludge Reduction System*, http://www.praxair.com, 2008.
155. Ovivo, *Sonolyzer™ Ultrasound Sludge Disintegrator*, http://www.ovivowater.com, 2015.
156. Rooksby, F., *Sonix™ A General Introduction*, http://www.biomedia.kr/sonix/pdf/g.pdf
157. VTA Technologie GmbH, *GSD: VTA's Patented Process for Disintegrating Sewage Sludge with Ultrasound*, http://www.vta.cc
158. *Paradigm Launches MicroSludge Portable Pilot System*, News on Water Online, August 20, 2009, https://www.wateronline.com/doc/paradigm-launches-microsludge-portable-pilot-0001
159. BIOGEST AG, *Crown – Disintegration System*, http://www.biogest.com/products/sludge-treatment/crown-disintegration-system
160. Vogelsang USA Inc., *BioCrack*, http://www.vogelsangusa.com, 2011.
161. Trojan Technologies, *OpenCel*, http://www.trojantechnologies.com, 2014.
162. Jeyanayagam, S., W. Khunjar, A. Pramanik, and C. Mehta, Accelerating the implementation of extractive nutrient recovery as an integral component of sustainable nutrient management, *Water e-Journal* (Online Journal of the Australian Water Association at http://www.awa.asn.au/AWA_MBRR/Publications/Water_e-Journal/02_-__RESOURCE_RECOVERY_-_Jeyanayagam.aspx), 1(2), 2016.
163. Cordell, D., J. Drangert, and S. White, The story of phosphorus: global food security and food for thought, *Global Environmental Change*, 19(2), 2009, 292–305.
164. CDM, *Charting the Future of Biosolids Management*, Water Environmental Federation and National Biosolids Partnership, 2011.
165. Wilson, C., F. Betancourt, and D. Hogan, How sidestream phosphate removal can benefit biosolids management programs in florida's nutrient-selective watersheds, *Florida Water Resources Journal*, http://fwrj.com/?page_id=163, 2013, 4–7.
166. Battistoni, P., P. Pavan, F. Cecchi, and J. Mata-Alvarez, Phosphate removal in real anaerobic supernatants: Modeling and performance of a fluidized bed reactor, *Water Science and Technology*, 38(1), 1998, 275–283.
167. P-REX, *AirPrex® Struvite Crystallization in Sludge*, http://www.p-rex.eu, February 2015.

168. Royal HaskoningDHV, *The Chrystalactor® Efficient Treatment without Waste*, http://www.crystalactor.com
169. Multiform Harvest, *Multiform Harvest*, http://www.multiformharvest.com
170. Ostara Nutrient Recovery Technologies Inc., *Nutrient Management Solution*, http://ostara.com/nutrient-management-solutions/, 2016.
171. Paques, *Phospaq™*, http://en.paques.nl/products/other/phospaq, 2016.
172. Zhang, Y., E. Desmidt, A. V. Looveren, L. Pinoy, B. Meesschaert, and B. Van der Bruggen, Phosphate separation and recovery from wastewater by novel electrodialysis, *Environmental. Science and Technology*, 47(11), 2013, 5888–5895.
173. Wang, X., Y. Wang, X. Zhang, H. Feng, C. Li, and T. Xu, Phosphorus recovery from excess sludge by conventional electrodialysis (CED) and electrodialysis with bipolar membranes (EDBM), *Industrial & Engineering Chemistry Research*, 52(45), 2013, 15896–15904.
174. Kelly, P. T. and H. Zhu, Nutrients removal and recovery in bioelectrochemical systems: A review, *Bioresource Technology*, 153, 2014, 351–360.
175. Fraunhofer IGB, *Electrochemical Process for Phosphorus Recovery*, http://www.igb.fraunhofer.de
176. Fraunhofer IGB, Fertilizer from Wastewater, *Fraunhofer Research News* at http://www.igb.fraunhofer.de, May 30, 2016.
177. Chishti, S. S., S. N. Hasnain, and M. A. Khan, Studies on the recovery of sludge protein, *Water Research*, 26(2), 1992, 241–248.
178. Hwang, J., L. Zhang, S. Seo, Y. Lee, and D. Jahng, Protein recovery from excess sludge for its use as animal feed, *Bioresource Technology*, 99(18), 2008, 8949–8954.
179. Pervaiz, M. and M. Sain, High-yield protein recovery from secondary sludge of paper mill effluent and its characterization, *BioResources*, 7(3), 2012, 2933–2947.
180. Lackner, S., E. M. Gilbert, S. E. Vlaeminck, A. Joss, H. Horn, and M. C. van Loosdrecht, Full-scale partial nitritation/anammox experiences - An application survey, *Water Research*, 55, 2014, 292–303.
181. Veolia Water Technologies, *ANITA™Mox Sustainably Treating Highly Concentrated Ammonia-Loaded Effluents*, http://www.veoliawatertechnologies.com, 2014.
182. Purac AB, *DeAmmon® Efficient Nitrogen Removal*, http://www.purac.se
183. DEMON GmbH, *DEMON® - Deammonification Process*, http://www.essdemon.com/en/demon/demon-process
184. Grontmij, *SHARON Nitrogen Removal over Nitrite*, http://www.grontmij.com
185. Paques, *Anammox™ Sustainable Nitrogen Removal*, http://www.paques.nl, 2016.
186. Turovskiy, I. S. and J. D. Westbrook, Recent advancements in wastewater sludge composting, *Florida Water Resources Journal*, http://fwrj.com/?page_id=163, 2003, 42–44.
187. U.S. Environmental Protection Agency, *Composting of Municipal Wastewater Sludges*, EPA/625/4-85/014, Seminar Publication, Center for Environmental Research Information, Office of Research and Development, Cincinnati, OH, August 1985.
188. Haug, P. T., *Compost Engineering: Principles and Practices*, Ann Arbor Science Publishers, Inc., Ann Arbor, MI, 1980.
189. U.S. Environmental Protection Agency, *Emerging Technologies for Wastewater Treatment and In-Plant Wet Weather Management*, EPA 832-R-06-006, Office of Wastewater Management, Washington, DC, February 2008.
190. U.S. Environmental Protection Agency, *Summary Report: In-Vessel Composting of Municipal Wastewater Sludge*, EPA/625/8-89/016, Center for Environmental Research Information, Cincinnati, OH, September 1989.
191. Veolia Water Technologies, *BioCon™ Biosolids Drying & Energy Recovery*, http://www.veoliawatertechnologies.com, 2015.
192. Tchobanoglous, G., H. Theisen, and S. Vigil, *Integrated Solid Waste Management: Engineering Principles and Management Issues*, McGraw-Hill, Inc., NY, 1993.

193. Corey, R. C., *Principles and Practices of Incineration*, Wiley-Interscience, NY, 1969.
194. Sulzle Kopf Anlagenbau GmbH, *SynGas Solutions*, http://suelzle-kopf.de/en/syngas/solutions/
195. Aries Clean Energy, *Gasification, Delivering Clean Electrical or Thermal Energy from Waste*, https://ariescleanenergy.com/gasification
196. Zemmermann, F. J., Wet air combustion, *Industrial Water and Wastes*, 6(4), 1961, 102–106.
197. Chauzy, J., J. C. Martin, D. Cretenot, and J. P. Rosiere, Wet air oxidation of municipal sludge: Return experience of the north brussels waste water treatment plant, *Water Practice & Technology*, 5(1), 2010, 1–8.
198. Siemens, *Zimpro® Wet Air Oxidation Systems: The Cleanest Way to Treat the Dirtiest Water*, http://www.siemens.com, 2011.
199. Galandak, J. and M. Racstain, Design considerations for pyrolysis of municipal sludge, *Journal of Water Pollution Control Federation*, 51(2), 1979, 370–377.
200. Chen, D., L. Yin, H. Wang, and P. He, Pyrolysis technologies for municipal solids waste: A review, *Waste Management*, 34(12), 2014, 2466–2486.
201. U.S. Environmental Protection Agency, 40 CFR Parts 527 and 258 – Solid Waste Disposal Facility Criteria; Final Rule, 56 *FR* 50978, Oct. 9, 1991, and as amended later.
202. Qasim, S. R., and W. W. Chiang, *Sanitary Landfill Leachate: Generation, Control and Treatment*, CRC Press LLC, Boca Raton, FL, 1994.
203. U.S. Environmental Protection Agency, 40 CFR Parts 86, 87, 89 et al. – Mandatory Reporting of Greenhouse Gases; Final Rule, 75 *FR* 56260, Oct. 30, 2009, and as amended later.
204. Koerner, R. M., *Designing with Geosynthetics*, 5th ed., Pearson/Prentice Hall, Upper Saddle River, NJ, 2005.
205. Goldman, L. J., L. I. Greenfield, A. S. Damle, G. L. Kingsbury, C. M. Northeim, and R. S. Truesdale, *Design, Construction, and Evaluation of Clay Liners for Waste Management Facilities*, EPA/530-SW-86-007-F, Office of Solid Waste and Emergency Response, U.S. Environmental Protection Agency, Washington, DC, 1986.
206. Haxo, H. E., R. S. Haxo, N. A. Nelson, P. D. Hexo, R. M. White, S. Vakessian, and M. A. Fong, *Liner Materials for Hazardous and Toxic Wastes and Municipal Solid Waste Leachate*, Noyes Publications, Park Ridge, NJ, 1985.
207. Matrecon, Inc., *Lining of Waste Containment and Other Impoundment Facilities*, USEPA/600/2-88/052, Risk Reduction Engineering Laboratory, Cincinnati, OH, September 1988.
208. U.S. Environmental Protection Agency, *Inventory of U.S. Greenhouse Gas Emissions and Sinks: 1990–2014*, EPA 430-R-16-002, Washington, DC, April 15, 2016.
209. Flanagan, K., Methane recovery does more than provide energy, *Solid Waste and Power*, 2(4), 1988, 30–33.
210. Knox, T. D., Landfill gas management technologies, *Waste Age*, 23(4), 1992, 233–244.

14

Plant Layout, Yard Piping, Plant Hydraulics, and Instrumentation and Controls

14.1 Chapter Objectives

Proper layout of treatment units, connecting conduits, roads and parking facilities, administration building, and maintenance shops is important for day to day plant operation and maintenance, and future expansion. Also, the elevations of treatment units and pipings are properly adjusted to give adequate hydraulic gradient for gravity flow, and to protect from submergence under peak flows. Proper use of instrumentation and automatic controls reduces labor requirements, chemical usages, and power consumptions; improves treatment process performance and reliability; and provides adequate data acquisition and recording. The objectives of this chapter are to:

- Present the basic considerations for plant layout, and in particular a compact and modular design concept
- Provide guidelines for selection of appurtenances and connecting pipings between the treatment units
- Present principles of designing yard piping and preparation of hydraulic profile
- Present basic process control strategies for physical and chemical measuring elements and instruments, and automatic control systems

14.2 Plant Layout

Plant layout is the physical arrangement of treatment units, pipings, buildings, and roads on the selected site. Proper plant layout can (1) enhance the attractiveness of the site; (2) minimize construction and operation costs; (3) offer flexibility in future process modifications and plant expansion; (4) fit the operational and maintenance needs of the processes and plant personnel; (5) integrate the functions of all components; and (6) maintain the landscaping and plant structure in perfect harmony with the environment.[1–3]

14.2.1 Factors Affecting Plant Layout and Site Development

Many factors are considered for plant layout on the selected site. Some of these factors are discussed below. A typical layout plan of a compact facility is presented in Example 6.24 and shown in Figure 6.12. Aerial views of several layouts of treatment plants are shown in Figure 14.1.

(a) (b) (c) (d)

FIGURE 14.1 Aerial views of wastewater treatment plants: (a) conventional activated sludge plant with nitrification, 22,800 m^3/d (6.0 MGD); (b) conventional activated sludge plant with nitrification, 28,500 m^3/d (7.5 MGD); (c) conventional activated sludge plant with nitrification, 22,000 m^3/d (5.8 MGD); and (d) oxidation ditch plant with nitrification, 3500 m^3/d (0.9 MGD) (Courtesy Freese and Nichols, Inc., Engineers, Architects, Planners, Fort Worth, Texas.)

14.2.2 Construction Requirements

Construction costs depend upon the site topography, geology and hydrology; foundation considerations; access roads; and building. These requirements are discussed below.[2–4]

Site Topography: A site on a sidehill with moderate slope will allow locating various treatment units in their normal sequence without excessive excavation or filling. This will provide gravity flow, reduce pumping requirement, cause the least disruption to the natural topography, and require the least erosion control measures. The sludge handling building can be located on the side hill providing direct access to the trucks from the ground or upper levels. Also, the chemical storage silos can have direct access both from the ground and from the upper levels of the chemical building. The sloping site development will not alter the existing site contours and drainage pattern.

Geology and Foundation: The ability of the ground to support structures without extensive piling is cost effective for supporting the basins, buildings, and permanent equipment. Thus, the load-bearing capacity, excessive and differential settlements, and floatation caused by high groundwater table are important considerations in plant layout. Separation of massive concrete structures such as basins, filters, and silos from other smal to medium size facilities is necessary to minimize uneven subsidence of the structures.

Access Roads: The plant layout should include access roads to all treatment units, buildings, equipment, and special areas. Railroads are helpful for the delivery of bulk chemicals and transport of residuals. Trucks and service traffic should be separated from visitors and employee traffic upon entry to the plant. Employee parking at the rear and visitor parking in the front of the administration building will enhance the traffic flow. Signs for visitor parking and other buildings should be posted. Additionally, turning radius and alternative access should be considered in the routing of chemical delivery trucks.

Building: Buildings are needed for plant personnel; major process equipment, such as pumps and compressors, chemicals, and instrumentation; maintenance and machine shops; and storage rooms and garages. These buildings should be provided in groups for maximum efficiency. The following considerations should be given to building design and location.

1. Locate process equipment with enclosures at the point of maximum usage.
2. The administration building should be located near the entrance and in public view. The administration building should contain offices, laboratory, instrumentation and control room, showers, lavatories, and locker rooms. In addition, the *visitor's lobby* should have education displays and other information for visitors.
3. Buildings may be used as barriers to block the undesired views of the treatment facility.
4. The climate of the area should be considered in building orientation and designed to minimize heating, ventilation, lighting, and air-conditioning costs. In cold regions, building should not shade trucking and parking areas. This will reduce snow and ice clearance.

14.2.3 Compact and Modular Site Development

A compact site plan minimizes buildup area and pipe requirement. Centralization of similar process units, process equipment, personnel, and facilities may reduce total staff size, as well as optimize plant supervision and operation needs. The traditional linear plant layout is not ideal for chemical feed and operation needs because of long distances. Many important design considerations of compact and modular design are given below.[2,4–6]

Modular Layout: Modular and compact plant layout is more suitable for future capacity expansion and upgrading to meet the stringent effluent quality limits after discharge permit renewal. The modular design uses parallel process trains with cross-connections for operational flexibility.

Construction and Structural Requirements: A compact design uses common-wall construction. This concept requires less excavation volumes, reduces concrete quantities because of common walls, and occupies less floor area because of multiple uses. Provisions for structural movement caused by concrete shrinkage, expansion, soil movement, and unequal settlement must be made. The location of isolation–expansion joints should be coordinated with all disciplines so that appropriate measures can be taken for potential movement.

Mechanical Equipment: The major mechanical equipment is consolidated into one central area near the support and maintenance facilities in the operations building. Equipment access in a compact design may be restricted. Trucks may not be able to get as close to the process equipment as they might with a dispersed plant layout. Provisions for monorails, bridge crane, access openings, and other measures should be provided for removal and placement of equipment.[5] Adequate clearance should also be provided around all equipment and accessories for maintenance and disassembly.

Elevations of Building Floors, Equipment, and Headroom: In a compact layout, the elevations of most treatment units, chemical storage and feeders, and building floors become interrelated. Thus, raising or lowering of one basin may affect the elevations of other units and building floors. Headroom clearance and operating elevations and other such details should be considered at an early stage of the design.

Electrical and Other Utility Lines: The electrical needs require special attention in a compact layout. Electrical conduits, chemical pipes, plumbing, and heating, ventilating, and air-conditioning (HVAC) will be concentrated into a smaller area and in two to three floors. Careful planning is needed to avoid conflict between disciplines in both vertical and horizontal directions, room layout, floor openings, and slab penetrations. In a compact design, the length of electrical and instrumentation conduits and cables are reduced. Cable sizes due to shorter lengths and voltage drop may also be reduced. Because the cables and other lines are routed indoors instead of in buried conduits between the structures, cost-effective indoor cable trays can be used, and they may be more accessible.[5]

Safety and Hazard Protection: In a compact layout, the plant staff will work close and around the storage, handling, and delivery of hazardous chemicals. Appropriate articles in the adopted fire code dealing with hazardous materials used in wastewater treatment should be applied.* Minimum distances from storage tank to building, openings in buildings, and air intakes are important design considerations. Other governing codes, and local regulations and permits should be checked at an early stage of the design. Related safety issues and compliance with relevant building codes for fire ratings, and access–egress because of the increased use of hallways and stairways must be met. Combining building use for administration and operation into a single structure may result in a building with *occupancy classifications*. Under such conditions, it can be difficult to meet all the requirements of the building code.[5,7,8]

General Operation and Maintenance: Compact and modular layout offers many other operational benefits. Some of these are:

1. Short walking distance to equipment and treatment processes.
2. Covered access by means of galleries and tunnels.
3. With fewer structures due to common uses, the number of internal access roads and paved pathways are reduced.
4. Shorter connecting piping reduces head losses.
5. Because of a compact design, the odors and other environmental control facilities are more effectively and economically installed and operated.
6. The number of plant operation staff may be reduced because the equipment needing regular attention is located close to the operations building.[5]

Underground and Multistory Treatment Plants: Many multistory, underground, and totally covered treatment plants have been built and operated successfully. Examples of these facilities may be found in Asia, Europe, the United States, and other countries.[2,9–13] These plants are examples of highly compact layouts. It is expected that more multistory, underground, and covered facilities will be built in the future as suitable sites become scarce; and emissions from treatment plants are strictly regulated. In the areas where the available land is limited, this trend will be especially necessary for sustainable devemopent.[14,15]

14.2.4 Plant Utilities

The utilities at a treatment plant include electrical power, natural gas, water lines, effluent lines, telephone lines, and intercommunication system. The digester gas at the plant is normally used for heating the buildings and digesters, and to produce hot water and steam. Power generation from digester gas may be an economical option for larger plants. All utility lines should conform to the applicable codes and regulations set by the municipality and to the operating rules of the concerned utilities. All utility lines should be grouped properly in a utility chase, marked on the site if exposed, and shown properly on the as-built layout plans. This will facilitate repairs, modifications, and future expansions.[1–3,5,8]

14.2.5 Environmental Considerations and Security

Many environmental issues at a plant are odors, noise, landscaping, shoreline planning, flood protection, lighting, and security.[1–3] These issues are briefly discussed below.

Odor and Aerosol Sources: At wastewater treatment facilities, the potential sources of odor and aerosol include such treatment processes as headworks (screens and grit removal), primary clarifiers, aeration basins, pumps stations, thickeners, and sludge holding tanks. The mostly concerned facilities should be located downwind from public spaces and residential developments near the site. Such units should be

* Originally, there were three fire codes: the Uniform Fire Code (UFC), the Standard Fire Code (SFC), and the National Fire Code (NFC). Combination of these codes occurred in late 1990s, resulting in only two fire codes: the NFPA® 1 Fire Code (also known as UFC) and the International Fire Code® (IFC®). The applicable fire code is one that is adopted in the city's ordinance.[7,8]

located near each other and covered if possible. Protective barriers such as large trees and sufficient buffer zones should also be considered.[16]

Noise Sources: Equipment such as pumps, ejectors, generators, blowers, and compressors produce disturbing sound levels. This equipment should be isolated to prevent discomfort to plant personnel and neighbors. Sound barriers and absorptive enclosures should be provided to control disturbing noise. Vehicular noise should also be considered in the layout of access roads.

Landscaping: Existing vegetation, trees, and shrubs on site should be assessed and retained if possible. Planting should be considered for control of slope erosion, surface runoff, enhance attractiveness, and provide barriers for noise, odor, and aerosol controls. Local soil, climatological, and biological conditions should be carefully reviewed to maximize the use of native plants and shrubs.

Lighting and Illumination: Proper lighting at the plant site promotes safer operation, safety and security, and efficiency. Both interior and exterior, and security lighting add to safer environment. Also, illumination to highlight the prominent structural and landscape features make the site more attractive and desirable.

Shoreline Planning: The waterfront is an important open area for recreation. Erosion of banks, siltation of waterways, and destruction of valuable ecological niches should be protected from any harm or damage. The treatment units and other structures should be kept as far back as possible from the water edge to permit other compatible uses of particularly recreation and open space. The structure should be properly oriented to least obstruct the shoreline view.

Security: Plant security is essential for safety of plant personnel and equipment. All accesses to the plant should be controlled. Fences and barriers should be provided to secure the facility. Security cameras, and audio and visual alarms should be installed to alert the plant personnel as needed.

Flood Plain Avoidance: Protection of treatment facility from flooding should be the top priority. The buildings and structures should be raised above the flood level, or flood protection levees should be constructed. The surface runoff within the plant site should be collected at a designated location. A stormwater management plan should be developed with outlet structure, flood gate, stormwater pumps, and outfall pipe for pumping the stormwater above the highest flood stage in the receiving water.

14.2.6 Occupational Health and Safety

A wastewater treatment facility has many types of occupational health hazards that must be considered as part of plant design and layout. Important factors for safeguard are chemicals and chemical handling, toxic gases, biological vectors, fire protection, explosions, burns, electric shocks, rotating machinery parts, material and equipment handling, falls and drowning, and the like.[2,4,7,8,17,18]

14.2.7 Future Expansion

Provision for future expansion and upgrading of the plant will be needed. Therefore, the considerations should be given to (1) keeping sufficient land in reserve for expansion and (2) least interruption to the existing plant operation during process modifications or capacity expansions.

EXAMPLE 14.1: CHECKLIST FOR A COMPACT AND MODULAR PLANT LAYOUT

Develop a checklist for physical layout and site development of a compact wastewater treatment plant.

Solution

The checklist of a compact layout site development plan of a wastewater treatment facility should include the necessary planning and design factors. Such a checklist is given below.

1. Site Characteristics.
   a. Land area: Select a site large enough for current development and future expansion.
   b. Topography: Select a site on a sidehill to facilitate gravity flow.
   c. Geotechnical data: Obtain soil data for foundation design, soil stability, and drainage.
   d. Review environmental constraints for site selection: (a) plant and animal communities, (b) ecosystem, (c) endangered and locally threatened species, (d) unique and valuable environmental features, and (e) unique archeological, historical, scientific, or cultural areas such as parks, wetlands, or stream corridors.
   e. Review permit requirements for construction: (a) zoning and land use, (b) NPDES permit, (c) Clean Water Act, Section 404, (d) River and Harbor Act, Section 10, (e) Clean Air Act, (f) U.S. Army Corps of Engineers, Section 404 and/or Section 10 Permits, and (g) building, electrical, plumbing, and fire codes.
2. Access Roads and Parking.
   a. Provide access roads to all treatment units, buildings, and equipment.
   b. Separate service traffic from employee and visitors.
   c. Provide separate employee parking at the back (or side) and visitors parking in the front of the administration building.
3. Unit Layout.
   a. Provide modular layout with parallel process trains.
   b. Utilize common walls and provide isolation–expansion joints between buildings and structures for shrinkage, expansion, and uneven settlements.
4. Equipment.
   a. Provide major mechanical equipment in the central area near support and maintenance facilities.
   b. Provide monorails, bridge crane, jib crane, and access openings for moving equipment for replacement.
   c. Provide sufficient space around equipment for maintenance.
5. Elevations.
   a. Check the compatibility of operating elevations of chemical storage, feeders, building floor levels, elevations of treatment units, and headroom clearances.
   b. Plan carefully to avoid conflicts between disciplines in both vertical and horizontal directions in room layout, floor openings, and slab penetration for electrical conduits, chemical pipes, plumping, and HVAC at different floors.
   c. Group properly all utility lines in a chase or conduit and show them on the layout plans.
6. Safety.
   a. Give prime consideration to safety and hazard protection because the plant personnel will work in multiple use areas around the delivery, storage, handling, and delivery and use of hazardous chemicals. Related safety codes for fire ratings, access–egress for storage tanks and air intakes will apply.
   b. Locate odor and aerosol sources downwind from public space. Provide protective barriers such as shrub and trees, and sound barriers and absorptive enclosures around noisy equipment.
7. Future Developments.
   a. Reserve adequate space for adding parallel process trains for future plant expansion.
   b. Reserve sufficient spaces at proper locations for future process improvements to meet potentially more stringent discharge permit, or to enhance plant performance.

EXAMPLE 14.2: EXAMPLE OF A CONCEPTUAL COMPACT AND MODULAR LAYOUT OF A WASTEWATER TREATMENT FACILITY

A compact and modular layout plan minimizes construction and operation costs, and offers flexibility in future process modification and plant expansion. A site is proposed for a wastewater treatment plant. The site is classified as a socially and ecologically sensitive suburban underdeveloped tract. The suburban developments are on the western and southwestern sides, and the main interceptor is coming from the west. Site access is available from the south or west of the track. The prevailing winds are northwesternly. In general, the site is down hilling from west to east, and a small stream meanders around the east and northeast edges of the property line. A levee is provided between the plant site and the stream for flood protection. Based on a watershed waste load allocation requirements, a BNR facility is recommended and the optimal effluent discharge point in the stream is near the northeastern corner. Develop a conceptual compact and modular layout plan of the proposed wastewater treatment plant. Show the layout plan with major process trains, and important components at the plant. Also describe the key features of the proposed facility.

Solution

The site development plan of the proposed secondary wastewater treatment plant with BNR process is shown in Figure 14.2. After a careful consideration of the treatment units, connecting pipings, functional buildings, access roads, parking and reserved spaces, and the basic principles presented in Section 14.2 and Example 14.1, a compact and modular layout plan is developed and shown in Figure 14.2. These components are arranged to integrate the functions of all components in perfect harmony with the environment. The following features of the layout plan may be noted.

1. Treatment processes, modular layout, and process trains.
   a. Wastewater treatment processes:

      The proposed wastewater treatment process trains are mainly arranged in the middle of the plant area. The reserved area for future expansion is on the south side of the track. The major process components are: (a) two mechanical bar screens with one bypass channel with a manual screen (Sections 7.2.1 and 7.2.2); (b) four fine screens (Sections 7.2.3 and 7.2.4); (c) one influent lift station (Section 6.3.7); (d) one peak flow pump station; (e) three-cell peak flow storage facility with spillways between cells and gate structures for returning flow from each cell (Section 6.3.7); (f) flow meter vaults for influent and return peak flow (Section 6.3.7); (g) degritting facility with four centrifugal vortex-induced grit chambers and grits disposal equipment (Chapter 8); (h) four primary clarifiers (Chapter 9) with a primary sludge pump station (Section 13.3.3); (i) four BNR process trains, including a three-chamber anaerobic zone, a six-chamber anoxic zone, and a plug flow aeration basin, and an internal recycle pump station in each train (Sections 10.6 through 10.8); (j) four final clarifiers with a RAS/WAS pump station (Section 10.9); (k) chemical building for storage and feed of coagulant for effluent polishing to meet phosphorus limit (Section 9.5.2) and sodium hypochlorite solution for cleaning, washing, deodorizing, and control slime growth, and other general housekeeping (Section 11.6); (l) five cloth media filters, including one redundant unit (Section 15.4.7); (m) UV disinfection system with two channels under a canopy (Section 11.9); (n) one Parshall flume for measurement of final discharge flow to the stream (Example 9.38); and (o) effluent outfall at the stream (Section 12.6).

   b. Solids handling processes:

      The sludge handling area is kept on the north side for both current and future facilities. The major solids handling process components with design references are: (a) two gravity thickeners for primary sludge thickening (Section 13.5.1); (b) a gravity belt thickener (GBT) building for thickening WAS (Section 13.5.4); (c) two sludge blending tanks (Example 13.7) with a digester feed pump station (Section 13.3.3); (d) four anaerobic digesters (Section 13.6.1); (e) a belt-filter

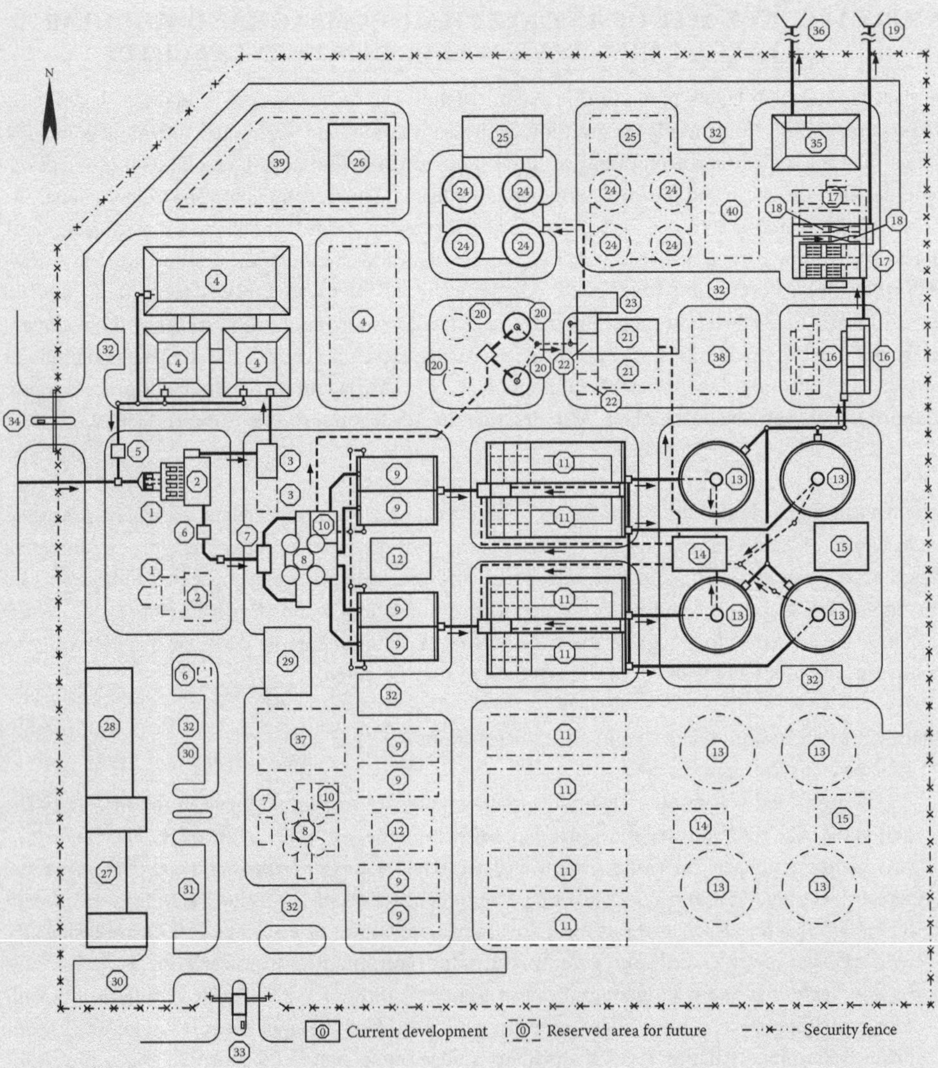

Facility code

1	Bar/fine screens	2	Influent lift station	3	Peak flow pump station	4	Peak flow storage facility
5	Returnflow meter vault	6	Influent flow meter vault	7	Flowsplitter	8	Degritting facility
9	Primary clarifiers	10	PS pump station	11	BNR basins	12	Blower building
13	Final clarifiers	14	RAS/WAS pump station	15	Chemical building	16	Cloth media tertiary filters
17	UV disinfection channels	18	Parshall flumes	19	Effluent discharge outfall	20	Gravity thickeners
21	GBT building	22	Sludge blending tank	23	Digester feed pump station	24	Anaerobic digesters
25	BFP building	26	Biosolids alkaline stabilization	27	Administration building	28	Maintenance building
29	Power supply system	30	Employees parking	31	Visitor parking	32	Service vehicle parking
33	Main entrance	34	Delivery entrance	35	Stormwater detention pond	36	Stormwater discharge outfall
37	Fermentation (future)	38	P-release tanks (future)	39	Emerging biosolids handing/disposalfacilities (future)		
40	Reclaimed water reuse (future)						

Abbreviation

BFP = Belt-filter press	BNR = Biological nutrient removal	GBT = Gravity belt filter
PS = Primary sludge	RAS = Return activated sludge	WAS = Waste activated sludge

FIGURE 14.2 Conceptual layout of the wastewater treatment plant (Example 14.2).

press (BFP) building with BFP feed pumps and a polymer feed system for sludge dewatering (Section 13.8.2); and (f) a biosolids alkaline stabilization system for producing Class A biosolids (Section 13.6.3 and Example 13.10).

2. Arrangement of treatment units.
   a. The pump station is located on the western side, and the wastewater is pumped to a flow splitter from where flow by gravity is achieved in normal sequence of treatment units.
   b. The physical arrangements of the pumping station and treatment units are such that it utilizes the existing features, site topography, and foundation requirements.
   c. The layout of various treatment units is compact with sufficient lengths of connecting conduits for process upgrading.
   d. The plant capacity expansion will be achieved by adding a parallel modular process trains. Such expansion will maintain a modular design and keep the operation similar to existing processing units.
3. Functional buildings, structures, and sites.

   Several auxiliary structures are also provided to support administrative and operational activities, and house maintenance shops. The important structures are described below.
   a. Administration building: The administration building is located near the entrance for public view. It serves as a barrier to undesired views of the treatment facilities. The administration building houses offices, instrumentation and control rooms, operational laboratory, visitors' lobby, showers, lavatories, and locker rooms.
   b. Maintenance building: A maintenance building equipped with machine shops, storage rooms, repair facilities, and garages is provided on the north and adjacent to the administration.
   c. Operational buildings: Several operational buildings are provided to house important equipment that may require indoor operating conditions. These buildings are also summarized as major process components in Step 1.
   d. Other structures: Miscellaneous structures include meter vaults, junction boxes, flow splitters, and gate structures.
   e. A power supply equipment site is also provided in the middle of the plant site. It is close to high-power-demand facilities, such as influent lift station, blower building, and administration and maintenance buildings.
4. Environmental considerations.
   a. The treatment plant is located away from the development. Existing trees and new landscaping is provided on the south side as buffers.
   b. As much as possible, the developed site does not alter the existing naturally stabilized site contour and drainage pattern. The landscaping reflects the rolling character of the surrounding area. The entire area is sodded to reduce erosion.
   c. The ground surface is drained away from the main structures, and the entire area is sloped toward a detention pond that is provided at the northeastern corner of the site plan. This location is the lowest point to permit natural drainage and to retain storm drainage. Storm water pumps are provided for discharging the surface runoff into the stream after proper retention in the pond.
   d. The treatment facility is located at the bank of the stream that meanders around the east and northeast edges of the property line. A recreation area is developed by building garden walks and esplanades, along the shoreline.
5. Future expansion and improvement.

   Sufficient land is provided for future construction of parallel modules south of the existing wastewater treatment facilities. These additional modules will be able to double the plant capacity as necessary in the future. Future sludge handling facilities will be added on the north side adjacent to the existing sludge handling facilities. Several areas are also reserved for future development that may include (a) fermentation of primary sludge for producing VFAs used to enhance EBPR (Section 10.7) and P-release from WAS (Section 13.10.3); (b) P-release tanks for P-recovery system (Section 13.10.3); (c) emerging biosolids handling processes that may include thermal hydrolysis (Section 13.10.1), P-recovery (Section 13.10.3), partial nitrification/Anammox (PN/A) process (Section 13.10.4), and heat drying (Section 13.11.2); and (d) reclaimed water reuse facility consisting of a pump station and a reclaimed water storage tank (Section 12.5). The sodium hypochlorite solution from existing chemical building will also be used for establishing a desired chlorine residual in the treated effluent for urban reuse.

6. Roads and parking.

   Paved service roads serve each treatment facility, building, and major equipment. Visitor parking is provided in the front of the administration building. Employee parking is provided in front of the maintenance building. Service vehicle parking is provided in front of the maintenance building and at several places near the major operational buildings.

7. Access gates and plant security.

   Access to the treatment facility is from the south and west. The main entrance at the south is in front of the administration building. The access for trucks is from the east gate. These truck transport residuals, bring septage, and deliver chemicals, parts, construction materials, and equipment. All gates are access controlled with an intercom system and card reader. In addition, video camera is provided to monitor the activity at each gate. The entire property is fenced with a security camera at each corner.

14.3 Yard Piping

Yard piping includes connecting conduits; collection and division boxes; valves, gates, and fittings; and appurtenances between various treatment units. The pipings with these appurtenances are important to transmit flows from one treatment unit to the other with least head losses. Often flows are collected from several treatment units operating in parallel, then divided again by splitter boxes. This allows flexibility of operation if one unit is taken out of service. A common practice is to use hydraulically similar inlet and splitter boxes for equal flow distribution. There are three basic considerations in preparing the piping layout: (1) convenience of construction and operation, (2) accessibility for maintenance, and (3) ease for new connections when future trains are added.

14.3.1 Organizing Yard Piping

The connecting conduits, including channels and piping may be above or below ground level. They must be marked clearly on layout plans with reference to the treatment unit. Provisions for future tie-ins should be made and properly indicated on the plans. Gate valves should be provided in the connecting piping to isolate or bypass units, and module for routine operation and maintenance. All connecting piping, valves, and fittings should be shown and marked properly on the layout plans, and identified on the ground with reference to permanent benchmarks. This will assist the designers and operators to fully understand the flow routes, unit isolations, and future connections, and to understand the operational capability of the plant.[2]

14.3.2 Pipe Galleries

In large plants, the yard pipings may be numerous and complex. For this reason, underground tunnels (also called pipe galleries or operating galleries) are constructed to locate piping and controls.

These galleries can be heated in cold regions to prevent freezing of pipes and valves, and provide easy access to the piping and valves. In small- and medium-sized plants, such investments may not be justified, and pipes may be located above and below ground. Plant construction photographs and videos showing yard piping are very helpful in reaching pipes and controls for operation and maintenance needs.

14.3.3 Considerations for Plant Upgrading

In a compact plant layout, the distances between adjacent treatment units are short to minimize the cost of connecting piping. However, the space between the treatment units may be the primary concerns of upgrading the plant by inserting a desired unit in the future. Plants with tight piping and hydraulics have resulted in serious upgrading problems. Often additional pumping may be needed to reroute the flow through the newly inserted treatment unit. This situation is unacceptable and must be avoided.

14.4 Plant Hydraulics

The principal factors that affect the flow in a treatment plant are differential elevations between the adjacent treatment units and the head losses in the connecting pipes. The head losses in a connecting pipe depend upon (1) slope, (2) cross-sectional area, (3) roughness of interior pipe surface, (4) flow conditions are full, partial full, steady, or varied flow, (5) presence or absence of obstructions, bends and valves, and (6) character, specific gravity, and viscosity of the liquid. The analysis of flow in both closed conduits and open channels is normally performed on the basis of three fundamental equations of fluid mechanics: (1) the Bernoulli's energy equation, Equation 6.16b, (2) momentum equation, Equation 8.14, and (3) continuity equation that is expressed by Equation 14.1.

$$Q = V_1 A_1 = V_2 A_2 \tag{14.1}$$

where

Q = flow rate through the conduit, m^3/s (ft^3/s)
A_1 and A_2 = areas at two different cross sections of the conduit, m^2 (ft^2)
V_1 and V_2 = velocities of fluid at sections A_1 and A_2, respectively, m/s (ft/s)

14.4.1 Hydraulic Profile

The hydraulic profile is a graphical representation of the hydraulic grade line through the treatment plant. It includes head losses through each treatment unit and the connecting piping. If the high water level in the receiving water is known, the head loss computations are done backwards through the plant using this known level as a control point. Sometimes the computations are started in the direction of the flow from the interceptor using the water surface as the reference point. In some plants, the hydraulic calculations are started somewhere in the middle using a benchmark or an arbitrary elevation. At the end, the elevations of water surface are adjusted in both directions.

The total available head at a plant is the difference between the water surface elevation in the junction box of the interceptor and the high flood level in the receiving water (typically at 100-year flood). If the total available head is less than the head loss through the plant, flow by gravity cannot be achieved. In such case, pumping ahead of the plant will be required to raise the head to achieve flow by gravity through the plant.

Hydraulic profile is an essential part of the plant design. The floor and foundation elevations of all treatment units and appurtenances are established with respect to the free water surface developed in the hydraulic profile. For this reason, the hydraulic profile is prepared at the peak and average design flows as well as at the minimum initial flow. Also, the hydraulic profile is prepared for all main paths of flow through the plant. Preparation of a typical hydraulic profile through a wastewater treatment plant is illustrated in Example 6.30.

The graphical representation of the hydraulic profile is also useful in showing the relative position of various treatment units with respect to the hydraulic grade line through the plant. In preparation of hydraulic grade line, the horizontal and vertical scales are intentionally distorted to show the location of treatment units and important elevations with respect to water surface. The ground surface is also indicated to establish the optimum elevations of the plant structures and the hydraulic controls.

14.4.2 Head Losses through Treatment Units

The total head loss through a treatment unit is the maximum value determined at peak design influent flow plus recirculation, if applicable. It is the sum of head losses and includes the following:

1. Head loss at the influent structure: influent pipe, channel, and inlet ports
2. Head loss at the effluent structure: effluent weir, launder, and collection and exit boxes
3. Head loss through the unit: baffle walls, turns, and obstruction due to equipment
4. Miscellaneous and surface free-fall allowances

The approximate head losses across different treatment units are summarized in Table 14.1.

14.4.3 Head Losses through Connecting Conduits

The connecting conduits are pipes, channels, junction and division boxes, and other appurtenances. The total head loss through these conduits is the sum of the following:

1. Head losses due to entrance and exit
2. Head loss due to contraction and enlargement
3. Head loss due to friction

TABLE 14.1 Approximate Head Loss in Various Treatment Units in a Wastewater Treatment Plant

Treatment Unit	Range of Head Losses, m (ft)
Bar screen	0.02–0.3 (0.07–1)
Grit removal	1–2.5 (3–8)
Velocity controlled grit channel	0.5–1.2 (1.5–4)
Aerated grit chamber	0.3–1.5 (1–5)
Vortex-type grit chamber	
Primary clarifier	0.5–1 (1.5–3)
Aeration basin	0.3–0.8 (1–2.5)
Trickling filter	
Low rate with dosing tank	2.5–5 (8–16)
High rate, single stage	2–5 (6–16)
Rotating biological contactor (RBC)	0.5–1 (1.5–3)
Combined attached and suspended growth process (dual)	4.5–6 (15–20)
Membrane biological reactor (MBR)	0.8–1.2 (2.5–4)
Final clarifier	0.5–1 (1.5–3)
Disinfection facility	0.2–2.5 (0.7–8)

Source: Adapted in part from Reference 2.

4. Head loss due to bends, fittings, gates, valves, and meters
5. Head required over the weir and other hydraulic controls
6. Free-fall surface allowances
7. Extra head allowance for future expansion of the treatment facility

There are some basic principles for calculating the head losses through the connecting conduits. These are listed below.

1. The velocity in the piping and conduits is kept high enough to keep the solids in suspension.
   a. Maintain a minimum velocity of 0.6 m/s (2 ft/s) at the peak design flow.
   b. Provide a minimum velocity of 0.3 m/s (1 ft/s) at the minimum initial flow to transport the organic solids.
   c. When the ratio of the maximum to minimum flows is very large, the self-cleaning velocity may not be achievable under the initial flows. Under such conditions, the flushing action and frequency at peak diurnal flows must be considered in design. In some cases, aeration is provided to keep solids in suspension. It may be desirable to provide smaller lines initially, and more lines are added in parallel as flow increases.
2. The minor head losses in channels, conduits, and appurtenances are calculated in terms of the velocity head from Equation 6.15. The minor head coefficients are provided in Sections C.1 and C.2 of Appendix C.
3. The friction loss in pressure conduits is calculated from either the Darcy–Weisbach or Hazen–Williams equation (Equation 6.13a or 6.13b). The Manning's equation (Equation 6.11) may also be used to calculate the grade of the water surface in pressure conduits.
4. In open channels, the depth of flow varies depending upon the flow conditions that may be uniform or nonuniform.
   a. The uniform flow occurs in channels that have constant cross section, and flow and velocity. In the design of channels, it is generally assumed that the flow is uniform at peak design flow. The Manning's equation is generally used to calculate the grade of water surface. The depth of flow in the channel and in the treatment unit at peak design flow should be used to set up a freeboard high enough to prevent overflow occurring from the channels or treatment units. Also, in sewers and connecting pipes, the water depth at low flow should at least match the hydraulic profile in downstream channel or unit to prevent surcharge. The hydraulic profile in sewers is prepared in Example 6.29.
   b. Nonuniform flow occurs in channels when cross section changes or the volume of water entering the channel is not constant. Nonuniform flow generally occurs in channels that have free-fall into effluent flume, launder, or channel junctions with surcharge (or backflow condition). Friction formulae do not apply to nonuniform flows. The computational techniques for nonuniform flows are covered in Section 8.4.2, and Examples 8.17 through 8.19. Readers are referred to References 19 through 22 on this topic.
5. Use of special methods is required to deal with variable flows at many locations in wastewater system. The variable flow distribution equations for calculating the discharges through multiple laterals are presented in Section 9.4.5, and Examples 9.13, 9.14, and 9.21. The design of effluent diffuser pipe at outfall structure is covered in Section 12.6.3, and Examples 12.40 and 12.41.
6. In wastewater treatment plants, sufficient allowances are made for transitions and nonuniform flows by providing invert drops. The head loss through the transitions is calculated by using the energy equation Equation 7.3a (Examples 7.3, 7.4, 7.8 through 7.10, and 8.13).
7. Most flow measuring devices operate on head loss measurements. Proper head loss calculations should be made for flow-measuring devices (Venturi tube, Parshall flume, orifice plate, weirs, etc.). The head loss through these flow measurement devices should be included in the hydraulic profile.

EXAMPLE 14.3: DESIGN OF A DIVISION STRUCTURE

Primary settled wastewater is collected from four primary sedimentation basins and discharged into a diversion structure. The flow is divided into three aeration basins. Develop the design of the division structure. The peak wet weather flow is 24,000 m^3/d (0.28 m^3/s). Provide sufficient free fall after the diversion.

Solution

1. Describe the hydraulic arrangement of the diversion structure.

   The influent enters an influent chamber (2-m long × 1-m wide) through an influent pipe. It flows into the diversion box through an opening near the bottom of the baffle wall. The box is a square with inside dimensions of 2 m × 2 m × 3 m high to the top of wall. Three identical rectangular weirs are provided on three sides of the box. The flow from each weir drops into an effluent chambers (2 m × 1 m wide) then transported by an effluent pipe to the designated aeration basin. Bottom slopes are also provided to prevent accumulating deposits in the dead zones. The conceptual plan and section view of the diversion structure are shown in Figure 14.3.

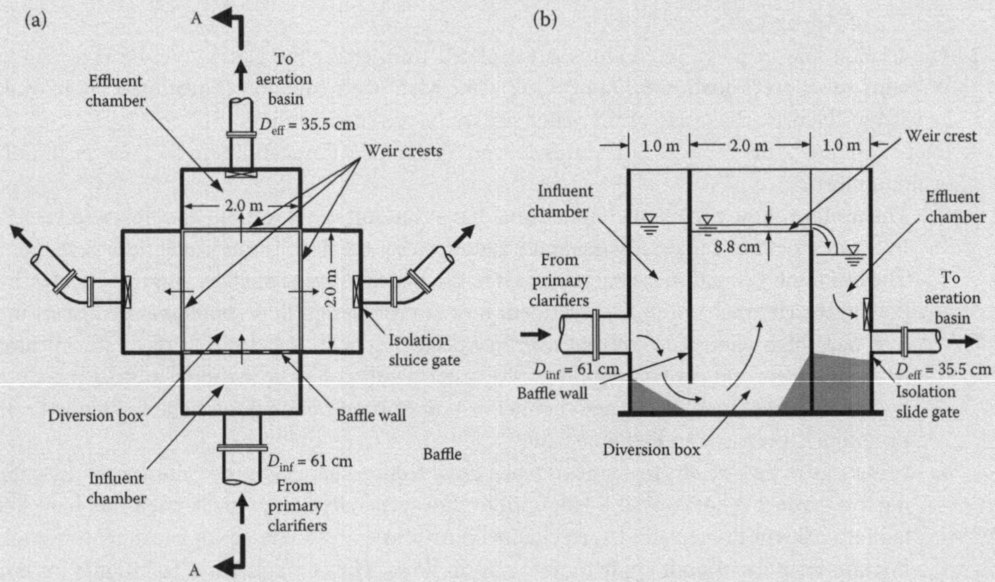

FIGURE 14.3 Definition sketch of a diversion structure: (a) plan view, and (b) sectional view (Section AA) (Example 14.3).

2. Determine the head over the weirs.

   The peak wet weather flow through each weir $Q = 8000\ m^3/d$ or $0.093\ m^3/s$.

   Calculated head over each weir from Equation 8.10 at $C_d = 0.6$, $L = 2$ m, and $n = 0$ under a free-fall condition.

$$L' = L - 0.1nH = L = 3\ \text{m}$$

$$H = \left(\frac{3}{2} \times \frac{Q}{C_d L' \sqrt{2g}}\right)^{2/3} = \left(\frac{3}{2} \times \frac{0.093\ \text{m}^3/\text{s}}{0.6 \times 2\ \text{m} \times \sqrt{2 \times 9.81\ \text{m/s}^2}}\right)^{2/3} = 0.088\ \text{m} \quad \text{or} \quad 8.8\ \text{cm}$$

3. Determine the diameters of influent and exit pipes.

   Assume that a velocity of $v = 1$ m/s is maintained in both influent and exit pipes at the peak wet weather flow.

$$\text{Influent pipe diameter, } D_{\text{inf}} = \left(\frac{4}{\pi} \times \frac{0.28\,\text{m}^3/\text{s}}{1\,\text{m/s}}\right)^{1/2} = 0.6\,\text{m} \quad \text{or} \quad 60\,\text{cm}$$

   Select standard diameter $D_{\text{inf}} = 61$ cm (24 in) pipe.

$$\text{Exit pipe diameter, } D_{\text{eff}} = \left(\frac{4}{\pi} \times \frac{0.093\,\text{m}^3/\text{s}}{1\,\text{m/s}}\right)^{1/2} = 0.34\,\text{m} \quad \text{or} \quad 34\,\text{cm}$$

   Select standard diameter $D_{\text{eff}} = 35.5$ cm (14 in) pipe.

   Note: The commercial pipe sizes are given in Section C.2 of Appendix C.

EXAMPLE 14.4: WATER SURFACE ELEVATION IN A DOWNSTREAM UNIT

The effluent from two secondary clarifiers are collected in a junction box, then conveyed to the combined influent channel of chlorine contact basins. The design details are shown in Figure 14.4. The connecting PVC pipe is 40.6 cm (16 in) in diameter and the linear length is 48 m. It carries a peak design flow of 0.15 m^3/s. The pipe has two 90° elbows in horizontal plane (not shown in the figure), four 45° elbows (two vertically and two horizontally (not shown in the figure)), and a slide gate. The values of minor head loss coefficients for 90° elbow $K = 0.3$, 45° elbow $K = 0.2$, slide gate $K = 0.8$ (including entrance loss), and exit $K = 1.0$. The coefficient of roughness $C = 130$ in the Hazen–William equation. Calculate the water surface elevation at the influent channel of the disinfection facility. The water surface elevation in the clarifier effluent junction box is 8.10 m. Draw the longitudinal section of the pipeline and indicate the elevation in the influent channel of chlorine contact basins.

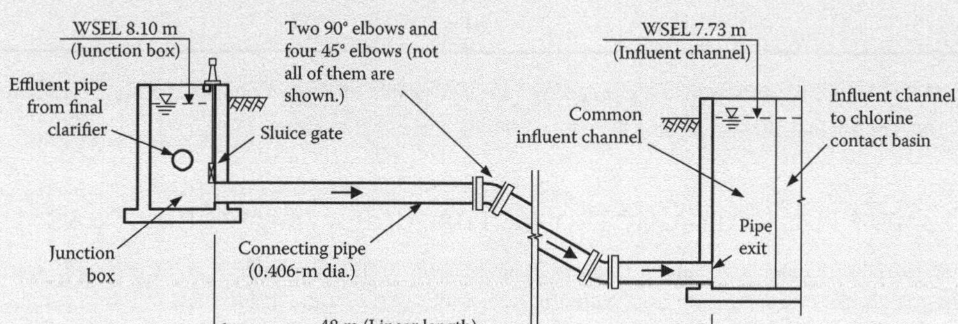

FIGURE 14.4 Longitudinal section of the connecting pipe (Example 14.4).

Solution

1. Calculate the velocity in the connecting pipe.

$$\text{Area of the pipe } (D = 40.6\,\text{cm} = 0.406\,\text{m}),\ A = \frac{\pi}{4} \times D^2 = \frac{\pi}{4} \times (0.406\,\text{m})^2 = 0.129\,\text{m}^2$$

Velocity in the pipe at $Q = 0.15\ \text{m}^3/\text{s}$, $V = \dfrac{Q}{A} = \dfrac{0.15\ \text{m}^3/\text{s}}{0.129\ \text{m}^2} = 1.16\ \text{m/s}$

2. Determine the friction head loss (h_f) through the pipe from the Hazen–Williams equation (Equation 6.13b).

$$h_f = 6.82\left(\frac{V}{C}\right)^{1.85} \times \frac{L}{D^{1.167}} = 6.82 \times \left(\frac{1.16\ \text{m/s}}{130}\right)^{1.85} \times \frac{48\ \text{m}}{(0.406\ \text{m})^{1.167}} = 0.15\ \text{m}$$

3. Calculate the minor head losses (h_m).

Apply Equation 6.15a to calculate the velocity head, $h_v = \dfrac{V^2}{2g} = \dfrac{(1.16\ \text{m/s})^2}{2 \times 9.81\ \text{m/s}^2} = 0.069\ \text{m}$

The overall minor head loss coefficient is sum of all minor head loss coefficients of all fittings and other losses.

$$K_{\text{overall}} = 2K_{90\ \text{elbow}} + 4K_{45\ \text{elbow}} + K_{\text{slide gate}} + K_{\text{exit}} = 2 \times 0.3 + 4 \times 0.2 + 0.8 + 1.0 = 3.2$$

Total minor head losses (Equation 6.15b), $h_m = K_{\text{overall}}\ h_v = 3.2 \times 0.069\ \text{m} = 0.22\ \text{m}$

Note: The minor head coefficients of commonly used fittings are provided in Sections C.1 and C.2 of Appendix C.

4. Determine the total head loss through the connecting pipe.

$$h_t = h_f + h_m = (0.15 + 0.22)\ \text{m} = 0.37\ \text{m}$$

5. Determine the water surface elevation at the influent channel of disinfection facility.
The water surface elevation,

$$WSEL_{\text{influent channel}} = WSEL_{\text{junction box}} - h_t = (8.1 - 0.37)\ \text{m} = 7.73\ \text{m}$$

6. The longitudinal section of the connecting pipe and water surface elevations at influent junction box and influent channel of chlorine contact basins are shown in Figure 14.4.

EXAMPLE 14.5: HYDRAULIC PROFILE THROUGH A TRAPEZOIDAL CHANNEL

A trapezoidal channel is designed to achieve post reaeration before discharge into the receiving water. The trapezoidal channel is 5-m long with a bottom width of 0.5 m and a side slope of 3 horizontal to 1 vertical. The channel has a smooth transition on each end. The influent end connects into a 2-m wide influent rectangular upstream channel. The effluent end connects into a square downstream effluent box (2 m × 2 m). The depth of the water surface in the influent rectangular channel is 0.96 m at peak design flow of $1.321\ \text{m}^3/\text{s}$. The invert elevation of the rectangular influent channel at the end is 9.85 m. The invert elevation of the trapezoidal channel at the upper end is 0.01 m lower than that of the influent rectangular channel. The invert elevation of effluent box is 0.01 m lower than that of the trapezoidal channel at the lower end. The plan view of the post reaeration channel is shown in Figure 14.5a. Draw the hydraulic profile and determine the invert and water surface elevations at the effluent end.

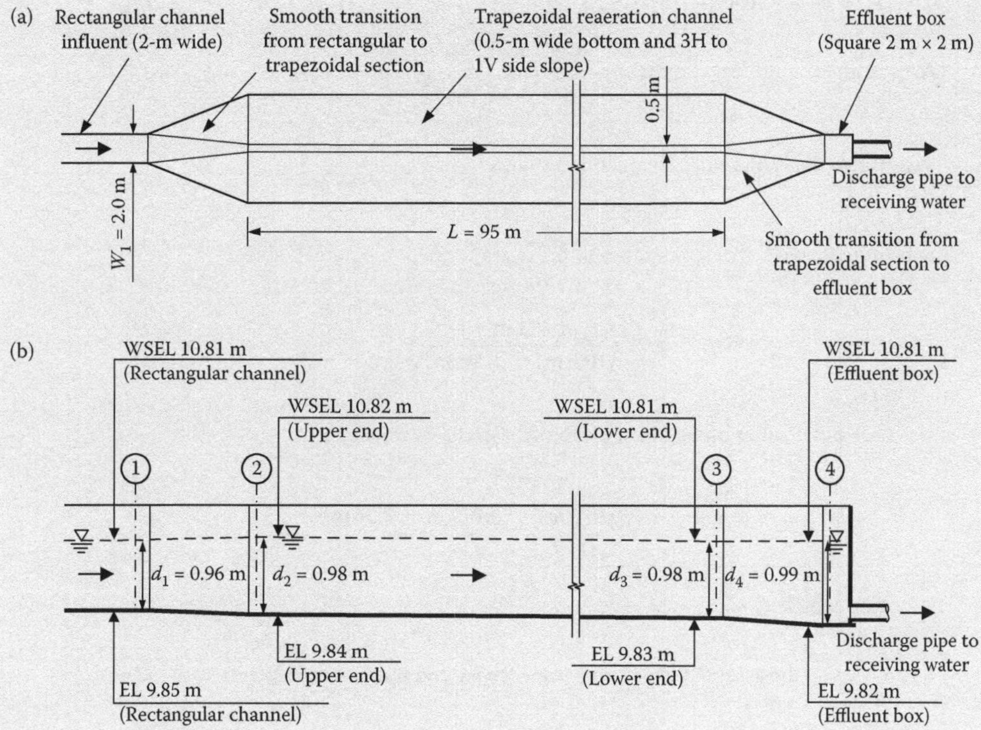

FIGURE 14.5 Sketches of post reaeration channel: (a) plan, and (b) longitudinal section (Example 14.5).

Solution

1. Determine the water depth and velocity in the trapezoidal channel at the upper end of the transition.

Applying the Bernoulli's energy equation (Equation 7.3a) at the rectangular and trapezoidal sections.

$$Z_1 + d_1 + \frac{v_1^2}{2g} = Z_2 + d_2 + \frac{v_2^2}{2g} + h_{L,1-2}$$

Drop in invert elevations, $Z_1 - Z_2 = 0.01$ m

Water surface elevation in the rectangular channel, $WSEL_1 = Z_1 + d_1 = (9.85 + 0.96)\text{ m} = 10.81$ m

The upper end transition is an expansion. The head loss at the transition is expressed by Equation 7.3b.

$h_{L,1-2} = \frac{K_e}{2g}(v_1^2 - v_2^2)$ Assume $K_e = 0.4$ for the transition (Section C.2 of Appendix C).

$$d_1 = 0.96\text{ m},\ A_1 = W_1 d_1 = 2\text{ m} \times 0.96\text{ m} = 1.92\text{ m, and } v_1 = \frac{Q}{A_1} = \frac{1.321\text{ m}^3/\text{s}}{1.92\text{ m}} = 0.688\text{ m/s}$$

Apply the energy equation at the rectangular and trapezoidal sections.

$$Z_1 + d_1 + \frac{v_1^2}{2g} = Z_2 + d_2 + \frac{v_2^2}{2g} + \frac{K_e}{2g}(v_1^2 - v_2^2) \quad \text{or} \quad (Z_1 - Z_2) + d_1 + (1 - K_e) \times \frac{v_1^2}{2g} = d_2 + (1 - K_e) \times \frac{v_2^2}{2g}$$

$$0.01\text{ m} + 0.96\text{ m} + (1 - 0.4) \times \frac{(0.688\text{ m/s})^2}{2 \times 9.81\text{ m/s}^2} = d_2 + (1 - 0.4) \times \frac{v_2^2}{2 \times 9.81\text{ m/s}^2}$$

$$0.984\,\text{m} = d_2 + (0.0306\,\text{s}^2/\text{m}) \times v_2^2$$

Area of the trapezoidal channel, $A_2 = \frac{1}{2} \times (0.5\,\text{m} + (0.5\,\text{m} + 6d_2)) \times d_2 = (0.5\,\text{m}) \times d_2 + 3d_2^2$

Velocity in the trapezoidal channel, $v_2 = \frac{Q}{A_2} = \frac{1.321\,\text{m}^3/\text{d}}{(0.5\,\text{m}) \times d_2 + 3d_2^2}$

Substitute the velocity expression into the energy equation obtained above.

$$0.984\,\text{m} = d_2 + (0.0306\,\text{s}^2/\text{m}) \times \left(\frac{1.321\,\text{m}^3/\text{d}}{(0.5\,\text{m}) \times d_2 + 3d_2^2}\right)^2$$

Use trial and error approach to solve for d_2. $d_2 = 0.979\,\text{m} \approx 0.98\,\text{m}$.

$$A_2 = \frac{1}{2} \times [0.5\,\text{m} + (0.5\,\text{m} + 6 \times 0.979\,\text{m})] \times 0.979\,\text{m} = 3.36\,\text{m}^2$$

$$v_2 = \frac{1.321\,\text{m}^3/\text{d}}{3.36\,\text{m}^2} = 0.393\,\text{m/s}$$

Since $v_1 > v_2$, the assumed expansion at the upper end transition is correct.

$$h_{L,1-2} = \frac{0.4}{2 \times 9.81\,\text{m/s}^2}[(0.688\,\text{m/s})^2 - (0.393\,\text{m/s})^2] = 0.007\,\text{m}$$

2. Determine the invert and water surface elevations at the upper end of the trapezoidal channel.
   Invert elevation at the upper end, $Z_2 = Z_1 - 0.01\,\text{m} = (9.85 - 0.01)\,\text{m} = 9.84\,\text{m}$
   Water surface elevation at the upper end, $WSEL_2 = Z_2 + d_2 = (9.84 + 0.979)\,\text{m} = 10.819\,\text{m} \approx 10.82\,\text{m}$
   Water surface elevation in the rectangular channel $= (9.85 + 0.96)\,\text{m} = 10.81\,\text{m}$
3. Verify the head loss at the upper end of the trapezoidal channel.

   Velocity head in the rectangular channel (Equation 6.15a), $h_{v,1} = \frac{V_1^2}{2g} = \frac{(0.688\,\text{m/s})^2}{2 \times 9.81\,\text{m/s}^2} = 0.024\,\text{m}$

   Hydraulic grade line in the rectangular channel, $HGL_1 = WSEL_1 + h_{v,1} = (10.81 + 0.024)\,\text{m} = 10.834\,\text{m}$
   Velocity head at the upper end of trapezoidal channel (Equation 6.15a),

$$h_{v,2} = \frac{V_2^2}{2g} = \frac{(0.390\,\text{m/s})^2}{2 \times 9.81\,\text{m/s}^2} = 0.008\,\text{m}$$

   Hydraulic grade line at the upper end, $HGL_2 = WSEL_2 + h_{v,2} = (10.819 + 0.008)\,\text{m} = 10.827\,\text{m}$
   Head loss at the upper end of the trapezoidal section,

$$h_{L,1-2} = HGL_1 - HGL_2 = (10.834 - 10.827)\,\text{m} = 0.007\,\text{m}$$

   This is the same as that calculated from Equation 7.3b at the end of Step 1.

4. Compute the slope of the trapezoidal channel.

   Assume that the uniform flow is achieved and the Manning's equation (Equation 6.11a) is applicable.

   Wetted perimeter, $p_2 = 0.5\,\text{m} + 2 \times \sqrt{(0.979\,\text{m})^2 + (3 \times 0.979\,\text{m})^2} = 0.5\,\text{m} + 2\times\, 3.1\,\text{m} = 6.7\,\text{m}$

   $$\text{Hydraulic mean radius, } R_2 = \frac{A_2}{p_2} = \frac{3.36\,\text{m}^2}{6.7\,\text{m}} = 0.50\,\text{m}$$

   Rearrange the Manning's equation (Equation 6.11a) using $n = 0.013$.

   $$S = \frac{(nV_2)^2}{R_2^{4/3}} = \frac{(0.013 \times 0.39\,\text{m/s})^2}{(0.50\,\text{m})^{4/3}} \times \frac{\text{s}^2}{\text{m}^{2/3}} = 0.000065\,\text{m/m}$$

5. Determine the invert and water surface elevations at the lower end of the trapezoidal channel.

   The differential elevation due to slope, $\Delta Z_{2-3} = LS = 95\text{ m} \times 0.000065\text{ m/m} = 0.006\text{ m}$

   Invert elevation at the lower end, $Z_3 = Z_2 - \Delta Z_{2-3} = (9.84 - 0.006)\text{ m} = 9.834\text{ m} \approx 9.83\text{ m}$

   Under uniform flow condition, $d_3 = d_2 = 0.979\text{ m} \approx 0.98\text{ m}$, and $v_3 = v_2 = 0.390\text{ m/s}$.

   Water surface elevation at the lower end, $WSEL_3 = Z_3 + d_3 = (9.834 + 0.979)\text{ m} = 10.813\text{ m} \approx 10.81\text{ m}$

6. Determine the depth of flow in the effluent box.

   The drop in invert elevations, $Z_4 - Z_3 = 0.01\text{ m}$

   Assume that the velocity in the effluent box is ignorable, $v_4 \approx 0$; and the head loss at the lower end is expressed by the minor head losses (Equation 6.15b) using $K_{\text{exit}} = 1.0$.

   $$h_{L,3-4} = h_{\text{m exit}} = K_{\text{exit}}\frac{v_3^2}{2g} = 1.0 \times \frac{(0.390\,\text{m/s})^2}{2 \times 9.81\,\text{m/s}^2} = 0.008\,\text{m}$$

   Note: The head loss at the lower end is equal to the velocity head in the trapezoidal channel.

7. Determine the invert and water surface elevations at the effluent box.

   Invert elevation of the effluent box, $Z_4 = Z_3 - 0.01\text{ m} = (9.834 - 0.01)\text{ m} = 9.824\text{ m} \approx 9.82\text{ m}$

   Water surface elevation in the effluent box, $WSEL_4 = WSEL_3 - h_{L,3-4} = (10.813 - 0.008)\text{ m} = 10.805\text{ m} \approx 10.81\text{ m}$

   Water depth in the effluent box, $d_4 = WSEL_4 - Z_4 = (10.81 - 9.82)\text{ m} = 0.99\text{ m}$

8. Draw the longitudinal section of the post reaeration channel in Figure 14.5b.

EXAMPLE 14.6: HYDRAULIC PROFILE THROUGH A SECONDARY TREATMENT PLANT

Hydraulic profile through a conventional activated sludge treatment plant is prepared at peak design flow of 0.35 m^3/s. The raw wastewater after screening is pumped into the influent channel of the aerated grit chamber. The wastewater flows by gravity through grit chamber, primary clarifier, aeration basin, secondary clarifier, and chlorination and dechlorination facilities, and then discharges into the receiving stream. The water surface elevation in the receiving stream at the 100-year flood is 100.00 m. The head losses across each unit from the influent channel to outlet box under the critical flow are: aerated grit chamber = 0.9 m, primary clarifier = 0.85 m, aeration basin = 0.45 m, final clarifier = 0.75 m, and chlorination and dechlorination facility = 0.6 m. The head loss through the outfall structure with Parshall flume followed by cascade aeration is 4.5 m. The head loss at the influent structure of every treatment unit is 0.15 m. Assume that (a) the head loss in the force main from the raw influent pump to the influent box of aerated grit chamber is 2 m; (b) the connecting pipes between the upstream effluent box and downstream influent structure of all adjacent treatment units including the outfall structure have a head loss of 0.5 m; (c) the head loss between the outlet box of disinfection facility and the influent box of outfall facility is 0.9 m. Draw the hydraulic profile with key elevations of the treatment units.

Solution

The hydraulic profile with key elevations is shown in Figure 14.6.

The hydraulic profile through the plant is developed based on the 100-year flood level in the receiving water. All WSELs are calculated using the head losses provided in the problem statement. The HGL

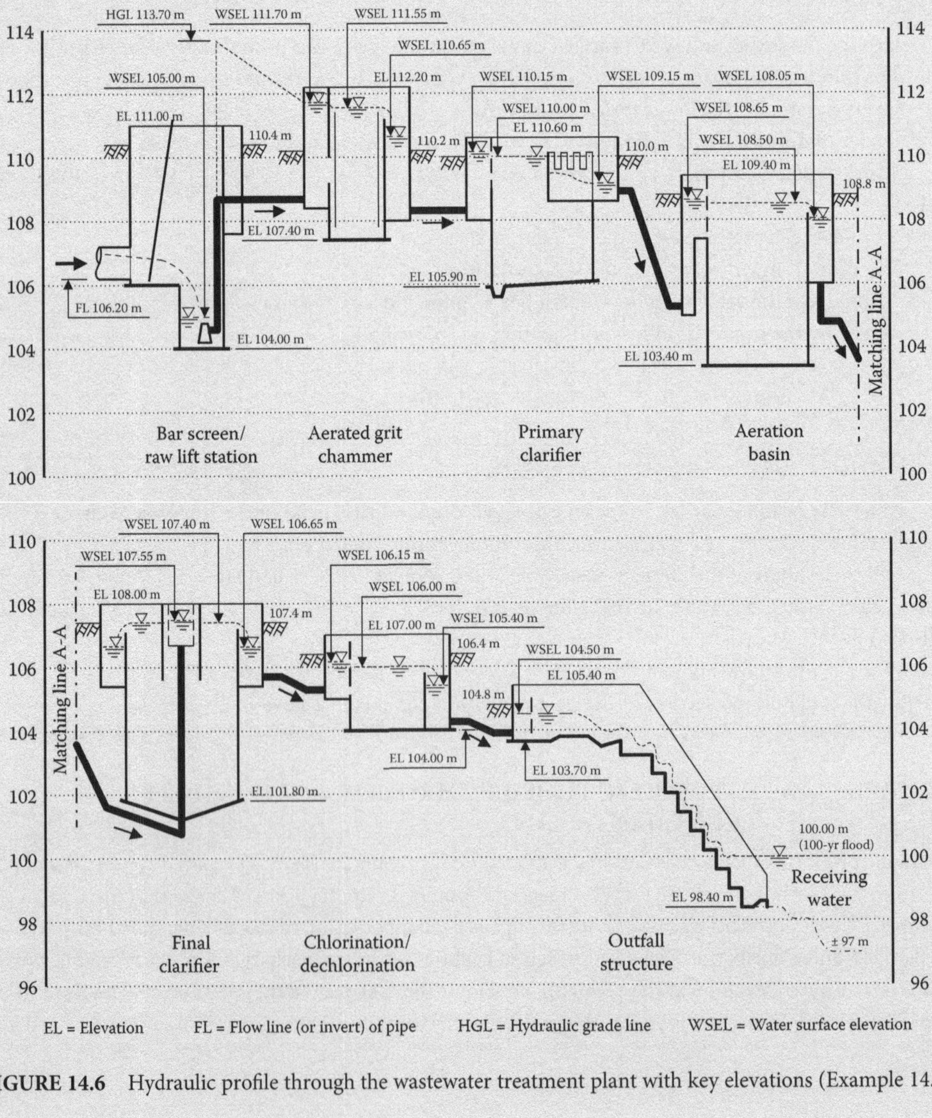

FIGURE 14.6 Hydraulic profile through the wastewater treatment plant with key elevations (Example 14.6).

elevation of 113.70 m at the pump discharge is obtained. At the peak design flow, the minimum water depth in the wet well is 105.00 m. The static head between the WSELs in the wet well and influent channel of the aerated grit chamber is (111.70 − 105.00) m = 6.70 m. The total head required for the pump may be obtained from two methods: (a) the sum of the static head of 6.70 m and the head loss of 2 m through the force main and (b) the difference between the HGL at the pump discharge and the WSEL in the wet well (113.70 − 105.00) m = 8.7 m. The invert elevation of the intercepting sewer is 106.20 m.

14.5 Instrumentation and Controls

The use of instrumentation and controls in wastewater treatment facilities is a common practice these days. This is because manual measurements of process control variables are infrequent and cannot be made within the short span of time and precision needed for operation and optimization of the facility. The subject of instrumentation and controls is very complex requiring an extensive background of measuring elements and instruments, computers, controllers and control elements, and other systems in the context of a wastewater treatment facility. Many factors that are considered in the selection of process control sophistication include: (1) plant size and plans for future expansion, (2) complexity of the treatment processes and treatability aspects, (3) number of control points, (4) interdependence of process elements and controls and process time delays, (5) education and training of operators and maintenance personnel, and (6) useful life and process obsolescence.[2,4,23,24] These factors are important for the selection of controls, data acquisition, and data display systems.

14.5.1 Benefits of Instrumentation and Control Systems

The instrumentation and control systems offer many benefits in process improvement, equipment performance, and personnel needs. The end result may be a reduction in labor, chemicals and energy costs; and overall improvement in process efficiency, reliability, and data acquisition and recording. Some of these benefits are summarized in Table 14.2.

14.5.2 Components of Control Systems

The process control systems utilize (1) *control variables* and (2) *associated controls*. The sensors of the control system measure the process variables and transmit, display, or record them so that the operators can make proper process adjustmenets.[25] The control variables and associated controls are discussed below.

Control Variables: The control variables are measured and adjusted for plant operation. They fall into three categories: physical, chemical, and biological. Common examples are given below.

- *Physical*: flow, pressure, temperature, level, weight, etc.
- *Chemical*: pH, turbidity, specific conductance, dissolved oxygen, chlorine residual, oxidation–reduction potential (ORP), and others
- *Biological*: oxygen consumption rate, TOC reduction rate, sludge growth rate, etc.

Associated Controls: The automatic control systems include (1) signal measurement or sensing devices, (2) signal-transmitting devices, (3) data display devices, (4) control systems and loops, (5) supervisory control and data acquisition (SCADA) system, and (6) central control room. These components are briefly discussed below.

TABLE 14.2 Benefits of Instrumentation and Control Systems in Wastewater Treatment

Purpose	Benefit
Process	• Improved process performance and better process results • Efficient use of energy • Efficient use of chemicals • Process changes detected in a timely manner • Automatic execution of corrective measures • Great ability to control complex processes
Equipment	• Immediate alert signal of malfunction • Ability to diagnose problems in remotely located equipment before malfunction occurs • Performance status monitored at all times • Automatic execution of corrective measures and automatic response to potentially disastrous situations • Automatic shutdown to prevent major damage • Increase in running time
Personnel	• Timely and accurate process information • Safer operation • Efficient use of labor • Capability to quickly solve analytical problems • Minimize the potential for human errors • Allows for an overview of plant operation • Decrease in manual paperwork • More complete records that may allow an overview of plant operation and behavior, and design of future expansion • Increased security

Source: Adapted in part from References 1, 2, 25, and 26.

Signal Measurement or Sensing Devices: The sensing devices measure signals. They may be online or offline and continuous or intermittent. These devices, measured signals, and common applications are summarized in Table 14.3.

Signal-Transmitting Devices: These devices transmit the process variable signals from the sensing devices to a controller or readout device. The transmission may be mechanically, pneumatically, or electronically by radio or microwave. Detailed discussion on mechanical, pneumatic, electric, and radio microwave transmissions may be found in References 29 through 31.

Data-Display Devices: The transmitted information from sensing device is displayed in a convenient location before use by the operating personnel. Most common types of readout devices are indicators, recorders, and totalizer on panels, computer screens, or monitors. The receiving instrument may display the movement, pressure, or current change signals directly, or it may be servo-operated.[27]

Control Loops or Systems: The time control systems are divided into three categories: (1) *digital*, (2) *analog*, and (3) *automatic* controls. The digital systems have two positions. Some examples are on/off, open/closed, high/low, or alarm/normal. They give a status change signal. The analog information has a range of values such as flow rate, water level, and concentration measurements. The automatic controls are discrete or continuous. The discrete control correlates equipment status (on/off) and status changes with a preset value or program of events such as pumping cycle based on wet well level. The operation may be initiated *manually* by an operator using a pushbutton or *automatically* by an internal process-generated event. The continuous controls may utilize *controller* with various *control loops* or *systems* to perform the control functions.[27–32] Examples of using these systems for control of a valve in a process are illustrated in Figure 14.7. The programmable logic controllers (PLCs) are commonly used to perform the discrete control tasks. A PLC may replace multiple components (relays and timers) in hard-wired logic networks.

There are two types of variables in continuous process control. These are (1) *uncontrolled* variable (sometimes also called the wild variable) such as plant flow, temperature, and BOD, TSS, nitrogen and

TABLE 14.3 Common Online Process Measurement Devices and Their Applications in Wastewater Treatment

Measured Variable	Primary Device	Measured Signal	Common Application
Flow	Venturi, flow nozzle and orifice meters	Differential pressure	Gas and liquid
	Electromagnetic meter	Magnetic field and voltage change	Liquid and sludge
	Turbine meter	Propeller rotation	Clean liquid
	Acoustic meter, including Doppler or transit time types	Sound waves	Liquid and sludge
	Parshall and Palmer-Bowlus flumes	Differential elevation of liquid surface	Liquid
	Weirs	Head over weir	Clean liquid
Liquid pressure	Liquid-to-air diaphragm	Balance pressure across a metal diaphragm	Pressure: 0–200 kN/m^2 (0–30 psi)
	Strain gauge	Dimensional change in sensor	Pressure: 0–350,000 kN/m^2 (0–50,000 psi)
	Bellows	Displacement of mechanical linkage	Pressure: 0–20,000 kN/m^2 (0–3000 psi)
	Bourdon tube	Uncurling motion of a curved tube	Pressure: 0–35,000 kN/m^2 (0–5000 psi)
Liquid level	Float	Movement of a float on liquid surface	Liquid head: 0–10 m (0–35 ft)
	Bubbler tube	Back pressure in the tube bubbling regulated air	Liquid head: 0–55 m (0–180 ft)
	Diaphragm bulb	Pressure change in air and liquid across a diaphragm	Liquid head: 0–15 m (0–50 ft)
	Ultrasonic	The sonic pulses reflected from the target surface	Liquid level in channel or basin
	Radar	The guided or noncontact radar waves reflected from the target surface	Liquid level in channel or basin
Sludge level	Photocell	Detection of light in a probe by a photocell across sludge blanket	Primary sedimentation, final clarifier, and gravity thickener
	Ultrasonic	Detection of the ultrasonic signal transmitted between two transducers	Primary clarifier, final clarifier, and gravity thickener
Temperature	Thermocouple	Current flows in a circuit made of two different metals	Anaerobic digester, thermal hydrolysis, heat drying, and hot-water boiler
	Thermal bulb	Absolute pressure of a confined gas	Sludge and water lines
	Resistance temperature detector	Change in electrical resistance of temperature-sensitive element	Bearing and winding temperatures of electrical machinery, anaerobic digesters, and hot-water boilers
Speed	Tachometer	Voltage and current	Variable-speed pump, blower, and mixer
Weight	Weight beam, hydraulic load cell, or strain gauge	Lever mechanism or spring, pressure across diaphragm, or dimensional change	Chemicals, sludge cake, and compost
Density	Gamma radiation	Absorption of gamma rays by the liquid between radiation source and the detector	MLSS concentration; returned, thickened, and digested sludge

(Continued)

TABLE 14.3 (*Continued*) Common Online Process Measurement Devices and Their Applications in Wastewater Treatment

Measured Variable	Primary Device	Measured Signal	Common Application
Suspended solids	Ultrasonic analyzer	Loss of ultrasonic signal across a liquid between transmitter and receiver	MLSS concentration; returned, thickened, and digested sludges
Turbidity	Turbidimeter	Amount of light scattered by particles suspended in a liquid sample	Reclaimed water
pH	Selective ion electrode	Voltage produced by hydrogen ion activity	Influent, chemical solution, anaerobic digester, dewatering, and effluent
Oxidation–reduction potential (ORP)	Electrode	Change in potential due to oxidation or reduction	Influent, aeration basin, BNR system, and anaerobic digester
TDS	Conductivity	Electrical current across a solution	Influent and effluent
Dissolved oxygen (DO)	Membrane electrode	Electric current across a membrane because of reduction of molecular oxygen	Influent, aeration basin, and effluent
Total organic carbon (TOC)	Carbon analyzer	CO_2 produced from combustion of a sample	Influent and effluent
Chlorine residual	Sensor	Amperometric change due to oxidation–reduction on two dissimilar electrodes	Chlorine contact basin, plant effluent, and reclaimed water
Gases (NH_3, CO, CO_2, H_2S, H_2, and CH_4)	Electrochemical, infrared or thermal conductivity sensors	Various types of sensors	Detection of hazardous condition in operating room or head space in covered treatment unit
Oxygen uptake rate (OUR)	Respirometer using sensor	Decrease in DO level with respect to time	Aeration basin and treatability
Biogas (CO_2, CH_4, and H_2)	Electrochemical, infrared or thermal conductivity sensors	CO_2, CH_4, and H_2 concentrations	Anaerobic digester
Volatile fatty acids (VFAs)	Infrared spectroscopic, automatic titrimetric, or GC-FIC- based analyzer	Various types of sensors	Anoxic zone, anaerobic digester, P-release of WAS, and fermentation
Orthophosphate (Ortho-P)	Spectrophotometer	Colorimetric measurement of stannous chloride and phosphate–molybdate complex	Influent, effluent, sidestream, and flow after P-release from WAS for phosphorus recovery
Nitrate/nitrite	Spectrophotometer	Colorimetric measurement of color complex	Influent, effluent, nitrification, and denitrification
Ammonia	Spectrophotometer	Colorimetric measurement of color complex	Influent, effluent, sidestream, and flow through partial nitritation/Anammox (PN/A) process

$1\ kN/m^2 = 0.145$ psi.

Source: Adapted in part from References 2, 4, and 27 through 29.

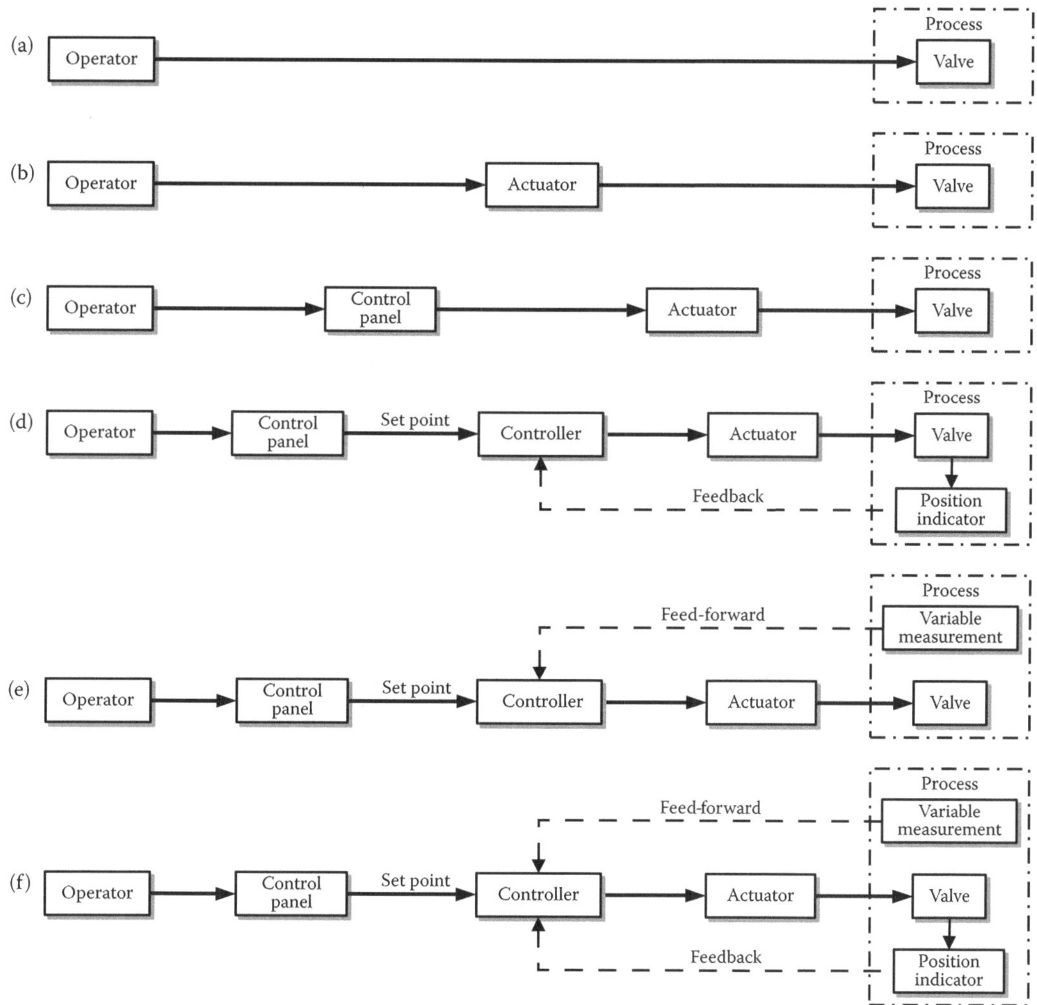

FIGURE 14.7 Examples of control systems: (a) manual control, (b) manual control made easier, (c) remote manual control, (d) automatic feedback control, (e) automatic feed-forward control, and (f) automatic feed-forward-feedback computer control. (Adapted in part from References 2, 4, and 25.)

phosphorus concentrations and (2) *controlled* variables such as chlorine dose in disinfection process, oxygen supply in aeration basin, and chemical feed in a rapid mix basin. In order to maintain a constant concentration of chemical in the wastewater, the chemical feed rate must change in proportion to the flow.[32] In a control system, a series of related or unrelated control loops are used for detection of variables that manipulate the process for the desired results. A typical control system and its components are illustrated in Figure 14.8.

The automatic control system consists of three basic parts: (1) a measurement device that detects the changes in the variables; (2) a reference source that compares the measured value of the variable with a preset reference; and (3) a controller or mechanism that manipulates the variable until the measured signal reaches the reference value.[27,29–32]

The control loop is defined by the Instrument Society of America (ISA) as a combination of one or more interconnected instruments to measure or control a process or both.[31,32] The control circuit or

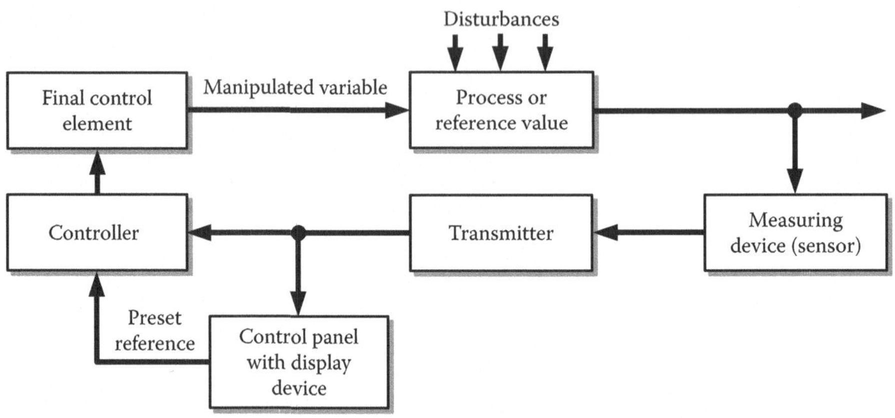

FIGURE 14.8 Typical control system components. (Adapted in part from References 2, 4, and 25.)

loop is composed of control elements and comes in two general types: (1) *feedback* or *closed loop* (Figure 14.7d) and (2) *feed-forward* or *open loop* (Figure 14.7e).

- In a feedback loop, the control is accomplished by "feeding back" the measured signal to the controller for the purpose of detecting the deviation from the set point. An example of closed feedback loop is an automatic chlorinator to maintain a desired chlorine residual as discussed in Chapter 11.
- In a feed-forward loop, the control is applied without the use of feedback information; the control action is initiated before a deviation in the process occurs. An example of such system is chemical-feeder output controlled in proportion to the plant flow at a manually preset dosage.

A series of relatively simple individual control loops may be combined into independent, interdependent, or dependent open and closed control loops to constitute an elaborate and complex system.[27,29–32]

The device that automatically performs the control function in regulating the control variable is a controller. The control function may be of different types, depending on the functional needs. Some of these controls are briefly described below.[27,29–31]

1. *Two-Position* or *On-Off*: An example is starting and stopping of a pump by a level controller.
2. *Floating Control*: It is a three-position control such as an air-conditioning system that has heating-off-cooling control action.
3. *Proportional Control*: The output of the controller is directly proportional to the deviation from the set point. This control is used for filter level control.
4. *Proportional-Integral Control*: The rate of output change is dependent on the amount and duration of the deviation (signal error). The integral function samples the difference between the fixed set point of the controller and the point at which the sensor stabilizes and adjusts the controller element to return the parameter valve to the set point.
5. *Proportional-Integral-Differential (PID) Control*: A differential (derivative or rate) function is added to modify the position of the controlled element in terms of the rate at which the parameter signal deviates from the fixed point.
6. *Sample Data Control*: The action is delayed to provide a time lag for the effect of a previously applied correction to be sensed. An excellent example is a chlorination facility, in which the chlorine dose is controlled by the residual measured after the contact period discussed in Chapter 11.
7. *Timer Control*: The timer control device operates on a set time schedule. An example is pumping of sludge at preset time intervals.

8. *Two- and Three-Mode Control*: Two or three modes of control may be combined with the basic controls to achieve variable and complex control operations. They are commonly known as compound loops. An example is chlorination followed by dechlorination discussed in Chapter 11.

Supervisory Control and Data Acquisition System: Most of the medium- to large-sized wastewater treatment plants utilize SCADA systems for data logging.[25] These systems (1) collect, format, record, and display large quantities of data effectively; (2) provide accurate, impartial documentation of all process measurements; (3) accumulate and display the process data; and (4) perform the necessary process corrections for optimum operation. These corrections may include control of chemical solutions, air supply, scheduling of pumps and blowers, and like. They can also be used to develop a maintenance schedule for the plant operator based on actual operating time of a particular equipment, such as a pump or blower.

Central Control Room: Central control is used to organize the plant operation in such a manner that all treatment information, important events, and alarms are displayed, indicated, and recorded at a centralized location. This location is called the *central control room*. In addition, most central facilities practice automatic or manual actuation of final control elements. Central control room reduces the number of personnel required to operate a large and complex treatment facility.

Application of the *expert systems* has also gained popularity in process control during recent years. This technology captures in a computer program the concepts, facts, and judgments used by an experienced operator in decision making. It assists the operators who have limited background in complex treatment processes.[2,4] The *neural network* techniques provide learning capabilities within the computer system. It allows the computer to recognize patterns and modify its own recognition criteria to improve its performance. Readers may find additional information on this topic in References 33 through 38.

Symbols and Identification of Process and Instrumentation Diagram (P&ID): The symbols and identification used in designing the P&IDs are based on those provided by the Instrumentation ISA, the Institute of Electrical and Electronic Engineers (IEEE), and the National Electrical Manufacturers' Association (NEMA). Commonly used symbols and identification systems may be found in Reference 39.

EXAMPLE 14.7: DESIGN CONSIDERATIONS FOR INSTRUMENTATION AND CONTROL (I&C) SYSTEMS

Design work on I&C systems requires coordinated effort of engineers, manufacturers, and plant personnel. The end result is several documents. Describe the coordinated effort and documents developed.

Solution

The coordinated effort of engineers, manufacturers, and plant personnel involves the following:

1. The design of I&C system is developed by an I&C engineer. It is, however, the responsibility of the project engineer to maintain close contact with the I&C engineer and vendors and check thoroughly the final design.
2. The system is used, operated, and maintained by the plant personnel. For this reason, plant operators and personnel, and owner must be involved during the design phase.
3. The process flow diagrams (PFDs) are prepared by the project engineer, and become the basis for the process and instrumentation diagrams (P&IDs).[25] The P&ID is a functional schematic presentation of the proposed treatment processes with the required instruments and controls. By using a consistent tagging system, the P&IDs can illustrate the functions of the treatment processes without referring to the actual hardware.

4. Based on the P&ID, the following design documents are further prepared by the I&C engineer:
   a. *An overall SCADA network architecture design*: It presents the basic concepts of how the plant SCADA system be interfaced with the proposed and existing PLCs, input/output (I/O) panels, and subnetwork panels.
   b. *Control loop diagrams (CLDs)*: The CLD specifies the detailed monitoring and control loops for each process equipment. This includes (a) monitoring interconnections from field measurement elements through PLC to the human machine interface (HMI) and (b) control commands from HMI through PLC to the field equipment. The CLDs may be arranged separately for each major process area with single or multiple treatment facilities.
   c. *PLC panel plans (P3s)*: These plans may include such details as typical panel dimensions, communication box plan, network details, and I/O panels.
   d. Instrument installation details (IIDs) represent final installation details.
   e. *Construction specifications under instrumentation section*: These documents describe in details about requirements for PLCs, HMIs, uninterruptible power supplies (UPSs), network panels, I/O list, instrumentation accessories, and commissioning. A control narrative should also be included in the specifications.

EXAMPLE 14.8: PRIMARY FLOW MEASURING DEVICES AND THEIR APPLICATIONS

Match the following primary flow-measuring devices with their applications. The primary flow-measuring devices include: orifice meter, flow nozzle, Parshall flume, ultrasonic flow meter (Doppler type), and electromagnetic flow meter. The applications are: (a) raw wastewater, (b) treated clean effluent, and (c) sludge.

Solution

Orifice meter	(b)
Flow nozzle	(b)
Parshall flume	(a) and (b)
Ultrasonic flow meter (Doppler type)	(a) and (c)
Electromagnetic flow meter	(a), (b), and (c)

EXAMPLE 14.9: MEASURED SIGNALS FOR VARIABLES

Write a measured signal for each of the following variables: conductivity, pH, density, oxidation-reduction potential (ORP), dissolved oxygen (DO), temperature, and pressure.

Solution

The list of measured variables and their measured signals is summarized in the table below.

Measured Variables	Measured Signals
Conductivity	Flow of electric current across the solution
pH	Voltage produced by hydrogen ion activity
Density	Absorption of gamma rays by the liquid between radiation source and the detector
ORP	Change in potential due to oxidation or reduction
DO	Electric current across a membrane due to reduction of molecular oxygen
Temperature	Current flow in a circuit made of two different metals
Pressure	Balance pressure across a diaphragm

EXAMPLE 14.10: SENSING DEVICES AND MONITORING OF MAJOR EQUIPMENT

At a wastewater treatment plant, monitoring and control of the process units are mostly done from the central control room. List the major sensing devices and summarize their purposes, display locations, and brief descriptions.

Solution

The treatment units, sensing devices, their purposes, locations, and brief descriptions at a wastewater treatment plant are summarized in Table 14.4.

TABLE 14.4 Major Sensing Devices Purpose, Display Location, and Description at Different Treatment Units[a]

Sensing Device	Purpose	Display Location	Description
Mechanical Bar Screens with a Manual Bar Screen for Emergency Bypass (Sections 7.2.1 and 7.2.2)			
Limit switches on gates in influent channel	Monitor gate open/close position	Central	Display green/red lights when gate is OPEN/CLOSE
Radar or ultrasonic level transmitter in influent and effluent channels	Monitor water level; and detect high water level due to excessive head loss through screen	Central	Display and record level continuously; and activate alarm per high level set point
Motor status monitors and data loggers for mechanical bar screens	Monitor screen operating status for run time, number of start, cycle on/off per duration timer set point, and emergency stop	Central and local	Display and record screen operating data; display green/red lights when screen is ON/OFF; and activate alarm for equipment malfunction
Temperature, pH, NH_3-N, Ortho-P, and TP sensors	Monitor raw influent quality parameters	Central	Display and record raw influent quality data; and activate alarm per high and low pH set points
Automatic sampler	Collect composite sample for routine chemical analysis by the laboratory	Central	Manually input analytical results for $CBOD_5$, TSS, VSS, total alkalinity, NH_3-N, TKN, and TP per analysis schedule
Raw influent lift station (Section 6.3.7)			
Limit switches on influent isolation gates	Monitor gate open/close position	Central	Display green/red lights when gate is OPEN/CLOSE
Radar or ultrasonic level transmitter in wet well	Monitor water level for controlling pump start/stop per level set points; and detect high or low level	Central	Display and record level continuously; and activate alarm per high and low level set points
Motor status monitors and data loggers for raw influent pumps	Monitor pump operating status for pump speed, run time, number of start, cycle on/off per duration timer set point, and emergency stop	Central and local	Display and record pump operating data; display green/red lights when pump is ON/OFF; activate alarm per high and low level set points, and for equipment malfunction
Temperature, amperage, vibration, and seal leakage sensors on motors	Monitor and provide motor protection	Central	Activate alarm for motor overheat, overload, and excessive vibration, and seal leakage
Pressure indicators on pump discharge pipes	Monitor discharge pressure of pumps	Local	Display pressure data

(Continued)

TABLE 14.4 (*Continued*) Major Sensing Devices Purpose, Display Location, and Description at Different Treatment Units[a]

Sensing Device	Purpose	Display Location	Description
Pressure switches on pump discharge pipes	Detect high discharge pressure due to potential blockage	Central	Activate alarm per high pressure set point
Limit switches on pump discharge valves	Monitor valve open/close position	Central	Display green/red lights when valve is OPEN/CLOSE
Float switch in a sump	Monitor liquid level in the sump in dry well	Central	Activate sump pump and alarm
Main circuit breakers and auxiliary contacts	Monitor power outage	Central	Activate alarm (standby power required)
Meter vault (Section 6.3.7)			
Magnetic or Venturi tube flow meter in force main	Monitor flow pumped	Central	Display, record, and totalize flow data continuously
Aerated grit chambers (Section 8.4.3)			
Limit switches on influent isolation valves	Monitor valve open/close position	Central	Display green/red lights when valve is OPEN/CLOSE
Limit switches on effluent isolation gates	Monitor gate open/close position	Central	Display green/red lights when gate is OPEN/CLOSE
Orifice plate with a square root extractor and display	Measure air flow to each chamber	Central and local	Display, record, and totalize air flow data continuously
Motor status monitors and data loggers for air blowers	Monitor blower operating status for run time, cycle on/off per duration timer set point, and emergency shutdown	Central and local	Display and record blower operating data; display green/red lights when blower is ON/OFF; and activate alarm for equipment malfunction
Pressure indicators on air piping	Monitor air pressure	Local	Display pressure data
Motor status monitors and torque sensors for screw conveyors	Monitor screw conveyor operating status	Central and local	Display green/red lights when conveyor is ON/OFF; and activate alarm for equipment malfunction
Motor status monitors for grit handling equipment	Monitor operating status of slurry pumps, grit classifiers, and grit storage hopper gates	Central and local	Display green/red lights when motor is ON/OFF; and activate alarm for equipment malfunction
Pressure indicators on grit pipes	Monitor discharge pressure	Local	Display pressure data
Magnetic flow meters on grit pipes	Monitor the grit slurry flow to each classifier	Central	Display, record, and totalize grit flow data
Centrifugal vortex-induced grit chambers (Section 8.4.4)			
Limit switches on influent and effluent isolation gates	Monitor gate open/close position	Central	Display green/red lights when gate is OPEN/CLOSE
Motor status monitors for center propellers	Monitor propeller operating status	Central and local	Display green/red lights when propeller is ON/OFF; and activate alarm for equipment malfunction
Radar or ultrasonic level transmitter in effluent channel	Monitor water level	Central	Display and record level continuously

(*Continued*)

TABLE 14.4 (*Continued*) Major Sensing Devices Purpose, Display Location, and Description at Different Treatment Units[a]

Sensing Device	Purpose	Display Location	Description
Motor status monitors for grit handling equipment	Monitor operating status of grit pumps and concentrators	Central and local	Similar to those for aerated grit chambers
Pressure indicators on grit pipes	Monitor discharge pressure	Local	Similar to those for aerated grit chambers
Magnetic flow meters on grit pipes	Monitor the grit flow to each concentrator	Central	Similar to those for aerated grit chambers
	Primary clarifiers (Chapter 9)		
Motor status monitors and torque sensors for sludge-raking mechanisms	Monitor sludge collector operating status, and emergency shutdown	Central and local	Display green/red lights when collector is ON/OFF; and activate alarm for equipment malfunction
Limit switches on sludge withdrawal valves	Monitor valve open/close position	Central	Display green/red lights when valve is OPEN/CLOSE
Radar or ultrasonic level transmitter in sludge wet well, and scum box	Monitor sludge level for controlling start/stop of primary sludge and scum pump, and air agitator per level set points	Central	Display and record level continuously
Motor status monitors for primary sludge and scum pumps	Monitor operating status of sludge and scum pumps	Central and local	Display green/red lights when pump is ON/OFF; and activate alarm for equipment malfunction
Pressure indicators on primary sludge and scum pump discharge pipes	Monitor discharge pressures	Local	Display pressure data
Magnetic flow meters on primary sludge piping	Monitor primary sludge flow	Central	Display, record, and totalize sludge flow data
	Biological nutrient removal (BNR) system –anaerobic zones (Sections 10.6 through 10.8)		
Motor status monitors and torque sensors for submersible mixers	Monitor mixer operating status	Central and local	Display green/red lights when mixer is ON/OFF; and activate alarm for equipment malfunction
Variable speed control device[b]	Monitor and control rotational speed of mixer blades	Central	Display and record rotational speed
VFA sensor in influent pipe	Monitor VFA concentration	Central	Display and record VFA data continuously
ORP and Ortho-P sensors	Monitor ORP and Ortho-P	Central	Display and record the monitoring data
	BNR system –anoxic zones (Sections 10.6 through 10.8)		
Motor status monitors and torque sensors for submersible mixers	Monitor mixer operating status	Central and local	Similar to those for anaerobic zones
Variable speed control device (similar to those for anaerobic zones)	Monitor and control rotational speed of mixer blades	Central and local	Similar to those for anaerobic zones

(*Continued*)

TABLE 14.4 (*Continued*) Major Sensing Devices Purpose, Display Location, and Description at Different Treatment Units[a]

Sensing Device	Purpose	Display Location	Description
ORP (or DO) and NO_3 sensors	Monitor ORP (or DO) and NO_3	Central	Display and record the monitoring data
BNR system -oxic (or aerobic) zones (Sections 10.6 through 10.8)			
Pressure transmitters on air piping	Monitor air pressure	Central	Display and record pressure continuously
Orifice plate with a square root extractor and display	Measure air flow to each aeration zone	Central and local	Display, record, and totalize air flow data continuously
DO sensors in multiple aeration zones	Monitor DO and control air supply rates	Central and local	Display and record DO continuously
Ultrasonic suspended solids analyzers in aeration basins	Measure MLSS concentrations	Central	Display and record MLSS data continuously
Final clarifiers (Section 10.9)			
Ultrasonic sludge blanket sensors in final clarifiers	Monitor the sludge blanket levels	Central	Display and record level continuously; and activate alarm per high or low level set point
Motor status monitors and torque sensors for sludge scrapper mechanisms	Monitor sludge scrapper operating status and emergency shutdown	Central and local	Similar to those for primary clarifiers
Radar or ultrasonic level transmitter in RAS/WAS wet well and scum box	Monitor sludge level for controlling start/stop of RAS, WAS, and scum pumps per level set points; and detect high level	Central	Display and record level continuously; and activate alarm per high or low level set point
Motor status monitors for RAS, WAS, and scum pumps	Monitor operating status of RAS, WAS, and scum pumps	Central and local	Similar to those for primary sludge pumps
Pressure indicators on RAS, WAS, and scum pump discharge pipes	Monitor discharge pressures	Local	Display pressure data
Magnetic flow meters on RAS and WAS pipes	Monitor RAS and WAS flows	Central	Display, record, and totalize sludge flow data
Ultrasonic suspended solids analyzers in RAS and WAS flows basins	Measure TSS concentrations	Central	Display and record TSS data continuously
DO, TSS, NH_3-N, Ortho-P, and TP sensors	Monitor effluent quality parameters	Central	Display and record clarifier effluent quality data; and activate alarm per high TSS and NH_3-N set points
Air supply by high-speed turbo (HST) blowers (Section 10.3.10)			
Monitoring and control packaged with HST blower system[c]	Monitor blower operating status and control of blower operation for rotation speed, run time; air flow; air, lube oil, and bearing temperatures; vibration, and emergency shutdown	Central and local	Display and record blower operating data; display green/red lights when blower is ON/OFF; and activate alarm for equipment malfunction
Pressure indicators on air piping	Monitor air pressure	Local	Display pressure data

(*Continued*)

TABLE 14.4 (*Continued*) Major Sensing Devices Purpose, Display Location, and Description at Different Treatment Units[a]

Sensing Device	Purpose	Display Location	Description
Flow meters on air headers, differential pressure type	Monitor air supply to each aeration zone	Central	Display, record, and totalize air flow data continuously
Limit switches on air piping isolation valves	Monitor valve open/close position	Central	Display green/red lights when valve is OPEN/CLOSE
Automatic backwash cloth media filters (Section 15.4.7)			
Limit switches on influent isolation gates	Monitor gate open/close position	Central	Display green/red lights when gate is OPEN/CLOSE
Radar or ultrasonic level transmitters in influent chambers	Monitor water level; and detect high level	Central	Display and record level continuously; and activate backwash cycle per high level set point
Motor status monitors for filter backwash pumps	Monitor operating status of backwash pumps	Central and local	Display green/red lights when pump is ON/OFF; and activate alarm for equipment malfunction
Pressure indicators on pump suction and discharge pipes	Monitor suction and discharge pressure	Local	Display pressure data
UV disinfection system (Section 11.9)			
Monitoring and control packaged with UV system[d]	Monitor and control of UV lamp module operation for run times for each lamp, row, module, and bank; UV light intensity; electrical power to lamps; temperature of UV lamp module; and burn out lamps	Central and local	Display and record UV system operating data; display green/red lights when UV module or bank is ON/OFF; and activate alarm for low UV dose and equipment malfunction
Radar or ultrasonic level transmitters at upstream and downstream of UV banks in UV channels	Monitor water levels for controlling effluent weir gates; and detect high and low levels	Central	Display and record level continuously; activate alarm per high or low set point
Limit switches on automatic effluent weir gates	Monitor gate open/close position	Central	Display green/red lights when gate is OPEN/CLOSE
Chlorine gas disinfection system (Section 11.6)			
Monitoring and control packaged with chlorination system[e]	Monitor and control of chlorination equipment operation for chlorine feed rate, weight of chlorine in storage	Central and local	Display and record chlorination system operating data; display green/red lights when chlorinator is ON/OFF; and activate alarm for equipment malfunction
Pressure indicator and low pressure switch on utility water line	Monitor water pressure	Central and local	Display pressure data and activate alarm per low pressure set point
Chlorine residual analyzer at end of chlorine contact basin	Measure chlorine residual to control chlorine and sodium bisulfite doses	Central	Display and record chlorine residual continuously

(*Continued*)

TABLE 14.4 (*Continued*) Major Sensing Devices Purpose, Display Location, and Description at Different Treatment Units[a]

Sensing Device	Purpose	Display Location	Description
Chlorine leak detector	Detection of chlorine leak in storage and feed facility	Central and local	Activate alarm when dangerous levels are reached
Sodium bisulfite dechlorination system (Section 11.7)			
Radar or ultrasonic level transmitters in storage and day tanks	Monitor solution level	Central	Display and record level continuously
Differential pressure transducers on metering pump suction pipes	Monitor pressure across solution strainer for potential line clogging	Central	Activate alarm per pressure differential set point
Pressure transmitters on metering pump discharge pipes	Monitor pump discharge pressure	Central	Display and record pressure continuously; and activate alarm per high pressure set point
Effluent flow and quality measurements (Section 12.6)			
Radar or ultrasonic level transmitters in Parshall flume (Example 9.38)	Measure flow	Central and local	Meter display, record, and totalize final discharge flow continuously
Chlorine residual analyzer after dechlorination	Measure chlorine residual for monitoring dechlorination	Central	Display and record residual continuously
DO, pH, and Ortho-P sensors	Monitor DO, pH, and Ortho-P	Central	Display and record the monitoring data
Automatic sampler	Collect composite sample for routine chemical analysis by the laboratory	Central	Manually input analytical results for $CBOD_5$, TSS, VSS, NH_3-N, and TP per analysis schedule
Gravity thickeners (Section 13.5.1)			
Ultrasonic sludge blanket sensors in thickeners	Monitor the sludge blanket levels	Central	Display and record level continuously
Motor status monitors and torque sensors for sludge-racking mechanisms	Monitor sludge rack operating status, and emergency shutdown	Central and local	Similar to those for clarifiers
Variable speed control device (a combination of PLC and VFD)	Monitor and control rotational speed of rack blades	Central and local	Display and record rotational speed
Limit switches on sludge withdraw valves	Monitor valve open/close position	Central	Display green/red lights when valve is OPEN/CLOSE
Ultrasonic solids analyzers in primary and thickened sludge pipes	Measure TSS concentrations	Central	Display and record TSS data continuously
Venturi meter on dilution water pipe	Flow measurement of dilution water	Central	Display, record, and totalize flow continuously
Pressure indicator on dilution water pipe	Monitor water pressure	Local	Display pressure data

(*Continued*)

TABLE 14.4 (*Continued*) Major Sensing Devices Purpose, Display Location, and Description at Different Treatment Units[a]

Sensing Device	Purpose	Display Location	Description
	Gravity belt thickeners (GBTs) (Section 13.5.4)		
Monitoring and control packaged with GBTs[f]	Monitor and control of GBT operation for run times for each unit; hydraulic/air pressure; wash water; and emergency shutdown	Central and local	Display and record GBT operating data; display green/red lights when GBT unit is ON/OFF; and activate alarm for belt misalignment and damage, low wash water pressure, low hydraulic/air pressure, and other equipment malfunction
Limit switches on thickened sludge valves	Monitor valve open/close position	Central	Display green/red lights when valve is OPEN/CLOSE
Ultrasonic solids analyzers in thickened sludge pipes	Measure TSS concentrations	Central	Display and record TSS data continuously
Monitoring and control packaged with polymer feed units[g]	Monitor and control of polymer unit operation for run times for each unit; polymer feed rate; dilution water; and emergency shutdown	Central and local	Display and record polymer unit operating data; display green/red lights when polymer feed unit is ON/OFF; and activate alarm for low dilution water pressure, and other equipment malfunction
Radar or ultrasonic level transmitters in emulsion polymer storage	Monitor polymer level	Central	Display and record level continuously; and activate alarm per low level set point
Pressure indicator transmitters and high pressure switches on polymer recirculation pump discharge pipes	Monitor water pressure	Central and local	Display and record pressure continuously; and activate alarm per high pressure set point
	Sludge blending tanks		
Radar or ultrasonic level transmitters in sludge blending tanks	Monitor sludge level	Central	Display and record level continuously; and activate alarm per high and low level set points
Motor status monitors and torque sensors for submersible mixers	Monitor mixer operating status	Central and local	Display green/red lights when mixer is ON/OFF; and activate alarm for equipment malfunction
Ultrasonic solids analyzers in sludge blending tanks	Measure TSS concentration	Central	Display and record TSS data continuously
	Anaerobic digesters (Section 13.6.1)		
Motor status monitors and flow control device for gas mixing compressors, and digester feed, and recirculation pumps	Monitor compressor and pump operating status and control flows	Central and local	Display green/red lights when compressor or pump is ON/OFF; display and record flow continuously; and activate alarm for equipment malfunction
Pressure indicator transmitters and pressure switches on the gas mixing lines; and digester feed, and recirculation pump discharge pipes	Monitor compressor suction and discharge pressures, and pump discharge pressures	Central and local	Display and record pressure continuously; and activate alarm per high and low pressure set points

(*Continued*)

TABLE 14.4 (*Continued*) Major Sensing Devices Purpose, Display Location, and Description at Different Treatment Units[a]

Sensing Device	Purpose	Display Location	Description
Pressure and temperature transmitters and pressure switches on hot water heating system	Monitor and control heating system operating status	Central and local	Display and record pressure and temperature; and activate alarm for equipment malfunction
VFA sensor in influent pipe	Monitor VFA concentration	Central	Display and record VFA data continuously; and activate alarm per high VFA set point
ORP, pH, and temperature sensors on digester content	Monitor ORP, pH, and temperature	Central	Display and record the monitoring data; and activate alarm per high ORP, low pH, or high/low temperature set points
CO_2, CH_4, and H_2 sensors and calorimeter of biogas piping	Monitor biogas composition and calorific value	Central	Display and record the monitoring data; and activate alarm per high H_2 set point
Magnetic flow meters on feed sludge and supernatant pipes	Monitor sludge and supernatant flows	Central	Display, record, and totalize flow data continuously
Ultrasonic gas flow meter	Monitor biogas flow	Central	Display, record, and totalize gas flow continuously
Ultrasonic solids analyzers in digesters	Measure SS concentrations	Central	Display and record SS data continuously
Radar or ultrasonic level transmitters in digesters	Monitor sludge level	Central	Display and record level continuously; and activate alarm per low level set point
Pressure and temperature transmitters and pressure switches in biogas storage tanks	Monitor biogas pressure	Central and local	Display and record pressure continuously; and activate alarm per high pressure set point
Flame, smoke, CO, H_2S, and CH_4 detectors	Monitor and detect flame and hazardous condition in building, and flare status	Central and local	Activate alarm when flame, smoke, or dangerous condition is detected
Belt filter press dewatering units (Section 13.8.2)			
Monitoring and control packaged with BFPs (similar to those for GBTs)	Monitor and control of BFP operation for run times for each unit; hydraulic/air pressure; wash water rate; sludge feed rate, and emergency shutdown	Central and local	Similar to those for GBTs
Limit switches on digested sludge valves	Monitor valve open/close position	Central	Display green/red lights when valve is OPEN/CLOSE
Motor status monitors for BFP feed pumps	Monitor pump operating status	Central and local	Similar to those for other sludge pumps
Magnetic flow meters on feed sludge pipes	Monitor sludge flow	Central	Display, record, and totalize flow data continuously
Ultrasonic solids analyzers in feed sludge pipes	Measure TSS concentrations	Central	Display and record TSS data continuously

(*Continued*)

TABLE 14.4 (*Continued*) Major Sensing Devices Purpose, Display Location, and Description at Different Treatment Units[a]

Sensing Device	Purpose	Display Location	Description
Monitoring and control packaged with polymer feed units (similar to those for GBTs)	Monitor and control of polymer unit operation for run times for each unit; polymer feed rate; dilution water; and emergency shutdown	Central and local	Similar to those for GBTs
Radar or ultrasonic level transmitters in emulsion polymer storage	Monitor polymer level	Central	Similar to those for GBTs
Pressure indicator transmitters and high pressure switches on polymer recirculation pump discharge pipes	Monitor discharge pressure	Central and local	Similar to those for GBTs
Motor status monitors for belt conveyers	Monitor belt conveyer operating status and emergency shutdown	Central and local	Display green/red lights when conveyer train is OPEN/CLOSE; and activate alarm for equipment malfunction
Load cell with weight transmitter	Monitor weight of sludge cake	Central and local	Display and record weight data
Ohm meter, lab test	Determine moisture content of sludge cake	Local	Manually input moisture data

[a] A control panel with PLC is typically provided for each treatment unit.
[b] It may include programmable logic controller (PLC) and variable frequency device (VFD).
[c] It may include VFD, PLC, and human machine interface (HMI) with multiple functional sensors and switches.
[d] It may include PLC and HMI with UV light intensity, temperature, and other sensors and switches.
[e] It may include vacuum regulator, pressure indicator transmitters, high pressure switch, load cell, rotameter, automatic shutoff valve, etc.
[f] It may include PLC and HMI with sensors, indicators, and switches for belt speed, belt tracking and tensioning, hydraulic/air pressure, belt wash water, etc.
[g] It may include PLC and HMI with sensors, indicators, and switches for emulsion, dilution water, and solution flows and pressures; polymer feed; etc.
Source: Adapted in part from Reference 2.

Discussion Topics and Review Problems

14.1 A design engineer is responsible to develop the plant layout, site plan, landscaping, and finished grading of a wastewater treatment facility. List the basic rules that the engineer must consider in these site developments.

14.2 Study the site plan given in Figure 14.2. List the basic features that make the site plan a modular layout and comment on the arrangement of the main treatment process components.

14.3 Obtain and review the layout plan, finished contours, and landscaping features of a wastewater treatment plant in your community. Compare the important features of this plant layout with that of the given in Figure 14.2.

14.4 Develop a comprehensive checklist used by a review committee for the design review of overall treatment process, plant layout, and site development. Arrange the checklist items under the following headings:

a. General layout
b. Process trains and components

c. Instrumentation and control
d. Piping and plant hydraulics
e. Administration building
f. Operation and maintenance flexibility
g. Access roads and plant security
h. Grading, drainage, and landscaping
i. Expansion flexibility

14.5 Discuss various design considerations that are necessary in developing the piping layout and hydraulic profile of a wastewater treatment plant.

14.6 A wastewater treatment plant has the following treatment units: (a) bar screen; (b) aerated grit channel; (c) primary clarifier; (d) aeration basin; (e) final clarifier; (f) chlorination facility; and (g) outfall structure with a Parshall flume. Select proper head losses across various treatment units using the approximate head loss information provided in Table 14.1. The outfall structure has a total head loss of 2 m. Assume that the connecting pipings between the units have head loss of approximately 20% of the average head loss across the upstream and downstream units. Prepare the hydraulic profile through the treatment plant where the lift station is located after the primary treatment. Also, justify your assumption with brief reasoning.

14.7 A wastewater treatment plant was designed to provide secondary treatment. The head losses in the primary clarifier, aeration basin and final clarifier complex, and chlorination facility at peak design flow were 0.8, 1.5, and 0.6 m, respectively. These units are directly connected by two connecting pipings with the following identical characteristics: diameter = 20 cm, straight length = 50 m, $K = 2.5$, and $C = 110$. Draw the hydraulic profile through the plant. Assume that the outfall pipe has a submerged discharge with a total head loss of 2 m. The high flood level in the receiving water is 100 m. The flow through all connecting piping is 0.02 m^3/s.

14.8 Obtain the hydraulic profile of the wastewater treatment plant in your community and summarize the following hydraulic conditions:
a. Head losses in each treatment unit
b. Head losses in each junction structure
c. Head losses in the connecting pipes
d. Water surface elevation in each treatment unit and junction structure

14.9 Two UV disinfection channels are provided after tertiary filters. The filtered effluent is received by an influent channel and divided by influent weir gates into two UV channels. The water surface elevation in the influent channel is 90.00 m at a design peak wet weather flow of 0.15 m^3/s. A 51-cm pipe connects the junction box of the filters and the influent channel of the UV channels. The pipe has two 45° and two 90° bends and a linear length of 30 m. Assume that the values of minor head loss coefficient (K) for 45° and 90° bends, entrance, and exit is 0.2, 0.3, 0.5, and 1.0, respectively and the coefficient of roughness $C = 110$. Determine the water surface elevation in the junction box of the tertiary filters.

14.10 A flow-monitoring float was installed in a junction box of a sewer interceptor prior to a wastewater treatment plant. The diameter, slope, and n of the interceptor are 30 cm, 0.005 m/m, and 0.013, respectively. At the average flow, the depth in the manhole is 12 cm while the float reads 26 cm at the peak flow. Assume that the depth in the junction box is the same as that of flow in the interceptor. Calculate the peak wet weather and average dry weather flows.

14.11 A sewer interceptor prior to a wastewater treatment plant is 100 cm in diameter. A mechanical regulator is installed that bypasses the excess flow into a flow equalization basin (FEB) during wet weather flow conditions. The maximum treatment capacity of the plant is reached at a flow depth of 48 cm. As the depth of flow increases above 48 cm, the regulator gate opens in proportion to the depth in the interceptor. The gate dimensions are 20 cm wide and 15 cm high. The differential elevation from the center of the gate to the 48-cm depth in the interceptor sewer is 1 m. Calculate the discharge into the FEB if the depth of flow in the interceptor sewer is 85 cm, and the gate opening is 75%. Assume that the coefficient of discharge of the gate is 0.8.

14.12 In Problem 14.11, the discharge flow to the FEB is calculated from the gate opening. Calculate the discharge that is bypassed from the interceptor sewer when the depth in the sewer is 85 cm. The sewer slope and n value are 0.0008 m/m and 0.013. Compare the result with that obtained in Problem 14.11.

14.13 A Venturi meter is used for continuous flow recording of industrial wastewater. The differential pressure is transmitted and recorded on a chart. Using Equation 7.4a with $C_d = 1.017$ for a throat diameter of 46 cm (18 in), calculate the hourly flows and totalize the flow. Use the following data:

Time	12 midnight	1	2	3	4	5	6	7	8	9	10	11
h, m H_2O	0.20	0.21	0.25	0.26	0.30	0.40	0.50	0.70	1.20	1.80	2.20	2.30
Time	12 noon	1	2	3	4	5	6	7	8	9	10	11
h, m H_2O	2.00	1.80	1.60	2.10	2.80	3.00	2.50	2.00	1.80	0.80	0.40	0.30

14.14 An industrial plant operates from 8 a.m. to 5 p.m. The hourly wastewater flow and COD concentrations are given below. Totalize the COD results and express in kilograms per operating day.

Time	7	8	9	10	11	12 noon	1	2	3	4	5	6	7
Flow, m^3/s	0	0.02	0.05	0.10	0.15	0.20	0.21	0.22	0.25	0.15	0.10	0.02	0
COD, mg/L	0	80	120	180	300	350	300	280	250	200	150	80	0

14.15 Using the typical control system components given in Figure 14.8, list various options available at each control system component for maintaining constant temperature in an anaerobic digester. The sludge is recirculated through external heat exchanger for heating purposes.

14.16 Match the primary flow measuring devices with their most suitable applications.

Orifice meter	(a) Sludge
Flow nozzle	(b) Clean liquid
Parshall flume	(c) Raw wastewater
Electromagnetic meter	

14.17 Write measured signals for the following variables:

a. Conductivity
b. pH
c. Density
d. Oxidation–reduction potential
e. Dissolved oxygen
f. Temperature
g. Pressure

14.18 Discuss various types of signal-transmitting devices. Give advantages and disadvantages and applications of each type.

14.19 What is the basic difference between feedback and feed-forward automatic control loops? Give three examples of each in wastewater treatment plants.

14.20 A simplified conventional activated sludge process diagram is shown in Figure 10.20. Draw an integrated process and instrumentation diagram (P&ID) for a process train that has a single aeration basin, final clarifier, return sludge pump, waste sludge pump, and blower. Also, include (a) DO sensor in aeration basin, (b) air flow meter and pressure sensor for air supply system, and (c) flow meters and pressure indicators for sludge pumping.

References

1. Task Force of the Water Environment Federation and the American Society of Civil Engineers/ Environmental and Water Resources Institute, *Design of Municipal Wastewater Treatment Plant*, WEF Manual of Practice No. 8 and ASCE Manuals and Reports on Engineering Practice No. 76, 5th ed., McGraw-Hill, New York, NY, 2009.
2. Qasim, S. R., *Wastewater Treatment Plants: Planning, Design, and Operation*, 2nd ed., CRC Press, Boca Raton, FL, 1999.
3. Metcalf & Eddy | AECOM, *Wastewater Engineering: Treatment and Resource Recovery*, 5th ed., McGraw-Hill, New York, NY, 2014.
4. Qasim, S. R., E. M. Motely, and G. Zhu, *Water Works Engineering: Planning, Design, and Operation*, Prentice Hall PTR, Upper Saddle River, NJ, 2000.
5. Corbin, D. J., R. D. G. Monk, C. J. Hoffman, and S. F., Crumb, Jr., Compact treatment plant layout, *Journal American Water Works Association*, 84 (8), 1992, 36–42.
6. Matsunaga, K., Design of multistory settling tanks in Osaka, *Journal Water Pollution Control Federation*, 52 (5), 1980, 950–954.
7. National Fire Protection Association (NFPA), *NFPA® 1: Fire Code*, 2015 ed., NFPA, Quincy, MA, 2014.
8. International Code Council, Inc. (ICC), *2015 International Fire Code®*, ICC, Country Club Hills, IL, 2014.
9. Nakai, K., Unique design features of Kobe's wastewater treatment plant, *Journal Water Pollution Control Federation*, 52 (5), 1980, 955–960.
10. Terada, K., Development of sewer systems and treatment plants in Tokyo, *Journal Water Pollution Control Federation*, 52 (5), 1980, 961–968.
11. Yuki, Y., E. Takayanagi, and T. Abe, Design of multi-story treatment in Osaka, Japan, *Water Science and Technology*, 23 (10–12), 1991, 1733–1742.
12. Isgard, E., Underground wastewater treatment plants, *Journal Water Pollution Control Federation*, 47 (4), 1975, 684–687.
13. Bishop, Jr., W. A., , and J. S. Fraser, Wastewater under home plate, *Civil Engineering, American Society of Civil Engineers*, 62 (10), 1992, 61–63.
14. Water World®, Largest Underground MBR Plant in Asia Completed, *Water World*, http://www.waterworld.com/articles/2010/09/largest-underground-mbr-plant-in-asia.html, September 30, 2010.
15. Water World®, China's Largest Underground Wastewater Recycling Plant Moves Ahead, *Water World*, http://www.waterworld.com/articles/wwi/2016/05/china-s-largest-underground-wastewater-recycling-plant-moves-ahead.html, May 26, 2016.
16. Ando, S., Odor control of wastewater treatment plants, *Journal Water Pollution Control Federation*, 52 (5), 1980, 906–913.
17. Occupational Safety and Health Administration (OSHA), U.S. Department of Labor, *OSHA Law & Regulations*, http://shop.iccsafe.org/codes/2015-international-codes-and-references.html
18. International Code Council, Inc. (ICC), *2015 International Codes and References*, https://www.osha.gov/law-regs.html
19. Chow, V. T., *Open-Channel Hydraulics*, McGraw-Hill Book Co., New York, 1959.
20. Henderson, F. M., *Open Channel Flow*, MacMillan, New York, 1966.
21. Finnemore, E. J., and J. B. Franzini, *Fluid Mechanics with Engineering Applications*, 10th ed., McGraw-Hill Education, New York, 2002.
22. Benefield, L. D., J. F. Judkins, and A. D. Parr, *Treatment Plant Hydraulics for Environmental Engineers*, Prentice-Hall Inc., Englewood Cliffs, NJ, 1984.
23. Montgomery Watson Harza, *Water Treatment: Principles and Design*, 2nd ed., John Wiley & Sons, Inc., Hoboken, NJ, 2005.

24. American Water Works Association and American Society of Civil Engineers, *Water Treatment Plant Design*, 5th ed., McGraw-Hill Professional, New York, NY, 2012.
25. Technical Practice Committee Control Group, *Process Instrumentation and Control Systems*, OM-6, Water Pollution Control Federation, Alexandria, VA, 1984.
26. Kawamura, S., *Integrated Design and Operation of Water Treatment Facilities*, 2nd ed., John Wiley & Sons, Inc., New York, 2000.
27. Molvar, A. E., J. F. Roesler, R. F. Wise, and R. H. Babcock, *Instrumentation and Automation Experiences in Wastewater-Treatment Facilities*, EPA-600/2-76-198, U.S. Environmental Protection Agency, Municipal Environmental Research Laboratory, Cincinnati, OH, October 1976.
28. Committee on Instrumentation, *Instrumentation in Wastewater Treatment Plants*, MOP 21, Water Pollution Control Federation, Washington, D.C., 1978.
29. U.S. Environmental Protection Agency, *Process Design Manual for Sludge Disposal*, Technology Transfer, EPA 625/1-79-011, Municipal Environmental Research Laboratory, Cincinnati, OH, September 1979.
30. Anderson, N. A., *Instrumentation for Process Measurement and Control*, 3rd ed., CRC Press, Boca Raton, FL, 1997.
31. Considine, D. M. (ed.), *Encyclopedia of Instrumentation and Control*, McGraw-Hill Book Co., New York, 1971.
32. Instrument Society of America, *Instrument Loop Diagrams, American National Standard*, ANSI/ISA-S5.4-1991, 1991.
33. Patry, G. G., and D. T. Chapman, *Dynamic Modeling and Expert Systems in Wastewater Engineering*, Lewis Publishers Inc., Chelsea, MI, 1989.
34. Waterman, D. A., *A Guide to Expert Systems*, Addison-Wesley, Reading, MA, 1986.
35. Krichten, D. J., K. D. Wilson, and K. D. Tracy, *Application of Expert System Technology to Control of Biological Phosphorus Removal Plant*, Air Products and Chemicals, Inc., Allentown, PA, August 1990.
36. Brown, R. J., An artificial neural network experiment, *Dr. Dobb's Journal*, 12 (4), 1987, 16–27.
37. Widrow, B. and R. Winter, Neural nets for adaptive filtering and adaptive pattern recognition, *Browse Journals & Magazines/Computer, IEEE*, 21 (3), 1988, 25–39.
38. Winston, P. H., *Artificial Intelligence*, Addison-Wesley, Reading, MA, 1984.
39. Cubberly, W. H. (ed.), *Comprehensive Dictionary of Measurement and Control*, 2nd ed., Instrument Society of America, Research Triangle Park, NC, 1991.

15

Advanced Wastewater Treatment and Upgrading Secondary Treatment Facility

15.1 Chapter Objectives

Upgrading of a secondary wastewater treatment plant may be necessary to meet the existing or stringent future effluent quality requirements. This may involve removal of suspended solids, organics, and nutrients beyond the capability of a secondary treatment plant. Advanced wastewater treatment is, therefore, removal of suspended and colloidal solids, nutrients, and many dissolved constituent remaining after conventional secondary treatment. The dissolved constituents may range from simple inorganic ions to complex synthetic organic compounds. The objectives of this chapter are to discuss the treatment processes used to upgrade the existing conventional secondary treatment facility and to achieve the advanced wastewater treatment. These treatment processes are:

- Natural systems
- Chemical precipitation
- Breakpoint chlorination for ammonia removal
- Air stripping for removal of dissolved gases
- Filtration, adsorption, and ion exchange
- Membranes and membrane bioreactors (MBRs)
- Distillation and other demineralization processes
- Treatment and disposal of concentrated brine

15.2 Application of Advanced Treatment Technologies and Upgrading of Secondary Effluent

The advanced treatment processes may be applied to secondary or primary treated effluents. They are used to remove four categories of constituents depending upon the degree of treatment desired. These are: (1) organic and inorganic suspended and colloidal solids, (2) dissolved organic constituents, (3) dissolved inorganic constituents, and (4) biological constituents, such as bacteria, viruses, and protozoan cysts and oocysts. The generalized process combinations are provided in Table 15.1.

15.3 Natural Treatment Systems

In the natural environment, physical, chemical, and biological processes occur when water, soil, plants, microorganisms, and atmosphere interact. The natural systems are designed to maximize these

TABLE 15.1 Process Combinations for Advanced Wastewater Treatment Systems Applied to Secondary and Primary Treated Effluents

Influent Quality and Treatment Process	Constituents Removed
Secondary Effluent	
Natural systems	TSS, dissolved organics, and nutrients
Coagulation, flocculation, sedimentation, filtration, and disinfection	TSS and colloidal solids (organic and inorganic), phosphorus, and biological constituents
Coagulation, flocculation, sedimentation, filtration, carbon adsorption, and disinfection	TSS and colloidal solids (organic and inorganic), refractory organics, phosphorus, and biological constituents
Lime precipitation, recarbonation, filtration, carbon adsorption, and disinfection	TSS colloidal solids (organic and inorganic), refractory organics, hardness, heavy metals, phosphorus, organic nitrogen, and biological constituents
Filtration, carbon adsorption, ultrafiltration, air stripping, and disinfection	TSS, refractory organics, ammonia, volatile organics, and biological constituents
Filtration, carbon adsorption, ultrafiltration, ion exchange (clinoptilite), and disinfection	TSS, refractory organics, ammonia, and biological constituents
Microfiltration, reverse osmosis, and disinfection	TSS, dissolved organics, demineralization, and biological constituents
Primary Effluent	
Natural systems and disinfection	TSS, dissolved organics, nutrients, and biological constituents
Membrane bioreactor and disinfection	TSS, dissolved organics, phosphorus, organic nitrogen, and biological constituents
Membrane bioreactor, reverse osmosis, and disinfection	TSS, dissolved organics, demineralization, and biological constituents

interactions. The treatment processes involved in the natural systems include sedimentation, filtration, gas transfer, adsorption, ion exchange, chemical precipitation, chemical oxidation and reduction, and biological conversions.

Additionally, some unique processes such as photosynthesis, photooxidation, and aquatic plants uptake also occur simultaneously.[1,2] In natural systems, these treatment processes occur simultaneously in a single ecosystem or reactor at the natural rate. In mechanical systems, these processes occur sequentially in separate reactors at accelerated rates due to energy input. The natural systems are divided into (a) terrestrial (or land) treatment systems and (b) aquatic treatment systems. Both systems are discussed below.

15.3.1 Terrestrial Treatment Systems

Terrestrial treatment systems remove contaminates by physical, chemical, and biological processes occurring within the soil–plant–water matrix. As a result, the terrestrial systems provide irrigation, nutrients reuse, crop production, recharge of groundwater, and reclaimed water for reuse. Basic information needed for developing a terrestrial treatment system is briefly listed below.[3,4]

1. Wastewater Characteristics: minimum, average, and peak design flows and concentration of constituents such as BOD_5, TSS, nitrogen, phosphorus, and heavy metals
2. Climatic Data: precipitation and frequency, evapotranspiration, temperature, growing season, occurrence and depth of frozen ground, storage requirement, and wind velocity and direction
3. Geological Data: seasonal depth of groundwater table and quality, bedrock type, depth, and permeability
4. Soil Characteristics: type and gradation, infiltration/permeability type and proportion of clay, cation exchange capacity, phosphorus and heavy-metal adsorption potentials, pH, and organic content

5. Plant Cover: indigenous plants, nutrients removal capacity, toxicity levels, moisture and shade tolerance, and marketability
6. Topography: ground slope, erosion hazard, and ease of ground cover management
7. Application: irrigation method, type of equipment, application rate, and type of drainage
8. Environmental Concerns: odors and socioeconomic and health aspects

There are three basic methods of land application: (1) slow-rate (SR), (2) rapid infiltration (RI) or soil aquifer treatment (SAT), and (3) overland flow (OF).[3,5,6] Each method has different overall objectives, adapted to different site conditions, and produces different quality of reclaimed water. In general, both SR and RI systems use the soil matrix for treatment after infiltration of wastewater at different application rates. In the OF systems, the soil surface and vegetation are mainly used for the treatment and only limited percolation is involved. Each method is briefly discussed below. Typical design features and performance expectations for three basic terrestrial systems are summarized in Table 15.2.

SR Systems: Slow-rate land treatment is the oldest and most common method of land treatment. The objective is irrigation and crop production. Surface runoff is not allowed. Water is lost by evapotranspiration and percolation. The design considerations include (1) availability of suitable site and needed land area, (2) wastewater characteristics and pretreatment, (3) climatic conditions and storage requirement, (4) soil type, (5) organic and hydraulic loadings, (6) crop production, (7) distribution method, (8) application cycle, and (9) groundwater quality criteria.[3–11]

Depending upon the application objective, the SR systems are usually categorized into two types: *Type 1*: slow infiltration and *Type 2*: crop irrigation. Type 1 systems are typically designed to use the least amount of land for treatment, while the crop irrigation requirements must be considered in design of Type 2 systems. Therefore, a Type 2 system may require a land area that is two to four times larger than that of a Type 1 system at the same wastewater flow.[6,11]

Hydraulic Loading Rate: The hydraulic loading rate is determined from a mass balance on the receiving soil. Runoff of applied wastewater is ignored. Also, wastewater should not be applied during wet weather. The hydraulic loading rates used in design of SR systems are usually expressed as the depth of water in cm (in) per time period: cm/week (in/week) for a week, cm/month (in/month) for a month, or cm/year (in/year) for a year.

The hydraulic loading rate for a Type 1 system is calculated from Equations 15.1a and 15.1b. It is normally determined by a water balance procedure on monthly basis using an unit of cm/month (in/month).[5,10,11] The procedure is shown in Example 15.1.

$$L_{SR1} = (ET - P) + K_d \tag{15.1a}$$

$$K_d = 24F_C T_A K \quad \text{or} \quad K_d = 24F_C(T_M - T_{NA})K \tag{15.1b}$$

where

L_{SR1} = hydraulic loading rate based on percolation limit for a Type 1 system, cm/month (in/month)
ET = evapotranspiration rate, cm/month (in/month)
P = precipitation rate, cm/month (in/month)
K_d = design percolation rate (or hydraulic conductivity), cm/month (in/month)
24 = hours per day (h/d)
F_C = factor accounting for soil clogging, fraction. Suggested value of $F_c = 0.02$–0.1.
T_A = days with application of wastewater in the month, d/month
T_M = total days in the month, d/month
T_{NA} = inoperative days for application of wastewater in the month, d/month. This includes the days such as rainy days, freezing temperature (below 0°C), and planting, cultivation, and harvesting periods.
K = soil permeability, m/h. The typical value of soil permeability may be in the range of 0.5–3 cm/h (0.2–1.2 in/h).

TABLE 15.2 Comparison of Design Features, Site Characteristics, and Effluent Quality from Land Treatment Processes

Condition	Land Treatment Process		
	Slow-Rate (SR)	Rapid Infiltration (RI)	Overland Flow (OF)
	Design Feature		
Application technique	Sprinkler, surface, or drip	Usually surface	Sprinkler or surface
Vegetation	Yes	Optional	Yes
Treatment goal	Secondary or AWT[a]	Secondary or ground water recharge	Secondary or AWT[a]
Field area required[b], ha (ac)	23–280 (60–700)	3–23 (7–60)	6.5–44 (16–110)
Weekly application rate, cm/week (in/week)	1.9–6.5 (0.75–2.5)	10–240 (4–95)	6–40 (2.5–16)[c]
Annual volumetric loading, m^3/m^2·year (gal/ft^2·year)	0.5–6 (12–150)	6–125 (150–3000)	3–20 (75–500)
Organic loading[d], kg BOD_5/ha·d (lb BOD_5/ac·d)	50–500 (45–450)	145–1000 (130–900)	40–110 (35–100)
Minimum pretreatment provided	Primary treatment	Primary treatment	Screening
Disposition of applied wastewater	Evapotranspiration and percolation	Mainly percolation	Surface runoff and evapotranspiration with some percolation
	Site Characteristics		
Slope	<20% on cultivated land and 35% on noncultivated land	Not critical; and excessive slopes require much earthwork	Finish slopes 2–8%
Soil permeability	Moderately slow	Rapid	Slow to none
Depth to groundwater, m (ft)	0.6–3 (2–10)	1–3 (3.3–10)[e]	Not critical
Climate restrictions	Storage for cold weather during winter	Not critical	Storage for cold weather during winter
	Effluent Quality From Land Treatment Process[f]		
BOD_5, mg/L	<2	5	10
TSS, mg/L	<1	2	10
Ammonia, mg/L as N	<0.5	0.5	<4
Total nitrogen, mg/L as N	3[g]	10	5[h]
Total phosphorus, mg/L as P	0.1	1	4
Fecal coliform, MPN/100 mL	≤1	10	>200

[a] Advanced wastewater treatment.
[b] Field areas are required for 44 L/s (1 mgd) flow. These areas exclude land required for buffer zone, roads, or ditches.
[c] This range includes screened wastewater to secondary effluent.
[d] Lower and upper ends of range are for municipal and industrial wastewater applications, respectively.
[e] Use underdrainage in locations with shallow groundwater.
[f] (1) All values for SR are obtained for percolation of primary or secondary effluent through 1.5 m (5 ft) of unsaturated soil; (2) All values for RI are obtained for percolation of primary or secondary effluent through 4.5 m (15 ft) of unsaturated soil; and (3) All values for OF are obtained for treatment of comminuted and screened wastewater over a length of 30–36 m (100–120 ft) of slope.
[g] Removal depends on loading rate, C:N ratio, and crop uptake.
[h] Higher removal may be achievable in moderately cold winter or using secondary effluent at high rates.

Note: 1 ha = 2.47 ac; 1 cm/week = 0.394 in/week; 1 m^3/m^2·year = 24.5 gal/ft^2·year; 1 kg BOD_5/ha·d = 0.892 lb BOD_5/ac·d; and 1 m = 3.28 ft.

Source: Adapted in part from References 2, and 3 through 11.

In Type 2 system, the crop irrigation requirements are normally considered on annual basis. The annual hydraulic loading rate can be calculated from Equation 15.1c at a unit of cm/year (ft/year).[5,11]

$$L_{SR2} = \frac{100\%\,(CET - P)}{E\,(1 - LR)} \quad (15.1c)$$

where

L_{SR2} = annual hydraulic loading rate for a Type 2 system, cm/year (in/year)
CET = crop evapotranspiration rate, cm/year (in/year)
E = irrigation efficiency, percent. The irrigation efficiencies are typically 70–80% and 65–75% for sprinkler and surface irrigation systems, respectively.
LR = factor for leaching requirement, fraction. The typical values of LR are 0.1–0.15 and 0.2–0.3 for low (<750 mg/L) and high (≥750 mg/L) TDS concentrations in the wastewater, respectively.

P has been defined previously. It must be in the unit of cm/year (in/year).

Nitrogen Loading Rate: The design hydraulic loading rate may be affected frequently by the nitrogen loading rate applied to a RS system. The allowable annual hydraulic loading based on nitrogen limit is calculated from Equations 15.1d.[5,10,11] This is applicable to both Type 1 and 2 systems.

$$L_{SRN} = \frac{C_P(P - ET) + F_N U_N}{(1 - f)C_N - C_P} \quad (15.1d)$$

where

L_{SRN} = allowable annual hydraulic loading rate based on nitrogen limit, cm/year (in/year)
C_P = nitrogen concentration in percolating water, mg/L as N. The value of C_P is normally 10 mg/L which is the limit for drinking water.
F_N = conversion factor, mg·ha-cm/kg·L (mg·ac-in/lb·L). The value of F_N is 10 mg·ha-cm/kg·L for SI units and 4.42 mg·ac-in/lb·L for U.S. customary units.
U_N = nitrogen uptake by crops, kg N/ha·year (lb N/ac·year). The nitrogen uptake depends upon the type of crop. Average nitrogen, phosphorus, and potassium uptakes for some forage, field, and forest crops are given in Table 15.3.[5,10]
f = fraction of applied nitrogen removed by nitrification/denitrification, volatilization, and soil storage, fraction. The values of f may be 0.1, 0.2, and 0.25 for tertiary, secondary, and primary effluents, respectively.
C_N = nitrogen concentration in applied wastewater, mg/L as N

P and *ET* have been defined previously. Both variables must be in the unit of cm/year (in/year).

Note: For evaluating storage requirement, Equation 15.1d may be applied on monthly basis (Example 15.1).

Organic Loading Rate: An organic loading rate up to 500-kg BOD_5/ha·d (450- lb BOD_5/ac·d) may be used in the design of SR systems (Table 15.2). Therefore, the design hydraulic loading rate for SR system is rarely limited by the organic loading rate for municipal wastewater application. However, the organic loading rate for SR system can be a limiting factor for industrial wastewater application.

Land Requirements: The land area required for SR irrigation is calculated from Equation 15.1e.[5,10,11]

$$A_{SR} = \frac{V}{F_{V1} L_W} \quad \text{or} \quad A_{SR} = \frac{365\,Q}{F_{V1} L_W} \quad (15.1e)$$

where

A_{SR} = total land area, ha (ac)
V = total annual volume of wastewater applied, m^3/year (Mgal/year)
F_{V1} = conversion factor, m^3/ha-cm (Mgal/ac-in). The value of F_{V1} is 100 m^3/ha-cm for SI units and 0.0272 Mgal/ac-in for U.S. customary units.

TABLE 15.3 Nutrients Uptake Rates for Different Crops

Crop	Nutrient Uptake, kg/ha·year (lb/ac·year)		
	Nitrogen	Phosphorus	Potassium
	Forage Crop		
Alfalfa	225–540 (200–480)	22–40 (20–35)	175–225 (155–200)
Coastal Bermuda grass	400–675 (360–600)	35–45 (30–40)	225 (200)
Ryegrass	200–280 (180–250)	60–85 (55–75)	270–325 (240–290)
Orchardgrass	250–350 (225–315)	20–50 (18–45)	225–315 (200–280)
	Field Crop		
Corn	175–200 (155–180)	20–30 (18–27)	110 (100)
Cotton	75–110 (65–100)	15 (13)	40 (35)
Grain, sorghum	135 (120)	15 (13)	70 (60)
Wheat	160 (145)	15 (13)	20–45 (18–40)
	Forest Crop		
Pine	110–320 (100–285)	–	–
Douglas fir	150–250 (135–225)	–	–

Note: 1 kg BOD_5/ha·year = 0.892 lb BOD_5/ac·year.
Source: Adapted in part from References 5 and 10.

L_w = design annual hydraulic loading rate, cm/year (in/year). L_w is the lower of L_{SR1} or L_{SR2} and L_{SRN}.
365 = days per year, d/year
Q = average daily flow rate, m^3/d (MGD)

Storage Requirements: Storage is required in most cases to balance the wastewater generation and allowable application rate. The procedure for designing the storage reservoir is covered in Section 12.4. Example 12.21 shows the step-by-step design calculations for a storage reservoir. Storage calculations are also given in Example 15.1.

RI Systems: The wastewater is percolated through the soil intermittently and treated effluent reaches the groundwater. The main objectives are (1) groundwater recharge, (2) achieve natural treatment while moving vertically through the soil and laterally along the groundwater, (3) recover treated water by pumping or through underdrain system, and (4) recharge a surface water resource.[4–6,8–11] Sandy and loamy soils improve BOD and TSS removals through RI systems. Nitrification/denitrification is achievable, but nitrogen and phosphorus removals are not sufficient. High groundwater quality criteria may limit the use of this method. The effluent quality and other design considerations for RI systems are also included and compared with other two systems in Table 15.2.

The cycle time for a site consists of periods of application and drying. The application cycle may vary from <1–5 days. The drying period may be in the range of 5–15 days. The land area is normally divided into sections or basins to achieve the required application and drying cycles. The design hydraulic loading rate, cycle time, number of cycles per year, loading rate per cycle, and the application rate per cycle are calculated from Equations 15.2a through 15.2e.[5,10,11]

$$L_{RI} = F_S K \quad (15.2a)$$

$$T_C = T_A + T_D \quad (15.2b)$$

$$N_C = \frac{365}{T_C} \quad (15.2c)$$

$$L_{\text{C}} = \frac{L_{\text{RI}}}{N_{\text{C}}} \tag{15.2d}$$

$$R_{\text{A}} = \frac{L_{\text{C}}}{T_{\text{A}}} \tag{15.2e}$$

where

L_{RI} = design hydraulic loading rate expressed on annual basis for RI systems, m/year (ft/year)
F_{S} = safety factor for soil permeability, fraction. The value of F_s may range from 0.02 to 0.1.
T_{C} = total time for one cycle, d/cycle
T_{A} = wastewater application days during one cycle, d/cycle. It may vary from 1 to 5 days.
T_{D} = drying days during one cycle, d/cycle. It may vary from 5 to 15 days.
N_{C} = number of cycles per year, cycle/year
L_{C} = loading rate per cycle, m/cycle (ft/cycle)
R_{A} = application rate applied during application cycle, m/d (ft/d)

K has been defined previously. It must be in the unit of m/year (ft/year).

The total land area required may be calculated from Equation 15.1e when the design hydraulic loading rate is determined.

Since the ponding in infiltrating zone is undesired, the application cycle (T_{A}) should also be checked against the criterion given by Equation 15.2f.[5,10] If the criterion is exceeded, then the T_{A} needs to be reduced, and the wastewater is delivered in infiltration zone sequentially. The area receiving flow during one application cycle and the total infiltration area are obtained from Equations 15.2f and 15.2h.

$$(R_{\text{A}} + P_{\text{max}} - K)T_{\text{A}} < D_{\text{C}} \tag{15.2f}$$

$$A_{\text{C}} = \frac{Q}{F_{\text{V2}}(R_{\text{A}} - P_{\text{max}})} \tag{15.2g}$$

$$A_{\text{RI}} = \frac{T_{\text{C}}}{T_{\text{A}}}A_{\text{C}} \quad \text{or} \quad A_{\text{RI}} = \frac{T_{\text{C}}Q}{F_{\text{V2}}T_{\text{A}}(R_{\text{A}} - P_{\text{max}})} \tag{15.2h}$$

where

P_{max} = maximum monthly precipitation once in 10 years, m/d (ft/d)
D_c = criterion for preventing ponding during application period, m (ft). A value of $D_c = 0.3$ m/cycle (1 ft/cycle) may be used.
A_{C} = average area that receives flow per application cycle, ha (ac)
Q = average daily flow rate, m^3/d (MGD)
F_{V2} = conversion factor, m^2/ha (Mgal/ac-ft). The value of F_{V1} is $10^4\ \text{m}^2$/ha for SI units and 0.326 Mgal/ac-ft for U.S. customary units.
A_{RI} = total infiltration area, ha (ac)

R_{A}, T_{A}, T_{C}, and K have been defined previously. It must be in the unit of m/d (ft/d).

It is desirable that the renovated water does not mix with the natural groundwater. If topography permits, the underdrains should lead to the surface water or pumped out from a collection sump. On account of these provisions, water quality requirements may become limiting design factors. Additional information about the potential removal efficiencies for BOD, TSS, nitrogen, and phosphorus in RI systems may be found in References 5, 10, and 11. A storage reservoir may also be needed for temporary storage (Section 12.4).

OF Systems: The OF system is used where soil permeability is too low and ground is sloping. The wastewater is applied over the upper reaches of the sloped terraces and allowed to flow overland. Land slope of 2–10% is desirable. The treated effluent is collected at the toe of the slope. Biochemical oxidation, filtration, and chemical adsorption are the primary mechanisms for removal of contaminants. Nitrogen removal is achieved by denitrification. Nutrients uptake is significant if crop harvesting is practiced. The design features, site characteristics, and effluent quality are summarized in Table 15.2.

The hydraulic loading is based on number of operative days. Storage is required during winter and wet weather periods (Section 12.4). The hydraulic loading rate, wastewater application rate, and sloped surface area are calculated from Equations 15.3a through 15.3d.[5,10,11]

$$L_{OF} = \frac{365Q + \Delta V}{F_{v1} T_{op} A_{OF}} \tag{15.3a}$$

$$N_{OF} = \frac{A_{OF}}{a_{OF}} \tag{15.3b}$$

$$a_{OF} = F_{v3} w_s l_s \tag{15.3c}$$

$$R_A = \frac{365Q + \Delta V}{F_{v4} w_s T_{op} \theta_{op}} \quad \text{or} \quad R_A = \frac{L_{OF} l_s}{F_{v5} \theta_{op}} \tag{15.3d}$$

where

L_{OF} = hydraulic loading based on operative days, cm/d (in/d)
ΔV = net change (gain or loss) in storage volume in sloped area due to precipitation, evaporation, and infiltration, m^3/year (Mgal/year)
T_{op} = number of days when wastewater is applied during a year, d/year
A_{OF} = total application area of OF system, ha (ac)
N_{OF} = number of application plots (plot)
a_{OF} = area of applied sloped surface in one application plot, ha/plot (ac/plot)
F_{v3} = conversion factor, ha/m^2 (ac/ft^2). The value of F_{v3} is 10^{-4} ha/m^2 for SI units and 2.3×10^{-5} ac/ft^2 for U.S. customary units.
w_s = width of sloped area in one application plot or length of distribution arm, m (ft)
l_s = length of sloped surface in one application plot, m (ft). The typical range is 30–60 m (100–200 ft).
R_A = application rate, m^3/m·h (gpm/ft). The application rate is expressed as flow in m^3/h (ft^3/h) per m (ft) of slope length.
θ_{op} = application period per day (h/d). The θ_{op} ranges 6–12 h/d with a typical value of 8 h/d.
F_{V4} = conversion factor, m^3/m^3 (Mgal·min/gal·h). The value of F_{V4} is 1 m^3/m^3 for SI units and 60×10^{-6} Mgal·min/gal·h for U.S. customary units.
F_{V5} = conversion factor, cm/m (ft^2·in·min/gal·h). The value of F_{V5} is 100 cm/m for SI units and 96.3 ft^2·in·min/gal·h for U.S. customary units.

Q and F_{V1} have been defined previously.

The general design guidance for OF systems are summarized in Table 15.4.[5,6,10,11] Additional information about the treatment performance for removals of BOD, TSS, nitrogen, and phosphorus in OF systems may be found in References 5, 10, and 11.

Other Environmental Concerns: There are many environmental concerns associated with the land treatment of wastewater. These concerns are accumulation of toxic chemicals in the soil and in the

TABLE 15.4 General Design Guidance for Overland Flow Systems

Effluent	Hydraulic Loading Rate, cm/d (in/d)	Application Rate, m^3/m·h (gpm/ft)
Primary	1.5–4 (0.6–1.6)	0.1–0.3 (0.13–0.4)
Stabilization pond	1.5–3 (0.6–1.2)	0.05–0.1 (0.07–0.13)
Secondary	3–6 (1.2–2.4)	0.2–0.45 (0.27–0.6)

Note: 1 cm/d = 0.394 in/d and 1 m^3/m·h = 1.34 gpm/ft.
Source: Adapted in part from References 5, 6, 10, and 11.

food chain, groundwater contamination, and odors problems. In all respects, the land treatment of wastewater must comply with the regulations promulgated under the Resource Conservation and Recovery Act (RCRA, as amended) to protect human health and the environment from the improper management of wastewater.[10,12,13]

EXAMPLE 15.1: LAND AREA REQUIRED FOR TYPE 1 SLOW-RATE (SR) SYSTEM

The monthly evapotranspiration (ET), precipitation (P), nitrogen uptake by crops (U_N), and the number of inoperative days in a month (T_{NA}) are given in Columns (2) through (5) of Table 15.5. The average wastewater flow from the community $Q = 1950\ m^3/d$. The average soil permeability K is 1.7 cm/h. The total nitrogen concentration in primary settled wastewater is 25 mg/L as N. The crop selected for the application field has annual nitrogen uptake rate of 320 kg N/ha·year. The soil permeability factor F_C due to clogging is 0.08, and loss of nitrogen by denitrification and volatilization is 25%. Determine (a) the area required for SR irrigation and (b) storage volume and area requirement for a Type 1 SR system.

Solution

1. Determine the design hydraulic loading rate on annual basis.

   The sample calculations for the month of January are given below. The procedure is repeated for other months, and the results are provided in Table 15.5.

   a. Determine from Equation 15.1b the hydraulic conductivity (K_d).

   $$K_d = 24F_C\,(T_M - T_{NA})K = 24\,h/d \times 0.08 \times (31 - 15)\,d/month \times 1.7\,cm/h = 52.2\,cm/month$$

   b. Determine from Equation 15.1a the hydraulic loading rate based on percolation limit for a Type 1 system (L_{SR1}).

   $$L_{SR1} = (ET - P) + K_d = (2.3 - 6.8)\,cm/month + 52.2\,cm/month = 47.7\,cm/month$$

   c. Determine from Equation 15.1d the allowable hydraulic loading rate based on nitrogen limit (L_{SRN}).

   $$L_{SRN} = \frac{C_P(P - ET) + F_N U_N}{(1 - f)C_N - C_P}$$

   $$= \frac{10\,mg/L \text{ as N} \times (6.8 - 2.3)\,cm/month + 10\,mg{\cdot}ha\text{-}cm/kg{\cdot}L \times 10.2\,kg\,N/ha{\cdot}month}{(1 - 0.25) \times 25\,mg/L \text{ as N} - 10\,mg/L \text{ as N}}$$

   $$= 16.8\,cm/month$$

   Note: The L_{SRN} is applied on monthly basis since it is needed for evaluating storage requirement later in Step 3. Normally, L_{SRN} is calculated on annual basis as illustrated below.

   $$L_{SRN} = \frac{10\,mg/L \text{ as N} \times (56.4 - 58.1)\,cm/year + 10\,mg{\cdot}ha\text{-}cm/kg{\cdot}L \times 320\,kg\,N/ha{\cdot}year}{(1 - 0.25) \times 25\,mg/L \text{ as N} - 10\,mg/L \text{ as N}}$$

   $$= 364\,cm/year$$

   It is the same as that obtained in the last row of Column (10) of Table 15.5.

   d. Tabulate the calculated monthly values of K_d, L_{SR1}, and L_{SRN} in Columns (8) through (10) of Table 15.5.

   e. Select the design hydraulic loading rate (L_w).

   The calculated monthly values of variables L_{SR1} and L_{SRN}, and total cumulative annual values are provided in Table 15.5. For SR Type 1 system, the design hydraulic loading rate (L_w) is the

TABLE 15.5 Monthly Data and Calculated Results for Determination of Design Hydraulic Loading Rate for a SR system (Example 15.1)

Month (1)	ET, cm/month (2)	P, cm/month (3)	U_N, kg N/ha·month (4)	T_{NA}, d/month (5)	T_M, d/month (6)	K_d, cm/month (8)	L_{SR1}, cm/year (9)	L_{SRN}, cm/month (10)
January	2.3	6.8	10.2	15	31	52.2	47.7	16.8
February	2.7	7.9	16.2	10	28	58.8	53.6	24.5
March	3.7	4.7	20.6	8	31	75.1	74.1	24.7
April	4.0	5.9	33.6	3	30	88.1	86.2	40.6
May	6.8	3.8	40.1	2	31	94.7	97.7	42.4
June	7.8	3.5	42.4	0	30	97.9	102.2	43.5
July	7.9	3.9	42.5	0	31	101.2	105.2	44.0
August	6.5	2.8	37.3	0	31	101.2	104.9	38.4
September	5.2	3.1	30.0	2	30	91.4	93.5	31.9
October	4.7	4.1	21.7	3	31	91.4	92.0	24.1
November	3.8	4.7	15.3	7	30	75.1	74.2	18.5
December	2.7	5.2	10.1	8	31	75.1	72.6	14.4
Total	58.1 cm/year	56.4 cm/year	320 kg N/ha·year	58 d/year	365 d/year	1002 cm/year	1004 cm/year	364 cm/year

Note: ET = evapotranspiration rate, P = Precipitation rate, U_N = uptake rate by crops, T_{NA} = number of wastewater application days in the month, T_M = total days in the month, K_d = hydraulic conductivity, L_{SR1} = hydraulic loading rate on Type 1, L_{SRN} = hydraulic loading rate on N limit.

lower value of L_{SR1} and L_{SRN} (whichever is less). In case L_{SR1} is lower for any month of the year, then lowest value of L_{SR1} should be used as the design hydraulic loading rate (L_w). In this example, L_{SRN} is lower than L_{SR1} for every month of the year and the design hydraulic loading rate $L_w = L_{SRN} = 364$ cm/year is used to determine the land area.

2. Determine the land area from Equation 15.1e.

$$A_{SR} = \frac{365Q}{F_{V1}L_W} = \frac{365\ \text{d/year} \times 1950\ \text{m}^3/\text{d}}{100\ \text{m}^3/\text{ha-cm} \times 364\ \text{cm/year}} = 19.56\ \text{ha}$$

3. Determine the storage volume requirement (V'_w).

   Using the same approach as in Step 1, the sample calculation of V'_w for the month of January are provided below.

   a. Design hydraulic loading rate ($L_w = L_{SRN}$).

      $L_w = 16.8$ cm/month obtained from Column (10) of Table 15.5.

   b. Wastewater volume expressed by the depth over the land area of 19.56 ha provided for the SR system.

$$V'_w = \frac{T_M Q}{F_{V1}A_{SR}} = \frac{31\ \text{d/month} \times 1950\ \text{m}^3/\text{d}}{100\ \text{m}^3/\text{ha-cm} \times 19.56\ \text{ha}} = 30.9\ \text{cm/month}$$

   c. Storage requirement V'_m for the month of January.

$$V'_m = V_w - L_w = (30.9 - 16.8)\ \text{cm/month} = 14.1\ \text{cm/month}$$

   Note: A positive value of V'_m indicates that the wastewater flow exceeds the treatment capacity of the SR system and storage volume for the month of January will be required. A negative value would have indicated that the SR system is capable of treating the entire flow generated in the month of January, and no storage would be required for the month. The excess treatment capacity would have allowed to treat only stored wastewater in previous months. Develop the above data for the remainder of the year.

4. The design hydraulic loading rate L_w, wastewater volume V'_w, and storage requirement V'_m for the months of February to December are calculated using the procedure in Step 3 and are summarized in Columns (2) through (4) of Table 15.6.

5. Determine the storage volume requirement.

   The positive value of V'_m from October until March in Table 15.6 indicates that the treated effluent is in surplus and storage volume will be filled. The negative value of V'_m from April until September in Table15.6 indicates that there is a deficiency of wastewater volume from the community and the storage volume will be emptied. At the end of September, the accumulative deficiency reaches maximum, and the storage reservoir is empty completely as the accumulative storage volume $V'_a = 0$. From October, the reservoir starts to fill, and at the end of March, the accumulative storage volume reaches maximum and $V'_{a,max} = 58.4$ cm. The reservoir start to empty from April and becomes empty at the end of September and cycle repeats.

$$V_{a,max} = V'_{a,max}A_{SR} = 58.4\ \text{cm} \times 19.56\ \text{ha} \times 10^2\ \text{m}^3/\text{ha-cm} = 114{,}200\ \text{m}^3$$

6. Determine the storage area requirement (A_s).

   Assume an average depth of 4 m in the storage basin.

   The seepage (bottom surface) area of the storage basin,

$$A_s = \frac{114{,}200\ \text{m}^3}{10^4\ \text{m}^2/\text{ha} \times 4\ \text{m}} = 2.9\ \text{ha}$$

TABLE 15.6 Calculated Results for Determination of Storage Requirement (Example 15.1)

Month (1)	L_w[a], cm/month (2)	V'_w, cm/month (3)	V'_m, cm/month (4)	V'_a, cm (5)
January	16.8	30.9	14.1	48.8
February	24.5	27.9	3.4	52.2
March	24.7	30.9	6.2	58.4[b]
April	40.6	29.9	−10.7	47.7
May	42.4	30.9	−11.5	36.2
June	43.5	29.9	−13.6	22.6
July	44.0	30.9	−13.1	9.5
August	38.4	30.9	−7.5	2.0
September	31.9	29.9	−2.0	0.0[c]
October	24.1	30.9	6.8	6.8[d]
November	18.5	29.9	11.4	18.2
December	14.4	30.9	16.5	34.7
Total	364 cm/year	364 cm/year	0.0 cm/year	–

[a] L_w is obtained from Column (10) of Table 15.5.
[b] The maximum accumulative storage volume $V'_{a,max} = 58.4$ cm is reached at the end of March.
[c] At the end of September, the accumulative storage volume $V'_a = 0$.
[d] (0.0 + 6.8) cm = 6.8 cm.

EXAMPLE 15.2: LAND AREA REQUIRED FOR TYPE 2 SR SYSTEM

Determine (a) the design hydraulic loading rate and (b) the area required for a Type 2 SR system. The annual difference between crop evapotranspiration and precipitation rates ($CEP - P$) = 29 in/year, leaching requirement $LR = 10\%$, the application efficiency of distribution system $E = 75\%$, and average wastewater flow rate = 0.5 MGD.

Solution

1. Calculate the hydraulic loading rate for a Type 2 system (L_{SR2}) from Equation 15.1c.

$$L_{SR2} = \frac{100\%\,(CET - P)}{E\,(1 - LR)} = \frac{100\% \times 29\text{ in/year}}{75\% \times (1 - 0.1)} = 43\text{ in/year (or 109 cm/year)}$$

2. Calculate the area required for Type 2 SR system (A_{SR}) from Equation 15.1e.

$$A_{SR} = \frac{365Q}{F_{V1}L_w} = \frac{365\text{d/year} \times 0.5\text{ MGD}}{0.0272\text{ Mgal/ac-in} \times 43\text{ in/year}} = 156\text{ ac}$$

EXAMPLE 15.3: DESIGN OF A RAPID INFILTRATION (RI) SYSTEM

A RI system is designed for a rural community. The average wastewater flow is 1580 m^3/d. The wastewater is primary settled then pumped into an equalization basin for temporary storage. The soil permeability K is 3.7 cm/h and permeability factor F_s is 0.08. The average wastewater application and drying periods to maximize the nitrogen removal are $T_A = 3$ d/cycle and $T_D = 10$ d/cycle, respectively. The maximum monthly precipitation over 10-year period P_{max} is 14.5 cm. Determine (a) total land area required, (b) total number of cycles per year, (c) area that receives flow per cycle, and (d) application rate.

Solution

1. Determine the design hydraulic loading (L_{RI}), cycle time (T_C), and number of cycles per year (N_C).

$$K = 24\,\text{h/d} \times 365\,\text{d/year} \times 3.7\,\text{cm/h} \times 10^{-2}\text{m/cm} = 324\,\text{m/year}$$

L_{RI} from Equation 15.2a, $L_{RI} = F_S K = 0.08 \times 324\,\text{m/year} = 25.9\,\text{m/year}$

T_C from Equation 15.2b, $T_C = T_A + T_D = (3 + 10)\,\text{d/cycle} = 13\,\text{d/cycle}$

N_C from Equation 15.2c, $N_C = \dfrac{365}{T_C} = \dfrac{365\,\text{d/year}}{13\,\text{d/cycle}} = 28\,\text{cycles/year}$

2. Determine the loading rate per cycle (L_C) and application rate (R_A).

L_C from Equation 15.2d, $L_C = \dfrac{L_{RI}}{N_C} = \dfrac{25.9\,\text{m/year}}{28\,\text{cycles/year}} = 0.925\,\text{m/cycle}$

R_A from Equation 15.2e, $R_A = \dfrac{L_C}{T_A} = \dfrac{0.925\,\text{m/cycle}}{3\,\text{d/cycle}} = 0.31\,\text{m/d}$ or $31\,\text{cm/d}$

3. Check for the ponding effect.
Calculate the 10-year maximum monthly precipitation on daily basis.

$$P_{max} = \frac{14.5\,\text{cm/month}}{30\,\text{d/month}} = 0.48\,\text{cm/d} \quad \text{or} \quad 0.0048\,\text{m/d}$$

$$K = 24\,\text{h/d} \times 3.7\,\text{cm/h} \times 10^{-2}\text{m/cm} = 0.888\,\text{m/d}$$

Apply Equation 15.2f, $(R_A + P_{max} - K)T_A = (0.31 + 0.0048 - 0.888)\,\text{m/d} \times 3\,\text{d/cycle}$
$= -1.7\,\text{m/cycle}$

Assume $D_c = 0.3$ m/cycle. Since $(R_A + P_{max} - K)\ T_A < D_c$, ponding problem will not occur.

4. Determine the average area (A_C) that receives flow per application cycle and total infiltration area (A_{RI}).

A_C from Equation 15.2g, $A_C = \dfrac{Q}{F_{V2}\,(R_A - P_{max})} = \dfrac{1580\,\text{m}^3/\text{d}}{10^4\,\text{m}^2/\text{ha} \times (0.31 - 0.0048)\,\text{m/d}} = 0.52\,\text{ha}$

A_{RI} from Equation 15.2h, $A_{RI} = \dfrac{T_C}{T_A} A_C = \dfrac{13\,\text{d/cycle}}{3\,\text{d/cycle}} \times 0.52\,\text{ha} = 2.25\,\text{ha}$

5. Summarize the sizing results.
Total land area required, $A_{RI} = 2.25$ ha
Area that receives flow per cycle, $A_C = 0.52$ ha
Total number of cycles per year, $N_C = 28$ cycles per year

Application rate, $R_A = 0.31$ m/d or $R_A = 0.31\,\text{m/d} \times 0.52\,\text{ha} \times 10^4\,\text{m}^2/\text{ha} = 1610\,\text{m}^3/\text{d}$

Note: The application rate of 1610 m^3/d is greater than the average daily flow of 1580 m^3/d. Therefore, the design capacity of RI system is sufficient to handle the wastewater flow.

EXAMPLE 15.4: LAND AREA REQUIRED FOR OVERLAND FLOW (OF) SYSTEM

Design an OF system for a rural community. The primary settled wastewater is stored in a basin. The wastewater flow is 1500 m^3/d and hydraulic loading rate is 3.5 cm/d. The operation is 10 h/d and 280 d/year excluding inoperative days. There is a net volume loss of 10,800 m^3/year from the OF system. Determine the land area required, application rate, and number of application plots if the dimension of sloped length is 50 m and the length of distributor arm is 20 m.

Solution

1. Determine the total land area (A_{OF}) required for the OF system.
   Hydraulic loading rate, $L_{OF} = 3.5\ \text{cm/d} \times 10^{-2}\text{m/cm} = 0.035\ \text{m/d}$
   Rearrange Equation 15.3a and calculate the total application area of OF system.

$$A_{OF} = \frac{365\,Q + \Delta V}{F_{v1} T_{op} L_{OF}} = \frac{365\ \text{d/year} \times 1500\ \text{m}^3/\text{d} + (-10{,}800\ \text{m}^3/\text{year})}{100\ \text{m}^3/\text{ha-cm} \times 280\ \text{d/year} \times 3.5\ \text{cm/d}} = 5.5\ \text{ha}$$

2. Determine the number of application plots (N_{OF}).
   Area covered by each application plot from Equation 15.3c,

$$a_{OF} = F_{v3} w_s l_s = 10^{-4}\ \text{ha/m}^2 \times 20\ \text{m} \times 50\ \text{m} = 0.1\ \text{ha/plot}$$

Number of application plots from Equation 15.3b, $N_{OF} = \dfrac{A_{OF}}{a_{OF}} = \dfrac{5.5\ \text{ha}}{0.1\ \text{ha/plot}} = 55\ \text{plots}$

3. Determine the application rate (R_A).

Apply Equation 15.3d, $R_A = \dfrac{L_{OF} l_s}{F_{v5}\theta_{op}} = \dfrac{3.5\ \text{cm/d} \times 50\ \text{m}}{100\ \text{cm/m} \times 10\ \text{h/d}} = 0.175\ \text{m}^3/\text{m}\cdot\text{h}$

15.3.2 Aquatic Treatment Systems

Aquatic systems utilize stabilization ponds (lagoons) and wetlands. In stabilization ponds algae and lower plants and animals provide treatment, while wetlands support rooted plants. Both systems are presented below.

Pond Systems: Stabilization ponds are shallow earthen basins to hold wastewater for extended periods. They are classified as aerobic, anaerobic, and facultative ponds. Treatment is mainly achieved by oxidation of organics due to natural aeration, microbial activities, and sedimentation. Algal growth provides oxygen and removes nutrients, but it creates serious suspended solids problem in the effluent from the ponds. The theory and design of stabilization ponds are extensively covered in Section 10.3.12 and Examples 10.82 through 10.91.

Wetland Treatment Systems: The natural wetlands are totally or intermittently inundated land areas in which the water table may reach above the ground. A wide variety of emergent vegetation and wildlife from fish to waterfowl thrive in the wet ecosystem. The wetlands are an important sink for nutrients and other contaminants. Plants also translocate oxygen from the leaves to the root zones. Submerged portions of vegetation provide surface area for growth of attached aerobic and facultative bacteria that demineralize the wastes for filtration and adsorption of constituents into the plant tissue. Also, growth of algae is restricted because penetration of sunlight is greatly reduced. The oxygen transport through the root hair of wetland plants is shown in Figure 15.1.[14–16]

The wetlands are designed to remove conventional pollutants such as BOD, TSS, and nutrients. Heavy metals are also removed to a significant extent. Primary settled wastewater, secondary effluent, storm water runoff, landfill leachate, acid mine drainage, and industrial wastes have been treated successfully; but the most common application is polishing of secondary effluent. Biomass typically contains 2% nitrogen and 0.4% phosphorus. Harvesting of emergent vegetation is labor intensive, and only a small portion of nutrients can be removed. However, water hyacinth and other floating wetland plants sequester significant amounts of nutrients and heavy metals, and their harvest is relatively less labor intensive.[15,16] The accumulation of nutrients and organics in a wetland occurs from the growth of vegetation, death or dieback, decay, and finally soil accretion.

Wetlands are of two types: *natural wetlands* and *constructed wetlands*. Both wetlands have been used for wastewater treatment, although the use of natural wetlands is generally limited to polishing or further treatment of secondary effluent.[2]

Natural Wetlands: The natural wetlands are considered a part of the receiving waters. Consequently, discharge to natural wetlands is subject to the NPDES permit requirements and only the effluent from secondary or advanced treatment facilities may be allowed (Section 6.2.2). The principal objective for effluent discharge is enhancement of the existing habitat.[17]

Constructed Wetlands: Constructed wetlands offer all the treatment capabilities of natural wetlands but without the NPDES permit requirement. Also, the constructed wetlands are not restricted to the special requirements on the influent quality. For these reasons, constructed wetlands offer more flexible operation

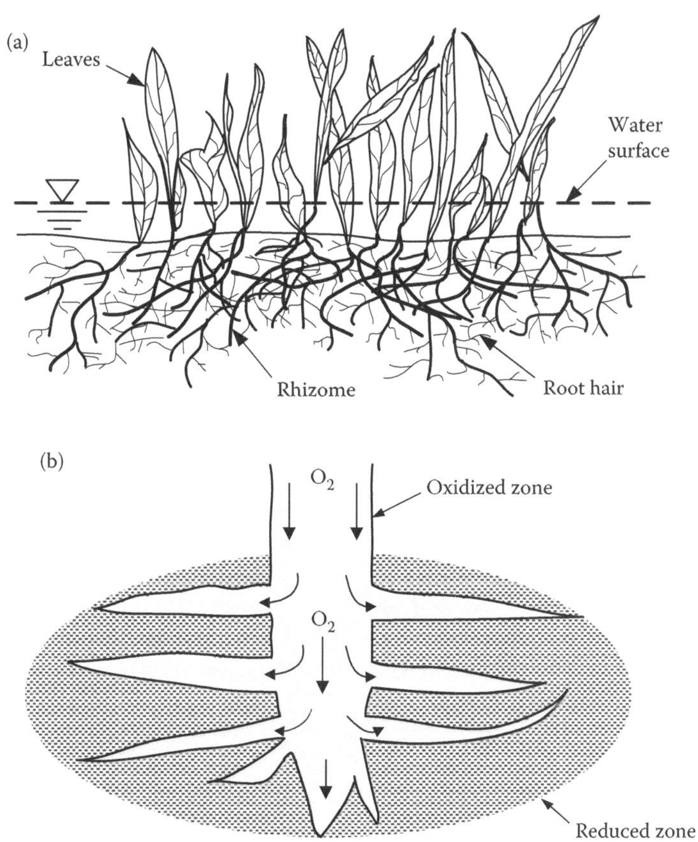

FIGURE 15.1 Wetland plants: (a) aquatic plants and (b) an enlarged root hair.

and perform more reliably than the natural wetlands.[17,18] Two primary types of constructed wetlands are developed for municipal wastewater treatment: (1) *free water surface* (*FWS*), and (2) *subsurface flow* (*SF*) wetlands. The SF constructed wetlands are also categorized by two subsystems: (1) *horizontal subsurface flow* (*HSF*), and (2) *vertical flow* (*VF*) systems. The general description is given below.

1. FWS Wetlands: The FWS wetlands typically consists of (a) a basin or channel with some type of barrier to prevent seepage, (b) soil or base media (up to 40 cm (15 in)) to support the root system of emergent vegetation, and (c) inlet and outlet structures. Water through the system is at a relatively shallow depth and large water surface is exposed to the atmosphere. The velocity is low and flow path is horizontal. Pretreated wastewater is applied continuously and treatment occurs as the water flows slowly through the support media, roots, and stems of emergent vegetation. The FWS wetlands may be designed to create new wildlife habitat and enhance the performance of existing nearby natural wetlands. The depth of ponding water is up to 80 cm (30 in).[19–21] The schematic flow diagram of FWS wetland is shown in Figure 15.2a.
2. HSF Wetlands: The HSF wetlands also have three components as in FWS wetlands. The major difference is no FWS, and the basin or channel is filled with rock or gravel. The medium supports the root system of emergent vegetation and provides horizontal flow through the pores. The depth of the bed varies from 30 to 60 cm (12–24 in). Deeper beds provide better support to roots zone and prevent beds from freezing.[3,10,21] The schematic flow diagram of HSF wetland is shown in Figure 15.2b.
3. VF Wetlands: The VF wetlands consist of a sand and gravel bed of 25–120-cm (12–48-in) deep. The wastewater is intermittently distributed from the top over the entire vegetated surface of the bed area to create a shallow ponding of 2.5–5 cm (1–2 in). Removal of contaminants is achieved when the wastewater percolates through the plant root zone. The bottom is normally sloped at 1–2% and perforated drainage pipes are provided to collect the treated effluent.[21,22] The schematic flow diagram of VF system is shown in Figure 15.2c. Reed drying bed is also an example of modified natural VF wetland system for sludge dewatering (Section 13.8.1 and Figure 13.23b).

The SF wetlands have several advantages over FWS wetlands: (1) little odor risk, public exposure, and insect vectors since no FWS is maintained; (2) medium provides a larger available surface area for attached organisms giving faster response and smaller surface area; (3) plant roots and porous medium improve oxygen transfer and favor the aerobic microbial activities in the biofilm; (4) accumulation of plant debris at the surface may give greater thermal protection in cold climate. The reported disadvantages of HSF or VF wetlands is clogging of the media and possible overflow.[2,3,10,20]

During recent years, the new developments in VF-constructed wetlands have gained more and more attentions. The VF wetlands are more efficient for nitrification than FWS or HSF systems due to a better oxygen transfer efficiency. Several hybrid processes, such as a combination of "VF + HSF" or "VF + FWS" wetlands in series, have also been investigated. The research results have demonstrated the benefits of enhancing nitrification/denitrification in these hybrid systems.[21–25]

Basic Design Considerations of Wetlands: The basic design considerations for wetlands include (1) site selection, (2) plants and vegetation, (3) physical facilities, (4) hydrologic and physical factors, (5) hydraulic and organic loading rates, and (6) performance expectations. The natural wetlands present a very sensitive ecosystem. Any modification to the existing wetlands to improve wastewater treatment capabilities may be very disruptive to natural ecosystem. For this reason, any attempt to modify the natural wetlands is not desired. Therefore, the basic design considerations apply to only the constructed wetlands and are presented below.

Site Selection: Constructed wetlands can be developed in most locations. Most desirable underlying soil permeability for FWS wetlands is 10^{-6}–10^{-7} m/s.[17] Compacted sandy clay and silty clay loams are suitable soils. Sites with sandy and other permeable soils may require clay or artificial liner. In heavy clay soils, addition of peat moss or top soil will improve permeability and accelerate initial plant growth.

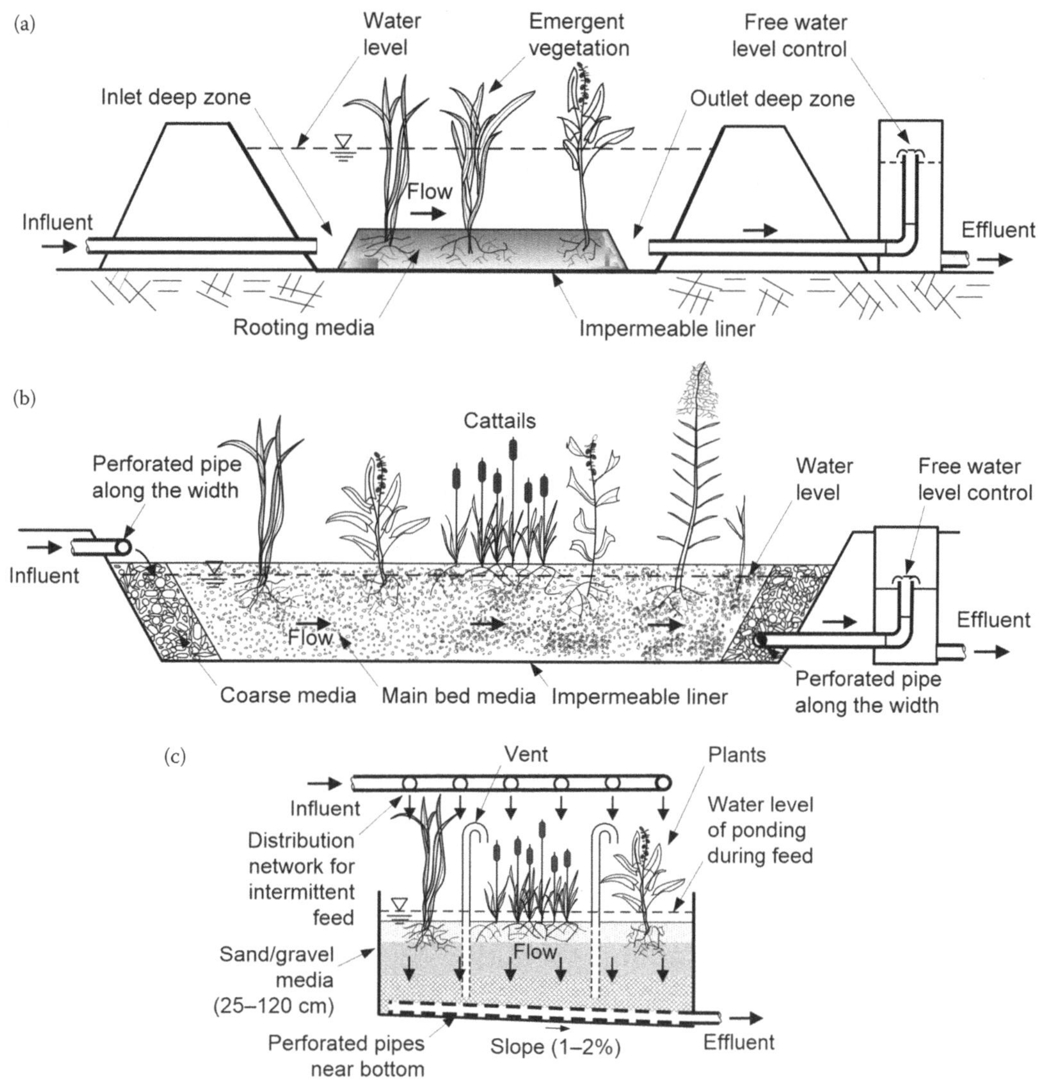

FIGURE 15.2 Constructed wetlands: (a) free water surface (FWS), (b) horizontal subsurface flow (HSF), and (c) vertical flow (VF). (Adapted in part from References 3, 10, 19, 21, and 22.)

Plants and Vegetation: Natural wetlands have wide diversity of plant life. Such diversity is not warranted in constructed wetlands, since eventually, only a few plants such as *cattail* alone or in combination with either *reeds* or *bulrushes* will dominate because of high nutrients levels in the wastewater. The most frequently found emergent plants in wastewater wetlands and their major environmental requirements are given in Table 15.7.[3,19]

Physical Facilities: The constructed wetlands typically behave like a plug-flow reactor. Many factors are considered in selection of proper system configuration for optimizing the performance. These factors are aspect ratio, compartmentalization, and locations of alternate and multiple influent feed points. The aspect ratio of length to width may range from 4:1 to 10:1, although a ratio of 1:1 may also be acceptable. Several alternative flow diagrams and configurations of constructed wetlands are shown in Figure 15.3.

Hydrologic and Physical Factors: Many hydrologic and physical factors are important in design and performance of constructed wetlands. Among these are precipitation, evapotranspiration, infiltration, water

TABLE 15.7 Emergent Aquatic Plants for Constructed Wetlands

Common Name (Scientific Name)	Temperature[a]		Maximum Salinity Tolerance, ppt[b]	Optimum pH, (Standard Unit)
	Survival	Desirable		
Cattail (*Typha* spp.)	10–30	12–24	30	4–10
Common reed (*Phragmites communis)*	12–33	10–30	45	2–8
Rush (*Juncus* spp.)	16–26	–	20	5–7.5
Bulrush (*Scirpus* spp.)	16–27	–	20	4–9
Sedge (*Carex* spp.)	14–32	–	–	5–7.5

[a] Temperature range is for seed germination; and roots and rhizomes can survive in frozen soils.
[b] ppt = parts per thousand.
Source: Adapted in part from References 3 and 26.

depth, flow velocity, and detention time. These factors affect the removal of organics, TSS, nutrients, and trace metals.[15–17] The water balance in a constructed wetland is expressed by Equations 15.4a.

$$\frac{dV}{dt} = Q_i + Q_P + Q_s - Q_e - Q_{ET} - Q_g \quad \text{or} \quad \frac{dV}{dt} = Q_i + PA_s + Q_s - Q_e - ET\,A_s - K_g A_s \tag{15.4a}$$

where

dV/dt = rate of change in volume with time, m^3/d
Q_i = influent wastewater flow, m^3/d (m^3/year)
Q_P = volume of precipitation, m^3/d (m^3/year). $Q_P = P\,A_s$.
P = precipitation, m/d (m/year)
A_s = surface area of the wetland, m^2
Q_s = surface drainage inflow, m^3/d (m^3/year)
Q_e = effluent flow, m^3/d (m^3/year)
Q_{ET} = volume of evapotranspiration, m^3/d (m^3/year). $Q_{ET} = (ET)\,A_s$.
ET = evapotranspiration, m/d (m/year)
Q_g = infiltration flow, m^3/d (m^3/year). $Q_g = K_g\,A_s$.
K_g = percolation rate, m/d (m/year)

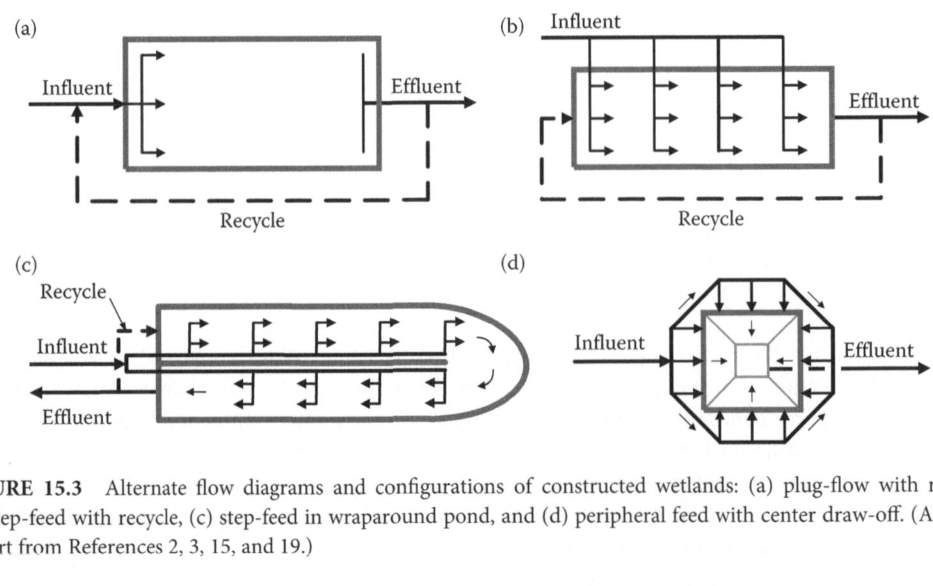

FIGURE 15.3 Alternate flow diagrams and configurations of constructed wetlands: (a) plug-flow with recycle, (b) step-feed with recycle, (c) step-feed in wraparound pond, and (d) peripheral feed with center draw-off. (Adapted in part from References 2, 3, 15, and 19.)

The nominal and actual volume, detention times, and water velocity through constructed wetlands are given by Equations 15.4b through 15.4g.[10,21]

$$V_n = lwh \quad \text{or} \quad V_n = A_s h \tag{15.4b}$$

$$V_a = eV_n \quad \text{or} \quad V_a = elwh \tag{15.4c}$$

$$t_n = \frac{V_n}{Q} \quad \text{or} \quad t_n = \frac{V_a}{eQ} \quad \text{or} \quad t_n = \frac{lwh}{Q} \quad \text{or} \quad t_n = \frac{t_a}{e} \tag{15.4d}$$

$$t_a = \frac{V_a}{Q} \quad \text{or} \quad t_a = \frac{eV_n}{Q} \quad \text{or} \quad t_a = \frac{elwh}{Q} \quad \text{or} \quad t_a = et_n \tag{15.4e}$$

$$u = \frac{Q}{wh} \quad \text{or} \quad u = ev \quad \text{or} \quad u = \frac{l}{t_n} \tag{15.4f}$$

$$v = \frac{Q}{ewh} \quad \text{or} \quad v = \frac{u}{e} \quad \text{or} \quad v = \frac{l}{t_n} \tag{15.4g}$$

where

V_n = nominal volume, m^3. It is based on the average length, width, and depth.
l and w = average length and width of the constructed wetlands, m
h = mean water depth, m. In SF wetlands, h is the depth of submergence within the support media. In FWS wetlands, h is the free water depth. The free water depth is normally <0.5 m (1.6 ft). As water depth increases to 0.7–1.0 m (2.3–3.3 ft), floating plants replace the emergent plants.
V_a = active volume, m^3. It is the available void volume for water to flow. In FWS wetlands, the V_a is obtained after subtracting the volume occupied by vegetative stems and litter. In SF wetlands, the V_a is the available void space through the stone and gravel base.
e = wetland volumetric availability factor, dimensionless. In FWS wetlands, the value of e ranges from 0.55 to 0.95 with typical value of 0.7–0.8 with cattails and other emergent vegetation. In SF wetlands, the value of e is 0.15–0.35 for coarse to fine gravel, respectively. The overall value of e for the sand/gravel media may vary from 0.25 to 0.4 with a typical value of 0.3.
t_n = nominal detention time, d
t_a = actual detention time, d
Q = flow rate, m^3/d
u = superficial water velocity, m/d. It is a virtual velocity through an empty wetland of average dimensions.
v = actual water velocity, m/d

A_s has been defined previously.

There are certain ambiguity about Q, e, and h in large wetlands. There may be a difference in influent and effluent flows due to water gain or loss. The wetland volumetric availability factor is only a gross assumption and may vary seasonally. The mean water depth in a large wetland may also vary greatly due to uneven bed and accumulation of solids in the bed.

The cross-sectional area of a bed is calculated from Darcy's law (Equation 15.4h).

$$A_c = \frac{1}{K}\frac{Q}{(\Delta h/\Delta l)} \quad \text{or} \quad A_c = wh \tag{15.4h}$$

where

A_c = cross-sectional area of the bed, m^2

K = hydraulic conductivity of the bed, m/s

$\Delta h/\Delta l$ = slope of the bed, fraction

In a wetland, the slope of the bed, FWS, and hydraulic gradient are practically the same. In a FWS system that has an overflow weir to control the water depth, the slope of the hydraulic gradient is normally <0.1%. In a gravel medium, the hydraulic conductivity is 10^{-3}–10^{-2} m/s (3.9×10^{-2} –3.9×10^{-1} in/s) and the slope of the bed is 4–5% or more.

Hydraulic and Organic Loading Rates: The hydraulic loading rate is the water flow rate per unit surface area of the wetland (m^3/m^2·d). Some wetlands, such as VF systems, are operated with intermittent flow. The loading rate during the feed portion of cycle is called *the instantaneous hydraulic loading rate* or *hydraulic application rate.*[21] The hydraulic loading rate for constructed wetlands is given by Equation 15.4i. The hydraulic loading rate may also equal the terminal velocity of the smallest particles removed, v_t (m/d). See Sections 8.3.2 and 9.4.5 for additional information about the discrete settling (Type I), terminal velocity, and hydraulic loading rate.

$$q_w = \frac{Q}{A_s} \quad \text{or} \quad q_w = \frac{Qh}{V_n} \quad \text{or} \quad q_w = \frac{h}{t_n} \tag{15.4i}$$

where q_w = hydraulic loading rate over the surface area of a wetland, m^3/m^2·d (m/d)

Other variables have been defined previously.

The hydraulic loading rate for FWS systems is closely tied to the hydrologic factors and conditions specific to the site. A typical value of 200 m^3/ha·d (15 gpm/ac) is considered efficient for optimum treatment efficiency.[3] Effective removal of nutrients may be achievable at a hydraulic loading rate of 100–150 m^3/ha·d (7.5–10 gpm/ac) or less. The hydraulic loading rates for preliminary design of FWS and SF wetlands are given in Table 15.8.

The organic loading is important for treatment efficiency and of emergent plants. High organic loading if not properly distributed may cause anaerobic condition, and plants may die. The organic loading also has an impact on nitrification and denitrification. Also, overloading reduces the oxygen transfer ability of the emergent plants. The maximum organic loading rate for both types of system should not exceed 110 kg BOD_5/ha·d (100 lb/ac·d).[2,3,15,17]

Performance Expectations: Wetlands effectively remove TSS, BOD, COD, nutrients, trace metals, organics, and pathogens. The basic mechanisms for removal are sedimentation, precipitation, microbial decomposition, adsorption, and uptake by vegetation. Plant harvesting may enhance the removal of contaminants.[26–28] Removal of TSS, BOD, nitrogen, phosphorus, and pathogens are presented below.

1. TSS Removal: Colloidal and suspended solids are effectively removed in the constructed wetlands. Most of the removal occurs within a few meters beyond the inlet zone. The major removal

TABLE 15.8 Suggested Hydraulic Loading Rates of Constructed Wetlands

Treatment Achieved	Hydraulic Loading Rate, m^3/ha·d (gpm/ac)	
	Free Water Surface (FWS)	Subsurface Flow (SF)
Basic	–	200–600 (15–45)
Secondary	135–450 (10–35)	450–1800 (35–140)
Polishing	200–900 (15–70)	450–1800 (35–140)

Source: Adapted in part from References 10 and 15.

mechanisms are sedimentation, straining, and impaction. Proper distribution of solids, low velocities, uniform cross-sectional loadings, and flow near the inlet zone enhance the removal of solids. The overall TSS removal may be 50–80%.[19] Kedlec and Wallace proposed a simplified equation for TSS removal in constructed wetlands. The relationship is given by Equation 15.5a.[21]

$$\frac{C_{TSS,e} - C^*_{TSS}}{C_{TSS,i} - C^*_{TSS}} = \exp\left(-\frac{q_w t_n}{h}\right) \quad \text{or} \quad \frac{C_{TSS,e} - C^*_{TSS}}{C_{TSS,i} - C^*_{TSS}} = \exp\left(-\frac{q_w l}{uh}\right) \tag{15.5a}$$

where

$C_{TSS,e}$ = effluent TSS concentration, mg/L (g/m^3)
$C_{TSS,i}$ = influent TSS concentration, mg/L (g/m^3)
C^*_{TSS} = background TSS concentration, mg/L. The value of C^*_{TSS} may range from 2 to 3 mg/L.

Other variables have been defined previously.

2. BOD Removal: The BOD_5 removal in constructed wetlands may vary from 60–85% depending upon the influent BOD_5.[19] Removal rate increases with higher influent concentration. Several relationships have been developed to estimate the effluent BOD_5 concentration. One such relationship that is also called *p-k-C** model is given by Equation 15.5b.[21,29–31]

$$\frac{C_{BOD,e} - C^*_{BOD}}{C_{BOD,i} - C^*_{BOD}} = \frac{1}{\left(1 + \dfrac{k}{365\,p\,q_w}\right)^p} \tag{15.5b}$$

where

$C_{BOD,e}$ = effluent BOD_5 concentration, mg/L (g/m^3)
$C_{BOD,i}$ = influent BOD_5 concentration, mg/L (g/m^3)
C^*_{BOD} = background BOD_5 concentration, mg/L. The value of C^* may range from 2 to 10 mg/L depending upon the decaying vegetation.
k = apparent *tanks-in-series* (TIS) rate constant, m/year[21]
p = apparent number of TIS, dimensionless
365 = days per year, d/year

q_w has been defined previously.

In SF wetlands, the major source of oxygen is that transmitted by the vegetation to the root zone. There is very little direct atmospheric reaeration because the water surface remains below the surface of the media (soil, gravel, rocks).[19,28] The plug-flow first-order kinetic equation with slight modifications has been applied for both FWS and SF wetlands. This relationship is given by Equation 15.5c.[16,17]

$$C_{BOD,e} = C_{BOD,i}\exp\left(-\frac{eK_w A_s h}{Q}\right) \quad \text{or} \quad C_{BOD,e} = C_{BOD,i}\exp\left(-\frac{eK_w h}{q_w}\right)$$
$$\text{or} \quad C_{BOD,e} = C_{BOD,i}\exp\left(-eK_w t_n\right) \tag{15.5c}$$

where K_w = reaction rate constant for wetlands, d^{-1}

Other variables have been defined previously.

3. Nitrogen Removal: Nitrogen removal is achievable in FWS and SF constructed wetlands. The removal is improved in SF systems, especially in VF wetlands. The nitrification–denitrification is the major path of nitrogen removal. If nitrification and nitrogen removal are desired, the hydraulic loading should be reduced to 100–150 m^3/ha·d (7.5–10 gpm/ac). Nitrogen removal may reach 80%

at nitrogen loading up to 45 kg N/ha·d (40 lb N/ac·d). Nitrogen removal is significantly higher if plant harvesting is praticed.[15,32]

4. Phosphorus Removal: Constructed wetlands are capable of phosphorus removal from wastewater. Phosphorus is an essential nutrient required for plant growth. The removal is higher in SF system where selected soils are used for support medium. Presence of clay, iron, and aluminum will enhance phosphorus removal.[16,33]
5. Pathogen Removal: The log removal of coliform organism in the wetlands is in the order of 1.5–3.5. The total coliform count in the effluent from wetlands may be 10^3/100 mL or lower if secondary effluent is treated by the wetlands. Such level may not be reached with primary effluent without disinfection.[10,15,19,21,31]
6. Mosquito Control and Plant Harvesting: Mosquito problem may occur when wetlands receive excessive organic loading and anaerobic condition develops. Biological control agents such as mosquito fish (*Gambusia affins*) may die due to low DO or hydrogen sulfide toxicity. Mosquito problem becomes severe if larvae are able to mature into adults. The possible control method include (1) effective pretreatment to reduce total organic loading, (2) step-feeding of the influent with proper influent distribution and effluent recycle, and (3) application of approved and environmentally safe chemical control agents. Vegetation management may also help control mosquito. Plant harvesting in FWS wetlands will reduce congestion at the water surface. It encourages oxygen transfer by reaeration and enhances the mosquito larvae control by fish. Single harvesting in a year may yield up to approximately 30,000 kg/ha (65,000 lb/ac) dry weight of *Phragmites* shoots. A higher yield is achieved with multiple harvestings.[3]

EXAMPLE 15.5: WATER BALANCE IN A WETLAND

A constructed wetland covers 10-ha area (A_s). The average annual precipitation and evapotranspiration are 88 and 64 cm, respectively. The percolation rate is 10^{-8} m/s. The average monthly surface drainage reaching the wetland is 1320 m^3/month. The average influent to the wetland is 1500 m^3/d. Estimate the effluent flow per day assuming that there is no net change in total storage volume of the wetland.

Solution

1. Determine the annual precipitation volume (Q_P) and evapotranspiration volume (Q_{ET}).

$$Q_P = PA_s = 88\ \text{cm/year} \times 10^{-2}\ \text{m/cm} \times 10\ \text{ha} \times 10^4\ \text{m}^2/\text{ha} = 88{,}000\ \text{m}^3/\text{year}$$

$$Q_{ET} = ET\ A_s = 64\ \text{cm/year} \times 10^{-2}\ \text{m/cm} \times 10\ \text{ha} \times 10^4\ \text{m}^2/\text{ha} = 64{,}000\ \text{m}^3/\text{year}$$

2. Determine the annual average surface drainage inflow (Q_s) and infiltration flow (Q_g).

$$Q_s = 1320\ \text{m}^3/\text{month} \times 12\ \text{month/year} = 15{,}840\ \text{m}^3/\text{year}$$

$$Q_g = K_g A_s = 10^{-8}\ \text{m/s} \times 86{,}400\ \text{s/d} \times 365\ \text{d/year} \times 10\ \text{ha} \times 10^4\ \text{m}^2/\text{ha} = 31{,}540\ \text{m}^3/\text{year}$$

3. Determine the annual wastewater influent flow (Q_i).

$$Q_i = 365\ \text{d/year} \times 1500\ \text{m}^3/\text{d} = 547{,}500\ \text{m}^3/\text{year}$$

4. Estimate the average effluent flow from the wetland (Q_e).

Rearrange Equation 15.4a and calculate Q_e at $dV/dt = 0$ since there is no net change in storage volume.

$$Q_e = Q_i + Q_P + Q_s - Q_{ET} - Q_g - \frac{dV}{dt} = (547{,}500 + 88{,}000 + 15{,}840 - 64{,}000 - 31{,}540 - 0)\,\text{m}^3/\text{year}$$

$$= 555{,}800\,\text{m}^3/\text{year}$$

$$\text{or} \quad Q_e = 555{,}800\,\text{m}^3/\text{year} \times \frac{\text{year}}{365\,\text{d}} = 1523\,\text{m}^3/\text{d}$$

EXAMPLE 15.6: NOMINAL AND ACTUAL DETENTION TIMES AND SUPERFICIAL AND ACTUAL FLOW VELOCITIES

A SF wetland is designed to treat 1500 m^3/d flow. The surface area $A_s = 10$ ha, width $w = 100$ m, and average submerged water depth $h = 0.6$ m. The overall volumetric availability factor of the support media $e = 0.38$. Determine the nominal and actual detention times and superficial and actual water velocities.

Solution

1. Determine the nominal and actual detention times.

Nominal volume from Equation 15.4b, $V_n = A_s h = 10\,\text{ha} \times 10^4\,\text{m}^2/\text{ha} \times 0.6\,\text{m} = 60{,}000\,\text{m}^3$

Actual volume from Equation 15.4c, $V_a = eV_n = 0.38 \times 60{,}000\,\text{m}^3 = 22{,}800\,\text{m}^3$

Nominal detention time from Equation 15.4d, $t_n = \dfrac{V_n}{Q} = \dfrac{60{,}000\,\text{m}^3}{1500\,\text{m}^3/\text{d}} = 40\,\text{d}$

Actual detention time from Equation 15.4e, $t_a = \dfrac{V_a}{Q} = \dfrac{22{,}800\,\text{m}^3}{1500\,\text{m}^3/\text{d}} = 15.2\,\text{d}$

2. Determine the superficial and actual velocities.

Superficial velocity u from Equation 15.4f, $u = \dfrac{Q}{wh} = \dfrac{1500\,\text{m}^3/\text{d}}{100\,\text{m} \times 0.6\,\text{m}} = 25\,\text{m/d}$

Actual velocity v from Equation 15.4g, $v = \dfrac{u}{e} = \dfrac{25\,\text{m/d}}{0.38} = 66\,\text{m/d}$

EXAMPLE 15.7: SLOPE OF BED AND HYDRAULIC CONDUCTIVITY OF A WETLAND

A SF wetland is designed for a flow of 1200 m^3/d. The slope of the bed is 5%. The width and mean depth through which the flow occurs are 100 and 0.6 m, respectively. The hydraulic loading rate is 250 m^3/ha·d. Determine the hydraulic conductivity, wetland area, and detention time.

Solution

1. Determine the cross-sectional area (A_C) through which the flow occurs from Equation 15.4h.

$A_c = wh = 100\,\text{m} \times 0.6\,\text{m} = 60\,\text{m}^2$

2. Determine the hydraulic conductivity (K).

Rearrange Equation 15.4h for calculating K from $\Delta h/\Delta l = 0.05$.

$$K = \frac{Q}{A_c(\Delta h/\Delta l)} = \frac{1200\,\text{m}^3/\text{d}}{60\,\text{m}^2 \times 0.05} \times \frac{\text{d}}{86{,}400\,\text{s}} = 4.6 \times 10^{-3}\,\text{m/s}$$

3. Determine the surface area of wetland (A_s).
   Rearrange Equation 15.4i to calculate A_s.

$$A_s = \frac{Q}{q_w} = \frac{1200\,\text{m}^3/\text{d}}{250\,\text{m}^3\text{ha}\cdot\text{d}}\,4.8\,\text{ha} \quad \text{or} \quad 48{,}000\,\text{m}^2$$

4. Determine the nominal detention time (t_n).

Hydraulic loading rate, $q_w = 250\,\text{m}^3/\text{ha·d} \times 10^{-4}\,\text{ha/m}^2 = 0.025\,\text{m/d}$

Rearrange Equation 15.4i and calculate the nominal detention time, $t_n = \dfrac{h}{q_w} = \dfrac{0.6\,\text{m}}{0.025\,\text{m/d}} = 24\,\text{d}$

EXAMPLE 15.8: TSS AND BOD$_5$ REMOVED IN A SF CONSTRUCTED WETLAND

A SF constructed wetland is designed to treat 1200 m^3/d primary settled wastewater. The total surface area (A_s) and mean water depth (h) are 10 ha and 0.5 m. The background TSS and BOD$_5$ concentrations are 2 and 3 mg/L, respectively. The influent TSS and BOD$_5$ concentrations are 60 and 80 mg/L. The apparent tank-in-series (TIS) rate constant $k = 100$ m/year, and p, the apparent number of TIS $= 1.2$. Determine the TSS and BOD$_5$ concentrations in the effluent.

Solution

1. Determine the nominal detention time (t_n) and hydraulic loading rate (q_w).

Nominal volume from Equation 15.4b, $V_n = A_s h = 10\,\text{ha} \times 10^4\,\text{m}^2/\text{ha} \times 0.5\,\text{m} = 50{,}000\,\text{m}^3$

Nominal detention time from Equation 15.4d, $t_n = \dfrac{V_n}{Q} = \dfrac{50{,}000\,\text{m}^3}{1200\,\text{m}^3/\text{d}} = 42\,\text{d}$

Hydraulic loading rate from Equation 15.4i, $q_w = \dfrac{h}{t_n} = \dfrac{0.5\,\text{m}}{42\,\text{d}} = 0.012\,\text{m/d}$ or $0.012\,\text{m}^3/\text{m}^2\text{d}$

2. Determine the effluent TSS concentration ($C_{TSS,e}$).
   Rearrange Equation 15.5a and calculate effluent TSS concentration.

$$\begin{aligned} C_{TSS,e} &= C_{TSS,i}\exp\left(-\frac{q_w t_n}{h}\right) + C^*_{TSS}\left(1 - \exp\left(-\frac{q_w t_n}{h}\right)\right) \\ &= 60\,\text{mg/L} \times \exp\left(-\frac{0.012\,\text{m/d} \times 42\,\text{d}}{0.5\,\text{m}}\right) + 2\,\text{mg/L} \times \left(1 - \exp\left(-\frac{0.012\,\text{m/d} \times 42\,\text{d}}{0.5\,\text{m}}\right)\right) \\ &= 23\,\text{mg/L} \end{aligned}$$

3. Determine the effluent BOD$_5$ concentration ($C_{BOD,e}$).

Rearrange Equation 15.5b and calculate effluent BOD_5 concentration.

$$C_{BOD,e} = \frac{C_{BOD,i}}{\left(1+\dfrac{k}{365\,p\,q_w}\right)^p} + C^*_{BOD}\left(1-\frac{1}{\left(1+\dfrac{k}{365\,p\,q_w}\right)^p}\right)$$

$$= \frac{80\text{ mg/L}}{\left(1+\dfrac{100\text{ m/year}}{365\text{ d/year}\times 1.2\times 0.012\text{ m/d}}\right)^{1.2}}$$

$$+ 3\text{ mg/L}\times\left(1-\frac{1}{\left(1+\dfrac{100\text{ m/year}}{365\text{ d/year}\times 1.2\times 0.012\text{ m/d}}\right)^{1.2}}\right) = 5\text{ mg/L}$$

EXAMPLE 15.9: BOD_5 REMOVAL IN FWS AND SF CONSTRUCTED WETLANDS

Compare the effluent BOD_5 concentrations from FWS and SF wetlands. The influent BOD_5, hydraulic loading, and water or submergence depth for both wetlands are the same, and they are 80 mg/L, 350 m^3/ha·d, and 0.5 m, respectively. The overall volumetric availability factors for FWS and SF wetlands are 0.8 and 0.4, respectively. The reaction rate constants for FSW and SF wetlands are $K_{w,FWS} = 0.2\text{ d}^{-1}$ and $K_{w,SF} = 0.5\text{ d}^{-1}$, respectively. Also, justify the reaction rate constants assumed for both wetlands.

Solution

1. Determine the hydraulic loading rate (qw) and detention time (t_n) for both wetlands.

   Hydraulic loading rate, $q_w = 350\text{ m}^3/\text{ha·d}\times 10^{-4}\text{ ha/m}^2 = 0.035\text{ m/d}$

   Rearrange Equation 15.4i and calculate the nominal detention time, $t_n = \dfrac{h}{q_w} = \dfrac{0.5\text{ m}}{0.035\text{ m/d}} = 14.3\text{ d}$

2. Determine the effluent BOD_5 concentration from FWS wetland ($C_{BOD,FWS,e}$) from Equation 15.5c.

   $$C_{BOD,FWS,e} = C_{BOD,i}\exp(-e_{FWS}K_{w,FWS}t_n) = 80\text{ mg/L}\times\exp(-0.8\times 0.2\text{ d}^{-1}\times 14.3\text{ d}) = 8.1\text{ mg/L}$$

3. Determine the effluent BOD_5 concentration from SF wetland ($C_{BOD,FWS,e}$) from Equation 15.5c.

   $$C_{BOD,SF,e} = C_{BOD,i}\exp(-e_{SF}K_{w,SF}t_n) = 80\text{ mg/L}\times\exp(-0.4\times 0.5\text{ d}^{-1}\times 14.3\text{ d}) = 4.6\text{ mg/L}$$

   Note: The effluent quality from SF wetland is lower than that from FWS system.

4. Justify the reaction rate constants assumed for FWS and SF wetlands.

   The flow through a single-cell FWS wetland will exhibit medium to low dispersion, while that in a SF wetland will be more like plug flow regime. See additional information about (a) nonideal flow with dispersion in reactors in Section 3.5.3 and Examples 3.45 through 3.47, and (b) determination of the reaction rate constants for stabilization ponds in Section 10.3.11 and Examples 10.82 through 10.90. The reaction rate constants of 0.2 and 0.5 d^{-1} for FWS and SF wetlands in the problem statement may be justified.

15.4 Advanced Wastewater Treatment Processes

Advanced wastewater treatment is applied to remove many constituents that are not adequately removed from secondary effluent. These constituents include suspended solids, nitrogen, phosphorus, and soluble organics and inorganic compounds such as heavy metals, hydrocarbons, and phenolic compounds. Many adverse effects of these constituents are accelerated plants growth, toxicity to fish, concentration into food chain, and taste and odor problems in water supplies. Removal of these constituents may be necessary to meet the stringent effluent discharge limits and water quality standards for reuse.

Commonly used advanced wastewater treatment processes and their applications are summarized in Table 15.9. In depth, discussion on effluent reuse may be found in Chapter 12. A detailed discussion about each of these processes is given below.

15.4.1 Biological Nutrient Removal (BNR) Processes

Biological nutrient removal (BNR) processes are generally integrated with conventional biological treatment processes. They provide enhanced removal of BOD, TSS, nitrogen, and phosphorus. The processes utilize anaerobic, anoxic, and aerobic zones in a wide variety of configuration to achieve phosphorus release and uptake, and nitrification and denitrification. The achievable concentrations of various components of nutrients in effluent from BNR processes are summarized in Table 10.58. The most commonly used modifications of BNR processes include A^2/O, modified Bardenpho, Johannesburg (JHB), SBR, and Orbal® processes. It is expected that in future, BNR technology will be utilized extensively to upgrade existing facilities and to build new plants.[34–37] In depth, coverage of BNR processes is given in Chapter 10 (Sections 10.6 through 10.8) and Examples 10.124 through 10.160.

TABLE 15.9 Advanced Wastewater Treatment Processes and Their Applications

Treatment Process	Common Application
Biological nutrient removal (BNR)	Enhanced removal of BOD, TSS, nitrogen, and phosphorus.
Coagulation, flocculation, and sedimentation	Enhanced removal of TSS, colloidal solids, and microorganisms; and precipitation of natural organic matter, phosphorus, and heavy metals.
Lime precipitation	Precipitation of phosphorus and heavy metals; and enhanced removal of TSS, colloidal solids, microorganisms, and BOD.
Breakpoint chlorination	Oxidation of ammonia nitrogen.
Stripping	Removal of NH_3, H_2S, CO_2, and volatile organic compounds (VOCs).
Filtration	Enhanced removal of TSS and colloidal solids, and microorganisms; polishing of secondary effluent; and providing pretreatment process prior to other advanced treatment processes.
Carbon absorption	Removal of refractory organics, organometallic compounds, and chlorine.
Ion exchange	Used for removal of ammonia and hardness; and demineralization of wastewater.
Membrane process and membrane bioreactor (MBR)	Microfiltration (MF) and ultrafiltration (UF) are used for removal of colloidal particles, color, and organic and inorganic contaminants. Nano filtration (NF), reverse osmosis (RO) and electrodialysis (ED) are exclusively used for demineralization. Membrane bioreactor (MBR) is a combination of aeration basin and final clarifier in a single reactor where membranes are used to separate clarified effluent from MLSS.
Distillation	Distillation is a demineralization process. It uses evaporation and subsequent condensation of water vapors.
Freeze process	It uses freezing and separation of ice crystals containing demineralized water. This process is suitable for cold climate.

15.4.2 Chemical Coagulation and Phosphorus Precipitation

Coagulation is reduction of surface charge of the particles and formation of complex hydrous oxides. Common chemicals used for this purpose are alum, ferric chloride, and ferrous and ferric sulfates. After rapid mixing of coagulant, the wastewater is slowly stirred to encourage formation of floc prior to settling. Enhanced removal of BOD and TSS is also achieved while phosphorus is precipitated. In depth, coverage of chemical coagulation is given in Sections 9.5.2 through 9.5.4, and Examples 9.25 through 9.29.

Chemical phosphorus removal can be applied to both suspended and attached growth biological reactors.[3] Depending on these chemical application locations, chemical phosphorus removal processes can be categorized as: (a) *preprecipitation*, (b) *coprecipitation*, (c) *postprecipitation*, and (d) *multiprecipitation*. These chemical feed locations are shown in Figure 15.4. In preprecipitation process, chemical is fed prior

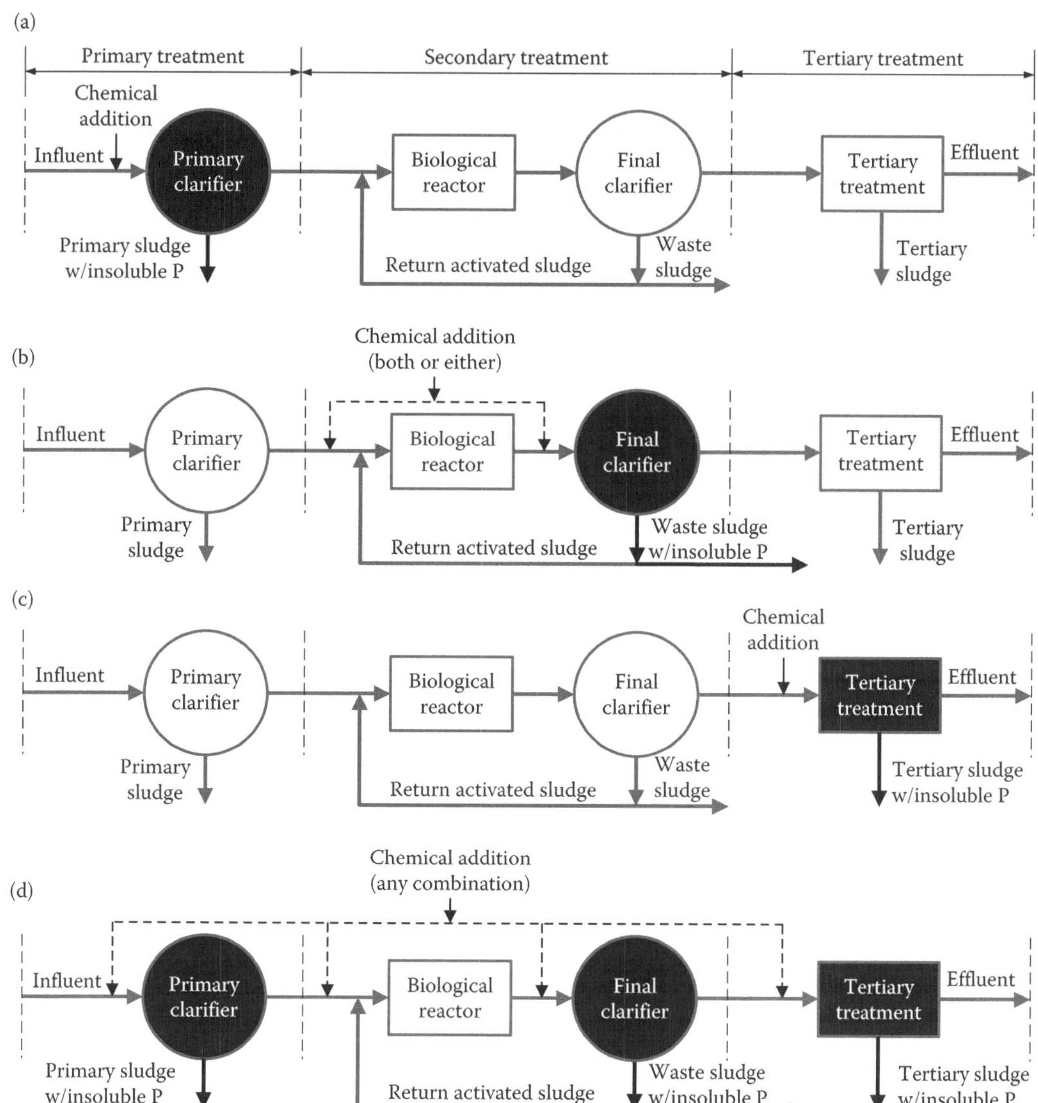

FIGURE 15.4 Alternate locations of chemical feed for chemical precipitation of phosphorus: (a) preprecipitation, (b) coprecipitation at pre- or postbiological reactor, (c) postprecipitation, and (d) multiprecipitations. (Adapted in part from References 3, 7, and 34 through 37.)

to the primary clarifier and phosphorus precipitates are mainly removed in primary clarifier (Figure 15.4a). Coprecipitation process involves chemical feed before and/or after the biological reactor. Phosphorus precipitates are embedded into the mixed liquor and separated in the final clarifier (Figure 15.4b). When post-precipitation process is applied, chemical is added into the final clarifier effluent and phosphorus precipitates are removed by tertiary treatment that consists of filtration with or without postclarification (Figure 15.4c). Postprecipitation process is also integrated into an EBPR process for effluent polishing and as a backup system when EBPR process is not meeting the effluent limit. In multiprecipitation process, chemical may be added in any combination at alternate locations for operational flexibility (Figure 15.4d). The chemical dose is normally optimized by jar test for desired floc formation to optimize BOD, TSS, and phosphorus removals. Polymer may be used in conjunction with the coagulants. The approximate alum and ferric chloride dosages for phosphorus removal from municipal wastewater are 170 and 80 mg/L, respectively.[38,39] The addition of coagulants in biological treatment process has marked influence upon the biota. The population of protozoans and higher animals is adversely affected. However, the BOD, TSS, and phosphorus removal is significantly improved. Overdosing of chemicals may cause toxicity to microorganisms.[3,37]

Rapid mixing of coagulant followed by flocculation is essential for maximum process effectiveness. The coagulants react with alkalinity to produce insoluble metal hydroxide for floc formation. The theory and design of rapid mix and flocculation units are presented in Sections 9.5.5 and 9.5.6, and Examples 9.32 through 9.47. Large quantities of sludge are produced from chemical precipitation. The chemical sludge may cause serious solids processing and disposal problems. The characteristics of sludge and quantities generated are covered in Section 9.5.4 and Examples 9.29 and 9.30.

15.4.3 Lime Precipitation

Lime precipitation enhances the removal of BOD, TSS, and phosphorus. Lime reacts with bicarbonate alkalinity and orthophosphate producing insoluble precipitates. The precipitates are flocculated and settled. Lime precipitation may also be carried out in a single- or two-stage lime treatment process.[40,41]

Single-Stage: Single-stage lime addition may be achieved by either pre- or postprecipitation application (Figures 15.4a or 15.4c). Normal lime dose is 100–180 mg/L as CaO to raise pH to about 10. The actual dosage may be affected by hardness, alkalinity, and phosphorus concentrations. The biological system is not adversely affected by lime addition in moderate amounts (80–120 mg/L as CaO). The carbon dioxide produced by microbial oxidation of organic matter is sufficient to maintain the pH near neutral. High lime dosages ahead of aeration basin may adversely affect the biological process.

Two-Stage: Two-stage lime treatment is applied prior to primary or after the secondary clarifiers in multiprecipitation option (Figure 15.4d). Overall lime dosage upto 450 mg/L as CaO may be necessary to raise the desired pH above 11. The effluent needs to be carbonated by adding CO_2 to lower the pH after flocculation and final sedimentation. High BOD, TSS, and phosphorus removals are achieved in two-stage lime treatment. However, excess lime causes scale formation in tanks, pipes, and other equipment. Also, handling and disposal of large quantities of lime sludge causes serious problem. Lime precipitation is presented in Section 9.5.2 and Examples 9.23, 9.24, 9.27, 9.30, and 9.31.

15.4.4 Breakpoint Chlorination for Ammonia Removal

Ammonia is oxidized to nitrogen gas by breakpoint chlorination. The desired ratio of chlorine (Cl_2) to ammonia (NH_3) is about 7:1 through 11:1, and the optimum pH range is 6–7 for breakpoint chlorination.[42,43] The advantages of breakpoint chlorination are (1) low capital cost, (2) high degree of efficiency and reliability to ammonia removal, (3) high degree of disinfection, and (4) insensitivity to cold weather. The disadvantages are formation of (a) high residual chloride concentration, (b) chlorinated organic compounds, and (c) dechlorination is necessary. In depth, discussion on breakpoint chlorination is provided in Section 11.6.2 and Examples 11.26 through 11.29.

15.4.5 Air Stripping for Removal of Dissolved Gases and VOCs

Process Description: Ammonia (NH_3), hydrogen sulfide (H_2S), carbon dioxide (CO_2), phenol (C_6H_5-OH), trichloroethylene (TCE, C_2HCl_3), toluene ($C_6H_5CH_3$), and many other volatile organic compounds (VOCs) are effectively removed from wastewater by air stripping. The stripping operation is carried out in a tower similar to a forced-air cooling tower. Water is trickled downward through the packing media of slates, rings, spheres, or corrugated surfaces. Air is blown countercurrent or crosscurrent through the dripping water. The surface-area-to-volume ratio of stripping material is 80–100 m^2/m^3, and porosity ratio is nearly 95%. Typical air velocity is 1–2 m/s and column height is 5–10 m.[44] The components of an air stripper are shown in Figure 15.5.

Ammonia Stripping: In an air stripper, ammonia is removed from an alkaline solution as expressed by Equations 15.6a.

$$NH_4^+ + OH^- \leftrightarrow NH_3 \uparrow + H_2O \tag{15.6a}$$

The ammonia stripping process requires the following major components: (1) pretreatment: raising the pH to above 9.3 and preheating to temperature above 80°C, (2) stripping: forming droplets in the stripping tower, and providing air–water contact and droplet agitation in a typical countercurrent flow, and (3) scrubber: capturing ammonia and other volatile compounds from the off-gas and forming liquid products. Ammonia stripping towers are simple to operate and are very effective except in cold climate. The major disadvantages are (a) need to neutralize the effluent and (b) prevent calcium carbonate scaling.

Air Requirement: In normal operation, the flow rates entering and leaving an air stripping tower may differ for both air and liquid because of entrainment, evaporation, and condensation. Also, loss due to aerosols may be significant. However, to simplify the analysis, the flow rates of air and water entering and leaving the stripper are assumed constant. Therefore, the air requirement for a countercurrent system is obtained from material balance within the tower. The exchange of volatile compounds and dissolved gases from water (H_2O) to air leaving from the stripper is expressed by Equation 15.6b.[36,44,45]

$$Lx_0 + Gy_0 = Lx_e + Gy_e \tag{15.6b}$$

where

L = liquid flow expressed in molar weight of H_2O per unit time, mole H_2O/s
G = air flow expressed in molar weight of air per unit time, mole air/s
x_0 = mole fraction of dissolved gas in liquid entering at the top of the tower, mole gas/mole H_2O

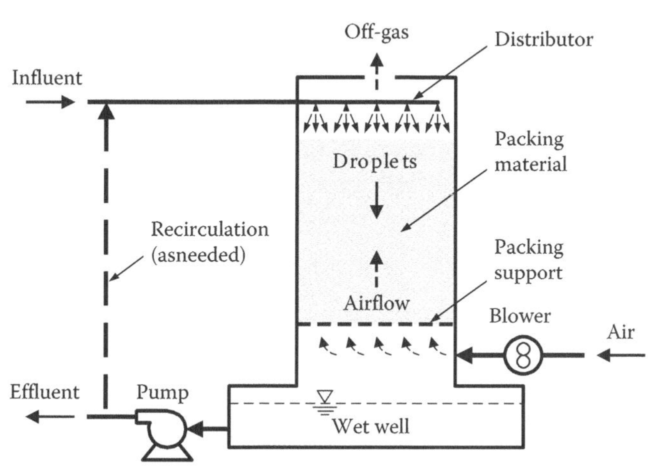

FIGURE 15.5 Components of a countercurrent gas stripping tower.

x_e = mole fraction of dissolved gas in liquid leaving from the bottom of the tower, mole gas/mole H_2O
y_0 = mole fraction of dissolved gas in air flow entering at the bottom of the tower, mole gas/mole air
y_e = mole fraction of dissolved gas in off-gas (or air) leaving from the top of the tower, mole gas/mole air

By assuming that the air entering the tower is relatively clean, the background concentration of dissolved gas is ignored ($y_0 \approx 0$); Equation 15.6b may be simplified by Equation 15.6c.[36,44,45]

$$Lx_0 = Lx_e + Gy_e \quad \text{or} \quad G = \left(\frac{x_0 - x_e}{y_e}\right)L \quad \text{or} \quad (G/L) = \frac{x_0 - x_e}{y_e} \quad \text{or} \quad y_e = \frac{x_0 - x_e}{(G/L)} \tag{15.6c}$$

where (G/L) = air requirement (or air-to-water ratio) expressed by the molar ratio of air to liquid flow, mole air/mole H_2O

Assume that Henry's law is applicable within air stripping tower and the presence of other dissolved gases is ignorable, the following relationships can be established: (a) the dissolved gas fraction in the air is equal to the ratio of partial pressure of the gas to total air pressure (Equation 15.6d) and (b) the partial pressure of the gas is proportional to the mole fraction of a dissolved gas in H_2O (Equation 15.6e).

$$y_g = \frac{P_g}{P_T} \tag{15.6d}$$

$$H = \frac{P_g}{x_g} \quad \text{or} \quad P_g = Hx_g \tag{15.6e}$$

where

y_g = mole fraction of gas in air, mole gas/mole air
P_g = partial pressure of the gas in air, atm
P_T = total air pressure, atm
H = mole-fraction-based Henry's law constant, atm·(mole gas/mole air)/(mole gas/mole H_2O), atm/mole fraction, atm·mole H_2O/mole air, or atm. The values of H for many gases in water are given in Table B.4 of Appendix B. The value of H may also be calculated from Equation 10.84b. A = 1887.12 and B = 6.315 for ammonia. These constants are available for many dissolved gases in Reference 36.
x_g = mole fraction of dissolved gas in H_2O that is at equilibrium with y_e in air, mole gas/mole H_2O

By combining these two equations and assuming $x_g \approx x_0$, y_e can be expressed by Equation 15.6f.[36]

$$y_e = \frac{H}{P_T}x_g \quad \text{or} \quad y_e \approx \frac{H}{P_T}x_0 \tag{15.6f}$$

This expression in the same form is given by Equation 10.84a.

By substituting Equation 15.6f into Equation 15.6c, the theoretical minimum air requirement ratios may be estimated from Equations 15.6g through 15.6i.[36]

$$(G/L)_m^{min} \approx \frac{P_T}{H}\frac{x_0 - x_e}{x_0} \tag{15.6g}$$

$$(G/L)_m^{min} \approx \frac{P_T}{H} \quad (\text{if} x_0 \gg x_e) \tag{15.6h}$$

$$(G/L)_v^{min} = f_{v/m}(G/L)_m^{min} \tag{15.6i}$$

where

$(G/L)_m^{min}$ = theoretical minimum molar air-to-water ratio, mole air/mole H_2O
$(G/L)_v^{min}$ = theoretical minimum volumetric air-to-water ratio, m^3 air/m^3 H_2O
$f_{v/m}$ = factor for converting from molar to volumetric ratio, (m^3/mole of air)/(m^3/mole as H_2O). The value of $f_{v/m}$ is 1340 (m^3/mole of air)/(m^3/mole as H_2O) at 20°C.

The Henry's law constant may be expressed in many forms and units. Other useful relationships of Henry's law constant are given by Equations 15.6j through 15.6m.

$$H_a = \frac{x_g}{P_g} \quad \text{or} \quad x_g = H_a P_g \tag{15.6j}$$

$$H_a = \frac{1}{H} \tag{15.6k}$$

$$x_g = \frac{n_a}{n_a + n_{H_2O}} \quad \text{or} \quad x_g \approx \frac{n_a}{n_{H_2O}} \quad (\text{if} n_a \ll n_{H_2O}) \tag{15.6l}$$

$$y_e = \frac{1}{H_a P_T} x_g \quad \text{or} \quad y_e \approx \frac{1}{H_a P_T} x_0 \quad (\text{if} x_0 \gg x_e) \tag{15.6m}$$

where

H_a = Henry's law constant reversed from H, (mole gas/mole H_2O)/(mole gas/mole air)·atm, mole fraction/atm, mole air/mole H_2O·atm, or atm^{-1}
n_a = mole concentration of gas, mole gas/L
n_{H_2O} = mole concentration of water, 55.6 mole H_2O/L

The Henry's law constant may also be determined from the solubility or saturation concentration of dissolved gas in liquid and the partial pressure (P_g) in the gaseous phase. Two solubility-based Henry's law constants are given by Equations 15.6n through 15.6q.

$$H_m = \frac{[C_s]}{P_g} \quad \text{or} \quad [C_s] = H_m P_g \tag{15.6n}$$

$$H_m = 10^{\left(\log(n_{H_2O}) - B + \frac{A}{T}\right)} \quad \text{or} \quad H_m = 10^{\left(-B' + \frac{A}{T}\right)} \tag{15.6o}$$

$$H_c = \frac{C_s}{P_g} \quad \text{or} \quad C_s = H_c P_g \tag{15.6p}$$

$$H_c = 1000\, mw_g H_m \tag{15.6q}$$

where

H_m = mole-concentration-based Henry's law constant, mole/L·atm
$[C_s]$ = mole saturation concentration of gas in liquid phase, mole/L
P_g = partial pressure in gaseous phase (atm). When a nearly pure gas is applied, the mole fraction of gas is near 1 so the $P_g \approx P_T$.
A and B = empirical constants. The constants are available for many dissolved gases in Reference 36.
B' = constant, $B' = B - \log(n_{H_2O})$
T = temperature, °K
H_c = mass-concentration-based Henry's law constant, mg/L·atm
C_s = solubility or saturation concentration of gas in liquid phase, mg/L
1000 = 1000 mg/g
mw_g = molecular weight of gas, g/mole

These two constants have been presented for dissolution of gaseous chlorine in Section 11.6.2 (Equations 11.10b through 11.10e). The mole-or mass-concentration-based Henry's law constant (H_m) can also be modified into a volumetric saturation-concentration-based constant that is expressed by Equations 15.6r and 15.6s.

$$H_v = \frac{C'_s}{P_g} \quad \text{or} \quad C'_s = H_v P_g \tag{15.6r}$$

$$H_v = 0.0224\, H_m \quad \text{or} \quad H_v = \frac{22.4\, H_c}{mw_g} \tag{15.6s}$$

where

H_v = volumetric saturation-concentration-based Henry's law constant, mL gas/L H_2O·atm. It is also called the absorption coefficient at standard temperature and pressure condition (20°C and 1 atm). The values of H_v for many gases in water are given in Table B.5 of Appendix B.

C_s' = volumetric saturation concentration of gas in liquid phase, mL/L. The relationship between C_s' and C_s is given by Equation B.5 in Example B.6 of Appendix B.

0.0224 = mL/mole

22.4 = mL·g/mole·mg

A dimensionless Henry's law constant is given by Equations 15.6t and 15.6u.

$$H_s = \frac{C_g}{C_s} \quad \text{or} \quad C_g = H_s\, C_s \tag{15.6t}$$

$$H_s = \frac{1}{1000\, H_m RT} \tag{15.6u}$$

where

H_s = dimensionless Henry's law constant, L H_2O/L gas, dimensionless. The value of H_s can be calculated from Equation 15.6u.

C_g = concentration of gas in gaseous phase, mg gas/L

1000 = 1000 L/m^3

R = universal gas constant, 8.206×10^{-5} atm·m^3/mole·°K

The pH and temperature are important for a gas to remain in equilibrium in the ionized forms. The fraction of unionized ammonia (NH_3) depends upon temperature. The Henry's law constant (H_a) and equilibrium constant for NH_3 at four temperatures, and the fraction of unionized ammonia at four different pH values and temperatures are summarized in Table 15.10.

Design Equation: The design equations and procedure for air stripping of dissolved gases and volatile organic compounds have been the subject of thorough development. Many of these equations and procedures may be found in References 36, and 44 through 48. The design equations for air stripping tower utilize a mass balance on a cross section. The basic assumptions are (a) plug-flow regime, (b) Henry's law applies, (c) negligible concentration of contaminants in the effluent, and (d) the liquid and air volumes remain constant. The depth of the air stripping tower is given by Equation 15.7a.[36,45]

$$Z = \left(\frac{L_s}{K_L a}\right)\left(\frac{R_s}{R_s - 1}\right)\ln\left(\frac{(C_0/C_e)(R_s - 1) + 1}{R_s}\right) \tag{15.7a}$$

TABLE 15.10 Henry's Law Constant (H_a), Equilibrium Constant (K_{NH3}), and Mole Fraction of Unionized Ammonia

Temperature, °C	Henry's Law Constant (H_a), atm^{-1}	Equilibrium Constant (K_{NH3}), 10^{-5} mole/L	Mole Fraction of Unionized Ammonia with Respect to Temperature and pH, $\frac{[NH_3]}{[NH_3]+[NH_4^+]}$			
			pH 9	pH 10	pH 11	pH 12
0	3.92	1.374	0.07	0.49	0.89	0
10	2.24	1.479	0.15	0.67	0.95	0
20	1.32	1.710	0.29	0.80	0.98	0
30	1.11	1.820	0.45	0.89	0.99	0

Source: Adapted in part from References 36, 39, and 41 through 48.

where

Z = depth of the packing medium, m (ft)
L_s = liquid loading rate, $m^3/m^2{\cdot}s$ ($ft^3/ft^2{\cdot}s$)
K_La = mass transfer coefficient, s^{-1}
R_s = stripping factor, dimensionless. The stripping factor R_s is a measure of the potential stripping of the contaminant from water to air. The practical range of R_s is 1.5–10.
C_0 = concentration of the contaminant in the influent, mg/L
C_e = concentration of the contaminant in the effluent, mg/L

When R_s is selected, the design volumetric air requirement (or air-to-water ratio) is determined by Equation 15.7b.

$$(G/L)_v^{design} = R_s(G/L)_v^{min} \tag{15.7b}$$

where $(G/L)_v^{design}$ = design volumetric air-to-water ratio, m^3 air/m^3 H_2O. The value of $(G/L)_v^{min}$ can be determined from either Equation 15.6i or 15.7c.

$$(G/L)_v^{min} = \frac{1}{H_s}\frac{C_0 - C_e}{C_0} \tag{15.7c}$$

All variables have been defined previously.

The air-to-water ratio is interrelated with *gas pressure drop* which describes the resistance the blower must overcome in the tower. Vendors should provide the pressure drop curves for the specific packing material with K_La for the targeted dissolved gas.[45]

EXAMPLE 15.10: AIRFLOWS REQUIRED FOR AMMONIA STRIPPING

A wastewater stream contains 26 mg/L ammonia nitrogen. The wastewater flow is 0.12 m^3/s and critical temperatures in the winter and summer are 0°C and 20°C. Determine the maximum and minimum air flow through the ammonia stripper required to produce an effluent ammonia nitrogen concentration of <3 mg/L. Assume that the total air pressure $P_T = 1$ atm.

Solution

1. Determine the operating pH for desired level of ammonia stripping.

$$\text{Ammonia removal required, } E = \frac{C_0 - C_e}{C_e} \times 100\% = \frac{(26-3)\text{ mg/L}}{26\text{ mg/L}} \times 100\% = 88\%$$

From Table 15.10, a pH of 11 is required to achieve a fraction of 0.89 for unionized ammonia at the temperature of 0°C. Select an operating pH of 11 for the ammonia stripper and the fractions of unionized ammonia are 0.89 and 0.98 at temperature of 0°C and 20°C, respectively.

2. Determine the mass concentrations of NH_3-N in the influent and effluent at pH = 11.

$$\text{At } 0°\text{C}, C_{0,0} = 0.89 \times 26 \text{ mg/L} = 23.1 \text{ mg/L and } C_{e,0} = 0.89 \times 3 \text{ mg/L} = 2.7 \text{ mg/L}$$

$$\text{At } 20°\text{C}, C_{0,20} = 0.98 \times 26 \text{ mg/L} = 25.5 \text{ mg/L and } C_{e,20} = 0.98 \times 3 \text{ mg/L} = 2.9 \text{ mg/L}$$

3. Determine the mole concentrations of NH_3-N in the influent and effluent.

$$\text{At } 0°\text{C, } n_{a,0,0} = \frac{C_{0,0}}{14 \text{ g/mole as } NH_3\text{-N}} = \frac{23.1 \text{ mg/L} \times 10^{-3} \text{ g/mg}}{14 \text{ g/mole as } NH_3\text{-N}} = 0.00165 \text{ mole/L as } NH_3\text{-N}$$

$$n_{a,e,0} = \frac{C_{e,0}}{14 \text{ g/mole as } NH_3\text{-N}} = \frac{2.7 \text{ mg/L} \times 10^{-3} \text{ g/mg}}{14 \text{ g/mole as } NH_3\text{-N}} = 0.00019 \text{ mole/L as } NH_3\text{-N}$$

$$\text{At } 20°\text{C, } n_{a,0,20} = \frac{C_{0,20}}{14 \text{ g/mole as } NH_3\text{-N}} = \frac{25.5 \text{ mg/L} \times 10^{-3} \text{ g/mg}}{14 \text{ g/mole as } NH_3\text{-N}} = 0.00182 \text{ mole/L as } NH_3\text{-N}$$

$$n_{a,e,20} = \frac{C_{e,20}}{14 \text{ g/mole as } NH_3\text{-N}} = \frac{2.9 \text{ mg/L} \times 10^{-3} \text{ g/mg}}{14 \text{ g/mole as } NH_3\text{-N}} = 0.00021 \text{ mole/L as } NH_3\text{-N}$$

4. Determine the mole fraction of NH_3-N in the influent and effluent.

$$\text{Number of moles of water in } V = 1 \text{ L, } n_{H_2O} = \frac{V}{18 \text{ g/mole as } H_2O} = \frac{10^3 \text{ g/L}}{18 \text{ g/mole as } H_2O}$$
$$= 55.6 \text{ mole/L as } H_2O$$

Since $n_a \ll n_{H_2O}$, estimate the mole fractions from Equation 15.6l.

$$\text{At } 0°\text{C, } x_{0,0} = \frac{n_{a,0,0}}{n_{H_2O}} = \frac{0.00165 \text{ mole/L as } NH_3\text{-N}}{55.6 \text{ mole/L as } H_2O} = 29.7 \times 10^{-6} \text{ mole } NH_3\text{-N/mole } H_2O$$

$$x_{e,0} = \frac{n_{a,e,0}}{n_{H_2O}} = \frac{0.00019 \text{ mole/L as } NH_3\text{-N}}{55.6 \text{ mole/L as } H_2O} = 3.4 \times 10^{-6} \text{ mole } NH_3\text{-N/mole } H_2O$$

$$\text{At } 20°\text{C, } x_{0,20} = \frac{n_{a,0,20}}{n_{H_2O}} = \frac{0.00182 \text{ mole/L as } NH_3\text{-N}}{55.6 \text{ mole/L as } H_2O} = 32.7 \times 10^{-6} \text{ mole } NH_3\text{-N/mole } H_2O$$

$$x_{e,20} = \frac{n_{a,e,20}}{n_{H_2O}} = \frac{0.00021 \text{ mole/L as } NH_3\text{-N}}{55.6 \text{ mole/L as } H_2O} = 3.8 \times 10^{-6} \text{ mole } NH_3\text{-N/mole } H_2O$$

5. Determine the Henry's law constant H for NH_3.
   Estimate H from Equation 10.84b using $A = 1887.12$ and $B = 6.315$.
   At 0°C, $T_0 = (273.15 + 0)°K = 273.15°K$.

$$H_0 = 10^{\left(-\frac{A}{T_0}+B\right)} = 10^{\left(-\frac{1887.12}{273.15°K}+6.315\right)}$$
$$= 0.255 \text{ atm·(mole } NH_3\text{-N/mole air)/(mole } NH_3\text{-N/mole } H_2O)$$
$$= 0.255 \text{ atm·mole } H_2O\text{/mole air}$$

At 20°C, $T_{20} = (273.15 + 20)°K = 293.15°K$.

$$H_{20} = 10^{\left(-\frac{A}{T_{20}}+B\right)} = 10^{\left(-\frac{1887.12}{293.15°K}+6.315\right)} = 0.754 \text{ atm·mole } H_2O\text{/mole air}$$

6. Determine the mole fraction of NH_3-N in the off-gas from Equation 15.6f.

$$\text{At } 0^\circ\text{C, } y_{e,0} \approx \frac{H_0}{P_T} x_{0,0} = \frac{0.255 \text{ atm·mole } H_2O/\text{mole air}}{1 \text{ atm}} \times 29.7 \times 10^{-6} \text{ mole } NH_3\text{-N/mole } H_2O$$

$$= 7.6 \times 10^{-6} \text{ mole } NH_3\text{-N/mole air}$$

$$\text{At } 20^\circ\text{C, } y_{e,20} \approx \frac{H_{20}}{P_T} x_{0,20} = \frac{0.754 \text{ atm·mole } H_2O/\text{mole air}}{1 \text{ atm}} \times 32.7 \times 10^{-6} \text{ mole } NH_3\text{-N/mole } H_2O$$

$$= 25 \times 10^{-6} \text{ mole } NH_3\text{-N/mole air}$$

7. Determine the molar flow rate of the wastewater (L).

$$\text{At } Q_w = 0.1 \text{ m}^3/\text{s, } L = \frac{Q_w}{18 \text{ g/mole as } H_2O} = \frac{0.12 \text{ m}^3/\text{s} \times 10^6 \text{ g/m}^3}{18 \text{ g/mole as } H_2O} = 6667 \text{ mole/s as } H_2O$$

8. Determine the molar flow rate of the air (Q) from Equation 15.6c.

$$\text{At } 0^\circ\text{C, } G_0 = \left(\frac{x_{0,0} - x_{e,0}}{y_{e,0}}\right) L = \left(\frac{(29.7 - 3.4) \times 10^{-6} \text{ mole } NH_3\text{-N/mole } H_2O}{7.6 \times 10^{-6} \text{ mole } NH_3\text{-N/mole air}}\right) \times 6667 \text{ mole } H_2O/\text{s}$$

$$= 23.1 \times 10^3 \text{ mole air/s}$$

$$\text{At } 20^\circ\text{C, } G_{20} = \left(\frac{x_{0,20} - x_{e,20}}{y_{e,20}}\right) L = \left(\frac{(32.7 - 3.8) \times 10^{-6} \text{ mole } NH_3\text{-N/mole } H_2O}{25 \times 10^{-6} \text{ mole } NH_3\text{-N/mole air}}\right)$$

$$\times 6667 \text{ mole } H_2O/\text{s} = 7.71 \times 10^3 \text{ mole air/s}$$

9. Determine the air requirements (Q_a) under standard conditions.
   Under standard conditions of 20°C and 1 atm, one mole of all gases occupies 22.4 L or 0.0224 m^3.
   At operating temperature of 0°C,

$$Q_{a,0} = 0.0224 \text{ m}^3 \times G_0 = 0.0224 \text{ m}^3 \times 23.1 \times 10^3 \text{ mole air/s} = 517 \text{ m}^3/\text{s as air}$$

   At operating temperature of 20°C,

$$Q_{a,20} = 0.0224 \text{ m}^3 \times G_{20} = 0.0224 \text{ m}^3 \times 7.71 \times 10^3 \text{ mole air/s} = 173 \text{ m}^3/\text{s as air}$$

Note: The air requirement at 0°C is approximately three times of that at 20°C. It is close to the ratio of Henry's Law constants (H) at temperatures of 20°C and 0°C ($H_{20}/H_0 \approx 3$).

EXAMPLE 15.11: MINIMUM AIR REQUIREMENT FOR AMMONIA REMOVAL

An air stripping process is use for ammonia removal from wastewater. Determine the theoretical minimum molar and volumetric air-to-water ratios for an efficient process. Assume that the operating temperature is 20°C and the total pressure of air $P_T = 1$ atm.

Solution

1. Determine the Henry's law constant H for NH_3.
   $H_{20} = 0.754$ atm·mole H_2O/mole air for NH_3 at 20°C has been calculated in Step 5 of Example 15.10.
2. Determine the theoretical minimum molar air-to-water ratio.

For an efficient ammonia stripping process, $x_0 \gg x_e$ and the ratio may be estimated from Equation 15.6h.

$$(G/L)_m^{min} \approx \frac{P_T}{H_{20}} = \frac{1\text{ atm}}{0.754\text{ atm·mole } H_2O/\text{mole air}} = 1.3\text{ mole air/mole } H_2O$$

3. Determine the theoretical minimum volumetric air-to-water ratio from Equation 15.6i.
   Use $f_{v/m} = 1340$ (m^3/mole of air)/(m^3/mole as H_2O) at 20°C.

$$\begin{aligned}(G/L)_v^{min} &= f_{v/m}(G/L)_m^{min} = 1340\,(m^3/\text{mole air})/(m^3/\text{mole } H_2O) \times 1.3\text{ mole air/mole } H_2O\\ &= 1740\,m^3\text{ air}/m^3\ H_2O\end{aligned}$$

Note: Significant amount of air is needed for effective removal of NH_3 from wastewater in an air stripper.

EXAMPLE 15.12: AIR REQUIREMENTS FOR STRIPPING TRICHLOROETHYLENE (TCE)

A contaminated ground water has 15 mg/L TCE. An air stripping tower is designed to treat the groundwater below 0.5 mg/L TCE. The percolation rate of groundwater is 0.15 m^3/s. The Henry's law constant H_a for TCE is 1.65×10^{-3} atm^{-1} (or mole air/mole H_2O·atm) and molecular weight for TCE is 132 g/mole TCE. Determine the minimum air flow (m^3/s) needed for stripping operation at $P_T = 1$ atm.

Solution

1. Determine the mole concentrations of TCE in the influent and effluent.

$$\text{Influent TCE concentration, } n_{a,0} = \frac{C_0}{132\text{ g/mole as TCE}} = \frac{15\text{ mg/L} \times 10^{-3}\text{ g/mg}}{132\text{ g/mole as TCE}} = 114 \times 10^{-6}\text{ mole/L as TCE}$$

$$\text{Effluent TCE concentration, } n_{a,e} = \frac{C_e}{132\text{ g/mole as TCE}} = \frac{0.5\text{ mg/L} \times 10^{-3}\text{ g/mg}}{132\text{ g/mole as TCE}} = 3.8 \times 10^{-6}\text{ mole/L as TCE}$$

2. Determine the mole fractions of TCE in the influent and effluent.
   Assume $n_a \gg n_{H_2O}$, estimate the mole fractions from Equation 15.6l using $n_{H_2O} = 55.6$ mole/L as H_2O.

$$x_0 = \frac{n_{a,0}}{n_{H2O}} = \frac{114 \times 10^{-6}\text{ mole/L as TCE}}{55.6\text{ mole/L as } H_2O} = 205 \times 10^{-8}\text{ mole TCE/mole } H_2O$$

$$x_e = \frac{n_{a,e}}{n_{H2O}} = \frac{3.8 \times 10^{-6}\text{ mole/LasTCE}}{55.6\text{ mole/LasH}_2\text{O}} = 6.8 \times 10^{-8}\text{ mole TCE/mole } H_2O$$

3. Determine the mole fraction of TCE in the off-gas.
   Assume $x_0 \gg x_e$, estimate the mole fraction of TCE in the off-gas from Equation 15.6m using $H_a = 1.65 \times 10^{-3}$ mole air/mole H_2O·atm.

$$\begin{aligned}y_e &\approx \frac{1}{H_aP_T}x_0 = \frac{1}{1.65 \times 10^{-3}\text{ mole air/mole } H_2O\text{·atm} \times 1\text{ atm}} \times 205 \times 10^{-8}\text{ mole TCE/mole } H_2O\\ &= 1.24 \times 10^{-3}\text{ mole TCE/mole air}\end{aligned}$$

4. Determine the molar flow rate of the wastewater (L).

$$\text{At } Q_w = 0.15, \text{m}^3/\text{s}, L = \frac{Q_w}{18\,\text{g/mole as } H_2O} = \frac{0.15\,\text{m}^3/\text{s} \times 10^6\,\text{g/m}^3}{18\,\text{g/mole as } H_2O} = 8333\,\text{mole/s as } H_2O$$

5. Determine the molar flow rate of the air (Q) from Equation 15.6c.

$$G = \left(\frac{x_0 - x_e}{y_e}\right)L = \left(\frac{(205 - 6.8) \times 10^{-8}\,\text{mole TCE/mole } H_2O}{1.24 \times 10^{-3}\,\text{mole TCE/mole air}}\right) \times 8333\,\text{mole } H_2O/\text{s}$$

$$= 13.3\,\text{mole air/s}$$

6. Determine the air requirements (Q_a) using 22.4 L/mole air under standard conditions (20°C and 1 atm).

$$Q_a = 22.4\,\text{L/mole} \times 10^{-3}\,\text{m}^3/\text{L} \times G = 22.4\,\text{L/mole} \times 10^{-3}\,\text{m}^3/\text{L} \times 13.3\,\text{mole air/s} = 0.30\,\text{m}^3/\text{s as air}$$

7. Determine the volumetric air-to-water ratio required under standard conditions (20°C and 1 atm).

$$(G/L)_v = \frac{Q_a}{Q_w} = \frac{0.30\,\text{m}^3/\text{s as air}}{0.15\,\text{m}^3/\text{s as water}} = 2$$

Note: Air stripping is an efficient process for removal of TCE from groundwater.

EXAMPLE 15.13: HENRY'S LAW CONSTANT OF TOLUENE IN DIFFERENT UNITS

Henry's law constant is expressed in different units. Estimate the mole-concentration-based Henry's law constant (H_m) for toluene in water at 20°C from partial pressure and solubility data, and check the value using regression constants of A = 1316 and B = 6.97. Also, estimate the Henry's law constant based on mass concentration (H_c) and dimensionless constants (H_s). The partial pressure and solubility of toluene at 20°C are 22.0 mmHg and 515 mg/L, respectively. Molecular weight of toluene is 92.13 g/mole.

Solution

1. Estimate the mole-concentration-based Henry's law constant (H_m) of toluene from partial pressure and solubility data.

$$\text{Partial pressure of toluene at 20°C, } p_g = \frac{22.0\,\text{mm Hg}}{760\,\text{mm Hg/atm}} = 0.0289\,\text{atm}$$

$$\text{Solubility of toluene in liquid, } [C_s] = \frac{C_s}{mw_g} = \frac{515\,\text{mg/L} \times 10^{-3}\,\text{g/mg}}{92.13\,\text{g/mole}} = 5.59 \times 10^{-3}\,\text{mole/L}$$

$$\text{Henry's law constant from Equation 15.6n, } H_m = \frac{[C_s]}{P_g} = \frac{5.59 \times 10^{-3}\,\text{mole/L}}{0.0289\,\text{atm}} = 0.193\,\text{mole/L·atm}$$

2. Check the H_m using regression constants.

$T_{20} = (273.15 + 20)°\text{K} = 293.15°\text{K}$

Henry's law constant from Equation 11.10d, $H_m = 10^{\left(\log(n_{H_2O}) - B + \frac{A}{T_{20}}\right)}$

$$= 10^{\left(\log(55.6\,\text{mole/L}) - 6.97 + \frac{1316}{293.15^\circ K}\right)}$$

$$= 0.184\,\text{mole/L·atm}$$

Note: The value of H_m obtained from regression constants is close to that obtained from partial pressure and solubility data.

3. Estimate the mass-concentration-based Henry's law constant (H_c).

   The value of H_c may be obtained from either one of the two procedures given below.

   Henry's law constant from Equation 15.6p, $H_c = \frac{C_s}{P_g} = \frac{515\,\text{mg/L}}{0.0289\,\text{atm}} = 1.78 \times 10^4\,\text{mg/L·atm}$

   Henry's law constant from Equation 15.6q, $H_c = 1000\, mw_g H_m$

$$= 1000\,\text{mg/g} \times 92.13\,\text{g/mole} \times 0.193\,\text{mole/L·atm}$$

$$= 1.78 \times 10^4\,\text{mg/L·atm}$$

4. Estimate the dimensionless Henry's law constant (H_s).

   $R = 8.206 \times 10^{-5}\,\text{atm·m}^3/\text{mole·°K}$

   Henry's law constant from Equation. 15.6u using $H_m = 0.193$ mole/L·atm in Step 1,

$$H_s = \frac{1}{1000\,H_m R T_{20}} = \frac{1}{1000\,\text{L/m}^3 \times 0.193\,\text{mole/L·atm} \times 8.206 \times 10^{-5}\,\text{atm·m}^3/\text{mole·°K} \times 293.15^\circ\text{K}}$$

$$= 0.22\,\text{(dimensionless)}$$

Note: The applications of Henry's law constants are presented in many examples. See Example 10.119 for saturation concentrations of CO_2 in anaerobic process, Examples 11.18 through 11.20 for dissolution of gaseous chlorine, and Examples 15.10 through 15.14 for air stripping.

EXAMPLE 15.14: DESIGN OF AN AIR STRIPPING TOWER

An air stripping tower is designed to strip TCE from a contaminated groundwater. The TCE concentration is 8.0 mg/L and treated water requirement is 0.5 mg/L. The mass transfer coefficient $K_L a$ of packing material is 0.02 s^{-1}. The flow rate of groundwater (Q_w) is 0.30 m^3/s. Determine (a) the column height and diameter, and (b) air–water flow rates. Operating temperature is 20°C and pressure is 1 atm. The regression coefficients A and B for TCE are 2076 and 9.7, respectively. Use a volumetric water loading rate of 0.04 m^3/m^2·s on the tower.

Solution

1. Determine the Henry's law constants H, H_m, and H_s for TCE.

   $T_{20} = (273.15 + 20)^\circ\text{K} = 293.15^\circ\text{K}$

   Henry's law constant from Equation 10.84b, $H = 10^{\left(-\frac{A}{T_{20}} + B\right)} = 10^{\left(-\frac{2076}{293.15^\circ K} + 9.7\right)}$

$$= 415\,\text{atm·mole H}_2\text{O/mole air}$$

Henry's law constant from Equation 11.10d, $H_m = 10^{\left(\log(n_{H2O}) - B + \frac{A}{T_{20}}\right)}$

$$= 10^{\left(\log(55.6\,\text{mole/L}) - 9.7 + \frac{2076}{293.15°K}\right)}$$

$$= 0.134\,\text{mole/L·atm}$$

Henry's law constant from Equation 15.6u at $R = 8.206 \times 10^{-5}$ atm·m^3/mole·°K,

$$H_s = \frac{1}{1000\,H_m R T_{20}} = \frac{1}{1000\,\text{L/m}^3 \times 0.134\,\text{mole/L·atm} \times 8.206 \times 10^{-5}\,\text{atm·m}^3/\text{mole·°K} \times 293.15°\text{K}}$$

$$= 0.31\,(\text{dimensionless})$$

2. Determine the theoretical minimum molar air-to-water ratio.

   Assume that the TCE stripping process is efficient ($x_0 \gg x_e$). Estimate the theoretical minimum molar air-to-water ratio from Equation 15.6h.

$$(G/L)_m^{\min} \approx \frac{P_T}{H} = \frac{1\,\text{atm}}{415\,\text{atm·mole H}_2\text{O/mole air}} = 0.0024\,\text{mole air/mole H}_2\text{O}$$

3. Determine the theoretical minimum volumetric air-to-water ratio.

   Use $f_{v/m} = 1340$ (m^3/mole of air)/(m^3/mole as H_2O) at 20°C.

$$(G/L)_v^{\min} = f_{v/m}(G/L)_m^{\min} = 1340\,(\text{m}^3/\text{mole air})/(\text{m}^3/\text{mole H}_2\text{O}) \times 0.0024\,\text{mole air/mole H}_2\text{O}$$
$$= 3.2\,\text{m}^3\,\text{air/m}^3\,\text{H}_2\text{O}$$

   Check the ratio from Equation 15.7c, $(G/L)_v^{\min} = \frac{1}{H_s}\frac{C_0 - C_e}{C_0} = \frac{1}{0.31} \times \frac{(8.0 - 0.5)\,\text{mg/L}}{8.0\,\text{mg/L}}$

$$= 3.0\,\text{m}^3\,\text{air/m}^3\,\text{H}_2\text{O}$$

   The ratios estimated from two methods are very close. Select the ratio of 3.2-m^3 air/m^3 H_2O for design.

4. Determine the design volumetric air-to-water ratio.

   Select a stripping ratio $R_s = 5$ and determine the design volumetric air-to-water ratio from Equation 15.7b.

$$(G/L)_v^{\text{design}} = R_s\,(G/L)_v^{\min} = 5 \times 3.2\,\text{m}^3\,\text{air/m}^3\,\text{H}_2\text{O} = 16\,\text{m}^3\,\text{air/m}^3\,\text{H}_2\text{O}$$

5. Determine the air flow and surface loading rate requirements.

   Air flow, $Q_a = (G/L)_v^{\text{design}}\,Q_w = 16\,\text{m}^3\,\text{air/m}^3\,\text{H}_2\text{O} \times 0.3\,\text{m}^3/\text{s as H}_2\text{O} = 4.8\,\text{m}^3/\text{s as air}$

   Air surface loading rate, $G_s = (G/L)_v^{\text{design}}\,L_s = 16\,\text{m}^3\,\text{air/m}^3\,\text{H}_2\text{O} \times 0.04\,\text{m}^3/\text{m}^2\text{·s as H}_2\text{O}$

$$= 0.64\,\text{m}^3/\text{m}^2\text{·s as air}$$

6. Determine the height of the packing medium from Equation 15.7a.

$$Z = \left(\frac{L_s}{K_L a}\right)\left(\frac{R_s}{R_s - 1}\right)\ln\left(\frac{(C_0/C_e)(R_s - 1) + 1}{R_s}\right)$$

$$= \left(\frac{0.04\,\text{m}^3/\text{m}^2\text{·s}}{0.02\,\text{s}^{-1}}\right)\left(\frac{5}{5 - 1}\right)\ln\left(\frac{(8.0\,\text{mg/L}/0.5\,\text{mg/L}) \times (5 - 1) + 1}{5}\right)$$

$$= 6.4\,\text{m}$$

   Select a design packing depth = 6.5 m.

7. Determine number of stripping towers, diameter, and air flow rate through each tower.

Total cross sectional area of stripping towers, $A_{\text{total}} = \dfrac{Q_w}{L_s} = \dfrac{0.30\,\text{m}^3/\text{s}}{0.04\,\text{m}^3/\text{m}^2\cdot\text{s}} = 7.5\,\text{m}^2$

Provide two air stripping towers.

Area of each tower, $A = \dfrac{A_{\text{total}}}{2\text{ towers}} = \dfrac{7.5\,\text{m}^2}{2\text{ towers}} = 3.75\,\text{m}^2/\text{tower}$

Required diameter of each tower, $d = \sqrt{\dfrac{4A}{\pi}} = \sqrt{\dfrac{4 \times 3.75\,\text{m}^2}{\pi}} = 2.2\,\text{m}$

Provide each tower 2.25 m in diameter and 6.5 m for packing depth.
Water flow rate through each tower $= 0.04\,\text{m}^3/\text{m}^2\cdot\text{s} \times 3.75\,\text{m}^2/\text{tower} = 0.15\,\text{m}^3/\text{s}$
Air flow rate through each tower $= 0.64\,\text{m}^3/\text{m}^2\cdot\text{s} \times 3.75\,\text{m}^2/\text{tower} = 2.4\,\text{m}^3/\text{s}$

15.4.6 Granular Filtration

Granular (or depth) filtration is categorized as a tertiary wastewater treatment process and mainly used to (1) polish secondary effluent, (2) meet the stringent discharge limits on pollutants that are in particular forms such as TSS, particular phosphorus, etc., and (3) provide pretreatment for UV disinfection and many advanced treatment processes that are affected by particulate matter. Examples of these advanced treatment processes are carbon adsorption, ion exchange, and membrane processes.

Process Description: The granular filtration involves passing of effluent through a *filter bed* or media that can strain out the colloidal particles. The filter bed may consist of such filtering media as fine sand, anthracite, mixed media, and/or diatomaceous earth. As solids accumulate in the filter bed, head loss increases and flow decreases. Backwashing is done periodically by pumping clean water in the opposite direction of the flow. The bed is expanded and trapped solids are washed off. The components and operation of a gravity filter are shown in Figure 15.6.

Filtration Mechanism: A number of mechanisms act simultaneously on solids removal by granular media. These are (1) *straining*, (2) *sedimentation*, (3) *impaction*, and (4) *interception*. The colloidal particles larger than the pore spaces in the media are trapped and removed by straining mechanism. The flow velocity through the filter bed is usually laminar. In low-velocity zones, some particles are removed by sedimentation. In a granular bed, there are sharp turns in the flow streamlines. The force of inertia moves many particles out of the flow stream causing them to strike the media and be held there. Thus, the removal is by impaction. At times, the particles following these stream lines touch a media grain and chemical bonding or electrostatic forces hold them there. Such a removal is due to interception.[36,37,49]

Filtration Stages: Typically, there are two operating stages during a complete filter operating cycle: (a) *filtration* and (b) *backwash* stages. The filtration stage (or a *filter run*) typically lasts 1–4 days while the backwash takes only 5–20 min. A filter run may also be broken into three different segments: (a) *ripening* (or *maturation*), (b) *effective*, and (c) *breakthrough* segments. Conceptual illustrations of turbidity and head loss through a filter run are provided in Figure 15.7.[49,50]

Types of Filters: The granular media depth filters are of many types depending upon the media, direction of flow, driving force, filtration rate, and filter operation. These filter types are briefly described below.[36,37,44,49]

1. Media: Filters may contain *monomedium*, and *dual*- and *multimedia*. The monomedium filters utilize one type of medium such as sand or anthracite. The media may be *uniform*, *nonuniform*, *stratified*, or *unstratified*. Uniform media bed remains unstratified after backwashing. Nonuniform media bed become stratified after backwashing. Also, the media may be shallow, medium, or

(a)

Operation cycle cycle	Valve				
	A	B	C	D	E
Normal filtration	Open	Open	Closed	Closed	Closed
Backwash	Closed	Closed	Open	Open	Closed
Rinse	Open	Closed	Closed	Closed	Open

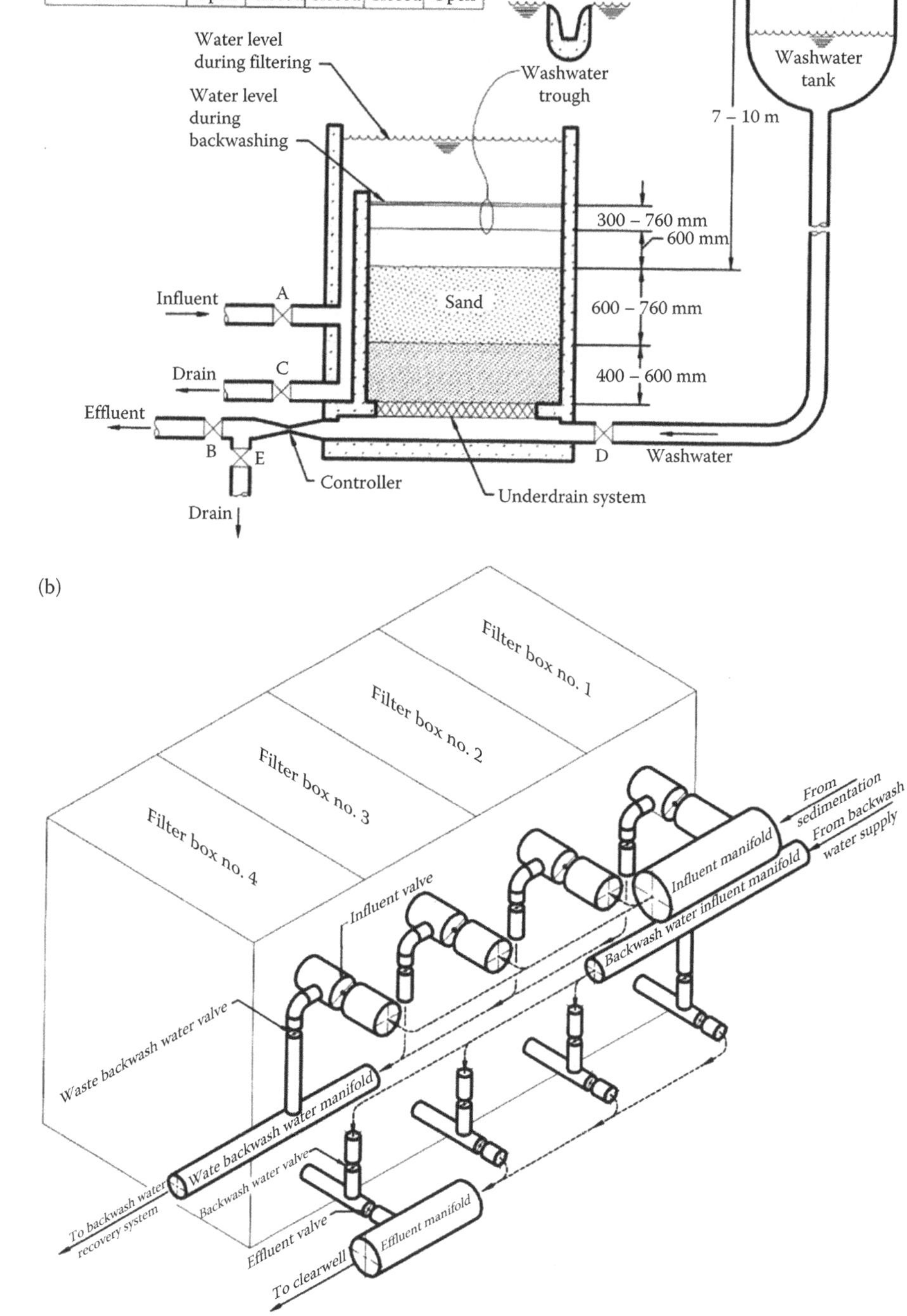

FIGURE 15.6 Filter components and details: (a) cross section of a rapid sand filter and operational features and (b) typical arrangement of four filter units with piping. (Adapted in part from References 3 and 49.)

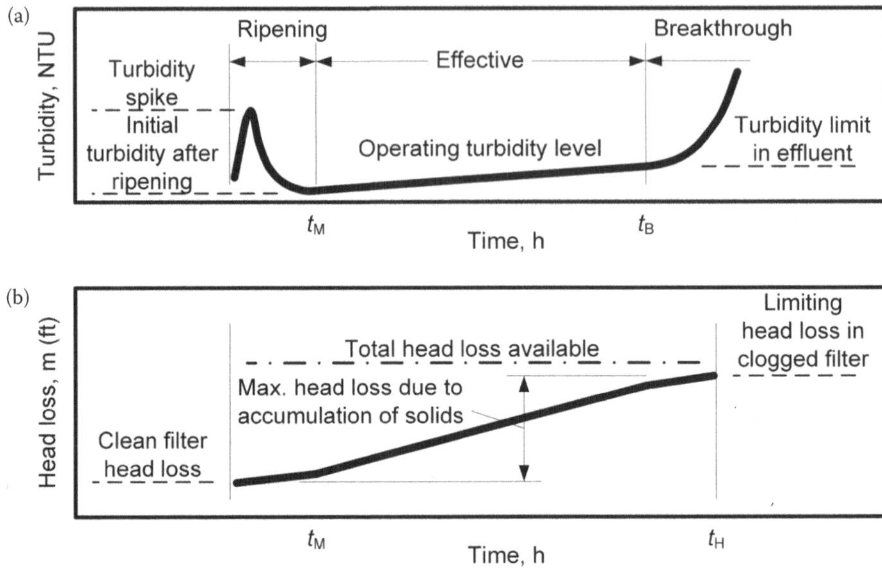

FIGURE 15.7 Typical filter performance curves during a filter run: (a) turbidity removal showing three segments of filter run and (b) head loss from clean to clogged filter. (Adapted in part from References 49 and 50.)

deep. Dual-media or multimedia filters are a combination of (a) anthracite and sand and (b) anthracite, sand, and garnet. These filters may be stratified or unstratified.

2. Direction of flow: Based on direction of flow the filters are *downflow*, *upflow*, and *center feed*.
3. Driving force: The filters are *gravity* or *pressure* feed.
4. Filtration rate: The filters are *slow* rate or *rapid* rate.
5. Filter operation: The filters are *declining* or *constant* rate; or *variable* level or *constant* level.

Media Design Parameters: The filter media are specified by *effective size* and *uniformity coefficient*. The effective size (d_{10}) is the size of the standard sieve opening that will pass 10% by weight of the media. The uniformity coefficient *UC* is the ratio of standard sieve opening that will pass 60% by weight of the medium to its effective size (d_{60}/d_{10}). Other parameters often specified are media *hardness* and its *resistance* to acid attack (acid solubility).[49]

Filter Components: Most filters have multiple units operating in parallel. The filtration system consists of *filter housing* (filter box); *underdrains*; and *influent*, *effluent*, and *washwater* piping. The filter box houses the media, underdrains, washwater troughs, and pipings. The underdrain system has gravel support, underdrains, and central flume called gullet. The filtered water flows out of the gullet into the effluent pipe. The influent piping brings the settled water to individual filter boxes. Each filter box has an influent valve to cut off the flow. The effluent pipe is connected to the gullet at the bottom of the filter. An effluent valve is provided to isolate the treated water pipe during backwash. The washwater piping conveys clean water to the bottom of the filter into the gullet and then to the underdrains to backwash the media.[36,37,49,51] The filter components and piping are also shown in Figure 15.6.

Filter Backwash System: The filter backwash is designed to effectively clean the filter and conserve the backwash water. Backwashing involves passing water upward through the filter media at a velocity sufficient to expand the bed. The particles become separated and porosity increases. The accumulated solids are washed out. The auxiliary surface washers break and disturb the media and air scouring may be used for enhanced cleaning and to conserve backwash water. The design of backwash system is covered later in this section. The backwash operation may be initiated by (a) reaching the turbidity breakthrough, which is determined by the turbidity limit in the treated effluent (see t_B in Figure 15.7a), (b) reaching the limiting head loss, which is lower by a safety margin than the total available head loss allowed in filter design (see t_H in Figure 15.7b), or c filter run timer (t_R), which is preset by the operator based on experience. Practically, these three times may not be identical and whichever is reached first triggers the backwash. The typical requirement of backwash water volume ranges from 4% to 8% of the filter production. However, such requirement may be doubled for deep-bed filters.[36,37,49]

Filter Design Parameters: Filtration has become a pretreatment process before carbon adsorption, ion exchange, and membrane processes. Proper design and operation of filtration system is therefore very essential. Most important design considerations are filter media selection and filter depth. There are two general approaches to determine the depth and size of the media: (1) pilot testing and (2) experience from the past.[51–54] Kawamura developed a relationship between the filter depth l and the combined weighted average of effective size d_e for dual- or multimedia bed. He found that the l/d_e ratio is constant in the range of 980–1020.[55–57] The conceptual relationship between the desired filter media depth and average effective grain size of the media as a function of the average porosity of the filter media is shown in Figure 15.8. The typical design parameters of granular filters are summarized in Table 15.11.

Filter Flow Control: Two basic strategies are typically used to control the flow rate through the granular media filters: (a) *constant-rate* and (b) *declining-rate* filtration.

Constant-Rate Filter: The constant-rate filters have a flow controller on the discharge end to maintain a *constant head* over the filter media. The throttling valve exerts head loss when the bed is clean. As the head loss in the filter increases, the valve opens to compensate for the increased head loss in the bed, and the flow rate remains constant. The total operating head on the filter is kept constant. A variation of constant-rate filter is *variable head*. The operating water level on the filter is allowed to rise as the bed is clogged. Without using a flow controller, a constant rate of flow is maintained from the individual filter over an effluent weir.[44,49]

Declining-Rate Filter: In the declining-rate filters, the rate of flow through the filter is allowed to decline as the filter head loss increases. A large influent channel or header supplies water to all filters keeping the water level essentially the same. Thus, the flow to each filter is equal to the rate that is allowed by the head loss in that filter. When a filter is taken out of service for backwashing, the water level in the remaining filters rises slightly to maintain the same total flow rate through the system. When the filter is brought back into service after backwash, extremely high filtration rate may occur through this filter because of least resistance to the flow. Therefore, a flow-restricting device such as an orifice may be required in the effluent discharge piping for limiting the excessive flow through an individual filter.[36,37,44,49]

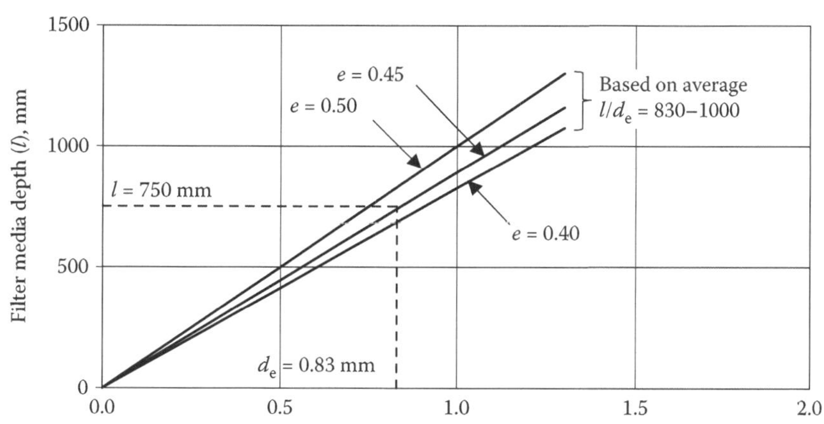

FIGURE 15.8 Relationship between the filter media and average effective size of media grain size as a function of average porosity of the media for dual- or multimedia bed. (Adapted in part from Reference 54.)

TABLE 15.11 Basic Design Values of Granular Filters

Filter Type	Design Value Range (Typical)			
	Total Depth, cm	Filtration Rate (*FR*), $m^3/m^2{\cdot}h$	Effective Size (d_{10}), mm	Uniformity Coefficient (*UC*)
Monomedium (stratified)				
Shallow bed of sand	25–35 (30)	5–15 (12)	0.45–0.65 (0.5)	1.2–1.6 (1.5)
Shallowbed of anthracite	30–50 (40)	5–15 (12)	0.8–1.5 (1.3)	1.3–1.8 (1.6)
Mediumbed of sand[a]	50–75 (60)	5–15 (12)	0.4–0.8 (0.65)	1.2–1.6 (1.5)
Mediumbed of anthracite[a]	60–90 (75)	5–24 (12)	0.8–2 (1.3)	1.3–1.8 (1.6)
Monomedium (unstratified)				
Deepbed of sand	90–180 (120)	5–24 (12)	2–3 (2.5)	1.2–1.6 (1.5)
Deepbed of anthracite	90–200 (150)	5–24 (12)	2–4 (2.7)	1.3–1.8 (1.6)
Dual-media				
Anthracite	35–90 (70)	5–24 (12)	0.8–2 (1.5)	1.3–1.8 (1.6)
Sand	20–40 (30)	–	0.4–0.8 (0.65)	1.2–1.6 (1.5)
Multimedia				
Anthracite[b]	40–120 (75)	5–15 (12)	1–2 (1.5)	1.3–1.8 (1.6)
Sand	25–45 (30)	–	0.4–0.8 (0.5)	1.3–1.8 (1.6)
Garnet	5–15 (10)	–	0.2–0.6 (0.35)	1.5–1.8 (1.6)

[a] Medium bed is also called conventional bed.
[b] Anthracite may be arranged into 2–3 layers.
Note: 1 cm = 0.394 in $1\ m^3/m^2{\cdot}h = 0.409\ gpm/ft^2$.
Source: Adapted in part from References 3, 8, 36, 49, 51, and 54.

Suspended Solids Removal by Granular Filters: Granular filters are efficient for removal of suspended solids and particulate forms of organics and nutrients. The typical effluent quality from granular filters used for tertiary treatment is given in Table 15.12.

The empirical correlations between TSS and turbidity in treated effluents are given by Equation 15.8.

$$C_{TSS} = f_{TSS/NTU} M_{Turbidity} \quad (15.8)$$

where

C_{TSS} = TSS concentration in effluent, mg/L
$M_{Turbidity}$ = turbidity measurement in effluent, NTU
$f_{TSS/NTU}$ = empirical factor, mg TSS/NTU. The value of $f_{TSS/NTU}$ may vary in the ranges of 2–2.4 and 1.3–1.6-mg TSS/NTU for settled secondary effluent and filtered effluent, respectively.

Proprietary Filtration Systems: Proprietary granular media filtration systems are commercially available from many manufacturers. These systems are developed for wastewater effluent filtration and may include: (a) traveling-bridge filter, (b) pulsed-bed filter, (c) deep-bed continuous backwash filter, and (d) synthetic medium filter (also call *Fuzzy* filter).[58–63]

TABLE 15.12 Typical Effluent Quality from Granular Filters for Tertiary Treatment

Process Application	Typical Effluent Quality	
	Turbidity, NTU	TSS, mg/L
Filtration of effluent from activated sludge process	0.5–4	2–8
Filtration of effluent with BNR facility	0.3–2	1–4

Source: Adapted in part from Reference 36.

Traveling-Bridge Filter: It is a proprietary continuous downflow, automatic backwash granular filter. The filter bed is divided into independent long horizontal cells. Each cell is typically 0.2–0.3 m (8–12 in) wide and backwashed individually by a washwater overhead mounted on a traveling bridge while other cells remain in service. Either monomedium of sand or dual media of sand and anthracite may be used in a traveling-bridge filter. The filtration rate is 5–12 $m^3/m^2{\cdot}h$ through a shallow filter bed that is <30 cm. An example of traveling-bridge continuous backwash filter is shown in Figure 15.9a.[36,37,54,58–60]

Pulsed-Bed Filter: The proprietary pulse-bed filter design uses a 25-cm shallow layer of fine-grained sand over an open atmosphere screened underdrain. The solids mat that builds up over the surface is pulsed with low-pressure diffused air bubbles that lift the accumulated solids from the surface, suspends the flocs into the liquid, and restores a clean surface for further filtration. The length of filter run is increased between the backwashing. During normal operation, the filter underdrains are not flooded as they are done in a conventional filter. High jet flows are also used in a short 3.5-min backwash duration to enhance effectiveness. The typical range of filtration rate for pulsed-bed filter is similar to that for the shallow monomedium sand filter. An example of pulsed-bed filter is shown in Figure 15.9b.[36,37,54,61]

Deep-Bed Continuous Backwash Filter: The wastewater is introduced at the bottom of the filter. It flows upward through the downward moving sand bed. Clean filtrate exits from the sand bed and discharges over an effluent weir. The sand and trapped particles move upward through a central pipe by airlift-pump action. The sand and solids are separated during upward flow. The liquid carries the solids into the central

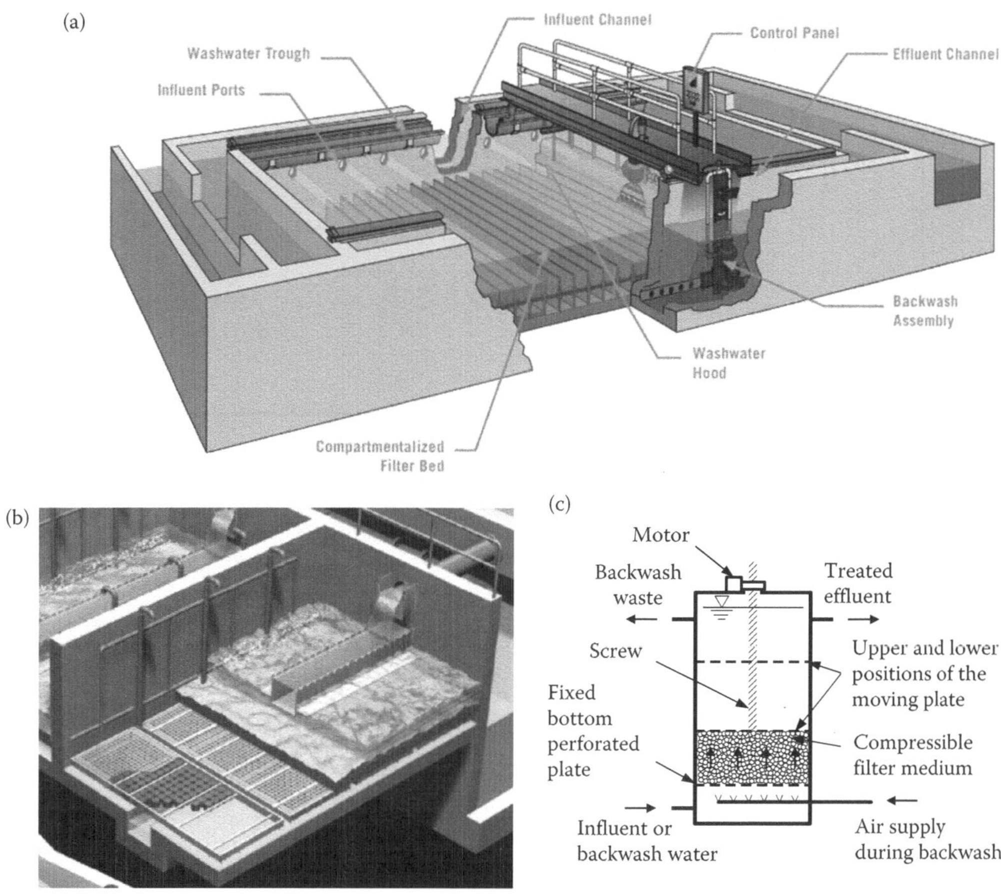

FIGURE 15.9 Proprietary filter systems: (a) traveling-bridge filter (Courtesy SUEZ), (b) pulsed-bed filter (Courtesy Evoqua Water Technologies), and (c) synthetic medium filter (or Fuzzy filter).

reject compartment for discharge, while the clean sand settles onto the top of the sand bed and is redistributed into the bed. This process has also been successfully modified for anoxic denitrification. Schematic of a continuous backwash denitrification filter (CBDNF) is shown in Figure 10.90a.[36,37,62]

Synthetic Medium Filter: This is an upflow filter using propriety synthetic medium. The basic features of the media are: (a) large size 2.5–3 cm, (b) low UC of 1.1–1.2, and (c) high porosity up to 90% (uncompressed medium). In this filter, a moving perforated plate is provided to compress the medium to reduce the porosity for filtration and uncompress the medium to expand the porosity for combined air/water backwashing. It may operate at a high filtration rate of 35–95 $m^3/m^2{\cdot}h$ in a filter bed depth of 60–108 cm.[36,37,63] Schematic of a synthetic medium filter is shown in Figure 15.9c.

Filter Hydraulics: The hydraulic design of filters is important but complex and not fully understood. Several semiempirical equations have been developed over 80 years to estimate the head loss through the clean filter bed. This is necessary for selection of media and depth. Likewise, numerous methods for estimating the head loss through clogged filter, duration of filter run between the backwashing, and quality of effluent have also been developed. The most reliable method for estimating the hydraulic characteristics of a filter is pilot testing with the flow to be treated.

Head Loss through a Clean Filter: Many empirical and semiempirical equations have been developed to predict the performance of a clean filter. The commonly used equations are Carman-Kozeny, Fair-Hatch, Hazen, and Rose equations. These relationships are given by Equations 15.9a through 15.9d.[36,37,44,50,57,64]

$$h_L = \frac{f}{\phi}\frac{(1-e)}{e^3}\frac{L}{d}\frac{v_s^2}{g} \quad \text{(Carman-Kozeny)} \quad (15.9a)$$

$$h_L = k\nu S^2 \frac{(1-e)^2}{e^3}\frac{L}{d^2}\frac{v_s}{g} \quad \text{(Fair-Hatch)} \quad (15.9b)$$

$$h_L = \frac{1}{C}\left(\frac{60}{T+10}\right)\frac{L}{d_{10}^2}v_s \quad \text{(Hazen)} \quad (15.9c)$$

$$h_L = \frac{1.067}{\phi}C_D\frac{1}{e^4}\frac{L}{d}\frac{v_s^2}{g} \quad \text{(Rose)} \quad (15.9d)$$

where

C = coefficient of compactness, dimensionless. It is usually 600–1200.
C_D = coefficient of drag, dimensionless. It is obtained from Equation 8.2.
d = medium grain diameter, m (ft)
d_{10} = effective medium size, mm
e = porosity ratio of medium, dimensionless. It is usually 0.4–0.5.
f = friction factor, dimensionless. It is calculated from Equations 15.9e and 15.9f.

$$f = 150\frac{(1-e)}{N_R} + 1.75 \quad (15.9e)$$

$$N_R = \phi\frac{\rho_w}{\mu}dv_s \quad \text{or} \quad N_R = \phi\frac{dv_s}{\nu} \quad (15.9f)$$

where

N_R = Reynolds number, dimensionless
ρ_w = density of water, kg/m^3 (lb/ft^3)
g = acceleration due to gravity, 9.81 m/s^2 (32.2 ft/s^2)
h_L = head loss through a clean filter bed, m (ft)
k = filtration constant, dimensionless. It is 5 and 6 for sieve openings and size of separation, respectively.

L = medium or layer depth, m (ft)
S = shape factor, dimensionless. It is usually 6.0–8.5.
S_v = specific surface area, dimensionless. It is $6/d$ and $5/(d{\cdot}\phi)$ for spheres and nonspherical particles, respectively.
T = temperature, °F
v_s = superficial (approach) filtration rate, m/s (ft/s)
μ = absolute viscosity, N·s/m^2 or kg/m·s (lb·s/ft^2)
ν = kinematic viscosity, m^2/s (ft^2/s)
ϕ = particle shape factor*, dimensionless. It is 1.0 for spheres, 0.94 for warn sand, 0.81 for sharp sand, 0.78 for angular sand, and 0.7 for crushed coal and sand.

Equation 15.9a is modified for a bed of nonuniform medium which becomes stratified after backwash. The modified equation for head loss in clean bed is given by Equation 15.9g.

$$h_L = \frac{L}{\phi}\frac{(1-e_s)}{e_s^3}\frac{v_s^2}{g}\sum\frac{f_{ij}x_{ij}}{d_{ij}} \tag{15.9g}$$

where
d_{ij} = average particle size between two adjacent sieve sizes (passing and retained), m (ft). The arithmetic or geometric mean may be used.
e_s = porosity of stratified layer of all fractions, dimensionless
f_{ij} = friction factor for different particle sizes, dimensionless. It is calculated from Equation 15.9f.
x_{ij} = fraction of particles retained on the sieve, fraction

The approximate head loss in a nonuniform bed can also be obtained from Equation 15.9a by using effective size (d_{10}). In a dual- or multimedia filter, the total head loss through the bed is the sum of head loss through each media layer. The procedure is shown in Example 15.16.

Head Loss in a Clogged Filter: During filtration, solids are captured in the pores of the media and the porosity decreases. The terminal head loss in the filter can be calculated if the porosity of the media at the end of the filter run is determined. The change in porosity can be estimated by assuming that the volume of the particles removed equals the reduction in pore volume.[65] This method is not very accurate in predicting the change in porosity. Another method is based on using several layers of the media in series. The increase in head loss for each layer is related to the amount of solid deposited in that layer. By considering each layer separately, a more reliable hydraulic model is obtained.[66] Extensive pilot testing data must be developed to calibrate the model.

Miscellaneous Head Losses: In addition to filter media, head losses are also encountered through (a) *gravel bed*, (b) *underdains*, and (c) *piping*. The head loss calculations through the supporting gravel bed are similar to that for filter media. The gravel bed is considered in several layers in series because of variability in gravel media. The underdrain systems are usually proprietary items and the manufacturers should provide the hydraulic characteristic data of their system. The system piping include influent, effluent, and washwater pipes, valves, flow rate controller, and complex manifolds with numerous fittings and specials. The friction head losses through such systems are calculated by Darcy–Weisbach equation (Equation 6.13a) or Hazan–Williams equation (Equation 6.13b). The minor losses through fittings and specials are calculated from Equation 6.15b. The minor loss coefficients are given in Appendix C. The applications of these equations to calculate head losses in filter piping system are shown in several examples given below.

* There is another approach to account for the effects of non-spherical particles. In this approach the settling velocity (v_s) or Reynolds number (N_R) is directly multiplied by a sphercity factor (ψ). This approach is practiced more commonly in filter media design. The relationship between ψ and ϕ is: $\psi = \frac{1}{\phi^{1/2}}$.

EXAMPLE 15.15: EFFECTIVE SIZE AND UNIFORMITY COEFFICIENT OF A FILTER MEDIUM

A sample of filter sand was sieved through the standard sieves. The results of sieve analysis are is given below. Determine the effective size d_{10} and uniformity coefficient (UC).

Sieve Size	Percent Retained, %	Percent Passing, %	Sieve Opening, mm
8	0	100	2.38
10	1	97	2.00
12	3	94	1.68
18	16	78	1.00
20	16	62	0.81
30	30	32	0.60
40	22	10	0.42
50	12	0	0.30

Solution

1. Draw the plot on probability paper of sieve opening versus percent passing by weight.
   The sieve analysis data are plotted in Figure 15.10.

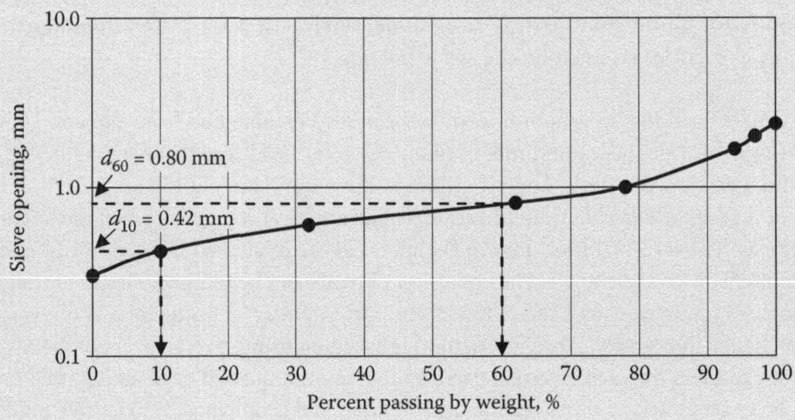

FIGURE 15.10 Plot of sieve opening versus percent passing (Example 15.15).

2. Read the values of d_{10} and d_{60} from the plot (Figure 15.10).
   $d_{10} = 0.42$ mm and $d_{60} = 0.80$ mm
3. Determine the value of UC.

$$\text{UC} = \frac{d_{60}}{d_{10}} = \frac{0.80\text{ mm}}{0.42\text{ mm}} = 1.90$$

EXAMPLE 15.16: DEPTHS OF THE MEDIA IN A DUAL-MEDIA FILTER

A dual-media filter is using anthracite coal and quartz sand. The specific gravity values of anthracite and sand are 1.55 and 2.65, and porosities are 0.48 and 0.40, respectively. The effective size of quartz sand is 0.5 mm. The depth of anthracite layer is twice the depth of sand layer. Determine (a) the effective size of the anthracite, (b) total depth and depth of each layer, and (c) overall l/d_e ratio.

Solution

1. Determine the effective size of anthracite.

   The relationship between two filter media with respect to their effective sizes is given by Equation 15.10a.

$$d_2 = d_1\left[\frac{Sg_1 - 1}{Sg_2 - 1}\right]^{2/3} \quad (15.10a)$$

   where

   d_1 = effective size of the medium with a specific gravity of Sg_1, mm

   d_2 = effective size of the medium with a specific gravity of Sg_2, mm

   Calculate the effective size of anthracite ($d_{anthracite}$) from Equation 15.10a,

$$d_{anthracite} = d_{sand}\left[\frac{(Sg_{sand} - 1)}{(Sg_{anthracite} - 1)}\right]^{2/3} = 0.5\ \text{mm} \times \left[\frac{2.65 - 1}{1.55 - 1)}\right]^{2/3} = 1.04\ \text{mm} \approx 1.0\ \text{mm}$$

2. Determine the weighted average size and porosity of the media.

$$\text{Weighted average size, } d_{average} = \frac{1}{3} \times d_{sand} + \frac{2}{3} \times d_{anthracite} = \frac{1}{3} \times 0.5\,\text{mm} + \frac{2}{3} \times 1.0\,\text{mm} = 0.83\,\text{mm}$$

$$\text{Weighted average porosity, } e_{average} = \frac{1}{3} \times e_{sand} + \frac{2}{3} \times e_{anthracite} = \frac{1}{3} \times 0.40 + \frac{2}{3} \times 0.48 = 0.45$$

3. Determine the total depth of the media and the depth of each layer.

   Read desired total depth l = 750 mm from Figure 15.8 at average media size = 0.83 mm and average porosity ratio = 0.45. Provide a total depth l = 75 cm (30 in).

$$\text{Depth of sand layer, } l_{sand} = \frac{1}{3}l = \frac{1}{3} \times 750\,\text{mm} = 250\,\text{mm} \quad \text{or} \quad 25\,\text{cm}\,(10\,\text{in})$$

$$\text{Depth of anthracite layer, } l_{anthracite} = \frac{2}{3}l = \frac{2}{3} \times 750\,\text{mm} = 500\,\text{mm} \quad \text{or} \quad 50\,\text{cm}\,(20\,\text{in})$$

4. Determine the overall l/d_e ratio in the filter bed with dual media.

$$l/d_e \text{ ratio provided by the sand layer, } (l/d_e)_{sand} = \frac{l_{sand}}{d_{sand}} = \frac{250\,\text{mm}}{0.5\,\text{mm}} = 500$$

$$l/d_e \text{ ratio provided by the anthracite layer, } (l/d_e)_{anthracite} = \frac{l_{anthracite}}{d_{anthracite}} = \frac{500\,\text{mm}}{1.0\,\text{mm}} = 500$$

Overall l/d_e ratio provided in dual media, $(l/d_e)_{overall} = (l/d_e)_{sand} + (l/d_e)_{anthracite} = 500 + 500 = 1000$

Note: Based on the weighted average size of the media, the $l/d_{average}$ = 750 mm/0.83 mm = 904. The overall l/d_e ratio based on each media l/d_e is 1000. This is equal to the typical ratio desired for the granular media filter design.

EXAMPLE 15.17: HEAD LOSS THROUGH A BED OF UNIFORM SAND

A filter bed of uniform sand is 0.42-m deep. Clear water at 20°C is passed through the bed at a filtering velocity of 4.5 m/h (1.25×10^{-3} m/s). The grain size is 0.45 mm, shape factor is 0.85, and porosity is 0.40. Determine the head loss through the clean bed.

Solution

1. Determine the Reynolds number (N_R) from Equation 15.9f.

At 20°C, $\mu = 1.002 \times 10^{-3}$ N·s/m^2 and $\rho_w = 998.2$ kg/m^3 are obtain from Table B.2 in Appendix B.

$d = 0.45$ mm $\times 10^{-3}$ m/mm $= 4.5 \times 10^{-4}$ m, $v_s = 1.25\ 10^{-3}$ m/s, and $\phi = 0.85$.

$$N_R = \phi \frac{\rho_w}{\mu} d v_s = 0.85 \times \frac{998.2\ \text{kg/m}^3}{1.002 \times 10^{-3}\ \text{kg/m·s}} \times 4.5 \times 10^{-4}\ \text{m} \times 1.25 \times 10^{-3}\ \text{m/s}$$

$$= 0.48 < 1.0\ \text{(laminar flow)}$$

2. Determine the friction factor (f) from Equation 15.9e.

$$f = 150 \frac{(1-e)}{N_R} + 1.75 = 150 \times \frac{(1-0.4)}{0.48} + 1.75 = 189$$

3. Determine the head loss from Equation 15.9a at $L = 0.42$ m.

$$h_L = \frac{f}{\phi} \frac{(1-e)}{e^3} \frac{L}{d} \frac{v_s^2}{g} = \frac{189}{0.85} \times \frac{(1-0.4)}{0.4^3} \times \frac{0.42\ \text{m}}{4.5 \times 10^{-4}\ \text{m}} \times \frac{(1.25 \times 10^{-3}\ \text{m/s})^2}{9.81\ \text{m/s}^2} = 0.31\ \text{m}$$

EXAMPLE 15.18: HEAD LOSS THROUGH A NONUNIFORM FILTER BED

A nonuniform filter bed is 0.7 m deep. The sieve analysis of the media is given below. The filtration rate is 4 m/h (1.11×10^{-3} m/s). The bed becomes fully stratified after backwash with smallest particles on the top. The specific gravity of sand is 2.65. The porosity and shape factor are 0.4 and 0.85, respectively. Determine the head loss through the clean bed. Water temperature is 20°C.

US Sieve No.		Mass Fraction Retained (x_{ij})	Sieve Opening, mm		Geometric Mean of Opening (d_{ij}), mm
Passing	Retained		Passing	Retained	
–	14	0.02	–	1.41	1.41
14	20	0.12	1.41	0.84	1.09
20	25	0.20	0.84	0.71	0.77
25	30	0.30	0.71	0.60	0.65
30	35	0.21	0.60	0.50	0.55
35	40	0.13	0.50	0.42	0.46
40	–	0.02	0.42	–	0.42
–	–	1.00 (Sum)	–	–	–

Solution

1. Develop the Reynolds number $(N_R)_{ij}$ as a function of d_{ij} from Equation 15.9f.
At 20°C, $\mu = 1.002 \times 10^{-3}$ N·s/m^2 and $\rho_w = 998.2$ kg/m^3 are obtain from Table B.2 in Appendix B.
$v_s = 1.11 \times 10^{-3}$ m/s and $\phi = 0.85$.

$$(N_R)_{ij} = \phi \frac{\rho_w}{\mu} d_{ij}\ v_s = 0.85 \times \frac{998.2\ \text{kg/m}^3}{1.002 \times 10^{-3}\ \text{kg/m·s}} \times d_{ij} \times 1.11 \times 10^{-3}\ \text{m/s} = 939.9\ d_{ij}$$

2. Determine the friction factor (f_{ij}) as a function of d_{ij} from Equation 15.9e.

$$f_{ij} = 150\frac{(1-e)}{(N_R)_{ij}} + 1.75 = 150 \times \frac{(1-0.4)}{939.9\,d_{ij}} + 1.75 = \frac{0.0958}{d_{ij}} + 1.75$$

3. Determine the value of $\Sigma f_{ij}\,x_{ij}/d_{ij}$ for the stratified layer of all fractions.
   a. Sample calculations for $d_{ij} = 1.41$ mm in the first row are presented below.

$$d_{ij} = 1.41\,\text{mm} \times 10^{-3}\,\text{m/mm} = 0.00141\,\text{m}$$

$$(N_R)_{ij} = 939.9\,d_{ij} = 939.9 \times 0.00141\,\text{m} = 1.33$$

$$f_{ij} = \frac{0.0958}{d_{ij}} + 1.75 = \frac{0.0958}{0.00141\,\text{m}} + 1.75 = 69.7$$

$$\frac{f_{ij}x_{ij}}{d_{ij}} = \frac{69.7 \times 0.02}{0.00141\,\text{m}} = 989\,\text{m}^{-1}$$

The calculation procedures for other grain sizes are the same as given above. The calculation results are tabulated in the table below.

Grain Size (d_{ij}), m	x_{ij}	$(N_R)_{ij}$	f_{ij}	$\frac{f_{ij}x_{ij}}{d_{ij}}$, m^{-1}
0.00141	0.02	1.33[a]	69.7	989
0.00109	0.12	1.02[a]	89.6	9864
0.00077	0.20	0.72	126	32,727
0.00065	0.30	0.61	149	68,769
0.00055	0.21	0.52	176	67,200
0.00046	0.13	0.43	210	59,348
0.00042	0.02	0.39	230	10,952
–	–	–	–	$\sum\frac{f_{ij}x_{ij}}{d_{ij}} = 249{,}849\,\text{m}^{-1}$

[a] The N_R values for the two largest grain sizes exceed 1. For nonuniform media, it is a common practice to discard the requirement of laminar flow ($N_R < 1$) and extend the application of Equations 15.9a, 15.9e, and 15.9g in transition flow regime (N_R up to 10,000). This will greatly simplify the calculation procedure. Also, the results estimated from these equations are normally within an acceptable range. This statement is also applicable to nonuniform or large-sized media.

4. Determine the head loss (h_L) from Equation 15.9g.

$$h_L = \frac{L}{\phi}\frac{(1-e_s)}{e_s^3}\frac{v_s^2}{g}\sum\frac{f_{ij}x_{ij}}{d_{ij}} = \frac{0.7\,\text{m}}{0.85} \times \frac{(1-0.4)}{0.4^3} \times \frac{(1.11\times10^{-3})^2\,\text{m}^2/\text{s}^2}{9.81\,\text{m/s}^2} \times 249{,}849\,\text{m}^{-1} = 0.24\,\text{m}$$

EXAMPLE 15.19: HEAD LOSS THROUGH A NONUNIFORM BED BASED ON EFFECTIVE SIZE OF THE MEDIA

The head loss through a nonuniform media bed is calculated in Example 15.18. Determine the effective size of the media from the sieve analysis, and calculate the head loss through the clean bed using the effective size. Compare with the head loss value obtained in Example 15.18.

Solution

1. Determine the effective size of the media from sieve analysis given in Example 15.18.
   a. Tabulate the data below.

Geometric Mean of Opening, mm	Mass Fraction Retained	Accumulative Mass Fraction	
		Retained	Passing
1.41	0.02	0.02	0.98
1.09	0.12	0.14	0.86
0.77	0.20	0.34	0.66
0.65	0.30	0.64	0.36
0.55	0.21	0.85	0.15
0.46	0.13	0.98	0.02
0.40	0.02	1.00	0.00
–	1.00 (Sum)	–	–

   b. Plot the sieve opening versus mass fraction of passing data in Figure 15.11.

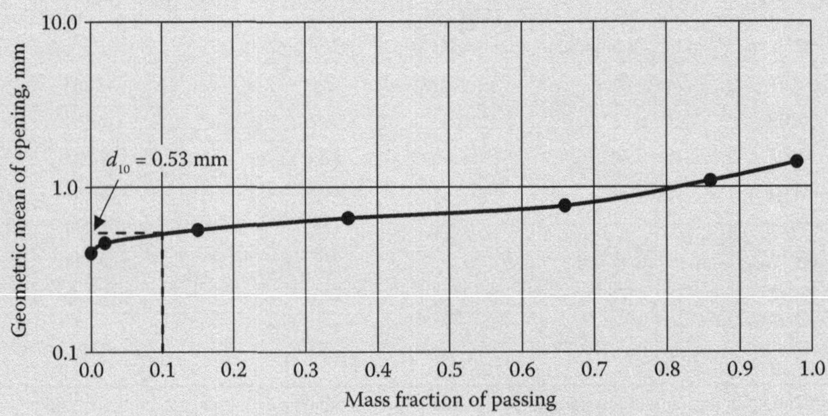

FIGURE 15.11 Plot of geometric mean of opening versus mass fraction of passing (Example 15.19).

   c. Obtain the effective size d_{10} from the plot.
      The effective size from the plot $d_{10} = 0.53$ mm.

2. Calculate the head loss through clean bed.
   Calculate the Reynolds number (N_R) from Equation 15.9f using $\mu = 1.002 \times 10^{-3}$ N·s/m^2, $\rho_w =$ 998.2 kg/m^3, $d_{10} = 0.58$ mm $= 5.3 \times 10^{-4}$ m, $v_s = 1.11 \times 10^{-3}$ m/s, and $\phi = 0.85$.

$$N_R = \phi \frac{\rho_w}{\mu} d\, v_s = 0.85 \times \frac{998.2\,\text{kg/m}^3}{1.002 \times 10^{-3}\,\text{kg/m·s}} \times 5.3 \times 10^{-4}\,\text{m} \times 1.11 \times 10^{-3}\,\text{m/s} = 0.50$$

Friction factor from (f) Equation 15.9e, $f = 150\dfrac{(1-e)}{N_R} + 1.75 = 150 \times \dfrac{(1-0.4)}{0.50} + 1.75 = 182$

Head loss from Equation 15.9a at $L = 0.7$ m.

$$h_L = \frac{f}{\phi}\frac{(1-e)}{e^3}\frac{L}{d}\frac{v_s^2}{g} = \frac{182}{0.85} \times \frac{(1-0.4)}{0.4^3} \times \frac{0.7\ \text{m}}{5.3 \times 10^{-4}\ \text{m}} \times \frac{(1.11 \times 10^{-3}\ \text{m/s})^2}{9.81\ \text{m/s}^2} = 0.33\ \text{m}$$

3. Compare the results.
   In Example 15.18, h_L of 0.24 m was obtained from Equation 15.9g. This is 0.09 m or 27% lower than 0.33 m obtained in Example 15.19. A conservative value of h_L is obtained from the effective size d_{10} data. The calculation procedure of h_L is similar to that of a bed of uniform size particles (Example 15.17).

EXAMPLE 15.20: HEAD LOSS IN A GRAVEL BED AND UNDERDRAIN

A rapid sand filter has two banks of four filters each. Each filter has a surface area of 67.5 m^2 and has *clay tile* underdrains, and four layers of gravel support. The gravel size, media depth, and porosity of each layer are given below. The specific gravity and shape factor of each layer are 2.65 and 1.0, respectively. The filtration rate is 10 m/h (0.0028 m/s) and the critical water temperature is 5°C. Calculate the head losses through the gravel bed and underdrain.

Gravel Layer	Media Size (mm)	Depth (cm)	Porosity	Reynolds Number (N_R)	Friction Fraction (f)	Head Loss (h_L) (m)
Top layer	2.5	4.5	0.45	4.6	19.7	0.0017
Second layer	4.5	5.5	0.45	8.3	11.7	0.0007
Third layer	10	7.0	0.50	17.5	6.0	0.0001
Bottom layer	15	8.0	0.50	28.6	4.4	0.0001
Total	–	25.0	–	–	–	0.0026

Solution

1. Determine the head loss through top layer.
   At water temperature of 5°C, $\mu = 1.519 \times 10^{-3}$ N·s/m^2 and $\rho_w = 999.9$ kg/m^3 from Table B.2 in Appendix B. Calculate the Reynolds number (N_R) from Equation 15.9f.

$$N_R = \phi\frac{\rho_w}{\mu} d\, v_s = 1.0 \times \frac{999.9\ \text{kg/m}^3}{1.519 \times 10^{-3}\ \text{kg/m·s}} \times 2.5 \times 10^{-3}\ \text{m} \times 2.8 \times 10^{-3}\ \text{m/s} = 4.6$$

Friction factor from Equation 15.9e, $f = 150\frac{(1-e)}{N_R} + 1.75 = 150 \times \frac{(1-0.45)}{4.6} + 1.75 = 19.7$

Head loss from Equation 15.9a,

$$h_L = \frac{f}{\phi}\frac{(1-e)}{e^3}\frac{L}{d}\frac{v_s^2}{g} = \frac{19.7}{1.0} \times \frac{(1-0.45)}{0.45^3} \times \frac{0.045\ \text{m}}{2.5 \times 10^{-3}\ \text{m}} \times \frac{(2.8 \times 10^{-3}\ \text{m/s})^2}{9.81\ \text{m/s}^2} = 0.0017\ \text{m}$$

2. Determine the head losses through other three layers.
   The procedure for head loss calculation through other layers is the same as that given in Step 1. The head loss values for other gravel layers and total head loss are also summarized in the above table.
3. Determine the head loss through the clay tile underdrain.

The head loss through the clay tile is given by Equation 15.10b.[65]

$$h_L = k_1 v_s^2 \tag{15.10b}$$

where

h_L = head loss through the clay tile underdrain, m

v_s = filtration rate (or hydraulic loading), m/h

k_1 = head loss constant that varies with the type of underdrain system and is given by the manufacturer, h^2/m. For the clay tile underdrain, $k_1 = 0.0005\ h^2/m$.

Head loss from Equation 15.10b, $h_L = k_1 v_s^2 = 0.0005\,h^2/m \times (10\,m/h)^2 = 0.05\,m$

4. Determine the total head loss through the gravel layers and clay tiles.

$$h_{L,total} = (0.0026 + 0.05)\,m = 0.053\,m \approx 0.05\,m$$

Note: The head loss through gravel media is small and may be ignored.

EXAMPLE 15.21: HEAD LOSS THROUGH A DUAL-MEDIA FILTER

The effective size and depth for sand and anthracite media layers of the dual-media filter are determined in Example 15.16. Determine the total head loss through the bed. The filtration rate is 10 m/h (0.0028 m/s) and the critical water temperature is 5°C. Assume that the shape factors for both media are 1.0.

Solution

1. Summarize the data given and developed in Examples 15.16 and 15.20.
   At 5°C, $\mu = 1.519 \times 10^{-3}$ N·s/m^2 and $\rho_w = 999.9$ kg/m^3.

Medium	Media Depth (L), cm	Effective Size ($d = d_{10}$), mm	Specific Gravity (S_s)	Porosity Ratio (e)
Quartz sand	25	0.5	2.65	0.40
Anthracite coal	50	1.0	1.55	0.48

2. Determine the head loss through the sand layer.
   N_R from Equation 15.9f,

$$N_R = \phi_{sand}\frac{\rho_w}{\mu}d_{sand}v_s = 1.0 \times \frac{999.9\,kg/m^3}{1.519 \times 10^{-3}\,kg/m{\cdot}s} \times 5 \times 10^{-4}\,m \times 2.8 \times 10^{-3}\,m/s = 0.92$$

f from Equation 15.9e, $f = 150\dfrac{(1 - e_{sand})}{N_R} + 1.75 = 150 \times \dfrac{(1 - 0.4)}{0.92} + 1.75 = 99.6$

h_L from Equation 15.9a,

$$h_L = \frac{f}{\phi}\frac{(1 - e_{sand})}{(e_{sand})^3}\frac{L_{sand}}{d_{sand}}\frac{v_s^2}{g} = \frac{99.6}{1.0} \times \frac{(1 - 0.4)}{0.4^3} \times \frac{0.25\,m}{5 \times 10^{-4}\,m} \times \frac{(2.8 \times 10^{-3}\,m/s)^2}{9.81\,m/s^2} = 0.373\,m$$

3. Determine the head loss through anthracite layer from Equations 15.9f, 15.9e, and 15.9a.

   $N_R = 1.84$, $f = 44.1$, and $h_L = 0.083$ m.

4. Determine the total head loss through clean dual-media filter.

$$h_L = (0.373 + 0.083)\,m = 0.456\,m \approx 0.46\,m$$

EXAMPLE 15.22: SIZING FILTERS FOR TERTIARY TREATMENT

Eight (8) dual-media filters are used for tertiary treatment at a water treatment facility. The design filtration rate is 10 $m^3/m^2{\cdot}h$ (4 gpm/ft^2) when one filter is out of service. The average daily flow is 0.63 m^3/s and the peaking factor for the maximum daily flow is 2.1. Determine (a) the filtration area for each filter at the design maximum daily flow, and (b) the overall dimensions of the filter cell. Also prepare a conceptual layout and brief description of the filter complex.

Solution

1. Determine the filtration area required for each filter cell.

   Design maximum daily flow, $Q_{max} = PF_{md}Q_{avg} = 2.1 \times 0.63\,m^3/s = 1.32\,m^3/s$

   Design flow to each filter, $Q = \dfrac{Q_{max}}{N-1} = \dfrac{1.32\,m^3/s}{8-1} = 0.19\,m^3/s$

   Filtration area required for each filter cell, $A_{req} = \dfrac{Q}{FR} = \dfrac{0.19\,m^3/s \times 3600\,s/h}{10\,m^3/m^2{\cdot}h} = 68\,m^2$

2. Select the overall dimensions of each filter cell.

   Using square filter units, the required dimensions, length(L) = width(W) = $\sqrt{68\,m^2} = 8.25\,m$

   Select dimensions of $L = W = 8.25$ m. The surface area $A = 68\,m^2$ per filter unit.

3. Describe the general arrange of the filter complex.

   A concept layout of the filter complex is shown in Figure 15.12. Eight filters are arranged into two rows of four filters each. A piping gallery is provided between two rows of filters for placing influent, effluent, and backwash pipes. Two identical pipes bring half of the design flow from the junction box of final clarifiers to each end of the gallery. The filter backwash waste flow is collected outside of each filter row.

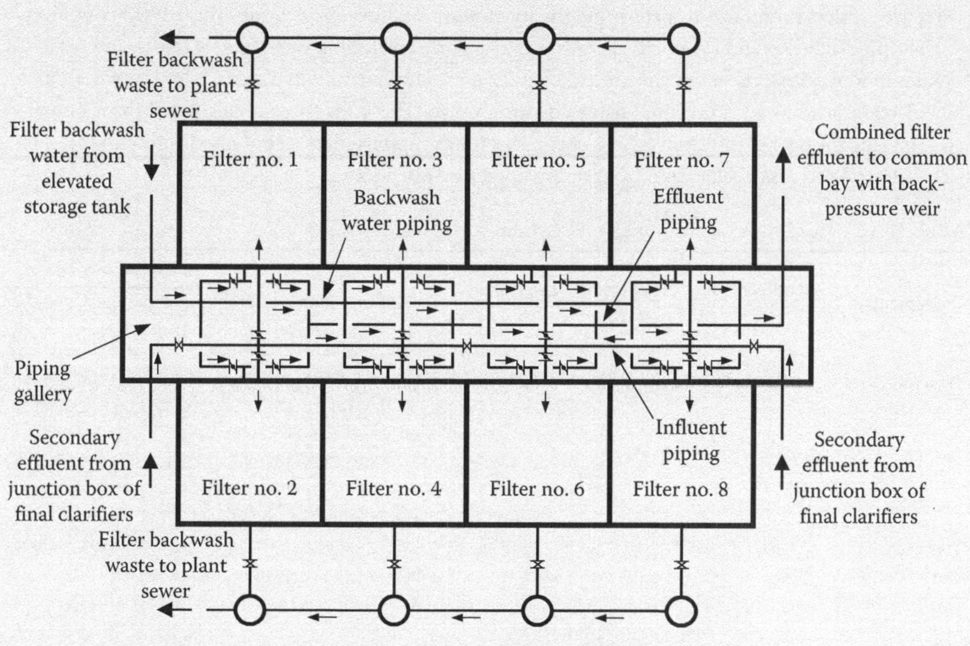

FIGURE 15.12 Conceptual layout of tertiary filter complex (Example 15.22).

EXAMPLE 15.23: HEAD LOSS THROUGH INFLUENT PIPING

Eight filters have been sized and arranged in a tertiary filter complex shown in Example 15.22. Calculate the head loss through the influent piping under the worst scenario. A simplified filter influent piping arrangement is shown in Figure 15.13. Use roughness coefficient $C = 120$.

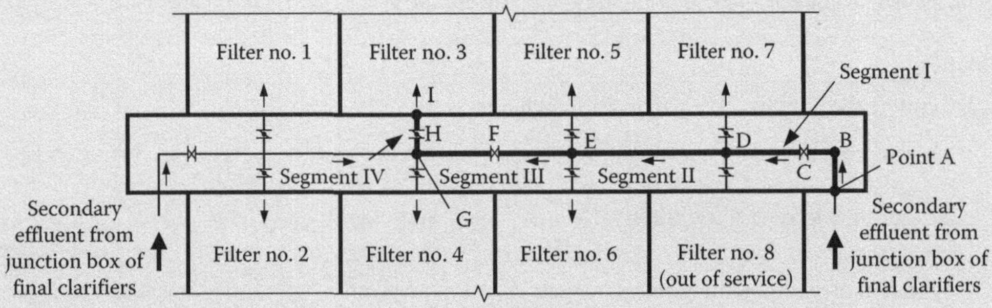

FIGURE 15.13 Filter influent piping details (Example 15.23).

Solution

1. Describe the critical condition for calculation of the maximum head loss in the influent piping.

   As described in Example 15.22, the filters are sized at the maximum daily flow when one filter is out of service for backwashing. The critical condition will occur when one of the four corner filter (Filter No. 1, 2, 7, or 8) is out of service. Assume that Filter No. 8 is out of service for backwashing; the maximum head loss through the influent piping will occur when either Filter No. 3 or 4 receives partial flow from the right side of the influent pipe. Therefore, the flow path from Point A–I is selected for calculating the maximum head loss under the critical condition. This flow path is highlighted in Figure 15.13.

2. Prepare a calculation table to determine the maximum total head loss under the critical condition.

   The pipe diameter, flow, velocity, length, appurtenance (including valves and fittings), and head loss values in four segments under the critical condition are summarized in Table 15.13. In each segment, the friction head loss is calculated from Equation 6.13b. The minor loss is calculated from Equation 6.15b using the overall minor head loss coefficient for the appurtenance contained within the segment. The minor head loss coefficients (K) are provided in Appendix C.

TABLE 15.13 Head Loss Calculations in Filter Influent Piping (Example 15.23)

Segment	Diameter, m	Flow, m^3/s	Velocity, m/s	Length, m	Appurtenance	Head Loss, m Friction (h_f)	Minor (h_m)	Total (h_L)
I (between Points A and D)	0.91	0.66[a]	1.01	9.00	$K = 0.4$ for a 90° elbow at Point B; $K = 1.0$ for an isolation gate valve at Point C; $K = 0.6$ for a cross (run-to-run) at Point D; and overall $K = 2.0$.	0.01[b]	0.10[c]	0.11[d]
II (between Points D and E)	0.91	0.47[e]	0.72	8.55	$K = 0.6$ for a cross (run-to-run) at Point E and overall $K = 0.6$.	0.01	0.02	0.03
	0.91	0.09[f]	0.14	8.55		0.00	0.00	0.00

(*Continued*)

TABLE 15.13 (*Continued*) Head Loss Calculations in Filter Influent Piping (Example 15.23)

Segment	Diameter, m	Flow, m^3/s	Velocity, m/s	Length, m	Appurtenance	Head Loss, m Friction (h_f)	Minor (h_m)	Total (h_L)
III (between Points E and G)					$K = 1.0$ for an isolation gate valve at Point F; $K = 1.8$ for a cross (run-to-branch) at Point G; and overall $K = 2.8$.			
IV (between Points G and I)	0.45	0.19[g]	1.19	4.80	$K = 1.2$ for a butterfly valve at Point H; $K = 1.0$ for the exit at Point I; and overall $K = 2.2$.	0.02	0.16	0.18
					Total maximum head loss in filter influent piping			0.32

[a] Flow in Segment I = secondary effluent flow in the feed pipe from each side, $Q_I = Q_{max} \div 2 = 1.32\,m^3/s \div 2 = 0.66\,m^3/s$.

[b] Friction head loss from Equation 6.13b, $h_f = 6.82\left(\frac{V}{C}\right)^{1.85} \times \frac{L}{D^{1.167}} = 6.82 \times \left(\frac{1.01\,m/s}{120}\right)^{1.85} \times \frac{9.0\,m}{(0.91\,m)^{1.167}} = 0.01\,m$

[c] Minor head loss from Equation 6.13b, $h_m = K\frac{V^2}{2g} = 2.0 \times \frac{(1.01\,m/s)^2}{2 \times 9.81\,m/s^2} = 0.10\,m$

[d] Total head loss in Segment I, $h_L = h_f + h_m = (0.01 + 0.10)\,m = 0.11\,m$.

[e] Flow in Segment II = flow in Segment I − flow to Filter No. 7, $Q_{II} = Q_I - Q = (0.66 - 0.19)\,m^3/s = 0.47\,m^3/s$.

[f] Flow in Segment III = flow in Segment II − flows to Filters No. 5 and 6, $Q_{III} = Q_{II} - 2Q = (0.47 - 2 \times 0.19)\,m^3/s = 0.09\,m^3/s$.

[g] Flow in Segment IV = flow to Filter No. 3, $Q_{IV} = Q = 0.19\,m^3/s$. (It equals the design flow to each filter cell.)

3. Determine the total maximum head loss through the influent piping under critical condition from Table 15.13.

   $h_L = 0.32\,m$

EXAMPLE 15.24: HEAD LOSS THROUGH THE EFFLUENT PIPING

The filter effluent piping collects filtered effluent from each filter and convey it into a common central effluent pipe. The combined filter effluent from the filter complex is conveyed by a conduit to a filter back-pressure weir common bay prior to the UV disinfection channels (see Example 15.25 for additional description). Use the flow data and conceptual filter layout given in Example 15.22. The simplified filter effluent pipings are shown in Figure 15.14.

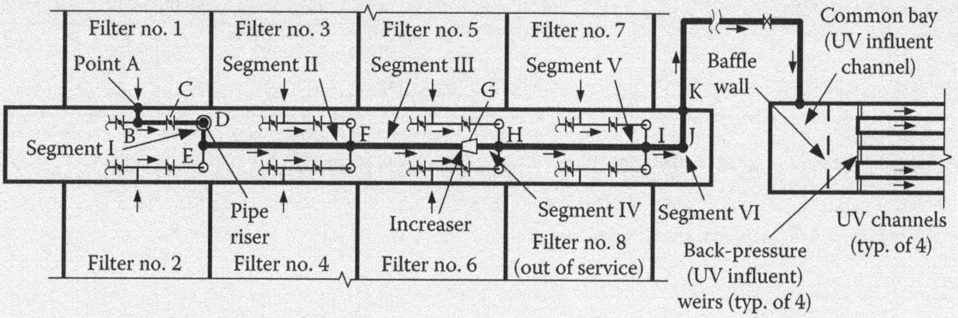

FIGURE 15.14 Details of filter effluent piping, including combined filter effluent conduit, and common bay with back-pressure weirs (Example 15.24).

Solution

1. Describe the critical condition for maximum head loss in the effluent piping.

   A procedure similar to that for influent piping in Example 15.23 is used. The flow path from Point A–K is selected for calculating the maximum head loss under the critical condition when Filter No. 8 is out of service for backwashing. This flow path is highlighted in Figure 15.14.

2. Prepare a calculation table to determine the maximum head loss through filter effluent piping.

   The calculation procedure is similar to that of influent piping. The head losses are calculated in Segments I–VI and the results are summarized in Table 15.14.

TABLE 15.14 Head Loss Calculations in Filter Effluent Piping (Example 15.24)

Segment	Diameter, m	Flow, m^3/s	Velocity, m/s	Length, m	Appurtenance	Head Loss, m		
						Friction (h_f)	Minor (h_m)	Total (h_L)
I (between Points A and E)	0.45	0.19[a]	1.19	10.80	$K=0.5$ for the entrance at Point A; $K=1.8$ for each of two tees (branch-to-run and run-to-branch) at Points B and E; $K=1.2$ for a (flow control) butterfly valve at Point C; $K=0.4$ for each of two 90° elbows (at two ends of the riser pipe) at Point D; and overall $K=6.1$.	0.04	0.44	0.48
II (between Points E and F)	0.91	0.38[b]	0.58	8.55	$K=0.6$ for a cross (run-to-run) at Point F and overall $K=0.6$.	0.00	0.01	0.01
III (between Points F and G)	0.91	0.76[c]	1.17	6.55	$K=0.2$ for the 91 cm × 122 cm increaser at Point G and overall $K=0.2$.	0.01	0.01	0.02
IV (between Points G and H)	1.22	0.76[c]	0.65	2.00	$K=0.6$ for a cross (run-to-run) at Point H and overall $K=0.6$.	0.00	0.01	0.01
V (between Points H and I)	1.22	1.14[d]	0.98	8.55	$K=0.6$ for a cross (run-to-run) at Point H and overall $K=0.6$.	0.01	0.03	0.04
VI (between Points I and K)	1.22	1.32[e]	1.13	4.50	$K=0.4$ for a 90° elbow at Point J and overall $K=0.4$.	0.00	0.03	0.03
					Total maximum head loss in filter effluent piping			0.59

[a] Flow in Segment I = flow from Filter No. 1, $Q_I = Q = 0.19\ m^3/s$. (It equals the design flow to each filter cell.)

[b] Flow in Segment II = combined flow from Filters No. 1 and 2, $Q_{II} = 2Q = 2 \times 0.19\ m^3/s = 0.38\ m^3/s$.

[c] Flows in Segments III and IV = combined flow from Filters No. 1–4, $Q_{III} = Q_{IV} = 4Q = 4 \times 0.19\ m^3/s = 0.76\ m^3/s$.

[d] Flow in Segment V = combined flow from Filters No. 1–6, $Q_V = 6Q = 6 \times 0.19\ m^3/s = 1.14\ m^3/s$.

[e] Flow in Segment VI = combined filter effluent flow from Filters No. 1–7, $Q_{VI} = 1.32\ m^3/s$. (It equals the design maximum daily flow.)

3. Determine the total maximum head loss through the effluent piping under critical condition from Table 15.14.

   $h_L = 0.59\ m$

EXAMPLE 15.25: HYDRAULIC PROFILE THROUGH A TERTIARY FILTER SYSTEM

The head losses in different components of a filtration system were calculated in Examples 15.20, 15.21, 15.23, and 15.24. The data used and results obtained are applicable for developing a hydraulic profile through the filtration system. In Examples 15.20 and 15.21, the head loss through the gravel support and underdrain system and through clean dual-media filter were calculated. Head losses in the influent and effluent pipings were separately calculated in Examples 15.23 and 15.24. As also shown in Figure 15.14, the combined filter effluent from the filter complex is conveyed by a single filtered effluent conduit to a filter back-pressure weir common bay that is also the influent chamber for the UV channels. Four (4) rectangular adjustable UV influent weir gates serve as the filter back-pressure weirs to control the water level in the filter units. A minimum water level at the weir must be selected to provide sufficient head over the filter media to prevent the development of negative pressure in the filter to avoid *air binding*. Calculate the total head loss in the effluent pipings, filtered water conduit, and head over the common bay weir. Draw the hydraulic profile from the secondary effluent junction box of the final clarifiers to the UV channels at the maximum design flow of 1.32 m^3/s. Also calculate the total head loss through the dirty and clean filter, and head exerted by the controller when the filter is clean. Assume that the constant-rate filter operates at a constant water surface in the filter unit. Additional information for preparing a complete hydraulic profile is also provided below: (a) the terminal head loss in the dirty filter media is 2.50 m; (b) the constant water depth above the media in the filter unit is 3.01 m; (c) the head loss in the influent pipe from the secondary effluent junction box to the filter influent header (Point A in Figure 15.13) is 0.64 m; (d) the head loss in the combined filter effluent conduit from the end of filter effluent pipe (Point K in Figure 15.14) to the filter back-pressure weir common bay is 0.52 m; (e) the length of back-pressure weir is 1.2 m each; and (f) the freeboard after the weir is 0.16 m. The water surface elevation in the upstream of UV influent channel is also assumed 100.00 m.

Solution

1. Determine the head over the back-pressure weirs.

   The combined filter effluent flow from the common bay is split over four rectangular back-pressure adjustable weir gates and drops into the UV channels (Figure 15.14).

   $$\text{Flow over each weir, } Q_{\text{weir}} = \frac{Q_{\max}}{N_{\text{weir}}} = \frac{1.32\ \text{m}^3/\text{s}}{4} = 0.33\ \text{m}^3/\text{s}$$

   Calculate the head over each rectangular weir from Equation 8.10 using $C_d = 0.6$ and $L = 1.2$ m at $n = 1$. Assume $L' = 1.17$ m.

   $$h_{\text{weir}} = \left(\frac{3}{2} \times \frac{Q_{\text{weir}}}{C_d L' \sqrt{2g}}\right)^{2/3} = \left(\frac{3}{2} \times \frac{0.33\ \text{m}^3/\text{s}}{0.6 \times 1.17\ \text{m} \times \sqrt{2 \times 9.81\ \text{m/s}^2}}\right)^{2/3} = 0.29\ \text{m}$$

   Check: $L' = L - 0.1n\,h_{\text{weir}} = 1.2\ \text{m} - 0.1 \times 1 \times 0.29\ \text{m} = 1.17\ \text{m}$

   Length of L' is same as the initial assumption.

2. Determine the top elevation (EL) of the filter back-pressure weir and the water surface elevation (WSEL) in the filter common bay.

   WSEL at the upstream of the UV channels = 100.00 m (given data)

   EL of the back-pressure weir crest = WSEL at the upstream of UV channels + freeboard after weir
   = 100.00 m (given data) + 0.16 m (given data) = 100.16 m

   WSEL in the common bay = top of weir + head over weir = 100.16 m + 0.29 m (Step 1) = 100.45 m

Note: A baffle wall is provided in front of the weirs. The head loss at the baffle wall is small and can be ignored.

3. Determine the hydraulic grade line (HGL) at the exit of filter effluent piping (Point K in Figure 15.14).

HGL at exit of filter effluent piping = WSEL in the common bay + head loss in filter effluent conduit
= 100.45 m + 0.52 m (given data) = 100.97 m

4. Determine the HGL at the entrance of filter effluent piping (Point A in Figure 15.14).

HGL at exit of filter effluent piping = HGL at exit of filter effluent piping + head loss in filter effluent piping
= 100.97 m + 0.59 m (Example 15.24, Table 15.14) = 101.56 m

5. Determine the constant water surface elevation in the filter unit.

WSEL in the filter = HGL at exit of filter effluent piping + head loss through underdrains and gravel support
+ terminal head loss in dirty filter media
= 101.56 m + 0.05 m (Example 15.20, Step 4) + 2.5 m (given data) = 104.11 m

6. Determine the HGL at the entrance of filter influent piping (Point A in Figure 15.13).

HGL at entrance of filter influent piping = WSEL in the filter + head loss in filter influent pipping
= 104.11 m + 0.32 m (Example 15.23, Table 15.13) = 104.43 m

7. Determine the WSEL in the secondary effluent junction box of the final clarifiers.

HGL in junction box = HGL at entrance of filter influent piping + head loss in influent pipe
= 104.43 m + 0.64 m (given data) = 105.07 m

8. Determine the elevations of filter media gravel support.

EL of top of filter media = WSEL in the filter − constant water depth above the media
= 104.11 m − 3.01 m (given data) = 101.10 m

EL of bottom of filter media = EL of top of filter media − depth of dual-media bed
= 101.10 m − 0.75 m (Example 15.16, Step 3) = 100.35 m

EL of bottom of gravel support = EL of bottom of filter media − depth of gravel layers
= 100.35 m − 0.25 m (Example 15.20) = 100.10 m

9. Draw the hydraulic profile in Figure 15.15.

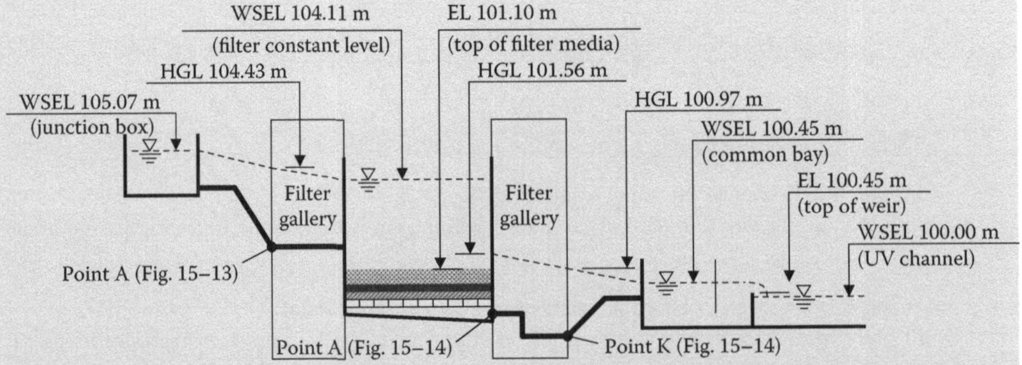

FIGURE 15.15 Hydraulic profile through the tertiary filter system (Example 15.25).

10. Determine the head exerted by the controller when filter is clean for maintaining a constant level in the filter unit.

Head exerted by the controller = terminal head loss in dirty filter media − head loss through clean filter media
= 2.5 m (given data) − 0.46 m (Example 15.21, Step 4) = 2.04 m

Note: Detailed information for preparation of hydraulic profile with step-by-step calculations of a dual-media filter system for drinking water treatment is available in Reference 49.

Backwash System for Granular Media: As filter accumulates solids, its porosity decreases, the head loss increases, and the filtration rate decreases. Under excessive solids accumulation, the *filter break through* may occur causing turbidity spike in the finished water. Before such condition is reached, the filter must be backwashed.

General Design Considerations of Backwash System: To ensure proper filter backwash, the design engineer must consider (1) the settling velocity of the media, (2) the backwash flow rate, (3) the head loss during backwash, (4) the supplemental cleaning methods, (5) the duration of backwash cycle, and (6) the quantity of water required to backwash each filter.

Settling Velocity of the Media after Backwash: The fluidized media settles after each backwash. Newton's or Stokes' equations (Equations 8.1a and 8.4) are used to calculate the settling velocity. In a dual- and mixed-media filter, the media selection should be such that the entire filter bed has approximately the same settling velocity. Equation 15.10a is used to calculate the size of the media grains of different specific gravities.[54,55]

Backwash Flow Rate: The backwash velocity must be higher than the settling velocity of the floc to lift it and wash it out, but it should be less than that of the media to prevent it from washing out. Generally, the settling velocity of the floc is in the range of 2.5 and 25 cm/min (1 and 10 in/min). For this reason, the minimum backwash rise rate is kept around 30 cm/min (12 in/min).[56,67] The settling velocity of the filter media is calculated by Newton's equation (Equation 8.1a). For water at 20°C (68°F), the Newton's equation reduces to Equation 15.11a for sand with specific gravity of 2.61 and to Equation 15.11b for anthracite with specific gravity of 1.5, respectively.[56,57]

$$v_t = 10d_{60} \quad \text{(sand)} \tag{15.11a}$$

$$v_t = 4.7d_{60} \quad \text{(anthracite)} \tag{15.11b}$$

where

v_t = terminal velocity of the media, m/min

d_{60} = size of the standard sieve opening that will pass 60% by weight of the media, mm. $d_{60} = d_{10} \times UC$ (uniformity coefficient). See media design parameters.

Many studies have shown that the most effective backwash rate is at approximately 10% of the terminal velocity of the media. Therefore, the backwash rates for sand and anthracite at 20°C are approximately expressed by Equations 15.12a and 15.12b, respectively.[56,57,68]

$$v_b = 0.1v_t \quad \text{or} \quad v_b = d_{60} \quad \text{(sand)} \tag{15.12a}$$

$$v_b = 0.1v_t \quad \textit{or} \quad v_b = 0.47d_{60} \quad \text{(anthracite)} \tag{15.12b}$$

where v_b = backwash rate, m/min

The above relationships apply at 20°C. If the water temperature is different, a temperature correction is applied to the backwash rate using Equation 15.12c.

$$v_{b,T} = v_{b,20}\,\mu^{1/3} \tag{15.12c}$$

where

$v_{b,T}$ = backwash rate at operating temperature T (°C), m/min
$v_{b,20}$ = backwash rate at temperature 20°C, m/min
μ = absolute viscosity of water at temperature T (°C), N·s/m^2 (kg/m·s)

The minimum backwash velocity needed to fluidize a 2-mm-size (d_{10}) granular sand media is about 2 m^3/m^2·min. The minimum velocities to fluidize the filter beds for different filter media of varying densities (garnet, sand, or anthracite) may be found in References 54, 56, and 57.

Head Loss during Backwash: During backwashing, the flow is reversed by forcing clean water upward through the media. The porosity and depth of expanded bed reach 65–70% and 120–155% of the packed bed. At this time, the head loss through the expanded bed must be at least equal to the buoyant mass of the particles in the fluid state. Also, the mass of packed bed media and fluidized bed media remains unchanged. The head loss due to backwash water flowing upward through the bed is given by Equation 15.13a.

$$h_{L,fb} = (1-e)L\frac{(\rho_m - \rho_w)}{\rho_w} \quad \text{or} \quad h_{L,fb} = (1-e)\,L\,(S_s - 1) \tag{15.13a}$$

where,

$h_{L,fb}$ = head loss through the expanded media during backwash, m (ft)
L = depth of stratified media at rest, m (ft)
e = porosity of the stratified bed at rest (fraction)
ρ_m = density of medium, kg/m^3 (lb/ft^3)
ρ_w = density of water, kg/m^3 (lb/ft^3)
S_s = specific gravity of the media, dimensionless

The porosity e_{fb} and depth of fluidized bed L_{fb} during backwashing are functions of the terminal velocity v_t and backwash velocity v_b. These relationships for uniform and nonuniform beds are given by Equations 15.13b through 15.13f.

$$e_{fb} = (v_b/v_t)^{0.22} \tag{15.13b}$$

$$L_{fb} = \frac{L(1-e)}{1-(v_b/v_t)^{0.22}} \quad \text{or} \quad L_{fb} = L\frac{(1-e)}{(1-e_{fb})} \tag{15.13c}$$

$$e_{fb,ij} = (v_b/v_{t,ij})^{0.22} \tag{15.13d}$$

$$e_{fb} = \sum x_{ij}(v_b/v_{t,ij})^{0.22} \quad \text{or} \quad e_{fb} = \sum x_{ij}e_{fb,ij} \tag{15.13e}$$

$$L_{fb} = L(1-e)\sum\frac{x_{ij}}{1-(v_b/v_{t,ij})^{0.22}} \quad \text{or} \quad L_{fb} = L(1-e)\sum\frac{x_{ij}}{1-e_{fb,ij}} \tag{15.13f}$$

where,

e_{fb} = porosity of fluidized bed, fraction
L_{fb} = depth of fluidized media during backwash, m (ft)
$e_{fb,ij}$ = porosity of fluidized bed of particles in the fraction of x_{ij}, fraction
x_{ij} = fraction of particles retained on the sieve, fraction
$v_{t,ij}$ = average terminal velocity of particles in the fraction of x_{ij} after backwash, m/s (ft/s). The arithmetic or geometric mean may be used.

Velocities v_b and v_t have been defined previously. Both variables are expressed in unit of m/s (ft/s).

Supplemental Cleaning Methods: Two basic methods for providing effective supplemental cleaning of clogged filter media are (a) *surface wash* and (b) *air-scour* systems.

1. Surface Wash System: Surface wash systems provide supplemental cleaning. They provide maximum agitation near the top of the filter bed where maximum accumulation is expected. In dual-media filter, this is done at the sand/anthracite interface. The agitation is provided by spraying water through nozzles located on either fixed piping manifolds or rotating arms. The turbulence created by surface wash system is expressed by Equation 15.14.

$$G_{sw} = \left[\frac{\Delta h\, g\, v_{sw}\, \rho_w}{\mu\, \alpha\, L_{fb}}\right]^{1/2} \qquad (15.14)$$

where

G_{sw} = velocity gradient created by water-jet-type surface wash, s^{-1}. Typical value of G_{sw} is in the range of 1100–1300 s^{-1}.[56]

v_{sw} = surface wash rate, m/s. The value of v_{sw} for a rotating-pipe discharge system is 1.2–2.4 m^3/m^2·h while that for a fixed-arm discharge system is 5–10 m^3/m^2·h.

Δh = head applied to the media by surface wash (m). The typical operating head is 35.69–53.03 m (350–520 kPa).

α = coefficient for surface wash, dimensionless. It is 0.25 and 0.5 for dual-arm surface and subsurface wash, respectively.

The surface wash systems are propriety units. They are normally started 1–3 min before the start of the backwash cycle and stopped 5–10 min before the end of backwash cycle.[68]

2. Air-Scour System: Air is also used for enhanced cleaning. Rising air bubbles increase turbulence throughout the bed and increase scouring. Typical air flow rates for different media are:[54,69–71]
   - For single-medium filters with effective size of 0.5 mm, airflow rate is 18–36 m^3/m^2·h followed by water backwash alone at a rate of 12–20 m^3/m^2·h.
   - For dual-media filters with effective sizes of 0.5 and 1.0 mm for sand and anthracite, air-flow rate is 55–91 m^3/m^2·h followed by water backwash at 37 to 49 m^3/m^2·h.
   - If air-scour and backwash are used simultaneously to clean larger sand with effective size of 1.0 mm, the air flow and water flow rates are 37–73 and 15 m^3/m^2·h, respectively.[72]
   - The filter bed can be made stratified or unstratified by backwashing. If air–water is used simultaneously during the entire backwashing cycle, the coarse and fine particles remain mixed in the settled bed. It is essential to turn off the air supply, while the water flow is continued for several minutes to fully stratify the mono-, dual-, and multimedia filter bed.

Duration of Backwash Cycle: The backwash cycle should extend long enough to clean the media bed. The time required to accomplish this depends on (a) the nature of the accumulated solids, (b) whether air-scour (or surface wash) is used, and (c) the geometry of the filter bed and the washwater troughs. Usually, the backwash cycle lasts from 5 to 20 min.[49,50]

Backwash Water Requirement: The backwash water requirement should be available at the plant site. The backwash water volume is typically in a range of 4–8% of the filter production. The total storage volume required is normally equal to the maximum amount needed to backwash each filter once during a 24-h period.[49,50]

Other Components of Backwash System: Other main components of a backwash system for media filtration are briefly presented below.

Washwater Collection Trough: The washwater troughs collect the backwash waste for removal. These troughs are designed to collect vertical backwash flow, have free fall over the lip of the trough, properly spaced, and vertically placed to carry dirty backwash water and not the media. Equation 8.13a or 8.13b is applied to design the wash water trough (Example 15.31).

Backwash Water Supply Systems: The backwash water is clean filter effluent that may be supplied by four methods for filter backwash: (1) a high-capacity pump to deliver the entire backwash flow directly to the manifold, (2) an elevated storage tank with a low capacity pump to fill the tank, (3) flow from high-service pumps, and (4) hydraulic gradient of the filter effluent flume, if available.

Backwash Water Return System: The filter backwash waste is quite large in volume and must be recycled for proper treatment prior to discharge or reuse.

Controlling Turbidity Spike after Backwashing: After backwashing, initially the filtered water will experience a turbidity spike. This may be detrimental to the water quality of filter effluent when low turbidity is desired for reuse. The following operating strategies may be used to control this turbidity spike as needed: (1) *filter-to-waste*, (2) *filter ripening*, and (3) *preconditioning filters*. Detailed information for effective control of turbidity spike after backwashing is available in Reference 67.

EXAMPLE 15.26: BACKWASHING OF A UNIFORM MEDIUM BED FILTER

The head loss calculations through a uniform medium filter bed are given in Example 15.17. It is backwashed at a backwash velocity of 9.5×10^{-3} m/s. Determine the porosity, depth of expanded bed, and head loss through the expanded bed. The specific gravity and porosity are 2.65 and 0.40, respectively.

Solution

1. Determine the terminal setting velocity of the media after backwash (v_t).

   The settling velocity of a particle is calculated from Newton's or Stokes' equations (Equation 8.1a or 8.4).

   Use the data given or obtained in Example 15.17: $d = 0.45\,\text{mm} \times 10^{-3}\,\text{m/mm} = 4.5 \times 10^{-4}\,\text{m}$, $\mu = 1.002 \times 10^{-3}\,\text{N·s/m}^2$, and $\rho_w = 998.2\,\text{kg/m}^3$ at 20°C.

   Apply Stokes' law (Equation 8.4) to determine the terminal velocity.

$$v_t = \frac{g(\rho_s - \rho_w)d^2}{18\mu} = \frac{9.81\,\text{m/s}^2 \times (2650 - 998.2)\,\text{kg/m}^3 \times (4.5 \times 10^{-4}\,\text{m})^2}{18 \times (1.002 \times 10^{-3}\,\text{kg/m·s})} = 0.18\,\text{m/s}$$

2. Check the Reynolds number (N_R) using Equation 15.9f.

$$N_R = \phi \frac{\rho_w}{\mu} d\, v_t = 0.85 \times \frac{998.2\,\text{kg/m}^3}{1.002 \times 10^{-3}\,\text{kg/m·s}} \times 4.5 \times 10^{-4}\,\text{m} \times 0.18\,\text{m/s} = 69$$

   N_R is in transition range ($1.0 < N_R < 10^4$) and Newton's equation is applicable.

3. Determine the terminal velocity (v_t) using the trial and error solution illustrated in Examples 8.4 and 8.5.

$$C_D \text{ from Equation 8.2, } C_D = \frac{24}{69} + \frac{3}{\sqrt{69}} + 0.34 = 1.05$$

$$v_t \text{ from Equation 8.1b, } v_t = \sqrt{\frac{4g(S_s - 1)d}{3C_D\phi}} = \sqrt{\frac{4 \times 9.81\,\text{m/s}^2 \times (2.65 - 1) \times (4.5 \times 10^{-4}\,\text{m})}{3 \times 1.05 \times 0.85}} = 0.10\,\text{m/s}$$

$$\text{Check } N_R,\ N_R = 0.85 \times \frac{998.2\,\text{kg/m}^3}{1.002 \times 10^{-3}\,\text{kg/m·s}} \times 4.5 \times 10^{-4}\,\text{m} \times 0.10\,\text{m/s} = 38$$

   Repeat calculations for C_D, v_t, and N_R until a stable value of v_t is reached after the fifth iteration. The final results are: $N_R = 32$, $C_D = 1.62$, and $v_t = 0.084$ m/s or 8.4×10^{-2}m/s.

4. Determine the porosity of the fluidized bed from Equation 15.13b.

$$e_{fb} = (v_b/v_t)^{0.22} = \left(\frac{9.5 \times 10^{-3}\ \text{m/s}}{8.4 \times 10^{-2}\ \text{m/s}}\right)^{0.22} = 0.62$$

5. Determine the depth of the fluidized media from Equation 15.13c using $L = 0.42$ m (from Example 15.17).

$$L_{fb} = L\frac{(1-e)}{(1-e_{fb})} = 0.42\ \text{m} \times \frac{1-0.40}{1-0.62} = 0.66\ \text{m}$$

Bed expansion based on depth ratio, $\frac{L_{fb}}{L} \times 100\% = \frac{0.66\ \text{m}}{0.42\ \text{m}} \times 100\% = 157\%$

Bed expansion based on porosity ratio, $\frac{e_{fb}}{e} \times 100\% = \frac{0.62}{0.40} \times 100\% = 155\%$

Note: The bed expansions obtained above based on the two ratios are virtually the same.

6. Determine the head loss through the fluidized bed from Equation 15.13a.

$$h_{L,fb} = (1-e)L(S_s - 1) = (1-0.40) \times 0.42\ \text{m} \times (2.65-1) = 0.42\ \text{m}$$

EXAMPLE 15.27: BACKWASHING OF A NONUNIFORM MEDIA BED FILTER

The head loss calculations through a nonuniform media bed filter are given in Example 15.18. Determine the porosity, depth of fluidized bed, and head loss through the expanded bed. The backwash velocity is 1.55×10^{-2} m/s. Assume that the media will become stratified after each backwash. The settling velocity of each fraction of the particles will be different depending upon the diameter, and must be calculated for each fraction as shown for uniform media in Example 15.18.

Solution

1. Determine the settling velocity of the largest size particle.

   Repeat Steps 1–3 in Example 15.26 to calculate the settling velocity ($v_{,t,ij}$) of the largest particle having $d_{ij} = 1.41$ mm (Example 15.18).

$v_{,t,ij}$ from Stokes; law (Equation 8.4), $v_{t,ij} = \frac{9.81\ \text{m/s}^2 \times (2650 - 998.2)\ \text{kg/m}^3 \times (1.41 \times 10^{-3}\ \text{m})^2}{18 \times (1.002 \times 10^{-3}\ \text{kg/m·s})}$

$= 1.79\ \text{m/s}$

N_R from Equation 15.9f, $N_R = 0.85 \times \frac{998.2\ \text{kg/m}^3}{1.002 \times 10^{-3}\ \text{kg/m·s}} \times 1.41 \times 10^{-3}\ \text{m} \times 1.79\ \text{m/s} = 2137$

C_D from Equation 8.2, $C_D = \frac{24}{2137} + \frac{3}{\sqrt{2137}} + 0.34 = 0.42$

$v_{,t,ij}$ from Equation 8.1b, $v_{t,ij} = \sqrt{\frac{4 \times 9.81\ \text{m/s}^2 \times (2.65-1) \times (1.41 \times 10^{-3}\ \text{m})}{3 \times 0.42 \times 0.85}} = 0.29\ \text{m/s}$

The final results are obtained after the third iteration: $N_R = 298$, $C_D = 0.59$, and $v_{t,ij} = 0.25$ m/s.

2. Determine the porosity of the fluidized layer of the largest size particle ($e_{fb,ij}$) from Equation 15.13d.

$$e_{fb,ij} = (v_b/v_{t,ij})^{0.22} = \left(\frac{1.55 \times 10^{-2}\,m/s}{0.25\,m/s}\right)^{0.22} = 0.54$$

3. Determine the factors of $x_{ij}\,e_{fb,ij}$ and $x_{ij}/(1 - e_{fb,ij})$ for the largest size particle using $x_{ij} = 0.02$ (from Example 15.18).

$$x_{ij}e_{fb,ij} = 0.02 \times 0.54 = 0.011$$

$$\frac{x_{ij}}{1 - e_{fb,ij}} = \frac{0.02}{1 - 0.54} = 0.043$$

4. Repeat all preceding steps for each layer and summarize the results in the table below.

Grain Size (d_{ij}), m	x_{ij}	$v_{t,ij}$	$e_{fb,ij}$	$x_{ij}e_{fb,ij}$	$\frac{x_{ij}}{1 - e_{fb,ij}}$
0.00141	0.02	0.250	0.54	0.011	0.043
0.00109	0.12	0.200	0.57	0.068	0.279
0.00077	0.20	0.150	0.61	0.122	0.513
0.00065	0.30	0.130	0.63	0.189	0.811
0.00055	0.21	0.110	0.65	0.137	0.600
0.00046	0.13	0.090	0.68	0.088	0.406
0.00042	0.02	0.080	0.70	0.014	0.067
–	–	–	–	$\sum x_{ij}e_{fb,ij} = 0.63$	$\sum \frac{x_{ij}}{1 - e_{fb,ij}} = 2.72$

5. Determine the porosity and depth of expanded media.

Porosity of expanded bed from Equation 15.13e, $e_{fb} = \sum x_{ij}e_{fb,ij} = 0.63$

Depth of expanded media from Equation 15.13f, $L_{fb} = L(1 - e)\sum \frac{x_{ij}}{1 - e_{fb,ij}}$

$$= 0.7\,m \times (1 - 0.40) \times 2.72 = 1.14\,m$$

Bed expansion based on depth ratio, $\frac{L_{fb}}{L} \times 100\% = \frac{1.14\,m}{0.7\,m} \times 100\% = 163\%$

Bed expansion based on porosity ratio, $\frac{e_{fb}}{e} \times 100\% = \frac{0.63}{0.40} \times 100\% = 158\%$

6. Determine the head loss through the expanded bed from Equation 15.13a.

$$h_{L,fb} = (1 - e)L(S_s - 1) = (1 - 0.40) \times 0.7\,m \times (2.65 - 1) = 0.69\,m$$

EXAMPLE 15.28: BACKWASHING OF A DUAL-MEDIA FILTER

A dual-media filter consists of anthracite coal and quartz sand. The media specifications are as follows.

Medium	Media Depth (L), cm	Effective Size (d_{10}), mm	Uniformity Coefficient (UC)	Specific Gravity (S_s)	Porosity Ratio (e)
Quartz sand	25	0.5	1.4	2.65	0.40
Anthracite coal	50	1.0	1.6	1.55	0.48

The effective size of the anthracite coal is such that the settling velocity of the anthracite and sand media are the same. The backwash rate is 10% of the settling velocity of the media or a minimum of 37 $m^3/m^2{\cdot}h$ (15 gpm/ft^2), whichever is higher. Determine (a) the average settling velocity of the media, (b) backwash velocity to fluidize the bed, (c) porosity ratio of the fluidized bed, (d) total depth of the expanded bed and bed expansion ratio, and (e) head loss through the expanded bed during the backwash. The shape factor $\phi = 1.0$. The critical operating temperature is 5°C.

Solution

1. Determine the media terminal velocity at 20°C.

   The approximate settling velocity (v_s) of each medium at 20°C is calculated from Equations 15.11a and 15.11b.

   Terminal velocity of sand, $v_t = 10\, d_{60} = 10(d_{10}UC) = 10 \times 0.5 \times 1.4 = 7\,\text{m/min} = 0.117\,\text{m/s}$

   Terminal velocity of anthracite, $v_t = 4.7\, d_{60} = 4.7(d_{10}UC) = 4.7 \times 1.0 \times 1.6 = 7.52\,\text{m/min} = 0.125\,\text{m/s}$

2. Determine the actual settling velocity of the media at 5°C.

   At 5°C, $\mu = 1.519 \times 10^{-3}\,\text{N·s/m}^2$ and $\rho_w = 999.9\,\text{kg/m}^3$ from Table B.2 in Appendix B.

   The settling velocity is calculated using trial and error procedure. Repeat Steps 1–3 in Example 15.26 to calculate the settling velocity. For initial trial, use v_t at 20°C calculated in Step 1 and d_{60} for particle size.

   a. Calculate the settling velocity of sand.

   $$N_R \text{ from Equation 15.9f, } N_R = 1.0 \times \frac{999.9\,\text{kg/m}^3}{1.519 \times 10^{-3}\,\text{kg/m·s}} \times (0.0005\,\text{m} \times 1.4) \times 0.117\,\text{m/s} = 54$$

   $$C_D \text{ from Equation 8.2, } C_D = \frac{24}{54} + \frac{3}{\sqrt{54}} + 0.34 = 1.19$$

   $$v_t \text{ from Equation 8.1b, } v_t = \sqrt{\frac{4 \times 9.81\,\text{m/s}^2 \times (2.65 - 1) \times (0.0005\,\text{m} \times 1.4)}{3 \times 1.19 \times 1.0}} = 0.113\,\text{m/s}$$

   The final results for sand are obtained after the third iteration: $N_R = 51$, $C_D = 1.23$, and $v_t = 0.111\,\text{m/s}$.

   b. Calculate the settling velocity of anthracite.

   $$N_R \text{ from Equation 15.9f, } N_R = 1.0 \times \frac{999.9\,\text{kg/m}^3}{1.519 \times 10^{-3}\,\text{kg/m·s}} \times (0.001\,\text{m} \times 1.6) \times 0.125\,\text{m/s} = 132$$

   $$C_D \text{ from Equation 8.2, } C_D = \frac{24}{132} + \frac{3}{\sqrt{132}} + 0.34 = 0.78$$

   $$v_t \text{ from Equation 8.1b, } v_t = \sqrt{\frac{4 \times 9.81\,\text{m/s}^2 \times (1.55 - 1) \times (0.001\,\text{m} \times 1.6)}{3 \times 0.78 \times 1.0}} = 0.121\,\text{m/s}$$

The final results for anthracite are obtained after the third iteration: $N_R = 126$, $C_D = 0.80$, and $v_t = 0.120$ m/s.

c. Determine the settling velocity of the media.

The calculated terminal velocities of sand and anthracite are quite close. It is safe to assume that the average settling velocity of the media $v_t = 0.12$ m/s.

3. Determine the backwash velocity to fluidize the media.

The average backwash velocity is approximately 10% of the settling velocity of the media (Equation 15.12a or 15.12b).

$$\text{Average backwash velocity, } v_b = 0.1\ v_t = 0.1 \times 0.12\ \text{m/s}$$
$$= 0\ .012\ \text{m/s} \quad \text{or} \quad 43.2\ \text{m}^3/\text{m}^2\cdot\text{h}\ (17.7\ \text{gpm/ft}^2)$$

The average backwash velocity of 43.2 m^3/m^2·h applies since it is >37 m^3/m^2·h (given data).

4. Calculate the porosity of the expanded bed from Equation 15.13b using $v_b/v_t = 0.1$.

$$e_{fb} = (v_b/v_t)^{0.22} = (0.1)^{0.22} = 0.60$$

5. Determine the depth of the fluidized media from Equation 15.13c.

$$\text{Depth of expanded sand, } L_{fb} = L\frac{(1-e)}{(1-e_{fb})} = 0.25\ \text{m} \times \frac{1-0.40}{1-0.60} = 0.38\ \text{m}$$

$$\text{Depth of expanded anthracite, } L_{fb} = 0.50\ \text{m} \times \frac{1-0.48}{1-0.60} = 0.65\ \text{m}$$

Total depth of expanded bed, $L_{fb} = 0.38\ \text{m} + 0.65\ \text{m} = 1.03$ m.

$$\text{Bed expansion based on depth ratio, } \frac{L_{fb}}{L} \times 100\% = \frac{1.03\ \text{m}}{0.75\ \text{m}} \times 100\% = 137\%$$

6. Determine the head loss through the expanded bed during the backwash cycle from Equation 15.13a.

Head loss of expanded sand, $h_{L,fb} = (1-e)L(S_s-1) = (1-0.40) \times 0.25\ \text{m} \times (2.65-1) = 0.25$ m

Depth of expanded anthracite, $h_{L,fb} = (1-e)L(S_s-1) = (1-0.48) \times 0.50\ \text{m} \times (1.55-1) = 0.14$ m

Total head loss of expanded bed, $h_{L,fb} = 0.25\ \text{m} + 0.14\ \text{m} = 0.39$ m.

EXAMPLE 15.29: REQUIREMENTS FOR SURFACE WASH AND BACKWASH

An effluent tertiary filtration facility is designed in Example 15.22. The surface wash velocity $v_{sw} = 0.06$ m^3/m^2·min and backwash velocity $v_b = 0.012$ m/s. The backwash cycle includes (a) performs surface wash 0–4 min ($t_{sw} = 4$ min), (b) backwash starts at $t_{b0} = 3$ min and reaches full rate in $t_{b1} = 1.5$ min, (c) backwash maintains full rate for $t_b = 6$ min, and (d) backwash drops from full rate to 0 in $t_{b2} = 1.5$ min. Determine (a) the surface wash and backwash flows, (b) total volume for surface wash and backwash, (c) velocity gradient created by the surface wash at 5°C, and (d) capacity of elevated storage tank required for backwashing two filters with additional 10% reserve. Also, select the pumps for filling the storage tank and prepare operational sequence for the pumps. Assume that each filter is backwashed once in 24 h; the applied total head to the media by surface wash is 42 m and the depth of expanded bed L_{fb} is 1.03 m.

Solution

1. Draw the filter backwash cycle, including surface wash and backwash in Figure 15.16.
2. Calculate the surface wash and backwash flows per filter.

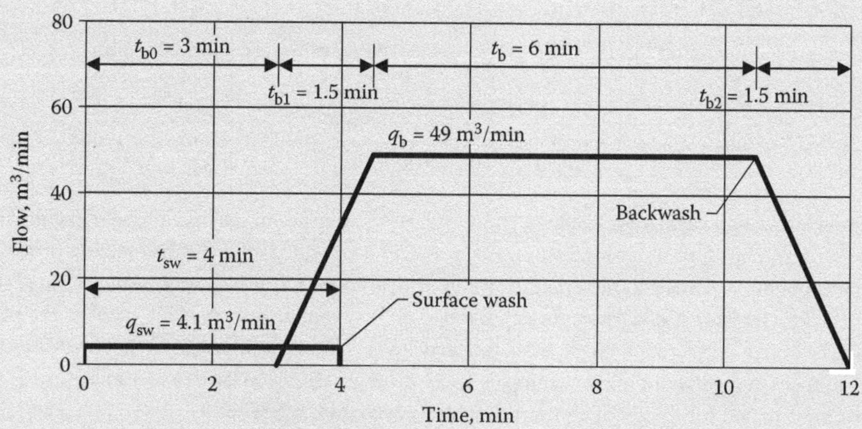

FIGURE 15.16 Filter backwash cycle (Example 15.29).

Surface area $A = 68\ \text{m}^2$ per filter (Example 15.22, Step 2).

Flow of surface wash per filter, $q_{sw} = v_{sw}A = 0.06\ \text{m}^3/\text{m}^2\cdot\text{min} \times 68\ \text{m}^2 = 4.1\ \text{m}^3/\text{min}$

Flow of backwash per filter, $q_b = v_b A = 0.012\ \text{m/s} \times 68\ \text{m}^2 \times 60\ \text{s/min} = 49\ \text{m}^3/\text{min}$

3. Determine the total volume of water for surface wash and backwash each cycle.

Volume of water for surface wash, $V_{sw} = q_{sw}t_{sw} = 4.1\ \text{m}^3/\ \text{min} \times 4\ \text{min} = 16.4\ \text{m}^3$

$$\text{Volume of water for backwash, } V_b = \frac{1}{2}q_b t_{b1} + q_b t_b + \frac{1}{2}q_b t_{b2} = q_b(0.5t_{b1} + t_b + 0.5t_{b2})$$
$$= 49\ \text{m}^3/\text{min} \times (0.5 \times 1.5\ \text{min} + 6\ \text{min} + 0.5 \times 1.5\ \text{min}) = 368\ \text{m}^3$$

Total volume of water per backwash cycle, $V_c = V_{sw} + V_b = (16.4 + 368)\ \text{m}^3 = 384\ \text{m}^3$

$$\text{Total volume of water backwash per day, } V_{daily} = 384\ \text{m}^3/\text{cycle} \times 1\ \text{cycle/filter}\cdot\text{d} \times 8\ \text{filters}$$
$$= 3072\ \text{m}^3/\text{d} \quad \text{or} \quad 0.81\ \text{MGD}$$

4. Determine the velocity gradient exerted by the surface wash from Equation 15.14.
At 5°C, $\mu = 1.519 \times 10^{-3}\ \text{N}\cdot\text{s/m}^2$ and $\rho_w = 999.9\ \text{kg/m}^3$ from Table B.2 in Appendix B.

$$\text{Velocity gradient, } G_{sw} = \left[\frac{\Delta h g v_{sw}\rho_w}{\mu\alpha L_{fb}}\right]^{1/2} = \left[\frac{42\ \text{m} \times 9.81\ \text{m/s}^2 \times 0.06\ \text{m/min} \times 999.9\ \text{kg/m}^3}{1.519 \times 10^{-3}\ \text{kg/m}\cdot\text{s} \times 0.25 \times 1.03\ \text{m} \times 60\ \text{s/min}}\right]^{1/2}$$
$$= 1030\ \text{s}^{-1}$$

5. Determine the capacity of elevated backwash storage tank.
Capacity of elevated backwash storage tank for two backwashes (including 10% reserve capacity),

$V_{tank} = 384\ \text{m}^3/\text{cycle} \times 2\ \text{cycles} \times 1.1 \approx 850\ \text{m}^3$

6. Size of the pumps to fill the elevated backwash storage tank.
Provide three pumps for filling the elevated backwash storage tank. The pump should be sized to fill the entire tank volume within 30 min by two pumps after backwashing the filter. The third pump will be a standby unit.

$$\text{Design flow for filling the tank, } q_{\text{fill}} = \frac{850\ \text{m}^3}{30\ \text{min}} = 28\ \text{m}^3/\text{min} \quad \text{or} \quad 28{,}000\ \text{L/min}$$

$$\text{Capacity required for each pump, } q_{\text{pump}} = \frac{28{,}000\ \text{L/min}}{2\ \text{pumps}} = 14{,}000\ \text{L/min per pump}$$

Provide a design capacity of 15,000 L/min (4000 gpm) per pump for operational flexibility.

Note: A reclaimed water ground storage tank (GST) is provided after the disinfection process. The backwash pumps will deliver the reclaimed water from this tank to the backwash water elevated storage tank (EST). The static head is the differential elevation between the water surface levels in the EST and GST. The head losses through the pipe can be calculated in a procedure similar to that presented for the filter influent and effluent piping (Examples 15.23 through 15.25 and Tables 15.13 and 15.14). Additional information for design of a pump station is covered in Reference 3.

7. Prepare the operational sequence of the pumps.

The pump operation will be automatically controlled by level switches in the elevated backwash storage tank. The conceptual operational sequence of the pumps is given in table below. The lead pump shall be alternated each day by SCADA. As an option, the third pump can be scheduled for use when the water surface elevation in the tank is below 25% full.

Elevation at Tank Volume	Lead Pump	Second Pump	Third Pump
Full tank	Off	Off	Off
80% full	On	Off	Off
50% full	On	On	Off
25% full	On	On	On (optional)
Empty	On	On	On (optional)

EXAMPLE 15.30: MINIMUM WATER HEAD FOR DELIVERING BACKWASH FLOW

The backwash system components of a tertiary filtration system are designed in Examples 15.28 and 15.29. The details of filter gravel and underdrain systems are given in Example 15.20. The average backwash velocity v_b is 0.012 m/s (Example 15.28, Step 3). Two filter rows with four filters in each row are separated by a filter piping gallery. The filter backwash feed pipes are connected to a backwash influent header that runs through the gallery (Figure 15.12). A backwash water connection pipe delivers the backwash water from the elevated storage tank (EST) to the header at one end of the gallery. Determine the minimum water head required to delivery an average backwash flow q_b of 49 m^3/min (Example 15.29, Step 2) from the EST to the filter unit at 5°C. The conceptual filter backwash system is shown in Figure 15.17. Assume that the following head losses have been obtained under the critical condition: (a) head loss in the backwash water connection pipe between Points A and B is 0.39 m and (b) maximum head loss in the backwash piping between Points B and C is 1.65 m (the critical condition occurs when Filter No. 8 is backwashed).

Note: The procedure to calculate losses through the backwash water piping is similar to that for calculating the head losses in filter influent and effluent piping (Examples 15.23 through 15.25 and Tables 15.13 and 15.14).

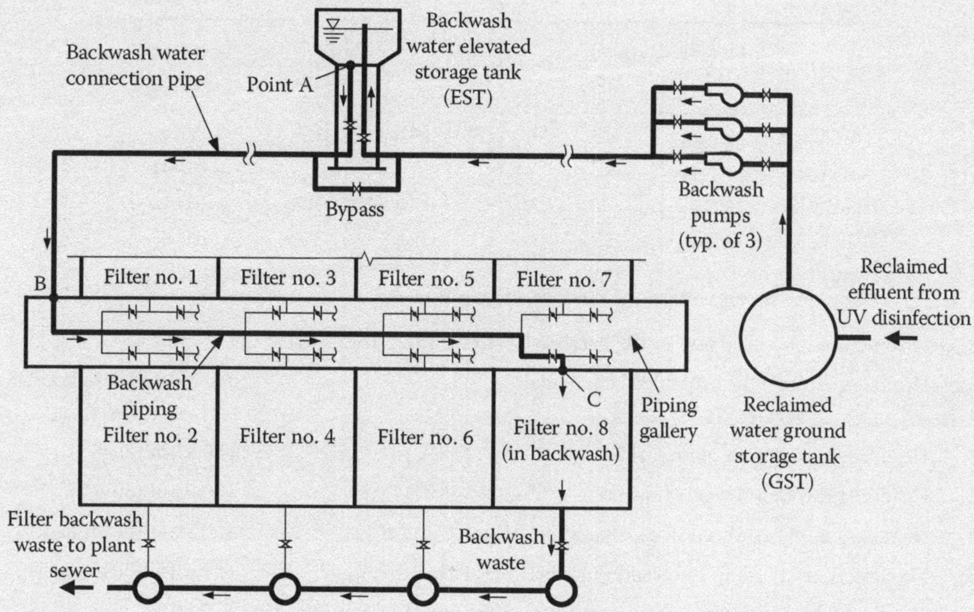

FIGURE 15.17 Conceptual layout of filter backwash system (Example 15.30).

Solution

1. Determine the head loss through the underdrain system during backwash.

   Average backwash velocity $v_b = 0.012$ m/s $= 43.2$ m^3/m^2·h is determined in Example 15.28, Step 3.

   Head loss through underdrain system from Equation 15.10b,

   $$h_L = k_1 v_b^2 = 0.0005\,\text{h}^2/\text{m} \times (43.2\,\text{m/h})^2 = 0.93\,\text{m}$$

2. Determine the head loss through the gravel layers during backwash.

   The details of gravel layers are given in Example 15.20.

   a. Determine the head loss through the top gravel layer.

   At 5°C, $\mu = 1.519 \times 10^{-3}$ N·s/m^2 and $\rho_w = 999.9$ kg/m^3 from Table B.2 in Appendix B.

   N_R from Equation 15.9f at d = 0.0025 m,

   $$N_R = \phi \frac{\rho_w}{\mu} d\, v_b = 1.0 \times \frac{999.9\,\text{kg/m}^3}{1.519 \times 10^{-3}\,\text{kg/m·s}} \times 0.0025\,\text{m} \times 0.012\,\text{m/s} = 19.7$$

   Friction factor from Equation 15.9e, $f = 150\frac{(1-e)}{N_R} + 1.75 = 150 \times \frac{(1-0.45)}{19.7} + 1.75 = 5.94$

   Head loss from Equation 15.9a,

   $$h_L = \frac{f}{\varphi}\frac{(1-e)}{e^3}\frac{L}{d}\frac{v_b^2}{g} = \frac{5.94}{1.0} \times \frac{(1-0.45)}{0.45^3} \times \frac{0.045\,\text{m}}{0.0025\,\text{m}} \times \frac{(0.012\,\text{m/s})^2}{9.81\,\text{m/s}^2} = 0.0095\,\text{m}$$

   b. Determine the head loss through other gravel layers.

   The procedure for head loss calculation through other layers is the same as given above. The head loss through other gravel layers and total head loss are summarized below.

Gravel Layer	Media Size, mm	Depth, cm	Porosity	Reynolds Number (N_R)	Friction Fraction (f)	Head Loss (h_L), m
Top layer	2.5	4.5	0.45	19.7	5.94	0.0095
Second layer	4.5	5.5	0.45	35.5	4.07	0.0044
Third layer	10	7.0	0.50	75.0	2.75	0.0012
Bottom layer	15	8.0	0.50	122	2.36	0.0007
Total	–	25.0	–	–	–	0.0158 ≈ 0.02

3. Determine the total head loss in the backwash water piping from EST to the filter unit.

Head loss through the expanded bed media = 0.39 m (Example 15.28, Step 6)

Head loss through the underdrain system = 0.93 m (Step 1)

Head loss through the gravel layers = 0.02 m (Step 2)

Head loss in the connection pipe between Points A and B = 0.39 m (given data)

Maximum head loss in the feed piping between Points B and C = 1.65 m (given data)

Total head loss in the backwash water piping from EST to the filter unit = 3.38 m

Add 25% allowance for operational flexibility, the overall head loss = 3.38 m × 1.25 = 4.23 m.

This is the minimum water head required to deliver the average backwash flow of 49 m^3/min from the EST to the filter unit.

EXAMPLE 15.31: WASHWATER TROUGH DESIGN

Design a filter backwash waste collection system for the tertiary filtration facility presented in Example 15.22. The conceptual layout of backwash waste collection system is shown in Figure 15.18. The backwash collection gullet is 0.75-m wide and is located in the center of the filter cell. Six washwater troughs are provided on both sides of the gullet at a center-to-center spacing of 2.75 m (4.51 ft). The length and width of the rectangular precast concrete troughs are 3.75 m (12.3 ft) and 0.5 m (1.6 ft), respectively. The designed backwash flow rate is 12 $L/m^2 \cdot s$ (18 gpm/ft^2), and the lip of the trough is 20 cm (8 in) above the maximum water depth in the trough (freeboard). The troughs have free fall into the gullet. The Manning's friction factor $n = 0.032$. Determine (a) the head over the weirs, (b) the water depth in the trough at the discharge end, (c) the water depth with and without the friction at the upstream end, and (d) the depth of the trough.

Solution

1. Determine the backwash flow per trough.

Total backwash flow, $q_b = 49\ m^3/min = 0.82\ m^3/s$ (Example 15.29, Step 2).

$$\text{Backwash flow per trough, } q_{b,trough} = \frac{q_{b,trough}}{N_{trough}} = \frac{0.82\ m^3/min}{6\ troughs} = 0.137\ m^3/s$$

2. Determine the head over the trough weirs.

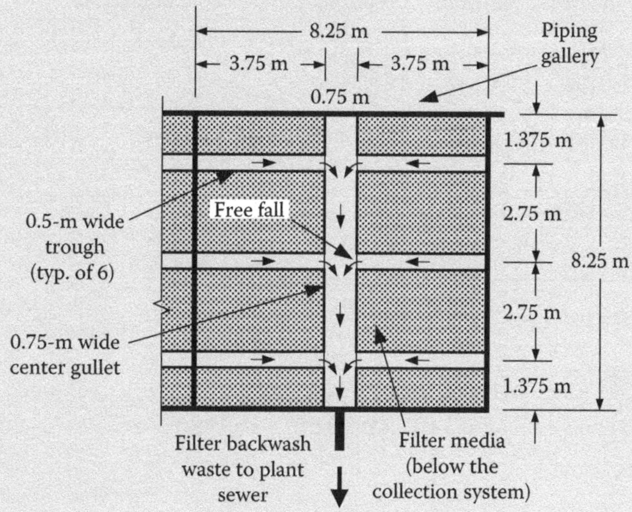

FIGURE 15.18 Conceptual layout of backwash waste collection system (Example 15.31).

The weirs are on both sides of the trough. The length of each weir, $L = 3.75$ m.

$$\text{Backwash flow per weir, } q_{b,weir} = \frac{q_{b,trough}}{N_{weir}} = \frac{0.137\,\text{m}^3/\text{s}}{2\text{ weirs}} = 0.0685\,\text{m}^3/\text{s}$$

Calculate the head over the rectangular weir from Equation 8.10 with $C_d = 0.6$ and $L' = L = 3.75$ m at $n = 0$.

$$h_{weir} = \left(\frac{3}{2} \times \frac{q_{b,weir}}{C_d L' \sqrt{2g}}\right)^{2/3} = \left(\frac{3}{2} \times \frac{0.0685\,\text{m}^3/\text{s}}{0.6 \times 3.75\,\text{m} \times \sqrt{2 \times 9.81\,\text{m/s}^2}}\right)^{2/3} = 0.05\,\text{m}$$

3. Determine the critical depth at the discharge end of the trough.
   Critical depth d_c is calculated from Equation. 7.10 with $C_w = 1.0$ and width of trough, $b = 0.5$ m.

$$d_c = \left(\frac{q_{b,trough}}{C_w b \sqrt{g}}\right)^{2/3} = \left(\frac{0.137\,\text{m}^3/\text{s}}{1.0 \times 0.5\,\text{m} \times \sqrt{9.81\,\text{m/s}^2}}\right)^{2/3} = 0.20\,\text{m}$$

   Critical depth d_c occurs at the discharge end. Assume $y_2 = d_c = 0.20$ m.
4. Determine the water depth at the upstream end of the trough.
   a. Upstream water depth without friction (Equation 8.13a).

$$y_1 = \sqrt{y_2^2 + \frac{2(q_{b,trough})^2}{gb^2 y_2}} = \sqrt{(0.20\,\text{m})^2 + \frac{2 \times (0.137\,\text{m}^3/\text{s})^2}{9.81\,\text{m/s}^2 \times (0.5\,\text{m})^2 \times 0.20\,\text{m}}} = 0.34\,\text{m}$$

   b. Upstream water depth with friction.
      For a level channel, the water surface profile is parabolic with an mean depth y_{mean} from Equation. 8.13c.

$$y_{mean} = y_1 - \frac{1}{3}(y_1 - y_2) = 0.34\,\text{m} - \frac{1}{3} \times (0.34\,\text{m} - 0.20\,\text{m}) = 0.29\,\text{m}$$

The mean hydraulic radius R_{mean} is estimated from Equation 8.13d.

$$R_{mean} = \frac{by_{mean}}{b + 2y_{mean}} = \frac{0.5\,m \times 0.29\,m}{0.5\,m + 2 \times 0.29\,m} = 0.13\,m$$

Upstream water depth with friction from Equation 8.13b with $c = 1.0\,m^{1/3}/s$.

$$y_1 = \sqrt{y_2^2 + \frac{2(q_{b,trough})^2}{gb^2 y_2} + \frac{2}{3}\frac{n^2 L (q_{b,trough})^2}{c^2 b^2 R_{mean}^{4/3} y_{mean}}}$$

$$= \sqrt{(0.20\,m)^2 + \frac{2 \times (0.137\,m^3/s)^2}{9.81\,m/s^2 \times (0.5\,m)^2 \times 0.20\,m} + \frac{2}{3}}$$

$$\times \sqrt{\frac{(0.032)^2 \times (3.75\,m) \times (0.137\,m^3/s)^2}{(1\,m^{1/3}/s)^2 \times (0.5\,m)^2 \times (0.13\,m)^{4/3} \times (0.29\,m)}}$$

$$= 0.35\,m$$

c. Select water depth.

The water depth with friction $y_1 = 0.36$ m is selected and used in design of trough.

5. Determine the depth of the trough.

The depth of trough, $D = y_1 + \text{Freeboard} = (0.35 + 0.2)\text{ m} = 0.55\text{ m}$

Provide a depth of 0.6 m.

Note: In depth, discussion on design of effluent troughs and launders is given in Section 8.4.2. Design details of effluent launders of different configurations are covered in Examples 8.17 through 8.19.

15.4.7 Surface Filtration

Surface filtration is an effective alternative to replace granular media filter for (1) polishing secondary effluent, (2) removing algae from stabilization pond effluent, and (3) providing pretreatment that is required by UV disinfection, microfiltration (MF), or ultrafiltration (UF). The main advantages of surface filters are acceptable filtrate quality, smaller footprint, low backwash flow rate, low backwash water volume, low labor and maintenance requirement, and less labor intensity. The surface filtration involves removal of suspended particles by sieving action through a thin filter medium (or septum). The typical pore size of the medium ranges from 5 to 40 μm. The filter medium may be cloth fabrics of different weaves, woven metal fabrics, and a variety of synthetic material. The surface filters may be categorized on the basis of direction of flow across the filter medium which may be *outside-in* or *inside-out* types. Key operational parameters and most commonly used systems are summarized in Table 15.15.[36] Both types are briefly discussed below.

Outside-In Surface Filters: The most popular examples of outside-in type filters are the cloth-media filtration systems manufactured by Aqua-Aerobic Systems, Inc. They are also called *cloth-media filters* (CMFs). These filters have tanks in which the filter media is completely submerged. The filter influent is distributed into the tanks, and filtration occurs through the fabric medium under a differential head. The filtrate is collected internally from the filter modules. The clogged solids are removed from the outside of the medium by a fixed or movable vacuum suction head.

Currently, there are three basic designs of CMFs: (a) *disk*, (b) *diamond*, and (c) *drum*. The filter medium may be acrylic pile, nylon pile, woven polyester pile, or needle felt fabrics. The operation and design features of these systems are summarized below. Important operating parameters of outside-in surface filters are also provided in Table 15.15.[36,73,74]

TABLE 15.15 Important Operating Parameters of Surface Filters

Parameter	Range	
	Outside-In Filter[a]	Inside-Out Filter[b]
Filtration rate (FR), $m^3/m^2{\cdot}h$ (gpm/ft^2)		
Average	5–12 (2–5)	5–20 (2–8)
Peak	14 (6)	14–40 (6–16)
Influent TSS concentration, mg/L	5–20	5–20
Diameter of disk filter[c], m (ft)	0.9–1.8 (3–6)	1.6–3 (5.25–10)
Normal pore size, μm	5–10	10–40
Submergence, percent of total disk area	100	45–70
Head loss, mm (in)	50–300 (2–12)	75–650 (3–26)
Backwash requirement, percent of throughput	2–5	2–4
Sample commercial systems[d]	AquaDisk®, AquaDiamond®, AquaDrum®, ISO-Disc™, and Five Star	Hydrotech®, DrumFilter, Ultrascreen®, and Forty-X™

[a] The performance data was developed from the available operating data on AquaDisk® and AquaDiamond® filters.
[b] The performance data were developed based on available operating data for Hydrotech®, DrumFilter, and Unltrascreen® filters.
[c] The dimensions are only applicable to disk filters.
[d] These filters were approved for reclaimed water reuse under California Water Recycling Criteria (Title 22) by the Division of Drinking Water (DDW) of State Water Resources Control Board, California Water Boards (September 2014).

Note: $1\ m^3/m^2{\cdot}h = 0.409\ gpm/ft^2$ 1 m = 3.28 ft 1 mm = 0.0394 in.

Source: Adapted in part from References 36, and 73 through 81.

Cloth-Media Disk Filter: These filters consist of several disks mounted on a horizontal shaft. In normal filtration mode, the disks stay still. The filtrate passes through the filter cloth on both sides of each disk, enters into a central collection header, and discharges over an effluent weir. The design details of disk type CMFs are shown in Figure 15.19a. The backwashing operation is initiated when a predetermined head loss is reached because of solids build up over the medium. The submerged disks start rotating at around 1 rpm. A reversed filtrate flow from the header is created when the cloth is passing through the fixed-type vacuum suction heads on each side of the disk. The accumulated solids are removed and carried away with the backwashing flow. The medium is fully cleaned and continues in operation. After prolong use the solids become lodged in the cloth media, and typical backwash cannot remove them. High-pressure spray wash may be required from the outside to flush the particles. The disk CMF system may be used in tertiary treatment for reuse of reclaimed water. These filters are also an attractive option for retrofitting existing granular filters for large plants.

Cloth-Media Diamond Filter: These filters consist of cloth filter elements mounted over the basin floor along the length of the filter. Each filter element has a diamond-shaped cross section. The filtrate is collected inside each element, released from one end, and discharged over an effluent weir. The design details of diamond shape filters are shown in Figure 15.19b. The cloth medium is cleaned by a vacuum sweep which moves back and forth along the element when the predetermined head loss is detected. The settled solids on the basin floor are also removed periodically by a movable vacuum device. The diamond CMF system is suitable for large scale applications because of long and skinny configuration. The diamond CMF system is also ideal in replacing existing traveling-bridge filters.

Cloth-Media Drum Filter: These filters are supplied as a packaged unit with a stationary cloth filter drum. The filtrate is collected inside the drum, and discharged through the center shaft into an effluent chamber. The major components of the AquaDrum® filter are shown in Figure 15.19c. During backwashing, the filter drum is rotated and the cloth medium is cleaned by a vacuum suction head. A sludge pump is also provided to remove the solids accumulated at the bottom of the tank. The drum CMF system is typically used at small plants.

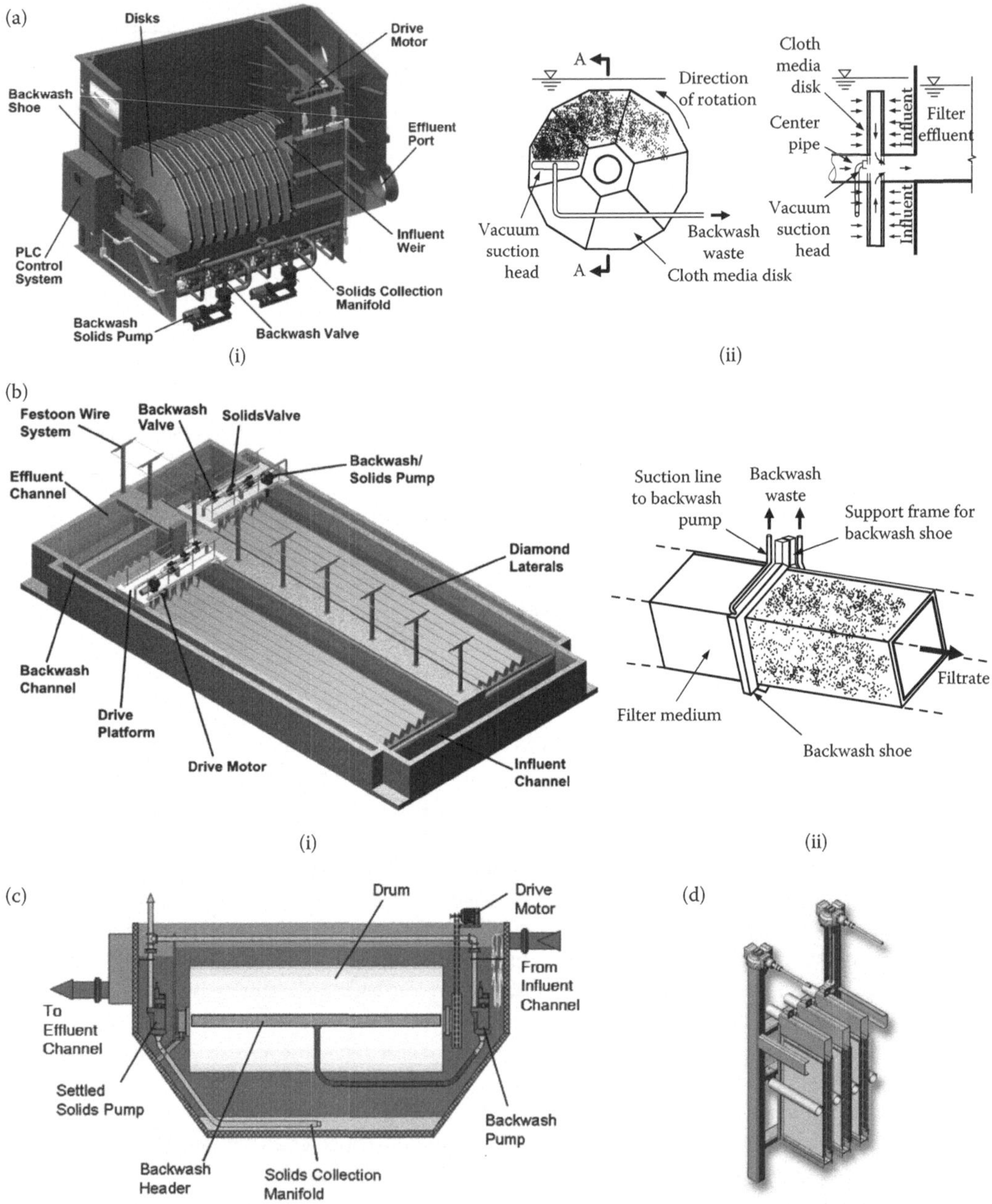

FIGURE 15.19 Outside-in surface filters: (a) cloth-media disk filter: (i) components (Courtesy Aqua-Aerobic Systems, Inc.) and (ii) conceptual filtration and backwashing mechanisms; (b) cloth-media diamond filter: (i) components (Courtesy Aqua-Aerobic Systems, Inc.) and (ii) solids removal mechanism; (c) components of cloth-media drum filter (Courtesy Aqua-Aerobic Systems, Inc.); and (d) components of cloth-media fixed-plate disk filter (Courtesy Alfa Laval Inc.)

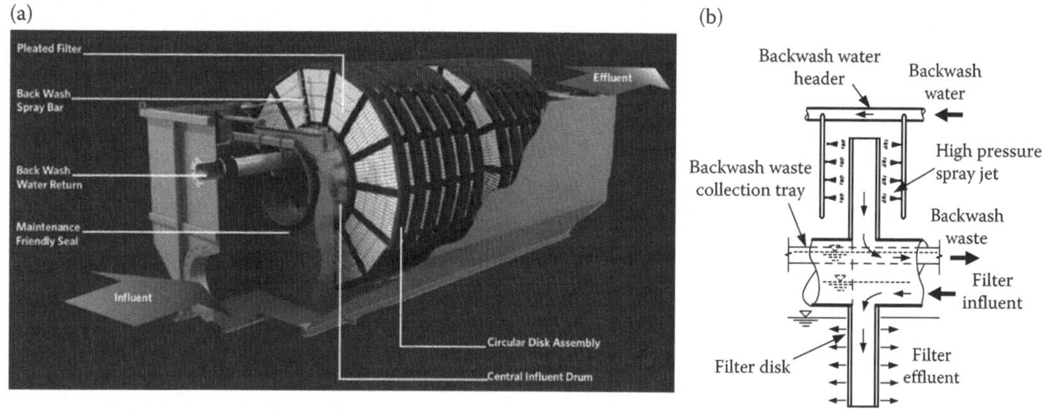

FIGURE 15.20 Inside-out disk surface filters: (a) components of Forty-X™ Disc Filter (Courtesy Evoqua Water Technologies) and (b) conceptual filtration and backwashing mechanisms.

Other innovative outside-in surface filter designs using fixed square or rectangular shaped disks are also commercially available. An example of such system (Iso-Disc™ by Alfa Laval Inc.) is shown in Figure 15.19d.[36,73–76]

Inside-Out Surface Filters: The inside-out surface filter consists of a single or multiple filter medium elements that are typically mounted on a horizontal shaft and submerged partially in a concrete basin or prefabricated stainless steel tank. The influent is fed through a central pipe into the element. As the filtrate flows from the inside to outside, solids are captured by the filter medium. The filtered effluent is collected in the basin, where an effluent weir is provided to maintain proper submergence of the filter elements. As the filter elements rotate slowly at 1–8.5 rpm, the filtration and backwashing may occur simultaneously or intermittently. The backwashing is achieved as the unsubmerged disk segment passes through the spray jets from the outside. The high-pressure spray penetrates the medium and separate the solids from the inside surface. The backwash waste is collected in an inside trough. The filter elements may be disks or drums. The medium is typically a polyester fabric cloth or stainless steel screen. The components of an example inside-out disc surface filter (Forty-X™ by Evoqua Water Technologies) is shown in Figure 15.20a. The conceptual filtration and backwash mechanisms of an inside-out surface filter are illustrated in Figure 15.20b. Important operating parameters of inside-out surface filters are also provided in Table 15.15.[36,77–80]

Performance of Surface Filters: The surface filters are very effective in removing TSS and turbidity. Recent operating data from full-scale installations have indicated that the cloth-media filters consistently produce effluent TSS concentration below 5 mg/L when the influent TSS concentrations ranged 15–20 mg/L. The turbidity in the filtrate is <2 NTU when the influent turbidity is up to 10 NTU. The log removal of coliform organisms may be 0–1, and that of virus is 0.0–0.5.[36,82–84]

EXAMPLE 15.32: CAPACITY, HEAD LOSS, AND BACKWASH VOLUME OF A SURFACE FILTER

An outside-in surface filter consists of five disks. The disk and central pipe have diameters of 3 and 0.5 m, respectively. The filter medium is stainless steel. The filtration rate (FR) = 12 $m^3/m^2{\cdot}h$. The effective opening area (efficiency factor) is 0.12. During the normal operation 60% of the disk surface area is submerged, and disk rotates at 5 rpm. The backwash volume is 3% of throughput volume. Determine (a) the total filtration capacity, (b) total head loss for clean and 80% clogged filter, and (c) daily volume of backwash water.

Solution

1. Determine the filtration capacity (or throughput volume) per filter.

$$\text{Total disk area on both sides of each disk, } A_{gross} = 2 \times \frac{\pi}{4} \times ((3\,\text{m})^2 - (0.5\,\text{m})^2) = 13.7\,\text{m}^2$$

$$\text{Total filtration area per disk at submergence of 60\%, } A_{filtration} = 0.6 \times A_{gross} = 0.6 \times 13.7\,\text{m}^2 = 8.2\,\text{m}^2$$

$$\text{Total filtration area per filter with 5 disks, } A_{filter} = 5 \times A_{filttration} = 5 \times 8.2\,\text{m}^2 = 41\,\text{m}^2$$

$$\text{Filtration capacity per filter, } Q_{filter} = FR \times A_{filter} = 12\,\text{m}^3/\text{m}^2{\cdot}\text{h} \times 41\,\text{m}^2 = 492\,\text{m}^3/\text{h}$$

$$\text{or} \quad Q_{filter} = 492\,\text{m}^3/\text{h} \times \frac{\text{h}}{3600\,\text{s}} = 0.14\,\text{m}^3/\text{s}\,(2200\,\text{gpm})$$

2. Determine the head loss when filter is clean and 80% clogged.

$$\text{Effective opening area per filter, } A_{opening} = 0.12 \times A_{filter} = 0.12 \times 41\,\text{m}^2 = 4.9\,\text{m}^2$$

Apply the head loss equation for fine screen (Equation 7.11a) to determine the head loss through the filter. Assume $C_d = 0.3$.

$$\text{Head loss } h_L \text{ for clean filter, } h_L = \frac{1}{2g} \times \left(\frac{Q_{filter}}{C_d \times A_{opening}}\right)^2 = \frac{1}{2 \times 9.81\,\text{m/s}^2} \times \left(\frac{0.14\,\text{m}^3/\text{s}}{0.3 \times 4.9\,\text{m}^2}\right)^2 = 0.0005\,\text{m}$$

$$\text{Head loss } h_L \text{ for 80\% clogged filter, } h_L = \frac{1}{2 \times 9.81\,\text{m/s}^2} \times \left(\frac{0.14\,\text{m}^3/\text{s}}{0.3 \times (1-0.8) \times 4.9\,\text{m}^2}\right)^2 = 0.012\,\text{m}$$

3. Determine the daily volume of backwash water per filter.

$$\text{Daily volume of backwash water per filter, } Q_{backwash} = 0.03 \times 0.14\,\text{m}^3/\text{s} = 0.004\,\text{m}^3/\text{s or 4 L/s (66 gpm)}$$

15.4.8 Carbon Adsorption

Carbon adsorption is used to remove soluble refractory organics. Many of these compounds are organometallic compounds, pesticides, chlorinated compounds, chlorine, and many others that are not removed by conventional secondary treatment and tertiary filtration. The process involves entrapping organic compounds onto the activated carbon surface. The activated carbon comes in two forms: *powdered* activated carbon (PAC) and *granular* activated carbon (GAC). The diameter of PAC particles is typically <0.074 mm (200 sieve) while that of GAC is typically >0.1 mm (~140 sieve). Both forms achieve removal of organics by absorption. Historically, PAC was first applied in the United States in 1929 for taste and odor control in public water supplies.[85]

Both PAC and GAC are used in wastewater applications. However, the most common method currently used is GAC column. The treated wastewater is percolated through the column until the GAC becomes saturated with the contaminants. The spent GAC is removed and sent for regeneration.

Manufacture and Regeneration of Activated Carbon: Activated carbon is manufactured from wood, coal, bone char, shell (coconut, pecan, and others), and some other carbon sources. The material is heated in absence of air to carbonize it and then activated by oxidation to remove noncarbon impurities. The activation burns away some carbon layers and produces a pore structure. Activation is accomplished by heating to a high temperature between 200°C and 1000°C in the presence of steam, carbon dioxide, or air; or by wet chemical treatment at lower temperatures. The chemical agents used are concentrated phosphoric acid, potassium hydroxide, or zinc chloride. The result is a highly porous carbon surface with high adsorptive properties. The pore-size distribution and adsorptive properties depend upon the carbon source and activation method employed.[49,72] The adsorptive capacity of carbon depends upon its porosity and surface area. The surface area in excess of 2000 m^2/g and dry weight density of 500 kg/m^3 can be achieved. The spent carbon is regenerated in special furnace by similar activation process. The adsorbed organic impurities are burnt off. Approximately 4–10% overall loss of carbon and reduced adsorptive capacity after regeneration is expected with each cycle; thus new material must be added.

Adsorption Mechanism: The large surface area enhances the physical adsorptive properties of carbon. The water borne solute molecules (*adsorbates*) pass through the maze of rough surfaces and irregular channels of activated carbon or *adsorbent*. The molecules are removed by *van der Waals* forces (physical adsorption) and by very weak bonding between adsorbent and adsorbate. For this reason, the adsorbate is approximately at equilibrium with the solute concentration in solution. Any change in solute concentration, therefore, affects the amount of chemicals that can be adsorbed.

The major considerations in selection of activated carbon are (1) adsorptive capacity, (2) the rate of adsorption, and (3) the form of activated carbon (PAC or GAC). Both PAC and GAC are marketed with specific adsorptive properties, and are available in different sieve-sizes.

Powdered Activated Carbon (PAC): PAC may be added in an aeration basin to achieve simultaneous biological oxidation and physical adsorption. Some of the benefits of PAC addition include (1) process stability during shock loadings, (2) reduction of refractory priority pollutants, (3) color and ammonia removal, and (4) improved sludge settleability. Addition of PAC may also improve the removal of some emerging constituents that are important for reuse of reclaimed water (Section 12.2.1). The PAC dose may range from 20 to 200 mg/L. PAC is also applied in the secondary effluent in a batch or continuous flow reactor. Tertiary coagulation/clarification or filtration is normally required to remove fine PAC particles.

The adsorption isotherms are normally developed to determine PAC dose needed for the desired removal. The *Freundlich* and *Langmuir isotherms* are commonly used.[86,87] The Freundlich isotherm and its linear corollary are expressed by Equations 15.15a and 15.15b. Similarly, the Langmuir isotherm and its linear relationships are given by Equations 15.15c through 15.15e.

Freundlich isotherm

$$\frac{X}{M} = kC^{1/n} \tag{15.15a}$$

$$\ln\left(\frac{X}{M}\right) = \frac{1}{n}\ln(C) + \ln(k) \tag{15.15b}$$

Langmuir Isotherm

$$\frac{X}{M} = \frac{abC}{1 + bC} \tag{15.15c}$$

$$\frac{C}{X/M} = \left(\frac{1}{a}\right)C + \frac{1}{ab} \tag{15.15d}$$

$$\left(\frac{1}{X/M}\right) = \left(\frac{1}{ab}\right)\left(\frac{1}{C}\right) + \frac{1}{a} \tag{15.15e}$$

where

X = adsorbate adsorbed by adsorbent ($X = C_0 - C$), mg adsorbate/L

M = carbon dose (g/L) as activated carbon

X/M = mass of adsorbate adsorbed per unit mass of adsorbent (mg adsorbate/g) activated carbon
C_0 = initial concentration of adsorbate, mg/L
C = final or equilibrium concentration of the adsorbate, mg/L
k = empirical constant or Freundlich capacity factor (mg adsorbate/g) activated carbon. The $\ln(k)$ is the Y-intercept when $\ln(X/M)$ is plotted with respect to $\ln(C)$.
$1/n$ = slope constant or Freundlich intensity parameter, dimensionless. It is the slope on the plot of $\ln(X/M)$ versus $\ln(C)$.
a = maximum mass of adsorbate adsorbed per unit mass of adsorbent (mg adsorbate/g) activated carbon
b = empirical constant (L/mg) activate carbon

Note: k and $1/n$ for chlorobenzene are 91 and 0.99, respectively. Both constants vary greatly for different organic compounds. These values for many compounds may be found in References 36 and 54.

The adsorption isotherms are experimentally developed by exposing a given concentration of adsorbate in a fixed volume of liquid to varying amounts of activated carbon. Typically, 10 or more batch reactors are used with different concentrations of activated carbon. The minimum time allowed for the samples to equilibrate is around 7 h for PAC. At the end of the test period, the concentration of adsorbate remaining is measured. The linear plots are developed and adsorption coefficients are determined. The procedure is given in Example 15.33.

Granular Activated Carbon (GAC): GAC is produced and regenerated in the same manner as PAC, and has similar adsorptive properties. GAC columns are often used instead of the conventional sand filter. Also, they are used in a postfiltration column. A continuous column study is normally needed to evaluate the feasibility and economics of carbon adsorption, and to develop the design parameters. The design parameters include the adsorption rate, area and depth of the carbon column, flow rate, run time, head loss, sieve size, service time, and regeneration frequency. The effectiveness of GAC in removing priority compounds can be express by the Freundlich and Langmuir isotherms discussed previously. Other important relationships for removal of contaminants in a GAC column are presented below.

Bohart–Adams Model: The Bohart–Adams model incorporates the reaction rate theory to predict the performance of a continuous flow GAC column. It is extensively applied for continuous flow carbon adsorption column design.[88] This relationship is expressed by Equation 15.16a.

$$\ln\left(\frac{C_0}{C_e} - 1\right) = \ln\left(e^{KN_0D/v_L} - 1\right) - KC_0t_B \tag{15.16a}$$

where

C_0 = influent concentration of adsorbate, mg/L or g/m³ (lb/ft³)
C_e = breakthrough or allowable concentration of adsorbate in the effluent, mg/L or g/m³ (lb/ft³)
K = rate constant, m³/g·h, (ft³/lb·h)
N_0 = adsorptive capacity, g adsorbate/m³ activated carbon (lb adsorbate/ft³ activated carbon)
D = depth of GAC bed, m (ft)
v_L = approaching velocity (or hydraulic loading), m/h or m³/m²·h (ft/h or ft³/ft²·h)
t_B = breakthrough (or service) time, h

The term e^{KN_0D/v_L} in most cases is much larger than unity, and Equation 15.16a is simplified and rearranged to give Equation 15.16b.

$$t_B = \frac{N_0}{C_0v_L}D - \frac{1}{KC_0}\ln\left(\frac{C_0}{C_e} - 1\right) \tag{15.16b}$$

A plot of t_B versus D in Equation 15.16b presents a linear relationship. The slope S and the intercept I are expressed by Equations 15.16c and 15.16d.

$$S = \frac{N_0}{C_0 v_L} \quad \text{or} \quad N_0 = C_0 v_L S \tag{15.16c}$$

$$I = \frac{1}{KC_0}\ln\left(\frac{C_0}{C_e} - 1\right) \quad \text{or} \quad K = \frac{1}{IC_0}\ln\left(\frac{C_0}{C_e} - 1\right) \tag{15.16d}$$

The critical depth of the bed is obtained from Equation 15.16e when $t_0 = 0$.

$$D_0 = \frac{v_L}{KN_0}\ln\left(\frac{C_0}{C_e} - 1\right) \tag{15.16e}$$

where D_0 = critical depth of GAC bed, m (ft)

The rate constant K and the adsorptive capacity N_0 are determined from a column test. Once the relationship is developed, this equation can be used to predict the service time and critical bed depth. Typical laboratory equipment recommended for this test is shown in Figure 15.21, and the laboratory procedure is explained below.[49,68,89,90]

1. Set up three or four carbon columns of known diameter in series. A set up of three columns is shown in Figure 15.21.
2. Select the adsorbate and prepare an influent feed solution of known concentration.
3. Start a constant flow rate through the first column and record time.
4. As the concentration of adsorbate reaches the predetermined breakthrough concentration, the time and depth of column are recorded.
5. The second column is added in series and the experiment is continued.
6. Finally, when the last column in series reaches the breakthrough, the first set of test is completed.
7. The columns are replaced with fresh GAC, and the experiment is repeated at a different flow rate while the feed concentrations are kept the same as that in the first column test.
8. Steps 3 through 6 are repeated to complete the second, third, and fourth rounds of test data (or as many flow rates are tested).

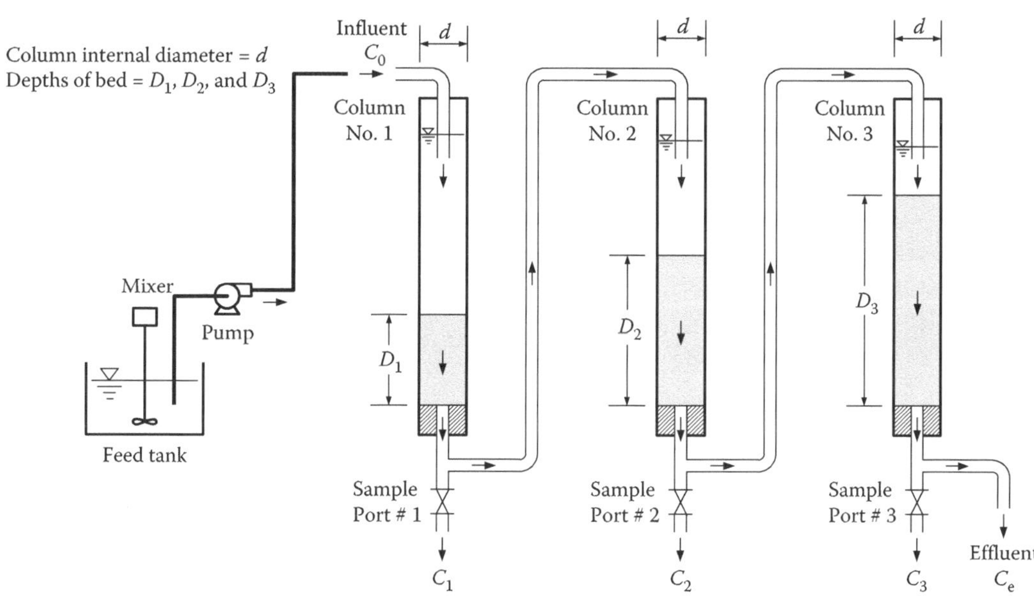

FIGURE 15.21 Equipment set up to develop design data from GAC column.

9. The resulting data give C_0, C_e, and V with time and depths of column. A plot of time versus depth (Equation 15.16b) produces straight lines for each flow rate.
10. The slope and the intercept of each line are obtained, and from them the constants K, N_0, and D_0 are calculated. Plotting these constants against the flow rate produces curves used for full-scale design of the carbon filter facility. Examples 15.35 and 15.36 illustrate GAC design procedures using the Bohart–Adams model.

Mass–Balance Relationship: A simplified mass–balance relationship for GAC column under steady-state condition is also used to design a GAC column. The mass–balance relationship is given by Equation 15.17.

$$\text{Accumulation} = \text{inflow} - \text{outflow} - \text{amount adsorbed} \tag{15.17a}$$

$$0 = QC_0t_B - QC_et_B - Mq_e \tag{15.17b}$$

$$\frac{M}{Qt_B} = \frac{M}{V_B} = \frac{(C_0 - C_e)}{q_e} \tag{15.17c}$$

$$\frac{M}{Qt_B} = \frac{M}{V_B} = \frac{C_0}{q_e} = \frac{C_0}{k(C_0)^{1/n}} \quad (\text{If } C_0 \gg C_e) \tag{15.17d}$$

$$q_e = k_m(C_0)^{1/m} \quad (\text{If } C_0 \gg C_e) \tag{15.17e}$$

$$M = EVB \times \rho_{GAC} \tag{15.17f}$$

$$EBV = Q \times EBCT \quad \text{or} \quad EBV = \frac{M}{\rho_{GAC}} \quad (\text{If } C_0 \gg C_e) \tag{15.17g}$$

$$V_B = Qt_B \quad \text{or} \quad V_B = \frac{Mq_e}{C_0} \quad \text{or} \quad V_B = \frac{M\,k_m\,(C_0)^{1/m}}{C_0} \quad (\text{If } C_0 \gg C_e) \tag{15.17h}$$

$$t_B = \frac{V_B}{Q} \tag{15.17i}$$

where

Q = flow rate, m^3/h (m^3/min)
M = mass of adsorbent (kg) activated carbon. The ratio of $M/Q\,t$ or M/V_B is the virtual GAC concentration in the volume of effluent treated per cycle in the unit of g/L or kg/m^3 of activated carbon.
V_B = breakthrough volume per cycle, m^3
q_e = mass of adsorbate adsorbed per unit mass of adsorbent (g adsorbate/kg) activated carbon. Also, $q_e = X/M$ (Equation 15.15a).
k_m = mass-balance-based rate constant, $(mg/g)(L/mg)^{1/m}$ or $(g/kg)(L/mg)^{1/m}$
$1/m$ = constant, dimensionless
EBV = empty bed volume that is occupied by the GAC, m^3
ρ_{GAC} = bulk density of GAC (g/L or kg/m^3) activated carbon. The typical range of ρ_{GAC} is 350–550 kg/m^3.
$EBCT$ = empty bed contact time, h (min)

Other variables have been defined previously in Equation 15.16.

Thomas Equation: A simplified equation is given by Thomas for an adsorption GAC column. It is given by Equation 15.18a.[91,92]

$$\frac{C_e}{C_0} = \frac{1}{1 + e^{k_1/Q(q_0M - C_0V_B)}} \quad \text{or} \quad e^{k_1/Q(q_0M - C_0V_B)} = \frac{C_0}{C_e} - 1 \tag{15.18a}$$

where

k_1 = rate constant, L/(mg adsorbate)·h or m^3/(g adsorbate)·h

q_0 = maximum solid-phase concentration of adsorbate adsorbed by adsorbent (g adsorbate/kg) activated carbon

Other variables have been defined previously in Equations 15.16 and 15.17.

The natural logarithms of both sides of Equation 15.18a give a linear equation (Equation 15.18b).

$$\ln\left(\frac{C_0}{C_e} - 1\right) = -\frac{k_1 C_0}{Q} V_B + \frac{k_1 q_0 M}{Q} \tag{15.18b}$$

A plot of $\ln((C_0/C_e) - 1)$ versus V_B gives a straight line. The slope (S) and intercept (I) of the line are expressed by Equations 15.18c and 15.18d.

$$S = -\frac{k_1 C_0}{Q} \tag{15.18c}$$

$$I = \frac{k_1 q_0 M}{Q} \tag{15.18d}$$

Scale-Up Approach: A simplified method was developed by Fornwalt and Hutchins for design of adsorption column.[93,94] A simple bench-scale test is needed to develop a breakthrough curve. The assumptions are that the breakthrough curve, liquid flow rate in terms of EBV per unit time, contact time, and the volume of liquid treated per unit mass of carbon for experimental and full-scale units are the same. Based on these assumptions several proportional relationships have been developed. These relationships are given by Equation 15.19.

$$q_b = \frac{Q}{EBV} \quad \text{or} \quad q_b = \frac{V_B}{EBVt_B} \tag{15.19a}$$

$$EBV = \frac{Q}{q_b} \quad \text{or} \quad EBV = \frac{V_B}{q_b t_B} \tag{15.19b}$$

$$V'_B = \frac{V_B}{M} \quad \text{or} \quad V'_B = \frac{Q}{M_t} \tag{15.19c}$$

$$M_t = \frac{Q}{V'_B} \quad \text{or} \quad M_t = \frac{QM}{V_B} \tag{15.19d}$$

$$t_B = \frac{V_B}{Q} \quad \text{or} \quad t_B = \frac{M}{M_t} \tag{15.19e}$$

$$V_B = Qt_B \quad \text{or} \quad V_B = q_b EBVt_B \tag{15.19f}$$

where

q_b = liquid flow rate in terms of fraction of EBV per unit time (h^{-1}). q_b is established based on breakthrough volume, the solute concentration, the maximum solid-phase concentration, and other data. Usually, the value of q_b is 0.2–3 h^{-1}.

V'_B = volume of liquid treated per unit mass of the adsorbent (m^3/kg) activated carbon

M_t = mass of adsorbent consumed per h, kg activated carbon/h

The other variables have been defined previously in Equations 15.16 and 15.17.

If the calculated breakthrough time t_B from Equation 15.19e or the calculated breakthrough volume V_B from Equation 15.19f are not within the acceptable range, another flow rate Q_b should be determined from the breakthrough curve. Another breakthrough test is conducted using new Q_b and process is repeated. The procedure is given in Example 15.39.

Moving-Bed Countercurrent Adsorption Column: The moving-bed adsorption column also called *pulsed-bed column* is a steady-state countercurrent operation. In this process, the adsorbent solids move downwards through the column while liquid moves upward. It is a common process for advanced wastewater treatment. The material balance relationship is expressed by Equation 15.20a.[92]

$$Q(C_0 - C_e) = M_s(X_m - X_i) \quad \text{or} \quad \frac{M_s}{Q} = \frac{C_0 - C_e}{X_m - X_i} \tag{15.20a}$$

where

Q = liquid flow rate, m^3/h

M_s = adsorbent flow rate (kg activated carbon/h). The ratio of M_s/Q is the carbon consumption in kg activated carbon per m^3 wastewater treated.

C_0 = influent concentration of adsorbate at the bottom of the column, mg/L (g/m^3)

C_e = effluent concentration of adsorbate at the top of the column, mg/L (g/m^3)

X_m = maximum solid-phase concentration of adsorbate adsorbed by adsorbent at the bottom the column (g adsorbate/kg) activated carbon. $X_m = q_0$ from Equation 15.18a.

X_i = initial solid-phase concentration of adsorbate adsorbed by adsorbent in the column (g adsorbate/kg) activated carbon. Typically, $X_i \approx 0$.

The minimum depth of the countercurrent adsorption column is obtained from the breakthrough curve. It is expressed by Equation 15.20b.[92]

$$Z_S = Z\left(\frac{V_T - V_B}{V_T - 0.5(V_T - V_B)}\right) \tag{15.20b}$$

where

Z_s = length of adsorption zone, m

Z = column depth, m

V_T = throughput volume to reach $C_e = 0.95\ C_0$ when the column is exhausted, m^3 (L)

V_B = breakthrough volume to reach the maximum allowable concentration in the effluent which is normally at $C_e = 0.05\ C_0$, m^3 (L)

Fluidized Bed: The advanced wastewater treatment is also achieved in a fluidized bed reactor. The liquid flows upward at a sufficient velocity to suspend the GAC. At the top of the column a distinct interface develops between the solids and the liquid. It is normally assumed that there is no interparticle contact and the bed operates without clogging. Usually, these beds operate as a continuous countercurrent column. The ratio of adsorbent required per unit volume of liquid (M_s/Q) is the same as that of a moving-bed countercurrent adsorption column.[95]

Empty Bed Contact Time Method: A more empirical approach to design a GAC column for removal of priority compounds is the *empty bed contact time* (EBCT) method. This method can be very inexpensive, yet still provide enough information to size and select carbon filters or columns. The basic design parameters are bed depth and approaching velocity (or hydraulic loading). Other parameters can also influence results. For a given short EBCT, a deep bed at high velocity removes organics better than shallow bed at low velocity.[96,97] The EBCT is calculated from Equation 15.21.

$$EBCT = \frac{D}{v_L} \quad \text{or} \quad EBCT = \frac{EBV}{Q} \tag{15.21}$$

All variables have been defined previously in Equations 15.16 and 15.17.

This equation is often used as the basis for determining contaminant removal and GAC regeneration rates. Example 20.41 shows a simple application of the EBCT method. Langmuir equations may also be applied to the EBCT to develop relationships.

Design Considerations of GAC Column: Experimental design of laboratory-scale and pilot testing of GAC column, and design of full-scale column requires considerable experience and expertise. Basic design considerations are listed below.[37,98]

1. GAC manufacturers have numerous publications on laboratory techniques. Side-wall effects due to channeling have significant effects on column operation. To reduce the side-wall effect both approaching velocity and column diameter are important. The maximum rates for different column diameters are as follows. The laboratory column should at least be 2.5 cm (1 in) in diameter and 60-cm (24-in) tall.

ID, cm (in)	Approaching Velocity, m/h or L/s·m^2 (gpm/ft^2)
4.5 (1¾)	1.2 or 0.34 (0.5)
7.0 (2¾)	2.4 or 0.68 (1.0)
9.5 (3¾)	3.6 or 1.02 (1.5)

2. The columns are unpressurized or pressurized, and downflow or upflow. However, downflow is more common.
3. The typical GAC size is 8 × 30 ·mesh in downflow column, and 12 × 40 mesh in upflow column, respectively.
4. The depth of GAC bed is 1.8–12 m (6–40 ft). Most common depth is in the range of 1.8–4.8 m (6–16 ft).
5. The diameter of the GAC bed ranges from 75 cm (30 in) to 6 m (20 ft). The depth to diameter ratio is 1.5:1 to 4:1.
6. The approaching velocity is 5–15 m/h or 1.4–4.2 L/s·m^2 (2–6 gpm/ft^2). The head loss is 2.3 m per meter (2.3 ft/ft) bed depth.
7. The effective contact time ranges from 2 to 10 min with an EBCT of 5–30 min. The operating cycle may range 100–600 d.
8. Backwash rate for downflow fixed beds with 8 × 30 mesh is 24–48 m/h or 6.8–13.5 L/s·m^2 (10–20 gpm/ft^2) and bed expansion is 25–500%. The backwash duration is 10–15 min. The backwash rate of upflow column (12 × 40 mesh) is 24–30 m/h or 6.8–8.5 L/s·m^2 (10–12.5 gpm/ft^2) and bed expansion is 10–15% and the backwash duration is 10–15 min.
9. The suspended solids blind the active sites of an activated carbon. It is therefore essential that the influent should have TSS concentration < 5 mg/L.
10. The filtered effluent from a POTW contains numerous organic compounds. For this reason, the design parameters are expressed in common terms. The examples are hydraulic loading in m/h, m^3/m^2·h or L/s·m^2 (gpm/ft^2), EBCT in min, organic (COD or TOC) loading in kg/kg (lb/lb), and effluent COD or TOC concentration in mg/L. For advanced treatment of filtered effluent, hydraulic loading rate = 5 to 15 m/h (2 to 6 gpm/ft2), EBCT = 30–45 min, organic loading = 0.2–0.8 kg COD/kg activated carbon (lb COD/lb activated carbon), and effluent COD = 0.5–15 mg/L.

EXAMPLE 15.33: DETERMINATION OF EMPIRICAL CONSTANTS IN FREUNDLICH AND LANGMUIR ISOTHERMS FOR PAC

A bench-scale study was conducted to determine the empirical constants in Freundlich and Langmuir equations. Filtered secondary effluent containing 40 μg/L of chlorobenzene was used. Effluent samples were placed in nine standard 2-L *jar testing* apparatus. A different dose of PAC was applied in each container. After 5 h of slow stirring the content was settled. The supernatant was centrifuged and analyzed to determine the equilibrium concentration of chlorobenzene. The results are given below. Determine (a) the empirical constants k and n in Freundlich equation, and a and b in Langmuir equation to describe the adsorption isotherm, and (b) the PAC dose required to achieve a chlorobenzene concentration of 0.5 μg/L if the influent concentration is 50 μg/L.

Carbon dose (M), mg/L	0	0.05	0.5	1	2	4	10	20	30
Equilibrium concentration (C), mg/L	40	35.10	15.85	9.50	4.91	2.41	0.95	0.45	0.34

Solution

1. Calculate the variables used for plotting the Freundlich and Langmuir isotherms.
   The ratios $\ln(C)$, $\ln(X/M)$, $C/(X/M)$, $1/C$, and $1/(X/M)$ are developed from the experimental data and are summarized in Table 15.16.

TABLE 15.16 Reduced Experimental Data for Plotting the Isotherms (Example 15.33)

j	C_0, µg/L	C, µg/L	$X = C_0 - C$, µg/L	X/M, µg/mg	Freundlich Isotherm		Langmuir Isotherm		
					$\ln(C)$	$\ln(X/M)$	$C/(X/M)$, mg/L	$1/C$, L/µg	$1/(X/M)$, mg/µg
0.05	40	35.10	4.90	98.0	3.56	4.58	0.36	0.028	0.010
0.5	40	15.85	24.15	48.3	2.76	3.88	0.33	0.063	0.021
1	40	9.50	30.50	30.5	2.25	3.42	0.31	0.11	0.033
2	40	4.91	35.09	17.5	1.59	2.86	0.28	0.20	0.057
4	40	2.41	37.59	9.4	0.88	2.24	0.26	0.41	0.11
10	40	0.95	39.05	3.91	−0.05	1.36	0.24	1.05	0.26
20	40	0.45	39.55	1.98	−0.8	0.68	0.23	2.22	0.51
30	40	0.34	39.66	1.32	−1.08	0.28	0.26	2.94	0.76

Note: 1 µg/mg = 1 mg/g.

2. Determine the empirical constants from isotherm plots.
   a. Freundlich adsorption isotherm.
      The plot of $\ln(X/M)$ versus $\ln(C)$ gives a linear relationship (Figure 15.22a). The Y-intercept or $\ln(k) = 1.37$, and the slope or $1/n = 0.913$. Therefore, $k = e^{1.37} = 3.94$, and $n = (1/0.913) = 1.10$.
   b. Langmuir adsorption isotherm.
      i. A plot of $C/(X/M)$ versus C (Figure 15.22b(i)) does not produce an ideal linear trend. However, a rough approximation gives a Y-intercept or $1/ab = 0.253$, and a slope or $1/a = 0.0035$. Solve for $ab = (1/0.253) = 3.95$, $a = (1/0.0035) = 286$, and $b = (1/0.253a) = (1/0.253 \times 286) = 0.0138$.
      ii. A plot of $1/(X/M)$ verses $1/C$ (Figure 15.22b(ii)) gives a clear linear relationship and much reliable values of the constant a and b are obtained. From this relationship Y-intercept $1/a = 0.0036$, and slope $1/ab = 0.247$.

$$\text{Solve for } a = \frac{1}{0.0036} = 278,\ ab = \frac{1}{0.247} = 4.045, \text{ and}$$

$$b = \frac{1}{0.247a} = \frac{1}{0.247 \times 278} = 0.0146.$$

These constants are used later to develop the Langmuir equation.

Note: Many times the plots may not show a clear linear trend on account of interference by presence of other compounds, inaccuracies in measurement, uncontrolled variables, and other causes. The results must be interpreted to represent the best situation.

3. Develop the Freundlich and Langmuir equations based on the empirical constants obtained in Step 2.

$$\text{Freundlich equation (Equation 15.15a): } \frac{X}{M} = 3.94C^{1/1.10} \quad \text{or} \quad \frac{X}{M} = 3.94C^{0.913}$$

$$\text{Langmuir equation (Equation 15.15c): } \frac{X}{M} = \frac{4.05C}{1 + 0.0146C}$$

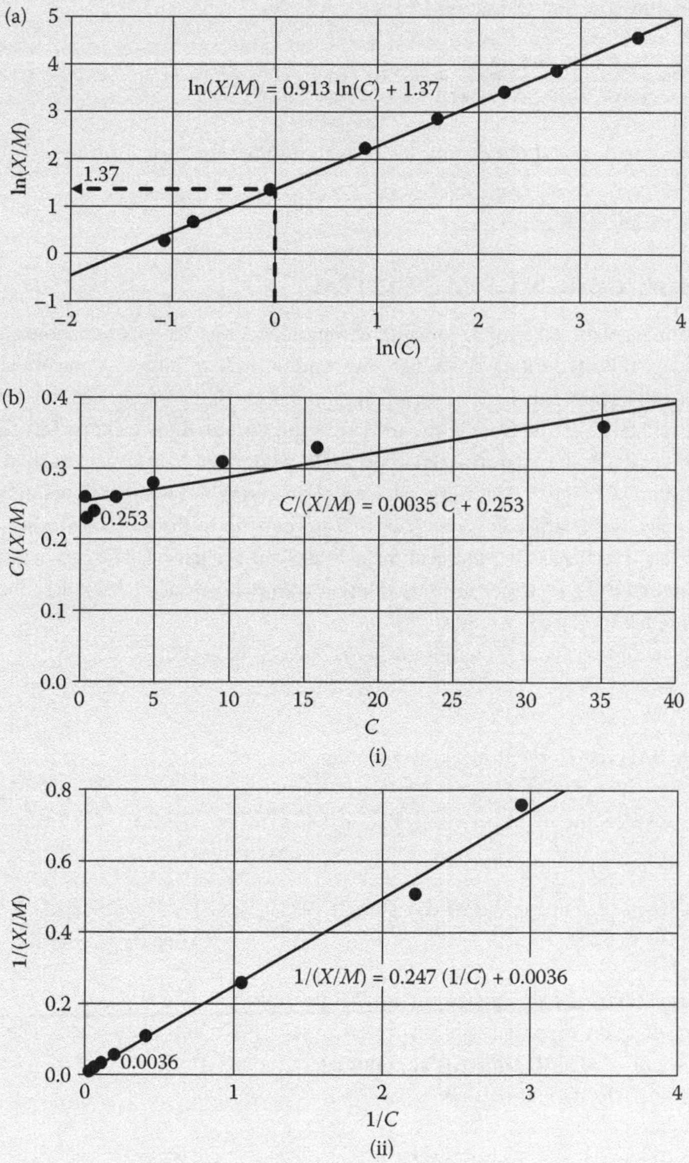

FIGURE 15.22 Plots of experimental data: (a) Freundlich adsorption isotherm plot: $\ln(X/M)$ versus $\ln(C)$; and (b) Langmuir adsorption isotherm plot: (i) $C/(X/M)$ versus C, and (ii) $1/(X/M)$ versus $1/C$ (Example 15.33).

4. Determine the carbon dose (M) if $C_0 = 50\ \mu g/L$ and $C_e = 0.5\ \mu g/L$.

$$X = C_0 - C_e = (50 - 0.5)\ \mu g/L = 49.5\ \mu g/L$$

Apply Freundlich equation (Equation 15.15a),

$$\frac{49.5\ \mu g/L}{M} = 3.94 \times (0.5\ \mu g/L)^{0.913} = 2.09 \quad \text{or} \quad M = 23.7\ mg/L$$

Apply Langmuir equation (Equation 15.15c),

$$\frac{49.5\ \mu g/L}{M} = \frac{4.05 \times 0.5\ \mu g/L}{1 + 0.0146 \times 0.5\ \mu g/L} = 2.01 \quad \text{or} \quad M = 24.6\ mg/L$$

Note: The carbon doses estimated from two equations are very close. Provide a design carbon dose of 25 mg/L.

EXAMPLE 15.34: DESIGN OF PAC SYSTEM

A wastewater treatment plant is serving a residential community. A toxicity test evaluation established toxicity level in the secondary effluent during wet weather flow conditions. A *toxicity reduction evaluation* (TRE) was conducted. This study involved *toxicity investigation examination* (TIE) which confirmed that initial storm water contained residual pesticides from lawns and landscaping areas. An *effluent toxicity treatability* (ETT) study was carried out under the corrective action plan. The recommended corrective action plan included a maximum application of 12 mg/L PAC in the aeration basin during wet weather flow conditions. The maximum weekly-average wet weather flow of 40,000 m^3/d occurs up to three events during a 90-day period annually. Calculate (a) total PAC requirement for a 90-day storage period. The carbon slurry is prepared at a concentration of 120-kg PAC per m^3 after dilution with plant effluent. Also, describe the equipment and components of the PAC feed system.

Note: The subject of toxicity and biomonitoring is covered in Section 5.7.

Solution

1. Determine the PAC requirement for 90-day storage.

$$\text{Maximum PAC requirement per event, } W_{event} = 12\ g/m^3 \times 40{,}000\ m^3/d \times 7\ d/event \times 10^{-3}\ kg/g = 3360\ kg/event$$

$$\text{Total PAC requirement for the 90-day storage, } W_{storage} = 3W_{event} = 3\ \text{events} \times 3360\ kg/event = 10{,}080\ kg \text{ for a 90-day storage period}$$

2. Determine the volume of a slurry tank for the 90-day storage.

The volume of slurry expands slightly after mixing the PAC with plant water. The volume ratio of dilution water and PAC in the slurry is approximately 1:1.2.

Volume of slurry for the 90-day storage,

$$V_{storage} = \frac{W_{storage}}{120\ kg/m^3} \times 1.2 = \frac{10{,}080\ kg}{120\ kg/m^3} \times 1.2 = 100.8\ m^3 \approx 100\ m^3$$

Provide a slurry tank 5 m × 5 m × 4.4 m (deep) The volume of the tank $V_{storage} = 110\ m^3$.

The storage volume of 110 m^3 gives an excess capacity of 10% over the 90-day storage requirement.

3. Describe the equipment and components for PAC feed system.
   a. *PAC bulk slurry storage tank*: The slurry tank has a level sensor with transmitter to monitor volume, and to activate the high- and low-level alarms for extreme conditions. Slurry mixer will have blades at two levels. It will operate at 30–50 rpm to create a vortex for keeping the slurry in suspension. The motor contacts will provide the status of mixer.
   b. *Carbon delivery system*: The PAC will be delivered pneumatically from a pressure differential (PD) truck. An air compressor will fluidize the PAC and deliver it at the vortex in the slurry

tank at a constant rate. Air scrubbers (both wet and dry) will provide carbon dust removal. Pressure indicators will monitor suction pressure, pressure loss across the filter, and the motor contact will monitor status of the fan.

c. *Air sparger for resuspension of PAC*: The settled PAC in the slurry tank will need a high pressure sparger system at the bottom to resuspend the settled PAC. A timer should start the mixer several minutes after air sparging.
d. *Recirculating pumps*: Two recirculation pumps should be provided for recirculation of the slurry. Progressive cavity pumps are the best suited for abrasive material such as PAC. Provide flushing water connections for cleaning the lines and pumps.
e. *Volumetric liquid feeder*: A rotating dipper type feeder with adjustable feed rate will transfer the required slurry volume by gravity into an adjacent constant-level tank.
f. *Constant-level tank*: A 400-L constant-level tank equipped with electric mixer will be provided. A float valve will maintain the liquid level in the tank.
g. *PAC feed pump*: Three variable speed progressive cavity pumps will be provided to feed the slurry in aeration basin. One pump will normally be in service, the second pump will be added for peak dosing, and the third pump will be a standby unit.

EXAMPLE 15.35: DEVELOPMENT OF BOHART–ADAMS EQUATIONS FROM GAC COLUMN TEST DATA

A continuous-flow bench-scale carbon adsorption study was conducted to develop the design data for full-scale post filtration GAC columns. A synthetic organic chemical (SOC) was to be removed from filtered effluent of an industrial plant. The Bohart–Adams model was used to develop the design data. The experimental setup included three 2.5-cm column arranged in series. The influent feed contained 25 mg/L SOC throughout the study. The maximum allowable concentration of SOC in the effluent was 1 mg/L. The temperature, pH, and other factors were relatively constant during the test. Four different approaching velocities (or hydraulic loading rates) were tested. The experimental results are summarized below. Prepare the necessary plots, and determine the relationships of K, N_0, and D_0 versus v_L.

Test Number	Approaching Velocity (v_L), $m^3/m^2 \cdot min$ (gpm/ft^2)	Bed Depth (D), m (ft)	Breakthrough Time (t_B), h	Breakthrough Volume (V_B), m^3 (gal)
1	0.081 (2)	$D_1 = 0.61$ (2)	600	1.48 (391)
		$D_2 = 1.22$ (4)	1520	3.77 (996)
		$D_3 = 1.83$ (6)	2450	6.07 (1604)
2	0.163 (4)	$D_1 = 1.22$ (4)	430	2.13 (563)
		$D_2 = 2.44$ (8)	1110	5.50 (1453)
		$D_3 = 3.66$ (12)	1780	8.82 (2330)
3	0.326 (8)	$D_1 = 1.52$ (5)	180	1.78 (470)
		$D_2 = 3.05$ (10)	530	5.25 (1387)
		$D_3 = 4.57$ (15)	900	8.92 (2357)
4	0.652 (16)	$D_1 = 1.52$ (5)	70	1.39 (367)
		$D_2 = 4.57$ (15)	430	8.52 (2251)
		$D_3 = 7.62$ (25)	800	15.85 (4188)

Note: $1\ m^3/m^2 \cdot min = 24.5\ gpm/ft^2$ $1\ m = 3.28\ ft$ $1\ m^3 = 264\ gal$.

Solution

1. Plot the breakthrough time (t_B) versus bed depth (D) for four approaching velocities (v_L) in Figure 15.23.

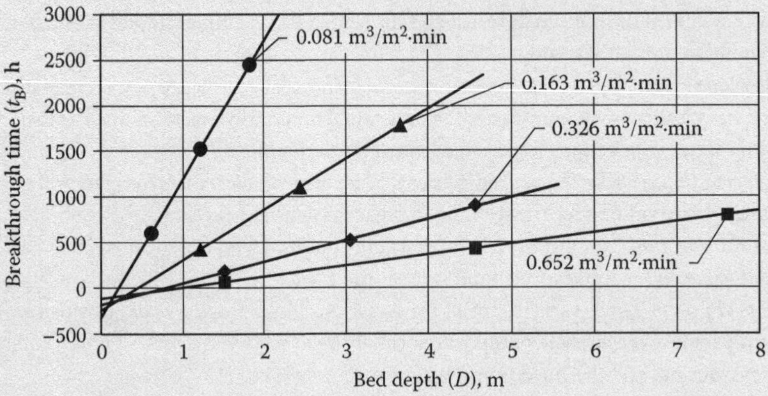

FIGURE 15.23 Plot of breakthrough time versus bed depth of a GAC column for four approaching velocities in Bohart–Adams model (Example 15.35).

2. Determine the intercept (I) and slope (S) for each of four velocities (v_L).

   Apply the Excel linear trendline function to the data (Figure 15.23), to obtain the values of I and S for each velocity of v_L, and summarize the results in the table below:

Test Number	Approaching Velocity (v_L)		Intercept (I), h	Slope (S), h/m
	$m^3/m^2{\cdot}min$	m/h		
1	0.081	4.86	327	1516
2	0.163	9.78	243	553
3	0.326	19.6	183	236
4	0.652	39.1	114	120

3. Determine the N_0, K, and D_0 of GAC column from the values of I and S.

   a. Determine the N_0, K, and D_0 of GAC column at approaching velocity $v_L = 0.081\ m^3/m^2{\cdot}min = 4.86\ m/h$.

      i. N_0 from Equation 15.16c.

$$N_0 = C_0 v_L S = 25\ g/m^3 \times 4.86\ m/h \times 1516\ h/m = 184{,}000\ g/m^3 \quad \text{or} \quad 184\ kg/m^3$$

      ii. K from Equation 15.16d.

$$K = \frac{1}{IC_0}\ln\left(\frac{C_0}{C_e} - 1\right) = \frac{1}{327\ h \times 25\ g/m^3} \times \ln\left(\frac{25\ g/m^3}{1\ g/m^3} - 1\right)$$
$$= 0.000389\ m^3/g{\cdot}h \quad \text{or} \quad 0.389\ m^3/kg{\cdot}h$$

      iii. D_0 from Equation 15.16e.

$$D_0 = \frac{v_L}{KN_0}\ln\left(\frac{C_0}{C_e} - 1\right) = \frac{4.86\ m/h}{0.000389\ m^3/g{\cdot}h \times 184{,}000\ g/m^3} \times \ln\left(\frac{25\ g/m^3}{1\ g/m^3} - 1\right)$$
$$= 0.22\ m$$

b. Determine the N_0, K, and D_0 of GAC column at each of other approaching velocities.
Similarly, calculate N_0, K, and D_0 of GAC column for tests 2, 3, and 4. These values are summarized in Table 15.17.

TABLE 15.17 Summary of N_0, K, and D_0 in Bohart–Adams Equations for GAC Columns in Four Test Runs (Example 15.35)

Test Number	Approaching Velocity (v_L)		Constant in Bohart–Adams Model		
	m^3/m^2·min (gpm/ft²)	m/h (ft/h)	N_0, kg/m³ (lb/ft³)	K, m³/kg·h (ft³/lb·h)	D_0, m (ft)
1	0.081 (2)	4.86 (16)	184 (11.5)	0.389 (6.23)	0.22 (0.72)
2	0.163 (4)	9.78 (32)	135 (8.43)	0.523 (8.38)	0.44 (1.44)
3	0.326 (8)	19.6 (64)	116 (7.24)	0.695 (11.1)	0.77 (2.53)
4	0.652 (16)	39.1 (128)	117 (7.30)	1.12 (17.9)	0.95 (3.12)

Note: 1 m³/m²·min = 24.5 gpm/ft²; 1 m/h = 3.28 ft/h; 1 kg/m³ = 0.0624 lb/ft³; 1 m³/kg·h = 16.02 ft³/lb·h; 1 m = 3.28 ft.

4. Prepare the plots of N_0, K, and D_0 versus approaching velocity (v_L) from four tests in Figure 15.24.

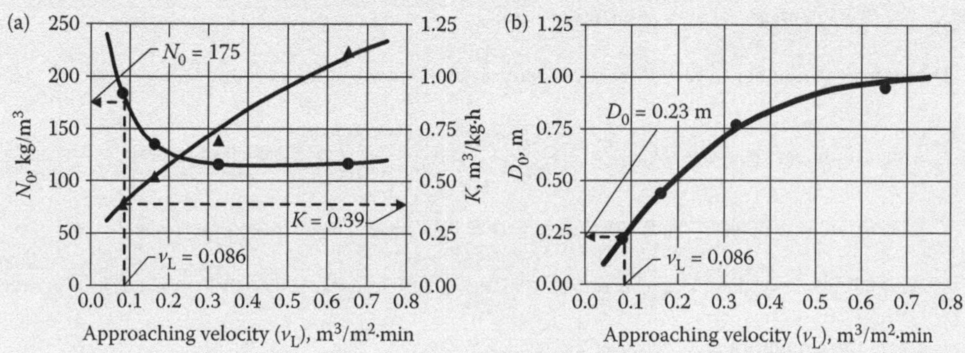

FIGURE 15.24 Plots of empirical constant versus approaching velocity (v_L) in Bohart–Adams model: (a) N_0 and K, and (b) D_0 (Example 15.35).

EXAMPLE 15.36: GAC BED DESIGN BASED ON CONSTANTS IN BOHART–ADAMS MODEL

Experimental data on GAC are developed in Example 20.35 to establish its performance on removal of a SOC. The Bohart–Adams plots of N_0, K, and D_0 for the target SOC concentrations of $C_0 = 25$ and $C_e = 1$ mg/L are shown in Figure 15.24. The GAC tested above is used to design an activated carbon bed of 0.75 m in diameter and 1.5-m deep. The influent and effluent concentrations of the target SOC are 25 mg/L and 1 mg/L. The flow rate through the GAC bed is 55 m³/d. Determine (a) the approaching velocity through the GAC bed, (b) the design values of N_0, K, and D_0, (c) service time per cycle, (d) number of carbon changes required per year, and annual carbon volume requirement, (e) amount of SOC (solute) removed per cycle, (f) adsorption capacity available per cycle, and (g) adsorptive efficiencies based on N_0 and D_0. Also, evaluate the potential error by assuming $e^{KN_0D/v_L} - 1 \approx e^{KN_0D/v_L}$.

Solution

1. Determine the approaching velocity through the GAC bed.

Flow rate through the GAC bed, $Q = 55\ \text{m}^3/\text{d} \times \dfrac{\text{d}}{1440\ \text{min}} = 0.038\ \text{m}^3/\text{min}$

Cross-sectional area of the GAC bed, $A = \dfrac{\pi}{4} \times (0.75\ \text{m})^2 = 0.44\ \text{m}^2$

Approaching velocity through the GAC bed, $v_L = \dfrac{Q}{A} = \dfrac{0.038\ \text{m}^3/\text{min}}{0.44\ \text{m}^2} = 0.086\ \text{m/min}$ or $5.2\ \text{m/h}$

2. Determine the values of N_0, K, and D_0 for design of the GAC bed.

Read K, N_0, and D_0 from Figure 15.24 and summarize the values below.

$N_0 = 175\ \text{kg/m}^3$ or $175{,}000\ \text{g/m}^3$, $K = 0.39\ \text{m}^3/\text{kg·h}$ or $0.00039\ \text{m}^3/\text{g·h}$, and $D_0 = 0.23\ \text{m}$.

3. Determine the service time (t_B) from Equation 15.16b at $D = 1.5$ m.

$$t_B = \frac{N_0}{C_0 v_L} D - \frac{1}{KC_0} \ln\left(\frac{C_0}{C_e} - 1\right)$$

$$= \frac{175{,}000\ \text{g/m}^3}{25\ \text{g/m}^3 \times 5.2\ \text{m/h}} \times 1.5\ \text{m} - \frac{1}{0.00039\ \text{m}^3/\text{g·h} \times 25\ \text{g/m}^3} \times \ln\left(\frac{25\ \text{g/m}^3}{1\ \text{g/m}^3} - 1\right)$$

$$= 1693\ \text{h/cycle} \quad \text{or} \quad \sim 70\ \text{d/cycle}$$

4. Determine the number of carbon changes per year and annual carbon volume requirement.

Annual number of cycles, $N_{annual} = \dfrac{365\ \text{d/year}}{t_B} = \dfrac{365\ \text{d/year}}{70\ \text{d/cycle}} = 5.2$ cycles per year

EBV(or volume of GAC bed), $EBV = A \times D = 0.44\ \text{m}^2 \times 1.5\ \text{m} = 0.66\ \text{m}^3/\text{cycle}$

Annual carbon volume, $V_{annual} = N_{annual} \times EBV = 5.2\ \text{cycles/year} \times 0.66\ \text{m}^3/\text{cycle} = 3.4\ \text{m}^3$ per year

5. Estimate the amount of SOC removed per cycle.

The actual SOC concentration during a cycle varies from zero at the start of the cycle to the breakthrough concentration of 1 mg/L at the end of the cycle. Assume the average SOC concentration $C_{e,avg} = 0.5$ mg/L and calculate the amount of SOC removed per cycle.

$$W_{SOC,removal} = t_B Q(C_0 - C_{e,avg}) = 70\ \text{d/cycle} \times 55\ \text{m}^3/\text{d} \times (25 - 0.5)\ \text{g/m}^3 \times 10^{-3}\ \text{kg/g} = 94\ \text{kg/cycle}$$

6. Determine the adsorption capacity available per cycle.

The constant of N_0 is the unit adsorption capacity of GAC, $N_0 = 175\ \text{kg/m}^3$ (Step 2). Calculate the adsorption capacity per cycle based on the volume of GAC bed.

$$W_{SOC,capacity} = N_0 \times EBV = 175\ \text{kg/m}^3 \times 0.66\ \text{m}^3/\text{cycle} = 116\ \text{kg/cycle}$$

7. Determine the adsorptive efficiencies.

a. Adsorption efficiency (E_{N0}) based on the unit adsorption capacity, N_0.

$$E_{N0} = \frac{W_{SOC,removal}}{W_{SOC,capacity}} \times 100\% = \frac{94\ \text{kg/cycle}}{116\ \text{kg/cycle}} \times 100\% = 81\%$$

b. Adsorption efficiency (E_D) based on the critical depth of the bed, $D_0 = 0.23$ m (Step 2).

$$E_D = \frac{D - D_0}{D} \times 100\% = \frac{(1.5 - 0.23)\text{ m}}{1.5\text{ m}} \times 100\% = 85\%$$

Note: The adsorptive efficiency estimated from N_0 or D_0 for the GAC bed is over 80 %. Therefore, the GAC process design is acceptable.

8. Evaluate the potential error by assuming $e^{KN_0D/v_L} - 1 \approx e^{KN_0D/v_L}$.

$$KN_0D/v_L = \frac{0.00039\text{ m}^3/\text{g·h} \times 175{,}000\text{ g/m}^3 \times 1.5\text{ m}}{5.2\text{ m/h}} = 19.7$$

$$e^{KN_0D/v_L} = e^{19.7} = 3.6 \times 10^8 \gg 1$$

Therefore, the potential error by assuming $e^{KN_0D/v_L} - 1 \approx e^{KN_0D/v_L}$ is extremely small, and the assumption is valid.

EXAMPLE 15.37: BREAKTHROUGH TIME OF A GAC BED FROM MASS–BALANCE RELATIONSHIP

A GAC column is used to remove carbon tetrachloride. The concentrations of carbon tetrachloride in the influent and effluent are 1.4 mg/L and 0.005 mg/L, respectively. The flow is 1800 m^3/d. The GAC in the column has density of 450 kg/m^3, and k_m and $1/m$ are 11 (g/kg)(L/mg)$^{0.83}$, and 0.83, respectively. The *EBCT* = 10 min. Determine (a) the amount of carbon required per cycle, (b) mass of carbon tetrachloride adsorbed per unit mass of GAC bed, (c) volume of effluent treated per cycle, and (d) breakthrough time (or useful life) of the GAC bed.

Solution

1. Determine the volume and mass of GAC per cycle.

Flow rate through the GAC bed, $Q = 1800\text{ m}^3/\text{d} \times \dfrac{\text{d}}{1440\text{ min}} = 1.25\text{ m}^3/\text{min}$

Volume of GAC bed (*EBV*) from Equation 15.17g, $EBV = Q \times EBCT = 1.25\text{ m}^3/\text{min} \times 10\text{ min} = 12.5\text{ m}^3\text{ GAC}$

Mass of GAC in the bed from Equation 15.17f, $M = EBV \times \rho_{GAC} = 1.25\text{ m}^3 \times 450\text{ kg/m}^3 = 5625\text{ kg GAC}$

2. Estimate the mass of carbon tetrachloride adsorbed per unit mass of GAC bed.
   Since $C_0 \gg C_e$, apply Equation 15.17e to estimate q_e.

$$q_e = k_m(C_0)^{1/m} = 11\text{ (g/kg)(L/mg)}^{0.83} \times (1.4\text{ mg/L})^{0.83} = 14.5\text{ g carbon tetrachloride/kg GAC}$$

3. Estimate the volume of effluent treated per cycle (V_B).

Apply Equation 15.17h, $V_B = \dfrac{Mq_e}{C_0} = \dfrac{5625\text{ kg} \times 14.5\text{ g/kg}}{1.4\text{ g/m}^3} = 58{,}260\text{ m}^3\text{ GAC per cycle}$

4. Determine the useful life of GAC bed per cycle.

Useful life from Equation 15.17i, $t_B = \dfrac{V_B}{Q} = \dfrac{58{,}260\text{ m}^3/\text{cycle}}{1800\text{ m}^3/\text{d}} = 32\text{ d/cycle} \approx 1\text{ month per cycle}$

EXAMPLE 15.38: GAC BED DESIGN USING THOMAS EQUATION

An industrial wastewater contains phenolic compounds. The TOC concentration in the influent is 200 mg/L. A bench-scale GAC column is operated to develop adsorption data. The column diameter and lengths are 9.5 cm and 1.05 m, and flow through the column is 1.6 *EBV* (empty bed volume) per hour. The packed carbon density in the column is 400 kg/m^3. Determine the kinetic constants k_1 and q_0 for the GAC. Also calculate (a) the mass of GAC required to treat the industrial wastewater, (b) volume and dimensions of the GAC bed, and (c) liquid flow rate in terms of EBV per unit time (q_b). The flow rate through the column is 5 m/h (1.4 L/m^2·s), the allowable breakthrough volume is 1050 m^3 at an effluent TOC concentration of 10 mg/L, and influent flow is 150 m^3/d (6.25 m^3/h). The breakthrough data are given below.

Parameter	Breakthrough Test Date										
Breakthrough volume (V_B), L	1000	1200	1400	1600	2000	2200	2400	2600	2800	3000	3200
Effluent concentration (C_e), mg/L	0.02	0.06	0.2	0.7	9.5	31	85	146	180	194	200

Note: Scale-up approach (Equation 15.19) can also be applied with Thomas equations in this example.

Solution

1. Determine the operating parameters of the experimental column.

Cross-sectional area of the GAC column, $A = \frac{\pi}{4} \times (0.095\ \text{m})^2 = 0.0071\ \text{m}^2$

EBV (or volume of GAC bed), $EBV = A \times D = 0.0071\ \text{m}^2 \times 1.05\ \text{m} = 0.0075\ \text{m}^3$

Rearrange Equation 15.19a to calculate the flow rate through the GAC column.

$$Q = q_b EBV = 1.6\ \text{h}^{-1} \times 0.0075\ \text{m}^3 = 0.012\ \text{m}^3/\text{h}$$

Mass of GAC in the column from Equation 15.17f, $M = EBV \times \rho_{GAC}$

$$= 0.0075\ \text{m}^3 \times 400\ \text{kg/m}^3 = 3\ \text{kg GAC}$$

2. Develop the breakthrough test data to plot linear relationship given by Thomas equation (Equation 15.18b).

At given $V_B = 1000\ \text{L} = 1\ \text{m}^3$, $C_0 = 200$ mg/L, and $C_e = 0.02$ mg/L, calculate $\ln(C_0/C_e - 1)$

$$\ln\left(\frac{C_0}{C_e} - 1\right) = \ln\left(\frac{200\ \text{mg/L}}{0.02\ \text{mg/L}} - 1\right) = 9.21$$

Similarly, calculate $\ln(C_0/C_e - 1)$ corresponding to all given volumes percolated from the GAC column.

Tabulate the results in Table 15.18.

3. Plot the linear relationship between $\ln(C_0/C_e - 1)$ and V_B in Figure 15.25.

TABLE 15.18 Breakthrough Test Data for Plotting Linear Relationship (Example 15.38)

Breakthrough Volume (V_B), m^3	Effluent Concentration (C_e), mg/L	$\ln(C_0/C_e - 1)$
1.0	0.02	9.21
1.2	0.06	8.11
1.4	0.2	6.91
1.6	0.7	5.65
2.0	9.5	3.00
2.2	31	1.70
2.4	85	0.30
2.6	146	−0.99
2.8	180	−2.20
3.0	194	−3.48
3.2	200	–

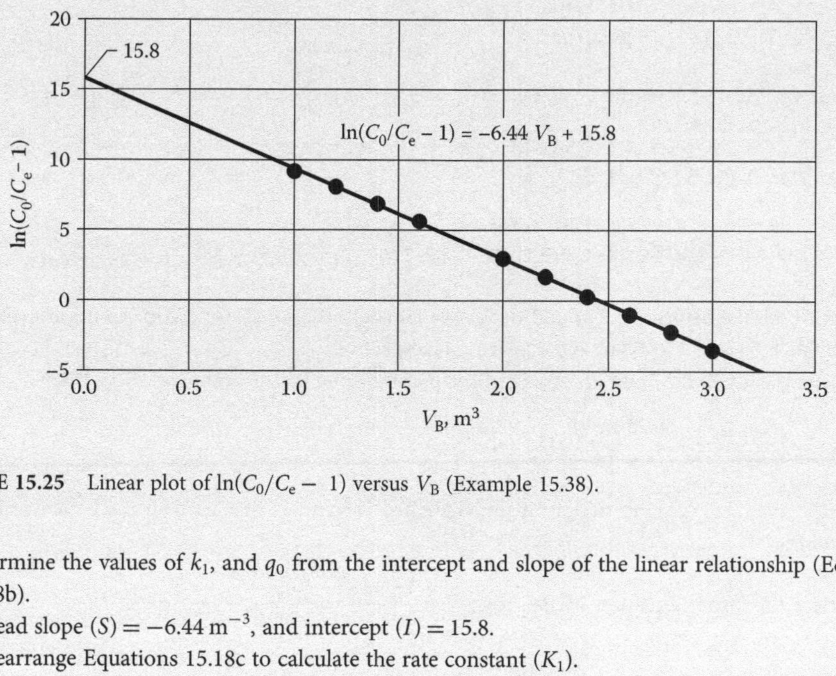

FIGURE 15.25 Linear plot of $\ln(C_0/C_e - 1)$ versus V_B (Example 15.38).

4. Determine the values of k_1, and q_0 from the intercept and slope of the linear relationship (Equation 15.18b).

   Read slope $(S) = -6.44\ \text{m}^{-3}$, and intercept $(I) = 15.8$.

   Rearrange Equations 15.18c to calculate the rate constant (K_1).

$$k_1 = -\frac{SQ}{C_0} = -\frac{(-6.44\ \text{m}^{-3}) \times 0.012\ \text{m}^3/\text{h}}{200\ \text{g/m}^3} = 0.00039\ \text{m}^3/\text{g·h} \quad \text{or} \quad 0.39\ \text{m}^3/\text{kg·h}$$

   Rearrange Equations 15.18d to calculate the maximum concentration of TOC on GAC (q_0).

$$q_0 = \frac{IQ}{k_1 M} = \frac{15.8 \times 0.012\ \text{m}^3/\text{h}}{0.00039\ \text{m}^3/\text{g·h} \times 3\ \text{kg}} = 162\ \text{g TOC/kg GAC}$$

5. Determine the mass of GAC (M_{design}) required in the full-scale column.

Apply Equation 15.18b at $Q_{design} = 150\ m^3/d = 6.25\ m^3/h$, and a breakthrough volume $V_{B,design} =$ $1000\ m^3$.

$$\ln\left(\frac{C_0}{C_e} - 1\right) = -\frac{k_1 C_0}{Q_{design}} V_{B,design} + \frac{k_1 q_0 M_{design}}{Q_{design}}$$

$$\ln\left(\frac{200\ mg/L}{10\ mg/L} - 1\right) = -\frac{0.00039\ m^3/g{\cdot}h \times 200\ g/m^3}{6.25\ m^3/h} \times 1050\ m^3$$

$$+\frac{0.00039\ m^3/g{\cdot}h \times 162\ g/kg \times M_{design}}{6.25\ m^3/h} 2.94 = -13.10 + 0.010\ kg^{-1} \times M_{design}$$

$$M_{design} = \frac{13.10 + 2.94}{0.010} kg = 1600\ kg$$

*EBV*Carbon requirement per unit wastewater, $\frac{M_{design}}{V_{B,design}} = \frac{1600\ kg}{1050\ m^3} = 1.5\ kg\ GAC/m^3$ wastewater

6. Calculate the volume and dimensions of the GAC bed.
   Rearrange Equation 15.17f to calculate the volume of GAC bed (EBV_{design}).

$$EBV_{design} = \frac{M_{design}}{\rho_{GAC}} = \frac{1600\ kg}{400\ kg/m^3} = 4\ m^3\ GAC$$

Cross-sectional area of the GAC bed at $v_L = 5\ m/h$, $A_{design} = \frac{Q_{design}}{v_L} = \frac{6.25\ m^3/h}{5\ m/h} = 1.25\ m^2$

Diameter of the GAC bed, $d_{design} = \sqrt{\frac{4A_{design}}{\pi}} = \sqrt{\frac{4 \times 1.25\ m^2}{\pi}} = 1.3\ m$

Depth of the GAC bed, $D_{design} = \frac{EBV_{design}}{A_{design}} = \frac{4\ m^3}{1.25\ m^2} = 3.2\ m$ GAC

Note: Both the depth of 3.2 m and depth to diameter ratio of 2.5 (or 3.2 m/1.3 m) are within the acceptable ranges of 1.8–4.8 m and 1.5:1 to 4:1, respectively.

7. Calculate the liquid flow rate in terms of EBV per unit time (q_b) from Equation 15.19a.

$$q_{b,design} \frac{Q_{design}}{EBV_{design}} = \frac{6.25\ m^3/h}{4\ m^3} = 1.6\ h^{-1}$$

Note: The value of q_b in design of the GAC bed is same as that used in the bench-scale GAC column test.

Service time from Equation 15.19e, $t_{B,design} = \frac{V_{B,design}}{Q_{design}} = \frac{1050\ m^3}{150\ m^3/d} = 7\ d$

EXAMPLE 15.39: DESIGN OF AN ADSORPTION COLUMN USING SCALE-UP APPROACH

An adsorption column is designed using scale-up approach. The experimental set up and data developed on industrial phenolic waste are given in Example 15.38. Design the GAC bed for the same flow of 150 m^3/d, and use the data developed in Example 15.38.

Solution

1. Draw the breakthrough curve.

Use the experimental data given in Table 15.18 (Example 15.38) to draw the breakthrough curve. The breakthrough curve is shown in Figure 15.26.

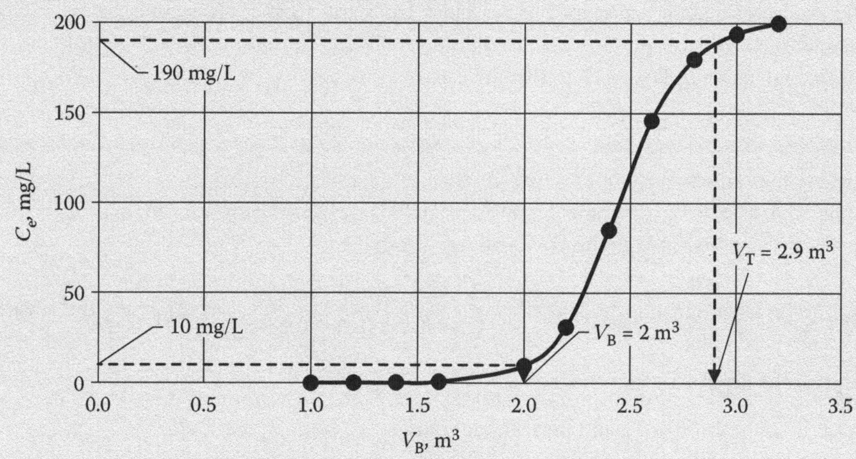

FIGURE 15.26 Breakthrough curve developed from experimental data (Example 15.39).

2. Determine the experimental V_B from the breakthrough curve in Figure 15.26.
   Read the experimental breakthrough volume V_B of 2 m^3 when the effluent TOC concentration $C_e =$ 10 mg/L is reached.
3. Determine the EBV and mass of GAC in the GAC bed.

   EBV_{design} (or volume of GAC bed) from Equation 15.19b, $EBV_{design} = \dfrac{Q_{design}}{q_{b,design}} = \dfrac{6.25\,m^3/h}{1.6\,h^{-1}} = 3.9\,m^3$

   Provide the GAC bed an EBV of 4 m^3. The volume calculated in this example is same as that presented in Example 15.38. Therefore, the dimensions of the GAC bed are also the same as those obtained in Step 6 of Example 15.38.
4. Verify the service time from scale-up approach.
   Obtain the mass of GAC in the experimental column $M = 3$ kg from Step 1 of Example 15.38.
   Calculate the volume of liquid treated per unit mass of GAC (V'_B) from Equation 15.19c using $V_B =$ 2 m^3 (Step 2).

   $$V'_B = \frac{V_B}{M} = \frac{2\text{ m}^3}{3\text{ kg}} = 0.67\text{ m}^3\text{ wastewater/kg GAC}$$

   Calculate the mass of GAC consumed per h (M_t) in the design GAC bed from Equation 15.19d.

   $$M_{t,design} = \frac{Q_{design}}{V'_B} = \frac{6.25\text{ m}^3/\text{h}}{0.67\text{ m}^3/\text{h}} = 9.3\text{ kg GAC/h}$$

   Obtain $M_{design} = 1600$ kg from Step 5 of Example 15.38.
   Service time from Equation 15.19e,

   Service time from Equation 15.19e, $t_{B,design} = \dfrac{M_{design}}{M_{t,design}} = \dfrac{1600\text{ kg}}{9.3\text{ kg/h}} = 172$ h

   or $t_{B,design} = 172\,h \times \dfrac{d}{24\,h} = 7.2\,d \approx 7\,d$

   The service time obtained in this example is same as that determined in Step 7 of Example 15.38.

EXAMPLE 15.40: DESIGN OF A MOVING-BED COUNTERCURRENT ADSORPTION COLUMN

Apply the data provided and developed in Examples 15.38 and 15.39 to determine the mass of carbon required and depth of adsorption zone. The allowable effluent concentration of TOC = 10 mg/L. Use the breakthrough curve in Figure 15.26 (Example 15.39).

Solution

1. Determine the carbon consumption (M_s/Q ratio).

   Apply the data from Example 15.38: $C_0 = 200$ mg/L, $C_e = 10$ mg/L, and $X_m = q_0 = 162$ g/kg (Step 4), and assume $X_i \approx 0$ in the GAC bed.

   M_s/Q ratio from Equation 15.20a, $\dfrac{M_s}{Q} = \dfrac{C_0 - C_e}{X_m - X_i} = \dfrac{(200-10)\text{g/m}^3}{(162-0)\text{g/kg}} = 1.2\text{ kg GAC/m}^3\text{ wastewater}$

2. Determine the depth of adsorption zone.

   Read the breakthrough V_b and throughput volumes V_T from Figure 15.26.

   $V_B = 2\text{ m}^3$ at $C_e = 0.05C_0 = 0.05 \times 200\text{ mg/L} = 10\text{ mg/L}$

   $V_T = 2.9\text{ m}^3$ at $C_e = 0.95C_0 = 0.95 \times 200\text{ mg/L} = 190\text{ mg/L}$

   Obtain the mass of GAC in the experimental column $M = 3$ kg from Step 1 of Example 15.38.

   Calculate the breakthrough volume of liquid treated per unit mass of GAC (V'_B) in the experimental column from Equation 15.19c.

   $$V'_B = \frac{V_B}{M} = \frac{2\text{ m}^3}{3\text{ kg}} = 0.67\text{ m}^3\text{ wastewater/kg GAC}$$

   Similarly, calculate the throughput volume of liquid treated per unit mass of GAC V'_T in the experimental column.

   $$V'_T = \frac{V_T}{M} = \frac{2.9\text{ m}^3}{3\text{ kg}} = 0.97\text{ m}^3\text{ wastewater/kg GAC}$$

   Obtain $M_{\text{design}} = 1600$ kg from Step 5 of Example 15.38.

   V_B for the design GAC bed, $V_{B,\text{design}} = V'_B M_{\text{design}} = 0.67\text{ m}^3/\text{kg} \times 1600\text{ kg} = 1070\text{ m}^3$ GAC

   V_T for the design GAC bed, $V_{T,\text{design}} = V'_T M_{\text{design}} = 0.97\text{ m}^3/\text{kg} \times 1600\text{ kg} = 1550\text{ m}^3$ GAC

   Obtain $Z_{\text{design}} = D_{\text{design}} = 3.2$ m in Step 6 of Example 15.38.

   Calculate the design depth of adsorption zone from Equation 15.20b.

   $$Z_{s,\text{design}} = Z_{\text{design}}\left(\frac{V_{T,\text{design}} - V_{B,\text{design}}}{V_{T,\text{design}} - 0.5\,(V_{T,\text{design}} - V_{B,\text{design}})}\right)$$

   $$= 3.2\text{ m} \times \left(\frac{(1550-1070)\text{ m}^3}{1550\text{ m}^3 - 0.5 \times (1550-1070)\text{ m}^3}\right) = 1.2\text{ m}$$

   Note: Comparing the carbon consumption, the moving bed countercurrent system is more efficient than the single fixed-bed column. Carbon consumption in countercurrent system is 1.2 kg GAC/m^3 wastewater (Step 1), and that in fixed-bed column is 1.5 kg GAC/m^3 wastewater from Step 5 of Example 15.38.

EXAMPLE 15.41: SELECTION OF A GAC BASED ON EMPTY BED CONTACT TIME (EBCT)

The filtered secondary effluent from a wastewater treatment plan contains 15 μg/L trichloroethane (TCE). Two samples of GAC were tested for removal of TCE from the effluent. Two identical bench-scale columns one for each sample were run in parallel to evaluate the performance of the GAC samples. The column height was 1.2 m (4 ft). The columns were run at different approaching velocities (or hydraulic loadings) and the effluent concentrations of TCE under equilibrium conditions determined. The results are given below. Compare the performance of both GAC and select the superior quality GAC.

Approaching Velocities (v_L), m^3/m^2·min (m/min)		0.12	0.16	0.22	0.33	0.41	0.57	0.81
Effluent TCE concentration (C_e), μg/L	Column 1	2.9	3.2	4.2	6.0	7.8	10.0	12.5
	Column 2	8.3	8.1	9.7	11.1	11.6	12.9	13.7

Solution

1. Determine the EBCT values at different values of throughput v_L.

$EBCT$ at $v_L = 0.12$ m/min fromEquation 15.21, $EBCT = \dfrac{D}{v_L} = \dfrac{1.2\text{ m}}{0.12\text{ m/min}} = 10\text{ min}$

Similarly, calculate the EBCT for all other approaching velocities and tabulate the data below.

Approaching Velocities (v_L), m^3/m^2·min	*EBCT*, min	Effluent TCE Concentration (C_e), mg/L	
		GAC in Column 1	GAC in Column 2
0.12	10	2.9	8.3
0.16	7.5	3.2	8.1
0.22	5.5	4.2	9.7
0.33	3.6	6.0	11.1
0.41	2.9	7.8	11.6
0.57	2.1	10	12.9
0.81	1.5	12.5	13.7
–	0	15	15

1 m^3/m^2·min = 24.5 gpm/ft^2

2. Plot the effluent TCE concentration (C_e) versus *EBCT* for both GAC samples in Figure 15.27.

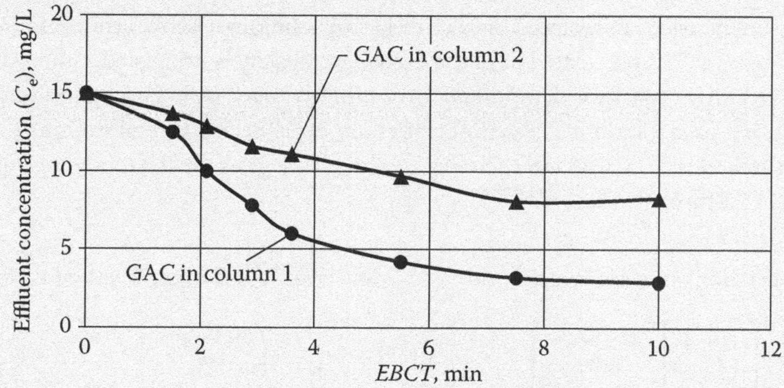

FIGURE 15.27 TCE concentration versus EBCT for two GAC samples (Example 15.41).

3. Select the GAC.

   The plot clearly shows that the GAC sample in Column 1 produces effluent of superior quality, and has superior performance. Therefore, GAC in column 1 is the obvious choice.

15.4.9 Ion Exchange

Ion exchange is the replacement of an ion in the aqueous phase by an ion in solid phase. The process is used for removal of hardness, ammonium and nitrate, heavy metals, total dissolved solids for demineralization, and some organic compounds. The ion exchange medium (or *resin*) consists of (1) a solid phase and (2) the mobile ions attached to an immobile functional group. The mobile ions in the resin are exchanged with ions in the water. The spent resin is regenerated and reused. The process may be operated in a batch or continuous mode. In a batch process, the resin is stirred in water until the reaction is complete, then the spent resin is settled, removed, and regenerated. In a continuous system, the water is passed through a packed column until the bed is exhausted and goes for regeneration cycle.

Ion Exchange Resin: An ion exchange resin is either *cationic* or *anionic*. The cationic resins exchange positive ions such as Ca^{2+}, Mg^{2+}, and Na^+ while the anionic resins exchange negative ions such as Cl^-, SO_4^{2-}, and NO_3^-. The ion exchange resins are either naturally occurring or synthetic materials. The naturally occurring materials are known as *zeolites*. The zeolites for water softening are *aluminosilicates* that has sodium as a mobile ion. They are also called sodium cation exchangers. The ammonium ion is removed by naturally occurring zeolite called *clinoptilolite*.[97,98]

The synthetic ion exchange resins are phenolic polymers. There are *sodium* (Na^+) and *hydrogen* (H^+) cation exchangers in which these positive ions are exchanged for various positive ions present in water. In anion exchange resins, the negative ions in water are exchanged by OH^- ions. The main types of anion exchangers are *weakly basic* and *strongly basic* anion exchangers. The weakly basic anion exchangers remove strongly ionized acids (HCl, H_2SO_4, etc.) but they will not remove weakly ionized acids such as H_2CO_3 and H_2SiO_3. Strongly basic anion exchangers remove both strongly and weakly ionized acids.

The important properties of ion exchange resins are *exchange capacity*, *grain size*, and *stability*. The exchange capacity (EC) is the quantity of exchangeable ions that can be removed per unit weight of the resin. It is expressed by weight (eq/kg or meq/g), or volume (eq/m^3, eq/L, or meq/L). Typical ion exchange capacity of zeolite resins is 0.05–0.1 eq/kg, and that of synthetic resins is 2–10 eq/kg of resin. The grain size of a resin influences the kinetics of ion exchange and hydraulics of the column. The rate of exchange is proportional to the inverse of the square of the particle diameter. The stability is important in determining the useful life of the resin.

The ion exchange capacity of an ion exchanger is determined by activating a known weight or volume of resin in a column. For example, a cationic resin in the column is washed with strong acid or NaCl solution until all active sites in the resin are replaced by Na^+ or H^+. A solution of known concentration of exchangeable ion (e.g., Ca^{2+} or Cu^{2+}) is percolated until the exchange capacity is exhausted. A breakthrough curve is prepared and the exchange capacity is determined. In case of hydrogen cation exchanger, the resin is titrated with a standard strong base solution. These procedures are explained in several examples.

Ion Exchanger Reactions: Typical ion exchange reactions are given by the following equilibrium equations (Equations 15.22 through 15.25).[49,97,98]

Sodium Cationic (Zeolite) Exchanger

1. Softening reactions involve removal of Ca^{2+}, Mg^{2+}, Fe^{2+}, and other bivalent cations by zeolite.

$$\underset{\text{(insoluble)}}{\underset{\text{Zeolite}}{Na_2Z}} + \begin{bmatrix} Ca \\ Mg \\ Fe \end{bmatrix} \cdot \underset{\text{(soluble)}}{\underset{\text{Hardness}}{\begin{bmatrix} (HCO_3)_2 \\ SO_4 \\ Cl_2 \end{bmatrix}}} \leftrightarrow \underset{\text{zeolite (insoluble)}}{\underset{\text{Hardness with}}{\begin{bmatrix} Ca \\ Mg \\ Fe \end{bmatrix} Z}} + \underset{\text{effluent (soluble)}}{\underset{\text{Residuals in treated}}{\begin{matrix} 2\ NaHCO_3 \\ Na_2SO_4 \\ 2\ NaCl \end{matrix}}} \qquad (15.22a)$$

2. Regeneration reactions with NaCl solution.

$$\begin{bmatrix} Ca \\ Mg \\ Fe \end{bmatrix} Z + \underset{\substack{\text{Regenerant NaCl} \\ \text{solution (soluble)}}}{2\,NaCl} \leftrightarrow \underset{\substack{\text{Regenerated} \\ \text{zeolite (insoluble)}}}{Na_2Z} + \underset{\substack{\text{Waste or brine} \\ \text{(soluble)}}}{\begin{matrix} CaCl_2 \\ MgCl_2 \\ FeCl_2 \end{matrix}} \tag{15.22b}$$

Hardness with zeolite (insoluble)

Hydrogen Cationic Exchanger

1. Reactions with storage acid anions such as SO_4^{2-} and Cl^-.

$$\underset{\substack{\text{Hydrogen cation} \\ \text{resin (insoluble)}}}{H_2R} + \underset{\substack{\text{Cations with strong} \\ \text{acidanions (soluble)}}}{\begin{bmatrix} Na_2 \\ Ca \\ Mg \end{bmatrix} \cdot \begin{bmatrix} SO_4 \\ Cl_2 \end{bmatrix}} \leftrightarrow \underset{\substack{\text{Cations with} \\ \text{resin (insoluble)}}}{\begin{bmatrix} Na_2 \\ Ca \\ Mg \end{bmatrix} R} + \underset{\substack{\text{Strong acids in treated} \\ \text{effluent (soluble)}}}{\begin{matrix} H_2SO_4 \\ 2\,HCl \end{matrix}} \tag{15.23a}$$

2. Reactions with weak acid anions such as HCO_3^-.

$$\underset{\substack{\text{Hydrogen cation} \\ \text{resin (insoluble)}}}{H_2R} + \underset{\substack{\text{Cations with} \\ \text{weak acid anions (soluble)}}}{\begin{bmatrix} Na_2 \\ Ca \\ Mg \end{bmatrix} \cdot (HCO_3)_2} \leftrightarrow \underset{\substack{\text{Cations with} \\ \text{resin (insoluble)}}}{\begin{bmatrix} Na_2 \\ Ca \\ Mg \end{bmatrix} R} + \underset{\substack{\text{Weak acids in treated} \\ \text{effluent (soluble)}}}{2\,CO_2 + H_2O} \tag{15.23b}$$

3. Regeneration reactions with strong acids such as H_2SO_4.

$$\underset{\substack{\text{Cations with} \\ \text{resin (insoluble)}}}{\begin{bmatrix} Na_2 \\ Ca \\ Mg \end{bmatrix} R} + \underset{\substack{\text{Regenerant} \\ H_2SO_4\text{(soluble)}}}{H_2SO_4} \leftrightarrow \underset{\substack{\text{Regenerated} \\ \text{resin (insoluble)}}}{H_2R} + \underset{\substack{\text{Waste or brine} \\ \text{(soluble)}}}{\begin{matrix} Na_2SO_4 \\ CaSO_4 \\ MgSO_4 \end{matrix}} \tag{15.23c}$$

Weak Base Anionic Exchanger

1. Reactions with strong acid anions such as SO_4^{2-}, Cl^-, and NO_3^-.

$$\underset{\substack{\text{Anions with} \\ \text{resin (insoluble)}}}{2\,(RNH_3)OH} + \underset{\substack{\text{Strong acids} \\ \text{(soluble)}}}{\begin{matrix} H_2SO_4 \\ 2\,HCl \\ 2\,HNO_3 \end{matrix}} \leftrightarrow \underset{\substack{\text{Anions with} \\ \text{resin (insoluble)}}}{\begin{matrix} (RNH_3)_2SO_4 \\ 2\,(RNH_3)Cl \\ 2\,(RNH_3)NO_3 \end{matrix}} + \underset{\substack{\text{Treated} \\ \text{effluent}}}{2\,H_2O} \tag{15.24a}$$

2. Regeneration reactions with sodium carbonate (Na_2CO_3).

$$\underset{\substack{\text{Anions} \\ \text{with resin (insoluble)}}}{\begin{matrix} (RNH_3)_2SO_4 \\ 2\,(RNH_3)Cl \\ 2\,(RNH_3)NO_3 \end{matrix}} + \underset{\substack{\text{Regenerant} \\ Na_2CO_3\text{ (soluble)}}}{Na_2CO_3} + H_2O \leftrightarrow \underset{\substack{\text{Regenerated} \\ \text{resin (insoluble)}}}{2\,(RNH_3)OH} + \underset{\substack{\text{Waste or brine} \\ \text{(soluble)}}}{\begin{matrix} Na_2SO_4 \\ 2\,NaCl \\ 2\,NaNO_3 \end{matrix}} + CO_2 \tag{15.24b}$$

Strong Base Anionic Exchanger

1. Reactions with strong and weak acid anions such as SO_4^{2-}, Cl^-, CO_3^-, and SiO_3^{2-}.

$$2\ (RN)OH + \begin{matrix} H_2SO_4 \\ 2\ HCl \\ 2\ H_2CO_3 \\ 2\ H_2SiO_3 \end{matrix} \leftrightarrow \begin{matrix} (RN)_2SO_4 \\ 2\ (RN)Cl \\ 2\ (RN)HCO_3 \\ 2\ (RN)HSiO_3 \end{matrix} + 2\ H_2O \tag{15.25a}$$

Anions with resin (insoluble) | Strong or weak acids (soluble) | Anions with resin (insoluble) | Treated effluent

2. Regeneration reactions with strong bases such as NaOH.

$$\begin{matrix} (RN)_2SO_4 \\ 2\ (RN)Cl \\ 2\ (RN)HCO_3 \\ 2\ (RN)HSiO_3 \end{matrix} + 2\ NaOH \leftrightarrow 2\ (RN)OH + \begin{matrix} Na_2SO_4 \\ 2\ NaCl \\ 2\ NaHCO_3 \\ 2\ NaHSiO_3 \end{matrix} \tag{15.25b}$$

Anions with resin (insoluble) | Regenerant NaOH (soluble) | Regenerated resin (insoluble) | Waste or brine (soluble)

Equilibrium Relationships for Ion Exchange: The law of mass action applies to chemical exchange reactions. The generalized equations of these reactions are expressed by Equation 15.26.

$$A^+ + B^+R_e^- \leftrightarrow B^+ + A^+R_e^- \tag{15.26a}$$

$$K_{B\rightarrow A} = \frac{[A^+R_e^-][B^+]}{[B^+R_e^-][A^+]} \tag{15.26b}$$

$$K_{B\rightarrow A} = \left[\frac{A^+}{B^+}\right]_{Resin}\left[\frac{B^+}{A^+}\right]_{Solution} \tag{15.26c}$$

where

$K_{B\rightarrow A}$ = mass action constant, or selectivity coefficient for replacing specie A by B in solution, dimensionless

$[A^+]$ or $[B^+]$ = equilibrium molar concentrations of cation species A or B in solution, mole/L

$[A^+R_e^-]$ or $[B^+R_e^-]$ = equilibrium molar content of cation species A or B on ion exchange resin, mole/g resin

$\left[\frac{A^+}{B^+}\right]_{Resin}$ = equilibrium molar content ratio of cation species A to B on ion exchange resin, dimensionless

$\left[\frac{B^+}{A^+}\right]_{Solution}$ = equilibrium molar concentration ratio of cation species B to A in solution, dimensionless

Larger value of $K_{B\rightarrow A}$ indicates greater preference for replacing the ion A by B in solution by the exchanger. An ion exchanger prefers (1) ions of higher valence, (2) ions with smaller solvated ions, (3) ions with greater ability to polarize, (4) ions that react strongly with the ion exchanger sites of the resins, and (5) ions that precipitate least with other ions to form complexes. The approximate selectivity scale on ion exchange resins has been established for many key cations and anions, and is summarized in Table 15.19. This selectivity scale is based on the assumption that the ion exchange capacity of chloride ion (Cl^-) is 1.0.

Breakthrough Curve: The breakthrough curve of an ion resin is similar to that of activated carbon medium, and is developed in the same way using a laboratory column. The simplified relationship proposed by Thomas (Equation 15.18) is applicable. The units of some variables in the equation are

TABLE 15.19 Approximate Selectivity Scales of Cations and Anions on Ion Exchange Resins

Parameter	Selectivity Scale									
				Cation on Strong-acid Resin						
Cation	Ag^+	Ca^{2+}	Mn^{2+}	Cu^{2+}	Mg^{2+}	K^+	NH_4^+	Na^+	H^+	Li^+
Selectivity	8.5	5.2	4.1	3.8	3.3	2.9	2.6	2.0	1.3	1.0
				Anion on Strong-base Anionic Resin						
Anions	NO_3^-	NO_2^-	CN^-	Cl^-	HCO_3^-	SO_4^{2-}	F^-	OH^-	CO_3^{2-}	HPO_4^{2-}
Selectivity	4	1.3	1.3	1.0	0.4	0.15	0.1	0.06	0.03	0.01

Source: Adapted in part from References 36, 37, and 99 through 101.

different. These units are: $k_1 = \text{L/s·kg}$, $\text{m}^3\text{/eq·d}$ or L/meq·d, $q_0 =$ eq/kg, C_0 and $C_e = \text{eq/m}^3$ or meq/L, and $Q = \text{m}^3\text{/d}$. The units in Equation 15.18 for M and V_B are compatible with these units. The use of breakthrough curve and Equation 15.18b are shown in Examples 15.38 and 15.47.

Applications of Ion Exchange: Common applications of ion exchange processes for wastewater treatment or effluent polishing are briefly described below.

Removal of Ammonia: Ammonium ion is removed by *clinoptilolite*, a naturally occurring zeolite. The selectivity of the resin is $K^+ > NH_4^+ > Na^+ > Ca^{2+} > Mg^{2+}$. The theoretical exchange capacity of clinoptilolite is 2 meq/L or eq/m^3, but it is lowered to 0.2–0.5 meq/L in wastewater application. Typically, the specific gravity of resins is 1.6 and the bulk density is 600–800 kg/m^3 (0.3–0.6 lb/ft^3). The media is 0.84 × 0.33 mesh. The volumetric hydraulic loading rate ranges from 5 to 10 m^3/m^3·h (2 to 4 gpm/ft^2), and is based on empty bed volume (EBV). As throughput volume increases the ammonium ion shows up in the effluent. At about 130 EBV, the concentration of ammonium ion increases 0.5–1.0 mg/L, and up to 5 mg/L at EBV of 170. Regeneration is achieved by adding lime ($Ca(OH)_2$) at a flow rate of about 10 EBV/h for 1–2 h, and followed by rinse cycle. Ammonium ion removed from zeolite is converted into ammonia because of high pH. Ammonia is then removed from the regenerant solution in a stripping tower, and captured in an acid scrubber solution. Calcium carbonate ($CaCO_3$) precipitates in the zeolite bed, stripping tower, pipings, and appurtenances. The zeolite bed is equipped with backwash system to remove the carbonate deposits.

Removal of Nitrate: Nitrate is removed by synthetic ion exchange resins. Most synthetic resins have greater affinity for sulfate over nitrate. As a result, the sulfate ion in the feedwater replaces nitrate ion from the resin into the effluent. This is called nitrate *leaching* or *dumping*.[36,54,102] Resins have been developed that overcome this problem. The procedure to calculate nitrate removal by ion exchange resin is quite involved. Calculation procedure without and with sulfate ion are different and the selectivity scale given in Table 15.19 are applied. An example to illustrate the procedure is given in Reference 36. Readers are referred to Example 11.18 in this source to study the procedure.

Removal of Heavy Metals: Natural and synthetic resins and chelating resins selectively remove heavy metals such as Cu, Ni, Cd, Zn, and others. Removal is highly depended upon pH, and most heavy metals bind better at higher pH than other cations.[36] Iron and manganese are removed by naturally occurring *greensand*. Oxidation by potassium permanganate, ozone, or chlorine is needed to convert iron and manganese to higher valence.[64]

Demineralization: Cation and anion exchangers are used in series to replace cations by H^+ and anion by OH^- that combine to produce water. The percolation rate is 0.2–0.4 m^3/m^2·min (5–10 gpm/ft^2) and bed depth is 0.75–2 m (2.5–6.5 ft). Split treatment is normally used to achieve a desired level of hardness and TDS (Example 15.45).[64]

Concentration of Specific Ions: Ion exchange beds are used to concentrate valuable ions such as silver, chromium, and radioactive contaminants. The low concentration of desired ions are removed by the resin. The regenerant solution provides high concentration of retained ions. The process is shown in Example 15.48.

Basic Design Considerations: The breakthrough curves, operating procedures, and design considerations for an ion exchange column are quite similar to those for a carbon adsorption column. Basic information is summarized below:

1. The exchange bed depth is 0.6–3 m (2–8.5 ft) and diameter is 0.75–3.5 m (2.5–11 ft). The depth to diameter ratio is 1.5:1–3:1. The operating flow is 2.5–20 m/h or 0.7–5.5 L/s·m^2 (1–8 gpm/ft^2).
2. The bed expansion may be 25–100% of bed depth. The maximum column height is normally 3.7 m (12 ft). If larger height is required two columns in series are used.
3. The regeneration flow rate is 2.5–5 m/h or 0.7–1.4 L/s·m^2 (1–2 gpm/ft^2). After resin regeneration a slow rinse followed by a short but fast rinse is used to flush out any remaining brine. Total volume of rinse water is 4000–20,000 L/m^3 (30–150 gal/ft^3).
4. The zeolite bed is regenerated by 5–10% salt solution, and salt requirement is 80–320 kg/m^3 (5–20 lb/ft^3). The hydrogen cation exchanger is regenerated by 2% H_2SO_4 solution followed by 10% solution. The lower concentration prevents $CaSO_4$ precipitation in the bed. The H_2SO_4 requirement is 96–192 kg/m^3 (6–12 lb/ft^3) of resin. If HCL is used for regeneration, the solution strength is 15% and quantity is 80–160 kg/m^3 (5–10 lb/ft^3).
5. The strong base anion exchanger uses 2–10% NaOH regenerant solution for regeneration, and base requirement is 48–96 kg/m^3 (3–6 lb/ft^3) resin.

EXAMPLE 15.42: ION EXCHANGE CAPACITY

A synthetic cation exchange resin has exchange capacity (*EC*) of 10 eq/kg. The density of resin $\rho_{resin} =$ 650 kg/m^3. Calculate exchange capacity of the resin in (a) eq/m^3, (b) g Cu^{2+}/kg and kg Cu^{2+}/m^3, and (c) g $CaCO_3$/kg and kg $CaCO_3$/m^3.

Solution

1. Compute the volumetric exchange capacity in eq/m^3.

$$\text{Volumetric exchange capacity} = EC \times \rho_{resin} = 10\ \text{eq/kg} \times 650\ \text{kg/m}^3 = 6500\ \text{eq/m}^3$$

2. Compute the cation exchange capacity as Cu^{2+}.

$$\text{Eq. wt. of Cu}^{2+} = \frac{\text{Molar wt. of Cu}}{\text{Eq. of Cu}^{2+}} = \frac{63.5\ \text{g/mole as Cu}^{2+}}{2\ \text{eq/mole as Cu}^{2+}} = 31.8\ \text{g/eq as Cu}^{2+}$$

Exchange capacity of the resin as Cu^{2+} per unit weight of resin (g Cu^{2+}/kg resin)

$$= EC \times \text{Eq. wt. of Cu}^{2+} = 10\ \text{eq/kg} \times 31.8\ \text{g/eq as Cu}^{2+} = 318\ \text{g Cu}^{2+}/\text{kg resin}$$

Exchange capacity of the resin as Cu^{2+} per unit volume of resin (kg Cu^{2+}/m^3 resin)

$$= \text{Volumetric exchange capacity} \times \text{Eq. wt. of Cu}^{2+}$$
$$= 6500\ \text{eq/m}^3 \times 31.8\ \text{g/eq as Cu}^{2+} \times 10^{-3}\ \text{kg/g} = 207\ \text{kg Cu}^{2+}/\text{m}^3\ \text{resin}$$

3. Compute the cation exchange capacity as $CaCO_3$.

$$\text{Eq. wt. of CaCO}_3 = \frac{\text{Molar wt. of CaCO}_3}{\text{Eq. of CaCO}_3} = \frac{100\ \text{g/mole}}{2\ \text{eq/mole}} = 50\ \text{g/eq as CaCO}_3$$

Exchange capacity of the resin as $CaCO_3$ per unit weight of resin (g $CaCO_3$/kg resin)

$$= EC \times \text{Eq. wt. of CaCO}_3 = 10\ \text{eq/kg} \times 50\ \text{g/eq as CaCO}_3 = 500\ \text{g CaCO}_3/\text{kg resin}$$

Exchange capacity of the resin as $CaCO_3$ per unit volume of resin (kg $CaCO_3/m^3$ resin)

$= \text{Volumetric exchange capacity} \times \text{Eq. wt. of } CaCO_3$

$= 6500 \text{ eq/m}^3 \times 50 \text{ g/eq as } CaCO_3 \times 10^{-3} \text{ kg/g} = 325 \text{ kg } CaCO_3/m^3 \text{ resin}$

EXAMPLE 15.43: CATION EXCHANGE CAPACITY DETERMINATION FROM BREAKTHROUGH CURVE

25 g of cation exchange resin is placed in a column. A standard 100-mg/L Cu^{2+} solution is percolated through the column and breakthrough data are developed. The allowable Cu^{2+} concentration in the effluent is 5 mg/L. The concentration of Cu^{2+} in the effluent with respect to throughput volume is given below. Determine the cation exchange capacity of the resin.

Breakthrough volume (V_B), L	10	14	16	18	20	22	25	27	30	32
Effluent concentration (C_e), mg Cu^{2+}/L	0.015	0.10	0.37	1.34	2.66	5.71	27	50	82	93

Solution

1. Plot the breakthrough curve.

   The plot of the concentration of Cu^{2+} in the effluent (C_e) versus throughput volume (V_B) is shown in Figure 15.28.

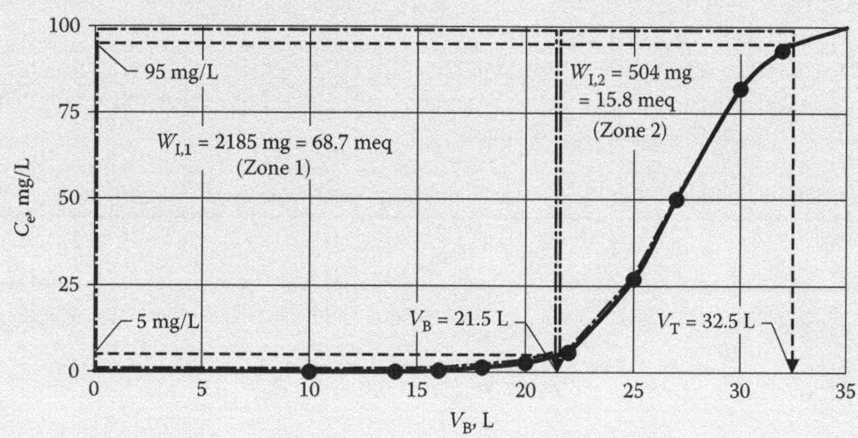

FIGURE 15.28 Breakthrough curve of Cu^{2+} of cationic exchange resin column (Example 15.43).

2. Determine the breakthrough and throughput volumes.

   Read the breakthrough and throughput volumes from Figure 15.28.

   $V_B = 21.5$ L at the allowable concentration $C_e = 0.05C_0 = 0.05 \times 100$ mg/L $= 5$ mg/L

   $V_T = 32.5$ L at the maximum concentration $C_e = 0.95C_0 = 0.95 \times 100$ mg/L $= 95$ mg/L

   The breakthrough curve has 2 distinct zones: (1) *Zone* 1 is between $V_B = 0$ and 21.5 L. This represents the volume to reach the breakthrough point at the allowable Cu^{2+} concentration $C_e = 5$ mg/L and (2) *Zone* 2 is between $V_B = 21.5$ and 32.5 L. This represents the additional volume needed after the breakthrough point to reach the throughput point (V_T) for bed exhaustion at the maximum Cu^{2+} concentration $C_e = 0.95\ C_0 = 95$ mg/L.

3. Determine the amount of Cu^{2+} removed by the resin.

The amount of Cu^{2+} ion removed by the resin is expressed by Equation 15.27a or Equation 15.27b.

$$M_I = V(C_0 - C_e) \tag{15.27a}$$

$$M_I = VC_0 \quad (\text{when } C_e \ll C_0) \tag{15.27b}$$

where

M_I = mg or g of ion removed by ion exchange. M_I may also be expressed in meq or eq of ion. It is obtained by dividing the mass concentration by the equivalent weight (Eq. wt.) of the ion.
V = volume of wastewater treated through the column, L (m^3)
C_0 = ionic concentration in the influent to the ion exchange column, mg/L (g/m^3)
C_e = ionic concentration in the effluent from the ion exchange column, mg/L (g/m^3)

The amount of Cu^{2+} removed in each zone is calculated from the tabulated data.

a. Determine the amount of Cu^{2+} ($M_{I,1}$) in Zone 1.

The amount of Cu^{2+} removed between $V_B = 0$ and 10 L.
At $V_B = 10$ L, effluent Cu^{2+} concentration $C_e = 0.015$ mg/L
Cu^{2+} concentration removed, $C_0 - C_e = (100 - 0.015)$ mg/L $= 99.985$ mg/L
Volume of wastewater treated, $V = (10 - 0)$ L $= 10$ L
Amount of Cu^{2+} removed from Equation 15.27a,

$$M_I = V(C_0 - C_e) = 10 \text{ L} \times 99.985 \text{ mg/L} = 999.85 \text{ mg} \approx 1000 \text{ mg}$$

Similarly, calculate M_I corresponding to all other volume segments treated by the resin column between $V_B = 10$ and 21.5 L. The results are given in Table 15.20. $M_{I,1} = 2185$ mg is the calculated sum of W_I between $V_B = 0$ and 21.5 L.

b. Determine the amount of Cu^{2+} ($M_{I,2}$) in Zone 2.

Similarly, calculate the values of M_I between $V_B = 21.5$ and 32.5 L. The results are also tabulated in Table 15.20. $M_{I,2} = 504$ mg is obtained.

TABLE 15.20 Breakthrough Test Data for Determining Exchange Capacity of Resin (Example 15.43)

Breakthrough Volume (V_B), L	10	14	16	18	20	21.5	22	25	27	30	32	32.5
Effluent concentration (C_e), mg Cu^{2+}/L	0.015	0.10	0.37	1.34	2.66	5	5.71	27	50	82	93	95
Concentration removed ($C_0 - C_e$), mg Cu^{2+}/L	99.985	99.9	99.63	98.66	97.34	95	94.29	73	50	18	7	5
Additional volume treated (V), L	10	4	2	2	2	1.5	0.5	3	2	3	2	0.5
Amount of Cu^{2+} removed (M_I), mg	1000	400	200	198	196	144	47	251	123	102	25	3
Amount of Cu^{2+} in each zone, mg	$M_{I,1} = 2185$ mg Cu^{2+} in Zone 1						$M_{I,2} = 504$ mg Cu^{2+} in Zone 2					

4. Calculate the ion exchange capacity (EC) of the resin.

The exchange capacity (CE) is expressed by Equation 15.27c.

$$EC = \frac{M_I}{M_R} \quad \text{or} \quad M_R = \frac{M_I}{EC} \tag{15.27c}$$

where

EC = ion exchange capacity, meq/g (eq/kg)

M_R = amount of resin in the column, g (kg)

$$\text{Eq. wt. of } Cu^{2+} = \frac{\text{Molar wt. of Cu}}{\text{Eq. of } Cu^{2+}} = \frac{63.5 \text{ g/mole as } Cu^{2+}}{2 \text{ eq/mole as } Cu^{2+}}$$

$$= 31.8 \text{ g/eq as } Cu^{2+} \quad \text{or} \quad 31.8 \text{ mg/meq as } Cu^{2+}$$

Calculate the amount of Cu^{2+} removed in the unit of meq when the breakthrough point is reached.

$$M_{I,1} = \frac{2185 \text{ mg } Cu^{2+}}{31.8 \text{ mg/meq as } Cu^{2+}} = 68.7 \text{ meq}$$

Calculate the amount of Cu^{2+} removed in the unit of meq between the breakthrough and throughput points.

$$M_{I,2} = \frac{504 \text{ mg } Cu^{2+}}{31.8 \text{ mg/meq as } Cu^{2+}} = 15.8 \text{ meq}$$

Calculate the amount of Cu^{2+} removed when the resin is exhausted at the throughput point ($M_{I,total}$).

$$M_{I,total} = M_{I,1} + M_{I,2} = (68.7 + 15.8) \text{ meq} = 84.5 \text{ meq}$$

Ion exchange capacity from Equation 15.27c at the weight of resin $M_R = 25$ g,

$$EC = \frac{M_{I,total}}{M_R} = \frac{84.5 \text{ meq}}{25 \text{ g}} = 3.4 \text{ meq/g resin} \quad \text{or} \quad 3.4 \text{ eq/kg resin}$$

Note: The amount of Cu^{2+} removed by the resin (M_I) may also be determined graphically from the total area above the curve between $V_B = 0$ and $V_T = 32.5$ L in Figure 15.28.

EXAMPLE 15.44: CATION EXCHANGE CAPACITY FROM RESIN TITRATION

A column titration study was conducted to determine the cation exchange capacity of a resin. The column was packed with 95 g of resin. The column was washed with NaCl solution to saturate the resin with Na^+ ion then rinsed to remove any remaining Cl^- ion. The column was titrated with 150 mg/L $CaCl_2$ standard solution. The breakthrough volume was measured for Cl^- and Ca^{2+} ions. The results are given below. Calculate (a) the cation exchange capacity of the resin in eq/kg and (b) mass and volume of resin required to treat 3800 m^3 of water containing 20 mg/L ammonium ion (NH_4^+). The density of resin is 680 kg/m^3.

Breakthrough volume (V_B), L	5	7	9	11	13	15	17	19	21	23
Breakthrough concentration (C_e), mg/L										
Ca^{2+}	0.0	0.0	0.0	1.1	5.4	12.4	32.4	48.6	54.0	54.0
Cl^-	0.0	4.8	19.2	57.6	86.4	96.0	96.0	96.0	96.0	96.0

Solution

1. Calculate the concentrations of Ca^{2+} and Cl^- in the standard titrant solution.

$$\text{For } Ca^{2+},\ C_{0,Ca} = \frac{\text{Molar wt. of Ca}}{\text{Molar wt. of } CaCl_2} \times 150 \text{ mg/L as } CaCl_2$$

$$= \frac{40 \text{ g/mole as Ca}}{111 \text{ g/mole as } CaCl_2} \times 150 \text{ mg/L as } CaCl_2 = 54 \text{ mg } Ca^{2+}/L$$

$$\text{For } Cl^-,\ C_{0,Cl} = \frac{2 \times \text{Molar wt. of Cl}}{\text{Molar wt. of } CaCl_2} \times 150 \text{ mg/L as } CaCl_2$$

$$= \frac{2 \times 35.5 \text{ g/mole as Cl}}{111 \text{ g/mole as } CaCl_2} \times 150 \text{ mg/L as } CaCl_2 = 96 \text{ mg } Cl^-/L$$

2. Tabulate the data for plotting C_e/C_0 ratios of Ca^{2+} and Cl^- with respect to the breakthrough volume (V_B).

Breakthrough Volume (V_B), L	Concentration (C_e), mg/L		C_e/C_0 Ratio	
	Ca^{2+}	Cl^-	Ca^{2+}	Cl^-
5	0.0	0.0	0.00	0.00
7	0.0	4.8	0.00	0.05
9	0.0	19.2	0.00	0.20
11	1.1	57.6	0.02	0.60
13	5.4	86.4	0.10	0.90
15	12.4	96.0	0.23	1.00
17	32.4	96.0	0.60	1.00
19	48.6	96.0	0.90	1.00
21	54.0	96.0	1.00	1.00
23	54.0	96.0	1.00	1.00

3. Plot the C_e/C_0 ratios of Ca^{2+} and Cl^- versus breakthrough volume (V_B) in Figure 15.29.

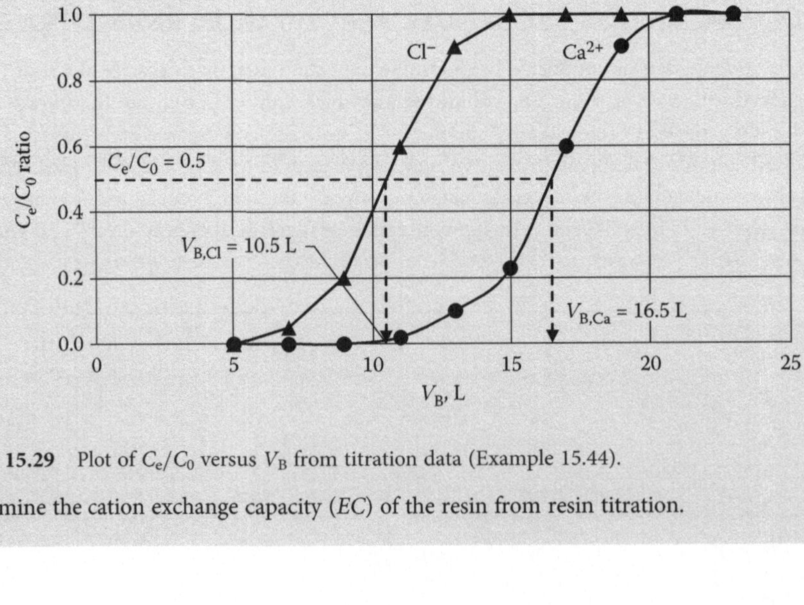

FIGURE 15.29 Plot of C_e/C_0 versus V_B from titration data (Example 15.44).

4. Determine the cation exchange capacity (*EC*) of the resin from resin titration.

The exchange capacity is expressed by Equation 15.27d using resin titration data.

$$EC = \frac{(V_{B,cationic} - V_{B,anionic})C_0}{M_R} \quad (15.27d)$$

where

EC = ion exchange capacity, meq/g (eq/kg)

$V_{B,cationic}$ and $V_{B,anionic}$ = breakthrough volume on cationic (Ca^{2+}) and anionic (Cl^-) curves at C_e/C_0 ratio = 0.5, L (m^3)

C_0 = ionic concentration in the influent to the ion exchange column, meq/L (eq/m^3)

Read $V_{B,cationic} = V_{B,Ca} = 16.5$ L and $V_{B,anionic} = V_{B,Cl} = 10.5$ L at C_e/C_0 ratio of 0.5 in Figure 15.29.

$$\text{Eq. wt. of } CaCl_2 = \frac{\text{Molar wt. of } CaCl_2}{\text{Eq. of } CaCl_2} = \frac{111 \text{ g/mole as } CaCl_2}{2 \text{ eq/mole as } CaCl_2}$$

$$= 55.5 \text{ g/eq as } CaCl_2 \quad \text{or} \quad 55.5 \text{ mg/meq as } CaCl_2$$

$$C_{0;CaCl_2} = \frac{150 \text{ mg/L as } CaCl_2}{55.5 \text{ mg/meq as } CaCl_2} = 2.7 \text{ meq/L} \quad \text{and} \quad W_R = 95 \text{ g of resin.}$$

Cation exchange capacity from Equation 15.27d at $M_R = 95$ g of resin,

$$EC = \frac{(V_{B,Ca} - V_{B,Cl})C_{0,CaCl_2}}{M_R} = \frac{(16.5 - 10.5) \times 2.7 \text{ meq/L}}{95 \text{ g}} = 0.17 \text{ meq/g} \quad \text{or} \quad 0.17 \text{ eq/kg}$$

5. Determine the volume of resin required to remove NH_4^+ ion.

$$\text{Eq. wt. of } NH_4^+ = \frac{\text{Molar wt. of } NH_4^+}{\text{Eq. of } NH_4^+} = \frac{18 \text{ g/mole as } NH_4^+}{1 \text{ eq/mole as } NH_4^+} = 18 \text{ g/eq or } 18 \text{ mg/meq as } NH_4^+$$

$$\text{Concentration of NH4+, } C_{0,NH4} = \frac{20 \text{ mg/L as } NH_4^+}{18 \text{ mg/meq as } NH_4^+} = 1.11 \text{ meq/L} \quad \text{or} \quad 1.11 \text{ eq/m}^3$$

Calculate the amount of NH_4^+ removed at the daily volume $V = 3800$ m^3 from Equation 15.27b,

$$M_I = VC_{0,NH4} = 3800 \text{ m}^3 \times 1.11 \text{ eq/m}^3 = 4220 \text{ eq}$$

$$\text{Mass of resin required from Equation 15.27c, } M_R = \frac{M_I}{EC} = \frac{4220 \text{ eq}}{0.17 \text{ eq/kg}} = 25{,}000 \text{ kg of resin}$$

The volume of resin required is calculated from Equation 15.27e.

$$V_R = \frac{M_R}{\rho_R} \quad \text{or} \quad M_R = V_R \times \rho_R \quad (15.27e)$$

where

V_R = volume of resin in the ion exchange column, L (m^3)

ρ_R = density of resin, g/L or kg/m^3

Volume of resin required from Equation 15.27e at $\rho_R = 680$ kg/m,

$$V_R = \frac{M_R}{\rho_R} = \frac{25{,}000 \text{ kg}}{680 \text{ kg/m}^3} = 37 \text{ m}^3$$

Provide a total volume $V_R = 45$ m^3. It is over 20% larger than the theoretical volume to account for leakage and short-circuiting.

EXAMPLE 15.45: SPLIT TREATMENT WITH ZEOLITE SOFTENER

Well water has total hardness (TH) of 300 mg/L as $CaCO_3$. Water is softened by split zeolite softener. The softener takes in three-fourth of total flow and produces water at a hardness of 3 mg/L as $CaCO_3$. One-fourth flow is directly mixed with softened water. The zeolite bed has 28-m^3 (1000-ft^3) resin capacity. The water softening capacity of the resin is 64 kg/m^3 (4 lb/ft^3) as $CaCO_3$. Exhausted resin is regenerated by 98% purity commercial salt. The salt consumption is two times the theoretical amount and used as 4% salt solution. The total flow treated by the split treat is 945 m^3/d (0.25 MGD). Calculate (a) the regeneration period in days, (b) hardness of finished water, (c) total salt consumption per generation cycle, and (d) total volume of regenerant solution. Ignore any water losses associated with the softener operation, such as backwashing, regeneration, etc.

Solution

1. Draw the definition sketch.

Flow to the softener $= 0.75Q = 0.75 \times 945\ m^3/d = 709\ m^3/d$

Flow bypassing the softener $= (945 - 709)\ m^3/d = 236\ m^3/d$

Draw the definition sketch in Figure 15.30.

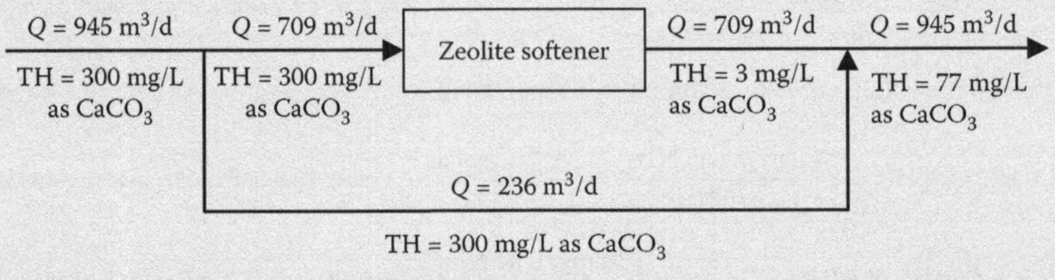

FIGURE 15.30 Definition sketch of split softening process diagram (Example 15.45).

2. Determine the regeneration period of the softener.

$$\text{Amount of TH removed per day} = 709\ m^3/d \times (300 - 3)\ g/m^3 \text{ as } CaCO_3 \times 10^{-3}\ kg/g = 211\ kg/d \text{ as } CaCO_3$$

$$\text{Regeneration capacity of the resin per cycle} = 64\ kg/m^3 \text{ as } CaCO_3 \times 28\ m^3 = 1790\ kg \text{ as } CaCO_3 \text{ per cycle}$$

$$\text{Regeneration period} = \frac{1790\ kg \text{ as } CaCO_3}{211\ kg/d \text{ as } CaCO_3} = 8.5\ d \text{ per cycle}$$

3. Calculate the hardness of finished water.

$$\text{Amount of TH in the well water} = 945\ m^3/d \times 300\ g/m^3 \text{ as } CaCO_3 \times 10^{-3}\ kg/g = 284\ kg/d \text{ as } CaCO_3$$

$$\text{Amount of TH remaining in finished water} = (284 - 211)\ kg/d \text{ as } CaCO_3 = 73\ kg/d \text{ as } CaCO_3$$

$$\text{TH concentration in finished water} = \frac{73\ kg/d \text{ as } CaCO_3}{945\ m^3/d} \times 10^3\ g/kg = 77\ g/m^3 \text{ or } 77\ mg/L \text{ as } CaCO_3$$

4. Determine the total commercial salt requirement per cycle.

$$\text{Pure NaCl required per cycle (Equation15.22b)} = 1790 \text{ kg } CaCO_3 \times \frac{58.5 \text{ g/eq as NaCl}}{50 \text{ g/eq as } CaCO_3}$$
$$= 2090 \text{ kg NaCl}$$

$$\text{Commercial salt at 98\% purity and twice amount} = 2 \times 2090 \text{ kg NaCl} \times \frac{100\%}{98\%} = 4270 \text{ kg salt per cycle}$$

5. Determine the volume of regenerant solution per cycle.

   Sodium chloride (NaCl) solution is used to regenerate geolite medium (Equation 15.22b). Assume the specific weight of regenerant solution at 4% of commercial salt = 1040 kg/m^3.

$$\text{Volume of regenerant solution per cycle} = \frac{4270 \text{ kg}}{0.04 \times 1040 \text{ kg/m}^3} = 103 \text{ m}^3 \text{ solution per cycle}$$

EXAMPLE 15.46: CATION AND ANION DEMINERALIZER

A hydrogen cation and strong base anion exchanger system is used to demineralize water. The flow is 50 m^3/d. The volumetric exchange capacities of cationic and anionic resins are 80 and 60 kg $CaCO_3$/m^3, respectively. The chemical quality of water is given below. Determine (a) volume and dimensions of both exchangers and (b) quantities of regeneration chemicals. The regeneration cycle is 4 d. Also, prepare a conceptual process diagram.

Parameter	Cation				Anion			
	Ca^{2+}	Mg^{2+}	Na^+	K^+	HCO_3^-	SO_4^{2-}	Cl^-	NO_3^-
Concentration, mg/L	145	18	130	50	258	220	214	50

Solution

1. Compute the total cation and anion concentrations in water.

   The concentrations of cations and anions in meq/L and mg/L as $CaCO_3$, and tabulated in the table below.

Ion	Cation Concentration				Ion	Anion Concentration			
	mg/L	Eq. wt.	meq/L	mg/L as $CaCO_3$		mg/L	Eq. wt.	meq/L	mg/L as $CaCO_3$
Ca^{2+}	145	20.0	7.25	362.5	HCO_3^-	258	61.0	4.23	211.5
Mg^{2+}	18	12.2	1.48	74.0	SO_4^{2-}	220	48.0	4.58	229.0
Na^+	130	23.0	5.65	282.5	Cl^-	214	35.5	6.03	301.5
K^+	50	39.1	1.28	64.0	NO_3^-	50	62.0	0.81	40.5
Total	–	–	15.66	783.0	Total	–	–	15.65	782.5

Note: The total concentrations of cations and anions are virtually the same. The water quality data is acceptable.

2. Determine the volume and dimensions of ion exchangers.

   a. Hydrogen cation exchangers.

      The volume of resin required may also be calculated from Equation 15.27f.

$$V_R = \frac{M_I}{EC'} \quad \text{or} \quad M_I = V_R \times EC' \qquad (15.27f)$$

where EC' = volumetric exchange capacity, meq/L (eq/m^3, g/L, or kg/m^3)

Total volume of water treated,

$$V = 50\ \text{m}^3/\text{d} \times 4\ \text{d/cycle} = 200\ \text{m}^3\ \text{per cycle}$$

The cation exchange reactions are expressed by Equations 15.23a and 15.23b.

Total quantity of cations exchanged per cycle from Equation 15.27b,

$$M_{\text{I,cation}} = VC_{0,\text{cation}} = 200\ \text{m}^3 \times 783\ \text{g/m}^3 \times 10^{-3}\ \text{kg/g} = 157\ \text{kg as CaCO}_3\ \text{per cycle}$$

Theoretical volume of hydrogen cation exchange resin from Equation 15.27f,

$$V_{\text{R,cation}} = \frac{M_{\text{I,cation}}}{EC'_{\text{cation}}} = \frac{157\ \text{kg as CaCO}_3}{80\ \text{kg CaCO}_3/\text{m}^3} = 1.96\ \text{m}^3 \approx 2\ \text{m}^3\ \text{of cationic resin per cycle}$$

In practice, because of short circuit and other operational and design limitations, the required volume of the bed is usually 1.2–1.5 times the theoretical volume. In this example, use a factor of 1.5. Total volume = 3 m^3. Provide two hydrogen cation exchange columns. The required volume of each column = 1.5 m^3.

Dimensions of each column: diameter = 1 m, surface area = 0.79 m^2, resin depth = 2 m, resin volume = 1.6 m^3, and total height of column = 3.5 m. This height will provide room for 50% bed expansion during backwash, and needed freeboard.

b. Strong base anion exchangers.

The anion exchange reactions are expressed by Equation 15.25a. Total quantity of anions exchanged per cycle from Equation 15.27b,

$$M_{\text{I,anion}} = VC_{0,\text{anion}} = 200\ \text{m}^3 \times 782.5\ \text{g/m}^3 \times 10^{-3}\ \text{kg/g} = 157\ \text{kg as CaCO}_3\ \text{per cycle}$$

Theoretical volume of anion exchange resin from Equation 15.27f,

$$V_{\text{R,anion}} = \frac{M_{\text{I,anion}}}{EC'_{\text{anion}}} = \frac{157\ \text{kg as CaCO}_3}{60\ \text{kg CaCO}_3/\text{m}^3} = 2.62\ \text{m}^3 \approx 2.6\ \text{m}^3\ \text{of anionic resin per cycle}$$

Provide a total resin volume = 3.2 m^3. It gives about 23% excess volume.

Provide two anion exchange columns with same dimensions as those for hydrogen cation exchangers. The total anionic resin volume = 1.6 m^3.

3. Determine the regenerant volume and chemicals per cycle.

a. Hydrogen cation exchangers.

Sulfuric acid (H_2SO_4) solution is used to regenerate the hydrogen cation exchanger (Equation 15.23c).

$$\begin{aligned}\text{Quantity of H}_2\text{SO}_4\ \text{required per cycle} &= M_{\text{I,cation}} \times \frac{\text{Eq. wt. of H}_2\text{SO}_4}{\text{Eq. wt. of CaCO}_3}\\ &= 157\ \text{kg as CaCO}_3 \times \frac{49\ \text{g/eq as H}_2\text{SO}_4}{50\ \text{g/eq as CaCO}_3}\\ &= 154\ \text{kg H}_2\text{SO}_4\ \text{per cycle}\end{aligned}$$

Assume that the purity of commercial acid is 93% and twice the theoretical amount is used.

$$\begin{aligned}\text{Quantity of commercial acid required per cycle} &= 2 \times 154\ \text{kg H}_2\text{SO}_4\ \text{per cycle} \times \frac{100\%}{93\%}\\ &= 331\ \text{kg acid per cycle}\end{aligned}$$

Normally, 2% acid solution is applied. The assumed density of 2% acid solution is 1005 kg/m^3.

$$\text{Volume of regenerant solution per cycle} = \frac{331\ \text{kg}}{0.02 \times 1005\ \text{kg/m}^3} = 16.5\ \text{m}^3\ \text{solution per cycle}$$

Volume of regenerant solution is 8.3 m^3 per column per cycle

b. Strong base anion exchangers.

Caustic (sodium hydroxide, NaOH) solution is used to regenerate the strongly basic anion exchanger (Equation 15.25b).

$$\text{Quantity of NaOH required per cycle} = M_{\text{I,anion}} \times \frac{\text{Eq. wt. of NaOH}}{\text{Eq. wt. of CaCO}_3}$$

$$= 157\,\text{kg as CaCO}_3 \times \frac{40\,\text{g/eq as NaOH}}{50\,\text{g/eq as CaCO}_3}$$

$$= 126\,\text{kg NaOH per cycle}$$

Assume that the purity of commercial caustic is 95% and twice the theoretical amount is used.

$$\text{Quantity of commercial caustic required per cycle} = 2 \times 126\ \text{kg NaOH per cycle} \times \frac{100\%}{95\%}$$

$$= 265\,\text{kg caustic per cycle}$$

Similarly 2% NaOH solution is applied. The assumed density of 2% caustic solution is 1002 kg/m^3.

$$\text{Volume of regenerant solution per cycle} = \frac{265\ \text{kg}}{0.02 \times 1002\ \text{kg/m}^3} = 13.2\ \text{m}^3\ \text{solution per cycle}$$

Volume of regenerant solution is 6.6 m^3 per column per cycle

4. Draw the conceptual process diagram in Figure 15.31.

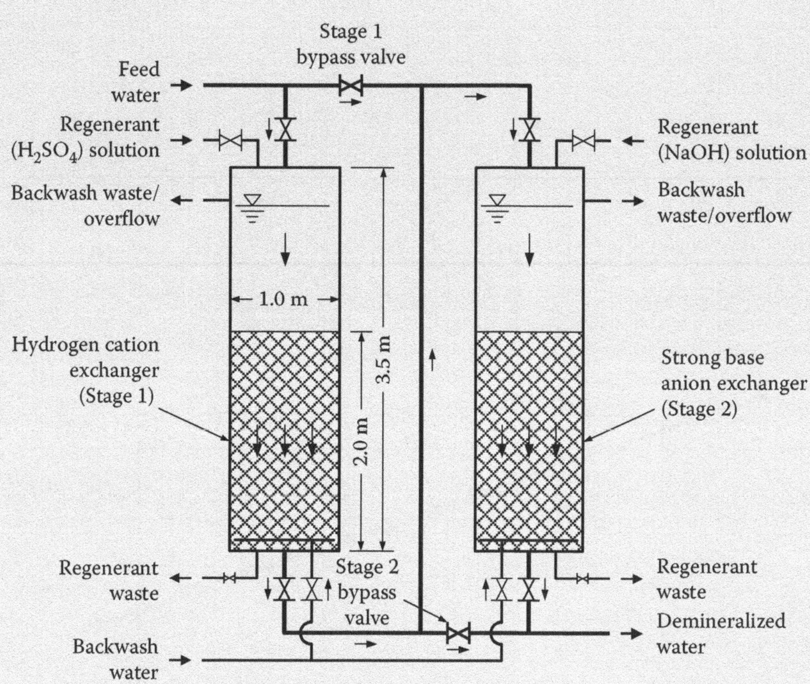

Note: Two process trains are provided in the system. One train is illustrated in this figure.

FIGURE 15.31 Conceptual process diagram of ion exchange demineralizers (Example 15.46).

EXAMPLE 15.47: DESIGN OF ION EXCHANGER USING THOMAS EQUATION

An industrial wastewater has Cu^{2+} concentration of 100 mg/L. The flow is 400 m^3/d. A bench-scale column test was conducted and the experimental data from the test are summarized in Example 15.43. Determine the kinetic constants k_1, and q_0 for the resin using the procedure similar to that developed in Example 15.38. The mass of resin in the column is 25 g and flow rate is 1.2 L/d. The density of resin is 650 kg/m^3. Also, calculate (a) kg of resin required per cycle, (b) dimensions of the column, and (c) the height of sorption zone for treating the industrial wastewater flow. Assume that the allowable concentration of Cu^{2+} in the effluent is 5% of C_0. The resin depth is twice the diameter. The breakthrough time is 2 d.

Solution

1. Develop the plotting data.

   Use the column test results in Example 15.43, and tabulate the breakthrough volume (V_B) and $\ln(C_0/C_e - 1)$ in Table 15.21.

TABLE 15.21 Breakthrough Test Data for Plotting Linear Relationship (Example 15.47)

Breakthrough Volume (V_B), L	Effluent Concentration (C_e), mg/L	$\ln(C_0/C_e - 1)$
10	0.015	8.80
14	0.10	6.91
16	0.37	5.60
18	1.34	4.30
20	2.66	3.60
22	5.71	2.80
25	27	0.99
27	50	0.00
30	82	−1.52
32	93	−2.59

2. Plot the linear relationship between $\ln(C_0/C_e - 1)$ and V_B in Figure 15.32.

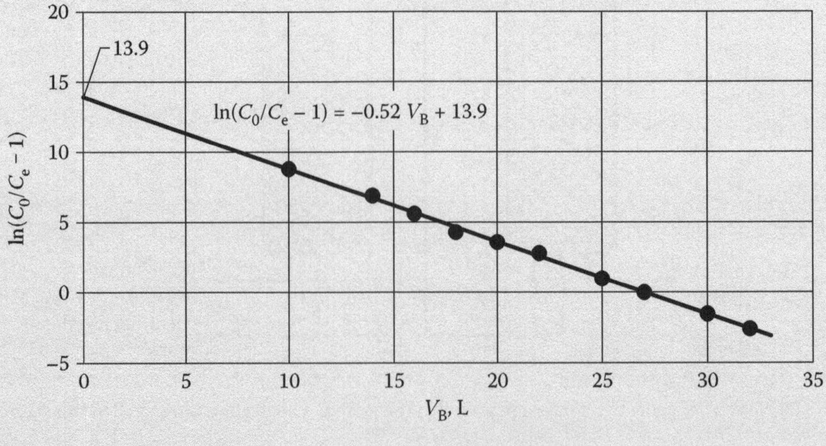

FIGURE 15.32 Linear plot of $\ln(C_0/C_e - 1)$ versus V_B (Example 15.47).

3. Determine the slope (S) and intercept (I) of the linear relationship (Equation 15.18b).
   Read slope $(S) = -0.52\ \text{L}^{-1}$, and intercept $(I) = 13.9$.
4. Determine the rate constant (k_1) and maximum solid-phase concentration of Cu^{2+} absorbed by resin.
   Rearrange Equation 15.18c to calculate K_1.

$$k_1 = -\frac{SQ}{C_0} = -\frac{(-0.52\ \text{L}^{-1}) \times 1.2\ \text{L/d}}{100\ \text{mg/L}} = 0.0062\ \text{L/mg·d}$$

eq. wt. = 31.8 g/eq or 31.8 mg/meg of Cu^{2+}
$k_1 = 0.0062\ \text{L/mg·d} \times 31.8\ \text{mg/meg} = 0.20\ \text{L/meg·d}$ or $0.20\ \text{m}^3/\text{eq·d}$
Rearrange Equation 15.18d to calculate q_0

$$q_0 = \frac{IQ}{k_1 M} = \frac{13.9 \times 1.2\ \text{L/d}}{0.20\ \text{L/meq·d} \times 25\ \text{g}} = 3.3\ \text{meq/g} \quad \text{or} \quad 3.3\ \text{eq/kg resin}$$

5. Determine the volume (V) and mass of resin needed for full-scale unit (M_{design}).
   Breakthrough volume at 2 d/cycle,

$$V_{\text{B,design}} = 2\ \text{d} \times 400\ \text{m}^3/\text{d} = 800\ \text{m}^3$$

Apply Equation 15.18b at $Q_{\text{design}} = 400\ \text{m}^3/\text{d}$, and $C_e = 0.05C_0 = 0.05 \times 100\ \text{mg/L} = 5\ \text{mg/L}$.

$$\ln\left(\frac{C_0}{C_e} - 1\right) = -\frac{k_1 C_0}{Q_{\text{design}}} V_{\text{B,design}} + \frac{k_1 q_0 M_{\text{design}}}{Q_{\text{design}}}$$

$$\ln\left(\frac{100\ \text{mg/L}}{5\ \text{mg/L}} - 1\right) = -\frac{0.20\ \text{m}^3/\text{eq·d} \times 100\ \text{g/m}^3}{400\ \text{m}^3/\text{d} \times 31.8\ \text{g/eq}} \times 800\ \text{m}^3 + \frac{0.20\ \text{m}^3/\text{eq·d} \times 3.3\ \text{eq/kg} \times M_{\text{design}}}{400\ \text{m}^3/\text{d}}$$

$$2.94 = -1.24 + 0.0017\ \text{kg}^{-1} \times M_{\text{design}}$$

$$M_{\text{design}} = \frac{1.24 + 2.94}{0.0017}\ \text{kg} = 2460\ \text{kg resin per cycle}$$

Resin requirement per unit wastewater, $\dfrac{M_{\text{design}}}{V_{\text{B,design}}} = \dfrac{2460\ \text{kg}}{800\ \text{m}^3} = 3.1\ \text{kg resin/m}^3\ \text{wastewater}$

6. Calculate the volume and dimensions of the full-scale resin unit.
   Rearrange Equation 15.17f to calculate the volume of resin bed (EBV_{design}).

$$EBV_{\text{design}} = \frac{M_{\text{design}}}{\rho_{\text{GAC}}} = \frac{2460\ \text{kg}}{650\ \text{kg/m}^3} = 3.8\ \text{m}^3 \approx 4\ \text{m}^3\ \text{resin per cycle}$$

It is given that the resin depth (D) is twice the diameter (d), $D = 2\ d$.

$$EBV_{\text{design}} = \frac{\pi}{4} d^2 D = \frac{\pi}{4} d^2 (2d) = \frac{\pi}{2} d^3$$

Diameter required, $d = \left(\dfrac{2EBV_{\text{design}}}{\pi}\right)^{1/3} = \left(\dfrac{2 \times 4\ \text{m}^3}{\pi}\right)^{1/3} = 1.37\ \text{m}$

Proved design diameter $d_{\text{design}} = 1.4\ \text{m}$, and resin bed depth $D_{\text{design}} = 2.8\ \text{m}$, and total height $H_{\text{design}} = 4.5\ \text{m}$ to accommodate a 150% bed expansion during backwash.

EXAMPLE 15.48: ION EXCHANGE SYSTEM TO RECOVER AND CONCENTRATE BY-PRODUCT

An electroplating shop is producing 182 m^3/d waste stream. The principal metals in the waste stream are: chromium = 130 mg/L as CrO_3, Zn = 20 /L, Ni = 30 mg/L, and Cu = 40 mg/L. Develop the process for metal removal and Cr recovery. The exchange capacity of the cation exchange bed is 3000 eq/m^3 (85 eq/ft^3), the H_2SO_4 regenerant solution is 5% and 30% excess is used. The rinse water is 16 m^3/m^3 (120 gal/ft^3) of resin volume. The anion exchanger has a capacity of 61 kg CrO_3/m^3 (3.8 lb/ft^3). The NaOH regenerant solution is 10% and 26% excess is used. The rinse water is 13.4 m^3/m^3 (100 gal/ft^3) of resin volume. The regeneration cycle of resin beds is 5 day.

Solution

1. Brief description of the treatment processes.
   The ion exchange system consists of a total of three ion exchangers. Two ion exchangers are arranged in series to achieve the desired wastewater treatment goals: (a) Exchange A: a hydrogen cationic exchanger for metal removal (Equation 15.23a) and (b) Exchange B: a strong base anionic exchanger for removal of chromium (Equation 15.25a). The third ion exchanger (Exchange C) is a hydrogen cationic exchanger for recovery of Cr in the regenerant waste from Exchange B (Equation 15.23a).
2. Design of the hydrogen cationic resin for metal removal in Exchanger A.
   The ion exchanger reactions of metals with the resin are described by Equation 15.23a.

$$H_2R + \begin{bmatrix} Zn \\ Ni \\ Cu \end{bmatrix} \cdot CrO_4 \leftrightarrow \begin{bmatrix} Zn \\ Ni \\ Cu \end{bmatrix} R + H_2CrO_4 \text{ (soluble)}$$

a. Determine the concentrations of Zn, Ni, and Cu as eq/d in the waste stream.

$$\text{Zn concentration, } C_{0,Zn} = \frac{20 \text{ mg/L as Zn}}{\text{Eq. wt. of Zn}^{2+}} = \frac{20 \text{ mg/L as Zn}}{32.7 \text{ mg/meq as Zn}^{2+}} = 0.61 \text{ meq/L}$$

$$\text{Ni concentration, } C_{0,Ni} = \frac{30 \text{ mg/L as Ni}}{\text{Eq. wt. of Ni}^{2+}} = \frac{30 \text{ mg/L as Ni}}{29.3 \text{ mg/meq as Ni}^{2+}} = 1.02 \text{ meq/L}$$

$$\text{Cu concentration, } C_{0,Cu} = \frac{30 \text{ mg/L as Cu}}{\text{Eq. wt. of Cu}^{2+}} = \frac{40 \text{ mg/L as Cu}}{31.8 \text{ mg/meq as Cu}^{2+}} = 1.26 \text{ meq/L}$$

$$\text{Total cationic concentration, } C_{0,cation} = C_{0,Zn} + C_{0,Ni} + C_{0,Cu}$$
$$= (0.61 + 1.02 + 1.26) \text{ meq/L} = 2.89 \text{ meq/L or } 2.89 \text{ eg/m}^3$$

b. Determine the quantity of cations discharged in the 5-d waste stream per cycle.

Total volume of waste stream treated per cycle, $V = 5\text{ d} \times Q = 5\text{ d} \times 182 \text{ m}^3/\text{d} = 910 \text{ m}^3$

Quantity of cations discharged per cycle from Equation 15.27b,

$$M_{I,cation} = VC_{0,cation} = 910 \text{ m}^3 \times 2.89 \text{ eq/m}^3 = 2630 \text{ eq per cycle}$$

c. Determine the theoretical volume of resin bed per cycle from Equation 15.27f.

$$V_{R,ExchangeA} = \frac{M_{I,cation}}{EC'_{cation}} = \frac{2630 \text{ eq}}{3000 \text{ eq/m}^3} = 0.88 \text{ m}^3 \approx 0.9 \text{ m}^3 \text{ of cationic resin per cycle}$$

d. Determine the dimensions of cationic exchange bed.
   Provide two hydrogen cationic exchange columns. The required volume of each column = 0.45 m^3.

Dimensions of each column: diameter = 0.61 m (2 ft), surface area = 0.29 m^2 (3.1 ft^2), resin depth = 1.55 m (5 ft), total resin volume = 0.45 m^3 (each), and total height of column = 3 m. The extra depth of 1.45 m (5 ft) above the resin bed gives 50% bed expansion and freeboard.

e. Determine the volume of acid regenerant solution required for regeneration.
Regeneration reactions with acid are described by Equation 15.23c.

$$\begin{bmatrix} Zn \\ Ni \\ Cu \end{bmatrix} R + H_2SO_4 \leftrightarrow H_2R + \begin{bmatrix} Zn \\ Ni \\ Cu \end{bmatrix} \cdot SO_4\,(\text{soluble})$$

$$\begin{aligned} \text{Quantity of H2SO4 required per cycle, } M_{H2SO4} &= M_{I,\text{cation}} \times \text{Eq. wt. of } H_2SO_4 \\ &= 2630\ \text{eq} \times 49\ \text{g/eq as } H_2SO_4 \\ &= 129{,}000\ \text{g}\,H_2SO_4 \text{ or } 129\ \text{kg}\,H_2SO_4 \text{ per cycle} \end{aligned}$$

Assume that the commercial acid is 98% pure H_2SO_4 and 30% excess is used per regeneration cycle.

$$\text{Quantity of commercial acid required per cycle} = 1.3 \times 129\ \text{kg} \times \frac{100\%}{98\%} = 171\ \text{kg acid per cycle}$$

Assumed density of 5% acid solution is 1008 kg/m^3.

$$\text{Volume of regenerant solution per cycle} = \frac{171\ \text{kg}}{0.05 \times 1008\ \text{kg/m}^3} = 3.4\ \text{m}^3 \text{ solution per cycle}$$

f. Determine the volume of rinse water and total volume of regenerant brine.

$$\text{Volume of rinse water} = 16\ \text{m}^3/\text{m}^3 \times V_{R,\text{ExchangeA}} = 16\ \text{m}^3/\text{m}^3 \times 0.9\ \text{m}^3 = 14.4\ \text{m}^3 \text{ per cycle}$$

$$\text{Total volume of brine (regenerant and rinse water)} = (3.4 + 14.4)\ \text{m}^3 = 17.8\ \text{m}^3 \text{ per cycle}$$

g. Suggest treatment method to precipitate Zn, Ni and Cu.
Use lime ($Ca(OH)_2$) to increase the pH of the regenerant brine in the range of 10–11. The metals combined with SO_4^{2-} will precipitate out as metal hydroxides.

3. Design of the strong base anionic resin for removal of chromium in Exchanger B.
The dilute chromic acid (H_2CrO_4) in the effluent from the Exchanger A is removed by the strong base anionic resin. The ion exchanger reaction of chromic acid (H_2CrO_4) with the resin is described by Equation 15.25a.

$$2\,(RN)OH + H_2CrO_4\,(\text{soluble}) \leftrightarrow (RN)_2CrO_4 + 2\,H_2O$$

a. Determine the volume and dimensions of the resin bed.
Quantity of chromium removed per cycle from Equation 15.27b,

$$M_{I,\text{anion}} = VC_{0,\text{cr}} = 910\ \text{m}^3 \times 130\ \text{g/m}^3 \times 10^{-3}\ \text{kg/g} = 118\ \text{kg as } CrO_3 \text{ per cycle}$$

Theoretical volume of resin bed per cycle from Equation 15.27f at 100% removal efficiency.

$$V_{R,\text{ExchangeB}} = \frac{M_{I,\text{anion}}}{EC'_{\text{anion}}} = \frac{118\ \text{kg}}{61\ \text{kg/m}^3} = 1.93\ \text{m}^3 \approx 2\ \text{m}^3 \text{ of anionic resin per cycle}$$

Provide two strong base anionic exchange columns. The required volume of each column = 1 m^3.
Dimensions of each column: diameter = 0.91 m (3 ft), surface area = 0.65 m^2 (7.1 ft^2), resin depth = 1.55 m (5 ft), total resin volume = 1 m^3 (each), and total height of column = 3 m. This height provides room for 50% bed expansion and freeboard.

b. Determine the volume of caustic regenerant solution or brine.
Regeneration reaction with caustic is described by Equation 15.25b.

$$(RN)_2CrO_4 + 2\,NaOH \leftrightarrow 2\,(RN)OH + Na_2CrO_4(\text{soluble})$$

$$\text{Quantity of NaOH required per cycle } M_{NaOH} = M_{I,anion} \times \frac{2 \times \text{Molar wt. of NaOH}}{\text{Molar wt. of } CrO_3}$$

$$= 118 \text{ kg as } CrO_3 \times \frac{2 \times 40 \text{ g/mole as NaOH}}{100 \text{ g/mole as } CrO_3}$$

$$= 94.4 \text{ kg NaOH per cycle}$$

If the commercial caustic contains 98% NaOH and 26% excess is used, the quantity is 121 kg caustic per cycle. Assumed density of 10% acid solution is 1009 kg/m^3, the volume of regenerant solution is 1.2 m^3 per cycle. The regenerant brine is discharge into Exchanger C for recovery of Cr.

c. Determine the volumes of rinse water and brine from the regeneration.

$$\text{Volume of rinse water} = 13.4 \text{ m}^3/\text{m}^3 \times V_{R,ExchangeB}$$
$$= 13.4 \text{ m}^3/\text{m}^3 \times 2 \text{ m}^3 = 26.8 \text{ m}^3 \text{ per cycle}$$

Daily average rinse water = 5.4 m^3/d during a 5-d cycle

This stream contains low concentrations of Na_2CrO_4. It is returned to Exchanger A for further removal of chromium and it passes through Exchanger B again. Therefore, the combined influent to Exchanger A is (182 + 5.4) m^3/d = 187.4 m^3/d (daily average). This is also the total flow discharged from Exchanger B.

4. Design of the hydrogen cationic resin for recovery of concentrated H_2CrO_4 in Exchanger C.

The purpose of Exchanger C is to recover H_2CrO_4 as a byproduct from Na_2CrO_4 in the regenerant waste discharged from the Exchanger B. The ion exchanger reaction of Na_2CrO_4 with the resin are described by Equation 15.23a.

$$H_2R + Na_2CrO_4 \text{ (dilute)} \leftrightarrow Na_2R + H_2CrO_4 \text{ (concentrate)}$$

a. Determine the volume and dimensions of the resin bed.

Quantity of NaOH required to regenerate anionic resin in Exchange B per cycle,

$$M_{NaOH} = \frac{94.4 \text{ kg as NaOH}}{40 \text{ g/eq as NaOH}} \times 10^3 \text{ g/kg} = 2360 \text{ eq per cycle}$$

Theoretical volume of hydrogen cationic resin bed per cycle from Equation 15.27f,

$$V_{R,ExchangeC} = \frac{M_{I,cation}}{EC'_{cation}} = \frac{2360 \text{ eq}}{3000 \text{ eq/m}^3} = 0.79 \text{ m}^3 \text{ of cationic resin per cycle}$$

Provide two columns of the same dimensions as the cationic resin bed and column in Exchanger A: diameter = 0.61 m (2 ft), resin depth = 1.55 m (5 ft), total resin volume = 0.45 m^3 (each), and total height of column = 3 m.

b. Determine the volume and concentration of H_2CrO_4 solution recovered.

The equivalent amount of Na_2CrO_4 in the regenerant waste from in Exchanger B = 118 kg as CrO_3 per cycle (Step 3a.), and the volume of regenerant waste is 1.2 m^3 per cycle (Step 3b.). This stream is passed through hydrogen cationic Exchanger C to recover H_2CrO_4.

$$\text{The quantity of } H_2CrO_4 \text{ recovered per cycle} = M_{I,anion} \times \frac{\text{Molar wt. of } H_2CrO_4}{\text{Molar wt. of } CrO_3}$$

$$= 118 \text{ kg as } CrO_3 \times \frac{118 \text{ g/mole as } H_2CrO_4}{100 \text{ g/mole as } CrO_3}$$

$$= 139 \text{ kg } H_2CrO_4 \text{ per cycle}$$

$$\text{Concentration of } H_2CrO_4 = \frac{139 \text{ kg as } H_2CrO_4}{1.2 \text{ m}^3 \times 1009 \text{ kg/m}^3}$$
$$= 0.115 \text{ kg } H_2CrO_4/\text{kg regenrant waste} \quad \text{or} \quad 11.5\%$$

c. Determine the volumes of acid regenerant solution, rinse water, and brine.

Regeneration reactions with acid are described by Equation 15.23c.

$$Na_2R + H_2SO_4 \leftrightarrow H_2R + Na_2SO_4(\text{soluble})$$

$$\text{Quantity of } H_2SO_4 \text{ required per cycle} = M_{I,NaOH} \times \text{Eq. wt. of } H_2SO_4$$
$$= 2360 \text{ eq} \times 49 \text{ g/eq as } H_2SO_4$$
$$= 116{,}000 \text{ g } H_2SO_4 \quad \text{or} \quad 116 \text{ kg } H_2SO_4 \text{ per cycle}$$

For the commercial acid of 98% H_2SO_4 when 30% excess is used, the quantity is 154 kg acid per cycle. Use density of 1008 kg/m^3 for 5% acid solution, the volume of regenerant solution is 3.1 m^3 per cycle.

$$\text{Volume of rinse water} = 16 \text{ m}^3/\text{m}^3 \times V_{R,\text{ExchangeC}} = 16 \text{ m}^3/\text{m}^3 \times 0.9 \text{ m}^3 = 14.4 \text{ m}^3 \text{ per cycle}$$

$$\text{Total volume of brine (regenerant and rinse water)} = (14.4 + 3.1) \text{ m}^3 = 17.5 \text{ m}^3 \text{ per cycle}$$

5. Draw the process schematic.

The conceptual process diagram of the ion exchange system with hydrogen cationic resin Exchanger A, strong base anionic resin Exchanger B and hydrogen cationic resin Exchanger C are shown in Figure 15.33.

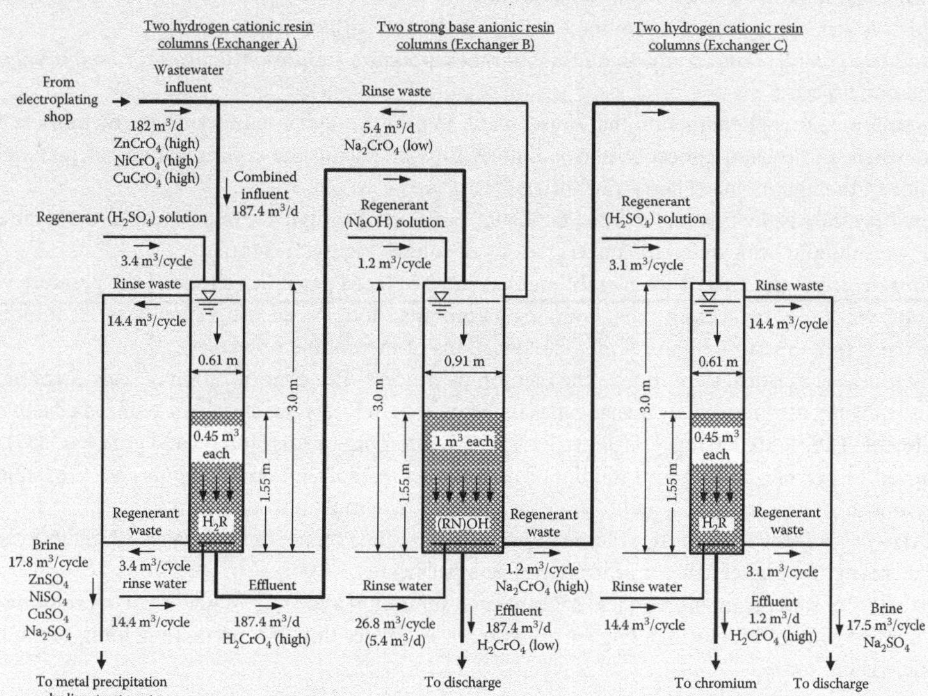

Note: Two identical ion exchanger treatment process trains are provided in the ion exchange system.

FIGURE 15.33 Process diagram of ion exchange system for metal removal and recovery of chromium (Example 15.48).

15.4.10 Membrane Processes

Membranes serve as a selective barrier that retains certain constituents by their sizes and allow the passage of others. The retention of constituents may vary from TSS, turbidity, and microorganisms to dissolved organics and ionized salt ions. In recent years, there have been rapid advancements in membrane technology. For this reason, the applications of membranes for treatment of water and wastewater will continue to grow rapidly in the future. These applications include removal of emerging constituents, recovery of resources, membrane bioreactor (MBR), biosolids thickening, renovation and production of high-quality reclaimed effluent for reuse, and demineralization/desalination.

Common Technical Terms: Common terms used in membrane processes are given below. Definition sketches are shown in Figure 15.34.

Cleaning is an operating procedure that is used periodically to clean membrane by depressurization, high-velocity water jet, air–water mixture, and chemical solutions.

Feedwater is the influent flow to the membrane system. *Permeate*, *filtrate*, or *product water* is the portion of feedwater that passes through the membrane. The fraction of feedwater that does not pass through the membrane is called *retentate*, *brine*, *concentrate*, or *waste stream*. Figure 15.34a shows the feedwater, permeate, and retentate streams in either inside-out or outside-in flow design.

Feedwater stream velocity is the velocity through the membrane. It must be high enough to create turbulence that is desired to prevent concentration polarization on the membrane surface.

Flux is the volume of product water recovered per unit area of the membrane. Flux decreases with length of run. Typically, the flux may decrease by 50% after 2 years of operation.

Fouling is deposition of solids on membrane surface that lowers the performance and shortens the life of membrane. Fouling can be either reversible or irreversible.

Lumen is the interior of a hallow fiber membrane.

Membrane element is a single membrane that provides the surface area.

Module is a complete unit comprised of membranes, pressure support structures for membranes, and inlet and outlet parts.

Osmosis is a natural phenomenon that causes water to pass through a semipermeable membrane into a solution where the solvent concentration is higher, thus balancing the concentrations of materials on either side of the membrane (Figure 15.34b(i)).

Osmotic pressure is the pressure exerted by the flow of water through a semipermeable membrane separating two solutions with different concentrations of solute (Figure 15.34b(ii)).

Packing density is the area of the membrane that can be placed per unit volume of the pressure vessel.

Pressure vessel is a metal casing that contains membrane modules, cartridges, or cassettes in different *arrangements* (see description later under classifications of membrane processes).

Pretreatment is a critical step prior to the membrane process. The concentration of scale-forming constituents (calcium, magnesium, iron, manganese, silicon, etc.) in feedwater should be reduced effectively by pretreatment. TDS $< 10{,}000$ mg/L is desired. Organics, microorganisms, and oil and grease (O&G) must be removed to prevent coating and fouling of membranes. Cellulose acetate membranes are subject to hydrolysis at high and low pH. The turbidity and particle size must also be controlled.

Recovery factor is the ratio of the volume of permeate to total volume of feed water. At higher recovery factor, there may be higher salt concentration in the permeate.

RO is a flow of water from more to less concentrated solution if a pressure higher than the osmotic pressure is applied (Figure 15.34b(iii)). Higher pressure gives greater flux, but pressure holding capability of membrane is limited.

Salt rejection is the ratio of salt in the retentate to that in the feed water. Salt rejection depends upon the type of membrane and salt concentration gradient.

Temperature is an important operating parameter that greatly affects the performance of membrane system. A greater flux is generally allowed at higher temperature. However, the life of membrane is reduced at a temperature $> 38°C$ (100°F).

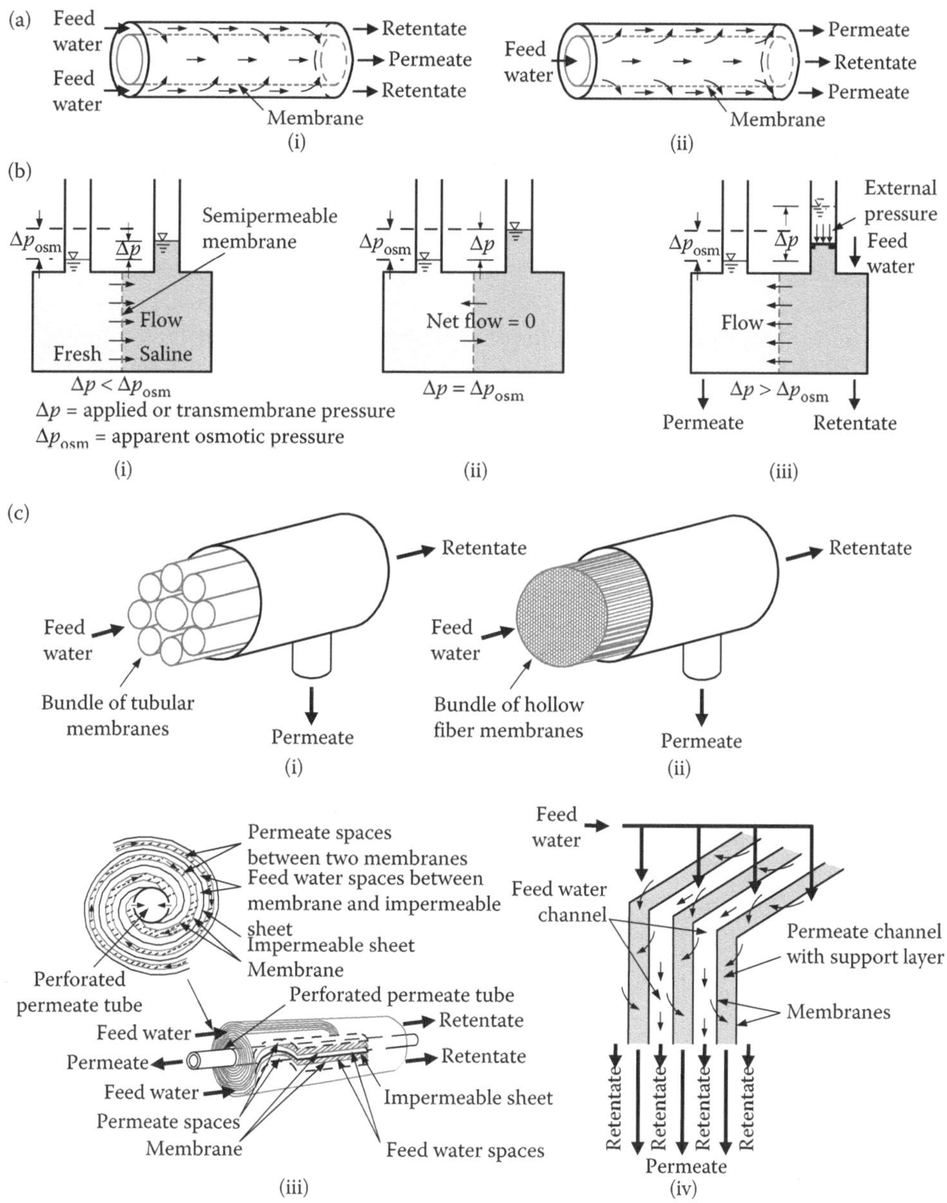

FIGURE 15.34 Definition sketches of membrane systems: (a) feedwater, permeate, and retentate streams: (i) outside-in system and (ii) inside-out system; (b) osmosis: (i) osmotic flow, (ii) osmotic equilibrium without flow, and (iii) reverse osmosis flow; and (c) membrane arrangements: (i) tubular, (ii) hollow fiber, (iii) spiral-wound, and (iv) flat plate (or sheet). (Adapted in part from References 10, 36, 37, 54, and 92.)

Transmembrane pressure (TMP) is the difference in pressure between two sides of the membrane. It is a valuable measurement because it describes how much force is needed to push water through the membrane. A low TMP indicates a clean and well-functioning membrane. A high TMP indicates a dirty or "fouled" membrane with reduced filtering ability.

Classifications of Membrane Processes: Membrane processes may be classified based on different characteristics of the membrane systems. These characteristics are (1) pore opening, (2) operating range, (3) driving force, (4) removal mechanism, (5) membrane material, and (6) membrane configuration.

General classifications of membrane systems based on these characteristics are identified in Table 15.22 and are briefly discussed below.

Pore Opening: Based on pore opening or size, the membrane processes are basically categorized as (1) MF, (2) UF, (3) nanofiltration (NF), (4) RO, and (5) electrodialysis (ED). The pores range from large to small, and may be recognized as macropores (>50 nm), messopores (2–50 nm), micropores (<2 nm), and denses (<2 nm). Detailed information about the pore sizes of these basic membrane systems are summarized in Table 15.22.

Operating Range: Because of the different opening sizes the membrane systems cover an extremely wide operating range from 0.0001 to 2 μm. The operating ranges of the basic membrane systems are identified in Table 15.22.

Driving Force: As identified in Table 15.22, the driving forces on membrane are (a) *hydrostatic pressure difference* or *submerged vacuum*, (b) *concentration difference*, and (c) *electromotive force*. The MF and UF membrane systems operate on differential pressure that may be created by either hydrostatic pressure in closed vessels or submerged vacuum condition in open vessels. The driving forces in NF and RO processes are both pressure and concentration differentials. In ED, the inorganic ions are transferred from an electrolyte by electromotive force across the selective permeable membrane.

Removal Mechanism: The removal mechanisms in membrane processes involve (a) sieving or straining action, (b) diffusion in solutions, (c) exclusion, and (d) ion exchange with selective membranes. The mechanisms that are utilized by the basic membrane processes are identified in Table 15.22.

Membrane Material: The membranes are manufactured from inorganic and organic synthetic materials. Typical membranes are approximately 100-μm thick having a skin of about 0.2-μm thick that serves as

TABLE 15.22 Membrane Processes and Classifications

Process	Pore Opening, nm	Operating Range, μm	Driving Force	Removal Mechanism	Material[a]	Configuration
Microfiltration (MF)	>50 (macropores)	0.07–2	Hydrostatic pressure difference or submerged vacuum	Sieve	Acrylonitrile, ceramic, PE, PES, PP, PS, PTFE, PVDF, and nylon	Tubular or hollow fiber, spiral wound, and flat plate (or sheet)
Ultrafiltration (UF)	2–50 (mesopores)	0.008–0.2	Hydrostatic pressure difference or submerged vacuum	Sieve	Aromatic polyamides, ceramic, CA, PE, PES, PP, PS, PVDF, and Teflon	Tubular or hollow fiber, spiral wound, and flat plate (or sheet)
Nanofiltration (NF)	<2 (micropores)	0.001–0.01	Hydrostatic pressure difference, or concentration difference	Sieve, diffusion, and exclusion	Cellulosic, aromatic polyamides, PS, PVDF, and TFC	Hollow fiber, spiral wound, and thin-film composite
Reverse osmosis (RO)	<2 (dense)[b]	0.0001–0.002	Hydrostatic pressure difference, or concentration difference	Diffusion and exclusion	Cellulosic, aromatic polyamides, and TFC	Hollow fiber, spiral wound, and thin-film composite
Electrodialysis (ED)	–	0.0003–0.002	Electromotive force	Ion exchange	Ion exchange sheet	Flat plate (or sheet)

[a] CA = cellulose acetate, PE = polyethylene, PES = polyethylsulfone, PP = polypropylene, PS = polysulfone, PTFE = polytetrafluorethylene, PVDF = polyvinylidene fluoride, and TFC = thin-film composite.

[b] The pore sizes of RO membrane are extremely small. This nonporous synthetic membrane is known as a *dense* membrane.

Note: 1 nm = 10^{-9} m and 1 μm = 10^{-6} m.

Source: Adapted in part from References 36, 37, 49, 54, and 64.

the rejecting surface. The remaining layer is porous and spongy, and serves as supporting material. The typical materials used for the basic membrane processes are summarized in Table 15.22.

Membrane Configuration: The common membrane configurations are (a) tubular membrane (Figure 15.34c(i)), (b) hollow fiber membrane (Figure 15.34c(ii)), (c) spiral wound thin-frame membrane (Figure 15.34c(iii)), and (d) flat plate (pleated-cartridge or rotating disk) membrane (Figure 15.34c(iv)).

The choice of membrane type and configuration depends upon (1) permeate rate, (2) rejection rate, (3) membrane fouling and cleaning, (4) useful life, (5) pretreatment of feedwater, (6) packing density, (7) power requirement, and (8) desired quality of finished product water. Readers are referred to several publications on these topics.[36,37,54,64,103–108] Also, membranes and equipment manufacturers should be consulted on this subject.

MF and UF Systems: MF or UF filtration is an alternative to solids removal by sedimentation and filtration. MF process removes TSS, turbidity, colloids, protozoan cysts, limited bacteria, some viruses, and large organic molecules. The high-quality permeate is suitable for reclaimed water reuse. UF system removes colloidal solids, most bacteria, some viruses, and large organic molecules and proteins. Permeate contains mostly small molecules. Both membranes operate under differential pressure or by vacuum under submerged condition. Discussions are provided below for both arrangements.

Components and Major Factors: Common equipment for operating MF and UF membranes include pretreatment facility with degritting and fine screening, feed pump, pressure vessel, membrane modules, and backwash system. The design factors are plant capacity, TSS and turbidity, flux rate, TMP, backwash quantity and quality, and average cleaning frequency. Some conceptual sketches of MF and UF membrane systems are shown in Figure 15.35. The basic design and operating parameters are summarized in Table 15.23.

Operation under Differential Pressure: Based on mass–balance relationships many equations have been developed to determine the operating pressure, permeate flow, recovery, and rejection of MF and UF systems (Figure 15.35a). These systems may also be in cross-flow or dead-end design (Figure 15.35b). These relationships are expressed by Equations 15.28 through 15.30.

$$P_t = \frac{P_f + P_r}{2} - P_p \quad \text{(Cross-flow model)} \quad (15.28a)$$

$$\Delta P = P_f - P_p \quad \text{(Cross-flow model)} \quad (15.28b)$$

$$P_t = P_f - P_p \quad \text{(Dead-end model)} \quad (15.28c)$$

$$Q_f = Q_p + Q_r \quad \text{(Flow balance)} \quad (15.29a)$$

$$Q_f C_f = Q_p C_p + Q_r C_r \quad \text{(Mass balance)} \quad (15.29b)$$

$$Q_p = F_t A \quad \text{(Permeate flow)} \quad (15.29c)$$

$$C_p = \frac{100\% - R}{r} C_f \quad \text{(Permeate concentration)} \quad (15.29d)$$

$$C_r = \frac{R}{100\% - r} C_f \quad \text{(Retentate concentration)} \quad (15.29e)$$

$$r = \frac{Q_p}{Q_f} \times 100\% \quad \text{or} \quad r = \frac{V_p}{V_f} \times 100\% \quad \text{(Recovery factor)} \quad (15.30a)$$

$$R = \left(\frac{C_f - C_p}{C_f}\right) \times 100\% = \left(1 - \frac{C_p}{C_f}\right) \times 100\% \quad \text{(Reject of contaminant)} \quad (15.30b)$$

$$R_{\log} = -\log\left(1 - \frac{R}{100\%}\right) = \log\left(\frac{C_f}{C_p}\right) \quad \text{(Reject of microoranisms)} \quad (15.30c)$$

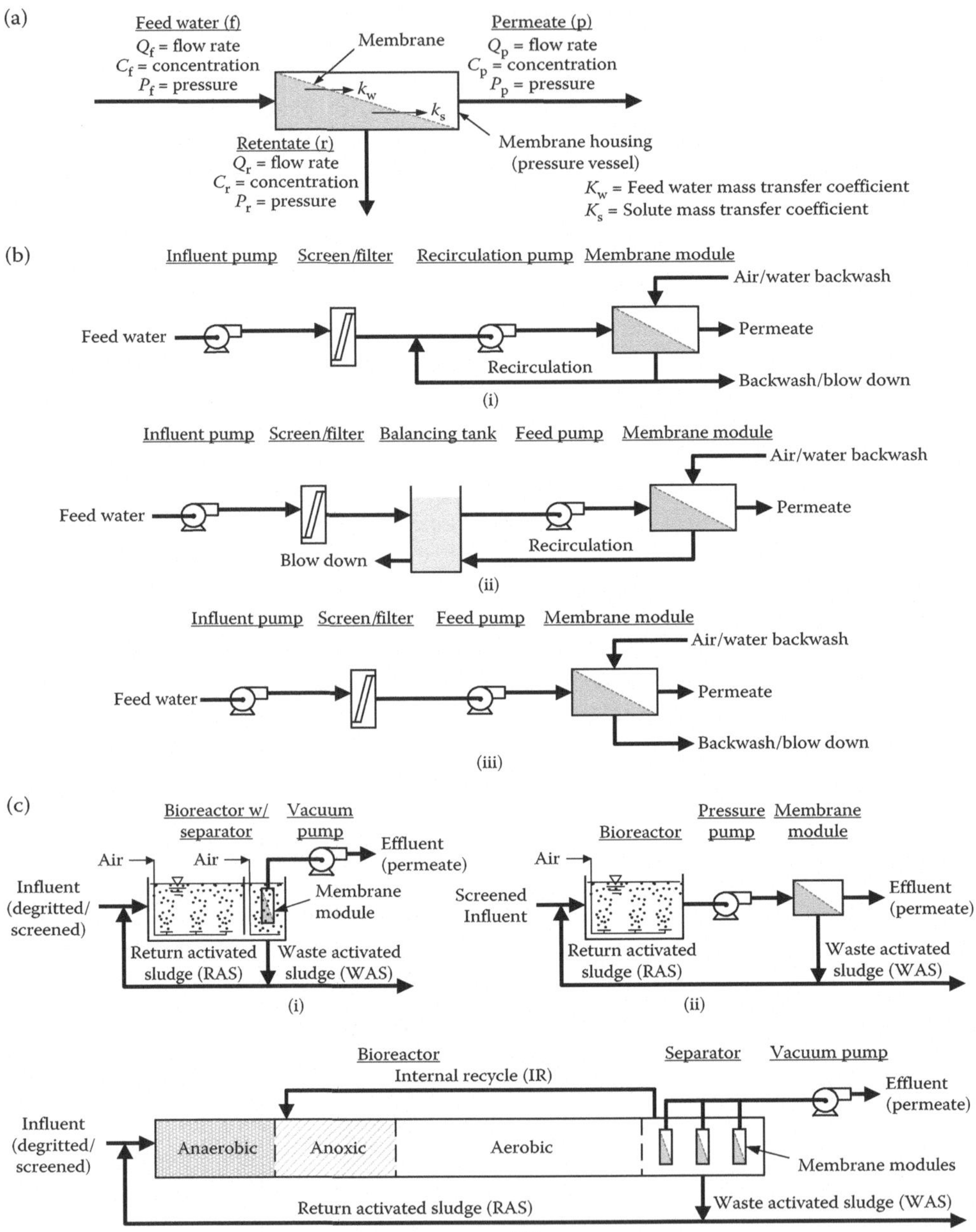

FIGURE 15.35 Definition sketches of MF and UF systems: (a) process operation; (b) process diagrams: (i) cross-flow, (ii) cross-flow with balancing tank, and (iii) dead-end; and (c) MBR processes: (i) immersed membrane module in a bioreactor, (ii) external membrane solids separator, and (iii) integrated BNR facility using A^2/O process. (Adapted in part from References 36, 37, 44, 54, 108, and 109.)

where

P_t = transmembrane pressure gradient, kPa
P_f = feed stream inlet pressure, kPa
P_p = permeate stream pressure, kPa
P_r = retentate stream pressure, kPa

TABLE 15.23 Basic Design and Operating Parameters of MF and UF Systems

Parameter	MF	UF
Flux rate, $L/m^2 \cdot d$ (gpd/ft^2)	600–1400 (15–35)	960–1400 (25–35)
Operating pressure or vacuum, kPa (psig)		
Pressure type	35–200 (5–30)	70–250 (10–35)
Vacuum type	−100 to −35 (−14 to −5)	−100 to −70 (−14 to −10)
Recovery factor, %	85–95	85–95
Energy requirement, kWh/m^3 (kWh/1000 gal)	0.2–0.3 (0.75–1)	0.2–0.3 (0.75–1)
TSS removal, %	95–98	96–99
BOD removal, %	75–90	80–90
COD removal, %	70–85	75–90
Total and fecal coliform, log removal	2–5	3–6

Note: $1\ L/m^2 \cdot d = 0.0245\ gpd/ft^2$; 1 kPa = 0.145 psig; $1\ kWh/m^3 = 3.785\ kWh/10^3$ gal.
Source: Adapted in part from References 36, 37, 54, 64, and 106.

ΔP = pressure drop across membrane module, kPa
Q_f = feedwater flow rate, m^3/h
Q_p = permeate flow rate, m^3/h
Q_r = retentate flow rate, m^3/h
C_f = concentration of contaminant in feed, mg/L (g/m^3)
C_p = concentration of contaminant in retentate, mg/L (g/m^3)
C_r = concentration of contaminant in permeate, mg/L (g/m^3)
F_t = transmembrane flux rate, $m^3/m^2 \cdot h$
A = total surface area of membrane, m^2
r = recovery factor, %
R = rejection factor, %
R_{log} = log rejection of microoranisms, dimensionless
V_p = net volume of permeate, m^3
V_f = volume of feed to the membrane, m^3

Operation under Submerged Vacuum: In these systems, the membranes remain totally submerged and effluent is withdrawn through the membranes under a vacuum. The transmembrane pressure is developed by a centrifugal pump (permeate pump) that creates a vacuum and permeate is removed. Owing to limitations of net positive suction head (NPSH) the maximum TMP is limited to 55 kPa (8 psi). This operation has received wide application in the design of MBR process. The MBRs combine suspended growth biological treatment with an integrated membrane separation. The membranes replace the clarifier and provide enhanced filtration. The process offers many benefits over conventional biological treatment processes.[109–112] Definition sketches of sample MBRs are shown in Figure 15.35c. Brief discussion on MBR system is provided below.[30,36,37,54,113–123]

1. The MBRs produce high-quality effluent (or permeate) that is low in BOD, TSS, turbidity and bacteria. The typical $CBOD_5$, TSS, turbidity, and total coliforms in the effluent from a MBR facility are: <5 and 1 mg/L, 0.2 NTU, and 100 MPN/100 mL, respectively. The MBR permeate quality is superior to filtered secondary effluent.
2. The kinetic equations and design procedures for biological reactors and BNR facilities presented in Chapter 10 also apply to MBRs. The major variations, however, are (a) higher MLSS concentration of 8000–12,000 mg/L resulting in a smaller reactor volume, higher volumetric loading of

0.8–1.5 kg BOD_5/m^3·d, and lower F/M < 0.1 kg BOD_5/kg VSS·d, (b) higher SRTs (10–30 d) to reduce the sludge production and enhance nitrification, (c) lower DO concentration in the reactor that may encourage simultaneous nitrification and denitrification, and (d) higher MLSS concentration may give more stable operation and less susceptibility to upset. Higher return sludge ratio of 300–500% of influent flow may be needed.

3. The MBR consists of MF or UF modules in flat sheet, flat plate (or rotating disk), hollow fiber, or tubular configurations. The typical pore sizes, specific surface areas, and TMPs are 0.04 μm, 300–330 m^2/m^3 (90–100 ft^2/ft^3), and 3–14 kPa (0.4–2 psi) for UF membranes, and 0.4 μm, 150–165 m^2/m^3 (45–50 ft^2/ft^3), and 14–55 kPa (2–8 psi) for MF membranes, respectively.
4. The additional basin volume is required to house the membrane modules is 3–10 m^3 per m^3 of membrane modules, and 0.015–0.05 m^3 per m^2 (0.05–0.16 ft^3 per ft^2) of membrane surface area.
5. The daily average flux rate is 400–600 L/m^2·d (10–15 gpd/ft^2), and the 6-h peak flux rate is < 1300 L/m^2·d (32 gpd/ft^2). The average surface area per module is 30 m^2, and up to 40 modules may be installed in a rack.
6. The membranes in MBR systems may be either vacuum or pressure driven. The vacuum-driven membranes are immersed in aeration basin or in a membrane separation tank (Figure 15.35c(i)). The pressure-driven membranes are external and are used less frequently (Figure 15.35(ii)).
7. MBRs can be used to retrofit or upgrade the existing activated sludge and BNR facilities (Figures 10.39, 10.103, and 15.35c(iii).
8. Membrane fouling is one of the most important concerns in MBR operations. Fouling can be caused by physical, chemical, and biological factors. Commonly used membrane fouling controls at MBR facilities include (a) fine screen (1–2 mm), (b) air scouring with intermittent or pulsed slug bubbling, (c) permeate backpulse by interrupting filtration every 15–30 min and backwashing for 30–45 s, and (d) maintenance cleaning at one to two times per week for 60–75 min by alternatively backflushing with sodium hypochlorite (200 mg/L as Cl_2) and citric acid (2000 mg/L). Less fouling risk has also been seen at higher SRT (30–50 d).
9. Recovery cleaning is also required every 3–6 months, or as needed when a sustained high TMP above the desired value is observed during MBR operations. The inorganic deposition on the membrane surface is removed by soaking membrane modules in sodium hypochlorite (1000 mg/L as Cl_2) and citric acid (2000 mg/L) solutions over a period of 4–6 h. Depending on the MBR design features, it may be performed by (a) cleaning in-place (CIP) without draining the tank or (b) cleaning after draining the tank to isolate the cell.
10. Many proprietary MBR systems are available in the market and the common ones include (a) *LEAPmbr** by GE Power/Water & Process Technologies, (b) *MEMPULSE®* by Evoqua Water Technologies, (c) *NEOSEP®* by Veolia Water Technologies, (d) *VRM®* by HUBER SE, (e) *KUBOTA Submerged Membrane Unit®* by KUBOTA Membrane USA Corporation, and (f) *SteraporeSADF™* by Mitsubishi Rayon Engineering Co., Ltd. Technical information of these MBRs are available in References 54, 113, and 118–123.

Computer Simulation, Modeling and Design of MBR: An accurate analysis of MBR process can be made by considering the mass of heterotrophic bacteria, cell debris, nitrifying bacteria, nonbiodegradable fractions of VSS in the influent and effluent, and many other parameters discussed in Section 10.8.4 and Example 10.161. The hand calculations become quite tedious. These parameters and many more are applied in the computer simulations. Many computer programs are commercially available.[124–127] They are used effectively for design of an MBR facility.

EXAMPLE 15.49: PERFORMANCE OF AN ULTRAFILTRATION (UF) MEMBRANE

An UF system is used to pretreat secondary effluent for an RO unit. After a flow equalization basin, the design average flow in across-flow mode of operation of the UF system is 4000 m^3/d. The feed, retentate, and permeate stream pressures are 70, 30, and 20 kPa. The flux rate is 800 L/m²·d and recovery factor is 80%. The turbidity values of feed and permeate are 5 and 0.5 NTU. The total coliform counts of feed and treated water are 10^4 and 10 MPN/100 mL, respectively. Determine (a) the transmembrane pressure gradient and overall pressure drop across the module, (b) membrane area required, (c) volume requirements, (d) net flow of permeate, and (e) turbidity rejection and log rejection of total coliform.

Solution

1. Determine the transmembrane pressure gradient and pressure drop across the cross-flow module.

   Transmembrane pressure gradient P_t from Equation 15.28a.

$$P_t = \frac{P_f + P_r}{2} - P_p = \frac{(70 + 30)\,\text{kPa}}{2} - 20\,\text{kPa} = 30\,\text{kPa}$$

   Pressure drop across the membrane module P from Equation 15.28b.

$$P = P_f - P_p = (70 - 20)\,\text{kPa} = 50\,\text{kPa}$$

2. Determine the net permeate flow.

   Rearrange Equation 15.30a and calculate net permeate flow rate Q_P.

$$Q_P = \frac{r}{100\%} \times Q_f = \frac{80\%}{100\%} \times 4000\,\text{m}^3/\text{d} = 3200\,\text{m}^3/\text{d}$$

3. Determine the total membrane area required from rearranged Equation 15.29c.

$$A_{mbr} = \frac{Q_P}{F_t} = \frac{3200\,\text{m}^3/\text{d} \times 1000\,\text{L/m}^3}{800\,\text{L/m}^2\cdot\text{d}} = 4000\,\text{m}^2$$

4. Determine the volume requirements.

   a. Hollow fiber membrane cassettes.

      Assume that the cassette contains 32 hollow fiber membrane modules, occupies 4.5 m³ volume, and provides 1000 m² of membrane surface area.

$$\text{Number of cassettes required, } N_{hf} = \frac{A_{mbr}}{1000\,\text{m}^2/\text{cassette}} = \frac{4000\,\text{m}^2}{1000\,\text{m}^2/\text{cassette}} = 4\,\text{cassettes}$$

$$\begin{aligned}\text{Volume occupied by 4 cassettes, } V_{hf} &= 4.5\,\text{m}^3/\text{cassette} \times N_{hf}\\ &= 4.5\,\text{m}^3/\text{cassette} \times 4\,\text{cassettes} = 18\,\text{m}^3\end{aligned}$$

      Provide 50% additional allowance for equipment and clearance. Total volume $V_{hf} = 27\,\text{m}^3$.

   b. Flat sheet membrane modules.

      Assume that the module contains 200 flat membrane sheets, occupies 1 m³ volume, and provides a membrane area of 200 m².

$$\text{Number of modules required, } N_{fs} = \frac{A_{mbr}}{200\,\text{m}^2/\text{module}} = \frac{4000\,\text{m}^2}{200\,\text{m}^2/\text{module}} = 20\,\text{modules}$$

$$\begin{aligned}\text{Volume occupied by 20 modules, } V_{fs} &= 1\,\text{m}^3/\text{module} \times N_{fs} = 1\,\text{m}^3/\text{module} \times 20\,\text{modules}\\ &= 20\,\text{m}^3\end{aligned}$$

      Assume a 50% additional allowance for equipment and clearance. Total volume $V_{fs} = 30\,\text{m}^3$.

5. Determine the rejection rate turbidity and total coliform.

Rejection factor R of turbidity from Equation 15.30b.

$$R = \left(1 - \frac{C_p}{C_f}\right) \times 100\% = \left(1 - \frac{0.5\text{ NTU}}{5\text{ NTU}}\right) \times 100\% = 90\%\text{ rejection}$$

Log rejection of total coliform from Equation 15.30c.

$$R_{\log} = \log\left(\frac{C_f}{C_p}\right) = \log\left(\frac{10^4\text{ MPN/100 mL}}{10\text{ MPN/100 mL}}\right) = \log(10^3) = 3\text{ log rejection}$$

EXAMPLE 15.50: REACTOR VOLUME, AERATION PERIOD, AND BIOMASS PRODUCTION OF CONVENTIONAL ACTIVATED SLUDGE (CAS) AND MBR PROCESSES

A municipal wastewater treatment plant is designed for an average flow of 20,000 m^3/d. Compare the reactor volume, aeration period, and biomass production in a CAS process, and those in an MBR process. The influent BOD_5 of primary treated effluent is 160 mg/L for the existing CAS. The MLVSS and SRT of conventional aeration basin are 2000 mg/L and 5 d, and those for MBR are 10,000 mg/L and 15 d. Assume that the kinetic coefficients are: $Y = 0.55$ g VSS/g BOD_5, $k = 5\text{ d}^{-1}$, $k_d = 0.05\text{ d}^{-1}$, and $K_s = 50$ mg/L BOD_5.

Solution

1. Determine the process parameters for the CAS process.
   a. Soluble BOD_5 (S) in the effluent from Equation 10.15h.

$$S = \frac{K_s(1 + k_d\theta_c)}{\theta_c(Yk - k_d) - 1} = \frac{50\text{ mg BOD}_5\text{/L} \times (1 + 0.05\text{ d}^{-1} \times 5\text{ d})}{5\text{ d} \times (0.55 \times 5\text{ d}^{-1} - 0.05\text{ d}^{-1}) - 1} = 5.0\text{ mg BOD}_5\text{/L}$$

   b. Observed yield coefficient (Y_{obs}) from Equation 10.15l.

$$Y_{obs} = \frac{Y}{1 + k_d\theta_c} = \frac{0.55\text{ mg VSS/mg BOD}_5}{1 + 0.05\text{ d}^{-1} \times 5\text{ d}} = 0.44\text{ mg VSS/mg BOD}_5 = 0.44\text{ g VSS/g BOD}_5$$

   c. Biomass growth (P_x) from Equation 10.15m.

$$P_x = Y_{obs}Q(S_0 - S) = 0.44\text{ g VSS/g BOD}_5 \times 20{,}000\text{ m}^3\text{/d} \times (160 - 5)\text{ g BOD}_5\text{/m}^3 \times 10^{-3}\text{ kg/g}$$
$$= 1364\text{ kg VSS/d}$$

   d. Volume (V) of aeration basin from Equation 10.15o.

$$V = \frac{\theta_c P_x}{X} = \frac{5\text{ d} \times 1364\text{ kg VSS/d} \times 10^3\text{ g/kg}}{2000\text{ g VSS/m}^3} = 3410\text{ m}^3$$

   e. Aeration period (θ) from rearranged Equation 10.15o.

$$\theta = \frac{V}{Q} = \frac{3410\text{ m}^3}{20{,}000\text{ m}^3\text{/d}} \times \frac{24\text{ h}}{\text{d}} = 4.1\text{ h}$$

2. Determine the process parameters for the MBR process.
   Similarly, calculate the parameters for the MBR process. The calculation results for both processes are summarized in the table below.

Parameter	CAS	MBR
Effluent soluble BOD_5 (S), mg/L	5.0	2.2
Biomass production (P_x), kg VSS/d	1364	978
Volume of aeration basin (V), m^3	3410	1467
Aeration period (θ), h	4.1	1.8

3. Compare the results of CAS and MBR processes.

The reactor volume of MBR is less than half of that of CAS. The biomass production in MBR is 28% less than that produced in a conventional aeration basin. In addition to these benefits, the effluent quality is significantly better due to a longer SRT in the MBR.

Note: A more accurate analysis can be made by using many established constants and coefficients. Hand calculations, however, become tedious. Such an example and solution may be found in Example 7.3 of Reference 37. Computer programs as mentioned earlier are extensively used for MBR design.

EXAMPLE 15.51: COMPARISON OF THE SURFACE AREA OF FINAL CLARIFIERS WITH THAT OF MEMBRANE CASSETTES

Final clarifiers occupy a large area as compared with that of MBR cassettes. Compare the surface area occupied by final clarifiers with that of external submerged basins for membrane cassettes. Apply the membrane information assumed in Example 15.49. The design surface overflow rate of final clarifier is $SOR = 18\ m^3/m^2{\cdot}d$ at the design average flow of $20{,}000\ m^3/d$.

Solution

1. Compute the total surface area of the final clarifiers (A_{fc}).

$$A_{fc} = \frac{Q}{SOR} = \frac{20{,}000\ m^3/d}{18\ m^3/m^2{\cdot}d} = 1100\ m^2$$

2. Determine the total membrane surface area required (A_{mbr}).

Rearrange Equation 15.30a and calculate net permeate flow rate Q_P at the recovery factor $r = 80\%$.

$$Q_P = \frac{r}{100\%} \times Q_f = \frac{80\%}{100\%} \times 20{,}000\ m^3/d = 16{,}000\ m^3/d$$

Calculate the total membrane area from rearranged Equation 15.29c at the flux rate $F_t = 800\ L/m^2{\cdot}d$.

$$A_{mbr} = \frac{Q_P}{F_t} = \frac{16{,}000\ m^3/d \times 1000\ L/m^3}{800\ L/m^2{\cdot}d} = 20{,}000\ m^2$$

3. Determine the surface areas required in the external submerged basins.

a. Hollow fiber membrane cassettes.

It was assumed that the membrane cassette occupies 4.5-m^3 volume, and provides membrane surface area of 1000 m^2 (Step 4a. of Example 15.49).

$$\text{Number of cassettes required, } N_{hf} = \frac{A_{mbr}}{1000\ m^2/\text{cassette}} = \frac{20{,}000\ m^2}{1000\ m^2/\text{cassette}} = 20\ \text{cassettes}$$

Volume occupied by the cassettes,

$$V_{hf} = 4.5\ m^3/\text{cassette} \times N_{hf} = 4.5\ m^3/\text{cassette} \times 20\ \text{cassettes} = 90\ m^3$$

Assume a 50% additional allowance for equipment and clearance. Total basin volume $V_{hf} = 135\ m^3$.

Assume that the depth of hollow fiber membrane cassette is 2.5 m.

Provide a submergence of 0.5 m. Total depth of external submerged basin $D_{hf} = 3$ m.

Surface area of the basin,

$$A_{hf} = \frac{V_{hf}}{V_{hf}} = \frac{135\ m^3}{3\ m} = 45\ m^2$$

b. Flat sheet membrane modules.

Similarly, calculate the requirements for the flat sheet membrane modules. The calculation results are: $N_{fs} = 100$ modules, total volume $V_{fs} = 150\ m^3$ (with 50% additional volume allowance), and basin surface area $A_{fs} = 100\ m^2$ at a depth of $D_{fs} = 1.5$ m.

4. Summarize and compare the results.

The total surface areas of conventional final clarifiers and external submerged basins with membranes are summarized below. The results indicate that the surface area required for solids separation using membrane technology may be <10% of that required by conventional final clarifiers.

Parameter	Final Clarifier	External Basin for MBR	
		Hollow Fiber Membrane	Flat Sheet Membrane
Surface Area, m^2	1100	45	100
Percent of final clarifier area, %	–	4.1	9.1

NF and RO Systems: The NF removes viruses, small molecules of compounds, and hardness. The permeate contains small molecules of compounds and ionic solute. It is commonly used for pretreatment of feed prior to an RO process that removes much smaller molecules, color, hardness, and sodium and nitrate ions. Permeate from RO is low in TDS and is demineralized. To achieve these treatment goals, the NF and RO systems are operated at much higher pressures.

Osmotic Pressure: If a solute solution in a semipermeable bag is placed into a container filled with distilled water, the water will diffuse into the bag. The pressure in the bag will eventually reach the osmotic pressure. The mathematical relationship between the osmotic pressure and the concentration is given by the *van't Hoff* equation (Equation 15.31).

$$P_{osm} = \Phi N \frac{nRT}{V} = \Phi NCRT \quad (15.31)$$

where

P_{osm} = osmotic pressure, atm

Φ = osmotic pressure coefficient, dimensionless

N = number of ions in each mole, ions/mole

n = moles of solute, mole

V = volume of solute, L

C = concentration, mole/L

R = ideal gas constant, 0.082 atm·L/mole·°K

T = absolute temperature, °K. °K = °C + 273.2.

RO, and Water and Solute Flux: If a pressure higher than the osmotic pressure is applied over a solution, the flow of water from higher to lower concentration will occur. This is called *RO*. The flux rates of feedwater and permeate flow through the membrane are a function of the pressure gradient. They are expressed by Equation 15.32a.

$$F_p = k_p(\Delta P_a - \Delta P_{osm}) = \frac{Q_p}{A} \quad \text{or} \quad F_f = k_f(\Delta P_a - \Delta P_{osm}) = \frac{Q_f}{A} \tag{15.32a}$$

where

F_p and F_f = permeate and feedwater flux rates, $L/m^2{\cdot}h$ ($kg/m^2{\cdot}h$)
k_p and k_f = permeate and feedwater mass transfer coefficients, $L/m^2{\cdot}h{\cdot}kPa$ ($kg/m^2{\cdot}h{\cdot}kPa$). The value of k_p or k_f depends upon type of membrane, solute characteristics, and temperature.
Q_p and Q_f = permeate and feedwater flow rates, L/h (kg/h)
A = area of membrane, m^2
ΔP_a = average applied or transmembrane pressure gradient, kPa. It is given by Equation 15.28a.
ΔP_{osm} = osmotic pressure gradient, kPa. It is expressed by Equation 15.32b.

$$\Delta P_{osm} = \frac{P_{osmf} + P_{osmr}}{2} - P_{osmp} \tag{15.32b}$$

where

P_{osmf} = osmotic pressure of feedwater, kPa
P_{osmr} = osmotic pressure of retentate, kPa
P_{osmp} = osmotic pressure of permeate, kPa

Some solutes pass through the membrane. The solute mass flux is expressed by Equations 15.33a through 15.33d.

$$F_s = k_s \Delta C_s = \frac{Q_p C_p}{A} \tag{15.33a}$$

$$\Delta C_s = \frac{C_f + C_r}{2} - C_p \quad \text{or} \quad \Delta C_s = \frac{Q_p}{k_s A} C_p \quad \text{or} \quad \Delta C_s = \frac{F_s}{k_s} \tag{15.33b}$$

$$C_p = \frac{k_s A}{2(Q_p + k_s A)}(C_f + C_r) \tag{15.33c}$$

$$C_r = \frac{2(Q_p + k_s A)}{k_s A} C_p - C_f \quad \text{or} \quad C_r = \frac{2F_s}{k_s} - (C_f - C_p) \tag{15.33d}$$

where

F_s = mass flux of solute, $kg/m^2{\cdot}h$
k_s = mass transfer coefficient of solute, m/h
ΔC_s = solute concentration gradient across membrane, mg/L (g/m^3)
Q_p = flow of permeate stream, kg/h
C_f = solute concentration in feed stream, mg/L (g/m^3)
C_r = solute concentration in retentate stream, mg/L (g/m^3)
C_p = solute concentration in permeate stream, mg/L (g/m^3)

The recovery and rejection factors for a NF or RO system are obtained from Equations 15.30a and 15.30b. In RO system, the range of recovery (r) is 80–95% depending upon feedwater quality. At increased recovery, the feedwater flow and brine volume are reduced, and the pressure is increased. The typical rejection of specific solute (R) in RO process is 90–98%. The rejection of microorganisms (R_{log}) is expressed by Equation 15.30c.

Design and O&M Considerations: The major components for NF and RO systems include (1) pretreatment system, (2) flow equalization, (3) membrane module, support system, and pressure vessel, (4) pumps, pipings, and controls for pressurizing membranes, recirculation, and hydraulic backflushing, and (5) handling and disposal of retenate (brine). The design parameters of a RO system are plant capacity, salinity of feed water, pretreatment needs, flux rate, recovery factor, rejection factor, applied pressure, feedwater temperature, method of retenate (brine) disposal, and energy requirements. The design procedures for a membrane plant are (1) determine feedwater quality, (2) select pretreatment processes, (3) select RO system after consultation with equipment manufacturers, (4) select operating parameters (flux rate, rejection factor, applied pressure, system life, performance level, and finished water quality), (5) calculate system size, (6) determine power requirement, (7) select brine disposal system, and (8) estimate system economics (amortization of capital cost, labor, supplies, energy consumption, and brine disposal). Important design and operational parameters of NF and RO systems are summarized in Table 15.24.

Fouling Control: Membrane fouling is caused by (1) buildup of solids carried in the feed, (2) chemical precipitation such as Fe, Mn, and hardness, (3) damage caused by chemicals such as chlorine, ozone, and strong acid and base that react with membranes, (4) fouling caused by organic compounds such as natural organic matter (NOM), and (5) slime growth due to colonization of microorganisms. The accumulation of particulates causes pore narrowing, pore plugging, and buildup of particulate matter close to membrane surface. This phenomenon is called gel/cake formation that increases the resistance to solvent transportation.[36,37,54] RO membrane fouling is controlled by (1) pretreatment of feedwater, (2) hydraulic flushing, and (3) chemical cleaning. The pretreatment of feedwater is achieved by precipitation of Fe, Mn, and hardness followed by filtration to remove colloidal matters. Commonly used filters are multimedia filters, cartridge filter, MF, and UF. In dual membrane system, coarse membranes such as MF or UF are integrated with RO. The hydraulic flushing is done periodically to clean the surface from solid accumulation. Chemical or physical treatment of membranes is done by ozone, chlorine or UV to prevent microbial growth. Oxygen removed from feedwater prevents oxidation of iron, manganese and hydrogen sulfide. The pH adjustment in the range of 4–7.5 prevents scale formation.

Fouling Indexes: Several fouling indexes are used to determine the pretreatment need of feedwater. These indexes are (1) *salt density index* (*SDI*), (2) *modified fouling index* (*MFI*), and (3) *mini-plugging factor index* (*MPFI*). To determine these indexes an experiment is conducted and the indices are calculated from the experimental data. The general experimental procedure is given below.[36,37,54,99,105,128]

1. A simple laboratory dead-end membrane filter apparatus is used.
2. The filtration apparatus consists of a 0.45-μm Millipore filter with a 47 mm internal diameter. A constant pressure of 207 kPa (30 psig) is applied.
3. The measurements of throughput volume (V) versus filtration time (t) are made at different time intervals. The total time to complete data collection may vary from 15 min to 2 h.

TABLE 15.24 Basic Design and Operating Parameters of NF and RO Systems

Parameter	NF	RO
Flux rate, $L/m^2{\cdot}d$ (gpd/ft^2)	340–480 (8–12)	340–480 (8–12)
Operating pressure, kPa (psig)	700–1400 (100–200)	800–1900 (125–230)
Recovery factor, %	85–90	80–85
Salt rejection, %	10–50	90–98
Energy requirements, kWh/m^3 ($kWh/10^3$ gal)	0.4–0.5 (1.5–1.9)	0.45–0.65 (1.7–2.5)
Molecular weight cut off (MWCO) (dalton)	300–1000	<300
Packing density, m^2/m^3 (ft^2/ft^3)		164–1640 (50–500)
Temperature, °C (°F)	16–38 (60–100)	16–38 (60–100)
pH (standard unit)	4–7.5	4–7.2

Note: 1 $L/m^2{\cdot}d$ = 0.0245 gpd/ft^2; 1 kPa = 0.145 psig; 1 kWh/m^3 = 3.785 $kWh/10^3$ gal.

Source: Adapted in part from References 36, 37, 49, 54, 64, and 106.

4. The relationships for *SDI*, *MFI*, and *MPEI* are given by Equations 15.34a through 15.34c.

$$SDI = \frac{P_{270}}{t} = \frac{1 - t_1/t_2}{t} \times 100\% \quad (15.34a)$$

$$\frac{1}{Q} = MFI \times V + a \quad (15.34b)$$

$$Q_{avg} = MPFI \times t + b \quad (15.34c)$$

where

SDI = silt density index, %
P_{207} = percent plugging at 207 kPa (30 psig), %
t_1 = time to collect the initial 500-mL sample, min
t_2 = time to collect the final 500-mL sample, min
t = total running time during the experiment, min
MFI = modified fouling index, s/L^2. It is the slope of linear portion of the plot of $(1/Q)$ versus V.
Q = average flow rate, L/s
V = throughput volume, L
a = intercept of linear portion of $1/Q$ versus V, s/L
$MPFI$ = mini-plugging factor index, L/s^2. It is the slope of linear portion of the plot of Q_{avg} versus t.
Q_{avg} = average flow at 30 s interval, L/s
t = time of operation, s
b = intercept of the linear portion of the plot of Q_{avg} versus t, L/s

The recommended values of three indexes are summarized in Table 15.25. The procedure to determine these indexes are given in Example 15.57.

Stability Indexes: Common methods of determining the stability of potable water in water distribution system are *Langelier saturation index* (*LSI*) and *Ryznar stability index* (*RSI*). These indexes can also be used as a measure of scaling, corrosive or balanced nature of feedwater for RO membranes. The LSI and RSI are expressed by Equations 15.35a through 15.35c.

$$LSI = pH - pH_s \quad (15.35a)$$

$$RSI = 2\,pH_s - pH \quad (15.35b)$$

$$pH_s = (pK_2 - pK_s) + pCa^{2+} + pAlk \quad (15.35c)$$

where

LSI = Langelier saturation index, dimensionless. In general a negative *LSI* indicates that the water is corrosive, and positive value indicates scaling property. The magnitude of the index does not give severity.
pH = measured pH of water, standard unit
pH_s = pH of $CaCO_3$ saturation, standard unit
RSI = Ryznar stability index, dimensionless. The severity in scaling and corrosive nature of feedwater based on *RSI* are given in Table 15.26.

TABLE 15.25 Recommendations of Fouling Indexes

Membrane Process	Fouling Index		
	SDI, %	MFI, s/L^2	MPFI, L/s^2
NF	0–3	0–10	0–1.5 × 10^{-4}
RO			
Hallow fiber	0–2	0–2	0–3 × 10^{-5}
Spiral wound	0–3	0–2	0–3 × 10^{-5}

Source: Adapted in part from References 36, 37, 54, and 108.

TABLE 15.26 Corrosive and Scaling Tendencies of Feedwater in Different RSI Range

RSI Range	Indication
<5.5	Heavy scale formation
5.5–6.2	Some scale will form
6.2–6.8	Nonscaling or noncorrosive
6.8–8.5	Corrosive water
>8.5	Very corrosive water

Source: Adapted in part from References 49 and 129.

pK_2 and pK_s = constants based on the ionic strength or the total dissolved solids (TDS) concentration of the feedwater at given temperature. The values of $(pK_2 - pK_s)$ are provided in Table B.9 of Appendix B.

pCa^{2+} = negative logarithm of calcium ion concentration in moles per liter ($pCa^{2+} = -\log[\text{Ca}^{2+}]$), dimensionless

$pAlk$ = negative logarithm of total alkalinity concentration in equivalents per liter ($pAlk = -\log[\text{Alk}]$), dimensionless

EXAMPLE 15.52: OSMOTIC PRESSURE OF SODIUM CHLORIDE SOLUTION

Determine the osmotic pressure of a 20,000 mg/L sodium chloride solution. Temperature is 25°C and $\Phi = 0.92$.

Solution

1. Calculate the molar concentration of NaCl.

   Molecular weight of NaCl = 58.5 g/mole NaCl

$$C = \frac{20{,}000\ \text{mg/L} \times 10^{-3}\ \text{g/mg}}{58.5\ \text{g/mole}} = 0.342\ \text{mole/L}$$

2. Determine the number of ions per mole of NaCl.

$$\text{NaCl} \leftrightarrow \text{Na}^+ + \text{Cl}^-$$

   Number of ions $N = 2$ per mole.

3. Estimate the osmotic pressure from Equation 15.31 at temperature of $T = 25°\text{C} + 273.2 = 298.2°\text{K}$.

$$P_{\text{osm}} = \Phi NCRT = 0.92 \times 2 \times 0.342\ \text{mole/L} \times 0.082\ \text{atm·L/mole·°K} \times 298.2\ \text{°K} = 15.4\ \text{atm}$$

or $P_{\text{osm}} = 15.4\ \text{atm} \times 10.33\ \text{m water/atm} = 159\ \text{m water}$

or $P_{\text{osm}} = 15.4\ \text{atm} \times 14.7\ \text{psi/atm} = 226\ \text{psi}$

EXAMPLE 15.53: OSMOTIC PRESSURE OF A SOLUTION

A 0.5 M solution of sucrose is separated from pure water by a semipermeable membrane. Calculate the osmotic pressure at 20°C and $\Phi = 1$.

Solution

Apply the *van't Hoff* equation (Equation 15.31) at $N = 1$ and $T = 20°\text{C} + 273.2 = 293.2°\text{K}$.

$$P_{\text{osm}} = \Phi NCRT = 1 \times 1 \times 0.5\ \text{mole/L} \times 0.082\ \text{atm·L/mole·°K} \times 293.2°\text{K} = 12.0\ \text{atm}$$

or $P_{\text{osm}} = 12.0\ \text{atm} \times 10.33\ \text{m water/atm} = 124\ \text{m water or } 176\ \text{psi}$

EXAMPLE 15.54: DEMINERALIZATION BY RO

Effluent from a POTW contains 1800 mg/L TDS. The effluent is demineralized for groundwater recharge. The desired TDS in the product water is $C_p = 250$ mg/L for recharge. The water mass transfer coefficient $k_f = 10^{-2}$ L/m^2·h·kPa and mass transfer coefficient for solute $k_s = 9.0 \times 10^{-4}$ m/h. The pressure differential $(\Delta P_a - \Delta P_{osm}) = 2500$ kPa, plant capacity $Q_f = 4000$ m^3/d (166.7 m^3/h), and flux recovery factor $r = 0.85$. Determine (a) the water flux rate (F_f), (b) product water flow rate (Q_p), (c) membrane surface area (A), (d) concentration of TDS in the retentate (C_r), (e) rejection factor and flow (R and Q_r), and (f) solute concentration gradient across membrane (ΔC_s) and mass flux of solute (F_s).

Solution

1. Calculate the feedwater flux rate (F_f) from Equation 15.32a.

$$F_f = k_f(\Delta P_a - \Delta P_{osm}) = 10^{-2}\,\text{L/m}^2\cdot\text{h}\cdot\text{kPa} \times 2500\,\text{kPa} = 25\,\text{L/m}^2\cdot\text{h} \quad \text{or} \quad 600\,\text{L/m}^2\cdot\text{d}$$

2. Calculate the product water (or permeate) rate (Q_p) from rearranged Equation 15.30a at the recovery factor $r = 85\%$.

$$Q_p = \frac{r}{100\%} \times Q_f = \frac{85\%}{100\%} \times 4000\,\text{m}^3/\text{d} = 3400\,\text{m}^3/\text{d} \quad \text{or} \quad 142\,\text{m}^3/\text{h}$$

3. Calculate the membrane area (A) from rearranged Equation 15.32a at the flux rate $F_f = 600$ L/m^2·d.

$$A = \frac{Q_f}{F_f} = \frac{4000\,\text{m}^3/\text{d} \times 1000\,\text{L/m}^3}{600\,\text{L/m}^2\cdot\text{d}} = 6670\,\text{m}^2$$

4. Calculate the TDS in the retentate (C_r) from Equation 15.33d at $C_f = 1800$ mg/L and $C_p = 250$ mg/L.

$$C_r = \frac{2(Q_p + k_s A)}{k_s A} C_p - C_f = \frac{2 \times (142\,\text{m}^3/\text{h} + 9 \times 10^{-4}\,\text{m/h} \times 6670\,\text{m}^2)}{9 \times 10^{-4}\,\text{m/h} \times 6670\,\text{m}^2} \times 250\,\text{mg/L} - 1800\,\text{mg/L}$$

$$= 10{,}530\,\text{mg/L}$$

5. Calculate the rejection factor (R) and rejected flow (Q_r).

   Rejection factor (R) from Equation 15.30b,

$$R = \left(1 - \frac{C_p}{C_f}\right) \times 100\% = \left(1 - \frac{250\,\text{mg/L}}{1800\,\text{mg/L}}\right) \times 100\% = 86\%$$

   Rejected flow (Q_r) from rearranged Equation 15.29a,

$$Q_r = Q_f - Q_p = (4000 - 3400)\,\text{m}^3/\text{d} = 600\,\text{m}^3/\text{d}$$

6. Estimate the solute concentration gradient across membrane (ΔC_s) and mass flux of solute (F_s).

   Solute concentration gradient (ΔC_s) across membrane from Equation 15.33b,

$$\Delta C_s = \frac{Q_p}{k_s A} C_p = \frac{142\,\text{m}^3/\text{h}}{9 \times 10^{-4}\,\text{m/h} \times 6670\,\text{m}^2} \times 250\,\text{mg/L} = 5910\,\text{mg/L}$$

   Mass flux of solute (Q_r) from Equation 15.33a,

$$F_s = k_s \Delta C_s = 9 \times 10^{-4}\,\text{m/h} \times 5910\,\text{g/m}^3 \times 10^{-3}\,\text{kg/g} = 5.3 \times 10^{-3}\,\text{kg/m}^2\cdot\text{h}$$

EXAMPLE 15.55: DESIGN OF AN RO PLANT FOR DESALINATION

Brackish water is used for municipal water supply. The TDS concentration of raw and finished water are 1000 and 300 mg/L. The plant capacity is 19,000 m^3/d (5 MGD). Develop the design and size of the various components of the RO system. Use the following data: recovery factor $r = 75\%$, salt rejection factor $R = 95\%$, permeate flux rate $F_p = 820$ L/m^2·d (20 gpd/ft^2), feed water temperature = 27°C, packing density of membrane = 1000 m^2/m^3, and design applied pressure $\Delta P_a = 4140$ kPa (600 psig).

Solution

1. Compute the TDS in permeate (C_p) from Equation 15.29d at $C_f = C_{raw} = 1000$ mg/L.

$$C_p = \frac{100\% - R}{r} C_f = \frac{100\% - 95\%}{75\%} \times 1000 \text{ mg/L} = 67 \text{ mg/L}$$

The TDS concentration in the RO permeate is considerably lower than that required in the finished water. Therefore, a split treatment will be needed to achieve the treatment goal.

2. Calculate the flows of permeate stream (Q_p) and feed stream (Q_f) to the RO system.
   Apply mass balance around Point A in Figure 15.36,

$$Q_{bypass} \times C_{bypass} + Q_p \times C_p = Q_{finished} \times C_{finished}$$

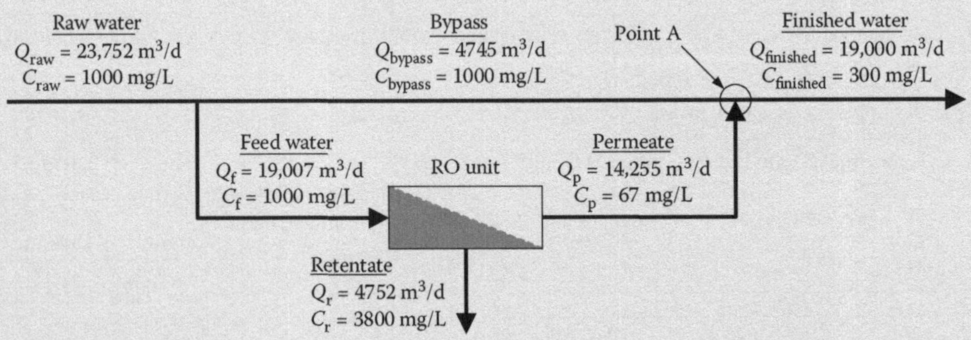

FIGURE 15.36 Flow and mass balances around the RO system (Example 15.55).

Since $Q_{bypass} = Q_{finished} - Q_p$ and $C_{bypass} = C_{raw}$, the following expression is obtained.

$$(Q_{finished} - Q_p) \times C_{raw} + Q_p \times C_p = Q_{finished} \times C_{finished}$$

$$\text{Permeate flow, } Q_p = \frac{C_{raw} - C_{finished}}{C_{raw} - C_p} \times Q_{finished} = \frac{(1000 - 300) \text{ mg/L}}{(1000 - 67) \text{ mg/L}} \times 19{,}000 \text{ m}^3/\text{d} = 14{,}255 \text{ m}^3/\text{d}$$

$$\text{Feed flow from rearranged Equation 15.30a, } Q_f = \frac{100\%}{r} \times Q_p = \frac{100\%}{75\%} \times 14{,}255 \text{ m}^3/\text{d} = 19{,}007 \text{ m}^3/\text{d}$$

3. Calculate the flows of bypass stream (Q_{bypass}) and raw water stream (Q_t).

$$\text{Bypass flow, } Q_{bypass} = Q_{finished} - Q_p = (19{,}000 - 14{,}255) \text{ m}^3/\text{d} = 4745 \text{ m}^3/\text{d}$$

$$\text{Total raw water flow, } Q_{raw} = Q_f + Q_{bypass} = (19{,}007 + 4745) \text{ m}^3/\text{d} = 23{,}752 \text{ m}^3/\text{d}$$

4. Determine the retentate (brine) flow (Q_r) and TDS concentration in retentate (C_r)

Retentate flow from rearranged Equation 15.29a, $Q_r = Q_f - Q_p = (19{,}007 - 14{,}255)\,m^3/d = 4752\,m^3/d$

TDS in retentate from Equation 15.29e, $C_r = \dfrac{R}{100\% - r} C_f = \dfrac{95\%}{100\% - 75\%} \times 1000\,mg/L = 3800\,mg/L$

The final results from flow and material mass balance analysis are shown in Figure 15.36.

5. Calculate the membrane area (A) and volume (V)

Membrane area from Equation 15.32a,

$$A = \frac{Q_p}{F_p} = \frac{14{,}255\,m^3/d \times 10^3\,L/m^3}{820\,L/m^2{\cdot}d} = 17{,}400\,m^2 \text{ in the RO system}$$

Provide a total membrane area $A = 20{,}000\,m^2$ that gives a 10% extra capacity in the RO system.

$$\text{Membrane volume, } V = \frac{A}{1000\,m^2/m^3} = \frac{20{,}000\,m^2}{1000\,m^2/m^3} = 20\,m^3 \text{ in the RO system}$$

6. Estimate the number of RO modules and pressure vessels.

Assume that the unit volume of each module is $0.04\,m^3$, and 10 modules are packed per pressure vessel.

$$\text{Number of modules, } N_{module} = \frac{V}{0.04\,m^3/\text{module}} = \frac{20\,m^3}{0.04\,m^3/\text{module}} = 500 \text{ modules in the RO system}$$

$$\text{Number of vessels, } N_{vessel} = \frac{N_{module}}{10\text{ modules/vessel}} = \frac{500\text{ modules}}{10\text{ modules/vessel}} = 50 \text{ vessels in the RO system}$$

7. Compute the power requirements in the RO system.

Power consumption of membrane system can be estimated from Equation 15.36.

$$P_w = \frac{\Delta P_a Q_f}{f_{power}} \tag{15.36}$$

where

P_w = water power required by the membrane unit, kW or kN·m/s (hp)
ΔP_a = applied pressure, kPa or kN/m^2 (psig)
Q_f = feedwater flow, m^3/s (gpm)
f_{power} = convention factor, 1 for SI unit, and 1715 for U.S. customary unit

Water power consumption from Equation 15.36,

$$P_w = \frac{\Delta P_a Q_f}{f_{power}} = \frac{4140\,kN/m^2 \times 19{,}007\,m^3/d}{1 \times 86{,}400\,s/d} = 911\,kN{\cdot}m/s \quad \text{or} \quad 911\,kW$$

Assume that the pump and motor efficiencies are 90% and 85%.

$$\text{Brake power of feed pumps, } P_p = \frac{P_w}{0.9} = \frac{911\,kW}{0.9} = 1010\,kW$$

$$\text{Motor power, } P_m = \frac{P_p}{0.85} = \frac{1010\,kW}{0.85} = 1190\,kW \text{ or } 1200\,kW$$

EXAMPLE 15.56: TWO-PASS RO SYSTEM

An RO system is designed to treat a feed flow of 200 L/min. The salinity in the feedwater is 2500 mg/L of sodium chloride (NaCl). An overall recovery factor of 75% is required. The temperature of feedwater is

25°C and $\Phi = 0.92$. An RO vendor has provided the following specifications of their system: permeate mass transfer coefficient $k_p = 2 \times 10^{-6}$ mole/cm^2·s·atm, area of the bundle = 28 m^2/module, permeate recovery factor $r = 50\%$, optimal applied pressure to the RO membrane = 35 atm, and NaCl rejection factor $R = 95\%$. Determine the following: (a) osmotic pressure of the solution, (b) permeate flow rate, (c) number of units required for 75% recovery factor, and (d) quality and quantity of final product.

Solution

1. Describe the RO system.

   The available RO unit has a recovery factor of 50%. To achieve 75% overall recovery factor, it is required to utilize a two-pass RO system where the reject stream from the first RO unit is sent to the second RO unit. In both passes, a rejection factor of 95% salt is expected.

2. Design the first RO unit.

   a. Calculate the osmotic pressure (P_{osm}).

   Molecular weight of NaCl = 58.5 g/mole NaCl.

   Molar concentration of NaCl in the feedwater to the first RO unit,

$$C_f = \frac{2500\,\text{mg/L} \times 10^{-3}\,\text{g/mg}}{58.5\,\text{g/mole}} = 0.0427\,\text{mole/L}$$

   Calculate P_{osm} from the *van't Hoff* equation (Equation 15.31) at $N = 2$ and $T = 25°\text{C} + 273.2 = 298.2°\text{K}$.

$$P_{osm} = \Phi N C_f RT = 0.92 \times 2 \times 0.0427\,\text{mole/L} \times 0.082\,\text{atm·L/mole·°K} \times 298.2°\text{K} = 1.92\,\text{atm}$$

   b. Calculate the permeate flux rate (F_p) from Equation 15.32a by assuming $\Delta P_{osm} \approx P_{osm} = 1.92$ atm.

$$F_p = k_p(\Delta P_a - \Delta P_{osm}) = 2 \times 10^{-6}\,\text{mole/cm}^2\text{·s·atm} \times (35 - 1.92)\,\text{atm}$$

$$= 6.62 \times 10^{-5}\,\text{mole/cm}^2\text{·s}$$

   c. Calculate the permeate flow (Q_p) and retentate flow (Q_r) in mole/s.

   Molecular weight of water = 18 g/mole H_2O.

$$\text{Feed flow rate, } Q_f = 200\,\text{L/min} \times 10^3\,\text{g/L} \times \frac{1\,\text{min}}{60\,\text{s}} \times \frac{1}{18\,\text{g/mole}} = 185\,\text{mole/s}$$

   Rearrange Equation 15.30a and calculate the permeate flow rate Q_P.

$$Q_p = \frac{r}{100\%} \times Q_f = \frac{50\%}{100\%} \times 185\,\text{mole/s} = 92.5\,\text{mole/s}$$

   Retentate flow, $Q_r = Q_f - Q_p = (185 - 92.5)\,\text{mole/s} = 92.5\,\text{mole/s}$

   Note: This is also the feedwater flow to the second RO unit.

   d. Calculate the membrane area (A) and number of bundles (N_{bundle}).

$$\text{Membrane area from Equation 15.32a, } A = \frac{Q_P}{F_p} = \frac{92.5\,\text{mole/s}}{6.62 \times 10^{-5}\,\text{mole/cm}^2\text{·s} \times 10^4\,\text{cm}^2/\text{m}^2} = 140\,\text{m}^2$$

$$\text{Number of bundles, } N_{bundle} = \frac{A}{28\,\text{m}^2/\text{bundle}} = \frac{140\,\text{m}^2}{28\,\text{m}^2/\text{bundle}} = 5 \text{ bundles in the first RO unit}$$

3. Design the second RO unit.

   a. Calculate the concentration of NaCl in the retentate or reject from the first RO unit.

NaCl concentration in retentate from Equation 15.29e,

$$C_r = \frac{R}{100\% - r} C_f = \frac{95\%}{100\% - 50\%} \times 2500\,\text{mg/L} = 4750\,\text{mg/L}$$

$$\text{Molar concentration of NaCl, } C_r = \frac{4750\,\text{mg/L} \times 10^{-3}\,\text{g/mg}}{58.5\,\text{g/mole}} = 0.0812\,\text{mole/L}$$

Note: This is also the NaCl concentration in the feedwater to the second RO unit.

b. Calculate the osmotic pressure, permeate flux rate, membrane area, and number of bundles required.

The calculation steps in Step 2 are repeated to determine the parameters in the second RO unit. The results are summarized below:

Osmotic pressure, $P_{osm} = 3.65$ atm; Permeate flux rate, $F_p = 6.27 \times 10^{-5}$ mole/cm²·s;

Permeate flow rate, $Q_p = 46.3$ mole/s; Retentate flow rate, $Q_r = 46.3$ mole/s;

Membrane area, $A = 70\ \text{m}^2$; and number of bundles, $N_{bundle} = 3$ bundles in the second RO unit

4. Determine the final product flow.

a. First RO unit.

Feed flow, $Q_f = 200$ L/min

$$\text{Permeate flow from rearranged Equation 15.30a, } Q_p = \frac{r}{100\%} \times Q_f = \frac{50\%}{100\%} \times 200\,\text{L/min} = 100\,\text{L/min}$$

$$\text{Retentate flow, } Q_r = Q_f - Q_p = (200 - 100)\,\text{L/min} = 100\,\text{L/min}$$

b. Second RO unit.

Similarly, the following flows are also calculated for the second RO unit:

Feed flow, $Q_f = 100$ L/min, $Q_p = 50$ L/min, and $Q_r = 50$ L/min.

Note: $Q_r = 50$ L/min is also the brine flow discharged from the two-pass RO system.

c. Final product water.

Final product flow, $Q_{final} = Q_p$ from the first RO unit + Q_p from the second RO unit = (100 + 50) L/min = 150 L/min.

Note: The overall recovery factor is 75%. It meets the requirement in the statement.

5. Determine the NaCl concentration in the final product stream.

a. First RO unit.

NaCl in the feed stream, $C_f = 2500$ mg/L

NaCl in the permeate stream from Equation 15.29d,

$$C_p = \frac{100\% - R}{r} C_f = \frac{100\% - 95\%}{50\%} \times 2500\,\text{mg/L} = 250\,\text{mg/L}$$

NaCl in the retentate stream, $C_r = 4750$ mg/L (Step 3.a)

b. Second RO unit.

Similarly, calculate the following NaCl concentrations in the second RO unit:

NaCl in feed flow, $C_f = 4750$ mg/L, $C_p = 475$ mg/L, and $C_r = 9025$ mg/L.

Note: $C_r = 9025$ mg/L is also the final NaCl concentration in the brine discharged from the two-pass RO system.

c. Final product water from the two-pass RO system.

NaCl concentration in the final product stream is obtained from mass balance.

$$C_{final} = \frac{250\,\text{mg/L} \times 100\,\text{L/min} + 475\,\text{mg/L} \times 50\,\text{L/min}}{150\,\text{L/min}} = 325\,\text{mg/L}$$

Overall rejection factor (R) from Equation [15-29(b)],

$$R = \left(1 - \frac{C_p}{C_f}\right) \times 100\% = \left(1 - \frac{325\ \text{mg/L}}{2500\ \text{mg/L}}\right) \times 100\% = 87\%$$

6. Draw the flows and concentrations in streams of the two-pass RO system in Figure 15.37.

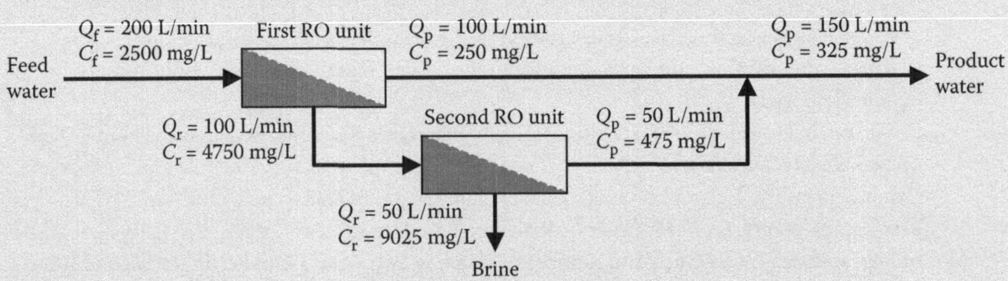

FIGURE 15.37 Flow and mass balances around the two-pass RO system (Example 15.56).

EXAMPLE 15.57: FOULING INDEX AND PRETREATMENT NEED FOR RO SYSTEM

A dead-end membrane filtration test was conducted to determine the pretreatment need of the feedwater to be demineralized by an RO system. A 0.45-μm Millipore filter with an internal diameter of 47 mm was used. A constant pressure of 207 kPa (30 psi) gauge was applied throughout the test. The test results are summarized in Table 15.27. Determine the *SDI*, *MFI*, and *MPFI* fouling indexes, and recommend pretreatment need.

TABLE 15.27 Filtration Test Results for Determination of Fouling Indexes (Example 15.57)

Filtration Time (t)		Throughput Volume (V)		Filtration Rate (Q), 10^{-3} L/s	Inverse Filtration Rate ($1/Q$), s/L
min	s	mL	L		
0.5	30	130	0.13	4.3	231
1	60	200	0.20	3.3	300
2	120	400	0.40	3.3	300
3	180	550	0.55	3.1	327
5	300	775	0.78	2.6	385
7	420	960	0.96	2.3	438
9	540	1080	1.08	2.0	500
11	660	1180	1.18	1.8	559
13	780	1270	1.27	1.6	614
15	900	1330	1.33	1.5	677
17	1020	1360	1.36	1.3	750
19	1140	1430	1.43	1.3	797
21	1260	1452	1.45	1.2	869
23	1380	1485	1.49	1.1	926
25	1500	1500	1.50	1.0	1000

Note: $Q = V/t$ and $1/Q = t/V$.

Solution

1. Determine the salt density index (*SDI*).
   a. Determine the time to filter first 500 mL (t_1), the last 500-mL filtrate (t_2), and the total running time (t).
   Filtrate volumes of 400 mL and 550 mL are obtained in 2 and 3 mins. By interpolation, 500-mL filtrate volume is obtained in 2.7 min. Likewise, 1000-mL sample is obtained in 7.5 min. The total running time $t = 25$ min is needed to reach the final sample volume of 1500 mL. Therefore, the time to collect first 500-mL sample $t_1 = 2.7$ min, and the time to collect last 500-mL sample $t_2 = (25 - 7.5)$ min $= 17.5$ min.
   b. Calculate the *SDI* value from Equation 15.34a.

$$SDI = \frac{1 - t_1/t_2}{t} \times 100\% = \frac{1 - (2.7\,\text{min}/17.5\,\text{min})}{25\,\text{min}} \times 100\% = 3.4\%$$

   c. Compare the calculated *SDI* value with the ranges in Table 15.25.
   The calculated *SDI* value of 3.4% is larger than the potentially allowable maximum value of 3 for RO configuration. Therefore, pretreatment is required for RO system.
2. Determine the modified fouling index (*MFI*).
   a. Plot the inverse flow rate ($1/Q$) versus throughput volume (V) in Figure 15.38.

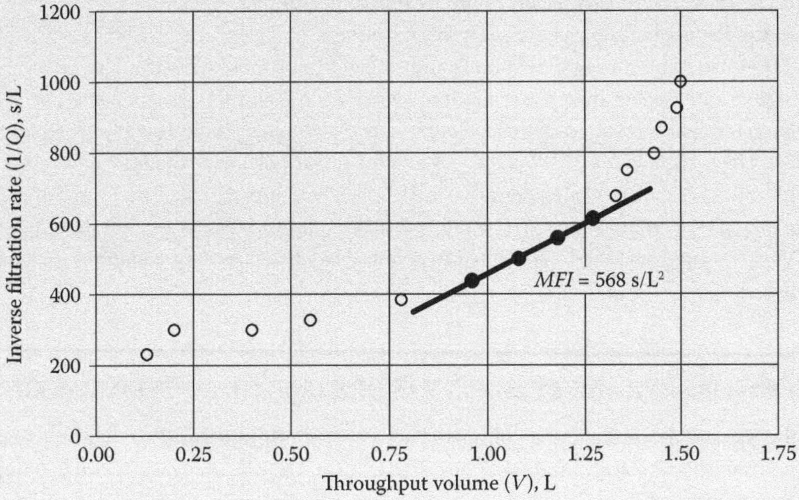

FIGURE 15.38 Plot of inverse flow rate versus throughput volume for determination of MFI (Example 15.57).

   b. Determine the *MFI* value.
   The slope of the linear portion of the line is *MFI*. Obtain $MFI = 568\ \text{s/L}^2$ from the plot.
   c. Compare the calculated *MFI* value with the range in Table 15.25.
   The calculated *MFI* value of $568\ \text{s/L}^2$ is significantly higher than the maximum value of 2 or 10 in Table 15.25. Therefore, pretreatment of feedwater prior to RO is needed.
3. Determine the mini-plugging factor index (*MPFI*).
   a. Plot the flow rate (Q) versus filtration time (t) in Figure 15.39.
   b. Determine the *MPFI* value.
   The slope of the linear portion of the line is *MPFI*. Obtain $MPFI = 3.1 \times 10^{-6}\ \text{L/s}^2$ from the plot.

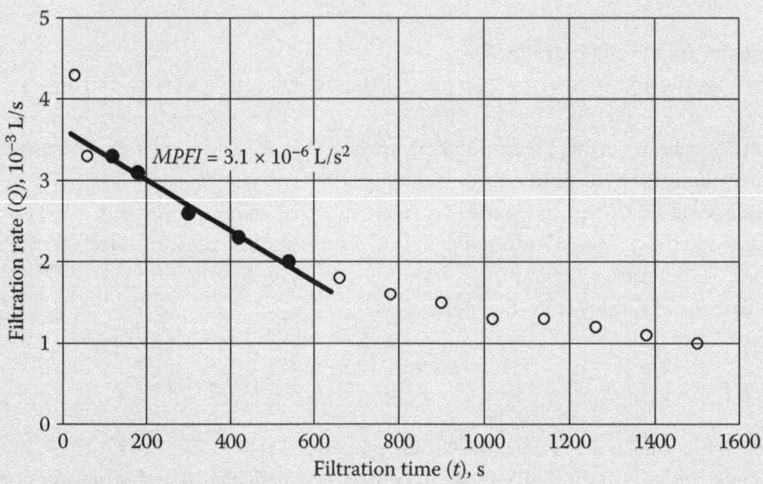

FIGURE 15.39 Plot of flow rate versus filtration time for determination of MPFI (Example 15.57).

c. Compare the calculated *MPFI* value with the range in Table 15.25.

The calculated *MPFI* value of 3.1×10^{-6} L/s^2 is within the acceptable range for RO system given in Table 15.25. This means no pretreatment is required.

4. Compare the fouling indexes calculated in this example.

The *MFI* is based on throughput volume and thought to be most sensitive. The *MFI* values of 568 s/L^2 is significantly higher than the maximum values for NF and RO systems given in Table 15.25. Clearly pretreatment is required. The calculated value of *SDI* is also indicating the requirement of pretreatment. The fouling index *MPFI* is the slope of the linear portion of the flow rate versus filtration time. This index is significantly less sensitive than *MFI*. Therefore, the calculated value of *MPFI* in this example may give false indication. To be safe, pretreatment of feedwater for RO should be provided. The RO system suppliers shall be consulted to obtain the detailed recommendations for pretreatment requirements.

EXAMPLE 15.58: RYZNAR STABILITY INDEX (RSI) OF RO FEEDWATER

Calculate the RSI and determine the stability of feedwater to an RO system. The chemical quality of the feedwater is given below: total alkalinity = 80 mg/L as $CaCO_3$, Ca^{2+} = 38 mg/L, TDS = 200 mg/L, pH = 7.5, and Temperature = 5°C.

Solution

1. Determine the *pAlk* and pCa^{2+}.

$$\text{Alkalinity} = \frac{80\,\text{mg/L as CaCO}_3 \times 10^{-3}\,\text{g/mg}}{50\,\text{g/eq as CaCO}_3} = 0.0016\,\text{eq/L}$$

$$pAlk = -\log(0.0016) = 2.8$$

$$[Ca^{2+}] = \frac{38\,\text{mg/L as Ca}^{2+} \times 10^{-3}\,\text{g/mg}}{40\,\text{g/mole as Ca}^{2+}} = 0.00095\,\text{mole/L}$$

$$pCa^{2+} = -\log(0.00095) = 3.0$$

2. Determine the value of $(pK_2 - pK_s)$ at 5°C, and TDS = 200 mg/L.
   Interpolate $(pK_2 - pK_s)$ from the data in Table B.9 of, Appendix B,

$$(pK_2 - pK_s) = \frac{2.71 + 2.49}{2} = 2.6$$

3. Calculate the value of *RSI*.
   pH_s from Equation 15.35c, $pH_s = (pK_2 - pK_s) + pCa^{2+} + pAlk = 2.6 + 3.0 + 2.8 = 8.4$
   RSI from Equation 15.35b, $RSI = 2\,pH_s - pH = 2 \times 8.4 - 7.5 = 9.3$
4. Determine the corrosive or scaling nature of feedwater.
   The *RSI* scale is provided in Table 15.26. $RSI > 8.5$ indicates very corrosive water. The *RSI* of 9.3 indicates the feedwater is highly corrosive. This tendency is not acceptable and the *RSI* must be adjusted by adding either lime ($Ca(OH)_2$), soda ash (Na_2CO_3), or caustic soda (NaOH). The acceptable range of *RSI* is 6.2–6.8 (nonscaling or noncorrosive).

ED Systems: The ED is also a demineralization process. It is an electrochemical separation process in which an electrical potential (DC current) is used as a driving force to transfer the ion through ion-selective membranes. Thus, the ions are gradually removed from the feed stream leaving behind the neutral species and particulate matters in the brine. It removes ionized salt ions effectively and produces a product with low ionic solute. The ED process is shown in Figure 15.40. The basic design features of ED systems are given below:[36,37,53,64,89,130–132]

1. The key element of an ED process is the ion selective membranes that are essentially ion exchange resins cast in sheet form.
2. The membranes that allow passage of positive ions (cations) such as Na^+ and K^+ are called *cation membranes* (Membrane C, Figure 15.40). Likewise, *anion membranes* (Membrane A, Figure 15.40) allow the passage of negative ions (anions) such as Cl^- and SO_4^{2-}.

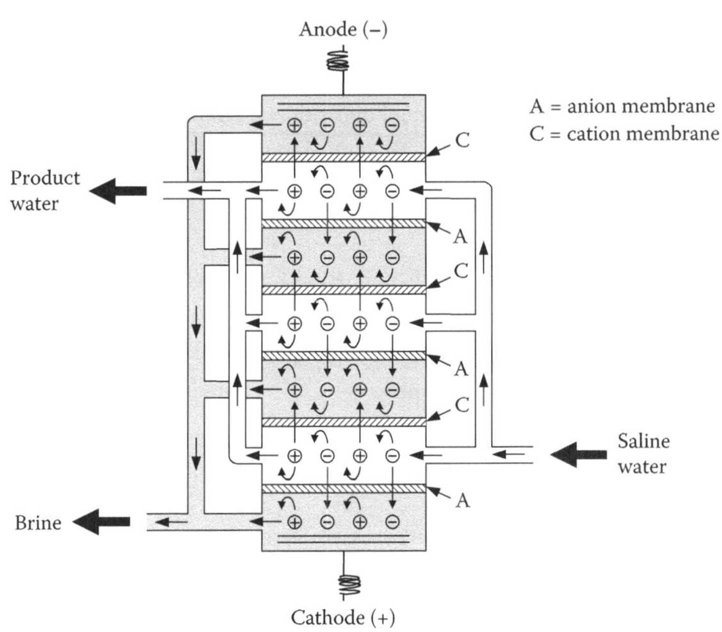

FIGURE 15.40 Schematic diagram and ion removal mechanism of electrodialysis process.

3. The ED membranes are alternately arranged between plastic spacers in a stalked configuration between cathode (positive) and anode (negative) electrodes. The assembly is called *stack*. A set of adjacent compartments is called a *cell pair*. There may be as many as 600 cell pairs in a stack.
4. As a DC voltage is applied the cations and anions migrate to the opposite electrodes, pass through the ion selective membrane then retained by the next membrane.
5. The typical flux rate is 0.8–1 $m^3/m^2{\cdot}d$ (20–25 gpd/ft^2). The typical salt rejection factor is 50–94% by weight; and the product water stream is 75–95% of feed stream by volume. The membrane resistance is 4–8 ohms and the current efficiency is 85–95%. The flux rate is 30–40 $L/m^2{\cdot}h$ (18–74 gpd/ft^2). The power consumptions is 1–2 kWh/m^3 (4–8 $kWh/10^3$ gal) of wastewater treated or 1–1.2 kWh/kg (0.45–0.5 kWh/lb) of salt removed.

Electrodialysis Reversal (EDR) Process: This process involves periodic reversal of DC current polarity so that the clean water compartment is switched with the waste stream compartment. The reversal of current effectively breaks up and removes scales, slime, and other deposits before they strongly build up. The product water is stopped for a short period during and after reversal.

Power Requirements: The Faraday's law of ED is used to estimate the current requirement for ED/EDR. One Faraday of electricity causes 1 g equivalent of a substance to migrate from one electrode to another. So the number of gram equivalent removed per unit time and current for stack of membranes are given by Equations 15.37a and 15.37b.

$$G_{eq} = Q_p(N_{inf} - N_{eff}) = Q_p\Delta N = Q_p N_{inf}\frac{E_r}{100\%} \quad (15.37a)$$

$$I = \frac{FQ_p(N_{inf} - N_{eff})}{n}\frac{100\%}{E_c} = \frac{FQ_p\Delta N}{n}\frac{100\%}{E_c} = \frac{FQ_p N_{inf}}{n}\frac{E_r}{E_c} \quad (15.37b)$$

where G_{eq} = gram equivalent per unit time, eq/s. The gram equivalent of salt is calculated from Equation 15.37c.

$$\text{gram equivalent of salt, eq} = \frac{\text{mass of salt, g}}{\text{equivalent weight of salt, g/eq}} \quad (15.37c)$$

Q_p = product flow rate, L/s
N_{inf} = normality of feedwater, eq/L
N_{eff} = normality of feedwater, eq/L
ΔN = change in normality between influent and effluent, eq/L
E_r = removal efficiency of salt (%). The value of E_r is expressed by Equation 15.37d.

$$E_r = \left(1 - \frac{C_p}{C_f}\right) \times 100\% \quad \text{or} \quad C_p = \left(1 - \frac{E_r}{100\%}\right) \times C_f \quad (15.37d)$$

C_f, C_p, and C_r have been previously defined in Equations 15.28 through 15.30.

I = current, A
F = Faraday's constant (A·s/eq). F = 96,485 A·s/eq.
n = number of cell pairs in the stack, pair
E_c = current efficiency, %

Recovery and Rejection: The recovery and rejection factors for an ED process are also obtained from Equations 15.30a and 15.30b, respectively.

The capacity of membrane to pass the electrical current is related to the current density (*CD*) and the normality of the feedwater (N_{inf}). The *CD* is defined as the current flowing per membrane area perpendicular to the direction of the current. It is expressed by Equation 15.37e.

$$CD = (CD/N_{inf}\text{ ratio})\, N_{inf} \quad \text{or} \quad CD = \frac{1000\, I}{A} \quad \text{or} \quad A = \frac{1000\, I}{CD} \quad (15.37e)$$

where

CD = current density, mA/cm^2

A = membrane area per cell pair, cm^2

(CD/N_{inf} ratio) = ratio of current density to normality of influent, $(mA/cm^2)/(eq/L)$

The normality of the feedwater (N_{inf}) is gram equivalent weight per liter of solution. The ratio of current density to normality (CD/N_{inf} ratio) is an important parameter and may vary from 400 to 700 $(mA/cm^2)/(eq/L)$. If the CD/N_{inf} ratio is too high, it will cause regions of low ionic concentration near the membranes, resulting in polarization causing high electrical consumption. The electrical resistance (R) of an ED stack is determined for a particular feedwater. Once the values of R and I are known, the voltage required (U) and power (P) are obtained from the Ohm's law (Equation 15.37f). The volumetric energy requirement is expressed by Equation 15.37g.

$$U = RI \quad \text{or} \quad P = RI^2 \quad \text{or} \quad P = UI \tag{15.37f}$$

$$En_v = \frac{P}{3600Q_p} \tag{15.37g}$$

where

En_v = volumetric energy requirement, kWh/m^3 of product water

P = power, W

U = voltage, V

R = resistance, ohm (O)

3600 = conversion factor, $(W{\cdot}s{\cdot}m^3)/(kW{\cdot}h{\cdot}L)$

EXAMPLE 15.59: CURRENT, POWER, AND MEMBRANE AREA REQUIREMENTS OF AN ED STACK

An ED stack has 400 pairs of cells. The TDS content of feedwater is 3000 mg/L. The equivalent cations or anion content in the feedwater $N_{inf} = 0.06$ eq/L. Pilot-scale tests show that current efficiency $E_c = 90\%$, salt removal efficiency $E_r = 50\%$, the electrical resistance $R = 4.5$ ohms, and CD/N_{inf} ratio is 400 $(mA/cm^2)/(eq/L)$. The feed rate $Q_f = 2000\ m^3/d$ (23 L/s) and product recovery factor $r = 90\%$. Determine (a) current (I), (b) volumetric energy consumption (P_v) of product water, (c) membrane area and dimensions of square configuration, (d) volume of product water and TDS, and (e) brine volume and TDS.

Solution

1. Calculate the current (I) requirements.

Product water from rearranged Equation 15.30a, $Q_p = \frac{r}{100\%} \times Q_f = \frac{90\%}{100\%} \times 23\ L/s = 21\ L/s$

Normality of feedwater,

Current from Equation 15.37b, $I = \frac{FQ_pN_{inf}}{n}\frac{E_r}{E_c} = \frac{96{,}485\ A{\cdot}s/eq \times 21\ L/s \times 0.06\ eq/L}{400\ \text{pairs of cell}} \times \frac{50\%}{90\%}$

$= 169$ A per cell pair

2. Calculate the volumetric energy requirement (En_v).

Voltage requirement from Equation 15.37f, $P = RI^2 = 4.5O \times (169A)^2 = 129{,}000\ W = 129\ kW$

Volumetric energy consumption from Equation 15.37g, $En_v = \dfrac{P}{3600\,Q_p}$

$$= \frac{129{,}000\ \text{W}}{3600\ (\text{W·s·m}^3)/(\text{kW·h·L}) \times 21\ \text{L/s}}$$

$= 1.7\ \text{kWh/m}^3$

3. Determine the membrane area per cell pair and dimensions.
   CD/N_{inf} ratio $= 400\ (\text{mA/cm}^2)/(\text{eq/L})$ and $N_{inf} = 0.06$ eq/L

   CD from Equation 15.37e, $CD = (CD/N_{inf}\text{ratio})N_{inf} = 400(\text{mA/cm}^2)/(\text{eq/L}) \times 0.06\ \text{eq/L}$
   $= 24\ \text{mA/cm}^2$

   Area of membrane per cell pair from Equation 15.37e, $A = \dfrac{1000I}{CD} = \dfrac{1000\ \text{mA/A} \times 169\text{A}}{24\ \text{mA/cm}^2}$

   $= 7040\ \text{cm}^2$ or $0.7\ \text{m}^2$

   Provide the dimensions of square configuration = 84 cm × 84 cm.
4. Determine the volume and TDS in product (or permeate) water.

   Product flow from rearranged Equation 15.30a, $Q_p = \dfrac{r}{100\%} \times Q_f = \dfrac{90\%}{100\%} \times 2000\ \text{m}^3/\text{d} = 1800\ \text{m}^3/\text{d}$

   TDS in product flow (C_p) from rearranged Equation 15.37d.

$$C_p = \left(1 - \frac{E_r}{100\%}\right) \times C_f = \left(1 - \frac{50\%}{100\%}\right) \times 3000\ \text{mg/L} = 1500\ \text{mg/L}$$

5. Determine the volume and TDS in brine (retentate) flow.
   Brine flow, $Q_r = Q_f - Q_p = (2000 - 1800)\ \text{m}^3/\text{d} = 200\ \text{m}^3/\text{d}$
   TDS in brine flow (C_r) is calculated from rearranged Equation 15.29b,

$$C_r = \frac{C_f Q_f - C_p Q_p}{Q_r} = \frac{3000\ \text{mg/L} \times 2000\ \text{m}^3/\text{d} - 1500\ \text{mg/L} \times 1800\ \text{m}^3/\text{d}}{200\ \text{m}^3/\text{d}} = 16{,}500\ \text{mg/L}$$

15.4.11 Concentration and Disposal of Waste Brine

Concentration and disposal of waste brine (or retentate) from ion exchange and membrane processes have created major problems. Many methods have been investigated and applied to deal with the brine concentration and disposal.

Concentration of Waste Brine: The brine generated from membrane and ion exchange processes contains high concentrations of dissolved salts and high-molecular weight organics. Most common concentration methods include (1) multiple stages RO, (2) solar evaporation, (3) controlled thermal distillation, and (4) freeze crystallization. These methods are briefly presented below.

Multiple-Stage RO: Multiple-stage RO systems are used to concentrate the brine and recover fresh water in several stages. After three-stage RO operation the TDS concentration in the retentate may reach well above 40,000 mg/L. The multiple stage membrane array is shown in Figure 15.41a. A three-stage RO system for brine concentration is designed in Example 15.60.

Solar Evaporation: Evaporation of brine in open and glass- or plastic-covered basins is effectively achieved if large land area is available. Annual evaporation from the surface may range from 90 to

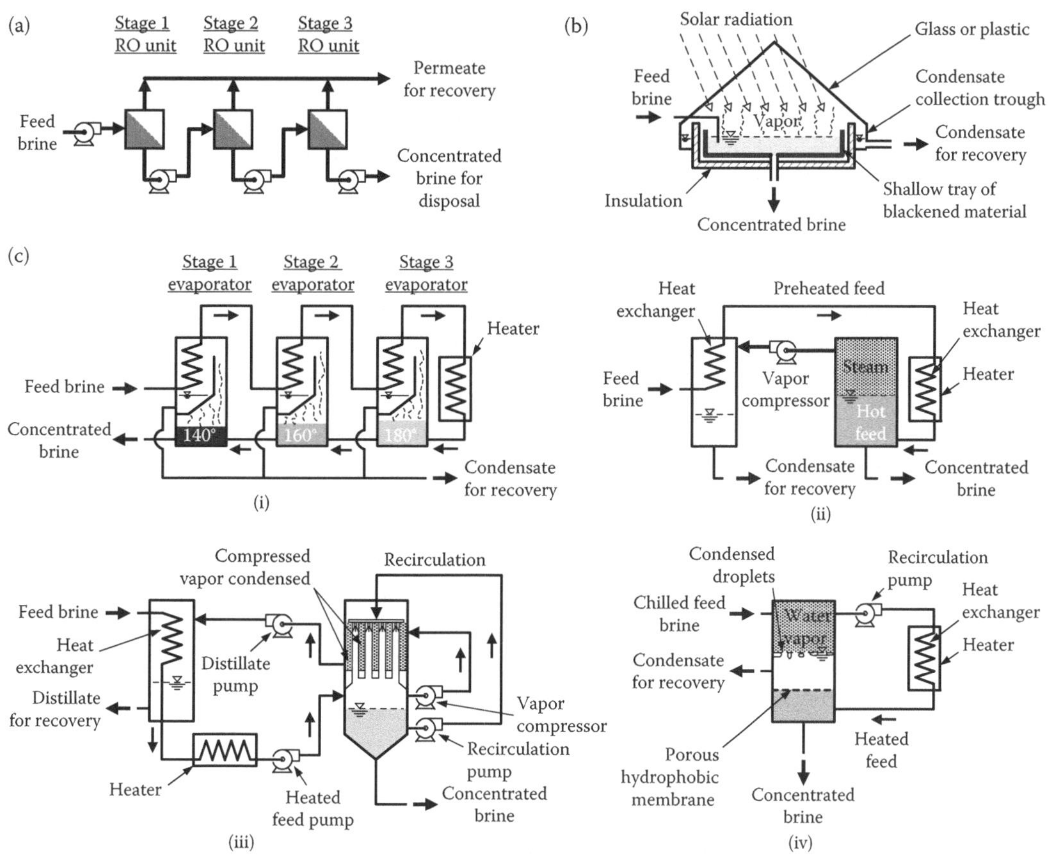

FIGURE 15.41 Concentration of brine: (a) multiple-stage RO; (b) solar evaporation; and (c) controlled thermal distillation: (i) multiple-effect distillation, (ii) distillation with vapor compression, (iii) falling-film evaporator, and (iv) membrane distillation.

250 cm depending upon the relative humidity, wind velocity, barometric pressure, and air temperature. An effective way of solar evaporation is *solar distillation*. The simplest solar distillation unit consists of a shallow blackened pan with sloping glass or plastic cover. The evaporated water condenses inside the sloping cover (since it is cooler) and runs down to the collecting channel for recovery of condensate. The temperature of water in the blackened pan may reach 60–80°C (140–175°C). With present technology about 3 kg of distillate is produced per day per m^2 of the evaporating pan area. This gives approximately 35% operating efficiency at average solar radiation of 47.5 calories/cm^2·d.[3,133,134] The design of a solar distillation unit is shown in Figure 15.41b.

Controlled Thermal Distillation: Thermal distillation is the oldest demineralization process. It consists of evaporating part or all of the water from a saline solution and subsequent condensation of mineral-free vapor. The energy requirements are very high. Several modifications are used to conserve energy. In a *multiple-effect distillation*, the water is evaporated in different stages. Incoming feedwater is preheated in each stage from the latent heat of condensation. A small vacuum is created by the condensed steam from the previous stage.[3,135] The process is shown in Figure 15.41c(i). Another modification is *distillation with vapor compression* (Figure 15.41c(ii)). In this process, the steam is compressed for condensation, and the latent heat of condensation is used to preheat the incoming flow to near boiling by passing through a heat exchanger. The preheated feed is further heated to produce steam. It is possible to operate the system with only 230 kJ of heat energy to evaporate each kg of water (100 Btu/lb).*

The TDS concentration in the product water is <10 mg/L, and that of concentrated brine is typically in the range of 2–10%. Another modification is *falling-film evaporator* in which the preheated feed is sprayed into the vapor compression chamber for condensation, and the heat of condensate is used to preheat the incoming feed (Figure 15.41c(iii)). As the TDS concentration in the brine increases the solids may crystalize resulting in zero discharge.[54,136] Another modification of controlled thermal distillation is *membrane distillation*. Heated feed is passed along the one side of a hydrophobic membrane. The water vapors pass through the membrane and are condensed over the surface of pipe carrying chilled influent feed.[53,136] A simplified process diagram is shown in Figure 15.41c(iv).

Freeze Crystallization: If saline water is cooled until freezing, the dissolved solids move to the bottom while pure water remains on the top. The concentration of TDS may vary from zero in the top layer to highest at the bottom layer of the slab of ice thus formed. This process has been used in cold climate where hard freeze occurs in the winter. The frozen slab is sliced to separate fresh water and brine portions. In the spring when ice melts, the fresh water and brine are recovered. It should be noted that the freezing point of a solution and quality of ice produced depend on the salinity of water.[137] Freezing temperature is lower for higher salinity water and ice crystals produced become coated with the brine at an excessively high salt content.

Disposal of Concentrated Waste Brine: Many disposal methods have been adapted depending upon the quantity, concentration, and availability of suitable site. Among the common methods are (1) disposal in wastewater collection system, (2) surface water discharge, (3) discharge into ocean or saline water, (4) deep-well injection, and (5) evaporation and land disposal of residues. These methods are briefly presented below.

Disposal in Wastewater Collection System: All discharges into municipal collection systems are governed by local regulations. This option is available for very small discharges. Conventional plants provide very little treatment except dilution. Increase in TDS level up to 100 mg/L in the wastewater is considered insignificant and should be allowed. Higher concentrations may interfere with the operation of the wastewater treatment plant, or effluent disposal in natural waters or reuse.

Surface Water Discharge: Discharge of small quantities of brine into surface water is the most common method of disposal. Extensive monitoring of the discharges and water body is needed to meet the requirements of the NPDES permit.

Discharge into Ocean or Saline Water: Coastal facilities extensively use this option. Larger facilities may require long pipeline for deep ocean outfall. Inland facilities may use rail haul, truck, or pipeline to transport concentrated brine.

Deep-Well Injection: Many oil producing states allow deep-well injection of brine. Among these states are California, Colorado, Florida, Louisiana, Pennsylvania, and Texas. Normally, brackish water or natural brines lie below a zone of fresh groundwater. In Louisiana, the fresh water reaches to a maximum depth of 240 m (800 ft). Most important considerations are site suitability, waste characteristics, and economics. Site suitability involves availability of saline water aquifer at a reasonable depth below an unfractured impermeable layer forming an artesian (confined) aquifer. The saline water aquifer should have good porosity and void space, permeability, and of sufficient thickness. The wastewater should be nontoxic and should not react with the natural formation to cause corrosion or precipitation of bivalent cations.

The economics of deep-well injection depend upon casing depth, type of formation, pretreatment need, waste volume, and injection pressure. Those states that allow deep-well injection have requirement for permit and monitoring. The permit is based on (1) geology of the area, (2) proposed casing depth, (3) brine volume, and (4) characteristics. Under the monitoring program the following data should be maintained

* Specific heat of water is 1 cal/g°C or 4200 J/kg°C (1 Btu/lb·°F). The heat of vaporization at 100°C (212°F) is 540 cal/g or 2260 kJ/kg (970 Btu/lb).

and reviewed every 6 months: (1) characteristics of waste, (2) continuous monitoring of pumping rate, (3) total volume injected in 6 months, (4) cumulative volume injection since start, (5) applied pressure and electrical log, and (6) monitoring disposal well injectivity and recording the *well index*. Well index is calculated from Equation 15.38.

$$WI = \frac{WQ}{\sqrt{WH}} \tag{15.38}$$

where
WI = well index
WQ = well flow, L/min (gpm)
WH = well-head pressure, kPa (psig)

The well-head pressure WH is the applied pressure at the bottom of the well. It includes driving forces, static head and pressure drop resulting from pipe friction, and minor losses. A sudden drop in driving force or electrical log or an increase in well flow or well index are clear indication of fracture occurring in the impermeable layer of the confined aquifer. The concept of deep-well injection may be looked upon as inverse of groundwater production. Same formula and equipment are used for determining the water-well productivity and deep-well injectivity.

Evaporation to Dryness and Land Disposal: Solar evaporation in ponds or covered basins or solar distillation may be used to completely dry the concentrated brine. The residue may be sent to a disposal facility. Spray dryer have also been developed for this purpose. A spray dryer has vertical spray drying chamber and a bag filter house. The concentrated brine is sprayed in the drying chamber through a centrifugal brine mixer. Heated air is also pumped into the dryer and drawn through the bag filters. Dry powder is collected from the bag house and sent to a residue disposal facility.

EXAMPLE 15.60: TREATMENT OF BRACKISH WATER AND CONCENTRATION OF RETENTATE BY RO

A RO facility is used to demineralize 4000 m^3/d brackish water for municipal water supply. The TDS of feed is 800 mg/L. Three RO systems are provided in two stages: one for raw water feed and two for concentration of the brine. The recovery and rejection factors of three RO systems are: (a) the first stage: 90% and 95%; (b) the second stage: 70 and 80; and (c) the third stage: 50% and 65%. Determine the volume and TDS of brine from the third stage and that of combined permeate stream. Also determine the percent solids in the concentrated brine if the bulk specific gravity $S_{b,brine} = 1.025$ (or 1025 kg/m^3).

Solution

1. Draw the definition sketch in Figure 15.42.
2. Determine the volumes and TDS concentrations in permeates and brines.
   a. First RO stage:

Permeate volume from rearranged Equation 15.30a, $Q_{p,1} = \frac{r_1}{100\%} \times Q_{f,1} = \frac{90\%}{100\%} \times 4000\,m^3/d$

$= 3600\,m^3/d$

TDS in permeate from rearranged Equation 15.30b, $C_{p,1} = \left(1 - \frac{R_1}{100\%}\right) C_{f,1}$

$= \left(1 - \frac{95\%}{100\%}\right) \times 800\,mg/L = 40\,mg/L$

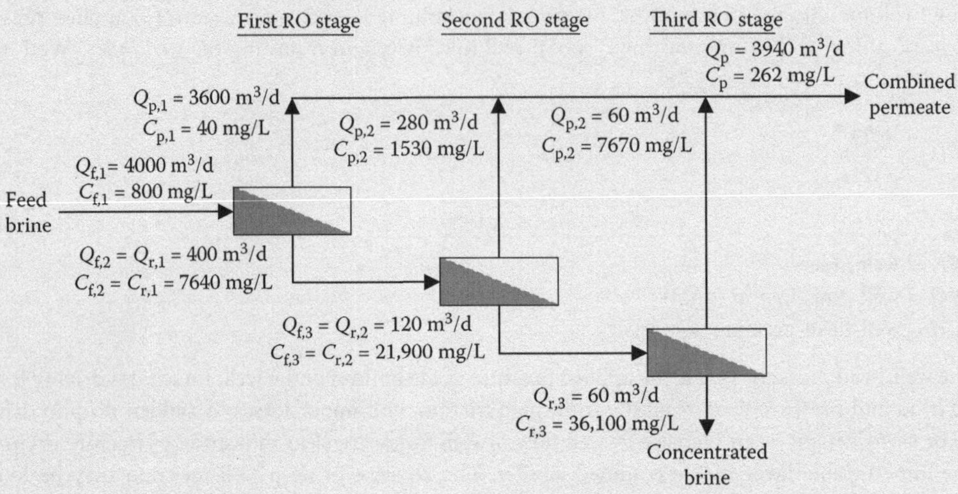

FIGURE 15.42 Definition sketch of three-stage RO concentration facility (Example 15.60).

Brine volume from rearranged Equation 15.29a, $Q_{r,1} = Q_{f,1} - Q_{p,1} = (4000 - 3600)\text{m}^3/\text{d} = 400\,\text{m}^3/\text{d}$

TDS in brine from rearranged Equation 15.29b, $C_{r,1} = \dfrac{C_{f,1}Q_{f,1} - C_{p,1}Q_{p,1}}{Q_{r,1}}$

$$= \frac{800\,\text{mg/L} \times 4000\,\text{m}^3/\text{d} - 40\,\text{mg/L} \times 3600\,\text{m}^3/\text{d}}{400\,\text{m}^3/\text{d}}$$

$$= 7640\,\text{mg/L}$$

Similarly, the flows and TDS concentrations are calculated below for the second and third RO stages.

b. Second RO stage:

From the calculated results in the first stage, $Q_{f,2} = Q_{r,1} = 400\,\text{m}^3/\text{d}$ and $C_{f,2} = C_{r,1} = 7640\,\text{mg/L}$.

Permeate volume, $Q_{p,2} = \dfrac{70\%}{100\%} \times 400\,\text{m}^3/\text{s} = 280\,\text{m}^3/\text{d}$

TDS concentration in permeate, $C_{p,2} = \left(1 - \dfrac{80\%}{100\%}\right) \times 7640\,\text{mg/L} = 1530\,\text{mg/L}$

Brine volume, $Q_{r,2} = Q_{f,2} - Q_{p,2} = (400 - 280)\,\text{m}^3/\text{d} = 120\,\text{m}^3/\text{d}$

$$\text{TDS concentration in brine, } C_{r,2} = \frac{7640\,\text{mg/L} \times 400\,\text{m}^3/\text{d} - 1530\,\text{mg/L} \times 280\,\text{m}^3/\text{d}}{120\,\text{m}^3/\text{d}} = 21{,}900\,\text{mg/L}$$

c. Third RO stage:

From the calculated results in the second stage, $Q_{f,3} = Q_{r,2} = 120\,\text{m}^3/\text{d}$ and $C_{f,3} = C_{r,2} = 21{,}900\,\text{mg/L}$.

Permeate volume, $Q_{p,3} = \dfrac{50\%}{100\%} \times 120\,\text{m}^3/\text{s} = 60\,\text{m}^3/\text{d}$

TDS concentration in permeate, $C_{p,3} = \left(1 - \dfrac{65\%}{100\%}\right) \times 21{,}900\,\text{mg/L} = 7670\,\text{mg/L}$

Brine volume, $Q_{r,3} = Q_{f,3} - Q_{p,3} = (120 - 60)\,m^3/d = 60\,m^3/d$

$$\text{TDS concentration in brine, } C_{r,3} = \frac{21{,}900\,mg/L \times 120\,m^3/d - 7670\,mg/L \times 60\,m^3/d}{60\,m^3/d}$$

$$= 36{,}100\,mg/L$$

3. Determine the volume and TDS in the combined permeate.

   Total volume, $Q_p = Q_{p,1} + Q_{p,2} + Q_{p,3} = (3600 + 280 + 60)\,m^3/d = 3940\,m^3/d$

   TDS concentration from mass balance,

$$C_p = \frac{C_{p,1}Q_{p,1} + C_{p,2}Q_{p,2} + C_{p,3}Q_{p,3}}{Q_p}$$

$$= \frac{40\,mg/L \times 3600\,m^3/d + 1530\,mg/L \times 280^3/d + 7670\,mg/L \times 60\,m^3/d}{3940\,m^3/d} = 262\,mg/L$$

4. Determine the percent solids in the brine.

   Flow and TDS concentration in the final brine, $Q_{brine} = Q_{r,3} = 60\,m^3/d$ and $C_{brine} = C_{r,3} = 36{,}100$ mg/L.

   Mass of TDS in the brine, $W_{brine} = C_{brine}Q_{brine} = 36{,}100\,g/m^3 \times 60\,m^3/d \times 10^{-3}\,kg/g = 2170\,kg/d$

   Rearrange Equation 13.1e to calculate the percent solids content in the brine ($p_{s,brine}$) assuming $\rho_w = 1000\,kg/m^3$.

$$p_{s,brine} = \frac{100\% \times W_{brine}}{Q_{brine}S_{s,brine}\rho_w} = \frac{100\% \times 2170\,kg/d}{60\,m^3/d \times 1.025 \times 1000\,kg/m^3} = 3.5\%$$

EXAMPLE 15.61: CONCENTRATION OF WASTE BRINE IN COVERED DRYING BEDS FOR LAND DISPOSAL

The concentrated brine obtained in Example 15.60 is further concentrated in a covered drying facility for land disposal of residues. The average evaporation rate is 2 cm/d. Determine (a) the volume of moisture evaporated, (b) the area required for covered drying facility, and (c) the volume and bulk specific gravity of the sludge concentrated for land disposal. Assume that the moisture content of the sludge is 70%.

Solution

1. Determine the volume of moisture evaporated.

   Total mass of liquid brine, $W_{brine} = Q_{brine}S_{s,brine}\rho_w = 60\,m^3/d \times 1.025 \times 1000\,kg/m^3 = 61{,}500\,kg/d$

   Total mass of wet sludge after drying to 70% moisture or $p_{s,sludge} = 30\%$ solids,

$$W_{sludge} = \frac{100\% \times W_{brine}}{p_{s,sludge}} = \frac{100\% \times 2170\,kg/d}{30\%} = 7230\,kg/d$$

   Mass of moisture evaporated, $W_{moisture} = W_{brine} - W_{sludge} = (61{,}500 - 7230)\,kg/d = 54{,}270\,kg/d$

$$\text{Volume of moisture evaporated, } Q_{moisture} = \frac{W_{moisture}}{\rho_w} = \frac{54{,}270\,kg/d}{1000\,kg/m^3} = 54\,m^3/d$$

2. Determine the area required for drying of concentrated brine.

For a covered facility, there is no impact of precipitation on drying process.

$$\text{Area required, } A_{\text{drying}} = \frac{Q_{\text{moisture}}}{E_{\text{p}}} = \frac{54\,\text{m}^3/\text{d}}{2\,\text{cm/d} \times 10^{-2}\,\text{m/cm}} = 2700\,\text{m}^2$$

Provide three covered areas each 40 m × 25 m to have a total drying area of 3000 m^2 at the facility.

3. Determine the volume and bulk specific gravity of sludge.

$$\text{Volume of sludge, } Q_{\text{sludge}} = Q_{\text{brine}} - Q_{\text{moisture}} = (60 - 54)\,\text{m}^3/\text{d} = 6\,\text{m}^3/\text{d}$$

Bulk specific gravity of sludge from rearranged Equation 13.1e,

$$S_{\text{s,sludge}} = \frac{100\% \times W_{\text{brine}}}{Q_{\text{sludge}}\rho_{\text{s,sludge}}\rho_{\text{w}}} = \frac{100\% \times 2170\,\text{kg/d}}{6\,\text{m}^3/\text{d} \times 30\% \times 1000\,\text{kg/m}^3} \approx 1.2$$

EXAMPLE 15.62: ENERGY RECOVERY IN A DISTILLATION SYSTEM WITH VAPOR COMPRESSION

The brine from an RO unit is concentrated using distillation with vapor compression. At an average 210 kJ of heat energy per kg is needed to bring the feed from 18°C to the boiling point. Approximately 40% heat of vaporization is lost by radiation, and 390 kJ/kg of equivalent mechanical energy is needed for vapor compression. Estimate the energy saving from distillation with vapor compression over conventional distillation. Heat energy for vaporization of water is 2260 kJ/kg (970 Btu/lb).

Solution

1. Calculate the total energy consumed for convention distillation.

Heat energy for heating feed sludge to digester is given by Equation 13.6a. This equation can also be used to calculate the heat requirement for heating concentrated brine. Assume that the specific heat of brine is same as that for water: 4200 J/kg·°C (1 Btu/lb·°F).

Heat energy required to raise the temperature of 1 kg brine from 18°C to 100°C

$$= 4200\,\text{J/kg·°C} \times (100-18)\text{°C} = 340{,}000\,\text{J/kg} \quad \text{or} \quad 340\,\text{kJ/kg}$$

Heat energy required for vaporization = 2260 kJ/kg (970 Btu/lb).

Total heat energy required for conventional distillation = (340 + 2260) kJ/kg = 2600 kJ/kg

2. Calculate the total energy consumed for distillation with vapor compression.

Heat energy required to reach boiling point = 210 kJ/kg (given)

Heat energy lost due to radiation during vaporization = 0.4 × 2260 kJ/kg = 900 kJ/kg

Equivalent heat energy for vapor compression = 390 kJ/kg (given)

Total heat energy required for distillation with vapor compression = (210 + 900 + 390) kJ/kg = 1500 kJ/kg

3. Estimate the energy saving from distillation with vapor compression over conventional distillation.

$$\text{Energy saving} = \frac{(2260 - 1500)\,\text{kJ/kg}}{2260\,\text{kJ/kg}} \times 100\% = 34\%$$

EXAMPLE 15.63: CONCENTRATION OF WASTE BRINE BY SINGLE-EFFECT EVAPORATOR

Waste brine from an RO facility is concentrated from 10% to 30% solids in a single-effect evaporator. The evaporator has 550 ft^2 of heating surface. The overall thermal coefficient of evaporator is 620 Btu/h·ft^2°F after considering heat loss. Dry steam at 10 psig is used at a vacuum of 26 in of mercury (gauge). The temperature of the brine entering and the condensate leaving the evaporator are 125°F and 180°F, respectively. The heat required for vaporization of water at 125° F is 1022 Btu/lb. The temperature of steam and enthalpy of dry steam at 10 psig (24.7 psia) obtained from the steam table are 240°F and 1160 Btu/lb. The heat release rate from the condensate at 180°F = 145 Btu/lb.[138] Calculate (a) the rate of feed brine to the evaporator and (b) the rate of steam utilization.

Solution

1. Conduct the mass balance of brine streams before or after the evaporator on basis of 1-lb feed brine.
   a. Mass in feed brine stream (10% solids) to the evaporator.
      Solids, $w_{s,feed} = 0.1$-lb solids/lb feed
      Water, $w_{w,feed} = 0.9$-lb water/lb feed
   b. Mass in concentrated brine stream (30% solids) from the evaporator.
      Solids, $w_{s,concentrate} = 0.1$-lb solids/lb feed

$$\text{Water}, w_{w,concentrate} = 0.1\text{ lb solids/lb feed} \times \frac{(1-0.3)\text{ lb water/lb concentrated brine}}{0.3\text{ lb solids/lb concentrated brine}}$$

$$= 0.23\text{-lb water/lb feed}$$

   c. Mass in vapor (pure water) from the evaporator.

$$\text{Water}, w_{w,vapor} = (0.9 - 0.23)\text{ lb water/lb feed} = 0.67\text{ lb water/lb feed}$$

2. Determine the overall evaporation capacity of the evaporator (H_c).

$$H_c = 550\text{ ft}^2 \times 620\text{ Btu/h·ft}^2\text{·°F} \times (180 - 125)\text{°F} = 1.9 \times 10^7\text{ Btu/h}$$

3. Determine the rate of feed brine to the evaporator (w_{feed}).
   At equilibrium, the heat required to evaporate the feed brine = overall evaporation capacity.

$$1022\text{ Btu/lb water} \times 0.67\text{ lb water/lb feed} \times w_{feed} = 1.9 \times 10^7\text{ Btu/h}$$

$$\text{Solve for } w_{feed} = \frac{1.9 \times 10^7\text{ Btu/h}}{1022\text{ Btu/lb water} \times 0.67\text{ lb water/lb feed}} = 28{,}000\text{ lb feed/h}$$

4. Determine the rate of steam utilization (w_{steam}).
   At equilibrium, the loss of enthalpy from the steam = overall evaporation capacity.

$$(1160 - 145)\text{ Btu/lb steam} \times w_{steam} = 1.9 \times 10^7\text{ Btu/h}$$

$$\text{Solve for } w_{steam} = \frac{1.9 \times 10^7\text{ Btu/h}}{(1160 - 145)\text{ Btu/lb steam}} = 19{,}000\text{ lb steam/h}$$

Discussion Topics and Review Problems

15.1 Describe the major design features, site characteristics, and effluent quality from land treatment systems.

15.2 List the environmental concerns of land treatment of municipal wastewater.

15.3 A SR irrigation system to treat 0.1 m^3/s primary treated municipal wastewater. The hydraulic loading rate is 1.5 m per year. Calculate (a) the total land area required for irrigation, (b) organic loading rate (kg BOD_5/ha·d) if influent has a BOD_5 concentration of 140 mg/L, and (c) overall land area required for. Assume that the additional land requirement is 350% of the theoretical land area for buffer zone, drainage ditches, and access roads.

15.4 A RI system is provided to treat 1000 m^3/d of primary effluent flow at an annual average hydraulic loading rate of 25 m/year. Determine (a) the average area that receives flow per application cycle, (b) total infiltration area, (c) annual average BOD_5 loading rate. Assume that the BOD_5 concentration is 100 mg/L in the primary effluent; the average application and drying periods are $T_A = 1$ d/cycle and $T_D = 7$ d/cycle, respectively; the 10-year maximum monthly precipitation is 0.4 cm/d; and the RI system is operated without ponding problem.

15.5 An OF system receives secondary effluent flow of 1000 m^3/d from a wastewater treatment plant. The OF system is operated for 300 d/year at an average application rate of 0.25 m^3/m·h for 8 h during each operative day. Determine the requirements of application area and sloped length at a hydraulic loading rate of 4 cm/d. Also estimate the number of plots required if the width of plot is 20 m.

15.6 A SF wetland is provided to treat a flow of 1000 m^3/d. Calculate the nominal detention time and volume required at a hydraulic loading rate of 200 m^3/ha·d and a mean depth of 0.5 m. Also, determine the width and length of the wetland at a bed slope of 5%. Assume that the hydraulic conductivity $K = 5 \times 10^{-3}$ m/s.

15.7 Use of FWS and SF constructed wetlands for removal of BOD_5 are evaluated. The influent flow and BOD_5 are 1000 m^3/d and 80 mg/L, respectively. A BOD_5 concentration of 15 mg/L is required in the effluents from both FWS and SF wetlands. Estimate (a) the nominal detention times, (b) hydraulic loading rates, and (c) surface areas required in both wetlands at a water or submergence depth of 0.5 m. Assume that the overall volumetric availability factors for FWS and SF wetlands are 0.8 and 0.4, respectively; and the reaction rate constants for FWS and SF wetlands are $K_{w,FWS} = 0.2\ d^{-1}$ and $K_{w,SF} = 0.5\ d^{-1}$, respectively.

15.8 Draw a process train of a secondary wastewater treatment plant. Identify the possible locations where suitable coagulant may be added for precipitation of phosphorus. List the advantages and disadvantages of coagulation at each location.

15.9 In a jar test, municipal wastewater was coagulated using aluminum sulfate $Al_2(SO_4)_3 \cdot 18H_2O$. The phosphorus concentration in the wastewater was 8 mg/L as P. At alum dosages of 89, 136, and 216 mg/L, soluble P levels remaining in the supernatants were 2, 1.5, and 0.4 mg/L, respectively. Calculate the molar ratio Al:P for these removals. Also prepare a plot of molar ratio of Al:P versus log of percent P removed.

15.10 What is the purpose of recarbonation after two-stage lime treatment? Give chemical equations to justify your reasoning.

15.11 Breakpoint chlorination is used for oxidation of 18 mg/L ammonia nitrogen in an effluent. Calculate chlorine dosage and how many mg/L chloride ions are added in the effluent.

15.12 Discuss the major limiting factors of ammonia stripping.

15.13 List the various types of filtration systems commonly used in the polishing of secondary effluent. Discuss the advantages and disadvantages of microscreens.

15.14 A filter is designed for a hydraulic loading rate of 5 m^3/m^2·h. The depth of sand bed is 0.25 m. The diameter and specific gravity of sand grains is 0.6 mm and 2.65, respectively. Assume that the porosity ratio is 0.45 and the shape factor is 0.9. Calculate the head loss through a clean bed of uniform sand at the water temperature of 15°C. Use the Carman–Kozeny equation (Equation 15.9a).

15.15 Develop the relationship between the head loss across a clean sand bed and the filtration rate between 2 and 8 m^3/m^2·h at an increment of 1 m^3/m^2·h. Assume that the filter bed is composed of uniform-size sand grains given in Problem 15.14. The water temperature is 15°C.

15.16 A sieve analysis of filter sand is performed to obtain the following experimental data.

Sieve Number (U.S. Sieve Series)	Mass Fraction of Sand Grains Retained, %	Geometric Mean Diameter, mm
14–20	0.8	1.09
20–25	5.25	0.77
25–30	23.02	0.65
30–35	32.65	0.54
35–40	28.81	0.46
40–50	5.15	0.35
50–60	2.65	0.27
60–70	1.13	0.23
70–100	0.54	0.18

Calculate the head loss across a clean stratified sand bed at a filtration rate of 150 m^3/m^2·d and a water temperature of 5°C. The sand bed is 60-cm deep. Assume that the porosity ratio is 0.4 and the shape factor is 1.0. Use the Carman–Kozeny equation (Equation 15.9a).

15.17 Use the sieve analysis data in Problem 15.16 to calculate (a) the effective size and uniformity coefficient of the sand and (b) the head loss through a clean stratified bed using the effective size value.

15.18 Calculate the head loss through a clean sand filter media, given the following:

Filtration rate	6 m^3/m^2·h
Media depth	30 cm
Media size	0.5 mm
Specific gravity	2.65
Porosity	0.4
Shape factor	0.9
Uniformity coefficient	1.4
Viscosity of water	1.3×10^{-3} N·s/m^2

15.19 Determine the size and depth of an anthracite coal filter medium to be compatible with the sand media given in Problem 15.18. The specific gravity of anthracite is 1.55 and media depth is 2/3 of total depth. Calculate also the head loss through the dual-media filter at a filtration rate of 6 m^3/m^2·h. Assume that the porosity ratio is 0.48, and the shape factor is 1.0 for the anthracite medium.

15.20 Use the filter sand bed information given in Problem 15.18 to calculate the backwash rates (m^3/m^2·h) required under the following conditions: (a) the media is just fluidized and (b) a bed expansion of 50% of the original bed depth is reached during backwashing.

15.21 Design a tertiary granular media filter system to treat 50,000 m^3/d of effluent. Determine (a) the number of filter units, (b) dual-media sizes and type, (c) head loss through media and gravel beds, (d) backwash rate, (e) backwash volume, and (f) backwash head requirements. Consult calculation procedures presented in Examples 15.20 through 22, and 28 through 30.

15.22 Calculate the turbulence (or velocity gradient, G_{sw}) required by a dual-arm surface wash system at a water temperature of 5°C. Use Equation 15.14 having the following parameters.

$\Delta h = 45$ m
$v_{sw} = 0.1\ m^3/m^2 \cdot min$
$L_{fb} = 60$ cm
$\alpha = 0.5$

15.23 Estimate the turbulence (velocity gradient, G) of an air-scour system based on the following given information:

Tank volume	$250\ m^3$
Water depth	4.0 m
Viscosity of water	$1.307 \times 10^{-3}\ N{\cdot}s/m^2$ (at 10°C)
Air density	$1.204\ kg/m^3$ (at 20°C)
Airflow	$50\ m^3/s$

15.24 Discuss the advantages and disadvantages of various demineralization processes (ion exchange, distillation, and membrane processes) if used for potable water reuse from a tertiary treatment effluent.

15.25 A tertiary effluent is treated by a split softening process for industrial reuse purpose. Partial tertiary effluent flow is softened by a zeolite resin bed and then mixed with the bypassed effluent to achieve a total hardness below 50 mg/L in the combined reuse water. The total hardness is 250 mg/L as $CaCO_3$ in the tertiary effluent. The hardness is 5 mg/L as $CaCO_3$ in the softened water from the resin bed. The exhausted resin is regenerated by 98% pure NaC1 solution. The salt consumption is three times the theoretical amount and used as 4% salt solution. The design reuse water flow is $1000\ m^3/d$. The softening capacity of the resin is $62\ kg/m^3$ total hardness as $CaCO_3$. Determine (a) the required flow rate through the resin bed, (b) volume of the zeolite bed at a regeneration cycle of 7 days, (c) salt consumption per regeneration cycle, and (d) total volume of regenerant solution.

15.26 Design an anion exchange system to treat $1000\ m^3/d$ of effluent. The resin bed has a capacity of 250 kg/m^3 as $CaCO_3$, and the regeneration cycle is 5 days. Calculate the theoretical amount of NaOH needed for regeneration. The concentrations of major anions in the effluent are: $SO_4^{2-} = 30$ mg/L, $Cl^- = 50$ mg/L, $HCO_3^- = 75$ mg/L, $CO_3^{2-} = 3$ mg/L, $NO_3^- = 15$ mg/L, $CO_2 = 15$ mg/L, and $SiO_2 = 4$ mg/L.

15.27 A hydrogen cation exchange bed has a capacity of $300\ kg/m^3$. Regeneration cycle is 5 days and effluent flow is $1000\ m^3/d$. Calculate (a) the required volume of resin bed and (b) theoretical amount of acid needed for regeneration. The concentrations of major cations are: $Ca^{2+} = 40$ mg/L, $Mg^{2+} = 15$ mg/L, $Na^+ = 60$ mg/L, and $K^+ = 3$ mg/L.

15.28 What are the general classifications of membrane processes? Describe (a) the pore opening and operating range and (b) driving force and removal mechanism of these different processes. Also explain the basic membrane configurations. Consult Table 15.22 and Figure 15.34 under Section 15.4.10.

15.29 Describe (a) the mechanism of RO process, (b) main application of the process, and (c) pretreatment normally required before a RO system. Also, draw process diagrams for a split RO system with bypass and a two-pass RO system.

15.30 Design an RO system to treat an industrial wastewater for demineralization. The pretreated industrial wastewater has TDS of 3000 mg/L at a flow rate of $1000\ m^3/d$. The following operating and performance data in provided by a RO system manufacturer: recovery factor $r = 80\%$, salt-rejection factor $R = 90\%$, and product water flux rate $= 0.65\ m^3/m^2{\cdot}d$ at the design pressure of 1800 kPa (kN/m^2).

15.31 Discuss the advantages and disadvantages of distillation process over membrane process.

References

1. Mitsch, W. J. and J. G. Gosselink, *Wetlands*, 5th ed., John Wiley & Sons, Inc., Hoboken, NJ, 2015.
2. Reed, S. C., R. W. Crites, and E. J. Middlebrooks, *Natural Systems for Waste Management and Treatment*, 2nd ed., McGraw-Hill Professional, New York, NY, 1998.
3. Qasim, S. R., *Wastewater Treatment Plants: Planning, Design, and Operation*, 2nd ed., CRC Press, Boca Raton, FL, 1999.

4. U.S. Environmental Protection Agency, *Land Treatment of Municipal Wastewater Effluents—Design Factors*, Vol. 1, EPA 625/4-76-010, Center for Environmental Research Information, Cincinnati, OH, January 1976.
5. U.S. Environmental Protection Agency, *Process Design Manual for Land Treatment of Municipal Wastewater*, EPA 625/1-81-013, Center for Environmental Research Information, Cincinnati, OH, October 1981.
6. U.S. Environmental Protection Agency, *Process Design Manual: Land Treatment of Municipal Wastewater Effluents*, EPA/625/R-06/016, Office of Research and Development, Cincinnati, OH, September 2006.
7. Sanks, R. L. and T. Asano (eds.), *Land Treatment and Disposal of Municipal and Industrial Wastewater*, Ann Arbor Science Publishers, Ann Arbor, MI, 1976.
8. Metcalf & Eddy, Inc., *Wastewater Engineering: Treatment, Disposal and Reuse*, 3rd ed., McGraw-Hill, Inc., New York, NY, 1991.
9. U.S. Environmental Protection Agency, *Applications of Sludges and Wastewaters on Agricultural Land: A Planning and Educational Guide*, MCD-35, Office of Water Program Operations, Washington, DC, March 1978.
10. Droste, R. L., *Theory and Practice of Water and Wastewater Treatment*, John Wiley & Sons, Inc., New York, NY, 1997.
11. Crites, R. W., E. J. Middlebrooks, R. K. Bastian, and S. C. Reed, *Natural Wastewater Treatment Systems*, 2nd ed., CRC Press, Boca Raton, FL, 2014.
12. U.S. Environmental Protection Agency, *Land Application of Wastewater and State Water Law: An Overview*, Vol. 1, EPA-600/2-77-232, Office of Research and Development, Ada, OK, November 1977.
13. U.S. Environmental Protection Agency, *Evaluation of Land Application Systems*, Technical Bulletin EPA-430/9-75-001, March 1975.
14. Kamal, M. S. and S. R. Qasim, Natural systems for wastewater-treatment, *The Encyclopedia of Environmental Science and Engineering*, 4th ed., J. R. Pfafflin and E. N. Ziegler (eds), Gordon and Breach, Science Publishers, New York, NY, 1998, pp. 821–831.
15. Hammer, D. A. (ed.), *Constructed Wetlands for Wastewater Treatment*, CRC Press, Boca Raton, FL, 1989.
16. U.S. Environmental Protection Agency, *Design Manual: Constructed Wetlands and Aquatic Plant Systems for Municipal Wastewater Treatment*, EPA/625/1-88-22, Office of Research and Development, Washington, DC, September 1988.
17. Reed, S. C. (ed.), *Natural Systems for Wastewater Treatment*, Manual of Practice No. FD-16, 2nd ed., Water Environment Federation, Alexandria, VA, 2001.
18. Reed, S. C., Constructed wetlands for industrial wastewaters, Paper presented at Purdue Industrial Waste Conference, Purdue University, West Lafayette, IN, May 1993.
19. Moshiri, G. A. (ed.), *Constructed Wetlands for Water Quality Improvement*, CRC Press, Boca Raton, FL, 1993.
20. U.S. Environmental Protection Agency, Subsurface flow constructed wetlands conference, Paper presented at the Conference of Subsurface Flow Constructed Wetlands at University of Texas at El Paso, August 16 and 17, El Paso, TX, 1993.
21. Kadlec, R. H. and S. D. Wallace, *Treatment Wetlands*, 2nd ed., CRC Press, Boca, Raton, FL, 2009.
22. Stefanakis, A., C. S. Akratos, and V. A. Tsihrintzis, *Vertical Flow Constructed Wetlands, Eco-Engineering Systems for Wastewater and Sludge Treatment*, Elsevier Inc., Amsterdam, Netherland, 2014.
23. Pandey, M. K., P. D. Jenssen, T. Krogstad, and S. Jonasson, Combination of horizontal and vertical flow constructed wetlands to improve nitrogen removal, *Water, Air, & Soil Pollution: Focus*, 9(3), 2009, 279–286.

24. Gaboutoeloe, G. K., S. Chen, M. E. Barber, and C. O. Stockle, Comparison of vertical and horizontal flow planted and unplanted subsurface flow wetlands treating municipal wastewater, *Water Science and Technology*, 68(1), 2013, 117–123.
25. Kim, S. H., J. S. Cho, J. H. Park, J. S. Heo, Y. S. Ok, R. D. Delaune, and D. C. Seo, Long-term performance of vertical-flow and horizontal-flow constructed wetlands as affected by season, N load, and operating stage for treating nitrogen from domestic sewage, *Environmental Science and Pollution Research International*, 23(2), 2016, 1108–1119.
26. Stephenson, M., G. Turner, P. Pope, J. Colt, A. Knight, and G. Tchobanoglous, The use and potential of aquatic species for wastewater treatment, Appendix A, *The Environmental Requirements of Aquatic Plants*, Publication No. 65, California State Water Resources Control Board, Sacramento, CA, 1980.
27. Tchobanoglous, G., Aquatic plant systems for wastewater treatment: Engineering considerations, *Aquatic Plants for Water Treatment and Resource Recovery*, K. R. Reedy and W. H. Smith (eds.), Magnolia Publishing, Orlando, FL, 1987, pp. 27–48.
28. U.S. Environmental Protection Agency, *Subsurface Flow Constructed Wetlands for Wastewater Treatment: A Technology Assessment*, EPA 832-R-93-008, Office of Water, Washington, DC, July 1993.
29. Kadlec, R. H., S. D. Wallace, and R. L. Knight, *Treatment Wetlands*, CRC Press, Boca Raton, FL, 1996.
30. Crites R. and G. Tchobanoglous, *Small and Decentralized Wastewater Management Systems*, McGraw-Hill Book Co., New York, NY, 1998.
31. U.S. Environmental Protection Agency, *Manual: Constructed Wetlands Treatment of Municipal Wastewaters*, EPA/625/R-99/010, Office of Research and Development, Cincinnati, OH, September 2000.
32. Dinges, R., *Natural Systems for Water Pollution Control*, Van Nostrand Reinhold, New York, NY, 1982.
33. Brix, H., Treatment of wastewater in the rhizosphere of wetland plants—The root-zone method, *Water Science and Technology*, 19, 1987, 107–118.
34. U.S. Environmental Protection Agency, *Nutrient Control Design Manual*, EPA/600/R-10/100, Office of Research and Development/National Risk Management Research Laboratory, Cincinnati, OH, August 2010.
35. U.S. Environmental Protection Agency, *Municipal Nutrient Removal Technologies Reference Document*, EPA-832-R-08-006, Office of Wastewater Management, Municipal Support Division, Washington, DC, September 2008.
36. Metcalf & Eddy|AECOM, *Wastewater Engineering: Treatment and Resource Recovery*, 5th ed., McGraw-Hill, New York, NY, 2014.
37. Metcalf & Eddy, Inc., *Wastewater Engineering, Treatment and Reuse*, 4th ed., McGraw-Hill Companies, Inc., New York, NY, 2003.
38. U.S. Environmental Protection Agency, *Process Design Manual for Upgrading Existing Wastewater Treatment Plants*, Technology Transfer EPA 625/1-71-004a, October 1974.
39. U.S. Environmental Protection Agency, *Handbook, Retrofitting POTWs*, Technology Transfer EPA/625/6-89/020, Center for Environmental Research Information, Cincinnati, OH, July 1989.
40. U.S. Environmental Protection Agency, *A Guide to the Selection of Cost-Effective Wastewater Treatment Systems*, Technical Report EPA-430/9-75-002, Office of Water Program Operations, Washington, DC, July 1975.
41. Culp, R. L. and G. L. Culp, *Advanced Wastewater Treatment*, Van Nostrand Reinhold Co., New York, NY, 1971.
42. U.S. Environmental Protection Agency, *Process Design Manual for Nitrogen Control*, Office of Technology Transfer, Washington, DC, October 1975.

43. White, G. C., *Disinfection of Wastewater and Water for Reuse*, Van Nostrand Reinhold Co., New York, NY, 1978.
44. Tchobanoglous, G. and E. D. Schroeder, *Water Quality: Characteristics, Modeling and Modification*, Addison-Wesley Publishing Co., Reading, PA, 1985.
45. Watts, R. J., *Hazardous Wastes: Sources, Pathways, Receptors*, John Wiley & Sons, Inc., New York, NY, 1998.
46. GaGrega, M. D., P. L. Buckingham, and J. C. Evans, *Hazardous Waste Management*, 2nd ed., Waveland Press, Inc., Long Grove, IL, 2001.
47. Green, D. W. and R. H. Perry (eds.), *Perry's Chemical Engineer' Handbook*, 8th ed., McGraw-Hill Professional, New York, NY, 2007.
48. Kavanaugh, M. C. and R. R. Trussell, Design of aeration towers to strip volatile contaminants from drinking water, *Water Pollution Control Federation*, 72(12), 1980, 684–692.
49. Qasim, S. R., E. M. Motley, and G. Zhu, *Water Works Engineering: Planning, Design, and Operation*, Prentice Hall PTR, Upper Saddle River, NJ, 2000.
50. Howe, K. J., D. W. Hand, J. C. Crittenden, R. R. Rrussell, and G. Tchobanoglous, Montgomery Watson Harza, *Principles of Water Treatment*, John Wiley & Sons, Inc., Hoboken, NJ, 2012.
51. U.S. Environmental Protection Agency, *Wastewater Filtration, Design Considerations*, Technology Transfer Seminar Publication, Washington, DC, July 1977.
52. Dahab, M. F. and J. C. Young, Unstratified-bed filtration of wastewater, *Journal of Environmental Engineering Division*, ASCE, 103(EE1), 1977, 21–36.
53. Tchobanoglous, G., Filtration of secondary effluent for reuse application, Paper presented at the 61st Annual Conference of the WPCF, Dallas TX, October 1988.
54. Metcalf & Eddy|AECOM, Asano, T., F. L. Burton, H. L. Leverenz, R. Tsuchihashi, and G. Tchobanoglous, *Water Reuse: Issues, Technologies, and Applications*, McGraw-Hill Professional, New York, NY, 2007.
55. Kawamura, S., Design and operation of high-rate filters: Part 1, *Journal of the American Water Works Association*, 67(10), 1975, 535–544.
56. Kawamura, S., Design and operation of high-rate filters: Part 2, *Journal of the American Water Works Association*, 67(11), 1975, 653–662.
57. Kawamura, S., Design and operation of high-rate filters: Part 3, *Journal of the American Water Works Association*, 67(12), 1975, 705–708.
58. SUEZ Treatment Solutions, Inc., *Infilco ABW® Automatic Backwash Filter*, http://www.suez-na.com, 2009.
59. Aqua-Aerobic Systems, Inc., *AquaABF® Automatic Backwash Filter*, http://www.aqua-aerobic.com, 2012.
60. Evoqua Water Technologies, *Gravisand® Filtration Systems, Davco™ Filtration System Technology*, http://www.evoqua.com, 2014.
61. Evoqua Water Technologies, *Hydro-Clear® Filters: Count on Them 24/7/365*, http://www.evoqua.com, 2014.
62. Parkson Corporation, *DynaSand® Continuous, Upflow, Granular Media Filter*, http://www.parkson.com, 2010.
63. Schreiber LLC/Bosman Water Management International, *Fuzzy Filter® Compressible Media Filter for High-Rate Suspended Solids Removal in Industrial, Municipal, Sewer Overflow and Reuse Applications*, http://www.bosman-water.com.
64. MWH, *Water Treatment: Principles and Design*, 2nd ed., John Wiley & Sons, Inc., Hoboken, NJ, 2005.
65. Hudson, H., *Water Clarification Processes: Practical Design and Evaluation*, Van Nostrand Reinhold Co., New York, NY, 1981.
66. Kawamura, S., Design and operation of high-rate filters, *Journal of the American Water Works Association*, 91(12), 1999, 77–90.

67. Monk, R. and J. F. Willis, Designing water treatment facilities, *Journal of the American Water Works Association*, 79(2), 1987, 45–57.
68. AWWA and J. K. Edzwald (ed.), *Water Quality and Treatment: A Handbook on Drinking Water*, 6th ed., McGraw-Hill Education, New York, NY, 2010.
69. ASCE and AWWA, *Water Treatment Plant Design*, 5th ed., McGraw-Hill Education, New York, NY, 2012.
70. Camp, J. R., Backwashing granular filters, *Proceedings of the American Society Civil Engineers Sanitary Division*, December 1971, p. 903.
71. U.S. Environmental Protection Agency, *Technologies for Upgrading Existing or Designing New Drinking Water Treatment Facilities*, EPA/625/4-89/023, Office of Drinking Water, Center for Environmental Research Information, Cincinnati, OH, March 1990.
72. Kawamura, S., *Integrated Design of Water Treatment Facilities*, 2nd ed., John Wiley & Sons, Inc., New York, NY, 2000.
73. Aqua-Aerobic Systems, Inc., *Cloth Media Filtration Featuring OptiFiber® Pile Cloth Media*, http://www.aqua-aerobic.com, 2013.
74. Aqua-Aerobic Systems, Inc., *AquaDrum® Cloth Media Filter*, http://www.aqua-aerobic.com, 2016.
75. Alfa Laval Inc., *AS-H Iso-Disc® Cloth Media Filter*, http://www.alfalaval.com, 2010.
76. Five Star Filtration LLC, *Five Star Disk Filter*, http://www.5starfiltration.com, 2014.
77. Veolia Water Technologies, *Flexible Microscreen Filter Hydrotech® Discfilterive*, http://www.krugerusa.com.
78. Xylem, Inc., *Sanitaire DrumFilter, Advanced Filtration System*, http://www.xyleminc.com, 2012.
79. Xylem, Inc., *Leopold® Ultrascreen® Disk Filter*, http://www.xyleminc.com, 2015.
80. Evoqua Water Technologies, *Forty-X™ Disc Filter*, http://www.evoqua.com, 2014.
81. California Water Boards, *Alternative Treatment Technology Report for Recycled Water*, State Water Resources Control Board, Division of Drinking Water, San Diego, CA, September 2014.
82. Batek, R. C., H. Lin, P. G. Baumann, and T. Reid, Cloth media filter retrofit of sand filters increases filtration capacity, *World Water: Water Reuse & Desalination*, Autumn, 3, 2011, 38–41.
83. Drzymala, I. E. and M. Verara, Life cycle analysis of tertiary filters for rerate expansions, *Texas WET*, May, 2010, 5–25.
84. Aqua-Aerobic Systems, Inc., *Success Stories: Trinity River Authority Increases Filtration Capacity and Decreases Backwash with AquaDiamond® Cloth Media Filters*, http://www.aqua-aerobic.com, 2011.
85. Lin, S. D., *Taste and Odors in Water Supplies—A Review*, ISWS/CIR-127/77, Illinois State Water Survey, Urbana, IL, 1977.
86. Kornegay, B. H., Activated carbon for taste and odor control, *Water Treatment Plant Design for the Practicing Engineer*, R. L. Sanks (ed.), Ann Arbor Science Publishers, Ann Arbor, MI, 1978.
87. Perrich, J. R., *Activated Carbon Adsorption for Wastewater Treatment*, CRC Press, Inc., Boca Raton, FL, 1981.
88. Bohart, G. S. and E. Q. Adams, Some aspects of the behavior of charcoal with respect to chlorine, *Journal of the American Chemical Society*, 42(3), 1920, 523–544.
89. Eckenfelder, W. W., D. L. Ford, and A. J., Englande, Jr., *Industrial Water Quality*, 4th ed., McGraw-Hill, New York, NY, 2009.
90. Ramalho, R. S., *Introduction to Wastewater Treatment Processes*, 2nd ed., Academic Press, New York, NY, 1983.
91. Thomas, H. C., Chromatography: A problem in kinetics, *Annals of the New York Academy of Sciences*, 49(2), 1948, 161–182.
92. Reynolas, T. D. and P. A. Richards, *Unit Operations and Processes in Environmental Engineering*, 2nd ed., PWS Publishing Co., Boston, MA, 1996.

93. Fornwalt, H. J. and R. A. Hutchins, Purifying liquids with activated carbon, Part I, *Chemical Engineering*, 73(8), 1966, 179–184.
94. Fornwalt, H. J. and R. A. Hutchins, Purifying liquids with activated carbon, Part II, *Chemical Engineering*, 73(10), 1966, 155–160.
95. Weber, W. J., Jr., *Physiochemical Processes for Water Quality Control*, Wiley-Interscience, New York, NY, 1972.
96. McGuire, M. J. and I. H. Suffet (eds.), *Activated Carbon Adsorption of Organics from the Aqueous Phase*, vol. 2, Ann Arbor Science Publishers, Ann Arbor, MI, 1980.
97. Dorfiner, K., *Ion Exchangers: Properties and Applications*, 3rd ed., Ann Arbor Science Publishers, Ann Arbor, MI, 1972.
98. U.S. Environmental Protection Agency, *Manual: Nitrogen Control*, Technical Report, EPA-625/R-93/010, Office of Research and Development, Washington, DC, September 1993.
99. Slater, M. J., *Principles of Ion Exchange Technology*, Butterworth Heinemann, New York, NY, 1991.
100. Peterson, S., Anion exchange processes, *Annual of the New York Academy of Science*, 57, 1953, 144.
101. Board, A. J., *Chemical Equilibrium*, Harper and Row Publishers, New York, 1966.
102. Anderson, R. E., Estimation of ion exchange process limits by selectivity calculations, *Adsorption and Ion Exchange*, I. M. Abrams, I. Zwiebel, and N. H. Sweed (eds.), Symposium Series, American Institute of Chemical Engineers, 71(152), 1975, 236–242.
103. Celenza, G. J., *Specialized Treatment Systems, Industrial Waste Treatment Process Engineering*, vol. 3, Technomic Publishing Co., Inc., Lancaster, PA, 2000.
104. Lien, L., Using membrane technology to minimize wastewater, *Pollution Engineering*, 30(5), 1998, 44–47.
105. Taylor, J. and W. Wiesner, Membranes, *Water Quality and Treatment*, 5th ed., R. D. Letterman (ed.), American Water Works Association and McGraw-Hill, Inc., New York, NY, 1999.
106. Paranjape, S., R. Reardon, and X. Foussereau, Pretreatment technology for reverse osmosis membrane used in wastewater reclamation application—Past, present and future—A literature review, *Proceedings of the 76th Annual WEFTEC®*, Water Environment Federation, Los Angeles, CA, October 2003.
107. Wong, J. M., A survey of advanced membrane technologies and their applications in water reuse projects, *Proceedings of the 76th Annual WEFTEC®*, Water Environment Federation, Los Angeles, CA, October 2003.
108. American Water Works Association Research Foundation, Lyonnaise des Eaux, and Water Research Commission of South Africa, *Water Treatment, Membrane Processes*, McGraw-Hill, New York, NY, 1996.
109. Hai, F. I., K. Yamamoto, and C. Lee (eds.), *Membrane Biological Reactors: Theory, Modeling, Design, Management and Applications to Wastewater Reuse*, IWA Publishing, London, UK, 2014.
110. Trussell, R. S., S. Adham, and R. R. Trussell, Process limits of municipal wastewater treatment with the submerged membrane bioreactor, *Journal of Environmental Engineering, ASCE*, 1313, 2005, 410–416.
111. Judd, S., *The MBR Book, Principles and Applications of Membrane Bioreactors for Water and Wastewater Treatment*, 2nd ed., Elsevier, Amsterdam, Netherland, 2010.
112. Chiemchaisri, C., K. Yamamoto, and S. Vigneswaran, Household membrane bioreactor in domestic wastewater treatment, *Water Science and Technology*, 27(1), 1993, 171–178.
113. Wallis-Lage, C., MBR systems: Similarities and differences between manufacturers, *Proceedings of the 76th Annual WEFTEC®*, Water Environment Federation, Los Angeles, CA, October 2003.
114. Crawford, G., D. Thompson, J. Lozier, G. Daigger, and E. Fleischer, Membrane reactors—A designer's perspective, *Proceedings of the 73rd Annual Conference & Exposition on Water Quality and Wastewater Treatment*, Water Environment Federation, Anaheim, CA, October 2000.

115. Cote, P., H. Buisson, and M. Praderie, Immersed membranes activated sludge process applied to the treatment of municipal wastewater, *Water Science and Technology*, 38(4–5), 1998, 437–442.
116. van Dijk, L. and G. C. Roncken, Membrane bioreactors for wastewater treatment: The state of the art and new developments, *Water Science and Technology*, 35(10), 1997, 35–41.
117. Stephenson, T., S. Judd, B. Jefferson, and K. Brindle, *Membrane Bioreactors for Wastewater Treatment*, IWA Publishing, London, UK, 2000.
118. GE Power/Water & Process Technologies, *LEAPmbr* Wastewater Membrane Bioreactor*, http://www.ge-energy.com, 2011.
119. Evoqua Water Technologies, *Efficiency Meets Simplicity, MEMPULSE® Membrane Bioreactor (MBR) System*, http://www.evoqua.com, 2014.
120. Veolia Water Technologies, *NEOSEP® Membrane Bioreactor (MBR)*, http://www.veoliawatertech.com, 2016.
121. HUBER SE, *HUBER Solution for MBR Applications*, http://www.huber.de, 2012.
122. KUBOTA Membrane USA Corporation, *KUBOTA Submerged Membrane Unit®*, http://www.kubota-membrane.com, 2009.
123. Mitsubishi Rayon Engineering Co., Ltd., *SteraporeSADFTM Hollow Fiber Membrane Element and Unit for Large Capacity Wastewater Treatment by Membrane Bioreactor*, http://www.mrc.co.jp, 2006.
124. Wintgues, T., J. Rosen, T. Melin, C. Brepols, K. Drensla, and N. Engelhardt, Modeling of a membrane bioreactor system for municipal wastewater treatment, *Journal of Membrane Science*, 216 (12–2), 2003, 55–65.
125. Matinia, J., *Mathematical Modelling and Computer Simulation of Activated Sludge Systems*, IWA Publishing, London, 2010.
126. EnviroSim Association, Ltd., *BioWin*, http://envirosim.com
127. Hydromantis Environmental Software Solutions, Inc., *GPS-X^{TM}*, http://www.hydromantis.com
128. American Society for Testing and Materials (ASTM), *Standard Test Method for Silt Density Index (SDI) of Water*, ASTM Procedure D4189-95, ASTM, Philadelphia, PA, 2002.
129. Snoeyink, V. L. and D. J. Jenkins, *Water Chemistry*, John Wiley & Sons, New York, 1980.
130. Cruver, J. E., Membrane processes, in *Physico-chemical Process for Water Quality Control*, W. J., Weber, Jr. (ed.), Wiley-Interscience, New York, NY, 1972.
131. Farrell, J. B. and R. N. Smith, Process applications of electrodialysis, *Industrial and Engineering Chemistry*, 54(6), 1962, 29–35.
132. Watson, I. C., O. J. Morin, and L. Henthorne, *Desalting Handbook for Planners*, 3rd ed., Desalination and Water Purification Research and Development Program Report No. 72, U.S. Department of the Interior, Bureau of Reclamation, Denver, CO, July 2003.
133. United Nations, *Solar Distillation as a Means of Meeting Small-Scale Water Demands*, Publication No. 70.II.B.1, United Nations Department of Economic and Social Affairs, New York, NY, 1970.
134. Qasim, S. R., Treatment of domestic sewage by using solar distillation and plant culture, *Journal of Environmental Science and Health*, 13(8), 1978, 615–627.
135. American Water Works Association, *Water Quality and Treatment: A Handbook of Public Water Supplies*, 4th ed., McGraw-Hill, Inc., New York, 1990.
136. Walton, J., H. Lu, C. Turner, S. Solis, and H. Hein, *Solar and Waste Heat Desalination by Membrane Distillation*, Desalination and Water Purification Research and Development Program Report No. 81, U.S. Department of the Interior, Bureau of Reclamation, Denver, CO, April 2004.
137. Beutel, J., *Water Recovery by Reiterative Freezing*, Wright-Patterson Air Force Base, Report No. NRL-TDR-62-14, Dayton, OH, 1962.
138. Perry, J. H., *Chemical Engineers Handbook*, 3rd ed., McGraw-Hill Book CO., New York, NY, 1950.

Appendix A: Abbreviations and Symbols, Basic Information about Elements, Useful Constants, Common Chemicals Used in Water and Wastewater Treatment, and the U.S. Standard Sieves and Size of Openings

In this appendix, many abbreviations and symbols, atomic numbers and weights, and valances of selected elements are listed in Tables A.1 and A.2. Many useful constants for unit conversions in different fields of engineering are summarized in Table A.3. Additionally, the alphabetical listing of commonly used chemicals in water and wastewater treatment and their properties are summarized in Table A.4. Also, the opening sizes of the U.S. Standard Sieves are summarized in Table A.5.[1–7]

TABLE A.1 Abbreviations and Symbols

Acceleration due to gravity (LT^{-2})	g	Gallon per minute (L^3T^{-1})	gal/min or gpm
Area (L^2)	A	Gallon per second (L^3T^{-1})	gal/s or gps
Atmosphere	atm	Horsepower	hp
British thermal unit	Btu	Horsepower-hour	hp·h
Calorie	cal	Hour (T)	h
Centimeter (L)	cm	Inch	in
Centipoise	cp	Joule	J
Cubic centimeter (L^3)	cm^3	Kilocalorie	kcal
Cubic feet per minute (L^3T^{-1})	cfm or ft^3/min	Kilogram (M)	kg
Cubic feet per second (L^3T^{-1})	cfs or ft^3/s	Kilonewton per cubic meter	kN/m^3
Cubic meter (L^3)	m^3	Kilowatt-hour	kWh
Cubic meter per day (L^3T^{-1})	m^3/d	Liter (L^3)	L
Cubic meter per second (L^3T^{-1})	m^3/s	Milligram (M)	mg
Degree Celsius	°C	Million gallons per day (L^3T^{-1})	MGD
Degree Fahrenheit	°F	Pound force	lb_f
Degree Kelvin	°K	Pound mass	lb_m
Degree Rankine	°R	Pounds per square inch	lb_f/in^2 or psi
Density (ML^{-3})	ρ	Power	P
Discharge or flow (L^3T^{-1})	Q	Second (T)	s
Energy	E	Square meter (L^2)	m^2
Feet per minute (LT^{-1})	fpm or ft/min	Standard temperature and pressure	STP (20°C and 760 mm Hg)
Feet per second (LT^{-1})	fps or ft/s	Volume (L^3)	V
Gallon (L^3)	gal	Watt	W

Note: M = mass, L = length, T = time.

Source: Adapted in part from References 1 through 5.

TABLE A.2 Basic Information of Selected Elements[a]

Element	Symbol	Atomic Number	Atomic Weight	Valence
Aluminum	Al	13	26.982	3
Antimony	Sb	51	121.760	3, 5
Argon	Ar	18	39.948	–
Arsenic	As	33	74.922	3, 5
Barium	Ba	56	137.327	2
Beryllium	Be	4	9.012	–
Bismuth	Bi	83	208.980	3, 5
Boron	B	5	10.811	3
Bromine	Br	35	79.904	1, 3, 5, 7
Cadmium	Cd	48	112.411	2
Calcium	Ca	20	40.078	2
Carbon	C	6	12.011	4

(*Continued*)

TABLE A.2 (*Continued*) Basic Information of Selected Elements[a]

Element	Symbol	Atomic Number	Atomic Weight	Valence
Chlorine	Cl	17	35.453	1, 3, 5, 7
Chromium	Cr	24	51.996	2, 3, 6
Cobalt	Co	27	58.933	2, 3
Copper	Cu	29	63.546	1, 2
Fluorine	F	9	18.998	1
Gold	Au	79	196.966	1, 3
Helium	He	2	4.003	–
Hydrogen	H	1	1.008	1
Iodine	I	53	126.904	1, 3, 5, 7
Iron	Fe	26	55.845	2, 3
Lead	Pb	82	207.190	2, 4
Lithium	Li	3	6.941	1
Magnesium	Mg	12	24.305	2
Manganese	Mn	25	54.938	2, 3, 4, 6, 7
Mercury	Hg	80	200.592	1, 2
Molybdenum	Mo	42	94.940	3, 4, 6
Neon	Ne	10	20.180	–
Nickel	Ni	28	58.693	2, 3
Nitrogen	N	7	14.007	3, 5
Oxygen	O	8	15.999	2
Phosphorus	P	15	30.974	3, 5
Platinum	Pt	78	195.084	2, 4
Potassium	K	19	39.098	1
Radium	Ra	88	226.05	–
Silicon	Si	14	28.086	4
Silver	Ag	47	107.868	1
Sodium	Na	11	22.990	1
Strontium	Sr	38	87.62	2
Sulfur	S	16	32.064	2, 4, 6
Tin	Sn	50	118.710	2, 4
Titanium	Ti	22	47.867	–
Tungsten	W	74	183.84	3, 4
Uranium	U	92	238.029	4, 6
Zinc	Zn	30	65.38	2

[a] For a complete list, consult a handbook of chemistry and physics.

TABLE A.3 Useful Constants

Acceleration due to gravity, $g = 9.807\ m/s^2$ ($32.174\ ft/s^2$)

pi, $\pi = 3.14159265$

$e = 2.71828$

U.S. standard atmosphere = 101.325 kN/m^2 (14.696 lb$_f$/in^2 or psi) = 101.325 kPa (1.013 bars) = 10.333 m (33.899 ft) of water = 760 mm Hg = 29.9213 in of Hg

Molecular weight of air (0.21 molar fraction of oxygen and 0.79 molar fraction of nitrogen) = 29.1 g/mole

Density of air (15°C and 1 atm) = 1.23 kg/m^3 (0.0765 lb$_m$/ft^3)

Dynamic viscosity of air, μ (15°C and 1 atm) = 17.8×10^{-6} N·s/m^2 (0.372×10^{-6} lb$_f$·s/ft^2)

Kinematic viscosity of air, v (15°C and 1 atm) = 14.4×10^{-6} m/s^2 (155×10^{-6} ft^2/s)

1 centipoise = 0.01 g/cm·s

1 stoke = 1 cm^2/s

Specific heat of air = 1005 J/kg·K (0.24 Btu/lb$_m$·°R)

Universal gas constant, R = 1.987 cal/mole·°K = 0.08206 L·atm/mole·°K = 8.314 J/mole·°K (R = 1.986 Btu/lb mole·°R = 0.7302 atm·ft3/lb mole·°R)

1 N = 1 kg·m/s^2 = 0.2248 lb$_f$

1 pascal (Pa) = 1 N/m^2

1 J = 1 N·m

1 J/s = 1 N·m/s = 1 w

1 kJ = 0.9478 Btu

1 amp = 1 coulomb/s

1 cal (20°C) = 0.003966 Btu

1 cal = 4.184 J

1°C = 1.8°F

$°C = \frac{5}{9}(°F - 32)$ or $°F = \frac{9}{5}°F + 32$

°K = °C + 273.15 or °R = °F + 446

1 Faraday = 96,485 coulombs/eq = 23,061 cal/volt eq

1 fathom = 1.829 m

1 grain = 0.06480 g

1 ha = 10,000 m^2 = 2.471 acres

1 knot = 1.852 km/h = 1.151 mph

1 hp = 745.7 w = 745.7 J/s = 550 ft·lb$_f$/s

1 lumen (at 5550 A) = 0.001471 W

1 lumen/cm^2 = 1 Lambert

1 lumen/m^2 = 0.0929 foot-candle

1 ton (U.S.) = 2000 lb

1 tonne (metric) = 1000 kg = 2204.6 lb

1 ton (U.K.) = 2240 lb

1 U.S. gallon = 3.78541 L = 8.34 lb of water

1 Imperial gallon (U.K.) = 4.546 L = 10 lb of water

1 ft^3 = 7.481 U.S. gallon = 62.43 lb$_m$ of water

Source: Adapted in part from References 1 through 5.

TABLE A.4 Common Chemicals Used in Water and Wastewater Treatment

Chemical Name and Formula	Common or Trade Name	Shipping Containers, and Suitable Storage Material[a]	Available Forms, Description, and Density	Solubility, lb/gal	Commercial Strength, %	Additional Characteristics and Properties
Activated carbon	Powdered carbon, PAC, Aqua Nuchor, Hydrodarco, Herite	Bags or bulk; dry: iron, steel; wet: rubber and silicon linings, or type 316 stainless steel	Black granules, powder, 5–30 lb/ft^3	Insoluble (suspension used)		1 lb/gal suspension used for storage and handling
Aluminum oxide, Al_2O_3	Activated alumina	Bags or drums; iron or steel	Powder granules (up to 1 in. in diameter); 95 lb/ft^3	Insoluble	100	
Aluminum sulfate, $Al_2(SO_4)_3{\cdot}14H_2O$ (dry)	Alum, filter alum, sulfate of alumina	100–200-lb bags, 300–400-lb barrels, bulk (carloads), or tank cars and trucks; dry: iron, or steel; wet: stainless steel, or rubber plastic	Ivory colored; powder (38–45 lb/ft^3), granule (60–63 lb/ft^3), or lump (62–70 lb/ft^3)	6.2 (60°F)	17 (Al_2O_3), or ~9 (Al)	pH of 1% solution: 3.5
Aluminum sulfate, $Al_2(SO_4)_3{\cdot}49\ H_2O$ (45-55% solution)	Alum solution	Tank cars and trucks; FRP, PE, type 316 stainless steel, rubber linings	Liquid; 10.8-11.2 lb/gal	–	8.5 (Al_2O_3), or 4.1-5 (Al)	pH: 1.9-2.3; freezing point: 4°F
Ammonium aluminum sulfate, $(NH_4)Al(SO_4)_2{\cdot}12H_2O$	Ammonium alum, crystal alum	100-lb bags, barrels, or bulk; FRP, PE, type 316 stainless steel, or rubber linings	Colorless crystals or white powder; 65-75 lb/ft^3	0.3 (32°F) 1.3 (68°F)	99	pH of 1% solution: 3.5
Ammonium hydroxide, NH_4OH	Ammonia water, ammonium hydrate, aqua ammonia	Carboys, 750-lb drums, or bulk; glass lining, steel, iron, FRP, or PE	Colorless liquid; 7.48 lb/gal	Complete	29.4 (NH_3) max 26° (Baumé)	pH 14; freezing point: −107°F
Ammonium silicofluoride, $(NH_4)_2SiF_6$	Ammonium fluorsilicate	100- and 400-lb drums; steel, iron, FRP, or PE	White crystals; 65-70 lb/ft^3	1.7 (63°F)	100	White, free flowing solid
Ammonium sulfate, $(NH_4)_2SO_4$	Sulfate of ammonia	50- and 100-lb bags, or 725-lb drums; FRP, PE, ceramic and rubber linings; or iron (dry)	White or brown crystal; 70 lb/ft^3	6.3 (68°F)	>99	Cakes in dry feed; add $CaSO_4$ for free flow
Anhydrous ammonia, NH_3	Ammonia	50-, 100-, and 150-lb cylinders, tank cars, trucks, or bulk; shipping containers	Colorless gas; 5.1 lb/gal (liquid at 68°F), or 0.048 lb/ft^3 (gas at 32°F and 1 atm)	3.9 (32°F) or 3.1 (60°F)	>99.9 (NH_3)	

(Continued)

TABLE A.4 (*Continued*) Common Chemicals Used in Water and Wastewater Treatment

Chemical Name and Formula	Common or Trade Name	Shipping Containers, and Suitable Storage Material[a]	Available Forms, Description, and Density	Solubility, lb/gal	Commercial Strength, %	Additional Characteristics and Properties
Bentonite	Colloidal clay, volclay, wilkinite	100-lb bags or bulk; iron, steel, FRP, or PE	Powder, pellet, or mixed sizes; 40–60 lb/ft^3	Insoluble (colloidal solution used)		Free flowing; nonabrasive
Calcium fluoride, CaF_2	Fluorspar	Bags, drums, barrels, hopper cars, or trucks; steel, iron, FRP, or PE	White crystalline solid or powder; 200 lb/ft^3 (crystal)	Very slight	85 (CaF_2) or < 5 (SiO_2)	
Calcium hydroxide, $Ca(OH)_2$	Hydrated lime, slaked lime	50-lb bags or bulk; FRP, PE iron, steel, or rubber lining	White powder, light, or dense; 28–36 lb/ft^3	0.14 (68°F), or 0.12 (90°F)	85–99 (Ca $(OH)_2$), or 63–73 (CaO)	Hopper agitation required for dry feed of light form
Calcium hypochlorite, $Ca(OCl)_2$	HTH, perchloron, pittchlor, bleaching power	5-lb cans; or 100-, 300-, and 800-lb drums; glass, plastic, and rubber linings, FRP, or PE	White granule, powder, or tablet; 50 lb/ft^3 (powder)	1.5 (25°F)	65 (available Cl_2)	1–3 (available Cl_2 solution used)
Calcium oxide, CaO	Burnt lime, chemical lime, quicklime, unslaked lime	80- and 100-lb bags, or bulk; FRP, PE, iron, steel, or rubber	Lump, pebble, or granule; 55–60 lb/ft^3	Slaked to form hydrated lime	75–95 (CaO)	pH of saturated solution depending on detention time and temperature; amount of water critical for efficient slaking
Carbon dioxide, liquid CO_2	Carbonic anhydride	Bulk; carbon steel (dry); or type 316 stainless steel (solution)	Liquid	0.012 (25°C)	99.5	Solution is acid
Chlorinated lime, CaO, $2CaOCl_2{\cdot}3H_2O$	Bleaching powder, chloride of lime	100-, 300-, and 800-lb drums; glass and rubber lining, FRP, PE	White powder; 48 lb/ft^3		25–37 (available Cl_2)	Deteriorates
Chlorine, Cl_2	Chlorine gas, liquid chlorine	100-lb and 150-lb cylinders, ton containers, or 16-, 30-, and 55-ton tank cars; shipping containers	Greenish-yellow, liquefied gas under pressure; 11.8 lb/gal (liquid at 68°F), or 0.201 lb/ft^3 (gas at 32°F and 1 atm)	0.07 (60°F), or 0.04 (100°F)	99.8 (Cl_2)	Forms HCl and HOCl when mixed with water

(*Continued*)

TABLE A.4 (*Continued*) Common Chemicals Used in Water and Wastewater Treatment

Chemical Name and Formula	Common or Trade Name	Shipping Containers, and Suitable Storage Material[a]	Available Forms, Description, and Density	Solubility, lb/gal	Commercial Strength, %	Additional Characteristics and Properties
Chlorine dioxide, ClO_2	Chlorine dioxide	Generated at site of application; glass, PVC, and rubber linings; FRP, or PE	Greening-yellow gas; 0.150 lb/ft^3 (gas at 32°F and 1 atm)	0.02 (30 mu)	263 (available Cl_2)	Explosive under certain conditions
Copper sulfate, $CuSO_4{\cdot}5H_2O$	Blue vitriol, blue stone	100-lb bags, 450-lb barrels, or drums; FRP, PE, silicon lining, iron, or stainless steel	Crystal, lump, or powder; 60–90 lb/ft^3	1.6 (32°F), 2.2 (68°F), or 2.6 (86°F)	99 ($CuSO_4$)	pH of 25% solution: ~3
Disodium phosphate, anhydrous, $Na_2HPO_4{\cdot}12H_2O$	Basic sodium phosphate, DSP, secondary sodium phosphate	100- and 300-lb drums, or 50- and 100-lb bags; cast iron, steel, FRP, or PE	White crystal, granular, or powder; 60–64 lb/ft^3	0.4 (32°F), 0.65 (86°F), or 1 (77°F)	64.3 (PO_4), or 48 (P_2O_5)	Precipitates with Ca and Mg; pH of 1% solution: 9.1
Ferric chloride, $FeCl_3$	Anhydrous ferric chloride	500-lb caska; 100-, 300-, 400-lb kegs; or 65-, 135-, and 250-lb drums; keep in original containers	Greenish-black powder, or crystals; 45-60 lb/ft^3		98 ($FeCl_3$), or 31-34 (Fe)	
Ferric chloride, $FeCl_3$ (solution)	Ferrichlor, iron chloride	55-gal drums, or bulk; Glass, PVC and rubber linings; FRP, or PE	Dark brown syrupy liquid; 10.5-12.4 lb/gal (25-47%)	Complete	37-45 ($FeCl_3$), or 8.6-16.3 (Fe)	
Ferric chloride, $FeCl_3{\cdot}6H_2O$	Crystal ferric chloride	300-lb barrels; keep in original containers	Yellow-brown lump; 60-65 lb/ft^3		59-61 ($FeCl_3$), or 20-21 (Fe)	Hygroscopic (store lumps and powder in tight container), no dry feed; optimum pH: 4-11
Ferric sulfate, $Fe_2(SO_4)_3{\cdot}9H_2O$	Ferrifloc, ferrisulfate	100–175-lb bags, or 400–425-lb drums; glass, plastic and rubber linings, FRP, PE, or type 316 stainless steel	Red-brown powder, or granule; 60–70 lb/ft^3	Soluble in 2–4 parts cold water	90-94 (Fe $(SO_4)_3$), or 25-26 (Fe)	Mildly hygroscopic coagulant at pH 3.5-11
Ferrous chloride, $FeCl_2{\cdot}4H_2O$	Iron(II) chloride tetrahydrate, ferrous dichloride tetrahydrate, ferrous chloride tetrahydrate	Bags, barrels, or bulk; glass, plastic, rubber linings, FRP, PE, or type 316 stainless steel	Light green crystal or lump; 56 lb/ft^3	~13 (50°F)	27-28 (Fe)	pH of 1% solution: <7

(*Continued*)

TABLE A.4 (*Continued*) Common Chemicals Used in Water and Wastewater Treatment

Chemical Name and Formula	Common or Trade Name	Shipping Containers, and Suitable Storage Material[a]	Available Forms, Description, and Density	Solubility, lb/gal	Commercial Strength, %	Additional Characteristics and Properties
Ferrous sulfate, $FeSO_4 \cdot 7H_2O$	Copperos, green vitriol	Bags, barrels, or bulk; Glass, plastic, rubber linings, FRP, PE, or type 316 stainless steel	Green crystal, granule, or lump; 45-75 lb/ft^3		55 ($FeSO_4$), or 20 (Fe)	Hygroscopic; cakes in storage; optimum pH: 8.5-11
Fluorosilicic acid, H_2SiF_6	Hexafluoro-silicic acid	Rubber-lined drums, trucks, or railroad tank cars; rubber-lined steel, or PE	Liquid; 11.5 lb/gal	~1.2 (68°F)	35 (approx.)	Corrosive, etches glass
Hydrogen fluoride, HF	Hydrofluoric acid	Steel drums, or tank cars; steel, FRP, PE	Liquid; 8.3 lb/gal		70 (HF)	Below 60% steel cannot be used
Oxygen, liquid	LOX	Dewars, cylinders, or truck and rail tankers; steel	Pale blue cryogenic liquid; 0.089 lb/ft^3 (gas at 68°F and 1 atm)	3.16% by volume (77°F)	>99.5	Prevent LOX from contacting grease, oil, asphalt, or other combustibles
Ozone, O_3	Ozone	Generated at site of application	Colorless gas; 0.125 lb/ft^3 (gas at 68°F and 1 atm)	12 mg/L (68°C)	1-20	Feed gas for ozone generation: air, oxygen-generated on-site, or liquid oxygen (LOX)
Phosphoric acid, H_3PO_4	Orthophosphoric acid, phosphoric(V) acid	Polyethylene drums or bulk; FRP, epoxy, rubber linings, polypropylene, or type 316 stainless steel	Odorless watery white liquid; 13-14 lb/gal		75, 80, or 85	Freezing point: 0.5°F at 75%, 40.2°F at 80%, and 70.0°F at 85%
Polyaluminum chloride, $Al_n(OH)_m Cl_{(3n-m)}$	PACl, SternPac	55-gal drums and bulk; FRP, PE, type 316 stainless steel, or rubber linings	Pale amber liquid; 10-11 lb/gal (bulk)	Soluble	8.5-18 (Al_2O_3)	Freezing point: −12°C; pH of 5% solution: 3-5
Potassium aluminum sulfate, $KAl(SO_4)_2 \cdot 12H_2O$	Potash alum, potassium alum	Bags, or lead-lined bulk (carloads); FRP, PE, or ceramic and rubber linings	Lump, granule, or powder; 60-67 lb/ft^3	0.5 (32°F), 1 (68°F), or 1.4 (86°F)	10-11 (Al_2O_3)	Low or even solubility; pH of 1% solution: 3.5
Potassium permanganate, $KMnO_4$	Purple salt	Bulk, barrels, or drums; iron, steel, FRP, or PE	Purple crystals; 170 lb/ft^3	0.54 (68°F)	100	Danger of explosion in contact organic matters

(*Continued*)

TABLE A.4 (*Continued*) Common Chemicals Used in Water and Wastewater Treatment

Chemical Name and Formula	Common or Trade Name	Shipping Containers, and Suitable Storage Material[a]	Available Forms, Description, and Density	Solubility, lb/gal	Commercial Strength, %	Additional Characteristics and Properties
Sodium aluminate, $Na_2OAl_2O_3$	Sodium aluminum oxide, Sodium metaaluminate, soda alum	100- and 150-lb bags, 250- and 440-lb drums, or 55-gal solution drums; iron, FRP, PE, rubber, or steel	Brown powder, or liquid; 50–65 lb/ft^3 (powder), or 12.7 lb/gal (30–45% solution)	3 (68°F), or 3.3 (86°F)	30–45 ($Na_2OAl_2O_3$), or 27° (Baumé)	Hopper agitation required for dry feed; very hygroscopic; and pH of 1% solution: 11.5–14
Sodium carbonate, Na_2CO_3	Soda ash, soda crystals, washing soda	Bags, barrels, bulk (carloads), or trucks; iron, rubber linings, steel, FRP, or PE	White powder; 31–56 lb/ft^3 (extra light or light), or 56–69 lb/ft^3 (dense)	1.5 (68°F) or 2.3 (86°F)	99.4 (Na_2CO_3) or 57.9 (Na_2O)	Hopper agitation required for dry feed of light and extra-light forms; pH of 1% solution: 11.3
Sodium chloride, NaCl	Common salt, salt	Bags, barrels, or bulk (carloads); bronze, FRP, PE, or rubber	Rock (50–60 lb/ft^3), or fine (58–78 lb/ft^3)	2.9 (32°F) or 3 (68°F)	98 (NaCl)	Absorbs moisture
Sodium chlorite, $NaClO_2$	ADOX dry	100-lb drums; metals (avoid cellulose materials)	Light orange powder, flake, or crystals; 53–56 lb/ft^3	3.5 (68°F)	80 ($NaClO_2$), or 30 (available Cl_2)	Generates ClO_2 at pH 3; explosive
Sodium fluoride, NaF	Fluoride	Bags, barrels, fiber drums, or kegs; iron, steel, FRP, or PE	Nile blue or white powder, light, or dense; 50–75 lb/ft^3	0.3–0.4 (32–68°F)	90–100 (NaF)	pH of 4% solution: 6.6
Sodium fluorosilicate, Na_2SiF_6	Sodium silicofluoride, sodium hexafluoro-silicate	Bags, barrels, fiber drums; cast iron, rubber lining, steel, FRP, PE	Nile blue or yellowish white powder; 72 lb/ft^3	0.03 (32°F), 0.06 (68°F), or 0.12 (140°F)	100	pH of 1% solution: 5.3
Sodium hexametaphosphate, $(NaPO_3)_6$	Calgon, glassy phosphate, metaphosphoric acid, sodium polyphosphate, SHMP	100-lb bags; rubber linings, plastics, or type 316 stainless steel	Crystal, flake, or powder; 47 lb/ft^3	1–4.2	90–100	pH of 1% solution: 7 (neutral)
Sodium hydroxide, NaOH	Caustic soda, soda lye, lye solution	100–700-lb drums, or bulk; carbon, steel, polypropylene, FRP, or rubber lining	Flake, lump, or liquid; 95.5 lb/ft^3 (solids), 9.6 lb/gal (25% solution), or 12.8 lb/gal (50% solution)	2.4 (32°F), 4.4 (68°F), or 4.8 (104°F)	25–100 (NaOH) or 74–76 (NaO_2)	Solid, hygroscopic; pH of 1% solution: 12.9; and freezing point of 50% solution: 54°F

(Continued)

TABLE A.4 (*Continued*) Common Chemicals Used in Water and Wastewater Treatment

Chemical Name and Formula	Common or Trade Name	Shipping Containers, and Suitable Storage Material[a]	Available Forms, Description, and Density	Solubility, lb/gal	Commercial Strength, %	Additional Characteristics and Properties
Sodium hypochlorite, NaOCl (6-12 trade %)	Sodium hypochlorite solution, bleach	5-, 13-, and 50-gal carboys; or 1300–2000-gal tank trucks; ceramic, glass, plastic, and rubber linings; FRP, or PE	Light yellow liquid; 8.8-10.1 lb/gal		3.8-13.2 (available Cl_2)	Unstable; pH of 1% solution: 7 (neutral)
Sodium pyrosulfite, $Na_2S_2O_5$	Sodium metabisulfite	Bags, drums, or barrels; iron, steel, FRP, or PE	White to yellow crystalline powder;	5.5 (68°F)	67 (SO_2 dry), or 33.3 (SO_2 solution)	Sulfurous odor
Sodium silicate, $Na_2O(SiO_2)_n$	Liquid glass, water glass	Drums or bulk (tank trucks and cars); cast iron, rubber lining, steel, FRP, or PE	Opaque, viscous liquid, 9.6 lb/gal (40% solution)	1.9 (77°F)	40, or 40° (Baumé)	Variable ratio of Na_2O to SiO_2; pH of 1% solution: 12.3
Sodium sulfite, Na_2SO_3	Sulfite	Bags, drums, or barrels; cast iron, rubber lining, steel, FRP, or PE	White crystalline powder, 80–90 lb/ft^3	2.3 (68°F)	23 (SO_2)	Sulfurous taste and odor
Sulfur dioxide, SO_2	Sulfurous acid anhydride	100–150-lb steel cylinders, ton-containers, or tank cars and trucks; shipping container	Colorless gas; 11.5 lb/gal (liquid at 68°F), or 0.201 lb/ft^3(gas at 32°F and 1 atm)	1.7 (68°F)	99 (SO_2)	Irritating gas
Sulfuric acid, H_2SO_4	Hydrogen sulfate, oil of vitriol, green vitriol	Bottles, carboys, drums, trucks, or tank cars; FRP, PE, porcelain, glass, or rubber linings	Solution; 15.4 lb/gal	Complete	95–98	pH of 0.5% solution: ~1.2
Tetrasodium pyrophosphate, $Na_4P_2O_7 \cdot 10H_2O$	Sodium pyrophosphate, tetrasodium phosphate, TSPP	125-lb kegs, 200-lb bags, or 300-lb barrels; cast iron, steel, or plastics	White powder, 68 lb/ft^3	0.6 (77°F), or 3.5 (212°F)	53 (P_2O_5)	pH of 1% solution: 10.8
Tricalcium phosphate, $Ca_{10}(OH)_2(PO_4)_6$	Tribasic calcium phosphate (TCP), bone phosphate of lime (BPL),	Bags, drums, bulk, or barrels; cast iron, steel, or plastics	Granular; ~200 lb/ft^3	Insoluble	100	Also available as white powder
Trisodium phosphate, $Na_3PO_4 \cdot 12H_2O$	Normal sodium phosphate, tertiary sodium phosphate, trisodium orthophosphate, TSP	125-lb kegs, 200-lb bags, or 325-lb barrels; cast iron, steel, or plastics	Crystal-course, medium; 56-61 lb/ft^3	1 (68°F) or 1.9 (104°F)	19 (P_2O_5)	pH of 1% solution: 11.9

[a] Always contact chemical suppliers to select best materials for handling.

Note: 1 lb = 0.4536 kg, 1 lb/ft^3 = 16.02 kg/m^3, 1 lb/gal = 119.8 kg/m^3, FRP = fiberglass-reinforced plastic, PE = polyethylene.

Source: Adapted in part from References 1 through 7.

TABLE A.5 U.S. Standard Sieves and Size of Openings

Sieve Size or Number	Size of Opening	
	in	mm
3/8	0.375	9.53
1/4	0.250	6.35
4	0.187	4.76
6	0.132	3.35
8	0.0937	2.36
10	0.0787	2.00
12	0.0661	1.68
14	0.0555	1.41
16	0.0469	1.19
18	0.0394	1.00
20	0.0331	0.841
25	0.0278	0.710
30	0.0232	0.595
35	0.0197	0.500
40	0.0165	0.420
45	0.0136	0.345
50	0.0117	0.300
60	0.0098	0.250
70	0.0083	0.210
80	0.0070	0.177
100	0.0059	0.149
120	0.0049	0.125
140	0.0041	0.105
170	0.0035	0.088
200 (75 μ)	0.0029	0.075
230 (63 μ)	0.0024	0.063
270 (53 μ)	0.0021	0.053
325 (45 μ)	0.0017	0.045
400 (38 μ)	0.0015	0.038
450 (32 μ)	0.0012	0.032
500 (25 μ)	0.0010	0.025
635 (20 μ)	0.0008	0.020
850 (10 μ)	0.0004	0.010

Source: Adapted in part from Reference 7.

References

1. Qasim, S. R., *Wastewater Treatment Plants: Planning, Design, and Operation*, 2nd ed., CRC Press, Boca Raton, FL, 1999.
2. Qasim, S. R., E. M. Motley, and G. Zhu, *Water Works Engineering: Planning, Design, and Operation*, Prentice Hall PTR, Upper Saddle River, NJ, 2000.

3. Droste, R. L., *Theory and Practice of Water and Wastewater Treatment*, John Wiley of Sons, Inc., New York, NY, 1997.
4. Metcalf & Eddy | AECOM, *Wastewater Engineering: Treatment and Resource Recovery*, 5th ed., McGraw-Hill, New York, NY, 2014.
5. Metcalf & Eddy, Inc., *Wastewater Engineering: Treatment and Reuse*, 4th ed., McGraw-Hill Companies, Inc., New York, NY, 2003.
6. AWWA and ASCE, *Water Treatment Plant Design*, 5th ed., McGraw-Hill Education, New York, NY, 2012.
7. Chen, W. F. and J. Y. Richard Liew (editors), *The Civil Engineering Handbook*, 2nd ed., CRC Press, Boca Raton, FL, 2002.

Appendix B: Physical Constants and Properties of Water, Solubility of Dissolved Gases in Water, and Important Constants for Stability and Sodicity of Water

The standard values of many useful physical constants and important properties of water, including specific weight, density, dynamic and kinetic viscosities, surface tension, and vapor pressure are provided in Section B.1. The definitions of these constants and properties, as well as the solubility and Henry's law constants of selected dissolved gases are provided in Section B.2. The relationships of these constants are also illustrated in many solved examples in Section B.2. Langelier saturation index (LSI) and Ryznar Stability Index (RSI) are used to determine the stability of feed water in design of membrane processes. The $(pK_2 - pK_s)$ values that are used to calculate these values are summarized in Section B.3. Also, the sodium adsorption ratio (SAR) is used to evaluate the sodicity of reclaimed effluent for irrigation. The adjusted concentrations of calcium ion $[Ca_X^{2+}]$ that are used to determine the SAR are listed in Section B.4.

B.1 Physical Constants

The useful physical constants of water are summarized in Tables B.1.[1–5] The principal physical properties of water are summarized in SI and the U.S. customary units in Tables B.2 and B.3, respectively.[3,4,6–8]

B.2 Definitions and Physical Relationships

The definitions and relationships of many physical constants and properties most commonly used in wastewater treatment process calculations are provided below.[3,4,6–8] The solubility of oxygen and other gases depends upon temperature, pressure, and TDS concentration. Many solved examples are included to illustrate these relationships and solubility calculations from Henry's law constants.

TABLE B.1 Useful Physical Constants of Water

Constant	Value
Molecular formula	= H_2O
Molecular weight	= 18 g/mole
Ionization constant (K_w) at 25°C	= 10^{-14}
Specific weight, γ at 4°C	= 9.81 kN/m^3 (62.4 lb$_f$/ft^3)
Weigh of 1 L of water (average)	= 1 kg of water (2.205 lb of water)
Weigh of 1 U.S. gal of water (average)	= 8.345 lb of water
1 U.S. gal	= 3.785 L = 0.003785 m^3 (0.1337 ft^3)
1 Imperial U.K. gal	= 4.546 L = 4.546 kg (10.02 lb) of water
1 ft^3	= 28.32 L (7.481 U.S. gal)
1 mg/L	= 1 g/m^3 ≈ 1 ppm
1 μg/L	= 1 mg/m^3 ≈ 1 ppb
1 g/U.S. gal	= 264.2 mg/L
1 lb/d	= 8.34 × 1 mg/L × 1 MGD
Density at 4°C, ρ_w	= 1 g/cm^3 = 1000 kg/m^3 (1.94 slug/ft^3)
Dynamic viscosity at 20°C (68°F), μ	= 1.002 × 10^{-3} N·s/m^2 (2.089 × 10^{-5} lb$_f$·s/ft^2)
Kinematic viscosity at 20°C (68°F), ν	= 1.003 × 10^{-6} m^2/s (1.078 × 10^{-5} ft^2/s)
1 atm	= 10.33 m (33.9 ft) of water = 760 mm Hg (14.7 psi)
1 psi	= 2.31 ft (0.703 m) of water = 0.0680 atm (51.71 mm Hg)
1 m head of water at 20°C	= 9.790 kN/m^2 (1.420 lb$_f$/in^2)
Boiling point at 1 atm	= 100°C (212°F)
Melting (or freezing) point at 1 atm	= 0°C (32°F)
Specific heat	= 1 cal/g·°C = 4200 J/kg·°C (1 Btu/lb·°F)
Heat of fusion at 0°C	= 80 cal/g = 335 kJ/kg (144 Btu/lb)
Heat of vaporization at 100°C	= 540 cal/g = 2260 kJ/kg (973 Btu/lb)
Modulus of elasticity at 15°C	= 2.15 kN/m^2 (312 × 10^3 lb$_f$/in^2)

Source: Adapted in part from References 1 through 5.

B.2.1 Density

The density (ρ) of a fluid is mass per unit volume. In SI units, density is expressed in g/cm^3 or kg/m^3. At 4°C, the density of water (ρ_w) is 1000 kg/m^3. In the U.S. customary units, the density is expressed in slug/ft^3. At 4°C, the density of water is 1.94 slug/ft^3.

B.2.2 Dynamic Viscosity

The dynamic viscosity (μ) is a measure of the fluid resistance to tangential or shear stress. Dynamic viscosity in SI units is expressed in Newton-second per square meter (N·s/m^2). In the U.S. customary units, it is expressed as lb$_f$·s/ft^2. The typical values of μ in both units at 10°C is 1.307 × 10^{-3} N·s/m^2 and 2.735 × 10^{-5} lb$_f$·s/ft^2.

B.2.3 Kinematic Viscosity

Kinematic viscosity (ν) is the ratio of dynamic viscosity and the density ($\nu = \mu/\rho$). The units for kinematic viscosity are m^2/s in SI units and ft^2/s in U.S. customary units. The typical values of kinematic viscosity at 10°C in SI and the U.S. customary units are 1.307 × 10^{-6} m^2/s and 1.410 × 10^{-5} ft^2/s.

TABLE B.2 Physical Properties of Water in SI Units

Temperature (T) °C	°F	Specific Weight (γ), kN/m^3	Density (ρ_w), kg/m^3	Dynamic Viscosity (μ), 10^{-3} $N{\cdot}s/m^2$	Kinematic Viscosity (ν), 10^{-6} m^2/s	Surface Tension against Air (σ), 10^{-3} N/m	Vapor Pressure (P_v) mm Hg	kN/m^2
0	32	9.805	999.8	1.787	1.786	75.6	4.579	0.61
3.98	39.2	9.806	1000	1.568	1.568	75.0	6.092	0.82
5	41	9.807	999.9	1.519	1.519	74.9	6.543	0.87
10	50	9.804	999.7	1.307	1.307	74.2	9.209	1.23
15	59	9.798	999.1	1.139	1.140	73.5	12.79	1.70
20	68	9.789	998.2	1.002	1.003	72.8	17.54	2.34
25	77	9.777	997.0	0.890	0.893	72.0	23.76	3.17
30	86	9.764	995.7	0.798	0.801	71.2	31.82	4.24
40	104	9.730	992.2	0.653	0.658	69.6	55.32	7.38
50	122	9.689	988.1	0.547	0.553	97.9	92.51	12.3
60	140	9.642	983.2	0.466	0.474	66.2	149.6	19.9
70	158	9.589	977.8	0.404	0.413	64.4	233.9	31.2
80	176	9.530	971.8	0.354	0.364	62.6	355.4	47.3
90	194	9.466	965.3	0.315	0.326	60.8	526.3	70.1
100	212	9.399	958.4	0.282	0.294	58.9	760.0	101

Source: Adapted in part from References 1 through 5.

TABLE B.3 Physical Properties of Water in U.S. Customary Units

Temperature (T) °F	°C	Specific Weight (γ), lb_f/ft^3	Density (ρ_w), $slug/ft^3$	Dynamic Viscosity (μ), 10^{-5} $lb_f{\cdot}s/ft^2$	Kinematic Viscosity (ν), 10^{-5} ft^2/s	Surface Tension against Air (σ), lb_f/ft	Vapor Pressure (P_v) in Hg	lb_f/in^2
32	0	62.42	1.940	3.746	1.931	0.00518	0.18	0.09
40	4	62.43	1.940	3.229	1.664	0.00614	0.24	0.12
50	10	62.41	1.940	2.735	1.410	0.00509	0.37	0.18
60	16	62.37	1.938	2.359	1.217	0.00504	0.53	0.26
70	21	62.30	1.936	2.050	1.059	0.00498	0.73	0.36
80	27	62.22	1.934	1.799	0.930	0.00492	1.04	0.51
90	32	62.11	1.931	1.595	0.826	0.00486	1.43	0.70
100	38	62.00	1.927	1.424	0.739	0.00480	1.93	0.95
110	43	61.86	1.923	1.284	0.667	0.00473	2.59	1.27
120	44	61.71	1.918	1.168	0.609	0.00467	3.44	1.69
130	54	61.55	1.913	1.069	0.558	0.00460	4.51	2.22
140	60	61.38	1.908	0.981	0.514	0.00454	5.88	2.89
150	66	61.20	1.902	0.905	0.476	0.00447	7.57	3.72
160	71	61.00	1.896	0.838	0.442	0.00441	9.65	4.74
170	77	60.80	1.890	0.780	0.413	0.00434	12.2	5.99
180	82	60.58	1.883	0.726	0.385	0.00427	15.3	7.51
190	88	60.36	1.876	0.678	0.362	0.00420	19.0	9.34
200	93	60.12	1.868	0.637	0.341	0.00413	23.5	11.5
212	100	59.83	1.860	0.593	0.319	0.00404	29.9	14.7

Source: Adapted in part from References 2 through 4.

B.2.4 Specific Gravity

The specific gravity (sp. gr.) is the relative density of a substance with respect to water at a given temperature. The density of water and other liquids change with temperature. The density of 1 g/cm^3 (or 1 g/mL, 1000 kg/m^3) is commonly used as the basis of specific gravity (water $= 1$).

B.2.5 Specific Weight

The specific weight (γ) of a fluid is its weight per unit volume. In SI units, it is expressed in kilonewton per cubic meter (kN/m^3). The commonly used relationship between specific weight (γ), density (ρ), and acceleration due to gravity (g) is $\gamma = \rho \cdot g$. The commonly used value of γ in SI units is 9.81 $kN/m^3(kg/m^2 \cdot s^2)$. In the U.S. customary, the value of γ is 62.4 lb_f/ft^3.

B.2.6 Surface Tension

Surface tension is a physical property that is created through the attraction of the molecules to each other in a liquid. Due to surface tension, a glass filled with water has surface slightly above the brim before it spills; a metal needle floats in the surface; or a drop of water is held in suspension at the tip of a pipette. The units of expression of surface tension are N/m (lb/ft). The surface tension of water at 10°C is 74.2 N/m (0.00509 lb_f/ft). The surface tension decreases as the temperature increases.

B.2.7 Vapor Pressure

Liquid molecules constantly change phase from liquid to gas due to sufficient kinetic energy they possess. The vapor pressure (p_v) is the pressure exerted by the vapors at the free water surface. The vapor pressure is normally expressed in mm mercury (Hg) or in Hg. In SI and the U.S. customary units, the vapor pressure is expressed in kilonewtons per square meter (kN/m^2) and lb_f/in^2. The vapor pressure of water at 10°C is 9.209 mm Hg, 1.23 kN/m^2, 0.37 in Hg, and 0.18 lb_f/in^2.

B.2.8 Henry's Law Constants

The saturation concentration of a dissolved gas in a liquid depends upon the partial pressure of the gas exerted over the liquid surface. Henry's law expresses the relationship between the partial pressure of the gas in the gas phase above the liquid and the concentration of gas in the liquid phase. The Henry's law constant may be expressed in a number of different ways and units. Some commonly used relationships are given by Equations 10.84a, 10.84b, 11.10, and 15.6, and explained in Example 10.119, and Sections 11.6.2 and 15.4.5. The values of Henry's law constants H (mole fraction based) and H_s (volumetric fraction based) for many gases are given in Tables B.4 and B.5.[2-4,6-8]

B.2.9 Solubility of Gases

The solubility of gases in natural waters depends upon four factors: (1) *temperature*, (2) *pressure*, (3) *gas content* in overlying atmosphere, and (4) salt concentration.[7,8]

Temperature: The solubility of any gas in water decreases with rising temperature. The solubility of gases including oxygen can be determined by Henry's law constant given in Table B.6. The saturation

TABLE B.4 Mole Fraction-Based Henry's Law Constant (H) for Common Gases

Temperature (T), °C	H (10^4 atm)							
	Air	O_2	CO_2	CO	H_2	H_2S	CH_4	N_2
0	4.32	2.55	0.0728	3.52	5.79	0.0268	2.24	5.29
10	5.49	3.27	0.104	4.42	6.36	0.0367	2.97	6.68
20	6.64	4.01	0.142	5.36	6.83	0.0483	3.76	8.04
30	7.71	4.75	0.186	6.20	7.29	0.0609	4.49	9.24
40	8.70	5.35	0.233	6.96	7.51	0.0745	5.20	10.4
50	9.46	5.88	0.283	7.61	7.65	0.0884	5.77	11.3
60	10.1	6.29	0.341	8.21	7.65	0.1030	6.26	12.0

Source: Adapted in part from Reference 6.

TABLE B.5 Volumetric Concentration-Based Henry's Law Constant (H_v) for Common Gases

Temperature (T)		Water Vapor Pressure (P_v), mm Hg	H_v at 0°C and 1 atm[a], mL/L·atm			
°C	°F		Air	O_2	CO_2	N_2
0	32	4.58	29.18	48.89	1713	23.54
2	36	5.29	27.69	46.33	1584	22.41
4	39	6.10	26.32	43.97	1473	21.35
6	43	7.01	25.06	41.80	1377	20.37
8	46	8.05	23.90	39.83	1282	19.45
10	52	9.21	22.84	38.02	1194	18.61
12	54	15.52	21.87	36.37	1117	17.86
14	57	11.99	20.97	34.86	1050	17.17
16	61	13.63	20.14	33.48	985	16.54
18	64	15.48	19.38	32.20	928	15.97
20	68	17.54	18.68	31.02	878	15.45
22	72	19.83	18.01	29.88	829	14.98
24	75	22.38	17.38	28.81	781	14.54
26	79	25.21	16.79	27.83	738	14.13
28	82	28.35	16.21	26.91	699	13.76
30	86	31.82	15.64	26.08	665	13.42
35	95	42.18	–	24.40	592	12.56
40	104	55.32	–	23.06	530	11.84
45	113	71.88	–	21.87	479	11.30
50	122	92.51	–	20.90	436	10.88
60	140	156.4	–	19.46	359	10.23
70	158	233.7	–	18.33	–	9.77
80	176	355.1	–	17.61	–	9.58
90	194	525.8	–	17.20	–	9.50
100	212	760.0	–	17.00	–	9.50

[a] It is also called the absorption *coefficient of gas* at 0°C and 1 atm.

Source: Adapted in part from References 6 through 8.

TABLE B.6 Solubility of Common Gases in Water

Temperature (T)		Solubility of Gases (C_s), mg/L			
°C	°F	Air	O_2	CO_2	N_2
0	32	37.50	14.62	1.00	22.81
5	41	32.94	12.80	0.83	20.16
10	50	29.27	11.33	0.70	17.93
15	59	26.25	10.15	0.59	16.15
20	68	23.74	9.17	0.51	14.72
25	77	21.58	8.38	0.43	13.55
30	86	19.60	7.63	0.38	12.53

Source: Adapted in part from References 6 through 8.

concentration of oxygen in water at 1 atm can also be calculated from several empirical equations (Equation B.1).[7,8]

$$C_s = 14.652 - 0.41022T + 0.00799T^2 - 0.0000777T^3 \tag{B.1a}$$

$$C_s = \left(0.68 - 6 \times 10^{-4}T\right)\left(\frac{760 - P_v}{T + 35}\right) \tag{B.1b}$$

$$C_s = 14.652 - 10.53(1 - e^{-0.03896T}) \tag{B.1c}$$

where

C_s = saturation concentration of oxygen in freshwater at 1 atm, mg/L
T = water temperature, °C
P_v = vapor pressure of water, mm Hg

The saturation concentrations of oxygen, nitrogen, carbon dioxide, and air in freshwater with respect to temperature and at 1 atm are summarized in Table B.6.

Pressure: The solubility of any gas increases at higher pressures and vice versa. The solubility of oxygen at any given pressure is calculated from Equation B.2.[9-11]

$$C_{s,P} = \frac{P - P_v}{760 - P_v} \times C_s \tag{B.2}$$

where

$C_{s,P}$ = solubility of oxygen at any given pressure (P), mg/L
C_s = solubility of oxygen at 1 atm, mg/L
P = prevailing barometric pressure, mm Hg
p_v = vapor pressure of water at 1 atm, mm Hg

At higher altitudes, the barometric pressure is lower, as a result the solubility of oxygen is also lower. The barometric pressure with altitude is given in Table B.7.

Gas Content in Air: The solubility of a gas in water increases if its content in the atmosphere increases because of its partial pressure above water–air interface. The oxygen concentration in dry air is 21% by volume, and 23.2% by weight.[9-11]

Salt Concentration: The solubility of a gas in water is inversely proportional to the salt concentration. The degree of salinity is normally expressed in concentration (mg/L) of total dissolved solid (TDS), chloride ions (Cl^-), or common salt (NaCl). The North Atlantic Ocean has chloride content of nearby 18,000 mg/L, and its DO saturation is about 82% of that in fresh water. The DO saturation of domestic wastewater is about 95% of that in fresh water. The TDS concentrations in strong, medium, and weak

TABLE B.7 Barometric Pressure with Altitude

Elevation above Sea Level		Absolute Pressure in Head of Water (H_{abs})		Barometric Pressure, mm Hg
m	ft	m	ft	
0	0	10.3	33.9	760
305	1000	10.0	32.8	736
457	1500	9.8	32.1	721
610	2000	9.6	31.5	706
1219	4000	8.9	29.2	655
1829	6000	8.3	27.2	611
2438	8000	7.7	25.2	566
3048	10,000	7.1	23.4	522
4572	15,000	5.9	19.2	434

Source: Adapted in part from References 1 and 12.

domestic wastewaters are 850, 500, and 250 mg/L, respectively. The saturation concentration of oxygen in saline water of different chloride concentrations and other gases in wastewater with respect to temperature are summarized in Table B.8.

The saturation concentrations of oxygen in saline water at different temperatures can be calculated from several methods. Three such methods are: (a) Equation B.3a based on the correction factor given in Table B.8, (b) the empirical relationship given by Equation B.3b based on chloride concentration, and (c) the empirical equation (Equation B.3c) based on chloride content, barometric pressure, and temperature.[12,13]

$$C_s'' = C_s - \frac{f_{Cl}}{100} S_{Cl} \qquad \text{(B.3a)}$$

$$C_s'' = C_s(1 - 9 \times 10^{-6} \times S_{Cl}) \qquad (T = 0\text{–}35°C) \qquad \text{(B.3b)}$$

$$C_s'' = \left(\frac{475 - 2.65 \times 10^{-3} \times S_{Cl}}{33.5 + T}\right)\left(\frac{P}{760}\right) \qquad \text{(B.3c)}$$

TABLE B.8 Saturation Concentration of Oxygen at Different Chloride Concentrations and Temperatures

Temperature (T)		Dissolved Oxygen Saturation (C_s), mg/L Chloride Concentration, mg/L					Decrease in Oxygen Concentration, mg O_2 per 100 mg Chloride
°C	°F	0	5000	10,000	15,000	20,000	
0	32	14.62	13.79	12.97	12.14	11.32	0.017
5	41	12.80	12.09	11.39	10.70	10.01	0.014
10	50	11.33	10.73	10.13	9.55	8.98	0.012
15	59	10.15	9.65	9.14	8.63	8.14	0.010
20	68	9.17	8.73	8.30	7.86	7.42	0.009
25	77	8.38	7.96	7.56	7.15	6.74	0.008
30	86	7.63	7.25	6.86	6.49	6.13	0.008

Source: Adapted in part from References 1, 3, 4, 8, 10 and 11.

where

C_s'' = saturation concentration of oxygen in saline water, mg/L
C_s = saturation concentration of oxygen in freshwater, mg/L
f_{Cl} = correction factor, mg/100 mg chloride. This factor is given in the last column of Table B.8.
S_{Cl} = salinity of chloride (Cl^-), mg/L
P = prevailing barometric pressure, mm Hg
T = temperature of saline water, °C

The solubility of oxygen at various temperatures, TDS concentrations and elevations are summarized in Table 10.16. The values in this table are extensively used in oxygen transfer relationships and solved examples in Section 10.3.8.

EXAMPLE B.1: SPECIFIC GRAVITY MEASUREMENT

An empty 100-mL graduated cylinder weighs 110.675 g. The cylinder filled with water at 20°C weighs 210.495 g. The same graduated cylinder filled with oil at 20°C weighs 209.475 g. Calculate the density of the oil and its specific gravity.

Solution

1. Calculate the density of water at 20°C.

$$\text{Weight of water} = (210.495 - 110.675)\text{ g} = 99.820\text{ g}$$

$$\text{Density of water at } 20^\circ\text{C} = \frac{\text{Mass}}{\text{Volume}} = \frac{99.820\text{ g}}{100\text{ mL}} = 0.9982\text{ g/mL or } 0.9982\text{ g/cm}^3.$$

2. Calculate the density of oil at 20°C.

$$\text{Weight of oil} = (209.475 - 110.675)\text{ g} = 98.800\text{ g}$$

$$\text{Density of oil at } 20^\circ\text{C} = \frac{98.800\text{ g}}{100\text{ mL}} = 0.9880\text{ g/mL or } 0.9880\text{ g/cm}^3.$$

3. Calculate the specific gravity of oil.

$$\text{The specific gravity of oil at } 20^\circ\text{C} = \frac{\text{Density of oil}}{\text{Density of water}} = \frac{0.9880\text{ g/cm}^3}{0.9982\text{ g/cm}^3} = 0.990.$$

Note: The specific gravity of oil can also be calculated directly from the weights.

$$\text{Specific gravity of oil} = \frac{\text{wt. of 100 mL of oil}}{\text{wt. of 100 mL of water}} = \frac{98.800\text{ g}}{99.820\text{ g}} = 0.990.$$

EXAMPLE B.2: WEIGHT OF A GIVEN VOLUME OF SOLUTION

The specific gravity of a salt solution is 1.15. How many lbs are in 55 gallons of the solution?

Solution

Calculate weight of 50 gal of the salt solution by assuming density of water is 1 g/cm^3.
Weight of 1 gallon of water = 8.345 lb/gal (Table B.1)
Density of salt solution = 1.15 × 8.345 lb/gal = 9.60 lb/gal of solution
Weight of 50 gal of salt solution = 55 gal × 9.60 lb/gal = 528 lbs

EXAMPLE B.3: RELATIVE WEIGHT OF A SUBSTANCE

The specific gravity of seawater is 1.25. How many percent it is heavier than that of water?

Solution

Calculate the relative weight of seawater heavier than the distilled water of a specific gravity of 1.

$$\begin{aligned}\text{Relative weight of seawater heavier than distilled water} &= \frac{\text{sp. gr. of seawater} - \text{sp. gr. of distilled water}}{\text{sp. gr. of distilled water}} \times 100\% \\ &= \frac{1.25 - 1.00}{1.00} \times 100\% = 25\%\end{aligned}$$

EXAMPLE B.4: EXPRESSION OF γ, μ, AND P_V IN g, s, AND m UNITS

The common unit for specific weight (γ) is N/m^3 (or kN/m^3); dynamic viscosity (μ) is $N{\cdot}s/m^2$; and vapor pressure (P_v) is N/m^2 (or kN/m^2). Express these units in g, s, and m units.

Solution

1. Convert the unit of specific weight (γ).
   Unit for $\gamma = N/m^3$
   N is a force and is equal to mass × acceleration, $N = g \times \frac{m}{s^2} = \frac{g{\cdot}m}{s^2}$

$$\gamma = \frac{N}{m^3} = \frac{g{\cdot}m}{s^2} \times \frac{1}{m^3} = g/m^2{\cdot}s^2$$

2. Convert the unit of dynamic viscosity (μ)

$$\mu = \frac{N{\cdot}s}{m^2} = \frac{g{\cdot}m}{s^2} \times \frac{s}{m^2} = g/m{\cdot}s$$

3. Convert the units of vapor pressure (P_v)

$$P_v = \frac{N}{m^2} = \frac{g{\cdot}m}{s^2} \times \frac{1}{m^2} = g/m{\cdot}s^2$$

EXAMPLE B.5: SOLUBILITY OF OXYGEN FROM MOLE-FRACTION-BASED HENRY'S LAW CONSTANT (H)

Determine the DO saturation concentration C_s in water at 10°C and 1 atm using the mole-fraction-based Henry's law constant, H (atm/mole fraction).

Solution

1. Select the applicable equation.
   Apply Equation 15.6e, $H = \frac{P_g}{x_g}$.
   The Henry's law constant, H at 10°C for oxygen = 3.27×10^4 atm/mole fraction (Table B.4).
2. Determine the partial pressure of oxygen (P_g) at 10°C and 1 atm.
   Vapor pressure of water on the surface of water at 10°C and 1 atm, $P_v = 9.21$ mm Hg (Table B.4).

Partial pressure of a gas can be calculated from Equation B.4, while the content of the gas in the air is known.

$$P_g = \frac{760 - P_v}{760} \times \frac{V_g}{100\%} \tag{B.4}$$

where

P_g = partial pressure of the gas in air, mm Hg
P_v = vapor pressure of water, mm Hg
V_g = content of gas in the air, % by volume

$$\text{Partial pressure of oxygen at 21\% (from Equation B.4)} = \frac{(760 - 9.21)\,\text{mm Hg}}{760\,\text{mm Hg/atm}} \times \frac{21\%}{100\%} = 0.207\,\text{atm}$$

3. Determine the mole fraction of oxygen in water (x_g) from rearranging Equation 15.6e.

$$x_g = \frac{P_g}{H} = \frac{0.207\,\text{atm}}{3.27 \times 10^4\,\text{atm/mole fraction}} = 6.33 \times 10^{-6}\,\text{mole fraction.}$$

4. Determine the mole concentration of O_2 in water (n_a).
   Since $n_a \ll n_{H_2O}$ calculate n_a from Equation 15.6l using mole concentration of water $n_{H_2O} = 55.6$ mole H_2O/L.

$$n_a \approx x_g n_{H_2O} = 6.33 \times 10^{-6}\,\text{mole fraction} \times 55.6\,\text{mole H}_2\text{O/L} = 3.52 \times 10^{-4}\,\text{mole O}_2\text{/L}$$

5. Determine the saturation concentration of oxygen (C_s) in mg/L.

$$C_s = 3.52 \times 10^{-4}\,\text{mole O}_2\text{/L} \times 32\,\text{g/mole} \times 103\,\text{mg/g} = 11.3\,\text{mg/L}$$

EXAMPLE B.6: SOLUBILITY OF OXYGEN FROM VOLUMETRIC SATURATION-CONCENTRATION-BASED HENRY'S LAW CONSTANT (H_V)

Determine the DO saturation C_s in water at 10°C and 1 atm using the volumetric saturation-concentration-based Henry's law constant at 0°C and 1 atm, H_v (mL/L·atm). Compare the result obtained in Example B.5.

Solution

1. Select the applicable equation.
   Apply Equation 15.6r, $C_s' = H_v P_g$
2. Determine the partial pressure of oxygen (P_g) at 10°C and 1 atm.
   Partial pressure of oxygen $P_g = 0.207$ atm (Example B.5, Step 2.)
3. Determine the absorption coefficient H_v at 10°C.
   From Table B.5, $H_v = 38.02$ mL/L·atm is obtained at 10°C after reducing to the standard condition, 0°C and 1 atm.
4. Determine the solubility of oxygen in water (volumetric saturation concentration, C_s') at 10°C and 1 atm from Equation 15.6r at $P_g = 0.207$ atm.

$$C_s' = H_v P_g = 38.02\,\text{mL/L·atm} \times 0.207\,\text{atm} = 7.87\,\text{mL/L}$$

5. Convert the solubility of oxygen from volumetric saturation concentration $\left(C'_s\right)$ in mL/L to mass concentration C_s in mg/L.

   By substituting Equations 15.6p and 15.6r into Equation 15.6s the relationship between these two concentrations is obtained in Equation B.5.

$$C'_s = \frac{22.4}{mw_g} C_s \quad \text{or} \quad C_s = \frac{mw_g}{22.4} C'_s \tag{B.5}$$

   where

   C'_s = volumetric saturation concentration of gas in liquid phase, mL/L
   C_s = solubility or saturation concentration of gas in liquid phase, mg/L
   mw_g = molecular weight of gas, g/mole
   22.4 = mL·g/mole·mg

   Mass concentration from Equation B.5, $C_s = \dfrac{32\,\text{g/mole}}{22.4\,\text{mL}\cdot\text{g/mole}\cdot\text{mg}} \times 7.87\,\text{mL/L} = 11.2\,\text{mg/L}$

6. Compare the result with that obtained in Example B.5.

   The C_s values obtained in Examples B.5 and B.6 are 11.3 and 11.2 mg/L, respectively.

EXAMPLE B.7: SATURATION CONCENTRATION OF OXYGEN FROM EMPIRICAL EQUATIONS

Determine the solubility of oxygen in distilled water at 10°C and 1 atm from Equations B.1a, B.1b, and B.1c, respectively.

Solution

1. Calculate the saturation concentration of oxygen (C_s) at 10°C and 1 atm from Equation B.1a.

$$\begin{aligned} C_s &= 14.652 - 0.41022T + 0.00799T^2 - 0.0000777T^3 \\ &= 14.652 - 0.41022 \times 10 + 0.00799 \times 10^2 - 0.0000777 \times 10^3 \\ &= 14.652 - 4.102 + 0.799 - 0.078 = 11.3\,\text{mg/L} \end{aligned}$$

2. Calculate the saturation concentration of oxygen (Cs) at 10°C and 1 atm from Equation B.1b.

   Vapor pressure at 10°C, $P_v = 9.21$ mm Hg (Table B.4).

$$\begin{aligned} C_s &= (0.68 - 6 \times 10^{-4}T)\left(\frac{760 - P_v}{T + 35}\right) = (0.68 - 6 \times 10^{-4} \times 10) \times \left(\frac{760 - 9.21}{10 + 35}\right) \\ &= 0.674 \times \frac{751}{45} = 11.2\,\text{mg/L.} \end{aligned}$$

3. Calculate the saturation concentration of oxygen (C_s) at 10°C and 1 atm from Equation B.1c.

$$\begin{aligned} C_s &= 14.652 - 10.53(1 - e^{-0.03896T}) = 14.652 - 10.53(1 - e^{-0.03896 \times 10} - 1) \\ &= 14.652 - 10.53 \times (1 - 0.677) = 11.3\,\text{mg/L.} \end{aligned}$$

EXAMPLE B.8: PREVAILING PRESSURE CORRECTION

Determine the saturation concentration of oxygen in surface water at an altitude of 6000 ft above sea level. The temperature of water is 15°C.

Solution

At an altitude of 6000 ft above sea level, the barometric pressure, $P = 611$ mm Hg (Table B.7).

At 15°C, the vapor pressure at 1 atm, $P_v = 12.788$ mm Hg (Table B.2).

The saturation concentration of oxygen at 15°C and 1 atm, $C_s = 10.15$ mg/L (Table B.6).

Calculate the saturation concentration of oxygen ($C_{s,P}$) at 15°C and 6000 ft above sea level from Equation B.2.

$$C_{s,P} = \frac{P - P_v}{760 - P_v} \times C_s = \frac{(611 - 12.788)\,\text{mm Hg}}{(760 - 12.788)\,\text{mm Hg}} \times 10.15\,\text{mg/L} = \frac{598\,\text{mm Hg}}{747\,\text{mm Hg}} \times 10.15\,\text{mg/L} = 8.13\,\text{mg/L}.$$

EXAMPLE B.9: SATURATION CONCENTRATION OF OXYGEN IN SALINE WATER

The saturation concentration of oxygen decreases at higher salt concentration. Calculate the DO saturation in saline water having chloride concentration of 5000 mg/L. The temperature and barometric pressure are 15°C and 1 atm. Use Equations B.3a, B.3b, and B.3c, respectively. Compare the results with the value given in Table B.8.

Solution

1. Calculate the DO saturation concentration using Equation B.3a.

   The saturation concentration of oxygen at 15°C and 1 atm, $C_s = 10.15$ mg/L (Table B.6).

   Correction factor, $f_{Cl} = 0.010$ mg O_2/100 mg chloride (Table B.8).

   DO saturation from Equation B.3a,

$$C_s'' = C_s - f_{Cl}\frac{S_{Cl}}{100} = 10.15\,\text{mg/L} - 0.010\,\text{mg/100 mg chloride} \times \frac{5000\,\text{mg chloride/L}}{100\,\text{mg chlorine}} = 9.65\,\text{mg/L}$$

2. Calculate the DO saturation concentration using Equation B.3b.

$$C_s'' = C_s(1 - 9 \times 10^{-6} \times S_{Cl}) = 10.15\,\text{mg/L} \times (1 - 9 \times 10^{-6} \times 5000)$$
$$= 10.15\,\text{mg/L} \times (1 - 0.045) = 9.69\,\text{mg/L}.$$

3. Calculate the DO saturation concentration using Equation B.3c.

$$C_s'' = \left(\frac{475 - 2.65 \times 10^{-3} \times S_{Cl}}{33.5 + T}\right)\left(\frac{P}{760}\right) = \left(\frac{475 - 2.65 \times 10^{-3} \times 5000\,\text{mg/L}}{33.5 + 15}\right) \times \left(\frac{760}{760}\right) = 9.52\,\text{mg/L}$$

4. Compare the results.

   The saturation concentration of oxygen in saline water containing 5000 mg/L chloride at 15°C and 1 atm is 9.65 mg/L (Table B.8). The values calculated from Equations B.3a and B.3b are close to the value given in the Table B.8 and acceptable. The value obtained from Equation B.3c is about 13% lower that obtained from Table B.8. Therefore, Equation B.3c should be used with caution.

B.3 Stability of Feed Water for Membrane Systems

The scaling or corrosive nature of feed water for treatment by membrane processes may be evaluated from LSI or RSI. These indexes are calculated by using ($pK_2 - pK_s$) values. These values with respect to temperature and TDS are summarized in Table B.9.[14] The application procedure is shown in Example 15.58.

TABLE B.9 Values of $pK_2 - pK_s$ with Respect to Temperature and Total Dissolved Solid (TDS)

TDS, mg/L	$pK_2 - pK_s$						
	0°C	10°C	20°C	30°C	40°C	50°C	80°C
0	2.45	2.23	2.02	1.86	1.68	1.52	1.08
40	2.58	2.36	2.15	1.99	1.81	1.65	1.21
80	2.62	2.40	2.19	2.03	1.85	1.69	1.25
120	2.66	2.44	2.23	2.07	1.89	1.73	1.29
160	2.68	2.46	2.25	2.09	1.91	1.75	1.31
200	2.71	2.49	2.28	2.12	1.94	1.78	1.34
240	2.74	2.52	2.31	2.15	1.97	1.81	1.37
280	2.76	2.54	2.33	2.17	1.99	1.83	1.39
320	2.78	2.56	2.35	2.19	2.01	1.85	1.41
360	2.79	2.57	2.36	2.20	2.02	1.86	1.42
400	2.82	2.59	2.38	2.22	2.04	1.88	1.44
440	2.83	2.61	2.40	2.24	2.06	1.90	1.46
480	2.84	2.62	2.41	2.25	2.07	1.91	1.47
520	2.86	2.64	2.43	2.27	2.09	1.93	1.49
560	2.87	2.65	2.44	2.28	2.10	1.94	1.50
600	2.88	2.66	2.45	2.29	2.11	1.95	1.51
640	2.90	2.68	2.47	2.31	2.13	1.97	1.53
680	2.91	2.69	2.48	2.32	2.14	1.98	1.54
720	2.92	2.70	2.49	2.33	2.15	1.99	1.55
760	2.92	2.70	2.49	2.33	2.15	1.99	1.55
800	2.93	2.71	2.50	2.34	2.16	2.00	1.56

Source: Adapted in part from Reference 14.

B.4 Salinity of Reclaimed Water for Irrigation

Sodicity is a condition when Na^+ ions build up in soil matrix from use of reclaimed water for irrigation. This condition reduces movement of water and air causing water logging.[15] The adjusted calcium concentration $[Ca_X^{2+}]$ value is needed to evaluate this situation. The $[Ca_X^{2+}]$ values based on different HCO_3^-/Ca^{2+} ratios and salinities are given in Table B.10. The procedure for determining this condition from $[Ca_X^{2+}]$ data is explained in Example 12.25.

TABLE B.10 Values of $[Ca_X^{2+}]$ as a Function of the HCO_3^--to-Ca^{2+} Ratio and Salinity

Ratio of HCO_3^-/Ca^{2+}, meq/L	Value of $[Ca_X^{2+}]$, meq/L											
	Salinity of Applied Water (EC_w), dS/m or mmhos/cm											
	0.1	0.2	0.3	0.5	0.7	1	1.5	2	3	4	6	8
0.05	13.2	13.6	13.9	14.4	14.8	15.3	15.9	16.4	17.3	18.0	19.1	19.9
0.1	8.31	8.57	8.77	9.07	9.31	9.62	10.0	10.4	10.9	11.3	12.0	12.6
0.15	6.34	6.54	6.69	6.92	7.11	7.34	7.65	7.90	8.31	8.64	9.17	9.58
0.2	5.24	5.40	5.52	5.71	5.87	6.06	6.31	6.52	6.86	7.13	7.57	7.91
0.25	4.51	4.65	4.76	4.92	5.06	5.22	5.44	5.62	5.91	6.15	6.52	6.82
0.3	4.00	4.12	4.21	4.36	4.48	4.62	4.82	4.98	5.24	5.44	5.77	6.04

Continued

TABLE B.10 (*Continued*) Values of $[Ca_X^{2+}]$ as a Function of the HCO_3^--to-Ca^{2+} Ratio and Salinity

Ratio of HCO_3^-/Ca^{2+}, meq/L	Value of $[Ca_X^{2+}]$, meq/L											
	Salinity of Applied Water (EC_w), dS/m or mmhos/cm											
	0.1	0.2	0.3	0.5	0.7	1	1.5	2	3	4	6	8
0.35	3.61	3.72	3.80	3.94	4.04	4.17	4.35	4.49	4.72	4.91	5.21	5.45
0.4	3.30	3.40	3.48	3.60	3.70	3.82	3.98	4.11	4.32	4.49	4.77	4.98
0.45	3.05	3.14	3.22	3.33	3.42	3.53	3.68	3.80	4.00	4.15	4.41	4.61
0.5	2.84	2.93	3.00	3.10	3.19	3.29	3.43	3.54	3.72	3.87	4.11	4.30
0.75	2.17	2.24	2.29	2.37	2.43	2.51	2.62	3.70	2.84	2.95	3.14	3.28
1	1.79	1.85	1.89	1.96	2.01	2.09	2.16	2.23	2.35	2.44	2.59	2.71
1.25	1.54	1.50	1.63	1.68	1.73	1.78	1.86	1.92	2.02	2.10	2.23	2.33
1.5	1.37	1.41	1.44	1.49	1.53	1.58	1.65	1.70	1.79	1.86	1.97	2.07
1.75	1.23	1.27	1.30	1.35	1.38	1.43	1.49	1.54	1.62	1.68	1.78	1.86
2	1.13	1.16	1.19	1.23	1.26	1.31	1.36	1.40	1.48	1.54	1.63	1.70
2.25	1.04	1.08	1.10	1.14	1.17	1.21	1.26	1.30	1.37	1.42	1.51	1.58
2.5	0.97	1.00	1.02	1.06	1.09	1.12	1.17	1.21	1.27	1.32	1.40	1.47
3	0.85	0.89	0.91	0.94	0.96	1.00	1.04	1.07	1.13	1.17	1.24	1.30
3.5	0.78	0.80	0.82	0.85	0.87	0.90	0.94	0.97	1.02	1.06	1.12	1.17
4	0.71	0.73	0.75	0.78	0.80	0.82	0.86	0.88	0.93	0.97	1.03	1.07
4.5	0.66	0.68	0.69	0.72	0.74	0.76	0.79	0.82	0.86	0.90	0.95	0.99
5	0.61	0.63	0.65	0.67	0.69	0.71	0.74	0.76	0.80	0.83	0.88	0.93
7	0.49	0.50	0.52	0.53	0.55	0.57	0.59	0.61	0.64	0.67	0.71	0.74
10	0.39	0.40	0.41	0.42	0.43	0.45	0.47	0.48	0.51	0.53	0.56	0.58
20	0.24	0.25	0.26	0.26	0.27	0.28	0.29	0.30	0.32	0.33	0.35	0.37

Source: Adapted in part from Reference 15.

References

1. Qasim, S. R., *Wastewater Treatment Plants: Planning, Design, and Operation*, 2nd ed., CRC Press, Boca Raton, FL, 1999.
2. Droste, R. L., *Theory and Practice of Water and Wastewater Treatment*, John Wiley of Sons, Inc., New York, NY, 1997.
3. Metcalf & Eddy | AECOM, *Wastewater Engineering: Treatment and Resource Recovery*, 5th ed., McGraw-Hill, New York, NY, 2014.
4. Metcalf & Eddy, Inc., *Wastewater Engineering: Treatment and Reuse*, 4th ed., McGraw-Hill Companies, Inc., New York, NY, 2003.
5. Qasim, S. R., E. M. Motley, and G. Zhu, *Water Works Engineering: Planning, Design, and Operation*, Prentice Hall PTR, Upper Saddle River, NJ, 2000.
6. Tchobanoglous, G., H. Theisen, and S. Vigil, *Integrated Solid Waste Management: Engineering Principles and Management Issues*, McGraw-Hill, Inc., New York, NY, 1993.
7. Lange, N. A. and J. A. Dean, *Lange's Handbook of Chemistry*, 13th ed., McGraw-Hill, Inc., New York, NY, 1985.
8. Tchobanoglous, G. and E. D. Schroeder, *Water Quality: Characteristics Modeling and Modification*, Addison-Welsey, Publishing Co., Reading, MA, 1985.
9. Elmore, H. L. and T. W. Hayes, Solubility of atmospheric oxygen in water, No. 29 Report of the Committee on Sanitary Engineering Research, *Journal of the Sanitary Engineering Division*, ASCE, 86(SA 4), 1960, 41–53.

10. Gameson, A. H. and K. G. Robertson, The solubility of oxygen in pure water and sea water, *Journal Chemical Technology and Biotechnology*, 5(9), 1955, 502.
11. APHA, AWWA, and WEF, *Standard Methods for Examination of Water and Wastewater*, 22nd ed., American Public Health Association, Washington, DC, 2012.
12. Hunter, J. S. and J. C. Ward, The effects of water temperature and elevation upon aeration, *Proceedings of International Symposium on Wastewater Treatment in Cold Climates*, Institute of Northern Studies, University of Saskatchewan, Canada, 1974.
13. Fair, G. M., J. C. Geyer, and D. A. Okum, *Water and Wastewater Engineering*, Vol. 2, John Wiley and Sons Inc., New York, NY, 1968.
14. Viessman, W., Jr., M. J. Hammer, E. M. Perez, and P. A. Chadik, *Water Supply and Pollution Control*, 8th ed., Prentice Hall PTR, Upper Saddle River, NJ, 2008.
15. Pettygrove, G. S. and T. Asano (editors), *Irrigation with Reclaimed Municipal Wastewater—A Guide Manual*, Lewis Publishers, Inc., Chelsea, MI, 1985.

Appendix C: Minor Head Loss Coefficients for Pressure Conduits and Open Channels, Normal Commercial Pipe Sizes, and Design Information of Parshall Flume

In this appendix, the constants commonly used to calculate the minor head losses in pressure conduits and open channels are provided in Sections C.1 and C.2. Standard commercial pipe sizes are listed in Table C.1 of Section C.3. Additionally, the basic information needed to design standard Parshall flumes are provided in Section C.4. The information provided in this appendix may be found in many hydraulics texts and handbooks. Many of these references are cited at the end of this appendix.

C.1 Minor Head Loss Coefficients (K) for Pressure Conduits

In a pressure conduit, the minor head losses are created at fittings, valves, bends, entrance, exit, etc. Each of these losses is calculated from Equation 6.15b by applying a proper minor head loss coefficient (K) to the velocity head at the application point. The typical values of K used in plant hydraulic calculations are presented below.[1–7]

C.1.1 Gate Valve

Full Open	One-Fourth Closed	One-Half Closed	Three-Fourths Closed	Typical Value	Equation 6.15b
0.19	1.15	5.6	24	1.0	$h_m = K\dfrac{V^2}{2g}$

Note: V = velocity in the pipe upstream or downstream of the valve.

C.1.2 Butterfly Valve

Open Full	Angle Closed			Typical Value	Equation 6.15b
	0°	40°	60°		
0.3	1.4	10	94	1.2	$h_m = K\frac{V^2}{2g}$

Note: V = velocity in the pipe upstream or downstream of the valve.

C.1.3 Other Valves

Check Valve	Plug Globe Valve (Full Open)	Diaphragm Valve (Full Open)	Equation 6.15b
1.5–2.5	1.0–4.0	2.3	$h_m = K\frac{V^2}{2g}$

Note: V = velocity in the pipe upstream or downstream of the valve.

C.1.4 Entrance and Exit

Entrance				Exit from Conduit to Still Water	Equation 6.15b
Pipe Projecting into Tank	End of Pipe Flushed with Tank	Slightly Rounded	Bell-Mouthed		
0.83	0.50	0.23	0.04	1.0	$h_m = K\frac{V^2}{2g}$

Note: V = velocity in the pipe after the entrance or before the exit.

C.1.5 Elbow (45–61 cm diameter)

22.5°	45°	90° Regular	90° Long	Equation 6.15b
0.1–0.2	0.15–0.3	0.20–0.40	0.14–0.23	$h_m = K\frac{V^2}{2g}$

Note: V = velocity in the pipe upstream or downstream of the elbow.

C.1.6 Tee

Run-to-Run	Branch-to-Run	Run-to-Branch	Equation 6.15b
0.25–0.6	0.6–1.8	0.6–1.8	$h_m = K\frac{V^2}{2g}$

Note: V = velocity in the pipe upstream or downstream of the elbow.

C.1.7 Increaser and Reducer

Increaser[a]			Reducer[b]			Equation 6.15b
$d/D = 0.25$	$d/D = 0.5$	$d/D = 0.75$	$d/D = 0.25$	$d/D = 0.5$	$d/D = 0.75$	
0.42	0.33	0.19	0.92	0.56	0.19	$h_m = K\frac{V^2}{2g}$

[a] d/D = ratio of pipe diameter upstream and downstream of increaser and V = velocity in the pipe upstream of the increaser.
[b] d/D = ratio of pipe diameters downstream and upstream of the reducer and V = velocity in the pipe downstream of the reducer.

C.2 Minor Head Loss Constants (K) for Open Channels

In an open channel, the minor head losses occur due to turbulences caused by changing the flow direction and cross sectional area. The minor head loss equation for the pressure conduit (Equation 6.15b) may also be applied to appurtenances such as manholes, junction and diversion boxes, syphons, and sluice gates (submerged or nonsubmerged). This equation is also modified to Equation 7.3b for enlargement (or outlet) losses, and Equation 7.3c for contraction (or inlet) losses, respectively. The typical minor head loss coefficients for these expressions are presented below.[1–7]

C.2.1 Manholes, and Junction and Diversion Boxes

Head losses in manholes, and junction and diversion boxes depend on the size of the appurtenances, change in direction, and the contour of the bottom.

1. Manhole, junction box, or diversion box with no change in channel or pipe size or direction.

Flow-Through Box	Terminal Box	Equation 6.15b
0.05	1.0	$h_m = K\frac{v^2}{2g}$

Note: v = velocity in the upstream or downstream channel or pipe of the manhole or box.

2. Manhole, junction box, or diversion box with no change in channel or pipe size, but a change in channel or pipe direction.

45° turn		90° turn		Equation 6.15b
Without Shaping	With Shaping	Without Shaping	With Shaping	
0.3–0.4	0.2–0.3	1.2–1.4	1.0–1.2	$h_m = K\frac{v^2}{2g}$

Note: v = velocity in the upstream or downstream channel or pipe of the manhole or box.

3. Large junction or diversion box in which the velocity is small.

Entrance	Exit	Equation 6.15b
0.5	1.0	$h_m = K\frac{v^2}{2g}$

Note: v = velocity in the channel or pipe after the entrance or before the exit.

C.2.2 Syphon

Syphon	Equation 6.15b
2.78	$h_m = K\frac{v^2}{2g}$

Note: v = velocity through the syphon.

C.2.3 Sluice Gate

Sluice Gate	Equation 6.15b
0.2–0.8	$h_m = K\frac{v^2}{2g}$

Note: v = velocity through the full gate opening (submerged or nonsubmerged).

C.2.4 Sudden Enlargement or Outlet Losses

Sharp-Cornered	Bell-Mouthed	Equation 7.3b
0.2–1.0	0.1	$h_L = \frac{K_e}{2g}(v_1^2 - v_2^2)$

Note: v_1 and v_2 = velocities upstream and downstream of the enlargement or outlet.

C.2.5 Sudden Contraction or Inlet Losses

Sharp-Cornered	Round-Cornered	Bell-Mouthed	Equation 7.3c
0.5	0.25	0.05	$h_L = \frac{K_c}{2g}(v_2^2 - v_1^2)$

Note: v_1 and v_2 = the velocities upstream and downstream of the contraction or inlet.

C.3 Standard Commercial Pipe Sizes

The normal diameters of pipes that may be commercially available in the U.S. market, and equivalent normal pipe sizes in SI units are summarized in Table C.1. Certain normal size may not be applicable for some pipe materials. The actual inside diameter of a pipe will also vary with the *pipe material*, and the *wall thickness* determined by the *pipe schedule* or *pipe class*. Readers should consult the pipe manufacturers to obtain the detailed technical data of the pipes used in the design.

C.4 Standard Design of Parshall Flume

A Parshall flume is a fixed hydraulic structure used for the flow measurement of water and wastewater. It has a converging section, throat, and the diverging section.[8–10] Under the free-flow condition, the depth of water at a specified location upstream of the throat is converted to flow rate by Equation 7.9. The constant C is a free-flow coefficient, and the exponent n varies with flume size. These constants for different throat width (W) are summarized in Table C.2.[9] The design components of a standard Parshall flume are shown in Figure C.1.[8] The standard dimensions are given in Table C.3.[9]

The free-flow condition is applied until the submergence ratio (submergence head to measuring head) is exceeded by a certain criteria (Table C.4).[9] Design procedures of Parshall flumes are presented in Examples 7.8, 8.13, and 9.38. The hydraulic profiles through these flumes are also illustrated in

Figures 7.14, 9.39, and 9.40. When a Parshall flume is submerged, the head loss through the flume depends upon the submergence.[8–10] The head losses through standard Parshall flumes for different percent submergences may be read from the graphics contained in Reference 9.

TABLE C.1 Commonly Available Commercial Pipe Sizes (4–240 in Diameter)

Normal Pipe Size Available in the U.S. Market		Equivalent Normal Pipe Size in SI Units, mm
in	mm	
4	101.6	100
5	127.0	–
6	152.4	150
8	203.2	200
10	254.0	250
12	304.8	300
14	355.6	350
15	381.0	–
16	406.4	400
18	457.2	450
20	508.0	500
21	533.4	–
24	609.6	600
27	685.8	700
30	762.0	750
36	914.4	900
42	1067	1050
48	1219	1200
54	1372	1400
60	1524	1500
66	1676	–
72	1829	1800
78	1981	2000
84	2134	2100
90	2286	2250
96	2438	2400
102	2591	–
108	2743	2700
114	2896	–
120	3048	3000
–	–	3300
144	3658	3600
180	4572	–
204	5182	–
240	6096	–

TABLE C.2 Coefficients (C) and Exponents (n) Used for Design of Parshall Flumes

Throat Width (W)	Coefficient (C)	Exponent (n)
1 in	0.338	1.55
2 in	0.676	1.55
3 in	0.992	1.55
6 in	2.06	1.58
9 in	3.07	1.53
1 ft	3.95	1.55
2 ft	8.00	1.55
3 ft	12.00	1.57
4 ft	16.00	1.58
5 ft	20.00	1.59
6 ft	24.00	1.59
7 ft	28.00	1.60
8 ft	32.00	1.61
10 ft	39.38	1.60
12 ft	46.75	1.60
15 ft	57.81	1.60
20 ft	76.25	1.60
25 ft	94.69	1.60
30 ft	113.13	1.60
40 ft	150.00	1.60
50 ft	186.88	1.60

Source: Adapted in part from Reference 9.

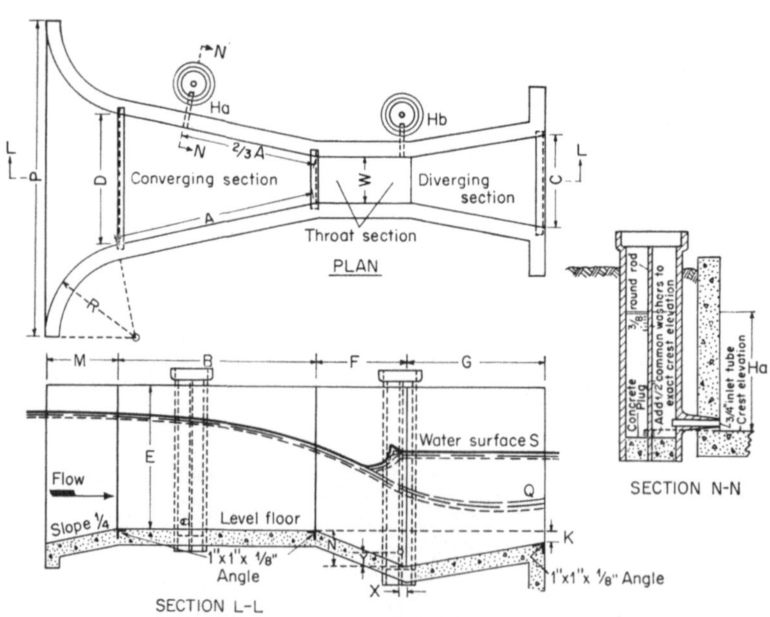

FIGURE C.1 Standard design details of Parshall flume. (Adapted in part from Reference 8. (Courtesy U.S. Department of the Interior, Bureau of Reclamation.))

TABLE C.3 Standard Dimensions of Selected Parshall Flumes ($W = 0.5-6$ ft)

Dimension and Capacity Parameter	Throat Size (W) in ft, Dimension in ft-in, and Capacity Parameter in ft^3/s									
	$W = 0.5$		$W = 1$		$W = 2$		$W = 4$		$W = 6$	
	ft	in	ft	in	ft	in	ft	in	ft	in
A	2	7/16	4	6	5	0	6	0	7	0
2/3 A	1	4 5/16	3	0	3	4	4	0	4	8
B	2	0	4	4 7/8	4	10 7/8	5	10 5/8	6	10 3/8
C	1	3½	2	0	3	0	5	0	7	0
D	1	3 5/8	2	9¼	3	11½	6	4¼	8	9
E	2	0	3	0	3	0	3	0	3	0
F	1	0	2	0	2	0	2	0	2	0
G	2	0	3	0	3	0	3	0	3	0
K	0	3	0	3	0	3	0	3	0	3
M	0	0	1	3	1	3	1	6	1	6
N	0	4¼	0	9	0	9	0	9	0	9
P	2	11½	4	10¾	6	1	8	10¾	11	3½
R	1	4	1	8	1	8	2	0	2	0
X	0	2	0	2	0	2	0	2	0	2
Y	0	3	0	3	0	3	0	3	0	3
Minimum capacity	0.05		0.11		0.42		1.3		2.6	
Maximum capacity	3.9		16.1		33.1		67.9		103.5	

Source: Adapted in part from Reference 9.

TABLE C.4 Determination of Free-Flow Condition for Parshall Flumes

Throat Width (W)	Submergence Ratio (H_b/H_a) Allowed under Free-Flow Condition[a], %
1–3 in	≤ 50
6–9 in	≤ 60
1–8 ft	≤ 70
10–50 ft	≤ 80

[a] H_a = water depth at the throat and H_b = submergence head downstream of the throat (Figure C.1).

Source: Adapted in part from Reference 9.

References

1. Qasim, S. R., E. M. Motley, and G. Zhu, *Water Works Engineering: Planning, Design, and Operation*, Prentice Hall PTR, Upper Saddle River, NJ, 2000.
2. Qasim, S. R., *Wastewater Treatment Plants: Planning, Design, and Operation*, 2nd ed., CRC Press, Boca Raton, FL, 1999.
3. FMC Corporation, *Hydraulics and Useful Information*, Chicago Pumps, Chicago, IL, 1963.
4. Benefield, L. D., J. F. Judkins, and A. D. Parr, *Treatment Plant Hydraulics for Environmental Engineers*, Prentice-Hall Inc., Englewood Cliffs, NJ, 1984.
5. Brater, E. F., H. W. King, J. E. Lindell, and C. Y. Wei, *Handbook of Hydraulics*, 7th ed., MaGraw-Hill, New York, NY, 1996.

6. Heald, C. C. (editor), *Cameron Hydraulic Data*, 18th ed., Ingersoll-Dresser Pumps, Liberty Corner, NJ, 1995.
7. Zipparro, V. J. (editor), *Davis's Handbook of Applied Hydraulics in Engineering*, 4th ed., McGraw-Hill Book Co., New York, 1993.
8. U.S. Department of the Interior, *Water Measurement Manual*, 2nd ed., Bureau of Reclamation, Denver, CO, 1967.
9. U.S. Department of the Interior, Bureau of Reclamation, *Water Measurement Manual, A Water Resources Technical Publication*, 3rd ed., 1997 (Revised Reprint, 2001), Bureau of Reclamation, http://www.usbr.gov, 2001.
10. Parshall, R. L., *Measuring Water in Irrigation Channels with Parshall Flumes and Small Weirs*, U.S. Soil Conservation Service, Circular 843, U.S. Department of Agriculture, Washington, DC, 1950.

Appendix D: Unit Conversions

In this appendix, the conversion factors between SI units and the U.S. customary units are provided in Table D.1. The conversion factors for mixed units are summarized in Table D.2.[1-3]

TABLE D.1 Conversion Factors between SI Units and the U.S. Customary Units

SI Units	Factor f Multiply by Factor f to Obtain → ← Divide by Factor f to Obtain	U.S. Customary Units
Length		
cm	0.3937	in
m	3.281	ft (1 yd = 3 ft)
km (1000 m)	0.6214	mile, mi (1760 yd or 5280 ft)
Area		
cm^2	0.1550	in^2
m^2	10.76	ft^2
m^2	1.196	yd^2 ($9\ ft^2$)
Hectare, ha ($10{,}000\ m^2$)	2.471	acre ($4840\ yd^2$ or $43{,}560\ ft^2$)
km^2	0.3861	mi^2
km^2	247.1	acre
Volume		
cm^3	6.102×10^{-2}	in^3
L	0.2642	gal
m^3 (1000 L)	35.31	ft^3 (7.481 gal)
m^3	1.308	yd^3 ($27\ ft^2$)
m^3	264.2	gal
m^3	8.13×10^{-4}	acre-ft ($43{,}560\ ft^3$ or 325,900 gal)
Mass		
g	3.527×10^{-2}	oz
g	2.205×10^{-3}	lb (1 lb = 16 oz)
g	15.43	grain (1 lb = 7000 grain)
kg	2.205	lb
Tonne (1000 kg)	2205	lb
Force		
N	0.2248	lb_f

(Continued)

TABLE D.1 (*Continued*) Conversion Factors between SI Units and the U.S. Customary Units

SI Units	Factor f Multiply by Factor f to Obtain → ← Divide by Factor f to Obtain	U.S. Customary Units
Velocity		
m/s (1 m/s = 86.4 km/d)	3.281	ft/s (16.36 mi/d)
m/s	2.237	mi/h
km/h	0.6214	mi/h
km/h	2.237	mi/h
Flow rate		
m^3/s	35.31	ft^3/s (7.481 gal/s or 448.9 gal/min)
m^3/d	264.2	gal/d
m^3/d	2.642×10^{-4}	Mgal/d
m^3/s	22.82	Mgal/d
L/s	22,820	gal/d
Acceleration		
m/s^2 ($g = 9.81\ m/s^2$)	3.281	ft/s^2 ($g = 32.2\ ft/s^2$)
Temperature		
°C	$1.8 \times (°C) + 32$ → ← $0.0555 \times (°F) - 32$	°F
°k	$1.8 \times (°k) - 459.7$ → ← $0.0555 \times (°F) + 459.7$	°F
Power		
kW	0.9478	Btu/s
kW	1.341	hp (550 ft·lb_f/s)
W	0.7376	ft·lb/s
Pressure		
kPa (Pascal)	0.1450	lb_f/in^2
Pa (N/m^2)	1.450×10^{-4}	lb_f/in^2
Pa (N/m^2)	2.089×10^{-2}	lb_f/in^2
Pa (N/m^2)	2.961×10^{-4}	in Hg
Pa (N/m^2)	4.019×10^{-3}	in of water
Pa (N/m^2)	9.869×10^{-3}	atm (14.7 lb/in^2 or 1 psi)
Dynamic viscosity (μ)		
N·s/m^2 (k/m·s)	0.021	lb_f·s/ft^2 (32.15 lb_m/ft·s)
N·s/m^2 (k/m·s)	0.671	lb_m/ft·s
Kinematic viscosity (ν)		
Centistoke	1.076×10^{-5}	ft^2/s
m^2/s	10.76	ft^2/s
Energy		
kW·h	1.341	hp·h
kW·h	3412	Btu

Source: Adapted in part from References 1 through 3.

TABLE D.2 Conversion Factors, Mixed Units

Length (L)					
Mile (mi)	yd	ft	in	m	cm
1	1760	5280	6.336×10^4	1.609×10^3	1.609×10^5
5.68×10^{-4}	**1**	3	36	0.9144	91.44
1.894×10^{-4}	0.333	**1**	12	0.3048	30.48
1.578×10^{-5}	0.028	0.083	**1**	0.0254	2.54
6.214×10^{-4}	1.094	3.281	39.37	**1**	100
6.214×10^{-6}	0.01094	0.03281	0.3937	0.01	**1**

Area (A)					
mi^2	acre	yd^2	ft^2	in^2	m^2
1	640[a]	3.098×10^6	2.788×10^7	4.014×10^9	2.59×10^6
1.563×10^{-3}	**1**	4840	43,560	6.27×10^6	4047
3.228×10^{-7}	2.066×10^{-4}	**1**	9	1296	0.836
3.587×10^{-8}	2.3×10^{-5}	0.111	**1**	144	0.093
2.491×10^{-10}	1.59×10^{-7}	7.716×10^{-4}	6.944×10^{-3}	**1**	6.452×10^{-4}
3.861×10^{-7}	2.5×10^{-4}	1.196	10.764	1550	**1**

Volume (V)						
acre·ft	U.S. gal	ft^3	in^3	L	m^3	cm^3
1	325,851	43,560	75.3×10^6	1.23×10^6	1230	1.23×10^9
3.07×10^{-6}	**1**	0.134	231.552	3.785	3.875×10^{-3}	3875.412
2.3×10^{-5}	7.481	**1**	1728	28.317	0.028	28,316.846
1.33×10^{-8}	4.329×10^{-3}	5.787×10^{-4}	**1**	0.016	1.639×10^{-5}	16.387
8.1×10^{-7}	0.264	0.035	61.024	**1**	1×10^{-3}	1000
8.13×10^{-4}	264.2	35.31	6.10×10^4	1000	**1**	10^6
8.13×10^{-10}	2.642×10^{-4}	3.531×10^{-5}	6.10×10^{-2}	10^{-3}	10^{-6}	**1**

Time (T)					
Year	Months	Days	h	min	s
1	12	365	8760	525,600	3.154×10^7

Velocity (L/T)				
ft/s	ft/min	m/s	m/min	cm/s
1	60	0.3048	18.29	30.48
0.017	**1**	5.08×10^{-3}	0.3048	0.5080

(*Continued*)

TABLE D.2 (*Continued*) Conversion Factors, Mixed Units

Velocity (L/T)				
ft/s	ft/min	m/s	m/min	cm/s
3.281	196.8	**1**	60	100
0.055	3.28	0.017	**1**	1.70
0.032	1.969	0.01	0.588	**1**

Discharge (L^3/T)					
mgd	gpm	ft^3/s	ft^3/min	L/s	m^3/d
1	694.4	1.547	92.82	43.75	3.78×10^3
1.44×10^{-3}	**1**	2.228×10^{-3}	0.134	0.063	5.45
0.646	448.9	**1**	60	28.32	2447
0.011	7.481	0.017	**1**	0.472	40.78
0.023	15.850	0.035	2.119	**1**	86.41
2.64×10^{-4}	0.183	4.09×10^{-4}	0.025	0.012	**1**

Mass (M)					
Ton	lb$_m$	Grain	Ounce (oz)	kg	g
1	2000	1.4×10^7	32,000	907.2	907,200
0.0005	**1**	7000	16	0.454	454
7.14×10^{-8}	1.429×10^{-4}	**1**	2.29×10^{-3}	6.48×10^{-5}	0.065
3.125×10^{-5}	0.0625	437.6	**1**	0.028	28.35
1.10×10^{-3}	2.205	1.54×10^4	35.27	**1**	1000
1.10×10^{-6}	2.20×10^{-3}	15.43	0.035	10^{-3}	**1**

Temperature (T)			
°F	°C	°K	°R
°F	$\frac{5}{9} \times (°\text{F} - 32)$	$\frac{5}{9} \times (°\text{F} - 32) + 273.2$	°F + 459.7
$1.8 \times (°\text{C} + 32)$	**°C**	°C + 273.2	$1.8 \times (°\text{C} + 273.2)$
$1.8 \times °\text{K} - 459.7$	°K − 273.2	**°K**	$1.8 \times °\text{K}$
°R − 459.7	$\frac{5}{9} \times (°\text{R} - 491.7)$	$\frac{5}{9} \times °\text{R}$	**°R**

Density (M/L^3)			
lb/ft^3	lb/gal (U.S.)	kg/m^3 (g/L)	kg/L (g/cm^3)
1	0.1337	16.02	0.01602
7.48	**1**	119.8	0.1198

(*Continued*)

TABLE D.2 (*Continued*) Conversion Factors, Mixed Units

Density (M/L^3)			
lb/ft^3	lb/gal (U.S.)	kg/m^3 (g/L)	kg/L (g/cm^3)
0.06243	8.345×10^{-3}	**1**	0.001
62.43	8.345	1000	**1**

Pressure (F/L^2)						
lb/in^2	ft water	in Hg	atm	mm Hg	kg/cm^2	N/m^2
1	2.307	2.036	0.068	51.71	0.0703	6895
0.4335	**1**	0.8825	0.0295	22.41	0.0305	2989
0.4912	1.133	**1**	0.033	25.40	0.035	3386.44
14.70	33.93	29.92	**1**	760	1.033	1.013×10^5
0.019	0.045	0.039	1.30×10^{-3}	**1**	1.36×10^{-3}	133.3
14.23	32.78	28.96	0.968	744.7	**1**	98,070
1.45×10^{-4}	3.35×10^{-4}	2.96×10^{-4}	9.87×10^{-6}	7.50×10^{-3}	1.02×10^{-5}	**1**

Force (F)		
lb$_f$	N	dyne
1	4.448	4.448×10^5
0.225	**1**	10^5
2.25×10^{-6}	10^{-5}	**1**

Energy (E)					
kW·h	hp·h	Btu	J	kJ	Calorie (cal)
1	1.341	3412	3.6×10^6	3600	8.6×10^5
0.7457	**1**	2545	2.684×10^6	2684	6.4×10^5
2.930×10^{-4}	3.929×10^{-4}	**1**	1054.8	1.055	252
2.778×10^{-7}	3.72×10^{-7}	9.48×10^{-4}	**1**	0.001	0.239
2.778×10^{-4}	3.72×10^{-4}	0.948	1000	**1**	239
1.16×10^{-6}	1.56×10^{-6}	3.97×10^{-3}	4.186	4.18×10^{-3}	**1**

Power (P)					
kW	Btu/min	hp	ft·lb/s	kg·m/s	cal/min
1	56.89	1.341	737.6	102	14,330
0.018	**1**	0.024	12.97	1.793	252
0.746	42.44	**1**	550	76.09	10,690

(*Continued*)

TABLE D.2 (*Continued*) Conversion Factors, Mixed Units

Power (P)					
kW	Btu/min	hp	ft·lb/s	kg·m/s	cal/min
1.35×10^{-3}	0.077	1.82×10^{-3}	1	0.138	19.43
9.76×10^{-3}	0.558	0.013	7.233	1	137.6
6.977×10^{-5}	3.97×10^{-3}	9.355×10^{-5}	0.0514	7.12×10^{-3}	1

Dynamic (or Absolute) Viscosity (μ)				
cp	$lb_f \cdot s/ft^2$	$lb_m/ft \cdot s$	g/cm·s	$N \cdot s/m^2$ (kg/m·s or dp)
1	2.09×10^{-5}	6.72×10^{-4}	0.01	1×10^{-3}
4.78×10^4	1	32.15	478.5	47.85
1488	0.031	1	14.88	1.488
100	2.09×10^{-3}	0.0672	1	0.10
1000	0.021	0.672	10	1

Kinematic Viscosity (ν)			
Centistoke	ft^2/s	cm^2/s	m^2/s (Myriastoke)
1	1.076×10^{-5}	0.01	10^{-6}
9.29×10^4	1	929.4	0.093
100	1.076×10^{-3}	1	10^{-4}
10^6	10.76	10^4	1

[a] 1 acre = 0.4047 hectare (ha), and 1 ha = 10,000 m^2.
Source: Adapted in part from References 1 through 3.

References

1. Qasim, S. R., *Wastewater Treatment Plants: Planning, Design, and Operation*, 2nd ed., CRC Press, Boca Raton, FL, 1999.
2. Qasim, S. R., E. M. Motley, and G. Zhu, *Water Works Engineering: Planning, Design, and Operation*, Prentice Hall PTR, Upper Saddle River, NJ, 2000.
3. Mechtly, E. A., *The International System of Units, Physical Constants and Conversion Factors*, 2nd rev., Scientific and Technical Information Office, National Aeronautics and Space Administration (NASA), Washington, D.C., 1973.

Appendix E: Summary of Design Parameters for Wastewater Treatment Processes

In this appendix, the practical ranges of selected major design parameters for wastewater treatment processes expressed in both SI and U.S. customary units, and the conversion factors between two units, are summarized in Table E.1.[1–3]

TABLE E.1 Design Parameter and Units of Expressions for Wastewater Treatment Processes

Design Parameter	Typical Range	Conversion Factor
	Coarse Screen, Pretreatment	
Opening size, mm (in)		
Manually cleaned	25–75 (1–3)	1 mm = 0.03937 in
Mechanically cleaned	6–50 (0.25–2)	1 mm = 0.03937 in
Velocity through clean screen, m/s (ft/s)		
Manually cleaned	0.3–0.6 (1–2)	1 m/s = 3.28 ft/s
Mechanically cleaned	0.6–1 (2–3.3)	1 m/s = 3.28 ft/s
Volume of screenings, $m^3/10^6\ m^3$ (ft^3/Mgal)	20–75 (2.5–10)	$1\ m^3/10^6\ m^3 = 0.1337\ ft^3$/Mgal
	Fine Screen, Pretreatment	
Opening size, mm (in)	1–6 (0.04–0.25)	1 mm = 0.03937 in
	Grit Removal	
Velocity in horizontal flow channel, m/s (ft/s)		1 m/s = 3.28 ft/s
Velocity controlled channel	0.25–0.4 (0.8–1.3)	
Aerated grit chamber	0.6–0.8 (2–2.5)	
Surface overflow rate (SOR) in horizontal flow chamber, m^3 flow/m^2 surface area·d (gpd/ft^2)	500–1000 (12,500–25,000)	$1\ m^3/m^2{\cdot}d = 24.54$ gpd/ft^2
Detention time, min		
Velocity controlled channel	0.75–1.5	
Aerated grit chamber	2–5	
Horizontal flow chamber	4–6	
Volume of grits, m^3 grit/$10^6\ m^3$ raw influent (ft^3/Mgal)	5–50 (0.7–7)	$1\ m^3/10^6\ m^3 = 0.1337\ ft^3$/Mgal

(Continued)

TABLE E.1 (*Continued*) Design Parameter and Units of Expressions for Wastewater Treatment Processes

Design Parameter	Typical Range	Conversion Factor
Flow Equalization		
Mixing power requirement, kW/10^3 m^3 volume (hp/Mgal)	4–8 (20–40)	1 kW/10^3 m^3 = 5.2 hp/Mgal
Air supply, m^3 air/10^3 m^3 tank volume·min (ft^3/Mgal·min)	10–20 (1350–2700)	1 m^3/10^3 m^3·min = 133.7 ft^3/Mgal·min
Primary Clarifier		
Surface overflow rate (SOR), m^3 flow/m^2 surface area·d (gpd/ft^2)		1 m^3/m^2·d = 24.54 gpd/ft^2
Average flow	24–48 (600–1200)	
Peak 2-h flow	40–100 (1000–2500)	
Detention time, h	1.5–4	
Weir loading rate, m^3 flow/m weir length·d (gpd/ft)	125–375 (10,000–30,000)	1 m^3/m·d = 80.52 gpd/ft
Volume of sludge, m^3 sludge/10^3 m^3 (ft^3/Mgal)	2.2–10 (300–1300)	1 m^3/10^3 m^3 = 133.7 ft^3/Mgal
Mass of sludge, kg dry solids/10^3 m^3 (lb/Mgal)	100–170 (850–1400)	1 kg/10^3 m^3 = 8.345 lb/Mgal
Solids content of sludge, % total solids	1–6	
Chemically Enhanced Primary Treatment (CEPT) for Phosphorus Removal		
Chemical dosage, practical molar ratio (mole/mole)		
Lime (Ca:P)	1.3:1–2:1	
Aluminum (Al:P)	1:1–2.5:1	
Iron, ferric (Fe:P)	1.5:1–3.5:1	
Iron, ferrous (Fe:P)	1.5:1–1.7:1	
High Rate Clarification for Primary Treatment		
Surface overflow rate (SOR), m^3 flow/m^2 surface area·h (gpm/ft^2)		1 m^3/m^2·h = 0.409 gpm/ft^2
Solids contact	2–3 (0.8–1.2)	
Inclined surface clarification	10–15 (4–6)	
Micro-sand ballasted flocculation	100–200 (40–80)	
Solids ballasted flocculation	24–36 (10–15)	
Overall detention time, h		
Solids contact	1–2	
Inclined surface clarification	0.15–0.5	
Micro-sand ballasted flocculation	0.15–0.2	
Solids ballasted flocculation	0.2–0.3	
Fine Mesh Screen for Primary Treatment		
Mesh size, mm (in)	0.15–0.5 (0.006–0.02)	1 mm = 0.03937 in
Surface overflow rate, m^3 flow/m^2 surface area·h (gpm/ft^2)	20–260 (8–106)	1 m^3/m^2·h = 0.409 gpm/ft^2
Rapid Mixing for Coagulation		
Detention time (t), min	0.5–2	
Hydraulic gradient (G), s^{-1}	300–1500	
Mixing opportunity parameter (Gt), dimensionless	30,00–90,000	

(*Continued*)

TABLE E.1 (*Continued*) Design Parameter and Units of Expressions for Wastewater Treatment Processes

Design Parameter	Typical Range	Conversion Factor
Flocculation after Coagulation		
Detention time (t), min	5–30	
Hydraulic gradient (G), s^{-1}	15–60	
Mixing opportunity parameter (Gt), dimensionless	10,000–150,000	
Suspended Growth Biological Treatment, Activated Sludge and Modifications		
Volumetric organic loading rate, kg BOD_5/m^3 aeration basin volume·d (lb/10^3 ft^3·d)		1 kg/m^3·d = 62.43 lb/10^3 ft^3·d
Conventional with operational modification	0.3–1 (20–60)	
High-rate aeration/high pure oxygen (HPO)	1–3 (60–185)	
Extended aeration/oxidation ditch/single nitrification/sequencing bioreactor (SBR)	0.08–0.3 (5–20)	
Fixed-film media	0.5–6 (30–375)	
Membrane bioreactor (MBR)	0.8–1.5 (50–95)	
F/M ratio, kg BOD_5/kg VSS·d (lb/lb·d)		1 kg/kg·d = 1 lb/lb·d
Conventional with operational modification	0.2–0.6 (0.2–0.6)	
High-rate aeration/HPO	0.5–2 (0.5–2)	
Extended aeration/oxidation ditch/single nitrification/SBR	0.04–0.2 (0.04–0.2)	
Fixed-film media	0.1–0.3 (0.1–0.3)	
MBR	<0.1 (0.1)	
Solids retention time (SRT), d		
Conventional with operational modification	3–15	
High-rate aeration/HPO	0.5–4	
Extended aeration/oxidation ditch/single nitrification/SBR	15–40	
Fixed-film media	10–25	
MBR	10–30	
Return activated sludge ratio (return flow/influent flow), dimensionless	0.2–1	
Air requirement by BOD_5 removal, m^3 air/kg BOD_5 removed (ft^3/lb)	20–100 (300–1500)	1 m^3/kg = 16.02 ft^3/lb
Air requirement by flow treated, m^3 air/m^3 wastewater treated (ft^3/gal)	3.75–15 (0.5–2)	1 m^3/m^3 = 0.1337 ft^3/gal
Oxygen requirement		
Organic, kg O_2/kg BOD_5 applied (lb/lb)	0.6–2.4 (0.6–2.4)	1 kg/kg = 1 lb/lb
Ammonia, kg O_2/kg NH_3-N applied·d (lb/lb·d)	4.57 (4.57)	
Aeration System		
Standard oxygen transfer efficiency (SOTE), %		
Fine bubble diffuser	25–35	
Coarse bubble diffuser/mechanical aerator	7–15	
Standard aeration efficiency (SAE), kg O_2 transferred/kW·h (lb/hp·h)		1 kg/kW·h = 1.644 lb/hp·h
Fine bubble diffuser	2.5–10 (4–16)	
Coarse bubble diffuser/mechanical aerator	1.2–2.5 (2–4)	

(*Continued*)

TABLE E.1 (*Continued*) Design Parameter and Units of Expressions for Wastewater Treatment Processes

Design Parameter	Typical Range	Conversion Factor
	Final Clarifier	
Surface overflow rate (SOR) at average flow, m^3 flow/m^2 surface area·d (gpd/ft^2)		1 m^3/m^2·d = 24.54 gpd/ft^2
Activated sludge	16–24 (400–600)	
Extended aeration/oxidation ditch	8–16 (200–400)	
High pure oxygen (HPO)	16–28 (400–700)	
Enhanced biological phosphorus removal (EBPR) with chemical addition/biological nutrient removal (BNR) or selector technologies	12–32 (300–800)	
Surface overflow rate (SOR) at peak 2-h flow, m^3 flow/m^2 surface area·d (gpd/ft^2)	40–64 (1000–1600)	1 m^3/m^2·d = 24.54 gpd/ft^2
Solids loading rate, kg solids/m^2 surface area·h (lb/ft^2·h)		1 kg/m^2·h = 0.2048 lb/ft^2·h
Average flow	1–8 (0.2–1.6)	
Peak 2-h flow	7–10 (1.4–2)	
	Stabilization Pond System	
Surface organic loading rate, kg BOD_5/ha pond surface area·d (lb/ac·d)		1 kg/ha·d = 0.8922 lb/ac·d
Aerobic (high rate)	70–225 (60–200)	
Facultative (aerobic–anaerobic)	20–90 (15–80)	
Anaerobic	225–1120 (200–1000)	
Volumetric organic loading rate for anaerobic, kg BOD_5/m^3 pond volume·d (lb/10^3 ft^3·d)	0.1–0.4 (6–25)	1 kg/m^3·d = 62.43 lb/10^3 ft^3·d)
Detention time, d		
Aerobic (high rate)	2–6	
Facultative (aerobic–anaerobic)	20–180	
Anaerobic	5–180	
Attached Growth Biological Treatment, Trickling Filter (TF) and Rotating Biological Contactor (RBC)		
Volumetric organic loading rate for TF, kg BOD_5/m^3 reactor volume·d (lb/10^3 ft^3·d)		1 kg/m^3·d = 62.43 lb/10^3 ft^3·d
Low rate/standard rate with nitrification TF	0.08–0.5 (5–30)	
Intermediate-rate/high-rate TF	0.24–2.4 (15–150)	
Super-rate/roughing TF	0.8–6 (50–375)	
Surface organic loading rate for RBC, g BOD_5/m^2·d (lb $BOD_5/10^3$ ft^2·d)		1 g/m^2·d = 0.2048 lb/10^3 ft^2·d
For BOD removal	8–20 (1.6–4)	
For BOD removal and nitrification	5–16 (1–3.3)	
For separate nitrification	1–2 (0.2–0.4)	
Hydraulic loading rate, m^3 flow/m^2 surface area·d (gpd/ft^2)		1 m^3/m^2·d = 24.54 gpd/ft^2
Low rate/standard rate with nitrification TF	1–16 (25–400)	
Intermediate-rate/high-rate TF	4–40 (100–1000)	
Super-rate/roughing TF	15–100 (350–2500)	
RBC for BOD removal	0.08–0.16 (2–4)	

(*Continued*)

TABLE E.1 (*Continued*) Design Parameter and Units of Expressions for Wastewater Treatment Processes

Design Parameter	Typical Range	Conversion Factor
RBC for BOD removal and nitrification	0.03–0.08 (0.75–2)	
RBC for separate nitrification	0.04–0.10 (1–2.5)	
Hydraulic retention time (HRT) in RBC, h		
BOD removal	0.75–1.5	
BOD removal and nitrification	1.5–4	
Separate nitrification	1.2–3	
Recirculation rate for TF, % of influent flow		
Low-rate/standard-rate with nitrification	0–2	
Intermediate rate/high rate	0.1–2	
Super rate/roughing	0–12	
Combined Attached and Suspended Growth Biological Treatment		
Organic loading rate in fixed film reactor, kg $BOD_5/m^3{\cdot}d$ ($lb/10^3\ ft^3{\cdot}d$)		$1\ kg/m^3{\cdot}d = 62.43\ lb/10^3\ ft^3{\cdot}d$
Activated biofilter (ABF)	0.24–1.2 (15–75)	
Trickling filter/solids contact (TF/SC)	0.4–1.6 (25–100)	
Biofilter/activated sludge (BF/AS)	1.4–4 (90–250)	
Trickling filter with intermediate clarifier/activated sludge (TF-IC/AS)	1–4.8 (60–300)	
F/M ratio for suspended growth, kg BOD_5/kg VSS·d (lb/lb·d)		1 kg/kg = 1 lb/lb
BF/AS	0.5–1.2 (0.5–1.2)	
TF-IC/AS	0.2–0.5 (0.2–0.5)	
Solids retention time (SRT), d	15–200	
Hydraulic retention time (HRT) for suspended growth, h		
TF/SC	0.75–2	
BF/AS, or TF-IC/AS	2–8	
Final clarifier surface overflow rate (SOR), $m^3/m^2{\cdot}h$ (gpm/ft^2)		$1\ m^3/m^2{\cdot}h = 0.409\ gpm/ft^2$
ABF	2–3 (0.8–1.2)	
TF/SC or BF/AS	1–2 (0.4–0.8)	
TF-IC/AS	2–3.5 (0.8–1.4)	
Integrated Fixed-Film Media Biological Treatment		
Volumetric organic loading rate, kg $BOD_5/m^3{\cdot}d$ (lb $BOD_5/10^3\ ft^3{\cdot}d$)		$1\ kg/m^3{\cdot}d = 62.43\ lb/10^3\ ft^3{\cdot}d$
Integrated fixed-film activated sludge (IFAS)	1.5–3 (95–190)	
Moving bed biofilm reactor (MBBR)	1–6 (60–380)	
Surface organic loading rate, g BOD_5/m^2 surface area·d ($lb/10^3\ ft^2{\cdot}d$)		$1\ g/m^2{\cdot}d = 0.2048\ lb/10^3\ ft^2{\cdot}d$
Surface area loading rate (SALR) for IFAS	10 (2)	
Surface area removal flux (SARF) for MBBR	5–20 (1–4)	
Hydraulic retention time (HRT), h	1.5–6	

(*Continued*)

TABLE E.1 (*Continued*) Design Parameter and Units of Expressions for Wastewater Treatment Processes

Design Parameter	Typical Range	Conversion Factor
Submerged Attached Growth Biological Treatment		
Volumetric organic loading rate, kg $BOD_5/m^3 \cdot d$ ($lb/10^3\ ft^3 \cdot d$)	2–10 (120–600)	1 $kg/m^3 \cdot d = 62.43\ lb/10^3\ ft^3 \cdot d$
Hydraulic loading rate, m^3 flow/m^2 reactor surface area·h ($ft^3/ft^2 \cdot h$)	2.4–40 (8–130)	1 $m^3/m^2 \cdot h = 3.281\ ft^3/ft^2 \cdot h$
Empty bed contact time (EBCT), min	5–60	
Anaerobic Suspended Growth Biological Treatment		
Volumetric organic loading rate, kg $COD/m^3 \cdot d$ ($lb/10^3\ ft^3 \cdot d$)	1–20 (60–1200)	1 $kg/m^3 \cdot d = 62.43\ lb/10^3\ ft^3 \cdot d$
Solids retention time (SRT), d	15–200	
Hydraulic loading rate, $m^3/m^2 \cdot h$ ($ft^3/ft^2 \cdot h$)		
Upflow velocity in upflow anaerobic sludge blanket process (UASB)	0.5–1.5 (1.5–5)	1 $m^3/m^2 \cdot h = 3.281\ ft^3/ft^2 \cdot h$
Surface overflow rate (SOR) in clarifier	0.5–1 (1.6–3.3)	
Hydraulic retention time (HRT), h	4–120	
Anaerobic Attached Growth Biological Treatment		
Volumetric organic loading rate, kg $COD/m^3 \cdot d$ ($lb/10^3\ ft^3 \cdot d$)	0.1–40 (6–1500)	1 $kg/m^3 \cdot d = 62.43\ lb/10^3\ ft^3 \cdot d$
Solids retention time (SRT), d	10–40	
Hydraulic loading, m^3/m^2 reactor surface area·h ($ft^3/ft^2 \cdot h$)	0.5–20 (1.5–60)	1 $m^3/m^2 \cdot h = 3.281\ ft^3/ft^2 \cdot h$
Hydraulic retention time (HRT), h	8–120	
Suspended Growth for Biological Nitrogen Removal		
Solids retention time (SRT), d	7–40	
Hydraulic retention time (HRT), h		
Anoxic	1–8	
Aerobic	4–24	
Overall	6–36	
Mixed liquor internal recycle ratio (recycle flow/influent flow), dimensionless	1–4	
Return activated sludge ratio (return flow/influent flow), dimensionless	0.5–1	
Suspended Growth for Biological Nutrient Removal (BNR)		
Solids retention time (SRT), d	5–40	
F/M ratio, kg BOD_5/kg VSS·d (lb/lb·d)	0.1–0.25 (0.1–0.25)	1 kg/kg·d = 1 lb/lb·d
Hydraulic retention time (HRT), h		
Anaerobic (AN)	0.5–6	
Anoxic (AX)	1–8	
Oxic (OX)	4–12	
Overall	6–24	
Mixed liquor internal recycle ratio (recycle flow/influent flow), dimensionless	1–4	
Return activated sludge ratio (return flow/influent flow), dimensionless	0.25–1	

(*Continued*)

TABLE E.1 (*Continued*) Design Parameter and Units of Expressions for Wastewater Treatment Processes

Design Parameter	Typical Range	Conversion Factor
Submerged Attached Growth for Biological Denitrification		
Volumetric organic loading rate, kg NO_5-N/m^3·d (lb/10^3 ft^3·d)	0.3–5 (20–300)	1 kg/m^3·d = 62.43 lb/10^3 ft^3·d
Hydraulic loading rate, m^3 flow/m^2 reactor surface area·h (ft^3/ft^2·h)	2.4–24 (1–10)	1 m^3/m^2·h = 3.281 ft^3/ft^2·h
Hydraulic retention time (HRT) or empty bed contact time (EBCT), min	5–30	
Filtration with Granular and Surface Filters		
Filtration rate, m^3 flow/m^2 surface area·h (gpm/ft^2)		1 m^3/m^2·h = 0.409 gpm/ft^2
Average flow	5–20 (2–8)	
Peak flow	14–40 (6–16)	
Land Treatment System		
Annual volumetric loading rate, m^3/m^2·yr (gal/ft^2·yr)		1 m^3/m^2·yr = 24.54 gal/ft^2·yr
Slow rate (SR)	0.5–6 (12–150)	
Rapid infiltration (RI)	6–125 (150–3000)	
Overland flow (OF)	3–20 (75–500)	
Organic loading rate[d], kg BOD_5/ha·d (lb/ac·d)		1 kg/ha·d = 0.8922 lb/ac·d
Slow rate (SR)	50–500 (45–450)	
Rapid infiltration (RI)	145–1000 (130–900)	
Overland flow (OF)	40–110 (35–100)	
Constructed Wetland Systems		
Hydraulic loading rate, m^3/ha·d (gpm/ac)		1 m^3/ha·d = 0.07424 gpm/ac
Free water surface (FWS)	135–900 (10–70)	
Subsurface flow (SF)	200–1800 (15–140)	
Adsorption by Granular Activated Carbon		
Operating velocity, L/s·m^2 (gpm/ft^2)	1.4–4.2 (2–6)	1 L/s·m^2 = 1.472 gpm/ft^2
Empty bed contact time (EBCT), min	5–30	
Ion Exchange Column		
Operating velocity, L/s·m^2 (gpm/ft^2)	0.7–5.5 (1–8)	1 L/s·m^2 = 1.472 gpm/ft^2
Empty bed contact time (EBCT), min	10–30	
Membrane System		
Flux rate, L/m^2·d (gpd/ft^2)		1 L/m^2·d = 0.0245 gpd/ft^2
Microfiltration (MF)	600–1400 (15–35)	
Ultrafiltration (UF)	960–1400 (25–35)	
Nanofiltration (NF) or reverse osmosis (RO)	340–480 (8–12)	
Recovery factor, %		
MF/UF	85–95	
NF/RO	80–90	
Disinfection		
Chlorine dosage, mg Cl_2 applied/L (lb/Mgal)	4–40 (33–330)	1 mg/L = 8.345 lb/Mgal

(*Continued*)

TABLE E.1 (*Continued*) Design Parameter and Units of Expressions for Wastewater Treatment Processes

Design Parameter	Typical Range	Conversion Factor
UV dose for inactivation of bacteria and virus in filtered secondary effluent, mJ/cm^2		
1- to 2-log	10–70	
3- to 4-log	20–190	
Gravity Sludge Thickening		
Solids loading rate, kg dry solids/m^2 surface area·d (lb/ft^2·d)	20–150 (4–30)	1 kg/m^2·d = 0.2048 lb/ft^2·d
Hydraulic loading rate, m^3 sludge/m^2 surface area·d (gpd/ft^2)	4–30 (100–750)	1 m^3/m^2·min = 24.54 gpm/ft^2
Chemical dosage		
Lime, mg/L as CaO (lb/Mgal)	6–9 (50–75)	1 mg/L = 8.345 lb/Mgal
Ferric chloride, mg/L as $FeCl_3$ (lb/Mgal)	1–3 (8–25)	
Potassium permanganate, mg/L $KMnO_4$ (lb/Mgal)	10–40 (80–330)	
Polymer, g polymer/kg dry solids (lb/ton)	2.5–6 (5–12)	1 g/kg = 2 lb/ton
Dissolved Air Flotation (DAF) for Sludge Thickening		
Solids loading rate with polymer addition, kg dry solids/m^2 surface area·d (lb/ft^2·d)	220–300 (45–60)	1 kg/m^2·d = 0.2048 lb/ft^2·d
Polymer dosage, g polymer/kg dry solids (lb/ton)	1–5 (2–10)	1 g/kg = 2 lb/ton
Gravity Belt Thickener (GBT)		
Solids loading rate, kg dry solids/h (lb/h) per m belt width	300–1350 (650–3000)	1 kg/h = 2.205 lb/h
Hydraulic loading rate, L wet sludge/min (gpm) per m belt width	400–1000 (100–250)	1 L/min = 0.2642 gpm
Polymer dosage, g polymer/kg dry solids (lb/ton)	1–7 (2–14)	1 g/kg = 2 lb/ton
Wash water requirement, L/min (gpm) per m belt width	60–80 (15–20)	1 L/min = 0.2642 gpm
Anaerobic Sludge Digestion		
Per capita capacity requirement (PCCR), m^3 (ft^3) per capita	0.03–0.18 (1–6)	1 m^3 = 35.31 ft^3
Volumetric solids loading rate, kg VSS/m^3 digester volume·d (lb/10^3 ft^3·d)		1 kg/m^3·d = 62.43 lb/10^3 ft^3·d
Standard rate or staged mesophilic	0.5–1.6 (30–100)	
High rate	1.6–4.8 (100–300)	
Staged thermophilic, acid/gas phased (A/GAnD), or temperature-phased (TPAnD)	4.8–6.4 (300–400)	
Solids retention time (SRT), d		
Standard rate mesophilic	30–60	
High rate mesophilic	15–20	
Staged mesophilic, 1st reactor (2nd reactor)	7–10 (variable)	
Staged thermophilic, 1st reactor (2nd or 3rd reactor)	17–22 (1.5–2)	
A/GAnD, 1st reactor (2nd reactor)	1–3 (>10)	
TPAnD, 1st reactor (2nd reactor)	3–10 (5–15)	

(*Continued*)

TABLE E.1 (*Continued*) Design Parameter and Units of Expressions for Wastewater Treatment Processes

Design Parameter	Typical Range	Conversion Factor
Digester heating loss, J/s·m^2 surface area·°C (Btu/h·ft^2·°F)	0.7–2.8 (0.12–0.5)	1 J/s·m^2·°C = 0.1763 Btu/ft^2·h·°F
Digester mixing requirement		
Gas injection, m^3 gas/10^3 m^3 volume·min (ft^3/10^3 ft^3·min)	4.5–7 (4.5–7)	1 m^3/10^3 m^3·min = 1 ft^3/10^3 ft^3·min
Mechanical mixing, kW/10^3 m^3 volume (hp/10^3 ft^3)	5–8 (0.2–0.3)	1 KW/10^3 m^3 = 0.03795 hp/10^3 ft^3
Hydraulic mixing (G and T), s^{-1} and min	50–80 and 20–30	
Aerobic Sludge Digestion		
Per capita capacity requirement (PCCR), m^3 (ft^3) per capita	0.06–0.2 (2–7)	1 m^3 = 35.31 ft^3
Volumetric solids loading rate, kg VSS/m^3 digester volume·d (lb/10^3 ft^3·d)		1 kg/m^3·d = 62.43 lb/10^3 ft^3·d
Conventional aerobic digestion	1.6–4.8 (100–300)	
Autothermal thermophilic aerobic digestion (ATAD)	3–4 (200–250)	
Solids retention time (SRT), d		
Conventional	10–60	
ATAD	4–15	
Oxygen requirement, kg O2/kg VSS reduced (lb/lb)	1.6–2.3 (1.6–2.3)	1 kg/kg = 1 lb/lb
Sludge Drying Beds		
Area requirement per capita for conventional sludge drying bed, m^2 bed area/person (ft^2/person)	0.09–0.23 (1–2.5)	1 m^2/person = 10.76 ft^2/person
Mass sludge applying rate, kg dry solids/m^2 bed area·yr (lb/ft^2·yr)		1 kg/m^2·yr = 0.2048 lb/ft^2·yr
Conventional sand drying bed	50–200 (10–40)	
Artificial-media drying bed	750–1800 (150–360)	
Solar drying bed	50–900 (10–180)	
Reed drying bed	30–100 (6–20)	
Mass sludge applying rate for vacuum assisted drying bed, kg dry solids/m^2 bed area·h (lb/ft^2·h)	6–60 (1.2–12)	1 kg/m^2·yr = 0.2048 lb/ft^2·yr
Belt-Filter Press (BFP) Dewatering		
Solids loading rate, kg dry solids/h (lb/h) per m belt width	200–1500 (450–3300)	1 kg/h = 2.205 lb/h
Hydraulic loading rate, L wet sludge/min (gpm) per m belt width	100–800 (25–200)	1 L/min = 0.2642 gpm
Polymer dosage, g polymer/kg dry solids (lb/ton)	1.5–12.5 (3–25)	1 g/kg = 2 lb/ton
Wash water requirement, L/min (gpm) per m belt width	150–300 (40–80)	1 L/min = 0.2642 gpm
Plate and Frame Filter Press Dewatering		
Solids loading rate, kg dry solids/m^2 press area·h (lb/ft^2·h)	~5 (1)	1 kg/m^2·h = 0.2048 lb/ft^2·h
Chemical dosage, g chemical/kg dry solids (lb/ton)		1 g/kg = 2 lb/ton
Ferric chloride	50 (100)	
CaO	100 (200)	

(Continued)

TABLE E.1 (*Continued*) Design Parameter and Units of Expressions for Wastewater Treatment Processes

Design Parameter	Typical Range	Conversion Factor
Vacuum Filter Dewatering		
Solids loading rate kg dry solids/m^2 filter area·h (lb/ft^2·h)	10–25 (2–5)	1 kg/m^2·h = 0.2048 lb/ft^2·h
Chemical dosage, g chemical/kg dry solids (lb/ton)		1 g/kg = 2 lb/ton
Ferric chloride	20–40 (40–80)	
CaO	100–150 (200–300)	
Rotary Press for Dewatering		
Solids loading rate, kg dry solids/m^2 filter area·h (lb/ft^2·h)	~250 (50)	1 kg/m^2·h = 0.2048 lb/ft^2·h
Volumetric sludge feed rate, m^3 feed wet sludge/m^2 filter area·h (ft^3/ft^2·h)	2.5–3.5 (1–1.5)	1 m^3/m^2·h = 0.409 gpm/ft^2
Polymer dosage, g polymer/kg dry solids (lb/ton)	5–15 (10–30)	1 g/kg = 2 lb/ton
Land Application of Biosolids		
Mass application rate, metric ton (mt) biosolids/ha field area·yr (ton/ac·yr)		1 mt/ha·yr = 0.4461 ton/ac·yr
Agricultural land utilization	2–70 (1–30)	
Forest land utilization	10–225 (4–100)	
Land reclamation utilization	7–450 (3–200)	
Dedicated land disposal	12–2250 (5–1000)	

Source: Adapted in part from References 1 through 3.

References

1. Qasim, S. R., *Wastewater Treatment Plants: Planning, Design, and Operation*, 2nd ed., CRC Press, Boca Raton, FL, 1999.
2. Metcalf & Eddy | AECOM, *Wastewater Engineering: Treatment and Resource Recovery*, 5th ed., McGraw-Hill, New York, NY, 2014.
3. Metcalf & Eddy, Inc., *Wastewater Engineering: Treatment and Reuse*, 4th ed., McGraw-Hill Companies, Inc., New York, NY, 2003.

Appendix F: List of Examples and Solutions

In this book, over 700 design examples with in-depth solutions are presented to cover the complete spectrum of wastewater treatment and reuse. A listing of these illustrative examples is provided in Table F.1.

Note: In this table, the examples of Chapters 1–10 are presented in Volume 1: Principles and Basic Treatment while the examples of Chapters 11–15 are presented in Volume 2: Post-Treatment, Reuse, and Disposal.

TABLE F.1 Summary of Examples

(*Continued*)

TABLE F.1 (*Continued*) Summary of Examples

(*Continued*)

TABLE F.1 (*Continued*) Summary of Examples

(*Continued*)

TABLE F.1 (*Continued*) Summary of Examples

(*Continued*)

TABLE F.1 (*Continued*) Summary of Examples

(*Continued*)

TABLE F.1 (*Continued*) Summary of Examples

(*Continued*)

TABLE F.1 (*Continued*) Summary of Examples

(*Continued*)

TABLE F.1 (*Continued*) Summary of Examples

(*Continued*)

TABLE F.1 (*Continued*) Summary of Examples

(*Continued*)

TABLE F.1 (*Continued*) Summary of Examples

No	Title	Page
Example 10.33	Biomass Production and Oxygen Utilization	**10**-48
Example 10.34	Oxygen Utilization Equation	**10**-48
Example 10.35	Profiles of Substrate Concentration in the Effluent and Process Removal Efficiency	**10**-49
Example 10.36	Design of Biological Process	**10**-50
Example 10.37	Estimate the Effect of Temperature on S	**10**-54
Example 10.38	Net Biomass Growth and Sludge Wasting	**10**-58
Example 10.39	Determination of Kinetic Coefficients from Continuous Flow Completely Mixed Bench-Scale Reactor Study	**10**-60
Example 10.40	Performance Comparison of Biological CFSTR and PFR	**10**-64
Example 10.41	Comparative Performance Curves of a Biological CFSTR and PFR	**10**-65
Example 10.42	Comparison between True and Dispersed PFRs	**10**-67
Example 10.43	Dispersion Number of a Dispersed Plug-Flow Aeration Basin	**10**-67
Example 10.44	Determination of SVI	**10**-70
Example 10.45	SVI, Settled Volume, and Return Sludge Relationships	**10**-70
Example 10.46	Maximum Solids Concentration in RAS Based on SVI	**10**-71
Example 10.47	RAS Flow Adjustment	**10**-73
Example 10.48	Oxygen Demand and Supply in Activated Sludge Process Modifications	**10**-86
Example 10.49	Advantages of a Completely Mixed Activated Sludge Process	**10**-86
Example 10.50	MLVSS Concentration in an Oxidation Ditch (Extended Aeration System) without Sludge Wasting	**10**-87
Example 10.51	Design of Oxidation Ditch with No Sludge Wasting	**10**-88
Example 10.52	Design of Deep-Shaft Biological Reactor	**10**-91
Example 10.53	Comparison of Basin Capacity and F/M Ratio of Contact Stabilization and Conventional Activated Sludge Processes	**10**-93
Example 10.54	Design of High-Purity Oxygen (HPO) Activated Sludge Process Reactor	**10**-95
Example 10.55	Describe Operational Details of the Sequencing Batch Reactor (SBR) process	**10**-96
Example 10.56	Cycle Time and Volume of a Sequencing Batch Reactor (SBR)	**10**-99
Example 10.57	Sodium Sulfite Dose for Deoxygenation	**10**-116
Example 10.58	Determination of Standard K_La of a Surface Aerator in Clean Water	**10**-117
Example 10.59	Equilibrium Concentration of Oxygen for Wastewater in Aeration Basin	**10**-119
Example 10.60	Determination of K_La_{ww} of a Diffused Aerator in Primary Settled Wastewater	**10**-121
Example 10.61	Estimation of K_La and Oxygen Transfer Efficiency of a Diffuser System	**10**-123
Example 10.62	Diffused Oxygen Transfer Rate at Higher Elevation	**10**-125
Example 10.63	Determination of DO Saturation Coefficient β, Oxygen Utilization Rate γ_m, and K_La_{ww}	**10**-126
Example 10.64	Oxygen Utilization Rate of Activated Sludge	**10**-129
Example 10.65	Impact of Water Depth on K_La	**10**-130
Example 10.66	Field Oxygen Transfer Efficiency of Diffused Aerator	**10**-130
Example 10.67	Aeration Efficiency of Diffused Aerator	**10**-132
Example 10.68	Total Horsepower and Number of Turbine Aerators	**10**-133
Example 10.69	Diffusers in a Basin of Known Width	**10**-136
Example 10.70	Theoretical and Standard Oxygen Requirements in a Single-Stage Nitrification Reactor	**10**-137
Example 10.71	Diffuser Configuration and Aeration Basin Design	**10**-158
Example 10.72	Sizing of Mechanical Aerator	**10**-165

(*Continued*)

TABLE F.1 (*Continued*) Summary of Examples

(*Continued*)

TABLE F.1 (*Continued*) Summary of Examples

No	Title	Page
Example 10.114	Biogas Generation from Industrial Waste	**10**-271
Example 10.115	Methane Generation from Mass COD Loading	**10**-272
Example 10.116	Biogas Generation from a Waste Composition	**10**-272
Example 10.117	Methane Generation from Substrate Utilization	**10**-275
Example 10.118	Derivation of Substrate Utilization and Biomass Growth Rate Equations	**10**-275
Example 10.119	Estimate Alkalinity Requirement in Anaerobic Process	**10**-276
Example 10.120	Sizing of an Anaerobic Contact Process Reactor	**10**-281
Example 10.121	Characteristics of Granules	**10**-286
Example 10.122	Design of a UASB Reactor	**10**-286
Example 10.123	Design of an Anaerobic Fluidized-Bed Reactor (AnFBR)	**10**-293
Example 10.124	Oxygen and Alkalinity Requirements in Nitrification	**10**-308
Example 10.125	Cell Synthesis Yields in Nitrification	**10**-310
Example 10.126	Determination of Nitrification Kinetic Coefficients under Field Operating Conditions	**10**-311
Example 10.127	Design SRT for Nitrification	**10**-312
Example 10.128	Design of a Single-Stage Suspended Growth Reactor for Combined CBOD Removal and Nitrification	**10**-313
Example 10.129	Heterotrophic and Autotrophic Biomass Growth	**10**-315
Example 10.130	Theoretical Oxygen Requirement (ThOR) in a Single Reactor with CBOD Removal and Nitrification	**10**-317
Example 10.131	Biomass Growth and Supplemental Alkalinity for Nitrification	**10**-317
Example 10.132	Combined CBOD Removal and Nitrification in an Attached Growth Reactor	**10**-320
Example 10.133	Tertiary Nitrification in a Fixed-Film Reactor	**10**-322
Example 10.134	Tertiary Nitrification in a RBC System	**10**-324
Example 10.135	Combined CBOD Removal and Nitrification in an IFAS System	**10**-326
Example 10.136	Tertiary Nitrification in a MBBR Reactor	**10**-332
Example 10.137	Denitrification Reactions with Carbon Source Other Than Methanol	**10**-355
Example 10.138	Cell Synthesis, Methanol and Equivalent BOD_LConsumptions, and Alkalinity Recovery Factors for Denitrification	**10**-355
Example 10.139	Determine Theoretical Oxygen Consumption Ratios	**10**-357
Example 10.140	Estimate Methanol Requirement for Postanoxic Denitrification	**10**-357
Example 10.141	Denitrification Kinetic Coefficients under Field Operating Conditions	**10**-358
Example 10.142	Kinetics-Based Nitrate Reduction and Specific Denitrification Rates	**10**-359
Example 10.143	Denitrification in Preanoxic Zone	**10**-360
Example 10.144	Denitrification in Postanoxic Zone with Substrate Driven Mode	**10**-363
Example 10.145	Denitrification in Postanoxic Zone with Endogenous Driven Mode	**10**-364
Example 10.146	AttachedGrowth Downflow Denitrification Filter	**10**-366
Example 10.147	Denitrification in an Anoxic Fluidized-Bed Bioreactor	**10**-369
Example 10.148	Effluent TP and HRT of Anaerobic Zone	**10**-377
Example 10.149	Phosphorus Release and Uptake, and COD Consumption in Anaerobic Zone	**10**-378
Example 10.150	Phosphorus Removal in an EBPR Process	**10**-381
Example 10.151	Mixer Sizing for Anaerobic or Anoxic Basins	**10**-384
Example 10.152	Concentration Profiles of $sBOD_5$, Ortho-P, and TN in a BNR Facility	**10**-391
Example 10.153	Phosphorus Content of Biomass and TP in the Effluent	**10**-393

(*Continued*)

TABLE F.1 (*Continued*) Summary of Examples

(*Continued*)

TABLE F.1 (*Continued*) Summary of Examples

(*Continued*)

TABLE F.1 (*Continued*) Summary of Examples

(*Continued*)

TABLE F.1 (*Continued*) Summary of Examples

(*Continued*)

TABLE F.1 (*Continued*) Summary of Examples

(*Continued*)

TABLE F.1 (*Continued*) Summary of Examples

(*Continued*)

TABLE F.1 (*Continued*) Summary of Examples

Index

Note: Volume 1—Chapters 1–10
Volume 2—Chapters 11–15

A

B

C

D

E

F

G

H

I

K

L

M

N

O

P

R

S

T

U

Y

Z